AF252048

PLATES, LAMINATES AND SHELLS

Asymptotic Analysis and Homogenization

Series on Advances in Mathematics for Applied Sciences – Vol. 52

PLATES, LAMINATES AND SHELLS

Asymptotic Analysis and Homogenization

T Lewiński
Warsaw University of Technology, Poland

J J Telega
Institute of Fundamental Technological Research,
Polish Academy of Sciences, Poland

World Scientific
Singapore • New Jersey • London • Hong Kong

Published by

World Scientific Publishing Co. Pte. Ltd.

P O Box 128, Farrer Road, Singapore 912805

USA office: Suite 1B, 1060 Main Street, River Edge, NJ 07661

UK office: 57 Shelton Street, Covent Garden, London WC2H 9HE

Library of Congress Cataloging-in-Publication Data
Lewiński, T.
    Plates, laminates, and shells : asymptotic analysis and
homogenization / T. Lewiński, J.J. Telega.
      p.   cm. -- (Series on advances in mathematics for applied
sciences : vol. 52)
    Includes bibliographical references and index.
    ISBN 9810232063
    1. Elastic plates and shells.   2. Homogenization (Differential
equations)   I. Telega, Józef Joachim.   II. Title.   III. Series.
QA935.L39   1999
531'.382--dc21
                                                99-40193
                                                  CIP

British Library Cataloguing-in-Publication Data
A catalogue record for this book is available from the British Library.

Copyright © 2000 by World Scientific Publishing Co. Pte. Ltd.

All rights reserved. This book, or parts thereof, may not be reproduced in any form or by any means, electronic or mechanical, including photocopying, recording or any information storage and retrieval system now known or to be invented, without written permission from the Publisher.

For photocopying of material in this volume, please pay a copying fee through the Copyright Clearance Center, Inc., 222 Rosewood Drive, Danvers, MA 01923, USA. In this case permission to photocopy is not required from the publisher.

Printed in Singapore by Regal Press (S) Pte. Ltd.

To the blessed memory of Pawełek Telega

and to

Ewa and Hanna

for their patience and understanding

Preface

Introduction of composite materials into many fields of engineering practice has made the stress analysis complex due to the presence of two levels of analysis. For instance, the structure as a whole may affect failure phenomena at the material level and these microdefects do have influence on the overall response of the whole structure. It is clear that the entire analysis has to be branched into two analyses: at the macro- and microlevel. A systematic, albeit formal, technique of such an analysis is provided by the multiscale expansion method. Its correctness is lucidly justified by the homogenization theory.

The theorems of homogenization apply to both media and structures of regular as well as irregular (random) layup. However, in the regular (or periodic) case the homogenization methods are constructive and thus applicable. Broad applications of the homogenization theory in investigation of effective properties of composite materials are the subject of the monographs by Bensoussan et al. (1978), Sanchez-Palencia (1980), Bakhvalov and Panasenko(1984) and Jikov et al. (1994); cf. also the comprehensive review paper by Ponte Castañeda and Suquet(1998). The effective properties characterize a hypothetic material with smeared-out inhomogeneities. From the mathematical standpoint the homogenization is equivalent to investigating G-limits of operators or Γ-limits of functionals which govern a given physical problem. In recent years new methods of finding these limits have been developed, which has made it possible to examine the overall behaviour of the composite materials in the physically and geometrically nonlinear range, cf. Müller (1987), Bouchitté and Suquet (1991), Sanchez-Palencia and Zaoui (1987), Messaoudi and Michaille (1994) and Sab (1994). It is also feasible to consider cracks, both unilateral and bilateral, with or without friction, cf. Telega (1990b). The dynamic overall behavior is less investigated.

The homogenization methods apply also to the plate and shell problems. The best results are obtained if the departure point is three-dimensional and an appropriate passage to a limit transforms the initial formulation to a two-dimensional one with the elastic moduli being smeared-out. Such an ingenious approach was proposed by Caillerie (1984) and Kohn and Vogelius (1984) in the papers on elastic plates. The monograph of Kalamkarov (1992) was inspired by this averaging method.

The aim of the present book is to synthesize applications of the homogenization methods to plate, shell and laminate problems in a possibly broad perspective, by encompassing various averaging methods, whether starting from the three-dimensional or two-dimensional mathematical models. The analysis is confined to the carefully selected both practically important and variationally consistent structural models. In almost all these models the dual formulations are given more or less explicitly.

The averaging by homogenization applied in the present book starts either from the three-dimensional or two-dimensional settings. In the latter case the homogenization process can go in various manners, depending on the assumed scaling, which leads to the formulae of various applicability ranges. One of the aims of the book is to investigate these discrepancies and reveal physical meaning of the homogenization formulae. On the other hand an emphasis is put on the mathematical rigour. The majority of the results found by the two-scale expansion technique are justified by the Γ-convergence method. If available, the theorems on correctors are reported.

The homogenization based on the two-dimensional models yields much simpler formulae than those obtained from the three-dimensional setting. But simplicity is not the only motivation for starting from the two-dimensional models. The other argument follows from the theory of relaxation of the layout optimization problems of plates and shells.

The layout optimization problems and the homogenization theory are linked with each other since the relaxed optimization problems are solvable and admit the presence of the composite domains, see Kohn and Strang (1986). This interesting interrelation is explained in Chapter VI by the fundamental layout problem for thin plates: mix two plate materials to form the stiffest plate of a given volume. The other optimum design problem: find the stiffest plate of varying thickness and given volume has found its partial solution with the help of the theory of Young measures.

Designers who use composite materials have to reckon with appearing the cracks. In fibrous composite laminates some cracks form regular patterns, which justifies using the homogenization method for predicting decay of stiffnesses and interrelate it with crack density parameters. In Chapter III some selected problems of such type are discussed in detail. The homogenization results turn out to be realistic only if an appropriate scaling is used. Some homogenization results, found by appropriate scalings, compare favourably with available experimental data.

The material for the present book has been collected during the joint work of the authors since 1985. A large part of the work was supported by the State Committee for Scientific Research (KBN, Poland) through the grants No. 3 P404 013 06 and No. 7 T07A 016 12 as well as through the Statutory Projects at the Faculty of Civil Engineering, Warsaw University of Technology. Moreover, the material for this book was partly collected by the first author during his stay at the Essen University (Germany) as an Alexander von Humboldt fellow. The support of the Alexander von Humboldt Foundation is gratefully acknowledged.

The present book would never be completed without stimulating discussions with our colleagues: W. Bielski, A. Gałka, B. Gambin, G. Jemielita, S. Jemioło, A. M. Othman, S. Tokarzewski, R. Wojnar, J. Bojarski, S. Bytner and S. Olszewski and without a creative atmosphere at the Institute of the Fundamental Technological Research of the Polish Academy of Sciences and the Institute of Structural Mechanics of the Faculty of Civil Engineering, Warsaw University of Technology.

Moreover, we would like to express our sincere thanks Mrs M. Rejmund and to Ms I. Malicka for their dedicated work of word-processing and editing the whole manuscript and to Mrs B. Sobolewska for drawing the figures.

We also express our thanks to the editors, and especially to Ms E.H.Chionh, for the invaluable editorial remarks and understanding, when the subsequent deadlines had gone.

Last but not least we wish to express our gratitude to Professor Nicola Bellomo, the Editor-in-Chief of the Series on Advances in Mathematics for Applied Sciences, for including our book into this collection.

Tomasz Lewiński and Józef Joachim Telega
Warsaw, March 1999

Contents

Chapter IV. ELASTIC-PERFECTLY PLASTIC PLATES
Introduction

Chapter V. ELASTIC AND PLASTIC SHELLS
Introduction

Chapter VI. APPLICATION OF HOMOGENIZATION METHODS IN OPTIMUM DESIGN OF PLATES AND SHELLS

CHAPTER I

MATHEMATICAL PRELIMINARIES

Introduction

Throughout the book rather advanced mathematical tools are used for the study of plates, laminates and shells. These tools comprise, grosso modo, functional analysis, modern variational, asymptotic and homogenization methods. Therefore to facilitate the reading of the book, in the first chapter we have gathered most of the relevant mathematical results. They include not only rather standard ones, though scattered throughout the literature, but also such that are either new or unattainable like those contained in Sections 1.3.2, 1.3.5, 1.3.6 and 1.4.

1. Function spaces, convex analysis, variational convergence

As already mentioned, the aim of this section is to introduce basic mathematical notions, definitions and results, which will be of primal importance throughout the book.

1.1. Function spaces: L^p and Sobolev spaces

In the present section we introduce the notion of weak (weak-$*$) convergence and most important, from our point of view, results concerning Lebesgue and Sobolev spaces.

1.1.1. Lebesgue spaces L^p

We start with an abstract result.

Definition 1.1.1. Let X be a Banach space, X^* its dual and $\langle \cdot, \cdot \rangle$ the bilinear canonical pairing over $X \times X^*$.
(a) We say that $\{v_n\}_{n \in \mathbb{N}} \subset X$ ($\mathbb{N}$ is the set of natural numbers) converges weakly to $v \in X$ and denote

$$v_n \rightharpoonup v \quad \text{in} \quad X , \quad \text{as} \quad n \to \infty$$

if

$$\forall v^* \in X^* \quad \langle v_n, v^* \rangle \to \langle v, v^* \rangle \quad \text{as} \quad n \to \infty .$$

(b) We say that $\{v_n^*\}_{n \in \mathbb{N}} \subset X^*$ converges weak-$*$ to $v^* \in X^*$ and denote

$$v_n^* \overset{*}{\rightharpoonup} v^* \quad \text{in} \quad X^* , \quad \text{as} \quad n \to \infty$$

if

$$\forall v \in X \quad \langle v, v_n^* \rangle \rightarrow \langle v, v^* \rangle \quad \text{as} \quad n \rightarrow \infty \, . \qquad \square$$

Weak convergence in Lebesgue spaces L^p

Let us first introduce the notion of L^p spaces.

Definition 1.1.2.

(i) Let $1 \le p < \infty$ and let $\Omega \subset \mathbb{R}^n$ be an open set. A measurable function $f : \Omega \rightarrow \mathbb{R}$ is said to be in $L^p(\Omega)$ if

$$||f||_{L^p(\Omega)} \equiv \left(\int_\Omega |f(x)|^p dx \right)^{1/p} < \infty \, .$$

(ii) Let $p = \infty$ and $\Omega \subset \mathbb{R}^n$ be an open set. A measurable function $f : \Omega \rightarrow \mathbb{R}$ is said to be in L^∞ if

$$||f||_{L^\infty(\Omega)} \equiv \inf\{\alpha : |f(x)| \le \alpha \text{ a.e. in } \Omega\} < \infty \, . \qquad \square$$

Sometimes we shall simply write

$$||f||_{L^p(\Omega)} = ||f||_{L^p} = ||f||_{0,p}$$

and similarly for $L^\infty(\Omega)$.

We recall that for $1 \le p < \infty$, the dual space of $L^p(\Omega)$ is $L^q(\Omega)$ where $1/p + 1/q = 1$. The dual space of $L^\infty(\Omega)$ contains strictly $L^1(\Omega)$.

Let us pass now to the notion of *weak convergence in $L^p(\Omega)$*.
Case 1. Let $1 \le p < \infty$, then $f_m \rightharpoonup f$ in $L^p(\Omega)$ if

$$\int_\Omega f_m(x)g(x)dx \rightarrow \int_\Omega f(x)g(x)dx \, , \quad \text{as} \quad m \rightarrow \infty$$

for every $g \in L^q(\Omega)$.
Case 2. Let $p = \infty$, then $f_m \overset{*}{\rightharpoonup} f$ in $L^\infty(\Omega)$ if

$$\int_\Omega f_m(x)g(x)dx \rightarrow \int_\Omega f(x)g(x)dx \, , \quad \text{as} \quad m \rightarrow \infty$$

for every $g \in L^1(\Omega)$.

We also have the following result.

Theorem 1.1.3. Let $\Omega \subset \mathbb{R}^n$ be an open set.
Case 1. Let $p = 1$ and let

$$f_m \rightharpoonup f \quad \text{in} \quad L^1(\Omega) \, .$$

Then there exists $K > 0$ such that

$$||f_m||_{L^1} \leq K .$$

Case 2. Let $1 < p < \infty$.
(i) Let

$$f_m \rightharpoonup f \quad \text{in} \quad L^p(\Omega) .$$

Then there exists $K > 0$ such that

$$||f_m||_{L^p} \leq K .$$

(ii) Conversely, if $||f_m||_{L^p} \leq K$, then there exists $f \in L^p(\Omega)$ and a subsequence $\{f_{m_j}\}$ of $\{f_m\}$ such that

$$f_{m_j} \rightharpoonup f \quad \text{in} \quad L^p(\Omega) .$$

Case 3. Let $p = \infty$.
(i) Let

$$f_m \overset{*}{\rightharpoonup} f \quad \text{in} \quad L^\infty(\Omega) .$$

Then there exists $K > 0$ such that

$$||f_m||_{L^\infty} \leq K .$$

(ii) Conversely, if $||f_m||_{L^\infty} \leq K$, then there exists $f \in L^\infty(\Omega)$ and a subsequence $\{f_{m_j}\}$ of $\{f_m\}$ such that

$$f_{m_j} \overset{*}{\rightharpoonup} f \quad \text{in} \quad L^\infty(\Omega) . \qquad \qquad \square$$

Case 3 (ii) is known in the literature as the Banach-Alaoglu theorem.

Corollary 1.1.4. Let χ_ε, $\varepsilon = \dfrac{1}{n}$, be a sequence of characteristic functions. Then there exist $\theta \in L^\infty(\Omega, [0, 1])$ and a subsequence $\{\chi_{\varepsilon'}\}$ of $\{\chi_\varepsilon\}$ such that

$$\chi_{\varepsilon'} \rightharpoonup \theta \quad \text{in} \quad L^\infty(\Omega) \quad \text{weak-}*$$

as $\varepsilon' \to 0$. $\qquad \qquad \square$

Throughout the whole book periodic homogenization will often be performed and the following theorem will play an important role.

Theorem 1.1.5. Let $Y = \prod\limits_{i=1}^{n} (a_i, b_i)$ and let $f \in L^p(Y)$, $1 \leq p \leq \infty$. Extend f by periodicity from Y to $\mathbb{R}^n$. Let $\Omega \subset \mathbb{R}^n$ be an open set and

$$f_\varepsilon(x) = f(x/\varepsilon) , \quad x \in \Omega , \quad \varepsilon = 1/m , \ m \in \mathbb{N} .$$

Then

$$f_\varepsilon \rightharpoonup \langle f \rangle = \frac{1}{|Y|} \int_Y f(y) dy \quad \text{in} \quad L^p(\Omega) , \quad \text{as} \quad \varepsilon \to 0$$

if $1 \le p < \infty$ and

$$f_\varepsilon \rightharpoonup \langle f \rangle \quad \text{in} \quad L^\infty(\Omega) \quad \text{weak-}* , \quad \text{as} \quad \varepsilon \to 0$$

if $p = \infty$. Here $|Y| = \text{meas } Y = \text{measure } Y$. $\qquad\square$

Let us provide two examples.

(a) *Riemann-Lebesgue lemma.* Let $Y = (0, 2\pi)$, $f(y) = \sin y$, and $\Omega \subset \mathbb{R}$. Then

$$\sin(x/\varepsilon) \rightharpoonup 0 \quad \text{in} \quad L^\infty(\Omega) \quad \text{weak-}* , \quad \text{as} \quad \varepsilon \to 0 .$$

(b) Let $Y = (0, 1), 0 < \lambda < 1, \alpha, \beta \in \mathbb{R}, \Omega \subset \mathbb{R}$, and

$$f(y) = \begin{cases} \alpha & \text{for} \quad y \in (0, \lambda) , \\ \beta & \text{for} \quad y \in (\lambda, 1) . \end{cases}$$

Then

$$f_\varepsilon(x) \rightharpoonup \lambda\alpha + (1 - \lambda)\beta \quad \text{in} \quad L^\infty(\Omega) \quad \text{weak-}* , \quad \text{as} \quad \varepsilon \to 0 . \qquad\square$$

Suppose now that $\Omega \subset \mathbb{R}^n$ is a domain and $f(x, y)$, $x \in \Omega$, $y \in Y$, is a Y-periodic function with respect to the second argument and sufficiently regular. Then

$$\lim_{\varepsilon \to 0} \int_\Omega f(x, x/\varepsilon) dx = \int_\Omega \langle f(x, y) \rangle dx , \tag{1.1.1}$$

where

$$g(x) := \langle f(x, y) \rangle = \frac{1}{|Y|} \int_Y f(x, y) dy .$$

1.1.2. Sobolev spaces and trace operators

Let $\Omega \subset \mathbb{R}^n$ be an open set and $f \in \mathbf{D}'(\Omega)$ a distribution on Ω. By $\alpha = (\alpha_1, \ldots, \alpha_n) \in \mathbb{N}^n$ we denote a multi-index and set

$$\partial^\alpha f = \frac{\partial^{|\alpha|} f}{\partial x_1^{\alpha_1} \ldots x_n^{\alpha_n}} = (\partial_1)^{\alpha_1} \ldots (\partial_n)^{\alpha_n} f , \tag{1.1.2}$$

where

$$|\alpha| = \alpha_1 + \ldots + \alpha_n \quad \text{(length of } \alpha) . \tag{1.1.3}$$

Here $\partial_k = \dfrac{\partial}{\partial x_k}$. For $\alpha = 2$ we write:

$$\partial_{ij} f = \frac{\partial^2 f}{\partial x_i \partial x_j} \quad (1 \le i, j \le n) .$$

Obviously, these notations are also used for regular functions on Ω.

We set $(1 \le p \le \infty)$

$$W^{m,p}(\Omega) = \{ v \in L^p(\Omega) | \partial^\alpha v \in L^p(\Omega) \quad \forall \alpha \in \mathbb{N}^n , |\alpha| \le m \} , \qquad (1.1.4)$$

where $\partial^\alpha v$ denotes the matrix of the αth derivative, in the sense of distributions, of the function v, that is

$$\forall \varphi \in \mathbf{D}(\Omega) , \quad \int_\Omega (\partial^\alpha v) \varphi dx = (-1)^{|\alpha|} \int_\Omega v \partial^\alpha \varphi dx , \qquad (1.1.5)$$

where $\mathbf{D}(\Omega)$ is the space of C^∞ functions with compact support in Ω. The space $W^{m,p}(\Omega)$ is endowed with the norm

$$\|v\|_{W^{m,p}(\Omega)} = \left(\sum_{|\alpha|=0}^m \|\partial^\alpha v\|_{L^p}^p \right)^{1/p} \quad \text{if} \quad 1 \le p < \infty , \qquad (1.1.6)$$

$$\|v\|_{W^{m,\infty}} = \max_{0 \le |\alpha| \le m} \{ \|\partial^\alpha v\|_{L^\infty} \} \quad \text{if} \quad p = \infty . \qquad (1.1.7)$$

Basic properties
(i) The space $W^{m,p}(\Omega)$ is a Banach space, separable if $1 \le p < \infty$ and reflexive if $1 < p < \infty$. The reflexivity is equivalent to weak compactness of the unit ball (and closed bounded sets).
(ii) If $1 \le p < \infty$, the $C^\infty(\Omega)$ functions are dense in $W^{m,p}(\Omega)$ endowed with its norm.
(iii) If Ω is bounded, then $W^{1,\infty}(\Omega)$ is the space of Lipschitz functions.
(iv) For $p = 2$ the space

$$W^{m,2}(\Omega) = H^m(\Omega) , \qquad (1.1.8)$$

is a Hilbert space with the scalar product

$$(u, v) \rightarrow \sum_{|\alpha| \le m} \int_\Omega (\partial^\alpha u) \partial^\alpha v dx . \qquad (1.1.9)$$

It is convenient to introduce the following notations:

$$\|v\|_{m,p,\Omega} = \|v\|_{W^{m,p}(\Omega)} , \quad 1 \le p \le \infty , \qquad (1.1.10)$$

$$||v||_{m,\Omega} = ||v||_{m,2,\Omega} \, . \tag{1.1.11}$$

For $1 \le p < \infty$ we introduce the spaces:

$$W_0^{m,p}(\Omega) = \overline{\mathbf{D}(\Omega)} \, , \tag{1.1.12}$$

where $\overline{\mathbf{D}(\Omega)}$ denotes the closure in $W^{m,p}(\Omega)$. For $p = 2$ we write:

$$W_0^{m,2}(\Omega) = H_0^m(\Omega) \, . \tag{1.1.13}$$

In general we have

$$W_0^{m,p}(\Omega) \subset W^{m,p}(\Omega)$$

and the inclusion is strict.

For $m \in \mathbb{N}$, by $H^{-m}(\Omega)$ we denote the dual space of the space $H_0^m(\Omega)$, endowed with the norm:

$$||f||_{-m,\Omega} = \sup\{|\langle f, v\rangle| : v \in H_0^m(\Omega) \, , ||v||_{m,\Omega} = 1\} \, . \tag{1.1.14}$$

It can be shown that each element $f \in H^{-m}(\Omega)$ can be written in the form:

$$f = \sum_{|\alpha| \le m} \partial^\alpha f_\alpha \, , \quad f_\alpha \in L^2(\Omega) \, , \quad |\alpha| \le m \, . \tag{1.1.15}$$

Such a representation is not uniquely determined.

The space $H^{-m}(\Omega)$ is a space of distributions. We observe that, in general, the dual space $(H^m(\Omega))^*$ of $H^m(\Omega)$ is not a space of distributions.

Regularity of the boundary $\Gamma = \partial\Omega$

From practical point of view the most important are domains $\Omega \subset \mathbb{R}^n$ of class $C^m(0 \le m \le \infty)$ or $C^{m,\alpha}$. Let $\mathbf{O}$ be an open set of $\mathbb{R}^n$. The functions of class $C^{m,\alpha}(\mathbf{O})(m \in \mathbb{N}, 0 < \alpha \le 1)$ are defined by:

$$v \in C^{m,\alpha}(\mathbf{O}) \Leftrightarrow \begin{cases} v \in C^m(\mathbf{O}) \, , \\ ||v||_{C^{m,\alpha}(\mathbf{O})} = ||v||_{m,\infty,\mathbf{O}} \\ \quad + \max_{|\beta|=m} \sup_{\substack{x,y \in \mathbf{O} \\ x \ne y}} \dfrac{|\partial^\beta v(x) - \partial^\beta v(y)|}{|x-y|^\alpha} < +\infty \, , \end{cases}$$

where $|\cdot|$ denotes the Euclidean norm in $\mathbb{R}^n$. The case $m = 0$, $\alpha = 1$ corresponds to Lipschitz functions. It is well known that $C^{0,1}(\mathbf{O})$ coincides with the space $W^{1,\infty}(\mathbf{O})$ of functions with bounded generalized derivatives.

Let us characterize more precisely the regularity of the boundary $\Gamma = \partial\Omega$ of $\Omega \subset \mathbb{R}^n$. To this end we assume that:

(i) Ω is bounded.

(ii) There exist $\alpha > 0$ and $\beta > 0$ and a finite number of maps $\mathbf{a}_r$, $1 \leq r \leq M$, associated with the local coordinate systems $(\mathbf{O}_r, x_1^r, \widehat{x}^r)$ such that

$$\Gamma = \bigcup_{r=1}^{M} \{(x_1^r, \widehat{x}^r) | x_1^r = \mathbf{a}_r(\widehat{x}^r), |\widehat{x}^r| < \alpha\}, \tag{1.1.16}$$

with the conditions:

$$\{(x_1^r, \widehat{x}^r) | \mathbf{a}_r(\widehat{x}^r) < x_1^r < \mathbf{a}_r(\widehat{x}^r) + \beta, |\widehat{x}^r| < \alpha\} \subset \Omega, 1 \leq r \leq M, \tag{1.1.17}$$

$$\{(x_1^r, \widehat{x}^r) | \mathbf{a}_r(\widehat{x}^r) - \beta < x_1^r < \mathbf{a}_r(\widehat{x}^r), |\widehat{x}^r| < \alpha\} \subset \mathbb{R}^n \backslash \overline{\Omega}, 1 \leq r \leq M. \tag{1.1.18}$$

Here $\widehat{x}^r = (x_2^r, \ldots, x_n^r)$ and the condition $|\widehat{x}^r| < \alpha$ has the following meaning: $|\widehat{x}_i^r| < \alpha, 2 \leq i \leq n$. The functions $\mathbf{a}_r$, $1 \leq r \leq M$ are of class C^m or $C^{m,\alpha}$ on the set Δ_r of points $\widehat{x}^r$ satisfying $|\widehat{x}_i^r| < \alpha$, $2 \leq i \leq n$, or $|\widehat{x}^r| \leq \alpha$.

Imbedding theorems and traces of functions on the boundary
Assume now that Ω is at least of class $C^{0,1}$.

Defining the space $C^\infty(\overline{\Omega})$ of restrictions to $\overline{\Omega}$ of the functions from the space $C^\infty(\mathbb{R}^n)$, we have the following density theorem:

$$\overline{C^\infty(\overline{\Omega})} = W^{m,p}(\Omega), \tag{1.1.19}$$

where the closure is to be taken in the sense of the space $W^{m,p}(\Omega)$.

We now quote the Sobolev and Rellich-Kondrashov imbedding theorems.

Theorem 1.1.6. Let $\Omega \subset \mathbb{R}^n$ be a bounded domain with Lipschitz boundary and let $1 \leq p \leq \infty$.
Case 1. If $1 \leq mp \leq n$, then

$$W^{m,p}(\Omega) \subset L^{p^*}(\Omega), \quad \frac{1}{p^*} = \frac{1}{p} - \frac{m}{n}$$

and the imbedding

$$W^{m,p}(\Omega) \subset L^q(\Omega)$$

is compact for every $1 \leq q < p^*$.
Case 2. If $mp = n$, then

$$W^{m,p}(\Omega) \subset L^q(\Omega) \quad \text{for every} \quad 1 \leq q < \infty$$

and the imbedding is compact.
Case 3. If $mp > n$, then

$$W^{m,p}(\Omega) \subset C(\overline{\Omega})$$

and the imbedding is compact.

Note that if a space $W^{1,p}$ is replaced by $W_0^{1,p}$, then no regularity of the boundary is required. If Ω is of class $C^{0,1}$, one can define the surface measure $d\Gamma$ on $\Gamma = \partial\Omega$ by

$$\int_\Gamma v\, d\Gamma = \sum_{r=1}^M \int_{\Delta_r} v(\widehat{x}^r, \mathbf{a}_r(\widehat{x}^r))\varphi_r(\widehat{x}^r, \mathbf{a}_r(\widehat{x}^r))\sqrt{1 + \sum_{i=1}^{n-1}(\partial_i \mathbf{a}_r(\widehat{x}^r))^2}\, d\widehat{x}^r \,, \quad (1.1.20)$$

where $\{\varphi_r\}_{r=1}^M$ is a partition of unity subordinate to the covering $\{\mathbf{a}_r(\Delta_r)\}_{r=1}^M$ of Γ. We observe that the partial derivatives $(\partial_i \mathbf{a}_r)$ of Lipschitz functions are in $L^\infty(\Delta_r)$. Consequently the r.h.s. of Eq. (1.1.20) is well defined.

The existence of a surface measure on Γ permits one to define the spaces $L^q(\Gamma)$, $1 \leq q \leq \infty$. We set

$$||v||_{0,q,\Gamma} = ||v||_{L^q(\Gamma)} \,. \qquad (1.1.21)$$

We are now in a position to formulate the first trace theorem.

Theorem 1.1.7. Let $\Omega \subset \mathbf{R}^n$ be a domain class of $C^{0,1}$.
Case 1. For $1 \leq p \leq n$, there exists a constant $K_p > 0$ such that:

$$\forall v \in C^\infty(\overline{\Omega}) \,, \quad ||v||_{0,q,\Gamma} \leq K_p||v||_{1,p,\Omega} \,, \quad q = \frac{(n-1)p}{n-p} \,. \qquad (1.1.22)$$

Case 2. For $p = n$ and $1 \leq q < \infty$, there exists a constant $K_q > 0$ such that:

$$\forall v \in C^\infty(\overline{\Omega}) \,, \quad ||v||_{0,q,\Gamma} \leq K_q||v||_{1,n,\Omega} \,. \qquad (1.1.23)$$

Case 3. For $p > n$, there exists a constant $K > 0$ such that:

$$\forall v \in C^\infty(\overline{\Omega}) \,, ||v||_{0,\infty,\Gamma} \leq K||v||_{1,p,\Omega} \,. \qquad (1.1.24)$$
$$\square$$

Corollary 1.1.8. By density result (1.1.19), the inequalities (1.1.22) - (1.1.24) remain valid for any $v \in W^{1,p}(\Omega)$. Consequently one can define the trace operator. For instance, if $p = 2$ then the trace mapping, denoted by tr, is an element of:

$$tr \in \mathbf{L}(H^1(\Omega), L^q(\Gamma)) \,,$$

(the space of linear and continuous operators mapping $H^1(\Omega)$ into $L^q(\Gamma)$) with

$$\left.\begin{array}{l} q = 2 + \dfrac{2}{n-2} > 2 \quad \text{for} \quad n > 2 \\[2mm] \forall q \quad \text{for} \quad n = 2 \,. \end{array}\right\} \qquad (1.1.25)$$
$$\square$$

Remark 1.1.9. In practice, we shall often use the notation $tr\, v = v_{|\Gamma}$ or even $trv = v$. $\quad\square$

Another surface measure, often used in variational analysis, is the so-called *Hausdorff measure*.

Definition 1.1.10. Let $O \subset \mathbb{R}^n$, $0 < k \leq \infty$ and $0 < \delta \leq \infty$. We set

$$H_k^\delta(O) = \frac{\omega_k}{2^k} \inf\{\sum_{j=1}^\infty (\text{diam } O_j)^k | O \subset \bigcup_{j=1}^\infty O_j , \text{ diam } O_j < \delta\} , \qquad (1.1.26)$$

and

$$H_k(O) = \lim_{\delta \to 0} H_k^\delta(O) = \sup_\delta H_k^\delta(O) , \qquad (1.1.27)$$

where ω_k is the measure of the unit ball in $\mathbb{R}^k$. The number $H_k(O)$ is called the k-dimensional Hausdorff measure of the set O. $\qquad\square$

In the specific case of $\Omega \subset \mathbb{R}^3$, $O = \partial\Omega$, $H_2(\partial\Omega)$ stands for the surface, 2-dimensional Hausdorff measure of the boundary $\partial\Omega$ of Ω.

If $\Omega \subset \mathbb{R}^n$ is of class $C^{0,1}$, the outward unit normal $n = (n_1, \ldots, n_n)$ is defined almost everywhere on $\Gamma = \partial\Omega$ for the surface measure $d\Gamma$. One can then define the normal derivative $\partial_n = \dfrac{\partial}{\partial n}$ for the functions of $C^\infty(\overline{\Omega})$ by:

$$v \in C^\infty(\overline{\Omega}) \implies \partial_n v = \sum_{i=1}^n n_i \partial_i v = \frac{\partial v}{\partial n} = \sum_{i=1}^n n_i \frac{\partial v}{\partial x_i} . \qquad (1.1.28)$$

Obviously, the last implication is also valid for $v \in C^1(\overline{\Omega})$. By using (1.1.25) we conclude that for $v = C^\infty(\overline{\Omega})$ one has

$$\left\|\frac{\partial v}{\partial n}\right\|_{0,q,\Gamma} \leq \sum_{i=1}^n \|\partial_i v\|_{0,q,\Gamma}, \quad 1 \leq q \leq \infty . \qquad (1.1.29)$$

On account of (1.1.19), (1.1.29) and applying Theorem 1.1.7 we infer that $\dfrac{\partial}{\partial n}$ is a continuous linear operator from the space $W^{2,p}(\Omega)$ into the space $L^q(\Gamma)$ with:

$$\begin{aligned}
q &= \frac{(n-1)p}{n-p} \quad \text{for} \quad 1 \leq p < \infty , \\
1 &\leq q < \infty \qquad \text{for} \quad p = n , \\
q &= \infty \qquad\quad \text{for} \quad p > n .
\end{aligned} \qquad (1.1.30)$$

Moreover, there exist positive constants K_p', K_q' and K', independent of $v \in W^{2,p}(\Omega)$ such that

$$\left\|\frac{\partial v}{\partial n}\right\|_{0,q,\Gamma} \begin{cases} \leq K_p'\|v\|_{2,p,\Omega} & \text{for} \quad 1 \leq p < n \quad \text{and} \quad q = \dfrac{(n-1)p}{n-p} , \\ \leq K_q'\|v\|_{2,n,\Omega} & \text{for} \quad p = n \quad \text{and} \quad 1 \leq q < \infty , \end{cases} \qquad (1.1.31)$$

$$\left\|\frac{\partial v}{\partial n}\right\|_{0,\infty,\Gamma} \leq K'\|v\|_{2,p,\Omega} \quad \text{for} \quad p > n, \tag{1.1.32}$$

for every $v \in W^{2,p}(\Omega)$.

By definition of the spaces $W_0^{m,p}(\Omega)$, it is evident that

$$W_0^{1,p}(\Omega) = \{v \in W^{1,p}(\Omega)|v = 0 \quad \text{on} \quad \Gamma\}, \tag{1.1.33}$$

$$W_0^{2,p}(\Omega) = \{v \in W^{2,p}(\Omega)|v = \frac{\partial v}{\partial n} = 0 \quad \text{on} \quad \Gamma\}. \tag{1.1.34}$$

Suppose now that the real numbers p and q satisfy:

$$\left. \begin{array}{l} 1 \leq p < n, 1 \leq q < n \quad \text{and} \quad \dfrac{1}{p} + \dfrac{1}{q} \leq 1 + \dfrac{1}{n}, \\[2mm] \text{or} \quad 1 < q \quad \text{and} \quad n \leq p, \\[1mm] \text{or} \quad 1 \leq q \quad \text{and} \quad n < p. \end{array} \right\} \tag{1.1.35}$$

Then for $u \in W^{1,p}(\Omega)$ and $v \in W^{1,q}(\Omega)$ we have the *Green formula*:

$$\int_\Omega v\partial_i u dx = -\int_\Omega u\partial_i v dx + \int_\Gamma uvn_i d\Gamma. \tag{1.1.36}$$

It can be verified that in all the three cases specified in (1.1.35), the product uv belongs to $L^1(\Gamma)$.

Fractional Sobolev spaces on Γ

Assume now that Ω is of class C^∞. For $0 < s < 1$ one can define the spaces $H^s(\Gamma)$ by:

$$u \in H^s(\Gamma) \Leftrightarrow \begin{cases} u \in L^s(\Gamma), \\ \displaystyle\int_\Gamma \int_\Gamma \frac{|u(x) - u(y)|^2}{|x - y|^{n-1+2s}} d\Gamma(x)d\Gamma(y) < +\infty. \end{cases} \tag{1.1.37}$$

The space $H^s(\Gamma)$ is a Hilbert space for the scalar product:

$$(u, v)_{H^s(\Gamma)} = \int_\Gamma uv d\Gamma + \int_\Gamma \int_\Gamma \frac{(u(x) - u(y))(v(x) - v(y))}{|x - y|^{n-1+2s}} d\Gamma(x)d\Gamma(y).$$

It can be shown that for $v \in H^1(\Omega)$, the trace mapping $tr \in \mathbf{L}(H^1(\Omega), L^q(\Gamma))$ with q as in (1.1.25), has the space $H^{1/2}(\Gamma)$ as the range. It means that
(i) $\forall v \in H^1(\Omega)$, $tr\ v \in H^{1/2}(\Gamma)$,
(ii) $\forall g \in H^1(\Gamma)$, $\exists\, v \in H^1(\Omega)$, $tr\ v = g$.

From (ii) it follows that

$$H^{1/2}(\Gamma) \subset L^q(\Gamma),$$

with q as in (1.1.25).

More precisely, the point (ii) means that there exist a lift operator $\mathbf{R}$ (not necessarily unique):

$$\mathbf{R} \in \mathbf{L}(H^{1/2}(\Gamma), H^1(\Omega)), \tag{1.1.38}$$

such that

$$\forall g \in H^{1/2}(\Gamma), tr(\mathbf{R}g) = g. \tag{1.1.39}$$

Similarly, we define:

$$\forall m \in \mathbb{N}\setminus\{0\}, \quad H^{m-\frac{1}{2}}(\Gamma) = \{tr\, v|\, v \in H^m(\Omega)\}. \tag{1.1.40}$$

Remark 1.1.11. The space $H^{m-\frac{1}{2}}(\Gamma)$, and more generally the space $H^s(\Gamma)$ $(s > 0)$, can be defined in an intrinsic manner. Then, however one has first to introduce the space $H^s(\mathbb{R}^n)$ by using the Fourier transformation for $s > 0$. $\qquad\square$

One can also define the following norm:

$$g \in H^{m-\frac{1}{2}}(\Gamma) \Longrightarrow ||g||_{m-\frac{1}{2},\Gamma} = \inf\{||v||_{m,\Omega}|\, v \in H^m(\Omega), tr\, v = g\}.$$

Equipped with this norm the space $H^{m-\frac{1}{2}}(\Gamma)$ is a Banach space. More precisely, the space $H^{m-\frac{1}{2}}(\Gamma)$ is a Hilbert space with the scalar product defined by

$$f, g \in H^{m-\frac{1}{2}}(\Gamma) \Longrightarrow (f, g)_{m-\frac{1}{2},\Gamma} = (u, v)_{m,\Omega}, \tag{1.1.41}$$

where u and v are the unique elements of the space $(H^m(\Omega) \cap H_0^1(\Omega))^\perp$ satisfying:

$$tr\, u = f, \quad tr\, v = g.$$

We recall that

$$H^{m-\frac{1}{2}}(\Gamma) = (H^m(\Omega) \cap H_0^1(\Omega))^\perp \quad (\text{in } H^m(\Omega)).$$

We set

$$H^\circ(\Gamma) = L^2(\Gamma). \tag{1.1.42}$$

Identifying the space $L^2(\Gamma)$ with its dual, we define the topological dual of $H^{m-\frac{1}{2}}(\Gamma)$:

$$H^{-m+\frac{1}{2}}(\Gamma) = (H^{m-\frac{1}{2}}(\Gamma))^*, \quad m \in \mathbb{N}\setminus\{0\}. \tag{1.1.43}$$

Equipped with the norm:

$$v \in H^{-m+\frac{1}{2}}(\Gamma) \Longrightarrow ||v||_{-m+\frac{1}{2},\Gamma}$$
$$= \max\{\langle v, \varphi\rangle|\varphi \in H^{m-\frac{1}{2}}(\Gamma), ||\varphi||_{m-\frac{1}{2},\Gamma} \le 1\}, \tag{1.1.44}$$

this space is a Banach space. In fact, the space $H^{-m+\frac{1}{2}}(\Gamma)$, the dual of a Hilbert space is also a Hilbert space since the norm (1.1.44) is generated by the scalar product:

$$u, v \in H^{-m+\frac{1}{2}}(\Gamma) \implies (u, v)_{-m+\frac{1}{2},\Gamma} = (\varphi, \psi)_{m-\frac{1}{2},\Gamma}, \qquad (1.1.45)$$

where, by the Riesz theorem, φ and ψ represent u and v in the space $H^{m-\frac{1}{2}}(\Gamma)$, respectively.

To avoid ambiguity, let us consider practically important case appearing in the study of mixed boundary conditions involving $\Gamma_0 \subset \Gamma$. For the sake of simplicity we take $m = 1$ and $p = 2$. By $H^{1/2}(\Gamma_0)$ we denote the usual fractional Sobolev space. The subspace consisting in the functions whose extension to Γ by zero lies in $H^{1/2}(\Gamma)$ is denoted by $H_{00}^{1/2}(\Gamma_0)$. The norm is taken as the graph norm of the extension by zero, which induces a finer topology than $H^{1/2}(\Gamma_0)$ norm. By $H^{-1/2}(\Gamma_0)$ we mean the normed dual of $H_{00}^{1/2}(\Gamma_0)$.

Theorem 1.1.12. (Riesz' representation theorem). Let X be a Hilbert space and f a bounded linear functional on X. Then there exists a *uniquely* determined element φ_f of X such that

$$f(u) = \langle f, u \rangle = (u, \varphi_f) \quad \text{for all } u \in X \text{ , and } \quad ||f|| = ||\varphi_f|| . \qquad (1.1.46)$$
$$\square$$

Three important inequalities

Theorem 1.1.13. (Poincaré's inequality). Let Ω be a bounded open set and $1 \le p < \infty$. There exists a constant $K = K(p, \Omega)$ such that

$$||v||_{m,p,\Omega} \le K \left(\sum_{|\alpha|=m} ||\partial^\alpha v||_{L^p}^p \right)^{1/p} , \qquad (1.1.47)$$

for every $v \in W_0^{m,p}(\Omega)$. Particularly, for $m = 1$ one has

$$||v||_{L^p} \le K ||\nabla u||_{L^p} , \qquad (1.1.48)$$

for every $v \in W_0^{1,p}(\Omega)$.

If $n = 1$, $\Omega = (0, 1)$ and $u \in H_0^1(0, 1)$, then

$$||v||_{L^2} \le \frac{1}{\pi} ||v'||_{L^2}$$

where $v' = \dfrac{dv}{dx}$. $\qquad\qquad\qquad\qquad\qquad\qquad\qquad\qquad\qquad \square$

Remark 1.1.14. Note that the constant $1/\pi$ is the best possible and it is attained whenever $v(x) = \sin \pi x$.

Theorem 1.1.15. (Poincaré-Wirtinger inequality). Let Ω be a connected domain of class C^1 and let $1 \le p < \infty$. There exists a constant $K > 0$ such that

$$||u - \langle u \rangle||_{L^p} \le K ||\nabla u||_{L^p} , \quad \langle u \rangle = \frac{1}{|\Omega|} \int_\Omega u\, dx , \qquad (1.1.49)$$

for any $u \in W_0^{1,p}(\Omega)$. $\qquad\qquad\qquad\qquad\qquad\qquad\qquad\qquad\qquad\qquad \square$

Let us pass now to *Korn's* inequality. Let Ω be a domain in $\mathbb{R}^3$. Define

$$\mathcal{R} = \{v \in H^1(\Omega)^3 | v = a + b \times x \, , x \in \Omega\} \, , \tag{1.1.50}$$

where $a, b \in \mathbb{R}^3$. The set $\mathcal{R}$ is a finite-dimensional (six-dimensional), closed linear subspace of $H^1(\Omega)^3 = [H^1(\Omega)]^3$.

Let V be a closed subspace of $H^1(\Omega)^3$ such that $H_0^1(\Omega)^3 \subset V \subset H^1(\Omega)^3$. Let $\mathcal{R}_V = \mathcal{R} \cap V$ and let $\mathcal{Q}_V$ be the orthogonal complement of $\mathcal{R}_V$ in V, i.e., $V = \mathcal{R}_V \oplus \mathcal{Q}_V$. The inequality

$$\sum_{i,j=1}^{3} \int_\Omega e_{ij}(v)e_{ij}(v)dx \geq K||v||_{1,\Omega}^2 \quad \text{for any } v \in \mathcal{Q}_V \, , \tag{1.1.51}$$

is called *Korn's inequality*. Here $K > 0$ is a constant independent of $v \in \mathcal{Q}_V$ and

$$e_{ij}(v) = v_{(i,j)} = \left(\frac{\partial v_i}{\partial x_j} + \frac{\partial v_j}{\partial x_i}\right)/2 \, . \tag{1.1.52}$$

The inequality

$$\sum_{i,j=1}^{3} \int_\Omega [e_{ij}(v)e_{ij}(v) + v_iv_i]dx \geq K_1||v||_{1,\Omega}^2 \quad \text{for any } H^1(\Omega)^3 \, , \tag{1.1.53}$$

defines the *coerciveness of strains*.

Let us provide an example of $\mathcal{Q}_V$. Let $\partial\Omega = \overline{\Gamma}_0 \cup \overline{\Gamma}_1$ with meas $\Gamma_0 > 0$. Then $\mathcal{R}_V = \{0\}$ and

$$V = \mathcal{Q}_V = \{v \in H^1(\Omega)^3 | v = 0 \text{ on } \Gamma_0\} \, . \tag{1.1.54}$$

Theorem 1.1.16. Let $\Omega \subset \mathbb{R}^3$ be a domain with a Lipschitz boundary. Then both the coerciveness of strains and Korn's inequality hold. $\square$

Remark 1.1.17. For more details on Lebesgue spaces the reader is referred to any standard textbook on functional analysis like Alexiewicz (1969), Edwards (1965) or Yosida (1978) as well as to more specialized books (Adams, 1975; Kufner et al., 1977). Sobolev spaces are investigated in Adams (1975), Brezis (1983), Kufner et al. (1977), cf. also Dautray and Lions (1990), Lions and Magenes (1968), Ciarlet and Rabier (Appendix, 1980). Theorem 1.1.5 is a slight modification of Dacorogna's Theorem 1.5 (1989, Chap. 2), who also provided a detailed proof, cf. also Appendix to the paper by Duvaut (1979) in the case of periodic L^2-functions. The properties of the space $H_{00}^{1/2}$ are studied by Lions and Magenes (1968).

The regularity of the boundary, including weaker assumptions and singular cases is studied by Adams (1975), cf. also Nečas (1976), Nazarov and Plamenevskii (1991) and Movchan and Movchan (1995). Definition 1.1.10 of the Hausdorff measure follows Giusti

(1984), cf. also Part II of the book by Morel and Solomini (1995). Riesz' Representation Theorem 1.1.12 is proved in standard textbooks on functional analysis, already mentioned.

For more details on Poincaré and Poincaré-Wirtinger inequalities the reader is referred to Brezis (1983) and Morrey (1966). Korn's inequality was proved by Duvaut and Lions (1976) and Nečas and Hlaváček (1981).

1.2. Elements of convex analysis and duality, minimization theorems, multivalued mappings

The present section is intended as a brief introduction to convex analysis and selected modern variational methods, including the theory of duality. For details the reader is referred to Castaing and Valadier (1977), Ekeland and Temam (1976), Hiriart-Urruty and Lemaréchal (1996), Ioffe and Tihomirov (1979), Moreau (1974), Rockafellar (1970), Rockafellar and Wets (1998), cf. also Panagiotopoulos (1993), Smith (1985), Struwe (1990). Topological vector space and locally convex spaces are expounded in Yosida (1978, Chap. I), cf. also Roubiček (1997). We recall that Banach spaces fall within this class of spaces.

1.2.1. Convex sets and functions

Let V be a real topological vector space and $f : V \to \overline{\mathbb{R}} = [-\infty, +\infty]$. Particularly, V may be a finite-dimensional space. In this specific case the strong and weak topologies coincide.

The *effective domain* of f is

$$dom\, f = \{v \in V | f(v) < \infty\} . \tag{1.2.1}$$

The *epigraph* of f is

$$epi\, f = \{(v, r) \in V \times \mathbb{R} | f(v) \leq r\} . \tag{1.2.2}$$

The set

$$[v_1, v_2] = \{v \in V | v = \lambda v_1 + (1 - \lambda)v_2 , \ \lambda \in [0, 1]\} ,$$

is said to be the *interval* joining the points v_1 and v_2. A subset C of the space V is said to be *convex* if it contains the interval joining any two of its points. The empty set is assumed to be convex by definition.

Let $C \subset V$. The intersection of all convex sets which contain C is a convex subset of the space V. This set is called the *convex hull* of the set C, and it is denoted by coC. The intersection of all closed convex sets which contain C is a closed convex set of V, which is called the *convex closure* of C and denoted by $\overline{co}C$.

Proposition 1.2.1. The closure of the convex hull of a set C coincides with its convex closure, $\overline{co}C = \overline{coC}$. $\square$

If $\{v_1, \ldots, v_k\}$ is a finite set of points of V, then every point $v \in V$ which can be represented in the form

$$v = \sum_{i=1}^{k} \lambda_i v_i \,,$$

where $\lambda_i \geq 0$, $i = 1, \ldots, k$, $\sum_i \lambda_i = 1$, is called a *convex combination* of the points $v_1, \ldots, v_k$.

Proposition 1.2.2. (Carathéodory's theorem). Let $C \subset \mathbb{R}^n$. Then every point of the set coC is a convex combination of no more than $n + 1$ distinct points of the set C. $\square$

A result from the theory of Hilbert spaces

Let V be a Hilbert space with the scalar product $(\cdot, \cdot)$, for instance the space $L^2(\Omega)$ or $L^2(\Omega)^n$. For an arbitrary set $C \subset V$ we denote

$$C^\perp = \{v | u \perp v \, \forall \, u \in C\} \,,$$

where $u \perp v$ means that $(u, v) = 0$. It is known that $C^\perp$ is a closed subspace of V.

To derive the homogenized complementary potential we shall frequently exploit the following result.

Proposition 1.2.3. Let $V_1, V_2 \subset V$ be closed subspaces of the Hilbert space V, then

$$(V_1 + V_2)^\perp = V_1^\perp \cap V_2^\perp \,.$$

Proof. Since $V_\alpha \subset V_1 + V_2 (\alpha = 1, 2)$ therefore $(V_1 + V_2)^\perp \subset V_\alpha^\perp$. Hence we conclude that

$$(V_1 + V_2)^\perp \subset V_1^\perp \cap V_2^\perp \,.$$

To obtain the inverse inclusion we take $u \in V_1^\perp \cap V_2^\perp$. Then $u \in V_\alpha^\perp (\alpha = 1, 2)$, which means that $u \perp V_\alpha$. Hence $u \perp V_1 + V_2$ and consequently $V_1^\perp \cap V_2^\perp \subset (V_1 + V_2)^\perp$. $\square$

We recall that $V_1 \oplus V_2$ denotes the *orthogonal* sum of subspaces V_1, V_2.

The function f is *convex* if for every $u, v \in V$, $\lambda \in [0, 1]$, $f[\lambda u + (1 - \lambda)v] \leq \lambda f(u) + (1 - \lambda)f(v)$ (with the convention $(+\infty) + (-\infty) = +\infty$). Convex functions assuming value $(-\infty)$ are very special.

Let $f : C \to \mathbb{R}$, $C \subset V$. With f we can associate the function $\overline{f}$ defined by

$$\overline{f}(v) = \begin{cases} f(v) & \text{if } v \in C \,, \\ +\infty & \text{if } v \notin C \,. \end{cases} \tag{1.2.3}$$

We observe that the function $\overline{f}$ is defined on the whole space V. This function is convex if and only if the set C is convex and the function f is convex on C. The convexity of f on C means that for each $u, v \in C$ the following inequality is satisfied:

$$f[\lambda u + (1 - \lambda)v] \leq \lambda f(u) + (1 - \lambda)f(v) \quad \forall \, \lambda \in [0, 1] \,. \tag{1.2.4}$$

Let C be a subset of V. The *indicator function* I_C is defined by:

$$I_C(v) = \begin{cases} 0 & \text{if } v \in C, \\ +\infty & \text{if } v \notin C. \end{cases} \qquad (1.2.5)$$

The set C is convex if and only if the function I_C is convex. In this way the study of convex sets can be reduced to the investigation of convex functions.

The inequality which follows is often used in proving homogenization theorems, see Dal Maso (1993).

Theorem 1.2.4. (Jensen's inequality). Let $\Omega \subset \mathbb{R}^n$ be a bounded open set, $u \in L^1(\Omega)$ and $f : \mathbb{R} \to \mathbb{R}$ be convex, then

$$f\left(\frac{1}{|\Omega|}\int_\Omega u(x)dx\right) \leq \frac{1}{|\Omega|}\int_\Omega f(u(x))dx . \qquad \square$$

Pedregal (1997) provides a general approach to Jensen's inequality in terms of parametrized measures, see also Sec. 21.6.

The next proposition will prove to be useful in the study of minimum compliance problems.

Proposition 1.2.5. If $\{f_i\}_{i\in\mathcal{I}}$ is a family of convex functions defined on a vector space V with values in $\overline{\mathbb{R}}$, then the function

$$f = \sup_{i\in\mathcal{I}} f_i$$

is also convex. $\qquad \square$

Let V be a Hausdorff locally convex topological vector space, for instance a Banach space or a finite-dimensional space. By V^* we denote the space of continuous linear functionals on V. This space is called the topological dual of V. The value of $v^* \in V^*$ on $v \in V$ will be denoted by $v^*(v)$ or $\langle v^*, v \rangle$. The triple $(V^*, V, \langle \cdot, \cdot \rangle)$ is an example of the so-called *dual pair*. Thus for each $v \in V$, $v \neq 0$, there exists $v^* \in V$ such that $\langle v^*, v \rangle \neq 0$ and for each $v^* \in V$, $v^* \neq 0$, there exists $v \in V$ such that $\langle v^*, v \rangle \neq 0$.

By $\sigma(V^*, V)$ we denote the weak topology on V^* generated by the duality between V and V^*. For instance, if $V = L^p(\Omega)$, then $V^* = L^q(\Omega)$ where $1/p + 1/q = 1$. Then the dual of $L^1(\Omega)$ is $L^\infty(\Omega)$ and vice versa. In contrast, the normed dual of $L^\infty(\Omega)$ is larger than $L^1(\Omega)$, cf. Rockafellar (1976), Yosida (1978).

Lower semicontinuity
A function $f : V \to \overline{\mathbb{R}}$ is *lower semicontinuous* on V (l.s.c.) if

$$\forall \bar{v} \in V \quad f(\bar{v}) \leq \liminf_{v \to \bar{v}} f(v) . \qquad (1.2.6)$$

For instance, the indicator function I_C of a set $C \subset V$ is l.s.c. if and only if C is closed.

A convex l.s.c. function is said to be *proper* if it is not the constant $+\infty$ and if it does not assume the value $-\infty$. The set of all such functions defined on V is denoted by $\Gamma_0(V)$.

Continuity of convex functions
The following theorem is very useful in proving homogenization theorems.

Theorem 1.2.6. Every l.s.c. convex function defined on a Banach space is continuous in interior points of its effective domain. $\quad\square$

In fact, this theorem holds in barrel spaces.

Polar functions
Let V be a topological vector space and $f : V \to \overline{\mathbb{R}}$. The function $f^* : V^* \to \overline{\mathbb{R}}$ defined by

$$f^*(v^*) = \sup\{\langle v^*, v\rangle - f(v)|v \in V\}$$

is called the *conjugate*, or *polar*, or *dual* function of f.
The function $f^{**} : V \to \overline{\mathbb{R}}$ defined by

$$f^{**}(v) = \sup\{\langle v^*, v\rangle - f^*(v^*)|v^* \in V^*\}$$

is called the *biconjugate* or *bipolar* of f.
The function $Cf : V \to \overline{\mathbb{R}}$ defined by

$$Cf = \sup\{g \le f|g \text{ convex}\}$$

is called the *(lower) convex envelope* of f.

Theorem 1.2.7. Let $f : V \to \mathbb{R} \cup \{+\infty\}$, then
(i) f^* is convex and lower semicontinuous.
(ii) If f is convex and lower semicontinuous, then $f^* \not\equiv +\infty$.
(iii) In general

$$f^{**} \le Cf \le f$$

and if f is convex and lower semicontinuous, then

$$f^{**} = Cf = f \ .$$

In particular if f takes only finite values then $f^{**} = Cf$.
(iv) In general

$$f^{***} = f^* \ .$$

(v) $f \le g$ implies $f^* \ge g^*$. $\quad\square$

Example 1.2.8. The function

$$I_C^*(v^*) = \sup\{\langle v^*, v\rangle | v \in C\}$$

is known as the *support function* of C. In the case of three-dimensional associated plasticity where $V = V^* = \mathbb{E}_s^3$ and $\mathbb{E}_s^3$ stands for the space of symmetric 3×3 matrices, the support function has clear physical meaning: it is the density of plastic dissipation provided that C stands for the elasticity convex, see Secs. 13 and 14.

Infimal-convolution

Let V be a vector space, f_1 and f_2 two functions from V to $\overline{\mathbb{R}}$. The *infimal-convolution* (or inf-convolution) is the function denoted by $f_1 \Box f_2$, from V to $\overline{\mathbb{R}}$, defined by

$$(f_1 \Box f_2)(v) = \inf\{f_1(v - u) \dotplus f_2(u) | u \in V\} . \tag{1.2.7}$$

Hence we conclude that

$$(f_1 \Box f_2)(v) = (f_2 \Box f_1)(v) = \inf\{f_1(v_1) \dotplus f_2(v_2) | v_1, \, v_2 \in V, \, v_1 + v_2 = v\} . \tag{1.2.8}$$

Examples 1.2.9.

1°. Let $C_\alpha \subset V$, $\alpha = 1, 2$. By I_{C_α} we denote their indicator functions. We calculate:

$$\begin{aligned}
(I_{C_1} \Box I_{C_2})(v) &= \inf\{I_{C_1}(v_1) + I_{C_2}(v_2) | v_1 + v_2 = v\} \\
&= \begin{cases} 0 & \text{if there exist } v_\alpha \in C_\alpha, \, \alpha = 1, 2 \text{ such that } v_1 + v_2 = v , \\ +\infty & \text{otherwise.} \end{cases}
\end{aligned}$$

Consequently we obtain

$$I_{C_1} \Box I_{C_2} = I_{(C_1 + C_2)} .$$

2°. Let $v \in V$ and

$$I_{\{v\}}(u) = \begin{cases} 0 & \text{if} \quad u = v , \\ +\infty & \text{otherwise.} \end{cases}$$

We find

$$(I_{\{v\}} \Box f)(u) = \inf\{I_{\{v\}}(v_1) + f(u - v_1) | v_1 \in V\} = f(u - v) .$$

3°. Let $C \subset V$ and $f : V \to \overline{\mathbb{R}}$. One has

$$(I_C \Box f)(v) = \inf\{I_C(v_1) + f(v - v_1) | v_1 \in V\} = \inf\{f(v - v_1) | v_1 \in C\} .$$

For instance, if V is a normed space and $f(v) = ||v||$, then

$$(I_C \Box || \cdot ||)(v) = \inf\{||v - v_1|| : v_1 \in C\} . \qquad \Box$$

Proposition 1.2.10. Let V be a vector space and $f_1, f_2 : V \to \overline{\mathbb{R}}$ convex functions. Then $f_1 \square f_2$ is also a convex function. $\qquad\square$

Theorem 1.2.11. Let V be a Hausdorff locally convex space and f_1 and f_2 two functions on V. Then

$$(f_1 \square f_2)^* = f_1^* + f_2^* , \qquad (1.2.9)$$

(with the convention that $(-\infty) + (+\infty) = -\infty$). Remark that f_α^* takes the value $(-\infty)$ if and only if f_α is the constant $(+\infty)$. $\qquad\square$

We shall say that the inf-convolution $f_1 \square f_2$ is *exact at* v, if $(f_1 \square f_2)(v) \in \mathbb{R}$ implies that there exist v_1 and v_2 such that $v_1 + v_2 = v$ and $f_1(v_1) + f_2(v_2) = (f_1 \square f_2)(v)$.

Theorem 1.2.12. Let V be a Hausdorff locally convex space and $f_1, f_2 \in \Gamma_0(V)$. Suppose there exists $v_0^* \in V^*$ such that f_1^* and f_2^* are finite at v_0^*, and that f_1^* is continuous at v_0^*. Then $f_1 \square f_2$ is exact and l.s.c. $\qquad\square$

Theorem 1.2.13. Let $f_1, f_2 \in \Gamma_0(V)$. Suppose there exists $v_0 \in V$ such that f_1 and f_2 are finite at v_0, and one of them (say f_1) is continuous at v_0. Then one has

$$(f_1 + f_2)^* = f_1^* \square f_2^* , \qquad (1.2.10)$$

and the inf-convolution $f_1^* \square f_2^*$ is exact. $\qquad\square$

Subdifferentiation

Let V be a topological vector space, $f : V \to \overline{\mathbb{R}}$, and $v_0 \in V$ such that $f(v_0) \in \mathbb{R}$. Then $v^* \in V^*$ is said to be a *subgradient* of f at v_0 if for every $v \in V$, $f(v) - f(v_0) \geq \langle v^*, v - v \rangle$.

The set of all subgradients of f at v_0 is called the *subdifferential* and denoted by $\partial f(v_0)$.

We observe that if $f : V \to (-\infty, +\infty]$ is not the constant $+\infty$, one can define a subgradient v^* at v_0 by the formula

$$\forall v , \quad f(v) \geq f(v_0) + \langle v^*, v - v_0 \rangle .$$

Then $f(v_0) = +\infty \Rightarrow \partial f(v_0) = \emptyset$.

Proposition 1.2.14. Let V be a Hausdorff locally convex space, $f : V \to \overline{\mathbb{R}}$, $v_0 \in V$ such that $f(v_0) \in \mathbb{R}$ and $v^* \in V^*$. Then the following properties are equivalent:
(i) $v^* \in \partial f(v_0)$,
(ii) $f^*(v^*) + f(v_0) = \langle v^*, v_0 \rangle$,
(iii) $f^*(v^*) + f(v_0) \leq \langle v^*, v_0 \rangle$.
Consequently $\partial f(v_0)$ is closed and convex. $\qquad\square$

Theorem 1.2.15. Let V be a Hausdorff locally convex space, f_1 and f_2 two convex l.s.c. proper functions, and $v_1 \in V$ such that f_1 and f_2 are finite at v_1 and f_1 is continuous at v_1.

Then for every $v \in V$,

$$\partial(f_1 + f_2)(v) = \partial f_1(v) + \partial f_2(v) \,. \qquad (1.2.11)$$

$\square$

We recall that for f_1, f_2 being only proper convex functions we have

$$\partial(f_1 + f_2)(v) \supset \partial f_1(v) + \partial f_2(v_2) \,.$$

Obviously at least for convex functions subdifferentiability is a generalization of differentiability. To this end we recall the definition of the Gâteaux derivative.

Let $f : V \to \overline{\mathbb{R}}$. The limit

$$\lim_{\lambda \to 0^+} \frac{f(u + \lambda v) - f(u)}{\lambda} \,,$$

if it exists, is called the derivative of f in the direction v and is denoted by $f'(u; v)$. If there exists $u^* \in V^*$ such that

$$\forall\, v \in V \,, \quad f'(u; v) = \langle u^*, v \rangle \,,$$

then f is said to be *Gâteaux differentiable* at u and u^* is called the *Gâteaux* derivative of f at u. Then we write $u^* = f'(u)$ or $u^* = Gf(u)$. In contrast to the subdifferential, which may consist of more than one elements, the Gâteaux derivative is unique. Deeper interrelationship is provided by the following proposition.

Proposition 1.2.16. Let $f : V \to \overline{\mathbb{R}}$ be a convex function. If f is Gâteaux differentiable at $u \in V$, then it is subdifferentiable at u and $\partial f(u) = \{f'(u)\}$. Inversely, if at $u \in V$ the function f is continuous, finite and possesses a unique subgradient, then f is Gâteaux differentiable at u and $\partial f(u) = \{f'(u)\}$. $\square$

The subdifferential as a multivalued mapping.

Let now $V = \mathbb{R}^n$ and let $f : \mathbb{R}^n \to \mathbb{R}$ be a convex function. Proposition 1.2.14 states that $\partial f(x)$ is a closed and convex set, thus $\partial f : x \to \partial f(x)$ is a *multifunction*.

Proposition 1.2.17. The subdifferential mapping is *monotone* in the sense that, for all x_1 and x_2 in $\mathbb{R}^n$

$$(x_2^* - x_1^*, x_2 - x_1) \geq 0 \quad \text{for all} \quad x_\alpha^* \in \partial f(x_\alpha) \,, \quad \alpha = 1, 2 \,. \qquad (1.2.12)$$

$\square$

In (1.2.12), (x^*, x) stands for the scalar product in $\mathbb{R}^n$.

In fact, we have more than monotonicity.

Proposition 1.2.18. The subdifferential mapping is *maximal monotone* in the sense that

$$(x^* - \eta, x - \xi) \geq 0 \quad \forall \eta \in \partial f(\xi) \,, \quad \text{implies} \quad x^* \in \partial f(x) \,. \qquad (1.2.13)$$

$\square$

Particularly, if f is differentiable, then $\eta = \dfrac{\partial f}{\partial \xi}$.

Remark 1.2.19. Propositions 1.2.17 and 1.2.18 can be extended to proper, lower semicontinuous functions defined on Hilbert spaces, cf. Brézis (1973, Chap. II).

1.2.2. Minimization theorems

Let (V, τ) be a topological space and let $f : V \to \overline{\mathbb{R}}$ be a function.

Definition 1.2.20. We say that f is τ-*coercive* if for every $t \in \mathbb{R}$ there exists a τ-compact (and τ-closed) subset $\mathcal{C}_t$ of V such that

$$\{v \in V \mid f(v) \le t\} \subset \mathcal{C}_t . \tag{1.2.14}$$

The next two propositions are due to Buttazzo (1989).

Proposition 1.2.21. Assume that
(i) f is τ-lower semicontinuous,
(ii) f is τ-coercive.
Then f admits a *minimum point* on V. $\qquad\qquad\square$

The last proposition is very general: the space V may be nonreflexive and the functional f nonconvex. We also observe that f may be a sum of two functionals: $f = f_1 + f_2$. Specifically, f_2 may be the indicator function of a set $C \subset V$. Then the minimization problem reads:

$$\left|\;\begin{aligned}&\text{find}\\&\inf\{f_1(v) + I_C(v)\mid v \in V\} = \inf\{f_1(v)\mid v \in C\} .\end{aligned}\right.$$

In dual Banach spaces, in particular in reflexive spaces, the coerciveness with respect to the weak-$*$ topology is phrased in the following proposition.

Proposition 1.2.22. Let $V = W^*$ be the dual of a Banach space W, and let τ be its weak-$*$ topology. Then a function $f : V \to \overline{\mathbb{R}}$ is τ-coercive if and only if

$$\lim_{\|v\|\to\infty} f(v) = +\infty . \tag{1.2.15}$$

Corollary 1.2.23. The condition (1.2.15) is equivalent to the following: there exists a function $\Phi : \mathbb{R} \to \mathbb{R}$ such that

$$\lim_{s\to+\infty} \Phi(s) = +\infty \quad \text{and} \quad f(v) \ge \Phi(\|v\|) . \qquad\qquad\square$$

We observe that the last two propositions apply to the following dual pairs: $(L^1(\Omega), L^\infty(\Omega),$ $\langle \cdot, \cdot \rangle_{L^1(\Omega)\times L^\infty(\Omega)})$ $(L^\infty(\Omega), L^1(\Omega), \langle \cdot, \cdot \rangle_{L^\infty(\Omega)\times L^1(\Omega)}),$
$(\mathbb{M}^1(\Omega), C_0(\Omega), \langle \cdot, \cdot \rangle_{\mathbb{M}^1(\Omega)\times C_0(\Omega)})$, where $\mathbb{M}^1(\Omega)$ stands for the space of bounded measures on Ω, see Sec. 13.1.

The proposition which follows is due to Cea (1971) and is, in fact, a corollary from Propositions 1.2.21 and 1.2.22.

Proposition 1.2.24. Let V be a reflexive Banach space and $f : V \to \mathbb{R}$ a weakly lower semicontinuous functional. If $C \subset V$ is a bounded and weakly closed set then there exists at least one minimizer of f in the set C. $\qquad\square$

The following proposition, due to Ekeland (1974, 1979, 1990), is useful in proving the existence of minimizing sequences, cf. also Ekeland and Temam (1976 Chap. I), Benaouda and Telega (1997).

Proposition 1.2.25. Let (V, d) be a complete metric space, and $f : V \to \mathbb{R} \cup \{+\infty\}$ a lower semicontinuous function, $\not\equiv \infty$, bounded from below. Let $\eta > 0$ be given, and a point $u \in V$ such that

$$f(u) \leq \inf_V f + \eta \ .$$

Then there exists some point $v \in V$ such that

$$f(v) \leq f(u) \leq \inf_V f + \eta \ , \tag{1.2.16}$$

$$d(u, v) \leq 1 \ , \tag{1.2.17}$$

$$\forall \, w \neq v \ , \quad f(w) > f(v) - \eta d(v, w) \ . \tag{1.2.18}$$

$$\square$$

Stronger version is formulated in the following form.

Proposition 1.2.26. Let (V, d) be a complete metric space, and $f : V \to \mathbb{R} \cup \{+\infty\}$ a l.s.c. function, $\not\equiv +\infty$, bounded from below. For any $\eta > 0$, there is some point $v \in V$ with

$$f(v) \leq \inf_V f + \eta \ , \tag{1.2.19}$$

$$\forall \, w \in V \ , \quad f(w) \geq f(v) - \eta d(v, w) \ . \tag{1.2.20}$$

$$\square$$

This relies on the fact that there always exists a point u with $f(u) \leq \inf_V f + \eta$. The inequality (1.2.19) then proceeds from (1.2.16) whilst (1.2.20) from (1.2.18).

The following theorem is also due to Ekeland (1974, 1979).

Theorem 1.2.27. Let f be a Gâteaux differentiable function on a Banach space V, bounded from below and satisfying the following condition:

(A) whenever $f'(u_n) \to 0$ in V^* and $\{f(u_n)\}$ is bounded, then either $f'(u_n) = 0$ for some n or the sequence $\{u_n\}$ has a cluster point in V.
Then the function f attains its minimum:

$$\exists \, \overline{u} \in V : f(\overline{u}) = \inf\{f(u) | u \in V\} \quad \text{and} \quad f'(\overline{u}) = 0 \ . \qquad\square$$

We recall that in the last theorem f' stands for the Gâteaux derivative of f.

For a large class of linear problems the following existence and uniqueness theorem is most appropriate, see Ciarlet (1988), Yosida (1978).

Theorem 1.2.28 (Lax-Milgram lemma). Let V be a Banach space with norm $\|\cdot\|$, let $L : V \to \mathbb{R}$ be a continuous linear form and let $a : V \times V \to \mathbb{R}$ be a symmetric and continuous bilinear form that is V-*elliptic (coercive)* in the sense that there exists a constant $K > 0$ such that

$$a(v,v) \geq K\|v\|^2 \quad \text{for all} \quad v \in V .$$

Then the problem:

$$\left|\begin{array}{l} \text{find} \quad u \in V \quad \text{such that} \\ \quad a(u,v) = L(v) \quad \text{for all} \quad v \in V , \end{array}\right.$$

has one and only one solution, which is also the unique solution of the equivalent minimization problem:

$$\left|\begin{array}{l} \text{find} \quad u \in V \quad \text{such that} \\ \quad J(u) = \inf\{J(v)|v \in V\} , \end{array}\right.$$

where

$$J(v) = \frac{1}{2}a(v,v) - L(v) . \qquad\qquad \square$$

A large class of unilateral contact problems *without friction* can be formulated in the form of the following *variational inequality*, cf. Telega (1987, 1988)

$$u \in C : \quad a(u, v-u) \geq \langle f, v-u \rangle \quad \text{for all} \quad v \in C \qquad (1.2.21)$$

where $a(u,v)$ is a bilinear form on a Hilbert space V, more precisely, $a : V \times V \to \mathbb{R}$ is continuous and linear in each of the variables u, v. The proof the theorem which follows is given in Kinderlehrer and Stampacchia (1980).

Theorem 1.2.29. Let $a(u,v)$ be a coercive bilinear form on V, $C \subset V$ closed and convex and $f \in V^*$. Then there exists a unique solutions to the problem (1.2.21). In addition, the mapping $f \to u$ is Lipschitz, that is, if u_1, u_2 are solutions to the problem (1.2.21) corresponding to $f_1, f_2 \in V^*$, then

$$\|u_1 - u_2\|_V \leq (1/K)\|f_1 - f_2\|_{V^*} . \qquad\qquad (1.2.22)$$
$$\square$$

We observe that the last theorem reduces to the Lax-Milgram lemma provided that $C = V$. The mapping $f \to u$ is linear if C is a subspace of V. If the bilinear form $a(u,v)$ is symmetric, i.e. $a(u,v) = a(v,u)$ for all $u,v \in V$, then the problem (1.2.21) is equivalent to the convex minimization problem

$$\frac{1}{2}a(u,u) - f(u) = \inf\{\frac{1}{2}a(v,v) - f(v)|v \in \mathcal{C}\}\,, \qquad (1.2.23)$$

where $f(v) = \langle f, v \rangle$.

Brezzi's theorem

Let now V and $\mathcal{S}$ be real Hilbert spaces, and let $a(\cdot,\cdot)$ and $b(\cdot,\cdot)$ be continuous bilinear forms on $\mathcal{S} \times \mathcal{S}$ and $\mathcal{S} \times V$ respectively. The search of the saddle point on $V \times \mathcal{S}$ of the functional

$$\mathcal{L}(v,\tau) = \frac{1}{2}a(\tau,\tau) + b(\tau,v) - \langle f,v \rangle - \langle g,\tau \rangle\,, \qquad (1.2.24)$$

is equivalent to the following problem, see Brezzi (1974)

$$\left|\begin{array}{lll}\text{Find } (u,\sigma) \in V \times \mathcal{S} & \text{such that} & \\ a(\sigma,\tau) + b(\tau,u) = \langle g,\tau \rangle & \text{for all } \tau \in \mathcal{S}\,, & (1.2.25) \\ b(\sigma,v) = \langle f,v \rangle & \text{for all } \tau \in V\,. & \end{array}\right.$$

Here f and g are given functions in V^* and $\mathcal{S}^*$ respectively.

Let $\mathcal{S}_0 = \{\tau \in \mathcal{S}|\; b(\tau,v) = 0 \text{ for all } v \in V\}$. One version of Brezzi's theorem was formulated by Arnold and Falk (1987).

Theorem 1.2.30. Suppose there is a constant $\gamma > 0$ such that

$$a(\tau,\tau) \geq \gamma \|\tau\|_{\mathcal{S}}^2 \quad \text{for all} \quad \tau \in \mathcal{S}\,,$$

and

$$\inf_{0 \neq v \in V} \sup_{0 \neq \tau \in \mathcal{S}} \frac{b(\tau,v)}{\|\tau\|_{\mathcal{S}}\|v\|_V} \geq \gamma\,.$$

Then for all $(f,g) \in V^* \times \mathcal{S}^*$, there is a unique solution $(u,\sigma) \in V \times \mathcal{S}$ of (1.2.24). Moreover,

$$\|u\|_V + \|\sigma\|_{\mathcal{S}} \leq K(\|f\|_{V^*} + \|g\|_{\mathcal{S}^*})\,,$$

where K depends only on γ and bounds for the bilinear forms $a(\cdot,\cdot)$ and $b(\cdot,\cdot)$. $\square$

Carathéodory function and the last general existence theorem

Let $\Omega \subset \mathbb{R}^n$ be an open set and let $f : \Omega \times \mathbb{R}^m \times \mathbb{R}^N \to \mathbb{R} \cup \{+\infty\}$. Then f is said to be a *Carathéodory function* if
(i) $f(x,\cdot,\cdot)$ is continuous for almost every $x \in \Omega$,
(ii) $f(\cdot,u,\xi)$ is measurable in x for every $(u,\xi) \in \mathbb{R}^m \times \mathbb{R}^N$.

Theorem 1.2.31. Let Ω be a bounded open set of $\mathbb{R}^n$ with Lipschitz boundary. Let $f :$ $\Omega \times \mathbb{R}^m \times \mathbb{R}^{nm} \to \mathbb{R} \cup \{+\infty\}$ be a Carathéodory function satisfying the coercivity condition

$$f(x, u, \xi) \geq \alpha(x) + \beta|\xi|^p ,$$

for almost every $x \in \Omega$, for every $(u, \xi) \in \mathbb{R}^m \times \mathbb{R}^{nm}$ and for some $\alpha \in L^1(\Omega)$, $\beta > 0$ and $p > 1$. Assume that $f(x, u, \cdot)$ is convex. Let

$$J(u) = \int\limits_{\Omega} f(x, u(x), \nabla u(x))dx .$$

Assume that there exists $\widetilde{u} \in u_0 + W_0^{1,p}(\Omega)^m$ such that $J(\widetilde{u}) < +\infty$, then

$$\inf\{J(u)|\ u \in u_0 + W_0^{1,p}(\Omega)^m\}$$

attains its minimum. $\hfill\square$

The proof of the last theorem can be found in Dacorogna (1989). For mixed boundary conditions where u_0 is prescribed on $\Gamma_0 \subset \partial\Omega$ with meas $\Gamma_0 > 0$ the minimization problem means evaluating

$$\inf\{J(u)|u \in W^{1,p}(\Omega)^m ,\ u = u_0 \quad \text{on} \quad \Gamma_0\} . \tag{1.2.26}$$

Under the assumptions of Theorem 1.2.31 the minimization problem (1.2.26) is solvable provided that $u_0 \in W^{1-\frac{1}{2},p}(\Gamma_0)$.

1.2.3.   Normal integrands, integral functionals and Rockafellar's theorem

Let us first recall the notion of a *Borel subset* in $\mathbb{R}^n$. Borel subsets are sets obtained from open and closed sets by performing the operation of taking countable unions, countable intersections, completion and any countable combination of these operations. A function f with values in $\overline{\mathbb{R}}$ is called the *Borel function* if $f^{-1}(F)$ is a Borel subset for any closed set F. For instance, continuous functions are Borel.

In the calculus of variations of great importance are *normal integrands*.

Definition 1.2.32. Let B be a Borel subset in $\mathbb{R}^m$. A function $f : \Omega \times B \to \overline{\mathbb{R}}$ is said to be a normal integrand if
(i) for almost every $x \in \Omega \subset \mathbb{R}^n$ the function $f(x, \cdot)$ is lower semicontinuous on B,
(ii) there is a Borel function $\widetilde{f} : \Omega \times B \to \overline{\mathbb{R}}$ such that $\widetilde{f}(x, \cdot) = f(x, \cdot)$ for almost every $x \in \Omega$.

It is convenient to call f a *proper integrand* if $f(x, \cdot)$ is proper for every $x \in \Omega$, i.e., if $f(x, \xi) > -\infty$ for all ξ and $f(x, \xi) \not\equiv +\infty$. Furthermore, f is said to be a *convex integrand* if $f(x, \xi)$ is convex in ξ for each $x \in \Omega$, i.e., if epi $f(x, \cdot)$ is convex-valued.

We observe that any Carathéodory function is a normal integrand. For more details the reader is referred to Ekeland and Temam (1976, Chap. VIII). An alternative definition can be found in Rockafellar (1976, p. 173), see also Ioffe and Tihomirov (1979). The definition given by Rockafellar involves the measurability of the *epigraph multifunction* epi $f : \Omega \to \mathbb{R}^{m+1}$ defined by

$$\text{epi } f(x, \cdot) = \{(\boldsymbol{\xi}, \alpha) \in R^m \times \mathbb{R} | f(x, \boldsymbol{\xi}) \leq \alpha\} \, .$$

Measurable multifunctions will be introduced in Section 1.2.6.

Let μ be a non-negative, σ-finite measure on $(\mathbf{O}, \mathbf{A})$ where $(\mathbf{O}, \mathbf{A}, \mu)$ is a measure space. For instance, we may have $\mu = dx$; if $\mathbf{O}$ stands for the boundary of a Lipschitz domain $\Omega \subset \mathbb{R}^n$ then μ denotes a surface measure, cf. Sec. 1.1.2.

For any normal integrand f on $\Omega \times \mathbb{R}^m$ and any measurable function $\boldsymbol{u} : \Omega \to \mathbb{R}^m$, we have $f(x, \boldsymbol{u}(x))$ measurable in x, and therefore the integral

$$J_f(\boldsymbol{u}) = \int_\Omega f(x, \boldsymbol{u}(x))dx \, , \tag{1.2.27}$$

has a well-defined value in $\overline{\mathbb{R}}$ under the following convention: if neither the positive nor the negative part of the function $x \to f(x, \boldsymbol{u}(x))$ is summable, we set $J(\boldsymbol{u}) = +\infty$. In particular, then,

$$J_f(\boldsymbol{u}) < +\infty \Rightarrow f(x, \boldsymbol{u}(x)) < +\infty \quad a.e.$$

J_f is called the *integral functional* associated with the integrand f. In practice, we are concerned with the restriction of J_f to a linear space X of measurable functions $\boldsymbol{u} : \Omega \to \mathbb{R}^m$. Obviously, J_f is a *convex functional* on X, if f is a *normal convex integrand*.

Among the linear spaces X of interest, besides the space of all measurable functions, are the various Lebesgue spaces, the space of constant functions, and in the case of topological or differentiable structure on Ω, spaces of continuous or differentiable functions. According to Rockafellar (1976), in their role in the theory of integral functionals these spaces fall into two different categories, distinguished by the presence or absence of a certain property of *decomposability*.

We shall say that X, a linear space of measurable functions $\boldsymbol{u} : \mathbf{O} \to \mathbb{R}^m$, is *decomposable* if $\mathbf{O}$ can be expressed as the union of a sequence of measurable subsets $\mathbf{O}_k (k = 1, 2, ...)$, such that for every $\mathbf{O}_k$ and bounded measurable function $\boldsymbol{u}_1 : \mathbf{O}_k \to \mathbb{R}^m$, and every $\boldsymbol{u}_2 \in X$, the (measurable) function

$$\boldsymbol{u}(x) = \begin{cases} \boldsymbol{u}_1(x) & \text{for} \quad x \in \mathbf{O}_k \, , \\ \boldsymbol{u}_2(x) & \text{for} \quad x \in \mathbf{O}/\mathbf{O}_k \, , \end{cases} \tag{1.2.28}$$

belongs to X. Since μ is σ-finite, the sets $\mathbf{O}_k$ can always be chosen with $\mu(\mathbf{O}_k)$ finite.

The space of all measurable functions, the Lebesgue spaces and Orlicz spaces, are all decomposable. However, the space of constant functions and spaces of continuous or differentiable functions furnish examples of nondecomposability.

The importance of the concept of decomposability results from the following celebrated theorem due to Rockafellar (1976).

Theorem 1.2.33. Let f be a normal integrand on $O \times \mathbb{R}^m$, and let X be a linear space of measurable functions $u : O \to \mathbb{R}^m$. For the relation

$$\inf_{u \in X} \int_O f(x, u(x))\mu(dx) = \int_O [\inf_{u \in \mathbb{R}^m} f(x, u)]\mu(dx) \qquad (1.2.29)$$

to hold, it is sufficient that X be decomposable and that the first infimum not be $+\infty$. (These conditions are superfluous in the case where X is the space of all measurable functions.) $\qquad \square$

We observe that a similar theorem can be formulated for a maximization problem, since $\sup \int f = -\inf \int (-f)$. In such form Rockafellar's theorem will be used in the case where X is a Lebesgue space.

An extension of this theorem to convex functionals defined on a space of measures will be given in Section 13.2.

Calculation of conjugate functionals

Let Ω be a bounded open set in $\mathbb{R}^n$, and f a *non-negative normal integrand* on $\Omega \times \mathbb{R}^m$. We define a non-negative functional by

$$J(u) = \int_\Omega f(x, u(x))dx ,$$

where $u \in L^p(\Omega)^m$, $1 \le p \le \infty$.

Proposition 1.2.34. Let J be a functional (1.2.30) defined on $L^p(\Omega)^m$, $1 \le p \le \infty$, and let $u_0 \in L^\infty(\Omega)^m$ be an element such that $J(u_0) < \infty$. Then for all $u^* \in L^q(\Omega)^m$ we have

$$J^*(u^*) = \int_\Omega f^*(x, u^*(x))dx . \qquad (1.2.30)$$

$$\square$$

We recall that $\dfrac{1}{p} + \dfrac{1}{q} = 1$, and if $p = 1$ then $q = \infty$ and vice versa. For the proof of the last proposition the reader is referred to the book by Ekeland and Temam (1976, Chap. IX).

1.2.4. Quasiconvexity and A-quasiconvexity

In optimal design problem the stored energy function may be nonconvex, cf. Chap. VI. Then the notion of quasiconvexity introduced by Morrey in 1952 proves very useful, cf. Morrey (1966). Here we follow Dacorogna (1982, 1989), cf. also Murat (1987).

Definition 1.2.35. A Borel measurable and locally integrable function $f : \mathbb{R}^{nm} \to \mathbb{R} -$ is said to be *quasiconvex* if

$$\int_D f(\boldsymbol{A} + \nabla\varphi(x))dx \geq |D|f(\boldsymbol{A}) , \qquad (1.2.31)$$

for every bounded domain $D \subset \mathbb{R}^n$, for every $\boldsymbol{A} \in \mathbb{R}^{nm}$ and for every $\varphi \in W_0^{1,\infty}(D)^m = [W_0^{1,\infty}(D)]^m$.

Suppose that $D -$ is a hypercube, φ is D-periodic (assumes equal values on the opposite faces of D) and set $\boldsymbol{\sigma} = \boldsymbol{A} + \nabla\varphi$. Then we readily get

$$\langle f(\boldsymbol{\sigma}) \rangle := \frac{1}{|D|}\int_D f(\boldsymbol{\sigma})dx \geq f(\langle \boldsymbol{\sigma} \rangle) , \qquad (1.2.32)$$

since

$$\langle \boldsymbol{\sigma} \rangle = \frac{1}{|D|}\int_D \boldsymbol{\sigma}dx = \frac{1}{|D|}\int_D (\boldsymbol{A} + \nabla\varphi)dx = \boldsymbol{A} + \frac{1}{|D|}\int_{\partial D} \varphi \otimes \boldsymbol{n}dx = \boldsymbol{A} .$$

Here $\boldsymbol{n} = (n_i)$ stands for the outward unit normal to D and

$$\int_D \varphi \otimes \boldsymbol{n}dx = 0 .$$

In the case of convex functions the inequality (1.2.31) is nothing but *Jensen's inequality*, see Th. 1.2.4.

We observe that the quasiconvexity condition (1.2.31) uses an infinite number of test functions φ and thus is by no means an easy condition to handle.

More generally, let $\Omega \subset \mathbb{R}^n$ be a bounded open set, and let $f : \Omega \times \mathbb{R}^m \times \mathbb{R}^{nm}$ be continuous and satisfy

$$|f(x, \boldsymbol{u}, \boldsymbol{A})| \leq a(x, |\boldsymbol{u}|, |\boldsymbol{A}|) , \qquad (1.2.33)$$

for every $(x, \boldsymbol{u}, \boldsymbol{A}) \in \Omega \times \mathbb{R}^m \times \mathbb{R}^{nm}$, where a is increasing with respect to $|\boldsymbol{u}|$ and $|\boldsymbol{A}|$ and locally integrable in x. Then the function f is said to be *quasiconvex* if

$$\int_D f(x_0, \boldsymbol{u}_0, \boldsymbol{A} + \nabla\varphi(z))dz \geq |D|f(x_0, \boldsymbol{u}_0, \boldsymbol{A}) , \qquad (1.2.34)$$

for every cube $D \subset \Omega$, for every $(x_0, \boldsymbol{u}_0, \boldsymbol{A}) \in \Omega \times \mathbb{R}^m \times \mathbb{R}^{nm}$ and for every $\varphi \in W_0^{1,\infty}(D)^m$.

The importance of the notion of quasiconvexity for the calculus of variations can be summarized in the form of the following result due to Morrey (1966), cf. also Dacorogna (1982, 1989).

Theorem 1.2.36. Let $f : \mathbb{R}^{nm} \to \mathbb{R}$ be continuous. A necessary and sufficient condition for f to be lower semicontinuous with respect to weak-$*$ convergence in $W^{1,\infty}(\Omega)^m$, i.e.,

$$\lim_{k \to \infty} \inf \int_D f(\nabla u^k(x))dx \geq \int_D f(\nabla u(x))dx \,, \tag{1.2.35}$$

whenever $u^k \rightharpoonup u$ weak-$*$ in $W^{1,\infty}(\Omega)^m$, is that f is quasiconvex. $\qquad\square$

Sverăk (1992) formulated a necessary and sufficient condition for quasiconvexity. Let us consider practically important case where $A \in \mathbb{E}^3$ is the space of 3×3 matrices. By $\mathbb{Z}$ we denote the set of integers.

Propositions 1.2.37. A continuous function $f : \mathbb{E}^3 \to \mathbb{R}$ is quasiconvex if and only if

$$\int_{[0,1]^3} f(A + \nabla\varphi(x))dx \geq f(A) \,, \tag{1.2.36}$$

for each $A \in \mathbb{E}^3$ and each smooth function $\varphi : \mathbb{R}^3 \to \mathbb{R}^3$ periodic with respect to $\mathbb{Z}^3$, i.e., such that for each $x \in \mathbb{R}^3$ and each $k \in \mathbb{Z}^3$ we have $\varphi(x + k) = \varphi(x)$. $\qquad\square$

Let us proceed to closely related notion of *A-quasiconvexity*, see Dacorogna (1982), Murat (1987). To this end we assume that

$$(\mathcal{H}) \left\{ \begin{array}{l} u^\nu \rightharpoonup u \quad \text{in} \quad L^\infty(\Omega)^m \quad \text{weak-}* \,, \\[2mm] (au^\nu)_i = \sum_{j=1}^{m} \sum_{k}^{n} a_{ijk}\dfrac{\partial u_j^\nu}{\partial x_k} \quad \text{bounded in} \quad L^2(\Omega)^q \ (i = 1, \ldots, q) \,, \\[2mm] f(u^\nu) \rightharpoonup l \quad \text{in} \quad L^\infty(\Omega) \quad \text{weak-}* \,, \end{array} \right.$$

where $a_{ijk} \in \mathbb{R}$ are constants.

Our aim now is to find for which f we have $l \geq f(u)$ provided that the hypothesis $(\mathcal{H})$ holds.

We observe that weak-$*$ convergence in L^∞ can be replaced by weak convergence in $L^p(p < \infty)$. However, in that case, in order to ensure that $f(u)$ is a distribution one has to impose a growth condition at infinity on f (e.g. $|f(\eta)| \leq \alpha + \beta|\eta|^p$ with $\beta > 0$) while in L^∞ the only requirement will be that f is continuous.

Definition 1.2.38. A function $f : \mathbb{R}^m \to \mathbb{R}$ is said to be *A-quasiconvex* if

$$\int_D f(b + \xi(x))dx \geq \int_D f(b)dx = |D|f(b) \,, \tag{1.2.37}$$

for every $b \in \mathbb{R}^m$, for every hypercube $D \subset \mathbb{R}^n$ and for every $\xi \in L(D)$ where

$$L(D) = \{\xi \in L^\infty(D)^m | \int_D \xi(x))dx = 0 \quad \text{and} \quad \xi \in Ker\,a\} \,,$$

(by $\xi \in Ker\,a$ we mean that $\displaystyle\sum_{j,k} a_{ijk}\frac{\partial \xi_j}{\partial x_k} = 0$).   □

Remark 1.2.39. In the Russian edition of the book by Dacorogna (1982) the set $L(D)$ is replaced by

$$L_1(D) = \{\xi \in L(D)|\ \xi \text{ is } D\text{-periodic}\}\,.$$

By D-periodicity of a function ξ is meant here the periodicity of ξ in $\mathbb{R}^n$ with the period D.

Theorem 1.2.40. (Necessary condition). Suppose that

$$\lim_{\nu\to\infty}\inf \int_\Omega f(u^\nu(x))dx \geq \int_\Omega f(u(x))dx\,, \tag{1.2.38}$$

holds (i.e., $l \geq f(u)$) for every sequence $\{u^\nu\}_{\nu\in\mathbb{N}}$ satisfying hypothesis $(\mathcal{H})$. Then f is A-quasiconvex.   □

Theorem 1.2.41. (Sufficiency condition). Suppose that $\{u^\nu\}_{\nu\in\mathbb{N}}, u$ satisfy hypothesis $(\mathcal{H})$ and

$$u^\nu - u \in Ker\,a\,. \tag{1.2.39}$$

If f is A-quasiconvex, then (1.2.38) holds for every bounded open set $\Omega \subset \mathbb{R}^n$.   □

Remark 1.2.42.
(i) A Borel measurable and locally integrable function $f : \mathbb{R}^{nm} \to \mathbb{R}$ is said to be *quasi-affine* if f and $-f$ are quasiconvex.
(ii) Examples of quasiconvex functions, which are not convex, are provided by Dacorogna (1989), Gibianskii and Cherkaev (1984) and Lurie and Cherkaev (1997). These authors also give examples of quasiaffine functions.
(iii) By Qf is denoted the *quasiconvex envelope* of f:

$$Qf = \sup\{g \leq f|g \text{ quasiconvex}\}.$$

We have

$$Cf \leq Qf \leq f\,,$$

where Cf is the convex envelope of f.

1.2.5. Elements of the duality theory

The stress approach to the analysis of solids and structures involves the complementary energy. Such an approach can be interrelated with the primal problem, formulated in terms of displacements, via the *duality theory*. The aim of this section is to present those features of this theory which will be exploited both in the asymptotic analysis and homogenization.

To expound the Rockafellar theory of duality we follow Ekeland and Temam (1976), see also Laurent (1972).

The *primal problem* means evaluating

$$(P) \qquad\qquad \inf\{J(u)|u \in V\}\,,$$

where V and V^* constitute a pair of dual linear topological spaces with canonical bilinear form $\langle \cdot, \cdot \rangle : V^* \times V \to \mathbb{R}$.

Let X and X^* be another pair of dual Hausdorff linear topological vector spaces with the canonical bilinear form $\langle \cdot, \cdot \rangle_{X^* \times X}$.

We introduce a function $\Phi : V \times X \to \overline{\mathbb{R}}$ such that

$$\Phi(u, 0) = J(u)\,. \qquad\qquad (1.2.40)$$

Consider the *perturbed minimization problem*

$$(P_p) \qquad\qquad \inf\{\Phi(u, p)|u \in V\}\,.$$

We observe that for $p = 0$ the problem P_0 coincides with P.

The *dual problem* means evaluating

$$(P^*) \qquad\qquad \sup\{-\Phi^*(0, p^*)|p^* \in X^*\}\,.$$

Proposition 1.2.43.

(i) $-\infty \le \sup P^* \le \inf P \le +\infty$.

(ii) If the problem P is nontrivial, then

$$\sup P^* \le \inf P < +\infty\,.$$

(iii) If the problem P^* is nontrivial, then

$$-\infty < \sup P^* \le \inf P\,.$$

(iv) If the problems P and P^* are nontrivial, then

$$-\infty < \sup P^* \le \inf P < +\infty\,. \qquad\qquad \square$$

Recalling that

$$\sup\{-\Phi^*(0, p^*)|p^* \in X^*\} = -\inf\{\Phi^*(0, p^*)|p^* \in X^*\}\,,$$

we can formulate the dual problem P^{**} of P^*:

$$(P^{**}) \quad \inf\{\Phi^{**}(u, 0)|u \in V\}\,.$$

Here Φ^{**} denotes the polar function of Φ^* or the so-called Γ-regularization of Φ, i.e., $\Phi^{**} \in \Gamma(V \times X)$. By $\Gamma(V \times X)$ we denote the space of *all* convex lower semicontinuous functions defined on $V \times X$. We recall that $\Gamma_0(V \times X)$ is the subset of proper functions belonging to $\Gamma(V \times X)$.

Since we always have $\Phi^{***} = \Phi^*$, therefore the dual problem P^{***} of P^{**} coincides with P^*.

The problem P^{**} coincides with $P(\Phi^{**}(u,0) = \Phi(u,0) \; \forall \, u \in V)$ provided that $\Phi^{**} = \Phi$, i.e., if

$$\Phi \in \Gamma_0(V \times X) \,. \tag{1.2.41}$$

Until the end of this section we assume that (1.2.41) is satisfied. For $p \in X$ we define the *marginal function* by

$$h(p) = \inf P_p = \inf\{\Phi(u,p) | u \in V\} \,. \tag{1.2.42}$$

Lemma 1.2.44. Under the assumption (1.2.41) the function $h : X \to \overline{\mathbb{R}}$ is convex. $\quad\square$

We observe that in general $h \notin \Gamma_0(X)$ though $\Phi \in \Gamma_0(V \times X)$.

Lemma 1.2.45.

(i) $\qquad\qquad \forall p^* \in X^* \qquad h^*(p^*) = \Phi^*(0, p^*).$

(ii) $\qquad\qquad \sup P^* = \sup\{-h^*(p^*) | p^* \in X^*\} = h^{**}(0).$ $\qquad\square$

Hence we conclude that the inequality $\sup P^* \leq \inf P$ is equivalent to $h^{**}(0) \leq h(0)$.

Definition 1.2.46.
(i) The problem P is called *normal* if h is finite and lower semicontinuous at zero.
(ii) The problem P is said to be *stable* if $h(0)$ is finite and h is subdifferentiable at zero.

Proposition 1.2.47. Under the assumption (1.2.41) the following conditions are equivalent:
(i) P and P^* are normal and possess solutions.
(ii) P and P^* are stable.
(iii) P is stable and has a solution. $\qquad\square$

The following result provides a stability criterion.

Proposition 1.2.48. Let Φ be a convex function, $\inf P < \infty$ and assume that

$$\begin{aligned} &\text{there exists } u_0 \in V \text{ such that} \\ &p \to \Phi(u_0, p) \text{ is finite and continuous at } 0 \in X \,. \end{aligned} \tag{1.2.43}$$

Then the problem P is stable. $\qquad\square$

Extremality relations

Proposition 1.2.49. If P and P^* possess solutions and if

$$-\infty < \inf P = \sup P^* < \infty \,, \tag{1.2.44}$$

then *all* solutions $\overline{u}$ of P and *all* solutions $\overline{p}^*$ of P^* are interrelated by the *extremality relation*:

$$\Phi(\overline{u}, 0) + \Phi^*(0, \overline{p}^*) = 0 , \qquad (1.2.45)$$

which is equivalent to

$$(0, \overline{p}^*) \in \partial\Phi(\overline{u}, 0) . \qquad (1.2.46)$$

Inversely, if $\overline{u} \in V$ and $\overline{p}^* \in X^*$ satisfy (1.2.45) then $\overline{u}$ is a solution to P and $\overline{p}^*$ is a solution to P and (1.2.44) is fulfilled. $\qquad\qquad\square$

Practically important specific case
Let $\Lambda : V \to X$ be a continuous linear operator, i.e., $\Lambda \in \mathbf{L}(V, X)$ and let Λ^* be the conjugate of Λ, $\Lambda^* \in \mathbf{L}(X^*, V^*)$ defined by, cf. Yosida (1978, Chap. VII)

$$\langle \Lambda v, x^* \rangle_{X \times X^*} = \langle v, \Lambda^* x^* \rangle_{V \times V^*} , \qquad (1.2.47)$$

for all $v \in D(\Lambda)$ and all $x^* \in D(\Lambda^*)$. Here $D(\Lambda)$ denotes the domain of the operator Λ; similarly $D(\Lambda^*)$ is the domain of Λ^*.

Let the functional J be given by

$$J(u) = G(\Lambda u) + F(u) . \qquad (1.2.48)$$

The perturbed functional $\Phi(u, p)$ may be assumed in the form

$$\Phi(u, p) = G(\Lambda u + p) + F(u) . \qquad (1.2.49)$$

It can easily be shown that the dual problem P^* means evaluating

$$\sup\{-G^*(p^*) - F^*(-\Lambda^* p^*) | p^* \in X^*\} . \qquad (1.2.50)$$

We observe that:
(i) if F and G are convex then Φ is also convex;
(ii) if $F \in \Gamma_0(V)$ and $G \in \Gamma_0(X)$ then $\Phi \in \Gamma_0(V \times X)$.
The condition (1.2.43) can be written as follows:

$$\begin{gathered} \text{there exists an element } u_0 \in V \text{ such that} \\ F(u_0) < +\infty , \ G(\Lambda u_0) < +\infty \text{ and } G \text{ is continuous at } \Lambda u_0 . \end{gathered} \qquad (1.2.51)$$

The extremality condition (1.2.45) yields

$$F(\overline{u}) + F^*(-\Lambda^* \overline{p}^*) = \langle -\Lambda^* \overline{p}^*, \overline{u} \rangle_{V^* \times V} , \qquad (1.2.52)$$

$$G(\Lambda \overline{u}) + G^*(\overline{p}^*) = \langle \overline{p}^*, \Lambda \overline{u} \rangle_{X^* \times X} . \qquad (1.2.53)$$

These relations are obviously equivalent to

$$-\Lambda^* \overline{p}^* \in \partial F(\overline{u}) , \qquad (1.2.54)$$

$$\overline{p}^* \in \partial G(\Lambda \overline{u}) . \qquad (1.2.55)$$

Remark 1.2.50. Ekeland and Temam (1976) assume the perturbed functional in the form

$$\Phi(u, p) = G(\Lambda u - p) + F(u) . \qquad (1.2.56)$$

1.2.6. Set-valued maps

As we already know, the subdifferential is a multivalued (set-valued) mapping. In plasticity, with a point x of a body is associated a so-called *elasticity convex*, being a closed and convex set of plastically admissible stresses (or moments in the case of plastic plates). Therefore in the present section we shall present basic notions related to set-valued mappings, cf. Rockaffellar (1976), Aubin and Cellina (1984), Aubin and Frankowska (1990).

Let X and Z be two sets. A *set-valued* map F from X to Z is a map that associates with any $x \in X$ a subset $F(x)$ of Z. The subsets $F(x)$ are the *images* or the *values* of F. Here we content ourselves with the case where $X = \Omega \subset \mathbb{R}^n$ and $(\Omega, \mathbf{A})$ is a measurable space. Also, the values $F(x)$ are in the finite-dimensional space $\mathbb{R}^m$, which usually will be identified with the space of symmetric $n \times n$ matrices. We set

$$dom\ F = \{x \in \Omega | F(x) \neq \emptyset\}\,,$$
$$gph\ F = \{(x, z) | z \in F(x)\}\,.$$

We shall denote by $F^{-1} : \mathbb{R}^m \to \Omega$ the multifunction given by

$$F^{-1}(z) = \{x \in \Omega | z \in F(x)\}\,.$$

A multifunction $F : \Omega \to \mathbb{R}^m$ is said to be a multivalued mapping with *closed-values* if $F(x)$ is a closed subset of $\mathbb{R}^m$ for every $x \in \Omega$. Such a multifunction is said to be *measurable*, relative to the σ-field $\mathbf{A}$, if for each closed set $C \subset \mathbb{R}^m$ the set $F^{-1}(C)$ is measurable, i.e., belongs to $\mathbf{A}$. Here

$$F^{-1}(C) = \bigcup_{z \in C} F^{-1}(z) = \{x \in \Omega | F(x) \cap C \neq \emptyset\}\,.$$

Proposition 1.2.51. For a closed-valued multifunction $F : \Omega \to \mathbb{R}^m$, the following properties are equivalent:
(a) F is measurable.
(b) $F^{-1}(C)$ is measurable for all open sets C.
(c) $F^{-1}(C)$ is measurable for all compact sets C.
(d) $F^{-1}(C)$ is measurable for all closed balls C.
(e)   dist $(z, F(x))$ is a measurable function of $x \in \Omega$ for each $z \in \mathbb{R}^m$, where
       dist $(z, F(x)| = \min\{|z - w|_{\mathbb{R}^m} : w \in F(x)\}$. $\square$

With a measurable closed-valued multifunction one can associate a function. Indeed, we have the following theorem.

Theorem 1.2.52. If $F : \Omega \to \mathbb{R}^m$ is a measurable closed-valued multimapping, there exists at least one measurable selection, i.e., a measurable function $f : dom\ F \to \mathbb{R}^m$ such that $f(x) \in F(x)$ for all $x \in dom\ F$. $\square$

The next issue is the problem of defining continuity of set-valued maps. In the case of single valued maps from Ω to $\mathbb{R}^m$, continuous functions are characterized by two equivalent properties:

(i) for any neighborhood $\mathcal{N}(f(x_0))$ of $f(x_0)$, there exists a neighborhood $\mathcal{N}(x_0)$ of x_0 such that $f(\mathcal{N}(x_0)) \subset \mathcal{N}(f(x_0))$;

(ii) for any sequence of elements $\{x^p\}_{p\in\mathbb{N}}$ converging to x_0, the sequence $f(x^p)$ converges to $f(x_0)$.

These two properties can be adapted to the case of set valued-maps from Ω to $\mathbb{R}^m$: they become:

(a) for any neighborhood $\mathcal{N}(F(x_0))$ of $F(x_0)$, there exists a neighborhood $\mathcal{N}(x_0)$ of x_0 such that $F(\mathcal{N}(x_0)) \subset \mathcal{N}(F(x_0))$;

(b) for any sequence of elements $\{x^p\}_{p\in\mathbb{N}}$ converging to x_0 and for any $z_0 \in F(x_0)$, there exists a sequence of elements $z^p \in F(x^p)$ that converges to z_0.

In the case of multivalued mappings, these two properties are no longer equivalent. We call *upper semicontinuous* maps those that satisfy property (a), *lower semicontinuous* maps are those that satisfy property (b). Obviously, continuous multivalued mappings are the ones that satisfy both properties (a) and (b).

The most famous continuous selection theorem is due to Michael, cf. Aubin and Cellina (1984).

Theorem 1.2.53. Let F from Ω into the closed convex subsets of $\mathbb{R}^m$ be lower semicontinuous. Then there exists $f : \Omega \to \mathbb{R}^m$, a continuous selection from F. $\qquad\qquad\square$

1.3. *Variational convergence of sequences of operators and functionals*

This section is intended as a brief introduction to the mathematical theory of homogenization. More precisely, we shall introduce the notion of G- and H-convergence of sequences of operators as well as the notion of Γ-convergence of sequences of functionals. With a sequence of functionals one can associate the sequence of conjugate functionals, which in our case will represent the functionals involved in the complementary energy principle. Thus naturally arises the problem of interrelationship between the Γ-convergence of the primal sequence of functionals and the Γ-convergence of the sequence of conjugate functionals. The dual homogenization will play an important role throughout the whole book. To derive effective models in the subsequent chapters of the book we shall often use the powerful method of two-scale asymptotic expansions. This method can be justified rigorously by using the method of two scale convergence. The books by Sanchez-Hubert and Sanchez-Palencia (1992, 1993) may serve as a good introduction to homogenization (and asymptotic methods), cf. also Persson et al. (1993).

As an application of the Γ-convergence theory we shall find the Γ-limit of a sequence of nonconvex functionals, yet convex with respect to the highest order derivatives. This case of nonuniform homogenization includes geometrically nonlinear elastic plates with a periodically nonuniform microstructure as well as geometrically linear and non-linear elastic shells.

To perform the homogenization of elastic-plastic plates we shall also need the notion of convergence in Kuratowski's sense of sequence of sets.

1.3.1. G-convergence

The notion of G-convergence was introduced by Spagnolo (1968), cf. also Attouch (1984), Dal Maso (1993) and Appendix A by Allaire to the book by Hornung (1997). The G-convergence is a notion of convergence associated with sequences of *symmetric*, second-order operators. The G means Green because this type of convergence corresponds to the convergence of inverse operators and thus to the convergence of the associated Green functions.

For the sake of simplicity, the notion of G-convergence is introduced in the case of the following equation:

$$\operatorname{div}\left(\boldsymbol{A}_\varepsilon \nabla u^\varepsilon\right) = f \quad \text{in} \quad \Omega\,,$$
$$u^\varepsilon = 0 \quad \text{on} \quad \partial\Omega\,. \tag{1.3.1}$$

Here $\Omega \subset \mathbb{R}^n$ is a bounded open set and $\varepsilon > 0$ a "small" parameter intended to tend to zero. More precisely, ε belongs to the set $\mathcal{E} = \{\varepsilon = 1/m \,|\, m \in \mathbb{N}\backslash\{0\}\}$. By $\mathcal{E}', \mathcal{E}'' \ldots$, we denote infinite subsequences of $\mathcal{E}$. The matrices $\boldsymbol{A}_\varepsilon$ belong to the following set of $n \times n$ symmetric matrices:

$$\mathcal{M}_s(\alpha,\beta,\Omega) = \{\boldsymbol{A}(x) \in L^\infty(\Omega, \mathbb{E}^n_s) : \ \alpha|\boldsymbol{\xi}|^2 \leq A(x)\boldsymbol{\xi}\cdot\boldsymbol{\xi} \leq \beta|\boldsymbol{\xi}|^2$$
$$\text{for any } \boldsymbol{\xi} \in \mathbb{R}^n \text{ and a.e. } x \text{ in } \Omega\}.$$

At this moment, we impose *no* periodicity assumption.

Definition 1.3.1. The sequence of matrices $\boldsymbol{A}_\varepsilon(x)$ is said to G-converge to a limit $\boldsymbol{A}_0(x)$, as ε tends to 0, if, for any right-hand side $f \in L^2(\Omega)$ in (1.3.1), the sequence of solutions u^ε converges weakly in $H_0^1(\Omega)$ to a limit u^0 which is the unique solution of the *homogenized* equation associated with $\boldsymbol{A}_0$:

$$-\operatorname{div}\left(\boldsymbol{A}_0 \nabla u^0\right) = f \quad \text{in} \quad \Omega\,,$$
$$u^0 = 0 \quad \text{on} \quad \partial\Omega\,. \tag{1.3.2}$$

The set $\mathcal{M}_s(\alpha,\beta,\Omega)$ is compact with respect to the Γ-convergence, as stated in the next theorem.

Theorem 1.3.2.
(i) If a sequence $\boldsymbol{A}_\varepsilon$ G-converges its G-limit is unique.
(ii) For any sequence $\boldsymbol{A}_\varepsilon$ in $\mathcal{M}_s(\alpha,\beta,\Omega)$, there exists a subsequence ε' and a homogenized limit $\boldsymbol{A}_0$, belonging to $\mathcal{M}_s(\alpha,\beta,\Omega)$ such that $\boldsymbol{A}_{\varepsilon'}$ G-converges to $\boldsymbol{A}_0$.
(iii) The G-limit of a sequence $\boldsymbol{A}_\varepsilon$ is independent of the term f and the boundary condition on $\partial\Omega$. $\qquad\qquad\square$

Similar properties will also hold for the H-convergence. The properties of H-limit will also be satisfied by the G-limit.

Remark 1.3.3. For a general study of G-convergence the reader is referred to the books by Attouch (1984) and Dal Maso (1993). These authors define G-convergence in terms of inverse operators. $\square$

1.3.2. H-convergence and the energy method

The H-convergence method was introduced by Murat (1977/78) and Tartar (1977). The abridged English version of these fundamental contributions has recently been published by Murat and Tartar (1997). We observe that "H" stems from "homogenization". The H-convergence is a generalization of the G-convergence to the case of nonsymmetric problems. Thus, symmetric problems, which are of main interest for us, are also covered by this theory.

As previously Ω is an open set of $\mathbb{R}^n$. By $\omega \subset\subset \Omega$ we denote a *bounded* open set of Ω such that $\overline{\omega} \subset \Omega$. By $\alpha, \beta, \alpha', \beta'$ we denote strictly positive real numbers such that $0 < \alpha < \beta < +\infty, 0 < \alpha' < \beta' < +\infty$. We set

$$\mathcal{M}(\alpha, \beta, \Omega) = \{A \in L^\infty(\Omega, \mathbb{E}^n) : \ \alpha|\xi|^2 \leq (A(x)\xi, \xi) \leq \beta|\xi|^2 \\ \text{for any } \xi \in \mathbb{R}^n \text{ and a.e. } x \text{ in } \Omega\} , \tag{1.3.3}$$

where $\mathbb{E}^n$ is the space of $n \times n$ matrices.

Definition of the H-convergence

Definition 1.3.4. A sequence A_ε, $\varepsilon \in \mathcal{E}$, of elements of $\mathcal{M}(\alpha, \beta, \Omega)$ H-converges to an element A_0 of $\mathcal{M}(\alpha,' \beta', \Omega)(A_\varepsilon \overset{H}{\rightharpoonup} A_0)$ if and only if, for any $\omega \subset\subset \Omega$ and any f in $H^{-1}(\omega)$ the solution u^ε of

$$-\text{div}\,(A_\varepsilon \nabla u^\varepsilon) = f \quad \text{in} \quad \omega\,, \\ u^\varepsilon \in H_0^1(\omega)\,, \tag{1.3.4}$$

is such that

$$u^\varepsilon \rightharpoonup u^0 \quad \text{weakly in} \quad H_0^1(\omega)\,, \\ A_\varepsilon \nabla u^\varepsilon \rightharpoonup A_0 \nabla u^0 \quad \text{weakly in} \quad L^2(\omega)^n\,, \tag{1.3.5}$$

for $\varepsilon \in \mathcal{E}$, where u^0 is the solution of

$$-\text{div}\,(A_0 \nabla u^0) = f \quad \text{in} \quad \omega\,, \\ u^0 \in H_0^1(\omega)\,. \tag{1.3.6}$$
$\square$

This definition can be extended to higher order elliptic equations. Let us consider the fourth-order plate equation[1] $(\Omega \subset \mathbb{R}^2)$:

$$(D_\varepsilon^{\alpha\beta\lambda\mu}(x)w_{,\lambda\mu}^\varepsilon)_{,\alpha\beta} = g \quad \text{in} \quad \Omega\,, \\ w \in H_0^2(\Omega)\,, \tag{1.3.7}$$

---

[1] Small Greek indices (except for ε) run over the set $\{1,2\}$. If repeated at different levels, they imply summation.

where $w_{,\alpha\beta} = \dfrac{\partial^2 w}{\partial x_\alpha \partial x_\beta}$. Here D_ε enjoys the usual symmetry property: $D_\varepsilon^{\alpha\beta\lambda\mu} = D_\varepsilon^{\lambda\mu\alpha\beta} = D_\varepsilon^{\beta\alpha\lambda\mu}$ and satisfies the following condition

$$K_1|\boldsymbol{\rho}|^2 \leq D_\varepsilon^{\alpha\beta\lambda\mu}(x)\rho_{\alpha\beta}\rho_{\lambda\mu} \leq K_2|\boldsymbol{\rho}|^2 , \tag{1.3.8}$$

for any $\rho \in \mathbb{E}_s^2$ and a.e. x in Ω ; here $|\boldsymbol{\rho}|^2 = \displaystyle\sum_{\alpha,\beta=1}^{2} \rho_{\alpha\beta}\rho_{\alpha\beta}$ and $g \in H^{-2}(\Omega)$. In this case H-convergence means that for any $\omega \subset\subset \Omega$ and g in $H^{-2}(\omega)$, the solution of

$$\begin{aligned}
(D_\varepsilon^{\alpha\beta\lambda\mu}(x)w_{,\lambda\mu}^\varepsilon)_{,\alpha\beta} &= g \quad \text{in} \quad \omega , \\
w &\in H_0^2(\omega) ,
\end{aligned} \tag{1.3.9}$$

is such that

$$\begin{aligned}
w^\varepsilon &\rightharpoonup w^0 \quad \text{weakly in} \quad H_0^2(\omega) , \\
\boldsymbol{D}_\varepsilon\nabla^2 w^\varepsilon &\rightharpoonup \boldsymbol{D}_0\nabla^2 w^0 \quad \text{weakly in} \quad L^2(\omega, \mathbb{E}_s^2) ,
\end{aligned} \tag{1.3.10}$$

as $\varepsilon \to 0$, where w^0 solves the *homogenized plate problem*:

$$\begin{aligned}
(D_0^{\alpha\beta\lambda\mu}(x)w_{,\lambda\mu}^0)_{,\alpha\beta} &= g \quad \text{in} \quad \omega , \\
w^0 &\in H_0^2(\omega) .
\end{aligned} \tag{1.3.11}$$

We observe that in the general case, the *homogenized coefficients* A_0^{ij}, $D_0^{\alpha\beta\lambda\mu}$ may still depend on the macroscopic variable $x \in \Omega$.

Example 1.3.5. Suppose that in $(1.3.7)_1$ only $D_\varepsilon^{1111}, D_\varepsilon^{2222}, D_\varepsilon^{1122} = D_\varepsilon^{2211}$ and $D_\varepsilon^{1212} = D_\varepsilon^{1221} = D_\varepsilon^{2112} = D_\varepsilon^{2121}$ are different from zero (orthotropic plate). We further assume that $D_\varepsilon^{\alpha\beta\lambda\mu} \in L^\infty(a,b)$ are functions of the variable x_1 only.

By w^0 we denote the solution to

$$\begin{aligned}
(D_0^{\alpha\beta\lambda\mu}w_{,\lambda\mu}^0)_{,\alpha\beta} &= g \quad \text{in} \quad \Omega , \\
w^0 &\in H_0^2(\Omega) ,
\end{aligned} \tag{1.3.12}$$

where

$$\begin{aligned}
(D_\varepsilon^{1111})^{-1} &\rightharpoonup (D_0^{1111})^{-1} , \\
D_\varepsilon^{1122}(D_\varepsilon^{1111})^{-1} &\rightharpoonup D_0^{1122}(D_0^{1111})^{-1} , \\
D_\varepsilon^{2222} - (D_\varepsilon^{1122})^2(D_\varepsilon^{1111})^{-1} &\rightharpoonup D_0^{2222} - (D_0^{1122})^2(D_0^{1111})^{-1} , \\
D_\varepsilon^{1212} &\rightharpoonup D_0^{1221} ,
\end{aligned} \tag{1.3.13}$$

weak-$*$ in $L^\infty(a,b)$ as $\varepsilon \to 0$.

We have, see Bonnetier and Vogelius (1987).

Proposition 1.3.6. Let $D_\varepsilon^{\alpha\beta\lambda\mu}$ and $D_0^{\alpha\beta\lambda\mu}$ denote orthotropic tensors such that (1.3.8) and (1.3.13) hold. If w^ε and $w^0 \in H_0^2(\Omega)$ denote the solutions to (1.3.7) and (1.3.12), respectively, then

$$w^\varepsilon \rightharpoonup w^0 \quad \text{weakly in} \quad H_0^2(\Omega) .$$

Proof. It is sufficient to prove that $w^{\varepsilon'} \rightharpoonup w^0$ for some subsequence of the original sequence. By using (1.3.8) we conclude that

$$||w^\varepsilon||_{H_0^2(\Omega)} \leq K||g||_{L^2(\Omega)} , \quad ||M^{\alpha\beta}||_{L^2(\Omega)} \leq K||g||_{L^2(\Omega)} ,$$

provided that $g \in L^2(\Omega)$. Here $M_\varepsilon^{\alpha\beta} = D_\varepsilon^{\alpha\beta\lambda\mu}\kappa_{\lambda\mu}(w^\varepsilon)$ stand for the components of the moment tensor and $\kappa_{\alpha\beta}(w^\varepsilon) = -w^\varepsilon_{,\alpha\beta}$. Consequently we may extract a subsequence ε' such that

$$\begin{aligned} w^{\varepsilon'} &\rightharpoonup w && \text{in} \quad H_0^2(\Omega) \quad \text{and} \\ M_{\varepsilon'}^{\alpha\beta} &\rightharpoonup M^{\alpha\beta} && \text{in} \quad L^2(\Omega) . \end{aligned} \tag{1.3.14}$$

Moreover we have $M^{\alpha\beta}{}_{,\alpha\beta} = \dfrac{\partial^2 M^{\alpha\beta}}{\partial x_\alpha \partial x_\beta} = g$. We shall now verify that

$$M^{\alpha\beta} = D_0^{\alpha\beta\lambda\mu}\kappa_{\lambda\mu}(w) \quad \text{in} \quad \Omega , \tag{1.3.15}$$

from which it follows that $w = w^0$ (the unique solution to (1.3.12)).
(i) Let $\alpha = 1, \beta = 2$. The definition of $M_{\varepsilon'}^{12}$ and the fact that $D_{\varepsilon'}^{\alpha\beta\lambda\mu}$ is independent of x_2 yields

$$M_{\varepsilon'}^{12} = \partial_2\left(2D_{\varepsilon'}^{1212}\partial_1(-w^{\varepsilon'})\right) . \tag{1.3.16}$$

Recalling that the imbedding $H_0^2(\Omega) \to H^1(\Omega)$ is compact, from $(1.3.14)_1$ we infer that

$$\partial_1 w^{\varepsilon'} \to \partial_1 w \quad \text{in} \quad L^2(\Omega) . \tag{1.3.17}$$

Combining it with $(1.3.13)_4$ we obtain

$$D_{\varepsilon'}^{1212}\partial_1 w^{\varepsilon'} \rightharpoonup D_0^{1212}\partial_1 w \quad \text{in} \quad L^2(\Omega) , \tag{1.3.18}$$

and consequently

$$\partial_2(2D_{\varepsilon'}^{1212}\partial_1(-w^{\varepsilon'})) \rightharpoonup \partial_2(2D_0^{1212}\partial_1(-w)) \quad \text{in} \quad H^{-1}(\Omega) .$$

Passing to the limit in (1.3.16) we get

$$M^{12} = 2D_0^{1212}\kappa_{12}(w) ,$$

as desired.

(ii) The cases $\alpha = \beta = 1$ and $\alpha = \beta = 2$ are more involved.

The local property of the H-limit implies that it suffices to prove (1.3.15) for any rectangle $R = (a_1, b_1) \times (a_2, b_2)$ contained in Ω, see below. Since

$$\partial_{11} M_{\varepsilon'}^{11} = g - \partial_{\alpha\beta} M_{\varepsilon'}^{\alpha\beta} = g - 2\partial_{12} M_{\varepsilon'}^{12} - \partial_{22} M_{\varepsilon'}^{22} \,,$$

it follows that

$$\partial_1 M_{\varepsilon'}^{11} = \int_{a_1}^{x_1} g\, dx - 2\partial_2 M_{\varepsilon'}^{12} - \partial_{22} \int_{a_1}^{x_1} M_{\varepsilon'}^{22}\, dx + k^{\varepsilon'}(x_2) \quad \text{in} \quad R\,. \tag{1.3.19}$$

The first three terms on the right-hand side of (1.3.19) are bounded in the space

$$L^2((a_1, b_1); H^{-2}(a_2, b_2))\,.$$

Next, integration of (1.3.19) with respect to x_1 gives $M_{\varepsilon'}^{11}$, and since these are bounded in $L^2(R)$, it follows that $k^{\varepsilon'}(x_2)$ are bounded in $H^{-2}(a_2, b_2)$. The relation (1.3.19) thus implies that $\partial_1 M_{\varepsilon'}^{11}$ are bounded in $L^2((a_1, b_1); H^{-2}(a_2, b_2))$. We conclude that $M_{\varepsilon'}^{11}$ are bounded in the space

$$S(R) = \{M \in L^2(R)|\ \partial_1 M \in L^2((a_1, b_1); H^{-2}(a_2, b_2))\}\,, \tag{1.3.20}$$

equipped with the natural norm.

By using Theorem (5.1) in Lions (1969) we conclude that $S(R)$ is compactly imbedded in $L^2((a_1, b_1); H^{-2}(a_2, b_2))$. Consequently we have

$$M_{\varepsilon'}^{11} \to M^{11} \quad \text{in} \quad L^2[(a_1, b_1); H^{-2}(a_2, b_2)]\,. \tag{1.3.21}$$

Further we have

$$(D_{\varepsilon''}^{1111})^{-1} M_{11}^{\varepsilon''} = \kappa_{11}(w^{\varepsilon''}) + \partial_{22}[D_{\varepsilon''}^{1122}(D_{\varepsilon''}^{1111})^{-1}(-w^{\varepsilon''})]\,, \tag{1.3.22}$$

and

$$M_{\varepsilon''}^{22} = D_{\varepsilon''}^{1122}(D_{\varepsilon''}^{1111})^{-1} M_{\varepsilon''}^{11} + \partial_{22}[(D_{\varepsilon''}^{2222} - (D_{\varepsilon''}^{1122})^2(D_{\varepsilon''}^{1111})^{-1})(-w^{\varepsilon''})]\,. \tag{1.3.23}$$

We recall that $D_{\varepsilon}^{2212} = 0$.

By virtue of (1.3.13) and (1.3.21) we get

$$(D_{\varepsilon''}^{1111})^{-1} M_{\varepsilon''}^{11} \rightharpoonup (D_0^{1111})^{-1} M^{11}$$

and

$$D_{\varepsilon''}^{1122}(D_{\varepsilon''}^{1111})^{-1} M_{\varepsilon''}^{11} \rightharpoonup D_0^{1122}(D_0^{1111})^{-1} M^{11} \tag{1.3.24}$$

in $L^2((a_1, b_1); H^{-2}(a_2, b_2))$.

Due to (1.3.13) and (1.3.14) we get

$$
\begin{aligned}
D_{\varepsilon''}^{1122}(D_{\varepsilon''}^{1111})^{-1}w^{\varepsilon''} &\rightharpoonup D_0^{1122}(D_0^{1111})^{-1}w \,, \\
[D_{\varepsilon''}^{2222} - (D_{\varepsilon''}^{1122})^2(D_{\varepsilon''}^{1111})^{-1}]\,w^{\varepsilon''} &\rightharpoonup [D_0^{2222} - (D_0^{1122})^2(D_0^{1111})^{-1}]w
\end{aligned}
\tag{1.3.25}
$$

in $L^2(R)$. A combination of (1.3.22) – (1.3.25) now leads to

$$
(D_0^{1111})^{-1}M^{11} = \partial_{11}(-w) + \partial_{22}[D_0^{1122}(D_0^{1111})^{-1}(-w)] \,,
\tag{1.3.26}
$$

and

$$
M^{22} = D_0^{1122}(D_0^{1111})^{-1}M^{11} + \partial_{22}[(D_0^{2222} - (D_0^{1122})^2(D_0^{1111})^{-1})(-w)] \,.
\tag{1.3.27}
$$

Multiplying (1.3.26) by D_0^{1111} we obtain the expression for M^{11}. Next, substituting (1.3.26) into (1.3.27) we obtain M^{22}. Recalling that $D_0^{\alpha\beta\lambda\mu}$ are independent of x_2, we get the desired constitutive relation for M^{11} and M^{22}. $\square$

Properties of the H-convergence and H-limit

Similarly to the G-convergence, H-convergence means, in essence, the convergence of the inverse operators $[-\mathrm{div}\,(A_\varepsilon\,\mathrm{grad}\,)]^{-1}$, which are bounded linear operators from $H^{-1}(\Omega)$ to $H_0^1(\Omega)$, when both spaces $H^{-1}(\Omega)$ and $H_0^1(\Omega)$ are equipped with their weak topologies. The underlying topology satisfies the property of *uniqueness of the H-limit* and the H-limit is *local* as the following proposition shows.

Proposition 1.3.7.
(i) A sequence A_ε , $\varepsilon \in \mathcal{E}$, of elements of $\mathcal{M}(\alpha, \beta, \Omega)$ has at most one H-limit.
(ii) Let A_ε and B_ε , $\varepsilon \in \mathcal{E}$, be two sequences in $\mathcal{M}(\alpha, \beta, \Omega)$ that satisfy

$$
A_\varepsilon \overset{H}{\rightharpoonup} A_0 \,, \quad B_\varepsilon \overset{H}{\rightharpoonup} B_0 \,,
$$

and are such that $A_\varepsilon = B_\varepsilon$ on an open set $\omega \subset \Omega$. Then $A_0 = B_0$ on ω. $\square$

Lemma 1.3.8 (compensated compactness). Let Ω be an open subset of $\mathbb{R}^n$ and $q^\varepsilon, v^\varepsilon, \varepsilon \in \mathcal{E}$, such that

$$
\begin{cases}
q^\varepsilon \in L^2(\Omega)^n \,, \\
q^\varepsilon \rightharpoonup q^0 \quad \text{weakly in} \quad L^2(\Omega)^n \,, \\
\mathrm{div}\, q^\varepsilon \to \mathrm{div} q^0 \quad \text{strongly in} \quad H^{-1}(\Omega) \,,
\end{cases}
$$

$$
\begin{cases}
v^\varepsilon \in H^1(\Omega) \,, \\
v^\varepsilon \rightharpoonup v^0 \quad \text{weakly in} \quad H^1(\Omega) \,.
\end{cases}
$$

Then

$$
(q^\varepsilon, \nabla v^\varepsilon) \rightharpoonup (q^0, \nabla v^0) \quad \text{weak-}* \quad \text{in} \quad \mathbf{D}'(\Omega) \,,
$$

i.e., for any $\varphi \in C_0^\infty(\Omega)$ we have

$$\int_\Omega (\boldsymbol{q}^\varepsilon, \nabla v^\varepsilon)\varphi dx \to \int_\Omega (\boldsymbol{q}^0, \nabla v^0)\varphi dx \,.$$

Here $(\cdot, \cdot)$ stands for the scalar product in $\mathbb{R}^n$. $\square$

We observe that the product $(\boldsymbol{q}^\varepsilon, \nabla v^\varepsilon)$ is that of two weakly and not strongly convergent sequences. This phenomenon is known as *compensated compactness*. Applying the last lemma we formulate the next one.

Lemma 1.3.9. Let Ω be an open subset of $\mathbb{R}^n$. Let $\boldsymbol{A}_\varepsilon$ belong to $\mathcal{M}(\alpha, \beta, \Omega)$ for $\varepsilon \in \mathcal{E}$. Assume that, for $\varepsilon \in \mathcal{E}$,

$$\begin{cases} u^\varepsilon \in H^1(\Omega) \,, \\ u^\varepsilon \rightharpoonup u^0 \quad \text{weakly in} \quad H^1(\Omega) \,, \\ \boldsymbol{q}^\varepsilon = \boldsymbol{A}_\varepsilon \nabla u^\varepsilon \rightharpoonup \boldsymbol{q}^0 \quad \text{weakly in} \quad L^2(\Omega)^n \,, \\ -\mathrm{div}\,(\boldsymbol{A}_\varepsilon \nabla u^\varepsilon) \to \mathrm{div}\boldsymbol{q}^0 \quad \text{strongly in} \quad H^{-1}(\Omega) \,, \end{cases}$$

$$\begin{cases} v^\varepsilon \in H^1(\Omega) \,, \\ v^\varepsilon \rightharpoonup v^0 \quad \text{weakly in} \quad H^1(\Omega) \,, \\ \boldsymbol{\eta}^\varepsilon = \boldsymbol{A}_\varepsilon^T \nabla v^\varepsilon \rightharpoonup \boldsymbol{\eta}^0 \quad \text{weakly in} \quad L^2(\Omega)^n \,, \\ -\mathrm{div}\,(\boldsymbol{A}_\varepsilon^T \nabla v^\varepsilon) \to -\mathrm{div}\boldsymbol{\eta}^0 \quad \text{strongly in} \quad H^{-1}(\Omega) \,. \end{cases}$$

Then

$$(\boldsymbol{q}^0, \nabla v^0) = (\nabla u^0, \boldsymbol{\eta}^0) \quad \text{a.e. in} \quad \Omega \,.$$ $\square$

Here $\boldsymbol{A}_\varepsilon^T$ stands for the transpose of the matrix $\boldsymbol{A}_\varepsilon$. Obviously, for symmetric matrices $\boldsymbol{A}_\varepsilon^T = \boldsymbol{A}_\varepsilon$.

The following two theorems are of primal importance.

Theorem 1.3.10. Assume that $\boldsymbol{A}_\varepsilon$, $\varepsilon \in \mathcal{E}$, belong to $\mathcal{M}(\alpha, \beta, \Omega)$ and H-converges to $\boldsymbol{A}_0 \in \mathcal{M}(\alpha', \beta', \Omega)$. Assume that

$$\begin{cases} u^\varepsilon \in H^1(\Omega) \,, \\ f^\varepsilon \in H^{-1}(\Omega) \,. \\ -\mathrm{div}\,(\boldsymbol{A}_\varepsilon \nabla u^\varepsilon) = f^\varepsilon \quad \text{in} \quad \Omega \,, \\ u^\varepsilon \rightharpoonup u^0 \quad \text{weakly in} \quad H^1(\Omega) \,, \\ f^\varepsilon \to f^0 \quad \text{strongly in} \quad H^{-1}(\Omega) \,, \end{cases}$$

for $\varepsilon \in \mathcal{E}$. Then

$$\boldsymbol{A}_\varepsilon \nabla u^\varepsilon \rightharpoonup \boldsymbol{A}_0 \nabla u^0 \quad \text{weakly in} \quad L^2(\Omega)^n \,,$$
$$(\boldsymbol{A}_\varepsilon \nabla u^\varepsilon, \nabla u^\varepsilon) \rightharpoonup (\boldsymbol{A}_0 \nabla u^0, \nabla u^0) \quad \text{weak-}* \quad \text{in} \quad \mathbf{D}'(\Omega) \,.$$

As previously, $(\cdot, \cdot)$ denotes the scalar product in $\mathbb{R}^n$. $\square$

Remark 1.3.11.
(i) Actually, it can be shown that the energy $(A_\varepsilon \nabla u^\varepsilon, \nabla u^\varepsilon)$ converges weakly in $L^1_{loc}(\Omega)$.
(ii) The boundary conditions have no influence on the H-limit.

Theorem 1.3.12. Let $A_\varepsilon, \varepsilon \in \mathcal{E}$, belong to $\mathcal{M}(\alpha, \beta, \Omega)$. There exists a subset $\mathcal{E}'$ of $\mathcal{E}$ and a matrix $A_0 \in \mathcal{M}(\alpha, \dfrac{\beta^2}{\alpha}, \Omega)$ such that A_ε H-converges to A_0 for $\varepsilon \in \mathcal{E}'$. $\qquad\square$

This theorem shows the sequential compactness of $\mathcal{M}(\alpha, \beta, \Omega)$ for the topology induced by the H-convergence.

Remark 1.3.13. By introducing the so-called *corrector matrix* it is possible to approximate ∇u^ε in the strong topology of a suitable space, cf. Murat (1977/78), Tartar (1977), Murat and Tartar (1997).

Remark 1.3.14. The *energy method* is a constructive proof for the compactness theorem of H-convergence, cf. Murat and Tartar (1997), Allaire (1997). This method, attributed to Tartar, has nothing to do with any kind of energy. It is sometimes more appropriately called the *oscillating test function method*, but it is most commonly referred to as the energy method.

Denseness of periodic composites
Dal Maso and Kohn (1991) provided a general characterization of H-limits and their approximation by periodic composites. At the moment of writing the book this important result has not yet been published. Therefore, we reproduce here the seminar given by Kohn during the 1st Workshop on Composite Media and Homogenization Theory in 1991, cf. also Francfort and Milton (1987). These results are limited to the scalar case. However, they can immediately be extended to the vector case (linear elasticity). The results which will now be presented are important, for instance, in optimal design problems.

Consider a mixture of two isotropic materials with conductivities a_1 and a_2, respectively. The volume fractions are θ and $1 - \theta$. We set

$$A_\varepsilon(x) = a_1 \chi_\varepsilon(x) + a_2(1 - \chi_\varepsilon(x)), \quad a.e. \ x \in \Omega, \tag{1.3.28}$$

where $\chi_\varepsilon(x)$ stands for the characteristic function of the first phase. Here $\Omega \subset \mathbb{R}^n$ is a fixed domain.

Let us introduce now the family of divergence operators

$$L_\varepsilon u = \mathrm{div}(A_\varepsilon \nabla u), \tag{1.3.29}$$

and consider all possible H-limits $A_0(x)$

$$A_\varepsilon I \overset{H}{\rightharpoonup} A_0(x). \tag{1.3.30}$$

Two fundamental problems arise naturally:

(1) characterize *all* possible H-limits $\{A_0(x)\} = \mathcal{L}$ (the notation A_h is preserved for periodic or non-uniformly periodic homogenization).

(2) Characterize *all* possible limits $\{A_0(x)\} = \mathcal{L}_{v_1}$ provided that vol $\{x \in \Omega | A_\varepsilon(x) = a_1\} = v_1$.

The answer to the first problem reads

$$\mathcal{L} = \{A(x) | A(x) \in G \text{ a.e.} \quad x \in \Omega\},$$

where G is a closed set of tensors. In fact, G is a closure of effective moduli of *periodic* composites, called G-closure of a_1 and a_2.

The answer to the second problem is

$$\mathcal{L}_{v_1} = \{A(x) | A(x) \in G_{\theta(x)} \text{ a.e. } x \in \Omega\},$$

where G_θ is a closed set of tensors and $0 \leq \theta(x) \leq 1, \displaystyle\int_\Omega \theta(x)dx = v_1$. G_θ is called the G_θ-closure of a_1 and a_2. More precisely, G_θ = the closure of effective moduli of *periodic* composites with the volume fraction θ. The last function is obviously the weak-$*$ limit of the sequence of the characteristic functions $\{\chi_\varepsilon\}_{\varepsilon>0}$.

Let us pass to the proof of the above assertions. First, however, we recall a result from real variables analysis, cf. Dal Maso (1993).

Lemma 1.3.15. Consider a set $S \subset L^1(\mathbb{R}^n, \mathbb{R}^m)$ such that it is:

(a) translation invariant or

$$f \in S \Rightarrow f(\cdot + a) \in S \qquad \forall\, a,$$

(b) closed under strong L^1 convergence on compact sets,

(c) S is *decomposable*, i.e. for any Borel set B, and $f_1, f_2 \in S$ the function

$$\widetilde{f}(x) = \begin{cases} f_1(x) & \text{on} \quad B, \\ f_2(x) & \text{on} \quad \mathbb{R}^n \setminus B, \end{cases}$$

also belongs to S. Then

$$S = \{f | f(x) \in \mathcal{S} \text{ a.e.}\}$$

for some *closed* subset $\mathcal{S} \subset \mathbb{R}^m$ and $\mathcal{S} = \{$ constant elements of $S\}$. $\square$

Let us return to the first problem. Now we have

$$S = H \text{ - closure of } \{A_\varepsilon(x)I\}. \tag{1.3.31}$$

The local character of H-convergence implies that we may assume that $A_\varepsilon(x)$ is defined on $\mathbb{R}^n$. The set S, given by (1.3.31), is translation invariant, closed for the strong L^1 convergence on compact sets and decomposable since H-convergence is local. To prove periodicity we take $A_0 \in G$. By the previous step there exist $A_\varepsilon(x)I \overset{H}{\to} A_0$. Consider now A_ε restricted to the unit cube Y and then extended by periodicity. Let us find the effective conductivity of periodic composites with this fine scale structure:

$$(\bar{A}_\varepsilon \boldsymbol{\xi}, \boldsymbol{\xi}) = \inf\{ \int_Y A_\varepsilon(y)(\boldsymbol{\xi} + \nabla\varphi, \boldsymbol{\xi} + \nabla\varphi) dy | \varphi \in H^1_{per}(Y) \} \, ,$$

where $Y = (0,1)^n$, $\boldsymbol{\xi} \in \mathbb{R}^n$ and

$$H^1_{per}(Y) = \{ v \in H^1(Y)) | v \text{ is } Y\text{-periodic} \} \, .$$

We claim that $\bar{A}_\varepsilon \to A_0$ as $\varepsilon \to 0$. Indeed, $(\bar{A}_\varepsilon \boldsymbol{\xi}, \boldsymbol{\xi})$ is determined by solving cell problems:

$$\operatorname{div}[A_\varepsilon(y)(\boldsymbol{\xi} + \nabla\varphi_\varepsilon)] = 0 \, , \quad \varphi_\varepsilon\text{-periodic}.$$

Then φ_ε tends to a solution associated to A_0, say φ_0, as $\varepsilon \to 0$. Moreover, we have

$$\lim_{\varepsilon \to 0} (\bar{A}_\varepsilon \boldsymbol{\xi}, \boldsymbol{\xi}) = \lim_{\varepsilon \to 0} [\text{energy of cell problem for } A_\varepsilon I]$$

$$= \text{energy of cell problem associated to } \{(H\text{-limit of } A_\varepsilon I) = A_0\} \, .$$

To justify the answer to the second problem we take

$$S = \text{ closure of } \{A_\varepsilon(x)I \, , \, \chi_\varepsilon(x)\} \, ,$$

in the topology of $(H\text{-convergence}) \times \sigma(L^\infty(\Omega), L^1(\Omega))$. This set is translation invariant, closed for the strong topology and decomposable. Now we have

$$\mathcal{S} = \{(A_0, \theta) = \text{closure of associated set for periodic composites}\} \, ,$$

and

$$G_\theta = \text{slice of } \mathcal{S} \text{ at given } \theta \, .$$

1.3.3. Two-scale convergence

Two-scale convergence was introduced by Nguesteng (1989) and developed by Allaire (1992). In the present section we shall expound the essential points of this method. To start with we observe that the two-scale convergence method is confined to periodic homogenization problems. The method of *two-scale asymptotic expansions* will often be used in our book. This is a formal method which enables us to find the homogenized problem. Let ε denote the size of periodic heterogeneities, ε is a small parameter which tends to zero in the asymptotic process. Assume that the sequence of solutions of the considered partial differential equation with microperiodically oscillating coefficients is denoted by u^ε. The two-scale asymptotic expansion is an *ansatz* of the form:

$$u^\varepsilon(x) = u^{(0)}\left(x, \frac{x}{\varepsilon}\right) + \varepsilon u^{(1)}\left(x, \frac{x}{\varepsilon}\right) + \varepsilon^2 u^{(2)}\left(x, \frac{x}{\varepsilon}\right) + \ldots \tag{1.3.32}$$

where each function $u^{(i)}(x,y)(y = \dfrac{x}{\varepsilon})$ in this series depends on two variables: the macroscopic (or slow) variable x and the microscopic (or fast) variable $y \in Y$. Here Y is a so-called basic cell. Substituting (1.3.32) into the equation satisfied by u^ε and identifying powers of ε leads to a chain of equations for each term $u^{(i)}(x,y)$. Next, averaging with respect to y yields the homogenized equation for $u^{(0)}$. From the mathematical point of view, the method of two-scale asymptotic expansions is only formal since, a priori, there is no reason for the ansatz (1.3.32) to hold true. Hence the need for a second step, a rigorous justification of the homogenized problem. The two-scale convergence combines these two steps into a single one. One may say that the two-scale convergence is a *rigorous* justification of the first term of the ansatz (1.3.32) for any bounded sequence u^ε.

Let us pass to the definition. To this end we introduce the space $C_{per}^\infty(Y)$ of infinitely differentiable function in $\mathbb{R}^n$ which are periodic of period Y. By $\mathbf{D}(\Omega; C_{per}^\infty(Y))$, we denote the space of infinitely differentiable functions with compact supports in Ω and with values in the space $C_{per}^\infty(Y)$.

Definition 1.3.16. A sequence of functions u^ε in $L^2(\Omega)$ is said to two-scale converge to a limit $u^{(0)}(x,y)$ belonging to $L^2(\Omega \times Y)$ if, for any function $\varphi(x,y)$ in $\mathbf{D}(\Omega; C_{per}^\infty(Y))$, it satisfies:

$$\lim_{\varepsilon \to 0} \int_\Omega u^\varepsilon(x)\varphi\left(x, \frac{x}{\varepsilon}\right) dx = \frac{1}{|Y|} \int_\Omega \int_Y u^{(0)}(x,y)\varphi(x,y)dxdy . \qquad (1.3.33)$$

$\square$

The compactness theorem which follows, justifies this notion of *two-scale convergence.*

Theorem 1.3.17. From each bounded sequence u^ε in $L^2(\Omega)$, one can extract a subsequence, and there exists a limit $u^{(0)}(x,y) \in L^2(\Omega \times Y)$ such that this subsequence two-scale converges to $u^{(0)}$. $\square$

The next two propositions provide more information on the two-scale limit.

Proposition 1.3.18. Let u^ε be a sequence of functions in $L^2(\Omega)$ which two-scale converges to a limit $u^{(0)} \in L^2(\Omega \times Y)$.
(i) Then, u^ε converges weakly in $L^2(\Omega)$ to

$$u(x) = \langle u^{(0)}(x,y)\rangle , \qquad (1.3.34)$$

where $\langle \cdot \rangle$ means averaging over Y and

$$\lim_{\varepsilon \to 0} ||u^\varepsilon||^2_{L^2(\Omega)} \geq ||u^{(0)}||^2_{L^2(\Omega \times Y)} \geq ||u||^2_{L^2(\Omega)} . \qquad (1.3.35)$$

(ii) Assume, further, that $u^{(0)}(x,y)$ is smooth and that

$$\lim_{\varepsilon \to 0} ||u^\varepsilon||^2_{L^2(\Omega)} = ||u^{(0)}||^2_{L^2(\Omega \times Y)} . \qquad (1.3.36)$$

Then

$$||u^\varepsilon(x) - u^{(0)}\left(x, \frac{x}{\varepsilon}\right)||^2_{L^2(\Omega)} \to 0 \quad \text{as } \varepsilon \to 0 . \qquad (1.3.37)$$

$\square$

Until now only bounded sequences in $L^2(\Omega)$ have been considered. The second proposition examines the case of a bounded sequence in $H^1(\Omega)$.

Proposition 1.3.19. Let u^ε be a bounded sequence in $H^1(\Omega)$. Then, up to a subsequence, u^ε two-scale converges to a limit $u \in H^1(\Omega)$, and ∇u^ε two-scale converges to $\nabla_x u(x) + \nabla_y u^{(1)}(x, y)$, where the function $u^{(1)}(x, y)$ belongs to $L^2(\Omega, H^1_{per}(Y)/\mathbb{R})$. $\square$

As an exercise, the reader is advised to study the two-scale convergence of a bounded sequence in $H^2(\Omega)$.

1.3.4. Γ-convergence

This type of convergence pertains to sequences of functionals. For instance, such sequences are generated by the variational principle of the total potential energy of solids or structures like plates or shells.

A detailed presentation of the theory of Γ-convergence is provided by Attouch (1984) and Dal Maso (1993). Attouch (1984) prefers to use the notion of epi-convergence, which in fact is a special case of Γ-convergence. In our specific case these notions coincide.

Definition 1.3.20. Let (X, τ) be a metrisable topological space and $\{G_\varepsilon\}_{\varepsilon>0}$ a sequence of functionals from X into $\overline{\mathbb{R}}$ – the extended reals.

(a) The $\Gamma(\tau)$-limit inferior, denoted also by G_i, is the functional on X defined by

$$G_i(u) = \Gamma(\tau) - \lim_{\varepsilon \to 0} \inf G_\varepsilon(u) = \min_{\{u_\varepsilon \overset{\tau}{\to} u\}} \lim_{\varepsilon \to 0} \inf G_\varepsilon(u_\varepsilon) .$$

(b) The $\Gamma(\tau)$-limit superior, denoted also by G_s, is the functional on X defined by

$$G_s(u) = \Gamma(\tau) - \lim_{\varepsilon \to 0} \sup G_\varepsilon(u) = \min_{\{u_\varepsilon \overset{\tau}{\to} u\}} \lim_{\varepsilon \to 0} \sup G_\varepsilon(u_\varepsilon) .$$

(c) The sequence $\{G_\varepsilon\}_{\varepsilon>0}$ is said to be $\Gamma(\tau)$-convergent if $G_i = G_s$; we then write

$$G = \Gamma(\tau) - \lim_{\varepsilon \to 0} G_\varepsilon .$$

Theorem 1.3.21. Let (X, τ) be a topological space with a countable base for τ. Then there exists a subsequence $\{G_{\varepsilon'}\}_{\varepsilon'>0}$ such that the limits G_i and G_s exist and $G_i(u) = G_s(u)$, for every $u \in X$. $\square$

Properties.
Let $G_\varepsilon : (X, \tau) \to \overline{\mathbb{R}}$ be a sequence of $\Gamma(\tau)$-convergent functionals and let $G = \Gamma(\tau) - \lim_{\varepsilon \to 0} G_\varepsilon$. Then the following properties hold:

(i) The functionals G_i and G_s are τ-lower semicontinuous (τ-l.s.c.).

(ii) If the functionals G_ε are convex, then $G_s = \Gamma(\tau)$ - $\lim\sup_{\varepsilon \to 0} G_\varepsilon$ is also a convex functional. Hence the limit $G = \Gamma(\tau) - \lim_{\varepsilon \to 0} G_\varepsilon$ is a τ-closed (τ-l.s.c.) convex functional.

(iii) If $\Phi : X \to \mathbb{R}$ is a τ-continuous functional, called a perturbation functional, then

$$\Gamma(\tau) - \lim_{\varepsilon \to 0} (G_\varepsilon + \Phi) = \Gamma(\tau) - \lim_{\varepsilon \to 0} G_\varepsilon + \Phi = G + \Phi \,.$$

(iv)

$$G(u) = \Gamma(\tau) - \lim_{\varepsilon \to 0} G_\varepsilon(u) \Leftrightarrow \begin{cases} \forall \{u_\varepsilon \xrightarrow{\tau} u\} \,, G(u) \leq \lim_{\varepsilon \to 0} \inf G_\varepsilon(u_\varepsilon) \,, u \in X \,; \\ \forall u \in X \quad \exists u_\varepsilon \xrightarrow{\tau} u, \text{ such that} \\ G(u) \geq \lim_{\varepsilon \to 0} \sup G_\varepsilon(u_\varepsilon) \,. \end{cases}$$

Further characterization is given by the convergence of minima.

Theorem 1.3.22. Let $G = \Gamma(\tau)$ - $\lim_{\varepsilon \to 0} G_\varepsilon$, and suppose that there exists a τ-relatively compact subset $X_0 \subset X$ such that $\inf_{X_0} G_\varepsilon = \inf_X G_\varepsilon$ ($\forall \varepsilon > 0$). Then $\inf_X G = \lim_{\varepsilon \to 0} (\inf_X G_\varepsilon)$. Moreover, if $\{u_\varepsilon\}_{\varepsilon > 0}$ is such that $G_\varepsilon(u_\varepsilon) - \inf_X G_\varepsilon \xrightarrow[\varepsilon \to 0]{} 0$, then every τ-cluster point of the sequence $\{u_\varepsilon : \varepsilon \to 0\}$ minimizes G on X.

Remark 1.3.23. From a practical point of view the following sufficient condition of existence of compact set X_0 is very useful. If $(X, \|\cdot\|)$ is a Banach space with τ-relatively compact balls, then a sufficient condition of existence of the compact set X_0 is that the sequence $\{G_\varepsilon\}_{\varepsilon > 0}$ satisfies the condition of *equi-coercivity*

$$\lim_{\varepsilon} \sup G_\varepsilon(u_\varepsilon) < +\infty \implies \lim_{\varepsilon} \sup \|u_\varepsilon\| < +\infty \,. \tag{1.3.38}$$

$\square$

The study of two-dimensional plate models obtained from three-dimensional ones will involve loading functionals dependent on a small parameter ε. Hence the need for perturbation functionals which depend on ε.

Let X be a general topological space.

Definition 1.3.24. We say that a sequence $\{\Phi_\varepsilon\}_{\varepsilon > 0}$ is *continuously convergent* in X to a function $\Phi : X \to \overline{\mathbb{R}}$ if for every $u \in X$ and for every neighborhood $\mathcal{V}$ of $\Phi(u)$ in $\overline{\mathbb{R}}$ there exist ε_0 and $\mathcal{U} \in \mathcal{N}(u)$ such that $\Phi_\varepsilon(v) \in \mathcal{V}$ for every $\varepsilon \leq \varepsilon_0$ and for every $v \in \mathcal{U}$. $\square$

We recall that $\mathcal{N}(u)$ stands for the set of all *open* neighborhoods of u in X.

Remark 1.3.25. It is clear that continuous convergence is stronger than pointwise convergence. Moreover, continuous convergence is stronger than Γ-convergence.

Proposition 1.3.26. Suppose that $\{\Phi_\varepsilon\}_{\varepsilon > 0}$ is continuously convergent to a functional Φ, and that Φ_ε and Φ are everywhere finite on X. Then

$$\Gamma(\tau) - \lim_{\varepsilon \to 0} \inf(G_\varepsilon + \Phi_\varepsilon) = \Gamma(\tau) - \lim_{\varepsilon \to 0} \inf G_\varepsilon + \Phi \,,$$

$$\Gamma(\tau) - \lim_{\varepsilon \to 0} \sup(G_\varepsilon + \Phi_\varepsilon) = \Gamma(\tau) - \lim_{\varepsilon \to 0} \sup G_\varepsilon + \Phi .$$

In particular, if $\{G_\varepsilon\}_{\varepsilon > 0}$ $\Gamma - converges$ to G in X, then $\{G_\varepsilon + \Phi_\varepsilon\}_{\varepsilon > 0}$ Γ-converges to $G + \Phi$ in X.   □

The following nonstandard diagonalization lemma is due to Attouch (1984).

Lemma 1.3.27. Let $\{a_{m,p}|m = 1, 2, \ldots; p = 1, 2, \ldots\}$ be a doubly indexed family in $\overline{\mathbb{R}}$. Then, there exists a mapping $m \to p(m)$, increasing to $+\infty$, such that

$$\lim_{m \to \infty} \sup a_{m,p(m)} \le \lim_{p \to \infty} \sup \left(\lim_{m \to \infty} \sup a_{m,p} \right) .$$
    □

1.3.5. Γ-convergence of sequence of nonconvex functionals convex in highest-order derivatives: non-uniform homogenization

The classes of homogenized models of geometrically nonlinear elastic plates as well as geometrically linear and nonlinear elastic shells, studied in the present book, can properly be modelled by Γ-limits of sequences of functionals of the following type (or possibly a straightforward extension), see Bielski and Telega (1999)

$$G_\varepsilon(\boldsymbol{u}, w) = \int_\Omega f[x, \frac{x}{\varepsilon}, \boldsymbol{u}(x), e(\boldsymbol{u}(x)); w(x), \nabla w(x), \nabla^2 w(x)]dx . \qquad (1.3.39)$$

Here Ω is a bounded and open set of $\mathbb{R}^2$, $\boldsymbol{u} \in W^{1,p}(\Omega)^2$, $w \in W^{2,q}(\Omega)$, $p, q \ge 2$ and $e_{\alpha\beta}(\boldsymbol{u}) = \left(\dfrac{\partial u_\alpha}{\partial x_\beta} + \dfrac{\partial u_\beta}{\partial x_\alpha} \right) /2$. Obviously we can consider a domain $\Omega \subset \mathbb{R}^3$. Having in mind the aforementioned application to plates and shells, the present section is confined to two-dimensional domains. The paper by Bielski and Telega (1999) has been inspired by Braides' paper (1983), who found the $\Gamma(L^p(\Omega))$-limit of the following sequence of functionals:

$$G_\varepsilon^1(w) = \int_\Omega f_1(x, \frac{x}{\varepsilon}, w(x), \nabla w(x))dx , \qquad (1.3.40)$$

where $\Omega \subset \mathbb{R}^n$.

Let us specify the assumptions on the integrand f appearing in (1.3.39). They are given by:

(A_1) $f = f(x, y, \boldsymbol{u}, \epsilon; w, \boldsymbol{\eta}, \boldsymbol{\rho}) : \mathbb{R}^2 \times \mathbb{R}^2 \times \mathbb{R}^2 \times \mathbb{E}_s^2 \times \mathbb{R} \times \mathbb{R}^2 \times \mathbb{E}_s^2 \to [0, \infty] ,$

which is measurable and Y-periodic in y, continuous in x, $\boldsymbol{u}$, w and $\boldsymbol{\eta}$, and convex in ϵ and $\boldsymbol{\rho}$.

(A_2) There exist Y-periodic function $a \in L_{loc}^1(\mathbb{R}^2)$, increasing function $g : \mathbb{R}^+ \to \mathbb{R}^+$ which is continuous at 0 and such that $g(0) = 0$, and function $b : \mathbb{R}^2 \to \mathbb{R}$, continuous and non-negative, for which the following inequalities are satisfied:

$$|f(x, y, \boldsymbol{u}, \epsilon; w, \boldsymbol{\eta}, \boldsymbol{\rho}) - f(x', y, \boldsymbol{u}', \epsilon; w', \boldsymbol{\eta}', \boldsymbol{\rho})| \le g(|x - x'| + |\boldsymbol{u}' - \boldsymbol{u}|$$
$$+|w' - w| + |\boldsymbol{\eta}' - \boldsymbol{\eta}|)(a(y) + f(x, y, \boldsymbol{u}, \epsilon; w, \boldsymbol{\eta}, \boldsymbol{\rho})) , \qquad (1.3.41)$$

$$0 \leq f(x, y, \boldsymbol{u}, \boldsymbol{\epsilon}; w, \boldsymbol{\eta}, \boldsymbol{\rho})$$
$$\leq b(x)[a(y) + |\boldsymbol{u}|^p + |\boldsymbol{\epsilon}|^p + |w|^q + |\boldsymbol{\eta}|^q + |\boldsymbol{\rho}|^q] , \qquad (1.3.42)$$

for all $x, x' \in \Omega$, $y \in Y$, $\boldsymbol{u}, \boldsymbol{u}', \boldsymbol{\eta}, \boldsymbol{\eta}' \in \mathbb{R}^2$, and $\boldsymbol{\epsilon}, \boldsymbol{\rho} \in \mathbb{E}_s^2$. Here $\mathbb{E}_s^2$ stands for the space of symmetric 2×2 matrices. Obviously, the norms in (1.3.41) and (1.3.42) are Euclidean norms.

Let us set $\tau = s - (L^p(\Omega)^2 \times W^{1,q}(\Omega))$. We are now in a position to formulate the homogenization theorem.

Theorem 1.3.28. Under the assumptions (A_1) and (A_2) the $\Gamma(\tau)$-limit of the sequence of functionals $\{G_\epsilon\}_{\epsilon>0}$ defined by (1.3.39) has the following form:

$$\Gamma(\tau) - \lim_{\epsilon \to 0} G_\epsilon(\boldsymbol{u}, w) = \int_\Omega f_h[x, \boldsymbol{u}(x), e(\boldsymbol{u}(x)); w(x), \nabla w(x), \nabla^2 w(x)]dx , \quad (1.3.43)$$

where $\boldsymbol{u} \in W^{1,p}(\Omega)^2$, $w \in W^{2,q}(\Omega)$ and

$$f_h(x, \boldsymbol{\xi}, \boldsymbol{\epsilon}; \psi, \boldsymbol{\eta}, \boldsymbol{\rho}) = \inf\{\frac{1}{|Y|}\int_Y f[x, y, \boldsymbol{\xi}, \boldsymbol{\epsilon} + e^y(\boldsymbol{v}(y)); \psi, \boldsymbol{\eta}, \boldsymbol{\rho} + \nabla_y^2 v(y)]dy$$

$$|\boldsymbol{v} \in W_{per}^{1,p}(Y)^2 , \ v \in W_{per}^{2,q}(Y)\} , \qquad (1.3.44)$$

for all $\psi \in \mathbb{R}$, $\boldsymbol{\xi}, \boldsymbol{\eta} \in \mathbb{R}^2$ and $\boldsymbol{\epsilon}, \boldsymbol{\rho} \in \mathbb{E}_s^2$. Here

$$W_{per}^{1,p}(Y)^2 = \{\boldsymbol{v} \in W^{1,p}(Y)^2| \ v \quad \text{is} \quad Y\text{-periodic}\} , \qquad (1.3.45)$$

$$W_{per}^{2,q}(Y) = \{\boldsymbol{v} \in W^{2,q}(Y)^2| \ v \quad \text{and} \quad \frac{\partial v}{\partial y_\alpha} \quad \text{are} \quad Y\text{-periodic}\} , \qquad (1.3.46)$$

and $e_{\alpha\beta}^y(\boldsymbol{v}) = \left(\dfrac{\partial v_\alpha}{\partial y_\beta} + \dfrac{\partial v_\beta}{\partial y_\alpha}\right)/2, \quad (\nabla_y^2 v)_{\alpha\beta} = \dfrac{\partial^2 v}{\partial y_\beta \partial y_\beta}.$

Proof. Detailed proof has been given in Bielski and Telega (1999). Let us sketch the main points.

(i) First we consider the simpler case where

$$f(x, y, \boldsymbol{u}, e(\boldsymbol{u}); w, \nabla w, \nabla^2 w) = \varphi(y, e(\boldsymbol{u}), \nabla^2 w) .$$

Following Braides (1983) and exploiting (A_1) and (A_2) we then prove that

$$\Gamma(\tau) - \lim_{\epsilon \to 0} \int_\Omega \varphi\left(\frac{x}{\epsilon}, e(\boldsymbol{u}), \nabla^2 w(x)\right) dx = \int_\Omega \varphi_h(e(\boldsymbol{u}(x)), \nabla^2 w(x))dx ,$$

where

$$\varphi_h(\boldsymbol{\epsilon}, \boldsymbol{\rho}) = \inf\{\frac{1}{|Y|}\int_Y \varphi[y, \boldsymbol{\epsilon} + e^y(\boldsymbol{v}(y)), \boldsymbol{\rho} + \nabla_y^2 v(y)]dy| \ \boldsymbol{v} \in W_{per}^{1,p}(Y)^2, v \in W_{per}^{2,q}(Y)\}.$$

(ii) Take now the following integrand

$$f(x, y, \boldsymbol{u}, e(\boldsymbol{u}); w, \nabla w, \nabla^2 w) = \varphi(x, y, e(\boldsymbol{u}), \nabla^2 w) \, .$$

Following Braides (1983) once again we prove that there exists a function $\varphi_h(x, e(\boldsymbol{u}), \nabla^2 w)$ such that

$$\Gamma(\tau) - \lim_{\varepsilon \to 0} \int_\Omega \varphi(x, \frac{x}{\varepsilon}, e(\boldsymbol{u}), \nabla^2 w) dx = \int_\Omega \varphi_h(x, e(\boldsymbol{u}), \nabla^2 w) dx \, ,$$

where

$$\varphi_h(x, \epsilon, \rho) = \inf\{\frac{1}{|Y|} \int_Y \varphi(x, y, \epsilon + e^y(v), \rho + \nabla^2 v) dy | \, v \in W_{per}^{1,p}(Y)^2 \, , \, v \in W_{per}^{2,q}(Y)\} \, .$$

(iii) In the final step we prove (1.3.43), by using the integral representation of Γ-limits of sequences of integral functionals due to Buttazzo and Dal Maso (1980). □

Remark 1.3.29. Let us fix $x \in \Omega$, $\psi \in \mathbb{R}$, and $\boldsymbol{\xi}, \boldsymbol{\eta} \in \mathbb{R}^2$. For each $\epsilon, \eta \in \mathbb{E}_s^2$ we have

$$f_h(x, \boldsymbol{\xi}, \epsilon; \psi, \boldsymbol{\eta}, \rho) = \inf\{\frac{1}{|Y|} \int_Y f_h(x, \boldsymbol{\xi}, \epsilon + e^y(v); \psi, \boldsymbol{\eta}, \rho + \nabla_y^2 w) dy$$

$$|v \in W_0^{1,p}(Y)^2 \, , \, v \in W_0^{2,q}(Y)\} \, , \qquad (1.3.47)$$

Indeed, since $0 \in W_0^{1,p}(Y)^2$ and $0 \in W_0^{2,q}(Y)$ therefore we can write

$$f_h(x, \boldsymbol{\xi}, \epsilon; \psi, \boldsymbol{\eta}, \rho) = \frac{1}{|Y|} \int_Y f_h(x, \boldsymbol{\xi}, \epsilon + 0; \psi, \boldsymbol{\eta}, \rho + 0) dy$$

$$\geq \inf\{\frac{1}{|Y|} \int_Y f_h[x, \boldsymbol{\xi}, \epsilon + e^y(v); \psi, \boldsymbol{\eta}, \rho + \nabla_y^2 w] dy | v \in W_0^{1,p}(Y)^2, \, v \in W_0^{2,q}(Y)\} \, .$$

On the other hand, Jensen's inequality yields

$$f_h(x, \boldsymbol{\xi}, \epsilon; \psi, \boldsymbol{\eta}, \rho) = f_h[x, \boldsymbol{\xi}, \frac{1}{|Y|} \int_Y (\epsilon + e^y(v)) dy; \psi, \boldsymbol{\eta}, \frac{1}{|Y|} \int_Y (\rho + \nabla_y^2 v) dy]$$

$$\leq \frac{1}{|Y|} \int_Y f_h(x, \boldsymbol{\xi}, \epsilon + e^y(v); \psi, \boldsymbol{\eta}, \rho + \nabla_y^2 v) dy$$

for each $v \in W_0^{1,p}(Y)^2$, $v \in W_0^{2,q}(Y)$.

1.3.6.   Γ-convergence and duality

With a sequence of functionals one can associate the sequence of conjugate or dual functionals. Suppose that the original sequence is Γ-convergent. In the present section we are going to study the interrelationship between the Γ-convergence of the primal sequence of functionals and the sequence of dual functionals. Here we follow the approach developed by Azé (1984), cf. also Azé (1986) and Jikov et al. (1994).

Let V and X be two separable Banach spaces and let V^* and X^* be their topological duals. In general, these spaces are not necessarily reflexive. In practice, however, V is often a reflexive space whilst X is a space of perturbation parameters, cf. Sec. 1.2.5.

Consider a sequence of functionals $\{G_\varepsilon\}_{\varepsilon>0}$ from $\Gamma_0(V^* \times X)$. This sequence is said to satisfy the assumption (A) if:

(A) There exists $r > 40$ such that for each sequence $\{p_\varepsilon\}_{\varepsilon>0}$ of elements from $B_r = \{p \in X : ||p|| \le r\}$, there exists a bounded sequence $\{v_\varepsilon^*\}_{\varepsilon>0}$ such that $\lim \sup_\varepsilon G_\varepsilon(v_\varepsilon^*, p_\varepsilon) < +\infty$.

We recall that $\varepsilon \in \mathcal{E} = \{\frac{1}{n} \mid n \in \mathbb{N}\setminus\{0\}\}$.

Similarly, the sequence $\{G_\varepsilon\}_{\varepsilon>0}$ satisfies the assumption (A*) if:

(A*) There exists $r^* > 0$ such that for each sequence $\{v_\varepsilon\}_{\varepsilon>0} \subset B_{r^*} = \{v \in V : ||v|| \le r^*\}$, there exists a bounded sequence $\{p_\varepsilon^*\}_{\varepsilon>0}$ such that $\lim \sup_\varepsilon G_\varepsilon^*(v_\varepsilon, p_\varepsilon^*) < +\infty$.

The assumption (A) is a universal qualification hypothesis whilst (A*) will be shown to play the role of a uniform coercivity hypothesis for the primal problems.

Let us first prove two auxiliary lemmas.

Lemma 1.3.30. Let $\{G_\varepsilon\}_{\varepsilon>0}$ be a sequence of functionals from $\Gamma_0(V^* \times X)$ satisfying (A). For $\lambda \ge 0$ we set:

$$G_{\varepsilon,\lambda}(v^*, p) = G_\varepsilon(v^*, p) + \frac{\lambda}{2}||v^*||^2 . \tag{1.3.48}$$

Let $\{\lambda_{1/k}\}_{k\in\mathbb{N}}$ be a sequence of positive numbers bounded from above. Then, for any subsequence $\{n_k\}$

$$\left\{ \begin{array}{l} \lim \sup_k (G_{1/n_k, \lambda_{1/k}})^*(v_{1/k}, p_{1/k}^*) < +\infty \\ \text{and } \{v_{1/k}\} \text{ bounded} \end{array} \right\} \Rightarrow \{p_{1/k}^*\} \text{ bounded.} \tag{1.3.49}$$

Proof. Let $r > 0$ be a constant appearing in (A) and let $p_{1/k} \in B_r$ be such that $r||p_{1/k}^*|| - 1 \le \langle p_{1/k}^*, p_{1/k}\rangle_{X^*\times X}$. Consider a sequence $\{p_{1/n}\}$ of elements from B_r such that $p_{1/n_k} = p_{1/k}$. According to (A) to this sequence corresponds a sequence $\{v_{1/n}^*\}$ bounded by a positive constant c such that:

$$\lim \sup_n G_{1/n}(v_{1/n}^*, p_{1/n}) < +\infty .$$

By using the definition of $(G_{1/n,\lambda})^*$ we get:

$$(G_{1/n_k,\lambda_k})^*(v_{1/k},p_{1/k}^*) + G_{1/n_k}(v_{1/n_k}^*,p_{1/n_k}) + \frac{\lambda_{1/k}}{2}||v_{1/n_k}^*||^2 \geq \langle v_{1/k}, v_{1/n_k}^*\rangle_{V\times V^*}$$
$$+\langle p_{1/k}^*, p_{1/n_k}\rangle_{X^*\times X} \geq -c||v_{1/k}|| + r||p_{1/k}^*|| - 1 .$$

The desired result follows by taking the limit superior with respect to k. $\qquad\square$

Corollary 1.3.31. If (A^*) is satisfied, then we have

$$\left\{ \begin{array}{l} \lim\sup_k G_{1/n_k}(v_{1/k}^*,p_{1/k}) < +\infty \\ \text{and } \{p_{1/k}\} \text{ bounded} \end{array} \right\} \Rightarrow \{v_{1/k}^*\} \text{ bounded.} \qquad (1.3.50)$$

Lemma 1.3.32. Let C be a convex set in $V \times X^*$. Then:

$$\overline{C}^{\sigma(V,V^*)\times\sigma(X^*,X)} = \overline{C}^{(s-V)\times\sigma(X^*,X)} . \qquad (1.3.51)$$

Here $(s - V)$ denotes the strong topology of the space V whilst the bar stands for the closure of C in the indicated topologies.

Proof. The topology $(s-V)\times(w^*-X^*)$ is compatible with the duality $\sigma(V\times X^*, V^*\times X)$. The same pertains to the topology $\sigma(V,V^*) \times \sigma(X^*,X)$ and (1.3.51) follows. $\qquad\square$

Hence we conclude that for these two topologies the lower semicontinuous regularizations coincide.

Let us pass to a result concerning the effect of duality on the Γ (strong $\times$ weak)-convergence.

Theorem 1.3.33. Let $\{G_\varepsilon\}_{\varepsilon>0}$ be a sequence of functionals belonging to $\Gamma_0(V^* \times X)$ and satisfying (A). Then we have

$$\Gamma(w^* \times s) - \lim\inf_{\varepsilon\to 0} G_\varepsilon \geq G \Rightarrow \Gamma(s \times w^*) - \lim\sup_{\varepsilon\to 0} G_\varepsilon^* \leq G^* . \qquad (1.3.52)$$

Proof. We can assume that $G > -\infty$. Otherwise the result is evident. We begin by showing the following lemma.

Lemma 1.3.34. Under the assumptions of Theorem 1.3.33 and $G > -\infty$ one has:
For each $B \subset X$ relatively strongly compact there exists $K \geq 0$ such that:

$$\forall v^* \in V^*, \quad \forall p \in B, \quad \forall \varepsilon \in \mathcal{E}, \quad G_\varepsilon(v^*,p) \geq -K(||v^*||+1) . \qquad (1.3.53)$$

Proof. Suppose that the conclusion were false. Then there exists $B \subset Y$, relatively strongly compact and such that

$$\forall k \in \mathbb{N}, \quad \exists v_{1/k}^* \in V^*, \quad \exists p_{1/k} \in B, \quad \exists n_k \in \mathbb{N} \quad \text{such that}$$
$$G_{1/n_k}(v_{1/k}^*,p_{1/k}) \leq -k(||v_{1/k}^*||+1) .$$

One may assume that the sequence $\{n_k\}$ is increasing.

Case 1: $\{v^*_{1/k}\}$ is bounded.

Since V is separable and B is relatively compact, therefore there exist subsequences still denoted by $\{v^*_{1/k}\}$, $\{p_{1/k}\}$ such that

$$v^*_{1/k} \xrightarrow{w^*} v \quad \text{and} \quad p_{1/k} \xrightarrow{s} p \, ,$$

as $k \to \infty$. Consequently we conclude that

$$\lim_k \inf G_{1/n_k}(v^*_{1/k}, p_{1/k}) = -\infty \, ,$$

and hence $G(v^*, p) = -\infty$, which contradicts the assumption on G.

Case 2: the sequence $\{v^*_{1/k}\}$ is unbounded.

One may then assume, up to the extraction of a subsequence, that $\lim_{k \to *\infty} ||v^*_{1/k}|| = +\infty$.

Consider a sequence $\widetilde{p}_{1/n} \xrightarrow{s} \widetilde{p}$, such that $\widetilde{p}_{1/n} \leq r$. According to the assumption (A), there exists a bounded sequence $\{\widetilde{v}^*_{1/n}\}$ such that $\lim_n \sup G_{1/n}(\widetilde{v}^*_{1/n}, \widetilde{p}_{1/n}) < +\infty$. We set

$$\widetilde{v}^*_{1/k} = \widetilde{v}^*_{1/n_k} \, , \quad \widetilde{p}_{1/k} = \widetilde{p}_{1/n_k} \, .$$

Since the Banach space V is separable, therefore the bounded sequence $\{\widetilde{v}^*_{1/k}\} \subset V^*$ admits a subsequence (still denoted by $\widetilde{v}^*_{1/k}$) convergent to a certain $\widetilde{v}^* \in V^*$ in the topology $\sigma(V^*, V)$.

We define

$$\xi^*_{1/k} = t_{1/k} v^*_{1/k} + (1 - t_{1/k})\widetilde{v}^*_{1/k} \, , \qquad \eta^*_{1/k} = t_{1/k} p_{1/k} + (1 - t_{1/k})\widetilde{p}_{1/k} \, ,$$

where

$$t_{1/k} = \frac{\sqrt{1/k}}{||v^*_{1/k} - \widetilde{v}^*_{1/k}||} \, .$$

For k sufficiently large, $t_{1/k}$ is well-defined and $\lim_{k \to +\infty} t_{1/k} = 0$. Moreover we have

$$||\xi^*_{1/k} - \widetilde{v}^*_{1/k}|| = t_{1/k}||v^*_{1/k} - \widetilde{v}^*_{1/k}|| = \frac{1}{\sqrt{k}} \, ,$$

and thus

$$\xi^*_{1/k} \xrightarrow{w^*} \widetilde{v}^* \quad \text{and} \quad \eta_{1/k} \xrightarrow{s} \widetilde{p} \quad \text{as} \quad k \to +\infty \, .$$

We also find

$$G_{1/n_k}(\xi^*_{1/k}, \eta_{1/k}) \leq t_{1/k} G_{1/n_k}(v^*_{1/k}, p_{1/k}) + (1 - t_{1/k}) G_{1/n_k}(\widetilde{v}^*_{1/k}, \widetilde{p}_{1/k})$$

$$\leq \frac{-\sqrt{k}(||v^*_{1/k}|| + 1)}{||v^*_{1/k} - \widetilde{v}^*_{1/k}||} + K_1 \, ,$$

and consequently

$$\lim_k \inf G_{1/n_k}(\xi^*_{1/k}, \eta_{1/k}) = -\infty \, ,$$

and

$$G(\widetilde{v}^*, \widetilde{p}) = -\infty \,,$$

which is impossible. $\square$

We now turn to the *proof of Theorem 1.3.33*. Since the space X is separable, therefore there exists a sequence $\{p_k\}_{k\geq 1}$ dense in it. By X_k we denote the subspace generated by $\{p_1, \ldots, p_k\}$ and B denotes the unit ball in X. We set $B_k = kB \cap X_k = \{p \in X_k : ||p|| \leq k\}$. B_k is a strongly compact subset of X. For $k \geq 1$ and $\lambda > 0$ we set:

$$G_{\varepsilon,k,\lambda}(v^*, p) = G_\varepsilon(v^*, p) + I_{B_k}(p) + \frac{\lambda}{2}||v^*||^2 \,,$$

$$G_{k,\lambda}(v^*, p) = G(v^*, p) + I_{B_k}(p) + \frac{\lambda}{2}||v^*||^2 \,,$$

where I_{B_k} stands for the indicator function of B_k.

It will now be shown that

$$\forall k \geq 1, \quad \forall \lambda > 0 , \quad \forall (v, p^*) \in V \times X^*$$
$$\lim_n \sup G^*_{1/n,k,\lambda}(v, p^*) \leq G^*_{k,\lambda}(v, p^*) \,. \tag{1.3.54}$$

Obviously, this is equivalent to:

$$\lim_n \inf \Big[\inf_{\substack{v^* \in V^* \\ p \in B_k}} \big(G_{1/n}(v^*, p) + \frac{\lambda}{2}||v^*||^2 - \langle v^*, v \rangle_{V^* \times V} - \langle p^*, p \rangle_{X^* \times X} \big) \Big]$$

$$\geq \inf_{\substack{v^* \in V^* \\ p \in B_k}} \big[G(v^*, p) + \frac{\lambda}{2}||v^*||^2 - \langle v^*, v \rangle - \langle p^*, p \rangle \big] \,. \tag{1.3.55}$$

On account of Lemma 1.3.34 there exists $K \geq 0$ such that

$$\forall v^* \in V^* , \quad \forall p \in B_k , \quad \forall n \in \mathbb{N} , \quad G_{1/n,\lambda}(v^*, p) \geq \frac{\lambda}{2}||v^*||^2 - K(||v^*|| + 1) \,. \tag{1.3.56}$$

Consider now an extracted sequence $\{n_l\}$ for which the limit inferior is just a limit. The infimum in $(v^*_{1/l}, p_{1/l})$ on the left-hand side of (1.3.54) is then attained.

Using now the sequence $\{\widetilde{v}^*_{1/n}\}$ appearing in the assumption (A) and associated with the sequence $p_{1/n} \equiv 0$ and taking into account the estimate (1.3.55) we conclude that $\{v^*_{1/l}\}$ is bounded. We can extract subsequences, still denoted by $\{v^*_{1/l}\}$ and $\{p_{1/l}\}$, such that

$$v^*_{1/l} \xrightarrow{w^*} v^* \in V^* \quad \text{and} \quad p_{1/l} \xrightarrow{s} p \in B_k \quad \text{as} \quad l \to +\infty \,.$$

Recalling the definition of $\Gamma(w^* \times s) \text{-} \lim_{\varepsilon \to 0} \inf G_\varepsilon$ and the semicontinuity of the norm one gets

$$\lim_l \inf \big[G_{1/n_l}(v^*_{1/l}, p_{1/l}) + \frac{\lambda}{2}||v^*_{1/l}||^2 \big] \geq G(v^*, p) + \frac{\lambda}{2}||v^*||^2 \,,$$

which establishes (1.3.54) and consequently (1.3.53). Thus we conclude that for each $k \geq 1$ and for any $\lambda > 0$ we have

$$\lim_{n} \sup G^*_{1/n,k,\lambda}(v,p^*) \leq G^*_{k,\lambda}(v,p^*) \leq G^*(v,p^*) .$$

Applying now diagonalization Lemma 1.3.27 we conclude that there exist $\lambda(n) \to 0^+$, $k(n) \to +\infty$ such that

$$\lim_{n} \sup G^*_{1/n,k(n),\lambda(n)}(v,p^*) \leq G^*(v,p^*) . \tag{1.3.57}$$

We may write $G_{1/n,k,\lambda} = G_{1/n,\lambda} + \Psi_k$ with $\Psi_k(v^*,p) = I_{B_k}(p)$. Hence, cf. Sec. 1.2

$$G^*_{1/n,k,\lambda} = (G^*_{1/n,\lambda}\square\Psi^*_k)^{**} ,$$

where $\square$ denotes the inf-convolution. Let us show that

$$G^*_{1/n,k,\lambda} = \overline{G^*_{1/n,\lambda}\square\Psi^*_k}^{(s-V)\times\sigma(X^*,X)} , \tag{1.3.58}$$

which represents the semicontinuous regularization. Indeed, $G^*_{1/n,k,\lambda} = (G_{1/n,\lambda} + \Psi_k)^*$ cannot assume the value $-\infty$ since in that case we would have $G_{1/n,\lambda} + \Psi_k \equiv +\infty$. After (A) there exists $\{\widetilde{v}_{1/n}\}$ such that $\lim_{n} \sup G_{1/n}(\widetilde{v}^*_{1/n},0) < +\infty$. Consequently, for n sufficiently large we have $(G_{1/n,\lambda} + \Psi_k)(v^*_{1/n},0) < +\infty$ and thus $G_{1/n,\lambda} + \Psi_k$ is not identically equal to $+\infty$. We deduce that

$$G^*_{1/n,k,\lambda} = (G^*_{1/n,\lambda}\square\Psi^*_k)^{**} = \overline{G^*_{1/n,\lambda}\square\Psi^*_k}^{\sigma(V,V^*)\times\sigma(X^*,X)} .$$

By Lemma 1.3.32 the last regularization can be taken in the sense of $(s-V) \times \sigma(X^*,X)$. Let us continue the proof of Theorem 1.3.33. We have:

$$\lim_{n} \sup G^*_{1/n,k(n),\lambda(n)}(v,p^*) \leq G^*(v,p^*) .$$

We may assume that $G^*(v,p^*) < +\infty$; otherwise the result is evident.

For each $n \in \mathbb{N}$ we introduce the following set:

$$\mathbf{U}^*_n = \{q^* \in X^* : |\langle q^* - p^*, p_i\rangle_{X^*\times X} \leq \frac{1}{n} , \; 1 \leq i \leq n\} . \tag{1.3.59}$$

We recall that $\{p_i\}_{1\leq i\leq n}$ constitute a subset of the sequence dense in X. $\mathbf{U}^*_n$ is a $\sigma(X^*,X)$-neigborhood of p^*.

Using the definition of the lim inf we write: there exists $(u_{1/n}, q^*_{1/n}) \in V \times X^*$ such that

$$\|u_{1/n} - v\| \leq \frac{1}{n} , \quad q^*_{1/n} \in \mathbf{U}^*_n ,$$

$$(G^*_{1/n,\lambda(n)}\square\Psi^*_{k(n)})(u_{1/n}, q^*_{1/n}) < \sup \{G^*_{1/n,k(n),\lambda(n)}(v,p^*) + \frac{1}{n}, -\frac{1}{n}\} . \tag{1.3.60}$$

Simple calculation yields

$$\Psi_k^*(v, p^*) = \begin{cases} kd(p^*, X_k^\perp) & \text{if} \quad v = 0 \,, \\ +\infty & \text{otherwise.} \end{cases} \tag{1.3.61}$$

Using the definition of inf-convolution we deduce the existence of $\eta_{1/n}^* \in X^*$ such that

$$G_{1/n,\lambda(n)}^*(u_{1/n}, \eta_{1/n}^*) + k(n)d(q_{1/n}^* - \eta_{1/n}^*, X_{k(n)}^\perp)$$

$$< \sup \left\{ G_{1/n,k(n),\lambda(n)}^*(v, p^*) + \frac{1}{n}, -\frac{1}{n} \right\} . \tag{1.3.62}$$

Since $\{u_{1/n}\}$ is bounded, therefore, by Lemma 1.3.30 the sequence $\{\eta_{1/n}^*\}$ is also bounded. Let us calculate $G_{1/n,\lambda}^*$. We have

$$G_{1/n,\lambda} = G_{1/n} + \gamma_\lambda \quad \text{with} \quad \gamma_\lambda(v^*, p) = \frac{\lambda}{2}\|v^*\|^2 .$$

Hence

$$G_{1/n,\lambda}^* = (G_{1/n}^* \square \gamma_\lambda^*)^{**} = \overline{G_{1/n}^* \square \gamma_\lambda^*} \,,$$

since $G_{1/n,\lambda}^* > -\infty$ $(G_{1/n,\lambda} \not\equiv +\infty)$.

By Lemma 1.3.32 the semicontinuous regularization may be taken in the sense of $(s - V) \times (w^* - X^*)$. Thus there exists $(w_{1/n}, p_{1/n}^*) \in V \times X^*$ such that

$$\|w_{1/n} - u_{1/n}\| \le \frac{1}{n} \,, \quad p_{1/n}^* \in W_n^* \,,$$
$$(G_{1/n}^* \square \gamma_{\lambda(n)}^*)(w_{1/n}, p_{1/n}^*) < G_{1/n,\lambda(n)}^*(u_{1/n}, \eta_{1/n}^*) + \frac{1}{n} \,, \tag{1.3.63}$$

where

$$W_n^* := \left\{ p^* \in X^* : |\langle p^* - \eta_{1/n}^*, p_i \rangle| \le \frac{1}{n}, \ 1 \le i \le n \right\} .$$

We note that

$$G_{1/n,\lambda(n)}^*(w_{1/n}, p_{1/n}^*) \le (G_{1/n}^* \square \gamma_{\lambda(n)}^*)(w_{1/n}, p_{1/n}^*) < G_{1/n,\lambda(n)}^*(u_{1/n}, \eta_{1/n}^*) + \frac{1}{n} \,.$$

Hence, using Lemma 1.3.30 we deduce that $\{p_{1/n}^*\}$ is bounded, since the sequence $\{w_{1/n}\}$ is bounded. Noting that

$$\gamma_\lambda^*(v, p^*) = \begin{cases} \dfrac{1}{2\lambda}\|v\|^2 & \text{if} \quad p^* = 0 \,, \\ +\infty & \text{otherwise,} \end{cases}$$

and exploiting the definition of the inf-convolution we obtain $v_{1/n} \in V$ such that

$$G_{1/n}^*(v_{1/n}, p_{1/n}^*) + \frac{1}{2\lambda(n)}\|v_{1/n} - w_{1/n}\|^2 \le G_{1/n,\lambda(n)}^*(u_{1/n}, \eta_{1/n}^*) + \frac{1}{n} \,. \tag{1.3.64}$$

From (1.3.56) – (1.3.59) we get

$$G^*_{1/n}(v_{1/n}, p^*_{1/n}) + \frac{1}{2\lambda(n)}\|v_{1/n} - w_{1/n}\|^2 + k(n)d(q^*_{1/n} - \eta^*_{1/n}, X^\perp_{k(n)})$$

$$\leq \sup\{G^*_{1/n,k(n),\lambda(n)}(v, p^*) + \frac{1}{n}, -\frac{1}{n}\} + \frac{1}{n}. \tag{1.3.65}$$

Taking the limit superior in n, by (1.3.57) we conclude that

$$\limsup_n G^*_{1/n}(v_{1/n}, p^*_{1/n})$$

$$\leq \limsup_n [\sup\{G^*_{1/n,k(n),\lambda(n)}(v, p^*) + \frac{1}{n}, -\frac{1}{n}\} + \frac{1}{n}] \leq G^*(v, p^*). \tag{1.3.66}$$

It thus remains to show that $v_{1/n} \xrightarrow{s} v$ and $p^*_{1/n} \xrightarrow{w^*} p^*$ as $n \to \infty$. We observe that due to the assumption (A) there exist bounded sequences $\{\widetilde{v}^*_{1/n}\}, \{\widetilde{p}_{1/n}\}$ such that

$$\limsup_n G_{1/n}(\widetilde{v}^*_{1/n}, \widetilde{p}_{1/n}) < +\infty.$$

Hence we deduce that there exists $K' \geq 0$ such that:

$$\forall v \in V, \forall p^* \in X^*, \forall n \in \mathbb{N}, G^*_{1/n}(v, p^*) \geq -K'(\|v\| + \|p^*\| + 1). \tag{1.3.67}$$

Substituting (1.3.67) into (1.3.64) and noting that the sequence $\{p^*_{1/n}\}$ is bounded, we conclude that there exist positive constants K'' and K''' such that:

$$-K'''(\|v_{1/n}\| + 1) + \frac{1}{2\lambda(n)}\|v_{1/n} - w_{1/n}\|^2 \leq K'''.$$

Since $\lambda(n) \to 0$ and $\{w_{1/n}\}$ is bounded, therefore $\{v_{1/n}\}$ is bounded and $\|v_{1/n} - w_{1/n}\|^2 \to 0$ as $n \to \infty$. Hence $\|w_{1/n} - v\| \leq \frac{2}{n}$ and consequently $v_{1/n} \to v$ strongly in V as $n \to \infty$. From (1.3.64) we get

$$\lim_{n\to\infty} d(q^*_{1/n} - \eta^*_{1/n}, X^\perp_{k(n)}) = 0.$$

We set

$$D = \bigcup_{k\geq 1} X_k.$$

Obviously, D is dense in X. Let $p \in D$, then beginning from a certain $k(n)$ we have $p \in X_{k(n)}$. Let us introduce $r^*_{1/n} \in X^\perp_{k(n)}$ such that $\|q^*_{1/n} - \eta^*_{1/n} - r^*_{1/n}\| \leq d(q^*_{1/n} - \eta^*_{1/n}, X^\perp_{k(n)} + \frac{1}{n})$. We have

$$\langle q^*_{1/n} - \eta^*_{1/n}, p\rangle_{X^*\times X} = \langle q^*_{1/n} - \eta^*_{1/n} - r^*_{1/n}, p\rangle$$

$$\leq \|p\|[d(q^*_{1/n} - \eta^*_{1/n}, X^\perp_{k(n)}) + \frac{1}{n}].$$

Hence

$$\lim_{n \to +\infty} \langle q^*_{1/n} - \eta^*_{1/n}, p \rangle = 0 . \qquad (1.3.68)$$

Moreover, from the definition of $\mathbf{U}^*_n$ one has

$$\lim_{n \to +\infty} \langle q^*_{1/n} - p^*, p \rangle = 0 . \qquad (1.3.69)$$

From (1.3.67) and (1.3.68) we conclude that

$$\lim_{n \to \infty} \langle \eta^*_{1/n}, -p^*, p \rangle = 0 ,$$

for each $p \in D$. Indeed, $p^*_{1/n} \in \mathbf{W}^*_n$ and thus for each $p \in D$ we have

$$\lim_{n \to \infty} \langle p^*_{1/n} - \eta^*_{1/n}, p \rangle = 0 .$$

Consequently we get

$$\forall\, p \in D , \quad \lim_{n \to \infty} \langle p^*_{1/n} - p^*, p \rangle = 0 .$$

Since $\{p^*_{1/n}\}$ is bounded and D is dense, therefore $p^*_{1/n} \rightharpoonup p^*$ weak-$*$ as $n \to \infty$ and the theorem is proved. $\qquad \square$

As a corollary we formulate the following result.

Theorem 1.3.35. Let V and X be two separable Banach spaces. Let $\{G_\varepsilon\}_{\varepsilon>0}$, G be functionals belonging to $\Gamma_0(V^* \times X)$, satisfying assumption (A) and

$$G = \Gamma(w^* \times s) - \lim_{\varepsilon \to 0} G_\varepsilon . \qquad (1.3.70)$$

Then

$$G^* = \Gamma(s \times w^*) - \lim_{\varepsilon \to 0} G^*_\varepsilon .$$

Proof. We conclude from Theorem 1.3.33 that

$$\Gamma(s \times w^*) - \limsup_{\varepsilon \to 0} G^*_\varepsilon \leq G^* .$$

The inequality

$$G^* \leq \Gamma(s \times w^*) - \liminf_{\varepsilon \to 0} G^*_\varepsilon$$

results immediately from (1.3.69) and the definition of the conjugate (dual) functional. $\quad \square$

We now pass to the central result of this section. This result concerns the Γ-convergence of sequence of marginal functionals and of their conjugate functionals.

Let $G_\varepsilon : V^* \times X \to \mathbb{R} \cup \{+\infty\}$ be a sequence of functionals from $\Gamma_0(V^* \times X), \varepsilon \in \mathcal{E}$. The marginal functional is given by, cf. the formula (1.2.42)

$$h_\varepsilon(p) = \inf \{G_\varepsilon(v^*, p) | v^* \in V^*\} .$$

This functional is convex, not necessarily lower semicontinuous, cf. Sec. 1.2. The dual problem will involve the functional $h_\varepsilon^*(p^*) = G_\varepsilon^*(0, p^*)$. We observe that the Γ-convergence of G_ε to G in a topology (τ, σ) does not necessarily imply the Γ-convergence of h_ε in the topology σ. Indeed, if $X = \{0\}$ and $G_\varepsilon(v, p) = J_\varepsilon(v)$, then the Γ-convergence of h_ε would imply the convergence of $\inf J_\varepsilon$, which is not always the case.

We now formulate the aforementioned central theorem.

Theorem 1.3.36. Let V and X be two separable Banach spaces. Let $\{G_\varepsilon\}_{\varepsilon>0}$ and G be functionals belonging to $\Gamma_0(V^* \times X)$, satisfying the assumptions (A) and (A*) and

$$G = \Gamma(w^* \times s) - \lim_{\varepsilon \to 0} G_\varepsilon . \tag{1.3.71}$$

Then

(i) $h = \Gamma(s) - \lim_{\varepsilon \to 0} h_\varepsilon,$

(ii) $h^* = \Gamma(w^*) - \lim_{\varepsilon \to 0} h_\varepsilon^*,$

(iii) $G(\cdot, 0) = \Gamma(w^*) - \lim_{\varepsilon \to 0} G_\varepsilon(\cdot, 0),$

(iv) if v_ε^* is a minimizer of P_ε up to δ_ε and if p_ε^* is a minimizer of P_ε^* up to δ_ε with $\varepsilon \to 0$, then the sequences $\{v_\varepsilon^*\}$ and $\{p_\varepsilon^*\}$ are bounded and if v^* and p^* are limits of subsequences, one has:

(a) the minimum of P is attained at v^*,
(b) the minimum of P^* is attained at p^*,
(c) $\inf P = - \inf P^*$,
(d) $\inf P_\varepsilon \to P$ and $\inf P_\varepsilon^* \to \inf P^*$
as $\varepsilon \to 0$. $\square$

Let us clarify the formulation of the primal and dual problems involved in the last theorem. We have:

(P_ε) $\inf\{G_\varepsilon(v, 0) | v \in V\} ,$

(P_ε^*) $\inf\{G_\varepsilon^*(0, p^*) | p^* \in X^*\} = \inf\{h_\varepsilon^*(p^*) | p^* \in X^*\}$
            $= - \sup\{-G_\varepsilon^*(0, p^*) | p^* \in X^*\} .$

We recall that the spaces V and X are not necessarily reflexive.

Proof of Theorem 1.3.36.
Step 1. Let $p_\varepsilon \to p$ strongly as $\varepsilon \to 0$. We want to show that

$$\lim_\varepsilon \inf h_\varepsilon(p_\varepsilon) \geq h(p) .$$

To this end we take a sequence of real numbers $\{M_{1/k}\}$ such that $M_{1/k} \downarrow \liminf_{\varepsilon} h_\varepsilon(p_\varepsilon)$. There exists a subsequence $\{\varepsilon_k\}$ and $v^*_{1/k} \in V^*$ such that

$$G_{\varepsilon_k}(v^*_{1/k}, p_{\varepsilon_k}) < M_{1/k} \;.$$

On account of Corollary 1.3.31 and the assumption (A^*) the sequence $\{v^*_{1/k}\}$ is bounded in V^*. Since the space V is separable therefore there exists a subsequence, still denoted by $v^*_{1/k}$, which converges to an element $v^* \in V^*$ in the topology $\sigma(V^*, V)$. Thus we have

$$\liminf_{\varepsilon} h_\varepsilon(p_\varepsilon) = \liminf_{k} M_{1/k} \geq \liminf_{k} G_{\varepsilon_k}(v^*_{1/k}, p_{\varepsilon_k}) \geq G(v^*, p) \geq h(p) \;.$$

Here we have used (1.3.70).

Step 2. Let us show that there exists $p_\varepsilon \xrightarrow{s} p$ as $\varepsilon \to 0$ such that $h_\varepsilon(p_\varepsilon) \leq h(p)$. We may assume that $h(p) < +\infty$. Let $\{M_{1/k}\}$ be a sequence of real numbers such that $M_{1/k} \downarrow h(p)$. There exists $v^*_{1/k} \in V^*$ such that $G(v^*_{1/k}, p) < M_{1/k}$. According to (1.3.70), there exist:

$$p_{\varepsilon, 1/k} \to p \;\; \text{in the strong topology of } X \text{ as } \varepsilon \to 0 \;,$$

and

$$v^*_{\varepsilon, 1/k} \rightharpoonup v^*_{1/k} \;\; \text{in the topology} \;\; \sigma(V^*, V) \;\; \text{as} \;\; \varepsilon \to 0 \;,$$

such that

$$\limsup_{\varepsilon} G_\varepsilon(v^*_{\varepsilon, 1/k}, p_{\varepsilon, 1/k}) \leq G(v^*_{1/k}, p) \;.$$

Hence

$$\limsup_{k} (\limsup_{\varepsilon} h_\varepsilon(p_{\varepsilon, 1/k})) \leq h(p) \;.$$

By diagonalization, there exists $k(n) \to \infty$ such that $(\varepsilon = 1/n)$

$$p_{\varepsilon, 1/k(n)} \xrightarrow{s} p \;, \qquad \limsup_{\varepsilon} h_\varepsilon(p_{\varepsilon, 1/k(n)}) \leq h(p)$$

as $n \to \infty$. Thus (i) is proved.

Step 3. The assumption (A) yields: for each sequence $\{p_\varepsilon\}_{\varepsilon \in \mathcal{E}}$ with $p_\varepsilon \in B_r$

$$\limsup_{\varepsilon} h_\varepsilon(p_\varepsilon) < +\infty \;.$$

We now apply Theorem 1.3.35 with $V = V^* = \{0\}$ and $\widetilde{G}_\varepsilon(v^*, p) = \overline{h}_\varepsilon(p)$, the lower semicontinuous regularization of h_ε. Since $h = \Gamma(s){-}\lim_{\varepsilon \to 0} h_\varepsilon$, therefore we also have $h = \Gamma(s){-}\lim_{\varepsilon \to 0} \overline{h}_\varepsilon$. It is obvious that $\overline{h}_\varepsilon \in \Gamma_0(X)$ and $\overline{h}^*_\varepsilon = h^*_\varepsilon$. Thus $h^* = \Gamma(w^*){-}\lim_{\varepsilon \to 0} h^*_\varepsilon$ and (ii) is shown.

Step 4. We observe that (iii) results directly from (i) and (ii) combined with Theorem 1.3.35. Indeed, from Theorem 1.3.35 we deduce that $G^* = \Gamma(s \times w^*) - \lim_{\varepsilon \to 0} G_\varepsilon^*$. Thus if we set $k_\varepsilon(v) = \inf \{G_\varepsilon^*(v, p^*) | p^* \in X^*\}$, by using (i) and (ii) we have

$$k = \Gamma(s) - \lim_{\varepsilon \to 0} k_\varepsilon \qquad \text{and} \qquad k^* = \Gamma(w^*) - \lim_{\varepsilon \to 0} k_\varepsilon^* \,.$$

Here $k_\varepsilon^*(v^*) = G_\varepsilon(v^*, 0)$ and $k^*(v^*) = G(v^*, 0)$.

Step 5. $\inf P = -\inf P^*$ results from the fact that h is lower semicontinuous as the strong Γ-limit of a sequence of functionals.

To terminate, it suffices to show that the sequence δ_ε is bounded, because the desired convergence results are then a consequence of variational properties of Γ-convergence. We observe that:

(A^*) implies $\lim_\varepsilon \sup \, (\inf P_\varepsilon^*) < +\infty$,

(A) implies $\lim_\varepsilon \sup \, (\inf P_\varepsilon) < +\infty$.

The duality theory yields, cf. Sec. 1.2

$$\inf P_\varepsilon^* \geq -\inf P_\varepsilon^* \,.$$

Hence, for ε sufficiently small we obtain

$$-\infty < \inf P_\varepsilon < +\infty \,, \qquad -\infty < \inf P_\varepsilon^* < +\infty \,.$$

Let $\{v_\varepsilon^*\}$ and $\{p_\varepsilon^*\}$ be, up to δ_ε, minimizing sequences for the problems P_ε and P_ε^*, respectively. We have

$$G_\varepsilon(v_\varepsilon^*, 0) \leq \inf P_\varepsilon + \delta_\varepsilon \,, \qquad G_\varepsilon^*(0, p_\varepsilon^*) \leq \inf P_\varepsilon^* + \delta_\varepsilon \,.$$

Hence we deduce, by using Lemma 1.3.30 and Corollary 1.3.31, that $\{v_\varepsilon^*\}$ and $\{p_\varepsilon^*\}$ are bounded. This completes the proof. $\qquad \square$

Remark 1.3.37. In applications the assumption (A) can easily be verified. The same, however, cannot be said about (A^*). We are going to indicate a situation where (A) and (A^*) hold.

Theorem 1.3.38. Under the following assumptions:

$$G = \Gamma(w^* \times s) - \lim_{\varepsilon \to 0} G_\varepsilon \,, \tag{1.3.72}$$

$$(A) \,, \tag{1.3.73}$$

$$\forall \varepsilon \in \mathcal{E} \,, \quad \forall v^* \in V^* \,, \quad G_\varepsilon(v^*, 0) \geq m(\|v^*\|) \,, \tag{1.3.74}$$

where m is a coercive, convex and even function, the conclusions of Theorem 1.3.36 are verified. $\qquad \square$

We shall need the following result.

Lemma 1.3.39. Under the assumption (A) one has:

$$\lim_{\varepsilon} \sup \left(\sup_{||p|| \leq r} h_\varepsilon(p) \right) < +\infty .$$

Proof. Suppose the assertion of the lemma is false. Then there would exist n_k and $p_{1/k} \in B_r$ such that $h_{1/n_k}(p_{1/k}) \geq k$. Thus we would have $\lim \sup_{1/k} h_{1/n_k}(p_{1/k}) = +\infty$, which on account of (A) is excluded. $\qquad\square$

The last lemma and (1.3.73) imply that beginning from a certain ε the function h_ε is finite and continuous at 0, cf. Sec. 1.2. Consequently we have $h_\varepsilon^{**} > -\infty$ and thus:

$$G_\varepsilon^*(0, \cdot) = h_\varepsilon^* \not\equiv +\infty . \tag{1.3.75}$$

Proof of Theorem 1.3.38. It is sufficient to show that (A*) is verified.

For $v \in V$ we set

$$k_\varepsilon(v) = \inf\{G_\varepsilon^*(v, p^*)|p^* \in X^*\} .$$

We have

$$k_\varepsilon^*(v^*) = G_\varepsilon(v^*, 0) \geq m(||v^*||) .$$

Suppose that $k_\varepsilon \in \Gamma_0(V)$, hence we would have $k_\varepsilon(v) \leq m^*(||v||)$. Since m is coercive, therefore there exist $M \in \mathbb{R}$ and r^* such that for each $v \in V$ with $||v|| \leq r^*$ one has $k_\varepsilon(v) < M$.

If $\{v_\varepsilon\}_{\varepsilon>0}$ is a sequence with $||v_\varepsilon|| \leq r^*$, there exists $\{p_\varepsilon^*\}_{\varepsilon>0}$ such that $\lim \sup_\varepsilon G^*(v_\varepsilon, p_\varepsilon^*) < +\infty$. According to Lemma 1.3.30, the sequence $\{p_\varepsilon^*\}_{\varepsilon>0}$ is bounded and (A*) is then obviously satisfied.

Thus it suffices to show that $k_\varepsilon \in \Gamma_0(V)$. We note that $k_\varepsilon \not\equiv +\infty$ because k_ε^* does not assume the value $-\infty$ and $k_\varepsilon > -\infty$ because k_ε^* is not identically equal to $+\infty$, at least for ε sufficiently small, due to (A). It remains to show that k_ε is lower semicontinuous. This will be shown if one proves that for each $r > 0$ the functional

$$l_\varepsilon(p^*) = \inf\{G_\varepsilon^*(v, p^*)| \, ||v|| \leq r\}$$

is coercive. In turn, this fact will be implied by $l_\varepsilon^* \in \Gamma_0(X)$ and its continuity at 0.

Let us calculate l_ε^*. To this end we introduce

$$\Phi_\varepsilon(v, p^*) = G_\varepsilon^*(v, p^*) + \psi(v, p^*)$$

with

$$\psi(v, p^*) = I_{B_r}(v) = \begin{cases} 0 & \text{if} \quad ||v|| \geq r , \\ +\infty & \text{otherwise} . \end{cases}$$

Obviously, ψ is convex and lower semicontinuous in the topology $(s - V) \times (w^* - X^*)$. Thus, by Lemma 1.3.32 it is lower semicontinuous in the topology $\sigma(V, V^*) \times \sigma(X^*, X)$.

Moreover, ψ is continuous at $(0, p^*)$ in $(s$ - $V) \times (w^*$ - $X^*)$, which is an admissible topology for the duality $\sigma(V \times X^*, V^* \times X)$.

According to (1.3.72) there exist elements of the form $(0, p^*)$ which belong to dom G^*_ε. Then we get:

$$\Phi^*_\varepsilon(v^*, p) = (G_\varepsilon \square \psi^*)(v^*, p) \,,$$

and

$$l_\varepsilon(p^*) = \inf\{\Phi_\varepsilon(v, p^*) | v \in V\} \,.$$

Hence

$$l^*_\varepsilon(p) = \Phi^*_\varepsilon(0, p) = (G_\varepsilon \square \psi^*)(0, p) \,.$$

We observe that

$$\psi^*(v^*, p) = \begin{cases} r\|v^*\| & \text{if} \quad p = 0 \,, \\ +\infty & \text{otherwise} \,. \end{cases}$$

Hence

$$l^*_\varepsilon(p) = \inf\{G_\varepsilon(v^*, p) + r\|v^*\| \mid v^* \in V^*\} \,.$$

It is evident that $l^*_\varepsilon \not\equiv +\infty$ and $l^*_\varepsilon > -\infty$, otherwise we would have $l^{**}_\varepsilon \equiv +\infty$ and $l_\varepsilon \equiv +\infty$, which is not the case.

According to Lemma 1.3.39, the functional l^*_ε is finite in a neighborhood of 0. Being lower semicontinuous on a Banach space this functional is continuous at 0. The proof is complete. $\qquad\square$

1.3.7. Convergence of sets in Kuratowski's sense

Let $\{C_\varepsilon\}_{\varepsilon>0}$ be a sequence of subsets of a topological space V.

Definition 1.3.40. The K-*lower limit* of the sequence $\{C_\varepsilon\}_{\varepsilon>0}$, denoted by K - $\lim\inf\limits_{\varepsilon\to 0} C_\varepsilon$, is the set of all points $v \in V$ with the following property: for every $U \in \mathcal{N}(v)$ there exists $\varepsilon_0 > 0$ such that $U \cap C_\varepsilon \neq \emptyset$ for every $\varepsilon \leq \varepsilon_0$. The K-*upper limit*, denoted by K - $\lim\sup\limits_{\varepsilon\to 0} C_\varepsilon$, is the set of all points $v \in V$ with the following property: for every $U \in \mathcal{N}(v)$, for every $\varepsilon_0 > 0$, $\varepsilon_0 \in \mathcal{E}$, there exists $\varepsilon \leq \varepsilon_0$ such that $U \cap C_\varepsilon \neq \emptyset$. If there exists a set $C \subseteq V$ such that $C = K$ - $\lim\inf\limits_{\varepsilon\to 0} C_\varepsilon = K$- $\lim\sup\limits_{\varepsilon\to 0} C_\varepsilon$ then we write $C = K$ - $\lim\limits_{\varepsilon\to 0} C_\varepsilon$, and we say that the sequence $\{C_\varepsilon\}_{\varepsilon>0}$ *converges to C in the sense of Kuratowski*, or *K-converges to C* (in V). $\qquad\square$

The K-lower limit is sometimes denoted by Lim inf, and similarly for K-upper limit, cf. Aubin and Frankowska (1990). Then $C = \operatorname*{Lim}\limits_{\varepsilon\to 0} C_\varepsilon$.

The above definition implies that

$$K-\lim\inf_{\varepsilon\to 0} C_\varepsilon \subset K-\lim\sup_{\varepsilon\to 0} C_\varepsilon \,,$$

hence $\{C_\varepsilon\}_{\varepsilon>0}$ K-converges to C if and only if

$$K-\limsup_{\varepsilon\to 0} C_\varepsilon \subseteq C \subseteq K-\liminf_{\varepsilon\to 0} C_\varepsilon .$$

Example 1.3.41. If C is a subset of V and $C_\varepsilon = C$ for every $\varepsilon \in \mathcal{E}$, then $\{C_\varepsilon\}_{\varepsilon>0}$ K-converges to $\overline{C}$, the closure of C in V. $\qquad\square$

The K-convergence of a sequence of sets is equivalent to the Γ-convergence of the corresponding indicator functions. Indeed, we have the following result.

Proposition 1.3.42. Let $\{C_\varepsilon\}_{\varepsilon>0}$ be a sequence of subsets of V and let

$$C' = K-\liminf_{\varepsilon\to 0} C_\varepsilon , \qquad C'' = K-\limsup_{\varepsilon\to 0} C_\varepsilon .$$

Then

$$I_{C'} = \Gamma-\limsup_{\varepsilon\to 0} I_{C_\varepsilon} \qquad I_{C''} = \Gamma-\liminf_{\varepsilon\to 0} I_{C_\varepsilon} .$$

In particular $\{C_\varepsilon\}_{\varepsilon>0}$ K-converges to C in V if and only if $\{I_{C_\varepsilon}\}$ Γ-converges to I_C in V.$\square$

The following statement reveals the connection between Γ-convergence of functions and K-convergence of their epigraphs. Due to this connection, the Γ-convergence is sometimes called epi-convergence, cf. Attouch (1984).

Theorem 1.3.43. Let $\{G_\varepsilon\}_{\varepsilon>0}$ be a sequence of functions from V int $\overline{\mathbb{R}}$, and let

$$G_i = \Gamma-\liminf_{\varepsilon\to 0} G_\varepsilon , \qquad G_s = \Gamma-\limsup_{\varepsilon\to 0} G_\varepsilon .$$

Then

$$\mathrm{epi}\, G_i = K-\limsup_{\varepsilon\to 0}\, \mathrm{epi}\, G_\varepsilon , \qquad \mathrm{epi}\, G_s = K-\liminf_{\varepsilon\to 0}\, \mathrm{epi}\, G_\varepsilon ,$$

where the K-limits are taken in the product topology $V \times \mathbb{R}$. In particular, $\{G_\varepsilon\}$ Γ-converges to G in V if and only if $\{\mathrm{epi}\, G_\varepsilon\}_{\varepsilon>0}$ K-converges to $\mathrm{epi}\, G$ in $V \times \mathbb{R}$. $\qquad\square$

Remark 1.3.44. The notion of K-convergence has been introduced after Dal Maso (1993, Chap. 4), cf. also Attouch (1984) and Aubin and Frankowska (1990).

Remark 1.3.45. In Section 13.2 we shall provide an example of K-convergence applicable to two-phase plastic composites.

1.4. *Two approximation results*

Most of the homogenization results presented in this book will be justified by using Γ-convergence. In turn, in the prevailing number of cases the density of continuous affine functions in the space $H^1(\Omega)$ will be exploited. Also, for most problems involving second-order derivatives, we shall use the fact that C^1-functions with piecewise constant second gradient are dense in $H^2(\Omega)$. In the last case Ω is a two-dimensional bounded domain.

Definition 1.4.1.
(a) Let $\Omega \subset \mathbb{R}^n (n \leq 3)$ be a bounded domain. A function $u : \Omega \to \mathbb{R}$ is affine if it is a restriction to Ω of an affine function on $\mathbb{R}^n$.
(b) A function $u : \Omega \to \mathbb{R}$ is called *piecewise affine* if it is continuous and there exists a partition of Ω on a set with zero measure and a finite number of open sets on which u is affine. $\qquad\qquad\square$

The first approximation result is formulated in the following form.

Proposition 1.4.2. Let Ω be a bounded domain in $\mathbb{R}^n (n \leq 3)$ with a Lipschitz boundary and $u \in H^1(\Omega)$. There exists a sequence $\{u^m\}_{m\in\mathbb{N}}$ of piecewise affine functions on Ω such that

$$u^m \to u \quad \text{in} \quad H^1(\Omega) \quad \text{when} \quad m \to \infty .$$

Proof. It is sufficient to prove the assertion for a function $u \in C^\infty(\Omega)$, continuous on $\bar{\Omega}$. Details are given in the book by Ekeland and Temam (1976, Chap. X). $\qquad\square$

Having in mind application to plate problems, we have formulated this nice approximation results for $n \leq 3$ only. It can be formulated for any finite n and spaces $W^{1,p}(\Omega)$.

The second result, this time concerning approximation of functions from the space $W^{2,p}(\Omega)$ $(1 \leq p < \infty, \Omega \subset \mathbb{R}^2)$ by means of C^1-functions with piecewise constant second gradient, seems not to be so well known. Here we shall give a detailed proof due to Descloux, cf. also Birman and Solomyak (1967), Neuman and Schmidt (1983), Orlov (1978), Powell and Sabin (1977) and Sablonnière (1989).

Let $\Omega \subset \mathbb{R}^2$ be a quadrangle represented in Fig. 1.4.1.

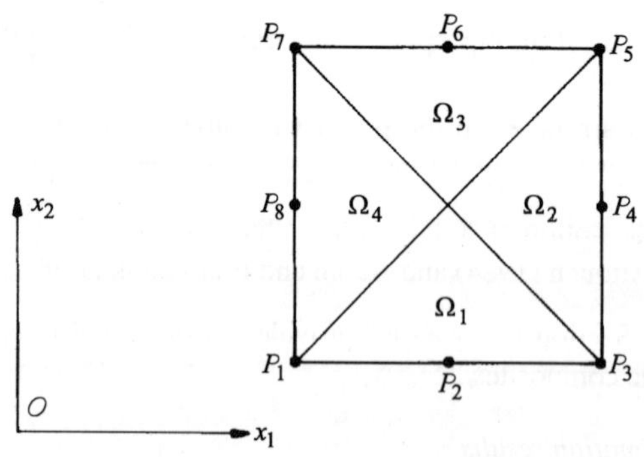

Fig. 1.4.1

Here P_2, P_4, P_6 and P_8 are mid-points of the corresponding sides of Ω.

Proposition 1.4.3. For each $f_k \in \mathbb{R}$, $1 \le k \le 8$, there exists one and only one function f such that

(a)
$$f \in C^1(\bar{\Omega}) \,,$$

(b)
$$f_{|\Omega_l} \in \mathcal{P}_2 \,, \quad 1 \le l \le 4 \,,$$

(c)
$$f(P_k) = f_k \,, \quad 1 \le h \le 8 \,,$$

where $\mathcal{P}_2$ denotes the set of polynomials of order not exceeding 2. $\qquad \square$

Prior to providing the proof, an auxiliary result will be given. By using rotation and homothety one can refer the results which follow to Fig. 1.4.2.

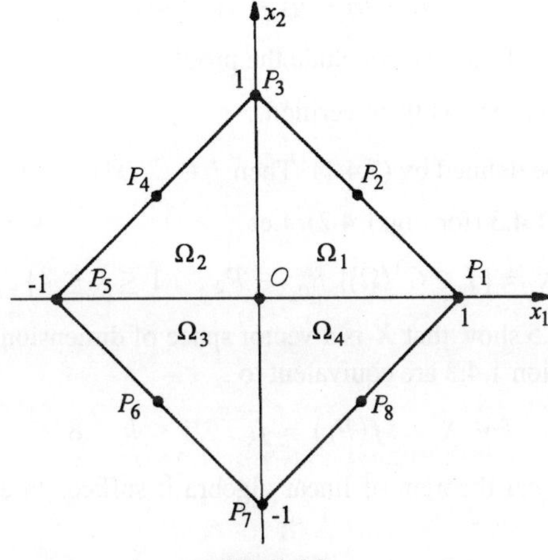

Fig. 1.4.2.

Lemma 1.4.4. Let $f \in C^1(\bar{\Omega})$ be such that $f_{|\Omega_k} \in \mathcal{P}_2$ for $1 \le k \le 4$. Then there exist constants $\alpha, \beta, \gamma, \delta, \omega, a, b, c$ such that

$$
\begin{aligned}
f(x_1, x_2) &= \alpha x_1^2 + \omega x_1 x_2 + \beta x_2^2 + a x_1 + b x_2 + c \quad &&\text{on} \quad \Omega_1 \,, \\
f(x_1, x_2) &= \gamma x_1^2 + \omega x_1 x_2 + \beta x_2^2 + a x_1 + b x_2 + c \quad &&\text{on} \quad \Omega_2 \,, \\
f(x_1, x_2) &= \gamma x_1^2 + \omega x_1 x_2 + \delta x_2^2 + a x_1 + b x_2 + c \quad &&\text{on} \quad \Omega_3 \,, \\
f(x_1, x_2) &= \alpha x_1^2 + \omega x_1 x_2 + \delta x_2^2 + a x_1 + b x_2 + c \quad &&\text{on} \quad \Omega_4 \,.
\end{aligned}
\tag{1.4.1}
$$

Proof. We set

$$a = \partial_1 f(0,0) \,, \quad b = \partial_2 f(0,0) \,, \quad c = f(0,0) \,, \tag{1.4.2}$$

where $\partial_1 = \partial/\partial x_1$, $\partial_2 = \partial/x_2$.
Let

$$g(x_1, x_2) = f(x_1, x_2) - a x_1 - b x_2 - c \,. \tag{1.4.3}$$

Hence

$$g_{|\Omega_k} \in \mathcal{P}_2 \,, \quad \partial_1 g_{|\Omega_k}(0,0) = \partial_2 g_{|\Omega_k}(0,0) = g_{|\Omega_k}(0,0) = 0$$

for $1 \leq k \leq 4$. Consequently there exist coefficients p_k, q_k and r_k such that

$$g_{|\Omega_k}(x_1, x_2) = p_k x_1^2 + q_k x_1 x_2 + r_k x_2^2 \,, \quad 1 \leq k \leq 4 \,. \tag{1.4.4}$$

The continuity of g along the axes x_1 and x_2 yields the following relations

$$p_1 = p_4 := \alpha, \quad r_1 = r_2 : +\beta, \quad p_2 = p_3 := \gamma \,, \quad r_3 = r_4 := \delta \,. \tag{1.4.5}$$

Next, from the continuity of the normal derivative of g along the same axes one gets

$$q_1 = q_2 = q_3 = q_4 := \omega \,. \tag{1.4.6}$$

On account of (1.4.3) – (1.4.6) we conclude the proof. $\square$

The following lemma can easily be verified.

Lemma 1.4.5. Let f be defined by (1.4.1). Then $f \in C^1(\bar{\Omega})$. $\square$

Proof of Proposition 1.4.3 (for Fig. 1.4.2). Let

$$X = \{f \in C^1(\bar{\Omega})|\ f_{|\Omega_k} \in \mathcal{P}_2 \,, \quad 1 \leq k \leq 4\} \,.$$

Lemmas 1.4.4 and 1.4.5 show that X is a vector space of dimension 8. The conditions (a), (b) and (c) of Proposition 1.4.3 are equivalent to

$$f \in X \,, \quad f(\mathcal{P}_k) = f_k \,, \quad 1 \leq k \leq 8 \,.$$

On account of a classical theorem of linear algebra it suffices to establish the following relation

$$f \in X \,, \quad f(\mathcal{P}_k) = 0 \,, \quad 1 \leq k \leq 8 \implies f = 0 \,. \tag{1.4.7}$$

Let $f \in X$ be such that $f(\mathcal{P}_k) = 0, 1 \leq k \leq 8$. Hence one immediately concludes that f vanishes on $\partial\Omega$ and that $\partial_1 f(\mathcal{P}_k) = \partial_2 f(\mathcal{P}_k) = 0$ for $k = 1, 3, 5, 7$. Thus we have

$$\begin{aligned}
f(1,0) &= \alpha + a + c = 0 &&\text{on}\quad \Omega_1 \,, &&(1.4.8)\\
f(0,1) &= \beta + b + c = 0 &&\text{on}\quad \Omega_1, &&(1.4.9)\\
\partial_1 f(1,0) &= 2\alpha + a = 0 &&\text{on}\quad \Omega_1 \,, &&(1.4.10)\\
\partial_2 f(1,0) &= \omega + b = 0 &&\text{on}\quad \Omega_1 \,, &&(1.4.11)\\
\partial_1 f(0,1) &= \omega + a = 0 &&\text{on}\quad \Omega_1 \,, &&(1.4.12)\\
\partial_2 f(0,1) &= 2\beta + b = 0 &&\text{on}\quad \Omega_1 \,, &&(1.4.13)\\
\partial_2 f(-1,0) &= -\omega + b = 0 &&\text{on}\quad \Omega_2 \,. &&(1.4.14)
\end{aligned}$$

Equations (1.4.11) and (1.4.14) yield $b = \omega = 0$, Eqs. (1.4.12), (1.4.10), (1.4.8) and (1.4.13) imply that $a = 0$, $\alpha = 0$, $c = 0$ and $\beta = 0$ respectively. We have thus established that $\alpha = \beta = \omega = a = b = c = 0$; moreover (1.4.1) readily gives $\gamma = \delta = 0$. The proof is complete. $\square$

Later on we shall need certain results on quadratic one-dimensional splines. Let N be a positive integer, $h = \dfrac{1}{N}$, $x_i = ih$, $0 \le i \le N$ and

$$S_h = \{ f \in C^1[0,1] \mid f_{|[x_{i-1},x_i]} \in \mathcal{P}_2 \, , \, 1 \le i \le N \} \, . \tag{1.4.15}$$

We employ here the conventional notation and "h" used in this section has obviously nothing in common with homogenized quantities. One immediately verifies the following result.

Lemma 1.4.6. Let $\beta, \alpha_0, \alpha_1, \ldots, \alpha_N \in \mathbb{R}$ be given numbers. Then there exists one and only one element $s \in S_h$ such that $s(x_i) = \alpha$, $0 \le i \le N$ and $s'(0) = \beta$. $\square$

We introduce the next definition.

Definition 1.4.7. Let $f \in C^1[0,1]$. An element $s \in S_h$ such that $s(x_i) = f(x_i)$, $0 \le i \le N$ and $s'(0) = f'(0)$ is called the interpolant of f with respect to S_h.

Lemma 1.4.8. Let $f \in C^2[0,1]$ and, for fixed i, $p \in \mathcal{P}_2$ be such that $f(x_i) = p(x_i)$, $f(x_{i+1}) = p(x_{i+1})$. Let $c \ge 0$ be such that

$$|f''(x) - p''(x)| \le ch \qquad \forall \, x \in [x_i, x_{i+1}] \, . \tag{1.4.16}$$

Then

(a) $\qquad\qquad |f'(x) - p'(x)| \le ch^2 \qquad \forall \, x \in [x_i, x_{i+1}] \, ,$

(b) $\qquad\qquad |f(x) - p(x)| \le ch^3 \qquad \forall \, x \in [x_i, x_{i+1}] \, .$

Proof. We set $r(x) = f(x) - p(x)$. By applying Rolle's theorem we conclude that there exists $\xi \in (x_i, x_{i+1})$ such that $r'(\xi) = 0$. We have

$$r'(x) = \int_{\xi}^{x} r''(\eta)d\eta \, .$$

Hence, by using (1.4.16) we conclude the assertion (a).

Next we write

$$r(x) = \int_{x_i}^{x} r'(\eta)d\eta \, .$$

Hence (b) readily follows. $\square$

Let us consider the function $\varphi : \mathbb{R} \to \mathbb{R}$ defined by

$$\varphi(x) = \begin{cases} 0 & \text{if } x \le -h \text{ and } x \ge 2h \, , \\[2mm] \left(\dfrac{x+h}{h} \right)^2 & \text{if } -h \le x \le 0 \, , \\[2mm] 1 - \dfrac{2}{h^2}x(x-h) & \text{if } 0 \le x \le h \, , \\[2mm] \left(\dfrac{x-2h}{h} \right)^2 & \text{if } h \le x \le 2h \, . \end{cases} \tag{1.4.17}$$

Properties of φ

$$\varphi \in C^1(\mathbb{R}) , \quad \varphi|_{[ih,(i+1)h]} \in \mathcal{P}_2 , \quad \forall\, i \in \mathbb{Z} \quad \varphi(0) = \varphi(h) = 1 , \tag{1.4.18}$$

$$\varphi''(x) = \begin{cases} \dfrac{2}{h} & \text{if } x \in [-h, 0] \cup [h, 2h], \\[2ex] -\dfrac{4}{h^2} & \text{if } x \in [0, h] , \end{cases} \tag{1.4.19}$$

where $\mathbb{Z}$ stands for the set of integers.

Performing simple calculation, from (1.4.18) and (1.4.19) one concludes

Lemma 1.4.9. Let $f \in C^1[0, 1]$ be such that $f(0) = f'(0) = 0$ and, for $0 \le x \le 1$, let s be the function given by

$$s(x) = f(0)\varphi(x) + (f(h) - f(0))\varphi(x - h) + (f(2h) - f(h) + f(0))\varphi(x - 2h)$$
$$+ \ldots + [f(Nh) - f((N-1)h) + f((N-2)h) + \ldots + (-1)^N f(0)]\varphi(x - Nh) . \tag{1.4.20}$$

Then

(a) s is the interpolant of f with respect to S_h;

(b) $s''(x) = \dfrac{2}{h^2}f(h) , \quad 0 \le x \le h;$ $\tag{1.4.21}$

(c) $s''(x) = \dfrac{2}{h^2}\{f(x_{2m}) + f(x_{2m-1}) - 4\sum_{l=0}^{m-1}[f(x_{2l+1}) - f(x_{2l})]\},$ $\tag{1.4.22}$

for x and m such that $h \le x_{2m-1} \le x \le x_{2m} \le 1$;

(d) $s''(x) = \dfrac{2}{h^2}\{-3f(x_{2m+1}) + f(x_{2m}) + 4\sum_{l=0}^{m}[f(x_{2l+1}) - f(x_{2l})]\},$ $\tag{1.4.23}$

for x and m such that $2h \le x_{2m} \le x \le x_{2m+1} \le 1.$ $\qquad\qquad \square$

Later on in this section we shall need the following result.

Proposition 1.4.10. Let $f \in C^4[0, 1]$ and denote by s_h the interpolant with respect to S_h. Then there exists a constant c, independent of f and h such that

$$||f - s_h|| + h||f' - s_h'|| + h^2||f'' - s_h''|| \le ch^3(||f'''|| + ||f^{iv}||) ,$$

where, for $g \in C^0[0, 1]$, we introduce the notation: $||g|| = \sup\limits_{0 \le x \le 1} |g(x)|.$

Proof. We note that if p is a polynomial of the second order on $[0,1]$, then the interpolant of $f + p$ is given by $s_h + p$. Thus, without loss of generality we may assume that

$$f(0) = f'(0) = 0 . \tag{1.4.24}$$

On account of Lemma 1.4.8, it is sufficient to establish existence of a constant c_1, independent of f and h, such that

$$||f'' - s_h''|| \leq c_1 h(||f'''|| + ||f^{\text{IV}}||) . \qquad (1.4.25)$$

Due to (1.4.24), the proof is reduced to evaluation of the expressions (1.4.21) – (1.4.23). In the sequel of this proof, we denote by r a *generic function* which may depend on x and h and such that there exists a constant c_2, independent of x, h and f, such that

$$|r(x, h)| \leq c_2(||f'''|| + ||f^{\text{IV}}||) .$$

Case 1. For $0 \leq x \leq h$, by (1.4.21), (1.4.24) and Taylor's formula one gets

$$s_h''(x) = \frac{2}{h^2} \left(\frac{h^2}{2} f''(0) + h^3 r \right) = f''(0) + hr .$$

This implies existence of c_1 such that

$$\max_{0 \leq x \leq h} |f''(x) - s_h''(x)| \leq c_1 h ||f'''|| . \qquad (1.4.26)$$

To treat the remaining two cases related to (1.4.22) and (1.4.23), from the trapezoid rule with correction one deduces, cf. Davis and Rabinowitz (1967, p. 53), Davis and Rabinowitz (1975)

$$2h\{f(a) + f(a + 2h) + f(a + 4h) + \ldots + f(a + 2mh)\}$$
$$= \quad h[f(a) + f(a + 2mh)] + \int_a^{a+2mh} f(x)dx + \frac{h^2}{3}[f'(a + 2mh) - f'(a)]$$
$$+ h^3 r , \quad 0 \leq a < a + 2mh \leq 1 . \qquad (1.4.27)$$

By applying successively the last formula with $a = 0$ and $a = h/2$, one gets

$$2h\sum_{l=0}^{m}[f(x_{2l+1}) - f(x_{2l})] = h[f(h) - f(0)] - \frac{h^2}{3}[f'(h) - f'(0)] - \int_0^h f(x)dx$$
$$+ h[f(x_{2m+1}) - f(x_{2m})] + \int_{x_{2m}}^{x_{2m+1}} f(x)dx + \frac{h^2}{3}[f'(x_{2m+1}) - f'(x_{2m})] + h^4 r . \quad (1.4.28)$$

Taking into account (1.4.24), we have

$$h[f(h) - f(0)] - \frac{h^2}{3}[f'(h) - f'(0)] - \int_0^h f(x)dx$$
$$= \frac{h^3}{2} f''(0) - \frac{h^3}{3} f''(0) - \frac{h^3}{6} f''(0) + h^4 r = h^4 r . \qquad (1.4.29)$$

Case 2. For $x_{2m-1} \le x \le x_{2m}$, on account of (1.4.22), (1.4.28) with m being replaced by $m-1$, (1.4.29) and setting $b = x_{2m-1}$ we obtain

$$s_h''(x) = \frac{2}{h^2}\{f(b+h) + f(b) - \frac{2}{h}[h(f(b) - f(b-h))$$

$$+ \int_{b-h}^{b} f(x)dx + \frac{h^2}{3}(f'(b) - f'(b-h))]\} + hr \, ,$$

$$s_h''(x) = \frac{2}{h^2}\{f(b+h) - f(b) + 2f(b-h) - \frac{2}{h}\int_{b-h}^{b} f(x)dx$$

$$- \frac{2h}{3}[f'(b) - f'(b-h)]\} + hr \, ,$$

$$s_h''(x) = \frac{2}{h^2}\{hf'(b) + \frac{h^2}{2}f''(b) + 2[f(b) - hf'(b) + \frac{h^2}{2}f''(b)]$$

$$- \frac{2}{h}[hf(b) - \frac{h^2}{2}f'(b) + \frac{h^3}{6}f''(b)] - \frac{2h^2}{3}f''(b) + h^3 r\}$$

$$= \frac{2}{h^2}[\frac{h^2}{2}f''(b) + h^3 r] \, ,$$

$$s_h''(x) = f''(x) + h^3 r \, , \quad x_{2m-1} \le x \le x_{2m} \, . \tag{1.4.30}$$

We recall that r is a generic function introduced earlier.

Case 3. For $x_{2m} \le x \le x_{2m+1}$, on account of (1.4.23), (1.4.28) and (1.4.29), putting $b = x_{2m}$ we get

$$s_h''(x) = \frac{2}{h^2}\{-3f(b+h) + f(b) + \frac{2}{h}[h(f(b+h) - f(b))$$

$$+ \int_{b}^{b+h} f(x)dx + \frac{h^2}{3}(f'(b+h) - f'(b))]\} + hr \, ,$$

$$s_h''(x) = \frac{2}{h^2}\{-f(b+h) - f(b)$$

$$+ \frac{2}{h}\int_{b}^{b+h} f(x)dx + \frac{2h}{3}[f'(b+h) - f'(b)]\} + hr \, ,$$

$$s_h''(x) = \frac{2}{h^2}\{-2f(b) - hf'(b)$$

$$- \frac{h^2}{2}f''(b) + 2f(b) + hf'(b) + \frac{h^2}{3}f''(b) + \frac{2h^2}{3}f''(b)\} + hr \, ,$$

$$s_h''(x) = \frac{2}{h^2}\frac{h^2}{2}f''(b) + hr \, ,$$

$$s_h''(x) = f''(x) + hr \, , \quad x_{2m} \le x \le x_{2m+1} \, . \tag{1.4.31}$$

From (1.4.26), (1.4.30) and (1.4.31) we eventually obtain (1.4.25) and thus the proof is complete. $\qquad\square$

Now we are in a position to pass to two-dimensional approximations.

Let $D = \{(x_1, x_2) | 0 \le x_1, x_2 \le 1\}$. By N we denote a positive integer and set $h = 1/N$. Next, we assume the following notations:

T_{ch} – the set of quadrangles $[ih, (i+1)h] \times [jh, (j+1)h]$, $0 \le i, j \le N-1$,
T_{th} – the set of triangles obtained by division of the quadrangles according to Fig. 1.4.3.
N_h – the set of the vertices and mid-points of the sides of the quadrangles of T_{ch},
$U_h = \{u \in C^1(D) | \ u_{|T} \in \mathcal{P}_2, T \in \mathsf{T}_{th}\}$.

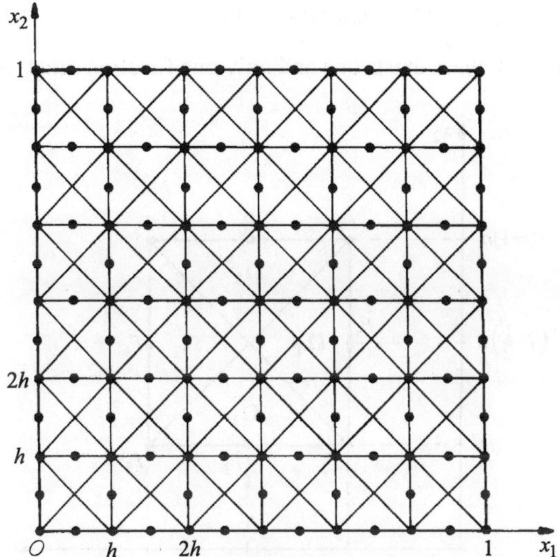

Fig. 1.4.3.

Let $f \in C^1(D)$. We are going to construct an approximation u of f. To this end we set:
$s_{hj} = s_{hj}(x_1)$ – the interpolant of $f(\cdot, jh)$ with respect to S_h in the sense of Definition 1.4.7, $1 \le j \le N$;
$s_{vi} = s_{vi}(x_2)$ – the interpolant of $f(ih, \cdot)$ with respect to S_h in the sense of Definition 1.4.7, $1 \le i \le N$.
By definition of s_{hj} and s_{vi} we have

$$s_{hj}(ih) = s_{vi}(jh) \quad 0 \le i, j \le N . \tag{1.4.32}$$

For $0 \le i, j \le N-1$, let $C_{ij} = [ih, (i+1)h] \times [jh, (j+1)h]$ and let $u_{ij} : C_{ij} \to \mathbb{R}$ satisfy the relations:

$$u_{ij|\Omega_k} \in \mathcal{P}_2 \; ; \quad k = 1, 2, 3, 4; \quad u_{ij} \in C^1(C_{ij}) \; (\Omega_k \text{ is defined in Fig. 1.4.4}),$$

$$u_{ij}(ih, jh) = s_{hj}(ih) \; , \; u_{ij}(i + \tfrac{1}{2}h, jh) = s_{hj}((i + \tfrac{1}{2})h) \; ,$$

$$u_{ij}((i + 1)h, jh) = s_{hj}((i + 1)h) \; ,$$

$$u_{ij}(ih, (j + \tfrac{1}{2})h) = s_{vi}((j + \tfrac{1}{2})h) \; ,$$

$$u_{ij}((i + 1)h, (j + \tfrac{1}{2})h) = s_{vi+1}((j + \tfrac{1}{2})h) \; , \tag{1.4.33}$$

$$u_{ij}(ih, (j + 1)h) = s_{hj+1}(ih) \; ,$$

$$u_{ij}((i + \tfrac{1}{2})h, (j + 1)h) = s_{hj+1}((i + \tfrac{1}{2})h) \; ,$$

$$u_{ij}((i + 1)h, (j + 1)h) = s_{hj+1}((i + 1)h) \; .$$

Due to Proposition 1.4.3, there exists one and only one function satisfying (1.4.33).

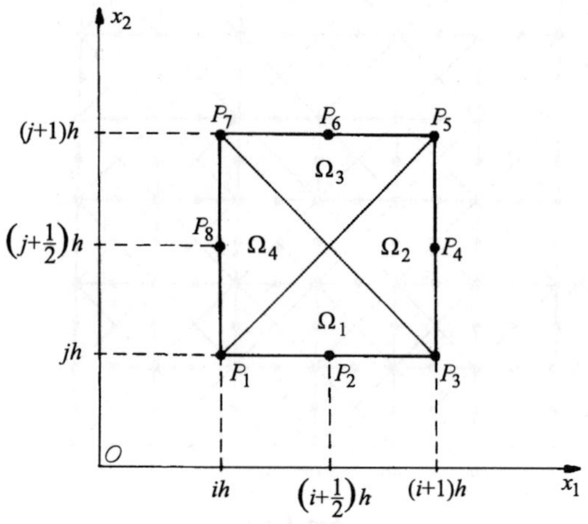

Fig. 1.4.4.

We define $u : D \to \mathbb{R}$ by setting

$$u_{|C_{ij}} = u_{ij} \qquad 0 \leq i, j \leq N - 1 \; . \tag{1.4.34}$$

Lemma 1.4.11. If u is given by (1.4.34) then $u \in U_h$.
Proof. One can easily verify that

$$u(x_1, jh) = s_{hj}(x_1) \; , \qquad 0 \leq x_1 \leq 1 \; , \qquad 0 \leq j \leq N \; ,$$

and

$$u(ih, x_2) = s_{vi}(x_2) \; , \qquad 0 \leq x_2 \leq 1 \; , \qquad 0 \leq i \leq N \; . \tag{1.4.35}$$

Hence one concludes that u is continuous in D. It remains to prove the continuity of the first derivatives $\partial_1 u$ and $\partial_2 u$ of u. To this end, let us consider Fig. 1.4.5 and verify the continuity of $\partial_2 u$ along the interval AC.

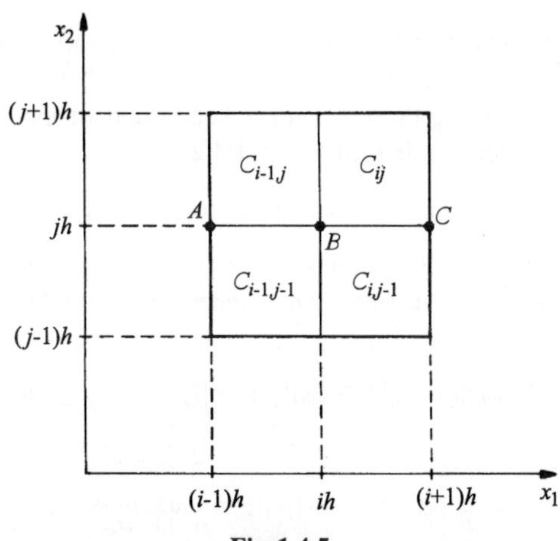

Fig. 1.4.5.

By virtue of (1.4.35), one has

$$\partial_2 u_{i-1,j-1}(A) = \partial_2 u_{i-1,j}(A) \,,$$
$$\partial_2 u_{i-1,j-1}(B) = \partial_2 u_{i,j-1}(B) = \partial_2 u_{i-1,j}(B) = \partial_2 u_{ij}(B) \,,$$
$$\partial_2 u_{i,j-1}(C) = \partial_2 u_{ij}(C) \,. \tag{1.4.36}$$

Along BC the derivatives $\partial_2 u_{i,j-1}$ and $\partial_2 u_{i,j}$ are the polynomials of the first order in the variable x_1. Hence, taking into account (1.4.36) we conclude that $\partial_2 u$ is continuous along AC. $\qquad\Box$

The last lemma shows that the following definition is meaningful.

Definition 1.4.12. The function u given by (1.4.34) is the interpolant of f with respect to U_h.

Lemma 1.4.13. Referring to Fig. 1.4.4, let

$$V_{ij} = \{v \in C^1(C_{ij}) \mid v_{|\Omega_k} \in \mathcal{P}_2 \,, 1 \le k \le 4\} \,, \quad 0 \le i,j \le N-1 \,.$$

Then there exists a constant c, independent of i, j, k and v, such that for $v \in V_{ij}$ one has

$$\|v\|_{i,j,0} + h\|v\|_{i,j,1} + h^2\|v\|_{i,j,2} \le c \max_{1 \le k \le 8} |v(P_k)| \,,$$

where

$$||v||_{i,j,0} = \max_{x \in C_{ij}} |v(x)| \, ,$$

$$||v||_{i,j,1} = \max(\max_{x \in C_{ij}} |\partial_1 v(x)|, \max_{x \in C_{ij}} |\partial_2 v(x)|) \, ,$$

$$||v||_{i,j,2} = \max(\sup_{x \in C_{ij}} |\partial_{11} v(x)|, \sup_{x \in C_{ij}} |\partial_{12} v(x)|, \sup_{x \in C_{ij}} |\partial_{22} v(x)|) \, .$$

Proof. It follows easily by using Proposition 1.4.3 and standard approach consisting in the passage to the reference element defined in Fig. 1.4.2. $\square$

We are now in a position to prove the following proposition.

Proposition 1.4.14. Let $f \in C^4(D)$ and denote by u_h the interpolant of f with respect to U_h (Def. 1.4.12). Then there exists a constant c, independent of h but dependent on f, such that

$$||f - u_h||_0 + h||f - u_h||_1 + h^2||f - u_h||_2 \le ch^3 \, ,$$

where

$$||v||_0 = \max_{x \in D} |v(x)| \, , \qquad ||v||_1 = \max_{\alpha=1,2}(\max_{x \in D} |\partial_\alpha v(x)|),$$

$$||v||_2 = \max_{1 \le \alpha, \beta \le 2}(\max_{x \in D} |\partial_\alpha \partial_\beta v(x)|) \, .$$

Proof. In the sequel c will denote a generic constant, independent of α, β, h, but which may depend on f. We shall only establish the relation

$$||f - u_h||_1 \le ch^2 \, . \tag{1.4.37}$$

For $0 \le i, j \le N - 1$, the reader is referred to Fig. 1.4.4. Let u_{ij} denote the restriction of u_h to C_{ij}. By virtue of construction of u_{ij} and due to Proposition 1.4.10, one has

$$\max_{1 \le k \le 8} |f(P_k) - u_{ij}(P_k)| \le ch^3 \, . \tag{1.4.38}$$

It is not difficult to establish (for instance, by using Taylor's expression) the existence of a second-order polynomial g such that

$$||f - g||_{i,j,0} + h||f - g||_{i,j,1} + h^2||f - g||_{i,j,2} \le ch^3 \, . \tag{1.4.39}$$

Here we have used notations of Lemma 1.4.13; g depends on i, j, h but not on c. We have

$$||f - u_h||_{i,j,1} \le ||f - g||_{i,j,1} + ||u_{ij} - g||_{i,j,1} \, . \tag{1.4.40}$$

We observe that $u_{ij} - g \in V_{ij}$, where V_{ij} is defined in Lemma 1.4.13. Due to this lemma one has

$$||u_{ij} - g||_{i,j,1} \le \frac{c}{h} \max_{1 \le k \le 8} |u_{ij}(P_k) - g(P_k)|$$

$$\le \frac{c}{h} \max_{1 \le k \le 8} |u_{ij}(P_k) - f(P_k)| + \frac{c}{h} \max_{1 \le k \le 8} |f(P_k) - g(P_k)| \le ch^2 \, , \tag{1.4.41}$$

where (1.4.38) and (1.4.39) have been taken into account. From (1.4.39) – (1.4.41) we deduce that

$$\|f - u_h\|_{i,j,1} \le ch^2 \,,$$

where c is independent of i, j. In this way we have obtained (1.4.37), which completes the proof. $\qquad\square$

Remark 1.4.15. The proof of Proposition 1.4.14 exploits the classical argument of "the inverse hypothesis" in the theory of finite elements, established by Lemma 1.4.13. We observe that in Proposition 1.4.14 it could not be difficult to provide explicit dependence of c on f.

We pass to proving the second density result. Let

$$\left.\begin{array}{l} \Omega \subset \mathbb{R}^2 \text{ be a bounded domain with Lipschitzian} \\ \text{boundary, } 1 \le p < \infty \,. \end{array}\right\} \tag{1.4.42}$$

Without loss of generality, we may assume that

$$\Omega \subset \{(x_1, x_2) | 0 < x_1, x_2 < 1\} \,. \tag{1.4.43}$$

Proposition 1.4.16. Let Ω satisfy (1.4.42) and (1.4.43) and let $f \in W^{2,p}(\Omega)$. Then there exists $u_h \in U_h$ such that

$$\lim_{h \to 0} \|f - u_h\|_{W^{2,p}(\Omega)} = 0 \,. \tag{1.4.44}$$

Proof. Due to Calderón's extension theorem (see Stein, 1970, Chap. VI), it suffices to consider the case where $\Omega = (0,1) \times (0,1)$. We note that $U_h \subset W^{2,p}(\Omega)$. From the density of $C^\infty(\bar{\Omega})$ in $W^{2,p}(\Omega)$ we deduce that one may assume that $f \in C^\infty(\bar{\Omega})$. Proposition 1.4.14 then implies the existence of $u_h \in U_h$ such that

$$\lim_{h \to 0} \|f - u_h\|_{W^{2,\infty}(\Omega)} = 0 \,.$$

For an $f \in W^{2,p}(\Omega)$ there exists a sequence $\{f_n\}_{n \in \mathbb{N}} \subset C^\infty(\bar{\Omega})$ such that

$$\|f - f_n\|_{W^{2,p}(\Omega)} \to 0 \quad \text{as} \quad n \to \infty \,.$$

In turn, for each n there exists $u_{n,h} \in U_h$ such that

$$\|f - u_{n,h}\|_{W^{2,\infty}(\Omega)} \to 0 \quad \text{when} \quad n \to \infty \,.$$

Applying the diagonalization method we infer the existence of a mapping $h \to n(h)$ with $\lim_{h \to 0} n(h) = +\infty$ such that denoting

$$u_h = u_{n(h),h} \,, \tag{1.4.45}$$

we obtain

$$\|f - u_h\|_{W^{2,p}(\Omega)} \leq \|f - f_{n(h)}\|_{W^{2,p}(\Omega)} + \|f_{n(h)} - u_h\|_{W^{2,p}(\Omega)}$$
$$\leq \|f - f_{n(h)}\|_{W^{2,p}(\Omega)} + c\|f_{n(h)} - u_h\|_{W^{2,\infty}(\Omega)} ,$$

where $c > 0$ depends on Ω only. The assertion is proved. $\qquad\square$

Remark 1.4.17. Sablonnière (1989) considered similar problem of approximation by using general triangular mesh.

1.5. *An augmented Lagrangian method for problems with unilateral constraints*

Augmented Lagrangian methods combine ordinary Lagrangian techniques and penalty methods without suffering from the disadvantages of these methods: slow convergence for the former and possible ill-conditioning as $\eta \to \infty$ for the latter. More precisely, augmented Lagrangian methods converge without requiring that the penalty parameter η tends to infinity.

Augmented Lagrangian methods have many applications in the analysis of optimization problems with constraints, cf. Glowinski and Le Tallec (1989), Ito and Kunish (1990, 1996) Telega and Gałka (1998, 1999).

Ito and Kunish (1990) proposed the augmented Lagrangian algorithm to solve the minimization problem (P) below. In Sections 8-11 we shall see that local problems for plates and laminates weakened by periodically distributed fissures are always of such a type provided that friction is neglected.

In the present section we shall present the main results obtained by Ito and Kunish (1990). To this end, we first introduce the following spaces and mappings:

V is a Hilbert space,

V_1 is a reflexive Banach space, continuously injected into V,

H is a Hilbert lattice with inner product $\langle \cdot, \cdot \rangle$,

$a(\cdot, \cdot) : V \times V \to \mathbb{R}$ is a bilinear, symmetric, continuous and V-elliptic form with $a(u, u) \geq c_0\|u\|_V^2$ for some constant $c_0 > 0$,

$l : V \to \mathbb{R}$ is a continuous linear functional,

$g : V_1 \to H$ is a convex, continuous and Gâteaux differentiable mapping.

We recall (Yosida, 1978) that H is a Hilbert lattice if it is a Hilbert space and a partially ordered set (with order denoted by $\leq$) such that every nonempty finite subset of H has a greatest lower and a least upper bound (denoted by sup), and such that

(i) $\mathbf{u} \leq \mathbf{v}$ implies $\mathbf{u} + \mathbf{w} \leq \mathbf{v} + \mathbf{w}$ for all $\mathbf{u}, \mathbf{v}, \mathbf{w} \in H$,

(ii) $\mathbf{u} \leq 0$ implies $\alpha\mathbf{u}$ for all $\mathbf{u} \in H, \alpha \in \mathbb{R}^+$,

(iii) $|\mathbf{u}| \leq |\mathbf{v}|$ implies $\|\mathbf{u}\|_H \leq \|\mathbf{v}\|_H$ for all $\mathbf{u}, \mathbf{v} \in H$, where $|\mathbf{u}| = \sup(\mathbf{u}, -\mathbf{u})$ and $\|\cdot\|_H$ denotes the norm in H.

In practice, $\leq$ stands for the pointwise almost everywhere ordering.

We identify H with its dual and assume that $\lambda \in H, \lambda \geq 0$, implies $\langle \lambda, \mathbf{u} \rangle \geq 0$ for all $\mathbf{u} \geq 0$.

The problem under investigation means evaluating

$$(P) \qquad \min \left\{ \tfrac{1}{2} a(u,u) - l(u) \,\middle|\, g(u) \leq 0, \quad u \in V_1 \right\}.$$

We need the following assumption:

(A) there exists $(\tilde{u}, \tilde{\lambda})| \in V_1 \times H$ such that $\tilde{u}$ is a solution to $(P), \tilde{\lambda} \geq 0$

and

(a) $\qquad \langle \tilde{\lambda}, g(\tilde{u}) \rangle = 0,$

(b) $\qquad a(\tilde{u}, v) - l(v) + \langle \tilde{\lambda}, g'(\tilde{u})v \rangle = 0$ for all $v \in V_1.$

Here g' denotes the Gâteaux derivative of g. In view of the ellipticity of the bilinear form $a(\cdot, \cdot)$ the solution $\tilde{u}$ of (P) is necessarily unique. The element $\tilde{\lambda}$ is referred to as a Lagrange multiplier for (P).

Now we are in a position to define a family of augmented Lagrangian problems associated with (P):

$$(P)_{\eta,\lambda} \qquad \min L_\eta(u, \lambda),$$

where

$$L_\eta(u, \lambda) = \frac{1}{2} a(u,u) - l(u) + \frac{1}{2\eta} \left(\|\sup(0, \lambda + \eta g(u))\|_H^2 - \|\lambda\|_H^2 \right),$$

and $\lambda \in H, \eta > 0, \eta \in \mathbb{R}^+$. We note that $L_\eta(u, \lambda)$ can be written in an alternative way. Let $\hat{g} : V_1 \times H \times \mathbb{R}^+ \to H$ be defined by

$$\hat{g}(u, \lambda, \eta) = \sup \left(g(u), -\frac{\lambda}{\eta} \right).$$

It can easily be shown that

$$L_\eta(u, \lambda) = \frac{1}{2} a(u,u) - l(u) + \langle \lambda, \hat{g}(u, \lambda, \eta) \rangle + \frac{\eta}{2} \|\hat{g}(u, \lambda, \eta)\|_H^2 . \qquad (1.5.1)$$

We now pass to presentation and examination of the augmented Lagrangian algorithm to solve (P). It consists of a sequence of unconstrained optimization problems $(P)_{\eta_n, \lambda_n}$ whose solutions (provided that they exist) converge to the solution of (P).

The Algorithm

(1) Choose $\lambda_1 \in H, \lambda_1 \geq 0$, and $\eta > 0$.
(2) Put $n = 1$.
(3) Solve $(P)_{\eta, \lambda_n}$ for u_n.
(4) Put $\lambda_{n+1} = \lambda_n + \eta \hat{g}(u_n, \lambda_n, \eta) = \sup(0, \lambda_n + \eta g(u_n))$.
(5) Put $n = n + 1$ and return to (3).

A stopping criterion must be added to the algorithm for actual computations.

Prior to studying the convergence of this algorithm we require two technical lemmas.

Lemma 1.5.1. Let $\lambda \in H, \lambda \geq 0, \eta > 0, \eta \in \mathbb{R}^+$ and assume that $u \to L_\eta(u, \lambda)$ is radially unbounded, i.e. $L_\eta(u, \lambda) \to \infty$ as $\|u\|_{V_1} \to \infty$. Then there exists a unique solution u_η of $(P)_{\eta, \lambda}$ which satisfies

$$a(u_\eta, v) - l(v) + \langle \sup(0, \lambda + \eta g(u_\eta)), g'(u_\eta)v \rangle = 0 \,, \tag{1.5.2}$$

for all $v \in V_1$.

Lemma 1.5.2. Let $\lambda \in H, \eta > 0, \eta \in \mathbb{R}^+$. Then the function

$$u \to \| \sup(0, \lambda + \eta g(u)) \|_H^2$$

from V_1 to $\mathbb{R}$ is convex.

Proof of Lemma 1.5.2. The convexity of g implies

$$\eta g(\mu u + (1 - \mu)v) \leq \mu \eta g(u) + (1 - \mu)\eta g(v)$$

for every $u, v \in V_1$ and $\mu \in (0, 1)$. Using properties of the Hilbert lattice structure of H (Yosida, 1978) we obtain

$$\lambda + \eta g(\mu u + (1 - \mu)v) \leq \mu(\lambda + \eta g(u)) + (1 - \mu)(\lambda + \eta g(v)) \,, \quad \lambda \in H$$

and

$$\begin{aligned}
&\sup(0, \lambda + \eta g(\mu u + (1 - \mu)v)) \\
&\leq \sup(0, \mu(\lambda + \eta g(u))) + \sup(0, (1 - \mu)(\lambda + \eta g(v))) \\
&= \sup(0, \lambda + \eta g(u)) + (1 - \mu) \sup(0, \lambda + \eta g(v)) \,.
\end{aligned}$$

Hence we deduce that

$$\begin{aligned}
&\| \sup(0, \lambda + \eta g(\mu u + (1 - \mu)v)) \|_H \\
&\leq \mu \| \sup(0, \lambda + \eta g(u)) \|_H + (1 - \mu) \| \sup(0, \lambda + \eta g(v)) \|_H \,,
\end{aligned}$$

and thus

$$\begin{aligned}
&\| \sup(0, \lambda + \eta g(\mu u + (1 - \mu)v)) \|_H^2 \\
&\leq \mu \| \sup(0, \lambda + \eta g(u)) \|_H^2 + (1 - \mu) \| \sup(0, \lambda + \eta g(v)) \|_H^2 \,,
\end{aligned}$$

from which the desired result follows. $\square$

Proof of Lemma 1.5.1. In view of the ellipticity of $a(\cdot, \cdot)$, the mapping $u \to \frac{1}{2}a(u, u) - l(u)$ from V_1 to $\mathbb{R}$ is strictly convex. Since also $u \to (1/2\eta) \| \sup(0, \lambda + \eta g(u)) \|_H^2$ is convex, it follows that $u \to L_\eta(u, \lambda)$ is strictly convex. Moreover, $u \to \sup(0, \lambda + \eta g(u))$ is continuous (Baiocchi and Capelo, 1984, Th. 19.8) and thus $u \to L_\eta(u, \lambda)$ is continuous as well. By assumption this mapping is also radially unbounded. Since V_1 is reflexive, the existence of a unique solution u_η of $(P)_{\eta,\lambda}$ follows. Thus characterization (1.5.2) can be proved by differentiation of $L_\eta(u, \lambda)$ with respect to u.

Theorem 1.5.3. Let hypothesis (A) be satisfied and assume that $u \to L_\eta(u, \lambda)$ is radially unbounded from V_1 to $\mathbb{R}$ for every $\lambda \geq 0$. Then

$$c_0||\tilde{u} - u_n||_V^2 + \frac{1}{2\eta}||\lambda_{n+1} - \tilde{\lambda}||_H^2 \leq \frac{1}{2\eta}||\lambda_n - \tilde{\lambda}||_H^2$$

for all $n \in \mathbb{N}$.

Proof. Let $J : V_1 \to \mathbb{R}$ be given by

$$J(u) = \frac{1}{2}a(u, u) - l(u),$$

and set $\theta(u, \lambda) = \sup(0, \lambda + \eta g(u))$ for $\lambda \in H$ and $u \in V$. We note that $\theta(u_n, \lambda_n) = \lambda_{n+1}$. With (A) we find

$$\begin{aligned}
J(u_\eta) - J(\tilde{u}) &= \frac{1}{2}a(u_n, u_n) - l(u_n) - \frac{1}{2}a(\tilde{u}, \tilde{u}) + l(\tilde{u}) \\
&= \frac{1}{2}a(u_n, u_n) - a(u_n, \tilde{u}) + \frac{1}{2}a(\tilde{u}, \tilde{u}) + a(\tilde{u}, u_n - \tilde{u}) + l(\tilde{u} - u_n) \\
&= \frac{1}{2}a(u_n - \tilde{u}, u_n - \tilde{u}) + \langle \tilde{\lambda}, g'(\tilde{u})(\tilde{u} - u_n) \rangle.
\end{aligned}$$

By using (1.5.2) we obtain

$$\begin{aligned}
J(\tilde{u}) - J(u_n) &= \frac{1}{2}a(\tilde{u}, \tilde{u}) - l(\tilde{u}) - \frac{1}{2}a(u_n, u_n) + l(u_n) \\
&= \frac{1}{2}a(\tilde{u} - u_n, \tilde{u} - u_n) - \langle \theta(u_n, \lambda_n), g'(u_n)(\tilde{u} - u_n) \rangle.
\end{aligned}$$

Adding these two expressions we get

$$0 = a(\tilde{u} - u_n, \tilde{u} - u) + \langle \tilde{\lambda}, g'(\tilde{u})(\tilde{u} - u) \rangle - \langle \theta(u_n, \lambda_n), g'(u_n)(\tilde{u} - u_n) \rangle. \qquad (1.5.3)$$

Next we define $G : V_1 \times H \to \mathbb{R}$

$$G(u, \lambda) = \frac{1}{2\eta}[||\sup(0, \eta g(u) + \lambda)||_H^2 - ||\lambda||_H^2],$$

for $\eta > 0$. It is easily seen that

$$G(\tilde{u}, \lambda) \leq 0 \quad \text{for all} \quad \lambda \in H, \ \lambda \geq 0. \qquad (1.5.4)$$

In fact, by (A) and since H is a Hilbert lattice we conclude that

$$\sup(0, \lambda + \eta g(\tilde{u})) \leq \lambda,$$

and

$$\| \sup(0, \lambda + \eta g(\tilde{u})) \|_H - \|\lambda\|_H \leq 0 \, .$$

Hence (1.5.4) follows. Moreover, for every $v \in V_1$ we find

$$G'_u(u, \lambda)v = \langle \sup(0, \eta g(u) + \lambda), g'(u)v \rangle \, .$$

By Lemma 1.5.2 the mapping $u \to G(u, \lambda)$ is convex, consequently we obtain the following subdifferential inequality

$$G(\tilde{u}, \lambda_n) \geq G(u_n, \lambda_n) + \langle \theta(u_n, \lambda_n), g'(u_n)(\tilde{u} - u_n) \rangle \, . \tag{1.5.5}$$

Next we find

$$\begin{aligned}
G(u_n, \lambda_n) + \langle \theta(u_n, \lambda_n), g'(u_n)(\tilde{u} - u_n) \rangle &= \frac{\eta}{2} \|\hat{g}(u_n, \lambda_n, \eta)\|_H^2 \\
&+ \langle \lambda_n, \hat{g}(u_n, \lambda_n, \eta) \rangle + \langle \theta(u_n, \lambda_n), g'(u_n)(\tilde{u} - u_n) \rangle = \frac{\eta}{2} \|\hat{g}(u_n, \lambda_n, \eta)\|_H^2 \\
&+ \langle \lambda_n - \tilde{\lambda}, \hat{g}(u_n, \lambda_n, \eta) \rangle + \langle \tilde{\lambda} \hat{g}(u_n, \lambda_n, \eta) \rangle + \langle \theta(u_n, \lambda_n), g'(u_n)(\tilde{u} - u_n) \rangle \\
&= \frac{1}{2\eta} \|\eta \hat{g}(u_n, \lambda_n, \eta) + \lambda_n - \tilde{\lambda}\|_H^2 - \frac{1}{2\eta} \|\lambda_n - \tilde{\lambda}\|_H^2 + \langle \tilde{\lambda}, \hat{g}(u_n, \lambda_n, \eta) \rangle \\
&+ \langle \lambda_{n+1}, g'(u_n)(\tilde{u} - u_n) \rangle = \frac{1}{2\eta} \|\theta(u_n, \lambda_n) - \tilde{\lambda}\|_H^2 - \frac{1}{2\eta} \|\lambda_n - \tilde{\lambda}\|_H^2 \\
&+ \langle \tilde{\lambda}, \hat{g}(u_n, \lambda_n, \eta) \rangle + a(\tilde{u} - u_n, \tilde{u} - u_n) + \langle \tilde{\lambda}, g'(\tilde{u})(\tilde{u} - u_n) \rangle \, , \tag{1.5.6}
\end{aligned}$$

where for the last equality we used (1.5.3). From (1.5.4) – (1.5.6) combined with the equality $\langle \tilde{\lambda}, g(\tilde{u}) \rangle = 0$ we obtain

$$\begin{aligned}
&\frac{1}{2\eta} \|\lambda_{n+1} - \tilde{\lambda}\|_H^2 - \frac{1}{2\eta} \|\lambda_n - \tilde{\lambda}\|_H^2 + a(\tilde{u} - u, \tilde{u} - u) \\
&+ \langle \tilde{\lambda}, \hat{g}(u_n, \lambda_n, \eta) + g'(\tilde{u})(\tilde{u} - u_n) - g(\tilde{u}) \rangle \leq 0 \, . \tag{1.5.7}
\end{aligned}$$

The convexity of g implies

$$g'(\tilde{u})(\tilde{u} - u_n) + \hat{g}(u_n, \lambda_n, \eta) - g(\tilde{u}) \geq g'(\tilde{u})(\tilde{u} - u_n) + g(u_n) - g(\tilde{u}) \geq 0 \, . \tag{1.5.8}$$

Combining (1.5.7) and (1.5.8) and using the fact that $\tilde{\lambda} \geq 0$ we obtain

$$\frac{1}{2\eta} \|\lambda_{n+1} - \tilde{\lambda}\|_H^2 - \frac{1}{2\eta} \|\lambda_n - \tilde{\lambda}\|_H^2 + a(\tilde{u} - u_n, \tilde{u} - u_n) \leq 0 \, .$$

This is the desired estimate in view of the ellipticity of $a(\cdot, \cdot)$.   □

Remark 1.5.4. The assumptions on the radial unboundedness of $u \to L_\eta(u, \lambda)$ and on the reflexivity of V_1 are only required in the proof of Lemma 1.5.1 and can be replaced by the assumption of existence of a solution of $(P)_{\eta,\lambda}$. If $V_1 = V$, then the radial unboundedness of $u \to L_\eta(u, \lambda)$ is implied by the V-ellipticity of $a(\cdot, \cdot)$. $\square$

The first-order necessary optimality condition associated with $(P)_{\eta,\lambda_n}$ of the algorithm has the following form:

$$a(u_n, v) - l(v) + \langle \sup(0, \lambda_n + \eta g(u_n)), g'(u_n)v \rangle = 0$$

for all $v \in V_1$. By definition of λ_{n+1} this condition is equivalent to

$$a(u_n, v) - l(v) + \langle \lambda_{n+1}, g'(u_n)v \rangle = 0 \quad \text{for all} \quad v \in V_1 . \tag{1.5.9}$$

Corollary 1.5.5. Under the assumptions of Theorem 1.5.3, the sequence $\{\lambda_n\}_{n \in \mathbb{N}}$ is bounded in H and $u_n \to \tilde{u}$ in V with

$$c_0 \sum_{n=1}^{\infty} ||\tilde{u} - u_n||_V^2 \le \frac{1}{2\eta} ||\lambda_1 - \lambda||_H^2 . \tag{1.5.10}$$

If, moreover, $g'(u_n)v \to g'(\tilde{u})v$ in H, for every $v \in V_1$, then every cluster point $\hat{\lambda}$ of $\{\lambda_n\}$ satisfies (A)(b) with $\tilde{\lambda}$ replaced by $\hat{\lambda}$ and $\hat{\lambda} \ge 0$.

Proof. Estimate (1.5.10) follows directly from Theorem 1.5.3. The second assertion is a consequence of (1.5.9). Finally, by step (4) of the algorithm $\langle \hat{\lambda}, v \rangle \ge 0$ for each $v \ge 0$ and hence $\hat{\lambda} \ge 0$. $\square$

The correspondence between (P) and $(P)_{\eta,\tilde{\lambda}}$ is formulated as

Corollary 1.5.6. Under the assumptions of Theorem 1.5.3, $\tilde{u}$ is the unique solution of the constrained problem (P) if and only if $\tilde{u}$ is the solution of the unconstrained problem $(P)_{\eta,\tilde{\lambda}}$.

Proof. The result follows immediately from Theorem 1.5.3 with $\lambda_1 = \tilde{\lambda}$. $\square$

Variable stepsize analysis

In the augmented Lagrangian algorithm with variable stepsize, step (4) is replaced by

$$(4') \qquad \text{put } \lambda_{n+1} = \lambda_n + \alpha \eta \hat{g}(u_n, \lambda_n, \eta) , \quad \text{where} \quad \alpha \in (0, 1] .$$

For numerical calculations it may be essential to take $\alpha < 1$ to obtain good performance of the algorithm.

Theorem 1.5.7. Assume that (A) is satisfied and that $u_n \to L_\eta(u, \lambda)$ is radially unbounded from V_1 to $\mathbb{R}$ for every $\lambda \ge 0$. Then we have

$$c_0 ||\tilde{u} - u_n||_V^2 + \frac{\eta}{2}(1 - \alpha)||\hat{g}(u_n, \lambda_n, \eta)||_H^2 + \frac{1}{2\eta\alpha}||\lambda_{n+1} - \tilde{\lambda}||_H^2$$

$$\le \frac{1}{2\eta\alpha} ||\lambda_n - \tilde{\lambda}||_H^2 , \tag{1.5.11}$$

for all $n \in \mathbb{N}$, and

$$c_0 \sum_{n=1}^{\infty} [||\tilde{u} - u_n||_V^2 + \frac{\eta}{2}(1-\alpha)||\hat{g}(u_n, \lambda_n, \eta)||_H^2] \leq \frac{1}{2\eta\alpha}||\lambda_1 - \tilde{\lambda}||_H^2 . \qquad (1.5.12)$$

Proof. The proof is left to the reader as an exercise, cf. also Ito and Kunisch (1990). $\qquad \square$

Remark 1.5.8. The same authors (Ito and Kunisch, 1996) developed the augmented Lagrangian algorithm for a more general class of convex optimization problems:

$$\left|\begin{array}{l} \text{find} \\ \quad \min\{f(u) + \varphi(\Lambda u)|\, u \in \mathbb{K}\} . \end{array}\right.$$

Here $f : V \to \mathbb{R}$ is a lower semicontinuous, continuously differentiable, convex functional, $\Lambda \in \mathbf{L}(V, \mathbf{H})$ and $\varphi : \mathbf{H} \to \mathbb{R}$ is a proper, lower semicontinuous convex functional. Moreover, V, $\mathbf{H}$ are real Hilbert spaces and $\mathbb{K}$ is a closed convex subset of V.

CHAPTER II

ELASTIC PLATES

Introduction

The subject of this chapter is a statical analysis of a linearly elastic plate of properties periodic in two orthogonal longitudinal directions. Such a plate can be constructed by repeating the basic periodicity cell $\mathcal{Z}$ along the directions of the sides of the reference rectangle $Z = (0, l_1^z) \times (0, l_2^z)$, see Fig. 2.0. Due to highly oscillating elastic properties, the solution of the elasticity problem is also oscillating. The aim of the analysis is to extract from this solution those parts which describe the overall plate response. It is the two-scale expansion method which makes it possible to extract the terms representing the overall plate behavior. This method has already been successfully applied to the overall analysis of periodic three-dimensional composites. Here this method should be appropriately modified to take into account that the transverse dimension of the plate is much smaller than its in-plane dimensions. This has two consequences: distribution of displacements is almost linear across the thickness and the in-plane stresses assume much greater values than other stress components. The latter property means that the plane-stress state prevails. It turns out that both these features can be disclosed by the two-scale expansion method if an appropriate scaling of the loading is adopted.

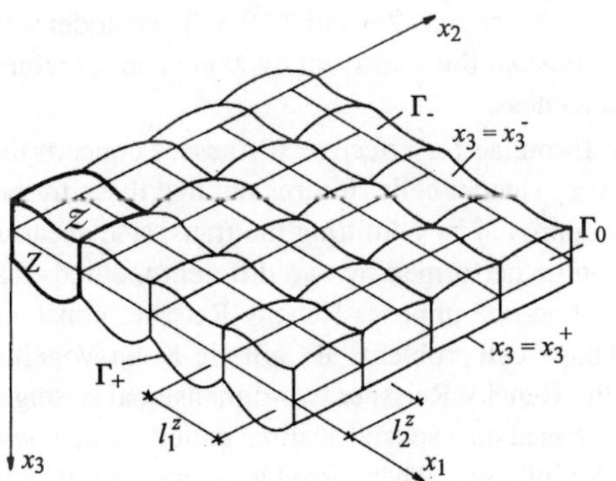

Fig. 2.0.

Let the plate thickness be scaled by the small parameter e and the dimensions l_1, l_2 – by the small parameter ε. A perfect effective plate model is constructed by fixing the ratio e/ε

and passing to zero with ε, cf. Caillerie (1982). This is equivalent to a simultaneous passage to the zero limit with ε and e, which can be performed formally by the asymptotic expansion method (Sections 2.2, 2.3) or rigorously by the Γ-convergence technique (Section 2.10.4). Both approaches lead to the Caillerie-Kohn-Vogelius plate model of periodic plates made from cells $\mathcal{Z}$ of arbitrary shape. Just this effective plate model will further be treated as a reference model and all other averaging methods will be better or worse approximations of this model, depending on the shape of the basic cell $\mathcal{Z}$. This model is, however, difficult to apply since it involves three-dimensional local problems. Difficulties that arise at any attempts to solve these local problems analytically justify further steps towards rational simplifications.

The first simplified model refers to the case of the basic cell $\mathcal{Z}$ being a thin plate. This model can be derived from by two methods. The first one was originated in 1976 by Duvaut and Metellus who homogenized the thin plate equations of Kirchhoff, without making any reference to the three-dimensional plate pattern. The second modelling consists in imposing Kirchhoff's constraints on the solutions of the Caillerie-Kohn-Vogelius local problems. This purely two-dimensional homogenization leads to relatively simple formulae for effective stiffnesses. In the case of plates with straight ribs these formulae can be put in one compact and algebraic formula, called the formula of Francfort and Murat, cf. Sec. 3.8.1. It will turn out clear in Chapter VI that this algebraic formula plays a crucial role in proving the extremal properties of ribbed plates. This shows that the homogenization results do not only provide us with rational averaging formulae for plates of repeated lay-up, but they yield complete characterizations of the sets of effective stiffness tensors of all plates constructed by mixing given constituents in fixed proportions. This creates a link between homogenization and optimization, a link that makes the homogenization results of major significance.

The second simplified model of periodic plates refers to the case of the basic cell $\mathcal{Z}$ being a slender column, see Sections 2.9 and 2.10.3. This model was proposed by Caillerie (1984) for the case of constant thickness and by Kohn and Vogelius (1984) for the case of periodically varying thickness.

Thus the analytical formulae for effective stiffnesses concern the periodic plates composed of very thin or very slender cells. It turns out that these former formulae (of Duvaut and Metellus) can be improved by admitting the transverse shear deformation within periodicity cells. This can be performed by two different methods leading to the same final results. In the first method one imposes Hencky-Reissner constraints on the solutions to the three-dimensional basic cell problems of Caillerie-Kohn-Vogelius, see Section 2.7. The second method takes the Hencky-Reissner two-dimensional setting as a point of departure. The homogenization is based on a special scaling, called refined, which preserves relations between all length scales involved. There are at least two such length scales. The first is the length of the basic cell. The second is the quantity $(D/H)^{1/2}$, where D and H represent the bending and shearing stiffnesses, respectively; see Sections 5.3 – 5.5.

If a thin plate is periodic with respect to a curvilinear parametrization, then homogenization is nonuniform and leads to an effective model with slowly varying effective stiffnesses.

This topic is discussed within the Kirchhoff framework in Section 3.10.

The homogenization analysis of periodic plates undergoing moderately large deflections is performed in Section 4.2 under the von Kármán assumptions and generalized in Section 5.7 to the case of transverse shear deformable plates. The influence of openings on the bifurcation analysis of thin plates is discussed in Section 4.3.

The Kirchhoff and Hencky-Reissner modelling cannot be applied to sandwich plates with stiff faces. The bending stiffness of the faces is taken into account in the model of Hoff. The formulae for effective stiffnesses of such sandwich plates of repeated lay-up (usually the core has a periodic structure) are derived in Section 6.

In the present chapter the following conventions are adopted. The small Latin indices, like: i, j, k, l, m, n, $s \ldots$ run over 1, 2, 3; the small Greek indices, like α, β, λ, $\mu, \ldots$ (expect for ε) assume the values 1, 2. The summation convention applies to the indices at different levels, e.g.

$$A^{\alpha\beta}b_\beta = \sum_{\beta=1}^{2} A^{\alpha\beta}b_\beta , \quad M^{ij}a_{jk} = \sum_{j=1}^{3} M^{ij}a_{jk} .$$

In the expressions like $A^{\alpha\beta}b^\beta$ or $M^{ij}a^{jk}$ the summation does not hold.

The three Cartesian orthogonal systems will be used: (x_i), (z_i), (y_i). The partial derivatives are denoted shortly as follows

$$\frac{\partial(\ )}{\partial x_i} = (\ )_{,i} , \qquad \frac{\partial(\ )}{\partial z_i} = (\ )_{;i} , \qquad \frac{\partial(\ )}{\partial y_i} = (\ )_{|i} . \tag{0.1}$$

We also introduce the following notation:

$$\begin{aligned} x &= (x_1, x_2) , & z &= (z_1, z_2), & y &= (y_1, y_2) , \\ \boldsymbol{x} &= (x_1, x_2, x_3) , & \boldsymbol{z} &= (z_1, z_2, z_3), & \boldsymbol{y} &= (y_1, y_2, y_3) . \end{aligned} \tag{0.2}$$

The symmetrized gradients are denoted as follows

$$\begin{aligned} e_{ij}(\boldsymbol{v}) &= \frac{1}{2}(v_{i,j} + v_{j,i}) , & e_{ij}(\boldsymbol{v}) &= e_{ij}^x(\boldsymbol{v}) , \\ e_{ij}^z(\boldsymbol{v}) &= \frac{1}{2}(v_{i;j} + v_{j;i}) , & e_{ij}^y(\boldsymbol{v}) &= \frac{1}{2}(v_{i|j} + v_{j|i}) , \end{aligned} \tag{0.3}$$

where $\boldsymbol{v} = (v_1, v_2, v_3)$; v_i depend on either $\boldsymbol{x}$ or $\boldsymbol{z}$ or $\boldsymbol{y}$.

Let $v = (v_1, v_2)$ and w be functions of x, z or y. We set

$$e_{\alpha\beta}(v) = \frac{1}{2}(v_{\alpha,\beta} + v_{\beta,\alpha}) ; \qquad \kappa_{\alpha\beta}(w) = -w_{,\alpha\beta} ,$$

$$e_{\alpha\beta}^z(v) = \frac{1}{2}(v_{\alpha;\beta} + v_{\beta;\alpha}) ; \qquad \kappa_{\alpha\beta}^z(w) = -w_{;\alpha\beta} , \tag{0.4}$$

$$e_{\alpha\beta}^y(v) = \frac{1}{2}(v_{\alpha|\beta} + v_{\beta|\alpha}) ; \qquad \kappa_{\alpha\beta}^y(w) = -w_{|\alpha\beta} .$$

The quantity e will be a small parameter scaling the plate thickness and should not be mistaken for the operators (0.3), (0.4).

The rescaled plane and spatial periodicity cells are denoted by Y and $\mathcal{Y}$ and are parametrized by (y_α) and (y_i) respectively. The averages of a function f defined on these cells are denoted by

$$\langle f \rangle = \frac{1}{|Y|} \int_Y f(y) dy , \qquad \prec f \succ = \frac{1}{|\mathcal{Y}|} \int_{\mathcal{Y}} f(\boldsymbol{y}) d\boldsymbol{y} , \tag{0.5}$$

where $|Y| = \mathrm{area}\, Y$, $|\mathcal{Y}| = \mathrm{vol}\, \mathcal{Y}$.

2.   Three-dimensional analysis and effective models of composite plates

The aim of this section is to put forward a derivation of these effective models of periodic plates which start from the three-dimensional setting. We consider two methods of modelling: (a) the process of reduction of the transverse dimension is indissolubly bonded with the process of smearing-out the stiffnesses; (b) both processes being subsequently performed.

A central role is played by model $(P_{\mathcal{H}})$ with three-dimensional local problems. This model is derived by two different methods: by asymptotic expansions and then independently derived and justified by Γ-convergence with all dimensions of the basic cell tending to zero. Other models discussed are approximations to model $(P_{\mathcal{H}})$ and concern plates composed of very slender or very thin periodicity cells. These approximate models are not derived here by an asymptotic expansion method; the method of Γ-convergence is here self-explanatory.

All convention given in the Introduction to the Chapter apply here.

2.1.   Equilibrium problem of a periodic plate

We consider a three-dimensional elastic body occupying the closure of a domain B lying between two surfaces of Z-periodic shape; $Z = (0, l_1^z) \times (0, l_2^z)$:

$$B = \{ \boldsymbol{x} \mid \boldsymbol{x} = (x, x_3) , \quad x = (x_\alpha) \in \Omega , \quad x_3^-(x) < x_3 < x_3^+(x) \} .$$

Here Ω is a plane, open reference domain parametrized by Cartesian coordinates $(x_\alpha)(\alpha = 1, 2)$; x_3-axis is directed normal to this plane and the (x_i) system forms a left-handed Cartesian system with orthonormal basis (e_1, e_2, e_3). The functions $x_3^\pm$ are assumed to be Z-periodic, i.e.

$$x_3^\pm(x_1 + m l_1^z,\ x_2 + n l_2^z) = x_3^\pm(x_1, x_2) , \quad m, n \in \mathbb{N} .$$

They determine the upper $(+)$ and lower $(-)$ faces of B :

$$\Gamma_\pm = \{ \boldsymbol{x} \mid x \in \Omega, \quad x_3 = x_3^\pm(x) \} .$$

Let $\Gamma = \partial\Omega$ be the boundary of Ω. The cylindrical surface

$$\Upsilon = \{x \mid x \in \Gamma, \quad x_3^-(x) < x_3 < x_3^+(x)\}$$

is referred to as a lateral surface of B. If the in-plane dimensions of Ω are much greater than $\max |x_3^+ - x_3^-|$, then such a body could be called a plate with varying thickness. The plate is composed of cells $\mathcal{Z}$

$$\mathcal{Z} = \{x \mid x \in Z, \quad x_3^-(x) < x_3 < x_3^+(x)\} \,,$$

except for a boundary layer around Υ.

The elastic moduli C_Z^{ijkl} of the plate material are assumed to be Z-periodic functions in x, hence index Z at the core letter C. Within the framework of linear elasticity the stresses σ^{ij} and strains ϵ_{ij}, both referred to the Cartesian system (x_i), are interrelated by Hooke's law

$$\sigma^{ij} = C_Z^{ijkl}(x)e_{kl} \,. \tag{2.1.1}$$

The strain-displacement relations are given by

$$e_{ij}(w) = w_{(i,j)} \tag{2.1.2}$$

where $w = (w_i)$.

The plate is subject to surface loads $r_i^\pm(x)$ on the $\Gamma_\pm$ faces, i.e.,

$$\sigma^{ij}(x, x_3^\pm)n_j^\pm(x) = r_\pm^i(x) \,; \tag{2.1.3}$$

here $n^\pm = (n_j^\pm)$ are versors normal to $\Gamma_\pm$. The body forces $b^i = b^i(x)$ are Z-periodic in x. Along Υ the plate is clamped. Thus the space of kinematically admissible displacement fields is given by

$$V_0(B) = \{v \in H^1(B)^3 \mid v = 0 \text{ on } \Gamma_0\} \,. \tag{2.1.4}$$

The equilibrium problem reads:

$$(P_{\mathcal{Z}}) \quad \left|
\begin{array}{l}
\text{find } w \in V_0(B) \text{ such that} \\[2mm]
\displaystyle\int_B \sigma^{ij}e_{ij}(v)dx = \int_{\Gamma_+} r_+^i(x)v_i(x, x_3^+)d\Gamma_+ \\[6mm]
\displaystyle + \int_{\Gamma_-} r_-^i(x)v_i(x, x_3^-)d\Gamma_- + \int_B b^i(x)v_i(x)dx \qquad \forall \, v \in V_0(B) \,.
\end{array}
\right. \tag{2.1.5}$$

Here σ^{ij} depend upon w according to (2.1.1) and (2.1.2). We assume that the matrix $C_Z = (C_Z^{ijkl})$ is positive definite: $\exists m > 0$ such that

$$\forall \gamma \in \mathbb{E}_s^3 \qquad C_Z^{ijkl}\gamma_{ij}\gamma_{kl} \geq m\sum_{i,j}(\gamma_{ij})^2 \,, \tag{2.1.6}$$

for almost every $x \in B$. Moreover, the following symmetry conditions hold

$$C_Z^{ijkl} = C_Z^{klij} = C_Z^{jikl} = C_Z^{ijlk} \ . \tag{2.1.7}$$

The area elements $d\Gamma_\pm$ are given by

$$d\Gamma_\pm = (G_Z^\pm(x))^{1/2}dx \ , \qquad dx = dx_1 dx_2 \tag{2.1.8}$$

with

$$G_Z^\pm(x) = 1 + (x_{3,1}^\pm)^2 + (x_{3,2}^\pm)^2 \ ; \tag{2.1.9}$$

here $(\cdot)_{,\alpha} = \partial/\partial x_\alpha$.

2.2. Family of problems (P_ε)

Problem (P_Z) involves three quantities: $h = \max \mid x_3^+ - x_3^- \mid$, l_1^z and l_2^z which are small in comparison with the global dimensions of B. The presence of small quantities makes the problem intractable by usual analytical and discretized methods. Thus it is reasonable to take advantage of these parameters being small and use an asymptotic method.

In the asymptotic method to be used we consider a family of problems (P_ε) such that for a certain $\varepsilon = \varepsilon_0$, $(P_{\varepsilon_0}) = (P_Z)$. A common feature of all problems (P_ε) is that the periodically cells for (P_ε) remain homothetic to the original cell $\mathcal{Z}$. Thus we substitute

$$x_3^\pm(x) \rightsquigarrow x_3^{\varepsilon\pm}(x) = \varepsilon c^\pm \left(\frac{x}{\varepsilon}\right) \ , \quad x_3^\pm(x) = \varepsilon_0 c^\pm \left(\frac{x}{\varepsilon_0}\right) \tag{2.2.1}$$

$$l_\alpha^z \rightsquigarrow \varepsilon l_\alpha \ , \quad l_\alpha^z = \varepsilon_0 l_\alpha \ ; \quad Z \rightsquigarrow \varepsilon Y \ , \quad Z = \varepsilon_0 Y \ .$$

Here $\rightsquigarrow$ means replacement and $Y = (0, l_1) \times (0, l_2)$. The functions $c^\pm$ are Y-periodic. Let

$$\mathcal{Y} = \{y \mid y = (y, y_3) \ , \ y = (y_1, y_2) \in Y \ , \ c^-(y) < y_3 < c^+(y)\} \ .$$

Thus replacements (2.2.1) mean

$$\mathcal{Z} \rightsquigarrow \mathcal{Z}_\varepsilon = \varepsilon \mathcal{Y} \ , \quad \mathcal{Z}_{\varepsilon_0} = \mathcal{Z} \ . \tag{2.2.2}$$

$\mathcal{Y}$ represents a rescaled cell of periodicity, cf. Fig. 2.2.1. Scaling (2.2.1) replaces domain B with domain B_ε.

To compensate for diminishing of the transverse dimensions of B_ε when $\varepsilon \to 0$ one must scale the loading

$$r_\pm^\alpha(x) \rightsquigarrow r_\pm^{\varepsilon,\alpha}(x) = \varepsilon^2 p_\pm^\alpha(x) \ , \quad r_\pm^{\varepsilon_0,\alpha}(x) = r_\pm^\alpha(x) \ ,$$

$$r_\pm^3(x) \rightsquigarrow r_\pm^{\varepsilon,3}(x) = \varepsilon^3 q_\pm(x) \ , \quad r_\pm^{\varepsilon_0,3}(x) = r_\pm^3(x) \ , \tag{2.2.3}$$

$$b^\alpha(x) \rightsquigarrow \varepsilon b^\alpha \left(\frac{x}{\varepsilon}\right) \ , \quad b^3(x) \rightsquigarrow \varepsilon^2 b^3 \left(\frac{x}{\varepsilon}\right) \ ,$$

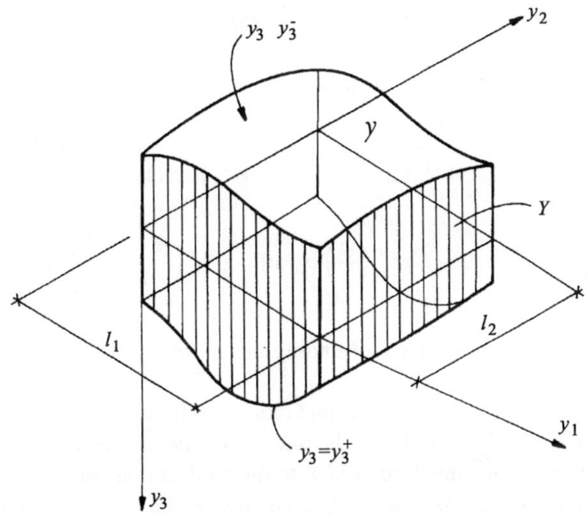

Fig. 2.2.1. Rescaled cell of periodicity

where $p_\pm^\alpha$, $q_\pm$ are ε-independent functions defined on Ω; $b^\alpha(\cdot, y_3)$, $b^3(\cdot, y_3)$ are Y-periodic. The unknown displacement field is denoted by $\boldsymbol{w}^\varepsilon = (w_i^\varepsilon)$ and the strains $\epsilon_{ij}^\varepsilon = e_{ij}(\boldsymbol{w}^\varepsilon)$ and stresses σ_ε^{ij} depend on ε. The boundary conditions (2.1.3) assume the form

$$\sigma_\varepsilon^{\alpha j}(x, x_3^{\varepsilon\pm}(x))n_j^{\varepsilon\pm}(x) = \varepsilon^2 p_\pm^\alpha(x) , \qquad \sigma_\varepsilon^{3j}(x, x_3^{\varepsilon\pm}(x))n_j^{\varepsilon\pm}(x) = \varepsilon^3 q_\pm(x) , \quad (2.2.4)$$

where $n^{\varepsilon\pm}$ represents the unit vector outward normal to $\Gamma_\pm^\varepsilon$ at $(x, x_3^{\varepsilon\pm})$. Assume that functions $c^\pm$ are differentiable. Then

$$\boldsymbol{n}_\varepsilon^\pm(x) = \left(G_\pm\left(\frac{x}{\varepsilon}\right)\right)^{-1/2} \boldsymbol{N}^\pm\left(\frac{x}{\varepsilon}\right) ,$$

$$\boldsymbol{N}^\pm(y) = [\mp c_{|1}^\pm, \ \mp c_{|2}^\pm, \ \pm 1] , \qquad (2.2.5)$$

$$G_\pm(y) = 1 + (c_{|1}^\pm)^2 + (c_{|2}^\pm)^2 ; \quad f_{|\alpha} = \frac{\partial f}{\partial y_\alpha} .$$

To make the scaling complete one should assume the following constitutive relation

$$\sigma_\varepsilon^{ij}(\boldsymbol{x}) = C^{ijkl}\left(\frac{x}{\varepsilon}, \frac{x_3}{\varepsilon}\right) \epsilon_{kl}^\varepsilon(\boldsymbol{x}) ; \qquad (2.2.6)$$

$$C_Z^{ijkl}(\boldsymbol{x}) = C^{ijkl}\left(\frac{x}{\varepsilon_0}, \frac{x_3}{\varepsilon_0}\right) ; \qquad (2.2.7)$$

$C^{ijkl}\left(\cdot, \frac{x_3}{\varepsilon}\right)$ are Y-periodic and $C^{ijkl}\left(\frac{x}{\varepsilon}, y_3\right)$ are defined for $y_3 \in (c^-(y), c^+(y))$. Hence $C^{ijkl}(y)$ are defined for $y \in \mathcal{Y}$.

The variational equilibrium equation (2.1.5) yields:

$$
\int_{B_\varepsilon} \sigma_\varepsilon^{ij} e_{ij}(v)\,dx = \int_{\Gamma_+^\varepsilon} [\varepsilon^2 p_+^\alpha v_\alpha(x,\varepsilon c^+) + \varepsilon^3 q^+ v_3(x,\varepsilon c^+)]\,d\Gamma_+^\varepsilon
$$

$$
+ \int_{\Gamma_-^\varepsilon} [\varepsilon^2 p_-^\alpha v_\alpha(x,\varepsilon c^-) + \varepsilon^3 q_- v_3(x,\varepsilon c^-)]\,d\Gamma_-^\varepsilon
$$

$$
+ \int_{B_\varepsilon} [\varepsilon b^\alpha\left(\frac{x}{\varepsilon}\right) v_\alpha(x) + \varepsilon^2 b^3\left(\frac{x}{\varepsilon}\right) v_3(x)]\,dx \tag{2.2.8}
$$

for all v vanishing on Υ^ε. The problem of finding w^ε, ϵ^ε, σ_ε satisfying (2.2.8) and (2.2.6) will be called (P_ε).

The asymptotic method applied in the sequel requires reformulation of problem (P_ε) so that the local variable $y \in \mathcal{Y}$ would play the role of an independent variable. To this end we shall extrapolate the fields involved in (P_ε) to the product domain $\Omega \times \mathcal{Y} \ni (x, y)$, as follows. It is assumed that stresses and displacements can be expressed in terms of new functions

$$
\sigma_\varepsilon^{ij}(x, x_3) = \tilde{\sigma}_\varepsilon^{ij}\left(x; \frac{x}{\varepsilon}, \frac{x_3}{\varepsilon}\right),
$$

$$
w_i^\varepsilon(x, x_3) = \tilde{w}_i^\varepsilon\left(x; \frac{x}{\varepsilon}, \frac{x_3}{\varepsilon}\right), \qquad v_i(x, x_3) = \tilde{v}_i\left(x; \frac{x}{\varepsilon}, \frac{x_3}{\varepsilon}\right), \tag{2.2.9}
$$

such that

$$
\tilde{\sigma}_\varepsilon^{ij}\left(x; \cdot, \frac{x_3}{\varepsilon},\right) \qquad \tilde{w}_i^\varepsilon\left(x; \cdot, \frac{x_3}{\varepsilon},\right), \qquad \tilde{v}_i\left(x; \cdot, \frac{x_3}{\varepsilon},\right)
$$

are Y-periodic functions. Let us rewrite (2.2.9) as follows

$$
f(x, x_3) = \tilde{f}(x_1, x_2, y_1, y_2, y_3)\big|_{y_\alpha = x_\alpha/\varepsilon,\ y_3 = x_3/\varepsilon} \tag{2.2.10}
$$

for $f = \sigma_\varepsilon^{ij}, u_i^\varepsilon, v_i^\varepsilon$. Derivatives of $\tilde{f}$ are denoted by

$$
\frac{\partial \tilde{f}}{\partial x_\alpha} = \tilde{f}_{,\alpha} \qquad \frac{\partial \tilde{f}}{\partial y_i} = \tilde{f}_{|i}\,. \tag{2.2.11}
$$

Hence

$$
\frac{\partial f}{\partial x_\alpha} = [\tilde{f}_{,\alpha} + \frac{1}{\varepsilon}\tilde{f}_{|\alpha}]_{|y=x/\varepsilon}\,. \tag{2.2.12}
$$

Using (2.2.10) – (2.2.12) one can express the integrand on the l.h.s. of (2.2.8) as follows

$$
\sigma_\varepsilon^{ij} e_{ij}(v) = [\tilde{\sigma}_\varepsilon^{i\alpha}\tilde{v}_{i,\alpha} + \frac{1}{\varepsilon}\tilde{\sigma}_\varepsilon^{ij}\tilde{v}_{i|j}]_{|y=x/\varepsilon}\,.
$$

Thus one can write Eq. (2.2.8) in the form

$$\varepsilon \int_{\Omega} \int_{c^-(\frac{x}{\varepsilon})}^{c^+(\frac{x}{\varepsilon})} \left[\tilde{\sigma}_\varepsilon^{i\alpha}\left(x, \frac{x}{\varepsilon}\right) \tilde{v}_{i,\alpha}\left(x, \frac{x}{\varepsilon}\right) + \frac{1}{\varepsilon}\tilde{\sigma}_\varepsilon^{ij}\left(x, \frac{x}{\varepsilon}\right) \tilde{v}_{i|j}\left(x, \frac{x}{\varepsilon}\right) \right] dx\, d\left(\frac{x_3}{\varepsilon}\right)$$

$$= \varepsilon^2 \int_{\Omega} \left[p_+^\alpha(x)\tilde{v}_\alpha\left(x, \frac{x}{\varepsilon}, c^+\right) G_+^{\frac{1}{2}}\left(\frac{x}{\varepsilon}\right) + p_-^\alpha(x)\tilde{v}_\alpha\left(x, \frac{x}{\varepsilon}, c^-\right) G_-^{\frac{1}{2}}\left(\frac{x}{\varepsilon}\right) \right] dx$$

$$+ \varepsilon^3 \int_{\Omega} \left[q^+(x)\tilde{v}_3\left(x, \frac{x}{\varepsilon}, c^+\right) G_+^{\frac{1}{2}}\left(\frac{x}{\varepsilon}\right) + q^-(x)\tilde{v}_3\left(x, \frac{x}{\varepsilon}, c^-\right) G_-^{\frac{1}{2}}\left(\frac{x}{\varepsilon}\right) \right] dx$$

$$+ \varepsilon^2 \int_{\Omega} \int_{c^-(\frac{x}{\varepsilon})}^{c^+(\frac{x}{\varepsilon})} \left[b^\alpha\left(\frac{x}{\varepsilon}\right) \tilde{v}_\alpha\left(x, \frac{x}{\varepsilon}\right) + \varepsilon b^3\left(\frac{x}{\varepsilon}\right) \tilde{v}_3\left(x, \frac{x}{\varepsilon}\right) \right] dx\, d\left(\frac{x_3}{\varepsilon}\right). \tag{2.2.13}$$

Now we set $y = \dfrac{x}{\varepsilon}$, $y_3 = \dfrac{x_3}{\varepsilon}$ or $y = \dfrac{x}{\varepsilon}$ and treat y as an independent variable. Integrating both sides over Y one arrives at

$$A^\varepsilon(\tilde{w}^\varepsilon, \tilde{v}) = F^\varepsilon(\tilde{v}) \tag{2.2.14}$$

where

$$A^\varepsilon(\tilde{w}^\varepsilon, \tilde{v}) = \varepsilon h_0 \int_{\Omega} \prec \tilde{\sigma}_\varepsilon^{i\alpha}(\tilde{w}^\varepsilon)\tilde{v}_{i,\alpha} + \frac{1}{\varepsilon}\tilde{\sigma}_\varepsilon^{ij}(\tilde{w}^\varepsilon)\tilde{v}_{i|j} \succ dx\,, \tag{2.2.15}$$

$$F^\varepsilon(\tilde{v}) = \varepsilon^2 \int_{\Omega} [h_0 \prec b^\alpha(y)\tilde{v}_\alpha(x, y) \succ + p_+^\alpha(x)\langle \tilde{v}_\alpha(x, y, c^+)(G_+(y))^{\frac{1}{2}}\rangle$$

$$+ p_-^\alpha(x)\langle \tilde{v}_\alpha(x, y, c^-)(G_-(y))^{\frac{1}{2}}\rangle] dx + \varepsilon^3 \int_{\Omega} [h_0 \prec b^3(y)\tilde{v}_3(x, y) \succ$$

$$+ q_+(x)\langle \tilde{v}_3(x, y, c^+)(G_+(y))^{\frac{1}{2}}\rangle + q_-(x)\langle \tilde{v}_3(x, y, c^-)(G_-(y))^{\frac{1}{2}}\rangle] dx\,. \tag{2.2.16}$$

The notations $\langle \cdot \rangle$ and $\prec \cdot \succ$ have been adopted in the Introduction. The quantity

$$h_0 = |\mathcal{Y}|/|Y| \tag{2.2.17}$$

represents the average thickness of the rescaled cell $\mathcal{Y}$.

The stress-displacement relations take the form

$$\tilde{\sigma}_\varepsilon^{ij}(x, y, y_3) = C^{ijkl}(y, y_3)\tilde{\epsilon}_{kl}^\varepsilon(x, y, y_3) \tag{2.2.18}$$

where

$$2\tilde{\epsilon}^{\varepsilon}_{\alpha\beta} = \tilde{w}^{\varepsilon}_{\alpha,\beta} + \tilde{w}^{\varepsilon}_{\beta,\alpha} + \frac{1}{\varepsilon}(\tilde{w}^{\varepsilon}_{\alpha|\beta} + \tilde{w}^{\varepsilon}_{\beta|\alpha})\,,$$

$$2\tilde{\epsilon}^{\varepsilon}_{\alpha 3} = \tilde{w}^{\varepsilon}_{3,\alpha} + \frac{1}{\varepsilon}(\tilde{w}^{\varepsilon}_{\alpha|3} + \tilde{w}^{\varepsilon}_{3|\alpha})\,, \qquad \tilde{\epsilon}^{\varepsilon}_{33} = \frac{1}{\varepsilon}\tilde{w}^{\varepsilon}_{3|3} \tag{2.2.19}$$

and $\tilde{w} \in H(\Omega \times \mathcal{Y})$, where

$$H(\Omega \times \mathcal{Y}) = \{v = (v_i(x,y)) \mid v(x,\cdot) \in W(\mathcal{Y}), v(\cdot,y) \in H^1_0(\Omega)^3\}\,,$$

$W(\mathcal{Y}) = \{v \in [H^1(\mathcal{Y})]^3 \mid v$ assumes equal values on the opposite lateral faces of $\mathcal{Y}\}$
and $H^1_0(\Omega)^3 = [H^1_0(\Omega)]^3$.

Now we are ready to formulate the boundary value problem on the product domain $\Omega \times \mathcal{Y}$. It will be called $(\mathcal{P}^{\varepsilon})$:

> find $\tilde{w}^{\varepsilon} \in H(\Omega \times \mathcal{Y})$ such that Eq. (2.2.14) holds for all $v \in H(\Omega \times \mathcal{Y})$.

The solution $\tilde{w}^{\varepsilon}$ will be approximated by the two-scale asymptotic expansion method.

2.3. Asymptotic analysis. Effective moduli and local problems

The solution $\tilde{w}^{\varepsilon}$ of problem $(\mathcal{P}^{\varepsilon})$ is sought in the form

$$\tilde{w}^{\varepsilon} = u^{(0)}(x) + \varepsilon u^{(1)}(x;y) + \varepsilon^2 u^{(2)}(x;y) + \dots \tag{2.3.1}$$

where $u^{(0)} \in H^1_0(\Omega)^3$ and $u^{(p)} \in H(\Omega \times \mathcal{Y})$, $p = 1, 2, \dots$. Consequently, the stresses assume the form of a similar expansion

$$\tilde{\sigma}^{ij}_{\varepsilon} = \sigma^{ij}_0 + \varepsilon\sigma^{ij}_1 + \varepsilon^2\sigma^{ij}_2 + \dots\,, \tag{2.3.2}$$

where $\sigma^{ij}_p = \sigma^{ij}_p(x,y)$, and

$$\sigma^{ij}_0 = C^{ijkl}u^{(1)}_{k|l} + C^{ijk\beta}u^{(0)}_{k,\beta}\,, \quad \sigma^{ij}_p = C^{ijkl}u^{(p+1)}_{k|l} + C^{ijk\beta}u^{(p)}_{k,\beta}\,;\quad p = 1, 2, \dots. \tag{2.3.3}$$

Substituting (2.3.2) into (2.2.14) and equating the terms of the same order with respect to ε results in the reformulation of problem $(\mathcal{P}^{\varepsilon})$ to the following sequence of problems:

Find $u^{(0)} \in H^1_0(\Omega)^3$ and $u^{(p)} \in H(\Omega \times \mathcal{Y})$ such that

$$\int_{\Omega} \prec \sigma^{ij}_0 \tilde{v}_{i|j} \succ dx = 0\,, \tag{2.3.4}$$

$$\int_{\Omega} \prec \sigma^{i\alpha}_0 \tilde{v}_{i,\alpha} + \sigma^{ij}_1 \tilde{v}_{i|j} \succ dx = 0\,, \tag{2.3.5}$$

$$h_0 \int_{\Omega} \prec \sigma^{i\alpha}_1 \tilde{v}_{i,\alpha} + \sigma^{ij}_2 \tilde{v}_{i|j} \succ dx = F_2(\tilde{v})\,, \tag{2.3.6}$$

$$h_0 \int_\Omega \prec \sigma_2^{i\alpha} \tilde{v}_{i,\alpha} + \sigma_3^{ij} \tilde{v}_{i|j} \succ dx = F_3(\tilde{v}) \,, \tag{2.3.7}$$

$$\int_\Omega \prec \sigma_p^{i\alpha} \tilde{v}_{i,\alpha} + \sigma_{p+1}^{ij} \tilde{v}_{i|j} \succ dx = 0 \,, \tag{2.3.8}$$

for each $\tilde{v} \in H(\Omega \times \mathcal{Y})$, where

$$F_2(\tilde{v}) = \int_\Omega [p_+^\alpha(x)\langle \tilde{v}_\alpha(x,y,c^+)(G_+(y))^{\frac{1}{2}}\rangle \tag{2.3.9}$$

$$+ p_-^\alpha(x)\langle \tilde{v}_\alpha(x,y,c^-)(G_-(y))^{\frac{1}{2}}\rangle]dx + h_0 \int_\Omega \prec b^\alpha(y)\tilde{v}_\alpha(x,y) \succ dx \,,$$

$$F_3(\tilde{v}) = \int_\Omega [q_+(x)\langle \tilde{v}_3(x,y,c^+)(G_+(y))^{\frac{1}{2}}\rangle$$

$$+ q_-(x)\langle \tilde{v}_3(x,y,c^-)(G_-(y))^{\frac{1}{2}}\rangle]dx + h_0 \int_\Omega \prec b^3(y)\tilde{v}_3(x,y) \succ dx \,. \tag{2.3.10}$$

The sequence of problems given by (2.3.4) – (2.3.8) can be subsequently solved, thus enabling a construction of the solution (2.3.1) and (2.3.2) of problem $(\mathcal{P}^\varepsilon)$.

To make this chapter self-contained, it is indispensable to follow at least first steps of the solution process.

Step 1. Substitute the representation $(2.3.3)_1$ into (2.3.4) and take $\tilde{v}_i = \varphi(x)w_i(y)$, $\varphi \in \mathbf{D}(\Omega)$, $w \in W(\mathcal{Y})$. Since φ is arbitrary, one finds

$$\prec [C^{ijkl}u_{k|l}^{(1)} + C^{ij\beta k}u_{k,\beta}^{(0)}]w_{i|j} \succ = 0 \,. \tag{2.3.11}$$

Hence $u^{(1)}$ can be expressed as

$$u_k^{(1)} = \Theta_k^{(j\beta)}(y)u_{j,\beta}^{(0)} + u_k(x) \,, \tag{2.3.12}$$

where $\Theta^{(j\beta)} \in W(\mathcal{Y})$ satisfies (for $(j\beta) = (kl)$)

$$(\mathcal{P}_\mathcal{Y}^1) \qquad | \quad a_\mathcal{Y}(\Theta^{(kl)}, w) = - \prec C^{ijkl}w_{i|j} \succ \qquad \forall w \in W(\mathcal{Y}) \tag{2.3.13}$$

and the bilinear form $a_\mathcal{Y}(\cdot,\cdot)$ is defined by

$$a_\mathcal{Y}(u,v) = \prec C^{ijkl}(y)u_{i|j}v_{k|l} \succ \,, \qquad u,v \in W(\mathcal{Y}) \,. \tag{2.3.14}$$

According to standard theorems problem $(\mathcal{P}_\mathcal{Y}^1)$ is uniquely solvable provided that, cf. Sec. 1.2.2

$$\prec \Theta^{(kl)} \succ = 0 \,. \tag{2.3.15}$$

Note that the function

$$\Theta^{(3\beta)} = (\Theta_k^{(3\beta)}) = (-\hat{y}_3\delta_k^\beta) , \tag{2.3.16}$$

where

$$\hat{y}_3 = y_3 - \prec y_3 \succ , \tag{2.3.17}$$

satisfies

$$C^{ijkh}\Theta_{k|h}^{(3\beta)} + C^{ij3\beta} = 0 , \tag{2.3.18}$$

along with (2.3.15). Thus $\Theta^{(3\beta)}$ solves problem $(\mathcal{P}_y^1)$. Substituting (2.3.16) into (2.3.12) one obtains

$$u_\sigma^{(1)} = \Theta_\sigma^{(\alpha\beta)}(\boldsymbol{y})u_{\alpha,\beta}^{(0)} - \hat{y}_3 w_{,\sigma} + u_\sigma(x) , \qquad u_3^{(1)} = \Theta_3^{(\alpha\beta)}(\boldsymbol{y})u_{\alpha,\beta}^{(0)} + u_3(x) , \tag{2.3.19}$$

where $w(x) = u_3^{(0)}(x)$.

Step 2. Substitute relation (2.3.19) into (2.3.3)$_1$ taking into account (2.3.18). One finds

$$\sigma_0^{ij} = A_\diamond^{ij\alpha\beta}u_{\alpha,\beta}^{(0)} , \tag{2.3.20}$$

$$A_\diamond^{ijhm} = C^{ijkl}a_{kl}^{(hm)} , \qquad a_{kl}^{(hm)} = \Theta_{k|l}^{(hm)} + \delta_k^{(h}\delta_l^{m)} . \tag{2.3.21}$$

Now let us substitute (2.3.20) into (2.3.5) with $\tilde{v}_\alpha = w_\alpha(x)$, $\tilde{v}_3 = 0$ and $w_\alpha \in H_0^1(\Omega)$. Then

$$\int_\Omega \prec \sigma_0^{\alpha\beta} \succ w_{\beta,\alpha}dx = 0 \qquad \forall \boldsymbol{w} \in H_0^1(\Omega)^2 , \tag{2.3.22}$$

where

$$\prec \sigma_0^{\alpha\beta} \succ = A_y^{\alpha\beta\lambda\mu}u_{\lambda,\mu}^{(0)} , \tag{2.3.23}$$

$$A_y^{\alpha\beta\lambda\mu} = \prec A_\diamond^{\alpha\beta\lambda\mu} \succ . \tag{2.3.24}$$

We shall prove that solution to the equation (2.3.22) is trivial: $u_\lambda^{(0)} = 0$, $\sigma_0^{\alpha\beta} = 0$ and hence $\sigma_0^{ij} = 0$. To this end we have to concentrate on properties of the tensor $\boldsymbol{A}_y$.

Let us take $(kl) = (\alpha\beta)$ and $\boldsymbol{w} = \Theta^{(\lambda\mu)}$ in (2.3.13). One easily finds an identity

$$a_y(\Theta^{(\alpha\beta)}, \Theta^{(\lambda\mu)}) + \prec C^{ij\alpha\beta}\Theta_{i|j}^{(\lambda\mu)} \succ = 0 . \tag{2.3.25}$$

Combining this identity with (2.3.24) and (2.3.21) one obtains

$$A_y^{\alpha\beta\lambda\mu} = \prec C^{ijkl}a_{ij}^{(\alpha\beta)}a_{kl}^{(\lambda\mu)} \succ , \tag{2.3.26}$$

which, by (2.1.7), implies the following symmetries

$$A_y^{\alpha\beta\lambda\mu} = A_y^{\lambda\mu\alpha\beta} , \qquad A_y^{\alpha\beta\lambda\mu} = A_y^{\beta\alpha\lambda\mu} = A_y^{\beta\alpha\mu\lambda} . \tag{2.3.27}$$

Moreover, one can prove that $(A_y^{\alpha\beta\lambda\mu})$ is positive definite. By (2.3.26) and (2.1.6) one can estimate

$$A_y^{\alpha\beta\lambda\mu}\gamma_{\alpha\beta}\gamma_{\lambda\mu} \geq c\sum_{i,j} \prec a_{ij}^{(\alpha\beta)} a_{ij}^{(\lambda\mu)} \succ \gamma_{\alpha\beta}\gamma_{\lambda\mu} = c\sum_{i,j} \prec \tilde{\gamma}_{ij}\tilde{\gamma}_{ij} \succ \geq 0 \qquad (2.3.28)$$

with

$$\tilde{\gamma}_{ij} = a_{ij}^{(\alpha\beta)}\gamma_{\alpha\beta} \ . \qquad (2.3.29)$$

On the other hand, the equality in (2.3.28) can only be attained if $\tilde{\gamma}_{ij} = 0$. Then $\langle\tilde{\gamma}_{ij}\rangle = 0$. Taking into account that $\langle\Theta_{\lambda|\mu}^{\alpha\beta}\rangle = 0$, one finds $\langle\tilde{\gamma}_{\lambda\mu}\rangle = \gamma_{\lambda\mu} = 0$. Hence A_y is positive definite, which implies that $u^0 = 0$ is the only solution to problem (2.3.22). Consequently, Eqs. (2.3.19) reduce to

$$u_\sigma^{(1)} = -\hat{y}_3 w_{,\sigma} + u_\sigma(x) \ , \qquad u_3^{(1)} = u_3(x) \ . \qquad (2.3.30)$$

Step 3. Let us substitute (2.3.30) into (2.3.3)$_2$ for $p = 1$. One obtains

$$\sigma_1^{ij} = C^{ijkl}u_{k|l}^{(2)} + C^{ij\alpha k}u_{k,\alpha} - \hat{y}_3 C^{ij\alpha\beta}w_{,\alpha\beta} \ . \qquad (2.3.31)$$

By inserting this formula into Eq. (2.3.5), taking into account that $\sigma_0^{ij} = 0$ and choosing $\tilde{v}_i = \varphi(x)w_i(y)$, $\varphi \in \mathbf{D}(\Omega)$, $w \in W(\mathcal{Y})$, one arrives at

$$\prec [C^{ijkl}u_{k|l}^{(2)} + C^{ij\beta k}u_{k,\beta} - \hat{y}_3 C^{ij\alpha\beta}w_{,\alpha\beta}]w_{i|j} \succ = 0 \qquad \forall \, w \in W(\mathcal{Y}) \ . \qquad (2.3.32)$$

By linearity of the problem (2.3.32) its solution can be represented in the form

$$u^{(2)} = \Theta^{(j\beta)}(y)u_{j,\beta} - \Xi^{(\alpha\beta)}(y)w_{,\alpha\beta} + \eta(x) \ , \qquad (2.3.33)$$

where $\eta = (\eta_\alpha)$ is at this moment undetermined while $\Theta^{(j\beta)}$ are solutions to problem (P_y^1) subject to condition (2.3.15) and the fields $\Xi^{(\alpha\beta)} \in W(\mathcal{Y})$ satisfy

$$(\mathcal{P}_y^2) \ \Big| \ a_y(\Xi^{(\alpha\beta)}, w) = - \prec \hat{y}_3 C^{ij\alpha\beta}w_{i|j} \succ \qquad \forall \, w \in W(\mathcal{Y}) \ . \qquad (2.3.34)$$

They are also normalized as follows

$$\prec \Xi^{(\alpha\beta)} \succ = 0 \ . \qquad (2.3.35)$$

Due to the last condition the local fields $\Xi^{(\alpha\beta)}$ are uniquely determined as solution to $(\mathcal{P}_y^2)$. The proof is similar to the proof of well posedness of the problem $(\mathcal{P}_y^1)$.

Substituting relation (2.3.16) into (2.3.33) one finds

$$u_\alpha^{(2)} = \Theta_\alpha^{(\gamma\beta)}(y)u_{\gamma,\beta} - \Xi_\alpha^{(\alpha\beta)}(y)w_{,\gamma\beta} - \hat{y}_3 u_{3,\alpha} + \eta_\alpha(x) \ ,$$
$$u_3^{(2)} = \Theta_3^{(\gamma\beta)}(y)u_{\gamma,\beta} - \Xi_3^{(\gamma\beta)}(y)w_{,\gamma\beta} + \eta_3(x) \ . \qquad (2.3.36)$$

Substitution of (2.3.33) into (2.3.31) results in

$$\sigma_1^{ij} = A_\diamond^{ij\alpha\beta}(\boldsymbol{y})u_{\beta,\alpha} - E_\diamond^{ij\alpha\beta}(\boldsymbol{y})w_{,\alpha\beta} \,, \tag{2.3.37}$$

where $\boldsymbol{A}_\diamond$ is defined by (2.3.21) and

$$E_\diamond^{ij\alpha\beta} = C^{ijkl}e_{kl}^{(\alpha\beta)} \,, \quad e_{kl}^{(\alpha\beta)} = \Xi_{k|l}^{(\alpha\beta)} + \hat{y}_3\delta_k^{(\alpha}\delta_l^{\beta)} \,. \tag{2.3.38}$$

Note that the field v_3 contributes to $u_3^{(1)}$ but does not affect σ_1^{ij}.
Step 4. One can easily prove that

$$\prec \sigma_1^{3j} \succ = 0 \,. \tag{2.3.39}$$

Let us substitute $w_i = \delta_{im}\hat{y}_3$ into (2.3.13). Hence

$$\prec A_\diamond^{m3kl} \succ = 0 \,. \tag{2.3.40a}$$

The same substitution into (2.3.34) results in

$$\prec E_\diamond^{m3\alpha\beta} \succ = 0 \,. \tag{2.3.40b}$$

Hence, by (2.3.37), one obtains (2.3.39).

Further, the following notation will be used

$$\mathcal{N}^{\alpha\beta} = h_o \prec \sigma_1^{\alpha\beta} \succ \,, \quad \mathcal{M}^{\alpha\beta} = h_o \prec \hat{y}_3\sigma_1^{\alpha\beta} \succ \,, \quad Q^\alpha = h_o \prec \sigma_2^{3\alpha} \succ \,. \tag{2.3.41}$$

Step 5. According to (2.3.1) and (2.3.19) we have

$$\tilde{w}_\sigma = \varepsilon(u_\sigma(x) - \hat{y}_3 w_{,\sigma}) + 0(\varepsilon^2) \,, \quad \tilde{w}_3 = w(x) + \varepsilon u_3(x) + 0(\varepsilon^2) \,. \tag{2.3.42}$$

The choice of test functions $\tilde{v}_i$ involved in (2.3.6) - (2.3.7) should be compatible with representations (2.3.42). Thus we assume

$$\tilde{v}_\alpha = \varepsilon[w_\alpha(x) - \hat{y}_3 v_{,\alpha}] \,, \quad \tilde{v}_3 = v(x) \,. \tag{2.3.43}$$

First, let us substitute (2.3.43) into (2.3.6), assuming that $w_\alpha \in H_0^1(\Omega)$ and $v \in H_0^2(\Omega)$. Taking into account (2.3.39) one finds

$$\int_\Omega \mathcal{N}^{\alpha\beta}w_{\alpha,\beta}dx = \int_\Omega \hat{p}^\alpha(x)w_\alpha(x)dx \,, \tag{2.3.44}$$

$$\int_\Omega (\mathcal{M}^{\alpha\beta}v_{,\alpha\beta} + Q^\alpha v_{,\alpha})dx = \int_\Omega \hat{m}^\alpha(x)v_{,\alpha}dx \,, \tag{2.3.45}$$

where the loadings are given by

$$
\begin{aligned}
\hat{p}^\alpha(x) &= h_0 \prec b^\alpha(\boldsymbol{y}) \succ + \langle (G_+)^{\frac{1}{2}}\rangle p_+^\alpha(x) + \langle (G_-)^{\frac{1}{2}}\rangle p_-^\alpha(x) , \\
\hat{m}^\alpha(x) &= \langle (c^+ - \prec y_3 \succ)(G_+)^{\frac{1}{2}}\rangle p_+^\alpha(x) \\
&\quad + \langle (c^- - \prec y_3 \succ)(G_-)^{\frac{1}{2}}\rangle p_-^\alpha(x) - \prec \hat{y}_3 b^\alpha(\boldsymbol{y}) \succ .
\end{aligned}
\tag{2.3.46}
$$

Now let us set $\tilde{v}_\alpha = 0$ and $\tilde{v}_3 = v(x)$ in (2.3.7). One obtains

$$
\int_\Omega Q^\alpha v_{,\alpha} dx = \int_\Omega \hat{q}(x) v(x) dx ,
\tag{2.3.47}
$$

where

$$
\hat{q}(x) = \langle (G_+)^{\frac{1}{2}}\rangle q_+(x) + \langle (G_-)^{\frac{1}{2}}\rangle q_-(x) + h_0 \prec b^3(\boldsymbol{y}) \succ .
\tag{2.3.48}
$$

If $v \in H_0^2(\Omega)$, then, on combining formulae (2.3.45) and (2.3.47), one finds

$$
\int_\Omega \mathcal{M}^{\alpha\beta} \kappa_{\alpha\beta}(v) dx = \int_\Omega (\hat{q}v - \hat{m}^\alpha v_{,\alpha}) dx ,
\tag{2.3.49}
$$

where $\kappa_{\alpha\beta}(v) = -v_{,\alpha\beta} = -\dfrac{\partial^2 v}{\partial x_\beta \partial x_\alpha}$. Let us define the effective moduli by

$$
\begin{aligned}
E_{\mathsf{y}}^{\alpha\beta\lambda\mu} &= \prec E_\diamond^{\alpha\beta\lambda\mu} \succ , \qquad & F_{\mathsf{y}}^{\alpha\beta\lambda\mu} &= \prec \hat{y}_3 A_\diamond^{\alpha\beta\lambda\mu} \succ , \\
D_{\mathsf{y}}^{\alpha\beta\lambda\mu} &= \prec \hat{y}_3 E_\diamond^{\alpha\beta\lambda\mu} \succ .
\end{aligned}
\tag{2.3.50}
$$

The stress and couple stress averages involved in Eqs (2.3.44) and (2.3.49) are interrelated with $e(\boldsymbol{u}), \kappa(w)$ by

$$
\begin{aligned}
\mathcal{N}^{\alpha\beta} &= h_0 A_{\mathsf{y}}^{\alpha\beta\lambda\mu} e_{\lambda\mu}(\boldsymbol{u}) + h_0 E_{\mathsf{y}}^{\alpha\beta\lambda\mu} \kappa_{\lambda\mu}(w) , \\
\mathcal{M}^{\alpha\beta} &= h_0 F_{\mathsf{y}}^{\alpha\beta\lambda\mu} e_{\lambda\mu}(\boldsymbol{u}) + h_0 D_{\mathsf{y}}^{\alpha\beta\lambda\mu} \kappa_{\lambda\mu}(w) ,
\end{aligned}
\tag{2.3.51}
$$

where A_{y} is defined by (2.3.24). Equations (2.3.51) follow from (2.3.37) and (2.3.41).

The variational equations (2.3.44) and (2.3.49) along with constitutive relationships (2.3.51) form the first homogenized problem

$$
(P_{hom}) \quad \left|
\begin{array}{l}
\text{Find}(\boldsymbol{u}, w) \in (H_0^1(\Omega))^2 \times H_0^2(\Omega) = V_K^0(\Omega) \text{ such that} \\
\text{Eqs. (2.3.44), (2.3.49), (2.3.51) hold for each } (\boldsymbol{w}, v) \in V_K^0(\Omega) .
\end{array}
\right.
$$

Step 6. We shall prove that problem (P_{hom}) is well posed.

Note first that tensors (2.3.50) can be written as follows

$$
\begin{aligned}
E_{\mathsf{y}}^{\alpha\beta\lambda\mu} &= \prec C^{ijkl} a_{ij}^{(\alpha\beta)} e_{kl}^{(\lambda\mu)} \succ , \qquad & F_{\mathsf{y}}^{\alpha\beta\lambda\mu} &= \prec C^{ijkl} e_{ij}^{(\alpha\beta)} a_{kl}^{(\lambda\mu)} \succ , \\
D_{\mathsf{y}}^{\alpha\beta\lambda\mu} &= \prec C^{ijkl} e_{ij}^{(\alpha\beta)} e_{ij}^{(\lambda\mu)} \succ ,
\end{aligned}
\tag{2.3.52}
$$

where $a_{ij}^{(\alpha\beta)}$ and $e_{ij}^{(\alpha\beta)}$ have been defined by Eqs. $(2.3.21)_2$ and $(2.3.38)_2$. The identities
(2.3.52) can be proved similarly as the identity (2.3.26) proved previously. Thus the symmetry properties (2.1.7) readily imply

$$E_y^{\alpha\beta\lambda\mu} = F_y^{\lambda\mu\alpha\beta} , \quad D_y^{\alpha\beta\lambda\mu} = D_y^{\lambda\mu\alpha\beta} . \tag{2.3.53a}$$

Note, that

$$E_y^{\alpha\beta\lambda\mu} \neq E_y^{\lambda\mu\alpha\beta} , \quad F_y^{\alpha\beta\lambda\mu} \neq F_y^{\lambda\mu\alpha\beta} , \tag{2.3.53b}$$

in general.

The elastic potential of the homogenized plate is expressed by

$$\mathcal{W}_y(\epsilon, \kappa) = (A_y^{\alpha\beta\lambda\mu}\epsilon_{\alpha\beta}\epsilon_{\lambda\mu} + E_y^{\alpha\beta\lambda\mu}\kappa_{\lambda\mu}\epsilon_{\alpha\beta} + F_y^{\alpha\beta\lambda\mu}\kappa_{\alpha\beta}\epsilon_{\lambda\mu} + D_y^{\alpha\beta\lambda\mu}\kappa_{\alpha\beta}\kappa_{\lambda\mu})/2 , \tag{2.3.54}$$

where

$$\epsilon_{\alpha\beta} = e_{\alpha\beta}(\boldsymbol{u}) , \quad \kappa_{\alpha\beta} = \kappa_{\alpha\beta}(w) . \tag{2.3.55}$$

By virtue of representations (2.3.26), (2.3.52) one can rearrange the potential $\mathcal{W}_y$ to the
form

$$\mathcal{W}_y(\epsilon, \kappa) = \frac{1}{2} \prec C^{ijkl}\bar{\gamma}_{ij}\bar{\gamma}_{kl} \succ , \tag{2.3.56}$$

where

$$\bar{\gamma}_{ij} = a_{ij}^{(\lambda\mu)}(\boldsymbol{y})\epsilon_{\lambda\mu} + e_{ij}^{(\lambda\mu)}(\boldsymbol{y})\kappa_{\lambda\mu} . \tag{2.3.57}$$

Due to (2.1.6) the potential $\mathcal{W}_y$ is non-negative. We shall prove more, that $\mathcal{W}_y = 0$ implies
$\epsilon = 0$ and $\kappa = 0$.

Let $\mathcal{W}_y = 0$. Then by (2.1.6) $\bar{\gamma}_{ij} = 0$. In particular $\bar{\gamma}_{\alpha\beta} = 0$. Hence $\langle\bar{\gamma}_{\alpha\beta}\rangle = 0$. Due to
Y-periodicity of $\Theta^{(ij)}$ and $\Xi^{(ij)}$, we have

$$\langle\bar{\gamma}_{\alpha\beta}\rangle = \epsilon_{\alpha\beta} + \hat{y}_3\kappa_{\alpha\beta} = 0 . \tag{2.3.58}$$

Hence $\epsilon_{\alpha\beta} = 0$ and $\kappa_{\alpha\beta} = 0$. Thus there exists a positive constant c such that

$$\mathcal{W}_y(\epsilon, \kappa) \geq c \sum_{\alpha,\beta}((\epsilon_{\alpha\beta})^2 + (\kappa_{\alpha\beta})^2) . \tag{2.3.59}$$

Therefore, the problem (P_{hom}) is uniquely solvable.

Thus the solution $(w^\varepsilon, \boldsymbol{\sigma}_\varepsilon)$ to problem (P_ε) is expressed by $(2.2.9)_{1,2}$ with $\tilde{\sigma}_\varepsilon^{ij}$, $\tilde{w}_i^\varepsilon$ given
by (2.3.1), (2.3.2). The iterative process can be continued. We stop at this stage since the
overall response of the periodic plate is described by problem (P_{hom}).

Step 7. Problem (P_{hom}) is referred to the rescaled plate. To come back to the original plate
problem (Sec. 2.1) one should introduce the following quantities

$$u_\alpha^h = \varepsilon_0 u_\alpha , \quad w^h = w ,$$
$$\epsilon_{\alpha\beta}^h = e_{\alpha\beta}(\boldsymbol{u}^h) , \quad \epsilon_{\alpha\beta}^h = \varepsilon_0\epsilon_{\alpha\beta} , \quad \kappa_{\alpha\beta}^h = \kappa_{\alpha\beta} \tag{2.3.60}$$

$$N_h^{\alpha\beta} = (\varepsilon_0)^2 \mathcal{N}^{\alpha\beta} , \qquad M_h^{\alpha\beta} = (\varepsilon_0)^3 \mathcal{M}^{\alpha\beta} , \tag{2.3.61}$$

$$A_{\mathcal{H}}^{\alpha\beta\lambda\mu} = h_0 \varepsilon_0 A_{\mathcal{y}}^{\alpha\beta\lambda\mu} , \quad E_{\mathcal{H}}^{\alpha\beta\lambda\mu} = h_0 (\varepsilon_0)^2 E_{\mathcal{y}}^{\alpha\beta\lambda\mu} ,$$
$$F_{\mathcal{H}}^{\alpha\beta\lambda\mu} = h_0 (\varepsilon_0)^2 F_{\mathcal{y}}^{\alpha\beta\lambda\mu} , \quad D_{\mathcal{H}}^{\alpha\beta\lambda\mu} = h_0 (\varepsilon_0)^3 D_{\mathcal{y}}^{\alpha\beta\lambda\mu} ; \tag{2.3.62}$$

$$q = (\varepsilon_0)^3 \hat{q} = \langle G_-^{\frac{1}{2}} \rangle r_-^3 + \langle G_+^{\frac{1}{2}} \rangle r_+^3 + \varepsilon_0 h_0 \prec b^3 \succ ,$$

$$p^\alpha = (\varepsilon_0)^2 \hat{p}^\alpha = \langle G_-^{\frac{1}{2}} \rangle r_-^\alpha + \langle G_+^{\frac{1}{2}} \rangle r_+^\alpha + \varepsilon_0 h_0 \prec b^\alpha \succ ,$$

$$m^\alpha = \varepsilon_0 \langle (c^- - \prec y_3 \succ) G_-^{\frac{1}{2}} \rangle r_-^\alpha \tag{2.3.63}$$
$$+ \varepsilon_0 \langle (c^+ - \prec y_3 \succ) G_+^{\frac{1}{2}} \rangle r_+^\alpha - \varepsilon_0 \prec \hat{y}_3 b^\alpha \succ .$$

The last three formulae express the equivalent loadings applied to the original, Z-periodic plate. Its effective membrane $(A_{\mathcal{H}})$, reciprocal $(E_{\mathcal{H}}, F_{\mathcal{H}})$ and bending $(D_{\mathcal{H}})$ stiffnesses are given by (2.3.62). The membrane forces N_h and moments M_h are interrelated with membrane strains $e_{\alpha\beta}(u^h)$ and changes of curvature $\kappa_{\alpha\beta}(w^h)$ by

$$N_h^{\alpha\beta} = A_{\mathcal{H}}^{\alpha\beta\lambda\mu} e_{\lambda\mu}(u^h) + E_{\mathcal{H}}^{\alpha\beta\lambda\mu} \kappa_{\lambda\mu}(w^h) ,$$
$$M_h^{\alpha\beta} = F_{\mathcal{H}}^{\alpha\beta\lambda\mu} e_{\lambda\mu}(u^h) + D_{\mathcal{H}}^{\alpha\beta\lambda\mu} \kappa_{\lambda\mu}(w^h) . \tag{2.3.64}$$

The effective problem for the original Z- periodic plate problem (P_Z) reads

$$(P_{\mathcal{H}}) \quad \left| \begin{array}{l} \text{Find } (u^h, w^h) \in H_0^1(\Omega)^2 \times H_0^2(\Omega) = V_K^0(\Omega) \text{ such that} \\[4pt] \displaystyle\int_\Omega [N_h^{\alpha\beta}(u^h, w^h) e_{\alpha\beta}(\tilde{w}) + M_h^{\alpha\beta}(u^h, w^h) \kappa_{\alpha\beta}(\tilde{w})] dx \\[4pt] = \displaystyle\int_\Omega (p^\alpha \tilde{w}_\alpha + q\tilde{w} - m^\alpha \tilde{w}_{,\alpha}) dx \qquad \forall\, (\tilde{w}, \tilde{w}) \in V_K^0(\Omega) \end{array} \right.$$

where relations $N_h^{\alpha\beta}(u^h, w^h)$, $M_h^{\alpha\beta}(u^h, w^h)$ are expressed by Eqs. (2.3.64). The displacement fields u^h, w^h represent averaged in-plane displacement and transverse deflection of the original plate.

Remark 2.3.1. Note that by Eqs. (2.3.2), $\sigma_0^{ij} = 0$, (2.3.37) and (2.3.57) one can rewrite

$$\sigma_\varepsilon^{ij} = \varepsilon[A_\diamond^{ij\alpha\beta} e_{\alpha\beta}(u) + E_\diamond^{ij\alpha\beta} \kappa_{\alpha\beta}(w)] + 0(\varepsilon^2) ,$$
$$\epsilon_{ij}^\varepsilon = \varepsilon \bar{\gamma}_{ij} + 0(\varepsilon^2) . \tag{2.3.65}$$

Thus by Eqs. (2.3.24), (2.3.50), (2.3.51) one finds

$$h_0 \prec \sigma_\varepsilon^{ij} \epsilon_{ij}^\varepsilon \succ = \varepsilon^2 [\mathcal{N}^{\alpha\beta} \epsilon_{\alpha\beta} + \mathcal{M}^{\alpha\beta} \kappa_{\alpha\beta}] + 0(\varepsilon^3) . \tag{2.3.66}$$

Such equivalence between average of the internal work of stresses and internal work of averaged stress – and couple stress resultants means that Hill's consistency criterion, written in the three-dimensional case as

$$\prec \sigma : \epsilon \succ = \prec \sigma \succ : \prec \epsilon \succ , \tag{2.3.67}$$

holds true in the case considered.

2.4.   Case of transverse symmetry

Assume now in addition that the geometry and elastic properties of the plate are symmetric with respect to the plane $x_3 = 0$, i.e. $C_Z^{ijkl}(x, x_3) = C_Z^{ijkl}(x, -x_3)$ and $x_3^+(x) = -x_3^-(x)$. Moreover the planes $x_3 = $ const are assumed to be planes of material symmetry, i.e.

$$C_Z^{3\alpha\beta\gamma} = C_Z^{3333\delta} = 0 \,.\tag{2.4.1}$$

The above assumptions can conveniently be put in the form

$$C^{ijkl}\left(\frac{x}{\varepsilon}, \frac{x_3}{\varepsilon}\right) = C^{ijkl}\left(\frac{x}{\varepsilon}, -\frac{x_3}{\varepsilon}\right)\,, \qquad C^{3\alpha\beta\gamma} = 0\,, \quad C^{3333\delta} = 0\,,$$

$$c^+(y) = -c^-(y) = c(y)\,, \quad y \in Y \tag{2.4.2}$$

$$G_+(y) = G_-(y) = G(y)\,, \quad \hat{y}_3 = y_3\,.$$

Under the assumptions (2.4.2) the fields $\Theta_\sigma^{(\alpha\beta)}$ and $\Xi_3^{(\alpha\beta)}$ are even functions of y_3, while the fields $\Theta_3^{(\alpha\beta)}$ and $\Xi_\sigma^{(\alpha\beta)}$ are odd in y_3. Hence fields $A_\diamond^{\gamma\delta\alpha\beta}$ are even in y_3 and $E_\diamond^{\gamma\delta\alpha\beta}$ are odd in y_3, see (2.3.21), (2.3.38). Consequently

$$\boldsymbol{E_y} = 0\,, \quad \boldsymbol{F_y} = 0 \tag{2.4.3}$$

or constitutive relations (2.3.51) decouple and problem (P_{hom}) (Section 2.3) is decomposed into two problems:

(i) the in-plane problem: find $\boldsymbol{u} \in (H_0^1(\Omega))^2$ such that

$$(P_{hom}^S) \quad \left| \begin{array}{l} h_0 \displaystyle\int_\Omega A_y^{\alpha\beta\lambda\mu} e_{\alpha\beta}(\boldsymbol{u}) e_{\lambda\mu}(\boldsymbol{w}) dx = \displaystyle\int_\Omega \hat{p}^\alpha w_\alpha dx \ \ \forall\, \boldsymbol{w} \in (H_0^1(\Omega))^3\,, \\[4mm] \text{where } \hat{p}^\alpha = \langle G^{\frac{1}{2}}\rangle(p_+^\alpha + p_-^\alpha) + h_o \prec b^\alpha \succ\,; \end{array} \right.$$

$$\tag{2.4.4}$$
$$\tag{2.4.5}$$

(ii) the bending problem:

$$(P_{hom}^A) \quad \left| \begin{array}{l} \text{find } w \in H_0^2(\Omega) \text{ such that} \\[2mm] h_0 \displaystyle\int_\Omega D_y^{\alpha\beta\lambda\mu} \kappa_{\alpha\beta}(w) \kappa_{\lambda\mu}(v) dx = \displaystyle\int_\Omega (\hat{q}v - \hat{m}^\alpha v_{,\alpha}) dx \ \ \forall\, v \in H_0^2(\Omega)\,, \\[2mm] \text{where} \\[2mm] \hat{q} = \langle G^{\frac{1}{2}}\rangle(q_+ + q_-) + h_o \prec b^3 \succ\,, \qquad \hat{m}^\alpha = \langle cG^{\frac{1}{2}}\rangle(p_+^\alpha - p_-^\alpha)\,. \end{array} \right.$$

$$\tag{2.4.6}$$
$$\tag{2.4.7}$$

The underlying constitutive relationships have the form

$$\mathcal{N}^{\alpha\beta} = h_0 A_y^{\alpha\beta\lambda\mu} e_{\lambda\mu}(\boldsymbol{u})\,, \qquad \mathcal{M}^{\alpha\beta} = h_0 D_y^{\alpha\beta\lambda\mu} \kappa_{\lambda\mu}(w)\,. \tag{2.4.8}$$

Similarly, Eqs. (2.3.64) concerning the original plate decouple and problem $(P_{\mathcal{H}})$ simplifies to membrane and bending problems.

2.5. *Centrosymmetry of the periodicity cell*

Apart from assumptions (2.4.2) assume in addition that the planes x_α =const are planes of material symmetry. Consequently the plate material is orthotropic with respect to the system (x_i). Let us shift the local coordinate system (y_i) to the centre of the cell, Fig. 2.5.1. Moreover, we assume that planes $y_\alpha = 0$ are planes of symmetry of the $\mathcal{Y}$ cell, i.e.

$$C^{ijkl}(y_1, y_2, y_3) = C^{ijkl}(\pm y_1, \pm y_2, \pm y_3) ,$$
$$G(y_1, y_2) = G(\pm y_1, \pm y_2) , \qquad c(y_1, y_2) = c(\pm y_1, \pm y_2) . \tag{2.5.1}$$

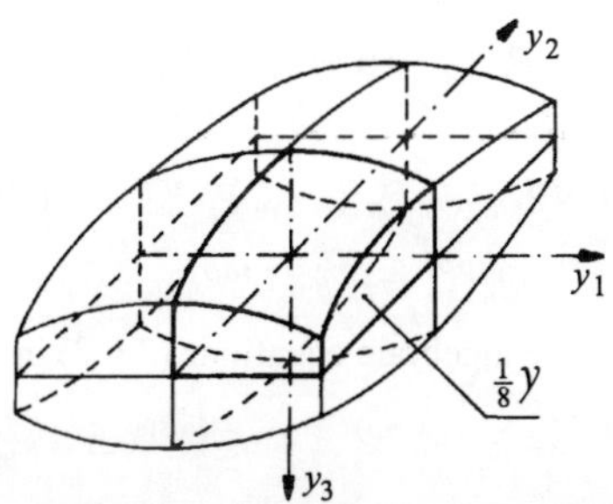

Fig. 2.5.1. A centrosymmetric cell of periodicity

Thus the cell $\mathcal{Y}$ is composed of eight identical segments made of the orthotropic material. Such symmetry, called centrosymmetry, results in the following simplification of the final expansion of the solution (2.3.1) and (2.3.2) of the problem $(\mathcal{P}_\varepsilon)$

$$u_\alpha^\varepsilon = \varepsilon(u_\alpha - y_3 w_{,\alpha}) + \varepsilon^2[\Theta_\alpha^{(\lambda\mu)} e_{\lambda\mu}(\boldsymbol{u}) + \Xi_\alpha^{(\lambda\mu)} \kappa_{\lambda\mu}(w)] + O(\varepsilon^3) ,$$
$$u_3^\varepsilon = w + \varepsilon^2[\Theta_3^{(\lambda\mu)} e_{\lambda\mu}(\boldsymbol{u}) + \Xi_3^{(\lambda\mu)} \kappa_{\lambda\mu}(v) + \eta_3] + O(\varepsilon^3) . \tag{2.5.2}$$

Here u_α, w, η_3 are functions of x, while $\Theta_i^{(\lambda\mu)}$, $\Xi_i^{(\lambda\mu)}$ depend on $\left(\dfrac{x}{\varepsilon}\right)$. The displacement field (u_α, w) are solutions to (P_{hom}^S) and (P_{hom}^A), Sec. 2.4, while η_3 solves a subsequent effective problem. The result (2.5.2) differs from results found in Section 2.3 in that now we have

$$\eta_\alpha = 0 , \qquad v_3 = 0 . \tag{2.5.3}$$

The stresses are given by the formula

$$\sigma_\varepsilon^{ij} = \varepsilon[A_\diamond^{ij\alpha\beta} e_{\alpha\beta}(\boldsymbol{u}) + E_\diamond^{ij\alpha\beta} \kappa_{\alpha\beta}(w)] + O(\varepsilon^2) \tag{2.5.4}$$

which does not differ from that valid in the general case, see (2.3.65).

2.6. *On computing effective stiffnesses*

The computational problem of finding the effective stiffnesses consists in solving the local problems (P_y^1) (Eq. (2.3.13)) and (P_y^2) (Eq. (2.3.34)). The unknown functions $\Theta^{(\alpha\beta)}$ and $\Xi^{(\alpha\beta)}$ can be approximated by the Galerkin method.

Let $(\phi_a)_{a=1}^N$ be a basis in N-dimensional subspace $W_N(\mathcal{Y})$ of $W(\mathcal{Y})$. Assume additionally that $\prec \phi_a \succ = 0$ to get rid of constants up to which solutions of problems (P_y^σ) are determined. We represent approximate solutions as

$$\Theta_k^{(\alpha\beta)} = \Theta_k^{(\alpha\beta)a}\phi_a(\boldsymbol{y})\,, \qquad \Xi_k^{(\alpha\beta)} = \Xi_k^{(\alpha\beta)a}\phi_a(\boldsymbol{y})\,, \quad a = 1,\dots,N \tag{2.6.1}$$

and take trial functions as $\boldsymbol{w} = \phi_b(\boldsymbol{y})$. One finds

$$K_{ba}^{ik}\Theta_k^{(\alpha\beta)a} + P_b^{i(\alpha\beta)} = 0\,, \qquad K_{ba}^{ik}\Xi_k^{(\alpha\beta)a} + Q_b^{i(\alpha\beta)} = 0 \tag{2.6.2}$$

where

$$K_{ab}^{ik} = \prec C^{ijkl}\phi_{a|l}\phi_{b|j} \succ\,, \qquad P_b^{i(\alpha\beta)} = \prec C^{ij\alpha\beta}\phi_{b|j} \succ\,,$$
$$Q_b^{i(\alpha\beta)} = \prec \hat{y}_3 C^{ij\alpha\beta}\phi_{b|j} \succ\,. \tag{2.6.3}$$

Let us represent solutions to the algebraic equations (2.6.2) in the form

$$\Theta_k^{(\alpha\beta)a} = -k_{ki}^{ab}P_b^{i(\alpha\beta)}\,, \qquad \Xi_k^{(\alpha\beta)a} = -k_{ki}^{ab}Q_b^{i(\alpha\beta)} \tag{2.6.4}$$

where $\boldsymbol{k} = \boldsymbol{K}^{-1}$. Substituting (2.6.4) into (2.3.24) and (2.3.50) one readily finds:

$$\begin{aligned}
A_y^{\gamma\delta\alpha\beta} &= \prec C^{\gamma\delta\alpha\beta} \succ - P_a^{k(\gamma\delta)}k_{ki}^{ab}P_b^{i(\alpha\beta)}\,, \\
E_y^{\gamma\delta\alpha\beta} &= \prec \hat{y}_3 C^{\gamma\delta\alpha\beta} \succ - Q_b^{i(\alpha\beta)}k_{ki}^{ab}P_a^{k(\gamma\delta)}\,, \\
F_y^{\gamma\delta\alpha\beta} &= \prec \hat{y}_3 C^{\gamma\delta\alpha\beta} \succ - Q_a^{k(\gamma\delta)}k_{ki}^{ab}P_b^{i(\alpha\beta)}\,, \\
D_y^{\gamma\delta\alpha\beta} &= \prec (\hat{y}_3)^2 C^{\gamma\delta\alpha\beta} \succ - Q_a^{k(\gamma\delta)}k_{ki}^{ab}Q_b^{i(\alpha\beta)}\,.
\end{aligned} \tag{2.6.5}$$

We see that Galerkin's method does not violate symmetry properties (2.3.27) and (2.3.53). The main difficulty in implementing the above algorithm lies in forming the basis (ϕ_a). First, ϕ_a should assume the same values at opposite lateral boundaries of $\mathcal{Y}$. This property can be realized by identifying appropriate degrees of freedom, if one uses the finite element method. We can, however, encounter difficulties in satisfying the condition $\prec \phi_a \succ = 0 -$ it is easy to satisfy if one uses the Fourier trigonometric representations and nontrivial to fulfil if one uses the finite element method.

The non-homogeneities of local problems (P_y^α) can be shifted to the displacement boundary conditions if both local problems are combined into the following form:

$$(\tilde{P}_y^\sigma) \quad \left| \begin{aligned} &\text{Find } \boldsymbol{u}_\sigma^{(\alpha\beta)} \text{ such that} \\ &a_y(\boldsymbol{u}_\sigma^{(\alpha\beta)}, \boldsymbol{u}_\sigma^{(\alpha\beta)}) = \min\{a_y(\boldsymbol{v}_\sigma^{(\alpha\beta)}, \boldsymbol{v}_\sigma^{(\alpha\beta)})\,| \\ &\boldsymbol{v}_\sigma^{(\alpha\beta)} - \frac{1}{2}(\hat{y}_3)^{\sigma-1}(y_\alpha\boldsymbol{e}_\beta + y_\beta\boldsymbol{e}_\alpha) \in W(\mathcal{Y})\}\,; \end{aligned} \right. \tag{2.6.6}$$

here $\sigma, \alpha, \beta = 1, 2$. The rescaled effective stiffnesses are given by

$$
\begin{aligned}
A_{\mathcal{Y}}^{\alpha\beta\lambda\mu} &= a_{\mathcal{Y}}(\boldsymbol{u}_1^{(\alpha\beta)}, \boldsymbol{u}_1^{(\lambda\mu)}) , & F_{\mathcal{Y}}^{\alpha\beta\lambda\mu} &= a_{\mathcal{Y}}(\boldsymbol{u}_2^{(\alpha\beta)}, \boldsymbol{u}_1^{(\lambda\mu)}) , \\
E_{\mathcal{Y}}^{\alpha\beta\lambda\mu} &= a_{\mathcal{Y}}(\boldsymbol{u}_1^{(\alpha\beta)}, \boldsymbol{u}_2^{(\lambda\mu)}) , & D_{\mathcal{Y}}^{\alpha\beta\lambda\mu} &= a_{\mathcal{Y}}(\boldsymbol{u}_2^{(\alpha\beta)}, \boldsymbol{u}_2^{(\lambda\mu)}) .
\end{aligned}
\tag{2.6.7}
$$

Note that nonhomogeneity in $(\tilde{P}_{\mathcal{Y}}^{\sigma})$ follows from imposing displacements along lateral edges of $\mathcal{Y}$. In this way we enforce stretching, in-plane shearing, bending as well as torsion of $\mathcal{Y}$.

This algorithm can be computerized with ease in the case where $\mathcal{Y}$ is centrosymmetric (Sec. 2.5), since symmetry conditions stabilize one quarter of $\mathcal{Y}$ subjected to boundary displacements.

2.7. Case of moderately thick periodicity cells

The modelling presented in Section 2.3 applies to thin plates of arbitrary periodicity cells $\mathcal{Z}$. The assumption of thinness has the form:
$\xi_{\Omega} = \max|x_3^+ - x_3^-|/\text{diam}(\Omega) \ll 1$ and is independent of the values of ratios $\xi_{\alpha} = \max|x_3^+ - x_3^-|/l_{\alpha}^z$ characterizing the shape of cell $\mathcal{Z}$. In general one cannot circumvent or avoid an analysis of three-dimensional local problems $(P_{\mathcal{Y}}^{\alpha})$. If, however, the cell $\mathcal{Z}$ has a specific shape: is flat or slender, then the local problems $(P_{\mathcal{Y}}^{\alpha})$ can be approximated by two-dimensional problems. In this section we discuss an approximate method that concerns the case of the cell $\mathcal{Z}$ (or $\mathcal{Y}$) being a moderately thick plate such that $\xi_{\alpha} \leq 1/3$, which represents a rough determination of the applicability range of the Hencky approach to transversely non-homogeneous moderately thick plates. On the other hand, one should assume $\xi_{\Omega} \leq 1/10$ to assure that the whole plate is thin and that results of Section 2.3 are applicable. Because of technical reasons we confine consideration to the case where

$$
c^{\alpha\beta} = c^{\alpha\beta}(y_3) , \qquad c^{\alpha\beta} = C^{33\alpha\beta}/C^{3333} ,
\tag{2.7.1}
$$

and the planes $y_3 = \text{const}$ are planes of material symmetry, i.e. (2.4.1) holds. The solutions to problems $(P_{\mathcal{Y}}^{\alpha})$ can be decomposed as follows

$$
\Theta^{(\alpha\beta)} = \bar{\Theta}^{(\alpha\beta)} + \tilde{\Theta}^{(\alpha\beta)} , \qquad \Xi^{\alpha\beta} = \bar{\Xi}^{(\alpha\beta)} + \tilde{\Xi}^{(\alpha\beta)} ,
\tag{2.7.2}
$$

with

$$
\tilde{\Theta}^{(\alpha\beta)} = (0, 0, -\int c^{\alpha\beta} dy_3) , \qquad \tilde{\Xi}^{(\alpha\beta)} = (0, 0, -\int \hat{y}_3 c^{\alpha\beta} dy_3) .
\tag{2.7.3}
$$

The local fields $\bar{\Theta}^{(\alpha\beta)}$ and $\bar{\Xi}^{(\alpha\beta)}$ are solutions to the following modified local problems:

$$
(\bar{P}_{\mathcal{Y}}^1) \quad \left|
\begin{aligned}
&\text{find } \bar{\Theta}^{(\alpha\beta)} \in W(\mathcal{Y}) \text{ such that} \\
&a(\bar{\Theta}^{(\alpha\beta)}, \boldsymbol{v}) + \prec \tilde{C}^{\gamma\delta\alpha\beta} v_{\gamma|\delta} \succ = 0 \quad \forall \, \boldsymbol{v} \in W(\mathcal{Y}) .
\end{aligned}
\right.
\tag{2.7.4}
$$

$$
(\bar{P}_{\mathcal{Y}}^2) \quad \left|
\begin{aligned}
&\text{Find } \bar{\Xi}^{(\alpha\beta)} \in W(\mathcal{Y}) \text{ such that} \\
&a(\bar{\Xi}^{(\alpha\beta)}, \boldsymbol{v}) + \prec \hat{y}_3 \tilde{C}^{\gamma\delta\alpha\beta} v_{\gamma|\delta} \succ = 0 \quad \forall \, \boldsymbol{v} \in W(\mathcal{Y}) .
\end{aligned}
\right.
\tag{2.7.5}
$$

Here

$$\tilde{C}^{\alpha\beta\lambda\mu} = C^{\alpha\beta\lambda\mu} - c^{\alpha\beta}C^{33\lambda\mu} \,. \tag{2.7.6}$$

The decomposition (2.7.2) follows from the following identities

$$C^{\alpha\beta ij}v_{i|j} = \tilde{C}^{\alpha\beta\gamma\delta}v_{\gamma|\delta} + c^{\alpha\beta}C^{33ij}v_{i|j},$$

$$\tilde{\Theta}^{(\alpha\beta)}_{3|3} = -c^{\alpha\beta}(y_3) \,, \qquad \tilde{\Xi}^{(\alpha\beta)}_{3|3} = -\hat{y}_3 c^{\alpha\beta}(y_3) \,. \tag{2.7.7}$$

The problems $(\bar{P}^\alpha_\mathcal{Y})$ can be interpreted as plate-type problems if $\mathcal{Y}$ has the shape of a plate. Thus the Hencky theory of plates can be applied. According to this theory we represent the solutions of $(\bar{P}^\alpha_\mathcal{Y})$ as follows

$$\bar{\Theta}^{(\alpha\beta)}_\lambda(\boldsymbol{y}) = T^{(\alpha\beta)}_\lambda(y) + \hat{y}_3 Z^{(\alpha\beta)}_\lambda(y) \,, \qquad \bar{\Theta}^{(\alpha\beta)}_3(\boldsymbol{y}) = \hat{X}^{\alpha\beta}(y) \,,$$

$$\bar{\Xi}^{(\alpha\beta)}_\lambda(\boldsymbol{y}) = \hat{U}^{(\alpha\beta)}_\lambda(y) + \hat{y}_3 \Phi^{(\alpha\beta)}_\lambda(y) \,, \qquad \bar{\Xi}^{(\alpha\beta)}_3(\boldsymbol{y}) = \chi^{\alpha\beta}(y) \,. \tag{2.7.8}$$

Similar constraints are imposed on the trial fields

$$v_\lambda(\boldsymbol{y}) = u_\lambda(y) + \hat{y}_3 \varphi_\lambda(y) \,, \qquad v_3(\boldsymbol{y}) = v(y) \,. \tag{2.7.9}$$

The unknown functions: $T^{(\alpha\beta)}_\lambda$, $Z^{(\alpha\beta)}_\lambda$, $\hat{X}^{\alpha\beta}$, $\hat{U}^{(\alpha\beta)}_\lambda$, $\Phi^{(\alpha\beta)}_\lambda$, $\chi^{\alpha\beta}$, u_λ, φ_λ, v belong to $H^1_{per}(Y)$.

In Hencky's theory of plates one also introduces the assumption of negligibility of transverse normal stresses, which modifies the constitutive relations. Similarly, it will be assumed here that the stress-type fields

$$\bar{\sigma}^{33}_{1(\lambda\mu)} := C^{33\gamma\delta}\bar{\Theta}^{(\lambda\mu)}_{\gamma|\delta} + C^{3333}\underline{\bar{\Theta}^{(\lambda\mu)}_{3|3}} \,, \qquad \bar{\sigma}^{33}_{2(\lambda\mu)} := C^{33\gamma\delta}\bar{\Xi}^{(\lambda\mu)}_{\gamma|\delta} + C^{3333}\underline{\bar{\Xi}^{(\lambda\mu)}_{3|3}} \tag{2.7.10}$$

are negligible. Eliminating the underlined quantities one can approximate the in-plane stress-type fields as follows:

$$\bar{\sigma}^{\alpha\beta}_{1(\lambda\mu)} := C^{\alpha\beta kl}\bar{\Theta}^{(\lambda\mu)}_{k|l} \approx \tilde{C}^{\alpha\beta\gamma\delta}\bar{\Theta}^{(\lambda\mu)}_{\gamma|\delta} \,, \qquad \bar{\sigma}^{\alpha\beta}_{2(\lambda\mu)} := C^{\alpha\beta kl}\bar{\Xi}^{(\lambda\mu)}_{k|l} \approx \tilde{C}^{\alpha\beta\gamma\delta}\bar{\Xi}^{(\lambda\mu)}_{\gamma|\delta} \,. \tag{2.7.11}$$

Hence

$$\sigma^{\alpha\beta}_{1(\lambda\mu)} := C^{\alpha\beta kl}\Theta^{(\lambda\mu)}_{k|l} \cong \tilde{C}^{\alpha\beta\gamma\delta}\bar{\Theta}^{(\lambda\mu)}_{\gamma|\delta} - c^{\alpha\beta}c^{\lambda\mu}C^{3333} \,,$$

$$\sigma^{\alpha\beta}_{2(\lambda\mu)} := C^{\alpha\beta kl}\Xi^{(\lambda\mu)}_{k|l} \approx \tilde{C}^{\alpha\beta\gamma\delta}\bar{\Xi}^{(\lambda\mu)}_{\gamma|\delta} - \hat{y}_3 c^{\alpha\beta}c^{\lambda\mu}C^{3333} \,, \tag{2.7.12}$$

which introduces simplifications to the problems $(P^\alpha_\mathcal{Y})$ and to the formulae for effective stiffnesses.

The kinematical assumptions (2.7.8) as well as stress assumptions (2.7.11), (2.7.12) make it possible to reduce the transverse dimension of $\mathcal{Y}$. Prior to formulating these reduced problems let us define the stiffnesses:

$$(\hat{A}^{\alpha\beta\lambda\mu}, \, \hat{E}^{\alpha\beta\lambda\mu}, \, \hat{D}^{\alpha\beta\lambda\mu})^T = \int\limits_{c^-(y)}^{c^+(y)} \tilde{C}^{\alpha\beta\lambda\mu}(\boldsymbol{y}) \begin{bmatrix} 1 \\ \hat{y}_3 \\ (\hat{y}_3)^2 \end{bmatrix} dy_3$$

$$\hat{H}^{\lambda\mu} = k \int\limits_{c^-(y)}^{c^+(y)} C^{\lambda 3\mu 3}(\boldsymbol{y}) dy_3 \,, \qquad k = 1 \,; \tag{2.7.13}$$

and the bilinear forms:

$$\hat{b}(\boldsymbol{u},\boldsymbol{v}) = \langle \hat{A}^{\lambda\mu\gamma\delta} u_{\gamma|\delta} v_{\lambda|\mu} \rangle\,, \qquad \hat{e}_H(\boldsymbol{u},\boldsymbol{v}) = \langle \hat{E}^{\lambda\mu\gamma\delta} u_{\gamma|\delta} v_{\lambda|\mu} \rangle\,,$$

$$\hat{d}_H(\boldsymbol{u},\boldsymbol{v}) = \langle \hat{D}^{\lambda\mu\gamma\delta} u_{\gamma|\delta} v_{\lambda|\mu} \rangle\,, \qquad \hat{g}_1(u,v) = \langle \hat{H}^{\alpha\beta} u_{\beta} v_{|\alpha} \rangle\,, \qquad (2.7.14)$$

$$\hat{g}_2(u,v) = \langle \hat{H}^{\alpha\beta} u_{|\alpha} v_{|\beta} \rangle\,, \qquad \hat{g}_3(\boldsymbol{u},\boldsymbol{v}) = \langle \hat{H}^{\alpha\beta} u_{\alpha} v_{\beta} \rangle\,,$$

where $\boldsymbol{u},\boldsymbol{v} \in H^1_{per}(Y)^2$, $u,v \in H^1_{per}(Y)$ and k is a transverse shear correction factor. We recall that angular brackets $\langle \cdot \rangle$ mean here the averaging over Y.

Let us define the rescaled membrane stress resultants

$$N^{\lambda\mu}_{1(\alpha\beta)} = \hat{A}^{\lambda\mu\gamma\delta} T^{(\alpha\beta)}_{\gamma|\delta} + \hat{E}^{\lambda\mu\gamma\delta} Z^{(\alpha\beta)}_{\gamma|\delta} + \hat{A}^{\lambda\mu\alpha\beta}\,,$$
$$N^{\lambda\mu}_{2(\alpha\beta)} = \hat{A}^{\lambda\mu\gamma\delta} \hat{U}^{(\alpha\beta)}_{\gamma|\delta} + \hat{E}^{\lambda\mu\gamma\delta} \Phi^{(\alpha\beta)}_{\gamma|\delta} + \hat{E}^{\lambda\mu\alpha\beta}\,, \qquad (2.7.15)$$

moments:

$$M^{\lambda\mu}_{1(\alpha\beta)} = \hat{E}^{\lambda\mu\gamma\delta} T^{(\alpha\beta)}_{\gamma|\delta} + \hat{D}^{\lambda\mu\gamma\delta} Z^{(\alpha\beta)}_{\gamma|\delta} + \hat{E}^{\lambda\mu\alpha\beta}\,,$$
$$M^{\lambda\mu}_{2(\alpha\beta)} = \hat{E}^{\lambda\mu\gamma\delta} \hat{U}^{(\alpha\beta)}_{\gamma|\delta} + \hat{D}^{\lambda\mu\gamma\delta} \Phi^{(\alpha\beta)}_{\gamma|\delta} + \hat{D}^{\lambda\mu\alpha\beta}\,, \qquad (2.7.16)$$

and transverse shear forces:

$$Q^{\lambda}_{1(\alpha\beta)} = \hat{H}^{\lambda\mu}(\hat{X}^{(\alpha\beta)}_{|\mu} + Z^{(\alpha\beta)}_{\mu})\,, \qquad Q^{\lambda}_{2(\alpha\beta)} = \hat{H}^{\lambda\mu}(\chi^{(\alpha\beta)}_{|\mu} + \Phi^{(\alpha\beta)}_{\mu})\,. \qquad (2.7.17)$$

Substituting (2.7.8) – (2.7.11) into (2.7.4) and (2.7.5) we reduce the transverse dimension and formulate the following two-dimensional problems

$$(\bar{P}^{\sigma}_Y) \left|
\begin{array}{l}
\text{find } \boldsymbol{X}^{(\alpha\beta)}_{\sigma} \in H_{H,per}(Y) = H^1_{per}(Y)^2 \times H^1_{per}(Y)^2 \times H^1_{per}(Y) \\[4pt]
\text{such that } \langle N^{\lambda\mu}_{\sigma(\alpha\beta)} u_{\lambda|\mu} + M^{\lambda\mu}_{\sigma(\alpha\beta)} \varphi_{\lambda|\mu} + Q^{\lambda}_{\sigma(\alpha\beta)}(w_{|\lambda} + \varphi_\lambda)\rangle = 0 \qquad (2.7.18) \\[4pt]
\hspace{6cm} \forall\,(\boldsymbol{u},\boldsymbol{\varphi},w) \in H_{H,per}(Y)\,.
\end{array}
\right.$$

Here

$$\boldsymbol{X}^{(\alpha\beta)}_1 = (\boldsymbol{T}^{(\alpha\beta)}, \boldsymbol{Z}^{(\alpha\beta)}, \hat{X}^{(\alpha\beta)})\,, \qquad \boldsymbol{X}^{(\alpha\beta)}_2 = (\hat{\boldsymbol{U}}^{(\alpha\beta)}, \boldsymbol{\Phi}^{(\alpha\beta)}, \chi^{(\alpha\beta)})\,. \qquad (2.7.19)$$

The local problems $(\bar{P}^{\sigma}_Y)$ are well posed. One can prove that fields $\boldsymbol{T}^{(\alpha\beta)}$, $\hat{\boldsymbol{U}}^{(\alpha\beta)}$, $\hat{X}^{(\alpha\beta)}$, $\chi^{(\alpha\beta)}$ are determined up to additive constants, while fields $\boldsymbol{Z}^{(\alpha\beta)}$, $\boldsymbol{\Phi}^{(\alpha\beta)}$ are determined uniquely. The effective stiffnesses: $h_0 A^{\alpha\beta\lambda\mu}_{\mathcal{Y}}$, $h_0 E^{\alpha\beta\lambda\mu}_{\mathcal{Y}}$, $h_0 F^{\alpha\beta\lambda\mu}_{\mathcal{Y}}$, $h_0 D^{\alpha\beta\lambda\mu}_{\mathcal{Y}}$, involved in Eqs. (2.3.51) are approximated by $\hat{A}^{\alpha\beta\lambda\mu}_{\hbar}$, $\hat{E}^{\alpha\beta\lambda\mu}_{\hbar}$, $\hat{F}^{\alpha\beta\lambda\mu}_{\hbar}$ and $\hat{D}^{\alpha\beta\lambda\mu}_{\hbar}$ according to the rules (2.7.8) – (2.7.12). One finds

$$\hat{A}^{\alpha\beta\lambda\mu}_{\hbar} = \langle N^{\lambda\mu}_{1(\alpha\beta)}\rangle\,, \quad \hat{E}^{\alpha\beta\lambda\mu}_{\hbar} = \langle N^{\lambda\mu}_{2(\alpha\beta)}\rangle\,,$$
$$\hat{F}^{\alpha\beta\lambda\mu}_{\hbar} = \langle M^{\lambda\mu}_{1(\alpha\beta)}\rangle\,, \quad \hat{D}^{\alpha\beta\lambda\mu}_{\hbar} = \langle M^{\lambda\mu}_{2(\alpha\beta)}\rangle\,. \qquad (2.7.20)$$

The homogenized problem assumes the form (P_{hom}), Section 2.3, with effective stiffnesses given above. The density of the elastic energy of the effective plate is given by

$$\mathcal{W}_{\hbar}(\boldsymbol{\epsilon},\boldsymbol{\kappa}) = (\hat{A}^{\alpha\beta\lambda\mu}_{\hbar}\epsilon_{\alpha\beta}\epsilon_{\lambda\mu} + \hat{E}^{\alpha\beta\lambda\mu}_{\hbar}\kappa_{\lambda\mu}\epsilon_{\alpha\beta} + \hat{F}^{\alpha\beta\lambda\mu}_{\hbar}\kappa_{\alpha\beta}\epsilon_{\lambda\mu} + \hat{D}^{\alpha\beta\lambda\mu}_{\hbar}\kappa_{\alpha\beta}\kappa_{\lambda\mu})/2\,. \qquad (2.7.21)$$

To prove the symmetry conditions required

$$
\hat{A}_\hbar^{\alpha\beta\lambda\mu} = \hat{A}_\hbar^{\lambda\mu\alpha\beta} , \quad
\hat{E}_\hbar^{\alpha\beta\lambda\mu} = \hat{F}_\hbar^{\lambda\mu\alpha\beta} , \quad
\hat{D}_\hbar^{\alpha\beta\lambda\mu} = \hat{D}_\hbar^{\lambda\mu\alpha\beta} ,
$$
$$
K_\hbar^{\alpha\beta\lambda\mu} = K_\hbar^{\beta\alpha\lambda\mu} = K_\hbar^{\beta\alpha\mu\lambda} , \qquad
K = \hat{A}, \hat{E}, \hat{F}, \hat{D} ,
\tag{2.7.22}
$$

one should form appropriate identities following from the variational equation of $(\bar{P}_Y^\sigma)$ and combining them with (2.7.20). Finally one arrives at

$$
\begin{aligned}
\hat{A}_\hbar^{\alpha\beta\lambda\mu} &= \langle \hat{A}^{\alpha\beta\lambda\mu} \rangle - \hat{b}(\boldsymbol{T}^{(\alpha\beta)}, \boldsymbol{T}^{(\lambda\mu)}) - [\hat{e}_H(\boldsymbol{Z}^{(\lambda\mu)}, \boldsymbol{T}^{(\alpha\beta)}) \\
&\quad + \hat{e}_H(\boldsymbol{T}^{(\lambda\mu)}, \boldsymbol{Z}^{(\alpha\beta)})] - \hat{d}_H(\boldsymbol{Z}^{(\alpha\beta)}, \boldsymbol{Z}^{(\lambda\mu)}) - \hat{g}_3(\boldsymbol{Z}^{(\lambda\mu)}, \boldsymbol{Z}^{(\alpha\beta)}) \\
&\quad + \hat{g}_2(\hat{X}^{(\alpha\beta)}, \hat{X}^{(\lambda\mu)}) , \\
\hat{F}_\hbar^{\lambda\mu\alpha\beta} &= \langle \hat{E}^{\alpha\beta\lambda\mu} \rangle - \hat{b}(\hat{\boldsymbol{U}}^{(\alpha\beta)}, \boldsymbol{T}^{(\lambda\mu)}) - [\hat{e}_H(\boldsymbol{\Phi}^{(\alpha\beta)}, \boldsymbol{T}^{(\lambda\mu)}) \\
&\quad + \hat{e}_H(\hat{\boldsymbol{U}}^{(\alpha\beta)}, \boldsymbol{Z}^{(\lambda\mu)})] - \hat{d}_H(\boldsymbol{\Phi}^{(\alpha\beta)}, \boldsymbol{Z}^{(\lambda\mu)}) - \hat{g}_3(\boldsymbol{\Phi}^{(\alpha\beta)}, \boldsymbol{Z}^{(\lambda\mu)}) \\
&\quad + \hat{g}_2(\hat{X}^{(\alpha\beta)}, \chi^{(\lambda\mu)}) , \\
\hat{E}_\hbar^{\lambda\mu\alpha\beta} &= \langle \hat{E}^{\alpha\beta\lambda\mu} \rangle - \hat{b}(\boldsymbol{T}^{(\alpha\beta)}, \hat{\boldsymbol{U}}^{(\lambda\mu)}) - [\hat{e}_H(\boldsymbol{T}^{(\alpha\beta)}, \boldsymbol{\Phi}^{(\lambda\mu)}) \\
&\quad + \hat{e}_H(\boldsymbol{Z}^{(\alpha\beta)}, \hat{\boldsymbol{U}}^{(\lambda\mu)})] - \hat{d}_H(\boldsymbol{Z}^{(\alpha\beta)}, \boldsymbol{\Phi}^{(\lambda\mu)}) - \hat{g}_3(\boldsymbol{Z}^{(\alpha\beta)}, \boldsymbol{\Phi}^{(\lambda\mu)}) \\
&\quad + \hat{g}_2(\chi^{(\alpha\beta)}, \hat{X}^{(\lambda\mu)}) , \\
\hat{D}_\hbar^{\alpha\beta\lambda\mu} &= \langle \hat{D}^{\alpha\beta\lambda\mu} \rangle - \hat{b}(\hat{\boldsymbol{U}}^{(\lambda\mu)}, \hat{\boldsymbol{U}}^{(\alpha\beta)}) - [\hat{e}_H(\hat{\boldsymbol{U}}^{(\lambda\mu)}, \boldsymbol{\Phi}^{(\alpha\beta)}) \\
&\quad + \hat{e}_H(\boldsymbol{\Phi}^{(\lambda\mu)}, \hat{\boldsymbol{U}}^{(\alpha\beta)})] - \hat{d}_H(\boldsymbol{\Phi}^{(\lambda\mu)}, \boldsymbol{\Phi}^{(\alpha\beta)}) - \hat{g}_3(\boldsymbol{\Phi}^{(\lambda\mu)}, \boldsymbol{\Phi}^{(\alpha\beta)}) \\
&\quad + \hat{g}_2(\chi^{(\alpha\beta)}, \chi^{(\lambda\mu)}) .
\end{aligned}
\tag{2.7.23}
$$

Due to symmetry of the bilinear forms $\hat{b}(\cdot,\cdot)$, $\hat{g}_2(\cdot,\cdot)$, $\hat{g}_3(\cdot,\cdot)$, $\hat{d}_H(\cdot,\cdot)$, and $\hat{e}_H(\cdot,\cdot)$, the stiffnesses (2.7.20) satisfy symmetry conditions (2.7.22). Note moreover, that the effective elastic potential $\mathcal{W}_\hbar$ can be rearranged to the following form

$$
\mathcal{W}_\hbar = (\prec \tilde{C}^{\alpha\beta\lambda\mu} \bar{\gamma}_{\alpha\beta} \bar{\gamma}_{\lambda\mu} \succ + 4 \prec \tilde{C}^{\alpha 3 \beta 3} \bar{\gamma}_{\alpha 3} \bar{\gamma}_{\beta 3} \succ)/2 ,
\tag{2.7.24}
$$

where

$$
\bar{\gamma}_{\alpha\beta} = a_{\alpha\beta}^{\lambda\mu}(\boldsymbol{y}) \epsilon_{\lambda\mu} + e_{\alpha\beta}^{\lambda\mu}(\boldsymbol{y}) \kappa_{\lambda\mu} , \qquad
2\bar{\gamma}_{\alpha 3} = a_\alpha^{\lambda\mu}(\boldsymbol{y}) \epsilon_{\lambda\mu} + e_\alpha^{\lambda\mu}(\boldsymbol{y}) \kappa_{\lambda\mu} ,
\tag{2.7.25}
$$

with

$$
a_{\alpha\beta}^{\lambda\mu} = \delta_{(\alpha}^\lambda \delta_{\beta)}^\mu + T_{(\alpha|\beta)}^{(\lambda\mu)} + \hat{y}_3 Z_{(\alpha|\beta)}^{(\lambda\mu)} , \qquad
a_\alpha^{\lambda\mu} = Z^{(\lambda\mu)} + \hat{X}_{|\alpha}^{(\lambda\mu)} ,
\tag{2.7.26}
$$
$$
e_{\alpha\beta}^{\lambda\mu} = \hat{y}_3 \delta_{(\alpha}^\lambda \delta_{\beta)}^\mu + \hat{U}_{(\alpha|\beta)}^{(\lambda\mu)} + \hat{y}_3 \Phi_{(\alpha|\beta)}^{(\lambda\mu)} , \qquad
e_\alpha^{\lambda\mu} = \phi^{(\lambda\mu)} + \chi_{|\alpha}^{(\lambda\mu)} .
\tag{2.7.27}
$$

The tensors $\tilde{C}^{\alpha\beta\lambda\mu}$ and $\tilde{C}^{\alpha 3 \beta 3}$ are positive definite. Hence $\mathcal{W}_\hbar$ is non-negative.

Assume that $\mathcal{W}_\hbar = 0$. Hence $\bar{\gamma}_{\alpha\beta} = 0$ and $\bar{\gamma}_{\alpha 3} = 0$ for all $\boldsymbol{y} \in \mathcal{Y}$ and consequently $\langle \bar{\gamma}_{\alpha\beta} \rangle = 0$, $\langle \bar{\gamma}_{\alpha 3} \rangle = 0$. Due to periodicity of $X_\sigma^{(\alpha\beta)}$ one concludes that $\epsilon_{\alpha\beta} + \hat{y}_3 \kappa_{\alpha\beta} = 0$ for almost every $y_3 \in (c^-(y), c^+(y))$. This implies that $\epsilon_{\alpha\beta} = 0$, $\kappa_{\alpha\beta} = 0$. Thus $\mathcal{W}_\hbar$ defines a positive definite quadratic form of (ϵ, κ).

The effective stiffnesses of the original plate of Section 2.1 are given by

$$A_\hbar^{\alpha\beta\lambda\mu} = \varepsilon_0 \hat{A}_\hbar^{\alpha\beta\lambda\mu} , \qquad F_\hbar^{\alpha\beta\lambda\mu} = \varepsilon_0^2 \hat{F}_\hbar^{\alpha\beta\lambda\mu} ,$$
$$E_\hbar^{\alpha\beta\lambda\mu} = \varepsilon_0^2 \hat{E}_\hbar^{\alpha\beta\lambda\mu} , \qquad D_\hbar^{\alpha\beta\lambda\mu} = \varepsilon_0^3 \hat{D}_\hbar^{\alpha\beta\lambda\mu} . \tag{2.7.28}$$

The homogenized constitutive relations assume the form

$$N_\hbar^{\alpha\beta\lambda\mu} = A_\hbar^{\alpha\beta\lambda\mu} \epsilon_{\lambda\mu}^h + E_\hbar^{\alpha\beta\lambda\mu} \kappa_{\lambda\mu}^h , \qquad M_\hbar^{\alpha\beta\lambda\mu} = F_\hbar^{\alpha\beta\lambda\mu} \epsilon_{\lambda\mu}^h + D_\hbar^{\alpha\beta\lambda\mu} \kappa_{\lambda\mu}^h , \tag{2.7.29}$$

where $\epsilon_{\lambda\mu}^h = \varepsilon_0 \epsilon_{\lambda\mu}$, $\kappa_{\lambda\mu}^h = \kappa_{\lambda\mu}$.

Let us examine consequences of transverse symmetry. Then $\hat{E}$ defined by (2.7.13) vanishes. Problems $(\bar{P}_Y^\sigma)$ decouple into membrane and bending-type problems:

$$(\bar{P}_{S,Y}^1) \quad \left| \begin{array}{l} \text{find } T^{(\alpha\beta)} \in H_{per}^1(Y)^2 \text{ such that} \\[2mm] \hat{b}(T^{(\alpha\beta)}, u) + \langle \hat{A}^{\gamma\delta\alpha\beta} u_{\gamma|\delta} \rangle = 0 , \qquad \forall\, u \in H_{per}^1(Y)^2 . \end{array} \right. \tag{2.7.30}$$

$$(\bar{P}_{S,Y}^2) \quad \left| \begin{array}{l} \text{Find } (\Phi^{(\alpha\beta)}, \chi^{(\alpha\beta)}) \in H_{per}^1(Y)^2 \times H_{per}^1(Y) \text{ such that} \\[2mm] \hat{d}_H(\Phi^{(\alpha\beta)}, \varphi) + \hat{g}_1(\varphi, \chi^{(\alpha\beta)}) + \hat{g}_3(\Phi^{(\alpha\beta)}, \varphi) + \langle \hat{D}^{\gamma\delta\alpha\beta} \varphi_{\gamma|\delta} \rangle = 0 , \\[2mm] \hat{g}_2(\chi^{(\alpha\beta)}, w) + \hat{g}_1(\Phi^{(\alpha\beta)}, w) = 0 , \\[2mm] \qquad\qquad\qquad \forall\, (\varphi, w) \in H_{per}^1(Y)^2 \times H_{per}^1(Y) , \end{array} \right. \tag{2.7.31}$$

and $Z^{(\alpha\beta)} = 0$, $\hat{U}^{(\alpha\beta)} = 0$, $\hat{X}^{(\alpha\beta)} = 0$.

The non-zero effective stiffnesses are:

$$\hat{A}_\hbar^{\alpha\beta\lambda\mu} = \langle \hat{A}^{\alpha\beta\lambda\mu} \rangle - \hat{b}(T^{(\alpha\beta)}, T^{(\lambda\mu)})$$
$$\hat{D}_\hbar^{\alpha\beta\lambda\mu} = \langle \hat{D}^{\alpha\beta\lambda\mu} \rangle - \hat{d}_H(\Phi^{(\lambda\mu)}, \Phi^{(\alpha\beta)}) - \hat{q}_3(\Phi^{(\lambda\mu)}, \Phi^{(\alpha\beta)}) + \hat{g}_2(\chi^{(\alpha\beta)}, \chi^{(\lambda\mu)}) . \tag{2.7.32}$$

The homogenized constitutive relations are decoupled:

$$N_\hbar^{\alpha\beta} = A_\hbar^{\alpha\beta\lambda\mu} \epsilon_{\lambda\mu}^h , \qquad M_\hbar^{\alpha\beta} = D_\hbar^{\alpha\beta\lambda\mu} \kappa_{\lambda\mu}^h . \tag{2.7.33}$$

2.8. *Case of thin periodicity cells. Derivation by imposing Kirchhoff's constraints*

If periodicity cells are thin plates themselves one can neglect their transverse shear deformations. Thus the transverse shear deformations associated with displacement fields $\bar{\Theta}^{(\alpha\beta)}$ and $\bar{\Xi}^{(\alpha\beta)}$ in problems $(\bar{P}_y^1)$ and $(\bar{P}_y^2)$ respectively can be neglected, i.e.

$$\bar{\Theta}_{\lambda|3}^{(\alpha\beta)} + \bar{\Theta}_{3|\lambda}^{(\alpha\beta)} = 0 , \qquad \bar{\Xi}_{\lambda|3}^{(\alpha\beta)} + \bar{\Xi}_{3|\lambda}^{(\alpha\beta)} = 0 .$$

The same assumption concerns the trial fields: $v_{\lambda|3} + v_{3|\lambda} = 0$. The above requirements are fulfilled if displacement assumptions (2.7.8) are modified by setting

$$Z_\lambda^{(\alpha\beta)} = -\hat{X}_{|\lambda}^{(\alpha\beta)} \,, \qquad \Phi_\lambda^{(\alpha\beta)} = -\chi_{|\lambda}^{(\alpha\beta)} \,, \qquad \varphi_\lambda = -v_{|\lambda} \,. \tag{2.8.1}$$

Consequently the displacement assumptions (2.7.8) are reduced to the form of Kirchhoff. As in Section 2.7 we assume that planes $y_3 = \text{const}$ are planes of material symmetry, i.e. (2.4.1) holds.

Let us introduce the strain measures

$$e_{\alpha\beta}^y(\boldsymbol{v}) = \frac{1}{2}(v_{\alpha|\beta} + v_{\beta|\alpha}) \,, \qquad \kappa_{\alpha\beta}^y(w) = -w_{|\alpha\beta} \,.$$

Next we define new bilinear forms

$$\hat{e}_K(w,\boldsymbol{v}) = \langle \hat{E}^{\alpha\beta\lambda\mu}(y)\kappa_{\lambda\mu}^y(w)e_{\alpha\beta}^y(\boldsymbol{v})\rangle \,,$$
$$\hat{d}_K(w,v) = \langle \hat{D}^{\alpha\beta\lambda\mu}(y)\kappa_{\lambda\mu}^y(w)\kappa_{\alpha\beta}^y(v)\rangle \,, \tag{2.8.2}$$

and introduce the function spaces

$$\tilde{H}_{K,per}(Y) = \tilde{H}_{per}^1(Y)^2 \times \tilde{H}_{per}^2(Y) \,, \qquad \tilde{H}_{per}^\alpha(Y) = \{v \in H_{per}^\alpha(Y) \mid \langle v \rangle = 0\} \,.$$

The space $H_{per}^1(Y)$ has been defined in Section 1.3 and $H_{per}^2(Y) = W_{per}^{2,2}(Y)$, cf. (1.3.46).

The local problems $(\bar{P}_Y^\sigma)$ are rearranged to the form:

$$(\hat{P}_{K,Y}^\sigma) \quad \left| \begin{array}{c} \text{find } \boldsymbol{X}_\sigma^{(\alpha\beta)} \in \tilde{H}_{K,per}(Y) \text{ such that} \\[1mm] \langle N_{\sigma(\alpha\beta)}^{\lambda\mu}e_{\lambda\mu}^y(\boldsymbol{u}) + M_{\sigma(\alpha\beta)}^{\lambda\mu}\kappa_{\lambda\mu}^y(w)\rangle = 0 \\[1mm] \forall\,(\boldsymbol{u},w) \in \tilde{H}_{K,per}(Y) \,, \end{array} \right. \tag{2.8.3}$$

where

$$\boldsymbol{X}_1^{(\alpha\beta)} = (\boldsymbol{T}^{(\alpha\beta)}, \hat{X}^{(\alpha\beta)}) \,, \qquad \boldsymbol{X}_2^{(\alpha\beta)} = (\hat{\boldsymbol{U}}^{(\alpha\beta)}, \chi^{(\alpha\beta)}) \,,$$

and quantities $N_{\sigma(\alpha\beta)}^{\lambda\mu}$, $M_{\sigma(\alpha\beta)}^{\lambda\mu}$ are changed accordingly. Let us write problems $(\hat{P}_{K,Y}^\sigma)$ more explicitly:

$$(\hat{P}_{K,Y}^1) \quad \left| \begin{array}{l} \text{find } (\boldsymbol{T}^{(\alpha\beta)}, \hat{X}^{(\alpha\beta)}) \in \tilde{H}_{K,per}(Y) \text{ such that} \\[1mm] \hat{b}(\boldsymbol{T}^{(\alpha\beta)}, \boldsymbol{u}) + \hat{e}_K(\hat{X}^{(\alpha\beta)}, \boldsymbol{u}) + \langle \hat{A}^{\alpha\beta\lambda\mu}(y)e_{\lambda\mu}^y(\boldsymbol{u})\rangle = 0 \,, \\[1mm] \hat{e}_K(w, \boldsymbol{T}^{(\alpha\beta)}) + \hat{d}_K(\hat{X}^{(\alpha\beta)}, w) + \langle \hat{E}^{\alpha\beta\lambda\mu}(y)\kappa_{\lambda\mu}^y(w)\rangle = 0 \\[1mm] \hfill \forall\,(\boldsymbol{u},w) \in \tilde{H}_{K,per}(Y) \,. \end{array} \right. \tag{2.8.4}$$

$$(\hat{P}_{K,Y}^2) \quad \left| \begin{array}{l} \text{Find } (\hat{\boldsymbol{U}}^{(\alpha\beta)}, \chi^{(\alpha\beta)}) \in \tilde{H}_{K,per}(Y) \text{ such that} \\[1mm] \hat{b}(\hat{\boldsymbol{U}}^{(\alpha\beta)}, \boldsymbol{u}) + \hat{e}_K(\chi^{(\alpha\beta)}, \boldsymbol{u}) + \langle \hat{E}^{\alpha\beta\lambda\mu}(y)e_{\lambda\mu}^y(\boldsymbol{u})\rangle = 0 \,, \\[1mm] \hat{e}_K(w, \hat{\boldsymbol{U}}^{(\alpha\beta)}) + \hat{d}_K(\chi^{(\alpha\beta)}, w) + \langle \hat{D}^{\alpha\beta\lambda\mu}(y)\kappa_{\lambda\mu}^y(w)\rangle = 0 \\[1mm] \hfill \forall\,(\boldsymbol{u},w) \in \tilde{H}_{K,per}(Y) \,. \end{array} \right. \tag{2.8.5}$$

Problems $(\hat{P}^\sigma_{K,Y})$ are uniquely solvable. The formulae (2.7.20) for effective stiffnesses assume the form (the subscript $\hbar$ is now replaced by h) :

$$\hat{A}_h^{\alpha\beta\lambda\mu} = \langle \hat{A}^{\lambda\mu\alpha\beta} + \hat{A}^{\lambda\mu\gamma\delta}T^{(\alpha\beta)}_{\gamma|\delta} - \hat{E}^{\lambda\mu\gamma\delta}\hat{X}^{(\alpha\beta)}_{|\gamma\delta} \rangle \, ,$$

$$\hat{F}_h^{\alpha\beta\lambda\mu} = \langle \hat{E}^{\lambda\mu\alpha\beta} + \hat{E}^{\lambda\mu\gamma\delta}T^{(\alpha\beta)}_{\gamma|\delta} - \hat{D}^{\lambda\mu\gamma\delta}\hat{X}^{(\alpha\beta)}_{|\gamma\delta} \rangle \, ,$$

$$\hat{E}_h^{\alpha\beta\lambda\mu} = \langle \hat{E}^{\lambda\mu\alpha\beta} + \hat{A}^{\lambda\mu\gamma\delta}\hat{U}^{(\alpha\beta)}_{\gamma|\delta} - \hat{E}^{\lambda\mu\gamma\delta}\chi^{(\alpha\beta)}_{|\gamma\delta} \rangle \, ,$$

$$\hat{D}_h^{\alpha\beta\lambda\mu} = \langle \hat{D}^{\lambda\mu\alpha\beta} + \hat{E}^{\lambda\mu\gamma\delta}\hat{U}^{(\alpha\beta)}_{\gamma|\delta} - \hat{D}^{\lambda\mu\gamma\delta}\chi^{(\alpha\beta)}_{|\gamma\delta} \rangle \, .$$

$$(2.8.6)$$

Taking $u = T^{(\lambda\mu)}$, $w = \hat{X}^{(\lambda\mu)}$ in $(\hat{P}^1_{K,Y})$ and $u = \hat{U}^{(\lambda\mu)}$, $w = \chi^{(\lambda\mu)}$ in $(\hat{P}^2_{K,Y})$ one obtains identities which combined with (2.8.6) lead to new formulae for effective stiffnesses

$$\hat{A}_h^{\alpha\beta\lambda\mu} = \langle \hat{A}^{\alpha\beta\lambda\mu} \rangle - \hat{b}(T^{(\alpha\beta)}, T^{(\lambda\mu)}) - [\hat{e}_K(\hat{X}^{(\alpha\beta)}, T^{(\lambda\mu)})$$
$$+\hat{e}_K(\hat{X}^{(\lambda\mu)}, T^{(\alpha\beta)})] - \hat{d}_K(\hat{X}^{(\alpha\beta)}, \hat{X}^{(\lambda\mu)}) \, ,$$

$$\hat{E}_h^{\lambda\mu\alpha\beta} = \langle \hat{E}^{\alpha\beta\lambda\mu} \rangle - \hat{b}(T^{(\alpha\beta)}, \hat{U}^{(\lambda\mu)}) - [\hat{e}_K(\hat{X}^{(\alpha\beta)}, \hat{U}^{(\lambda\mu)})$$
$$+\hat{e}_K(\chi^{(\alpha\beta)}, T^{(\lambda\mu)})] - \hat{d}_K(\hat{X}^{(\alpha\beta)}, \chi^{(\lambda\mu)}) \, ,$$

$$\hat{F}_h^{\alpha\beta\lambda\mu} = \langle \hat{E}^{\alpha\beta\lambda\mu} \rangle - \hat{b}(\hat{U}^{(\alpha\beta)}, T^{(\lambda\mu)}) - [\hat{e}_K(\hat{X}^{(\lambda\mu)}, \hat{U}^{(\alpha\beta)})$$
$$+\hat{e}_K(\chi^{(\alpha\beta)}, T^{(\lambda\mu)})] - \hat{d}_K(\chi^{(\alpha\beta)}, \hat{X}^{(\lambda\mu)}) \, ,$$

$$\hat{D}_h^{\alpha\beta\lambda\mu} = \langle \hat{D}^{\alpha\beta\lambda\mu} \rangle - \hat{b}(\hat{U}^{(\alpha\beta)}, \hat{U}^{(\lambda\mu)}) - [\hat{e}_K(\chi^{(\lambda\mu)}, \hat{U}^{(\alpha\beta)})$$
$$+\hat{e}_K(\chi^{(\alpha\beta)}, \hat{U}^{(\alpha\beta)})] - \hat{d}_K(\chi^{(\alpha\beta)}, \chi^{(\lambda\mu)}) \, .$$

$$(2.8.7)$$

By virtue of symmetry properties of bilinear forms $\hat{b}(\cdot, \cdot)$, $\hat{d}_K(\cdot, \cdot)$ it is readily seen that the effective stiffnesses (2.8.7) satisfy symmetry conditions (2.7.22). The effective potential (2.7.24) (set h instead of $\hbar$) can be written in the form

$$\mathcal{W}_h = \prec \tilde{C}^{\alpha\beta\lambda\mu}\bar{\gamma}_{\alpha\beta}\bar{\gamma}_{\lambda\mu} \succ /2 \, , \tag{2.8.8}$$

and $\bar{\gamma}_{\alpha\beta}$ are given by $(2.7.25)_1$ with

$$a^{\lambda\mu}_{\alpha\beta} = \delta^\lambda_{(\alpha}\delta^\mu_{\beta)} + T^{(\lambda\mu)}_{(\alpha|\beta)} - \hat{y}_3\hat{X}^{(\lambda\mu)}_{|\alpha\beta} \, , \qquad e^{\lambda\mu}_{\alpha\beta} = \hat{y}_3\delta^\lambda_{(\alpha}\delta^\mu_{\beta)} + \hat{U}^{(\lambda\mu)}_{(\alpha|\beta)} - \hat{y}_3\chi^{(\lambda\mu)}_{|\alpha\beta} \, . \tag{2.8.9}$$

As previously the effective elastic potential is strictly convex.

The effective constitutive relations assume the form (2.7.29), where the subscript $\hbar$ should be replaced by h. The effective stiffnesses of the original plate are

$$A_h = \varepsilon_0 \hat{A}_h \, , \quad F_h = (\varepsilon_0)^2 \hat{F}_h \, , \quad E_h = (\varepsilon_0)^2 \hat{E}_h \, , \quad D_h = (\varepsilon_0)^3 \hat{D}_h \, . \tag{2.8.10}$$

Let us consider consequences of transverse symmetry. Problem $(\hat{P}^1_{K,Y})$ turns out to be the same as $(\bar{P}^1_{S,Y})$ of Section 2.7. Problem $(\hat{P}^2_{K,Y})$ reduces to:

$$(\hat{P}^2_{KS,Y}) \quad \left|\begin{array}{l} \text{find } \chi^{(\alpha\beta)} \in H^2_{per}(Y) \text{ such that} \\[2mm] \hat{d}_K(\chi^{(\alpha\beta)}, v) + \langle \hat{D}^{\alpha\beta\lambda\mu}(y)\kappa^y_{\lambda\mu}(v)\rangle = 0 \\[2mm] \hspace{3cm} \forall\, v \in H^2_{per}(Y)\,. \end{array}\right. \qquad (2.8.11)$$

The homogenized constitutive relations have the form (2.7.33) with $\hbar$ replaced by h; $A_\hbar = A_h$ and D_h is given by

$$D^{\alpha\beta\lambda\mu}_h = \varepsilon_0^3 \left[\langle \hat{D}^{\alpha\beta\lambda\mu}\rangle - \hat{d}_K(\chi^{(\alpha\beta)}, \chi^{(\alpha\beta)})\right]\,, \qquad (2.8.12\text{a})$$

or

$$D^{\alpha\beta\lambda\mu}_h = \varepsilon_0^3\, \hat{d}_K(\chi^{(\alpha\beta)} - p^{(\alpha\beta)}, \chi^{(\lambda\mu)} - p^{(\lambda\mu)})\,, \qquad (2.8.12\text{b})$$

where

$$p^{(\alpha\beta)}{}_{|\sigma\kappa} = \delta^\alpha_\sigma \delta^\beta_\kappa\,.$$

Hence $p^{(\alpha\beta)}$ are polynomials of second order in y_1, y_2. In other words

$$D^{\alpha\beta\lambda\mu}_h = \langle D^{\gamma\delta\rho\sigma}[\delta^\alpha_\gamma \delta^\beta_\delta + \kappa^y_{\gamma\delta}(\chi^{(\alpha\beta)})][\delta^\lambda_\rho \delta^\mu_\sigma + \kappa^y_{\rho\sigma}(\chi^{(\lambda\mu)})]\rangle\,, \qquad (2.8.12\text{c})$$

where $D = (\varepsilon_0)^3 \hat{D}$ represents the bending stiffness tensor of the original plate.

2.9. Case of transversely slender periodicity cells of constant thickness

We consider a plate of constant thickness with Z-periodic variation of elastic moduli. We assume in this section that the plate thickness $|x_3^+ - x_3^-|$ is much greater than in-plane dimensions l^z_α. The periodicity cell $\mathcal{Z}$ is said to be transversely slender. Planes $x_3 = $ constant are not necessarily planes of material symmetry. Although this case is comprised by the asymptotic method of Sections 2.2, 2.3 it seems reasonable to derive approximate formulae for effective stiffnesses in order to avoid difficulties associated with solving the three-dimensional local problems $(P^\alpha_{\mathcal{Y}})$ of Section 2.3. The results of this section will be justified in Section 2.10.3 by the Γ-convergence method.

Step 1. To find effective stiffnesses we first introduce a small parameter ε by considering a sequence of plate problems with εY-periodic elastic moduli. Transverse geometry is kept ε-independent. Thus we apply the scaling, cf. (2.2.1) – (2.2.3),

$$x_3^\pm(x) = x_3^\pm \rightsquigarrow x_3^\pm\,, \qquad l^z_\alpha \rightsquigarrow \varepsilon l_\alpha \quad l^z_\alpha = \varepsilon_0 l_\alpha\,,$$

$$Z \rightsquigarrow \varepsilon Y\,, \quad Z = \varepsilon_0 Y\,; \qquad (2.9.1)$$

$$r^i_\pm(x) \rightsquigarrow r^i_\pm(x)\,, \qquad b^i \rightsquigarrow b^i\,.$$

The elastic moduli are εY- periodic in x

$$C^{ijkl}_Z(x, x_3) \rightsquigarrow C^{ijkl}\left(\frac{x}{\varepsilon}, x_3\right)\,, \qquad C^{ijkl}_Z(x, x_3) = C^{ijkl}\left(\frac{x}{\varepsilon_0}, x_3\right) \qquad (2.9.2)$$

and $C^{ijkl}(\cdot, x_3)$ are Y- periodic.

The first step of the algorithm is to perform homogenization of the material based on the scaling (2.9.1), (2.9.2). In the next step we reduce the third dimension either by the asymptotic method of Friedrichs-Dressler-Goldenveizer or by imposing stress-displacement constraints of Kirchhoff type.

Let us pay attention to the first step. We shall outline briefly the process of smearing out elastic properties in x_1, x_2 directions.

The variational equilibrium equation has the form

$$\int_B \sigma_\varepsilon^{ij} e_{ij}(\boldsymbol{v})\,d\boldsymbol{x} = \int_\Omega [p_+^\alpha v_\alpha(x, x_3^+) + q_+ v_3(x, x_3^+)]\,dx$$

$$+ \int_\Omega [p_-^\alpha v_\alpha(x, x_3^-) + q_- v_3(x, x_3^-)]\,dx + \int_B b^i\left(\frac{x}{\varepsilon}\right) v_i(\boldsymbol{x})\,d\boldsymbol{x} \quad \forall\, \boldsymbol{v} \in V_0(B) \quad (2.9.3)$$

where

$$\sigma_\varepsilon^{ij} = C^{ijkl}\left(\frac{x}{\varepsilon}, x_3\right) e_{kl}(\boldsymbol{w}^\varepsilon) . \qquad (2.9.4)$$

To find effective behavior of the plate considered we apply the method of two-scale expansions:

$$\boldsymbol{w}^\varepsilon = \boldsymbol{w}^{(0)}(x, x_3) + \varepsilon \boldsymbol{w}^{(1)}(x, x_3, y) + \varepsilon^2 \boldsymbol{w}^{(2)}(x, x_3, y) + \cdots ,$$

$$\boldsymbol{v} = \boldsymbol{v}^{(0)}(x, x_3) + \varepsilon \boldsymbol{v}^{(1)}(x, x_3, y) + \varepsilon^2 \boldsymbol{v}^{(2)}(x, x_3, y) + \cdots ,$$

where $y = x/\varepsilon$ and $\boldsymbol{w}^{(k)}(x, x_3, \cdot)$, $\boldsymbol{v}^{(k)}(x, x_3, \cdot)$ are Y-periodic functions for $k \geq 1$. Expansions of deformations read:

$$e_{\alpha\beta}(\boldsymbol{w}^\varepsilon) = e_{\alpha\beta}(\boldsymbol{w}^{(0)}) + w_{(\alpha|\beta)}^{(1)} + 0(\varepsilon) ,$$

$$e_{33}(\boldsymbol{w}^\varepsilon) = e_{33}(\boldsymbol{w}^{(0)}) + 0(\varepsilon) , \qquad 2e_{\alpha 3}(\boldsymbol{w}^\varepsilon) = 2e_{\alpha 3}(\boldsymbol{w}^{(0)}) + w_{3|\alpha}^{(1)} . \qquad (2.9.5)$$

Let us substitute $\boldsymbol{v} = \boldsymbol{v}^{(0)}(\boldsymbol{x})$ into Eq. (2.9.3) and pass to zero with ε. One obtains

$$\int_B \sigma_\mathfrak{h}^{ij} e_{ij}(\boldsymbol{v}^{(0)})\,d\boldsymbol{x} = \int_\Omega [p_+^\alpha v_\alpha^{(0)}(x, x_3^+) + p_-^\alpha v_\alpha^{(0)}(x, x_3^-) \qquad (2.9.6)$$

$$+ q_+ v_3(x, x_3^+) + q_- v_3(x, x_3^-)]\,dx + \int_B \langle b^i\rangle v_i(\boldsymbol{x})\,d\boldsymbol{x} ,$$

with

$$\sigma_\mathfrak{h}^{ij} = \langle \sigma_0^{ij}\rangle , \qquad (2.9.7)$$

$$\sigma_0^{ij} = C^{ijkl} e_{kl}(\boldsymbol{w}^{(0)}) + C^{ij\lambda\mu} w_{\lambda|\mu}^{(1)} + C^{ij\lambda 3} w_{3|\lambda}^{(1)} . \qquad (2.9.8)$$

Note that

$$\sigma_\varepsilon^{ij} e_{ij}(\boldsymbol{v}) = \sigma_0^{ij} e_{ij}(\boldsymbol{v}^{(0)}) + \sigma_0^{\alpha\beta} v_{\alpha|\beta}^{(1)} + \sigma_0^{\alpha 3} v_{3|\alpha}^{(1)} + 0(\varepsilon) . \qquad (2.9.9)$$

Hence

$$\langle \sigma_\varepsilon^{ij} e_{ij}(\boldsymbol{v})\rangle = \sigma_\mathfrak{h}^{ij} e_{ij}(\boldsymbol{v}^{(0)}) + \langle \sigma_0^{\alpha\beta} v_{\alpha|\beta}^{(1)} + \sigma_0^{\alpha 3} v_{3|\alpha}^{(1)}\rangle + 0(\varepsilon) . \qquad (2.9.10)$$

Passing to zero with ε in (2.9.3) and taking into account (2.9.6) one finds

$$\int_B \langle \sigma_0^{\alpha\beta} v_{\alpha|\beta}^{(1)} + \sigma_0^{\alpha3} v_{3|\alpha}^{(1)} \rangle dx = 0 \qquad (2.9.11)$$

for any $v^{(1)}(x,\cdot) \in H_{per}^1(Y)^3$.

By taking $v^{(1)} = \phi(x) \cdot v(y)$, $\phi \in \mathbf{D}(B)$ one can localize Eq. (2.9.11):

$$\langle \sigma_0^{\alpha\beta} v_{\alpha|\beta}^{(1)} + \sigma_0^{\alpha3} v_{3|\alpha}^{(1)} \rangle = 0 \qquad \forall\, v \in H_{per}^1(Y)^3\,. \qquad (2.9.12)$$

Hence we may write

$$w_\lambda^{(1)} = \Theta_\lambda^{(kl)} e_{kl}(w^{(0)})\,, \qquad w_3^{(1)} = \vartheta^{(kl)} e_{kl}(w^{(0)})\,, \qquad (2.9.13)$$

where functions $\Theta_\lambda^{(kl)}$, $\vartheta^{(kl)}$ are solutions to the following local problem

$$(P_Y^{(kl)}) \quad \left| \begin{array}{l} \text{find } (\Theta_\lambda^{(kl)}\,,\ \vartheta^{(kl)}) \in H_{per}^1(Y)^3 \text{ such that} \\[6pt] \langle [C^{\alpha\beta\lambda\mu}\Theta_{\lambda|\mu}^{(kl)} + C^{\alpha\beta\lambda3}\vartheta_{|\lambda}^{(kl)} + C^{\alpha\beta kl}]v_{\alpha|\beta} \rangle = 0\,, \\[6pt] \langle [C^{\alpha3\lambda\mu}\Theta_{\lambda|\mu}^{(kl)} + C^{\alpha3\lambda3}\vartheta_{|\lambda}^{(kl)} + C^{\alpha3 kl}]v_{3|\alpha} \rangle = 0\,, \\[6pt] \text{for any } v \in H_{per}^1(Y)^3\,. \end{array} \right. \qquad (2.9.14)$$

The solution of the above problem is determined up to an additive constant. Substituting representations (2.9.13) into (2.9.8) one finds:

$$\sigma_0^{ij} = C_0^{ijkl} e_{kl}(w^{(0)})\,, \qquad (2.9.15)$$

with

$$C_0^{ijkl} = C^{ijkl} + C^{ij\lambda\mu}\Theta_{\lambda|\mu}^{(kl)} + C^{ij\lambda3}\vartheta_{|\lambda}^{(kl)}\,. \qquad (2.9.16)$$

Hence

$$\sigma_\mathfrak{h}^{ij} = C_\mathfrak{h}^{ijkl} e_{kl}(w^{(0)})\,, \qquad C_\mathfrak{h}^{ijkl} = \langle C_0^{ijkl} \rangle\,. \qquad (2.9.17)$$

The variational equilibrium equation (2.9.6) along with the constitutive relation (2.9.17) constitute the homogenized three-dimensional elasticity problem. Note that if $x_3 =$const are planes of material symmetry (cf. conditions (2.4.1)) then problem $(P_Y^{(kl)})$ decouples into:

– the in-plane problem: find $\Theta^{(kl)} \in H_{per}^1(Y)^2$ such that

$$(\check{P}_Y^1) \quad \langle [C^{\alpha\beta\lambda\mu}\Theta_{\lambda|\mu}^{(kl)} + C^{\alpha\beta kl}]v_{\alpha|\beta} \rangle = 0\,, \qquad \forall\, v \in H_{per}^1(Y)^2\,; \qquad (2.9.18)$$

– and the transverse shear problem

$$\text{find } \vartheta^{(kl)} \in H_{per}^1(Y) \text{ such that}$$

$$(\check{P}_Y^2) \quad \langle (C^{\alpha3\lambda\mu}\vartheta_{|\lambda}^{(kl)} + C^{\alpha3 kl})v_{3|\beta} \rangle = 0\,, \qquad \forall\, v_3 \in H_{per}^1(Y)\,. \qquad (2.9.19)$$

Consequently $\Theta_\lambda^{(\alpha 3)} = 0$, $\vartheta^{(\alpha\beta)} = 0$, $\vartheta^{(33)} = 0$. Hence

$$C_0^{\alpha\beta\gamma\delta} = C^{\alpha\beta\gamma\delta} + C^{\alpha\beta\lambda\mu}\Theta_{\lambda|\mu}^{(\alpha\beta)}\,, \qquad C_0^{\alpha\beta33} = C^{\alpha\beta33} + C^{\alpha\beta\lambda\mu}\Theta_{\lambda|\mu}^{(33)}\,,$$

$$C_0^{\alpha3\beta3} = C^{\alpha3\beta3} + C^{\alpha3\lambda3}\vartheta_{|\lambda}^{(\beta3)}\,, \qquad C_0^{3333} = C^{3333} + C^{33\lambda\mu}\vartheta_{\lambda|\mu}^{(33)}\,, \qquad (2.9.20)$$

$$C_0^{\alpha3\beta\lambda} = C_0^{333\alpha} = 0\,,$$

and $\sigma_\mathfrak{h}^{\alpha\beta}$, $\sigma_\mathfrak{h}^{33}$ depend upon $\Theta^{(kl)}$, while $\sigma_\mathfrak{h}^{\alpha3}$ depends on $\vartheta^{(\beta3)}$.

Equations (2.9.6) and (2.9.17) constitute a $3D$ problem $(P_\mathfrak{h}^{3D})$.

Step 2. Having found the problem $(P_\mathfrak{h}^{3D})$ one can now impose the condition of the thickness being small as compared with diam Ω. This process of reduction of the transverse dimension can be performed in two ways:

a) by imposing Kirchhoff constraints,

b) by asymptotic analysis.

Both approaches lead to the same plate equations with effective constitutive relations of the form (2.7.29) and with effective stiffnesses given by

$$\begin{bmatrix} A_\mathfrak{h}^{\alpha\beta\lambda\mu} \\ E_\mathfrak{h}^{\alpha\beta\lambda\mu} \\ D_\mathfrak{h}^{\alpha\beta\lambda\mu} \end{bmatrix} = \int_{x_3^-}^{x_3^+} \tilde{C}_\mathfrak{h}^{\alpha\beta\lambda\mu}(x_3) \begin{bmatrix} 1 \\ \hat{x}_3 \\ (\hat{x}_3)^2 \end{bmatrix} dx_3\,, \quad \hat{x}_3 = x_3 - \frac{1}{2}(x_3^+ + x_3^-)\,, \qquad (2.9.21)$$

since the conditions $c^\pm = \text{const}$ imply $\prec y_3 \succ = \dfrac{1}{2}(c^+ + c^-)$. The tensor $\tilde{C}_\mathfrak{h}$ refers to the plane-stress state and is constructed as follows. Let us write the homogenized constitutive relations in the form

$$\sigma_\mathfrak{h}^{\alpha\beta} = C_\mathfrak{h}^{\alpha\beta\lambda\mu}\epsilon_{\lambda\mu} + C_\mathfrak{h}^{\alpha\beta k3}\gamma_{k3}\,, \qquad \sigma_\mathfrak{h}^{k3} = C_\mathfrak{h}^{k3\lambda\mu}\epsilon_{\lambda\mu} + C_\mathfrak{h}^{k3\lambda3}\gamma_{k3}\,, \qquad (2.9.22)$$

where $\gamma_{\lambda3} = 2\epsilon_{\lambda3}$, $\gamma_{33} = \epsilon_{33}$. Let us require that $\sigma_\mathfrak{h}^{k3} = 0$, which makes it possible to eliminate γ_{k3} :

$$\gamma_{k3} = -\tilde{c}_{ki}^{\mathfrak{h}} C_\mathfrak{h}^{i3\lambda\mu}\epsilon_{\lambda\mu}\,, \qquad (2.9.23)$$

with

$$\tilde{c}^{\mathfrak{h}} = [C_\mathfrak{h}']^{-1}\,, \qquad C_\mathfrak{h}' = (C_\mathfrak{h}^{i3j3})\,. \qquad (2.9.24)$$

Thus

$$\sigma_\mathfrak{h}^{\alpha\beta} = \tilde{C}_\mathfrak{h}^{\alpha\beta\lambda\mu}\epsilon_{\lambda\mu}\,, \qquad (2.9.25)$$

where

$$\tilde{C}_\mathfrak{h}^{\alpha\beta\lambda\mu} = C_\mathfrak{h}^{\alpha\beta\lambda\mu} - C_\mathfrak{h}^{\alpha\beta i3}\tilde{c}_{ij}^{\mathfrak{h}} C_\mathfrak{h}^{j3\lambda\mu}\,. \qquad (2.9.26)$$

In the case of planes $x_3 = \text{const}$ being planes of material symmetry, the formula for $\tilde{C}_\mathfrak{h}$ reduces to

$$\tilde{C}_\mathfrak{h}^{\alpha\beta\lambda\mu} = C_\mathfrak{h}^{\alpha\beta\lambda\mu} - C_\mathfrak{h}^{\alpha\beta33}(C_\mathfrak{h}^{3333})^{-1}C_\mathfrak{h}^{33\lambda\mu}\,, \qquad (2.9.27)$$

which coincides with formula (2.7.6). Note that in this case tensor $\tilde{C}_\mathfrak{h}$ depends only on $\Theta^{(kl)}$. A rigorous justification of formulae (2.9.21) will be given in Section 2.10.3.

2.10. Γ-convergence and justification of three models of thin, transversely inhomogeneous and anisotropic plates with constant thickness

The method of assessing effective stiffnesses presented in Sections 2.2, 2.3 as well as the formulation of the effective problem concern a general shape of periodicity cell $\mathcal{Z}$. Thus these formulae can be viewed as reference formulae. Further models were concerned with specific shapes of $\mathcal{Z}$. Results of Sections 2.7, 2.8 are applicable to flat cells, while results of Section 2.9 refer to cells $\mathcal{Z}$ that are transversely slender.

The aim of this section is to explain the meaning of the effective models of Sections 2.2, 2.3, 2.8 and 2.9 approximate the overall properties of periodic plates. It turns out that these models describe a limit behavior at dimensions of $\mathcal{Z}$ tending to zero. To show it two small parameters ε and e will be introduced, the latter controls the transverse dimensions of $\mathcal{Z}$. In the present section the plate thickness will be assumed as constant: x_3^- and x_3^+ are fixed constants. The plane $x_3 = 0$ determines the middle plane of the plate; $x_3^- = -x_3^+$. We scale the thickness

$$x_3^- = -x_3^+ = -ec, \qquad e > 0,$$

and consider the plate with periodicity cells

$$\mathcal{Z}_{e\varepsilon} = \varepsilon Y \times (-ec, ec), \quad Y = (0, l_1) \times (0, l_2).$$

The present section is arranged as follows. In Section 2.10.2 we prove that the effective model of Section 2.8 can be obtained by passing to zero: first $e \to 0$ and next $\varepsilon \to 0$. In Section 2.10.3 we show that model of Section 2.9 follows also from passing to zero: first $\varepsilon \to 0$ and next $e \to 0$. A simultaneous passage to zero: $(e, \varepsilon) \to (0, 0)$ is considered in Section 2.10.4, which results in formulae of Section 2.3. The convergence proofs given here justify the asymptotic expansion methods used in Sections 2.1 – 2.9.

2.10.1.   Basic relations

We shall assume that the plate thickness is constant but do not assume any material symmetry; in particular (2.4.1) is not imposed here. The domain of the plate in its undeformed configuration is denoted by $B_e = \Omega \times (-ec, ec)$. Plate faces are $\Gamma_\pm^e = \Omega \times \{\pm ec\}$ and the lateral boundary is: $\Upsilon^e = \Gamma \times (-ec, ec)$, $\Gamma = \partial\Omega$. The elastic moduli are here scaled as follows,

$$C_{e\varepsilon}^{ijkl}(\boldsymbol{x}) = \frac{1}{e^3} C^{ijkl}\left(\frac{x_\alpha}{\varepsilon}, \frac{x_3}{e}\right), \tag{2.10.1}$$

where $C^{ijkl}(\,\cdot\,, x_3/e)$ are Y-periodic; $\varepsilon > 0$ scales the dimensions of the periodicity rectangle εY. To work with an e-independent domain of the plate we introduce new variables

$$z_\alpha = x_\alpha, \qquad z_3 = x_3/e, \tag{2.10.2}$$

hence $z_3 \in (-c, c)$. Now we set

$$C_\varepsilon^{ijkl}(\boldsymbol{z}) = C^{ijkl}\left(\frac{z_\alpha}{\varepsilon}, z_3\right), \tag{2.10.3}$$

$$C_{e\varepsilon}^{ijkl}(\boldsymbol{z}) = \frac{1}{e^3} C^{ijkl}(\boldsymbol{z}). \tag{2.10.4}$$

As usual, we make the following assumptions, cf. (2.1.7)

$$C^{ijkl}(\boldsymbol{y}) = C^{jikl}(\boldsymbol{y}) = C^{klij}(\boldsymbol{y}) \tag{2.10.5}$$

for a.e. $\boldsymbol{y} \in \mathcal{Y} = Y \times (-c, c)$;

$$C^{ijkl} \in L^\infty(\mathcal{Y}) ; \tag{2.10.6}$$

there exists a constant $m > 0$ such that

$$\forall \boldsymbol{\epsilon} \in \mathbb{E}_s^3 \qquad C^{ijkl}(\boldsymbol{y})\epsilon_{ij}\epsilon_{kl} \geq m \sum_{i,j=1}^{3} \epsilon_{ij}\epsilon_{ij} , \tag{2.10.7}$$

for a.e. $\boldsymbol{y} \in \mathcal{Y}$.

The plate is subject to body forces $\tilde{\boldsymbol{b}} = (\tilde{b}^i)$ and tractions $(p_+^\alpha/e, q^+) = \boldsymbol{g}^+$, $(p_-^\alpha/e, q^-) = \boldsymbol{g}^-$ on Γ_+^e and Γ_-^e, respectively. The plate is assumed to be clamped on Υ^e . Let $\tilde{\boldsymbol{b}}(\boldsymbol{x}) = \boldsymbol{b}(x_\alpha, x_3/e)$ belong to $L^2(B_e)^3 = [L^2(B_e)]^3$ while $\boldsymbol{g}^+ = (g_+^i)$ and $\boldsymbol{g}^- = (g_-^i)$ are elements of $L^2(\Omega)^3$.

The space of kinematically admissible displacement fields is defined by

$$V_0(B_e) = \{\boldsymbol{v} \in H^1(B_e)^3 \mid \boldsymbol{v} = 0 \text{ on } \Upsilon^e\} . \tag{2.10.8}$$

If $\tilde{\boldsymbol{v}} \in H^1(B_e)^3$ then the strain tensor is given by

$$e_{ij}^x(\tilde{\boldsymbol{v}}) = \left(\frac{\partial \tilde{v}_i}{\partial x_j} + \frac{\partial \tilde{v}_j}{\partial x_i}\right)/2 .$$

For fixed $e > 0$ and $\varepsilon > 0$ the functional of the total potential energy assumes the form

$$\tilde{J}_{e\varepsilon}(\tilde{\boldsymbol{v}}) = \frac{1}{2}\int_{B_e} C_{e\varepsilon}^{ijkl} e_{ij}^x(\tilde{\boldsymbol{v}})e_{kl}^x(\tilde{\boldsymbol{v}})d\boldsymbol{x} - \int_{B_e} \tilde{b}^i\tilde{v}_i d\boldsymbol{x} - \int_{\Gamma_\pm^e} \frac{1}{e}p_\pm^\alpha\tilde{v}_\alpha d\Gamma - \int_{\Gamma_\pm^e} q^\pm\tilde{v}_3 d\Gamma , \tag{2.10.9}$$

where

$$\int_{\Gamma_\pm^e} p_\pm^\alpha\tilde{v}_\alpha d\Gamma = \int_{\Gamma_+^e} p_+^\alpha\tilde{v}_\alpha d\Gamma + \int_{\Gamma_-^e} p_-^\alpha\tilde{v}_\alpha d\Gamma ,$$

and similarly for the integral of $q^\pm$.

The minimum principle of the total potential energy means evaluating

$$\tilde{J}_{e\varepsilon}(\tilde{\boldsymbol{v}}) = \inf\{\tilde{J}_{e\varepsilon}(\tilde{\boldsymbol{v}}) \mid \tilde{\boldsymbol{v}} \in V_0(B_e)\} . \tag{2.10.10}$$

On account of (2.10.5) – (2.10.7), $\tilde{\boldsymbol{v}}^{e\varepsilon}$ exists and is unique.

From now on until the end of this section we shall work with the domain $B = \Omega \times (-c, c)$. To achieve this we proceed in the following fashion. Let $\tilde{\boldsymbol{v}} \in V_0(B_e)$, then the function $\boldsymbol{v}$ defined by

$$v_\alpha(z_i) = \frac{1}{e}\tilde{v}_\alpha(z_\beta, ez_3) , \qquad v_3(z_i) = \tilde{v}_3(z_\alpha, ez_3) , \tag{2.10.11}$$

belongs to

$$V_0(B) = \{v \in H^1(B)^3 \mid v = 0 \text{ on } \Upsilon\}, \qquad \Upsilon = \Gamma \times (-c, c). \tag{2.10.12}$$

Moreover we have

$$e^z_{\alpha\beta}(v) = \frac{1}{e} e^x_{\alpha\beta}(\tilde{v}), \qquad e^z_{\alpha3}(v) = e^x_{\alpha3}(\tilde{v}), \qquad e^z_{33}(v) = e e^x_{33}(\tilde{v}), \tag{2.10.13}$$

where

$$e^z_{ij}(v) = \left(\frac{\partial v_i}{\partial z_j} + \frac{\partial v_j}{\partial z_i}\right)/2.$$

By taking into account (2.10.1) – (2.10.4), (2.10.11) and (2.10.13) in (2.10.9) we obtain the rescaled functional of the total potential energy

$$J_{e\varepsilon}(v) := \tilde{J}_{e\varepsilon}(ev_\alpha, v_3) = \frac{1}{2} \int_B C^{ijkl}_\varepsilon (\mathbf{Q}^e e^z(v))_{ij} (\mathbf{Q}^e e^z(v))_{kl} dz - L_e(v), \tag{2.10.14}$$

where $v \in V_0(B)$ while $\mathbf{Q}^e : \mathbb{E}^3_s \to \mathbb{E}^3_s$ is defined by

$$(\mathbf{Q}^e \epsilon)_{\alpha\beta} = \epsilon_{\alpha\beta}, \quad (\mathbf{Q}^e \epsilon)_{\alpha3} = \frac{1}{e}\epsilon_{\alpha3}, \quad (\mathbf{Q}^e \epsilon)_{33} = \frac{1}{e^2}\epsilon_{33}. \tag{2.10.15}$$

The rescaled loading functional has the following form

$$L_e(v) = \int_B e(eb^\alpha v_\alpha + b^3 v_3) dz + \int_{\Gamma_+} (p^\alpha_+ v_\alpha + q^+ v_3) d\Gamma + \int_{\Gamma_-} (p^\alpha_- v_\alpha + q^- v_3) d\Gamma. \tag{2.10.16}$$

We observe that the rescaled stress tensor $\sigma_{e\varepsilon}$ is interrelated with the strain $\epsilon^{e\varepsilon}$ by

$$\sigma^{ij}_{e\varepsilon} = \frac{\partial j_\varepsilon(z, \mathbf{Q}^e \epsilon^{e\varepsilon})}{\partial(\mathbf{Q}^e \epsilon^{e\varepsilon})_{ij}}, \tag{2.10.17}$$

or

$$\sigma^{ij}_{e\varepsilon} = C^{ij\alpha\beta}_\varepsilon \epsilon^{e\varepsilon}_{\alpha\beta} + \frac{2}{e} C^{ij\alpha3}_\varepsilon \epsilon^{e\varepsilon}_{\alpha3} + \frac{1}{e^2} C^{ij33}_\varepsilon \epsilon^{e\varepsilon}_{33}, \tag{2.10.17a}$$

where

$$j_\varepsilon(z, \mathbf{Q}^e \epsilon) = j(\frac{z_\alpha}{\varepsilon}, z_3, \mathbf{Q}^e \epsilon) = \frac{1}{2} C^{ijkl}(\frac{z_\alpha}{\varepsilon}, z_3)(\mathbf{Q}^e \epsilon)_{ij}(\mathbf{Q}^e \epsilon)_{kl}. \tag{2.10.18}$$

Prior to passing to zero with the "small" parameters e and ε we shall formulate a lemma playing a crucial role in our subsequent developments.

Lemma 2.10.1. Let $v^{e\varepsilon} \in V_0(B)$ be a sequence weakly convergent to $v^\varepsilon \in V_0(B)$ in $H^1(B)^3$ as $e \to 0$. Suppose that there exists a constant $K > 0$, independent of e, such that

$$\int_B j\left[\frac{z_\alpha}{\varepsilon}, z_3, \mathbf{Q}^e e^z(v^{e\varepsilon})\right] dz \leq K. \tag{2.10.19}$$

Then $e^z_{i3}(\boldsymbol{v}^\varepsilon) = 0$ and

$$v^\varepsilon_\alpha = u^\varepsilon_\alpha(z_\beta) - z_3\frac{\partial w^\varepsilon_3}{\partial z_\alpha}\,,\quad v^\varepsilon_3 = w^\varepsilon_3(z_\alpha)\,,\qquad\qquad (2.10.20)$$

where $u^\varepsilon_\alpha \in H^1_0(\Omega)$ and $w^\varepsilon_3 \in H^2_0(\Omega)$.
Proof. From (2.10.7), (2.10.18) and (2.10.19) we obtain

$$m\,\|\,\mathbf{Q}^e e^z(\boldsymbol{v}^{e\varepsilon})\,\|^2_{L^2(B,\mathbf{E}^3_s)} \le K\,,$$

where

$$\|\,\mathbf{Q}^e e^z(\boldsymbol{v}^{e\varepsilon})\,\|^2_{L^2(B,\mathbf{E}^3_s)} = \|\,(e^z_{\alpha\beta}(\boldsymbol{v}^{e\varepsilon}))\,\|^2_{L^2(B,\mathbf{E}^2_s)}$$
$$+\frac{1}{e^2}\,\|\,(e^z_{\alpha3}(\boldsymbol{v}^{e\varepsilon}))\,\|^2_{L^2(B)^2} +\frac{1}{e^4}\,\|\,e^z_{33}(\boldsymbol{v}^{e\varepsilon})\,\|^2_{L^2(B)}\,.$$

Hence

$$\frac{1}{e}\,\|\,e^z_{\alpha3}(\boldsymbol{v}^{e\varepsilon})\,\|_{L^2(B)^2} \le K_1\,,\qquad \frac{1}{e^2}\,\|\,e^z_{33}(\boldsymbol{v}^{e\varepsilon})\,\|_{L^2(B)} \le K_1\,,$$

where K_1 does not depend on e. Consequently we have

$$\frac{\partial v^\varepsilon_3}{\partial z_3} = 0\,,\qquad e^z_{\alpha3}(\boldsymbol{v}^\varepsilon) = 0\,.\qquad\qquad (2.10.21)$$

Now we readily obtain (2.10.20) by applying Theorem 1.4-1 and Theorem 3.3-1 due to Ciarlet (1990), cf. also Destuynder (1986a). $\qquad\square$

Suppose that $\boldsymbol{v}^{e\varepsilon}$ solves the following minimization problem

$$(P_{e\varepsilon})\qquad J_{e\varepsilon}(\boldsymbol{v}^{e\varepsilon}) = \inf\{J_{e\varepsilon}(\boldsymbol{v}) \mid \boldsymbol{v} \in V_0(B)\}\,.\qquad\qquad (2.10.22)$$

In other words we may write

$$J_{e\varepsilon}(\boldsymbol{v}^{e\varepsilon}) \le J_{e\varepsilon}(\boldsymbol{v})\qquad \forall \boldsymbol{v} \in V_0(B)\,.$$

For $\boldsymbol{v} = 0$ we have

$$\frac{1}{2}\int_B j\left[\frac{z_\alpha}{\varepsilon}, z_3, \mathbf{Q}^e e^z(\boldsymbol{v}^{e\varepsilon})\right]dz \le L_e(\boldsymbol{v}^{e\varepsilon})$$

$$\le |L_e(\boldsymbol{v}^{e\varepsilon})| \le K_2\left(\|\,\boldsymbol{v}^{e\varepsilon}\,\|_{L^2(B)^3} + \|\,\boldsymbol{v}^{e\varepsilon}\,\|_{L^2(\Gamma_+)^3} + \|\,\boldsymbol{v}^{e\varepsilon}\,\|_{L^2(\Gamma_-)^3}\right)\,.\qquad (2.10.23)$$

Thus

$$\frac{m}{2}\,\|\,\mathbf{Q}^e e^z(\boldsymbol{v}^{e\varepsilon})\,\|^2_{L^2(B,\mathbf{E}^3_s)} \le \tilde{K}_2\left(\|\,\boldsymbol{v}^{e\varepsilon}\,\|_{L^2(B)^3} + \|\,\boldsymbol{v}^{e\varepsilon}\,\|_{L^2(\Gamma_+)^3} + \|\,\boldsymbol{v}^{e\varepsilon}\,\|_{L^2(\Gamma_-)^3}\right)\,,$$

where $\tilde{K}_2 > 0$ depends on Ω, $\boldsymbol{b}$ and $\boldsymbol{g}_\pm$ but not on e. As in the previous case, for $0 < e < 1$ we write

$$\frac{m}{2} \parallel \boldsymbol{e}^z(\boldsymbol{v}^{e\varepsilon}) \parallel^2_{L^2(B,\mathbf{E}^3_s)} \leq \tilde{K}_2 \left(\parallel \boldsymbol{v}^{e\varepsilon} \parallel_{L^2(B)^3} + \parallel \boldsymbol{v}^{e\varepsilon} \parallel_{L^2(\Gamma_+)^3} + \parallel \boldsymbol{v}^{e\varepsilon} \parallel_{L^2(\Gamma_-)^3} \right)$$
$$\leq K_2 \parallel \boldsymbol{v}^{e\varepsilon} \parallel_{H^1(B)^3} .$$

Korn's inequality finally yields

$$K_3 \parallel \boldsymbol{v}^{e\varepsilon} \parallel^2_{H^1(B)^3} \leq K_2 \parallel \boldsymbol{v}^{e\varepsilon} \parallel_{H^1(B)^3} .$$

Thus the sequence $\boldsymbol{v}^{e\varepsilon}$ is bounded and there exists a subsequence, still denoted by $\boldsymbol{v}^{e\varepsilon}$, weakly convergent to $\boldsymbol{v}^\varepsilon$ in $H^1(B)^3$ and thus strongly convergent in $L^2(B)^3$. The boundedness of the sequence $\boldsymbol{v}^{e\varepsilon}$ implies now the existence of a positive constant K appearing in (2.10.19).

Lemma 2.10.1 introduces Kirchhoff's rescaled displacement fields. It is thus natural to introduce the following subspaces of the space $H^1(B)^3$:

$$V_K(B) = \{\boldsymbol{v} \in H^1(B)^3 \mid e^z_{i3}(\boldsymbol{v}) = 0\}$$
$$= \{\boldsymbol{v} \in H^1(B)^3 \mid v_\alpha = u_\alpha - z_3 \frac{\partial w}{\partial z_\alpha}, \ v_3 = w, \ u_\alpha \in H^1(\Omega), \ w \in H^2(\Omega)\} \quad (2.10.24)$$

and

$$V_K^0(B) = \left\{\boldsymbol{v} \in V_K(B) \mid u_\alpha \in H^1_0(\Omega), w \in H^2_0(\Omega)\right\} . \quad (2.10.25)$$

2.10.2.   Justification of the effective plate model of Sec. 2.8 by passing to zero:
$e \to 0$ and then $\varepsilon \to 0$

The aim of this section is to put forward a rigorous justification of the homogenized plate model derived in Section 2.8.1 by imposing displacement constraints (2.8.1). Having this in mind we pass now to the study of Γ-convergence of the sequence of functionals $\{J_{e\varepsilon}\}_{e>0,\varepsilon>0}$ defined by (2.10.14). We shall first pass to zero with the thickness parameter ($e \to 0$) and next with the parameter ε characterizing the periodic structure of the plate. The result of this passage to the limit is formulated in the following form.

Theorem 2.10.2. For any fixed $\varepsilon > 0$ the sequence of functionals $\{J_{e\varepsilon}\}_{e>0}$ is Γ-convergent in the strong topology of the space $L^2(B)^3$ (weak topology of $H^1(B)^3$) to the functional

$$J_\varepsilon(\boldsymbol{v}) = \begin{cases} \frac{1}{2} \int_B \tilde{C}^{\alpha\beta\lambda\mu}_\varepsilon(\boldsymbol{z}) e^z_{\alpha\beta}(\boldsymbol{v}) e^z_{\lambda\mu}(\boldsymbol{v}) d\boldsymbol{z} - L(\boldsymbol{v}) & \text{if } \boldsymbol{v} \in V_K(B) \\ +\infty & \text{otherwise} , \end{cases} \quad (2.10.26)$$

where

$$\tilde{C}^{\alpha\beta\lambda\mu}_\varepsilon(\boldsymbol{z}) = \tilde{C}^{\alpha\beta\lambda\mu}\left(\frac{z_\alpha}{\varepsilon}, z_3\right) , \qquad \tilde{C}^{\alpha\beta\lambda\mu} = C^{\alpha\beta\lambda\mu} - C^{\alpha\beta3i} \tilde{c}_{ij} C^{j3\lambda\mu} , \quad (2.10.27)$$

$$L(\boldsymbol{v}) = \int_{\Gamma^e_\pm} p^\alpha_\pm v_\alpha d\Gamma + \int_{\Gamma^e_\pm} q^\pm v_3 d\Gamma$$

$$= \int_\Omega [(p^\alpha_+ + p^\alpha_-)u_\alpha + (q^+ + q^-)w - (p^\alpha_+ - p^\alpha_-)c\frac{\partial w}{\partial z_\alpha}]d\Omega$$

$$= \int_\Omega [(p^\alpha_+ + p^\alpha_-)u_\alpha + (q^+ + q^-)w + cw\frac{\partial}{\partial z_\alpha}(p^\alpha_+ - p^\alpha_-)]d\Omega , \quad (2.10.28)$$

and $\tilde{\boldsymbol{c}} = (\tilde{c}_{ij})$ is the inverse of the invertible (3×3)-matrix (C^{i3j3}), $\boldsymbol{v} = (u_\alpha - z_3\frac{\partial w}{\partial z_\alpha}, w)$, $u_\alpha \in H^1_0(\Omega)$, $w \in H^2_0(\Omega)$.

Proof. We divide it into several steps.

(i) The sequence of functionals $\{L_e\}_{e>0}$ is continuously convergent. In fact, for any sequence $\{v^e\}_{e>0} \subset L^2(B)^3$ strongly convergent to $v \in L^2(B)^3$ we have $|L_e(v^e) - L(v)| \to 0$ as $e \to 0$. By applying Proposition 1.3.26, we conclude that it suffices to study the Γ-limit of the sequence $J^1_{e\varepsilon}$ defined by

$$J^1_{e\varepsilon}(\boldsymbol{v}) = \int_B j_\varepsilon(z, \mathbb{Q}^e e^z(\boldsymbol{v}))dz , \quad \boldsymbol{v} \in V^0_K(B) . \quad (2.10.29)$$

Since the space $(H^1(B)^3, L^2)$ has a countable base, therefore the Γ-limit of the sequence of functionals $\{J^1_{e\varepsilon}\}_{e>0}$ exists, cf. Th. 1.3.21. We claim that it is given by the functional

$$J^1_\varepsilon(\boldsymbol{v}) = \frac{1}{2}\int_B \tilde{C}^{\alpha\beta\lambda\mu}_\varepsilon(z)e^z_{\alpha\beta}(\boldsymbol{v})e^z_{\lambda\mu}(\boldsymbol{v})dz , \quad \boldsymbol{v} \in V_K(B) . \quad (2.10.30)$$

(ii) It will now be shown that for any $\boldsymbol{v} \in H^1(B)^3 \backslash V_K(B)$ we have

$$J^1_\varepsilon(\boldsymbol{v}) = +\infty . \quad (2.10.31)$$

In fact, for any such $\boldsymbol{v}$, at least one of the strain measures $e^z_{3i}(\boldsymbol{v})$ does not vanish. We may assume that $e^z_{33}(\boldsymbol{v}) \neq 0$; the two remaining cases can be treated similarly.

There exists a sequence $\{v^e\}_{e>0} \subset H^1(B)^3$ such that $v^e \to v$ in $L^2(B)^3$ and $\lim_{e\to 0} J^1_{e\varepsilon}(v^e) = J^1_\varepsilon(\boldsymbol{v})$. On account of (2.10.7), $\| e^z(v^e) \|_{L^2(B,,\mathbf{E}^3_s)} \leq$ const. Consequently, $e^z_{33}(v^e) \rightharpoonup e^z_{33}(\boldsymbol{v})$ weakly in $L^2(B)$ when $e \to 0$ and, by the lower-semicontinuity of the norm

$$0 < \| e^z_{33}(\boldsymbol{v}) \|_{L^2(B)} \leq \lim_{e\to 0} \inf \| e^z(v^e) \|_{L^2(B,\mathbf{E}^3_s)} .$$

Thus we obtain

$$J^1_\varepsilon(\boldsymbol{v}) = \lim_{e\to 0} J^1_{e\varepsilon}(\boldsymbol{v}) \geq \lim_{e\to 0} \| \mathbb{Q}^e e^z(v^e) \|^2_{L^2(B,\mathbf{E}^3_s)} \geq \lim_{e\to 0} \frac{m}{e^2} \| e^z(v^e) \|^2_{L^2(B,\mathbf{E}^3_s)} = +\infty ,$$

where m is a constant appearing in (2.10.7).

(iii) Let $c := C^{-1}$. It means that

$$c^{\varepsilon}_{ijkl}C^{klmn}_{\varepsilon} = \frac{1}{2}(\delta^m_i\delta^n_j + \delta^n_i\delta^m_j) , \qquad (2.10.32)$$

$$C^{ijkl}_{\varepsilon}c^{\varepsilon}_{klmn} = \frac{1}{2}(\delta^i_m\delta^j_n + \delta^i_n\delta^j_m) . \qquad (2.10.33)$$

Now we are in a position to prove that for any $v \in V_K(B)$ we have

$$J^1_{\varepsilon}(v) \le \frac{1}{2}\int_B \tilde{C}^{\alpha\beta\lambda\mu}_{\varepsilon}e^z_{\alpha\beta}(v)e^z_{\lambda\mu}(v)dz . \qquad (2.10.34)$$

Let $v = (u_\alpha - z_3\dfrac{\partial w}{\partial z_\alpha}, w) \in V_K(B)$ and take a sequence $v^e = (u_\alpha - z_3\dfrac{\partial w}{\partial z_\alpha}, w + e^3\xi)$, where $\xi \in H^1(B)$. If $v \in V^0_K(B)$, then ξ is a function from the space $V^0_K(B)$. It is evident that $v^e \to v$ in $L^2(B)^3$ when $e \to 0$. Moreover

$$\mathbf{Q}^e e^z(v^e) = \begin{bmatrix} & & e^2\dfrac{\partial\xi}{\partial z_1} \\ & e^z_{\alpha\beta}(v) & \\ & & e^2\dfrac{\partial\xi}{\partial z_2} \\ e^2\dfrac{\partial\xi}{\partial z_1} & e^2\dfrac{\partial\xi}{\partial z_2} & e\dfrac{\partial\xi}{\partial z_3} \end{bmatrix} .$$

The definition of the Γ-limit yields

$$J^1_{\varepsilon}(v) \le \lim_{e\to 0} J^1_{e\varepsilon}(\mathbf{Q}^e e^z(v^e))$$

$$= \lim_{e\to 0}\ \sup_{\sigma\in C^{\infty}(\bar{B},\mathbf{E}^s_3)}\int_B [(\mathbf{Q}^e e^z(v^e))_{ij}\sigma^{ij} - \frac{1}{2}c^{\varepsilon}_{ijkl}\sigma^{ij}\sigma^{kl}]dz$$

$$= \lim_{e\to 0}\ \sup_{\sigma\in C^{\infty}(\bar{B},\mathbf{E}^s_3)}\int_B [(\mathbf{Q}^e e^z(v^e))_{ij}\sigma^{ij} - \frac{1}{2}c^{\varepsilon}_{ijkl}\sigma^{ij}\sigma^{kl} - \frac{1}{2}c^{\varepsilon}_{i3j3}\sigma^{i3}\sigma^{j3}]dz$$

$$\le \lim_{e\to 0}\ \sup_{\sigma\in C^{\infty}(\bar{B},\mathbf{E}^s_3)}\int_B [\sigma^{\alpha\beta}e^z_{\alpha\beta}(v) + 2e^2\frac{\partial\xi}{\partial z_\alpha}\sigma^{\alpha3}$$

$$+ e\frac{\partial\xi}{\partial z_3}\sigma^{33} - \frac{1}{2}c^{\varepsilon}_{\alpha\beta\lambda\mu}\sigma^{\alpha\beta}\sigma^{\lambda\mu}]dz = \lim_{e\to 0}\ \sup_{(\sigma^{\alpha\beta})\in C^{\infty}(\bar{B},\mathbf{E}^s_2)}\int_B [\sigma^{\alpha\beta}e^z_{\alpha\beta}(v)$$

$$- \frac{1}{2}c^{\varepsilon}_{\alpha\beta\lambda\mu}\sigma^{\alpha\beta}\sigma^{\lambda\mu}]dz + \lim_{e\to 0}\ \sup_{\sigma^{i3}\in C^{\infty}(\bar{B})}\int_B \left(2e^2\frac{\partial\xi}{\partial z_\alpha}\sigma^{\alpha3} + e\frac{\partial\xi}{\partial z_3}\sigma^{33}\right)dz$$

$$= \begin{cases} \dfrac{1}{2}\displaystyle\int_B \tilde{C}^{\alpha\beta\lambda\mu}_{\varepsilon}e^z_{\alpha\beta}(v)e^z_{\lambda\mu}(v)dz & \text{if } \sigma^{i3} = 0 , \\ +\infty & \text{otherwise,} \end{cases} \qquad (2.10.35)$$

where $\tilde{c}^{\varepsilon}_{\alpha\beta\lambda\mu}$ is the inverse of the matrix $\tilde{C}^{\alpha\beta\lambda\mu}_{\varepsilon}$, see step **(v)** below.

(iv) We shall prove the inverse inequality to that specified by (2.10.34). For any $\{v^e\}_{e>0} \subset H^1(B)^3$ we have

$$
\begin{aligned}
J^1_{e\varepsilon}(v^e) &= \sup_{\sigma \in C^\infty(\bar{B}, \mathbf{E}^s_3)} \int_B [(\mathbf{Q}^e e^z(v^e))_{ij}\sigma^{ij} - \frac{1}{2}c^\varepsilon_{ijkl}\sigma^{ij}\sigma^{kl}]dz \\
&\geq \sup_{\substack{\sigma \in C^\infty(\bar{B}, \mathbf{E}^s_3) \\ \sigma^{i3}=0}} \int_B [(\mathbf{Q}^e e^z(v^e))_{ij}\sigma^{ij} - \frac{1}{2}c^\varepsilon_{ijkl}\sigma^{ij}\sigma^{kl}]dz \\
&= \sup_{\substack{\sigma \in C^\infty(\bar{B}, \mathbf{E}^s_3) \\ \sigma^{i3}=0}} \int_B [\sigma^{ij}e^z_{ij}(v^e) - \frac{1}{2}c^\varepsilon_{ijkl}\sigma^{ij}\sigma^{kl}]dz .
\end{aligned}
\tag{2.10.36}
$$

Suppose that $\{v^e\}_{e>0} \subset H^1(B)^3$ is such that $v^e \to v$ in $L^2(B)^3$ strongly when $e \to 0$ and $\lim_{e \to 0} J^1_{e\varepsilon}(v^e) = J^1_\varepsilon(v)$. Then, by using (2.10.36) we infer that

$$
J^1_\varepsilon(v) \geq \sup_{(\sigma^{\alpha\beta}) \in C^\infty(\bar{B}, \mathbf{E}^s_2)} \int_B [\sigma^{\alpha\beta}e^z_{\alpha\beta}(v) - \frac{1}{2}c^\varepsilon_{\alpha\beta\lambda\mu}\sigma^{\alpha\beta}\sigma^{\lambda\mu}]dz .
\tag{2.10.37}
$$

(v) It remains to prove that the matrix $\tilde{C}_\varepsilon$, given by (2.10.27), is the inverse of $(c^\varepsilon_{\alpha\beta\lambda\mu})$, where $(c^\varepsilon_{ijkl}) = (C_\varepsilon)^{-1}$. Firstly, however, let us justify (2.10.27).
For stresses and strains interrelated by the constitutive relationship:

$$
\sigma^{ij} = C^{ijkl}_\varepsilon \epsilon_{kl} ,
$$

and knowing that $\sigma^{i3} = 0$ we get

$$
\sigma_{\alpha\beta} = \tilde{C}^{\alpha\beta\lambda\mu}_\varepsilon \epsilon_{\lambda\mu} ,
\tag{2.10.38}
$$

with $\tilde{C}_\varepsilon$ defined by (2.10.27). The derivation of this formula is similar to the derivation of formula (2.9.26) and hence is omitted here. We observe that (2.10.7) implies that both $(C^{\alpha\beta\lambda\mu})$ and (C^{i3j3}) are invertible matrices.

To prove that $(\tilde{C}^{\alpha\beta\lambda\mu})$ is the inverse of $(c_{\alpha\beta\lambda\mu})$, we calculate:

$$
\begin{aligned}
\tilde{C}^{\alpha\beta\gamma\delta}c_{\gamma\delta\lambda\mu} &= (C^{\alpha\beta\gamma\delta} - C^{\alpha\beta i3}\tilde{c}_{ij}C^{j3\gamma\delta})c_{\gamma\delta\lambda\mu} \\
&= C^{\alpha\beta\gamma\delta}c_{\gamma\delta\lambda\mu} - C^{\alpha\beta i3}\tilde{c}_{ij}(C^{j3kl}c_{kl\lambda\mu} - 2C^{j3\gamma3}c_{\gamma3\lambda\mu} - C^{j333}c_{33\lambda\mu}) \\
&= C^{\alpha\beta\gamma\delta}c_{\gamma\delta\lambda\mu} - \frac{1}{2}C^{\alpha\beta i3}\tilde{c}_{ij}(\delta^j_\lambda\delta^3_\mu + \delta^j_\mu\delta^3_\lambda) + 2C^{\alpha\beta\gamma3}c_{\gamma3\lambda\mu} + C^{\alpha\beta33}c_{33\lambda\mu} \\
&= C^{\alpha\beta ij}c_{ij\lambda\mu} = \frac{1}{2}(\delta^\alpha_\lambda\delta^\beta_\mu + \delta^\alpha_\mu\delta^\beta_\lambda) ,
\end{aligned}
$$

where we have used (2.10.33) and the relations $\tilde{c} = (C^{i3j3})^{-1}$, $\delta^3_\mu = 0$. Similarly we find

$$
c_{\alpha\beta\gamma\delta}\tilde{C}^{\gamma\delta\lambda\mu} = \frac{1}{2}(\delta^\lambda_\alpha\delta^\mu_\beta + \delta^\mu_\alpha\delta^\lambda_\beta) ,
$$

which concludes the proof. $\qquad\qquad\square$

The properties of the matrix (C^{ijkl}) imply that there exists a positive constant M such that for a.e. $\boldsymbol{y} \in \mathcal{Y}$ we have

$$C^{ijkl}\epsilon_{ij}\epsilon_{kl} \leq M \sum_{i,j=1}^{3} \epsilon_{ij}\epsilon_{ij} , \qquad (2.10.39)$$

for each $\epsilon \in \mathbb{E}_s^3$. Both sides of the last inequality represent (strictly) convex functions. Hence the following coercivity condition readily follows, cf. Th. 1.2.7$_{(v)}$

$$c_{ijkl}(\boldsymbol{y})T^{ij}T^{kl} \geq \frac{1}{M} \sum_{i,j=1}^{3} (T^{ij})^2 \qquad (2.10.40)$$

for each $T \in \mathbb{E}_3^s$ and a.e. $\boldsymbol{y} \in \mathcal{Y}$. Thus the matrix $(c_{\alpha\beta\lambda\mu})$ is also coercive:

$$c_{\alpha\beta\lambda\mu}(\boldsymbol{y})T^{\alpha\beta}T^{\lambda\mu} \geq \frac{1}{M} \sum_{\alpha,\beta=1}^{2} (T^{\alpha\beta})^2 , \qquad (2.10.41)$$

for each $(T^{\alpha\beta}) \in \mathbb{E}_2^s$ and a.e. $\boldsymbol{y} \in \mathcal{Y}$. Similarly we obtain

$$c_{ijkl}(\boldsymbol{y})T^{ij}T^{kl} \leq \frac{1}{m} \sum_{i,j=1}^{3} (T^{ij})^2 . \qquad (2.10.42)$$

Hence

$$c_{\alpha\beta\lambda\mu}(\boldsymbol{y})T^{\alpha\beta}T^{\lambda\mu} \leq \frac{1}{m} \sum_{\alpha,\beta=1}^{2} (T^{\alpha\beta})^2 . \qquad (2.10.43)$$

The last relation readily yields

$$\tilde{C}^{\alpha\beta\lambda\mu}(\boldsymbol{y})\epsilon_{\alpha\beta}\epsilon_{\lambda\mu} \geq m \sum_{\alpha,\beta=1}^{2} (\epsilon_{\alpha\beta})^2 , \qquad (2.10.44)$$

for each $(\epsilon_{\alpha\beta}) \in \mathbb{E}_s^2$ and a.e. $\boldsymbol{y} \in \mathcal{Y}$. It is worth noting that m is the constant appearing in (2.10.7).

Problem (P_ε) and its dual

Having determined the Γ-limit of J_ε ($\varepsilon > 0$) we find it useful to study its mechanical properties. Towards this end we formulate the following minimization problem.
Problem (P_ε) ($\varepsilon > 0$ and fixed)

$$\left| \begin{aligned} &\text{Find} \\ &J_\varepsilon(\boldsymbol{v}^\varepsilon) = \inf\{J_\varepsilon^1(\boldsymbol{v}) - L(\boldsymbol{v}) \mid \boldsymbol{v} \in V_K^0(B)\} . \end{aligned} \right. \qquad (2.10.45)$$

The property (2.10.44) and the continuity of the functional L ensure that the minimizer $\boldsymbol{v}^\varepsilon = (u_\alpha^\varepsilon - z_3 \partial w^\varepsilon / \partial z_\alpha, w^\varepsilon)$ exists, where $u_\alpha^\varepsilon \in H_0^1(\Omega)$ and $w^\varepsilon \in H_0^2(\Omega)$. Then we have

$$J_\varepsilon^1(\boldsymbol{v}^\varepsilon) = \frac{1}{2}\int_B \tilde{C}_\varepsilon^{\alpha\beta\lambda\mu}(\boldsymbol{z})[e_{\alpha\beta}^z(\boldsymbol{u}^\varepsilon) + z_3\kappa_{\alpha\beta}^z(w^\varepsilon)][e_{\lambda\mu}^z(\boldsymbol{u}^\varepsilon) + z_3\kappa_{\lambda\mu}^z(w^\varepsilon)]d\boldsymbol{z}$$

$$= \frac{1}{2}\int_\Omega [\hat{N}_\varepsilon^{\alpha\beta}e_{\alpha\beta}^z(\boldsymbol{u}^\varepsilon) + \hat{M}_\varepsilon^{\alpha\beta}\kappa_{\alpha\beta}^z(w^\varepsilon)]dz \, ,$$

or

$$J_\varepsilon^1(\boldsymbol{v}^\varepsilon) = \int_\Omega \hat{\mathcal{W}}_\varepsilon[z_\alpha, z_3, \boldsymbol{e}^z(\boldsymbol{u}^\varepsilon), \boldsymbol{\kappa}^z(w^\varepsilon)]dz \, , \tag{2.10.46}$$

where $\boldsymbol{u}^\varepsilon = (u_\alpha^\varepsilon)$, $\kappa_{\alpha\beta}^z(w) = -\partial^2 w/\partial z_\alpha \partial z_\beta$. Moreover

$$\hat{N}_\varepsilon^{\alpha\beta} = \hat{A}_\varepsilon^{\alpha\beta\lambda\mu}e_{\lambda\mu}^z(\boldsymbol{u}^\varepsilon) + \hat{E}_\varepsilon^{\alpha\beta\lambda\mu}\kappa_{\lambda\mu}^z(w^\varepsilon) \, , \tag{2.10.47}$$

$$\hat{M}_\varepsilon^{\alpha\beta} = \hat{F}_\varepsilon^{\alpha\beta\lambda\mu}e_{\lambda\mu}^z(\boldsymbol{u}^\varepsilon) + \hat{D}_\varepsilon^{\alpha\beta\lambda\mu}\kappa_{\lambda\mu}^z(w^\varepsilon) \, , \tag{2.10.48}$$

where $\hat{F}_\varepsilon^{\alpha\beta\lambda\mu} = \hat{E}_\varepsilon^{\lambda\mu\alpha\beta}$ and

$$\hat{X}_\varepsilon^{\alpha\beta\lambda\mu}(z) = \hat{X}^{\alpha\beta\lambda\mu}(z/\varepsilon) \, , \quad \hat{\boldsymbol{X}} \in \{\hat{\boldsymbol{A}}, \hat{\boldsymbol{E}}, \hat{\boldsymbol{F}}, \hat{\boldsymbol{D}}\} \, ,$$

with

$$\begin{bmatrix} \hat{A}_\varepsilon^{\alpha\beta\lambda\mu} \\ \hat{E}_\varepsilon^{\alpha\beta\lambda\mu} \\ \hat{D}_\varepsilon^{\alpha\beta\lambda\mu} \end{bmatrix} (z_1, z_2) = \int_{-c}^{c} \begin{bmatrix} 1 \\ z_3 \\ z_3^2 \end{bmatrix} \tilde{C}_\varepsilon^{\alpha\beta\lambda\mu}(\boldsymbol{z})dz_3 \, . \tag{2.10.49}$$

Note that $\hat{\boldsymbol{A}}$, $\hat{\boldsymbol{E}}$, $\hat{\boldsymbol{D}}$ have already appeared, see (2.7.13); moreover

$$e_{\lambda\mu}(\boldsymbol{u}^\varepsilon) = e_{\lambda\mu}^z(\boldsymbol{u}^\varepsilon) \, , \quad \kappa_{\lambda\mu}(w^\varepsilon) = \kappa_{\lambda\mu}^z(w^\varepsilon) \, .$$

The potential $\hat{\mathcal{W}}_\varepsilon$ has the form

$$\hat{\mathcal{W}}_\varepsilon(z_\alpha, \boldsymbol{\epsilon}, \boldsymbol{\kappa}) = \frac{1}{2}(\hat{A}_\varepsilon^{\alpha\beta\lambda\mu}\epsilon_{\alpha\beta}\epsilon_{\lambda\mu} + 2\hat{E}_\varepsilon^{\alpha\beta\lambda\mu}\epsilon_{\alpha\beta}\kappa_{\lambda\mu} + \hat{D}_\varepsilon^{\alpha\beta\lambda\mu}\kappa_{\alpha\beta}\kappa_{\lambda\mu}) \, ; \quad \boldsymbol{\epsilon}, \boldsymbol{\kappa} \in \mathbb{E}_s^2 \, . \tag{2.10.50}$$

Properties of the matrices $\hat{\boldsymbol{A}}, \hat{\boldsymbol{E}}, \hat{\boldsymbol{F}}, \hat{\boldsymbol{D}}$

(i) $\hat{\boldsymbol{E}} = \hat{\boldsymbol{F}}^T$, where "T" stands for the transpose of a matrix.

(ii) Their components belong to $L^\infty(\Omega)$.

(iii) $\hat{X}^{\alpha\beta\lambda\mu} = \hat{X}^{\beta\alpha\lambda\mu} = \hat{X}^{\lambda\mu\alpha\beta}$, $\hat{\boldsymbol{X}} \in \{\hat{\boldsymbol{A}}, \hat{\boldsymbol{E}}, \hat{\boldsymbol{F}}, \hat{\boldsymbol{D}}\}$.

(iv)

$$\hat{A}^{\alpha\beta\lambda\mu}\epsilon_{\alpha\beta}\epsilon_{\lambda\mu} + \hat{E}^{\alpha\beta\lambda\mu}(\epsilon_{\alpha\beta}\rho_{\lambda\mu} + \epsilon_{\lambda\mu}\rho_{\alpha\beta})$$

$$+ \hat{D}^{\alpha\beta\lambda\mu}\rho_{\alpha\beta}\rho_{\lambda\mu} \geq \frac{2}{3}m \sum_{\alpha,\beta=1}^{2} (\epsilon_{\alpha\beta}\epsilon_{\alpha\beta} + \rho_{\alpha\beta}\rho_{\alpha\beta}) \, , \tag{2.10.51}$$

provided that $c = 1$; $\boldsymbol{\epsilon}, \boldsymbol{\rho} \in \mathbb{E}_s^2$ and m is the constant.

Proof. The first three properties are evident. To prove the last one, we calculate

$$\int_{-c}^{c} \tilde{C}^{\alpha\beta\lambda\mu}(\epsilon_{\alpha\beta} + z_3\rho_{\alpha\beta})(\epsilon_{\lambda\mu} + z_3\rho_{\lambda\mu})dz_3 \geq m \sum_{\alpha,\beta=1}^{2} \int_{-c}^{c} (\epsilon_{\alpha\beta} + z_3\rho_{\alpha\beta})^2 dz_3$$

$$= 2mc \sum_{\alpha,\beta=1}^{2} (\epsilon_{\alpha\beta})^2 + \frac{2}{3}mc^3 \sum_{\alpha,\beta=1}^{2} (\rho_{\alpha\beta})^2 \ .$$

For $c = 1$ one obtains **(iv)**. $\square$

It is instructive now to derive the dual problem $(P_\varepsilon)^*$ or the maximum principle of the total complementary energy. Firstly, however, we shall formulate the dual problem $(P_{e\varepsilon})^*$ of $(P_{e\varepsilon})$. To apply Rockafellar's theory of duality presented in Section 1.2.5, we set

$$\Lambda v = \mathbf{Q}^e e^z(v) \ , \quad v \in V_0(B) \ .$$

For $p^* \in L^2(B, \mathbb{E}_s^3)$ we find

$$\langle \Lambda v, p^* \rangle_{L^2 \times L^2} = \int_B (\mathbf{Q}^e e^z(v))_{ij} p^{*ij} dz$$

$$= \int_B [p^{*\alpha\beta} e_{\alpha\beta}^z(v) + \frac{2}{e} p^{*\alpha 3} e_{\alpha 3}^z(v) + \frac{1}{e^2} p^{*33} e_{33}^z(v)] dz$$

$$= \int_B (\mathbf{Q}^e p^*)^{ij} e_{ij}^z(v) dz = -\int_B (\mathbf{Q}^e p^*)^{ij}{}_{;j} v_i dz$$

$$+ \int_{\Gamma_\pm} (\mathbf{Q}^e p^*)^{ij} n_j v_i d\Gamma = \langle \Lambda^* p^*, v \rangle_{V_0^*(B) \times V_0(B)} \ . \qquad (2.10.52)$$

We denote: $(\ )_{;j} = \partial(\ )/\partial z_j$. The last relation yields

$$\Lambda^* p^* = \begin{cases} -div\ \mathbf{Q}^e p^* & \text{in } B \ , \\ (\mathbf{Q}^e p^*) n & \text{on } \Gamma_\pm \ . \end{cases} \qquad (2.10.53)$$

We assume the following notation: $\mathbf{Q}^{-e} = (\mathbf{Q}^e)^{-1}$; hence $\mathbf{Q}^{-e}\mathbf{Q}^e = \mathbf{I}$, where $\mathbf{I}$ is the identity matrix.

Let us set

$$j_{e\varepsilon}(z, \epsilon) = j_\varepsilon(z, \mathbf{Q}^e\epsilon) = j\left(\frac{z_\alpha}{\varepsilon}, z_3, \mathbf{Q}^e\epsilon\right) \ , \quad \epsilon \in \mathbb{E}_s^3 \ . \qquad (2.10.54)$$

For $\epsilon^* \in \mathbb{E}_3^s$ we calculate

$$j_{e\varepsilon}^*(z, \epsilon^*) = \sup\{\epsilon^* : \epsilon - j_{e\varepsilon}(z, \epsilon) \mid \epsilon \in \mathbb{E}_s^3\}$$

$$= \sup\{\epsilon^* : (\mathbf{Q}^{-e}\mathbf{Q}^e\epsilon) - j_\varepsilon(z, \mathbf{Q}^e\epsilon) \mid \epsilon \in \mathbb{E}_s^3\}$$

$$= \sup\{((\mathbf{Q}^{-e})^T\epsilon^*) : \mathbf{Q}^e\epsilon - j_\varepsilon(z, \mathbf{Q}^e\epsilon) \mid \epsilon \in \mathbb{E}_s^3\}$$

$$= \sup\{(\mathbf{Q}^{-e}\epsilon^*) : \epsilon - j_\varepsilon(z, \epsilon) \mid \epsilon \in \mathbb{E}_s^3\}$$

$$= j_\varepsilon^*(z, \mathbf{Q}^{-e}\epsilon^*) = \frac{1}{2}c_{ijkl}^\varepsilon(\mathbf{Q}^{-e}\epsilon^*)^{ij}(\mathbf{Q}^{-e}\epsilon^*)^{kl} , \qquad (2.10.55)$$

because $(\mathbf{Q}^{-e})^T = \mathbf{Q}^{-e}$ and $c^\varepsilon = C_\varepsilon^{-1}$. One can easily verify that

$$(\mathbf{Q}^{-e}\epsilon^*)^{\alpha\beta} = \epsilon^{*\alpha\beta} , \quad (\mathbf{Q}^{-e}\epsilon^*)^{\alpha 3} = e\epsilon^{*\alpha 3} , \quad (\mathbf{Q}^{-e}\epsilon^*)^{33} = e^2\epsilon^{*33} . \qquad (2.10.56)$$

Now we must find the conjugate functional of $(-L_e)$. By taking into account (2.10.53), one obtains

$$(-L_e)^*(\sigma) = \sup\{\langle -\Lambda^*\sigma, v\rangle + L_e(v) \mid v \in V_0(B)\}$$

$$= \sup\{\langle \operatorname{div} \mathbf{Q}^e\sigma, v\rangle - \int_{\Gamma_\pm} (\mathbf{Q}^e\sigma)^{ij} n_j v_i d\Gamma + L_e(v) \mid v \in V_0(B)\}$$

$$= \sup\{\int_B [(\mathbf{Q}^e\sigma)^{\alpha j}{}_{;j} v_\alpha + (\mathbf{Q}^e\sigma)^{3j}{}_{;j} v_3 + e^2 b^\alpha v_\alpha + e b^3 v_3] dz$$

$$- \int_{\Gamma_\pm} [(\mathbf{Q}^e\sigma)^{\alpha 3} v_\alpha + (\mathbf{Q}^e\sigma)^{33} v_3] d\Gamma + \int_{\Gamma_\pm} (p_\pm^\alpha v_\alpha + q^\pm v_3) d\Gamma \mid v \in V_0(B)\}$$

$$= \begin{cases} 0 & \text{if } \begin{cases} \sigma^{\alpha\beta}{}_{;\beta} + \dfrac{1}{e}\sigma^{\alpha 3}{}_{;3} + e^2 b^\alpha = 0 \ \text{in } B , & (2.10.57) \\[2mm] \dfrac{1}{e}\sigma^{3\alpha}{}_{;\alpha} + \dfrac{1}{e^2}\sigma^{33}{}_{;3} + e b^3 = 0 \ \text{in } B , & (2.10.58) \\[2mm] \dfrac{1}{e}\sigma^{\alpha 3} = p_\pm^\alpha , \ \dfrac{1}{e^2}\sigma^{33} = q^\pm \ \text{on } \Gamma_\pm , & (2.10.59) \end{cases} \\[2mm] +\infty & \text{otherwise} , \end{cases}$$

because $n = (0, 0, \pm 1)$ on $\Gamma_\pm$. Multiplying Eq. (2.10.58) by $e > 0$ and next adding to Eq. (2.10.58) we get

$$\sigma^{i\alpha}{}_{;\alpha} + \frac{1}{e}\sigma^{i3}{}_{;3} + e^2 b^i = 0 \qquad \text{in } B . \qquad (2.10.60)$$

Let us set

$$\mathcal{S}_e = \{\sigma \in L^2(B, \mathbb{E}_3^s) \mid \operatorname{div} \sigma \in L^2(B)^3$$

$$\text{and Eqs. } (2.10.60), (2.10.59) \text{ are satisfied}\} , \qquad (2.10.61)$$

$$\mathcal{G}_{e\varepsilon}(\sigma) = -\int_B j_\varepsilon^*(z, \mathbf{Q}^{-e}\sigma) dz - I_{\mathcal{S}_e}(\sigma) . \qquad (2.10.62)$$

Now we are in a position to formulate the dual problem $(P_{e\varepsilon})^*$ of $(P_{e\varepsilon})$ and exhibit their interrelationship.

Problem $(P_{e\varepsilon})^*$

$$\left|\begin{array}{l} \text{Find} \\ \mathcal{G}_{e\varepsilon}(\boldsymbol{\sigma}^{e\varepsilon}) = \sup\{\mathcal{G}_{e\varepsilon}(\boldsymbol{\sigma}) \mid \boldsymbol{\sigma} \in L^2(B, \mathbb{E}_3^s)\} \, . \end{array}\right.$$

Corollary 2.10.3. The solution $\boldsymbol{\sigma}^{e\varepsilon}$ of the dual problem $(P_{e\varepsilon})^*$ exists and is unique. Moreover, we have

$$\inf(P_{e\varepsilon})^* = \min(P_{e\varepsilon})^* = \sup(P_{e\varepsilon})^* = \max(P_{e\varepsilon})^* \, . \tag{2.10.63}$$

Proof. The properties of the complementary energy j_ε^* and the regularity of (b^i) and $(p_\pm^\alpha, q^\pm)$ imply the existence and uniqueness of $\boldsymbol{\sigma}^{e\varepsilon}$. The duality relation (2.10.63) results then from Proposition 1.2.49. $\qquad\qquad\qquad\square$

We proceed to the derivation of the dual problem $(P_\varepsilon)^*$. Now the operator Λ involved in the theory of duality (Sec. 1.2.5) has the form

$$\Lambda(\boldsymbol{u}, w) = (\Lambda_1 \boldsymbol{u}, \Lambda_2 w) = (e(\boldsymbol{u}), \kappa(w)) \, , \tag{2.10.64}$$

where $\boldsymbol{u} \in H_0^1(\Omega)$, $w \in H_0^2(\Omega)$. Proceeding similarly to the derivation of (2.10.53) one obtains $\Lambda^* = (\Lambda_1^*, \Lambda_2^*)$, where

$$\Lambda_1^* \boldsymbol{N} = -\text{div}\boldsymbol{N} \text{ in } \Omega \, , \tag{2.10.65}$$

$$\Lambda_2^* \boldsymbol{M} = -\text{div div}\boldsymbol{M} \text{ in } \Omega \, . \tag{2.10.66}$$

Here $\boldsymbol{N}, \boldsymbol{M} \in L^2(\Omega, \mathbb{E}_2^s)$. Next one finds

$$\hat{\mathcal{W}}_\varepsilon^*(z_\alpha, \epsilon^*, \rho^*) = \sup\{\epsilon^* : \epsilon + \rho^* : \rho - \hat{\mathcal{W}}_\varepsilon(z_\alpha, \epsilon, \rho) \mid \epsilon, \rho \in \mathbb{E}_s^2\}$$
$$= \frac{1}{2}(\hat{a}_{\alpha\beta\lambda\mu}^\varepsilon \epsilon^{*\alpha\beta} \epsilon^{*\lambda\mu} + 2\hat{e}_{\alpha\beta\lambda\mu}^\varepsilon \epsilon^{*\alpha\beta} \rho^{*\lambda\mu} + \hat{d}_{\alpha\beta\lambda\mu}^\varepsilon \rho^{*\alpha\beta} \rho^{*\lambda\mu}) \, , \tag{2.10.67}$$

where

$$\hat{\boldsymbol{b}}^\varepsilon(z_\alpha) = \begin{bmatrix} \hat{\boldsymbol{a}}^\varepsilon & \hat{\boldsymbol{e}}^\varepsilon \\ \hat{\boldsymbol{f}}^\varepsilon & \hat{\boldsymbol{d}}^\varepsilon \end{bmatrix} = \hat{\boldsymbol{\mathcal{B}}}_\varepsilon^{-1}(z_\alpha) \, , \qquad \hat{e}_{\alpha\beta\lambda\mu}^\varepsilon = \hat{f}_{\lambda\mu\alpha\beta}^\varepsilon \, , \tag{2.10.68}$$

and

$$\hat{\boldsymbol{\mathcal{B}}}_\varepsilon(z_\alpha) = \begin{bmatrix} \hat{\boldsymbol{A}}_\varepsilon & \hat{\boldsymbol{E}}_\varepsilon \\ \hat{\boldsymbol{F}}_\varepsilon & \hat{\boldsymbol{D}}_\varepsilon \end{bmatrix} \, . \tag{2.10.69}$$

Further, we have:

$$(-L)^*(-\Lambda^*(\boldsymbol{N}, \boldsymbol{M})) = \sup\{\langle -\Lambda_1^* \boldsymbol{N}, \boldsymbol{u}\rangle_{H^{-1}(\Omega) \times H_0^1(\Omega)}$$
$$+ \langle -\Lambda_2^* \boldsymbol{M}, w\rangle_{H^{-2}(\Omega) \times H_0^2(\Omega)} + L(\boldsymbol{u}, w) \mid (\boldsymbol{u}, w) \in H_0^1(\Omega)^2 \times H_0^2(\Omega)\}$$

$$= \begin{cases} 0 & \text{if} \begin{cases} \text{div}\boldsymbol{N} + \boldsymbol{p}_+ + \boldsymbol{p}_- = 0 & \text{in } \mathbf{D}'(\Omega), \tag{2.10.70} \\ \text{div div}\boldsymbol{M} + c\,\text{div}(\boldsymbol{p}_+ - \boldsymbol{p}_-) + (q^+ + q^-) = 0 \text{ in } \mathbf{D}'(\Omega), \tag{2.10.71} \end{cases} \\ +\infty, \quad \text{otherwise} \, . \end{cases}$$

Here $p_\pm = (p_\pm^\alpha)$. We set

$$\mathcal{S} = \{(N, M) \in [L^2(\Omega, \mathbb{E}_S)]^2 \mid \mathrm{div}N \in L^2(\Omega)^2 , \mathrm{div\,div}M \in L^2(\Omega) ,$$
$$\text{and Eqs } (2.10.70), (2.10.71) \text{ are satisfied}\}. \tag{2.10.72}$$

We formulate the dual problem.

Problem $(P_\varepsilon)^*$

Find

$$\sup\{-\int_\Omega \hat{\mathcal{W}}_\varepsilon^*[z_\alpha, N(z), M(z)]dz - I_S(N, M) \mid N, M \in L^2(\Omega, \mathbb{E}_2^s)\} .$$

By applying the results presented in Section 1.2.5 we conclude that solution $(\hat{N}_\varepsilon, \hat{M}_\varepsilon) \in \mathcal{S}$ of (P_ε^*) exists and is unique. Further, the minimizers $(u^\varepsilon, w^\varepsilon) \in H_0^1(\Omega)^2 \times H_0^2(\Omega)$ and $(\hat{N}_\varepsilon, \hat{M}_\varepsilon)$ are interrelated by the extremality condition, which now takes the form of Eqs. (2.10.47), (2.10.48). The second extremality condition means that in this case $\hat{N}_\varepsilon$ satisfies Eq. (2.10.70) while $\hat{M}_\varepsilon$ satisfies Eq. (2.10.71).

Remark 2.10.4. For $(\sigma_{e\varepsilon}^{ij})$ the weak form of Eqs. (2.10.57) – (2.10.59) is given by

$$\forall v \in V_0(B) , \quad \int_B \sigma_{e\varepsilon}^{i\alpha} \frac{\partial v_i}{\partial z_\alpha}dz + \frac{1}{e}\int_B \sigma_{e\varepsilon}^{i3} \frac{\partial v_i}{\partial z_3}dz$$
$$= e^2 \int_B b^i v_i dz + \int_{\Gamma_\pm} p_\pm^\alpha v_\alpha d\Gamma + e \int_{\Gamma_\pm} q^\pm v_3 d\Gamma . \tag{2.10.73}$$

Caillerie (1984) has shown that by a proper choice of test functions in the last equation, after the passage to zero with e, the equilibrium equations assume the following form:

$$\mathrm{div}\hat{N}_\varepsilon + p_+ + p_- = 0 \quad \text{in } \Omega ,$$
$$\mathrm{div}\hat{M}_\varepsilon - \hat{Q}_\varepsilon + c(p_+ - p_-) = 0 \quad \text{in } \Omega , \tag{2.10.74}$$
$$\mathrm{div}\hat{Q}_\varepsilon + (q^+ + q^-) = 0 \quad \text{in } \Omega .$$

Note that Eqs. (2.10.70), (2.10.71) are formally equivalent to Eqs. (2.10.74).

Let us set

$$N_{e\varepsilon}^{\alpha\beta} = \int_{-c}^{c} \sigma_{e\varepsilon}^{\alpha\beta}dz_3 ; \quad \alpha, \beta = 1, 2,$$

$$M_{e\varepsilon}^{\alpha\beta} = \int_{-c}^{c} z_3 \sigma_{e\varepsilon}^{\alpha\beta}dz_3 , \qquad Q_{e\varepsilon}^{\alpha} = \frac{1}{e}\int_{-c}^{c} \sigma_{e\varepsilon}^{\alpha3}dz_3 , \tag{2.10.75}$$

where $(\sigma_{e\varepsilon}^{ij})$ is the solution of $(P_{e\varepsilon})^*$. Caillerie (1984) proves that for subsequences of $N_{e\varepsilon}$, $M_{e\varepsilon}$ and $Q_{e\varepsilon}$, still denoted by the same subscripts, one has

$$N_{e\varepsilon}^{\alpha\beta} \rightharpoonup \hat{N}_{\varepsilon}^{\alpha\beta} \qquad \text{weakly in } L^2(\Omega) \,,$$

$$M_{e\varepsilon}^{\alpha\beta} \rightharpoonup \hat{M}_{\varepsilon}^{\alpha\beta} \qquad \text{weakly in } L^2(\Omega) \,, \qquad (2.10.76)$$

$$Q_{e\varepsilon}^{\alpha} \rightharpoonup \hat{Q}_{\varepsilon}^{\alpha} \qquad \text{weak-} * \text{ in } H^{-1}(\Omega) \,,$$

when $e \to 0$. A similar convergence takes place when e and ε tend to zero. Then the limits in (2.10.76) do not depend on ε. $\qquad\qquad\square$

Now we are in a position to formulate the dual Γ-convergence theorem.
Theorem 2.10.5. For a fixed $\varepsilon > 0$ the sequence of functionals:

$$\mathcal{G}_{e\varepsilon}(\boldsymbol{\sigma}) = -\int_B j_\varepsilon^*(z, \mathbf{Q}^{-e}\boldsymbol{\sigma})dz - I_{S_e}(\boldsymbol{\sigma}) \,, \qquad (2.10.77)$$

Γ-converges in the weak topology of $L^2(B, \mathbb{E}_s^3)$, when $e \to 0$, to the functional

$$\mathcal{G}_\varepsilon(\boldsymbol{N}, \boldsymbol{M}) = -\int_\Omega \mathcal{W}_\varepsilon^*[z_\alpha, \boldsymbol{N}(z), \boldsymbol{M}(z)]dz - I_S(\boldsymbol{N}, \boldsymbol{M}) \,, \qquad (2.10.78)$$

where S is defined by (2.10.72). Moreover, for convergent subsequences we have:
(i)

$$\sigma_{e\varepsilon}^{ij} \rightharpoonup \sigma_\varepsilon^{ij} \qquad \text{weakly in } L^2(B) \,,$$

$$N_{e\varepsilon}^{\alpha\beta} \rightharpoonup \hat{N}_\varepsilon^{\alpha\beta} = \int_{-c}^c \sigma_\varepsilon^{\alpha\beta} dz_3 \qquad \text{weakly in } L^2(\Omega) \,,$$

$$M_{e\varepsilon}^{\alpha\beta} \rightharpoonup \hat{M}_\varepsilon^{\alpha\beta} = \int_{-c}^c z_3\sigma_\varepsilon^{\alpha\beta} dz_3 \qquad \text{weakly in } L^2(\Omega) \,,$$

when $e \to 0$, where $\sigma_\varepsilon^{i3} = 0$ for $i = 1, 2, 3$.
(ii)

$$\inf P_{e\varepsilon} \to \inf P_\varepsilon \,, \qquad \sup P_{e\varepsilon}^* \to \sup P_\varepsilon^* \,,$$

when $e \to 0$.
Proof. The proof is based on Theorem 1.3.36. It is sufficient to show that for a fixed $\varepsilon > 0$ the sequence of functionals:

$$J_{e\varepsilon}^{(2)}(\boldsymbol{v}, \boldsymbol{p}) = \int_B j_\varepsilon[z, \mathbf{Q}^e(e^z(\boldsymbol{v}) + \boldsymbol{p})]dz \,, \quad \boldsymbol{v} \in H^1(B)^3 \,, \quad \boldsymbol{p} \in L^2(B, \mathbb{E}_s^3), \qquad (2.10.79)$$

is Γ-convergent in the topology $(s - L^2(B)^3) \times (s - L^2(B, \mathbb{E}_s^3))$ to

$$J_\varepsilon^{(2)}(\boldsymbol{v}, \boldsymbol{p}) = \frac{1}{2} \int\limits_B \tilde{C}_\varepsilon^{\alpha\beta\lambda\mu}(\boldsymbol{z})[e_{\alpha\beta}^z(\boldsymbol{v}) + p_{\alpha\beta}]\,[e_{\lambda\mu}^z(\boldsymbol{v}) + p_{\lambda\mu}]d\boldsymbol{z} \ , \quad \boldsymbol{v} \in V_K(B) \ ,$$

where $p_{i3} = 0$. Details are left to the reader, cf. also Sec. 5.5. $\hfill \square$

Homogenization: $\varepsilon \to 0$

The elastic moduli specified by (2.10.49) are εY-periodic. In order to derive an effective plate model one has to perform homogenization; according to our procedure a Γ-limit has to be found when $\varepsilon \to 0$. The loading functional L, given by (2.10.28), still plays the role of a perturbation functional.

Effective elastic potential and its properties

Let us denote by $\hat{\mathcal{W}}_h$ the elastic potential describing the effective plate behavior. By applying the general homogenization theorem, namely Theorem 1.3.28 one has

$$\hat{\mathcal{W}}_h(\boldsymbol{\epsilon}, \boldsymbol{\rho}) = \inf\{\langle \hat{\mathcal{W}}(y_\alpha, e^y(\boldsymbol{v}) + \boldsymbol{\epsilon}, \boldsymbol{\kappa}^y(v) + \boldsymbol{\rho}) \rangle \mid (\boldsymbol{v}, v) \in \tilde{H}_{K,per}(Y)\} \ , \quad (2.10.80)$$

where $\boldsymbol{\epsilon}, \boldsymbol{\rho} \in \mathbb{E}_s^2$; $e_{\alpha\beta}^y(\boldsymbol{v})$, $\kappa_{\alpha\beta}^y(v)$ are defined by

$$e_{\alpha\beta}^y(\boldsymbol{v}) = (v_{\alpha|\beta} + v_{\beta|\alpha})/2 \ , \quad \kappa_{\alpha\beta}^y(v) = -v_{|\alpha\beta} \ ,$$

$(\ )_{|\alpha} = \partial/\partial y_\alpha$, and $\tilde{H}_{K,per}(Y)$ is defined in Section 2.8; moreover

$$\hat{\mathcal{W}}(y_\sigma, \boldsymbol{\epsilon}, \boldsymbol{\rho}) = \frac{1}{2}\hat{A}^{\alpha\beta\lambda\mu}(y_\sigma)\epsilon_{\alpha\beta}\epsilon_{\lambda\mu} + \hat{E}^{\alpha\beta\lambda\mu}(y_\sigma)\epsilon_{\alpha\beta}\rho_{\lambda\mu} + \frac{1}{2}\hat{D}^{\alpha\beta\lambda\mu}(y_\sigma)\rho_{\alpha\beta}\rho_{\lambda\mu} \ .$$

Properties of $\hat{\mathcal{W}}_h$

(i)

$$\hat{\mathcal{W}}_h(\boldsymbol{\epsilon}, \boldsymbol{\rho}) \le \langle \hat{\mathcal{W}}(y_\alpha, \boldsymbol{\epsilon}, \boldsymbol{\rho}) \rangle \le M_1 \sum_{\alpha,\beta=1}^{2} [(\epsilon_{\alpha\beta})^2 + (\rho_{\alpha\beta})^2] \ , \qquad (2.10.81)$$

for each $\boldsymbol{\epsilon}, \boldsymbol{\rho} \in \mathbb{E}_s^2$ and $c = 1$. Here M_1 is a positive constant. To prove this property it suffices to take $\boldsymbol{v} = \boldsymbol{0}$ and $v = 0$ in (2.10.80).

(ii)

$$\hat{\mathcal{W}}_h(\boldsymbol{\epsilon}, \boldsymbol{\rho}) \ge \frac{m}{3|Y|} \sum_{\alpha,\beta=1}^{2} [(\epsilon_{\alpha\beta})^2 + (\rho_{\alpha\beta})^2] \ , \qquad (2.10.82)$$

for each $\boldsymbol{\epsilon}, \boldsymbol{\rho} \in \mathbb{E}_s^2$ and $c = 1$. In fact, by taking into account (2.10.51) one obtains

$$\hat{\mathcal{W}}_h(\boldsymbol{\epsilon}, \boldsymbol{\rho}) = \langle \hat{\mathcal{W}}(y_\alpha, e^y(\bar{\boldsymbol{v}}) + \boldsymbol{\epsilon}, \boldsymbol{\kappa}^y(\bar{v}) + \boldsymbol{\rho}) \rangle$$

$$\ge \frac{m}{3}\langle (\ |e^y(\bar{\boldsymbol{v}}) + \boldsymbol{\epsilon}|^2 + |\boldsymbol{\kappa}^y(\bar{v}) + \boldsymbol{\rho}|^2) \rangle \ge \frac{m}{3|Y|}(|\boldsymbol{\epsilon}|^2 + |\boldsymbol{\rho}|^2) \ , \quad \text{for } c = 1 \ ,$$

because

$$\sum_{\alpha,\beta=1}^{2} \langle \epsilon_{\alpha\beta} e_{\alpha\beta}^{y}(\bar{\boldsymbol{v}}) \rangle = 0 \,, \qquad \sum_{\alpha\beta=1}^{2} \langle \rho_{\alpha\beta} \kappa_{\alpha\beta}^{y}(\bar{v}) \rangle = 0 \,.$$

By $(\bar{\boldsymbol{v}}, \bar{v})$ we have denoted a minimizer of the functional appearing on the r.h.s. of (2.10.80). Let us discuss now the problem of existence of $(\bar{\boldsymbol{v}}, \bar{v})$. To this end we write

$$\hat{\mathcal{W}}_h(\boldsymbol{\epsilon}, \boldsymbol{\rho}) = \inf\{J_Y(\boldsymbol{v}, v) + \ell(\boldsymbol{v}, v) + m_1 \mid (\boldsymbol{v}, v) \in \tilde{H}_{K,per}\}$$

where

$$J_Y(\boldsymbol{v}, v) = \langle \hat{\mathcal{W}}(y_\alpha, \boldsymbol{\epsilon}^y(\boldsymbol{v}), \boldsymbol{\kappa}^y(v)) \rangle \,,$$

$$\ell(\boldsymbol{v}, v) = \frac{1}{2} \langle 2\hat{A}^{\alpha\beta\lambda\mu} e_{\alpha\beta}^{y}(\boldsymbol{v}) \epsilon_{\lambda\mu} + \hat{E}^{\alpha\beta\lambda\mu} \epsilon_{\alpha\beta} \kappa_{\lambda\mu}^{y}(v)$$

$$+ \hat{E}^{\alpha\beta\lambda\mu} e_{\alpha\beta}^{y}(\boldsymbol{v}) \rho_{\lambda\mu} + \hat{F}^{\alpha\beta\lambda\mu} \rho_{\alpha\beta} e_{\lambda\mu}^{y}(\boldsymbol{v}) + \hat{F}^{\alpha\beta\lambda\mu} \kappa_{\alpha\beta}^{y}(v) \epsilon_{\lambda\mu} + 2\hat{D}^{\alpha\beta\lambda\mu} \rho_{\alpha\beta} \kappa_{\lambda\mu}^{y}(v) \rangle \,,$$

$$m_1 = \frac{1}{2} \langle \hat{A}^{\alpha\beta\lambda\mu} \epsilon_{\alpha\beta} \epsilon_{\lambda\mu} + \hat{E}^{\alpha\beta\lambda\mu} \epsilon_{\alpha\beta} \rho_{\lambda\mu} + \hat{F}^{\alpha\beta\lambda\mu} \rho_{\alpha\beta} \epsilon_{\lambda\mu} + \hat{D}^{\alpha\beta\lambda\mu} \rho_{\alpha\beta} \rho_{\lambda\mu} \rangle \,.$$

The functional J_Y is strictly convex and coercive on the space $\tilde{H}_{K,per}(Y)$. In fact by using (2.10.82) we easily obtain

$$J_Y(\boldsymbol{v}, v) \geq K(\| \boldsymbol{v} \|_{H^1(Y)^2}^2 + \| v \|_{H^2(Y)}^2) \,, \quad (\boldsymbol{v}, v) \in \tilde{H}_{K,per}(Y) \,.$$

Here K is a positive constant. In the space $H^1(Y)^2$, the set of rigid body motion is given by

$$\mathcal{R} = \{\boldsymbol{v} \mid v_\alpha(y_1, y_2) = a_{\alpha\beta} y_\beta + K_\alpha\} \,, \tag{2.10.83}$$

where $(a_{\alpha\beta})$ is a skew-symmetric matrix and $(K_\alpha) \in \mathbb{R}^2$. To satisfy the periodicity requirement, $a_{\alpha\beta}$ $(\alpha, \beta = 1, 2)$ should vanish. Thus in the space $H^1_{per}(Y)^2$, rigid body motions reduce to the set of constant vectors, i.e. to $\mathbb{R}^2$. It implies that in $\tilde{H}^1_{per}(Y)^2$ the kernel of the operator e^y is equal to the null vector. Similarly, in $\tilde{H}^2_{per}(Y)^2$ the kernel of the operator κ^y reduces to zero.

It is evident that the functional ℓ is continuous on the space $H^1(Y)^2 \times H^2(Y)$, and thus also on $\tilde{H}_{K,per}(Y)$.

Summarizing, we conclude that the following minimization problem:

$$J_Y(\bar{\boldsymbol{v}}, \bar{v}) + \ell(\bar{\boldsymbol{v}}, \bar{v}) + m_1$$
$$= \inf\{J_Y(\boldsymbol{v}, v) + \ell(\boldsymbol{v}, v) + m_1 \mid (\boldsymbol{v}, v) \in \tilde{H}_{K,per}(Y)\} \,, \tag{2.10.84}$$

is uniquely solvable. We observe that for the minimization problem:

$$\left| \begin{array}{l} \text{find} \\ \inf\{J_Y(\boldsymbol{v}, v) + \ell(\boldsymbol{v}, v) + m_1 \mid (\boldsymbol{v}, v) \in H^1_{per}(Y)^2 \times H^2_{per}(Y)\} \end{array} \right.$$

a minimizer exists and is determined up to a constant vector $(\boldsymbol{K}, K) \in \mathbb{R}^2 \times \mathbb{R}$.

Effective constitutive relationships

Due to linearity of the problem under investigation one can write

$$\bar{v} = T^{(\alpha\beta)}\epsilon_{\alpha\beta} + \hat{U}^{(\alpha\beta)}\rho_{\alpha\beta} , \qquad (2.10.85)$$

$$\bar{v} = \hat{X}^{(\alpha\beta)}\epsilon_{\alpha\beta} + \chi^{(\alpha\beta)}\rho_{\alpha\beta}, \qquad (2.10.86)$$

where $T^{(\alpha\beta)}, \hat{U}^{(\alpha\beta)} \in \tilde{H}^1_{per}(Y)^2$ while $\hat{X}^{(\alpha\beta)}, \chi^{(\alpha\beta)} \in \tilde{H}^2_{per}(Y)$. The effective constitutive relationships are given by

$$\hat{N}_h^{\alpha\beta} = \frac{\partial \hat{\mathcal{W}}_h}{\partial \epsilon_{\alpha\beta}} = \hat{A}_h^{\alpha\beta\lambda\mu}\epsilon_{\lambda\mu} + \hat{E}_h^{\alpha\beta\lambda\mu}\rho_{\lambda\mu} , \qquad (2.10.87)$$

$$\hat{M}_h^{\alpha\beta} = \frac{\partial \hat{\mathcal{W}}_h}{\partial \rho_{\alpha\beta}} = \hat{F}_h^{\alpha\beta\lambda\mu}\epsilon_{\lambda\mu} + \hat{D}_h^{\lambda\mu\alpha\beta}\rho_{\lambda\mu} , \qquad (2.10.88)$$

where $\hat{F}_h^{\alpha\beta\lambda\mu} = \hat{E}_h^{\lambda\mu\alpha\beta}$ and

$$\hat{A}_h^{\alpha\beta\gamma\delta} = \frac{\partial^2 \hat{\mathcal{W}}_h}{\partial \epsilon_{\gamma\delta}\partial \epsilon_{\alpha\beta}} , \qquad \hat{E}_h^{\alpha\beta\gamma\delta} = \frac{\partial^2 \hat{\mathcal{W}}_h}{\partial \rho_{\gamma\delta}\partial \epsilon_{\alpha\beta}} , \qquad (2.10.89)$$

$$\hat{F}_h^{\alpha\beta\gamma\delta} = \frac{\partial^2 \hat{\mathcal{W}}_h}{\partial \epsilon_{\gamma\delta}\partial \rho_{\alpha\beta}} , \qquad \hat{D}_h^{\alpha\beta\gamma\delta} = \frac{\partial^2 \hat{\mathcal{W}}_h}{\partial \rho_{\alpha\beta}\partial \rho_{\gamma\delta}} . \qquad (2.10.90)$$

The formulae above are fully equivalent to the formula (2.8.6) found previously. The symmetry properties (2.7.22) can easily be deduced from relations (2.10.89) and (2.10.90).

The Y-periodic functions $T^{(\alpha\beta)}$, $\hat{U}^{(\alpha\beta)}$, $\hat{X}^{(\alpha\beta)}$, $\chi^{(\alpha\beta)}$ are solutions to local problems $(\hat{P}^\alpha_{K,Y})$ of Section 2.8. Those local problems are here derived as necessary conditions for the minimization problem involved on the r.h.s. of (2.10.80). As we know, this minimization problem is a convex optimization problem. Thus local problems $(\hat{P}^\alpha_{K,Y})$ are equivalent to solving the minimization problem just mentioned.

Now we can formulate the next important result of the present section. We recall that $x_\alpha = z_\alpha$ while the integrand in (2.10.91) below is specified by relation (2.10.50).

Theorem 2.10.6. The sequence of functionals

$$\tilde{J}^1_\varepsilon(u, w) = J^1_\varepsilon(v) = \int_\Omega \hat{\mathcal{W}}\left[\frac{x_\alpha}{\varepsilon}, \, e(u), \kappa(w)\right] dx , \qquad (2.10.91)$$

where $u \in H^1(\Omega)^2$ and $w \in H^2(\Omega)$ is Γ-convergent in the strong topology of $L^2(\Omega)^2 \times H^1(\Omega)$ to the limit functional

$$\tilde{J}^1_h(u, w) = \int_\Omega \hat{\mathcal{W}}_h\left[e(u), \kappa(w)\right] dx . \qquad (2.10.92)$$

Here $\hat{\mathcal{W}}_h$ is given by (2.10.80) and $v = (u_\alpha - z_3 \partial w / \partial x_\alpha, w)$.

For any fixed $\varepsilon > 0$ the problem

$$(P_\varepsilon) \qquad \tilde{J}_\varepsilon^1(u^\varepsilon, w^\varepsilon) - L(u^\varepsilon, w^\varepsilon)$$
$$= \inf\{\tilde{J}_\varepsilon^1(u, w) - L(u, w) \mid (u, w) \in H_0^1(\Omega)^2 \times H_0^2(\Omega)\}$$

and the homogenized problem

$$(P_h) \qquad \tilde{J}_h^1(\bar{u}, \bar{w}) - L(\bar{u}, \bar{w})$$
$$= \inf\{\tilde{J}_h^1(u, w) - L(u, w) \mid (u, w) \in H_0^1(\Omega)^2 \times H_0^2(\Omega)\}$$

are uniquely solvable. Moreover, we have

$$(u^\varepsilon, w^\varepsilon) \rightharpoonup (u, w) \text{ weakly in } H^1(\Omega)^2 \times H^2(\Omega)$$
$$\inf P_\varepsilon \to \inf P_h \,. \tag{2.10.93}$$

Proof. Γ-convergence is left to the reader as an exercise. In fact, it can be performed along the same lines as for the scalar case or three-dimensional elasticity, cf. also Sec. 2.10.4 where a more complicated case will be studied.

The existence and uniqueness of $(\bar{u}, \bar{w})$ is ensured by the properties of the effective potential $\hat{\mathcal{W}}_h$. Problem (P_ε) has already been discussed, cf. (2.10.45). Finally, (2.10.93) results immediately by an application of Theorem 1.3.22.

Dual effective potential

The effective complementary energy $\hat{\mathcal{W}}_h^*$ can be found as Fenchels's conjugate of $\hat{\mathcal{W}}_h$:

$$\hat{\mathcal{W}}_h^*(\epsilon^*, \rho^*) = \sup\{\epsilon^* : \epsilon + \rho^* : \rho - \hat{\mathcal{W}}_h(\epsilon, \rho) \mid \epsilon, \rho \in \mathbb{E}_s^2\} \,,$$

where $\epsilon^*, \rho^* \in \mathbb{E}_s^2$. By using (2.10.80) we find

$$\hat{\mathcal{W}}_h^*(\epsilon^*, \rho^*) = \sup\{\langle \epsilon^* : (e^y(v) + \epsilon) + \rho^* : (\kappa^y(v) + \rho)$$
$$-\hat{\mathcal{W}}(y_\alpha, e^y(v) + \epsilon, \kappa^y(v) + \rho)\rangle \mid \epsilon, \rho \in \mathbb{E}_s^2 \,, (v, v) \in \tilde{H}_{K,per}(Y)\} \,, \tag{2.10.94}$$

because

$$\langle \epsilon^* : e^y(v) \rangle = 0 \text{ and } \langle \rho^* : \kappa^y(v) \rangle = 0 \,, \text{ where } v \in H_{per}^1(Y)^2, \, v \in H_{per}^2(Y) \,.$$

We set

$$\mathbf{H}(Y) = [e^y(H_{per}^1(Y)^2) \times \kappa^y(H_{per}^2(Y))] \oplus (\mathbb{E}_s^2 \times \mathbb{E}_s^2) \,,$$
$$\mathbf{K}(p_1, p_2) = |Y| \langle \hat{\mathcal{W}}(y_\alpha, p_1, p_2) \rangle \,, \quad p_\alpha \in L^2(Y, \mathbb{E}_s^2) \,.$$

Then we can write, cf. Sec. 1.2.1,

$$\hat{\mathcal{W}}_h^*(\epsilon^*, \rho^*) = |Y|^{-1}(\mathbf{K} + I_{\mathbf{H}(Y)})^*(\epsilon^*, \rho^*) = |Y|^{-1}(\mathbf{K}^* \square I_{S_{per}(Y)})(\epsilon^*, \rho^*) \,, \tag{2.10.95}$$

where ϵ^*, ρ^* are identified with elements of $L^2(Y, \mathbb{E}_2^s)$; moreover

$$\mathbf{K}^*(\mathbf{n}, \mathbf{m}) = |Y| \langle \hat{\mathcal{W}}^*(y_\alpha, \mathbf{n}, \mathbf{m}) \rangle , \quad \mathcal{S}_{per}(Y) = [\mathbf{H}(Y)]^\perp$$

and, cf. (2.10.68), (2.10.69),

$$\hat{\mathcal{W}}^*(y_\alpha, \mathbf{n}, \mathbf{m}) = \frac{1}{2}[\mathbf{n}, \mathbf{m}]\mathcal{B}^{-1}(y_\alpha)\,[\mathbf{n}, \mathbf{m}]^T .$$

The symbol $\square$ represents here the inf-convolution.

Let us find the orthogonal complement of $[\mathbf{H}(Y)]^\perp$ in the sense of L^2:

$$\mathcal{S}_{per}(Y) = [\mathbf{H}(Y)]^\perp = [e^y(H^1_{per}(Y)^2) \times \kappa^y(H^2_{per}(Y))]^\perp \cap (\mathbb{E}_s^2 \times \mathbb{E}_s^2)^\perp$$

while $I_{\mathcal{S}_{per}(Y)}$ is the indicator function of the set $\mathcal{S}_{per}(Y)$.

We calculate

$$\begin{aligned}
(\mathbb{E}_s^2 \times \mathbb{E}_s^2)^\perp &= \{(\mathbf{n}, \mathbf{m}) \in L^2(Y, \mathbb{E}_2^s)^2 \mid \langle \mathbf{n}(y) : \epsilon \rangle + \langle \mathbf{m}(y) : \rho \rangle = 0 , \quad \forall \epsilon, \rho \in \mathbb{E}_s^2\} \\
&= \{(\mathbf{n}, \mathbf{m}) \in L^2(Y, \mathbb{E}_2^s)^2 \mid \langle \mathbf{n}(y) \rangle = 0 , \ \langle \mathbf{m}(y) \rangle = 0\} , \qquad (2.10.96)
\end{aligned}$$

$$\begin{aligned}
[e^y(H^1_{per}(Y)^2) &\times \kappa^y(H^2_{per}(Y))]^\perp \\
&= \{(\mathbf{n}, \mathbf{m}) \in L^2(Y, \mathbb{E}_2^s)^2 \mid \langle \mathbf{n}(y) : e^y(\boldsymbol{v}) + \mathbf{m}(y) : \kappa^y(v) \rangle = 0 , \\
&\qquad\qquad\qquad \forall(\boldsymbol{v}, v) \in H^1_{per}(Y)^2 \times H^2_{per}(Y)\} . \qquad (2.10.97)
\end{aligned}$$

Performing integration by parts in the following variational equations:

$$\int_Y \mathbf{n}(y) : e^y(\boldsymbol{v}) dy = 0 , \quad \forall \boldsymbol{v} \in H^1_{per}(Y)^2 ,$$

$$\int_Y \mathbf{m}(y) : \kappa^y(\boldsymbol{v}) dy = 0 , \quad \forall \boldsymbol{v} \in H^2_{per}(Y) ,$$

and taking into account (2.10.96) we finally obtain

$$\begin{aligned}
\mathcal{S}_{per}(Y) = \{(\mathbf{n}, \mathbf{m}) &\in L^2(Y, \mathbb{E}_2^s)^2 \mid \langle \mathbf{n}(y) \rangle = 0, \quad \langle \mathbf{m}(y) \rangle = 0 , \ \mathrm{div}_y \mathbf{n} = 0 , \\
&\mathrm{div}_y \mathrm{div}_y \mathbf{m} = 0 \text{ in } Y ; \ \mathbf{m}_\mu \text{ assume equal values and } \mathbf{n}\mu \text{ and } q \\
&\text{opposite values on the opposite sides of } Y\} , \qquad (2.10.98)
\end{aligned}$$

where

$$\mathbf{n}\mu = (\mathbf{n}^{\alpha\beta}\mu_\beta) , \quad \mathbf{m}_\mu = \mathbf{m}^{\alpha\beta}\mu_\alpha\mu_\beta ,$$

$$q = \mu_\alpha \frac{\partial \mathbf{m}^{\alpha\beta}}{\partial y_\beta} + \frac{\partial \mathbf{m}_\tau}{\partial s} , \quad \mathbf{m}_\tau = \mathbf{m}^{\alpha\beta}\mu_\alpha\tau_\beta . \qquad (2.10.99)$$

Here $\tau = (\tau_\alpha)$ is the tangent unit vector and s denotes an abscissa on ∂Y measured positively in the direction of τ. We observe that notation such as (2.10.99) is typical for the mechanical setting. In general, $\mathbf{m}_\mu$ and q are to be understood in the sense traces, cf. Temam (1985).

According to the definition of inf-convolution, from (2.10.95) one obtains

$$\hat{\mathcal{W}}_h^*(\boldsymbol{\epsilon}^*, \boldsymbol{\rho}^*) = \frac{1}{|Y|} \inf\{\mathbf{K}^*(\mathbf{n}_1, \mathbf{m}_1) + I_{\mathcal{S}_{per(Y)}}(\mathbf{n}_2, \mathbf{m}_2) \mid \boldsymbol{\epsilon}^* = \mathbf{n}_1 + \mathbf{n}_2\,,$$

$$\boldsymbol{\rho}^* = \mathbf{m}_1 + \mathbf{m}_2\,,\ (\mathbf{n}_\alpha, \mathbf{m}_\alpha) \in \mathcal{S}_{per}(Y)\}$$

$$= \inf\{\frac{1}{|Y|}\mathbf{K}^*(\boldsymbol{\epsilon}^* - \mathbf{n}_2, \boldsymbol{\rho}^* - \mathbf{m}_2) \mid (\mathbf{n}_2, \mathbf{m}_2) \in \mathcal{S}_{per}(Y)\}$$

$$= \inf\{\langle \hat{\mathcal{W}}^*(y_\alpha, \mathbf{n}(y) + \boldsymbol{\epsilon}^*, \mathbf{m}(y) + \boldsymbol{\rho}^*)\rangle \mid (\mathbf{n}, \mathbf{m}) \in \mathcal{S}_{per}(Y)\}\,, \quad (2.10.100)$$

because the set $\mathcal{S}_{per}(Y)$ of admissible generalized local stresses $(\mathbf{n}, \mathbf{m})$ is a linear space.

Having derived the complementary effective potential $\hat{\mathcal{W}}_h^*$, given by (2.10.100) one can formulate the dual homogenization theorem.

Theorem 2.10.7. The sequence of functionals $\{\mathcal{G}_\varepsilon\}_{\varepsilon>0}$, given by (2.10.78) is Γ-convergent in the weak topology of $L^2(\Omega, \mathbb{E}_s^2) \times L^2(\Omega, \mathbb{E}_s^2)$ to the functional

$$\mathcal{G}_h(\boldsymbol{N}, \boldsymbol{M}) = -\int_\Omega \hat{\mathcal{W}}_h[\boldsymbol{N}(x_\alpha), \boldsymbol{M}(x_\alpha)]dx - I_{\mathcal{S}}(\boldsymbol{N}, \boldsymbol{M})\,. \qquad (2.10.101)$$

If $(\hat{\boldsymbol{N}}_\varepsilon, \hat{\boldsymbol{M}}_\varepsilon)$ is a solution of the problem (P_ε^*) and

$$(P_h^*) \qquad \mathcal{G}_h(\hat{\boldsymbol{N}}_h, \hat{\boldsymbol{M}}_h) = \sup\{\mathcal{G}_h(\boldsymbol{N}, \boldsymbol{M}) \mid (\boldsymbol{N}, \boldsymbol{M}) \in L^2(\Omega, \mathbb{E}_2^s)^2\}\,,$$

then

$$(\hat{\boldsymbol{N}}_\varepsilon, \hat{\boldsymbol{M}}_\varepsilon) \rightharpoonup (\hat{\boldsymbol{N}}_h, \hat{\boldsymbol{M}}_h) \text{ weakly in } L^2(\Omega, \mathbb{E}_2^s)^2\,,$$

$$\sup(P_\varepsilon^*) \to \sup(P_h^*)\,,$$

when $\varepsilon \to 0$.

Proof. The proof follows by applying Theorem 1.3.36 and Theorem 2.10.6, cf. also Sec. 5.5. $\qquad\qquad \square$

2.10.3. Justification of the effective plate model of Section 2.9 by passing to zero: $\varepsilon \to 0$ and next $e \to 0$

For any fixed $e > 0$ the Γ-convergence when $\varepsilon \to 0$ for a thin body $\bar{B}_e$ with εY-periodic microstructure is a particular case of the homogenization of the equations of the three-dimensional elasticity. It is thus sufficient to find the homogenized potential when $\varepsilon \to 0$.

Homogenized elastic potential

For our particular case of a two-dimensional homogenization, this potential, denoted now by j_0, is given by

$$j_0(y_3, \boldsymbol{\epsilon}) = \inf\{\frac{1}{2|Y|} \int_Y C^{ijkl}(y_\alpha, y_3)(e_{ij}^y(\boldsymbol{v})$$

$$+\epsilon_{ij})(e_{kl}^y(\boldsymbol{v}) + \epsilon_{kl})dy \mid \boldsymbol{v} \in \tilde{H}_{per}^1(Y)^3\}\,, \qquad (2.10.102)$$

where $\epsilon \in \mathbb{E}_s^3$, and

$$\tilde{H}^1_{per}(Y)^3 = \{v \in H^1(Y)^3 \mid v_i \text{ is } Y\text{-periodic} , \ i = 1, 2, 3; \ \langle v \rangle = 0\} .$$

Now we have

$$e^y_{\alpha 3}(v) = \frac{1}{2}\frac{\partial v_3}{\partial y_\alpha} , \quad e^y_{33}(v) = 0 .$$

It is evident that a minimizer $\bar{v} \in \tilde{H}^1_{per}(Y)^3$ of the minimization problem on the r.h.s. of (2.10.102) exists and is unique, cf. Sec. 2.10.2. Moreover, we may write

$$\bar{v} = \Theta^{(ij)}\epsilon_{ij} , \quad \Theta^{(ij)} \in \tilde{H}^1_{per}(Y)^3. \tag{2.10.103}$$

The homogenized constitutive equation has the form (2.9.18), which will be written as

$$\sigma^{ij}_{\mathbf{h}} = \frac{\partial j_0}{\partial \epsilon_{ij}} = C^{ijkl}_{\mathbf{h}}(z_3)\epsilon_{kl} , \tag{2.10.104}$$

where $C_{\mathbf{h}}$, already defined in Section 2.9, can be put in the form

$$C^{rsmn}_{\mathbf{h}}(z_3) = \langle C^{ijkl}(y_\alpha, y_3)[e^y_{ij}(\Theta^{(mn)}) + \delta^m_i\delta^n_j] \ [e^y_{kl}(\Theta^{(rs)}) + \delta^r_k\delta^s_l]\rangle , \tag{2.10.105}$$

which can further be simplified since $e^y_{33}(\Theta^{(mn)}) = 0$. We note that $\Theta^{(mn)}_\lambda$ are defined as in Section 2.9 and $\Theta^{(mn)}_3 = \vartheta^{(mn)}$, see Sec. 2.9, problem $(P^{(kl)}_Y)$.

The following properties of j_0 may easily be verified:

(i) there exist positive constants $m' \leq M'$ such that for a.e. $z_3 \in (-1, 1)$

$$m'|\epsilon|^2 \leq j_0(z_3, \epsilon) \leq M'|\epsilon|^2 ,$$

for each $\epsilon \in \mathbb{E}_s^3$ $(c = 1)$.

(ii)

$$C^{ijkl}_{\mathbf{h}} = C^{jikl}_{\mathbf{h}} = C^{klij}_{\mathbf{h}} .$$

Now we are in a position to formulate the Γ-convergence theorem.

Theorem 2.10.8. For an $e > 0$ and fixed, the sequence of functionals $\{J^1_{e\varepsilon}\}_{\varepsilon>0}$ is Γ-convergent in the weak topology of $H^1(B)^3$ (strong of $L^2(B)^3$) to the functional

$$J^1_e(v) = \int_B j_0(z_3, \mathbb{Q}^e e^z(v))dz , \qquad v \in H^1(B)^3 .$$

The minimization problem

$$(P_e) \qquad J^1_e(v^e) - L_e(v^e) = \inf\{J^1_e(v) - L_e(v) \mid v \in V_0(B)\} ,$$

is uniquely solvable and

$$v^{e\varepsilon} \rightharpoonup v^e \quad \text{weakly in} \quad H^1(B)^3 , \ \inf(P_{e\varepsilon}) \to \inf(P_e) ,$$

when $\varepsilon \to 0$, where $(P_{e\varepsilon})$ is defined by (2.10.22). $\qquad\qquad\square$

Plate model: e tends to zero

Now the situation is similar to that for the functional $J_{e\varepsilon}$ when $e \to 0$. The only difference is in the elastic moduli; in the present case they are specified by (2.10.105). Thus we can formulate the limit theorem.

Theorem 2.10.9. The sequence of functionals $\{J_e^1 - L_e\}_{e>0}$ is Γ-convergent in the strong topology of $L^2(B)^3$ to the functional

$$J_{\mathbf{h}}(v) = \frac{1}{2} \int_B \tilde{C}_{\mathbf{h}}^{\alpha\beta\lambda\mu}(z_3)(e_{\alpha\beta}^z(v)e_{\lambda\mu}^z(v))dz - L(v) , \quad v \in V_K^0(B) \qquad (2.10.106)$$

where $\tilde{C}_{\mathbf{h}}$ is given by (2.9.27).   □

Corollary 2.10.10. The constant two-dimensional elastic moduli are calculated by the formula (2.9.22), with $x_3^- = -c$, $x_3^+ = c$, $\hat{x}_3 = z_3$, $x_3 = z_3$.

Remark 2.10.11. Dual formulations are left to the reader as an exercise.

2.10.4.   Justification of the effective plate model of Section 2.3 by passing to zero:
$e \to 0$ and $\varepsilon \to 0$ simultaneously

Now we have to deal with two small parameters which tend to zero simultaneously. The study of Γ-convergence of the sequence of functionals $\{J_{e\varepsilon}\}_{e>0,\ \varepsilon>0}$ is different from the previous two cases. In fact, our proof essentially exploits some tools elaborated in homogenization of microperiodic bodies. Prior to passing to the study of Γ-convergence we shall describe the macroscopic or effective (homogenized) elastic potential of the plate and derive its dual or the complementary energy density.

We shall frequently refer to formulae of Sections 2.2 and 2.3. Here, however, the plate of constant thickness is considered, cf. Introduction to Sec. 2.10.

Effective potential

Due to the assumption of the thickness being constant, the space $W(\mathcal{Y})$ is defined by, see Sec. 2.2

$$W(\mathcal{Y}) = \{v \in [H^1(\mathcal{Y})]^3 \mid v(\cdot, y_3) \text{ is } Y\text{-periodic for } y_3 \in (-c, c)\} , \qquad (2.10.107)$$

where $\mathcal{Y} = Y \times (-c, c)$.

Let us define also the space

$$\tilde{W}(\mathcal{Y}) = \{v \in W(\mathcal{Y}) \mid \prec v \succ = 0\} , \qquad (2.10.108)$$

where averaging over $\mathcal{Y}$ is denoted by $\prec \cdot \succ$, see Sec. 2.2.

The effective elastic potential has now the following form

$$j_{\mathcal{H}}(\epsilon, \rho) = \inf\{\tfrac{h_0}{2} \prec C^{ijkl}(y)[e_{ij}^y(v) + \epsilon_{ij} + y_3\rho_{ij}][e_{kl}^y(v) + \epsilon_{kl}$$
$$+ y_3\rho_{kl}] \succ \mid v \in W(\mathcal{Y})\} , \qquad (2.10.109)$$

where $h_0 = vol\ \mathcal{Y}/\text{area}\ Y = 2c;\ \boldsymbol{\epsilon}, \boldsymbol{\rho} \in \mathbb{E}_s^3;\ \epsilon_{i3} = 0,\ \rho_{i3} = 0;\ e_{ij}^y(\boldsymbol{v}) = (v_{i|j} + v_{j|i})/2$ and $(\ )_{|i} = \partial(\ )/\partial y_i$.

The properties of the elasticity tensor C immediately imply that the minimization problem on the r.h.s. of (2.10.109) is uniquely solvable in the space $\tilde{W}(\mathcal{Y})$ and up to a constant vector in the space $W(\mathcal{Y})$. Let us denote by $\bar{v} \in \tilde{W}(\mathcal{Y})$ the minimizer.

Properties of $j_{\mathcal{H}}$

For each $\boldsymbol{\epsilon}, \boldsymbol{\rho} \in \mathbb{E}_s^3$ with $\epsilon_{i3} = 0,\ \rho_{i3} = 0$ we have

$$1.\ j_{\mathcal{H}}(\boldsymbol{\epsilon}, \boldsymbol{\rho}) \leq \frac{h_0}{2} \prec C^{ijkl}(\boldsymbol{y})(\epsilon_{ij} + y_3\rho_{ij})(\epsilon_{kl} + y_3\rho_{kl}) \succ \leq M(|\boldsymbol{\epsilon}|^2 + |\boldsymbol{\rho}|^2), \quad (2.10.110)$$

where M is the constant appearing in (2.10.39).

$$2.\ j_{\mathcal{H}}(\boldsymbol{\epsilon}, \boldsymbol{\rho}) \geq \frac{mh_0}{2} \prec |e^y(\boldsymbol{v}) + \boldsymbol{\epsilon} + y_3\boldsymbol{\rho}|^2 \succ \geq \frac{m}{3}(|\boldsymbol{\epsilon}|^2 + |\boldsymbol{\rho}|^2)\,, \ \text{for}\ c = 1\,, \quad (2.10.111)$$

since

$$\sum_{i,j=1}^{3} \prec \epsilon_{ij} e_{ij}^y(\bar{v}) \succ = 0.$$

$\square$

The minimizer $\bar{v}$ depends linearly on $\boldsymbol{\epsilon}$ and $\boldsymbol{\rho}$. Therefore we may write:

$$\bar{v} = \Theta^{(\alpha\beta)}(\boldsymbol{y})\epsilon_{\alpha\beta} + \Xi^{(\alpha\beta)}(\boldsymbol{y})\rho_{\alpha\beta}\,. \quad (2.10.112)$$

Taking into account (2.10.112) in (2.10.109) we derive the effective constitutive relationships

$$\mathcal{N}^{\alpha\beta} = \frac{\partial j_{\mathcal{H}}}{\partial \epsilon_{\alpha\beta}}\,, \qquad \mathcal{M}^{\alpha\beta} = \frac{\partial j_{\mathcal{H}}}{\partial \rho_{\alpha\beta}}\,, \quad (2.10.113)$$

that assume the form (2.3.51) with $e_{\lambda\mu}(v)$ replaced with $\epsilon_{\lambda\mu}$ and $\kappa_{\lambda\mu}(w)$ replaced with $\rho_{\lambda\mu}$. The effective stiffnesses involved in (2.3.51) satisfy the expected symmetry conditions (Sec. 2.3). It is worth noting that property (2.10.111) of $j_{\mathcal{H}}$ is equivalent to the positive definiteness of the matrix

$$\mathbf{B}_{\mathcal{H}} = \begin{bmatrix} \boldsymbol{A}_{\mathcal{H}} & \boldsymbol{E}_{\mathcal{H}} \\ \boldsymbol{F}_{\mathcal{H}} & \boldsymbol{D}_{\mathcal{H}} \end{bmatrix}\,, \quad (2.10.114)$$

cf. (2.3.62). The minimization problem on the r.h.s. of (2.10.109) is a convex optimization problem. The necessary condition for a minimum leads naturally to two local problems (P_y^α) (see Sec. 2.3) for the determination of the functions $\Theta^{(\alpha\beta)}$ and $\Xi^{(\alpha\beta)}$; here one should replace $\hat{y}_3$ with y_3 in (P_y^α) , since $\prec y_3 \succ = 0$.

Dual effective potential

To derive the dual elastic potential $j_{\mathcal{H}}^*$, which represents the density of the complementary energy of the plate, we shall apply the theory of duality outlined in Section 1.2.5.

By definition of the conjugate function one has

$$j_{\mathcal{H}}^*(\epsilon^*, \rho^*) = \sup\{\epsilon^* : \epsilon + \rho^* : \rho - j_{\mathcal{H}}(\epsilon, \rho) \mid \epsilon, \rho \in \mathbb{E}_s^2\}$$

$$= \sup_{\substack{\epsilon,\rho \in \mathbb{E}_s^3 \\ \epsilon_{i3}=0, \rho_{i3}=0}} \{\epsilon^* : \epsilon + \rho^* : \rho - \inf_{v \in \tilde{\mathcal{W}}(\mathcal{Y})} \frac{1}{|Y|} \int_{\mathcal{Y}} j(\boldsymbol{y}, e^y(v) + \epsilon + y_3\rho) d\boldsymbol{y}\} . \quad (2.10.115)$$

For fixed $\epsilon, \rho \in \mathbb{E}_s^3$ with $\epsilon_{i3} = 0, \rho_{i3} = 0$ we first consider the following minimization problem:

$$(P_{(\epsilon,\rho)}) \qquad \inf\{\int_{\mathcal{Y}} j(\boldsymbol{y}, e^y(v) + \epsilon + y_3\rho) d\boldsymbol{y} \mid v \in \mathcal{W}(\mathcal{Y})\} .$$

Let $\Lambda v = e^y(v)$, $\Lambda : H^1(\mathcal{Y})^3 \to V(\mathcal{Y}) = L^2(\mathcal{Y}, \mathbb{E}_s^3)$. For any $p^* \in V^*(\mathcal{Y}) = L^2(\mathcal{Y}, \mathbb{E}_3^s)$ one finds

$$\langle \Lambda v, \boldsymbol{p}^* \rangle_{L^2(\mathcal{Y}, \mathbb{E}_s^3) \times L^2(\mathcal{Y}, \mathbb{E}_3^s)} = \int_{\mathcal{Y}} \boldsymbol{p}^* : e^y(v) d\boldsymbol{y}$$

$$= -\int_{\mathcal{Y}} (\mathrm{div}_y \boldsymbol{p}^*) \cdot \boldsymbol{v} d\boldsymbol{y} + \int_{\partial \mathcal{Y}} p^{*ij} \nu_j v_i dS = \langle \Lambda^* \boldsymbol{p}^*, \boldsymbol{v} \rangle_{[H^1(\mathcal{Y})^3]^* \times H^1(\mathcal{Y})^3} .$$

Hence

$$\Lambda^* \boldsymbol{p}^* = \begin{cases} -\mathrm{div}_y \boldsymbol{p}^* & \text{in } \mathcal{Y}, \\ \boldsymbol{p}^* \nu & \text{on } \partial \mathcal{Y} . \end{cases} \quad (2.10.116)$$

Further we set

$$j_{(\epsilon,\rho)}(\boldsymbol{y}, \boldsymbol{\Delta}) = j(\boldsymbol{y}, \boldsymbol{\Delta} + \epsilon + y_3\rho) , \quad \boldsymbol{\Delta} \in \mathbb{E}_s^3 ,$$

$$G_{(\epsilon,\rho)}(\boldsymbol{p}) = \int_{\mathcal{Y}} j_{(\epsilon,\rho)}(\boldsymbol{y}, \boldsymbol{p}(\boldsymbol{y})) d\boldsymbol{y} , \quad \boldsymbol{p} \in L^2(Y, \mathbb{E}_s^3) .$$

By using Proposition 1.2.34, one has

$$G_{(\epsilon,\rho)}^*(\boldsymbol{p}^*) = \int_{\mathcal{Y}} j_{(\epsilon,\rho)}^*(\boldsymbol{y}, \boldsymbol{p}^*(\boldsymbol{y})) d\boldsymbol{y} , \quad \boldsymbol{p}^* \in L^2(Y, \mathbb{E}_3^s) . \quad (2.10.117)$$

Here, for $\boldsymbol{\Delta}^* \in \mathbb{E}_3^s$, we calculate

$$j_{(\epsilon,\rho)}^*(\boldsymbol{y}, \boldsymbol{\Delta}^*) = \sup\{\boldsymbol{\Delta}^* : \boldsymbol{\Delta} - j_{(\epsilon,\rho)}(\boldsymbol{y}, \boldsymbol{\Delta}) \mid \boldsymbol{\Delta} \in \mathbb{E}_s^3\}$$

$$= \sup\{\boldsymbol{\Delta}^* : \boldsymbol{\Delta} - j(\boldsymbol{y}, \boldsymbol{\Delta}') \mid \boldsymbol{\Delta}' \in \mathbb{E}_s^3\} , \quad (2.10.118)$$

where $\boldsymbol{\Delta}' = \boldsymbol{\Delta} + \epsilon + y_3\rho$. Hence one readily obtains

$$j_{(\epsilon,\rho)}^*(\boldsymbol{y}, \boldsymbol{\Delta}^*) = \frac{1}{2} c_{ijkl}(\boldsymbol{y}) \Delta^{*ij} \Delta^{*kl} - \boldsymbol{\Delta}^* : (\epsilon + y_3\rho) , \quad (2.10.119)$$

where $c = \boldsymbol{C}^{-1}$.

If $F \equiv 0$ then

$$F^*(-\Lambda^* \boldsymbol{p}^*) = \sup\{ \langle -\Lambda^* \boldsymbol{p}^*, \boldsymbol{v} \rangle - 0 \mid \boldsymbol{v} \in W(\mathcal{Y})\}$$
$$= \begin{cases} 0 & \text{if } \boldsymbol{p}^* \in \mathcal{S}^0(\mathcal{Y}) \\ +\infty & \text{otherwise} \end{cases}, \qquad (2.10.120)$$

where

$$\mathcal{S}^0(\mathcal{Y}) = \{\boldsymbol{p}^* \in L^2(\mathcal{Y}, \mathbb{E}_3^s) \mid \mathrm{div}_y \boldsymbol{p}^* = 0 \text{ in } \mathcal{Y}, \ \boldsymbol{p}^* \boldsymbol{\nu} \text{ takes opposite values}$$
$$\text{on the opposite faces of } \partial Y \times (-c, c) , \ \boldsymbol{p}^* \boldsymbol{\nu} = 0 \text{ on } Y \times \{\pm c\}\}. \qquad (2.10.121)$$

Taking into account (2.10.118) – (2.10.120) one can formulate the dual problem of $(P_{(\boldsymbol{\epsilon}, \boldsymbol{\rho})})$:

$$(P_{(\boldsymbol{\epsilon}, \boldsymbol{\rho})})^* \qquad \sup\{- \int_{\mathcal{Y}} j^*_{(\boldsymbol{\epsilon}, \boldsymbol{\rho})}(\boldsymbol{y}, \boldsymbol{p}^*(\boldsymbol{y})) dy \mid \boldsymbol{p}^* \in \mathcal{S}^0(\mathcal{Y})\} .$$

Our assumptions imply

$$\inf(P_{(\boldsymbol{\epsilon}, \boldsymbol{\rho})}) = \sup(P_{(\boldsymbol{\epsilon}, \boldsymbol{\rho})})^*$$
$$= - \inf\{ \int_{\mathcal{Y}} j^*_{(\boldsymbol{\epsilon}, \boldsymbol{\rho})}(\boldsymbol{y}, \boldsymbol{p}^*(\boldsymbol{y})) dy \mid \boldsymbol{p}^* \in \mathcal{S}^0(\mathcal{Y})\} . \qquad (2.10.122)$$

Thus from (2.10.115) and (2.10.122) one obtains

$$j^*_{\mathcal{H}}(\boldsymbol{\epsilon}^*, \boldsymbol{\rho}^*) = \sup_{\substack{\boldsymbol{\epsilon}, \boldsymbol{\rho} \in \mathbb{E}_s^3 \\ \epsilon_{i3}=0, \rho_{i3}=0}} \{\boldsymbol{\epsilon}^* : \boldsymbol{\epsilon} + \boldsymbol{\rho}^* : \boldsymbol{\rho}$$
$$+ h_0 \inf_{\boldsymbol{p}^* \in \mathcal{S}^0(\mathcal{Y})} \prec j^*(\boldsymbol{y}, \boldsymbol{p}^*(\boldsymbol{y})) - \boldsymbol{p}^* : (\boldsymbol{\epsilon} + y_3 \boldsymbol{\rho}) \succ \}$$
$$= \inf_{\boldsymbol{p}^* \in \mathcal{S}^0(\mathcal{Y})} \sup_{\substack{\boldsymbol{\epsilon}, \boldsymbol{\rho} \in \mathbb{E}_s^3 \\ \epsilon_{i3}=0, \rho_{i3}=0}} \{(\boldsymbol{\epsilon}^* - h_0 \prec \boldsymbol{p}^*(\boldsymbol{y}) \succ) : \boldsymbol{\epsilon}$$
$$+ (\boldsymbol{\rho}^* - h_0 \prec y_3 \boldsymbol{p}^*(\boldsymbol{y}) \succ) : \boldsymbol{\rho} + h_0 \prec j^*(\boldsymbol{y}, \boldsymbol{p}^*(\boldsymbol{y})) \succ \}$$
$$= \inf\{h_0 \prec j^*(\boldsymbol{y}, \boldsymbol{p}^*(\boldsymbol{y})) \succ \mid \boldsymbol{p}^* \in \mathcal{S}^1(\mathcal{Y})\} , \qquad (2.10.123)$$

where

$$\mathcal{S}^1(\mathcal{Y}) = \{\boldsymbol{p}^* \in \mathcal{S}^0(\mathcal{Y}) \mid h_0 \prec \boldsymbol{p}^*(\boldsymbol{y}) \succ = \boldsymbol{\epsilon}^* , \ h_0 \prec y_3 \boldsymbol{p}^* \succ = \boldsymbol{\rho}^*\} . \qquad (2.10.124)$$

An equivalent form of (2.10.123) is given by

$$j^*_{\mathcal{H}}(\boldsymbol{\epsilon}^*, \boldsymbol{\rho}^*) = \inf\{h_0 \prec j^*(\boldsymbol{y}, \frac{1}{h_0} \boldsymbol{\epsilon}^* + \boldsymbol{q}^*(\boldsymbol{y})) \succ \mid \boldsymbol{q}^* \in \mathcal{S}(\mathcal{Y})\} , \qquad (2.10.125)$$

where

$$\mathcal{S}(\mathcal{Y}) = \{\boldsymbol{q}^* \in L^2(\mathcal{Y}, \mathbb{E}_3^s) \mid \mathrm{div}_y \boldsymbol{q}^* = 0 \text{ in } Y , \ \prec \boldsymbol{q}^* \succ = 0 , \quad h_0 \prec y_3 \boldsymbol{q}^* \succ = \boldsymbol{\rho}^* ,$$
$$\boldsymbol{q}^* \boldsymbol{\nu} \text{ takes opposite values on the opposite faces of}$$
$$\partial Y \times (-c, c) , \ \boldsymbol{q}^* \boldsymbol{\nu} = 0 \text{ on } Y \times \{\pm c\}\} . \qquad (2.10.126)$$

We recall that $\epsilon^{*i3} = 0$. To corroborate (2.10.125) and (2.10.126) we note that

$$h_0 \prec p^* \succ = \epsilon^* \iff \prec (p^* - \frac{1}{h_0}\epsilon^*) \succ = 0 \ .$$

Then, for $q^*(y) = p^*(y) - \frac{1}{h_0}\epsilon^*$ one has $\prec q^* \succ = 0$. Now the proof is straightforward and is left to the reader. $\square$

Below, in the proof of a Γ-convergence theorem an important role will be played by the following lemma.

Lemma 2.10.12. Assume that $T \in \mathcal{S}(\mathcal{Y})$ and let $\{v^\varepsilon\}_{\varepsilon>0} \subset H^1(B)^3$ be such that $\{\mathbf{Q}^\varepsilon e^z(v^\varepsilon)\}_{\varepsilon>0}$ is bounded in $L^2(B, \mathbb{E}_s^3)$ and v^ε converges strongly to $v \in H^1(B)^3$. Then one has

$$\lim_{\varepsilon \to 0} \int_B \psi(z_\alpha) T^{ij}\left(\frac{z_\beta}{\varepsilon}, z_3\right) (\mathbf{Q}^\varepsilon e^z(v^\varepsilon))_{ij} dz = \int_\Omega \psi \rho^* : \kappa^z(w)dz \ , \qquad (2.10.127)$$

where $\psi \in \mathbf{D}(\Omega)$, $v = (u_\alpha - z_3\partial w/\partial z_\alpha, w)$ and $\rho^* \in \mathbb{E}_2^s$ appears in (2.10.126).
Proof. Let us set

$$R_\varepsilon = \int_B \psi(z_\alpha) T^{ij}(\frac{z_\beta}{\varepsilon}, z_3)(\mathbf{Q}^\varepsilon e^z(v^\varepsilon))_{ij} dz \ , \quad T_\varepsilon^{ij} = (\frac{z_\beta}{\varepsilon}, z_3) \ .$$

Integration by parts yields

$$R_\varepsilon = - \int_B (\psi \mathbf{Q}^\varepsilon T_\varepsilon)_{;j}^{ij} v_i^\varepsilon dz + \int_{\Gamma_\pm \cup \Gamma_0} \psi T_\varepsilon^{ij} n_j v_i^\varepsilon d\Gamma \ . \qquad (2.10.128)$$

Since $\psi = 0$ on Γ and $T_\varepsilon^{ij} n_j = 0$ on $\Gamma_\pm$, therefore the last integral vanishes. We know that $\mathrm{div}_y T = 0$ in $\mathcal{Y}$. After a rescaling $y \to \left(\frac{z_\alpha}{\varepsilon}, z_3\right)$ we obtain

$$\left(\varepsilon T^{i\alpha}{}_{;\alpha} + T^{i3}{}_{;3}\right) \left(\frac{z_\beta}{\varepsilon}, z_3\right) = 0 \quad \text{in } B \ ,$$

because $\dfrac{\partial}{\partial y_\alpha} = \varepsilon \dfrac{\partial}{\partial z_\alpha}$ and $\dfrac{\partial}{\partial y_3} = \dfrac{\partial}{\partial z_3}$. For $\varepsilon > 0$ the last relation is equivalent to

$$T_{\varepsilon;\alpha}^{i\alpha} + \frac{1}{\varepsilon}T_{\varepsilon;3}^{i3} = 0 \quad \text{in} \quad B \ .$$

On the other hand we have

$$(\mathbf{Q}^\varepsilon T_\varepsilon)_{;j}^{\alpha j} = T_{\varepsilon;\beta}^{\alpha\beta} + \frac{1}{\varepsilon}T_{\varepsilon;3}^{\alpha3} \ , \qquad (\mathbf{Q}^\varepsilon T_\varepsilon)_{;j}^{3j} = \frac{1}{\varepsilon}T_{\varepsilon;\alpha}^{3\alpha} + \frac{1}{\varepsilon^2}T_{\varepsilon;3}^{33} \ .$$

Thus

$$(\mathbf{Q}^\varepsilon T_\varepsilon)_{;j}^{ij} = 0 \ ,$$

and consequently

$$R_\varepsilon = - \int\limits_{\Omega} \int\limits_{-c}^{c} \psi_{;j}(\mathbf{Q}^\varepsilon \boldsymbol{T}_\varepsilon)^{ij} v_i^\varepsilon d\boldsymbol{z} = - \int\limits_{\Omega} \psi_{;\alpha} \int\limits_{-c}^{c} (\mathbf{Q}^\varepsilon \boldsymbol{T}_\varepsilon)^{i\alpha} v_i^\varepsilon d\boldsymbol{z}$$

$$= - \int\limits_{\Omega} \psi_{;\alpha} \int\limits_{-c}^{c} (T_\varepsilon^{\alpha\beta} v_\beta^\varepsilon + \frac{1}{\varepsilon} T_\varepsilon^{3\alpha} v_3^\varepsilon) d\boldsymbol{z} \ .$$

We observe that

$$|R_\varepsilon| = ||\psi_{;j}(\mathbf{Q}^\varepsilon \boldsymbol{T}_\varepsilon)^{ij} e_{ij}^z(\boldsymbol{v}^\varepsilon)||_{L^1(B)} = |\int\limits_{\Omega} \psi_{;\alpha} \int\limits_{-c}^{c} (T_\varepsilon^{\alpha\beta} v_\beta^\varepsilon + \frac{1}{\varepsilon} T_\varepsilon^{3\alpha} v_2^\varepsilon) d\boldsymbol{z}|$$

$$\leq K ||\boldsymbol{T}_\varepsilon||_{L^2} ||\mathbf{Q}^\varepsilon e^z(\boldsymbol{v}^\varepsilon)||_{L^2} \leq K_1 \ ,$$

where K_1 is a positive constant independent of ε. Finally we obtain

$$\lim_{\varepsilon \to 0} R_\varepsilon = - \lim_{\varepsilon \to 0} \int\limits_{\Omega} \psi_{;\alpha} \int\limits_{-c}^{c} T_\varepsilon^{\alpha\beta} v_\beta^\varepsilon d\boldsymbol{z} - \lim_{\varepsilon \to 0} \int\limits_{\Omega} \psi_{;\alpha} \int\limits_{-c}^{c} \frac{1}{\varepsilon} T_\varepsilon^{3\alpha} v_3^\varepsilon d\boldsymbol{z}$$

$$= - \int\limits_{\Omega} \psi_{;\alpha} \int\limits_{-c}^{c} \langle T^{\beta\alpha}(y_\lambda, y_3)\rangle \left(u_\beta - z_3 \frac{\partial w}{\partial z_\beta}\right) d\boldsymbol{z} = \int\limits_{\Omega} \psi \rho^{*\alpha\beta} \kappa_{\alpha\beta}^z(w) d\boldsymbol{z} \ ,$$

because $\psi_{;3} = 0$, $\prec \boldsymbol{T} \succ = 0$, $h_0 \prec y_3 \boldsymbol{T} \succ = \rho^*$, and

$$\lim_{\varepsilon \to 0} \int\limits_{\Omega} \psi_{;\alpha} \int\limits_{-c}^{c} \frac{1}{\varepsilon} T_\varepsilon^{3\alpha} v_3^\varepsilon d\boldsymbol{z} = 0 \ .$$

$\square$

Γ-convergence of the sequence $\{J_{e\varepsilon}\}_{e>0, \ \varepsilon>0}$

As we know from the proof of Theorem 2.10.2 the sequence of loading functionals $\{L_e\}_{e>0}$ is continuously convergent. The main result of Γ-convergence when $e \to 0$ and $\varepsilon \to 0$ simultaneously is formulated in the form of the following theorem.

Theorem 2.10.13. The sequence of functionals $\{J_{e\varepsilon}^1\}_{e>0, \ \varepsilon>0}$ defined by (2.10.29) is Γ-convergent in the strong topology of $L^2(B)^3$ (weak topology of $H^1(B)^3$) to the limit functional $J_{\mathcal{H}}^1$ given by

$$J_{\mathcal{H}}^1(u_\alpha, w) = \int\limits_{\Omega} j_{\mathcal{H}}(e^z(\boldsymbol{u}), \boldsymbol{\kappa}^z(w)) d\boldsymbol{z} \ , \tag{2.10.129}$$

where $u_\alpha \in H^1(\Omega)$, $w \in H^2(\Omega)$.

Proof. We divide it into two major parts.

I. We shall prove that for any $v = (v_\alpha, v_3) \in V_K(B)$, $v_\alpha = u_\alpha - z_3 w_{;\alpha}$, $v_3 = w$, there exists a sequence $\{v^\varepsilon\}_{\varepsilon>0} \subset H^1(B)^3$ strongly convergent to v in $L^2(B)^3$ and such that

$$J_{\mathcal{H}}^1(u, w) \geq \limsup_{\varepsilon \to 0} J_{\varepsilon\varepsilon}^1(v^\varepsilon) \, , \tag{2.10.130}$$

where $J_{\varepsilon\varepsilon}^1 = J_{e\varepsilon|e=e}^1$. Tacitly we assume that v^ε stands for $v^{e\varepsilon}$ where $e = \varepsilon$.

Step 1. We take $v = (v_\alpha, v_3)$ in the form

$$v_\alpha = \sum_{\beta=1}^{2} [\epsilon_{\alpha\beta} z_\beta + z_3(\rho_{\alpha\beta} z_\beta + a_\alpha)] + c_\alpha \, , \qquad y_3 = z_3 \, ,$$

$$v_3 = \frac{1}{2} \sum_{\alpha,\beta=1}^{2} (-\rho_{\alpha\beta} z_\alpha z_\beta) - \sum_{\alpha=1}^{2} a_\alpha z_\alpha + a \, , \tag{2.10.131}$$

where $\epsilon, \rho \in \mathbb{E}_s^3$ with $\epsilon_{i3} = 0$, $\rho_{i3} = 0$ and $a_\alpha, c_\alpha, a \in \mathbb{R}$. Obviously, v belongs to $V_K(B)$. Let $\bar{v}$ be a function at which the functional appearing on the r.h.s. of (2.10.109) achieves a minimum value. For v specified by (2.10.131) we set

$$v_\alpha^\varepsilon = v_\alpha + \varepsilon \bar{v}_\alpha \left(\frac{z_\beta}{\varepsilon} , z_3 \right) \, , \qquad v_3^\varepsilon = v_3 + \varepsilon^2 \bar{v}_3 \left(\frac{z_\alpha}{\varepsilon} , z_3 \right) \, . \tag{2.10.132}$$

The sequence $\left\{ \bar{v}_i \left(\frac{\cdot}{\varepsilon}, z_3 \right) \right\}_{\varepsilon>0}$ is bounded for a.e. $z_3 \in (-c, c)$. Hence we conclude that

$$v^\varepsilon \to v \quad \text{in} \quad L^2(B)^3 \quad \text{when} \quad \varepsilon \to 0 \, .$$

Next we calculate

$$(\mathbb{Q}^\varepsilon e^z(v^\varepsilon))_{\alpha\beta} = \epsilon_{\alpha\beta} + z_3 \rho_{\alpha\beta} + e_{\alpha\beta}^z(\bar{v}) \left(\frac{z_\delta}{\varepsilon} , z_3 \right) \, ,$$

$$(\mathbb{Q}^\varepsilon e^z(v^\varepsilon))_{\alpha3} = e_{\alpha3}^z(\bar{v}) \left(\frac{z_\beta}{\varepsilon} , z_3 \right) \, , \tag{2.10.133}$$

$$(\mathbb{Q}^\varepsilon e^z(v^\varepsilon))_{33} = \frac{\partial v_3}{\partial z_3} \left(\frac{z_\alpha}{\varepsilon} , z_3 \right) \, .$$

Taking into account (2.10.133) we obtain

$$J_{\varepsilon\varepsilon}^1(v^\varepsilon) = \int_B j_\varepsilon(z, \mathbb{Q}^\varepsilon e^z(v^\varepsilon)) dz$$

$$= \int_\Omega \{ \int_{-c}^{c} \frac{1}{2} C^{ijkl} \left(\frac{z_\alpha}{\varepsilon} , z_3 \right) \left[e_{ij}^z(\bar{v}) \left(\frac{z_\beta}{\varepsilon} , z_3 \right) + \epsilon_{ij} + z_3 \rho_{ij} \right]$$

$$\times \left[e_{kl}^z(\bar{v}) \left(\frac{z_\delta}{\varepsilon} , z_3 \right) + \epsilon_{kl} + z_3 \rho_{kl} \right] dz_3 \} dz \, . \tag{2.10.134}$$

We recall that $\epsilon_{i3} = 0$, $\rho_{i3} = 0$ and $z_3 = y_3$. The integrand in (2.10.134) is an εY-periodic function and therefore tends weakly in $L^1(B)$ to $j_{\mathcal{H}}(\epsilon, \rho)$ when $\varepsilon \to 0$, cf. Th. 1.1.5. Consequently we write

$$\lim_{\varepsilon \to 0} J^1_{\varepsilon\varepsilon}(v^\varepsilon) = \int_\Omega j_{\mathcal{H}}(\epsilon, \rho) dz = \int_\Omega j_{\mathcal{H}}[e^z_{\alpha\beta}(u), \kappa^z(w)] dz ,$$

where, on account of (2.10.131), $u_\alpha = \sum_{\beta=1}^2 \epsilon_{\alpha\beta} z_\beta + c_\alpha$, $w = v_3$, $\kappa^z_{\alpha\beta}(w) = -\dfrac{\partial^2 w}{\partial z_\beta \partial z_\alpha} = \rho_{\alpha\beta}$.

Step 2. Let $\{\Omega_K\}_{K\in\mathcal{K}}$ be a finite partition of the domain Ω formed by polygonal sets. We set

$$\Omega^\delta_K = \{z \in \Omega \mid dist\,(z, \partial\Omega_K) > \delta\} , \quad \delta > 0.$$

We take a function $v = (v_\alpha, v_3) \in [C(B)^2 \times C^1(\Omega)] \cap V_K(B)$, given by

$$v_\alpha(z) = \sum_{\beta=1}^2 [\epsilon^K_{\alpha\beta} z_\beta + z_3(\rho^K_{\alpha\beta} z_\beta + a^K_\alpha)] + c^K_\alpha , \; z_3 = (z_\alpha, z_3) \in \Omega \times (-c, c),$$

$$v_3(z_1, z_2) = \sum_{\alpha,\beta=1}^2 \frac{1}{2}(-\rho^K_{\alpha\beta}) z_\alpha z_\beta - \sum_{\alpha=1}^2 a^K_\alpha z_\alpha + a^K ,$$

$$\tag{2.10.135}$$

where $\epsilon^K, \rho^K \in \mathbb{E}^3_s$ with $\epsilon^K_{i3} = 0$ and $\rho^K_{i3} = 0$; $a^K_\alpha, c^K_\alpha, a^K \in \mathbb{R}$, $K \in \mathcal{K}$.

We observe that $v_3 \in C^1(\bar\Omega)$ and $\dfrac{\partial v_3}{\partial z_\alpha} \in C(\bar\Omega)$. Further, we assume that the function

$$u_\alpha(z_1, z_2) = \sum_{\beta=1}^2 \epsilon^K_{\alpha\beta} z_\beta + c^K_\alpha, \; (z_\alpha) \in \Omega_K,$$

being piecewise affine is continuous. Hence we conclude that $v_\alpha \in C(\bar\Omega)$ and for a fixed $z_3 \in (-c, c)$ it is also piecewise affine.

The partition just introduced enables us to exploit the local character of the functionals $J^1_{\varepsilon\varepsilon}$. Let $\psi^\delta_K \in \mathbf{D}(\Omega_K)$ be such that $0 \le \psi^\delta_K \le 1$ and $\psi^\delta_K(z_\alpha) = 1$ for $(z_\alpha) \in \Omega^\delta_K$. With every family of functions $v^K \in \mathcal{W}(\mathcal{Y})$ we associate the following sequences

$$v^{\varepsilon,\delta}_\alpha(z) = v_\alpha(z) + \varepsilon \sum_{K\in\mathcal{K}} \psi^\delta_K(z_\beta) v^K_\alpha \left(\frac{z_\gamma}{\varepsilon}, z_3\right) ,$$

$$v^{\varepsilon,\delta}_3(z) = v_3(z_\alpha) + \varepsilon^2 \sum_{K\in\mathcal{K}} \psi^\delta_K(z_\beta) v^K_3 \left(\frac{z_\gamma}{\varepsilon}, z_3\right) . \tag{2.10.136}$$

It is evident that for $\varepsilon \to 0$ one has $v^{\varepsilon,\delta} \to v$ strongly in $L^2(B)^3$.

Let us take $t \in (0, 1)$. It is not difficult to show that

$$t\mathbf{Q}^\varepsilon e^z\left(v^{\varepsilon,\delta}\left(\frac{z_\alpha}{\varepsilon}, z_3\right)\right) = t\psi^\delta_K(z_\alpha)[e^z(v^K)\left(\frac{z_\beta}{\varepsilon}, z_3\right) + \epsilon^K + z_3\rho^K]$$

$$+ t(1 - \psi^\delta_K(z_\alpha))(\epsilon^K + z_3\rho^K) + (1-t)\frac{\varepsilon t}{1-t}\left(\psi^\delta_{K,(i}(z_\alpha) v^K_{j)}\left(\frac{z_\beta}{\varepsilon}, z_3\right)\right) ,$$

where

$$\psi^\delta_{K,(i}(z_\alpha)v^K_{j)}\left(\frac{z_\beta}{\varepsilon},z_3\right)=\frac{1}{2}\left[\frac{\partial\psi^\delta_K}{\partial z_i}v^K_j\left(\frac{z_\beta}{\varepsilon},z_3\right)+\frac{\partial\psi^\delta_K}{\partial z_j}v^K_i\left(\frac{z_\beta}{\varepsilon},z_3\right)\right],$$

and $(z_\alpha,z_3)\in\Omega_K\times(-c,c)$.

For instance, let us find $t(\mathbb{Q}^\varepsilon e(v^{\varepsilon,\delta}))_{\alpha3}$. We calculate

$$t(\mathbb{Q}^\varepsilon e^z(v^{\varepsilon,\delta}))_{\alpha3}=t\frac{1}{\varepsilon}e_{\alpha3}(v^{\varepsilon,\delta})=t\frac{1}{\varepsilon}\frac{1}{2}\left(\frac{\partial v^{\varepsilon,\delta}_\alpha}{\partial z_3}+\frac{\partial v^{\varepsilon,\delta}_3}{\partial z_\alpha}\right)$$

$$=t\frac{1}{\varepsilon}\frac{1}{2}\sum_{\beta=1}^{2}\left(\rho^K_{\alpha\beta}z_\beta+a^K_\alpha+\varepsilon\psi^\delta_K\frac{\partial v^K_\alpha}{\partial z_3}-\rho^K_{\alpha\beta}z_\beta-a^K_\alpha+\varepsilon\psi^\delta_K\frac{\partial v^K_3}{\partial z_\alpha}+\varepsilon^2\frac{\partial\psi^\delta_K}{\partial z_\alpha}v^K_3\right)$$

$$=t\psi^\delta_K e_{\alpha3}(v^K)\left(\frac{z_\beta}{\varepsilon},z_3\right)+(1-t)\frac{\varepsilon t}{2(1-t)}\frac{\partial\psi^\delta_K}{\partial z_\alpha}v^K_3\left(\frac{z_\beta}{\varepsilon},z_3\right)$$

$$=t\psi^\delta_K\left[e_{\alpha3}(v^K)\left(\frac{z_\beta}{\varepsilon},z_3\right)+\epsilon^K_{\alpha3}+z_3\rho^K_{\alpha3}\right]+t(1-\psi^\delta_K)(\epsilon^K_{\alpha3}+z_3\rho^K_{\alpha3})$$

$$+(1-t)\frac{\varepsilon t}{1-t}\psi^\delta_{K,(\alpha}(z_\beta)v_{3)}\left(\frac{z_\gamma}{\varepsilon},z_3\right),\quad(z_\alpha,z_3)\in\Omega_K\times(-c,c),$$

because $\epsilon^K_{\alpha3}=0$, $\rho^K_{\alpha3}=0$ and $\partial\psi^K/\partial z_3=0$. By using (2.10.7), the convexity of the function $j_\varepsilon(z,\cdot)$ and noting that $t\psi^\delta_K+t(1-\psi^\delta_K)+(1-t)=1$, one gets

$$J^1_{\varepsilon\varepsilon}(tv^{\varepsilon\delta})=\sum_{K\in\mathcal{K}}\int_{\Omega_K}\int_{-c}^{c}j\left[\frac{z_\alpha}{\varepsilon},z_3,t\mathbb{Q}^\varepsilon e(v^{\varepsilon\delta})\right]dz$$

$$=\sum_{K\in\mathcal{K}}\int_{\Omega_K}\int_{-c}^{c}j[\frac{z_\alpha}{\varepsilon},z_3,t\psi^\delta_K(z_\alpha)\left(e(v^K)\left(\frac{z_\alpha}{\varepsilon},z_3\right)+\epsilon^K+z_3\rho^K\right)$$

$$+t(1-\psi^\delta_K(z_\alpha))(\epsilon^K+z_3\rho^K)+(1-t)\frac{\varepsilon t}{1-t}\left(\psi^\delta_{K,(i}v_{j)}\left(\frac{z_\beta}{\varepsilon},z_3\right)\right)]dz$$

$$\leq\sum_{K\in\mathcal{K}}\int_{\Omega_K}\left\{\int_{-c}^{c}j\left[\frac{z_\alpha}{\varepsilon},z_3,e(v^K)\left(\frac{z_\alpha}{\varepsilon},z_3\right)+\epsilon^K+z_3\rho^K\right]dz_3\right\}dz$$

$$+\sum_{K\in\mathcal{K}}m\int_{\Omega_K}(1-\psi^\delta_K)dz\int_{-c}^{c}\mid\epsilon^K+z_3\rho^K\mid^2dz_3$$

$$+\sum_{K\in\mathcal{K}}m(1-t)\int_{\Omega_K}\int_{-c}^{c}\left(\frac{\varepsilon t}{1-t}\right)^2\left|\left(\psi^\delta_{K,(i}(z_\alpha)v^K_{j)}\left(\frac{z_\beta}{\varepsilon},z_3\right)\right)\right|^2dz,$$

because $j\geq0$. Now let ε tend to zero. By using Step 1, we arrive at

$$\limsup_{\varepsilon\to0}J^1_{\varepsilon\varepsilon}(tv^{\varepsilon\delta})\leq\sum_{K\in\mathcal{K}}\{h_0\,|\Omega_K|\prec j[y,e^y(v^K)+\epsilon^K+y_3\rho^K]\succ$$

$$+ m(1-t) \int_{\Omega_K} (1 - \psi_K^\delta) dz \int_{-c}^{c} \mid \boldsymbol{\epsilon}^K + z_3 \boldsymbol{\rho}^K \mid^2 dz_3 \} \,,$$

where $|\Omega_K| = $ the Lebesgue measure of Ω_K. Next, let $t \to 1$ and $\delta \to 0$, then

$$\limsup_{\substack{t \to 1 \\ \delta \to 0}} \; \limsup_{\varepsilon \to 0} J^1_{\varepsilon\varepsilon}(t\boldsymbol{v}^{\varepsilon\delta}) \le \sum_{K \in \mathcal{K}} \{ h_0 \, |\Omega_K| \; \prec j[\boldsymbol{y}, e^y(\boldsymbol{v}^K) + \boldsymbol{\epsilon}^K + y_3 \boldsymbol{\rho}^K] \succ \} \,.$$

By applying Lemma 1.3.27, one can construct a mapping $\varepsilon \to (t(\varepsilon), \delta(\varepsilon))$ with $(t(\varepsilon), \delta(\varepsilon)) \to (1, 0)$ such that setting $\boldsymbol{v}^\varepsilon = t(\varepsilon) \boldsymbol{v}^{\varepsilon, \delta(\varepsilon)}$, we conclude that

$$\boldsymbol{v}^\varepsilon \to \boldsymbol{v} \text{ strongly in } L^2(B)^3 \,,$$

and

$$\limsup_{\varepsilon \to 0} J^1_{\varepsilon\varepsilon}(\boldsymbol{v}^\varepsilon) \le \sum_{K \in \mathcal{K}} \{ h_0 \, |\Omega_K| \; \prec j[\boldsymbol{y}, e^y(\boldsymbol{v}^K) + \boldsymbol{\epsilon}^K + y_3 \boldsymbol{\rho}^K] \succ \} \,.$$

By taking the infimum on the r.h.s. of the last inequality for $\boldsymbol{v}^K$ running over $W(\mathcal{Y})$ one obtains

$$J^{1s}_{\mathcal{H}}(\boldsymbol{u}, w) \le \limsup_{\varepsilon \to 0} J^1_{\varepsilon\varepsilon}(\boldsymbol{v}^\varepsilon) \le \sum_{K \in \mathcal{K}} |\Omega_K| \, j_{\mathcal{H}}(\boldsymbol{\epsilon}^K, \boldsymbol{\rho}^K)$$

$$= \int_\Omega j_{\mathcal{H}}[e^z(\boldsymbol{u}), \boldsymbol{\kappa}^z(w)] dz = J^1_{\mathcal{H}}(\boldsymbol{u}, w) \,,$$

where (2.10.135) has been taken into account and $J^{1s}_{\mathcal{H}}$ stands for the Γ-limit superior.

The properties of the effective elastic potential imply that the functional $J^1_{\mathcal{H}}$ is convex and finite on $H^1(\Omega)^2 \times H^2(\Omega)$, thus it is also continuous on this space. By applying density Propositions 1.4.2, 1.4.16, we conclude the proof of (2.10.130).

II. The second part consists in proving that for any sequence $\{\boldsymbol{v}^\varepsilon\}_{\varepsilon > 0} \subset H^1(B)^3$ converging to $\boldsymbol{v} \in V_K(B)$ strongly in $L^2(B)^3$, the following inequality is satisfied:

$$J^1_{\mathcal{H}}(\boldsymbol{u}, w) \le \liminf_{\varepsilon \to 0} J^1_{\varepsilon\varepsilon}(\boldsymbol{v}^\varepsilon) \,, \tag{2.10.137}$$

where, as in the previous case, $v_\alpha = u_\alpha - z_3 \dfrac{\partial w}{\partial z_\alpha}$, $v_3 = w$, $u_\alpha \in H^1(\Omega)$ and $w \in H^2(\Omega)$.

As we know from Section 2.10.2, if $\boldsymbol{v} \in H^1(B)^3 \backslash V_K(B)$, then

$$\liminf_{\varepsilon \to 0} J^1_{\varepsilon\varepsilon}(\boldsymbol{v}^\varepsilon) = +\infty \,,$$

and inequality (2.10.137) is trivially satisfied.

By using duality arguments we claim that

$$\liminf_{\varepsilon \to 0} J^1_{\varepsilon\varepsilon}(\boldsymbol{v}^\varepsilon) \ge J^1_{\mathcal{H}}(\boldsymbol{u}, w) = \int_\Omega j_{\mathcal{H}}(e^z(\boldsymbol{u}), \boldsymbol{\kappa}^z(w)) dz = \sup \{ \int_\Omega [\boldsymbol{N} \, : \, e^z(\boldsymbol{u})$$

$$+ \boldsymbol{M} \, : \, \boldsymbol{\kappa}^z(w) - j^*_{\mathcal{H}}(\boldsymbol{N}(z_\alpha), \boldsymbol{M}(z_\alpha))] dz \, | \, \boldsymbol{N}, \boldsymbol{M} \in L^2(\Omega, \mathbb{E}^s_2) \} \,.$$

Step 3. First we take

$$N(z_\alpha) = \chi^K(z_\alpha) N_K , \qquad N_K \in \mathbb{E}_s^2 ,$$

where

$$\chi^K(z_\alpha) = \begin{cases} 1 & \text{if} \quad (z_\alpha) \in \Omega_K , \\ 0 & \text{if} \quad (z_\alpha) \in \Omega_K . \end{cases}$$

Now $\{\Omega_K\}_{K \in \mathcal{K}}$ is a finite family of open disjoint sets such that $\bar{\Omega} = \bigcup_K \bar{\Omega}_K$.

Let $T_K \in \mathcal{S}(\mathcal{Y})$, $K \in \mathcal{K}$. By applying Lemma 2.10.12 and recalling that $j \geq 0$ we write

$$\liminf_{\varepsilon \to 0} J_{\varepsilon\varepsilon}^1(v^\varepsilon) \geq \liminf_{\varepsilon \to 0} \sum_{K \in \mathcal{K}} \int_{\Omega_K} \psi_K^\delta(z_\alpha) \{ \int_{-c}^{c} [j\left(\frac{z_\beta}{\varepsilon}, z_3, \mathbb{Q}^\varepsilon e^z(v^\varepsilon)\right)$$

$$- T_K^{ij}\left(\frac{z_\alpha}{\varepsilon}, z_3\right) : (\mathbb{Q}^\varepsilon e_{ij}^z(v^\varepsilon))_{ij}] dz_3 + M : \kappa^z(w) \} dz$$

$$= \liminf_{\varepsilon \to 0} \sum_{K \in \mathcal{K}} \int_{\Omega_K} \psi_K^\delta \{ M : \kappa^z(w) + \int_{-c}^{c} [j\left(\frac{z_\alpha}{\varepsilon}, z_3, \mathbb{Q}^\varepsilon e^z(v^\varepsilon)\right)$$

$$- T_{K\varepsilon} : \mathbb{Q}^\varepsilon e^z(v^\varepsilon)] dz_3 \} dz , \tag{2.10.138}$$

where $T_{K\varepsilon}(z) = T_K\left(\frac{z_\alpha}{\varepsilon}, z_3\right)$.

We set

$$j_{T_{K\varepsilon}}\left(\frac{z_\alpha}{\varepsilon}, z_3, \epsilon\right) = j\left(\frac{z_\alpha}{\varepsilon}, z_3, \mathbb{Q}^\varepsilon \epsilon\right) - T_{K\varepsilon} : \mathbb{Q}^\varepsilon \epsilon . \tag{2.10.139}$$

Fenchel's conjugate of $j_{T_{K\varepsilon}}\left(\frac{z_\alpha}{\varepsilon}, z_3, \cdot\right)$ is calculated as follows:

$$j_{T_{K\varepsilon}}^*\left(\frac{z_\alpha}{\varepsilon}, z_3, \epsilon^*\right) = \sup\{\epsilon^* : \epsilon + T_{K\varepsilon} : \mathbb{Q}^\varepsilon \epsilon - j\left(\frac{z_\alpha}{\varepsilon}, z_3, \mathbb{Q}^\varepsilon \epsilon\right) \mid \epsilon \in \mathbb{E}_s^3\}$$

$$= \sup\{(\mathbb{Q}^{-\varepsilon}\epsilon^* + T_{K\varepsilon}) : \mathbb{Q}^\varepsilon \epsilon - j\left(\frac{z_\alpha}{\varepsilon}, z_3, \mathbb{Q}^\varepsilon \epsilon\right) \mid \epsilon \in \mathbb{E}_s^3\}$$

$$= \sup\{(\mathbb{Q}^{-\varepsilon}\epsilon^* + T_{K\varepsilon}) : \epsilon' - j\left(\frac{z_\alpha}{\varepsilon}, z_3, \epsilon'\right) \mid \epsilon' \in \mathbb{E}_s^3\}$$

$$= j^*\left(\frac{z_\alpha}{\varepsilon}, z_3, \mathbb{Q}^{-\varepsilon}\epsilon^* + T_{K\varepsilon}\right) , \tag{2.10.140}$$

where $\epsilon^* \in \mathbb{E}_3^s$. By applying Fenchel's inequality to $j_{T_{K\varepsilon}}\left(\frac{z_\alpha}{\varepsilon}, z_3, \cdot\right)$ at the point $[\mathbb{Q}^\varepsilon e^z(v^\varepsilon(z)), \frac{1}{2c} N_K]$, where $N_K^{i3} = 0$, we get

$$j_{T_{K\varepsilon}}\left(\frac{z_\alpha}{\varepsilon}, z_3, \mathbb{Q}^\varepsilon e^z(v^\varepsilon)\right) \geq \frac{1}{2c} N_K : \mathbb{Q}^\varepsilon e^z(v^\varepsilon) - j_{T_{K\varepsilon}}^*\left(\frac{z_\alpha}{\varepsilon}, z_3, \frac{1}{2c} N_K\right) \tag{2.10.141}$$

$$= \frac{1}{2c} N_K : \mathbb{Q}^\varepsilon e^z(v^\varepsilon) - j^*\left(\frac{z_\alpha}{\varepsilon}, z_3, \frac{1}{2c} \mathbb{Q}^{-\varepsilon} N_K + T_{K\varepsilon}\right) .$$

Taking into account (2.10.139) – (2.10.141) in (2.10.138) one obtains

$$\liminf_{\varepsilon \to 0} J^1_{\varepsilon\varepsilon}(v^\varepsilon) \geq \liminf_{\varepsilon \to 0} \sum_{K \in \mathcal{K}} \{ \int_{\Omega_K} \psi^\delta_K M : \kappa^z(w)$$

$$+ \int_{-c}^{c} \left[\frac{1}{2c} N^{\alpha\beta}_K e^z_{\alpha\beta}(v^\varepsilon) - j^* \left(\frac{z_\alpha}{\varepsilon}, z_3, \frac{1}{2c} Q^{-\varepsilon} N_K + T_{K\varepsilon} \right) \right] dz_3 \} dz . \qquad (2.10.142)$$

It is evident that

$$\lim_{\varepsilon \to 0} \int_\Omega \psi^\delta_K \int_{-c}^{c} \frac{1}{2c} N^{\alpha\beta}_K e^z_{\alpha\beta}(v^\varepsilon) dz$$

$$= \int_\Omega \psi^\delta_K \int_{-c}^{c} \frac{1}{2c} N^{\alpha\beta}_K e^z_{\alpha\beta}(u - z_3 \nabla w) dz = \int_\Omega \psi^\delta_K N^{\alpha\beta}_K e^z_{\alpha\beta}(u) dz . \qquad (2.10.143)$$

The sequence of εY-periodic functions $j^* \left[\frac{\cdot}{\varepsilon}, z_3, \frac{1}{2c} Q^{-\varepsilon} N_K + T_K \left(\frac{\cdot}{\varepsilon}, z_3 \right) \right]$ is bounded in $L^1(B)$ and weakly convergent to

$$\langle j^* [y_\alpha, y_3, \frac{1}{2c} N_K + T_K(y)] \rangle \quad \text{as} \quad \varepsilon \to 0 . \qquad (2.10.144)$$

We recall that $z_3 = y_3$, $2c = h_0$ and $N^{i3}_K = 0$. From (2.10.142) – (2.10.144) we obtain

$$\liminf_{\varepsilon \to 0} J^1_{\varepsilon\varepsilon}(v^\varepsilon) \geq \sum_{K \in \mathcal{K}} \{ \int_{\Omega_K} \psi^\delta_K [N^{\alpha\beta}_K e^z_{\alpha\beta}(u) + M^{\alpha\beta} \kappa^z_{\alpha\beta}(w)] dz$$

$$- h_0 \int_{\Omega_K} \psi^\delta_K dz \prec j^* [y, \frac{1}{2c} N_K + T_K(y)] \succ \} .$$

Passing to the supremum on the r.h.s. of the last inequality when T_K runs over $\mathcal{S}(\mathcal{Y})$, where $\rho^* = M$, we arrive at

$$\liminf_{\varepsilon \to 0} J^1_{\varepsilon\varepsilon}(v^\varepsilon) \geq \sum_{K \in \mathcal{K}} \int_{\Omega_K} \psi^\delta_K [N^{\alpha\beta}_K e^z_{\alpha\beta}(u) + M^{\alpha\beta} \kappa^z_{\alpha\beta}(w)] dz$$

$$- \int_{\Omega_K} \psi^\delta_K j^*_\mathcal{H}(N_K, M) dz , \qquad (2.10.145)$$

because $\sup(-\int) = -\inf \int$. Recalling that N is a step function, we obtain

$$\liminf_{\varepsilon \to 0} J^1_{\varepsilon\varepsilon}(v^\varepsilon) \geq \int_\Omega \sum_{K \in \mathcal{K}} \psi^\delta_K(z_\alpha) [N(z_\beta) : e^z(u(z_\gamma))$$

$$+ M(z_\alpha) \kappa^z(w(z_\beta))] dz - \int_\Omega \sum_{K \in \mathcal{K}} \psi^\delta_K(z_\alpha) j^*_\mathcal{H}(N(z_\beta), M(z_\beta)) dz . \qquad (2.10.146)$$

The inequality $0 \leq \sum\limits_{K \in \mathcal{K}} \psi_K^\delta \leq 1$ implies

$$0 \leq \sum_{K \in \mathcal{K}} \psi_K^\delta(z_\alpha) j_{\mathcal{H}}^*(N(z_\beta), M(z_\beta)) \leq j_{\mathcal{H}}^*(N(z_\alpha), M(z_\alpha)) \,,$$

because $j_{\mathcal{H}}^* \geq 0$. Thus we get

$$\lim_{\varepsilon \to 0} \inf J_{\varepsilon\varepsilon}^1(v^\varepsilon) \geq \int_\Omega \sum_{K \in \mathcal{K}} \psi_K^\delta(z_\alpha)[N(z_\beta) : e^z(u(z_\gamma))$$

$$+M(z_\beta) : \kappa^z(w(z_\gamma))]dz - \int_\Omega j_{\mathcal{H}}^*(N(z_\alpha), M(z_\alpha))dz \,.$$

When $\delta \to 0$, $\sum\limits_{K \in \mathcal{K}} \psi_K^\delta$ tends to 1 for a.e. $(z_\alpha) \in \Omega$ and (2.10.146) yields

$$\lim_{\varepsilon \to 0} \inf J_{\varepsilon\varepsilon}^1(v^\varepsilon) \geq \int_\Omega [N(z_\alpha) : e^z(u(z_\beta))$$

$$+M(z_\alpha) : \kappa^z(z_\beta) - j_{\mathcal{H}}^*(N(z_\alpha), M(z_\alpha))]dz \,.$$

Step 4. For each $N \in L^2(\Omega, \mathbb{E}_s^2)$ there exists a sequence of step functions $\{N_A\}_{A \in \mathbf{N}} \subset L^2(\Omega, \mathbb{E}_s^2)$ such that

$$N_A \to N \quad \text{strongly in} \quad L^2(\Omega, \mathbb{E}_s^2) \text{ as } A \to \infty \,.$$

Each function N_A can be represented in the form

$$N_A(z_\alpha) = \sum_{K(A)} \chi_{K(A)}^{\delta_A}(z_\alpha) N_{K(A)} \,,$$

where

$$\chi_{K(A)}^{\delta_A}(z_\alpha) = \begin{cases} 1, & \text{if} \quad (z_\alpha) \in \Omega_{K(A)} \,, \\ 0, & \text{otherwise} \,, \end{cases}$$

$\delta_A = 1/A$, $\operatorname{diam} \Omega_{K(A)} \leq \delta_A$ and $\bar{\Omega} = \bigcup\limits_{K(A)} \bar{\Omega}_{K(A)}$. By using the previous step one has

$$\lim_{\varepsilon \to 0} \inf J_{\varepsilon\varepsilon}^1(v^\varepsilon) \geq \int_\Omega [N_A(z_\alpha) : e^z(u(z_\beta)) + M(z_\alpha) : \kappa^z(z_\beta) - j_{\mathcal{H}}^*(N_A(z_\alpha), M(z_\alpha))]dz \,.$$

Passing to the limit on the r.h.s. of the last inequality we eventually obtain

$$\lim_{\varepsilon \to 0} \inf J_{\varepsilon\varepsilon}^1(v^\varepsilon) \geq \int_\Omega [N(z_\alpha) : e^z(u(z_\beta)) + M(z_\alpha) : \kappa^z(z_\beta) - j_{\mathcal{H}}^*(N(z_\alpha), M(z_\alpha))]dz \,,$$

which concludes the proof. $\square$

Remark 2.10.14. In the second part of the above proof $M(z_\alpha)$ can be approximated similarly as $N(z_\alpha)$, since M also belongs to $L^2(\Omega, \mathbb{E}_s^2)$. The proof is not influenced by such an approximation.

Remark 2.10.15. The reader will now be able to perform dual homogenization and formulate a counterpart of Theorem 2.10.5. $\qquad\square$

Remark 2.10.16. The scaling $(2.2.3)_2$ concerning the body forces adopted in Section 2.2 has balanced the fact that the volume tends faster to zero than the area. Consequently the body forces entered into the homogenized problem, see $(2.3.44) - (2.3.46)$. In this section we have proceeded differently and have not treated body forces in such a special way. Thus, by necessity, they do not enter the functional L given by $(2.10.28)$ and do not affect the homogenized solutions. The scaling $(2.10.1)$ was proposed by Caillerie (1984). We observe that one can either scale the elastic moduli or the loading.

Remark 2.10.17. The method of determination of the effective stiffnesses of plates presented in Sections 2.3 and 2.10.4 refers also to the case of hollow plates, widely used in the building industry as ceiling plates. In the special case of hollows going in one direction the local problems $(2.3.13)$ and $(2.3.34)$ are posed on a plane domain, which simplifies the numerical algorithm, see Fig. 2.10.1. If the openings have slender cross-sections and are located transversely symmetric with respect to the middle plane one can apply the simplified formulae of Section 3.7. In the case of transverse asymmetry one should use more complicated formulae derived in Lewiński (1995).

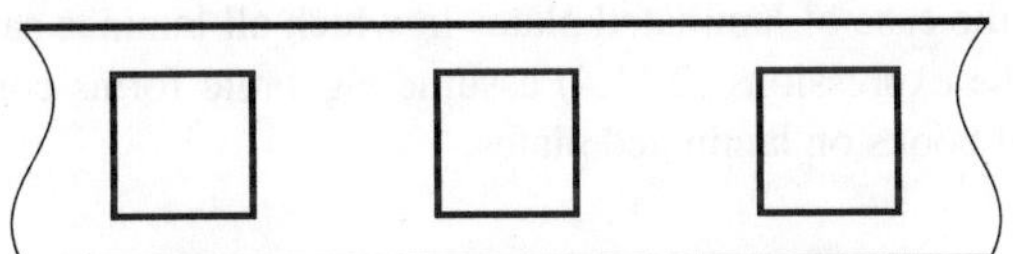

Fig. 2.10.1. A plate hollowed in one direction

2.11. *Effective stiffnesses of longitudinally homogeneous plates*

Consider a plate of constant thickness whose elastic moduli C^{ijkl} vary only in transverse direction. Thus c^- and c^+ are constant and

$$C^{ijkl} = C^{ijkl}(y_3), \qquad c^{\alpha\beta} = c^{\alpha\beta}(y_3),$$
$$h_0 = c^+ - c^-, \qquad \langle y_3 \rangle = \frac{1}{2}(c^- + c^+). \tag{2.11.1}$$

The planes $y_3 =$ const are assumed to be planes of material symmetry, i.e. $(2.4.1)$ holds. Under these conditions the decompositions $(2.7.2)$ take place with

$$\overline{\Theta}^{(\alpha\beta)} = 0, \qquad \overline{Xi}^{\alpha\beta}(y_3) = 0. \tag{2.11.2}$$

Thus by (2.7.7) we find

$$\Theta^{(\alpha\beta)}_{3|3} = -c^{\alpha\beta}(y_3) \,, \qquad \Xi^{(\alpha\beta)}_{3|3} = -\hat{y}_3 c^{\alpha\beta}(y_3) \,,$$
$$\Theta^{(\alpha\beta)}_{3|\delta} = \Theta^{(\alpha\beta)}_{\sigma|\gamma} = 0 \,, \qquad \Xi^{(\alpha\beta)}_{3|\delta} = \Xi^{(\alpha\beta)}_{\sigma|\gamma} = 0 \,. \tag{2.11.3}$$

Substitution of (2.11.3) into (2.3.21) and (2.3.38) gives

$$A^{\alpha\beta\lambda\mu}_y = \frac{1}{h_0}\int_{c^-}^{c^+} \widetilde{C}^{\alpha\beta\lambda\mu}(y_3)dy_3 \,,$$

$$E^{\alpha\beta\lambda\mu}_y = \frac{1}{h_0}\int_{c^-}^{c^+} \hat{y}_3\widetilde{C}^{\alpha\beta\lambda\mu}(y_3)dy_3 \,, \qquad F^{\alpha\beta\lambda\mu}_y = E^{\lambda\mu\alpha\beta}_y \,, \tag{2.11.4}$$

$$D^{\alpha\beta\lambda\mu}_y = \frac{1}{h_0}\int_{c^-}^{c^+} (\hat{y}_3)\widetilde{C}^{\alpha\beta\lambda\mu}(y_3)dy_3 \,,$$

where $\hat{y}_3 = y_3 - \frac{1}{2}(c^- + c^+)$. These formulae coincide with the conventional formulae for the stiffnesses of transversely nonhomogeneous plates. Note that the tensor $\widetilde{C}$ defined by (2.7.3), representing the reduced moduli of the plane stress state, appears here without extra stress-type assumptions.

Remark 2.11.1 In the case of laminated plates in which all laminae satisfy the conditions mentioned above the expressions (2.11.4) assume algebraic forms coinciding with those reported in classical books on laminated plates.

3. Thin plates in bending and stretching

3.1. Kirchhoff type description

Consider once again the plate problem of Section 2.1 with two additional simplifications: the body forces will be omitted and planes $x_3 = $ const will be assumed as planes of material symmetry, i.e. relations (2.4.1) hold here. In the approach presented here, the asymptotic analysis is preceded by the Kirchhoff construction of a two-dimensional plate model. This modelling is based on:

i) kinematic assumptions:

$$w_\alpha(x, x_3) = u_\alpha(x) - \hat{x}_3 w_{,\alpha} , \quad \hat{x}_3 = x_3 - x_3^0;$$
$$w_3(x, x_3) = w(x); \tag{3.1.1}$$

$$x_3^0 = \frac{1}{2} \int\limits_Z [(x_3^+)^2 - (x_3^-)^2] dx / \int\limits_Z (x_3^+ - x_3^-) dx, \tag{3.1.2}$$

accordingly to the definition of $\hat{y}_3$, see Eq. (2.3.17).

ii) Stress assumptions: stress-strain relations have the form

$$\sigma^{\alpha\beta} = \tilde{C}_Z^{\alpha\beta\lambda\mu} \epsilon_{\lambda\mu}, \qquad \epsilon_{\lambda\mu} = e_{\lambda\mu}(\boldsymbol{u}) ; \tag{3.1.3}$$

where

$$\tilde{C}_Z^{\alpha\beta\lambda\mu} = C_Z^{\alpha\beta\lambda\mu} - c_Z^{\lambda\mu} C_Z^{33\alpha\beta}, \tag{3.1.4}$$

$$c_Z^{\lambda\mu} = C_Z^{33\lambda\mu} / C_Z^{3333}. \tag{3.1.5}$$

The formula given by Eqs. (3.1.4), (3.1.5) determines so-called reduced moduli of the generalized plane-stress state. They follow from elimination of ϵ_{33} by assumption: $\sigma^{33} = 0$. As usual, the elasticity tensor C_Z is assumed to be positive definite.

Kinematic assumptions (3.1.1) are simultaneously imposed on trial kinematic fields $v = \tilde{v}$, Eq. (2.1.5). Thus

$$\sigma^{ij} e_{ij}(\tilde{\boldsymbol{v}}) = \sigma^{\alpha\beta} e_{\alpha\beta}(\tilde{\boldsymbol{v}}) = \sigma^{\alpha\beta} [e_{\alpha\beta}(\boldsymbol{v}) + \hat{x}_3 \kappa_{\alpha\beta}(\boldsymbol{v})], \tag{3.1.6}$$

where $\tilde{v}_\alpha = v_\alpha(x) - \hat{x}_3 v_{,\alpha}$, $\tilde{v}_3 = v$. Consequently, Eq. (2.1.5) assumes the form

$$\int\limits_\Omega [N^{\alpha\beta} e_{\alpha\beta}(\boldsymbol{v}) + M^{\alpha\beta} \kappa_{\alpha\beta}(v)] dx = \int\limits_\Omega [p_z^\alpha v_\alpha - m_z^\alpha v_{,\alpha} + q_z v] dx , \tag{3.1.7}$$

where the loadings are

$$p_z^\alpha(x) = (G_Z^+(x))^{\frac{1}{2}} r_+^\alpha(x) + (G_Z^-(x))^{\frac{1}{2}} r_-^\alpha(x),$$

$$m_z^\alpha(x) = (G_Z^+(x))^{\frac{1}{2}} (x_3^+(x) - x_3^0) r_+^\alpha(x) + (G_Z^-(x))^{\frac{1}{2}} (x_3^-(x) - x_3^0) r_-^\alpha(x), \tag{3.1.8}$$

$$q_z(x) = (G_Z^+(x))^{\frac{1}{2}} r_+^3(x) + (G_Z^-(x))^{\frac{1}{2}} r_-^3(x) .$$

The membrane forces and moments

$$N^{\alpha\beta} = \int_{x_3^-(x)}^{x_3^+(x)} \sigma^{\alpha\beta} dx_3, \qquad M^{\alpha\beta} = \int_{x_3^-(x)}^{x_3^+(x)} \hat{x}_3 \sigma^{\alpha\beta} dx_3 \tag{3.1.9}$$

are interrelated with deformations by

$$\begin{aligned}
N^{\alpha\beta} &= A_z^{\alpha\beta\lambda\mu} e_{\lambda\mu}(\boldsymbol{u}) + E_z^{\alpha\beta\lambda\mu} \kappa_{\lambda\mu}(w), \\
M^{\alpha\beta} &= F_z^{\alpha\beta\lambda\mu} e_{\lambda\mu}(\boldsymbol{u}) + D_z^{\alpha\beta\lambda\mu} \kappa_{\lambda\mu}(w) .
\end{aligned} \tag{3.1.10}$$

where

$$\begin{bmatrix} A_z^{\alpha\beta\lambda\mu}(x) \\ E_z^{\alpha\beta\lambda\mu}(x) \\ D_z^{\alpha\beta\lambda\mu}(x) \end{bmatrix} = \int_{x_3^-(x)}^{x_3^+(x)} \begin{bmatrix} 1 \\ \hat{x}_3 \\ (\hat{x}_3)^2 \end{bmatrix} \tilde{C}_Z^{\alpha\beta\lambda\mu}(x, x_3) dx_3 ; \tag{3.1.11}$$

and $F_z^{\alpha\beta\lambda\mu} = E_z^{\alpha\beta\lambda\mu} = E_z^{\lambda\mu\alpha\beta}$.

The equilibrium problem of a plate clamped along its edge $\Gamma = \partial\Omega$ reads:

$$(P^{(K)}) \quad \left| \begin{aligned} &\text{find } (\boldsymbol{u}, w) \in V_K^0(\Omega) = [H_0^1(\Omega)]^2 \times H_0^2(\Omega) \\ &\text{such that Eqs. (3.1.7) and (3.1.10) are satisfied} \\ &\text{for all } (\boldsymbol{v}, v) \in V_K^0(\Omega). \end{aligned} \right.$$

One can show that the problem $(P^{(K)})$ is uniquely solvable provided that $p_Z^\alpha, m_Z^\alpha, q_Z^\alpha$ are elements of the space $L^2(\Omega)$. The outline of the proof reads as follows. The positive definiteness condition (2.1.6) implies positivity of the energy density $N^{\alpha\beta} e_{\alpha\beta}(\boldsymbol{u}) + M^{\alpha\beta} \kappa_{\alpha\beta}(w)$. Consequently the form

$$\begin{aligned}
2\mathcal{W}(e_{\alpha\beta}, \kappa_{\alpha\beta}) = {}& A_z^{\alpha\beta\lambda\mu} e_{\alpha\beta} e_{\lambda\mu} + E_z^{\alpha\beta\lambda\mu} \kappa_{\lambda\mu} e_{\alpha\beta} + F_z^{\alpha\beta\lambda\mu} e_{\lambda\mu} \kappa_{\alpha\beta} \\
& + D_z^{\alpha\beta\lambda\mu} \kappa_{\alpha\beta} \kappa_{\lambda\mu}
\end{aligned} \tag{3.1.12}$$

is non-negative: $\mathcal{W} \geq 0$. Since the problem is finite-dimensional one can find $c_\alpha > 0$ such that

$$2\mathcal{W}(e, \kappa) \geq c_1 \sum_{\alpha,\beta} (e_{\alpha\beta})^2 + c_2 \sum_{\alpha,\beta} (\kappa_{\alpha\beta})^2 . \tag{3.1.13}$$

If we assume that all quantities are dimensionless then we can put $c_1 = c_2$.

By applying the Korn inequalities we note that the bilinear form of the left-hand side of (3.1.7) is V_K^0-elliptic. Since this form as well as the linear form on the right-hand side of (3.1.7) are continuous, we invoke the Lax-Milgram lemma to conclude that the problem $(P^{(K)})$ is uniquely solvable.

Let us pass now to the local formulation of the problem $(P^{(K)})$. We shall draw all local consequences of the variational equilibrium equation (3.1.7). First, note that $N^{\alpha\beta}e_{\alpha\beta}(v) = N^{\alpha\beta}v_{\alpha,\beta}$ by symmetry of N and hence

$$N^{\alpha\beta}e_{\alpha\beta}(\boldsymbol{v}) = (N^{\alpha\beta}v_\alpha)_{,\beta} - N^{\alpha\beta}{}_{,\beta}v_\alpha \ . \tag{3.1.14}$$

Moreover

$$M^{\alpha\beta}\kappa_{\alpha\beta}(v) = -(M^{\alpha\beta}v_{,\alpha})_{,\beta} - M^{\alpha\beta}{}_{,\alpha\beta}\,v + (M^{\alpha\beta}{}_{,\beta}\,v)_{,\alpha} \tag{3.1.15}$$

and consequently

$$\int_\Omega [N^{\alpha\beta}e_{\alpha\beta}(\boldsymbol{v}) + M^{\alpha\beta}\kappa_{\alpha\beta}(v)]dx$$

$$= -\int_\Omega [N^{\alpha\beta}_{,\beta}v_\alpha + M^{\alpha\beta}{}_{,\alpha\beta}v]dx + \int_\Gamma (N^{\alpha\beta}n_\beta v_\alpha - M^{\alpha\beta}n_\beta v_{,\alpha} + M^{\alpha\beta}_{,\beta}n_\alpha v)ds \ . \tag{3.1.16}$$

The right-hand side of (3.1.7) can be put in the form

$$\int_\Omega [p_z^\alpha v_\alpha + (q + m^\alpha_{z,\alpha})v]dx - \int_\Gamma m_z^\alpha n_\alpha v ds \ . \tag{3.1.17}$$

Here $\boldsymbol{n} = (n_\alpha)$ represents a unit vector outward normal to $\Gamma = \partial\Omega$. The clamping condition implies $v = 0$ and $v_{,\alpha=0}$ on Γ, and the contour integrals in (3.1.16) – (3.1.17) vanish. Then, equating these formulae and using the classical lemma of variational calculus leads to the local equilibrium equations in Ω:

$$-N^{\alpha\beta}_{,\beta} = p_z^\alpha \ , \qquad -M^{\alpha\beta}_{,\alpha\beta} = q + m^\alpha_{z,\alpha} \ . \tag{3.1.18}$$

Thus the local (or strong) formulation of the equilibrium problem of a clamped plate amounts to finding the fields $\boldsymbol{v}, v, e_{\alpha\beta}(\boldsymbol{v}), \kappa_{\alpha\beta}(v), N^{\alpha\beta}, M^{\alpha\beta}$ satisfying
– boundary conditions on Γ:

$$\boldsymbol{v} = \boldsymbol{0} \ , \qquad v = 0 \ , \qquad \frac{\partial v}{\partial \boldsymbol{n}} = 0 \text{ on } \Gamma \ , \tag{3.1.19}$$

– equilibrium equation (3.1.18) in Ω,
– constitutive relations (3.1.10).

Remark 3.1.1. The Kirchhoff theory of plates admits other types of boundary conditions on Γ. To disclose them let us substitute the expressions

$$v_{,\alpha} = \frac{\partial v}{\partial \boldsymbol{n}}n_\alpha + \frac{\partial v}{\partial s}\tau_\alpha \ , \qquad v_\alpha = v_n n_\alpha + v_\tau \tau_\alpha \ , \qquad v_n = v_\alpha n^\alpha \ , \qquad v_\tau = v_\alpha \tau^\alpha \ , \tag{3.1.20}$$

into (3.1.16) and introduce the notation

$$N_n = N^{\alpha\beta} n_\alpha n_\beta \,, \qquad N_\tau = N^{\alpha\beta} n_\alpha \tau_\beta \,,$$
$$M_n = M^{\alpha\beta} n_\alpha n_\beta \,, \qquad M_\tau = M^{\alpha\beta} n_\alpha \tau_\beta \,,$$
$$Q = M^{\alpha\beta}{}_{,\beta} n_\alpha + \frac{\partial M_\tau}{\partial s} \,. \tag{3.1.21}$$

Here τ represents a unit vector tangent to Γ; $\partial/\partial\tau = \partial/\partial s$. Q is called the effective Kirchhoff force. Thus we find

$$\int_\Omega [N^{\alpha\beta} e_{\alpha\beta}(\boldsymbol{v}) + M^{\alpha\beta} \kappa_{\alpha\beta}(v)] dx = -\int_\Omega [N^{\alpha\beta}_{,\beta} v_\alpha + M^{\alpha\beta}{}_{,\alpha\beta} v] dx$$
$$+ \int_\Gamma \left(N_n v_n + N_\tau v_\tau + M_n \left(-\frac{\partial v}{\partial \boldsymbol{n}}\right) + Q v \right) ds + L_R \,, \tag{3.1.22}$$

where

$$L_R = -\int_\Gamma \frac{\partial}{\partial s}(M_\tau v) ds \,. \tag{3.1.23}$$

Assume that Γ has m corner points O_k at $s = s_k$, $k = 1, \ldots, m$, where n and τ jump. Then

$$L_R = -\sum_{k=1}^{m} [M_\tau(s_k - 0) v(s_k - 0) - M_\tau(s_{k-1} + 0) v(s_{k-1} + 0)] \tag{3.1.24}$$

with $s_0 = s_m$. Assuming that v is continuous at s_k we represent L_R in the form

$$L_R = \sum_{k=1}^{m} R_k v(s_k) \,, \tag{3.1.25}$$

where

$$R_k = M_\tau(s_k + 0) - M_\tau(s_k - 0) \tag{3.1.26}$$

or $R_k = [\![M_\tau]\!](s_k)$.

The theory of thin plates admits $2^4 = 16$ types of boundary conditions, since one can assign the values of N_n or v_n, N_τ or v_τ, M_n or $(-\partial v/\partial n)$, Q or v. The most important boundary conditions are

(i) free edge; N_n, N_τ, M_n, Q are given,

(ii) simply supported sliding edge; N_n, N_τ, M_n, v are given,

(iii) immovable supported edge; v_n, v_τ, M_n, v are given,

(iv) clamped edge; v_n, $v_\tau - \partial v/\partial n$, v are given.

At free corner points O_k, $R_k = 0$. If $v(s_k) = 0$, then R_k are concentrated reactions and are a priori unknown.

Assume that the plate is rectangular with corners at points $O_1(0,0)$, $O_2(a,0)$, $O_3(a,b)$, $O_4(0,b)$. If the field $M^{12}(x)$ is continuous in $\bar{\Omega}$, then

$$R_i = 2(-1)^i \cdot M^{12}(O_i) \, . \tag{3.1.27}$$

3.2. Asymptotic homogenization. In-plane scaling approach

If dimensions of $Z = (0, l_1^z) \times (0, l_2^z)$ are much smaller than $\mathrm{diam}(\Omega)$ the $(P^{(K)})$ problem becomes intractable even by numerical methods. Thus it is thought helpful to use the two-scale asymptotic method in order to separate the local analysis within the periodicity cells from the global analysis of the plate behavior. The asymptotic analysis used in this section will be called in-plane scaling approach, since, in contrast to the scaling used in Section 2.2, the scaling concerns only the in-plane dimensions l_α^z of the cell of periodicity. Instead of considering problem $(P^{(K)})$ we shall deal with a family of problems $(P_\varepsilon^{(K)})$ indexed by $\varepsilon > 0$. This family is constructed by replacing

$$Z \rightsquigarrow \varepsilon Y, \quad l_\alpha^z \rightsquigarrow \varepsilon l_\alpha, \quad l_\alpha^z = \varepsilon_0 l_\alpha \, ,$$

$$x_3^\pm(x) \rightsquigarrow x^\pm \left(\frac{x}{\varepsilon}\right) = x_3^\pm \left(\frac{x\varepsilon_0}{\varepsilon}\right) = \varepsilon_0 c^\pm \left(\frac{x}{\varepsilon}\right) \, ,$$

$$G_Z^\pm(x) \rightsquigarrow G_\pm \left(\frac{x}{\varepsilon}\right) = G_Z^\pm \left(\frac{x\varepsilon_0}{\varepsilon}\right) \, , \tag{3.2.1a}$$

$$C_Z^{ijkl}(x, x_3) \rightsquigarrow \bar{C}^{ijkl} \left(\frac{x}{\varepsilon}, x_3\right) = C_Z^{ijkl} \left(\frac{\varepsilon_0}{\varepsilon} x, x_3\right) \, .$$

Consequently we replace

$$\boldsymbol{K}_z(x) \rightsquigarrow \boldsymbol{K} \left(\frac{x}{\varepsilon}\right) = \boldsymbol{K}_z \left(\frac{x\varepsilon_0}{\varepsilon}\right) \, ; \quad \boldsymbol{K} = \boldsymbol{A}, \boldsymbol{E}, \boldsymbol{D} \, ;$$

$$p_z^\alpha(x) \rightsquigarrow \tilde{p}^\alpha \left(x, \frac{x}{\varepsilon}\right) \, , \quad m_z^\alpha(x) \rightsquigarrow \tilde{m}^\alpha \left(x, \frac{x}{\varepsilon}\right) \, , \tag{3.2.1b}$$

$$q_z(x) \rightsquigarrow \tilde{q} \left(x, \frac{x}{\varepsilon}\right) \, , \quad r_\pm^k(x) \rightsquigarrow r_\pm^k(x) \, ,$$

where

$$\tilde{p}^\alpha(x, y) = (G_+(y))^{\frac{1}{2}} r_+^\alpha(x) + (G_-(y))^{\frac{1}{2}} r_-^\alpha(x) \, ,$$

$$\tilde{m}^\alpha(x, y) = (G_+(y))^{\frac{1}{2}} (x^+(y) - x_3^0) r_+^\alpha(x) + (G_-(y))^{\frac{1}{2}} (x^-(y) - x_3^0) r_-^\alpha(x) \, , \tag{3.2.1c}$$

$$\tilde{q}(x, y) = (G_+(y))^{\frac{1}{2}} r_+^3(x) + (G_-(y))^{\frac{1}{2}} r_-^3(x) \, .$$

Thus moduli $K^{\alpha\beta\lambda\mu}$, $K = A, E, D$, are interrelated with moduli $\hat{K}^{\alpha\beta\lambda\mu}$ (cf. (2.7.13)) by

$$\boldsymbol{A}(y) = \varepsilon_0 \hat{\boldsymbol{A}}(y), \quad \boldsymbol{E}(y) = (\varepsilon_0)^2 \hat{\boldsymbol{E}}(y) \, , \quad \boldsymbol{D}(y) = (\varepsilon_0)^3 \hat{\boldsymbol{D}}(y) \, . \tag{3.2.1d}$$

The scaling (3.2.1a) will be named in-plane, since the transverse plate dimension remains intact under this scaling. The ε-dependent counterpart of problem $(P^{(K)})$ reads:

$$(P_\varepsilon^{(K)}) \quad \left|\begin{array}{l} \text{find } (\boldsymbol{u}^\varepsilon, w^\varepsilon) \in V_K^0(\Omega) \text{ such that} \\[2mm] \displaystyle\int_\Omega [N_\varepsilon^{\alpha\beta}(\boldsymbol{u}^\varepsilon, w^\varepsilon)e_{\alpha\beta}(\boldsymbol{v}) + M_\varepsilon^{\alpha\beta}(\boldsymbol{u}^\varepsilon, w^\varepsilon)\kappa_{\alpha\beta}(v)]dx = \\[2mm] \displaystyle= \int_\Omega [\tilde{p}^\alpha\left(x, \frac{x}{\varepsilon}\right)v_\alpha - \tilde{m}^\alpha\left(x, \frac{x}{\varepsilon}\right)v_{,\alpha} + \tilde{q}\left(x, \frac{x}{\varepsilon}\right)v]dx \\[2mm] \hphantom{xxxxxxxxxxxxxxxxxxxxxxx} \forall\ (\boldsymbol{v}, v) \in V_K^0(\Omega)\,, \end{array}\right. \tag{3.2.2}$$

where

$$N_\varepsilon^{\alpha\beta}(\boldsymbol{u}^\varepsilon, w^\varepsilon) = A^{\alpha\beta\lambda\mu}\left(\frac{x}{\varepsilon}\right)e_{\lambda\mu}(\boldsymbol{u}^\varepsilon) + E^{\alpha\beta\lambda\mu}\left(\frac{x}{\varepsilon}\right)\kappa_{\lambda\mu}(w^\varepsilon)\,,$$

$$M_\varepsilon^{\alpha\beta}(\boldsymbol{u}^\varepsilon, w^\varepsilon) = E^{\alpha\beta\lambda\mu}\left(\frac{x}{\varepsilon}\right)e_{\lambda\mu}(\boldsymbol{u}^\varepsilon) + D^{\alpha\beta\lambda\mu}\left(\frac{x}{\varepsilon}\right)\kappa_{\lambda\mu}(w^\varepsilon)\,. \tag{3.2.3}$$

We see that now $(P_{\varepsilon 0}^{(K)}) = (P^{(K)})$. We assume that $\tilde{p}^\alpha, \tilde{m}^\alpha, \tilde{q} \in L^2(\Omega \times Y)$.

Let us proceed now to construct asymptotic process of solving the $(P_\varepsilon^{(K)})$ problem. We use a two-scale asymptotic expansion method. The solution $\boldsymbol{u}^\varepsilon$, w^ε is represented in the form

$$u_\alpha^\varepsilon = u_x^{(0)}(x) + \varepsilon u_\alpha^{(1)}\left(x, \frac{x}{\varepsilon}\right) + \varepsilon^2 u_\alpha^{(2)}\left(x, \frac{x}{\varepsilon}\right) + \dots\,,$$

$$w^\varepsilon = w^{(0)}(x) + \varepsilon^2 w^{(2)}\left(x, \frac{x}{\varepsilon}\right) + \varepsilon^3 w^{(3)}\left(x, \frac{x}{\varepsilon}\right) + \dots\,. \tag{3.2.4}$$

The trial fields are expanded similarly

$$v^\alpha = v_\alpha^{(0)}(x) + \varepsilon v_\alpha^{(1)}\left(x, \frac{x}{\varepsilon}\right) + \varepsilon^2 v_\alpha^{(2)}\left(x, \frac{x}{\varepsilon}\right) + \dots\,,$$

$$v = v^{(0)}(x) + \varepsilon^2 v^{(2)}\left(x, \frac{x}{\varepsilon}\right) + \varepsilon^3 v^{(3)}\left(x, \frac{x}{\varepsilon}\right) + \dots\,. \tag{3.2.5}$$

It is assumed that $(\boldsymbol{u}^{(0)}, w^{(0)}) \in V_K^0$, $(\boldsymbol{v}^{(0)}, v^{(0)}) \in V_K^0$,

$$u_\alpha^{(1)}(x, \cdot), v_\alpha^{(1)}(x, \cdot) \in H_{per}^1(Y)\,; \qquad w^{(2)}(x, \cdot), v^{(2)}(x, \cdot) \in H_{per}^2(Y). \tag{3.2.6}$$

The spaces $H_{per}^\alpha(Y)$ have been defined in Sections 1.3 and 2.8.

The first-order terms of the deformation measures assume the form

$$e_{\alpha\beta}(\boldsymbol{u}^\varepsilon) = e_{\alpha\beta}^0 + O(\varepsilon)\,, \qquad e_{\alpha\beta}^0 = e_{\alpha\beta}(\boldsymbol{u}^{(0)}) + e_{\alpha\beta}^y(\boldsymbol{u}^{(1)}),$$

$$\kappa_{\alpha\beta}(w^\varepsilon) = \kappa_{\alpha\beta}^0 + O(\varepsilon^2), \qquad \kappa_{\alpha\beta}^0 = \kappa_{\alpha\beta}(w^{(0)}) + \kappa_{\alpha\beta}^y(w^{(2)})\,, \tag{3.2.7}$$

where the strain measures $e_{\alpha\beta}^y$, $\kappa_{\alpha\beta}^y$ are defined in Section 2.8.

The first-order terms of the stress and stress couple resultants read

$$N_\varepsilon^{\alpha\beta} = N_0^{\alpha\beta} + O(\varepsilon)\,, \qquad M_\varepsilon^{\alpha\beta} = M_0^{\alpha\beta} + O(\varepsilon)\,, \tag{3.2.8}$$

where

$$N_0^{\alpha\beta} = A^{\alpha\beta\lambda\mu}\left(\frac{x}{\varepsilon}\right) e_{\lambda\mu}^0 + E^{\alpha\beta\lambda\mu}\left(\frac{x}{\varepsilon}\right) \kappa_{\lambda\mu}^0, \tag{3.2.9}$$

$$M_0^{\alpha\beta} = E^{\alpha\beta\lambda\mu}\left(\frac{x}{\varepsilon}\right) e_{\lambda\mu}^0 + D^{\alpha\beta\lambda\mu}\left(\frac{x}{\varepsilon}\right) \kappa_{\lambda\mu}^0. \tag{3.2.10}$$

Further analysis is based on the averaging result (1.1.1) for Y-periodic functions.

Let us substitute (3.2.8) and $v_\alpha = v_\alpha^{(0)}$, $v = v^{(0)}$ into Eq. (3.2.2) and pass to zero with ε using (1.1.1). One finds

$$\int_\Omega [N_h^{\alpha\beta} e_{\alpha\beta}(\boldsymbol{v}^{(0)}) + M_h^{\alpha\beta}\kappa_{\alpha\beta}(v^{(0)})]dx = \int_\Omega (p^\alpha v_\alpha^{(0)} - m^\alpha v_{,\alpha}^{(0)} + qv^{(0)})dx \tag{3.2.11}$$

$$\forall\, (\boldsymbol{v}^{(0)}, v^{(0)}) \in V_K^0(\Omega)\,,$$

where

$$N_h^{\alpha\beta} = \langle N_0^{\alpha\beta}\rangle\,, \qquad M_h^{\alpha\beta} = \langle M_0^{\alpha\beta}\rangle\,; \tag{3.2.12}$$

$$p^\alpha(x) = \langle\tilde{p}^\alpha(x,y)\rangle\,,\ m^\alpha(x) = \langle\tilde{m}^\alpha(x,y)\rangle\,,\ q(x) = \langle\tilde{q}(x,y)\rangle\,,$$

and $\langle\cdot\rangle$ means averaging over Y. The loadings p^α, m^α, q coincide with effective loadings (2.3.63) introduced in Section 2.3.

Now let us substitute Eqs. (3.2.4) and the whole expansions given by Eqs. (3.2.5) (first two terms are sufficient) into Eq. (3.2.2), pass with ε to 0 and combine equation thus obtained with Eq. (3.2.11). Then one arrives at

$$\int_\Omega \langle N_0^{\alpha\beta} e_{\alpha\beta}^y(\boldsymbol{v}^{(1)}) + M_0^{\alpha\beta}\kappa_{\alpha\beta}^y(v^{(2)})\rangle dx = 0\,. \tag{3.2.13}$$

Let us assume $v_\alpha^{(1)} = \xi_\alpha(x)\tilde{v}_\alpha(y)$, $v^{(2)} = \eta(x)\tilde{v}(y)$, $\xi_\alpha, \eta \in \mathbf{D}(\Omega)$ and $(\tilde{\boldsymbol{v}}, \tilde{v}) \in H_{K,per}(Y)$ where

$$H_{K,per}(Y) = [H_{per}^1(Y)]^2 \times H_{per}^2(Y)\,. \tag{3.2.14}$$

Due to ξ_α, η being arbitrary functions of $\mathbf{D}(\Omega)$ one finds the equation

$$\langle N_0^{\alpha\beta} e_{\alpha\beta}^y(\tilde{\boldsymbol{v}}) + M_0^{\alpha\beta}\kappa_{\alpha\beta}^y(\tilde{v})\rangle = 0 \qquad \forall(\tilde{\boldsymbol{v}}, \tilde{v}) \in H_{K,per}(Y)\,, \tag{3.2.15}$$

for the determination of unknown functions $(\boldsymbol{u}^{(1)}(x,\cdot), w^{(2)}(x,\cdot)) \in H_{K,per}(Y)$.

Since this problem is linear one can represent its solution as follows

$$\boldsymbol{u}^{(1)} = \boldsymbol{T}^{(\lambda\mu)}(y)\epsilon_{\lambda\mu}^h + \boldsymbol{U}^{(\lambda\mu)}(y)\kappa_{\lambda\mu}^h\,, \qquad w^{(2)} = X^{(\lambda\mu)}(y)\epsilon_{\lambda\mu}^h + \chi^{(\lambda\mu)}(y)\kappa_{\lambda\mu}^h\,. \tag{3.2.16}$$

We recall that

$$\epsilon_{\lambda\mu}^h = e_{\lambda\mu}(\boldsymbol{u}^{(0)})\,, \qquad \kappa_{\lambda\mu}^h = \kappa_{\lambda\mu}(w^{(0)})\,. \tag{3.2.17}$$

New functions in $y \in Y$ involved in Eqs. (3.2.16) are defined as follows.

Let us introduce the bilinear forms

$$b(\boldsymbol{u}, \boldsymbol{v}) = \langle A^{\alpha\beta\lambda\mu}(y) e^y_{\lambda\mu}(\boldsymbol{u}) e^y_{\alpha\beta}(\boldsymbol{v})\rangle,$$
$$e_K(w, \boldsymbol{v}) = \langle E^{\alpha\beta\lambda\mu}(y) \kappa^y_{\lambda\mu}(w) e^y_{\alpha\beta}(\boldsymbol{v})\rangle, \tag{3.2.18a}$$
$$d_K(w, v) = \langle D^{\alpha\beta\lambda\mu}(y) \kappa^y_{\lambda\mu}(w) \kappa^y_{\alpha\beta}(v)\rangle$$

for $\boldsymbol{u}, \boldsymbol{v} \in [H^1_{per}(Y)]^2$, $v, w \in H^2_{per}(Y)$. Note that

$$b(\boldsymbol{u}, \boldsymbol{v}) = \varepsilon_0 \hat{b}(\boldsymbol{u}, \boldsymbol{v}) , \qquad e_K(w, \boldsymbol{v}) = (\varepsilon_0)^2 \hat{e}_K(w, \boldsymbol{v}) ,$$
$$d_K(w, v) = (\varepsilon_0)^3 \hat{d}_K(w, v) , \tag{3.2.18b}$$

cf. (2.7.14), (2.8.2). The new fields involved in Eqs. (3.2.16) are solutions to the following problems:

$$(P^1_{K,Y}) \quad \left|
\begin{array}{l}
\text{find } (\boldsymbol{T}^{\alpha\beta}, X^{\alpha\beta}) \in H_{K,per}(Y) \text{ such that} \\[4pt]
b(\boldsymbol{T}^{\alpha\beta}), \boldsymbol{u}) + e_K(X^{(\alpha\beta)}, \boldsymbol{u}) + \langle A^{\alpha\beta\lambda\mu}(y) e^y_{\lambda\mu}(\boldsymbol{u})\rangle = 0 , \\[8pt]
e_K(w, \boldsymbol{T}^{(\alpha\beta)}) + d_K(X^{(\alpha\beta)}, w) + \langle E^{\alpha\beta\lambda\mu}(y) \kappa^y_{\lambda\mu}(w)\rangle = 0 \\[8pt]
\hfill \forall \ (\boldsymbol{u}, w) \in H_{K,per}(Y) ;
\end{array}
\right. \tag{3.2.19a}$$

$$(P^2_{K,Y}) \quad \left|
\begin{array}{l}
\text{find } (\boldsymbol{U}^{(\alpha\beta)}, \chi^{(\alpha\beta)}) \in H_{K,per}(Y) \text{ such that} \\[4pt]
b(\boldsymbol{U}^{(\alpha\beta)}, u) + e_K(\chi^{(\alpha\beta)}, \boldsymbol{u}) + \langle E^{\alpha\beta\lambda\mu}(y) e^y_{\lambda\mu}(\boldsymbol{u})\rangle = 0, \\[8pt]
e_K(w, \boldsymbol{U}^{(\alpha\beta)}) + d_K(\chi^{(\alpha\beta)}, w) + \langle D^{\alpha\beta\lambda\mu}(y) \kappa^y_{\lambda\mu}(w)\rangle = 0 , \\[8pt]
\hfill \forall \ (\boldsymbol{u}, w) \in H_{K,per}(Y) .
\end{array}
\right. \tag{3.2.19b}$$

Compare problems $(P^\sigma_{K,Y})$ with $(\hat{P}^\sigma_{K,Y})$ of Section 2.8. On taking into account relations (3.2.18b) one concludes that functions $T^{(\alpha\beta)}$, $\chi^{(\alpha\beta)}$ coincide with those defined by $(\hat{P}^\sigma_{K,Y})$; moreover

$$X^{(\alpha\beta)} = \hat{X}^{(\alpha\beta)}/\varepsilon_0 , \qquad U^{(\alpha\beta)} = \varepsilon_0 \hat{U}^{(\alpha\beta)} . \tag{3.2.20}$$

One can prove that solutions to problems $(P^\sigma_{K,Y})$ exist and are determined up to additive constants. Since x is viewed here as a parameter, these constants can be functions of $x \in \Omega$.

Assume that functions $\boldsymbol{T}^{(\alpha\beta)}$, $X^{(\alpha\beta)}$, $\boldsymbol{U}^{(\alpha\beta)}$, $\chi^{(\alpha\beta)}$ have been found. Then,

$$e^0_{\alpha\beta} = S^{\lambda\mu}_{\alpha\beta} \epsilon^h_{\lambda\mu} + R^{\lambda\mu}_{\alpha\beta} \kappa^h_{\lambda\mu} , \qquad \kappa^0_{\alpha\beta} = W^{\lambda\mu}_{\alpha\beta} \epsilon^h_{\lambda\mu} + P^{\lambda\mu}_{\alpha\beta} \kappa^h_{\lambda\mu} , \tag{3.2.21}$$

where

$$S^{\lambda\mu}_{\alpha\beta} = \delta^{(\lambda}_\alpha \delta^{\mu)}_\beta + e^y_{\alpha\beta}(\boldsymbol{T}^{(\lambda\mu)}) , \qquad W^{\lambda\mu}_{\alpha\beta} = \kappa^y_{\alpha\beta}(X^{(\lambda\mu)}) ,$$
$$R^{\lambda\mu}_{\alpha\beta} = e^y_{\alpha\beta}(\boldsymbol{U}^{(\lambda\mu)}) , \qquad P^{\lambda\mu}_{\alpha\beta} = \delta^{(\lambda}_\alpha \delta^{\mu)}_\beta + \kappa^y_{\alpha\beta}(\chi^{(\lambda\mu)}) ; \tag{3.2.22}$$

here $\delta^{(\lambda}_\alpha \delta^{\mu)}_\beta = \dfrac{1}{2}(\delta^\lambda_\alpha \delta^\mu_\beta + \delta^\mu_\alpha \delta^\lambda_\beta)$. Thus the quantities (3.2.9) can be interrelated with deformations (3.2.17) as follows

$$N^{\alpha\beta}_0 = A^{\alpha\beta\lambda\mu}_0 \epsilon^h_{\lambda\mu} + E^{\alpha\beta\lambda\mu}_0 \kappa^h_{\lambda\mu} , \qquad M^{\alpha\beta}_0 = F^{\alpha\beta\lambda\mu}_0 \epsilon^h_{\lambda\mu} + D^{\alpha\beta\lambda\mu}_0 \kappa^h_{\lambda\mu} , \tag{3.2.23}$$

where

$$A_0^{\alpha\beta\lambda\mu} = A^{\alpha\beta\gamma\delta} S_{\gamma\delta}^{\lambda\mu} + E^{\alpha\beta\gamma\delta} W_{\gamma\delta}^{\lambda\mu}, \qquad E_0^{\alpha\beta\lambda\mu} = A^{\alpha\beta\gamma\delta} R_{\gamma\delta}^{\lambda\mu} + E^{\alpha\beta\gamma\delta} P_{\gamma\delta}^{\lambda\mu},$$
$$F_0^{\alpha\beta\lambda\mu} = E^{\alpha\beta\gamma\delta} S_{\gamma\delta}^{\lambda\mu} + D^{\alpha\beta\gamma\delta} W_{\gamma\delta}^{\lambda\mu}, \qquad D_0^{\alpha\beta\lambda\mu} = E^{\alpha\beta\gamma\delta} R_{\gamma\delta}^{\lambda\mu} + D^{\alpha\beta\gamma\delta} P_{\gamma\delta}^{\lambda\mu}. \qquad (3.2.24)$$

According to the definition (3.2.12) the constitutive relations of the effective plate assume the form

$$N_h^{\alpha\beta} = A_h^{\alpha\beta\lambda\mu} \epsilon_{\lambda\mu}^h + E_h^{\alpha\beta\lambda\mu} \kappa_{\lambda\mu}^h, \qquad M_h^{\alpha\beta} = F_h^{\alpha\beta\lambda\mu} \epsilon_{\lambda\mu}^h + D_h^{\alpha\beta\lambda\mu} \kappa_{\lambda\mu}^h, \qquad (3.2.25)$$

where the effective stiffnesses are given by

$$A_h^{\alpha\beta\lambda\mu} = \langle A_0^{\alpha\beta\lambda\mu} \rangle, \qquad E_h^{\alpha\beta\lambda\mu} = \langle E_0^{\alpha\beta\lambda\mu} \rangle,$$
$$F_h^{\alpha\beta\lambda\mu} = \langle F_0^{\alpha\beta\lambda\mu} \rangle, \qquad D_h^{\alpha\beta\lambda\mu} = \langle D_0^{\alpha\beta\lambda\mu} \rangle. \qquad (3.2.26)$$

Let us substitute (3.2.1a) into (3.2.24) and take into account (3.2.20). Then comparing (3.2.26) with (2.8.10), (2.8.6) we conclude that stiffnesses given by (3.2.26) coincide with stiffnesses determined by (2.8.10). Consequently the stiffnesses (3.2.26) possess the required symmetry and the matrix

$$\mathbf{A}^h = \begin{bmatrix} \mathbf{A}_h & \mathbf{E}_h \\ \mathbf{F}_h & \mathbf{D}_h \end{bmatrix}$$

is positive definite, cf. Sec. 2.8. The homogenized problem:

$$(P_h) \quad \left| \begin{array}{l} \text{find } (\boldsymbol{u}^{(0)}, w^{(0)}) \in V_K^0(\Omega) \text{ such that Eq. (3.2.11)} \\[4pt] \text{holds for each } (\boldsymbol{v}^{(0)}, v^{(0)}) \in V_K^0(\Omega), \\[4pt] \text{the effective constitutive relations being given by (3.2.25)}, \end{array} \right.$$

is well posed. The homogenized potential is given by

$$\mathcal{W}_h = (N_h^{\alpha\beta} \epsilon_{\alpha\beta}^h \mid M_h^{\alpha\beta} \kappa_{\alpha\beta}^h)/2, \qquad (3.2.27)$$

or

$$\mathcal{W}_h = (\langle N_0^{\alpha\beta} \rangle \langle e_{\alpha\beta}^0 \rangle + \langle M_0^{\alpha\beta} \rangle \langle \kappa_{\alpha\beta}^0 \rangle)/2. \qquad (3.2.28)$$

Obviously, $\mathcal{W}_h$ is a strictly convex function of ϵ^h and κ^h. By making use of all identities produced by substituting $\boldsymbol{u} = \boldsymbol{T}^{(\lambda\mu)}$ or $\boldsymbol{U}^{(\lambda\mu)}$ and $w = X^{(\alpha\beta)}$ or $\chi^{(\alpha\beta)}$ into Eqs. (3.2.19), (3.2.20) one can rearrange formula (3.2.28) to the form:

$$\mathcal{W}_h = \langle N_0^{\alpha\beta} e_{\alpha\beta}^0 + M_0^{\alpha\beta} \kappa_{\alpha\beta}^0 \rangle/2. \qquad (3.2.29)$$

Equivalence of expressions (3.2.28) and (3.2.29) confirms that the homogenization process satisfies the consistency criterion (2.3.67) of Hill.

Remark 3.2.1. By exploiting the properties of the elasticity tensor $\tilde{C}$ it can be shown that the matrix

$$\mathbf{A}(y) = \begin{bmatrix} \mathbf{A}(y) & \mathbf{E}(y) \\ \mathbf{F}(y) & \mathbf{D}(y) \end{bmatrix}$$

is also positive definite.

Problem $(P_\varepsilon^{(K)})$ is equivalent to the following minimization problem:

$$\left| \begin{array}{l} \text{find} \\ J_v(\boldsymbol{u}^\varepsilon, w^\varepsilon) = \inf\{J_\varepsilon(\boldsymbol{v}, v)|(\boldsymbol{v}, v) \in V_K^0(\Omega)\} \,, \end{array} \right.$$

where

$$J_\varepsilon(\boldsymbol{v}, v) = \frac{1}{2}a^\varepsilon(\boldsymbol{v}, v; \boldsymbol{v}, v) - \mathcal{L}_\varepsilon(\boldsymbol{v}, v)$$

and

$$a^\varepsilon(\boldsymbol{v}, v; \boldsymbol{v}, v) = \int_\Omega [(A_\varepsilon^{\alpha\beta\lambda\mu} e_{\lambda\mu}(\boldsymbol{v}) + E_\varepsilon^{\alpha\beta\lambda\mu} \kappa_{\lambda\mu}(\boldsymbol{v})) e_{\alpha\beta}(\boldsymbol{v})$$
$$+ (E_\varepsilon^{\alpha\beta\lambda\mu} e_{\lambda\mu}(\boldsymbol{v}) + D_\varepsilon^{\alpha\beta\lambda\mu} \kappa_{\lambda\mu}(\boldsymbol{v})) \kappa_{\alpha\beta}(\boldsymbol{v})] dx \,,$$

$$\mathcal{L}_\varepsilon(\boldsymbol{v}, v) = \int_\Omega [\tilde{p}^\alpha \left(x, \frac{x}{\varepsilon}\right) v_\alpha - \tilde{m}^\alpha \left(x, \frac{x}{\varepsilon}\right) v_{,\alpha} + \tilde{q}\left(x, \frac{x}{\varepsilon}\right) v] dx \,.$$

We recall that $A_\varepsilon^{\alpha\beta\lambda\mu}(x) = A^{\alpha\beta\lambda\mu}\left(x, \frac{x}{\varepsilon}\right)$, etc. It is not difficult to prove the following convergence result.

Theorem 3.2.2. The sequence of functionals $\{J_\varepsilon\}_{\varepsilon>0}$ is Γ-convergent in the weak topology of $H_0^1(\Omega)^2 \times H_0^2(\Omega)$ to, cf. (3.2.11) and (3.2.27)

$$J_h(\boldsymbol{v}, v) = \int_\Omega \mathcal{W}_h(e(\boldsymbol{v}), \kappa(\boldsymbol{v})) dx - \mathcal{L}_h(\boldsymbol{v}, v)$$

and $\mathcal{L}_h(\boldsymbol{v}, v)$ denotes the loading functional appearing on the r.h.s. of (3.2.11).

Moreover, at least for a subsequence still denoted by $\{\boldsymbol{u}^\varepsilon, w^\varepsilon\}_{\varepsilon>0}$ we have

$$(\boldsymbol{u}^\varepsilon, w^\varepsilon) \rightharpoonup (\boldsymbol{u}^{(0)}, w^{(0)}) \text{ weakly in } H_0^1(\Omega)^2 \times H_0^2(\Omega) \,,$$

where

$$J_h(\boldsymbol{u}^{(0)}, w^{(0)}) = \inf\{J_h(\boldsymbol{v}, v)|(\boldsymbol{v}, v) \in V_K^0(\Omega)\} \,. \qquad \square$$

Consider the transverse symmetry case. Then $X^{(\alpha\beta)} = 0$, $U^{(\alpha\beta)} = 0$. Problems $(P^\sigma_{K,Y})$ assume the form

$$(P^1_{KS,Y}) \quad \left| \begin{array}{l} \text{find } T^{(\alpha\beta)} \in H^1_{per}(Y)^2 \text{ such that} \\[4pt] b(T^{(\alpha\beta)}, u) + \langle A^{\alpha\beta\lambda\mu}(y)e^y_{\lambda\mu}(u)\rangle = 0 \qquad \forall\ u \in H^1_{per}(Y)^2\ ; \end{array} \right. \qquad (3.2.30)$$

$$(P^2_{KS,Y}) \quad \left| \begin{array}{l} \text{find } \chi^{(\alpha\beta)} \in H^2_{per}(Y) \text{ such that} \\[4pt] d_K(\chi^{(\alpha\beta)}, v) + \langle D^{\alpha\beta\lambda\mu}(y)\kappa^y_{\lambda\mu}(v)\rangle = 0 \qquad \forall\ v \in H^2_{per}(Y)\ . \end{array} \right. \qquad (3.2.31)$$

The homogenized stiffnesses are given by

$$\begin{aligned} A_h^{\alpha\beta\lambda\mu} &= \langle A^{\alpha\beta\lambda\mu} + A^{\alpha\beta\gamma\delta}e^y_{\gamma\delta}(T^{(\lambda\mu)})\rangle, \\[4pt] D_h^{\alpha\beta\lambda\mu} &= \langle D^{\alpha\beta\lambda\mu} + D^{\alpha\beta\gamma\delta}\kappa^y_{\gamma\delta}(\chi^{(\lambda\mu)})\rangle, \qquad E_h^{\alpha\beta\lambda\mu} = 0. \end{aligned} \qquad (3.2.32)$$

The alternative formulae will be given in the sequel.

Let us consider the strong form of the local problems $(P^\sigma_{KS,Y})$. By standard arguments of the variational calculus one arrives at:

A. The strong formulation of $(P^1_{KS,Y})$: find $T^{(\alpha\beta)} \in C^1(Y)$ such that $T^{(\alpha\beta)}$ assume equal values on the opposite sides of Y and functions

$$N^{\lambda\mu}_{(\alpha\beta)} = A^{\lambda\mu\gamma\delta}(y)e^y_{\gamma\delta}(T^{(\alpha\beta)}) + A^{\alpha\beta\lambda\mu}(y) \qquad (3.2.33)$$

satisfy:

(i) differential equations

$$N^{\lambda\mu}_{(\alpha\beta)|\mu} = 0 \qquad \text{in } Y\ , \qquad (3.2.34)$$

where $(\ )_{|\mu} = \partial/\partial y_\mu$,

(ii) boundary conditions:

$$N^{\lambda\mu}_{(\alpha\beta)}\mu_\mu\mu_\lambda \quad \text{and} \quad N^{\lambda\mu}_{(\alpha\beta)}\mu_\mu\tau_\lambda\ ,$$

assume equal values on the opposite sides of Y; μ and τ are unit vectors normal and tangent to ∂Y.

(iii) Continuity conditions along line $\tilde\gamma$ of discontinuity of stiffnesses $A^{\alpha\beta\lambda\mu}(y)$:

$$\sum_\lambda T^{(\alpha\beta)}_\lambda \nu_\lambda\ , \ \sum_\lambda T^{(\alpha\beta)}_\lambda \tau_\lambda\ , \qquad N^{\lambda\mu}_{(\alpha\beta)}\nu_\lambda\nu_\mu \text{ and } N^{\lambda\mu}_{(\alpha\beta)}\nu_\lambda\tau_\mu\ ,$$

assume equal values on both sides of $\tilde\gamma$; here the vectors ν and τ are outward normal and tangent to $\tilde\gamma$ and are common to both sides of $\tilde\gamma$.

B. The strong formulation of $(P^2_{KS,Y})$: find $\chi^{(\alpha\beta)} \in C^2(Y)$ such that

(i) $\chi^{(\alpha\beta)}$ assume equal values and $\dfrac{\partial\chi^{(\alpha\beta)}}{\partial\mu}$ – opposite values on the opposite sides of Y.

(ii) The functions

$$M^{\lambda\mu}_{(\alpha\beta)} = D^{\lambda\mu\gamma\delta}(y)\kappa^{y}_{\gamma\delta}(\chi^{(\alpha\beta)}) + D^{\alpha\beta\lambda\mu}(y)\,, \tag{3.2.35}$$

satisfy:

– differential equation

$$M^{\lambda\mu}_{(\alpha\beta)|\lambda\mu} = 0 \qquad \text{in } Y\,; \tag{3.2.36}$$

– boundary conditions:

$$M^{(\alpha\beta)}_{n} = M^{\lambda\mu}_{(\alpha\beta)}\mu_\lambda\mu_\mu \text{ assume equal values on the opposite sides of } Y\,; \tag{3.2.37}$$

$$Q^{(\alpha\beta)} = M^{\lambda\mu}_{(\alpha\beta)}|_\lambda\,\mu_\mu + \frac{\partial}{\partial\tau}(M^{\lambda\mu}_{(\alpha\beta)}\mu_\lambda\tau_\mu) \tag{3.2.38}$$

assume opposite values on the opposite sides of Y;

– equilibrium of point reactions at the corners O_i of Y:

$$\sum_{i=1}^{4}(-1)^i M^{12}_{(\alpha\beta)}(\mathrm{O}_i) = 0; \tag{3.2.39}$$

this condition is identically satisfied.

– continuity conditions along line $\tilde{\gamma}$ of discontinuity of stiffnesses $D^{\alpha\beta\lambda\mu}(y)$:

$$\chi^{(\alpha\beta)}, \qquad M^{(\alpha\beta)}_{n}\,, \qquad \frac{\partial\chi^{(\alpha\beta)}}{\partial\nu} \quad \text{and} \quad Q^{(\alpha\beta)} \tag{3.2.40}$$

assume equal values on both sides of $\tilde{\gamma}$; ν and τ are common to both sides of $\tilde{\gamma}$.

3.3. Refined scaling approach

Within the framework of the in-plane scaling based approach of Section 3.2 thickness of the plate is viewed as ε-independent. Thus this analysis runs counter to the asymptotic analysis of Section 2.2, based upon the scaling (2.2.1) according to which the transverse dimension of the periodicity cell goes to zero simultaneously with in-plane dimensions. The aim of this section is to show that the latter scaling applied to Kirchhoff plate equations leads to the homogenization formulae found by the in-plane scaling. Scaling (2.2.1) will be from now onward called refined scaling, since it preserves three-dimensional shape of the $\mathcal{Z}$ cell, see Fig. 1.1.

Similarly to Section 2.2 let us assume the refined scaling defined by Eqs. (2.2.1) and (2.2.2). Consequently we replace

$$A^{\alpha\beta\lambda\mu}_z(x) \rightsquigarrow \varepsilon \hat{A}^{\alpha\beta\lambda\mu}\left(\frac{x}{\varepsilon}\right)\,,$$

$$E^{\alpha\beta\lambda\mu}_z(x) \rightsquigarrow \varepsilon^2 \hat{E}^{\alpha\beta\lambda\mu}\left(\frac{x}{\varepsilon}\right)\,, \qquad D^{\alpha\beta\lambda\mu}_z(x) \rightsquigarrow \varepsilon^3 \hat{D}^{\alpha\beta\lambda\mu}\left(\frac{x}{\varepsilon}\right)\,, \tag{3.3.1}$$

and the arrows become equalities for $\varepsilon = \varepsilon_0$. Hence $\hat{A} = A/\varepsilon_0$, $\hat{E} = E/(\varepsilon_0)^2$, $\hat{D} = D/(\varepsilon_0)^3$.

To compensate for the loss of stiffnesses for $\varepsilon \to 0$ we scale the loadings

$$\tilde{p}^\alpha\left(x,\frac{x}{\varepsilon}\right) \rightsquigarrow \varepsilon^s \, \overset{\approx}{p}{}^\alpha\left(x,\frac{x}{\varepsilon}\right), \qquad \tilde{q}\left(x,\frac{x}{\varepsilon}\right) \rightsquigarrow \varepsilon^{s+1} \, \overset{\approx}{q}\left(x,\frac{x}{\varepsilon}\right),$$

$$\tilde{m}^\alpha\left(x,\frac{x}{\varepsilon}\right) \rightsquigarrow \varepsilon^{s+1} \, \overset{\approx}{m}{}^\alpha\left(x,\frac{x}{\varepsilon}\right),$$

$$\tag{3.3.2}$$

s is an integer, $s \geq 2$.

Problem $P_\varepsilon^{(K)}$ of Section 3.3 is replaced with:

$$(\hat{P}_\varepsilon^{(K)}) \quad \left|\begin{aligned} &\text{find } (\hat{\boldsymbol{u}}^\varepsilon, \hat{w}^\varepsilon) \in V_K^0(\Omega) \text{ such that}\\ &\int_\Omega [\hat{N}_\varepsilon^{\alpha\beta}(\hat{\boldsymbol{u}}^\varepsilon,\hat{w}^\varepsilon)e_{\alpha\beta}(\boldsymbol{v}) + \hat{M}_\varepsilon^{\alpha\beta}(\hat{\boldsymbol{u}}^\varepsilon,\hat{w}^\varepsilon)\kappa_{\alpha\beta}(v)]dx\\ &= \int_\Omega [\varepsilon^s \, \overset{\approx}{p}{}^\alpha\left(x,\frac{x}{\varepsilon}\right)v_\alpha - \varepsilon^{s+1}\,\overset{\approx}{m}{}^\alpha\left(x,\frac{x}{\varepsilon}\right)v_{,\alpha} + \varepsilon^{s+1}\,\overset{\approx}{q}\left(x,\frac{\varepsilon}{x}\right)v]dx\\ &\hspace{6cm}\forall\,(\boldsymbol{v},v) \in V_K^0(\Omega)\,, \end{aligned}\right. \tag{3.3.3}$$

where

$$\hat{N}_\varepsilon^{\alpha\beta}(\hat{\boldsymbol{u}}^\varepsilon,\hat{w}^\varepsilon) = \varepsilon \hat{A}^{\alpha\beta\lambda\mu}\left(\frac{x}{\varepsilon}\right)e_{\lambda\mu}(\hat{\boldsymbol{u}}^\varepsilon) + \varepsilon^2 \hat{E}^{\alpha\beta\lambda\mu}\left(\frac{x}{\varepsilon}\right)\kappa_{\lambda\mu}(w^\varepsilon)\,,$$

$$\hat{M}_\varepsilon^{\alpha\beta}(\hat{\boldsymbol{u}}^\varepsilon,\hat{w}^\varepsilon) = \varepsilon^2 \hat{E}^{\alpha\beta\lambda\mu}\left(\frac{x}{\varepsilon}\right)e_{\lambda\mu}(\hat{\boldsymbol{u}}^\varepsilon) + \varepsilon^3 \hat{D}^{\alpha\beta\lambda\mu}\left(\frac{x}{\varepsilon}\right)\kappa_{\lambda\mu}(\hat{w}^\varepsilon).$$

$$\tag{3.3.4}$$

By virtue of relations (3.3.1) – (3.3.4) solutions to problems $(P_\varepsilon^{(K)})$ of Section 3.3 and $(\hat{P}_\varepsilon^{(K)})$ above are interrelated by

$$\hat{u}_\alpha^\varepsilon = \left(\frac{\varepsilon}{\varepsilon_0}\right)^{s-1} u_\alpha^\varepsilon, \qquad \hat{w}^\varepsilon = \left(\frac{\varepsilon}{\varepsilon_0}\right)^{s-2} w^\varepsilon,$$

$$\hat{N}_\varepsilon^{\alpha\beta} = \left(\frac{\varepsilon}{\varepsilon_0}\right)^s N_\varepsilon^{\alpha\beta}, \qquad \hat{M}_\varepsilon^{\alpha\beta} = \left(\frac{\varepsilon}{\varepsilon_0}\right)^{s+1} M_\varepsilon^{\alpha\beta}\,.$$

$$\tag{3.3.5}$$

Moreover, the following homogeneity relation for the plate potential $\mathcal{W}_\varepsilon$ holds:

$$\mathcal{W}_\varepsilon(\boldsymbol{w},w) = \left(\frac{\varepsilon_0}{\varepsilon}\right)^{2s-1} \hat{\mathcal{W}}_\varepsilon\left(\left(\frac{\varepsilon}{\varepsilon_0}\right)^{s-1}\boldsymbol{u},\left(\frac{\varepsilon}{\varepsilon_0}\right)^{s-2}w\right)\,. \tag{3.3.6}$$

Just this homogeneity relation shows that both in-plane and refined scaling approaches are equivalent in the considered case of Kirchhoff plate modelling. It will turn out that, if applied to different plate modelling, both scalings yield different results.

Owing to the linearity of the model, the parameter $s \geq 2$ was indetermined. This parameter assumes the fixed value $s = 3$ if the asymptotic analysis is applied to the geometrically nonlinear von Kármán equations of thin plates; this will be discussed in Section 4.

3.4. *Variational formulae for effective stiffnesses*

We say that strain measures $\epsilon_{\alpha\beta}$ and $\kappa_{\alpha\beta}$ are kinematically admissible and write

$$\epsilon \in \mathcal{K}_{\boldsymbol{\epsilon}}(\Omega), \qquad \kappa \in \mathcal{K}_{\boldsymbol{\kappa}}(\Omega)\,,$$

if they satisfy the compatibility equations. Thus

$$\mathcal{K}_{\boldsymbol{\epsilon}}(\Omega) = \{\epsilon = (\epsilon_{\beta\alpha}) \in L^2(\Omega, \mathbb{E}_s^2)|\ \int_{\Omega} s^{\alpha\beta}\epsilon_{\beta\alpha}dx = 0\,, \quad \forall\ s \in \mathbf{D}(\Omega, \mathbb{E}_s^2),\ \mathrm{div}\ s = 0\}\,,$$

$$\mathcal{K}_{\boldsymbol{\kappa}}(\Omega) = \{\kappa = (\kappa_{\alpha\beta}) \in L^2(\Omega, \mathbb{E}_s^2)|\ \int_{\Omega} M^{\alpha\beta}\kappa_{\beta\alpha}dx = 0\,,$$

$$\forall\ M \in \mathbf{D}(\Omega, \mathbb{E}_s^2),\ \mathrm{div\ div}\ M = 0\}\,,$$

Ω being arbitrary, even multiconnected.

Kinematically admissible strains $\epsilon_{\beta\alpha}$ are associated with a displacement field $u = (u_\alpha)$ such that $\epsilon_{\beta\alpha} = \dfrac{1}{2}(u_{\alpha,\beta} + u_{\beta,\alpha})$. Similarly kinematically admissible strains $\kappa_{\alpha\beta}$ are associated with a scalar field w such that $\kappa_{\alpha\beta} = -w_{,\alpha\beta}$.

Having in mind application to the determination of the homogenized potential we introduce two spaces:

$$\mathcal{K}_{\boldsymbol{\epsilon}}^{per}(Y) = \{\epsilon^y \in L^2(Y, \mathbb{E}_s^2)|\epsilon^y \in \mathcal{K}_{\boldsymbol{\epsilon}}(Y) \text{ and } \epsilon^y \text{ is associated with}$$
$$\text{a } Y\text{-periodic displacement field}\},$$

$$\mathcal{K}_{\boldsymbol{\kappa}}^{per}(Y) = \{\kappa^y \in L^2(Y, \mathbb{E}_s^2)|\kappa^y \in \mathcal{K}_{\boldsymbol{\kappa}}(Y) \text{ and } \kappa^y \text{ is associated with}$$
$$\text{a } Y\text{-periodic displacement field}\}.$$

The local potential is given by

$$\mathcal{W}(\epsilon^y, \kappa^y) = \frac{1}{2}[A^{\alpha\beta\lambda\mu}(y)\epsilon_{\alpha\beta}^y\epsilon_{\lambda\mu}^y + E^{\alpha\beta\lambda\mu}(y)\kappa_{\lambda\mu}^y\epsilon_{\alpha\beta}^y$$
$$+E^{\alpha\beta\lambda\mu}(y)\epsilon_{\lambda\mu}^y\kappa_{\alpha\beta}^y + D^{\alpha\beta\lambda\mu}(y)\kappa_{\lambda\mu}^y\kappa_{\alpha\beta}^y]\,, \tag{3.4.1}$$

and the homogenized potential (3.3.27) has the form

$$\mathcal{W}_h(\epsilon^h, \kappa^h) = \frac{1}{2}[A_h^{\alpha\beta\lambda\mu}\epsilon_{\alpha\beta}^h\epsilon_{\lambda\mu}^h + E_h^{\alpha\beta\lambda\mu}\kappa_{\lambda\mu}^h\epsilon_{\alpha\beta}^h$$
$$+F_h^{\alpha\beta\lambda\mu}\epsilon_{\lambda\mu}^h\kappa_{\alpha\beta}^h + D_h^{\alpha\beta\lambda\mu}\kappa_{\lambda\mu}^h\kappa_{\alpha\beta}^h]\,. \tag{3.4.2}$$

Thus we have

$$\mathcal{W}_h(\epsilon^h, \kappa^h) = \min\{\langle\mathcal{W}(\epsilon^y, \kappa^y)\rangle|\epsilon^y \in \mathcal{K}_{\boldsymbol{\epsilon}}^{per}(Y),$$
$$\kappa^y \in \mathcal{K}_{\boldsymbol{\kappa}}^{per}(Y) \text{ and } \langle\epsilon^y\rangle = \epsilon^h, \langle\kappa^y\rangle = \kappa^h\}\,. \tag{3.4.3}$$

To prove (3.4.3) let us note that the formula (3.2.29) can be rearranged as follows

$$\mathcal{W}_h = \langle \mathcal{W}(e^0, \kappa^0) \rangle \,, \tag{3.4.4}$$

where e^0, κ^0 are given by (3.2.7), or

$$e^0 = \epsilon^h + e^y(u^{(1)}) \,, \quad \kappa^0 = \kappa^h + \kappa^y(w^{(2)}) \,, \tag{3.4.5}$$

with $(u^{(1)}, w^{(2)}) \in H_{K,per}(Y)$. By definition of $(u^{(1)}, w^{(2)})$, cf. Eq. (3.2.15), the expression (3.4.4) can be represented by

$$\mathcal{W}_h = \min\{\langle \mathcal{W}(\epsilon^h + e^y(v) \,, \ \kappa^h + \kappa^y(v)) \rangle | (v, v) \in H_{K,per}(Y)\} \,. \tag{3.4.6}$$

Let us define a set

$$A = \{(\epsilon, \kappa) \in L^2\left(Y, \mathbb{E}_s^2\right)^2 | \ \epsilon = e^y(u), \ \kappa = \kappa^y(w) \,, u_\alpha = \epsilon_{\alpha\beta}^h y^\beta + a\varepsilon_{\alpha\beta} y^\beta + v_\alpha \,,$$
$$w = \frac{1}{2}\kappa_{\alpha\beta}^h y^\alpha y^\beta + b_\alpha y^\alpha + v \,, \quad (v, v) \in H_{K,per}(Y)\} \,, \tag{3.4.7}$$

where $a, b_\alpha \in \mathbb{R}$ and $\varepsilon_{\alpha\beta}$ stand for the components of Ricci's pseudotensor.

Note that $(\epsilon, \kappa) \in A$ means that

$$\epsilon = \epsilon^h + \epsilon^y(v) \,, \quad \kappa = \kappa^h + \kappa^y(v) \,, \tag{3.4.8}$$

where $(v, v) \in H_{K,per}(Y)$. Thus potential $\mathcal{W}_h$ given by (3.4.6) can be put in the form

$$\mathcal{W}_h = \min\{\langle \mathcal{W}(\epsilon, \kappa) \rangle | (\epsilon, \kappa) \in A\} \,, \tag{3.4.9}$$

and since $\langle \epsilon \rangle = \langle e^y(u) \rangle = \epsilon^h$, $\langle \kappa \rangle = \langle \kappa^y(w) \rangle = \kappa^h$ we arrive at the representation (3.4.3).

In the case of transverse symmetry the formula (3.4.3) splits into
– formula for D_h:

$$\kappa_{\alpha\beta}^h D_h^{\alpha\beta\lambda\mu} \kappa_{\lambda\mu}^h = \min\{\langle \kappa_{\alpha\beta}^y D^{\alpha\beta\lambda\mu} \kappa_{\lambda\mu}^y \rangle \mid \kappa^y \in \mathcal{K}_\kappa^{per}(Y), \langle \kappa^y \rangle = \kappa^h\} \,, \tag{3.4.10}$$

– formula for A_h:

$$\epsilon_{\alpha\beta}^h A_h^{\alpha\beta\lambda\mu} \epsilon_{\lambda\mu}^h = \min\{\langle \epsilon_{\alpha\beta}^y A^{\alpha\beta\lambda\mu} \epsilon_{\lambda\mu}^y \rangle \mid \epsilon^y \in \mathcal{K}_\epsilon^{per}(Y), \langle \epsilon^y \rangle = \epsilon^h\} \,. \tag{3.4.11}$$

Remark 3.4.1. If Ω is a connected domain and $s^{\alpha\beta}$, $M^{\alpha\beta}$ involved in definitions of $\mathcal{K}_\epsilon(\Omega)$, $\mathcal{K}_\kappa(\Omega)$ are twice differentiable, then

$$\text{div } s = 0, \quad \text{div div } M = 0 \,, \tag{3.4.12}$$

provided that s, M are represented as follows

$$s^{11} = \kappa_{22}(\varphi) , \qquad s^{22} = \kappa_{11}(\varphi) \qquad s^{12} = -\kappa_{12}(\varphi) ,$$
$$M^{11} = e_{22}(\boldsymbol{\varphi}) , \qquad M^{22} = e_{11}(\boldsymbol{\varphi}) , \qquad M^{12} = -e_{12}(\boldsymbol{\varphi}) . \tag{3.4.13}$$

In such a case the sets of kinematically admissible strain measures can be defined as follows

$$\mathcal{K}'_{\boldsymbol{\epsilon}}(\Omega) = \{\boldsymbol{\epsilon} \in L^2(\Omega, \mathbb{E}^2_s)| \int_\Omega [\epsilon_{11}\kappa_{22}(\varphi) + \epsilon_{22}\kappa_{11}(\varphi) - 2\epsilon_{12}\kappa_{12}(\varphi)]dx = 0$$
$$\forall \; \varphi \in \mathbf{D}(\Omega)\} , \tag{3.4.14}$$

$$\mathcal{K}'_{\boldsymbol{\kappa}}(\Omega) = \{\boldsymbol{\kappa} \in L^2(\Omega, \mathbb{E}^2_s)| \int_\Omega [\kappa_{11}e_{22}(\varphi) + \kappa_{22}e_{11}(\varphi) - 2\kappa_{12}e_{12}(\varphi)]dx = 0$$
$$\forall \; \varphi \in [\mathbf{D}(\Omega)]^2\} . \tag{3.4.15}$$

Here the primes indicate that additional regularity assumptions are imposed.

3.5. Correctors

In Theorem 3.2.2 the convergence of $(\boldsymbol{u}^\varepsilon, w^\varepsilon)$ to $(\boldsymbol{u}^{(0)}, w^{(0)})$ is only a weak one. Consequently the first gradient of u_α^ε and second gradients of w^ε are, in general, poorly approximated. To achieve strong convergence we make the following assumptions:

$$u_\alpha^{(0)} \in H^3(\Omega) , \qquad w^{(0)} \in H^4(\Omega) , \tag{3.5.1}$$
$$T_\alpha^{(\beta\gamma)} \in W^{1,\infty}(Y) , \qquad U_\alpha^{(\beta\gamma)} \in W^{1,\infty}(Y) , \tag{3.5.2}$$
$$X^{(\alpha\beta)} \in W^{2,\infty}(Y) , \qquad \chi^{(\alpha\beta)} \in W^{2,\infty}(Y) . \tag{3.5.3}$$

Let us introduce the functions

$$\boldsymbol{v}^\varepsilon = \boldsymbol{u}^\varepsilon - \boldsymbol{u}^{(0)} - \varepsilon\boldsymbol{u}^{(1)} , \qquad z^\varepsilon = w^\varepsilon - w^{(0)} - \varepsilon^2 w^{(2)} , \tag{3.5.4}$$

where $\boldsymbol{u}^{(1)}, w^{(2)}$ are given by (3.2.16). The functions $\varepsilon\boldsymbol{u}^{(1)}$ and $\varepsilon^2 w^{(2)}$ are called *correctors*. As we shall see, the functions $\boldsymbol{u}^{(0)} + \varepsilon\boldsymbol{u}^{(1)}$ and $w^{(0)} + \varepsilon^2 w^{(2)}$ provide much better approximations of $\boldsymbol{u}^\varepsilon$ and w^ε respectively than $\boldsymbol{u}^{(0)}, w^{(0)}$ alone. Below we tacitly assume that loading functions are square integrable.

Theorem 3.5.1. Under the assumptions (3.5.1) – (3.5.3) we have

$$\boldsymbol{v}^\varepsilon \longrightarrow 0 \quad \text{strongly in } H^1(\Omega)^2 , \qquad z^\varepsilon \longrightarrow 0 \quad \text{strongly in } H^2(\Omega) ,$$

when $\varepsilon \to 0$.

Proof. From (3.2.16), (3.5.1)$_1$ and (3.5.2) we conclude that $v_\alpha^\varepsilon \to 0$ strongly in $L^2(\Omega)$ as $\varepsilon \to 0$; similarly $z^\varepsilon \to 0$ in $H^1(\Omega)$. It is thus sufficient to show that $a^\varepsilon(\boldsymbol{v}^\varepsilon, z^\varepsilon; \boldsymbol{v}^\varepsilon, z^\varepsilon)$

converges to zero as $\varepsilon \to 0$. We calculate, cf. Remark 3.2.1

$$a^\varepsilon(\boldsymbol{v}^\varepsilon, z^\varepsilon; \boldsymbol{v}^\varepsilon, z^\varepsilon)$$
$$= a^\varepsilon(\boldsymbol{u}^\varepsilon - \boldsymbol{u}^{(0)} - \varepsilon\boldsymbol{u}^{(1)}, w^\varepsilon - w^{(0)} - \varepsilon^2 w^{(2)}; \boldsymbol{u}^\varepsilon - \boldsymbol{u}^{(0)} - \varepsilon\boldsymbol{u}^{(1)}, w^\varepsilon - w^{(0)} - \varepsilon^2 w^{(2)})$$
$$= \mathcal{L}_\varepsilon(\boldsymbol{u}^\varepsilon, w^\varepsilon) - 2\mathcal{L}_\varepsilon(\boldsymbol{u}^{(0)} + \varepsilon\boldsymbol{u}^{(1)}, w^{(0)} + \varepsilon^2 w^{(2)}) + \Delta^\varepsilon \,, \qquad (3.5.5)$$

where

$$\Delta^\varepsilon = a^\varepsilon(\boldsymbol{u}^{(0)} + \varepsilon\boldsymbol{u}^{(1)}, w^{(0)} + \varepsilon^2 w^{(2)}; \boldsymbol{u}^{(0)} + \varepsilon\boldsymbol{u}^{(1)}, w^{(0)} + \varepsilon^2 w^{(2)}) \,. \qquad (3.5.6)$$

We recall that the loading functional $\mathcal{L}_\varepsilon$ is defined in Remark 3.2.1. Let us assume for the moment that

$$\lim_{\varepsilon \to 0} \Delta^\varepsilon = \mathcal{L}_h(\boldsymbol{u}^{(0)}, w^{(0)}) \,, \qquad (3.5.7)$$

where $\mathcal{L}_h(\ )$ has been defined in Remark 3.2.1. Then, by virtue of (3.5.5), we obtain

$$\lim_{\varepsilon \to 0} a^\varepsilon(\boldsymbol{v}^\varepsilon, z^\varepsilon; \boldsymbol{v}^\varepsilon, z^\varepsilon) = 0 \,,$$

and the theorem follows because

$$\lim_{\varepsilon \to 0} \mathcal{L}_\varepsilon(\boldsymbol{u}^\varepsilon, w^\varepsilon) = \mathcal{L}_h(\boldsymbol{u}^{(0)}, w^{(0)}) \,.$$

In this way it remains to prove (3.5.7). To this end we calculate

$$e_{\alpha\beta}(\boldsymbol{u}^{(0)} + \varepsilon\boldsymbol{u}^{(1)}) = e_{\alpha\beta}(\boldsymbol{u}^{(0)}) + \varepsilon e_{\alpha\beta}(\boldsymbol{u}^{(1)}) + e_{\alpha\beta}^y(\boldsymbol{u}^{(1)}(y)) \,, \ y = x/\varepsilon$$
$$\kappa_{\alpha\beta}(w^{(0)} + \varepsilon^2 w^{(2)}) = \kappa_{\alpha\beta}(w^{(0)}) + \varepsilon^2 \kappa_{\alpha\beta}(w^{(2)}) + \kappa_{\alpha\beta}^y(w^{(2)}(y)) \,, \ y = x/\varepsilon \,.$$

We recall that both $\boldsymbol{u}^{(1)}$ and $w^{(2)}$ are functions of x and x/ε. Hence

$$\Delta^\varepsilon = \int_\Omega \{ [A_\varepsilon^{\alpha\beta\lambda\mu}(e_{\lambda\mu}(\boldsymbol{u}^{(0)}) + \varepsilon e_{\lambda\mu}(\boldsymbol{u}^{(1)}) + e_{\lambda\mu}^y(\boldsymbol{u}^{(1)})(\tfrac{x}{\varepsilon}))$$
$$+ E_\varepsilon^{\alpha\beta\lambda\mu}(\kappa_{\lambda\mu}(w^{(0)}) + \varepsilon^2 \kappa_{\lambda\mu}(w^{(2)}) + \kappa_{\lambda\mu}^y(w^{(2)})(\tfrac{x}{\varepsilon}))](e_{\alpha\beta}(\boldsymbol{u}^{(0)})$$
$$+ \varepsilon e_{\alpha\beta}(\boldsymbol{u}^{(1)}) + e_{\alpha\beta}^y(\boldsymbol{u}^{(1)})(\tfrac{x}{\varepsilon}))$$
$$+ [E_\varepsilon^{\alpha\beta\lambda\mu}(e_{\lambda\mu}(\boldsymbol{u}^{(0)}) + \varepsilon e_{\lambda\mu}(\boldsymbol{u}^{(1)}) + e_{\lambda\mu}^y(\boldsymbol{u}^{(1)})(\tfrac{x}{\varepsilon}))$$
$$+ D_\varepsilon^{\alpha\beta\lambda\mu}(\kappa_{\lambda\mu}(w^{(0)}) + \varepsilon^2 \kappa_{\lambda\mu}(w^{(2)}) + \kappa_{\lambda\mu}^y(w^{(2)})(\tfrac{x}{\varepsilon}))(\kappa_{\alpha\beta}(w^{(0)})$$
$$+ \varepsilon^2 \kappa_{\alpha\beta}(w^{(0)}) + \kappa_{\alpha\beta}^y(w^{(2)})\tfrac{x}{\varepsilon}))\}dx$$
$$= \Delta_1^\varepsilon + \varepsilon \int_\Omega \{ [A_\varepsilon^{\alpha\beta\lambda\mu}(e_{\lambda\mu}(\boldsymbol{u}^{(0)}) + e_{\lambda\mu}^y(\boldsymbol{u}^{(1)})(\tfrac{x}{\varepsilon}))$$

$$
\begin{aligned}
&+ E_\varepsilon^{\alpha\beta\lambda\mu}(\kappa_{\lambda\mu}(w^{(0)}) + \kappa^y_{\lambda\mu}(w^{(2)})(\tfrac{x}{\varepsilon}))]e_{\alpha\beta}(\boldsymbol{u}^{(1)}) \\
&+ E_\varepsilon^{\alpha\beta\lambda\mu}(\kappa_{\alpha\beta}(w^{(0)}) + \kappa^y_{\alpha\beta}(w^{(2)})(\tfrac{x}{\varepsilon}))e_{\lambda\mu}(\boldsymbol{u}^{(1)})\}dx \\
&+ \varepsilon^2 \int_\Omega [A_\varepsilon^{\alpha\beta\lambda\mu}e_{\lambda\mu}(\boldsymbol{u}^{(1)})e_{\alpha\beta}(\boldsymbol{u}^{(1)}) + E_\varepsilon^{\alpha\beta\lambda\mu}\kappa_{\lambda\mu}(w^{(2)})(e_{\alpha\beta}(\boldsymbol{u}^{(0)}) + e^y_{\alpha\beta}(\boldsymbol{u}^{(1)})(\tfrac{x}{\varepsilon})) \\
&+ E_\varepsilon^{\alpha\beta\lambda\mu}(e_{\lambda\mu}(\boldsymbol{u}^{(0)}) + e^y_{\lambda\mu}(\boldsymbol{u}^{(1)})(\tfrac{x}{\varepsilon}))\kappa_{\alpha\beta}(w^{(2)}) \\
&+ 2D_\varepsilon^{\alpha\beta\lambda\mu}(\kappa_{\lambda\mu}(w^{(0)}) + \kappa^y_{\lambda\mu}(w^{(2)})(\tfrac{x}{\varepsilon}))\kappa_{\alpha\beta}(w^{(2)})]dx \\
&+ \varepsilon^3 \int_\Omega [E_\varepsilon^{\alpha\beta\lambda\mu}\kappa_{\lambda\mu}(w^{(2)})e_{\alpha\beta}(\boldsymbol{u}^{(1)}) + E_\varepsilon^{\alpha\beta\lambda\mu}e_{\lambda\mu}(\boldsymbol{u}^{(1)})\kappa_{\alpha\beta}(w^{(2)})]dx \\
&+ \varepsilon^4 \int_\Omega D_\varepsilon^{\alpha\beta\lambda\mu}\kappa_{\alpha\beta}(w^{(2)})\kappa_{\lambda\mu}(w^{(2)})dx \;,
\end{aligned}
$$

where

$$
\begin{aligned}
\Delta_1^\varepsilon = \int_\Omega &[A_\varepsilon^{\alpha\beta\lambda\mu}(e_{\lambda\mu}(\boldsymbol{u}^{(0)}) + e^y_{\lambda\mu}(\boldsymbol{u}^{(1)})(\tfrac{x}{\varepsilon}))(e_{\alpha\beta}(\boldsymbol{u}^{(0)}) + e^y_{\alpha\beta}(\boldsymbol{u}^{(1)})(\tfrac{x}{\varepsilon})) \\
&+ E_\varepsilon^{\alpha\beta\lambda\mu}(\kappa_{\lambda\mu}(w^{(0)}) + \kappa^y_{\lambda\mu}(w^{(2)})(\tfrac{x}{\varepsilon}))(e_{\alpha\beta}(\boldsymbol{u}^{(0)}) + e^y_{\alpha\beta}(\boldsymbol{u}^{(1)})(\tfrac{x}{\varepsilon})) \\
&+ E_\varepsilon^{\alpha\beta\lambda\mu}(e_{\lambda\mu}(\boldsymbol{u}^{(0)}) + e^y_{\lambda\mu}(\boldsymbol{u}^{(1)})(\tfrac{x}{\varepsilon}))(\kappa_{\alpha\beta}(w^{(0)}) + \kappa^y_{\alpha\beta}(w^{(2)})(\tfrac{x}{\varepsilon})) \\
&+ D_\varepsilon^{\alpha\beta\lambda\mu}(\kappa_{\alpha\beta}(w^{(0)}) + \kappa^y_{\lambda\mu}(w^{(2)})(\tfrac{x}{\varepsilon}))(\kappa_{\lambda\mu}(w^{(0)}) + \kappa^y_{\lambda\mu}(w^{(2)})(\tfrac{x}{\varepsilon}))]dx \;. \quad (3.5.8)
\end{aligned}
$$

The integrands associated with ε, ε^2, ε^3 and ε^4 are functions bounded in $L^1(\Omega)$. Thus it remains to find the limit of Δ_1^ε when $\varepsilon \to 0$. Equation (3.2.16) yields

$$
\begin{aligned}
e^y_{\alpha\beta}(\boldsymbol{u}^{(1)}) &= e^y_{\alpha\beta}(\boldsymbol{T}^{(\lambda\mu)})e_{\lambda\mu}(\boldsymbol{u}^{(0)}) + e^y_{\alpha\beta}(\boldsymbol{U}^{(\lambda\mu)})\kappa_{\lambda\mu}(w^{(0)}) \;, \\
\kappa^y_{\alpha\beta}(w^{(2)}) &= \kappa^y_{\alpha\beta}(X^{(\lambda\mu)})e_{\lambda\mu}(\boldsymbol{u}^{(0)}) + \kappa^y_{\alpha\beta}(\chi^{(\lambda\mu)})\kappa_{\lambda\mu}(w^{(0)}) \;.
\end{aligned} \tag{3.5.9}
$$

In accordance with (3.5.2) and (3.5.3) we infer that $e^y_{\alpha\beta}(\boldsymbol{u}^{(1)}) \in L^2(\Omega)$ and $\kappa^y_{\alpha\beta}(w^{(0)}) \in L^2(\Omega)$. Also, we observe that

$$
\kappa_{\alpha\beta}(w^{(2)}) = X^{(\lambda\mu)}\kappa_{\alpha\beta}(e_{\lambda\mu}(\boldsymbol{u}^{(0)})) + \chi^{(\lambda\mu)}\kappa_{\alpha\beta}(\kappa_{\lambda\mu}(w^{(0)})) \;.
$$

Hence, on account of (3.5.1) and (3.5.3) we deduce that $\kappa_{\alpha\beta}(w^{(2)}) \in L^2(\Omega)$. Similarly, (3.5.1) and (3.5.3) yield $e_{\alpha\beta}(\boldsymbol{u}^{(1)}) \in L^2(\Omega)$. Consequently, by using (3.2.9) – (3.2.13) and Theorem 1.1.5 we find

$$
\lim_{\varepsilon\to 0}\Delta_1^\varepsilon = \int_\Omega \frac{1}{|Y|}\int_Y [A^{\alpha\beta\lambda\mu}(y)(e_{\alpha\beta}(\boldsymbol{u}^{(0)}) + e^y_{\alpha\beta}(\boldsymbol{u}^{(1)}))(e_{\lambda\mu}(\boldsymbol{u}^{(0)}) + e^y_{\lambda\mu}(\boldsymbol{u}^{(1)}))
$$

$$+ E^{\alpha\beta\lambda\mu}(y)(\kappa_{\lambda\mu}(w^{(0)}) + \kappa_{\lambda\mu}^y(w^{(2)}))(e_{\alpha\beta}(w^{(0)}) + e_{\alpha\beta}^y(\boldsymbol{u}^{(1)}))$$

$$+ E^{\alpha\beta\lambda\mu}(y)(e_{\lambda\mu}(\boldsymbol{u}^{(0)}) + e_{\lambda\mu}^y(\boldsymbol{u}^{(1)}))(\kappa_{\alpha\beta}(w^{(0)}) + \kappa_{\alpha\beta}^y(w^{(2)}))$$

$$+ D^{\alpha\beta\lambda\mu}(y)(\kappa_{\alpha\beta}(w^{(0)}) + \kappa_{\alpha\beta}^y(w^{(2)}))(\kappa_{\lambda\mu}(w^{(0)}) + \kappa_{\lambda\mu}^y(w^{(2)}))]dydx$$

$$= \int_\Omega [N_h^{\alpha\beta} e_{\alpha\beta}(\boldsymbol{u}^{(0)}) + M_h^{\alpha\beta}\kappa_{\alpha\beta}(w^{(0)})]dx$$

$$+ \int_\Omega \langle N_0^{\alpha\beta} e_{\alpha\beta}^y(\boldsymbol{u}^{(1)}) + M_0^{\alpha\beta}\kappa_{\alpha\beta}^y(w^{(2)})\rangle dx$$

$$= \int_\Omega [N_h^{\alpha\beta} e_{\alpha\beta}(\boldsymbol{u}^{(0)}) + M_h^{\alpha\beta}\kappa_{\alpha\beta}(w^{(0)})]dx = \mathcal{L}_h(\boldsymbol{u}^{(0)}, w^{(0)}) \ .$$

Thus the theorem has been proved. $\qquad\square$

Remark 3.5.2. In the last theorem the strong convergence takes place in $H^1(\Omega)^2 \times H^2(\Omega)$. Since $\boldsymbol{u}^{(0)} \in H_0^1(\Omega)$ and $w^{(0)} \in H_0^2(\Omega)$, a natural question arises: how to construct correctors leading to strong convergence in $H_0^1(\Omega)^2 \times H_0^2(\Omega)$? To this effect we introduce cut-off functions m_ε satisfying the following properties

(i) $m_\varepsilon \in \mathbf{D}(\Omega)$,

(ii) $m_\varepsilon(x) = 0$ if $d(x, \Gamma) = $ (distance from x to $\partial\Omega) \leq \varepsilon$,

(iii) $m_\varepsilon(x) = 1$ if $d(x, \Gamma) \geq 2\varepsilon$,

(iv) $\varepsilon^{|\alpha|}|D^\alpha m_\varepsilon(x)| \leq c_\alpha$, $\forall\alpha$, where c_α depends on α but is independent of ε .

Here $D^1 = \nabla = $ gradient, $D^2 = \nabla^2$ ($|\alpha| = 2$), etc. Such functions m_ε exist, provided Γ is smooth enough.

In our case it would be sufficient to take $m_\varepsilon \in C^2(\Omega)$ and to take the α's with $|\alpha| = 1$ and $|\alpha| = 2$. Obviously the above conditions still have to be satisfied.

The correctors are now defined by

$$\boldsymbol{u}_\varepsilon^{(1)} = \varepsilon m_\varepsilon \boldsymbol{u}^{(1)} , \qquad w_\varepsilon^{(2)} = \varepsilon^2 m_\varepsilon w^{(2)} . \tag{3.5.10}$$

We are in a position to formulate the second result on correctors.

Theorem 3.5.3. Under the assumption (3.5.1) – (3.5.3), if $\boldsymbol{u}_\varepsilon^{(1)}$ and $w_\varepsilon^{(2)}$ are defined by (3.5.10), then

$$\boldsymbol{v}^\varepsilon = \boldsymbol{u}^\varepsilon - \boldsymbol{u}^{(0)} - \boldsymbol{u}_\varepsilon^{(1)} \to 0 \quad \text{in } H_0^1(\Omega)^2 \quad \text{strongly,}$$
$$z^\varepsilon = w^\varepsilon - w^{(0)} - w_\varepsilon^{(2)} \to 0 \quad \text{in } H_0^2(\Omega)^2 \quad \text{strongly.}$$

Proof. The proof runs precisely along the lines of Theorem 3.5.1. $\qquad\square$

3.6. *Variational formulae for effective compliances. Dual effective potential*

We say that local membrane forces $\mathbf{n}^{\alpha\beta}$ are statically admissible, and write $\mathbf{n} \in \mathcal{S}_1^{per}(Y)$ if they satisfy:

$$\mathcal{S}_1^{per}(Y) = \{\mathbf{n} \in L^2(Y, \mathbb{E}_2^s) | \langle \mathbf{n}^{\alpha\beta} e_{\alpha\beta}^y(\boldsymbol{\varphi}) \rangle = 0 \quad \forall\, \boldsymbol{\varphi} \in \mathbf{D}(Y, \mathbb{E}_s^2)$$

$$\text{and } \mathbf{n}^{\alpha\beta}\mu_\beta \text{ take opposite values on opposite sides of } Y\}\,.$$

Here $\boldsymbol{\mu}$ represents a unit vector outward normal to ∂Y.

Let $\boldsymbol{\tau}$ represent a unit vector tangent to ∂Y. Let us define

$$\mathbf{m}_n = \mathbf{m}^{\alpha\beta}\mu_\alpha\mu_\beta\,, \qquad q = \mu_\alpha \mathbf{m}^{\alpha\beta}{}_{|\beta} + \frac{\partial \mathbf{m}_\tau}{\partial s}\,, \qquad \mathbf{m}_\tau = \mathbf{m}^{\alpha\beta}\mu_\alpha\tau_\beta \qquad (3.6.1)$$

for $\mathbf{m} = (\mathbf{m}^{\alpha\beta}) \in L^2(Y, \mathbb{E}_2^s)$; s parametrizes ∂Y and $\partial/\partial s = \partial/\partial\tau$.

Statically admissible local moments are such that $\mathbf{m} \in \mathcal{S}_2^{per}(Y)$, where

$$\mathcal{S}_2^{per}(Y) = \{\mathbf{m} \in L^2(Y, \mathbb{E}_2^s) | \langle \mathbf{m}^{\alpha\beta}\kappa_{\alpha\beta}^y(\varphi) \rangle = 0 \qquad \forall\, \varphi \in \mathbf{D}(Y, \mathbb{E}_s^2)\,;$$

$$\mathbf{m}_n \text{ takes equal and } q \text{ takes opposite values on opposite sides of } Y\}\,.$$

Let us invert the constitutive relations (3.2.3):

$$e_{\lambda\mu}(\boldsymbol{u}^\varepsilon) = a_{\lambda\mu\alpha\beta}\left(\frac{x}{\varepsilon}\right) N_\varepsilon^{\alpha\beta} + e_{\lambda\mu\alpha\beta}\left(\frac{x}{\varepsilon}\right) M_\varepsilon^{\alpha\beta}\,,$$

$$\kappa_{\lambda\mu}(w^\varepsilon) = e_{\lambda\mu\alpha\beta}\left(\frac{x}{\varepsilon}\right) N_\varepsilon^{\alpha\beta} + d_{\lambda\mu\alpha\beta}\left(\frac{x}{\varepsilon}\right) M_\varepsilon^{\alpha\beta}\,. \qquad (3.6.2)$$

Consequently the dual potential assumes the form

$$\mathcal{W}^*(\boldsymbol{N}, \boldsymbol{M}) = \frac{1}{2}[a_{\alpha\beta\lambda\mu}(y)N^{\alpha\beta}N^{\lambda\mu} + e_{\alpha\beta\lambda\mu}(y)M^{\lambda\mu}N^{\alpha\beta}$$

$$+ e_{\alpha\beta\lambda\mu}(y)M^{\alpha\beta}N^{\lambda\mu} + d_{\alpha\beta\lambda\mu}(y)M^{\lambda\mu}M^{\alpha\beta}]\,. \qquad (3.6.3)$$

Its effective counterpart is given by

$$\mathcal{W}_h^*(\boldsymbol{N}_h, \boldsymbol{M}_h) = \frac{1}{2}(a_{\alpha\beta\lambda\mu}^h N_h^{\alpha\beta}N_h^{\lambda\mu} + e_{\alpha\beta\lambda\mu}^h M_h^{\lambda\mu}N_h^{\alpha\beta}$$

$$+ f_{\alpha\beta\lambda\mu}^h N_h^{\lambda\mu}M_h^{\alpha\beta} + d_{\alpha\beta\lambda\mu}^h M_h^{\lambda\mu}M_h^{\alpha\beta})\,. \qquad (3.6.4)$$

The following variational formula

$$\mathcal{W}_h^*(\boldsymbol{N}_h, \boldsymbol{M}_h) = \inf\{\langle \mathcal{W}^*(\mathbf{n} + \boldsymbol{N}_h, \mathbf{m} + \boldsymbol{M}_h)\rangle | \mathbf{n} \in \mathcal{S}_1^{per}(Y)\,,$$

$$\mathbf{m} \in \mathcal{S}_2^{per}(Y),\ \langle \mathbf{n} \rangle = 0,\ \langle \mathbf{m} \rangle = 0\} \qquad (3.6.5)$$

determines the effective flexibilities a^h, e^h, f^h, d^h. The tensors e^h and f^h are linked by: $e_{\alpha\beta\lambda\mu}^h = f_{\lambda\mu\alpha\beta}^h$. Formula (3.6.5) can be expressed as follows

$$\mathcal{W}_h^*(\boldsymbol{N}_h, \boldsymbol{M}_h) = \inf\{\langle \mathcal{W}^*(\mathbf{n}, \mathbf{m})\rangle | \mathbf{n} \in \mathcal{S}_1^{per}(Y)\,,$$

$$\mathbf{m} \in \mathcal{S}_2^{per}(Y),\ \langle \mathbf{n} \rangle = \boldsymbol{N}_h,\ \langle \mathbf{m} \rangle = \boldsymbol{M}_h\}\,. \qquad (3.6.6)$$

The formula (3.6.5) is a particular case of the analogous formula for the effective dual potential for thin periodic shells. The formula for shells will be rigorously derived in Section 17.2 by using the duality theory. Therefore a less complicated derivation of Eq. (3.6.5) is omitted here.

In the case of transverse symmetry the formula (3.6.6) splits into
– formula for d^h

$$M_h^{\alpha\beta} d_{\alpha\beta\lambda\mu}^h M_h^{\lambda\mu} = \inf\{\langle \mathbf{m}^{\alpha\beta} d_{\alpha\beta\lambda\mu}(y) \mathbf{m}^{\lambda\mu}\rangle \mid \mathbf{m} \in \mathcal{S}_2^{per}(Y),\ \langle \mathbf{m}\rangle = M_h\} \quad (3.6.7)$$

– formula for a^h

$$N_h^{\alpha\beta} a_{\alpha\beta\lambda\mu}^h N_h^{\lambda\mu} = \inf\{\langle \mathbf{n}^{\alpha\beta} a_{\alpha\beta\lambda\mu}(y) \mathbf{n}^{\lambda\mu}\rangle \mid \mathbf{n} \in \mathcal{S}_1^{per}(Y),\ \langle \mathbf{n}\rangle = N_h\}\,. \quad (3.6.8)$$

3.7. Transversely symmetric plates periodic in one direction

Assume that the moduli $A^{\alpha\beta\lambda\mu}$, $D^{\alpha\beta\lambda\mu}$ are l_1-periodic functions in y_1 and independent of y_2. The aim of this section is to derive the formulae for $D_h^{\alpha\beta\lambda\mu}$. The formulae for $A_h^{\alpha\beta\lambda\mu}$ will be given without derivation.

Our task is to solve $(P_{KS,Y}^2)$ in its strong formulation, cf. Sec. 3.2. The functions $M_{(\alpha\beta)}^{\lambda\mu}$ are y_2-independent, hence Eq. (3.2.37) reduces to

$$\frac{d^2}{d(y_1)^2}\left[-D^{1111}(y_1)\frac{d^2}{d(y_1)^2}\chi^{(\alpha\beta)} + D^{11\alpha\beta}(y_1)\right] = 0\,. \quad (3.7.1)$$

Conditions (3.2.28) imply

$$\frac{d^2\chi^{(\alpha\beta)}}{d(y_1)^2}(0) = \frac{d^2\chi^{(\alpha\beta)}}{d(y_1)^2}(l_1)\,, \quad (3.7.2)$$

and (3.2.35) give

$$\frac{d\chi^{(\alpha\beta)}}{dy_1}(0) = \frac{d\chi^{(\alpha\beta)}}{dy_1}(l_1)\,. \quad (3.7.3)$$

By (3.7.1) – (3.7.3) we get

$$\kappa_{11}^y(\chi^{(\alpha\beta)}) = -\frac{D^{11\alpha\beta}(y_1)}{D^{1111}(y_1)} + \frac{\langle \frac{D^{11\alpha\beta}}{D^{1111}}\rangle}{\langle \frac{1}{D^{1111}}\rangle}\cdot\frac{1}{D^{1111}(y_1)}\,, \quad (3.7.4)$$

$$\kappa_{12}^y(\chi^{(\alpha\beta)}) = \kappa_{22}^y(\chi^{(\alpha\beta)}) = 0$$

where

$$\langle\cdot\rangle = \frac{1}{l_1}\int_0^{l_1}(\cdot)\,dy_1\,. \quad (3.7.5)$$

Note that Eq. (3.7.4) is sufficient to evaluate the effective stiffnesses. The complete characterization of $\chi^{(\alpha\beta)}$ is not necessary. By applying the formula $(3.2.32)_2$ one finds

$$D_h^{1111} = \langle (D^{1111})^{-1} \rangle^{-1} ,$$

$$D_h^{1112} = \left\langle \frac{D^{1112}}{D^{1111}} \right\rangle D_h^{1111} ,$$

$$D_h^{1122} = \left\langle \frac{D^{1122}}{D^{1111}} \right\rangle D_h^{1111} ,$$

$$D_h^{1222} = \langle D^{1222} \rangle - \left\langle \frac{D^{1112} D^{2211}}{D^{1111}} \right\rangle + \left\langle \frac{D^{1122}}{D^{1111}} \right\rangle \left\langle \frac{D^{1112}}{D^{1111}} \right\rangle D_h^{1111} ,$$

$$D_h^{2222} = \langle D^{2222} \rangle - \left\langle \frac{(D^{2211})^2}{D^{1111}} \right\rangle + \left\langle \frac{D^{2211}}{D^{1111}} \right\rangle^2 D_h^{1111} ,$$

$$D_h^{1212} = \langle D^{1212} \rangle - \left\langle \frac{(D^{1211})^2}{D^{1111}} \right\rangle + \left\langle \frac{D^{1211}}{D^{1111}} \right\rangle^2 D_h^{1111} .$$

$$(3.7.6)$$

The remaining stiffnesses can be found using symmetries (2.7.23).

Assume now that the plate material is orthotropic with orthotropy axes coinciding with y_1, y_2. Then $D^{1112} = D^{2221} = 0$ and formulae (3.7.6) reduce to

$$D_h^{1111} = \langle (D^{1111})^{-1} \rangle^{-1} , \qquad D_h^{1112} = D_h^{2221} = 0 ,$$

$$D_h^{1122} = \left\langle \frac{D^{1122}}{D^{2222}} \right\rangle D_h^{1111} ,$$

$$D_h^{2222} = \langle D^{2222} \rangle - \left\langle \frac{(D^{2211})^2}{D^{1111}} \right\rangle + \left\langle \frac{D^{2211}}{D^{1111}} \right\rangle^2 D_h^{1111} ,$$

$$D_h^{1212} = \langle D^{1212} \rangle .$$

$$(3.7.7)$$

To find similar formulae for effective compliances let us introduce functions:

$$f(x) = \langle x^{-1} \rangle^{-1} , \quad j(x) = \langle x \rangle , \qquad g(x,y) = \left\langle \frac{y}{x} \right\rangle f(x) ,$$

$$h(x,y,z) = \langle z \rangle - \left\langle \frac{y^2}{x} \right\rangle + \left\langle \frac{y}{x} \right\rangle^2 f(x) .$$

$$(3.7.8)$$

Note that

$$D_h^{1111} = f(D^{1111}) , \quad D_h^{1212} = j(D^{1212}) ,$$

$$D_h^{1122} = g(D^{1111}, D^{1122}) , \qquad D_h^{2222} = h(D^{1111}, D^{1122}, D^{2222}) .$$

$$(3.7.9)$$

In the orthotropic case the inverted constitutive relations have the form

$$\kappa_{11} = d_{1111} M^{11} + d_{1122} M^{22} ,$$

$$\kappa_{22} = d_{2211} M^{11} + d_{2222} M^{22} , \qquad \kappa_{12} = 2 d_{1212} M^{12} .$$

$$(3.7.10)$$

The effective compliances for the bending problem are given by

$$d_{1111}^h = h(d_{2222}, d_{1122}, d_{1111}) , \qquad d_{1122}^h = g(d_{2222}, d_{1122}) ,$$

$$d_{2222}^h = f(d_{2222}) , \quad d_{1212}^h = f(d_{1212}) .$$

$$(3.7.11)$$

These formulae follow from (3.7.9) by a direct algebraic computation.

Let us report now the counterparts of formulae (3.7.6) for effective in-plane stiffnesses; the derivation will be omitted to save space, cf. Section 5.6.1 where a short outline of a similar derivation is reported.

Define

$$d^2 = A^{1111} \cdot A^{1212} - (A^{1112})^2 \,, \qquad d_{11} = \left\langle \frac{A^{1212}}{d^2} \right\rangle \,, \qquad d_{12} = - \left\langle \frac{A^{1211}}{d^2} \right\rangle \,,$$

$$d_{22} = \left\langle \frac{A^{1111}}{d^2} \right\rangle \,, \qquad \Delta = d_{11}d_{22} - (d_{12})^2 \,; \qquad (3.7.12)$$

$$d_1 = \left\langle \frac{(A^{1212}A^{1122} - A^{1112}A^{1222})}{d^2} \right\rangle \,, \qquad d_2 = \left\langle \frac{(A^{1111}A^{1222} - A^{1112}A^{1122})}{d^2} \right\rangle \,.$$

The effective stiffnesses read

$$A_h^{1111} = \frac{d_{22}}{\Delta} \,, \qquad A_h^{1112} = -\frac{d_{12}}{\Delta} \,, \qquad A_h^{1212} = \frac{d_{11}}{\Delta} \,,$$

$$A_h^{1122} = \frac{(d_{22}d_1 - d_{12}d_2)}{\Delta} \,, \qquad A_h^{2212} = \frac{(d_{11}d_2 - d_{12}d_1)}{\Delta} \,,$$

$$A_h^{2222} = \left\langle A^{2222} - \frac{[A^{1111}(A^{1222})^2 + A^{1212}(A^{1122})^2 - 2A^{1112}A^{1122}A^{2212}]}{d^2} \right\rangle$$

$$+ \frac{[d_{22}(d_1)^2 + d_{11}(d_2)^2 - 2d_{12}d_1d_2]}{\Delta} \,. \qquad (3.7.13)$$

For the case of orthotropy the effective membrane stiffnesses $A_h^{\alpha\beta\lambda\mu}$ are given by

$$A_h^{1111} = f(A^{1111}) \,, \qquad A_h^{1212} = f(A^{1212}) \,, \qquad A_h^{1122} = g(A^{1111}, A^{2211}) \,,$$

$$A_h^{2222} = h(A^{1111}, A^{1122}, A^{2222}) \,, \qquad A_h^{1112} = A_h^{1222} = 0 \,. \qquad (3.7.14)$$

Let $a_{\alpha\beta\lambda\mu}$ be compliances for the membrane problem. The effective compliances in the orthotropic case are given by

$$a_{1111}^h = h(a_{2222}, a_{1122}, a_{1111}) \,, \qquad a_{1122}^h = g(a_{2222}, a_{1122}) \,,$$

$$a_{1112}^h = a_{2122}^h = 0 \,, \qquad a_{2222}^h = f(a_{2222}) \,, \qquad a_{1212}^h = j(a_{1212}) \,. \qquad (3.7.15)$$

Let us emphasize an analogy between formulae in the orthotropic case for:
(a) effective bending stiffnesses and effective compliances due to stretching,
(b) effective bending compliances and effective stretching stiffnesses.

The mutual formulae can be found by replacing indices: $(1, 2) \to (2, 1)$.

3.8. Ribbed plates. Bending problem

Let us consider a transversely symmetric plate made from strips of constant alternate properties. Our aim is to find effective bending stiffnesses $D_h^{\alpha\beta\lambda\mu}$. This problem has already been solved and its solution is given by (3.7.6). These formulae, however, do not take into account that $D^{\alpha\beta\lambda\mu}$ are piece-wise constant. Moreover these formulae are not explicitly dependent on the direction of stiffeners. The objective of this section is to derive the formulae for effective stiffnesses $D_h^{\alpha\beta\lambda\mu}$ as explicit functions of direction of stiffeners and area fractions θ_α of both constituents.

3.8.1.   Formula of Francfort and Murat for stiffnesses

Consider a transversely symmetric plate made from strips of stiffnesses D_1 and D_2 with area fractions θ_1 and θ_2 respectively; $\theta_1 + \theta_2 = 1$. Assume here that the quadratic form

$$f(\kappa) = \kappa_{\alpha\beta}(D_2^{\alpha\beta\lambda\mu} - D_1^{\alpha\beta\lambda\mu})\kappa_{\lambda\mu} \,, \tag{3.8.1}$$

is positive definite, which will be briefly written: $D_2 > D_1$ and called the "ordered case". The directions of ribs $n = (n_\alpha)$ is not correlated with anisotropy directions of D_α, see Fig. 3.8.1.

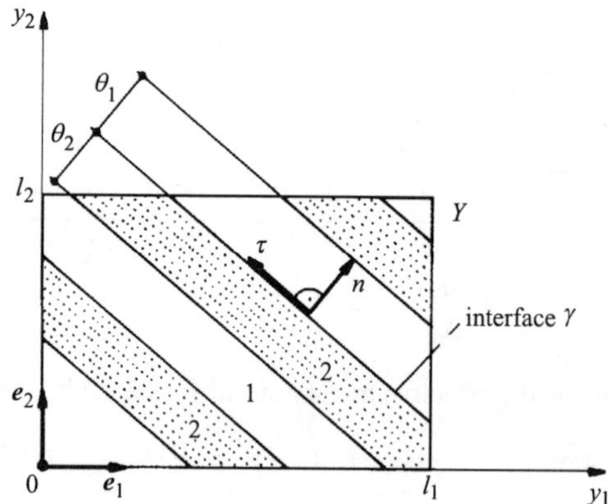

Fig. 3.8.1. Ribbed plate of first rank

The effective tensor D_h is determined by the following implicit formula of Francfort and Murat

$$\theta_1(D_2 - D_h)^{-1} = (D_2 - D_1)^{-1} - \theta_2\Gamma_D \,, \tag{3.8.2}$$

where

$$\Gamma_D = (n_\alpha n_\beta D_2^{\alpha\beta\lambda\mu} n_\lambda n_\mu)^{-1}\Gamma_n \,, \quad (\Gamma_n)_{\alpha\beta\lambda\mu} = n_\alpha n_\beta n_\lambda n_\mu \,, \tag{3.8.3}$$

and operation $(\cdot)^{-1}$ is understood as follows.
Let

$$M^{\alpha\beta} = D^{\alpha\beta\lambda\mu}\kappa_{\lambda\mu} \,, \quad \kappa_{\lambda\mu} = d_{\lambda\mu\alpha\beta}M^{\alpha\beta} \,. \tag{3.8.4}$$

Then we write

$$d = D^{-1} \,. \tag{3.8.5}$$

Formula (3.8.2) is compatible with $(3.2.32)_2$ and with variational formula (3.4.4). Because of algebraic nature of formula (3.8.2), its direct (algebraic) derivation is easier than a rigorous derivation based on variational formula (3.4.4). Thus we have found it appropriate to report below this easiest method of derivation. We shall use only some selected facts resulting from the very definition of D_h given in the previous sections.

Proof of (3.8.2). We refer to Section 3.7 and omit indices $(\alpha\beta)$ assigned to $\chi^{(\alpha\beta)}$. According to (3.7.4) $\kappa_{11}^y(\chi)$ is a piece-wise constant function. Consequently the strain tensor κ^y within Y is a constant tensor $\overset{1}{\kappa}$ in phase 1 and a constant tensor $\overset{2}{\kappa}$ within phase 2; we omit index y. Upon averaging we find

$$\kappa_{\alpha\beta}^h = \theta_1 \overset{1}{\kappa}_{\alpha\beta} + \theta_2 \overset{2}{\kappa}_{\alpha\beta} , \tag{3.8.6}$$

where $\kappa_{\alpha\beta}$ are referred to the $e_\alpha \otimes e_\beta$ basis. Similarly

$$M_h^{\alpha\beta} = \theta_1 \overset{1}{M}{}^{\alpha\beta} + \theta_2 \overset{2}{M}{}^{\alpha\beta} ,$$
$$\overset{\sigma}{M}{}^{\alpha\beta} = D_\sigma^{\alpha\beta\lambda\mu} \overset{\sigma}{\kappa}_{\alpha\beta} \quad \text{(do not sum over } \sigma \text{ !)}. \tag{3.8.7}$$

The continuity conditions (3.2.41), (3.2.42) on both sides of γ (see Fig. 3.8.1) imply

$$[\![\chi]\!]_\gamma = 0 , \quad \left[\!\!\left[\frac{\partial^2\chi}{\partial\tau\partial n}\right]\!\!\right]_{|\gamma} = 0 , \quad \left[\!\!\left[\frac{\partial\chi}{\partial n}\right]\!\!\right]_{|\gamma} = 0 , \quad \left[\!\!\left[\frac{\partial^2\chi}{\partial\tau^2}\right]\!\!\right]_{|\gamma} = 0 . \tag{3.8.8}$$

We observe that here $|\gamma$ does not denote partial differentiation with respect to y_γ. Hence

$$\overset{1}{\kappa}_{n\tau} = \overset{2}{\kappa}_{n\tau} , \qquad \overset{1}{\kappa}_{\tau\tau} = \overset{2}{\kappa}_{\tau\tau} , \tag{3.8.9}$$

where

$$\kappa_{n\tau} = \kappa_{\alpha\beta} n^\alpha \tau^\beta , \qquad \kappa_{\tau\tau} = \kappa_{\alpha\beta} \tau^\alpha \tau^\beta . \tag{3.8.10}$$

Let us represent both these tensors $\overset{\sigma}{\kappa}$ in the basis generated by (n, τ) :

$$\overset{\sigma}{\kappa} = \overset{\sigma}{\kappa}_{nn}\, n \otimes n + \overset{\sigma}{\kappa}_{n\tau}\, (n \otimes \tau + \tau \otimes n) + \overset{\sigma}{\kappa}_{\tau\tau}\, \tau \otimes \tau . \tag{3.8.11}$$

Hence

$$\overset{1}{\kappa} - \overset{2}{\kappa} = (\overset{1}{\kappa}_{nn} - \overset{2}{\kappa}_{nn}) n \otimes n , \tag{3.8.12a}$$

or

$$\overset{1}{\kappa}_{\alpha\beta} - \overset{2}{\kappa}_{\alpha\beta} = k n_\alpha n_\beta . \tag{3.8.12b}$$

The constant k will be specified in the sequel.

Let us define the homogenized stiffnesses by postulating a linear relation between M_h and κ^h:

$$M_h^{\alpha\beta} = D_h^{\alpha\beta\lambda\mu} \kappa_{\lambda\mu}^h \ . \tag{3.8.13}$$

By (3.8.6) and (3.8.13) we have

$$D_h^{\alpha\beta\lambda\mu} \kappa_{\lambda\mu}^h = \theta_1 D_1^{\alpha\beta\lambda\mu} \overset{1}{\kappa}_{\lambda\mu} + \theta_2 D_2^{\alpha\beta\lambda\mu} \overset{2}{\kappa}_{\lambda\mu} \ . \tag{3.8.14}$$

Let us substitute

$$\theta_2 \overset{2}{\kappa}_{\lambda\mu} = \kappa_{\lambda\mu}^h - \theta_1 \overset{1}{\kappa}_{\lambda\mu} \tag{3.8.15}$$

into (3.8.14). Hence

$$(D_2^{\alpha\beta\lambda\mu} - D_h^{\alpha\beta\lambda\mu})\kappa_{\lambda\mu}^h = \theta_1(D_2^{\alpha\beta\lambda\mu} - D_1^{\alpha\beta\lambda\mu}) \overset{1}{\kappa}_{\lambda\mu} \ . \tag{3.8.16}$$

On the other hand (3.8.6) and (3.8.12b) imply

$$\kappa_{\alpha\beta}^h = \theta_1 \overset{1}{\kappa}_{\alpha\beta} + \theta_2[\overset{1}{\kappa}_{\alpha\beta} - k\, n_\alpha n_\beta] \ ,$$

or

$$\kappa_{\alpha\beta}^h = \overset{1}{\kappa}_{\alpha\beta} - k\theta_2 n_\alpha n_\beta \ . \tag{3.8.17}$$

The continuity condition

$$\overset{1}{M}{}^{\alpha\beta} n_\alpha n_\beta = \overset{2}{M}{}^{\alpha\beta} n_\alpha n_\beta \ , \tag{3.8.18}$$

along with (3.8.6)$_2$ and (3.8.12b) implies

$$(D_2^{\alpha\beta\lambda\mu} - D_1^{\alpha\beta\lambda\mu}) \overset{1}{\kappa}_{\lambda\mu}\, n_\alpha n_\beta = (D_2^{\alpha\beta\lambda\mu} n_\alpha n_\beta n_\lambda n_\mu)k \ , \tag{3.8.19}$$

which determines k involved in (3.8.12b).

Let us define

$$\wp^{\alpha\beta} = (D_2^{\alpha\beta\lambda\mu} - D_1^{\alpha\beta\lambda\mu}) \overset{1}{\kappa}_{\lambda\mu} \ . \tag{3.8.20}$$

Upon substituting k given by (3.8.19) into (3.8.17) one finds

$$\kappa_{\alpha\beta}^h = \overset{1}{\kappa}_{\alpha\beta} - \frac{\theta_2 n_\lambda \wp^{\lambda\mu} n_\mu}{n_\gamma n_\delta D_2^{\gamma\delta\lambda\mu} n_\lambda n_\mu} n_\alpha n_\beta \ . \tag{3.8.21}$$

Combining (3.8.16) with (3.8.20) one obtains

$$(D_2^{\alpha\beta\lambda\mu} - D_h^{\alpha\beta\lambda\mu})\kappa_{\lambda\mu}^h = \theta_1 \wp^{\alpha\beta} \ . \tag{3.8.22}$$

Let us invert (3.8.20) and (3.8.22)

$$\overset{1}{\kappa}_{\lambda\mu} = (D_2 - D_1)^{-1}_{\lambda\mu\alpha\beta}\wp^{\alpha\beta} \qquad \kappa^h_{\lambda\mu} = \theta_1(D_2 - D_h)^{-1}_{\lambda\mu\alpha\beta}\wp^{\alpha\beta} , \tag{3.8.23}$$

and insert these expressions into (3.8.21). One arrives at the formula

$$\theta_1(D_2 - D_h)^{-1}_{\lambda\mu\alpha\beta} = (D_2 - D_1)^{-1}_{\lambda\mu\alpha\beta} - \frac{\theta_2}{n_\gamma n_\delta D_2^{\gamma\delta\nu\rho} n_\nu n_\rho} n_\lambda n_\mu n_\alpha n_\beta \tag{3.8.24}$$

equivalent to (3.8.2) – (3.8.3). This completes the proof. $\qquad\square$

Note that relations (3.8.22) are invertible, since condition (3.8.1) implies that the quadratic form

$$f_1(\kappa) = \kappa_{\alpha\beta}(D_2^{\alpha\beta\lambda\mu} - D_h^{\alpha\beta\lambda\mu})\kappa_{\lambda\mu} \tag{3.8.25}$$

is positive definite; it means that the strongest phase is stronger than the homogenized phase. This intuitive property can be deduced from (3.4.4). Note that

$$\kappa^y_{\alpha\beta} D^{\alpha\beta\lambda\mu}(y)\kappa^y_{\lambda\mu} \leq \kappa^y_{\alpha\beta} D_2^{\alpha\beta\lambda\mu}\kappa^y_{\lambda\mu} . \tag{3.8.26}$$

Hence if $\theta_1 \neq 0$ we can estimate

$$\kappa^h_{\alpha\beta} D_h^{\alpha\beta\lambda\mu}\kappa^h_{\lambda\mu} < \min\{\langle\kappa^y_{\alpha\beta} D_2^{\alpha\beta\lambda\mu}\kappa^y_{\lambda\mu}\rangle \,|\, \kappa^y \in K_\kappa(Y)\,,\, \langle\kappa^y\rangle = \kappa^h\}$$

$$= \kappa^h_{\alpha\beta} D_2^{\alpha\beta\lambda\mu}\kappa^h_{\lambda\mu} , \tag{3.8.27}$$

which completes the proof.

Remark 3.8.1. The ordering assumption $D_2 - D_1 > 0$ implies that the matrix $D_2 - D_1$ is invertible and the formula (3.8.24) makes sense. The same formula still holds if we assume only that the matrix $D_2 - D_1$ is invertible.

3.8.2.   Ribbed plates of higher rank with the stronger phase taken as an envelope

Assume that both stiffness tensors D_σ are isotropic. Thus they possess the following representations

$$D_\sigma^{\alpha\beta\lambda\mu} = 2k_\sigma I_1^{\alpha\beta\lambda\mu} + 2\mu_\sigma I_2^{\alpha\beta\lambda\mu} , \qquad \sigma \in \{1,2\} , \tag{3.8.28}$$

where

$$I_1^{\alpha\beta\lambda\mu} = \frac{1}{2}\delta^{\alpha\beta}\delta^{\lambda\mu} , \qquad I_2^{\alpha\beta\lambda\mu} = \frac{1}{2}(\delta^{\alpha\lambda}\delta^{\beta\mu} + \delta^{\alpha\mu}\delta^{\beta\lambda} - \delta^{\alpha\beta}\delta^{\lambda\mu}) . \tag{3.8.29}$$

Thus we have

$$D_\sigma^{1111} = D_\sigma^{2222} = k_\sigma + \mu_\sigma , \qquad D_\sigma^{1122} = k_\sigma - \mu_\sigma , \qquad D_\sigma^{1212} = \mu_\sigma . \tag{3.8.30}$$

The condition $D_2 > D_1$ means that

$$k_2 > k_1, \qquad \mu_2 > \mu_1 . \tag{3.8.31}$$

Due to the isotropy of D_2 one notes that the quantity

$$n_\gamma n_\delta D_2^{\gamma\delta\alpha\beta} n_\alpha n_\beta = k_2 + \mu_2 \tag{3.8.32}$$

is independent of n and the formula of Francfort and Murat (3.8.2) reduces to

$$\theta_1 (D_2 - D_h)^{-1} = (D_2 - D_1)^{-1} - \frac{\theta_2}{k_2 + \mu_2}\Gamma_\mathbf{n} , \tag{3.8.33}$$

or

$$\theta_1 (D_2 - D_h)^{-1} = (D_2 - D_1)^{-1} - \theta_2 \Gamma_\mathbf{D} ,$$

where

$$\Gamma_\mathbf{D} = \frac{1}{k_2 + \mu_2}\Gamma_\mathbf{n} ,$$

and

$$\Gamma_{\mathbf{D}_{\alpha\beta\lambda\mu}} = \frac{1}{2\mu_2}\left\{ n_\alpha n_\beta n_\lambda n_\mu - \frac{k_2 - \mu_2}{k_2 + \mu_2} n_\alpha n_\beta n_\lambda n_\mu \right\} .$$

The formula (3.8.33) holds irrespective of the isotropy of D_1. That is why this formula turns out to be very helpful in evaluating effective stiffnesses of so-called ribbed plates of higher rank. To be more specific, let us consider a ribbed plate of 2nd rank made by stacking together material 2 with the homogenized material of stiffness D_h determined by (3.8.33). The area fractions of materials "2" and "h" are α_2 and α_1 respectively. Let m be the direction of stiffeners, see Fig. 3.8.2.

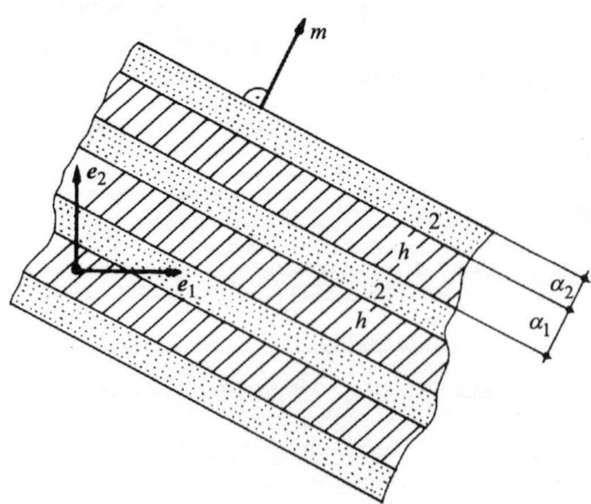

Fig. 3.8.2. Ribbed plate of second rank

On using formula (3.8.33) one finds the effective tensor D_{hh} of the new composite plate:

$$\alpha_1(D_2 - D_{hh})^{-1} = (D_2 - D_1)^{-1} - \frac{\alpha_2}{k_2 + \mu_2}\Gamma_m . \qquad (3.8.34)$$

Let us multiply both sides of (3.8.34) by θ_1 and combine this identity with (3.8.33). One finds the formula for effective stiffnesses of a ribbed plate of second rank:

$$\alpha_1\theta_1(D_2 - D_{hh})^{-1} = (D_2 - D_1)^{-1} - \frac{1}{k_2 + \mu_2}\Gamma_{n,m} , \qquad (3.8.35)$$
$$\Gamma_{n,m} = \theta_2\Gamma_n + \theta_1\alpha_2\Gamma_m .$$

Note that $\alpha_1\theta_1$ means the resulting area fraction of material "1".

This process of mixing material "2" with ribbed plates of subsequent ranks can be easily continued. We shall prove in Chapter VI that three subsequent laminations (or introducing ribs) done in an appropriate way lead to an isotropic effective plate of maximal stiffness. The idea of subsequent laminations is crucial in understanding the extremal properties of optimal plate designs.

3.8.3. Formula of Lurie-Cherkaev-Fedorov for stiffnesses

Let us consider once again the problem of Section 3.8.1 but without assumption (3.8.1). We shall derive the following formula for D_h:

$$D_h^{\alpha\beta\delta\gamma} = \langle D^{\alpha\beta\delta\gamma}\rangle_\theta - \theta_1\theta_2\frac{n_\lambda n_\mu(D_2^{\lambda\mu\alpha\beta} - D_1^{\lambda\mu\alpha\beta})(D_2^{\varphi\psi\delta\gamma} - D_1^{\varphi\psi\delta\gamma})n_\varphi n_\psi}{n_\kappa n_\omega[D^{\kappa\omega\rho\nu}]_\theta n_\rho n_\nu} , \qquad (3.8.36)$$

where

$$\langle f\rangle_\theta = \theta_1 f_1 + \theta_2 f_2 , \qquad [f]_\theta = \theta_1 f_2 + \theta_2 f_1 . \qquad (3.8.37)$$

First, let us solve (3.8.6), (3.8.12b) with respect to $\overset{\sigma}{\kappa}$:

$$\overset{1}{\kappa}_{\alpha\beta} = \kappa_{\alpha\beta}^h + \theta_2 k n_\alpha n_\beta , \qquad \overset{2}{\kappa}_{\alpha\beta} = \kappa_{\alpha\beta}^h - \theta_1 k n_\alpha n_\beta . \qquad (3.8.38)$$

Upon substituting (3.8.38) into (3.8.6) one finds

$$M_h^{\alpha\beta} = \theta_1 D_1^{\alpha\beta\lambda\mu}(\kappa_{\lambda\mu}^h + \theta_2 k n_\lambda n_\mu) + \theta_2 D_2^{\alpha\beta\lambda\mu}(\kappa_{\lambda\mu}^h - \theta_1 k n_\lambda n_\mu) ,$$

or

$$M_h^{\alpha\beta} = \langle D^{\alpha\beta\lambda\mu}\rangle_\theta \kappa_{\lambda\mu}^h + \theta_1\theta_2 k(D_1^{\alpha\beta\lambda\mu} - D_2^{\alpha\beta\lambda\mu})n_\lambda n_\mu . \qquad (3.8.39)$$

Let us multiply $(3.8.38)_1$ by $D_1^{\alpha\beta\lambda\mu}$ and $(3.8.38)_2$ by $D_2^{\alpha\beta\lambda\mu}$. Upon subtracting these identities we find

$$M_1^{\lambda\mu} - M_2^{\lambda\mu} = -(D_2^{\lambda\mu\alpha\beta} - D_1^{\lambda\mu\alpha\beta})\kappa_{\alpha\beta}^h + k[D^{\lambda\mu\alpha\beta}]_\theta n_\alpha n_\beta . \qquad (3.8.40)$$

Let us multiply both sides of (3.8.40) by $n_\lambda n_\mu$ and take into account (3.8.18). Hence

$$k = \frac{n_\lambda n_\mu (D_2^{\lambda\mu\alpha\beta} - D_1^{\lambda\mu\alpha\beta})\kappa_{\alpha\beta}^h}{n_\kappa n_\nu [D^{\kappa\nu\rho\omega}]_\theta n_\rho n_\omega} . \tag{3.8.41}$$

Now it is sufficient to substitute k given by (3.8.41) into (3.8.39) and we arrive at (3.8.36). The denominator of (3.8.41) is always positive definite, irrespective of materials being ordered or not.

3.8.4. Formula of Francfort-Murat-type for compliances

Let us consider the same problem as in Section 3.8.1. Our aim is to find a counterpart of formula (3.8.2) for the compliance tensors: d^σ, d^h. The following set of equations is at our disposal, cf. Sec. 3.8.1

$$M_h^{\alpha\beta} = \theta_1 M_1^{\alpha\beta} + \theta_2 M_2^{\alpha\beta} , \tag{3.8.42}$$

$$d_{\alpha\beta\lambda\mu}^h M_h^{\lambda\mu} = \theta_1 \overset{1}{\kappa}_{\alpha\beta} + \theta_2 \overset{2}{\kappa}_{\alpha\beta} , \tag{3.8.43}$$

$$\overset{\sigma}{\kappa}_{\alpha\beta} = d_{\alpha\beta\lambda\mu}^\sigma M_\sigma^{\lambda\mu} \qquad \text{(do not sum over } \sigma) , \tag{3.8.44}$$

$$(M_1^{\alpha\beta} - M_2^{\alpha\beta})n_\alpha n_\beta = 0 , \tag{3.8.45}$$

$$d_{\alpha\beta\lambda\mu}^1 M_1^{\lambda\mu} - d_{\alpha\beta\lambda\mu}^2 M_2^{\lambda\mu} = k\, n_\alpha n_\beta . \tag{3.8.46}$$

The derivation runs as follows. Let us define

$$N^{\alpha\beta} = M_1^{\alpha\beta} - M_2^{\alpha\beta} , \tag{3.8.47}$$

$$\rho_{\alpha\beta} = (d_{\alpha\beta\lambda\mu}^1 - d_{\alpha\beta\lambda\mu}^2)M_2^{\lambda\mu} . \tag{3.8.48}$$

Equation (3.8.42) can be rearranged to the form

$$M_h^{\alpha\beta} = M_2^{\alpha\beta} + \theta_1 N^{\alpha\beta} . \tag{3.8.49}$$

Functions (3.8.42) -(3.8.44) imply

$$d_{\alpha\beta\lambda\mu}^h M_h^{\lambda\mu} = d_{\alpha\beta\lambda\mu}^1 (M_h^{\lambda\mu} - \theta_2 M_2^{\lambda\mu}) + \theta_2 d_{\alpha\beta\lambda\mu}^2 M_2^{\lambda\mu} , \tag{3.8.50}$$

hence

$$(d_{\alpha\beta\lambda\mu}^1 - d_{\alpha\beta\lambda\mu}^2)M_h^{\lambda\mu} = \theta_2 \cdot \rho_{\alpha\beta} . \tag{3.8.51}$$

Thus we can express $M_h^{\lambda\mu}$, $M_2^{\lambda\mu}$ in terms of ρ

$$M_h^{\lambda\mu} = \theta_2 [(\boldsymbol{d}^1 - \boldsymbol{d}^h)^{-1}]^{\lambda\mu\alpha\beta} \rho_{\alpha\beta} , \qquad M_2^{\lambda\mu} = [(\boldsymbol{d}^1 - \boldsymbol{d}^2)^{-1}]^{\lambda\mu\alpha\beta} \rho_{\alpha\beta} . \tag{3.8.52}$$

Our task now is to express $N^{\alpha\beta}$ in terms of ρ. We use (3.8.46) and (3.8.49) to get

$$d_{\alpha\beta\lambda\mu}^1 (N^{\lambda\mu} + M_2^{\lambda\mu}) - d_{\alpha\beta\lambda\mu}^2 M_2^{\lambda\mu} = k\, n_\alpha n_\beta$$

or

$$\rho_{\alpha\beta} + d^1_{\alpha\beta\lambda\mu} N^{\lambda\mu} = k\, n_\alpha n_\beta$$

and

$$N^{\lambda\mu} = D_1^{\lambda\mu\alpha\beta}(k\, n_\alpha n_\beta - \rho_{\alpha\beta}) \,. \tag{3.8.53}$$

Equation (3.8.45) implies $n_\lambda N^{\lambda\mu} n_\mu = 0$, hence

$$k(n_\lambda n_\mu D_1^{\lambda\mu\alpha\beta} n_\alpha n_\beta) = n_\lambda n_\mu D_1^{\lambda\mu\alpha\beta} \rho_{\alpha\beta} \,. \tag{3.8.54}$$

We go back to (3.8.53) and find

$$N^{\lambda\mu} = \Gamma_{\mathbf{d}}^{\lambda\mu\alpha\beta} \rho_{\alpha\beta} \,, \tag{3.8.55}$$

$$\Gamma_{\mathbf{d}}^{\lambda\mu\alpha\beta} = \frac{n_\gamma n_\delta D_1^{\gamma\delta\alpha\beta} D_1^{\lambda\mu\varphi\psi} n_\varphi n_\psi}{n_\rho n_\nu D_1^{\rho\nu\kappa\omega} n_\kappa n_\omega} - D_1^{\lambda\mu\alpha\beta} \,. \tag{3.8.56}$$

Now we insert (3.8.55), (3.8.52) into (3.8.49) to obtain

$$\theta_2[(\boldsymbol{d}^1 - \boldsymbol{d}^h)^{-1}]^{\lambda\mu\alpha\beta} = [(\boldsymbol{d}^1 - \boldsymbol{d}^2)^{-1}]^{\lambda\mu\alpha\beta} + \theta_1 \Gamma_{\mathbf{d}}^{\lambda\mu\alpha\beta} \,, \tag{3.8.57}$$

or

$$\theta_2(\boldsymbol{d}^1 - \boldsymbol{d}^h)^{-1} = (\boldsymbol{d}^1 - \boldsymbol{d}^2)^{-1} + \theta_1 \Gamma_{\mathbf{d}} \,. \tag{3.8.58}$$

The ordering assumption (3.8.1) decides that inversion operations in (3.8.57) make sense. In the case of material "1" being isotropic one can compute operations in (3.8.56) to find

$$\frac{1}{2\mu_1} \Gamma_{\mathbf{d}}^{\lambda\mu\alpha\beta} = n^\lambda n^\mu n^\alpha n^\beta - \frac{1}{2}(\delta^{\lambda\alpha}\delta^{\mu\beta} + \delta^{\alpha\mu}\delta^{\beta\lambda})$$

$$- \frac{k_1 - \mu_1}{k_1 + \mu_1}(\delta^{\lambda\mu} - n^\lambda n^\mu)(\delta^{\alpha\beta} - n^\alpha n^\beta) \,. \tag{3.8.59}$$

3.9. *Ribbed plates. Plane elasticity problem*

We consider the same composite plate as in Section 3.8. Our aim is to find closed formulae for effective in-plane stiffnesses $A_h^{\alpha\beta\lambda\mu}$ and compliances $a^h_{\alpha\beta\lambda\mu}$.

3.9.1. Formula of Francfort and Murat for stiffnesses

Our considerations will be confined to the ordered case: the matrix $\boldsymbol{A}_2 - \boldsymbol{A}_1$ is assumed to be positive definite; $\boldsymbol{A}_\sigma$ represent the in-plane stiffness tensors of strips; the area fractions are denoted by θ_α; the lamination direction is determined by a versor $\boldsymbol{n}$, see Fig. 3.8.1 which remains valid here.

On analyzing problem $(P^1_{KS,Y})$, see Sec. 3.2.A, we conclude that $e^y_{\alpha\beta}(\boldsymbol{T})$ (indices $(\alpha\beta)$ are suppressed here) are piece-wise constants; they will be denoted by $\overset{1}{\epsilon}_{\alpha\beta}$, $\overset{2}{\epsilon}_{\alpha\beta}$; the indices

1 and 2 refer to materials "1" and "2". Let us refer these strains to the (n, τ) basis; n, τ being normal and tangent to the discontinuity line γ, see Fig. 3.8.1.

$$\overset{\sigma}{\epsilon}_{\alpha\beta} = \overset{\sigma}{\epsilon}_{nn}\, n_\alpha n_\beta + \overset{\sigma}{\epsilon}_{n\tau}\, (n_\alpha \tau_\beta + n_\beta \tau_\alpha) + \overset{\sigma}{\epsilon}_{\tau\tau}\, \tau_\alpha \tau_\beta \,, \tag{3.9.1}$$

where

$$\overset{\sigma}{\epsilon}_{nn} = \frac{\partial T_n}{\partial n}\Big|_\sigma\,, \qquad \overset{\sigma}{\epsilon}_{\tau\tau} = \frac{\partial T_\tau}{\partial s}\Big|_\sigma\,, \qquad 2\,\overset{\sigma}{\epsilon}_{n\tau} = \frac{\partial T_n}{\partial s}\Big|_\sigma + \frac{\partial T_\tau}{\partial n}\Big|_\sigma\,. \tag{3.9.2}$$

Note that $T_{n|_1} = T_{n|_2}$ and $T_{\tau|_1} = T_{\tau|_2}$, hence $\frac{\partial T_n}{\partial s}\big|_1 = \frac{\partial T_n}{\partial s}\big|_2$ and $\frac{\partial T_\tau}{\partial s}\big|_1 = \frac{\partial T_\tau}{\partial s}\big|_2$. Consequently

$$\overset{1}{\epsilon}_{\alpha\beta} - \overset{2}{\epsilon}_{\alpha\beta} = k\, n_\alpha n_\beta + \frac{1}{2}l(n_\alpha \tau_\beta + n_\beta \tau_\alpha)\,, \quad k = -\left[\!\!\left[\frac{\partial T_n}{\partial n}\right]\!\!\right]_{|\gamma}\,, \quad l = -\left[\!\!\left[\frac{\partial T_\tau}{\partial n}\right]\!\!\right]_{|\gamma}$$

and we can write

$$\overset{1}{\epsilon}_{\alpha\beta} - \overset{2}{\epsilon}_{\alpha\beta} = p_\alpha n_\beta + p_\beta n_\alpha\,, \quad p_\alpha = \frac{1}{2}(kn_\alpha + l\tau_\alpha)\,. \tag{3.9.3}$$

The quantities p_1, p_2 will be found in the sequel.

Moreover, the following relationships, that can be inferred from (i) – (iii) of part A of Section 3.2, are at our disposal:

$$\epsilon^h_{\alpha\beta} = \theta_1 \overset{1}{\epsilon}_{\alpha\beta} + \theta_2 \overset{2}{\epsilon}_{\alpha\beta}\,, \tag{3.9.4}$$

$$N^{\alpha\beta}_h = \theta_1 \overset{1}{N}{}^{\alpha\beta} + \theta_2 \overset{2}{N}{}^{\alpha\beta}\,, \tag{3.9.5}$$

$$\overset{1}{N}{}^{\alpha\beta} = A_1^{\alpha\beta\lambda\mu} \overset{1}{\epsilon}{}^{\lambda\mu}\,, \quad \overset{2}{N}{}^{\alpha\beta} = A_2^{\alpha\beta\lambda\mu} \overset{2}{\epsilon}{}^{\lambda\mu}\,, \tag{3.9.6}$$

$$\overset{1}{N}{}^{\alpha\beta} n_\alpha = \overset{2}{N}{}^{\alpha\beta} n_\alpha\,. \tag{3.9.7}$$

Proceeding similarly to Section 3.8 one finds, see Eqs. (3.8.16), (3.8.17)

$$(A_2^{\alpha\beta\lambda\mu} - A_h^{\alpha\beta\lambda\mu})\epsilon^h_{\lambda\mu} = \theta_1 \mathcal{N}^{\alpha\beta}\,, \tag{3.9.8}$$

where

$$\mathcal{N}^{\alpha\beta} = (A_2^{\alpha\beta\lambda\mu} - A_1^{\alpha\beta\lambda\mu})\,\overset{1}{\epsilon}_{\lambda\mu}\,, \tag{3.9.9}$$

and

$$\epsilon^h_{\alpha\beta} = \epsilon^1_{\alpha\beta} - \theta_2(n_\alpha p_\beta + n_\beta p_\alpha)\,. \tag{3.9.10}$$

To find p_α we use (3.9.7), substitute there (3.9.6) and eliminate $\overset{2}{\epsilon}$ by (3.9.3) to obtain

$$\mathcal{N}^{\alpha\beta} n_\alpha = G^{\beta\sigma} p_\sigma\,, \tag{3.9.11}$$

$$G^{\beta\mu} = 2A_2^{\alpha\beta\lambda\mu} n_\lambda n_\alpha\,. \tag{3.9.12}$$

Let g be the matrix inverse to G ; then

$$p_\alpha = g_{\alpha\beta}\mathcal{N}^{\sigma\beta}n_\sigma , \qquad g_{\lambda\beta}G^{\beta\alpha} = \delta^\alpha_\lambda . \tag{3.9.13}$$

Now let us substitute $(3.9.13)_1$ into (3.9.10) to obtain

$$\epsilon^h_{\alpha\beta} = \overset{1}{\epsilon}_{\alpha\beta} - \theta_2(n_\alpha g_{\beta\gamma}n_\sigma + n_\beta g_{\alpha\gamma}n_\sigma)\mathcal{N}^{\sigma\gamma} . \tag{3.9.14}$$

On inverting (3.9.8) and (3.9.9) and substituting into (3.9.14) one finds

$$\theta_1[(\boldsymbol{A}_2 - \boldsymbol{A}_h)^{-1}]_{\alpha\beta\lambda\mu} = [(\boldsymbol{A}_2 - \boldsymbol{A}_1)^{-1}]_{\alpha\beta\lambda\mu} - \theta_2(\boldsymbol{\Gamma_A})_{\alpha\beta\lambda\mu} , \tag{3.9.15}$$

$$(\boldsymbol{\Gamma_A})_{\alpha\beta\lambda\mu} = \frac{1}{2}(n_\alpha g_{\beta\lambda}n_\mu + n_\beta g_{\alpha\lambda}n_\mu + n_\alpha g_{\beta\mu}n_\lambda + n_\beta g_{\alpha\mu}n_\lambda) . \tag{3.9.16}$$

In the case of $\boldsymbol{A}_2$ being isotropic, Eq. (3.9.11) can be explicitly inverted. Assume that

$$A_2^{\alpha\beta\lambda\mu} = 2k_2 I_1^{\alpha\beta\lambda\mu} + 2\mu_2 I_2^{\alpha\beta\lambda\mu} , \tag{3.9.17}$$

where I_σ are given by (3.8.29). Then tensor $\boldsymbol{\Gamma_A}$ assumes the form

$$\Gamma_{\boldsymbol{A}\alpha\beta\lambda\mu} = \frac{1}{2\mu_2}\left\{ [\frac{1}{2}(\delta_{\beta\mu}n_\alpha n_\lambda + \delta_{\beta\lambda}n_\alpha n_\mu + \delta_{\alpha\mu}n_\beta n_\lambda + \delta_{\alpha\lambda}n_\beta n_\mu) \right.$$
$$\left. - n_\alpha n_\beta n_\lambda n_\mu] - \frac{k_2 - \mu_2}{k_2 + \mu_2}n_\alpha n_\beta n_\lambda n_\mu \right\} . \tag{3.9.18}$$

3.9.2. Formula of Francfort and Murat-type for compliances

Let us consider the same ribbed plate as in the previous section. We shall prove that the effective compliance tensor $\boldsymbol{a}^h$ is determined by

$$\theta_2(\boldsymbol{a}_1 - \boldsymbol{a}^h)^{-1} = (\boldsymbol{a}^1 - \boldsymbol{a}^2)^{-1} + \theta_1 \boldsymbol{\Gamma_a} , \tag{3.9.19}$$

where

$$\Gamma_{\boldsymbol{a}}^{\alpha\beta\lambda\mu} = -\overset{1}{A}{}^{\alpha\beta\lambda\mu}$$
$$+ \frac{1}{2} \overset{1}{A}{}^{\alpha\beta\sigma\rho} \overset{1}{A}{}^{\delta\gamma\lambda\mu}(n_\delta n_\rho h_{\sigma\gamma} + n_\sigma n_\delta h_{\gamma\rho} + n_\delta n_\sigma h_{\rho\gamma} + n_\sigma n_\gamma h_{\delta\rho}) , \tag{3.9.20}$$

where

$$h_{\lambda\beta}H^{\beta\mu} = \delta^\mu_\lambda , \qquad H^{\beta\mu} = 2 \overset{1}{A}{}^{\alpha\beta\lambda\mu}n_\alpha n_\lambda \tag{3.9.21}$$

or $\boldsymbol{h} = \boldsymbol{H}^{-1}$.

Indeed, the following relations are at our disposal:

$$N_h^{\alpha\beta} = \theta_1 \overset{1}{N}{}^{\alpha\beta} + \theta_2 \overset{2}{N}_{\alpha\beta} , \qquad \overset{\sigma}{\epsilon}_{\alpha\beta} = \overset{\sigma}{a}_{\alpha\beta\lambda\mu}\overset{\sigma}{N}{}^{\lambda\mu} \quad \text{(do not sum over } \sigma\text{)}$$
$$\epsilon^h_{\alpha\beta} = a^h_{\alpha\beta\lambda\mu} N_h^{\lambda\mu} , \qquad (\overset{1}{N}{}^{\alpha\beta} - \overset{2}{N}{}^{\alpha\beta})n_\alpha = 0 , \qquad \overset{1}{\epsilon}_{\alpha\beta} - \overset{2}{\epsilon}_{\alpha\beta} = n_\alpha p_\beta + n_\beta p_\alpha . \tag{3.9.22}$$

Let us define

$$\rho_{\alpha\beta} = (\overset{1}{A}_{\alpha\beta\lambda\mu} - \overset{2}{A}_{\alpha\beta\lambda\mu})\, \overset{2}{N}{}^{\lambda\mu}, \qquad N^{\alpha\beta} = \overset{1}{N}{}^{\alpha\beta} - \overset{2}{N}{}^{\alpha\beta}. \tag{3.9.23}$$

Hence

$$N_h^{\alpha\beta} = \overset{2}{N}{}^{\alpha\beta} + \theta_1 N_{\alpha\beta}, \tag{3.9.24}$$

$$N_h^{\alpha\beta} = \theta_2[(\boldsymbol{a}^1 - \boldsymbol{a}^h)^{-1}]^{\alpha\beta\lambda\mu}\rho_{\lambda\mu}, \qquad \overset{2}{N}_{\alpha\beta} = [(\boldsymbol{a}^1 - \boldsymbol{a}^2)^{-1}]^{\alpha\beta\lambda\mu}\rho_{\lambda\mu}. \tag{3.9.25}$$

The main problem is to express $\boldsymbol{N}$ in terms of $\boldsymbol{\rho}$. To find this relation we first write

$$\overset{1}{a}_{\alpha\beta\lambda\mu}\,(N^{\lambda\mu} + \overset{2}{N}{}^{\lambda\mu}) - \overset{2}{a}_{\alpha\beta\lambda\mu}\overset{2}{N}{}^{\lambda\mu} = n_\alpha p_\beta + n_\beta p_\alpha,$$

which gives

$$\overset{1}{a}_{\alpha\beta\lambda\mu}\,N^{\lambda\mu} + \rho_{\alpha\beta} = n_\alpha p_\beta + n_\beta p_\alpha,$$

and

$$N^{\alpha\beta} = \overset{1}{A}{}^{\alpha\beta\lambda\mu}(n_\lambda p_\mu + n_\mu p_\lambda - \rho_{\lambda\mu}). \tag{3.9.26}$$

Since $N^{\alpha\lambda}n_\alpha = 0$ we find the set of equations

$$H^{\beta\mu}\, p_\mu = q^\beta \tag{3.9.27}$$

with

$$H^{\beta\mu} = 2\,\overset{1}{A}{}^{\alpha\beta\lambda\mu}n_\alpha n_\lambda, \qquad q^\beta = \overset{1}{A}{}^{\alpha\beta\lambda\mu}\rho_{\lambda\mu}n_\alpha.$$

Let $\boldsymbol{h} = \boldsymbol{H}^{-1}$ or $h_{\lambda\beta}H^{\beta\mu} = \delta_\lambda^\mu$. Thus

$$p_\lambda = h_{\lambda\beta}\,\overset{1}{A}{}^{\alpha\beta\rho\mu}\rho_{\rho\mu}n_\alpha. \tag{3.9.28}$$

Now we return to (3.9.26) to find

$$N^{\alpha\beta} = \Gamma^{\alpha\beta\lambda\mu} - \rho_{\lambda\mu}, \tag{3.9.29}$$

with Γ given by (3.9.20). Upon substituting (3.9.25), (3.9.29) into (3.9.24) one obtains (3.9.19). $\qquad\qquad\qquad\qquad\qquad\qquad\qquad\qquad\qquad\qquad\qquad\qquad\qquad\qquad\qquad\quad\square$

In the case when $\boldsymbol{a}^1$ is isotropic, i.e.,

$$\boldsymbol{a}^1 = \frac{1}{2k_1}\boldsymbol{I}_1 + \frac{1}{2\mu_1}\boldsymbol{I}_2, \tag{3.9.30}$$

and

$$\overset{1}{A} = 2k_1 I_1 + 2\mu_1 I_2 \,, \tag{3.9.31}$$

one can express the tensor Γ_a as follows

$$\frac{1}{2\mu_1}\Gamma_a^{\alpha\beta\lambda\mu} = \frac{1}{2}(\delta^{\beta\mu}n_\lambda n_\alpha + \delta^{\beta\lambda}n_\mu n_\alpha + \delta^{\alpha\mu}n_\lambda n_\beta + \delta^{\alpha\lambda}n_\mu n_\beta)$$

$$-n_\alpha n_\beta n_\lambda n_\mu - \frac{k_1 - \mu_1}{k_1 + \mu_1}(\delta^{\alpha\beta} - n^\alpha n^\beta)(\delta^{\lambda\mu} - n^\lambda n^\mu) - \frac{1}{2}(\delta^{\alpha\lambda}\delta^{\beta\mu} + \delta^{\alpha\mu}\delta^{\beta\lambda}) \,. \tag{3.9.32}$$

Remark 3.9.1 Let us recall all the Francfort-Murat formulae derived in Sections 3.8 and 3.9:

$$\theta_1(\boldsymbol{X}_2 - \boldsymbol{X}_h)^{-1} = (\boldsymbol{X}_2 - \boldsymbol{X}_1)^{-1} - \theta_2\Gamma_X \,,$$
$$\theta_2(\boldsymbol{x}^1 - \boldsymbol{x}^h)^{-1} = (\boldsymbol{x}^1 - \boldsymbol{x}^2)^{-1} + \theta_1\Gamma_x \,, \tag{3.9.33}$$

for $\boldsymbol{X} \in \{\boldsymbol{A}, \boldsymbol{D}\}$, $\boldsymbol{x} \in \{\boldsymbol{a}, \boldsymbol{d}\}$, $\boldsymbol{x} = \boldsymbol{X}^{-1}$.

Despite the disclosed analogies between formulae for $\boldsymbol{A}_h$, $\boldsymbol{d}^h$, $\boldsymbol{a}^h$ and $\boldsymbol{D}_h$ (see Sec. 3.7) we see no similarity between the tensors Γ_A and Γ_d and between Γ_a and Γ_D. This lack of visible similarity can be justified as follows. The analogy between formulae: $\boldsymbol{A}_h(\boldsymbol{A})$ and $\boldsymbol{d}^h(\boldsymbol{d})$ has involved changing indices $1 \leftrightarrow 2$. Such a change becomes unclear if the axes y_α are deviated from directions of orthotropy, which is the case here.

One can, however, disclose analogies between formulae for $\boldsymbol{X}_h$ and $\boldsymbol{x}_h$ in the case of isotropic constituents. To this end let us recall the projection tensors of Hill for plane elasticity:

$$\mathcal{E}^A_{\alpha\beta\lambda\mu} = \frac{1}{2}(\delta_{\beta\mu}n_\lambda n_\alpha + \delta_{\beta\lambda}n_\mu n_\alpha + \delta_{\alpha\mu}n_\lambda n_\beta + \delta_{\alpha\lambda}n_\mu n_\beta) - n_\alpha n_\beta n_\lambda n_\mu \tag{3.9.34}$$

and for plate theory:

$$\mathcal{E}^D_{\alpha\beta\lambda\mu} = n_\alpha n_\beta n_\lambda n_\mu \,; \qquad \mathcal{E}^D = \Gamma_n \,. \tag{3.9.35}$$

Let us introduce the tensors

$$(\boldsymbol{\Pi}_n)_{\alpha\beta\lambda\mu} = (\delta_{\alpha\beta} - n_\alpha n_\beta)(\delta_{\lambda\mu} - n_\lambda n_\mu) \,, \quad I_{\alpha\beta\lambda\mu} = \frac{1}{2}(\delta_{\alpha\lambda}\delta_{\beta\mu} + \delta_{\alpha\mu}\delta_{\beta\lambda}) \tag{3.9.36}$$

and the ratios

$$\sigma_\alpha = \frac{k_\alpha - \mu_\alpha}{k_\alpha + \mu_\alpha} \,. \tag{3.9.37}$$

Then one can express the tensors Γ_X and Γ_x as follows

$$\Gamma_X = \frac{1}{2\mu_2}(\mathcal{E}^X - \sigma_2\Gamma_n) \,, \qquad \Gamma_x = 2\mu_1(\mathcal{E}^X - \boldsymbol{I} - \sigma_1\boldsymbol{\Pi}_n) \,. \tag{3.9.38}$$

These formulae (for $\boldsymbol{X} = \boldsymbol{A}$) have been discovered by Norris.

3.10. Plates periodic with respect to a curvilinear parametrization. Non-uniform homogenization

Assume that domain Ω is parametrized by a curvilinear coordinate system (ξ^α). Thus Ω can be viewed as an image of a certain domain Ω_0: $\Omega = \Phi(\Omega_0)$.

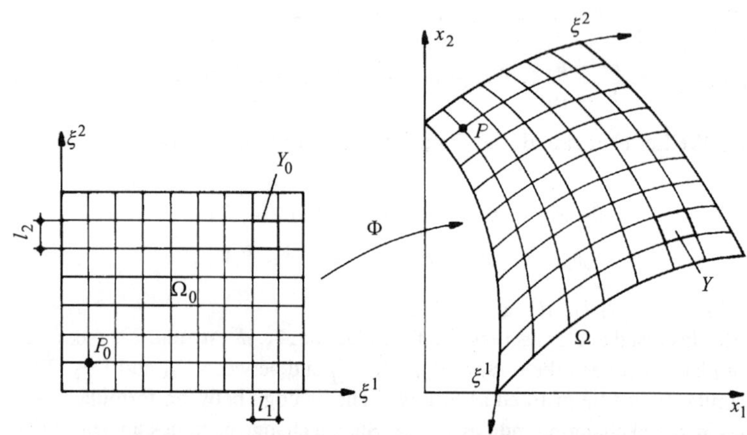

Fig. 3.10.1. Periodicity with respect to a curvilinear parametrization

The mapping Φ is determined by relations: $x_\alpha = x_\alpha(\xi^1, \xi^2)$. The metric tensor is defined by $g_{\alpha\beta} = \boldsymbol{r}_\alpha \cdot \boldsymbol{r}_\beta$, $\boldsymbol{r}_\alpha = \partial r / \partial \xi^\alpha$, $\boldsymbol{r} = [x_1(\xi), x_2(\xi)]$, $\xi = (\xi^\alpha)$. The Christoffel symbols are denoted by $\Gamma^\gamma_{\alpha\beta}$ and the covariant derivative assumes the form

$$f_{\alpha\|\beta} = f_{\alpha,\beta} - \Gamma^\gamma_{\alpha\beta} f_\gamma \,, \tag{3.10.1}$$

where $(\ )_{,\beta} = \partial(\ )/\partial\xi^\beta$; in this section comma means differentiation with respect to ξ^β.

Similarly to Section 3.8 we focus attention on the bending (anti-plane) problem of a transversely symmetric plate. The only unknown is a scalar function w representing deflection of the plate. Since scalars are invariants, the same quantity represents deflection of the plate with the middle plane parametrized by (ξ^α). Consider the bending stiffness tensor of the form

$$\boldsymbol{D}_\varepsilon = D^{\alpha\beta\lambda\mu}\left(\xi, \frac{\xi}{\varepsilon}\right) \boldsymbol{r}_\alpha \otimes \boldsymbol{r}_\beta \otimes \boldsymbol{r}_\lambda \otimes \boldsymbol{r}_\mu \,, \tag{3.10.2}$$

with $D^{\alpha\beta\lambda\mu}(\xi, \cdot)$ being Y_0-periodic functions, $Y_0 = (0, l_1) \times (0, l_2)$. The domain Ω is formed by curvilinear rectangles $Y = \Phi(Y_0)$. On the other hand the functions $D^{\alpha\beta\lambda\mu}(\cdot, \xi/\varepsilon)$ determine slow variations of stiffnesses.

Assume that the plate is subject to transverse loading $q = q(\xi)$, the quantity being ε-independent. Moreover, assume for simplicity that the plate is clamped along $\Gamma = \partial\Omega$.

Thus

$$w^\varepsilon = 0 \quad \text{and} \quad \frac{\partial w^\varepsilon}{\partial n} = 0 \quad \text{on } \Gamma . \tag{3.10.3}$$

Here w^ε represents the unknown deflection; ε indicates that this deflection depends on the size of periodicity cells. The moments $M_\varepsilon^{\alpha\beta}$ are interrelated with changes of curvature

$$\kappa_{\alpha\beta}(w^\varepsilon) = -w^\varepsilon_{\|\alpha\beta} \tag{3.10.4}$$

by the constitutive relationship

$$M_\varepsilon^{\alpha\beta} = D^{\alpha\beta\lambda\mu}\left(\xi, \frac{\xi}{\varepsilon}\right) \kappa_{\lambda\mu}(w^\varepsilon) . \tag{3.10.5}$$

Let us define the bilinear form

$$d_K^\varepsilon(w, v) = \int_{\Omega_0} D^{\alpha\beta\lambda\mu}\left(\xi, \frac{\xi}{\varepsilon}\right) \kappa_{\lambda\mu}(w)\kappa_{\alpha\beta}(v)\sqrt{g}\, d\xi , \qquad g = \det(g_{\alpha\beta}) , \tag{3.10.6}$$

for $w, v \in H^2(\Omega_0)$ and the linear form

$$f_K(v) = \int_{\Omega_0} qv\sqrt{g}\, d\xi . \tag{3.10.7}$$

The equilibrium problem reads:

$$(P_{KS}^\varepsilon) \quad \left|\begin{array}{l} \text{find } w^\varepsilon \in H_0^2(\Omega_0) \text{ such that} \\[4pt] d_K^\varepsilon(w^\varepsilon, v^\varepsilon) = f_K(v^\varepsilon) \qquad \forall\ v^\varepsilon \in H_0^2(\Omega) . \end{array}\right. \tag{3.10.8}$$

To find the effective characteristics of the plate considered we discuss the behavior of w^ε, $M_\varepsilon^{\alpha\beta}$, $\kappa_{\alpha\beta}(w^\varepsilon)$ as $\varepsilon \to 0$. To this end we represent w^ε and v^ε by the two-scale asymptotic form

$$z^\varepsilon(\xi) = z^{(0)}(\xi) + \varepsilon^2 z^{(2)}(\xi, y) + \varepsilon^3 z^{(3)}(\xi, y) + \ldots\Big|_{y=\frac{\xi}{\varepsilon}} \tag{3.10.9}$$

with $z = w$ or v and assume that

$$z^{(0)} \in H_0^2(\Omega_0) , \quad z^{(k)}(\xi, \cdot) \in H_{per}^2(Y_0) ; \quad k \geq 2 , \ z = w, v .$$

Hence we have

$$\kappa_{\alpha\beta}(z^\varepsilon) = \kappa_{\alpha\beta}(z^{(0)}) + \kappa_{\alpha\beta}^y(z^{(2)}) + O(\varepsilon)\Big|_{y=\xi/\varepsilon} \tag{3.10.10}$$

with $\kappa_{\alpha\beta}^y(z) = -z_{|\alpha\beta}$.

We shall use the following averaging lemma:. Let $F(\xi, \cdot)$ be Y_0-periodic. Then, cf. the relation (1.1.1)

$$\lim_{\varepsilon \to 0} \int_{\Omega_0} F\left(\xi, \frac{\xi}{\varepsilon}\right) \sqrt{g(\xi)}\, d\xi = \int_{\Omega_0} \langle F \rangle \sqrt{g(\xi)}\, d\xi \,,$$

with

$$\langle F \rangle = \frac{1}{|Y_0|} \int_{Y_0} F(\xi, y)\, dy \,.$$

(3.10.11)

By using the same homogenization technique as in Section 3.2 one finds:
(a) the local problem:

$$(P^2_{KS,Y_0}) \quad \left| \begin{array}{l} \text{find } \chi^{(\delta\gamma)}(\xi, \cdot) \in H^2_{per}(Y_0) \text{ such that} \\[2mm] \langle D^{\alpha\beta\lambda\mu}(\xi, y)[\kappa^y_{\lambda\mu}(\chi^{(\delta\gamma)}) + \delta^\delta_\lambda \delta^\gamma_\mu] \kappa^y_{\alpha\beta}(v) \rangle = 0 \quad \forall\, v^\varepsilon \in H^2_{per}(Y_0) \,, \end{array} \right.$$

(3.10.12)

(b) the homogenized stiffnesses

$$D_h^{\alpha\beta\lambda\mu}(\xi) = \langle D^{\alpha\beta\lambda\mu}(\xi, y) + D^{\alpha\beta\delta\gamma}(\xi, y) \kappa^y_{\delta\gamma}(\chi^{(\lambda\mu)}(\xi, y)) \rangle \,,$$

(3.10.13)

(c) the homogenized problem

$$(P_h) \quad \left| \begin{array}{l} \text{find } w^{(0)} \in H^2_0(\Omega_0) \text{ such that} \\[2mm] \int_{\Omega_0} M_h^{\alpha\beta} \kappa_{\alpha\beta}(v) \sqrt{g}\, d\xi = \int_{\Omega_0} qv \sqrt{g}\, d\xi \quad \forall\, v \in H^2_0(\Omega_0) \end{array} \right.$$

(3.10.14)

where

$$M_h^{\alpha\beta} = D_h^{\alpha\beta\lambda\mu} \kappa_{\lambda\mu}(w^{(0)}) \,.$$

(3.10.15)

Justification of (a) – (c) will be given in Chapter V in a broader context of periodic shells. A new approach to non-periodic homogenization is offered by the method of $\theta - 2$ convergence developed by Alexandre (1997).

3.11. *Effective bending stiffnesses of plates with quadratic inclusions*

The subject of this section is a transversely symmetric plate stiffened (or weakened) by quadratic inclusions of doubly periodic layout, see Fig. 3.11.1.

The cell Y consists of the inclusion $Y_1 = (-r/2, r/2) \times (-r/2, r/2)$ and the matrix $Y_2 = Y \backslash Y_1$; $Y = (-s/2, s/2) \times (-s/2, s/2)$. The distribution of bending stiffnesses is assumed as follows

$$D^{\alpha\beta\lambda\mu}(y) = \begin{cases} D_0^{\alpha\beta\lambda\mu} & \text{for } y \in Y_2 \,, \\ \lambda D_0^{\alpha\beta\lambda\mu} & \text{for } y \in Y_1 \,, \end{cases}$$

(3.11.1)

and $D_0^{\alpha\beta\lambda\mu}$ are constants; λ is a positive parameter. For $\lambda = 1$ the plate is homogeneous.

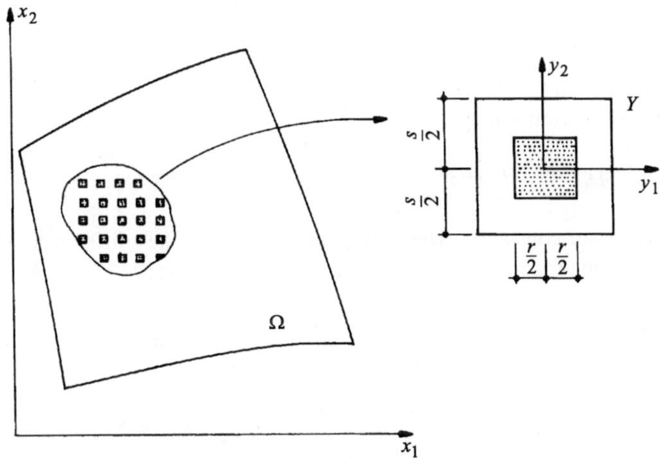

Fig. 3.11.1. A plate with quadratic inclusions. Geometry of the basic cell

Solutions to the local problems $(P^2_{KS,Y})$ of Sec. 3.2 can be found by the Galerkin method with trigonometric basis functions. We represent the $(\alpha\beta)$ solution

$$\chi^{(\alpha\beta)} = \sum_{a=1}^{N} \chi^{(\alpha\beta)a}\phi_a(y) , \qquad (3.11.2)$$

with the help of Y-periodic functions ϕ_a taken from the set

$$\{C_{m,n}(y_1, y_2) , \ S_{m,n}(y_1, y_2)\} , \qquad (3.11.3)$$

where

$$C_{m,n}(y) = c_m(y_1)c_n(y_2) , \quad S_{m,n}(y) = s_m(y_1)s_n(y_2) ,$$

and

$$c_n(t) = \cos(2\pi nt) , \quad s_n(t) = \sin(2\pi nt) .$$

We have put $s = 1$ for simplicity. On substituting (3.11.2) into (3.2.31) and choosing $v = \phi_b$ one finds

$$D_{ab}\chi^{(\alpha\beta)a} + D_b^{(\alpha\beta)} = 0 , \qquad (3.11.4)$$

where $a, b \in \{1, \ldots, N\}$ and

$$D_{ab} = d_K(\phi_a, \phi_b) , \quad D_b^{(\alpha\beta)} = \langle D^{\alpha\beta\gamma\delta}(y)\kappa^y_{\gamma\delta}(\phi_b)\rangle . \qquad (3.11.5)$$

The homogenized stiffnesses are given by $(3.2.32)_2$, which results in

$$D_h^{\alpha\beta\lambda\mu} = \langle D^{\alpha\beta\lambda\mu}\rangle - D_a^{(\alpha\beta)}(\boldsymbol{D}^{-1})^{ab} D_b^{(\lambda\mu)} \ . \tag{3.11.6}$$

The main problem is to solve (3.11.4).

Assume that $\boldsymbol{D}_0$ is isotropic or

$$D_0^{\alpha\beta\lambda\mu} = D\left[\nu\delta^{\alpha\beta}\delta^{\lambda\mu} + \frac{1-\nu}{2}(\delta^{\alpha\lambda}\delta^{\beta\mu} + \delta^{\alpha\mu}\delta^{\beta\lambda})\right] \ , \tag{3.11.7}$$

and that D and ν are constants. The only parameter that makes difference between the matrix and inclusion properties is λ. To be specific let us assume $N = 12$. We choose the basis functions as follows

$$\phi_1 = C_{1,0}\ , \quad \phi_2 = C_{0,1}\ , \quad \phi_3 = C_{1,1}\ , \quad \phi_4 = C_{2,2}\ ,$$
$$\phi_5 = C_{2,1}\ , \quad \phi_6 = C_{1,2}\ , \quad \phi_7 = C_{2,0}\ , \quad \phi_8 = C_{0,2}\ ,$$
$$\phi_9 = S_{1,1}\ , \quad \phi_{10} = S_{1,2}\ , \quad \phi_{11} = S_{2,1}\ , \quad \phi_{12} = S_{2,2}\ .$$

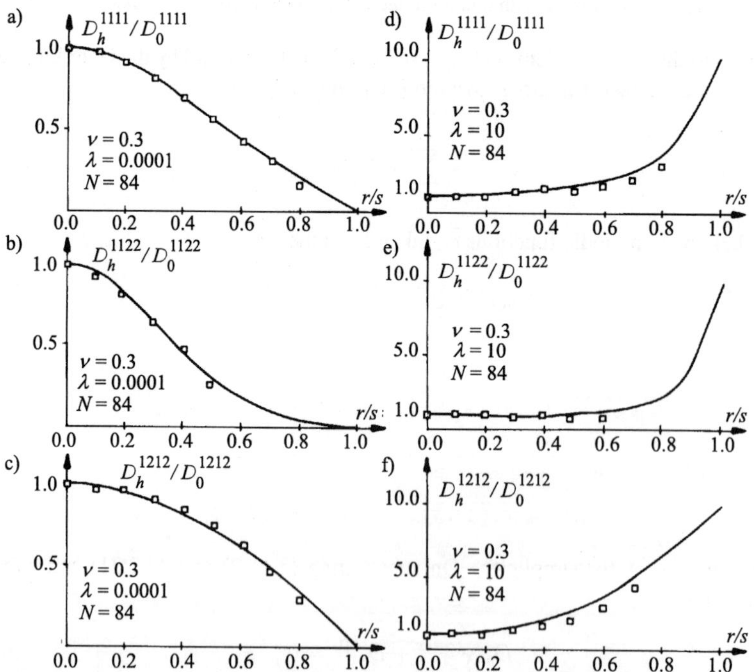

Fig. 3.11.2. Effective bending stiffnesses of plate of Fig. 3.11.1 versus the ratio r/s. The squares "□" represent results of Grigoliuk and Filshtinskii (1970). $\lambda = 0.0001$ for a) – c), $\lambda = 10.0$ for d) – f).

Note that the function $\phi_0 = C_{0,0}$ is excluded, thus making the matrix D invertible. In this manner we get rid of an additive constant up to which the solutions of $(P^2_{KS,Y})$ are determined. It is also clear how to define basis functions for $N \geq 12$. Note moreover, that due to symmetries of the periodicity cell the basis functions: $c_m(y_\sigma)s_n(y_{3-\sigma})$ are redundant.

All components of matrices (3.11.5) can be explicitly expressed in terms of ν and λ. One observes that for $\lambda \gg 1$ the convergence as N increases becomes poor, see Tables 3.11.1a-c.

Table 3.11.1

a) D_h^{1111}/D $\nu = 1/6$ $r/s = 1/3$

λ \ N	4	12	24	40	60	84
10^{-3}	0.853	0.821	0.811	0.807	0.803	0.800
10^{-2}	0.854	0.822	0.811	0.804	0.804	0.801
10^{-1}	0.856	0.824	0.814	0.812	0.808	0.805
10^0	0.873	0.850	0.846	0.844	0.841	0.839
10^1	1.000	1.000	1.000	1.000	1.000	1.000
10^2	1.374	1.357	1.322	1.298	1.297	1.285
10^3	1.947	1.755	1.559	1.547	1.489	1.464
10^4	5.261	1.969	1.847	1.684	1.605	1.585
10^5	37.473	2.354	2.121	1.801	1.767	1.650

b) D_h^{1122}/D $\nu = 1/6$ $r/s = 1/3$

λ \ N	4	12	24	40	60	84
10^{-3}	0.127	0.101	0.091	0.089	0.089	0.090
10^{-2}	0.127	0.101	0.091	0.089	0.089	0.090
10^{-1}	0.127	0.103	0.094	0.092	0.092	0.093
10^0	0.133	0.117	0.112	0.112	0.111	0.110
10^1	0.157	0.167	0.167	0.167	0.167	0.167
10^2	0.142	0.146	0.140	0.142	0.141	0.142
10^3	0.132	0.054	0.089	0.083	0.094	0.099
10^4	0.194	0.044	0.048	0.061	0.077	0.074
10^5	0.306	0.053	0.006	0.061	0.047	0.070

c) D_h^{1212}/D $\nu = 1/6$ $r/s = 1/3$

λ \ N	4	12	24	40	60	84
10^{-3}	0.366	0.366	0.362	0.361	0.361	0.361
10^{-2}	0.366	0.363	0.362	0.361	0.361	0.361
10^{-1}	0.366	0.363	0.362	0.362	0.362	0.362
10^0	0.371	0.369	0.369	0.368	0.368	0.368
10^1	0.417	0.417	0.417	0.417	0.417	0.417
10^2	0.674	0.669	0.653	0.649	0.645	0.642
10^3	1.821	1.223	1.131	1.052	1.005	0.981
10^4	12.070	2.144	1.667	1.366	1.294	1.194
10^5	114.338	6.603	2.147	1.849	1.512	1.392

The algorithm works fairly well for very small λ, which means that results of $D_h^{\alpha\beta\lambda\mu}$ tend uniformly to the case of quadratic openings – see Fig. 3.11.2, where the results of Filshtinskii and Grigoliuk are additionally depicted. The latter results were found for circular holes and here they are put by recalculating the r/s ratio from the condition of the square and circular areas being equal. Let us look lastly at global behavior of a clamped square plate with 25 inclusions, as in Fig. 3.11.3; $s = a/5$. Poisson's ratio ν is constant and equals $1/6$. The stiffness D in the matrix and inclusions equal D and λD respectively.

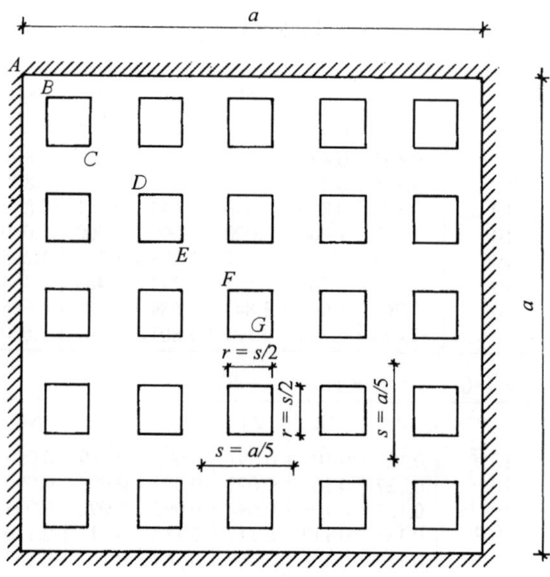

Fig. 3.11.3. Plate with quadratic inclusions

The plate is subject to a constant uniformly distributed transverse loading q. The nondimensional deflection is defined by $\bar{w} = Dw/qa^4$. In case of $\lambda \ll 1$ the plate deflection grows rapidly within the inclusions, see Fig. 3.11.4a ($\lambda = 10^{-4}$).

In the matrix region $\bar{w}^{(0)}$ approximates $\bar{w}$ from the upper side fairly well. In case of $\lambda > 1$ function $\bar{w}^{(0)}$ is a lower estimate of $\bar{w}$, see Fig. 3.11.4b.

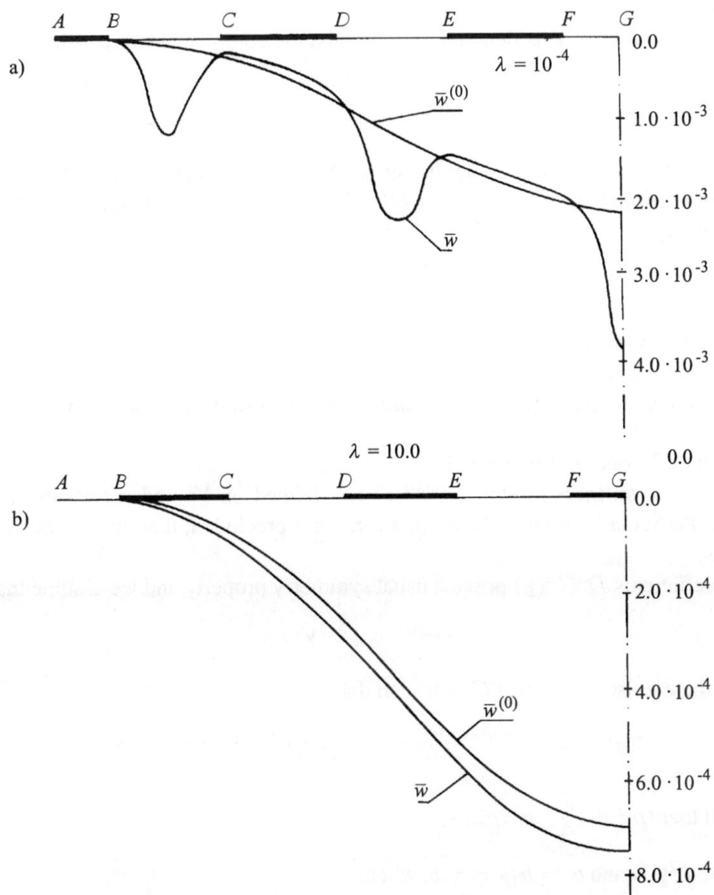

Fig. 3.11.4. The homogenized delfections $\bar{w}^{(0)}$ versus the finite element approximation $\bar{w}$.
$\lambda = 10^{-4}$ and 10.0 for a) and b) respectively

3.12.   *Perforated plates*

In the present section we shall study homogenization problem for periodically perforated plates. The proof of homogenization Theorem 3.12.1 is based on penalization, which consists in replacing the voids with a soft material.

Comments on the derivation of the effective dynamic plate model from the three-dimensional dynamic equations of a thin perforated solid are given in Section 7.

Let us consider a periodically perforated plate, clamped along $\Gamma = \partial\Omega$. Now the basic cell Y contains a finite number of holes. The voids are cylindrical and traverse all the thickness of the plate. The vertical axes of the voids are perpendicular to the mid-plane of the plate.

The sum of holes in Y is denoted by Y^0 whilst the complementary part by Y^c. Thus we have

$$Y = Y^c \cup Y^0 .$$

By P_i, $i = 1, \ldots, \mathcal{I}(\varepsilon)$ we denote the set of εY-cells which intersect Ω. We observe that $\mathcal{I}(\varepsilon)$ is proportional to ε^{-2}, i.e., $\mathcal{I}(\varepsilon) = O(\varepsilon^{-2})$. The intersection of voids with Ω is denoted by T_i^ε and

$$\Omega^\varepsilon = \Omega \setminus \bigcup_i T_i^\varepsilon . \tag{3.12.1}$$

We make the following assumptions:

(A_1) Ω^ε is connected.

(A_2) The holes T_i^ε have regular boundaries and are locally located on one side of their boundary.

(A_3) Each of T_i^ε does not intersect Γ.

The bending stiffness moduli $D^{\alpha\beta\lambda\mu}(y)$ are defined on Y^c and are assumed to be Y-periodic. Particularly, homogeneous plates are not precluded, then the moduli $D^{\alpha\beta\lambda\mu}$ are constants.

The coefficients $D^{\alpha\beta\lambda\mu}(y)$ possess usual symmetry property and we assume that

$$D^{\alpha\beta\lambda\mu} \in L^\infty(Y^c) , \tag{3.12.2}$$

and that there exists a constant $C > 0$ such that

$$\forall \, \boldsymbol{\rho} \in \mathbb{E}_s^2 \quad D^{\alpha\beta\lambda\mu}(y)\rho_{\alpha\beta}\rho_{\lambda\mu} \geq C|\boldsymbol{\rho}|^2 \quad \text{for a.e. } y \in Y^c . \tag{3.12.3}$$

We recall that $|\boldsymbol{\rho}|^2 = \displaystyle\sum_{\alpha,\beta=1}^{2} \rho_{\alpha\beta}\rho_{\alpha\beta}$.

Let $b \in L^2(\Omega)$ and $b^\varepsilon = b_{|\Omega^\varepsilon} = \chi_\varepsilon b$, where

$$\chi_\varepsilon(x) = \begin{cases} 1 & \text{if} \quad x \in \Omega^\varepsilon , \\ 0 & \text{if} \quad x \in \Omega \setminus \Omega^\varepsilon . \end{cases} \tag{3.12.4}$$

Obviously, the function χ_ε is εY-periodic. We set

$$V_\varepsilon = \{ v \in H^2(\Omega^\varepsilon) | v = 0 , \quad \frac{\partial v}{\partial n} = 0 \quad \text{on} \quad \Gamma \} . \tag{3.12.5}$$

For any fixed $\varepsilon > 0$ the functional of the total potential energy of the plate is given by

$$J_\varepsilon(w) = \frac{1}{2} a^\varepsilon(w, w) - \int_{\Omega^\varepsilon} b^\varepsilon w dx = \frac{1}{2} a^\varepsilon(w, w) - \int_\Omega \chi_\varepsilon(x) b(x) w(x) dx , \tag{3.12.6}$$

where

$$a^\varepsilon(u,v) = \int_{\Omega^\varepsilon} D_\varepsilon^{\alpha\beta\lambda\mu}(x)\kappa_{\alpha\beta}(u)\kappa_{\lambda\mu}(v)dx \,, \qquad u,v \in V^\varepsilon \,. \tag{3.12.7}$$

The minimum principle of the total potential energy means evaluating

$$(P_\varepsilon) \qquad J_\varepsilon(w^\varepsilon) = \inf\{J_\varepsilon(w)|w \in V_\varepsilon\} \,. \tag{3.12.8}$$

The assumptions $(A_1) - (A_2)$ and (3.12.3) imply the existence and uniqueness of w^ε, cf. Sec. 1.2.2.

Homogenization

To obtain the effective plate model we let ε tend to zero. Let τ stand for the strong topology of $H^1(\Omega)$.

Theorem 3.12.1. The sequence of functionals $\{J_\varepsilon\}_{\varepsilon>0}$, defined by (3.12.6), is $\Gamma(\tau)$-convergent to

$$J_h(w) = \int_\Omega \mathcal{W}_h(\kappa(w))dx - \int_\Omega \theta bw dx \,, \tag{3.12.9}$$

where $w \in H_0^2(\Omega)$ and

$$W_h(\rho) = \inf\{\frac{1}{2|Y|}\int_{Y^c} D^{\alpha\beta\lambda\mu}(y)(\kappa_{\alpha\beta}^y(\xi) + \rho_{\alpha\beta})(\kappa_{\lambda\mu}^y(\xi) + \rho_{\lambda\mu})dy|\xi \in H_{per}^2(Y^c)\} \,. \tag{3.12.10}$$

Here $\rho \in \mathbb{E}_s^2$ and $\theta = |Y^c|/|Y|$.

Proof. Since $\{\chi_\varepsilon\}_{\varepsilon>0}$ is a sequence of εY-periodic functions and $\chi_\varepsilon \in L^2(\Omega^\varepsilon)$, therefore we have, cf. Sec. 1.1.1

$$\chi_\varepsilon \rightharpoonup \frac{1}{|Y|}\int_Y \chi_{Y^c}(y)dy = \frac{|Y^c|}{|Y|} = \theta \,,$$

where χ_{Y^c} stands for the characteristic function of the set Y^c.

A solution $\bar{\xi}$ of the minimization problem appearing on the r.h.s. of (3.12.10) exists and is unique up to a rigid motion or a polynomial of the first order.

Suppose now that the holes are replaced by soft inclusions with the elastic moduli $\eta D_{1\varepsilon}^{\alpha\beta\lambda\mu}(x)$ $(\eta > 0)$. The loading is still applied to Ω^ε only. The functional of the total potential energy now takes the form

$$J_\varepsilon^\eta(v) = J_\varepsilon(v^c) + \eta \int_{\Omega\backslash\Omega^\varepsilon} D_{1\varepsilon}^{\alpha\beta\lambda\mu}(x)\kappa_{\alpha\beta}(v^0)\kappa_{\lambda\mu}(v^0)dx \,, \tag{3.12.11}$$

where $v = (v^c, v^0) \in H_0^2(\Omega)$, $v^c = v_{|\Omega^\varepsilon}$, $v^0 = v_{|\Omega \setminus \Omega^\varepsilon}$. For a fixed $\eta > 0$ the homogenization $(\varepsilon \to 0)$ yields the limit functional

$$J_h^\eta(v) = \int_\Omega \mathcal{W}_h^\eta(\kappa(v))dx - \int_\Omega \theta bv dx \,, \tag{3.12.12}$$

where $v \in H_0^2(\Omega)$ and

$$\mathcal{W}_h^\eta(\rho) = \inf\{\frac{1}{2|Y|} \int_Y D_\eta^{\alpha\beta\lambda\mu}(y)(\kappa_{\alpha\beta}^y(\xi) + \rho_{\alpha\beta})(\kappa_{\lambda\mu}^y(\xi) + \rho_{\lambda\mu})dy | \xi \in H_{per}^2(Y)\} \,. \tag{3.12.13}$$

Here

$$D_\eta^{\alpha\beta\lambda\mu}(y) = \begin{cases} D^{\alpha\beta\lambda\mu}(y) & \text{if} \quad y \in Y^c \,, \\ \eta D_1^{\alpha\beta\lambda\mu}(y) & \text{if} \quad y \in Y^0 \,. \end{cases} \tag{3.12.14}$$

To complete the proof it is thus sufficient to show that for any fixed $\rho \in \mathbb{E}_s^2$ one has

$$\mathcal{W}^\eta(\rho) \to \mathcal{W}_h(\rho) \qquad \text{as} \quad \eta \to 0 \,. \tag{3.12.15}$$

To corroborate (3.12.15), let us denote by ξ^η a solution to the minimization problem on the r.h.s. of (3.12.13). Then

$$\begin{aligned}
\mathcal{W}_h^\eta(\rho) &= \frac{1}{2|Y|} \int_Y D_\eta^{\alpha\beta\lambda\mu}(y)(\kappa_{\alpha\beta}^y(\xi^\eta) + \rho_{\alpha\beta})(\kappa_{\lambda\mu}^y(\xi^\eta) + \rho_{\lambda\mu})dy \\
&= \frac{1}{2|Y|} \int_{Y^c} D^{\alpha\beta\lambda\mu}(y)(\kappa_{\alpha\beta}^y(\xi_c^\eta) + \rho_{\alpha\beta})(\kappa_{\lambda\mu}^y(\xi_c^\eta) + \rho_{\lambda\mu})dy \\
&\quad + \frac{\eta}{2|Y|} \int_{Y^0} D_1^{\alpha\beta\lambda\mu}(y)(\kappa_{\alpha\beta}^y(\xi_0^\eta) + \rho_{\alpha\beta})(\kappa_{\lambda\mu}^y(\xi_0^\eta) + \rho_{\lambda\mu})dy \\
&\leq \frac{1}{2|Y|} \int_{Y^c} D^{\alpha\beta\lambda\mu}(y)\rho_{\alpha\beta}\rho_{\lambda\mu}dy + \frac{\eta}{2|Y|} \int_{Y^0} D_1^{\alpha\beta\lambda\mu}(y)\rho_{\alpha\beta}\rho_{\lambda\mu}dy \\
&\leq C_1(|\rho|^2 + \eta|\rho|^2) \leq C_2|\rho|^2 \,, \tag{3.12.16}
\end{aligned}$$

since $0 < \eta < 1$. The positive constant C_2 is independent of η and we have set $\xi^\eta = (\xi_c^\eta, \xi_0^\eta) \in H_{per}^2(Y)$, $\xi_c^\eta = \xi_{|Y^c}^\eta$, $\xi_0^\eta = \xi_{|Y^0}^\eta$. From (3.12.16) we conclude that

$$||\kappa^y(\xi_c^\eta) + \rho||_{L^2(Y^c, \mathbb{E}_s^2)}^2 \leq C_3|\rho|^2 \,,$$

where $C_3 > 0$ is independent of η. Consequently, the sequence ξ_c^η is bounded in $H_{per}^2(Y)$ and there exists a subsequence, still denoted by ξ_c^η, such that

$$\xi_c^\eta \rightharpoonup \bar{\xi}_c \quad \text{in} \quad H^2(Y^c) \quad \text{weakly as} \quad \eta \to 0 \,.$$

Similarly

$$||\boldsymbol{\kappa}^y(\xi_0^\eta) + \boldsymbol{\rho}||^2_{L^2(Y^0,\mathbf{E}_s^2)} \le C_4|\boldsymbol{\rho}|^2$$

and

$$\eta \int_{Y^0} D_1^{\alpha\beta\lambda\mu}(\kappa^y_{\alpha\beta}(\xi_0^\eta) + \rho_{\alpha\beta})(\kappa^y_{\lambda\mu}(\xi_0^\eta) + \rho_{\lambda\mu})dy \to 0 \quad \text{as} \quad \eta \to 0 .$$

Hence

$$\mathcal{W}_h^\eta(\boldsymbol{\rho}) \to \frac{1}{2|Y|} \int_{Y^c} D^{\alpha\beta\lambda\mu}(y)(\kappa^y_{\alpha\beta}(\bar{\xi}_c) + \rho_{\alpha\beta})(\kappa^y_{\lambda\mu}(\bar{\xi}_c) + \rho_{\lambda\mu})dy$$

$$= \inf\{\frac{1}{2|Y|} \int_{Y^c} D^{\alpha\beta\lambda\mu}(y)(\kappa^y_{\alpha\beta}(\xi) + \rho_{\alpha\beta})(\kappa^y_{\lambda\mu}(\xi) + \rho_{\lambda\mu})dy|\xi \in H^2_{per}(Y^c)\}$$

$$= \mathcal{W}_h(\boldsymbol{\rho}) \quad \text{as} \quad \eta \to 0, \tag{3.12.17}$$

since the functionals involved in (3.12.17) are convex, finite, lower semicontinuous and equi-coercive with respect to η. Being convex and finite they are continuous. This completes the proof. $\square$

Corollary 3.12.2. The effective (homogenized) moduli are now given by

$$\widehat{D}^{\gamma\delta\nu\rho} = \frac{1}{|Y|} \int_{Y^c} D^{\alpha\beta\lambda\mu}(y)(\kappa^y_{\alpha\beta}(\psi^{(\gamma\delta)}) + \delta_{\alpha\gamma}\delta_{\beta\delta})(\kappa^y_{\lambda\mu}(\psi^{(\nu\rho)}) + \delta_{\lambda\nu}\delta_{\mu\rho})dy . \tag{3.12.18}$$

Indeed, $\widehat{D}^{\gamma\delta\nu\rho} = \dfrac{\partial^2 \mathcal{W}_h}{\partial\rho_{\gamma\delta}\partial\rho_{\nu\rho}}$ and due to linearity of the problem we may set $\bar{\xi} = -\rho_{\alpha\beta}\psi^{(\alpha\beta)}$.

The Y-periodic functions $\psi^{(\gamma\delta)} \in H^2_{per}(Y)$ are solutions to

$$\mathcal{W}_h(\boldsymbol{I}^{\gamma\delta}) = \inf\{\frac{1}{2|Y|} \int_{Y^c} D^{\alpha\beta\lambda\mu}(y)(\kappa^y_{\alpha\beta}(\xi) + I^{\gamma\delta}_{\alpha\beta})(\kappa^y_{\lambda\mu}(\xi) + I^{\gamma\delta}_{\lambda\mu})dy|\xi \in H^2_{per}(Y^c)\} ,$$

where $I^{\gamma\delta}_{\alpha\beta}$ are the components of the identity tensor, i.e., $I^{\lambda\mu}_{\alpha\beta} = \delta_\alpha^\lambda\delta_\beta^\mu = \delta_{\alpha\lambda}\delta_{\beta\mu}$, because the coordinate system is Cartesian. More precisely, one should consider the identity tensor with the components given by $\frac{1}{2}(\delta_{\alpha\lambda}\delta_{\beta\mu} + \delta_{\alpha\mu}\delta_{\beta\lambda})$. However, due to the symmetry of the tensor D, it suffices to deal with $I^{\gamma\delta}_{\alpha\beta}$.

Remark 3.12.2. To derive the dual principle P_ε^* one can exploit the theory of duality outlined in Section 1.2.5. Similarly, to perform dual homogenization one can use Azé's theory presented in Section 1.3.6.

### 3.13.	Plates stiffened with rigid inclusions

Assume now that voids are filled with rigid inclusions. We retain the notations of the previous section. The essential difference is that now instead of voids we deal with rigid inclusions. The assumptions $(A_1) - (A_3)$, (3.12.2) and (3.12.3) are assumed to be satisfied. Obviously, ∂T_i^ε are boundaries of rigid inclusions. As before, the loading b is of class $L^2(\Omega)$.

For any fixed $\varepsilon > 0$ the space of kinematically admissible transverse displacements is given by

$$V_r^\varepsilon = \{w \in H_0^2(\Omega)|v_{|\Omega\setminus\Omega^\varepsilon} \in \mathcal{R}\} \tag{3.13.1}$$

where $\mathcal{R}$ stands for the set of rigid displacements. Since $\kappa(w) = 0$ if and only if $w(x) = a_1x_1 + a_2x_2 + b$, where $a_1, a_2, b \in \mathbb{R}$, therefore the set $\mathcal{R}$ coincides with the space of first order polynomials in x_1 and x_2.

The functional of the total potential energy is given by

$$G_\varepsilon(w) = \frac{1}{2} \int\limits_{\Omega^\varepsilon} D_\varepsilon^{\alpha\beta\lambda\mu}(x)\kappa_{\alpha\beta}(w)\kappa_{\lambda\mu}(w)dx - \int\limits_\Omega bw\ dx \ . \tag{3.13.2}$$

The minimum principle of the total potential energy means evaluating

$$(Q_\varepsilon) \qquad G_\varepsilon(w^\varepsilon) = \inf\{G_\varepsilon(w)|w \in V_r^\varepsilon\} \ . \tag{3.13.3}$$

For any $\varepsilon > 0$, a solution w^ε exists and is unique. Indeed, for $w \in V_r^\varepsilon$ we write

$$\int\limits_{\Omega^\varepsilon} D_\varepsilon^{\alpha\beta\lambda\mu}(x)\kappa_{\alpha\beta}(w)\kappa_{\lambda\mu}^y(w)dx \geq C\|\kappa(w)\|_{L^2(\Omega^\varepsilon)}^2 \geq C_1\|w\|_{H_0^2(\Omega)}^2 \ , \tag{3.13.4}$$

since

$$\sum_{\alpha,\beta=1}^2 \int\limits_\Omega \kappa_{\alpha\beta}(w)\kappa_{\alpha\beta}(w)dx \tag{3.13.5}$$

is a norm on $H_0^2(\Omega)$ equivalent to the natural norm of this space. Moreover, the linear form

$$L(v) = \int\limits_\Omega bw\ dx \tag{3.13.6}$$

is continuous on V_r^ε. Here C and C_1 are positive constants independent of ε.

Let us pass now to homogenization.

Theorem 3.13.1. The sequence of functionals $\{G_\varepsilon\}_{\varepsilon>0}$ is $\Gamma(w - H^2(\Omega))$ convergent to

$$G_h(w) = \int\limits_\Omega W_h^r(\kappa(w))dx - L(w) \ , \qquad w \in H_0^2(\Omega) \ , \tag{3.13.7}$$

where

$$\mathcal{W}_h^r(\boldsymbol{\rho}) = \inf\{\frac{1}{2|Y|}\int_{Y^c} D^{\alpha\beta\lambda\mu}(y)(\kappa_{\alpha\beta}^y(v)+\rho_{\alpha\beta})(\kappa_{\lambda\mu}^y(v)+\rho_{\lambda\mu})dy|v \in H_{r,per}(p)\}, \quad (3.13.8)$$

where $\rho \in \mathbb{E}_s^2$ and

$$H_{r,per}(p) = \{v \in H_{per}^2(Y)|(v-p)_{|Y\backslash Y^c} \in \mathcal{R}\}. \qquad (3.13.9)$$

Here $p = \dfrac{1}{2}\displaystyle\sum_{\alpha,\beta=1}^{2} \rho_{\alpha\beta}y_\alpha y_\beta$.

Proof. The notation $(v-p)_{|Y\backslash Y^c} \in \mathcal{R}$ means that on the part $Y\backslash Y^c$ occupied by the rigid inclusion (or inclusions) a function $(v-p)$ coincides with a rigid displacement.

Let us replace the rigid inclusions with deformable ones for which the elastic moduli are given by $\lambda D_r^{\alpha\beta\lambda\mu}(y)$ with λ being a positive number intended to tend to infinity. Then the functional of the total potential energy takes the following form

$$G_\varepsilon^\lambda(w) = G_\varepsilon(w) + \lambda \int_{\Omega\backslash\Omega^\varepsilon} D_{r,\varepsilon}^{\alpha\beta\lambda\mu}(x)\kappa_{\alpha\beta}(w)\kappa_{\lambda\mu}(w)dx. \qquad (3.13.10)$$

For any fixed $\lambda > 0$ the $\Gamma(s - H^1(\Omega))$-limit of $\{G_\varepsilon^\lambda\}_{\varepsilon>0}$ is given by

$$G_\varepsilon^\lambda(w) = \int_\Omega \mathcal{W}_h^{r,\lambda}(\kappa(w))dx - L(w), \quad w \in H_0^2(\Omega), \qquad (3.13.11)$$

where

$$\mathcal{W}_h^{r,\lambda}(\boldsymbol{\rho}) = \inf\{\frac{1}{2|Y|}\int_{Y^c} D^{\alpha\beta\lambda\mu}(y)(\kappa_{\alpha\beta}^y(v) + \rho_{\alpha\beta})(\kappa_{\lambda\mu}^y(v) + \rho_{\lambda\mu})dy$$

$$+ \frac{\lambda}{2|Y|}\int_{Y\backslash Y^c} D_r^{\alpha\beta\lambda\mu}(y)(\kappa_{\alpha\beta}^y(v) + \rho_{\alpha\beta})(\kappa_{\lambda\mu}^y(v) + \rho_{\lambda\mu})dy|v \in H_{per}^2(Y)\}. \quad (3.13.12)$$

It thus remains to show that for any fixed $\rho \in \mathbb{E}_s^2$ we have

$$\mathcal{W}_h^{r,\lambda}(\boldsymbol{\rho}) \to \mathcal{W}_h^r(\boldsymbol{\rho}) \quad \text{as} \quad \lambda \to +\infty. \qquad (3.13.13)$$

In fact, let v^λ be a minimizer of the minimization problem appearing on the r.h.s. of (3.13.12). Take $v \in H_{per}^2(Y)$ such that $\kappa^y(v) + \rho = \kappa^y(v-p) = 0$, where p is the same as in (3.13.9). Such a function v can always be chosen provided that the inclusion is contained in Y, i.e. does not intersect ∂Y. Then

$$C_0||\kappa^y(v^\lambda) + \rho||_{L^2(Y^c,\mathbb{E}_s^2)}^2 + C_1\lambda||\kappa^y(v^\lambda) + \rho||_{L^2(Y\backslash Y^c,\mathbb{E}_s^2)}^2$$

$$\leq C_2||\kappa^y(v) + \rho||_{L^2(Y^c,\mathbb{E}_s^2)}^2 = C_3. \qquad (3.13.14)$$

The positive constants $C_0, \ldots, C_3$ do not depend on λ. Hence

$$C_4||\kappa^y(v^\lambda) + \rho||^2_{L^2(Y,\mathbf{E}_s^2)} \leq C_0||\kappa^y(v^\lambda) + \rho||^2_{L^2(Y^c,\mathbf{E}_s^2)}$$
$$+C_1\lambda||\kappa^y(v^\lambda) + \rho||^2_{L^2(Y\setminus Y^c,\mathbf{E}_s^2)} \leq C_3 \,,$$

since $\lambda > 1$. Consequently the sequence $\{v^\lambda\}$ is bounded in $H^2_{per}(Y)$ and there exists a subsequence which converges weakly to $\bar{v} \in H^2_{per}(Y)$. Moreover, from (3.13.14) we conclude that

$$||\kappa^y(v) + \rho||^2_{L^2(Y\setminus Y^c,\mathbf{E}_s^2)} \leq \frac{C_5}{\lambda} \,.$$

Hence

$$||\kappa^y(\bar{v}) + \rho||^2_{L^2(Y\setminus Y^c,\mathbf{E}_s^2)} \leq \lim_{\lambda\to\infty} \inf ||\kappa^y(v^\lambda) + \rho||^2_{L^2(Y\setminus Y^c,\mathbf{E}_s^2)} \leq 0$$

and thus $\kappa^y(\bar{v} - p) = 0$ on $Y\setminus Y^c$. It means that $\bar{v} \in H_{r,per}(Y)$.

Eventually we find

$$\lim_{\lambda\to\infty} \mathcal{W}_h^{r,\lambda}(\rho) = \frac{1}{2|Y|} \int_{Y^c} D^{\alpha\beta\lambda\mu}(y)(\kappa^y_{\alpha\beta}(\bar{v}) + \rho_{\alpha\beta})(\kappa^y_{\lambda\mu}(\bar{v}) + \rho_{\lambda\mu})dy = \mathcal{W}_h^r(\rho)$$

and the proof is complete.   □

Remark 3.13.2. Derivation of the dual problem Q_ε^* as well as dual homogenization are left to the reader. Rigid inclusions make the problem interesting in itself.

4. Nonlinear behavior of plates

In this section we shall deal with two homogenization problems for von Kármán plates. In Sec. 4.2, by exploiting the results of Sec. 1.3.5, we shall perform (two-dimensional) homogenization of such plates exhibiting a periodic structure. Next, in Sec. 4.3, perforated von Kármán plates will be investigated. Thus the homogenization results of Sec. 3.12 will be extended to this nonlinear plate model. Also, we shall discuss the problem of convergence of the spectrum and of the bifurcating branches of the perforated von Kármán plate to the corresponding elements of the homogenized plate. Primarily, however, in Sec. 4.1 basic equations of von Kármán plate model will be introduced in a standard manner, cf. also Sec. 5.7.

4.1. Von Kármán equations

Let us consider a transversely symmetric thin plate of elastic moduli C^{ijkl} satisfying conditions $(2.4.2)_{1,2}$. Assume first that variability of C in $x \in \Omega$ is arbitrary and the plate is of constant thickness. We introduce the following notation: $B = \Omega \times (-\frac{h}{2}, \frac{h}{2})$, $\Gamma = \partial\Omega$, $\Gamma = \overline{\Gamma}_0 \cup \overline{\Gamma}_1$, $\Gamma_0 \cap \Gamma_1 = \emptyset$, $\Upsilon = \Gamma \times (-\frac{h}{2}, \frac{h}{2})$, $\Upsilon = \overline{\Upsilon}_0 \cup \overline{\Upsilon}_1$, $\Upsilon_0 \cap \Upsilon_1 = \emptyset$, $\Upsilon_0 = \Gamma_0 \times (-\frac{h}{2}, \frac{h}{2})$, $\Upsilon_1 = \Gamma_1 \times (-\frac{h}{2}, \frac{h}{2})$, $\Gamma_\pm = \Omega \times \{\pm\frac{h}{2}\}$. The plate is subjected to transverse loading $q(x)$ on Γ_+, in-plane and transversely symmetric loading: $t = (t^\alpha, 0)$, $t^\alpha(x, z) = t^\alpha(x, -z)$, $z = x^3$, on Υ_1. Displacements $w(x, z)$ vanish on Υ_0. The nonlinear strain tensor $\gamma_{ij}(w)$ is given by

$$\gamma_{ij}(w) = e_{ij}(w) + \frac{1}{2}\delta^{kl}w_{k,i}w_{l,j} \tag{4.1.1}$$

where $e_{ij}(w) = w_{(i,j)}$. Its variation $\delta\gamma_{ij}$ is denoted by $\Delta_{ij}(w, v)$:

$$\Delta_{ij}(w, v) = e_{ij}(v) + \frac{1}{2}\delta^{kl}(w_{l,j}v_{k,i} + w_{k,i}v_{l,j}), \tag{4.1.2}$$

where v vanishes on Υ_0. In the traditional notation we write

$$\delta\gamma_{ij} = \frac{1}{2}(\delta w_{i,j} + \delta w_{j,i}) + \frac{1}{2}\delta^{kl}(w_{l,j}\delta w_{k,i} + w_{k,i}\delta w_{l,j}). \tag{4.1.3}$$

Denoting by S^{ij} the components of the second (symmetric) Piola-Kirchhoff stress tensor, the constitutive relationship is taken in the linear from:

$$S_{ij} = C^{ijkl}\gamma_{kl}. \tag{4.1.4}$$

The functional of the total potential energy is now given by

$$J(w) = \frac{1}{2}\int_B C^{ijkl}\gamma_{ij}(w)\gamma_{kl}(w)dx - \int_\Omega q(x)w_3(x, h/2)dx - \int_{\Upsilon_1} t^\alpha(x)w_\alpha(x)dS, \tag{4.1.5}$$

where $x = (x_\alpha, x_3) = (x_\alpha, z)$.

The equilibrium problem means evaluating

$$J(\overline{w}) = \inf\{J(w)|w \in W^{1,4}(B)^3 , \qquad w = 0 \text{ on } \Upsilon_0\} . \tag{4.1.6}$$

The reader is advised to formulate an existence theorem, assuming that the loading functional is continuous, cf. Sec. 1.2.2. We observe that the nonlinearity of the strain measure $\gamma(w)$ implies that the functional (4.1.5) is nonconvex.

The condition of vanishing of the first variation of J yields

$$\int_B C^{ijkl}\gamma_{ij}(\overline{w})\Delta_{kl}(\overline{w}, v)dx = \int_\Omega q(x)v_3(x, h/2)dx + \int_{\Upsilon_1} t^\alpha v_\alpha dS , \tag{4.1.7}$$

for each v vanishing on Υ_0. To solve Eq. (4.1.7) one can use the Faedo-Galerkin method. All minimizers solving (4.1.6) satisfy Eq. (4.1.7), but not necessarily vice versa.

Let us pass to a concise presentation of the von Kármán plate model. This modelling is based on:

(i) kinematic assumptions

$$w_\alpha(x, x_3) = u_\alpha(x) - zw_{,\alpha} , \qquad w_3(x, z) = w(x) . \tag{4.1.8}$$

(ii) Stress assumptions which replace (4.1.4) with

$$S^{\alpha\beta} = \widetilde{C}^{\alpha\beta\lambda\mu}\gamma_{\lambda\mu} , \qquad S^{\alpha 3} = 2C^{\alpha 3\lambda 3}\gamma_{\lambda 3} , \qquad S^{33} = 0 , \tag{4.1.9}$$

where $\widetilde{C}$ is defined similarly to $\widetilde{C}_z$, cf. Sec. 3.1.

(iii) Strain assumption: nonlinear terms with respect to u_α are neglected in the definition of the strain measure γ, hence it is assumed that

$$\gamma_{\alpha\beta}(w_\alpha, w) = e_{\alpha\beta}(u) + \frac{1}{2}w_{,\alpha}w_{,\beta} + z\kappa_{\alpha\beta}(w) ,$$

$$\gamma_{\alpha 3}(w_\beta, w) = 0 , \quad \gamma_{33}(w_\alpha, w) = 0 , \tag{4.1.10}$$

where $\kappa_{\alpha\beta}(w) = -w_{,\alpha\beta}$. Consequently

$$\Delta_{\alpha\beta}(w_\gamma, w; v_\gamma, v) = e_{\alpha\beta}(v_\gamma) + \frac{1}{2}(w_{,\beta}v_{,\alpha} + w_{,\beta}v_{,\beta}) + z\kappa_{\alpha\beta}(v) ,$$

$$\Delta_{k3} = 0 , \tag{4.1.11}$$

and

$$\int_\Omega \int_{-h/2}^{h/2} S^{ij}\Delta_{ij}(w_\alpha, w; v_\alpha, v)dxdz = \int_\Omega [N^{\alpha\beta}e_{\alpha\beta}(v_\gamma)$$

$$+ N^{\alpha\beta}w_{,\beta}v_{,\alpha} + M^{\alpha\beta}\kappa_{\alpha\beta}(v)]dx , \tag{4.1.12}$$

where

$$N^{\alpha\beta} = \int_{-h/2}^{h/2} S^{\alpha\beta}dz , \qquad M^{\alpha\beta} = \int_{-h/2}^{h/2} zS^{\alpha\beta}dz . \tag{4.1.13}$$

Similarly, by using (4.1.8) the virtual work of the loading functional L reduces to

$$L(v_\alpha, v) = \int_\Omega qv dx + \int_{\Gamma_1} \widetilde{N}^\alpha v_\alpha d\Gamma , \qquad (4.1.14)$$

where

$$\widetilde{N}^\alpha = \int_{-h/2}^{h/2} t^\alpha dz . \qquad (4.1.15)$$

According to (4.1.9) and (4.1.13) the constitutive relationships assume the following form

$$N^{\alpha\beta} = A^{\alpha\beta\lambda\mu} E_{\lambda\mu}(\boldsymbol{u}, w) , \qquad M^{\alpha\beta} = D^{\alpha\beta\lambda\mu} \kappa_{\lambda\mu}(w) , \qquad (4.1.16)$$

where

$$E_{\alpha\beta}(\boldsymbol{u}, w) = e_{\alpha\beta}(\boldsymbol{u}) + \frac{1}{2} w_{,\alpha} w_{,\beta} . \qquad (4.1.17)$$

Under the hypotheses (i) - (iii) the variational Eq. (4.1.7) takes the form

$$\int_\Omega [N^{\alpha\beta} e_{\alpha\beta}(\boldsymbol{v}) + N^{\alpha\beta} w_{,\beta} v_{,\alpha} + M^{\alpha\beta} \kappa_{\alpha\beta}(v)] dx = L(\boldsymbol{v}, v) , \qquad (4.1.18)$$

valid for $\boldsymbol{v} = (v_\alpha)$, v , $\partial v/\partial n$ vanishing on Γ_0. Here $\boldsymbol{n} = (n_\alpha)$ denotes the unit outward normal vector to Γ.

The equilibrium problem: find $(\overline{u}, \overline{w})$ satisfying (4.1.18), with $N^{\alpha\beta}$ and $M^{\alpha\beta}$ given by (4.1.16), is solvable in the space $H^1(\Omega)^2 \times H^2(\Omega)$. Obviously $(\overline{u}, \overline{w})$ is to be such that $\overline{u}$, $\overline{w}$ and $\partial\overline{w}/\partial n$ vanish on Γ_0.

Remark 4.1. Duvaut and Lions (1974) used the Faedo-Galerkin method to show the existence of solutions to a large class of boundary value problems, including unilateral boundary conditions. Minimization methods were used by Ciarlet and Rabier (1980), cf. also Ciarlet (1997).

4.2. Homogenization

Consider now a von Kármán plate with an εY-periodic structure, cf. Secs. 2 and 3. It means that the elastic moduli

$$A_\varepsilon^{\alpha\beta\lambda\mu}(x) = A^{\alpha\beta\lambda\mu}\left(\frac{x}{\varepsilon}\right) , \qquad D_\varepsilon^{\alpha\beta\lambda\mu}(x) = D^{\alpha\beta\lambda\mu}\left(\frac{x}{\varepsilon}\right) , \qquad (4.2.1)$$

are εY-periodic.

We assume that $A^{\alpha\beta\lambda\mu} \in L^\infty(Y)$ and $D^{\alpha\beta\lambda\mu} \in L^\infty(Y)$ and that there exist constants $c_1 \geq c_0 > 0$ such that

$$\begin{aligned}
\forall\, \boldsymbol{\epsilon} \in \mathbb{E}_s^2 \qquad & c_0|\boldsymbol{\epsilon}|^2 \leq A^{\alpha\beta\lambda\mu}(y)\epsilon_{\alpha\beta}\epsilon_{\lambda\mu} \leq c_1|\boldsymbol{\epsilon}|^2 , \\
\forall\, \boldsymbol{\rho} \in \mathbb{E}_s^2 \qquad & c_0|\boldsymbol{\rho}|^2 \leq D^{\alpha\beta\lambda\mu}(y)\rho_{\alpha\beta}\rho_{\lambda\mu} \leq c_1|\boldsymbol{\rho}|^2 .
\end{aligned} \qquad (4.2.2)$$

The functional of the total potential is now given by

$$
J_\varepsilon(\boldsymbol{u}, w) = \frac{1}{2} \int\limits_\Omega [A_\varepsilon^{\alpha\beta\lambda\mu}(x) E_{\alpha\beta}(\boldsymbol{u}, w) E_{\lambda\mu}(\boldsymbol{u}, w)
$$
$$
+ D_\varepsilon^{\alpha\beta\lambda\mu}(x) \kappa_{\alpha\beta}(w) \kappa_{\lambda\mu}(w)] dx - L(\boldsymbol{u}, w) , \qquad (4.2.3)
$$

where $(\boldsymbol{u}, w) \in H^1(\Omega)^2 \times H^2(\Omega)$. We recall that in the two-dimensional case $H^2(\Omega) \subset W^{1,4}(\Omega)$. The loading functional L is not necessarily of the form (4.1.14). It suffices to assume its continuity in the weak topology of $H^1(\Omega)^2 \times H^2(\Omega)$. Then this functional is a perturbation functional. We observe that the virtual work of boundary bending moments can readily be included into the functional L.

We set

$$
V(\Omega) = \{(\boldsymbol{u}, w) \in H^1(\Omega)^2 \times H^2(\Omega) | \boldsymbol{u} = \boldsymbol{0} , \ w = \frac{\partial w}{\partial \boldsymbol{n}} \quad \text{on} \quad \Gamma_0\} . \qquad (4.2.4)
$$

For a fixed $\varepsilon > 0$ the minimization problem:

$$
(P_\varepsilon) \quad J_\varepsilon(\boldsymbol{u}^\varepsilon, w^\varepsilon) = \inf\{J_\varepsilon(\boldsymbol{u}, w) | (\boldsymbol{u}, w) \in V(\Omega)\}
$$

is solvable.

Let us pass to finding the limit functional $J_h = G_h - L$. To this end we set

$$
G_\varepsilon(\boldsymbol{u}, w) = \frac{1}{2} \int\limits_\Omega [A_\varepsilon^{\alpha\beta\lambda\mu}(x) E_{\alpha\beta}(\boldsymbol{u}, w) E_{\lambda\mu}(\boldsymbol{u}, w) + D_\varepsilon^{\alpha\beta\lambda\mu}(x) \kappa_{\alpha\beta}(w) \kappa_{\lambda\mu}(w)] dx . \qquad (4.2.5)
$$

Now we are in a position to formulate the homogenization theorem.

Theorem 4.2.1. Under the assumption (4.2.2) the sequence of functionals $\{G_\varepsilon\}_{\varepsilon>0}$ is $\Gamma(\tau)$-convergent $(\tau = s - (L^2(\Omega)^2 \times H^1(\Omega)))$ to

$$
G_h(\boldsymbol{u}, w) = \frac{1}{2} \int\limits_\Omega [A_h^{\alpha\beta\lambda\mu} E_{\alpha\beta}(\boldsymbol{u}, w) E_{\lambda\mu}(\boldsymbol{u}, w) + D_h^{\alpha\beta\lambda\mu} \kappa_{\alpha\beta}(w) \kappa_{\lambda\mu}(w)] dx , \qquad (4.2.6)
$$

where $(\boldsymbol{u}, w) \in H^1(\Omega)^2 \times H^2(\Omega)$.

The homogenized elastic potential is given by

$$
\mathcal{W}_h(\boldsymbol{E}^h, \boldsymbol{\kappa}^h) = \inf\{\frac{1}{2|Y|} \int\limits_Y A^{\alpha\beta\lambda\mu}(y)(E_{\alpha\beta}^h + e_{\alpha\beta}^y(\boldsymbol{v}))(E_{\lambda\mu}^h + e_{\lambda\mu}^y(\boldsymbol{v})) dy | \boldsymbol{v} \in \tilde{H}_{per}^1(Y)^2\}
$$
$$
+ \inf\{\frac{1}{2|Y|} \int\limits_Y D^{\alpha\beta\lambda\mu}(y)(\kappa_{\alpha\beta}^h + \kappa_{\alpha\beta}^y(v))(\kappa_{\lambda\mu}^h + \kappa_{\lambda\mu}^y(v)) dy | v \in \tilde{H}_{per}^2(Y)\} ,
$$

$$
(4.2.7)
$$

where $E^h, \kappa^h \in \mathbb{E}_s^2$. The homogenized moduli are calculated from

$$A_h^{\alpha\beta\lambda\mu} = \frac{\partial^2 \mathcal{W}_h}{\partial E_{\alpha\beta}^h \partial E_{\lambda\mu}^h} = \frac{1}{|Y|} \int_Y A^{\alpha\beta\sigma\tau}(y)(\delta_{\sigma\lambda}\delta_{\tau\mu} + e_{\sigma\tau}^y(\chi^{(\lambda\mu)}))dy , \qquad (4.2.8)$$

$$D_h^{\alpha\beta\lambda\mu} = \frac{\partial^2 \mathcal{W}_h}{\partial \kappa_{\alpha\beta}^h \partial \kappa_{\lambda\mu}^h} = \frac{1}{|Y|} \int_Y D^{\alpha\beta\sigma\tau}(y)(\delta_{\sigma\lambda}\delta_{\tau\mu} + \kappa_{\sigma\tau}^y(\Theta^{(\lambda\mu)}))dy , \qquad (4.2.9)$$

respectively.

The functions $\chi^{(\alpha\beta)}$ and $\Theta^{(\alpha\beta)}$ are solutions to the following local problems.

Problem P_{loc}^1. Find $\chi^{(\sigma\tau)} \in \widetilde{H}_{per}^1(Y)^2$ such that

$$\int_Y A^{\alpha\beta\lambda\mu}(y)(\delta_{\lambda\sigma}\delta_{\mu\tau} + e_{(\lambda\mu)}^y(\chi^{(\sigma\tau)}))e_{\alpha\beta}^y(v)dy = 0 , \quad \forall\, v \in \widetilde{H}_{per}^1(Y)^2 . \quad (4.2.10)$$

Problem P_{loc}^2. Find $\Theta^{(\sigma\tau)} \in \widetilde{H}_{per}^2(Y)^2$ such that

$$\int_Y D^{\alpha\beta\lambda\mu}(y)(\delta_{\lambda\sigma}\delta_{\mu\tau} + \kappa_{(\lambda\mu)}^y(\Theta^{(\sigma\tau)}))\kappa_{\alpha\beta}^y(v)dy = 0 , \quad \forall\, v \in \widetilde{H}_{per}^2(Y) . \quad (4.2.11)$$

Moreover

$$\inf\{J_h(\boldsymbol{u}, w)|\, (\boldsymbol{u}, w) \in V(\Omega)\} = \lim_{\varepsilon \to 0}(\inf\{J_\varepsilon(\boldsymbol{u}, w)|\, (\boldsymbol{u}, w) \in V(\Omega)\}) , \quad (4.2.12)$$

and there exists a subsequence ε' such that

$$(\boldsymbol{u}^{\varepsilon'}, w^{\varepsilon'}) \rightharpoonup (\overline{\boldsymbol{u}}, \overline{w}) \quad \text{weakly in} \quad H^1(\Omega)^2 \times H^2(\Omega) \quad \text{as} \quad \varepsilon' \to 0 .$$

The element $(\overline{\boldsymbol{u}}, \overline{w}) \in V(\Omega)$ is a minimizer of the homogenized plate problem

$$(P_h) \quad J_h(\overline{\boldsymbol{u}}, \overline{w}) = \inf\{J_h(\boldsymbol{u}, w)|(\boldsymbol{u}, w) \in V(\Omega)\} .$$

Proof. Applying Theorem 1.3.28 we conclude that $\Gamma(\tau) - \lim_{\varepsilon \to 0} G_\varepsilon - G_h$ and the homogenized potential is given by (4.2.7), where $E_{\alpha\beta}^h = \epsilon_{\alpha\beta} + \frac{1}{2}w,_\alpha w,_\beta$, $\epsilon \in \mathbb{E}_s^2$. In the local minimization problem appearing on the r.h.s. of (4.2.7) the macroscopic gradient $\nabla w(x) = (w,_\alpha(x))$ plays the role of a vector parameter. Therefore it is reasonable to treat $\frac{1}{2}(\nabla w) \otimes \nabla w$ as a part of E^h.

Let us denote by $(\overline{\boldsymbol{v}}, \overline{v}) \in \widetilde{H}_{per}^1(Y)^2 \times \widetilde{H}_{per}^2(Y)$ a solution to the local minimization problem. We obviously have

$$\overline{v}_\lambda = E_{\alpha\beta}^h \chi_\lambda^{(\alpha\beta)}(y) , \qquad \overline{v} = \kappa_{\alpha\beta}^h \Theta^{(\alpha\beta)} , \qquad (4.2.13)$$

and the formulae (4.2.9), (4.2.10) follow.

The local problems (4.2.10), (4.2.11) are equivalent to the two minimization problems involved in (4.2.7).

The remaining part of the proof is based on Theorem 1.3.22 and Remark 1.3.23. First, we have to find a τ-relatively compact subset $X_0 \subset V(\Omega)$ such that

$$\inf\{J_\varepsilon(\boldsymbol{u}, w)|(\boldsymbol{u}, w) \in X_0\} = J_\varepsilon(\boldsymbol{u}^\varepsilon, w^\varepsilon) , \tag{4.2.14}$$

cf. also the problem P_ε. We may write

$$J_\varepsilon(\boldsymbol{u}^\varepsilon, w^\varepsilon) \leq J_\varepsilon(\boldsymbol{u}, w) \qquad \forall (\boldsymbol{u}, w) \in V(\Omega) .$$

Hence, for $\boldsymbol{u} = 0$ and $w = 0$ we get

$$J_\varepsilon(\boldsymbol{u}^\varepsilon, w^\varepsilon) \leq 0 , \tag{4.2.15}$$

and consequently

$$\|\boldsymbol{u}^\varepsilon\|_{1,\Omega} + \|w^\varepsilon\|_{2,\Omega} \leq c , \tag{4.2.16}$$

where $c > 0$ is a constant, independent of ε. Indeed, the following result has been proved by Bielski and Telega (1996).

Lemma 4.2.2. Let $\eta > 0$ and $K > 0$ be given constants. There exists a constant $c_1 = c_1(\eta, K)$ such that

$$\sum_{\alpha,\beta=1}^{2} \int_\Omega |e_{\alpha\beta}(\boldsymbol{u}) + \frac{1}{2}w_{,\alpha}w_{,\beta}|^2 dx \geq c_1[\|\boldsymbol{u}\|_{1,\Omega}^2 + \|w\|_{2,\Omega}^2] , \tag{4.2.17}$$

for each couple $(\boldsymbol{u}, w) \in V(\Omega)$ satisfying

$$w \in T_M , \quad \|\boldsymbol{u}\|_{1,\Omega}^2 + \|w\|_{2,\Omega}^2 \geq \eta^2 , \tag{4.2.18}$$

where

$$T_M = \{w \in H^2_{\Gamma_0}(\Omega)| \ \|w\|_{2,\Omega} \leq Mp(w)\} . \tag{4.2.19}$$

Here

$$H^2_{\Gamma_0} = \{w \in H^2(\Omega)| \ w = \frac{\partial w}{\partial \boldsymbol{n}} = 0 \qquad \text{on } \Gamma_0\} , \tag{4.2.20}$$

$M > 0$ is a given constant and

$$p(w) = \{\int_\Omega (w_{,\alpha}w_{,\alpha})^2 dx\}^{1/4} \tag{4.2.21}$$

is a seminorm on $W^{1,4}(\Omega)$. The set T_M is a weakly closed one. $\square$

Let us continue the proof of Theorem 4.2.1. To prove (4.2.16) we use (4.2.2), (4.2.15), (4.2.17) and the continuity of the loading functional. On account of (4.2.16) we conclude that there exists a τ-relatively compact set $X_0 \subset V(\Omega)$ such that (4.2.14) is satisfied. Thus, by Theorem 1.3.22, (4.2.12) follows.

We can extract a subsequence $\{u^{\varepsilon'}, w^{\varepsilon'}\}_{\varepsilon'>0}$ such that

$$(u^{\varepsilon'}, w^{\varepsilon'}) \rightharpoonup (\overline{u}, \overline{w}) \quad \text{weakly in} \quad H^1(\Omega)^2 \times H^2(\Omega) \quad \text{as} \quad \varepsilon' \to 0 \,,$$

where $(\overline{u}, \overline{w}) \in V(\Omega)$. Moreover we have

$$\lim_{\varepsilon' \to 0} J_{\varepsilon'}(u^{\varepsilon'}, w^{\varepsilon'}) = \lim_{\varepsilon' \to 0} \left(\inf\{J_{\varepsilon'}(u, w) | (u, w) \in V(B)\}\right) = \lim_{\varepsilon' \to 0} J_{\varepsilon'}(u^{\varepsilon'}, w^{\varepsilon'}) \,.$$

Consequently $(\overline{u}, \overline{w})$ is a minimizer of J_h on $V(B)$. This completes the proof. $\qquad\square$

Remark 4.2.3. Ciarlet and Rabier (1980) defined the set T_M as a subset of $H_0^2(\Omega)$. The formula (4.2.19) extends their definition provided that Γ_0 constitutes only a part of Γ and $\mathrm{meas}\Gamma_0 > 0$. Also, the characterization of T_M due to Ciarlet and Rabier (1980, p. 69) remains valid in our more general case since the imbedding $H^2(\Omega) \subset W^{1,4}(\Omega)$ is compact.

$\qquad\square$

4.3.　Bifurcation and homogenization of perforated von Kármán plates

The main aim of the present section is to combine homogenization and bifurcation of a periodically perforated von Kármán plate subjected to in-plane boundary loading. For such plate it is natural to study the relationships between the critical values of the homogenized plate and those characterizing the perforated one. The same pertains to the bifurcating branches.

4.3.1.　Homogenization of perforated von Kármán plates

Consider now a von Kármán plate perforated similarly to the linear Kirchhoff plate, cf. Section 3.12. We assume that the perforated von Kármán plate is made of a homogeneous material. The consideration which follow can readily be extended to periodic and even non-uniformly periodic elastic moduli. Obviously, the inequalities (4.2.2) hold now in the part Y^c of the basic cell. The boundary Γ of Ω is assumed to consist of four parts: $\Gamma_{00}, \Gamma_0, \Gamma_1$ and Γ_2 and $\Gamma = \overline{\Gamma}_{00} \cup \overline{\Gamma}_0 \cup \overline{\Gamma}_1 \cup \overline{\Gamma}_2$. On Γ_{00} the plate is clamped: $w = \dfrac{\partial w}{\partial n} = 0$ and the in-plane displacement vanish: $u = 0$. On Γ_0 the plate is clamped: $w = \dfrac{\partial w}{\partial n} = 0$ whilst on Γ_1 it is simply supported: $w = 0$. On $\Gamma_0 \cup \Gamma_1$ the plate is subjected to a boundary loading described by the loading functional $L_1(u, w)$, which in the homogenization procedure plays the role of a perturbation functional. No boundary conditions are imposed on Γ_2.

The functional of the total potential energy is given by

$$J_\varepsilon^1(u, w) = G_\varepsilon(u, w) - \int_\Omega \chi_\varepsilon(x) q(x) w(x) dx - L_1(u, w) \,. \tag{4.3.1}$$

Here χ_ε is defined by (3.12.4) and G_ε is given by (4.2.5) with Ω being replaced by Ω^ε. The minimum principle of the total potential energy means evaluating

$$(Q_\varepsilon) \qquad J_\varepsilon^1(\boldsymbol{u}^\varepsilon, w^\varepsilon) = \inf\{J_\varepsilon^1(\boldsymbol{u}, w)|(\boldsymbol{u}, w) \in V_1(\Omega^\varepsilon)\}\,,$$

where

$$V_1(\Omega^\varepsilon) = \{(\boldsymbol{u}, w) \in H^1(\Omega^\varepsilon)^2 \times H^2(\Omega^\varepsilon)|\ \boldsymbol{u} = \boldsymbol{0}\,, \quad w = \frac{\partial w}{\partial \boldsymbol{n}} = 0 \quad \text{on} \quad \Gamma_{00}\,;$$

$$w = \frac{\partial w}{\partial \boldsymbol{n}} = 0 \quad \text{on} \quad \Gamma_0\,,\ w = 0 \quad \text{on} \quad \Gamma_1\}\,. \tag{4.3.2}$$

Korn's type inequalities are preserved for the domain Ω^ε, cf. Oleinik et al. (1992), Duvaut (1977b) and Secs. 8.3 and 9.3. Consequently a solution $(\boldsymbol{u}^\varepsilon, w^\varepsilon) \in V_1(\Omega^\varepsilon)$ exists since minimization methods still apply.

Prior to performing homogenization we shall introduce two extension operators for functions defined on Ω^ε.

Lemma 4.3.1. There exists an extension operator:

$$\mathbb{P}_1^\varepsilon :\ H^1(\Omega^\varepsilon)^2 \to H^1(\Omega)^2$$

such that

$$\exists\, C > 0\,, \qquad \forall\, \varepsilon > 0\,, \forall\, \boldsymbol{v} \in H^1(\Omega^\varepsilon)^2\,,$$
$$||\mathbb{P}_1^\varepsilon \boldsymbol{v}||_{0,\Omega} \leq C||\boldsymbol{v}||_{0,\Omega^\varepsilon}\,, \qquad ||e(\mathbb{P}_1^\varepsilon \boldsymbol{v})||_{0,\Omega} \leq C||e(\boldsymbol{v})||_{0,\Omega^\varepsilon}\,.$$

The constant C does not depend on ε. $\qquad\qquad \square$

The proof of the above lemma is left to the reader. We note, however, that similar problem is dealt with below and in Sec. 9.3.

Let us introduce the functional space V, which in the present section will play an important role

$$V = V(\Omega) = \{w \in H^2(\Omega)|\ w = 0 \quad \text{on} \quad \Gamma_{00} \cup \Gamma_0 \cup \Gamma_1\,,$$
$$\frac{\partial w}{\partial \boldsymbol{n}} = 0 \quad \text{on} \quad \Gamma_{00} \cup \Gamma_0\}\,. \tag{4.3.3}$$

The space $V_\varepsilon = V(\Omega^\varepsilon)$ is defined similarly.

Lemma 4.3.2. There exists an extension operator:

$$\mathbb{P}^\varepsilon :\ V_\varepsilon \to V$$

such that

$$\exists\, K > 0\,, \qquad \forall\, \varepsilon > 0\,, \forall\, w \in V^\varepsilon\,,$$
$$||\mathbb{P}^\varepsilon w||_{0,\Omega} \leq K||w||_{0,\Omega^\varepsilon}\,,$$
$$||\mathbb{P}^\varepsilon w||_{2,\Omega} \leq K||w||_{2,\Omega^\varepsilon}\,,$$
$$||\nabla^2(\mathbb{P}^\varepsilon w)||_{0,\Omega} \leq K||\nabla^2 w||_{0,\Omega^\varepsilon}\,. \qquad\qquad \square$$

Before proceeding to the proof of the last lemma we comment on the homogenization Theorem 3.12.1. In this case $V = H_0^2(\Omega)$ and V_ϵ is defined by (3.12.5). Let $\overline{w} \in H_0^2(\Omega)$ solve the following homogenized problem:

$$J_h(\overline{w}) = \inf\{J_h(w)|\ w \in H_0^2(\Omega)\}\,, \qquad (4.3.4)$$

with J_h given by (3.12.9). Then $\mathbf{P}^\varepsilon w^\varepsilon$ converges to $\overline{w}$ in $H_0^2(\Omega)$ weakly.

The proof of Lemma 4.3.2 exploits the following result, cf. Duvaut (1977b).

Lemma 4.3.3. There exists an extension operator:

$$\mathbf{P}:\ H^2(Y^c) \to H^2(Y)$$

such that

$$\begin{aligned}
&\exists\, K_1 > 0\,, \qquad \forall\, w \in H^2(Y^c)\,, \\
&||\mathbf{P}w||_{0,Y} \le K_1||w||_{0,Y^c}\,, \\
&||\mathbf{P}w||_{2,Y} \le K_1||w||_{2,Y^c}\,, \\
&||\nabla_y^2(\mathbf{P}w)||_{0,Y} \le K_1||\nabla_y^2 w||_{0,Y^c}\,,
\end{aligned} \qquad (4.3.5)$$

and the constant K_1 is independent of w.

Proof. Let $\mathcal{P}$ be the set of first order polynomials. Any function $w \in H^2(Y^c)$ can be written as follows

$$w = w_1 + p\,,$$

where $p \in \mathcal{P}$ and $w_1 \perp \mathcal{P}$ in $L^2(Y^c)$. Then we have

$$\exists\, K > 0\,,\ ||w_1||_{2,Y^c} \le K \sum_{|\alpha|=2} ||\partial_y^{|\alpha|} w_1||_{0,Y^c}\,, \qquad \forall\, w \in H^2(Y^c)\,. \qquad (4.3.6)$$

We recall that $\partial_y^{|\alpha|} = \nabla_y^2$, $|\alpha| = 2$. Now we choose an extension of w to Y which leaves p invariant. To this end w_1 is extended to Y by continuous lifting of the traces of w_1 and $\dfrac{\partial w_1}{\partial \mu}$ on Y^0, where μ is the inward unit normal vector to ∂Y^0. If this extension is denoted by $\widetilde{w}_1$, one has

$$\sum_{|\alpha|=2} ||\partial_y^{|\alpha|} \widetilde{w}_1||_{0,Y} \le K_2||w_1||_{2,Y^c}\,. \qquad (4.3.7)$$

We set

$$\mathbf{P}w = \widetilde{w}_1 + p\,. \qquad (4.3.8)$$

Now the inequalities (4.3.5) readily follow. $\qquad\qquad\qquad\qquad\qquad\qquad\qquad\square$

Proof of Lemma 4.3.2. It suffices to show that:
(i) if $w \in V_\varepsilon$, one can find an extension $\widetilde{w} \in V$ of w such that there exists a constant $K_3 > 0$ independent of ε such that

$$||\nabla^2 \widetilde{w}||_{0,\Omega} \leq K_3 ||\nabla^2 w||_{0,\Omega^\varepsilon} . \tag{4.3.9}$$

(ii) If $w \in V_\varepsilon$, there exists a constant $K_4 > 0$ independent of ε and w such that

$$||w||_{0,\Omega^\varepsilon} \leq K_4 ||\nabla^2 w||_{0,\Omega^\varepsilon} . \tag{4.3.10}$$

To prove the last two inequalities we observe that since Ω^ε is connected and the distribution of holes is periodic we may write

$$\Omega = (\bigcup_i T_i^\varepsilon) \cup A_0^\varepsilon , \tag{4.3.11}$$

where A_0^ε contains no hole. By using the homothetic transformation: $y = x/\varepsilon$ each domain T_i^ε is mapped into the same domain $Y^0 \subset \mathbb{R}^2$. Here, for the sake of simplicity, we assume that Y^0 contains only one hole. We set

$$v(y) = w(\varepsilon y) , \tag{4.3.12}$$

and apply Lemma 4.3.3. Hence we deduce the existence of extension $\widetilde{w}(x)$ to εY given by

$$\widetilde{w}(x) = \widetilde{v}(x/\varepsilon). \tag{4.3.13}$$

Here we tacitly assume that εY has been suitably shifted in the plane $\mathbb{R}^2$, cf. Sec. 9.3. Moreover we have

$$\sum_{\alpha,\beta} \int_{\varepsilon Y} (\widetilde{w}_{,\alpha\beta})^2 dx = \frac{1}{\varepsilon^2} \sum_{\alpha,\beta} \int_Y (\widetilde{v}_{|\alpha\beta})^2 dy$$

$$\leq \frac{1}{\varepsilon^2} K_1 \sum_{\alpha,\beta} \int_{Y^c} (v_{|\alpha\beta})^2 dy = K_1 \sum_{\alpha,\beta} \int_{\varepsilon Y^c} (w_{,\alpha\beta})^2 dx . \tag{4.3.14}$$

We recall that $v_{|\alpha\beta} = \dfrac{\partial^2 v}{\partial y_\alpha \partial y_\beta}$. Summing over all cells εY covering Ω and noting that the function w remains unchanged in a neighborhood of Γ we arrive at (4.3.9).

To prove (4.3.10) we denote by $\widetilde{w}$ the extension of $w \in V_\varepsilon$ given by the previous step. Since $\widetilde{w} \in V$ therefore there exists a constant K_5 independent of ε and $\widetilde{v}$ and such that

$$||\widetilde{w}||_{0,\Omega}^2 \leq K_5 \sum_{|\alpha|=2} ||\partial_y^{|\alpha|} \widetilde{w}||_{0,\Omega}^2 , \tag{4.3.15}$$

provided that $\text{meas}(\Gamma_{00} \cup \Gamma_0) > 0$. From (4.3.9) and (4.3.15) we immediately get (4.3.10), because $||w||_{0,\Omega^\varepsilon} \leq ||\widetilde{w}||_{0,\Omega}$. $\qquad\square$

Now we are in a position to formulate the homogenization theorem.

Theorem 4.3.4. The sequence of functionals $\{J_\varepsilon^1\}_{\varepsilon>0}$ is Γ-convergent in the weak topology of $H^1(\Omega)^2 \times H^2(\Omega)$ to

$$J_h^1(\boldsymbol{u}, w) = G_h^1(\boldsymbol{u}, w) - \int_\Omega \theta q(x)w(x)dx - L_1(\boldsymbol{u}, w), \quad (\boldsymbol{u}, w) \in V_1(\Omega) , \qquad (4.3.16)$$

where $\theta = |Y^c|/|Y|$ and G_h^1 has the form (4.2.6). The homogenized moduli $D_h^{\alpha\beta\lambda\mu}$, $A_h^{\alpha\beta\lambda\mu}$ are given by (4.2.9), (4.2.10), whilst the local problems have the form (4.2.10), (4.2.11) with Y being replaced by Y^c.

Let

$$J_h^1(\overline{\boldsymbol{u}}, \overline{w}) = \inf\{J_h^1(\boldsymbol{u}, w)|\,(\boldsymbol{u}, w) \in V_1(\Omega)\} . \qquad (4.3.17)$$

There is a subsequence $\{\mathbf{P}_1^{\varepsilon'}\boldsymbol{u}^{\varepsilon'}, \mathbf{P}^{\varepsilon'}w^{\varepsilon'}\}_{\varepsilon'>0}$ weakly convergent in $H^1(\Omega^2) \times H^2(\Omega)$ to $(\overline{\boldsymbol{u}}, \overline{w}) \in V_1(\Omega)$, being the solution of (4.3.17). $\qquad\square$

4.3.2. Bifurcation of von Kármán plates: basic results

Prior to the discussion of bifurcation problems of perforated von Kármán plates we introduce the relevant operators and results on bifurcation of von Kármán plates without holes. For the sake of simplicity we assume that on $\Gamma_0 \cup \Gamma_1$ the plate is subjected to one-parameter in-plane loading, i.e.,

$$N^{\alpha\beta}n_\alpha = \lambda\widetilde{N}^\alpha \text{ on } \Gamma_0 \cup \Gamma_1 . \qquad (4.3.18)$$

Let

$$V_{\Gamma_{00}} = \{\boldsymbol{u} \in H^1(\Omega)^2|\boldsymbol{u} = \boldsymbol{0} \quad \text{on} \quad \Gamma_{00}\} . \qquad (4.3.19)$$

Since

$$\mathrm{div}\boldsymbol{N} = \boldsymbol{0} \quad \text{in} \quad \Omega , \qquad (4.3.20)$$

therefore the following variational equation is readily obtained

$$\int_\Omega N^{\alpha\beta}e_{\alpha\beta}(\boldsymbol{v})dx = \lambda \int_{\Gamma_0\cup\Gamma_1} \widetilde{N}^\alpha v_\alpha d\Gamma , \quad \forall\, \boldsymbol{v} \in V_{\Gamma_{00}} . \qquad (4.3.21)$$

Taking into account the constitutive equation $(4.1.16)_1$ we conclude that the solution $\boldsymbol{N} = (N^{\alpha\beta})$ of (4.3.21) can be written in the following form

$$N^{\alpha\beta} = \lambda N_0^{\alpha\beta} + S^{\alpha\beta}(w) , \qquad (4.3.22)$$

where

$$N_0^{\alpha\beta} = A^{\alpha\beta\lambda\mu}e_{\lambda\mu}(\boldsymbol{u}^0) , \qquad (4.3.23)$$

and $\boldsymbol{u}^0 \in V_{\Gamma_{00}}$ is a solution to

$$\int_\Omega A^{\alpha\beta\lambda\mu}e_{\lambda\mu}(\boldsymbol{u}^0)e_{\alpha\beta}(\boldsymbol{v})dx = \int_{\Gamma_0\cup\Gamma_1} \widetilde{N}^\alpha v_\alpha d\Gamma , \qquad \forall\, \boldsymbol{v} \in V_{\Gamma_{00}} . \qquad (4.3.24)$$

$S = S(w)$ is given by

$$S^{\alpha\beta}(w) = A^{\alpha\beta\lambda\mu}[e_{\lambda\mu}(\widehat{u}(w)) + \frac{1}{2}w,_\lambda w,_\mu] \,, \tag{4.3.25}$$

and $\widehat{u}(w) \in V_{\Gamma_{00}}$ solves the following problem

$$\int_\Omega A^{\alpha\beta\lambda\mu} e_{\lambda\mu}(\widehat{u}(w)) e_{\alpha\beta}(v) dx = -\int_\Omega A^{\alpha\beta\lambda\mu} f_{\lambda\mu}(w) e_{\alpha\beta}(v) dx \,, \quad \forall\, v \in V_{\Gamma_{00}} \,. \tag{4.3.26}$$

Here $f_{\alpha\beta}(w) = \frac{1}{2}w,_\alpha w,_\beta$. We observe that because $w,_\alpha \in L^p(\Omega)$ for each finite p, therefore $f_{\alpha\beta}(w) \in L^2(\Omega)$.

The reader is advised to find the strong form of the variational equations (4.3.24), (4.3.26). Taking into account (4.3.22) in (4.1.18) we arrive at

$$\int_\Omega D^{\alpha\beta\lambda\mu} \kappa_{\lambda\mu}(w) \kappa_{\alpha\beta}(v) dx + \lambda \int_\Omega N_0^{\alpha\beta} \frac{\partial w}{\partial x_\alpha} \frac{\partial v}{\partial x_\beta} dx + \int_\Omega S_{\alpha\beta}(w) \frac{\partial w}{\partial x_\alpha} \frac{\partial v}{\partial x_\beta} dx = 0 \,,$$

$$w \in V \,, \quad v \in V \,. \tag{4.3.27}$$

We recall that the space V is now defined by (4.3.3). Equation (4.3.27) involves three operators. The first operator

$$A : V \to V^* \,, \tag{4.3.28}$$

is defined by

$$\langle Aw, v \rangle = \int_\Omega D^{\alpha\beta\lambda\mu} \kappa_{\lambda\mu}(w) \kappa_{\alpha\beta}(v) dx \,. \tag{4.3.29}$$

Since $D = (D^{\alpha\beta\lambda\mu})$ is positive definite, therefore A is an isomorphism of V on V^*.

The second operator

$$B : V \to V^* \,,$$

is defined by

$$\langle Bw, v \rangle = -\int_\Omega N_0^{\alpha\beta} w,_\beta v,_\alpha dx \,, \tag{4.3.30}$$

Lemma 4.3.5. The operator B is linear, self-adjoint and compact from V into V^*.

Proof. Indeed, $w \in V$ and thus $w,_\alpha \in L^p(\Omega)$ for any finite p. On the other hand, $N_0^{\alpha\beta} \in L^2(\Omega)$. Hence we conclude that $N_0^{\alpha\beta} w,_\beta \in L^{2-\eta}(\Omega)$, $\eta > 0$. Consequently B is compact. The remaining properties of B are evident.

Remark 4.3.6. Let us recall the definition of completely continuous and compact operators, cf. Yosida (1978).

Definition 4.3.7. Let X be a reflexive Banach space and let X_1 be a Banach space. A mapping

$$T : X \to X_1 ,$$

is called *completely continuous* if for each sequence $\{v_n\}_{n\in\mathbb{N}} \subset X$ weakly convergent to $v \in X$, the sequence $\{T(v_n)\}_{n\in\mathbb{N}}$ converges in the norm to $T(v)$ in the space X_1. $\qquad\square$

It is evident that each completely continuous operator is continuous.

To define a *compact operator* we require: (i) the continuity in the sense of norms, (ii) that the image of each set bounded in X be relatively compact in X_1.

We observe that due to the reflexivity of the space X, a completely continuous operator from X to X_1 is compact. The converse statement is not true in general, the exception being the linear operators. $\qquad\square$

The third operator

$$C : V \to V^* ,$$

is defined by

$$\langle C(w), v \rangle = - \int_\Omega S^{\alpha\beta}(w) w_{,\beta} v_{,\alpha} dx . \tag{4.3.31}$$

Lemma 4.3.8. The operator $C : V \to V^*$ is completely continuous. It is homogeneous of degree 3 and is derivable from a positive potential Q :

$$C(w) = \frac{1}{4} Q'(w) , \tag{4.3.32}$$

where

$$Q(w) = \int_\Omega S^{\alpha\beta}(w) w_{,\beta} w_{,\alpha} dx . \tag{4.3.33}$$

The functional $\widehat{Q}$ defined on $L^2(\Omega, \mathbb{E}_s^2)$ by

$$\widehat{Q}(\boldsymbol{f}) = 2 \int_\Omega \widehat{S}^{\alpha\beta}(\boldsymbol{f}) f_{\alpha\beta} dx , \tag{4.3.34}$$

is convex and positive. Here $S^{\alpha\beta}(w) = \widehat{S}^{\alpha\beta}(\boldsymbol{f}(w))$, $f_{\alpha\beta}(w) = \frac{1}{2} w_{,\alpha} w_{,\beta}$.

Proof. For $\varphi, \xi \in V$, by $(\boldsymbol{T}(\varphi,\xi), \boldsymbol{u}(\varphi,\xi))$ we denote the solution in $L^2(\Omega, \mathbb{E}_2^s) \times V_{\Gamma_{00}}$ of the following problem

$$\begin{aligned}
\mathrm{div}\boldsymbol{T} &= \boldsymbol{0} &&\text{in } \Omega , \\
T^{\alpha\beta} n_\beta &= 0 &&\text{on } \Gamma\backslash\Gamma_{00} , \\
\boldsymbol{u} &= \boldsymbol{0} &&\text{on } \Gamma_{00} , \\
T^{\alpha\beta} &= A^{\alpha\beta\lambda\mu}[e_{\lambda\mu}(\boldsymbol{u}) + \frac{1}{2}\varphi_{,\lambda}\xi_{,\mu}] . &&
\end{aligned} \tag{4.3.35}$$

It is evident that $T^{\alpha\beta}(\varphi,\xi)$ is bilinear in φ and ξ; moreover

$$S^{\alpha\beta}(w) = T^{\alpha\beta}(w,w) \ . \tag{4.3.36}$$

We set

$$\widetilde{Q}(\varphi,\xi,\eta,\theta) \leq \int_\Omega T^{\alpha\beta}(\varphi,\xi)\eta_{,\beta}\theta_{,\alpha}dx \ . \tag{4.3.37}$$

We have

$$|\widetilde{Q}(\varphi,\xi,\eta,\theta)| \leq K||\varphi||_V||\xi||_V||\eta||_V||\theta||_V \ , \tag{4.3.38}$$

where K is a positive constant. Moreover, the mapping $(\varphi,\xi,\eta,\theta) \rightarrow \widetilde{Q}(\varphi,\xi,\eta,\theta)$ is completely continuous from V^4 to $\mathbb{R}$.

It can be shown that

$$\widetilde{Q}(\varphi,\xi,\eta,\theta) = \widetilde{Q}(\xi,\varphi,\eta,\theta) = \widetilde{Q}(\eta,\theta,\varphi,\xi) = \widetilde{Q}(\varphi,\xi,\theta,\eta) \ . \tag{4.3.39}$$

By using (4.3.37) and (4.3.38) we conclude that $Q(w) = \widetilde{Q}(w,w,w,w)$ is Fréchet continuously differentiable on V and

$$\langle Q'(w),\theta\rangle = 4\widetilde{Q}(w,w,w,\theta) = 4\langle C(w),\theta\rangle \ , \tag{4.3.40}$$

It is not difficult to show that $Q(w)$ and $C(w)$ are completely continuous.

To show that the functional (4.3.34) is positive we write

$$f_{\alpha\beta} = -e_{\alpha\beta}(\boldsymbol{u}(\boldsymbol{f})) + a_{\alpha\beta\lambda\mu}\widehat{S}^{\lambda\mu}(\boldsymbol{f}) \ ,$$

where $\boldsymbol{a} = \boldsymbol{A}^{-1}$. Hence

$$\int_\Omega \widehat{S}^{\alpha\beta}(\boldsymbol{f})f_{\alpha\beta}dx = \int_\Omega a_{\alpha\beta\lambda\mu}\widehat{S}^{\alpha\beta}(\boldsymbol{f})\widehat{S}^{\lambda\mu}(\boldsymbol{f})dx - \int_\Omega e_{\alpha\beta}(\boldsymbol{u}(\boldsymbol{f}))\widehat{S}^{\alpha\beta}(\boldsymbol{f})dx \ .$$

On account of $(4.3.35)_1$, (4.3.36) and recalling the definition of $\widehat{S}(\boldsymbol{f})$ we conclude that the last term vanishes. Since the matrix $\boldsymbol{a}$ is positive definite, the functional $\widehat{Q}$ is positive. The convexity follows from the fact that it is quadratic. $\qquad\square$

Remark 4.3.9. It is worth noting that the solutions of $Q(w) = 0$ are solutions to the Monge Ampère equation, cf. Mignot et al. (1981)

$$\frac{\partial^2 w}{\partial x_\alpha^2} \cdot \frac{\partial^2 w}{\partial x_\beta^2} - \left(\frac{\partial^2 w}{\partial x_\alpha \partial x_\beta}\right)^2 = 0 \ . \tag{4.3.41}$$

$\qquad\square$

Having introduced the operators A, B and C we may write the von Kármán equations in the form

$$(P_\lambda) \qquad w \in V , \quad Aw - \lambda Bw + C(w) = 0 . \tag{4.3.42}$$

An equivalent form is given by

$$w - \lambda B_1 w + C_1(w) = 0 , \tag{4.3.43}$$

where $B_1 = A^{-1}B$ and $C_1 = A^{-1}C$. If V is equipped with the scalar product $(\varphi, \theta) = \langle A\varphi, \theta \rangle$, the operator B_1 is compact and self-adjoint, but not necessarily positive.

The study of the nonlinear problem (P_λ) is closely related to the linearized eigenvalue problem:

$$Aw - \lambda Bw = 0 . \tag{4.3.44}$$

Indeed, the properties of the operators A, B and C imply that the following general theorem applies, cf. Krasnosel'skii (1964).

Theorem 4.3.10. Let $\mathbf{H}$ be a Hilbert space and $N : \mathbf{H} \to \mathbf{H}$ a nonlinear completely continuous operator with $N(0) = 0$. It is assumed that N is the gradient of a weakly continuous functional F, $F(0) = 0$, and F is uniformly differentiable in a neighborhood of 0. Let N be Fréchet differentiable at 0 and set $N'(0) = D$. Let D be a completely continuous self-adjoint operator.

Then every characteristic number of the linear operator D is a bifurcation point of the nonlinear operator N. $\qquad\qquad\square$

Remark 4.3.11. Consider the following nonlinear equation

$$\varphi = \mu N(\varphi) , \tag{4.3.45}$$

where $N(0) = 0$. A number μ_0 is said to be a *bifurcation point* if for any $\eta, \delta > 0$ there exists a characteristic number μ of the operator N such that $|\mu - \mu_0| < \eta$ and to this characteristic number corresponds at least one eigenfunction φ satisfying (4.3.45) with $\|\varphi\| < \delta$. $\qquad\qquad\square$

It is evident that the linearized problem (4.3.44) admits a countable family of eigenvalues λ_j, $j \in (n, m)$, $n \leq 0$, $m \geq 0$, and at least one of the integers is infinite:

$$\ldots \leq \lambda_{-2} \leq \lambda_{-1} < 0 < \lambda_1 \leq \lambda_2 \leq \ldots .$$

The spectrum admits a positive part provided that there exists w such that $\langle Bw, w \rangle > 0$. Then

$$\frac{1}{\lambda_1} = \max_{w \in V} \frac{\langle Bw, w \rangle}{\langle Aw, w \rangle} , \tag{4.3.46}$$

and

$$E(\lambda_1) = \{ w \in V | \ Aw - \lambda_1 Bw = 0 \} = \{ w \in V | \ \langle Aw, w \rangle - \lambda_1 \langle Bw, w \rangle = 0 \} . \tag{4.3.47}$$

The eigenspace corresponding to the first positive (or negative) eigenvalue is not necessarily of dimension 1, since the problem is of fourth order.

Let us pass to the study of bifurcation points corresponding to the eigenvalues λ_{-1} and λ_1. We assume that λ_1 exists ($\lambda_1 < +\infty$). If the negative part of the spectrum is empty, we set $\lambda_{-1} = -\infty$.

Theorem 4.3.12.

(i) If $\lambda \in (\lambda_{-1}, \lambda_1)$ then the problem (P_λ) admits only the trivial solution $w = 0$.

(ii) For $\lambda = \lambda_1$ the set of solutions to (P_λ) consists of the cone of eigenvectors corresponding to λ_1, which verify $Q(w) = 0$.

Proof.

(i) Let $\lambda \in (\lambda_{-1}, \lambda_1)$ and let w be a solution to (P_λ). Then we have

$$\langle Aw, w \rangle - \lambda \langle Bw, w \rangle + \langle C(w), w \rangle = 0 \,. \tag{4.3.48}$$

By using (4.3.46) we get

$$\langle Aw, w \rangle - \lambda \langle Bw, w \rangle > \left(1 - \frac{\lambda}{\lambda_1} \right) \langle Aw, w \rangle > 0 \,,$$

for $\lambda \in [0, \lambda_1)$, and

$$\langle Aw, w \rangle - \lambda \langle Bw, w \rangle > \left(1 - \frac{\lambda}{\lambda_{-1}} \right) \langle Aw, w \rangle > 0 \,,$$

for $\lambda \in (\lambda_{-1}, 0]$. Since

$$\langle C(w), w \rangle = \frac{1}{4} Q(w) \geq 0 \,,$$

therefore $w = 0$.

(ii) If w is a solution to (4.3.44) for $\lambda = \lambda_1$ then w satisfies

$$\langle Aw, w \rangle - \lambda_1 \langle Bw, w \rangle = 0 \,, \qquad Q(w) = 0 \,. \tag{4.3.49}$$

Hence, by (4.3.46), we conclude that $w \in E(\lambda_1)$. Moreover, we know that $Q(w) = 0$ is equivalent to $S^{\alpha\beta}(w) = 0$, which implies $C(w) = 0$. The solutions of (4.3.49) are thus also solutions to (P_λ). It remains to observe that if w solves (4.3.49), then ηw, $\eta \in \mathbb{R}$, is also a solution. $\square$

Remark 4.3.13.

(a) For $\lambda = \lambda_{-1}$ and λ_{-1} finite the part (ii) of the last theorem is formulated similarly.

(b) If the equation of Monge Ampère with the boundary conditions:

$$w = \frac{\partial w}{\partial \boldsymbol{n}} = 0 \qquad \text{on } \Gamma_{00} \cup \Gamma_0 \,, \qquad w = 0 \qquad \text{on } \Gamma_1 \,,$$

admits only the solution $w = 0$, then for $\lambda = \lambda_1$ the problem (P_λ) also admits only the solution $w = 0$.

Theorem 4.3.14. Let λ_1 be finite. Then the point $(\lambda_1, w = 0)$ is a bifurcation point for the von Kármán plate considered. More precisely, there exists a bifurcating branch $(\lambda_1(R), w_1(R))$ defined for every $R > 0$ such that

$$Aw_1(R) - \lambda_1(R)Bw_1(R) + C(w_1(R)) = 0 \,,$$

$$\|w_1(R)\| = \sqrt{2R(1 + O(R))} \,, \qquad \lim_{R \to 0} \lambda_1(R) = \lambda_1 \,, \tag{4.3.50}$$

where $\|w_1(R)\|^2 = \langle Aw_1, w_1 \rangle$.

Proof. Let us introduce the manifold Σ_R in V:

$$\Sigma_R = \{w \in V \mid \frac{1}{2}\langle Aw, w \rangle + \frac{1}{4}\langle C(w), w \rangle = R\} \,, \qquad R > 0 \,. \tag{4.3.51}$$

By using the Lagrange multiplier method to solve the maximization problem:

$$(P_R) \quad \left| \begin{array}{l} \text{Find} \\ \qquad \max\{J(w) \mid w \in \Sigma_R\} \end{array} \right.$$

and denoting the relevant multiplier by μ we have $\lambda = 1/\mu$. Here $J(w) = \dfrac{1}{2}\langle Bw, w \rangle$. We observe that the set Σ_R is bounded and star-shaped with respect to the origin. Indeed, for each $w \in V$, there exists unique t such that $tw \in \Sigma_R$. It can be shown that problem (P_R) possesses a maximizer $\overline{w} \in \Sigma_R$. To this end one can use the direct method of the calculus of variations. Consequently, for each $R > 0$ there exists $(\lambda_1(R), w_1(R))$ being a solution to $(4.3.50)_1$.

To prove $(4.3.50)_2$ we observe that for each $w \in \Sigma_R$ one has

$$2R - \frac{1}{2}\langle C(w), w \rangle = \langle Aw, w \rangle = \|w\|^2 \le 2R \,.$$

On the other hand

$$0 \le \langle C(w), w \rangle \le K\|w\|^4 \le 4KR^2 \,, \qquad K > 0 \,.$$

Thus, for each $w \in \Sigma_R$ we obtain

$$2R - 2KR^2 \le \|w\|^2 \le 2R \,, \tag{4.3.52}$$

and $(4.3.50)_2$ follows.

Let us pass to proving $(4.3.50)_3$. Theorem 4.3.12 implies that $\lambda_1(R) \ge \lambda_1$. Let t_R be such that $t_R w_1 \in \Sigma_R$, $\|w_1\| = 1$, $Aw_1 - \lambda_1 Bw_1 = 0$. From the relation (4.3.52) we conclude that t_R^2 is of the order of $2R$ or $t_R^2 \sim 2R$. Hence

$$\frac{1}{\lambda_1} \ge \frac{1}{\lambda_1(R)} = \frac{\langle Bw_1(R), w_1(R) \rangle}{\langle Aw_1(R), w_1(R) \rangle + \langle C(w_1(R)), w_1(R) \rangle}$$

$$\ge \frac{\langle Bw_1(R), w_1(R) \rangle}{2R + 4KR^2} \ge \frac{t_R^2 \langle Bw_1, w_1 \rangle}{2R + 4KR^2} = \frac{1}{\lambda_1}(1 + O(R)) \,,$$

which establishes the formula. Here we have exploited $(4.3.50)_1$ and (4.3.46). $\qquad \square$

Remark 4.3.15. To prove the last theorem we made no assumption on the solutions of the Monge Ampère equation. In fact a vertical bifurcating branch is not precluded, i.e., $\lambda_1(R) \equiv \lambda_1$ in the case where (4.3.41) admits non-trivial solutions.

Remark 4.3.16. Consider now the case of von Kármán plate subjected additionally to transverse loading q. Written in the weak form the equilibrium equations consists of (4.3.21) and, cf. (4.1.18)

$$\int_\Omega (N^{\alpha\beta} w_{,\beta} v_{,\alpha} + M^{\alpha\beta} \kappa_{\alpha\beta}(v)) dx = \int_\Omega q(x) v(x) dx , \quad \forall\ v \in V . \tag{4.3.53}$$

We recall that now V is defined by (4.3.3). In this case Eq. (4.3.42) is to be replaced by

$$(\widetilde{P}_\lambda) \quad w \in V , \qquad Aw - \lambda Bw + C(w) = q . \tag{4.3.54}$$

We have the following result, provided that previous assumptions are satisfied.

Theorem 4.3.17. If Eq. (4.3.41) admits only null solution, then for each λ the problem $(\widetilde{P}_\lambda)$ possesses at least one solution w.

Proof. The solutions of $(\widetilde{P}_\lambda)$ are obviously the critical points of the functional

$$J(v) = \frac{1}{2}\langle Av, v\rangle - \frac{\lambda}{2}\langle Bv, v\rangle + Q(v) - \int_\Omega qv dx . \tag{4.3.55}$$

This functional is lower semicontinuous. Thus it suffices to show that

$$\lim J(v) \to +\infty \quad \text{as} \quad ||v|| \to \infty .$$

To obtain a contradiction, suppose that there exist K and a sequence $\{v^n\}_{n\in\mathbb{N}}$, $||v^n|| \to +\infty$ such that

$$J(v^n) < K .$$

Taking

$$\xi^n = \frac{v^n}{||v^n||} ,$$

one easily shows that

$$\frac{1}{2} - \frac{\lambda}{2}\langle B\xi^n, \xi^n\rangle - \frac{1}{||v^n||}\langle q, \xi^n\rangle \le \frac{K}{||v^n||^2} .$$

Hence

$$\frac{1}{2} - \frac{\lambda}{2}\langle B\xi, \xi\rangle \le 0 , \tag{4.3.56}$$

where ξ is a weak limit of $\{\xi^n\}$. Then $Q(\xi) = 0$, and since the equation of the Monge Ampère admits only null solution we conclude that $\xi = 0$ which contradicts (4.3.56). $\qquad\square$

4.3.3. Bifurcation points of the homogenized plate and the linearized problem

In this section it is shown that the bifurcation points of the homogenized plate are the limits of the bifurcation points of the perforated plate when $\varepsilon \to 0$.

For the perforated plates characterized by a small parameter $\varepsilon > 0$ the relevant operators are denoted by $A^\varepsilon, B^\varepsilon, C^\varepsilon$. The space $V_\varepsilon = V(\Omega^\varepsilon)$ is defined by (4.3.3) with Ω being replaced by Ω^ε.

Consider the linearized problem for the perforated plate

$$w^\varepsilon \in V^\varepsilon, \qquad A^\varepsilon w^\varepsilon - \lambda^\varepsilon B^\varepsilon w^\varepsilon = 0,$$
$$\langle B^\varepsilon w^\varepsilon, w^\varepsilon \rangle = \pm 1. \tag{4.3.57}$$

We observe that the operators A^ε are uniformly coercive. Consequently the space V_ε can be endowed with the norm

$$||v||^2 = \langle A^\varepsilon v, v \rangle. \tag{4.3.58}$$

Similarly to the previous section the problem (4.3.57) admits a countable family of eigenvalues $\{\lambda_j^\varepsilon\}$, $j \in (n^\varepsilon, m^\varepsilon)$, where at least one of the two integers $n^\varepsilon, m^\varepsilon$ is infinite, with

$$\ldots \le \lambda_{-j}^\varepsilon \le \lambda_{-j+1}^\varepsilon \le \ldots \le \lambda_{-1}^\varepsilon < 0 < \lambda_1^\varepsilon \le \lambda_2^\varepsilon \le \ldots \le \lambda_k^\varepsilon \le \ldots.$$

It is convenient to repeat the eigenvalues λ_j^ε with their multiplicities. With each λ_i^ε is associated an eigenvector w_i^ε normalized by $\langle B^\varepsilon w_i^\varepsilon, w_i^\varepsilon \rangle = \operatorname{sign} \lambda_i^\varepsilon$. In this way the system $\{w_i^\varepsilon\}$ constitutes a complete orthogonal system for the space V_ε equipped with the norm $||v||^2 = \langle A^\varepsilon v, v \rangle$. By $E^\varepsilon(\lambda)$ we denote the subspace of eigenvectors associated with the eigenvalue λ.

Similarly, the homogenized von Kármán plate with smeared out holes is described by the operators A^h, B^h, C^h. These operators involve the homogenized moduli $A_h^{\alpha\beta\lambda\mu}, D_h^{\alpha\beta\lambda\mu}$ given by Theorem 4.3.4. With the linearized eigenvalue problem is associated the system $\{(\lambda_i^h, w_i^h)\}$, $i \in (n_0, m_0)$,

$$w_i^h \in V, \qquad A^h w_i^h = \lambda_i^h B^h w_i^h,$$
$$\langle B^h w_i^h, w_i^h \rangle = \operatorname{sign} \lambda_i^h. \tag{4.3.59}$$

Here the space V is defined by (4.3.3).

The following theorem specifies the spectral properties of the sequence $\{A^\varepsilon - \lambda B^\varepsilon\}_{\varepsilon>0}$ when ε tends to zero.

Theorem 4.3.18. Let $\{\lambda_{k_0}^\varepsilon\}_{\varepsilon>0}$ (resp. $\{\lambda_{-j_0}^\varepsilon\}_{\varepsilon>0}$) remain bounded independently of ε. Then there exists a subsequence ε' of ε such that

$$\forall\, i \in (0, k_0] \ (\text{resp. } \forall\, j \in [-j_0, 0)) \quad \lambda_i^{\varepsilon'} \to \lambda_i^h$$

and

$$\mathbb{P}^{\varepsilon'} w_i^{\varepsilon'} \rightharpoonup w_i^h \quad \text{weakly in} \quad V$$

as $\varepsilon' \to 0$.

Proof. The extension operator $\mathbb{P}^\varepsilon$ and its properties are specified by Lemma 4.3.2. The proof of the last theorem is a consequence of the compactness of B^h and of the lemma, which follows, cf. Sec. 1.3.2. $\qquad\square$

Lemma 4.3.19. Let $\{g^\varepsilon\}_{\varepsilon>0}$ be a sequence in V^* strongly converging to g^h. Consider the sequence ζ^ε of the solutions in V_ε of

$$A^\varepsilon \zeta^\varepsilon = g^\varepsilon \ .$$

Then

$$\mathbb{P}^\varepsilon \zeta^\varepsilon \rightharpoonup \zeta^\varepsilon \quad \text{in} \quad V \quad \text{weakly} \ ,$$

where ζ^h is the solution of the homogenized problem

$$A^h \zeta^h = \theta g^h \ .$$

Here $\theta = |Y^c|/|Y|$, cf. Sec. 3.12. $\qquad\square$

We also have the following result.

Theorem 4.3.20. Let λ^h be an eigenvalue of multiplicity p of the problem

$$w \in V \ , \qquad A^h w = \lambda^h B^h w \ ,$$
$$\lambda^h_{i-1} < \lambda^h_i = \lambda^h_{i+1} = \ldots \lambda^h_{i+p-1} < \lambda^h_{i+p} \ . \tag{4.3.60}$$

Then, for every $\eta > 0$ there exist $\varepsilon_0 > 0$ such that for $\varepsilon \le \varepsilon_0$, $\lambda^\varepsilon_i, \ldots, \lambda^\varepsilon_{i+p-1}$ lie in the interval $(\lambda^h - \eta, \lambda^h + \eta)$.

Proof. For ε sufficiently small there exists an eigenvalue λ^ε in $(\lambda^h - \eta, \lambda^h + \eta)$. To corroborate this statement suppose that it were not true. Then we could find a subsequence ε' tending to zero such that $(A^{\varepsilon'} - \lambda^h B^{\varepsilon'})^{-1}: V^*_{\varepsilon'} \to V_{\varepsilon'}$ has a norm bounded by $K/\eta, K > 0$. Moreover, for $g \in L^2(\Omega)$ the solution $\zeta^{\varepsilon'}$ of

$$A^{\varepsilon'} \zeta^{\varepsilon'} - \lambda^h B^{\varepsilon'} \zeta^{\varepsilon'} = g_{|_{\Omega^{\varepsilon'}}} \ ,$$

verifies

$$||\zeta^{\varepsilon'}||_{V_{\varepsilon'}} \le \frac{K}{\eta} ||g||_{0,\Omega^{\varepsilon'}} \le K_1 \ .$$

Hence

$$\mathbb{P}^{\varepsilon'} \zeta^{\varepsilon'} \rightharpoonup \zeta \quad \text{in} \quad V \text{ weakly.}$$

By using Lemma 4.3.19 we conclude that

$$A^h \zeta - \lambda^h B^h \zeta = \theta g \ .$$

This contradicts the Fredholm alternative, since

$$(A^h - \lambda^h B^h)(V) \cap L^2(\Omega) = \{g \in L^2(\Omega)| \int_\Omega g\varphi dx = 0 \ , \ \forall \, \varphi \in E(\lambda^h)\} \ .$$

The case of multiple eigenvalue can be treated similarly. $\qquad\square$

4.3.4.   Bifurcating branches of perforated and homogenized plates

The results of Sec. 4.3.2 can be applied to the perforated von Kármán plate. Thus for each $\varepsilon > 0$ the point $(\lambda_1^\varepsilon, 0)$ is a bifurcation point: there exists a bifurcating branch $(\lambda_1^\varepsilon(R),\ w_1^\varepsilon(R))$ defined for $R > 0$, such that

$$A^\varepsilon w_1^\varepsilon(R) - \lambda_1^\varepsilon(R)B^\varepsilon w_1^\varepsilon(R) + C^\varepsilon(w_1^\varepsilon(R)) = 0 \,,$$
$$\langle A^\varepsilon w_1^\varepsilon(R), w_1^\varepsilon(R)\rangle = 2R(1 + O(R)) \,, \qquad \lim_{R\to 0} \lambda_1^\varepsilon(R) = \lambda_1^\varepsilon \,. \tag{4.3.61}$$

The parameter $\lambda_1^\varepsilon(R)$ is the Lagrange multiplier associated with the maximization problem

$$\max\{\langle B^\varepsilon v, v\rangle|\, v \in \Sigma_R^\varepsilon\} \,,$$

where

$$\Sigma_R^\varepsilon = \{v \in V_\varepsilon|\frac{1}{2}\langle A^\varepsilon v, v\rangle + \frac{1}{4}\langle C^\varepsilon v, v\rangle = R\} \,.$$

The passage with ε to zero yields the following result.

Theorem 4.3.21. Let the homogenized problem $A^h w - \lambda B^h w$ admit a positive eigenvalue. Then the bifurcating branch $(\lambda_1^\varepsilon(R), w_1^\varepsilon(R))$ converges in $\mathbb{R}\times V$ weakly to $(\lambda_1^h(R), w_1^h(R))$, being the non-trivial solution of

$$A^h w_1^h(R) - \lambda_1^h(R)B^h w_1^h(R) + C^h(w_1^h(R)) = 0 \,.$$

Moreover,

$$\langle B^h w_1^h(R), w_1^h(R)\rangle = \max\{\langle B^h v, v\rangle|\, v \in \Sigma_R^h\} \,,$$

where

$$\Sigma_R^h = \{v \in V|\frac{1}{2}\langle A^h v, v\rangle + \frac{1}{4}\langle C^h v, v\rangle = R\} \,.$$

Proof. Let us fix R and set $w_1^\varepsilon(R) = w_1^\varepsilon$.

A priori estimates

Since $w_1^\varepsilon \in \Sigma_R^\varepsilon$ and $\langle C^\varepsilon w_1^\varepsilon, w_1^\varepsilon\rangle \geq 0$, we conclude that

$$\|w_1^\varepsilon\|_{V_\varepsilon} \leq 2R \,. \tag{4.3.62}$$

To find an estimate of $\lambda_1^\varepsilon(R)$ we write

$$\lambda_1^\varepsilon(R) = \frac{\langle A^\varepsilon w_1^\varepsilon, w_1^\varepsilon\rangle + \langle C^\varepsilon(w_1^\varepsilon), w_1^\varepsilon\rangle}{\langle B^\varepsilon w_1^\varepsilon, w_1^\varepsilon\rangle} \,. \tag{4.3.63}$$

Since $\langle A^\varepsilon w_1^\varepsilon, w_1^\varepsilon\rangle \leq 2R$ and $\langle C^\varepsilon(w_1^\varepsilon), w_1^\varepsilon\rangle \leq 4R$, to estimate $\lambda_1^\varepsilon(R)$ from above it suffices to minorize $\langle B^\varepsilon w_1^\varepsilon, w_1^\varepsilon\rangle$.

According to the assumption of the theorem, the problem $(A^h - \lambda B^h)$ admits a positive eigenvalue λ_1^h :

$$A^h \varphi_1^h - \lambda_1^h B^h \varphi_1^h = 0 \,, \qquad \langle A^h \varphi_1^h, \varphi_1^h \rangle = 0 \,.$$

Let t_ε be such that $t_\varepsilon \varphi_1^h \in \Sigma_R^\varepsilon$. We set $\varphi_1^{h\varepsilon} = \varphi_{1|\Omega^\varepsilon}^h$. Then

$$\frac{t_\varepsilon^2}{2} \langle A^\varepsilon \varphi_1^{h\varepsilon}, \varphi_1^{h\varepsilon} \rangle + \frac{t_\varepsilon^4}{4} \langle C^\varepsilon(\varphi_1^{h\varepsilon}), \varphi_1^{h\varepsilon} \rangle = R \,,$$

$$0 \leq \langle C^\varepsilon(\varphi_1^{h\varepsilon}), \varphi_1^{h\varepsilon} \rangle \leq K_1 \,, \qquad K_2 \leq \langle A^\varepsilon \varphi_1^{h\varepsilon}, \varphi_1^{h\varepsilon} \rangle \leq K_3 \,,$$

where K_1 is independent of ε. It can easily be shown that

$$\frac{2R}{K_2 + K_4 R} \leq t_\varepsilon^2 \leq \frac{2R}{K_2} \,, \qquad K_4 = K_1/K_2 \,.$$

Since

$$\langle B^\varepsilon \varphi_1^\varepsilon, \varphi_1^\varepsilon \rangle \geq t_\varepsilon^2 \langle B^\varepsilon \varphi_1^{h\varepsilon}, \varphi_1^{h\varepsilon} \rangle \,,$$

and

$$\lim_{\varepsilon \to 0} \langle B^\varepsilon \varphi_1^{h\varepsilon}, \varphi_1^{h\varepsilon} \rangle = \frac{1}{\lambda_1^h} \,,$$

we deduce that

$$\langle B^\varepsilon \varphi_1^\varepsilon, \varphi_1^\varepsilon \rangle \geq K_5 \frac{2R}{K_2 + K_4 R} \,.$$

Thus we eventually obtain

$$0 < \lambda_1^\varepsilon(R) < K_6 \,. \tag{4.3.64}$$

Passage to the limit
(i) On account of (4.3.62) and (4.3.64) one can extract a subsequence, still denote by ε, such that

$$\lim_{\varepsilon \to 0} \lambda_1^\varepsilon(R) = \mu \,, \qquad \lim_{\varepsilon \to 0} \mathbf{P}^\varepsilon w_1^\varepsilon = w_1^h \qquad \text{in } V \text{ weakly} \,.$$

(ii) Let $\mathbf{R}^\varepsilon$ be the restriction operator from V to V_ε, $\mathbf{R}^\varepsilon v = v_{|\Omega^\varepsilon}$. The dual operator, sometimes called the adjoint operator, $\mathbf{R}_*^\varepsilon : V_\varepsilon^* \to V^*$ has the following properties:
1. If $f \in L^p(\Omega^\varepsilon)$, $\mathbf{R}_*^\varepsilon(f) = \widetilde{f}$ where

$$\widetilde{f}(x) = \begin{cases} f(x) & \text{for } x \in \Omega^\varepsilon \,, \\ 0 & \text{for } x \in \Omega \backslash \Omega^\varepsilon \,. \end{cases}$$

2. $\mathbf{R}_*^\varepsilon(f_{,\alpha}) = \dfrac{\partial \widetilde{f}}{\partial x_\alpha}.$

3. $\mathbf{R}_*^\varepsilon(f_{,\alpha\beta}) = \dfrac{\partial^2 \widetilde{f}}{\partial x_\alpha \partial x_\beta}.$

4. There exists a constant $K > 0$ such that

$$||\mathbb{R}^\varepsilon_*||_{L(V^*_\varepsilon, V^*)} \leq K \, ,$$

for each $\varepsilon > 0$.

When ε tends to zero, we have

$$\lim_{\varepsilon \to 0} \mathbb{R}^\varepsilon_*(B^\varepsilon w_1^\varepsilon) = B^h w_1^h \, , \qquad \text{in} \quad V^* \text{ strongly.}$$

Indeed, for $v \in V$ we write

$$\langle \mathbb{R}^\varepsilon_* B^\varepsilon w_1^\varepsilon, v \rangle = -\int_{\Omega^\varepsilon} N_\varepsilon^{\alpha\beta} \frac{\partial w_1^\varepsilon}{\partial x_\beta} \cdot \frac{\partial v}{\partial x_\alpha} dx = -\int_\Omega \widetilde{N}_\varepsilon^{\alpha\beta} \left[\frac{\partial}{\partial x_\beta}(\mathbb{P}^\varepsilon w_1^\varepsilon) \right] \frac{\partial v}{\partial x_\alpha} dx \, .$$

Since

$$\widetilde{N}_\varepsilon^{\alpha\beta} \rightharpoonup N_h^{\alpha\beta} \quad \text{in } L^2(\Omega) \text{ weakly} \, ,$$

and

$$\frac{\partial}{\partial x_\beta}(\mathbb{P}^\varepsilon w_1^\varepsilon) \to \frac{\partial w_1^h}{\partial x_\beta} \qquad \text{in } L^p(\Omega) \text{ strongly (for each } p < \infty) \, ,$$

therefore

$$\widetilde{N}_\varepsilon^{\alpha\beta} \frac{\partial}{\partial x_\beta}(\mathbb{P}^\varepsilon w_1^\varepsilon) \rightharpoonup N_h^{\alpha\beta} \frac{\partial w_1^h}{\partial x_\beta} \quad \text{in } L^q(\Omega) \text{ weakly, } q < 2 \, .$$

Consequently, $\dfrac{\partial}{\partial x_\alpha} \left[\widetilde{N}_\varepsilon^{\alpha\beta} \dfrac{\partial}{\partial x_\beta}(\mathbb{P}^\varepsilon w_1^\varepsilon) \right]$ converges to $B^h w_1^h$ in $W^{-1,q}(\Omega)$ weakly and hence in V^* strongly.

(iii) Similarly, it can be shown that

$$\mathbb{R}^\varepsilon_* C^\varepsilon(w^\varepsilon) \to C^h(w_1^h) \qquad \text{in } V^* \text{ strongly as } \varepsilon \to 0 \, .$$

We recall that

$$\langle \mathbb{R}^\varepsilon_* C^\varepsilon(w_1^\varepsilon), v \rangle = \int_{\Omega^\varepsilon} S_\varepsilon^{\alpha\beta}(w_1^\varepsilon) \frac{\partial w_1^\varepsilon}{\partial x_\beta} \frac{\partial v}{\partial x_\alpha} dx$$

$$= \int_\Omega \widetilde{S}_\varepsilon^{\alpha\beta} \left[\frac{\partial}{\partial x_\beta}(\mathbb{P}^\varepsilon w_1^\varepsilon) \right] \frac{\partial v}{\partial x_\alpha} dx \, , \quad v \in V \, , \qquad (4.3.65)$$

where $S_\varepsilon^{\alpha\beta}(w_1^\varepsilon)$ is the solution of the following problem:

$$\boldsymbol{S}_\varepsilon \in L^2(\Omega^\varepsilon, \mathbb{E}_2^s) : \int_{\Omega^\varepsilon} S_\varepsilon^{\alpha\beta}(w_1^\varepsilon) e_{\alpha\beta}(\boldsymbol{v}) dx = 0 \, , \quad \forall \, \boldsymbol{v} \in V^\varepsilon_{\Gamma_{00}} \, ,$$

$$S_\varepsilon^{\alpha\beta}(w_1^\varepsilon) = A^{\alpha\beta\gamma\mu} \left[e_{\lambda\mu}(\boldsymbol{u}^\varepsilon) + \frac{1}{2} \frac{\partial w_1^\varepsilon}{\partial x_\lambda} \frac{\partial w_1^\varepsilon}{\partial x_\mu} \right] \, . \tag{4.3.66}$$

The space $V_{\Gamma_{00}}^\varepsilon$ is defined by (4.3.19) with Ω being replaced by Ω^ε. We already knew that $\mathbb{P}^\varepsilon w_1^\varepsilon$ tends to w_1^h weakly in V, cf. step (i). Hence

$$w_{1,\alpha}^\varepsilon w_{1,\beta}^\varepsilon - w_{1,\alpha}^h w_{1,\beta}^h \to 0 \qquad \text{in } L^2(\Omega^\varepsilon) \text{ when } \varepsilon \to 0 .$$

Homogenization yields

$$\widetilde{S}_\varepsilon^{\alpha\beta}(w_1^\varepsilon) \rightharpoonup S^{\alpha\beta}(w_1^h) \qquad \text{in } L^2(\Omega, \mathbb{E}_2^s) \text{ weakly as } \varepsilon \to 0 ,$$

where $\boldsymbol{S}(w_1^h)$ is the solution of the homogenized problem:

$$\boldsymbol{S}_h(w_1^h) \in L^2(\Omega, \mathbb{E}_2^s) : \int_\Omega S_h^{\alpha\beta}(w_1^h) e_{\alpha\beta}(\boldsymbol{v}) dx = 0 , \qquad \forall\, \boldsymbol{v} \in V_{\Gamma_{00}} ,$$

$$S_h^{\alpha\beta}(w_1^h) = A_h^{\alpha\beta\lambda\mu}[e_{\lambda\mu}(\boldsymbol{u}^{ho}) + w_{1,\lambda}^h w_{1,\mu}^h] . \tag{4.3.67}$$

Here $\boldsymbol{u}^{ho}$ is the solution of

$$\int_\Omega A_h^{\alpha\beta\lambda\mu} e_{\lambda\mu}(\boldsymbol{u}^{ho}) e_{\alpha\beta}(\boldsymbol{v}) dx = \int_{\Gamma_0 \cup \Gamma_1} \widetilde{N}^\alpha v_\alpha d\Gamma , \qquad \forall\, \boldsymbol{v} \in V_{\Gamma_{00}} , \tag{4.3.68}$$

and Eq. (4.3.23) is to be replaced by

$$N_{ho}^{\alpha\beta} = A_h^{\alpha\beta\lambda\mu} e_{\lambda\mu}(\boldsymbol{u}^{ho}) . \tag{4.3.69}$$

(iv) The previous steps imply that

$$\mathbb{R}_*^\varepsilon[-\lambda_1^\varepsilon(R) B^\varepsilon w_1^\varepsilon + C^\varepsilon(w_1^\varepsilon)] \to -\nu B^h w_1^h + C^h(w_1^h)$$

in V^* strongly as $\varepsilon \to 0$. Hence we conclude that

$$A^h w_1^h - \nu B^h w_1^h + C^h(w_1^h) = 0 .$$

It thus remains to show that $w_1^h \neq 0$, $w_1^h \in \Sigma_R^h$, and that

$$\langle B^h w_1^h, w_1^h \rangle = \max\{\langle B^h w, w \rangle | w \in \Sigma_R^h\} . \tag{4.3.70}$$

This will imply that $\nu = \lambda_1^h(R)$ is the Lagrange multiplier corresponding to w_1^h.

Since $\mathbb{R}_*^\varepsilon A^\varepsilon w_1^\varepsilon$ converges to $A^h w_1^h$ in V^* strongly, we have

$$\lim_{\varepsilon \to 0} \langle A^\varepsilon w_1^\varepsilon, w_1^\varepsilon \rangle = \lim_{\varepsilon \to 0} \langle \mathbb{R}_*^\varepsilon A^\varepsilon w_1^\varepsilon, \mathbb{P}^\varepsilon w_1^\varepsilon \rangle = \langle A^h w_1^h, w_1^h \rangle .$$

Passing to the limit in

$$\frac{1}{2} \langle A^\varepsilon w_1^\varepsilon, w_1^\varepsilon \rangle + \frac{1}{4} \langle C^\varepsilon w_1^\varepsilon, w_1^\varepsilon \rangle = R , \tag{4.3.71}$$

we get

$$\frac{1}{2} \langle A^h w_1^h, w_1^h \rangle + \frac{1}{4} \langle C^h w_1^h, w_1^h \rangle = R . \tag{4.3.72}$$

Hence $w_1^h \in \Sigma_R^h$, and particularly $w_1^h \neq 0$.

It remains to prove (4.3.70). Let (λ, w) be the solution of

$$A^h w^0 - \lambda^h B^h w^0 + C^h(w^0) = 0 \,,$$
$$\langle B^h w^0, w^0 \rangle = \sup\{\langle B^h w, w \rangle | w \in \Sigma_R^h\} \,. \tag{4.3.73}$$

We define $T^h \in V^*$ by

$$\langle T^h, v \rangle = \lambda^h \int_\Omega N_0^{\alpha\beta} w^0_{,\beta} v_{,\alpha} dx - \int_\Omega S^{\alpha\beta}(w^0) w^0_{,\beta} v_{,\alpha} dx \,, \qquad \forall\, v \in V \,.$$

Similarly, $T^\varepsilon \in V_\varepsilon^*$ is defined by

$$\langle T^h, v \rangle = \frac{1}{\theta}[\lambda^h \int_{\Omega^\varepsilon} N_0^{\alpha\beta} w^0_{,\beta} v_{,\alpha} dx - \int_{\Omega^\varepsilon} S^{\alpha\beta}(w^0) w^0_{,\beta} v_{,\alpha} dx \,, \qquad \forall\, v \in V_\varepsilon \,.$$

Here $N_0^{\alpha\beta} = A_h^{\alpha\beta\lambda\mu} e_{\lambda\mu}(u^0)$, and u^0 is the solution of (4.3.24) with A being replaced by A_h. It is clear that $\mathbb{R}_*^\varepsilon T^\varepsilon$ tends to T^h in V^* strongly when $\varepsilon \to 0$.

Let ξ^ε be the solution of

$$A^\varepsilon \xi^\varepsilon = T^\varepsilon \,.$$

It is obvious that $\mathbb{P}^\varepsilon \xi^\varepsilon$ tends to ξ^0, where ξ^0 is the solution of

$$A^h \xi^0 = T^h \,.$$

Hence $\xi^0 = w^0$.

There exists $t_\varepsilon > 0$ such that $t_\varepsilon \xi^\varepsilon \in \Sigma_R^\varepsilon$:

$$\frac{t_\varepsilon^2}{2}\langle A^\varepsilon \xi^\varepsilon, \xi^\varepsilon \rangle + \frac{t_\varepsilon^4}{4}\langle C^\varepsilon(\xi^\varepsilon), \xi^\varepsilon \rangle = R \,. \tag{4.3.74}$$

We observe that

$$\lim_{\varepsilon \to 0}\langle A^\varepsilon \xi^\varepsilon, \xi^\varepsilon \rangle = \langle A^h w^0, w^0 \rangle \,, \qquad \lim_{\varepsilon \to 0}\langle C^\varepsilon(\xi^\varepsilon), \xi^\varepsilon \rangle = \langle C^h(w^0), w^0 \rangle \,.$$

It is not difficult to show that the positive solution of (4.3.74) tends to the positive solution of

$$\frac{t_0^2}{2}\langle A^h w^0, w^0 \rangle + \frac{t_0^4}{4}\langle C^h(w^0), w^0 \rangle = R \,.$$

We conclude that $t_0 = 1$ and $w^0 = w_1^h$. This completes the proof. $\qquad\square$

Remark 4.3.22.
(i) The spectrum of $(A^h - \lambda B^h)$ always has a positive or negative eigenvalue and the bifurcating branch at this point depends on ε in the manner described by the last theorem. The geometry of the plate and the anisotropic properties of the homogenized moduli strongly influence this spectrum.

(ii) The method based on maximization of the functional $\frac{1}{2}\langle B, w, w\rangle$ over Σ_R (or $\frac{1}{2}\langle B^h, w, w\rangle$ over Σ_R^h) is due to Berger (1967). $\square$

Kikuchi's method

Berger's method just presented permits one to obtain only the bifurcation branch corresponding to the first positive or negative eigenvalue. We observe that the solution thus obtained realizes the minimum of the energy and is stable in the sense of Lyapunov.

Our aim now is to show that the bifurcating branch in a neighborhood of a *simple eigenvalue* of the homogenized problem is the limit of the bifurcating branch (at the corresponding simple eigenvalue) of the perforated plate. The approach used exploits an idea due to Kikuchi (1976), cf. also Kesavan (1974). It is based on a parametrization of the bifurcating branch.

Let λ_h be a simple eigenvalue of the problem $(A^h - \lambda B^h)$ and w^h an associated eigenvector:

$$A^h \varphi^h - \lambda_h B^h \varphi^h = 0 , \qquad \langle B^h \varphi^h, \varphi^h \rangle = \operatorname{sign} \lambda_h . \tag{4.3.75}$$

We introduce the following spaces

$$U_h = \{v \in V | \langle B^h v, \varphi^h \rangle = 0\} , \tag{4.3.76}$$
$$U_h^* = \{f \in V^* | \langle f, \varphi^h \rangle = 0\} . \tag{4.3.77}$$

By Fredholm's alternative the operator $(A^h - \lambda_h B^h)$ realizes an isomorphism between U_h and U_h^*.

On account of Theorem 4.3.20 there exist $\eta > 0$ and $\varepsilon_0 > 0$ such that for each ε, $0 < \varepsilon < \varepsilon_0$, the problem $(A^\varepsilon - \lambda B^\varepsilon)$ admits only one simple eigenvalue λ_ε, $\lambda_\varepsilon \in (\lambda_h - \eta, \lambda_h + \eta)$ associated with an eigenvector φ^ε such that

$$A^\varepsilon \varphi^\varepsilon - \lambda_\varepsilon B^\varepsilon \varphi^\varepsilon = 0 , \qquad \langle B^\varepsilon \varphi^\varepsilon, \varphi^\varepsilon \rangle = \operatorname{sign} \lambda_\varepsilon = \operatorname{sign} \lambda_h , \tag{4.3.78}$$

and

$$\lim_{\varepsilon \to 0} \lambda_\varepsilon = \lambda_h , \qquad \lim_{\varepsilon \to 0} \mathbb{P}^\varepsilon \varphi^\varepsilon = \varphi^h \quad \text{in } V \text{ weakly} . \tag{4.3.79}$$

We also introduce the following spaces

$$U_\varepsilon = \{v \in V_\varepsilon | \langle B^\varepsilon v, \varphi^\varepsilon \rangle = 0\} , \qquad U_\varepsilon^* = \{f \in V_\varepsilon^* | \langle f, \varphi^\varepsilon \rangle = 0\} .$$

Thus we have the orthogonal decomposition of V_ε for the scalar product generated by $\langle A^\varepsilon \xi, \varphi \rangle$:

$$V_\varepsilon = \mathbb{R} w^\varepsilon \oplus U_\varepsilon ,$$

or

$$w = t\varphi^\varepsilon + v , \qquad t \in \mathbb{R} .$$

For $\delta > 0$ we set

$$\mathcal{N}_\varepsilon(\delta) = \{(\lambda, w) \in \mathbb{R} \times V_\varepsilon | \ w = t\varphi^\varepsilon + v \,, \ |\lambda - \lambda_\varepsilon| < \delta \,, \ |t| < \delta \,, \ ||v||_{V_\varepsilon} < \delta \} \,.$$

It is evident that $\mathcal{N}_\varepsilon(\delta)$ is a neighborhood of $(\lambda_\varepsilon, 0)$ in $\mathbb{R} \times V_\varepsilon$. Similar interpretation holds for $\mathcal{N}_h(\delta)$ in $\mathbb{R} \times V$. We recall that the problem (P_ε) refers to the perforated plate whilst (P_h) to the homogenized one. The following convergence result shows that the bifurcating branch of (P_ε) converges to the bifurcating branch of the homogenized problem. This convergence takes place in $C^{0,\alpha}(\Omega)$ $(\alpha < 1)$.

Theorem 4.3.23. There exist $\delta > 0$ and $\varepsilon_0 > 0$ such that for each ε, $0 < \varepsilon < \varepsilon_0$, the bifurcating branch of the problem (P_ε) at $(\lambda_\varepsilon, 0)$ admits in the neighborhood $\mathcal{N}_\varepsilon(\delta)$ of this point the following continuous parametrization:

$$t \to (\lambda^\varepsilon(t) \,, \ w^\varepsilon(t) = t\varphi^\varepsilon + v^\varepsilon(t)) \,,$$

the function $t \to v^\varepsilon(t)$ assumes its values in U_ε and $(\lambda^\varepsilon, w^\varepsilon)$ satisfies

$$A^\varepsilon w^\varepsilon(t) - \lambda^\varepsilon(t) B^\varepsilon w^\varepsilon(t) + C^\varepsilon(w^\varepsilon(t)) = 0 \,,$$
$$\lambda^\varepsilon(0) = \lambda_\varepsilon \,, \quad w^\varepsilon(0) = 0 \,. \tag{4.3.80}$$

When ε tends to zero, for each t in $(-\delta, \delta)$, $(\lambda^\varepsilon(t), \mathbb{P}^\varepsilon v^\varepsilon(t))$ tends to $(\lambda(t), v(t))$ in $\mathbb{R} \times V$ weakly. The function $t \to (\lambda(t), t\varphi^h + v(t) = w(t))$ is the continuous parametrization of the bifurcating branch of the problem (P_h) at $(\lambda_h, 0)$ in the neighborhood $\mathcal{N}_h(\delta)$ of this point:

$$A^h w^h(t) - \lambda^h(t) B^h w^h(t) + C^h(w^h(t)) = 0 \,.$$

Moreover, the above representations give all non-zero solutions of (P_ε) (resp. (P_h)) in $\mathcal{N}_\varepsilon(\delta)$ (resp. $\mathcal{N}_h(\delta)$). $\qquad\square$

Prior to passing to the proof we formulate four lemmas. We assume the following notation: an assertion valid for $\varepsilon \in [0, \varepsilon_0]$ will mean that it is true for the problems (P_ε), $\varepsilon \in (0, \varepsilon_0]$ as well as for the homogenized problem (P_h). To simplify our subsequent considerations we assume that $\lambda_h > 0$, and consequently $\langle B^\varepsilon \varphi^\varepsilon, \varphi^\varepsilon \rangle = 1$.

Lemma 4.3.24. For each $q < +\infty$ there exists a constant K_q such that for each $\varepsilon \in [0, \varepsilon_0]$, and each $w \in H^2(\Omega^\varepsilon)$ one has

$$||w||_{W^{1,q}(\Omega^\varepsilon)} \leq K_q ||w||_{H^2(\Omega^\varepsilon)} \,.$$

Proof. By using the properties of the extension operator $\mathbb{P}^\varepsilon$ we get, see Lemma 4.3.2

$$||w||_{W^{1,q}(\Omega^\varepsilon)} \leq ||\mathbb{P}^\varepsilon w||_{W^{1,q}(\Omega^\varepsilon)} \leq K'_q(\Omega) ||\mathbb{P}^\varepsilon w||_{H^2(\Omega)} \leq K_q ||w||_{H^2(\Omega^\varepsilon)} \,. \qquad\square$$

The nonlinear behavior of C^ε is made precise by the next lemma.

Lemma 4.3.25. There exist a constant $K > 0$ independent of $\varepsilon \in [0, \varepsilon_0]$ such that

$$||C^\varepsilon(w) - C^\varepsilon(v)||_{V_\varepsilon^*} \leq K(||w||_{V_\varepsilon} + ||v||_{V_\varepsilon})^2||w - v||_{V_\varepsilon} , \quad \forall\, v, w \in V_\varepsilon .$$

Proof. It suffices to consider the case $\varepsilon > 0$. Recalling that, cf. (4.3.31)

$$\langle C^\varepsilon(w), \xi \rangle = \int_{\Omega^\varepsilon} S_\varepsilon^{\alpha\beta}(w)w_{,\beta}v_{,\alpha}dx , \quad \forall\, w, v \in V_\varepsilon$$

we get

$$||C^\varepsilon(w) - C^\varepsilon(v)||_{V_\varepsilon^*} \leq \sup_{\substack{||\xi|| \leq 1 \\ \xi \in V_\varepsilon}} ||(S_\varepsilon^{\alpha\beta}(w)w_{,\beta} - S_\varepsilon^{\alpha\beta}(v)v_{,\beta})||_{L^q(\Omega^\varepsilon)^2}||\nabla\xi||_{L^{q^*}(\Omega^\varepsilon)^2} ,$$

where q is fixed, $q \in (1, 2)$, $q^* = \dfrac{q}{q-1}$.

The continuous embedding $H^2(\Omega^\varepsilon) \subset W^{1,q^*}(\Omega^\varepsilon)$ and Lemma 4.3.24 yield

$$||C^\varepsilon(w) - C^\varepsilon(v)||_{V_\varepsilon^*} \leq K_1||(S_\varepsilon^{\alpha\beta}(w)w_{,\beta} - S_\varepsilon^{\alpha\beta}(v)v_{,\beta})||_{L^q(\Omega^\varepsilon)^2} .$$

Next we have

$$||(S_\varepsilon^{\alpha\beta}(w)w_{,\beta} - S_\varepsilon^{\alpha\beta}(v)v_{,\beta})||_{L^q(\Omega^\varepsilon)^2} \leq ||[(S_\varepsilon^{\alpha\beta}(w) - S_\varepsilon^{\alpha\beta}(v))w_{,\beta}]||_{L^q(\Omega^\varepsilon)^2}$$
$$+ ||[(S_\varepsilon^{\alpha\beta}(v)(w_{,\beta} - v_{,\beta})]||_{L^q(\Omega^\varepsilon)^2} .$$

By the previous lemma we get

$$||(S_\varepsilon^{\alpha\beta}(v)(w_{,\beta} - v_{,\beta}))||_{L^q(\Omega^\varepsilon)^2} \leq K_2||[S_\varepsilon^{\alpha\beta}(v)]||_{L^2(\Omega^\varepsilon, \mathbf{E}_2^s)}||w - v||_{V_\varepsilon} .$$

Moreover

$$||\boldsymbol{S}_\varepsilon(v)||_{L^2(\Omega^\varepsilon, \mathbf{E}_2^s)} \leq K_3||v||_{V_\varepsilon} ,$$
$$||\boldsymbol{S}_\varepsilon(w) - \boldsymbol{S}_\varepsilon(v)||_{L^2(\Omega^\varepsilon, \mathbf{E}_2^s)} \leq K_4||(w_{,\alpha}w_{,\beta} - v_{,\alpha}v_{,\beta})||_{L^2(\Omega^\varepsilon, \mathbf{E}_s^2)}$$
$$\leq K_5(||w||_{V_\varepsilon} + ||v||_{V_\varepsilon})||w - v||_{V_\varepsilon} ,$$

since $\nabla w \otimes \nabla w - \nabla v \otimes \nabla v = (\nabla w + \nabla v) \otimes (\nabla w - \nabla v)$. This completes the proof. $\square$

To exhibit the nontrivial solutions of (P_ε) in the neighborhood of $(\lambda_\varepsilon, 0)$ we may set

$$w^\varepsilon = t\varphi^\varepsilon + v^\varepsilon , \quad v^\varepsilon \in U_\varepsilon .$$

We are going to show that these solutions are parametrized by t.

Recalling that $A^\varepsilon\varphi^\varepsilon - \lambda_\varepsilon B^\varepsilon\varphi^\varepsilon = 0$, we write the problem (P_ε) in the following form

$$A^\varepsilon v^\varepsilon - \lambda_\varepsilon B^\varepsilon v^\varepsilon = -C^\varepsilon(t\varphi^\varepsilon + v^\varepsilon) + (\lambda^\varepsilon - \lambda_\varepsilon)B^\varepsilon(t\varphi^\varepsilon + v^\varepsilon) .$$

The Fredholm alternative implies that this problem admits a solution if and only if

$$(\lambda^\varepsilon - \lambda_\varepsilon)\langle B^\varepsilon(t\varphi^\varepsilon + v^\varepsilon), \varphi^\varepsilon \rangle - \langle C^\varepsilon(t\varphi^\varepsilon + v^\varepsilon), \varphi^\varepsilon \rangle = 0 .$$

Hence

$$\lambda^\varepsilon - \lambda_\varepsilon = \frac{\langle C^\varepsilon(t\varphi^\varepsilon + v^\varepsilon), \varphi^\varepsilon\rangle}{t} \,,$$

since $\langle B^\varepsilon \varphi^\varepsilon, \varphi^\varepsilon\rangle = \text{sign}\, \lambda_\varepsilon = +1$, cf. (4.3.78). Thus the problem (P_ε) is equivalent to

$$A^\varepsilon v^\varepsilon - \lambda_\varepsilon B^\varepsilon v^\varepsilon = -C^\varepsilon(t\varphi^\varepsilon + v^\varepsilon) + \frac{1}{t}\langle C^\varepsilon(t\varphi^\varepsilon + v^\varepsilon), \varphi^\varepsilon\rangle B^\varepsilon(t\varphi^\varepsilon + v^\varepsilon) \,,$$

$$(4.3.81)$$

$$\lambda^\varepsilon = \lambda_\varepsilon + \frac{1}{t}\langle C^\varepsilon(t\varphi^\varepsilon + v^\varepsilon), \varphi^\varepsilon\rangle \,.$$

Suppose that t is fixed. Below it is shown that Eq. $(4.3.81)_1$ admits a unique solution v^ε in a neighborhood of 0. To this end a fixed point method is used. Then Eq. $(4.3.81)_2$ yields λ^ε as a function of t. We set

$$H_{\varepsilon,t}(v) = -C^\varepsilon(t\varphi^\varepsilon + v) + \frac{1}{t}\langle C^\varepsilon(t\varphi^\varepsilon + v), \varphi^\varepsilon\rangle B^\varepsilon(t\varphi^\varepsilon + v) \,. \qquad (4.3.82)$$

Since $(A^\varepsilon - \lambda_\varepsilon B^\varepsilon)$ is an isomorphism between U_ε and U_ε^*, Eq. $(4.3.81)_1$ can be written as follows

$$v = (A^\varepsilon - \lambda_\varepsilon B^\varepsilon)^{-1} H_{\varepsilon,t}(v) = T_{\varepsilon,t}(v) \,. \qquad (4.3.83)$$

The first estimate of v^ε is given by the following lemma.

Lemma 4.3.26. Let $k \geq 1$. There exists $\tau(k) > 0$, independent of $\epsilon \in [0, \varepsilon_0]$, such that each solution of $(P_\varepsilon) : (\lambda^\varepsilon, w^\varepsilon)$, $w^\varepsilon = t\varphi^\varepsilon + v^\varepsilon$, which belongs to

$$\mathcal{N}(\tau(k), k) = \{(\lambda^\varepsilon, w^\varepsilon)|\ |\lambda^\varepsilon - \lambda_\varepsilon| \leq \tau(k),\ |t| \leq \tau(k),\ ||v^\varepsilon||_{V_\varepsilon} \leq k\tau(k)\} \,,$$

satisfies

$$||v^\varepsilon||_{V_\varepsilon} \leq t \,.$$

Proof. There exists a constant $M > 0$ independent of ε such that

$$||(A^\varepsilon - \lambda_\varepsilon B^\varepsilon)^{-1}||_{\mathbf{L}(U_\varepsilon^*, U_\varepsilon)} \leq M \,, \quad ||B^\varepsilon||_{\mathbf{L}(V_\varepsilon, V_\varepsilon^*)} \leq M \,,$$
$$||C^\varepsilon(w^\varepsilon)||_{V_\varepsilon^*} \leq M||w^\varepsilon||_{V_\varepsilon}^3 \,.$$

By Lemma 4.3.25 and Eqs. (4.3.82), (4.3.83) we write

$$||v^\varepsilon||_{V_\varepsilon} \leq M(||t\varphi^\varepsilon + v^\varepsilon||_{V_\varepsilon}^2 + |\lambda^\varepsilon - \lambda_\varepsilon|)(|t| + ||v^\varepsilon||_{V_\varepsilon}) \,,$$

or

$$||v^\varepsilon||_{V_\varepsilon} - M(||t\varphi^\varepsilon + v^\varepsilon||_{V_\varepsilon}^2 + |\lambda^\varepsilon - \lambda_\varepsilon|)||v^\varepsilon||_{V_\varepsilon}$$
$$\leq M(||t\varphi^\varepsilon + v^\varepsilon||_{V_\varepsilon}^2 + |\lambda^\varepsilon - \lambda_\varepsilon|)|t| \,.$$

Hence, for $|t| \leq \tau$, $||v^\varepsilon||_{V_\varepsilon} \leq k\tau$, $|\lambda^\varepsilon - \lambda_\varepsilon| \leq \tau$ one has

$$||v^\varepsilon||_{V_\varepsilon} \leq \frac{M[(1+k)^2\tau^2 + \tau]}{1 - M[(1+k)^2\tau^2 + \tau]}|t| \,. \tag{4.3.84}$$

It is thus sufficient to take for $\tau(k)$ the positive root of

$$2M(1+k)^2\tau^2 + 2M\tau - 1 = 0 \,,$$

and the lemma follows. $\square$

Remark 4.3.27. Analysis similar to that in the last proof shows that

$$\lim_{t \to 0} \frac{||v^\varepsilon||_{V_\varepsilon}}{t} = 0 \,, \tag{4.3.85}$$

the limit being uniform in ε: the bifurcating branches are uniformly tangent to φ^ε at $t = 0$. $\square$

For the remaining part of the present section we take

$$k = \sup_{\varepsilon \in [0,\varepsilon_0]} ||\mathbf{P}^\varepsilon||_{\mathbf{L}(V_\varepsilon, V)} \,. \tag{4.3.86}$$

This choice will be justified by the study of the homogenized bifurcating branch.

We now turn to giving the contraction property of the operator $T_{\varepsilon,t}$ defined by (4.3.83).

Lemma 4.3.28. There exists a constant $K > 0$, independent of ε ($\varepsilon < \varepsilon_0$), such that for each t and for each $v, \widehat{v} \in U_\varepsilon$, $||v||_{V_\varepsilon} \leq t$, $||\widehat{v}||_{V_\varepsilon} \leq t$ one has

$$||T_{\varepsilon,t}(v) - T_{\varepsilon,t}(\widehat{v})||_{V_\varepsilon} \leq Kt^2||v - \widehat{v}||_{V^\varepsilon} \,. \tag{4.3.87}$$

Proof. By (4.3.82) we have

$$H_{\varepsilon,t}(v) - H_{\varepsilon,t}(\widehat{v}) = -C^\varepsilon(t\varphi^\varepsilon + v) + C^\varepsilon(t\varphi^\varepsilon + \widehat{v})$$

$$+\frac{1}{t}[\langle C^\varepsilon(t\varphi^\varepsilon + v), \varphi^\varepsilon\rangle B^\varepsilon(t\varphi^\varepsilon + v) - \langle C^\varepsilon(t\varphi^\varepsilon + \widehat{v}), \varphi^\varepsilon\rangle B^\varepsilon(t\varphi^\varepsilon + \widehat{v})] \,.$$

Application of Lemma 4.3.25 gives

$$||C^\varepsilon(t\varphi^\varepsilon + v) - C^\varepsilon(t\varphi^\varepsilon + \widehat{v})||_{V_\varepsilon^*} \leq K_1 t^2||v - \widehat{v}||_{V_\varepsilon} \,, \tag{4.3.88}$$

where $K_1 > 0$ is independent of ε. Moreover

$$X_\varepsilon = \langle C^\varepsilon(t\varphi^\varepsilon + v), \varphi^\varepsilon\rangle B^\varepsilon(t\varphi^\varepsilon + v) - \langle C^\varepsilon(t\varphi^\varepsilon + \widehat{v}), \varphi^\varepsilon\rangle B^\varepsilon(t\varphi^\varepsilon + \widehat{v})$$

$$= \langle C^\varepsilon(t\varphi^\varepsilon + v), \varphi^\varepsilon\rangle B^\varepsilon(v - \widehat{v}) + \langle C^\varepsilon(t\varphi^\varepsilon + v) - C^\varepsilon(t\varphi^\varepsilon + \widehat{v}), \varphi^\varepsilon\rangle B^\varepsilon(t\varphi^\varepsilon + \widehat{v}) \,.$$

Hence

$$||X_\varepsilon||_{V_\varepsilon^*} \leq K_2 t^3 ||v - \widehat{v}||_{V_\varepsilon} + K_3 t^3 ||v - \widehat{v}||_{V_\varepsilon} . \tag{4.3.89}$$

We conclude from (4.3.88) and (4.3.89) that

$$||H_{\varepsilon,t}(v) - H_{\varepsilon,t}(\widehat{v})||_{V_\varepsilon^*} \leq K_4 t^2 ||v - \widehat{v}||_{V_\varepsilon} . \tag{4.3.90}$$

Since the operators $(A^\varepsilon - \lambda_\varepsilon B^\varepsilon)^{-1}$ are uniformly bounded in $\mathbf{L}(U_\varepsilon^*, U_\varepsilon)$, the inequality (4.3.87) results from (4.3.90). $\qquad\square$

Proof of Theorem 4.3.23. By Lemmas 4.3.26 and 4.3.28, one can choose τ independent of ε such that if $(\lambda^\varepsilon, w^\varepsilon = t\varphi^\varepsilon + v^\varepsilon)$ belongs to $\mathcal{N}_\varepsilon(\tau)$ and is the solution to (P_ε) then $||v^\varepsilon||_{V_\varepsilon} \leq t$. Taking $\tau^2 \leq 1/2M$, v^ε belongs to the contraction domain of $T_{\varepsilon,t}$. Consequently Equation $(4.3.81)_1$ admits a unique solution $v^\varepsilon(t)$, $||v^\varepsilon(t)||_{V_\varepsilon} < \tau$, for each $t \in (-\tau, +\tau)$. Moreover, the contraction $T_{\varepsilon,t}$ being uniform with respect to t, the function $v^\varepsilon(t)$ is continuous. Equation $(4.3.81)_2$ gives $\lambda^\varepsilon(t)$. This proves the first part of the theorem provided that $\delta \leq \tau$.

It remains to investigate the behavior of the bifurcating branch when ε tends to zero. Let us fix $t \in (-\tau, +\tau)$. By Lemma 4.3.26 we have

$$||v^\varepsilon(t)||_{V_\varepsilon} \leq |t| , \quad ||\mathbf{P}^\varepsilon v^\varepsilon(t)||_V \leq k|t| ,$$
$$||w^\varepsilon(t)||_{V_\varepsilon} \leq 2|t| , \quad ||\mathbf{P}^\varepsilon w^\varepsilon(t)||_V \leq 2k|t| . \tag{4.3.91}$$

Since

$$\mathbf{P}^\varepsilon w^\varepsilon(t) = t\mathbf{P}^\varepsilon\varphi^\varepsilon + \mathbf{P}^\varepsilon v^\varepsilon(t) , \tag{4.3.92}$$

for a subsequence, still denoted by ε, one has

$$\mathbf{P}^\varepsilon\varphi^\varepsilon \rightharpoonup \varphi^h , \qquad \mathbf{P}^\varepsilon v^\varepsilon(t) \rightharpoonup v^h(t) , \qquad \mathbf{P}^\varepsilon w^\varepsilon(t) \rightharpoonup t\varphi^h + w^h(t) ,$$

in V weakly as $\varepsilon \to 0$.

Let us show that

$$\mathbb{R}_*^\varepsilon B^\varepsilon w^\varepsilon \to B^h w^h \quad \text{in } V^* \text{ strongly} ,$$
$$\mathbb{R}_*^\varepsilon(C^\varepsilon(w^\varepsilon)) \to C^h(w^h) \quad \text{in } V^* \text{ strongly} \tag{4.3.93}$$

as $\varepsilon \to 0$.

Indeed, since $\mathbf{P}^\varepsilon w^\varepsilon$ tends weakly to w^h in V, it is not difficult to show that (4.3.93) really holds true. Consequently, application of Lemma 4.3.19, now slightly modified, to

$$A^\varepsilon w^\varepsilon(t) = \lambda^\varepsilon(t)B^\varepsilon w^\varepsilon(t) - C^\varepsilon(w^\varepsilon(t)) = T^\varepsilon ,$$

with $\mathbb{R}_*^\varepsilon T^\varepsilon$ strongly converging to $\lambda^h B^h w^h(t) - C^h(w^h(t))$, yields

$$A^h w^h(t) - \lambda^h(t)B^h w^h(t) + C^h(w^h(t)) = 0 .$$

The last relation is valid for $t \in (-\tau, +\tau)$.

Let us show that $w^h(t) \neq 0$ (the limit solution *is not* the branch $w^h(t) \equiv 0$). To this end it suffices to obtain

$$\langle B^h \varphi^h, v^h \rangle = 0 \,, \tag{4.3.94}$$

because then $w^h(t) = t\varphi^h + v^h$ is the decomposition of $w^h(t)$ in $\mathbb{R}\varphi^h \oplus U_h$ and its first component is different from 0. Indeed, one has

$$\langle B^h \varphi^h, v^h \rangle = \lim_{\varepsilon \to 0} \langle \mathbb{R}_*^\varepsilon B^\varepsilon \varphi^\varepsilon, \mathbb{P}^\varepsilon v^\varepsilon \rangle = \lim_{\varepsilon \to 0} \langle B^\varepsilon \varphi^\varepsilon, w^\varepsilon \rangle = 0 \,.$$

It remains to show that the branch obtained is really the bifurcating branch of the homogenized problem at $(\lambda_h, 0)$. First, $\lambda^h(0) = \lambda_h$ and $w^h(0) = 0$. By Lemma 4.3.26, the choice of k given by (4.3.86) and estimate (4.3.91)$_2$, now implying that $||v^h(t)||_V \leq k|t|$ we conclude that $||v^h(t)||_V \leq |t|$. Consequently $v^h(t)$ belongs to the contraction domain of $T_{0,t}$: it is thus the unique solution of the problem (4.3.81)$_1$ satisfying $||v^h(t)||_V \leq |t|$ and coinciding with the bifurcating branch. This proves the second part of the theorem. $\square$

5.   Moderately thick transversely symmetric plates

A plate is called thin if a characteristic lengthwave of a deformation pattern is much longer than the plate thickness. On the other hand, a plate is viewed as thick if this lengthwave is comparable with the plate thickness. Plates of intermediate thickness to deformation lengthwave ratios are called moderately thick. In the case of in-plane periodic plates the deformation lengthwaves are comparable with in-plane dimensions of the basic (the smallest) periodicity cell. Consequently it is the shape of $\mathcal{Z}$ (or $\mathcal{Y}$) that decides which plate model should be applied: the model of thin (Kirchhoff) plates or one of those plate models in which transverse shear (or possibly also-normal) deformations are taken into account. The global plate dimensions are irrelevant here.

In the present section we focus attention on transversely symmetric plates subject to transverse loading. Transverse symmetry conditions are given by (2.4.2). Not all available models of plates with transverse shear and normal deformation can be applied to transversely non-homogeneous plates. For instance, the most important theory of Reissner developed in the years 1944-1950 applies only to transversely homogeneous plates of constant thickness. This is the simplest theory in which σ^{33} effect is taken into account. Thus we have to give up considering σ^{33} effect if we prefer to use a theory the degree of complexity of which is comparable to that of Reissner's theory. In this section the thin plate model will be improved by considering transverse shear deformations in the simplest possible way, as proposed by Hencky in 1947.

5.1.   Reissner-Hencky model

Consider the transversely symmetric $\mathcal{Z}$-periodic plate described in Section 2.4. The two-dimensional modelling of Hencky is based upon:
(i) kinematic assumptions

$$w_\alpha(x, x_3) = u_\alpha(x) + x_3\varphi_\alpha(x) , \qquad w_3(x, x_3) = w(x) ; \qquad (5.1.1)$$

here $x_3^0 = 0$ due to transverse symmetry condition, cf. Eqs. (3.1.1), (3.1.2).
(ii) Stress assumptions

$$\sigma^{\alpha\beta} = \tilde{C}_Z^{\alpha\beta\lambda\mu}e_{\lambda\mu}(\boldsymbol{w}) , \qquad \sigma^{\alpha3} = 2C_Z^{\alpha3\lambda3}e_{\lambda3}(\boldsymbol{w}) , \qquad (5.1.2)$$

cf. (3.1.3)-(3.1.5). Note that the assumption: $\varphi_\alpha = -\partial w/\partial x_\alpha$ or $e_{\lambda3}(\boldsymbol{w}) = 0$ reduces hypotheses of Hencky to those of Kirchhoff, see Sec. 3.1. Further steps of constructing the plate theory are similar to those described in Section 3.1.

The variational equilibrium equation of the clamped plate assumes the form

$$\int_\Omega [N^{\alpha\beta}e_{\alpha\beta}(\boldsymbol{v}) + M^{\alpha\beta}e_{\alpha\beta}(\boldsymbol{\psi}) + Q^\alpha(\psi_\alpha + v_{,\alpha})]dx$$

$$= \int_\Omega (p_z^\alpha v_\alpha + m_z^\alpha \psi_\alpha + q_z v)dx \qquad \forall\, v_\alpha, \psi_\alpha, v \in H_0^1(\Omega) , \qquad (5.1.3)$$

where $\tilde{v}_\alpha = v_\alpha(x) + x_3\psi_\alpha(x)$, $\tilde{v}_3 = v(x)$ and $\tilde{v}_i$ are trial displacements of problem (P_Z) and p_z^α, m_z^α, q_z are given by (3.1.8) with $G_Z^+ = G_Z^-$, $x_3^0 = 0$, $x_3^+ = -x_3^-$. Stress resultants $N^{\alpha\beta}$ and $M^{\alpha\beta}$ are defined by (3.1.9) while the transverse shear force is given by

$$Q^\alpha = \int\limits_{-x_3^+(x)}^{x_3^+(x)} \sigma^{\alpha 3} dx_3 \; . \tag{5.1.4}$$

Due to transverse symmetry we have $E_z^{\alpha\beta\lambda\mu} = F_z^{\alpha\beta\lambda\mu} = 0$. The constitutive relations assume the form:

$$N^{\alpha\beta} = A_z^{\alpha\beta\lambda\mu}e_{\lambda\mu}(\boldsymbol{u}) \; , \qquad M^{\alpha\beta} = D_z^{\alpha\beta\lambda\mu}e_{\lambda\mu}(\boldsymbol{\varphi}) \; ,$$
$$Q^\alpha = H_z^{\alpha\beta}(\varphi_\beta + w_{,\beta}) \; , \tag{5.1.5}$$

where $\boldsymbol{A}_z$, $\boldsymbol{D}_z$ are defined by (3.1.11) and the transverse shear stiffnesses are calculated from

$$H_z^{\alpha\beta}(x) = k \int\limits_{-x_3^+(x)}^{x_3^+(x)} C_Z^{\alpha 3\beta 3}(x, x_3) dx_3 \; . \tag{5.1.6}$$

Here k represents a shear – correction factor; assumptions (5.1.1) imply $k = 1$.

One notes that the problem of finding $\boldsymbol{u}$, $\boldsymbol{\varphi}$, w that satisfy (5.1.3), with (5.1.5), for all $\boldsymbol{v}$, $\boldsymbol{\psi}$, v vanishing on $\partial\Omega$ decouples into: a) the in-plane problem with u_α as unknowns, b) the anti-plane problem involving $(\boldsymbol{\varphi}, w)$. Problem (a) does not differ from that of the Kirchhoff theory and that is why will not be discussed here. The anti-plane or bending/transverse shearing problem reads:

$$(P^{(H)}) \quad \left|
\begin{aligned}
&\text{find } (\boldsymbol{\varphi}, w) \in V_H^0(\Omega) = H_0^1(\Omega)^2 \times H_0^1(\Omega) \text{ such that}\\
&\int\limits_\Omega [D_z^{\alpha\beta\lambda\mu}e_{\alpha\beta}(\boldsymbol{\varphi})e_{\lambda\mu}(\boldsymbol{\psi}) + H_z^{\alpha\beta}(\varphi_\alpha + w_{,\alpha})(\psi_\beta + v_{,\beta})]dx\\
&= \int\limits_\Omega (q_z v + m_z^\alpha \psi_\alpha)dx \qquad\qquad \forall\ (\boldsymbol{\psi}, v) \in V_H^0(\Omega) \; .
\end{aligned}
\right. \tag{5.1.7}$$

Problem $(P^{(H)})$ is elliptic and uniquely solvable. To save space this standard proof is omitted.

5.2. The in-plane scaling-based asymptotic homogenization

The method of the in-plane scaling has been discussed in Sec. 3.2 in the context of Kirchhoff's plate theory. This method can equally well be applied to solving problem $(P^{(H)})$. The range of applicability of results obtained in this way will be discussed in the sequel.

Proceeding similarly as in Sec. 3.2 we replace problem $(P^{(H)})$ with a family of problems:

$$(P_\varepsilon^{(H)}) \quad \left|\begin{array}{l} \text{find } (\varphi^\varepsilon, w^\varepsilon) \in V_H^0(\Omega) \text{ such that} \\[2mm] \displaystyle\int_\Omega [M_\varepsilon^{\alpha\beta}(w^\varepsilon, \varphi^\varepsilon)e_{\alpha\beta}(\psi^\varepsilon) + Q_\varepsilon^\alpha(v_\alpha^\varepsilon + \psi_\alpha^\varepsilon)]dx \\[2mm] \displaystyle = \int_\Omega [\tilde{m}^\alpha\left(x, \frac{x}{\varepsilon}\right)\psi_\alpha^\varepsilon + \tilde{q}\left(x, \frac{x}{\varepsilon}\right)v^\varepsilon]dx \qquad \forall\, (\psi^\varepsilon, v^\varepsilon) \in V_H^0(\Omega) \end{array}\right. \tag{5.2.1}$$

where

$$M_\varepsilon^{\alpha\beta} = D^{\alpha\beta\lambda\mu}\left(\frac{x}{\varepsilon}\right)e_{\lambda\mu}(\varphi^\varepsilon)\,, \qquad Q_\varepsilon^\alpha = H^{\alpha\beta}\left(\frac{x}{\varepsilon}\right)(\varphi_\beta^\varepsilon + w_{,\beta}^\varepsilon) \tag{5.2.2}$$

and $H^{\alpha\beta}(x/\varepsilon_0) = H_z^{\alpha\beta}(x)$; $H^{\alpha\beta}(\cdot)$ are Y-periodic. Thus $(P_{\varepsilon_0}^{(H)}) = (P^{(H)})$. Note moreover, that the transverse dimensions of the plate remain constant as ε tends to zero. On account of differences in dimensions of bending $(D^{\alpha\beta\lambda\mu})$ and transverse shearing $(H^{\alpha\beta})$ stiffnesses problem $(P_\varepsilon^{(H)})$ involves length scales characterizing the transverse dimensions of the plate and just with these length scales the vanishing (with $\varepsilon \to 0$) in-plane dimensions of cells εY can be compared. Consequently, the homogenized stiffnesses that follow from asymptotic analysis of $(P_\varepsilon^{(H)})$ concern plates composed of transversely slender periodicity cells $\varepsilon\mathcal{Y}$. One can say that the asymptotic homogenization will result in a "transversely fibrous" microstructure.

The asymptotic process itself is similar to that known from plane elasticity homogenization. Thus an outline of the technique seems to be sufficient.

The solution $(w^\varepsilon, \varphi^\varepsilon)$ and trial fields $(v^\varepsilon, \psi^\varepsilon)$ are represented in the form $(3.2.4)_1$ or

$$\eta^\varepsilon = \eta^{(0)}(x) + \varepsilon\eta^{(1)}(x, y) + \varepsilon^2\eta^{(2)}(x, y) + \cdots|_{y=x/\varepsilon} \tag{5.2.3}$$
$$\eta \in \{w, \varphi_\alpha, v, \psi_\alpha\}\,.$$

It is assumed that $\eta^{(0)} \in H_0^1(\Omega)$ and $\eta^{(k)}(x, \cdot) \in H_{per}^1(Y)$, the space $H_{per}^1(Y)$ being defined in Sec. 2.7. Local problems are derived in the standard manner. They read as follows:

(i) bending local problem

$$(P_{H,Y}^1) \quad \left|\begin{array}{l} \text{find } \Upsilon^{(\gamma\delta)} \in H_{per}^1(Y)^2 \text{ such that} \\[2mm] d_H(\Upsilon^{(\gamma\delta)}, v) + \langle D^{\alpha\beta\gamma\delta}(y)e_{\alpha\beta}^y(v)\rangle = 0 \qquad \forall\, v \in H_{per}^1(Y)^2\,, \end{array}\right. \tag{5.2.4}$$

where

$$d_H(u, v) = \langle D^{\alpha\beta\lambda\mu}u_{\lambda|\mu}v_{\alpha|\beta}\rangle\,; \tag{5.2.5}$$

(ii) transverse shear local problem:

$$(P_{H,Y}^2) \quad \left|\begin{array}{l} \text{find } \Pi^{(\lambda)} \in H_{per}^1(Y) \text{ such that} \\[2mm] g_2(\Pi^{(\lambda)}, v) + \langle H^{\alpha\lambda}(y)v_{|\alpha}\rangle = 0 \qquad \forall\, v \in H_{per}^1(Y)\,, \end{array}\right. \tag{5.2.6}$$

where

$$g_2(u, v) = \langle H^{\alpha\beta}u_{|\alpha}v_{|\beta}\rangle\,. \tag{5.2.7}$$

Solutions to problems $(P^{\alpha}_{H,Y})$ are determined up to additive constants. Note that

$$d_H(\boldsymbol{u}, \boldsymbol{v}) = (\varepsilon_0)^3 \hat{d}_H(\boldsymbol{u}, \boldsymbol{v}) \,, \qquad g_2(u, v) = \varepsilon_0 \hat{g}_2(u, v) \,, \tag{5.2.8}$$

where $\hat{d}_H(\cdot, \cdot)$ and $\hat{g}_2(\cdot, \cdot)$ were defined by (2.7.14).

The effective stiffnesses are determined by

$$\begin{aligned} D^{\alpha\beta\gamma\delta}_H &= \langle D^{\alpha\beta\gamma\delta}(y) + D^{\alpha\beta\sigma\mu}(y) e^y_{\sigma\mu}(\boldsymbol{\Upsilon}^{(\gamma\delta)}) \rangle \,, \\ H^{\alpha\delta}_H &= \langle H^{\alpha\delta}(y) + H^{\alpha\beta}(y) \Pi^{(\delta)}_{|\beta} \rangle \,, \end{aligned} \tag{5.2.9}$$

or

$$\begin{aligned} D^{\sigma\rho\delta\gamma}_H &= \langle D^{\alpha\beta\lambda\mu}[\delta^\sigma_\alpha \delta^\rho_\beta + e^y_{\alpha\beta}(\boldsymbol{\Upsilon}^{(\sigma\rho)})][\delta^\gamma_\lambda \delta^\delta_\mu + e^y_{\lambda\mu}(\boldsymbol{\Upsilon}^{(\delta\gamma)})] \rangle \,, \\ H^{\sigma\delta}_H &= \langle H^{\alpha\lambda}[\delta^\sigma_\alpha + \Pi^{(\sigma)}_{|\alpha}][\delta^\delta_\lambda + \Pi^{(\delta)}_{|\lambda}] \rangle \,. \end{aligned} \tag{5.2.10}$$

Assume that $\boldsymbol{p}^{(\sigma\rho)}(y)$ and $p^{(\sigma)}(y)$ are polynomials in y_1, y_2 that satisfy

$$e^y_{\alpha\beta}(\boldsymbol{p}^{\sigma\rho}) = \delta^\sigma_{(\alpha} \delta^\rho_{\beta)} \,, \qquad p^{(\sigma)}_{|\alpha} = \delta^\sigma_\alpha \,. \tag{5.2.11}$$

Then

$$\begin{aligned} D^{\sigma\rho\delta\gamma}_H &= d_H(\boldsymbol{p}^{(\sigma\rho)} + \boldsymbol{\Upsilon}^{(\sigma\rho)} \,, \boldsymbol{p}^{(\delta\gamma)} + \boldsymbol{\Upsilon}^{(\delta\gamma)}) \,, \\ H^{\sigma\delta}_H &= g_2(p^{(\sigma)} + \Pi^{(\sigma)} \,, p^{(\delta)} + \Pi^{(\delta)}) \,. \end{aligned} \tag{5.2.12}$$

Having found the formulae (5.2.12) one can easily prove that the quadratic form

$$\epsilon_{\alpha\beta} D^{\alpha\beta\lambda\mu}_H \epsilon_{\lambda\mu} + \beta_\lambda H^{\lambda\mu}_H \beta_\mu$$

is positive definite and the tensors $\boldsymbol{D}_H, \boldsymbol{H}_H$ possess desired symmetry properties of (2.1.7) – type for $\boldsymbol{D}_H$ and $H^{\alpha\beta}_H = H^{\beta\alpha}_H$.

The tensor $\boldsymbol{D}_H$ can also be defined by the formula (3.4.5), where $\boldsymbol{A}$ should be replaced with $\boldsymbol{D}$. A similar formula for $\boldsymbol{H}_H$ can be written as follows.

We say that $\boldsymbol{\beta} = (\beta_1, \beta_2)$, where $\beta_\alpha \in L^2(Y, \mathbb{R})$, is associated with a scalar field Π defined on Y if $\beta_\alpha = \partial\Pi/\partial y_\alpha$. Then

$$\langle \varphi_{|2}\beta_1 - \varphi_{|1}\beta_2 \rangle = 0 \qquad \forall \varphi \in \mathbf{D}(Y) \,.$$

Let us introduce the space

$$\mathcal{K}^{per}_{\beta}(Y) = \{\boldsymbol{\beta} \in L^2(Y, \mathbb{R}^2) \,|\, \boldsymbol{\beta} \text{ is associated with a } Y\text{-periodic field}\} \,.$$

One can prove that tensor $\boldsymbol{H}_H$ is determined by

$$\beta^h_\lambda H^{\lambda\mu}_H \beta^h_\mu = \min\{\langle \beta_\lambda H^{\lambda\mu}(y) \beta_\mu \rangle \,|\, \boldsymbol{\beta} \in \mathcal{K}^{per}_{\beta}(Y) \,, \langle \boldsymbol{\beta} \rangle = \boldsymbol{\beta}^h\} \,. \tag{5.2.13}$$

The properties of $\boldsymbol{D}_H$ and $\boldsymbol{H}_H$ mentioned above suffice to prove the homogenized problem:

$$(P_h^{(H)}) \quad \left| \begin{aligned} &\text{find } (\varphi^{(0)}, w^{(0)}) \in V_H^0(\Omega) \text{ such that} \\[2mm] &\int_\Omega [D_H^{\alpha\beta\lambda\mu} \varphi_{\alpha,\beta}^{(0)} \psi_{\lambda,\mu} + H_H^{\lambda\mu}(\varphi_\lambda^{(0)} + w_{,\lambda}^{(0)})(\psi_\mu + v_{,\mu})] dx \\[2mm] &= \int_\Omega (m^\alpha \psi_\alpha + qv) dx \qquad \forall \, (\psi, v) \in V_H^0(\Omega) \,; \\[2mm] &m^\alpha(x) = \langle \tilde{m}^\alpha(x,y) \rangle \,, q(x) = \langle \tilde{q}(x,y) \rangle \end{aligned} \right. \qquad (5.2.14)$$

is uniquely solvable. Note that after homogenization the plate remains a plate of Hencky-Reissner type.

Solutions to problems $(P_{H,Y}^\alpha)$ determine fields $w^{(1)}$ and $\varphi_\alpha^{(1)}$ such that

$$w^\varepsilon = w^{(0)}(x) + \varepsilon \Pi^{(\alpha)} \left(\frac{x}{\varepsilon} \right) \underline{\left(\frac{\partial w^{(0)}}{\partial x_\alpha}(x) + \varphi_\alpha^{(0)}(x) \right)} + \varepsilon \tilde{w}(x) + O(\varepsilon^2) \,,$$

$$\varphi_\alpha^\varepsilon = \varphi_\alpha^{(0)}(x) + \varepsilon \Upsilon_\alpha^{(\lambda\delta)} \left(\frac{x}{\varepsilon} \right) \underline{\frac{\partial \varphi_\lambda^{(0)}}{\partial x_\delta}(x)} + \varepsilon \tilde{\varphi}_\alpha(x) + O(\varepsilon^2) \,. \qquad (5.2.15)$$

The underlined terms are called correctors. They have the following interesting property; assume that $w^{(0)}, \varphi_\alpha^{(0)} \in H_0^2(\Omega)$, $\Pi^{(\alpha)} \in W^{1,\infty}(Y)$, $\Upsilon^{(\gamma\delta)} \in W^{1,\infty}(Y)^2$. Then

$$w^\varepsilon - w^{(0)}(x) - \varepsilon \Pi^{(\alpha)} \left(\frac{x}{\varepsilon} \right) \left(w_{,\alpha}^{(0)} + \varphi_\alpha^{(0)} \right) \to 0 \,,$$

$$\varphi_\alpha^\varepsilon - \varphi_\alpha^{(0)} - \varepsilon \Upsilon_\alpha^{(\lambda\delta)} \left(\frac{x}{\varepsilon} \right) \varphi_{\lambda\delta}^{(0)} \to 0 \,, \qquad (5.2.16)$$

strongly in $H_0^1(\Omega)$ when ε tends to zero.

Remark 5.2.1. Problem $(P_\varepsilon^{(H)})$ is obviously equivalent to a convex minimization problem. It is not difficult to find the homogenized (effective) potential, which yields the local problems $(P_{H,Y}^\alpha)$ $(\alpha = 1, 2)$. Next, one can justify the homogenization results (except correctors) obtained in the present section by using the method of Γ-convergence. The proof is quite similar to the one of linear elasticity.

5.3. The refined scaling analysis

Problem $(P^{(H)})$ of Sec. 5.1 involves natural length scales defined by $(D_z^{\alpha\beta\lambda\mu}/H_z^{\sigma\rho})^{1/2}$ for $H_z^{\sigma\rho} \neq 0$ and the length scales l_α^z (dimension of Z) induced by the periodicity of variation of the moduli C_Z^{ijkl} and the transverse dimension $2x_3^+$ representing the varying thickness. Recall that our attention is confined to transversely symmetric plates, i.e.: $x_3^+ = -x_3^-$, $G_Z^+ = G_Z^-$, $C_Z(x, x_3) = C_Z(x, -x_3)$, $x_3^0 = 0$.

The ε-indexed family of $(P_{\varepsilon,ref}^{(H)})$ problems will be formed so that $(P_{\varepsilon_0,ref}^{(H)}) = (P^{(H)})$, $(P^{(H)})$ given in Sec. 5.1 and for each ε the ratios of all length scales will be the same. This can be achieved by assuming the following scaling

$$Z \rightsquigarrow \varepsilon Y \ , \quad l_\alpha^z \rightsquigarrow \varepsilon l_\alpha \ , \quad G_Z^+(x) \rightsquigarrow G\left(\frac{x}{\varepsilon}\right) = G_Z^+\left(\frac{x\varepsilon_0}{\varepsilon}\right) \ ,$$

$$x_3^+(x) \rightsquigarrow \varepsilon c\left(\frac{x}{\varepsilon}\right) = \frac{\varepsilon}{\varepsilon_0}x_3^+\left(\frac{x\varepsilon_0}{\varepsilon}\right) \ ,$$

$$C_Z^{ijkl}(x, x_3) \rightsquigarrow C^{ijkl}\left(\frac{x}{\varepsilon}, \frac{x_3}{\varepsilon}\right) = C_Z^{ijkl}\left(\frac{x\varepsilon_0}{\varepsilon}, \frac{x_3\varepsilon_0}{\varepsilon}\right) \ , \tag{5.3.1}$$

$$q_z(x) \rightsquigarrow \varepsilon^3 \hat{q}\left(x, \frac{x}{\varepsilon}\right) \ ; \quad \hat{q}(x,y) = (q^+ + q^-)(x,y)[G(y)]^{1/2} \ ,$$

$$m_z^\alpha(x) \rightsquigarrow \varepsilon^3 \hat{m}^\alpha\left(x, \frac{x}{\varepsilon}\right) \ ; \quad \hat{m}^\alpha(x,y) = (G(y))^{1/2}c(y)[p_\alpha^+(x) - p_\alpha^-(x)] \ ,$$

where $q^\pm, p_\alpha^\pm$ have been introduced in Sec. 2.2. Functions: $G(\cdot), c(\cdot), C^{ijkl}(\cdot, x_3/\varepsilon), \hat{q}(x, \cdot),$ $\hat{m}^\alpha(x, \cdot)$ are Y-periodic. Scaling (5.3.1) implies the following scaling of stiffnesses

$$D_z^{\alpha\beta\lambda\mu}(x) \rightsquigarrow \varepsilon^3 \hat{D}^{\alpha\beta\lambda\mu}\left(\frac{x}{\varepsilon}\right) \qquad H_z^{\alpha\beta} \rightsquigarrow \varepsilon \hat{H}^{\alpha\beta}\left(\frac{x}{\varepsilon}\right) \ , \tag{5.3.2}$$

where $\hat{D}, \hat{H}$ have been defined by (2.7.13).

The scaling introduced above can be understood in two manners. On the one hand, scaling (5.3.1) is such that the three-dimensional periodicity cell $\mathcal{Z} = Z \times (-x_3^+, x_3^+)$ of the original plate is for each ε homothetic to the three-dimensional cells $\varepsilon\mathcal{Y} = \varepsilon Y \times (-\varepsilon c, \varepsilon c)$ of the problem with the small parameter ε, cf. asymptotic analysis of Sec. 2.2. On the other hand, scaling (5.2.3) implies that the length scales: $l_{\alpha\beta\lambda\mu\sigma\rho}^\varepsilon = (\varepsilon^3 \hat{D}^{\alpha\beta\lambda\mu}/\varepsilon\hat{H}^{\sigma\rho})^{1/2},$ for $\hat{H}^{\sigma\rho} \neq 0$, are proportional to ε, hence all length scales of the problem: $l_{\alpha\beta\lambda\mu\sigma\rho}^\varepsilon$ and εl_α tend to zero with $\varepsilon \rightarrow 0$ with the same speed. The scaling of the loading compensates the stiffness loss when $\varepsilon \rightarrow 0$.

Thus problem $(P^{(H)})$ is now replaced with

$$(P_{\varepsilon,ref}^{(H)}) \quad \left| \begin{array}{l} \text{find } (\hat{\varphi}^\varepsilon, \hat{w}^\varepsilon) \in V_H^0(\Omega) \text{ such that} \\[2mm] \displaystyle\int_\Omega [\hat{D}^{\alpha\beta\lambda\mu}\left(\frac{x}{\varepsilon}\right) e_{\alpha\beta}(\hat{\varphi}^\varepsilon)e_{\lambda\mu}(\psi) + \frac{1}{\varepsilon^2}\hat{H}^{\alpha\beta}\left(\frac{x}{\varepsilon}\right)(\hat{\varphi}_\alpha^\varepsilon + \hat{w}_{,\alpha}^\varepsilon)(\psi_\beta + v_{,\beta})]dx \\[4mm] \displaystyle = \int_\Omega [\hat{m}^\alpha\left(x, \frac{x}{\varepsilon}\right)\psi_\alpha + \hat{q}\left(x, \frac{x}{\varepsilon}\right)v]dx \qquad \forall\, (\psi, v) \in V_H^0(\Omega) \ . \end{array} \right. \tag{5.3.3}$$

To find effective properties of the plate considered we represent the solution as well as trial fields in the form (5.2.3). Upon substituting these representations into (5.3.3) and demanding that the left-hand side does not tend to infinity when $\varepsilon \rightarrow 0$ one concludes that, cf. Sec. 5.4 for a rigorous justification

$$\varphi_\alpha^{(0)} = -w_{,\alpha}^{(0)} \ , \qquad \psi_\alpha^{(0)} = -v_{,\alpha}^{(0)} \ , \qquad w^{(1)} = w^{(1)}(x) \ , \qquad v^{(1)} = v^{(1)}(x) \ . \tag{5.3.4}$$

Thus the solution of the problem $P_{\varepsilon,ref}^{(H)}$ is represented in the form

$$\hat{\varphi}_\alpha^\varepsilon = -w_{,\alpha}^{(0)} + \varepsilon\varphi_\alpha^{(1)}\left(x,\frac{x}{\varepsilon}\right) + \varepsilon^2\varphi_\alpha^{(2)}\left(x,\frac{x}{\varepsilon}\right) + \ldots,$$

$$\hat{w}^\varepsilon = w^{(0)}(x) + \varepsilon w^{(1)}(x) + \varepsilon^2 w^{(2)}\left(x,\frac{x}{\varepsilon}\right) + \ldots.$$

Let us define

$$\tilde{\varphi}_\alpha = \varphi_\alpha^{(1)} - w_{,\alpha}^{(1)}, \qquad \tilde{\psi}_\alpha = \psi_\alpha^{(1)} - v_{,\alpha}^{(1)}. \tag{5.3.5}$$

The main terms of Eq. (5.3.3) assume the following form

$$\int_\Omega [\hat{D}^{\alpha\beta\lambda\mu}\left(\frac{x}{\varepsilon}\right)[\kappa_{\lambda\mu}^h + \tilde{\varphi}_{\lambda|\mu}][\kappa_{\alpha\beta}(v^{(0)}) + \tilde{\psi}_{\alpha|\beta}]$$

$$+ \hat{H}^{\alpha\beta}\left(\frac{x}{\varepsilon}\right)(\tilde{\varphi}_\alpha + w_{|\alpha}^{(2)})(\tilde{\psi}_\beta + v_{|\beta}^{(2)})]dx = \int_\Omega(\hat{q}v^{(0)} - \hat{m}^\alpha v_{,\alpha}^{(0)})dx + 0(\varepsilon), \tag{5.3.6}$$

where $\kappa_{\alpha\beta}^h = -w_{,\alpha\beta}^{(0)}$.

Let us put $\tilde{\psi} = 0$ in (5.3.6) and let $\varepsilon \to 0$. Making use of the averaging formula (1.1.1) gives

$$\int_\Omega \hat{M}_h^{\alpha\beta}\kappa_{\alpha\beta}(v^{(0)})dx = \int_\Omega(\check{q}v^{(0)} - \check{m}^\alpha v_{,\alpha}^{(0)})dx, \tag{5.3.7}$$

where

$$\hat{M}_h^{\alpha\beta} = \langle\hat{D}^{\alpha\beta\lambda\mu}(y)(\kappa_{\lambda\mu}^h + \tilde{\varphi}_{\lambda|\mu})\rangle, \quad \check{q}(x) = \langle\hat{q}(x,y)\rangle, \quad \check{m}^\alpha(x) = \langle\hat{m}^\alpha(x,y)\rangle. \tag{5.3.8}$$

Let us pass to zero with ε in Eq. (5.3.6) and combine the resulting equation with (5.3.7). One arrives at

$$\int_\Omega \langle\hat{D}^{\alpha\beta\lambda\mu}(y)[\kappa_{\lambda\mu}^h(x) + \tilde{\varphi}_{\lambda|\mu}(x,y)]\tilde{\psi}_{\alpha|\beta}(x,y)$$

$$+ \hat{H}^{\alpha\beta}(y)[\tilde{\varphi}_\alpha(x,y) + w_{|\alpha}^{(2)}(x,y)][\tilde{\psi}_\beta(x,y) + v_{|\beta}^{(2)}(x,y)]\rangle dx = 0. \tag{5.3.9}$$

Let us put $\tilde{\psi}_\alpha = \bar{\psi}(x)\psi_\alpha(y)$, $v^{(2)} = \bar{v}(x)v(y)$, where $\psi_\alpha, v \in H_{per}^1(Y)$; $\bar{\psi}, \bar{v} \in \mathbf{D}(\Omega)$. Hence one obtains the local equations

$$\langle\hat{D}^{\alpha\beta\lambda\mu}(\kappa_{\lambda\mu}^h + \tilde{\varphi}_{\lambda|\mu})\psi_{\alpha|\beta} + \hat{H}^{\alpha\beta}(\tilde{\varphi}_\alpha + w_{|\alpha}^{(2)})\psi_\beta\rangle = 0 \qquad \forall\,\psi \in H_{per}^1(Y)^2, \tag{5.3.10}$$

$$\langle\hat{H}^{\alpha\beta}(\tilde{\varphi}_\alpha + w_{|\alpha}^{(2)})v_{|\beta}\rangle = 0 \qquad \forall\,v \in H_{per}^1(Y). \tag{5.3.11}$$

Solutions to equations given above can be represented as follows

$$w^{(2)} = \chi^{(\gamma\delta)}(y)\kappa_{\gamma\delta}^h(x), \qquad \tilde{\varphi}_\lambda = \Phi_\lambda^{(\gamma\delta)}(y)\kappa_{\gamma\delta}^h(x). \tag{5.3.12}$$

Let us denote

$$M^{\lambda\mu}_{2(\alpha\beta)} = \hat{D}^{\lambda\mu\gamma\delta}\Phi^{\alpha\beta}_{(\gamma|\delta)} + \hat{D}^{\lambda\mu\alpha\beta} , \qquad Q^{\lambda}_{2(\alpha\beta)} = \hat{H}^{\lambda\mu}(\chi^{\alpha\beta}_{|\mu} + \Phi^{(\alpha\beta)}_{\mu}) . \qquad (5.3.13)$$

The functions $\chi^{(\gamma\delta)}$, $\Phi^{(\gamma\delta)}$ satisfy the following variational equation

$$\langle M^{\lambda\mu}_{2(\alpha\beta)}\psi_{\lambda|\mu} + Q^{\lambda}_{2(\alpha\beta)}(v_{|\lambda} + \psi_{\lambda})\rangle = 0 \qquad \forall \; (\psi, v) \in H^1_{per}(Y)^2 \times H^1_{per}(Y) . \qquad (5.3.14)$$

Let us compare Eqs. (5.3.13) – (5.3.14) with (2.7.16) – (2.7.19) in the case of transverse symmetry ($\hat{E} = 0$). We conclude that these equations coincide. Problem of finding ($\Phi^{(\gamma\delta)}$, $\chi^{(\gamma\delta)}$) $\in H^1_{per}(Y)^2 \times H^1_{per}(Y)$ satisfying (5.3.14), (5.3.13) assumes the form of ($\bar{P}^2_{S,Y}$), cf. Eqs. (2.7.31).

Upon substituting $(5.3.12)_2$ into $(5.3.8)_1$ and multiplying both sides by $(\varepsilon_0)^3$ one finds

$$M^{\alpha\beta}_{\hbar} = D^{\alpha\beta\lambda\mu}_{\hbar}\kappa^h_{\lambda\mu} , \qquad M^{\alpha\beta}_{\hbar} = (\varepsilon_0)^3\hat{M}^{\alpha\beta}_{\hbar} ,$$
$$D^{\alpha\beta\lambda\mu}_{\hbar} = (\varepsilon_0)^3\langle\hat{D}^{\alpha\beta\lambda\mu} + \hat{D}^{\alpha\beta\gamma\delta}\Phi^{(\lambda\mu)}_{\gamma|\delta}\rangle \qquad\qquad (5.3.15a)$$

or by $(3.2.1d)_3$

$$D^{\alpha\beta\lambda\mu}_{\hbar} = \langle D^{\alpha\beta\lambda\mu} + D^{\alpha\beta\gamma\delta}\Phi^{(\lambda\mu)}_{\gamma|\delta}\rangle . \qquad (5.3.15b)$$

In this way we have arrived at the homogenized constitutive relations of the original plate, cf. Eq. $(2.7.33)_2$. The homogenized problem reads:

$$(P_\hbar) \quad \left|\begin{array}{l} \text{find } w^{(0)} \in H^2_0(\Omega) \text{ such that} \\[2mm] \displaystyle\int_\Omega D^{\alpha\beta\lambda\mu}_{\hbar}\kappa_{\alpha\beta}(w^{(0)})\kappa_{\lambda\mu}(v)dx = \int_\Omega (qv - m^\alpha v_{,\alpha})dx \quad \forall\, v \in H^2_0(\Omega) ; \end{array}\right. \qquad (5.3.16)$$

where

$$q = (\varepsilon_0)^3\breve{q} , \qquad m^\alpha = (\varepsilon_0)^3\breve{m}^\alpha . \qquad (5.3.17)$$

Due to analogy between the present analysis and that of Sec. 2.7 there is no need to prove independently that the local problems for finding $\chi^{(\gamma\delta)}$ and $\Phi^{(\gamma\delta)}$ and homogenized problem for $w^{(0)}$ are well-posed. It is, however, necessary to prove the correctness of the asymptotic derivation starting from $(P^{(H)}_{\varepsilon,ref})$. Such proof of convergence justifying the present asymptotic derivation is put forward in the next section.

5.4. Justification of the refined scaling approach

The homogenized plate model derived in Sec. 5.2 can be justified rigorously by the Γ-convergence method. The justification procedure of the in-plane scaling is, however, much simpler than that of the refined scaling. Therefore in this and in the next section we shall focus on the homogenized plate model developed in Sec. 5.3. The presence of the singular term, associated with ε^{-2}, makes the justification procedure worth of being studied in more detail.

5.4.1. Basic relations and auxiliary results

Problem $(P^{(H)}_{\varepsilon,ref})$ is equivalent to the following convex minimization problem

$$J_\varepsilon(\widetilde{\boldsymbol{\varphi}}^\varepsilon, \widetilde{w}^\varepsilon) - L_\varepsilon(\widetilde{\boldsymbol{\varphi}}^\varepsilon, \widetilde{w}^\varepsilon) = \inf\{J_\varepsilon(\boldsymbol{\varphi}, w) - L_\varepsilon(\boldsymbol{\varphi}, w) | (\boldsymbol{\varphi}, w) \in V_H^0(\Omega)\}\,, \qquad (5.4.1)$$

where

$$J_\varepsilon(\boldsymbol{\varphi}, w) = \int_\Omega k\left[\frac{x}{\varepsilon}, e(\boldsymbol{\varphi}), \boldsymbol{\gamma}(\boldsymbol{\varphi}, w)\right] dx\,, \qquad (5.4.2)$$

$$L_\varepsilon(\boldsymbol{\varphi}, w) = \int_\Omega \left[\widehat{m}^\alpha\left(x, \frac{x}{\varepsilon}\right)\varphi_\alpha + \widehat{q}\left(x, \frac{x}{\varepsilon}\right)w\right] dx\,. \qquad (5.4.3)$$

Here

$$k\left(\frac{x}{\varepsilon}, \boldsymbol{\epsilon}, \boldsymbol{\rho}\right) = \frac{1}{2}\left(\widehat{D}^{\alpha\beta\lambda\mu}\left(\frac{x}{\varepsilon}\right)\epsilon_{\alpha\beta}\epsilon_{\lambda\mu} + \frac{1}{\varepsilon^2}\widehat{H}^{\alpha\beta}\left(\frac{x}{\varepsilon}\right)\rho_\alpha\rho_\beta\right)\,, \qquad \boldsymbol{\epsilon} \in \mathbb{E}^2_s, \boldsymbol{\rho} \in \mathbb{R}^2\,, \qquad (5.4.4)$$

$$\gamma_\alpha(\boldsymbol{\varphi}, w) = \varphi_\alpha + w_{,\alpha}\,. \qquad (5.4.5)$$

We recall that $V_H^0(\Omega) = H_0^1(\Omega)^2 \times H_0^1(\Omega)$. There exist constants $m_1 \geq m_0 > 0$ such that

$$\forall\, \boldsymbol{\epsilon} \in \mathbb{E}^2_s \quad m_0|\boldsymbol{\epsilon}|^2 \leq \widehat{D}^{\alpha\beta\lambda\mu}(y)\epsilon_{\alpha\beta}\epsilon_{\lambda\mu} \leq m_1|\boldsymbol{\epsilon}|^2\,, \qquad (5.4.6)$$
$$\forall\, \boldsymbol{a} \in \mathbb{R}^2 \quad m_0|\boldsymbol{a}|^2 \leq \widehat{H}^{\alpha\beta}(y)\rho_\alpha\rho_\beta \leq m_1|\boldsymbol{a}|^2\,,$$

for a.e. $y \in Y$.

Straightforward calculation yields the dual potential:

$$k^*\left(\frac{x}{\varepsilon}, \boldsymbol{\epsilon}^*, \boldsymbol{\rho}^*\right) = \sup\{\boldsymbol{\epsilon}^* : \boldsymbol{\epsilon} + \boldsymbol{a}^* \cdot \boldsymbol{a} - k\left(\frac{x}{\varepsilon}, \boldsymbol{\epsilon}, \boldsymbol{\rho}\right) | \boldsymbol{\epsilon} \in \mathbb{E}^2_s, \boldsymbol{\rho} \in \mathbb{R}^2\}$$
$$= \frac{1}{2}\left[\widehat{d}_{\alpha\beta\lambda\mu}\left(\frac{x}{\varepsilon}\right)\epsilon^{*\alpha\beta}\epsilon^{*\lambda\mu} + \varepsilon^2\widehat{h}_{\alpha\beta}\left(\frac{x}{\varepsilon}\right)\rho^{*\alpha}\rho^{*\beta}\right]\,; \boldsymbol{\epsilon}^* \in \mathbb{E}^s_2, \boldsymbol{\rho}^* \in \mathbb{R}^2\,, \qquad (5.4.7)$$

where $\widehat{d} = \widehat{D}^{-1}, \widehat{h} = \widehat{H}^{-1}$. Formally, the last complementary potential can be obtained from the potential

$$j^*\left(\frac{x}{\varepsilon}, \boldsymbol{\epsilon}^*, \boldsymbol{\rho}^*\right) = \frac{1}{2}\left[\widehat{d}_{\alpha\beta\lambda\mu}\left(\frac{x}{\varepsilon}\right)\epsilon^{*\alpha\beta}\epsilon^{*\lambda\mu} + \widehat{h}_{\alpha\beta}\left(\frac{x}{\varepsilon}\right)\rho^{*\alpha}\rho^{*\beta}\right]\,. \qquad (5.4.8)$$

Indeed, it is sufficient to make the following replacement in (5.4.8)

$$\rho^* \rightsquigarrow \varepsilon\rho^*\,. \qquad (5.4.9)$$

Macroscopic potential and its dual

The macroscopic stored energy or elastic potential is here denoted by $\mathcal{W}_h$. It is given by

$$\mathcal{W}_h(\boldsymbol{\kappa}^h) = \inf\{\frac{1}{2}\langle \widehat{D}^{\alpha\beta\lambda\mu}(y)(\kappa_{\alpha\beta}^h + e_{\alpha\beta}^y(\boldsymbol{\varphi}))(\kappa_{\lambda\mu}^h + e_{\lambda\mu}^y(\boldsymbol{\varphi}))$$

$$+\widehat{H}^{\alpha\beta}(y)\gamma_\alpha^y(\boldsymbol{\varphi},w)\gamma_\beta^y(\boldsymbol{\varphi},w)\rangle|\boldsymbol{\varphi} \in H_{per}^1(Y)^2\,, w \in H_{per}^1(Y)\}\,, \qquad (5.4.10)$$

where $\boldsymbol{\kappa}^h \in \mathbb{E}_s^2$ and $\gamma_\alpha^y(\boldsymbol{\varphi},w) = \varphi_\alpha + w_{|\alpha}$.

Properties of $\mathcal{W}_h$

(1)
$$\mathcal{W}_h(\boldsymbol{\kappa}^h) \leq \frac{1}{2|Y|}\int_Y \widehat{D}^{\alpha\beta\lambda\mu}(y)\kappa_{\alpha\beta}^h\kappa_{\lambda\mu}^h dy \leq \frac{m_1}{2}|\boldsymbol{\kappa}^h|^2\,. \qquad (5.4.11)$$

To obtain (5.4.11) we take in (5.4.10) $\boldsymbol{\varphi} = \mathbf{0}$ and $w = 0$.

(2)
$$\mathcal{W}_h(\boldsymbol{\kappa}^h) \geq \frac{m_0}{2|Y|}\int_Y |e^y(\widetilde{\boldsymbol{\varphi}}) + \boldsymbol{\kappa}^h|^2 dy \geq \frac{m_0}{2}|\boldsymbol{\kappa}^h|^2\,, \quad \boldsymbol{\kappa}^h \in \mathbb{E}_s^2\,, \qquad (5.4.12)$$

because

$$\sum_{\alpha,\beta=1}^2 \int_Y \kappa_{\alpha\beta}^h e_{\alpha\beta}^y(\widetilde{\boldsymbol{\varphi}}(y))dy = 0\,.$$

Here $(\widetilde{\boldsymbol{\varphi}},\widetilde{w})$ solves the minimization problem appearing in (5.4.10). Obviously, $(\widetilde{\boldsymbol{\varphi}},\widetilde{w})$ coincides with $(\widetilde{\boldsymbol{\varphi}},w^2)$, cf. (5.3.10) and (5.3.11).

(3)
$$\mathcal{W}_h(\boldsymbol{\kappa}^h) = \frac{1}{2}\widehat{D}^{\alpha\beta\lambda\mu}\kappa_{\alpha\beta}^h\kappa_{\lambda\mu}^h\,, \quad \boldsymbol{\kappa}^h \in \mathbb{E}_s^2 \qquad (5.4.13)$$

where, cf. (5.3.15a)$_2$

$$\widehat{D}^{\alpha\beta\lambda\mu} = \langle \widehat{D}^{\alpha\beta\lambda\mu} + \widehat{D}^{\alpha\beta\lambda\mu}\Phi_{\lambda|\mu}^{(\lambda\mu)}\rangle\,. \qquad (5.4.14)$$

The last relation can be obtained by differentiating $\mathcal{W}_h$ twice and taking into account (5.3.12).

Comparing (5.3.15a)$_2$ with (5.4.14) we have

$$D_{\hbar}^{\alpha\beta\lambda\mu} = \varepsilon_0^3 \widehat{D}_h^{\alpha\beta\lambda\mu}\,.$$

The dual or complementary potential $\mathcal{W}_h^*$ can be derived similarly to Sec. 2.10 by applying Rockafellar's theory of duality outlined in Sec. 1.2.5. Finally we arrive at

$$\mathcal{W}_h^*(\boldsymbol{\epsilon}^h) = \inf\{\langle j^*(y,\boldsymbol{\epsilon}^* + \boldsymbol{p}^*(y),\boldsymbol{q}^*(y))\rangle|(\boldsymbol{p}^*,\boldsymbol{q}^*) \in \mathcal{S}_{per}\}\,, \qquad (5.4.15)$$

where

$$\mathcal{S}_{per} = \{(\boldsymbol{p}^*,\boldsymbol{q}^*) \in L^2(Y,\mathbb{E}_2^s) \times L^2(Y)^2| -\operatorname{div}_y\boldsymbol{p}^* + \boldsymbol{q}^* = 0\,, \operatorname{div}_y\boldsymbol{q}^* = 0\,, \text{ in } Y\,;$$

$$\boldsymbol{p}^*\mu \text{ and } \boldsymbol{q}^*\mu \text{ take opposite values on the opposite sides of } Y\,;$$

$$\int_Y \boldsymbol{p}^*(y)dy = 0\}\,. \qquad (5.4.16)$$

Remark 5.4.1. We also have $\langle q^* \rangle = 0$. Indeed, for $(p^*, q^*) \in \mathcal{S}_{per}$ we write

$$\langle q^* \rangle = \langle \operatorname{div}_y p^* \rangle = \frac{1}{|Y|} \int\limits_{\partial Y} p^* \mu ds = 0 \ .$$

Remark 5.4.2. Fields $(p^*, q^*) \in \mathcal{S}_{per}$ are local or microscopic moments and shear forces. Therefore we shall prefer the notation (m, t) instead of (p^*, q^*).

Suppose that $(m, t) \in \mathcal{S}_{per}$ and $\operatorname{div}_y \operatorname{div}_y m \in L^2(Y)$ and $\operatorname{div}_y t \in L^2(Y)$. Then one has

$$\operatorname{div}_y \operatorname{div}_y m = \operatorname{div}_y t = 0 \quad \text{in} \quad Y \ ,$$

and

$$0 = \int\limits_Y w \operatorname{div}_y \operatorname{div}_y m \, dy = \int\limits_{\partial Y} w(\operatorname{div}_y m) \cdot \mu ds \qquad \forall \, w \in H^1_{per}(Y) \ .$$

Hence we conclude that $(\operatorname{div}_y m) \cdot \mu$ takes opposite values on the opposite sides of Y. We know that for $(m, t) \in \mathcal{S}_{per}$ the following local equilibrium equation is satisfied

$$-\operatorname{div}_y m + t = 0 \quad \text{in} \quad Y \ .$$

After rescaling $y \rightsquigarrow x/\varepsilon$ we get

$$-\varepsilon \operatorname{div} m\left(\frac{\cdot}{\varepsilon}\right) + t\left(\frac{\cdot}{\varepsilon}\right) = 0 \quad \text{in} \ \Omega \ , \tag{5.4.17}$$

because $\partial/\partial y_\alpha = \varepsilon \partial/\partial x_\alpha$.

We now pass to the formulation of a counterpart of Lemma 2.10.12.

Lemma 5.4.3. Let $(m, t) \in \mathcal{S}_{per}$, $\psi \in \mathbf{D}(\Omega)$ and let $\{\varphi^\varepsilon, w^\varepsilon\}_{\varepsilon > 0} \subset H^1(\Omega)^2 \times H^1(\Omega)$ be a sequence strongly convergent in $L^2(\Omega)^2 \times L^2(\Omega)$. Then

$$\lim_{\varepsilon \to 0} \int\limits_\Omega \psi(x)[m^{\alpha\beta}\left(\frac{x}{\varepsilon}\right) e_{\alpha\beta}(\varphi^\varepsilon) + \frac{1}{\varepsilon} t^\alpha\left(\frac{x}{\varepsilon}\right) \gamma_\alpha(\varphi^\varepsilon, w^\varepsilon)] dx = 0 \ . \tag{5.4.18}$$

Proof. Assume that

$$\varphi^\varepsilon \to \varphi \qquad \text{strongly in } L^2(\Omega)^2 \ , \qquad w^\varepsilon \to w \qquad \text{strongly in } L^2(\Omega) \ ,$$

and denote by R_ε the integral over Ω in (5.4.18). Using integration by parts we obtain

$$\lim_{\varepsilon \to 0} R_\varepsilon = \lim_{\varepsilon \to 0} \int\limits_\Omega \{\psi(x)[-m^{\alpha\beta}_{,\beta}\left(\frac{x}{\varepsilon}\right) + \frac{1}{\varepsilon} t^\alpha\left(\frac{x}{\varepsilon}\right)] \varphi^\varepsilon_\alpha(x) - m^{\alpha\beta}\left(\frac{x}{\varepsilon}\right) \psi_{,\beta}(x) \varphi^\varepsilon_\alpha(x)$$

$$- \frac{1}{\varepsilon} t^\alpha_{,\alpha}\left(\frac{x}{\varepsilon}\right) w^\varepsilon(x) - \frac{1}{\varepsilon} t^\alpha\left(\frac{x}{\varepsilon}\right) \psi_{,\alpha}(x) w^\varepsilon(x)\} dx = 0 \ ,$$

because, cf. (5.4.17) and Remark 5.4.1

$$-\operatorname{div} m\left(\frac{x}{\varepsilon}\right) + \frac{1}{\varepsilon} t\left(\frac{x}{\varepsilon}\right) = 0 \ , \qquad \operatorname{div} t\left(\frac{x}{\varepsilon}\right) = 0 \qquad \text{in} \ \Omega \ ,$$

$$\lim_{\varepsilon \to 0} \int_\Omega m^{\alpha\beta}\left(\frac{x}{\varepsilon}\right)\psi_{,\beta}(x)\varphi_\alpha^\varepsilon(x)dx = \langle m^{\alpha\beta}(y)\rangle \int_\Omega \psi_{,\beta}(x)\varphi_\alpha(x)dx = 0 \,,$$

$$\lim_{\varepsilon \to 0} \int_\Omega \frac{1}{\varepsilon}t^\alpha\left(\frac{x}{\varepsilon}\right)\psi_\alpha w^\varepsilon(x)dx = \lim_{\varepsilon \to 0} \int_\Omega m_{,\beta}^{\alpha\beta}\left(\frac{x}{\varepsilon}\right)\psi_\alpha(x)w^\varepsilon(x)dx$$

$$= \langle m_{,\beta}^{\alpha\beta}(y)\rangle \int_\Omega \psi_\alpha(x)w(x)dx = 0 \,.$$

This establishes the formula (5.4.18). $\qquad\qquad\qquad\qquad\qquad\qquad\qquad\qquad\qquad\qquad$ $\square$

Lemma 5.4.4. Let $\{\varphi^\varepsilon, w^\varepsilon\}_{\varepsilon>0} \subset H^1(\Omega)^2 \times H^1(\Omega)$ be weakly convergent to (φ, w) in $H^1(\Omega)^2 \times H^1(\Omega)$. Assume that there exists a constant $c > 0$ independent of ε and such that

$$J_\varepsilon(\varphi^\varepsilon, w^\varepsilon) \le c \,.$$

Then $\varphi = -\nabla w$ and $w \in H^2(\Omega)$.

Proof. Applying (5.4.6) we obtain

$$m_0(\|e(\varphi^\varepsilon)\|_{L^2}^2 + \frac{1}{\varepsilon^2}\|\varphi^\varepsilon + \nabla w^\varepsilon\|_{L^2}^2) \le c \,.$$

Hence

$$\|\varphi^\varepsilon + \nabla w^\varepsilon\|_{L^2} \to 0 \quad \text{as} \quad \varepsilon \to 0 \,.$$

On the other hand

$$\varphi^\varepsilon + \nabla w^\varepsilon \rightharpoonup \varphi + \nabla w \quad \text{in} \quad L^2(\Omega)^2 \quad \text{weakly} \,.$$

Thus $\|\varphi + \nabla w\|_{L^2} = 0$ and $\varphi(x) = -\nabla w(x)$ for a.e. $x \in \Omega$. Consequently, $w \in H^2(\Omega)$ because $\varphi \in H^1(\Omega)^2$. $\qquad\qquad\qquad\qquad\qquad\qquad$ $\square$

Lemma 5.4.5. Assume that $\widehat{m}^\alpha \in L^2(\Omega \times Y)$, $\widehat{q} \in L^2(\Omega \times Y)$. A solution $(\widetilde{\varphi}^\varepsilon, \widetilde{w}^\varepsilon)$ to the problem (5.4.1) exists and is unique. Moreover, there exists a subsequence $\{\widetilde{\varphi}^{\varepsilon'}, \widetilde{w}^{\varepsilon'}\}_{\varepsilon'>0} \subset V_H^0(\Omega)$ satisfying the previous lemma with $c = c(\widehat{m}, \widehat{q}, \Omega)$.

Proof. For $\varepsilon \in (0, 1)$ and $(\varphi, w) \in V_H^0(\Omega)$ we have

$$J_\varepsilon(\varphi, w) - L_\varepsilon(\varphi, w) \ge \frac{m_0}{\varepsilon^2}\|\varphi + \nabla w^\varepsilon\|_{L^2}^2 + m_0\|e(\varphi)\|_{L^2}^2 - \|\widehat{m}\|_{L^2}\|\varphi\|_{L^2}$$

$$- \|\widehat{q}\|_{L^2}\|w\|_{L^2} \ge m_0(\|\varphi + \nabla w\|_{L^2}^2 + \|e(\varphi)\|_{L^2}^2)$$

$$- \|\widehat{m}\|_{L^2}\|\varphi\|_{L^2} - \|\widehat{q}\|_{L^2}\|w\|_{L^2} \,.$$

Applying the elementary inequality:

$$\forall\, f_1, f_2 \in L^2(\Omega) \,, \quad \forall \rho \in \mathbb{R} \,, \quad \rho > 0$$

$$\|f_1 + f_2\|_{L^2}^2 \ge \frac{\rho}{1+\rho}\|f_1\|_{L^2}^2 - \rho\|f_2\|_{L^2}^2 \,, \qquad\qquad\qquad (5.4.19)$$

combined with the Korn and Poincaré inequalities we readily obtain

$$J_\varepsilon(\varphi, w) - L_\varepsilon(\varphi, w) \geq m_2(||\varphi||_{H^1}^2 + ||w||_{H^1}^2) - m_3(||\widehat{m}||_{L^2}||\varphi||_{H^1} + ||\widehat{q}||_{L^2}||w||_{H^1}),$$

where m_2 and m_3 are positive constants.

For a fixed $\varepsilon > 0$ ($\varepsilon \in (0,1)$) the functional $J_\varepsilon - L_\varepsilon$ is strictly convex. Thus $(\widetilde{\varphi}^\varepsilon, \widetilde{w}^\varepsilon)$, which solves (5.4.1), exists and is unique.

The minimization problem (5.4.1) may be written in the following way

$$\forall\,(\varphi, w) \in V_H^0(\Omega)\;\; J_\varepsilon(\widetilde{\varphi}^\varepsilon, \widetilde{w}^\varepsilon) - L_\varepsilon(\widetilde{\varphi}^\varepsilon, \widetilde{w}^\varepsilon) \leq J_\varepsilon(\varphi, w) - L_\varepsilon(\varphi, w)\,. \quad (5.4.20)$$

For $(\varphi, w) = (0, 0)$ we have

$$m_0(||e(\varphi^\varepsilon)||_{L^2}^2 + \frac{1}{\varepsilon^2}||\widetilde{\varphi}^\varepsilon + \nabla\widetilde{w}^\varepsilon||_{L^2}^2) \leq ||\widehat{m}||_{L^2}||\widetilde{\varphi}^\varepsilon||_{L^2} + ||\widehat{q}||_{L^2}||\widetilde{w}^\varepsilon||_{L^2}\,. \quad (5.4.21)$$

Applying (5.4.18) to (5.4.21) for $\rho = \varepsilon^2$ we obtain

$$||\nabla\widetilde{w}^\varepsilon||_{L^2}^2 \leq m_4(1 + \varepsilon^2)(||\widehat{m}||_{L^2}||\widetilde{\varphi}^\varepsilon||_{L^2} + ||\widehat{q}||_{L^2}||\widetilde{w}^\varepsilon||_{L^2})\,.$$

Thus the sequence $\{\widetilde{\varphi}^\varepsilon, \widetilde{w}^\varepsilon\}_{\varepsilon>0}$ is bounded in $V_H^0(\Omega)$ and there exists a subsequence, say $\{\widetilde{\varphi}^{\varepsilon'}, \widetilde{w}^{\varepsilon'}\}_{\varepsilon'>0}$ converging weakly. Moreover, (5.4.20) yields

$$J_{\varepsilon'}(\widetilde{\varphi}^{\varepsilon'}, \widetilde{w}^{\varepsilon'}) \leq L_\varepsilon(\widetilde{\varphi}^{\varepsilon'}, \widetilde{w}^{\varepsilon'}) \leq c(\widehat{m}, \widehat{q}, \Omega)\,.$$

This completes the proof. $\qquad\qquad\qquad\qquad\qquad\qquad\qquad\qquad\qquad\qquad\square$

5.4.2. Γ-convergence of the sequence $\{J_\varepsilon - L_\varepsilon\}_{\varepsilon>0}$

The main result of Sec. 5.4 is formulated as the following theorem.

Theorem 5.4.6. The sequence of functionals $\{J_\varepsilon - L_\varepsilon\}_{\varepsilon>0}$, defined by (5.4.1) and (5.4.2), is Γ-convergent in the strong topology of $L^2(\Omega)^2 \times H^1(\Omega)$ to the functional $\bar{J}_h - \bar{L}_h$ specified by

$$(\bar{J}_h - \bar{L}_h)(w) = \begin{cases} (J_h - L_h)(w) & \text{if }\;\; w \in H_0^2(\Omega)\,, \\ +\infty & \text{otherwise}, \end{cases} \quad (5.4.22)$$

where

$$J_h(w) = \int_\Omega \mathcal{W}_h(\kappa(w(x)))dx\,, \quad w \in H^2(\Omega) \quad (5.4.23)$$

$$L_h(w) = \int_\Omega (\breve{q}(x)w(x) - \breve{m}^\alpha(x)w_{,\alpha}(x))dx\,, \quad w \in H^2(\Omega)\,, \quad (5.4.24)$$

and the loading $(\breve{q}, \breve{m})$ is given by $(5.3.8)_{2,3}$.

Proof. First, we note that

$$L_\varepsilon(\varphi, w) \rightharpoonup \int_\Omega [\check{m}^\alpha(x)\varphi_\alpha(x) + \check{q}(x)w]dx ,$$

weakly in $L^1(\Omega)$, since according to our assumptions the integrand $\widehat{m}^\alpha\left(x, \frac{\cdot}{\varepsilon}\right) \times \varphi_\alpha(x) +$ $\widehat{q}\left(x, \frac{\cdot}{\varepsilon}\right) w(x)$ is εY-periodic. It is thus sufficient to study the Γ-convergence of the sequence $\{J_\varepsilon\}_{\varepsilon>0}$. Now the proofs falls naturally into two parts.

I. For any $w \in H^2(\Omega)$ we have to demonstrate the existence of a sequence $\{\phi^\varepsilon, w^\varepsilon\}_{\varepsilon>0} \subset H^1(\Omega)^2 \times H^1(\Omega)$,

$$(\phi^\varepsilon, w^\varepsilon) \overset{\tau}{\to} (-\nabla w, w) , \quad \text{as} \quad \varepsilon \to 0 , \tag{5.4.25}$$

such that

$$J_h(w) \geq \limsup_{\varepsilon \to 0} J_\varepsilon(\varphi^\varepsilon, w^\varepsilon) , \tag{5.4.26}$$

where $\tau = s - [L^2(\Omega)^2 \times H^1(\Omega)]$.

Step 1. We take

$$w(x) = \sum_{\alpha,\beta=1,2} \frac{1}{2}(\epsilon_{\alpha\beta}x_\alpha x_\beta + a_\alpha x_\beta) + b , \quad \epsilon \in \mathbb{E}_s^2 , \ a \in \mathbb{R}^2 , \ b \in \mathbb{R} , \tag{5.4.27}$$

and set

$$w^\varepsilon(x) = w(x) + \varepsilon^2 \widetilde{w}\left(\frac{x}{\varepsilon}\right) , \qquad \varphi^\varepsilon(x) = -\nabla w(x) + \varepsilon\widetilde{\varphi}\left(\frac{x}{\varepsilon}\right) ,$$

where $\widetilde{w}$ and $\widetilde{\varphi}$ are solutions to the minimization problem appearing in (5.4.10).

One sees that such a choice is admissible since $\{\varphi^\varepsilon, w^\varepsilon\}_{\varepsilon>0} \subset H^1(\Omega)^2 \times H^1(\Omega)$ and (5.4.25) is satisfied. Moreover we have

$$\frac{1}{\varepsilon^2}\widehat{H}^{\alpha\beta}\left(\frac{x}{\varepsilon}\right)(\varphi_\alpha^\varepsilon + w_{,\alpha}^\varepsilon)(\varphi_\beta^\varepsilon + w_{,\beta}^\varepsilon)$$
$$= \widehat{H}^{\alpha\beta}\left(\frac{x}{\varepsilon}\right)\left[\widetilde{\varphi}_\alpha\left(\frac{x}{\varepsilon}\right) + \widetilde{w}_{,\alpha}\left(\frac{x}{\varepsilon}\right)\right]\left[\widetilde{\varphi}_\beta\left(\frac{x}{\varepsilon}\right) + \widetilde{w}_{,\beta}\left(\frac{x}{\varepsilon}\right)\right] .$$

Hence

$$J_\varepsilon(\varphi^\varepsilon, w^\varepsilon) = \frac{1}{2}\int_\Omega \left\{ \widehat{D}^{\alpha\beta\lambda\mu}\left(\frac{x}{\varepsilon}\right)\left[-\epsilon_{\alpha\beta} + e_{\alpha\beta}(\widetilde{\varphi})\left(\frac{x}{\varepsilon}\right)\right]\left[-\epsilon_{\lambda\mu} + e_{\lambda\mu}(\widetilde{\varphi})\left(\frac{x}{\varepsilon}\right)\right] \right.$$
$$\left. + \widehat{H}^{\alpha\beta}\left(\frac{x}{\varepsilon}\right)\left[\widetilde{\varphi}_\alpha\left(\frac{x}{\varepsilon}\right) + \widetilde{w}_{,\alpha}\left(\frac{x}{\varepsilon}\right)\right]\left[\widetilde{\varphi}_\beta\left(\frac{x}{\varepsilon}\right) + \widetilde{w}_{,\beta}\left(\frac{x}{\varepsilon}\right)\right] \right\} dx . \tag{5.4.28}$$

Employing Theorem 1.1.5 we readily obtain

$$\lim_{\varepsilon \to 0} J_\varepsilon(\varphi^\varepsilon, w^\varepsilon) = \int_\Omega \mathcal{W}_h(-\epsilon)dx = \int_\Omega \mathcal{W}_h(\kappa(w))dx ,$$

because w is given by (5.4.27) and $\kappa_{\alpha\beta}(w) = -\partial^2 w/\partial x_\beta \partial x_\alpha$.

More precisely, the integrand in (5.4.28) is an εY-periodic function and tends weakly in $L^1(\Omega)$ as $\varepsilon \to 0$ to

$$\frac{1}{|Y|}\int_Y j[y, -\boldsymbol{\epsilon} + e^y(\widetilde{\boldsymbol{\varphi}}(y)), \widetilde{\boldsymbol{\varphi}}(y) + \nabla_y \widetilde{w}(y)]dy = \mathcal{W}_h(-\boldsymbol{\epsilon}) \ . \tag{5.4.29}$$

Step 2. Let $\{\Omega_K\}_{K \in \mathcal{K}}$ be a finite partition of the domain Ω formed by polygonal sets. We take a function $w \in C^1(\Omega)$ given by

$$w(x) = \sum_{\alpha,\beta=1}^{2} (\epsilon_{\alpha\beta}^K x_\alpha x_\beta + a_\alpha^K x_\beta) + b^K \ , \quad x \in \Omega_K \ , \tag{5.4.30}$$

where $\epsilon^K \in \mathbb{E}_s^2$, $a^K \in \mathbb{R}^2$ and $b^K \in \mathbb{R}$, $K \in \mathcal{K}$. Such a partition enables one to exploit the local character of the functionals J_ε. We set

$$\Omega_K^\delta = \{x \in \Omega_K | \ \mathrm{dist}(x, \partial\Omega_K) > \delta\} \ , \quad \delta > 0 \ .$$

Let $\psi_K^\delta \in \mathbf{D}(\Omega_K)$ be such that $0 \le \psi_K^\delta \le 1$ and $\psi_K^\delta(x) = 1$ for $x \in \Omega_K^\delta$. With every family of functions $w^K \in H_{per}^1(Y)$ and $\varphi^K \in H_{per}^1(Y)^2$ we associate the following sequences

$$w^{\varepsilon,\delta}(x) = w(x) + \varepsilon^2 \sum_{K \in \mathcal{K}} \psi_K^\delta w^K\left(\frac{x}{\varepsilon}\right) \ ,$$

$$\varphi^{\varepsilon,\delta}(x) = -\nabla w(x) + \varepsilon \sum_{K \in \mathcal{K}} \psi_K^\delta(x)\varphi^K\left(\frac{x}{\varepsilon}\right) \ . \tag{5.4.31}$$

For $\varepsilon \to 0$ we obviously have

$$w^{\varepsilon,\delta} \to w \ , \ \text{strongly in } H^1(\Omega) \ , \qquad \phi^{\varepsilon,\delta} \to -\nabla w \ , \ \text{strongly in } L^2(\Omega)^2 \ .$$

Let us take $0 < t < 1$. By convexity of the function $j(x/\varepsilon, \cdot, \cdot)$ and noting that $t\psi_K^\delta + t(1 - \psi_K^\delta) + (1 - t) = 1$ we obtain

$$J_\varepsilon(t\boldsymbol{\varphi}^{\varepsilon,\delta}, tw^{\varepsilon,\delta}) = \sum_{K \in \mathcal{K}} \int_{\Omega_K} k\{\frac{x}{\varepsilon}, t\psi_K^\delta(x)\left[-\boldsymbol{\epsilon}^K + e(\varphi^K)\left(\frac{x}{\varepsilon}\right)\right] + t(1 - \psi_K^\delta(x))(-\boldsymbol{\epsilon}^K)$$

$$+ \left((1 - t)\frac{\varepsilon t}{1 - t}\psi_{K,(\beta}^\delta(x)\varphi_{\alpha)}^K\left(\frac{x}{\varepsilon}\right)\right), \ t\psi_K^\delta(x)\left[\varepsilon\varphi^K\left(\frac{x}{\varepsilon}\right)\right.$$

$$\left. + \varepsilon(\nabla w^K)\left(\frac{x}{\varepsilon}\right)\right] + t(1 - \psi_K^\delta(x))0 + (1 - t)\frac{\varepsilon^2 t}{1 - t}w^K\left(\frac{x}{\varepsilon}\right)\nabla\psi_K^\delta(x)\}dx$$

$$\le \sum_{K \in \mathcal{K}} \left\{ \int_{\Omega_K} j\left[\frac{x}{\varepsilon}, -\boldsymbol{\epsilon}^K + e(\varphi^K)\left(\frac{x}{\varepsilon}\right), (\varphi^K)\left(\frac{x}{\varepsilon}\right) + (\nabla w^K)\left(\frac{x}{\varepsilon}\right)\right] dx \right.$$

$$+ m_1|\boldsymbol{\epsilon}^K|^2 \int_{\Omega_K}(1 - \psi_K^\delta)dx + m_1(1 - t)\int_{\Omega_K}\left[\left|\frac{\varepsilon t}{1 - t}\psi_{K,(\alpha}^\delta(x)\varphi_{\beta)}^K\left(\frac{x}{\varepsilon}\right)\right|^2\right.$$

$$\left.\left. + \left|\frac{\varepsilon^2 t}{1 - t}w^K\left(\frac{x}{\varepsilon}\right)\nabla\psi_K^\delta(x)\right|^2\right]dx \right\} \ ,$$

since $j \geq 0$. Here m_1 is the constant appearing in (5.4.6) and

$$\psi^\delta_{K,(\alpha}(x)\varphi^K_{\beta)}\left(\frac{x}{\varepsilon}\right) = \frac{1}{2}\left[\frac{\partial\psi^\delta_K}{\partial x_\alpha}\varphi^K_\beta\left(\frac{x}{\varepsilon}\right) + \frac{\partial\psi^\delta_K}{\partial x_\beta}\varphi^K_\alpha\left(\frac{x}{\varepsilon}\right)\right] .$$

Let now ε tend to zero in the last inequality. By using the previous step we obtain

$$\limsup_{\varepsilon\to 0} J_\varepsilon(t\varphi^{\varepsilon,\delta} tw^{\varepsilon\delta}) \leq \sum_{K\in\mathcal{K}}\{|\Omega_K|\frac{1}{|Y|}\int_Y j[y, -\epsilon^K$$

$$+e^y(\varphi^K(y)), \varphi^K(y) + \nabla_y w^K(y)]dy + m_1|\epsilon^K|^2\int_{\Omega_K}(1-\psi^\delta_K)dx\} ,$$

where $|\Omega_K| = $ the Lebesgue measure of Ω_K. Next, let $t \to 1$ and $\delta \to 0$. Then one has

$$\limsup_{\substack{t\to 1 \\ \delta\to 0}}\limsup_{\varepsilon\to 0} J_\varepsilon(t\varphi^{\varepsilon,\delta}, tw^{\varepsilon,\delta})$$

$$\leq \sum_{K\in\mathcal{K}}|\Omega_K|\frac{1}{|Y|}\int_Y j[y, -\epsilon^K + e^y(\varphi^K(y)), \varphi^K(y) + \nabla_y w^K(y)]dy .$$

Applying Lemma 1.3.27 one can construct a mapping $\varepsilon \to (t(\varepsilon), \delta(\varepsilon))$ with $(t(\varepsilon), \delta(\varepsilon)) \to (1,0)$ such that setting

$$v^\varepsilon = t(\varepsilon)w^{\varepsilon,\delta(\varepsilon)} , \qquad \phi^\varepsilon = t(\varepsilon)\varphi^{\varepsilon,\delta(\varepsilon)} ,$$

we deduce that

$$\limsup_{\varepsilon\to 0} J_\varepsilon(\phi^\varepsilon, v^\varepsilon)$$

$$\leq \sum_{K\in\mathcal{K}}|\Omega_K|\frac{1}{|Y|}\int_Y j[y, -\epsilon^K + e^y(\varphi^K(y)), \varphi^K(y) + \nabla_y w^K(y)]dy . \tag{5.4.32}$$

It is obvious that if $\varepsilon \to 0$ then

$$v^\varepsilon \to w , \text{ strongly in } H^1(\Omega) , \qquad \phi^\varepsilon \to -\nabla w , \text{ strongly in } L^2(\Omega)^2 .$$

Taking now the infimum on the right-hand side of (5.4.32) when (φ^K, w^K) run over $H^1_{per}(Y)^2 \times H^1_{per}(Y)$ one obtains

$$J^s(-\nabla w, w) \leq \limsup_{\varepsilon\to 0} J_\varepsilon(\phi^\varepsilon, v^\varepsilon) \leq \sum_{K\in\mathcal{K}}|\Omega_K|\mathcal{W}_h(-\epsilon^K)$$

$$= \int_Y \mathcal{W}_h(\kappa(w))dx = J_h(w) . \tag{5.4.33}$$

Step 3. The property (ii) of the Γ-limit (epi-limit) implies that the convexity of J_ε is preserved by the epi-limit superior J^s. By virtue of (5.4.8) we have

$$J^s(-\nabla w, w) \le \frac{m_1}{2} \int_\Omega |\kappa(w)(x)|^2 dx , \quad w \in H^2(\Omega) .$$

Being convex and finite, the functional J^s is continuous on the space $H^2(\Omega)$.

Exploiting the properties of the homogenized potential $\mathcal{W}_h$ we readily conclude that the functional J_h is also convex and continuous on this space. Due to the density of functions of class $C^1(\Omega)$ with piecewise constant second order derivatives in $H^2(\Omega)$, cf. Sec. 1.4, the inequality $J^s(-\nabla w, w) \le J_h(w)$ can be extended to this space.

II. We have to show that for every sequence $\{\varphi^\varepsilon, w^\varepsilon\}_{\varepsilon>0} \subset H^1(\Omega)^2 \times H^1(\Omega)$ such that $(\varphi^\varepsilon, w^\varepsilon) \rightharpoonup (\varphi, w)$ weakly in $H^1(\Omega)^2 \times H^1(\Omega)$ the following inequality holds true

$$J_h(w) \le \lim_{\varepsilon \to 0} \inf J_\varepsilon(\varphi^\varepsilon, w^\varepsilon) . \tag{5.4.34}$$

If $\phi \ne -\nabla w$ then obviously

$$\lim_{\varepsilon \to 0} \inf J_\varepsilon(\varphi^\varepsilon, w^\varepsilon) = +\infty ,$$

and the desired inequality follows. Therefore, for the sequel of the proof we assume that Lemma 5.4.4 is satisfied, see also Lemma 5.4.5.

To prove (5.4.34) some duality arguments are exploited in an essential manner. In fact, it has to be shown that

$$\lim_{\varepsilon \to 0} \inf J_\varepsilon(\varphi^\varepsilon, w^\varepsilon) \ge J_h(w) = \int_\Omega \mathcal{W}_h(\kappa(w(x)))dx$$

$$= \sup\{ \int_\Omega \boldsymbol{M} : \kappa(w(x))dx - \int_\Omega \mathcal{W}_h^*(\boldsymbol{M}(x))dx | \boldsymbol{M} \in L^2(\Omega, \mathbb{E}_2^s)\} . \tag{5.4.35}$$

Step 4. First we take

$$\boldsymbol{M}(x) = \sum_{K \in \mathcal{K}} \chi^K(x) \boldsymbol{M}_K , \qquad \boldsymbol{M}_K \in \mathbb{E}_2^s , \tag{5.4.36}$$

where

$$\chi^K(x) = \begin{cases} 1, \text{ if } x \in \Omega_K , \\ 0, \text{ if } x \notin \Omega_K . \end{cases}$$

Here $\{\Omega_K\}_{K \in \mathcal{K}}$ is a finite family of open disjoint sets such that $\bar{\Omega} = \bigcup_{K \in \mathcal{K}} \bar{\Omega}_K$.

Let $(\boldsymbol{m}_K, \boldsymbol{t}_K) \in \mathcal{S}_{per}$, $K \in \mathcal{K}$. By using Lemma 5.4.3 and recalling that $k \geq 0$ we write

$$
\begin{aligned}
\liminf_{\varepsilon \to 0} \ & J_\varepsilon(\boldsymbol{\varphi}^\varepsilon, w^\varepsilon) \\
&\geq \liminf_{\varepsilon \to 0} \sum_{K \in \mathcal{K}} \int_{\Omega_K} \psi_K^\delta(x) \{ k \left[\frac{x}{\varepsilon}, e(\boldsymbol{\varphi}^\varepsilon), \boldsymbol{\gamma}(\boldsymbol{\varphi}^\varepsilon, w^\varepsilon) \right] \\
&\qquad - \left[\boldsymbol{m}_K \left(\frac{x}{\varepsilon} \right) : e(\boldsymbol{\varphi}^\varepsilon) + \frac{1}{\varepsilon} \boldsymbol{t}_K \left(\frac{x}{\varepsilon} \right) \cdot \boldsymbol{\gamma}(\boldsymbol{\varphi}^\varepsilon, w^\varepsilon) \right] \} dx \\
&= \liminf \sum_{K \in \mathcal{K}} \int_{\Omega_K} \psi_K^\delta(x) k_{(\boldsymbol{m}_K, \boldsymbol{t}_K)} \left[\frac{x}{\varepsilon}, e(\boldsymbol{\varphi}^\varepsilon), \boldsymbol{\gamma}(\boldsymbol{\varphi}^\varepsilon, w^\varepsilon) \right] dx \, ,
\end{aligned}
$$

where

$$
k_{(\boldsymbol{m}_K, \boldsymbol{t}_K)} \left(\frac{x}{\varepsilon}, \boldsymbol{\epsilon}, \boldsymbol{a} \right) = k \left(\frac{x}{\varepsilon}, \boldsymbol{\epsilon}, \boldsymbol{a} \right) - \boldsymbol{m}_K : \boldsymbol{\epsilon} - \frac{1}{\varepsilon} \boldsymbol{t}_K \cdot \boldsymbol{a} \, .
$$

Applying Fenchel's inequality to the function $k_{(\boldsymbol{m}_K, \boldsymbol{t}_K)}(x/\varepsilon, \cdot, \cdot)$ at $[\{e(\boldsymbol{\varphi}^\varepsilon(x)), \boldsymbol{\gamma}(\boldsymbol{\varphi}^\varepsilon(x), w^\varepsilon(x))\}; (\boldsymbol{M}_K, \boldsymbol{0})]$ one obtains:

$$
\begin{aligned}
k^*_{(\boldsymbol{m}_K, \boldsymbol{t}_K)} \left(\frac{x}{\varepsilon}, \boldsymbol{M}_K, \boldsymbol{0} \right) &\geq \boldsymbol{M}_K : e(\boldsymbol{\varphi}^\varepsilon(x)) \\
&\quad - k_{(\boldsymbol{m}_K, \boldsymbol{t}_K)} \left[\frac{x}{\varepsilon}, e(\boldsymbol{\varphi}^\varepsilon(x)), \boldsymbol{\gamma}(\boldsymbol{\varphi}^\varepsilon(x), w^\varepsilon(x)) \right] \, .
\end{aligned}
$$

Hence

$$
\begin{aligned}
\liminf_{\varepsilon \to 0} J_\varepsilon(\boldsymbol{\varphi}^\varepsilon, w^\varepsilon) &\geq \liminf_{\varepsilon \to 0} \sum_{K \in \mathcal{K}} \int_{\Omega_K} \{ \psi_K^\delta(x) [\boldsymbol{M}_K : e(\boldsymbol{\varphi}^\varepsilon(x))] \\
&\quad - k^*_{(\boldsymbol{m}_K, \boldsymbol{t}_K)} \left(\frac{x}{\varepsilon}, \boldsymbol{M}_K, \boldsymbol{0} \right) \} dx \, .
\end{aligned}
$$

According to our assumptions we have

$$
e(\boldsymbol{\varphi}^\varepsilon) \rightharpoonup \boldsymbol{\kappa}(w), \quad \text{weakly in} \quad L^2(\Omega, \mathbb{E}_s^2) \, .
$$

Thus

$$
\lim_{\varepsilon \to 0} \int_\Omega \psi_K^\delta(x) \boldsymbol{M}_K : e(\boldsymbol{\varphi}^\varepsilon(x)) dx = \int_\Omega \psi_K^\delta(x) \boldsymbol{M}_K : \boldsymbol{\kappa}(w(x)) dx \, .
$$

Let us find the explicit form of $k^*_{(\boldsymbol{m}_K, \boldsymbol{t}_K)}(x/\varepsilon, \cdot, \boldsymbol{0})$. For $\boldsymbol{T} \in \mathbb{E}_2^s$ standard calculation yields

$$
\begin{aligned}
k^*_{(\boldsymbol{m}_K, \boldsymbol{t}_K)} \left(\frac{x}{\varepsilon}, \boldsymbol{T}, \boldsymbol{0} \right) &= \sup \left\{ \boldsymbol{T} : \boldsymbol{\epsilon} - k_{(\boldsymbol{m}_K, \boldsymbol{t}_K)} \left(\frac{x}{\varepsilon}, \boldsymbol{\epsilon}, \boldsymbol{a} \right) \mid (\boldsymbol{\epsilon}, \boldsymbol{a}) \in \mathbb{E}_s^2 \times \mathbb{R}^2 \right\} \\
&= \frac{1}{2} \widehat{d}^\varepsilon_{\alpha\beta\lambda\mu} (T^{\alpha\beta} + m_K^{\alpha\beta})(T^{\lambda\mu} + m_K^{\lambda\mu}) + \frac{1}{2} \frac{1}{\varepsilon^2} \widehat{h}^\varepsilon_{\alpha\beta} t_K^\alpha t_K^\beta \, ,
\end{aligned}
$$

where $\widehat{d}^\varepsilon = (\widehat{D}^\varepsilon)^{-1}, \widehat{h}^\varepsilon = (\widehat{H}^\varepsilon)^{-1}$.

The sequence of periodic functions $k^*_{(m_K,t_K)}(\cdot/\varepsilon, M_K, 0)$ is bounded in $L^1(\Omega)$ and converges weakly to

$$\frac{1}{|Y|}\int_Y j^*[y, M_K + m_K(y), t_K(y)]dy \ .$$

To corroborate this statement it is sufficient to take into account (5.4.17) and Remark 5.4.2.
Hence

$$\liminf_{\varepsilon\to 0} J_\varepsilon(\varphi^\varepsilon, w^\varepsilon) \geq \sum_{K\in\mathcal{K}} \{ \int_{\Omega_K} \psi_K^\delta(x) M_K : \kappa(w(x))dx$$

$$-(\int_{\Omega_K} \psi_K^\delta(x)dx)\frac{1}{|Y|}\int_Y j^*[y, M_K + m_K(y), t_K(y)]dy\} \ . \qquad (5.4.37)$$

Taking now the supremum on the right-hand side of (5.4.37) when (m_K, t_K) runs over $\mathcal{S}_{per}$ one obtains

$$\liminf_{\varepsilon\to 0} J_\varepsilon(\varphi^\varepsilon, w^\varepsilon) \geq \sum_{K\in\mathcal{K}} \{ \int_{\Omega_K} \psi_K^\delta(x) M_K : \kappa(w(x))dx - (\int_{\Omega_K} \psi_K^\delta(x)dx)\mathcal{W}^*(M_K)\} \ ,$$

because $\sup(-\int) = -\inf\int$ and $\mathcal{W}_h^*$ is given by (5.4.15). Recalling that $\psi_K^\delta \geq 0$ and $M(x)$ has the form (5.4.36) we obtain

$$\liminf_{\varepsilon\to 0} J_\varepsilon(\varphi^\varepsilon, w^\varepsilon) \geq \int_\Omega \sum_{K\in\mathcal{K}} \psi_K^\delta(x) M : \kappa(w(x))dx - \int_\Omega \sum_{K\in\mathcal{K}} \psi_K^\delta(x)dx \mathcal{W}_h^*(M(x))dx \ .$$

The inequality $0 \leq \sum_{K\in\mathcal{K}} \psi_K^\delta \leq 1$ implies

$$0 \leq \sum_{K\in\mathcal{K}} \psi_K^\delta(x)\mathcal{W}^*(M(x)) \leq \mathcal{W}^*(M(x)) \ ,$$

because $\mathcal{W}_h^* \geq 0$. Consequently

$$\liminf_{\varepsilon\to 0} J_\varepsilon(\varphi^\varepsilon, w^\varepsilon) \geq \int_\Omega \sum_{K\in\mathcal{K}} \psi_K^\delta(x) M(x) : \kappa(w(x))dx - \int_\Omega \mathcal{W}_h^*(M(x))dx \ .$$

We now pass to the limit when $\delta \to 0$; $\sum_{K\in\mathcal{K}} \psi_K^\delta(x)$ tends to 1 a.e. and thus we have

$$\liminf_{\varepsilon\to 0} J_\varepsilon(\varphi^\varepsilon, w^\varepsilon) \geq \int_\Omega M(x) : \kappa(w(x))dx - \int_\Omega \mathcal{W}_h^*(M(x))dx \ .$$

Step 5. For each $M \in L^2(\Omega, \mathbb{E}_2^s)$ there exists a sequence $\{M_n\}_{n\in\mathbb{N}} \subset L^2(\Omega, \mathbb{E}_2^s)$ of simple functions such that

$$M_n \to M \text{ strongly in } L^2(\Omega, \mathbb{E}_2^s) \text{ as } n \to \infty \ .$$

Here

$$M_n(x) = \sum_{K(n)} \chi_{\delta_n}^{K(n)}(x) M_{K(n)} ,$$

$$\chi_{\delta_n}^{K(n)}(x) = \begin{cases} 1 , & \text{if } x \in \Omega_{K(n)} , \\ 0 , & \text{otherwise,} \end{cases}$$

and $\delta_n = 1/n$, $\operatorname{diam} \Omega_{K(n)} \le \delta_n$ and $\bar{\Omega} = \underset{K(n)}{\cup} \bar{\Omega}_{K(n)}$.

The previous step yields

$$\liminf_{\varepsilon \to 0} J_\varepsilon(\varphi^\varepsilon, w^\varepsilon) \ge \int_\Omega M_n(x) : \kappa(w(x))dx - \int_\Omega \mathcal{W}_h^*(M_n(x))dx . \qquad (5.4.38)$$

Passing to the limit on the right-hand side of (5.4.38) as $n \to \infty$ we finally obtain

$$\liminf_{\varepsilon \to 0} J_\varepsilon(\varphi^\varepsilon, w^\varepsilon) \ge \int_\Omega M(x) : \kappa(w(x))dx - \int_\Omega \mathcal{W}_h^*(M(x))dx ,$$

and the proof is complete. $\qquad\qquad\qquad\qquad\qquad\qquad\qquad\qquad\qquad\qquad\qquad\qquad\square$

Remark 5.4.7. The proof remains valid for $w \in H_0^2(\Omega)$ because then the approximating sequences are in $H_0^1(\Omega)^2 \times H_0^1(\Omega)$.

5.5. Dual homogenization

In this section we shall perform dual homogenization, which means Γ-convergence of the sequence of the complementary energy functionals, being dual to $J_\varepsilon - L_\varepsilon$, cf. (5.4.1), (5.4.2). For a fixed $\varepsilon > 0$ the primal problem means evaluating

$$(P_\varepsilon) \qquad \inf\{J_\varepsilon(\varphi, w) - L_\varepsilon(\varphi, w) | \varphi \in H_0^1(\Omega)^2, \ w \in H_0^1(\Omega)\} .$$

We now pass to the formulation of the dual problem (P_ε^*). To this end, the theory of duality presented in Section 1.2.5 applied. We set

$$\Lambda(\varphi, w) = (\Lambda_1 \varphi, \Lambda_2(\varphi, w)) = (e(\varphi), \gamma(\varphi, w)); \ (\varphi, w) \in H_0^1(\Omega)^2 \times H_0^1(\Omega) ,$$
$$\Lambda_1 : H_0^1(\Omega)^2 \to L^2(\Omega, \mathbb{E}_s^2), \ \Lambda_2 : H_0^1(\Omega)^2 \times H_0^1(\Omega) \to L^2(\Omega)^2 , \qquad (5.5.1)$$
$$\mathcal{S}(\Omega) = L^2(\Omega, \mathbb{E}_2^s) \times L^2(\Omega)^2 .$$

Proceeding similarly to Section 2.10 we obtain

$$\Lambda_1^* M = -\operatorname{div} M , \qquad\qquad\qquad\qquad\qquad\qquad (5.5.2)$$

$$\Lambda_2^* Q = \begin{cases} Q , & \text{in } \Omega , \ (\varphi) , \\ -\operatorname{div} Q , & \text{in } \Omega , \ (w) , \end{cases} \qquad\qquad\qquad (5.5.3)$$

where $(M, Q) \in \mathcal{S}(\Omega)$. Moreover, simple calculation yields

$$(-L_\varepsilon)^*[-\Lambda^*(M, Q)] = \begin{cases} 0, & \text{if } \operatorname{div} M - Q + \widehat{m} = 0, \ \operatorname{div} Q + \widehat{q} = 0, \text{ in } \Omega , \\ +\infty & \text{otherwise.} \end{cases} \qquad (5.5.4)$$

We recall that the equilibrium equations are to be understood in the sense of distributions.
Next, we set

$$G_\varepsilon(\boldsymbol{p}_1, \boldsymbol{p}_2) = \int_\Omega k\left[\frac{x}{\varepsilon}, \boldsymbol{p}_1(x), \boldsymbol{p}_2(x)\right] dx \,, \qquad (5.5.5)$$

where $\boldsymbol{p}_1 \in L^2(\Omega, \mathbb{E}_s^2)$, $\boldsymbol{p}_2 \in L^2(\Omega)^2$. According to Proposition 1.2.34 we have

$$G_\varepsilon^*(\boldsymbol{M}, \boldsymbol{Q}) = \int_\Omega k^*\left[\frac{x}{\varepsilon}, \boldsymbol{M}(x), \boldsymbol{Q}(x)\right] dx \,, \quad (\boldsymbol{M}, \boldsymbol{Q}) \in S(\Omega) \,. \qquad (5.5.6)$$

Taking into account (5.5.4) and (5.5.6) we can formulate the dual problem (P_ε^*), which means evaluating

$$(P_\varepsilon^*) \qquad \sup\left\{ -\int_\Omega k^*\left[\frac{x}{\varepsilon}, \boldsymbol{M}(x), \boldsymbol{Q}(x)\right] dx \,\big|\, (\boldsymbol{M}, \boldsymbol{Q}) \in S_\varepsilon(\Omega) \right\} \,,$$

where

$$S_\varepsilon = \{(\boldsymbol{M}, \boldsymbol{Q}) \in S(\Omega)|\ \mathrm{div}\,\boldsymbol{M} - \boldsymbol{Q} + \widehat{\boldsymbol{m}} = 0,\ \mathrm{div}\,\boldsymbol{Q} + \widehat{q} = 0,\ \text{in } \Omega\} \,. \qquad (5.5.7)$$

Using Proposition 1.2.50 we obtain

$$\inf P_\varepsilon = \sup P_\varepsilon^* \,. \qquad (5.5.8)$$

Appliction of Theorem 1.3.36 to dual homogenization
We shall now deal with the Γ-limit of the sequence of functionals

$$\mathcal{G}_\varepsilon(\boldsymbol{M}, \boldsymbol{Q}) = -\int_\Omega k^*\left[\frac{x}{\varepsilon}, \boldsymbol{M}(x), \boldsymbol{Q}(x)\right] dx - I_{S_\varepsilon(\Omega)}(\boldsymbol{M}, \boldsymbol{Q}) \,, \qquad (5.5.9)$$

where $I_{S_\varepsilon(\Omega)}$ is the indicator function of $S_\varepsilon(\Omega)$.
We set

$$G_\varepsilon^1(\boldsymbol{\varphi}, w, \boldsymbol{p}, \boldsymbol{q}) = \int_\Omega k\left[\frac{x}{\varepsilon}, e(\boldsymbol{\varphi}) + \boldsymbol{p}, \gamma(\boldsymbol{\varphi}, w) + \boldsymbol{q}\right] dx \,, \qquad (5.5.10)$$

where $\boldsymbol{\varphi} \in H^1(\Omega)^2$, $w \in H^1(\Omega)$, $(\boldsymbol{p}, \boldsymbol{q}) \in [L^2(\Omega, \mathbb{E}_s^2) \times L^2(\Omega)^2]$.
We observe that the functional L_ε, given by (5.4.3), is not influenced by perturbations $\boldsymbol{p}, \boldsymbol{q}$.
Let $\{\boldsymbol{q}^\varepsilon\} \subset L^2(\Omega)^2$ be a bounded sequence. Under the assumptions of Lemma 5.4.4 we have $1/\varepsilon \|\boldsymbol{q}^\varepsilon\|_{L^2} \leq m$ and $\boldsymbol{q} = \lim_{\varepsilon \to 0} \boldsymbol{q}^\varepsilon = 0$ a.e. in Ω. Here m is a constant independent of ε.

Further, we put

$$G_h^1(w, \boldsymbol{p}) = \int\limits_\Omega \mathcal{W}_h[\boldsymbol{\kappa}(w(x)) + \boldsymbol{p}(x)]dx , \tag{5.5.11}$$

where $w \in H^2(\Omega)$ and $\boldsymbol{p} \in L^2(\Omega, \mathbb{E}_s^2)$.

To exploit Theorem 1.3.36 we shall first prove that

$$G_h^1 = \Gamma(\tau \times s_{\mathcal{S}(\Omega)}) - \lim_{\varepsilon \to 0} G_\varepsilon^1 , \tag{5.5.12}$$

where $s_{\mathcal{S}(\Omega)}$ denotes the strong topology of $\mathcal{S}(\Omega)$. The topology τ has been defined in the previous section. We assume that Theorem 5.4.6 holds true.

Step 1. Let $\{\Omega_i\}_{i \in \mathcal{I}}$ be a finite partition of Ω into open disjoint sets. We take $\boldsymbol{p}(x) = \sum\limits_{i \in I} \chi_{\Omega_i}(x)\boldsymbol{p}^i, \boldsymbol{p}^i \in \mathbb{E}_s^2$. Since Ω is a bounded domain, hence $\boldsymbol{p} \in L^2(\Omega, \mathbb{E}_s^2)$.

The Γ-limit exhibits the local character, cf. Sec. 1.3.4. This property will be used in our subsequent considerations.

Let $\{\boldsymbol{\varphi}^\varepsilon, w^\varepsilon\}_{\varepsilon > 0} \subset H_0^1(\Omega)^2 \times H_0^1(\Omega)$ be a sequence such that $(\boldsymbol{\varphi}^\varepsilon, w^\varepsilon) \to (\boldsymbol{\varphi} = -\nabla w, w^\varepsilon)$ strongly in $L^2(\Omega)^2 \times H^1(\Omega)$. We set

$$\theta^i(x) = \frac{1}{2}\sum_{\alpha,\beta} p_{\alpha\beta}^i x_\alpha x_\beta , \qquad v(x) = w(x) - \theta^i(x) , \qquad x \in \Omega_i ,$$

$$v^\varepsilon = w^\varepsilon - \theta^i , \qquad \omega^\varepsilon = \boldsymbol{\varphi}^\varepsilon + \nabla \theta^i .$$

Hence $v^\varepsilon \to v = w - \theta^i$ strongly in $H^1(\Omega_i)$, and $\omega^\varepsilon \to -\nabla w + \nabla \theta^i$ strongly in $L^2(\Omega^2)$ when $\varepsilon \to 0$. Moreover, $\gamma(\omega^\varepsilon, v^\varepsilon) = \boldsymbol{\varphi}^\varepsilon + \nabla w^\varepsilon$. Due to Theorem 5.4.6 and the local property of the Γ-limit we write

$$\int\limits_{\Omega_i} \mathcal{W}_h[\boldsymbol{\kappa}(v)]dx = \int\limits_{\Omega_i} \mathcal{W}_h[\boldsymbol{\kappa}(w) + \boldsymbol{p}^i]dx \le \lim_{\varepsilon \to 0} \inf \int\limits_{\Omega_i} k\left[\frac{x}{\varepsilon}, e(\omega^\varepsilon), \gamma(\omega, \varepsilon^\varepsilon)\right] dx$$

$$= \lim_{\varepsilon \to 0} \inf \int\limits_{\Omega_i} k\left[\frac{x}{\varepsilon}, e(\boldsymbol{\varphi}^\varepsilon) + \boldsymbol{p}^i, \gamma(\boldsymbol{\varphi}^\varepsilon, w^\varepsilon)\right] dx .$$

Hence

$$G_h^1(w, \boldsymbol{p}) \le \lim_{\varepsilon \to 0} \inf G_\varepsilon^1(\boldsymbol{\varphi}^\varepsilon, w^\varepsilon, \boldsymbol{p}, 0) .$$

Step 2. The properties of the stiffnesses $\widehat{D}$ and $\widehat{H}$ (Sec. 5.4) imply that there exists a constant $m > 0$ such that

$$|k(y, \boldsymbol{\epsilon}_1, \boldsymbol{a}_1) - k(y, \boldsymbol{\epsilon}_2, \boldsymbol{a}_2)| \le m[(|\boldsymbol{\epsilon}_1| + |\boldsymbol{\epsilon}_2|)|\boldsymbol{\epsilon}_1 - \boldsymbol{\epsilon}_2| + \frac{1}{\varepsilon^2}(|\boldsymbol{a}_1| + |\boldsymbol{a}_2|)|\boldsymbol{a}_1 - \boldsymbol{a}_2|] .$$

Hence

$$|G_\varepsilon^1(\varphi, w, \boldsymbol{p}_1, \boldsymbol{q}_1) - G_\varepsilon^1(\varphi, w, \boldsymbol{p}_2, \boldsymbol{q}_2)| \leq m \int_\Omega (2|e(\varphi)| + |\boldsymbol{p}_1| + |\boldsymbol{p}_2|)|\boldsymbol{p}_1 - \boldsymbol{p}_2| dx$$

$$+ \frac{m}{\varepsilon^2} \int_\Omega (2|\gamma(\varphi, w)| + |\boldsymbol{q}_1| + |\boldsymbol{q}_2|)|\boldsymbol{q}_1 - \boldsymbol{q}_2| dx \,.$$

Consequently, there exists a positive constant m_R such that if

$$\varphi \in H^1(\Omega)^2\,, \quad w \in H^1(\Omega)\,, \quad (\boldsymbol{p}_\alpha, \boldsymbol{q}_\alpha) \in \mathcal{S}(\Omega), \quad ||e(\varphi)||_{L^2} \leq R\,,$$
$$||\gamma(\varphi, w)||_{L^2} \leq \varepsilon\,, \quad ||\boldsymbol{p}_\alpha||_{L^2} \leq R\,, \quad ||\boldsymbol{q}_\alpha||_{L^2} \leq \varepsilon R \quad (\alpha = 1, 2)\,,$$
$$\tag{5.5.13}$$

then

$$|G_\varepsilon^1(\varphi, w, \boldsymbol{p}_1, \boldsymbol{q}_1) - G_\varepsilon^1(\varphi, w, \boldsymbol{p}_2, \boldsymbol{q}_2)| \leq m_R(||\boldsymbol{p}_1 - \boldsymbol{p}_2||_{L^2} + \frac{1}{\varepsilon}||\boldsymbol{q}_1 - \boldsymbol{q}_2||_{L^2})\,. \tag{5.5.14}$$

Let us now take $\{\varphi^\varepsilon, w^\varepsilon\}_{\varepsilon>0} \subset H^1(\Omega)^2 \times H^1(\Omega)$ and $\{\boldsymbol{p}^\varepsilon, \boldsymbol{q}^\varepsilon\}_{\varepsilon>0} \subset L^2(\Omega, \mathbb{E}_s^2) \times L^2(\Omega)^2$ such that $(\varphi^\varepsilon, w^\varepsilon) \rightharpoonup (\varphi, w)$ weakly in $H^1(\Omega)^2 \times H^1(\Omega)$, $\boldsymbol{p}^\varepsilon \to \boldsymbol{p}$ strongly in $L^2(\Omega, \mathbb{E}_s^2)$, $1/\varepsilon||\boldsymbol{q}^\varepsilon||_{L^2} \leq$ const.$< +\infty$,

$$\liminf_{\varepsilon \to 0} G_\varepsilon^1(\varphi^\varepsilon, w^\varepsilon; \boldsymbol{p}^\varepsilon, \boldsymbol{q}^\varepsilon) < +\infty\,. \tag{5.5.15}$$

Obviously, $\varphi = -\nabla w$ and $\boldsymbol{q}^\varepsilon \to 0$ in $L^2(\Omega)^2$.

Further, let $\{\boldsymbol{p}^\delta, \boldsymbol{q}^\delta\}_{\delta>0} \subset \mathcal{S}(\Omega)$ be a sequence of simple functions such that

$$\boldsymbol{p}^\delta \to \boldsymbol{p} \text{ strongly in } L^2(\Omega, \mathbb{E}_s^2) \text{ when } \delta \to 0\,,$$
$$\frac{1}{\delta}||\boldsymbol{q}^\delta||_{L^2} \leq \text{ const. } < +\infty\,.$$

We assume that $\delta \leq \varepsilon$. Let R and m_R be positive constants such that, cf. (5.5.13)

$$||e(\varphi^\varepsilon)||_{L^2} \leq R\,, \quad ||\gamma(\varphi^\varepsilon, w^\varepsilon)||_{L^2} \leq \varepsilon\,, \quad ||\boldsymbol{p}^\varepsilon||_{L^2} \leq R\,,$$
$$\frac{1}{\varepsilon}||\boldsymbol{q}^\varepsilon||_{L^2} \leq R\,, \quad ||\boldsymbol{p}^\delta||_{L^2} \leq R\,, \quad \frac{1}{\delta}||\boldsymbol{q}^\delta||_{L^2} \leq R\,,$$
$$\tag{5.5.16}$$

implies, cf. (5.5.14)

$$|G_\varepsilon^1(\varphi^\varepsilon, w^\varepsilon, \boldsymbol{p}^\delta, \boldsymbol{q}^\delta) - G_\varepsilon^1(\varphi^\varepsilon, w^\varepsilon, \boldsymbol{p}^\varepsilon, \boldsymbol{q}^\varepsilon)|$$
$$\leq m_R(||\boldsymbol{p}^\varepsilon - \boldsymbol{p}^\delta||_{L^2} + \frac{1}{\varepsilon}||\boldsymbol{q}^\varepsilon||_{L^2} + \frac{1}{\delta}||\boldsymbol{q}^\delta||_{L^2})\,.$$

By virtue of the previous step we have

$$G_\varepsilon^1(\varphi^\varepsilon, w^\varepsilon, \boldsymbol{p}^\varepsilon, \boldsymbol{q}^\varepsilon) \geq G_\varepsilon^1(\varphi^\varepsilon, w^\varepsilon, \boldsymbol{p}^\delta, \boldsymbol{q}^\delta)$$
$$- m_R(||\boldsymbol{p}^\varepsilon - \boldsymbol{p}^\delta||_{L^2} + \frac{1}{\varepsilon}||\boldsymbol{q}^\delta||_{L^2} + \frac{1}{\delta}||\boldsymbol{q}^\delta||_{L^2})\,,$$

and consequently

$$\liminf_{\varepsilon \to 0} G^1_\varepsilon(\boldsymbol{\varphi}^\varepsilon, w^\varepsilon, \boldsymbol{p}^\varepsilon, \boldsymbol{q}^\varepsilon)$$

$$\geq \int_\Omega \mathcal{W}_h(\boldsymbol{\kappa}(w) + \boldsymbol{p}^\delta)dx - m_R(||\boldsymbol{p} - \boldsymbol{p}^\delta||_{L^2} + \frac{1}{\delta}||\boldsymbol{q}^\delta||_{L^2}) \,.$$

Since $(\boldsymbol{p}^\delta, \boldsymbol{q}^\delta) \to (\boldsymbol{p}, \boldsymbol{0})$ strongly in $L^2(\Omega, \mathbb{E}^2_s) \times L^2(\Omega)^2$ as $\delta \to 0$, we conclude that

$$\liminf_{\varepsilon \to 0} G^1_\varepsilon(\boldsymbol{\varphi}^\varepsilon, w^\varepsilon, \boldsymbol{p}^\varepsilon, \boldsymbol{q}^\varepsilon) \geq \int_\Omega \mathcal{W}_h(\boldsymbol{\kappa}(w) + \boldsymbol{p})dx \,.$$

Step 3. For given $w \in H^2(\Omega)$, $\boldsymbol{p} \in L^2(\Omega, \mathbb{E}^2_s)$ and $\boldsymbol{q} = 0$ we need to find $\{\boldsymbol{\varphi}^\varepsilon, w^\varepsilon\}_{\varepsilon > 0} \subset H^1(\Omega)^2 \times H^1(\Omega)$ such that

$$(\boldsymbol{\varphi}^\varepsilon, w^\varepsilon) \xrightarrow[\varepsilon \to 0]{\tau} (\boldsymbol{\varphi} - \nabla w, w) \,,$$

and

$$\limsup_{\varepsilon \to 0} G^1_\varepsilon(\boldsymbol{\varphi}^\varepsilon, w^\varepsilon, \boldsymbol{p}^\varepsilon, \boldsymbol{q}^\varepsilon) \leq G^1_h(w, \boldsymbol{p}) \,, \tag{5.5.17}$$

provided that $\boldsymbol{p}$ is a simple function, cf. the first step. For the sake of simplicity we take $I = \{1, 2\}$ and set

$$\Sigma = \partial(\Omega_1 \cup \Omega_2) \,, \qquad \Sigma_\delta = \{x \in \Omega, \ \mathrm{dist}\,(x, \Sigma) < \delta\} \,, \delta > 0 \,.$$

Theorem 5.4.6 implies the existence of $\{\boldsymbol{\varphi}^{\varepsilon,i}, w^{\varepsilon,i}\} \subset H^1(\Omega_i)^2 \times H^1(\Omega_i)$ such that

$$(\boldsymbol{\varphi}^{\varepsilon,i}, w^{\varepsilon,i}) \to (\boldsymbol{\varphi}^i, w^i) = (-\nabla w, w)_{|\Omega_i} \,,$$

strongly in $H^1(\Omega_i)^2 \times H^1(\Omega_i)$ and

$$\limsup_{\varepsilon \to 0} \int_\Omega k\left[\frac{x}{\varepsilon}, e(\boldsymbol{\varphi}^{\varepsilon,i}) + \boldsymbol{p}, \boldsymbol{\gamma}(\boldsymbol{\varphi}^{\varepsilon,i}, w^{\varepsilon,i})\right] dx \leq \int_{\Omega_i} \mathcal{W}_h[\boldsymbol{\kappa}(w^i) + \boldsymbol{p}]dx \,. \tag{5.5.18}$$

Let $\psi^\delta \in \mathbf{D}(\Omega)$ be such that $0 \leq \psi^\delta \leq 1$, $\psi^\delta = 1$ on Σ_δ and $\psi^\delta = 0$ on $\Omega \backslash \Sigma_{2\delta}$.
To combine the functions $\boldsymbol{\varphi}^{\varepsilon,i}$ and $w^{\varepsilon,i}(i = 1, 2)$ we set

$$\left.\begin{array}{l} \boldsymbol{\varphi}^{\varepsilon,\delta} = (1 - \psi^\delta)\boldsymbol{\varphi}^{\varepsilon,i} + \psi^\delta \boldsymbol{\varphi} \\ w^{\varepsilon,\delta} = (1 - \psi^\delta)w^{\varepsilon,i} + \psi^\delta w \end{array}\right\} \text{ on } \Omega_i \,.$$

Hence we conclude that $(\boldsymbol{\varphi}^{\varepsilon,\delta}, w^{\varepsilon,\delta}) \in H^1(\Omega)^2 \times H^1(\Omega)$, because $\boldsymbol{\varphi}^{\varepsilon,\delta} = \boldsymbol{\varphi}$ and $w^{\varepsilon,\delta} = w$ on Σ_δ.

Taking $t \in (0,1)$ and performing simple calculations one gets

$$e(t\boldsymbol{\varphi}^{\varepsilon,\delta}) + \boldsymbol{p} = t\psi^{\delta}[e(\boldsymbol{\varphi}) + \boldsymbol{p}] + t(1 - \psi^{\delta})[e(\boldsymbol{\varphi}^{\varepsilon,i}) + \boldsymbol{p}]$$

$$+\frac{(1-t)t}{2(1-t)}[(\nabla\psi^{\delta}) \otimes (\boldsymbol{\varphi} - \boldsymbol{\varphi}^{\varepsilon,i}) + (\boldsymbol{\varphi} - \boldsymbol{\varphi}^{\varepsilon,i}) \otimes \nabla\psi^{\delta}] + (1-t)\boldsymbol{p} \,,$$

$$\boldsymbol{\gamma}(t\boldsymbol{\varphi}^{\varepsilon,\delta}, tw^{\varepsilon,\delta}) = t\psi^{\delta}\boldsymbol{\gamma}(\boldsymbol{\varphi}, w) + t(1 - \psi^{\delta})\boldsymbol{\gamma}(\boldsymbol{\varphi}^{\varepsilon,i}, w^{\varepsilon,i}) + \frac{(1-t)t}{1-t}(w - w^{\varepsilon,i})\nabla\psi^{\delta} \,,$$

where $\otimes$ denotes the tensor product.

Recalling that $\boldsymbol{\gamma}(\boldsymbol{\varphi}, w) = \boldsymbol{0}$, $t\psi^{\delta} + t(1 - \psi^{\delta}) + 1 - t = 1$ and exploiting convexity of the function $k(x/\varepsilon, \cdot, \cdot)$ we obtain

$$G_{\varepsilon}^{1}\left(t\boldsymbol{\varphi}^{\varepsilon,\delta}, tw^{\varepsilon,\delta}, \boldsymbol{p}, 0\right) \leq t\int_{\Sigma_{2\delta}} \psi^{\delta}k\left[\frac{x}{\varepsilon}, e(\boldsymbol{\varphi}) + \boldsymbol{p}, 0\right] dx$$

$$+t\sum_{i=1}^{2}\int_{\Omega_{i}}(1 - \psi^{\delta})k\left[\frac{x}{\varepsilon}, e(\boldsymbol{\varphi}^{\varepsilon,i}) + \boldsymbol{p}, \boldsymbol{\gamma}(\boldsymbol{\varphi}^{\varepsilon,i}, w^{\varepsilon,i})\right] dx$$

$$+(1-t)\sum_{i=1}^{2}\int_{\Omega_{i}}k\{\frac{x}{\varepsilon}, \frac{t}{2(1-t)}[(\nabla\psi^{\delta}) \otimes (\boldsymbol{\varphi} - \boldsymbol{\varphi}^{\varepsilon,i}) + (\boldsymbol{\varphi} - \boldsymbol{\varphi}^{\varepsilon,i}) \otimes \nabla\psi^{\delta}]$$

$$+\boldsymbol{p}, \frac{t}{1-t}(w - w^{\varepsilon,i})\nabla\psi^{\delta}\}dx \leq m_{1}\int_{\Sigma_{2\delta}}|e(\boldsymbol{\varphi}) + \boldsymbol{p}|^{2}dx$$

$$+\sum_{i=1}^{2}\int_{\Omega_{i}}k\left[\frac{x}{\varepsilon}, e(\boldsymbol{\varphi}^{\varepsilon,i}) + \boldsymbol{p}, \boldsymbol{\gamma}(\boldsymbol{\varphi}^{\varepsilon,i}, w^{\varepsilon,i})\right] dx$$

$$+2(1-t)m_{1}\sum_{i=1}^{2}\int_{\Omega_{i}}\left[\left(\frac{t}{1-t}\right)^{2}(|\nabla\psi^{\delta}|^{2}|\boldsymbol{\varphi} - \boldsymbol{\varphi}^{\varepsilon,i}|^{2} + |\boldsymbol{p}|^{2})\right] dx$$

$$+\frac{m_{1}(1-t)}{\varepsilon^{2}}\sum_{i=1}^{2}\int_{\Omega_{i}}\left(\frac{t}{1-t}\right)^{2}(|\nabla\psi^{\delta}|^{2}|w - w^{\varepsilon,i}|^{2})dx \,.$$

Because $w^{\varepsilon,i} \to w$ strongly in $L^{2}(\Omega_{i})$, therefore the last term is bounded from above by

$$\frac{m_{1}t^{2}}{1-t}\int_{\Omega}|\nabla\psi^{\delta}|^{2}dx \,. \tag{5.5.19}$$

Taking into account (5.5.18) and (5.5.19) one has

$$\limsup_{\varepsilon \to 0} G_{\varepsilon}^{1}(t\boldsymbol{\varphi}^{\varepsilon,\delta}, tw^{\varepsilon,\delta}, \boldsymbol{p}, 0) \leq m_{1}\int_{\Sigma_{2\delta}}|e(\boldsymbol{\varphi}) + \boldsymbol{p}|^{2}dx$$

$$+2(1-t)m_{1}\int_{\Omega}|\boldsymbol{p}|^{2}dx + \int_{\Omega}\mathcal{W}_{h}[\boldsymbol{\kappa}(w) + \boldsymbol{p}]dx + \frac{m_{1}t^{2}}{1-t}\int_{\Omega}|\nabla\psi^{\delta}|^{2}dx \,.$$

Noting that

$$\int_\Omega |\nabla\psi^\delta|^2 dx = \int_{\Sigma_{2\delta}} |\nabla\psi^\delta|^2 dx \to 0 , \qquad \text{when } \delta \to 0 ,$$

we obtain

$$\limsup_{t\to 1^-}[\limsup_{\delta\to 0}(\limsup_{\varepsilon\to 0} G^1_\varepsilon(t\varphi^{\varepsilon,\delta}, tw^{\varepsilon,\delta}, \boldsymbol{p}, 0))] \le G^1_h(w,\boldsymbol{p}) .$$

By using Lemma 1.3.27 we conclude that there exists $(t(\delta(\varepsilon)), \delta(\varepsilon)) \to (1^-, 0)$ such that

$$\limsup_{\varepsilon\to 0} G^1_\varepsilon[t(\delta(\varepsilon))\varphi^{\varepsilon,\delta(\varepsilon)}, t(\delta(\varepsilon))w^{\varepsilon,\delta(\varepsilon)}, \boldsymbol{p}, 0] \le G^1_h(w,\boldsymbol{p}) .$$

Setting $\varphi^\varepsilon = t(\delta(\varepsilon))\varphi^{\varepsilon,\delta(\varepsilon)}$ and $w^\varepsilon = t(\delta(\varepsilon))w^{\varepsilon,\delta(\varepsilon)}$ we have

$$\varphi^\varepsilon \to \varphi = -\nabla w \quad \text{strongly in } L^2(\Omega)^2 , \qquad w^\varepsilon \to w \quad \text{strongly in } \;\; H^1(\Omega) ,$$

as $\varepsilon \to 0$.

Step 4. Let $(\boldsymbol{p}, 0) \in \mathcal{S}(\Omega)$, otherwise arbitrary. There exists a sequence of simple functions $\{\boldsymbol{p}^k\}_{k\in\mathbb{N}}$ such that $\boldsymbol{p}^k \to \boldsymbol{p}$ strongly in $L^2(\Omega, \mathbb{E}^2_s)$ when $k \to \infty$. Further, for each $k \in \mathbb{N}$ there exists a sequence $\{\varphi^{\varepsilon,k}, w^{\varepsilon,k}\}_{\varepsilon>0} \subset H^1(\Omega)^2 \times H^1(\Omega)$ such that

$$\varphi^{\varepsilon,k} \to \varphi = -\nabla w \quad \text{strongly in } L^2(\Omega)^2 \quad \text{as} \quad \varepsilon \to 0 ,$$
$$w^{\varepsilon,k} \to w \quad \text{strongly in} \;\; H^1(\Omega) \quad \text{as} \quad \varepsilon \to 0 ,$$
$$\limsup_{\varepsilon\to 0} G^1_\varepsilon(\varphi^{\varepsilon,k}, w^{\varepsilon,k}, \boldsymbol{p}^k, 0) \le G^1_h(w, \boldsymbol{p}^k) .$$

Hence

$$\limsup_{k\to 0}\limsup_{\varepsilon\to 0} G^1_\varepsilon(\varphi^{\varepsilon,k}, w^{\varepsilon,k}, \boldsymbol{p}^k, 0) \le G^1_h(w,\boldsymbol{p}) .$$

Applying Lemma 1.3.27 once again we deduce existence of $k(\varepsilon) \underset{\varepsilon\to 0}{\to} \infty$ such that

$$\limsup_{\varepsilon\to 0} G^1_\varepsilon(\varphi^{\varepsilon,k(\varepsilon)}, w^{\varepsilon,k(\varepsilon)}, \boldsymbol{p}^{k(\varepsilon)}, 0) \le G^1_h(w,\boldsymbol{p}) ,$$

and

$$\varphi^{\varepsilon,k(\varepsilon)} \to \varphi = -\nabla w \quad \text{strongly in } L^2(\Omega)^2 \quad \text{as} \quad \varepsilon \to 0 ,$$
$$w^{\varepsilon,k(\varepsilon)} \to w \quad \text{strongly in} \;\; H^1(\Omega) \quad \text{as} \quad \varepsilon \to 0 .$$

Thus the Γ-convergence (5.5.12) is proved.

The assumption (A) of Theorem 1.3.36 is satisfied for arbitrary $0 < r < +\infty$. If $B_r = \{(\boldsymbol{p}, \boldsymbol{q}) \in L^2(\Omega, \mathbb{E}^2_s) \times L^2(\Omega)^2 \,|\,(\|\boldsymbol{p}\|_{L^2} + \|\boldsymbol{q}\|_{L^2})| \le r\}$ and $\{\boldsymbol{p}^\varepsilon, \boldsymbol{q}^\varepsilon\}_{\varepsilon>0} \subset B_r$ then taking $(\varphi^\varepsilon, w^\varepsilon)$ such that $e(\varphi^\varepsilon) = -\boldsymbol{p}^\varepsilon$ and $\varphi^\varepsilon + \nabla w^\varepsilon = -\boldsymbol{q}^\varepsilon$, we deduce

$$\limsup_\varepsilon \{G^1_\varepsilon(\varphi^\varepsilon, w^\varepsilon, \boldsymbol{p}^\varepsilon, \boldsymbol{q}^\varepsilon) - L_\varepsilon(\varphi^\varepsilon, w^\varepsilon)\} = \limsup_\varepsilon(-L_\varepsilon(\varphi^\varepsilon, w^\varepsilon)) < +\infty .$$

Let us now verify (1.3.75). For each $(\varphi, w) \in H_0^1(\Omega)^2 \times H_0^1(\Omega)$ one has, cf. the proof of Lemma 5.2

$$G_\varepsilon^1(\varphi, w, 0, 0) - L_\varepsilon(\varphi^\varepsilon, w^\varepsilon) \geq \widetilde{m}(\|\varphi\|_{H^1}^2 + \|w\|_{H^1}^2 - \|\widehat{q}\|_{L^2}\|w\|_{H^1} - \|\widehat{m}\|_{L^2}\|\varphi\|_{L^2}),$$

where $\widetilde{m}$ is a positive constant. Having satisfied the assumptions of Theorem 1.3.36 we conclude that

$$\mathcal{G}_h = \Gamma(w_{S(\Omega)}) - \lim_{\varepsilon \to 0} \mathcal{G}_\varepsilon,$$

where

$$\mathcal{G}_h(M) = -\int_\Omega \mathcal{W}_h^*(M(x))dx - I_{\mathcal{S}_h(\Omega)}(M),$$

and

$$\mathcal{S}_h(\Omega) = \{M \in L^2(\Omega, \mathbb{E}_2^s)|\ \mathrm{div\ div}M + \check{q} - \check{m}^\alpha{}_{,\alpha} = 0,\ \mathrm{in}\ \Omega\}.$$

Further, let $(\widetilde{\varphi}^\varepsilon, \widetilde{w}^\varepsilon)$ be a minimizer of the problem (P_ε) and $\{\varphi^{\varepsilon'}, w^{\varepsilon'}\}_{\varepsilon>0}$ a convergent subsequence:

$$(\varphi^{\varepsilon'}, w^{\varepsilon'}) \xrightarrow{\tau} (-\nabla\bar{w}, \bar{w})$$

as $\varepsilon \to 0$. Then

$$(P_h) \qquad J_h(\bar{w}) - L(\bar{w}) = \inf\{J_h(w) - L(w)|w \in H_0^2(\Omega)\},$$
$$\inf P_{\varepsilon'} \to \inf P_h \quad \mathrm{when}\ \varepsilon' \to 0.$$

Moreover, if $(\widetilde{M}^\varepsilon, \widetilde{Q}^\varepsilon)$ is a solution to the problem (P_ε^*) and $\{M^{\varepsilon'}, Q^{\varepsilon'}\}_{\varepsilon'>0}$ a convergent subsequence:

$$(M^{\varepsilon'}, Q^{\varepsilon'}) \xrightarrow{w_{S(\Omega)}} (\bar{M}, 0) \quad \mathrm{as}\ \varepsilon' \to 0,$$

then

$$(P_h^*) \qquad \mathcal{G}_h(\bar{M}) = \sup\{-\int_\Omega W^*(M(x))dx|M \in \mathcal{S}_h(\Omega)\},$$
$$\sup P_{\varepsilon'}^* \to \sup P_h^* \quad \mathrm{as}\ \varepsilon' \to 0, \qquad \inf P_h^* = \sup P_h^*. \qquad \square$$

5.6.   *Orthotropic plates periodic in one direction*

Assume that both elastic moduli and external surfaces defining faces of the plate vary periodically in one direction, say x_1. Thus stiffnesses D and H are $l_1 = a - $ periodic functions in y_1;

$$D^{\alpha\beta\lambda\mu}(y) = D^{\alpha\beta\lambda\mu}(y_1), \qquad H^{\alpha\beta}(y) = H^{\alpha\beta}(y_1). \qquad (5.6.1)$$

Averaging over Y assumes a simplified form:

$$\langle f \rangle = \frac{1}{a} \int_0^a f(y_1) dy_1 \,, \tag{5.6.2}$$

but notation $\langle \cdot \rangle$ will still be used here. The aim of the present section is to find effective stiffnesses according to procedures of Secs. 5.2 and 5.3 subsequently.

5.6.1. Effective stiffnesses according to the in-plane scaling approach

To find closed formulae for $D_H^{\alpha\beta\lambda\mu}$ and $H_H^{\alpha\beta}$ (see Eqs. (5.2.9)) one should solve the local problems $(P_{H,Y}^\alpha)$ of Sec. 5.2. Let us start with solving problem $(P_{H,Y}^2)$.

The strong solution of $(P_{H,Y}^2)$ is sought in the space: $\mathcal{C}^2[0,a] \cap H_{per}^1(0,a)$. We conjecture that $\Pi^{(\lambda)}$ do not depend on y_2. The local equation for $\Pi^{(\lambda)}$ reads

$$-\frac{d}{dy_1}\left[H^{11}(y_1)\frac{d\Pi^{(\lambda)}}{dy_1} \right] = \frac{dH^{1\lambda}}{dy_1} \tag{5.6.3}$$

and by integrating one finds

$$\frac{d\Pi^{(\lambda)}}{dy_1} = -\frac{H^{1\lambda}(y_1)}{H^{11}(y_1)} + \frac{C^{(\lambda)}}{H_{11}(y_1)} \,. \tag{5.6.4}$$

Periodicity of $\Pi^{(\lambda)}$ implies $\langle d\Pi^{(\lambda)}/dy_1 \rangle = 0$, hence

$$C^{(\lambda)} = \frac{\langle H^{1\lambda}/H^{11} \rangle}{\langle (H^{11})^{-1} \rangle} \tag{5.6.5}$$

and the quantities $d\Pi^{(\lambda)}/dy_1$ are now determined. Substituting them into formula $(5.2.9)_2$ one arrives at

$$H_H^{11} = \langle (H^{11})^{-1} \rangle^{-1} \,, \qquad H_H^{12} = H_H^{21} = \langle \frac{H^{12}}{H^{11}} \rangle H_H^{11} \,,$$

$$H_H^{22} = \left\langle H^{22} - \frac{(H^{12})^2}{H^{11}} \right\rangle + \left\langle \frac{H^{12}}{H^{11}} \right\rangle^2 H_H^{11} \,. \tag{5.6.6}$$

Note that finding $\Pi^{(\lambda)}$ was unnecessary. In the orthotropic case when y_α axes coincide with orthotropy axes $(H^{12} = 0)$ formulae (5.6.6) reduce to

$$H_H^{11} = \langle (H^{11})^{-1} \rangle^{-1} \,, \qquad H_H^{22} = \langle H^{22} \rangle \,, \qquad H_H^{12} = 0 \,. \tag{5.6.7}$$

Let us pass to finding stiffnesses $D_H^{\alpha\beta\lambda\mu}$. The strong solution of $(P_{H,Y}^1)$ is sought in the space $[\mathcal{C}^2[0,a] \cap H_{per}^1(Y)]^2$. We conjecture that solutions $\Upsilon^{(\lambda\mu)}$ are y_2-independent. Upon integrating, the local equations of $(P_{H,Y}^1)$ assume the form

$$D^{1111}(y_1)\frac{d\Upsilon_1^{(\lambda\mu)}}{dy_1} + D^{1112}(y_1)\frac{d\Upsilon_2^{(\lambda\mu)}}{dy_1} + D^{11\lambda\mu}(y_1) = C_1^{(\lambda\mu)} \,,$$

$$D^{2111}(y_1)\frac{d\Upsilon_1^{(\lambda\mu)}}{dy_1} + D^{2112}(y_1)\frac{d\Upsilon_2^{(\lambda\mu)}}{dy_1} + D^{21\lambda\mu}(y_1) = C_2^{(\lambda\mu)} \,, \tag{5.6.8}$$

where $C_\alpha^{(\lambda\mu)}$ are unknown constants. On solving these equations with respect to $d\Upsilon_1^{(\lambda\mu)}/dy_1$ and $d\Upsilon_2^{(\lambda\mu)}/dy_1$ and applying periodicity conditions:

$$\left\langle \frac{d\Upsilon_\alpha^{(\lambda\mu)}}{dy_1} \right\rangle = 0 , \qquad \alpha = 1, 2 \tag{5.6.9}$$

one finds $C_\alpha^{(\lambda\mu)}$, $\alpha = 1, 2$. Consequently one finds closed formulae for $d\Upsilon_\alpha^{(\lambda\mu)}/dy_1$ and substitute them into $(5.2.9)_1$. We have

$$D_H^{11\lambda\mu} = C_1^{(\lambda\mu)} , \qquad D_H^{12\lambda\mu} = C_2^{(\lambda\mu)} ,$$

$$D_H^{22\lambda\mu} = \left\langle D^{22\lambda\mu} + D^{2211}\frac{d\Upsilon_1^{(\lambda\mu)}}{dy_1} + D^{2212}\frac{d\Upsilon_2^{(\lambda\mu)}}{dy_1} \right\rangle . \tag{5.6.10}$$

After performing algebraic manipulations one finds formulae of the same form as those for plane elasticity case: $(3.7.12) - (3.7.13)$; one should replace there A with D and subscript "h" with "H". In the orthotropic case when y_α are orthotropy axes the formulae for $D_H^{\alpha\beta\lambda\mu}$ reduce to formulae $(3.7.14)$ with A replaced by D and the subscript "h" with "H". Consequently, the formulae $(3.7.15)$ can be used to find effective compliances $d_{\alpha\beta\lambda\mu}^H$; $d^H = (D_H)^{-1}$.

Let us compare formulae for $D_h^{\alpha\beta\lambda\mu}$ and $D_H^{\alpha\beta\lambda\mu}$ for the orthotropic case (cf. $(3.7.7)$ and $(3.7.14)$). The non-zero stiffnesses are

$$D_h^{\alpha\alpha\beta\beta} = D_H^{\alpha\alpha\beta\beta} ,$$
$$D_h^{1212} = \langle D^{1212}\rangle , \qquad D_H^{1212} = \langle D_H^{1212}\rangle = \langle (D^{1212})^{-1}\rangle^{-1} , \tag{5.6.11}$$

hence $D_h^{1212} \geq D_H^{1212}$ and equality holds only if the plate is homogeneous. This discrepancy will be discussed in Sec. 5.6.2 in more detail.

5.6.2. Effective stiffnesses according to the refined scaling approach

According to the in-plane scaling approach the homogenized Reissner-Hencky plate remains a Reissner-Hencky plate but with new stiffnesses. Quite differently the refined scaling approach works. Upon performing homogenization one obtains a Kirchhoff (thin) plate characterized by one stiffness tensor $D_{\hbar}$. The aim of this section is to find its components for the plate of stiffnesses $(5.6.1)$ with orthotropy axes coinciding with y_1, y_2. The orthotropy assumption is not indispensable.

The strong solution to problem $(5.3.14)$ (cf. $\bar{P}_{S,Y}^2$, Sec. 2.7): $(\Phi^{(\alpha\beta)}(y_1), \chi^{(\alpha\beta)}(y_1))$ is sought in the space $[C^2[0, a] \cap H_{per}^1(0, a)]^3$. The variational equation $(5.3.14)$ implies the following local differential equations

$$-\frac{dM_{(\alpha\beta)}^{11}}{dy_1} + Q_{(\alpha\beta)}^1 = 0 , \qquad \frac{dQ_{(\alpha\beta)}^1}{dy_1} = 0 , \qquad -\frac{dM_{(\alpha\beta)}^{12}}{dy_1} + Q_{(\alpha\beta)}^2 = 0 , \tag{5.6.12}$$

where

$$M_{(\alpha\beta)}^{11} = \hat{D}^{1111}\frac{d\Phi_1^{(\alpha\beta)}}{dy_1} + \hat{D}^{11\alpha\beta} , \qquad M_{(\alpha\beta)}^{12} = \hat{D}^{1212}\frac{d\Phi_2^{(\alpha\beta)}}{dy_1} + \hat{D}^{12\alpha\beta} ,$$

$$Q_{(\alpha\beta)}^1 = \hat{H}^{11}\left(\frac{d\chi^{(\alpha\beta)}}{dy_1} + \Phi_1^{(\alpha\beta)}\right) , \qquad Q_{(\alpha\beta)}^2 = \hat{H}^{22}\Phi_2^{(\alpha\beta)} .$$

$$(5.6.13)$$

The periodicity conditions implied by (5.3.14) have the form

$$\Phi_\sigma^{(\alpha\beta)}(0) = \Phi_\sigma^{(\alpha\beta)}(a) , \quad \chi^{(\alpha\beta)}(0) = \chi^{(\alpha\beta)}(a) ,$$

$$M_{(\alpha\beta)}^{1\sigma}(0) = M_{(\alpha\beta)}^{1\sigma}(a) , \quad Q_{(\alpha\beta)}^1(0) = Q_{(\alpha\beta)}^1(a) .$$

$$(5.6.14)$$

By integrating the first two of equations (5.6.12) and using periodicity conditions $(5.6.14)_{3,4}$ one finds

$$\frac{d\Phi_1^{(\alpha\beta)}}{dy_1} = \frac{1}{\hat{D}^{1111}(y_1)}\left(\langle(\hat{D}^{1111})^{-1}\rangle^{-1} \cdot \left\langle\frac{\hat{D}^{\alpha\beta 11}}{\hat{D}^{1111}}\right\rangle - \hat{D}^{\alpha\beta 11}(y_1)\right) . \qquad (5.6.15)$$

On substituting this result into (5.3.15a) and taking into account $(3.2.1d)_4$ one concludes that

$$D_\hbar^{\alpha\alpha\beta\beta} = D_h^{\alpha\alpha\beta\beta} = D_H^{\alpha\alpha\beta\beta} , \qquad (5.6.16)$$

cf. $(5.6.11)_1$. Moreover

$$D_\hbar^{\alpha\alpha\alpha\beta} - D_\hbar^{\alpha\alpha\beta\alpha} - D_h^{\alpha\beta\alpha\alpha} - D_\hbar^{\beta\alpha\alpha\alpha} = 0 , \qquad (5.6.17)$$

for $\beta = 3 - \alpha$.

Much more effort is required to find a closed formula for $D_\hbar^{1212}$. This stiffness is given by

$$D_\hbar^{1212} = \left\langle \hat{D}^{1212}(y_1)\left(1 + \frac{d\Phi_2^{(12)}}{dy_1}\right)\right\rangle , \qquad (5.6.18)$$

where $\Phi_2^{(12)}$ is a solution to the differential equation $(5.6.12)_3$ or

$$\frac{d}{dy_1}\left[\hat{D}^{1212}(y_1)\frac{d\Phi_2^{(12)}}{dy_1}\right] - \hat{H}^{22}(y_1)\Phi_2^{(12)} = -\frac{d\hat{D}^{1212}}{dy_1} \qquad (5.6.19)$$

with boundary conditions

$$\Phi_2^{(12)}(0) = \Phi_2^{(12)}(a) , \qquad \frac{d\Phi_2^{(12)}}{dy_1}(0) = \frac{d\Phi_2^{(12)}}{dy_1}(a) . \qquad (5.6.20)$$

The variational formulation of this one-dimensional problem reads

$$\left|\begin{array}{l} \text{find } \Phi_2^{(12)} \in H_{per}^1(0, a) \text{ such that} \\[2mm] \langle\hat{D}^{1212}\dfrac{d\Phi_2^{(12)}}{dy_1}\dfrac{dv}{dy_1} + \hat{H}^{22}\Phi_2^{(12)}v\rangle = -\langle\hat{D}^{1212}\dfrac{dv}{dy_1}\rangle \qquad \forall\, v \in H_{per}^1(0, a) . \end{array}\right. \qquad (5.6.21)$$

The problem (5.6.19 – 5.6.20) or (5.6.21) is in general too complex to be explicitly solved and to find $D_{\hbar}^{1212}$ one can resort to solving problem (5.6.21) numerically. An explicit formula for $D_{\hbar}^{1212}$ can be found if $\hat{D}^{1212}$ and $\hat{H}^{22}$ are piece-wise constant functions:

$$\hat{D}^{1212} = \begin{cases} \hat{D}_1^{1212} \,, \; y_1 \in I_1 \\ \hat{D}_2^{1212} \,, \; y_1 \in I_2 \end{cases}, \qquad \hat{H}^{22} = \begin{cases} \hat{H}_1^{22} \,, \; y_1 \in I_1 \\ \hat{H}_2^{22} \,, \; y_1 \in I_2 \end{cases}, \qquad (5.6.22)$$

where $I_1 = (0, b_1)$, $I_2 = (b_1, a)$, $b_2 = a - b_1$.

In the same way the unknown function is represented

$$\Phi_2^{12} = \begin{cases} \phi_1 \,, \; y_1 \in I_1 \\ \phi_2 \,, \; y_1 \in I_2 \end{cases}. \qquad (5.6.23)$$

Let us denote

$$\frac{1}{d_\sigma} = \left(\frac{\hat{H}_\sigma^{22}}{\hat{D}_\sigma^{1212}} \right)^{1/2} , \qquad \gamma_\alpha = \frac{b_\alpha}{a} \,; \qquad \chi_\sigma = \frac{a}{d_\sigma} \,,$$

$$\lambda = \frac{b_1}{d_1} \,, \qquad \lambda\sigma = \frac{b_1}{d_2} \,, \qquad \sigma = \frac{d_1}{d_2} \qquad \xi = \frac{y_1}{b_1} \,, \qquad \omega = \frac{a}{b_1} \,. \qquad (5.6.24)$$

Quantities d_σ are of length dimension; other coefficients are non-dimensional. The governing equation (5.6.19) assumes the form of two equations

$$\frac{d^2\phi_1}{d\xi^2} - \lambda^2 \phi_1 = 0 \,, \qquad \frac{d^2\phi_2}{d\xi^2} - \lambda^2\sigma^2 \phi_2 = 0 \,, \qquad (5.6.25)$$

for intervals I_1 and I_2 respectively. Their solutions have the form

$$\phi_1 = C_1 e^{-\lambda\xi} + B_1 e^{-\lambda(1-\xi)} \,, \qquad \xi \in (0, 1)$$

$$\phi_2 = C_2 e^{-\lambda\sigma(\xi-1)} + B_2 e^{-\lambda\sigma(\omega-\xi)} \,, \qquad \xi \in (1, \omega) \,. \qquad (5.6.26)$$

The switching conditions

$$\phi_1(1) = \phi_2(1) \,, \qquad M_{(12)}^{12}(1 - 0) = M_{(12)}^{12}(1 + 0) \qquad (5.6.27)$$

and periodicity conditions

$$\phi_1(0) = \phi_2(\omega) \,, \qquad M_{(12)}^{12}(0) = M_{(12)}^{12}(\omega) \qquad (5.6.28)$$

make it possible to find the constants C_α, B_α; it turns out that $C_\alpha = -B_\alpha$. Omitting a lengthy derivation one eventually finds

$$D_{\hbar}^{1212} = \gamma_1 P_1 + \gamma_2 P_2 - \frac{(P_1 - P_2)^2}{Z} \,, \qquad (5.6.29)$$

where

$$P_\sigma = (\varepsilon_0)^3 \hat{D}_\sigma^{1212} = D_\sigma^{1212} \,,$$

$$Z = \frac{1}{2}\chi_1 P_1 \coth\left(\frac{1}{2}\gamma_1\chi_1\right) + \frac{1}{2}\chi_2 P_2 \coth\left(\frac{1}{2}\gamma_2\chi_2\right) \qquad (5.6.30)$$

and $\coth(x) := \cosh(x)/\sinh(x)$.

5.6.3. Effective torsional stiffness of plates of step-wise varying thickness

Consider a plate made of an isotropic elastic material, the thickness of which varies periodically in x_1 direction and alternately assumes the values $2h_1$ and $2h_2$, see Fig. 5.6.1.

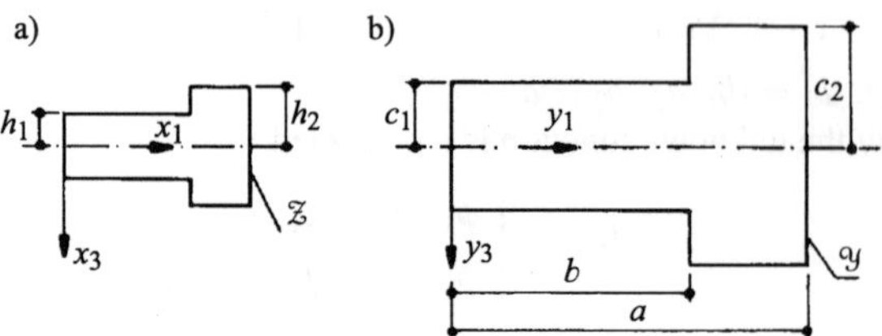

Fig. 5.6.1. Original and rescaled cells of periodicity

The dimensions of the rescaled cell $\mathcal{Y}$ are given in Fig. 5.6.1b.

The rescaled torsional and transverse shearing stiffnesses are, cf. (5.6.22)

$$\hat{D}_\sigma^{1212}(y_1) = \frac{2}{3}G_s(c_\sigma)^3 , \qquad \hat{H}_\sigma^{22} = 2kc_\sigma G_s , \tag{5.6.31}$$

where $2G_s = E/(1+\nu)$; k is a shear correction factor; E represents Young's modulus, ν is Poisson's ratio. Thus $d_\sigma = (1/3k)^{1/2}c_\sigma$ and $\sigma = c_1/c_2$.

The following formulae determine the effective torsional stiffness of the plate of Fig. 5.6.1 according to:

– thin plate-based homogenization:

$$\frac{D_h^{1212}}{D_2} = \frac{\sigma^3 + \omega - 1}{\omega} , \tag{5.6.32}$$

– in-plane scaling homogenization of Reissner-Hencky equations

$$\frac{D_H^{1212}}{D_2} = \frac{\omega\sigma^3}{\sigma^3(\omega - 1) + 1} , \tag{5.6.33}$$

– refined scaling homogenization of Reissner-Hencky equations

$$\frac{D_h^{1212}}{D_2} = \frac{\sigma^3 + \omega - 1}{\omega} - \frac{(1 - \sigma^3)^2}{\omega\sigma\lambda}g(\sigma, \omega, \lambda) , \tag{5.6.34}$$

where

$$g(\sigma, \omega, \lambda) = \frac{2}{\sigma^2 \coth\left(\dfrac{\lambda}{2}\right) + \coth\left(\dfrac{\lambda(\omega - 1)\sigma}{2}\right)} , \qquad D_2 = \frac{2}{3}G_s(h_2)^3 . \tag{5.6.35}$$

Consider a family of plates of constant c_α and ω. Then $\lambda \sim a$. Note that D_h^{1212} and D_H^{1212} do not depend on λ. One can prove that

$$D_H^{1212} = D_h^{1212}(\lambda = 0) \leq D_h^{1212} \leq D_h^{1212}(\lambda = \infty) = D_h^{1212} . \tag{5.6.36}$$

If c_α and ω are fixed then the mean thickness is fixed. Hence all plates have the same volume, see Fig. 5.6.2.

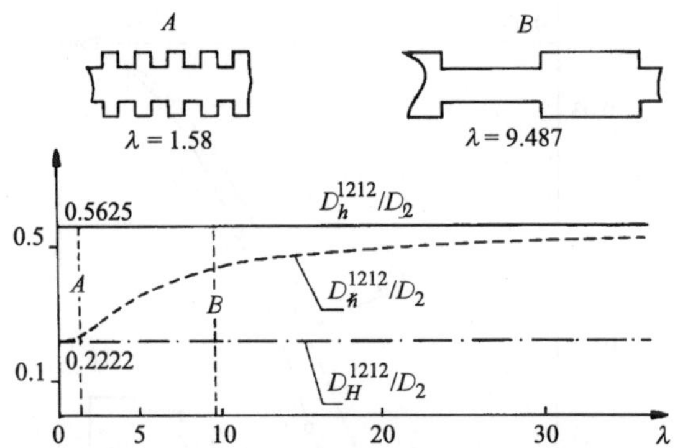

Fig. 5.6.2. The effective torsional stiffness versus λ.

Comparison of three approaches for the data: $k = 5/6$, $\sigma = 0.5$, $\omega = 2$

The speed of convergence of $D_h^{1212}(\lambda)$ to D_h^{1212} if $\lambda \to \infty$ depends on $\sigma = c_1/c_2$. For $\sigma \gg 1$ or $\sigma \ll 1$ this convergence is very slow.

Consider now another family of constant volume plates for which a, ω and the mean thickness

$$c = \gamma_1 c_1 + \gamma_2 c_2$$

are kept fixed. This family is indexed by $\sigma = c_1/c_2$. Let us introduce the non-dimensional stiffnesses:

$$(\bar{D}_h, \ \bar{D}_{\bar{h}}, \ \bar{D}_H) = \frac{(D_h^{1212}, \ D_{\bar{h}}^{1212}, \ D_H^{1212})}{E(l_1^z)^3} \tag{5.6.37}$$

which are functions of σ. Assume the data: $\nu = 0.25$, $\omega = 2$, $a/c = 2$, $c_1 + c_2 = a$. Functions $\bar{D}_h(\sigma)$, $\bar{D}_{\bar{h}}(\sigma)$, $\bar{D}_H(\sigma)$ for $\sigma \in [0, 1]$ are given in Fig. 5.6.3.

In the same figure two results due to Caillerie-Kohn-Vogelius approach of Section 2.3 (formulae (2.3.62)) are presented; these results are taken from Kohn and Vogelius (1984). It is obvious that $\bar{D}_h$ overestimates the effective torsional stiffness.

Let us consider now the effective torsional stiffness of a gridwork. To this end let us fix a, b_1 and c_2. Then D_2 and ω are fixed. Let $c_1 \to 0$ ($\sigma \to 0$), hence $\lambda \to \infty$, but $\lambda\sigma$ is fixed. We find the limits:

$$D_h^{1212}/D_2 \to \frac{\omega - 1}{\omega}, \qquad D_H^{1212}/D_2 \to 0,$$
$$D_{\bar{h}}^{1212}/D_2 \to \frac{\omega - 1}{\omega} - \frac{1}{\beta}\tanh\left(\frac{\omega - 1}{\omega}\beta\right) \tag{5.6.38}$$

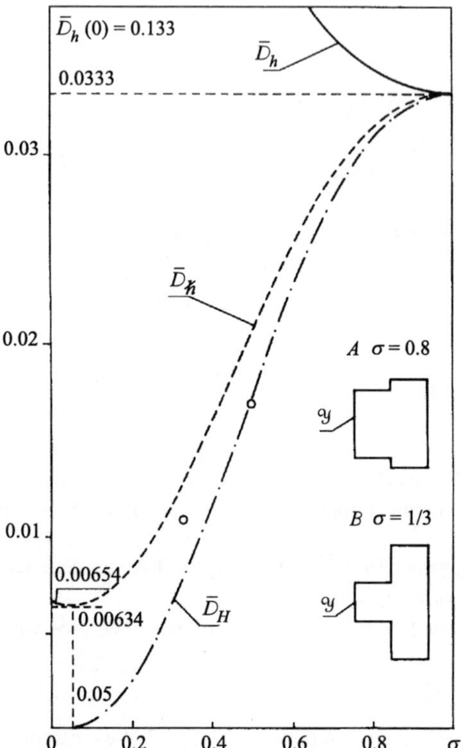

Fig. 5.6.3. The effective torsional stiffness versus σ.
Comparison for the data: $\nu = 0.25$, $\omega = 2$, $a/c = 2$, $c_1 + c_2 = a$.
The circles "o" represent the results found by the method of Section 2.3

where $\beta = \dfrac{1}{2}\left(\dfrac{5}{2}\right)^{1/2} \cdot \dfrac{a}{c_2}$ for $k = \dfrac{5}{6}$. These formulae describe the torsional stiffnesses of a gridwork of Fig. 5.6.4a.

Note that D_h^{1212}/D_2 does not depend on the ratio $(a - b_1)/c_2$.

Let us keep the parameters a and c_2 fixed and let $b \to 0$ ($\omega \to \infty$). Then

$$D_h^{1212}/D_2 \to 1 \qquad \text{and} \qquad D_h^{1212}/D_2 \to 1 - \frac{\tanh\beta}{\beta}. \qquad (5.6.39)$$

The above results refer to the case of a plate cut transversely with period a, see Fig. 5.6.4b. The formula for D_h^{1212} disregards this cutting.

The discussion presented above shows that the refined scaling approach is superior to the former two-dimensional methods of assessing the effective stiffnesses of plates. By virtue of the refined scaling the formulae for $D_\hbar^{1212}$ in the one-dimensional case, and all $D_\hbar^{\alpha\beta\lambda\mu}$ in general, become sensitive to the transverse shape of the periodicity cell.

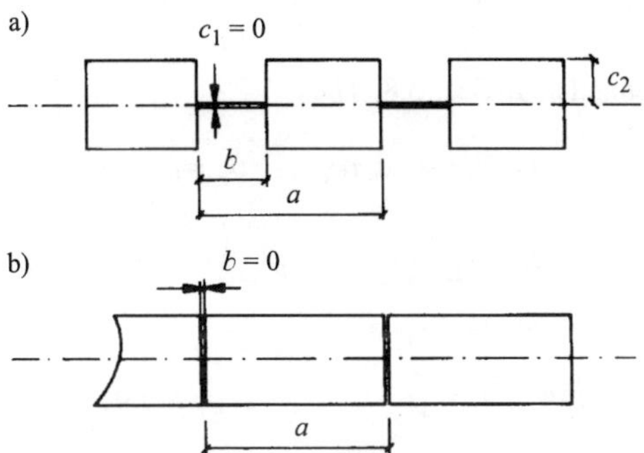

Fig. 5.6.4. A limiting case: gridwork. In case (b) the distance between the beams is zero

5.6.4. Formula of Tartar-Francfort-Murat type for effective stiffnesses of ribbed plates. In-plane scaling approach

Consider a transversely symmetric plate made from ribs of stiffnesses D_1, H_1 and D_2, H_2 with area fractions θ_1 and θ_2 respectively; $\theta_1 + \theta_2 = 1$, see Fig. 3.8.1 which remains relevant here now, however, the plate undergoes transverse shear deformations. Assume that condition (3.8.1) holds as well as the quadratic form $q_\alpha(H_2^{\alpha\beta} - H_1^{\alpha\beta})q_\beta$ is positive definite. The direction of stiffening $n = (n_\alpha)$ is not correlated with anisotropy directions of D_α and H_α.

The effective tensors D_H and H_H are determined by the following formulae of Francfort and Murat

$$\theta_1(D_2 - D_H)^{-1} = (D_2 - D_1)^{-1} - \theta_2\Gamma_A , \tag{5.6.40}$$

where Γ_A is given by (3.9.16) with $(g_{\beta\mu})$ defined by a matrix inverse to

$$2D_2^{\alpha\beta\lambda\mu}n_\lambda n_\alpha , \tag{5.6.41}$$

and of Tartar:

$$\theta_1(H_2 - H_H)^{-1} = (H_2 - H_1)^{-1} - \frac{\theta_2}{n_\alpha H_2^{\alpha\beta} n_\beta}n \otimes n . \tag{5.6.42}$$

Formula (5.6.40) has the same form as formula (3.9.15) derived in Sec. 3.9, because of analogy between the local problems $(P^1_{H,Y})$ of Sec. 5.2 and $(P^1_{KS,Y})$ of Sec. 3.2. Hence its derivation is unnecessary.

Formula (5.6.42) can be justified as follows. By (5.6.4) the quantities

$$\beta_\alpha = \partial\Pi/\partial y_\alpha , \tag{5.6.43}$$

are piece-wise constant; they are equal to $\overset{1}{\beta}$ and $\overset{2}{\beta}$ in phases 1 and 2 respectively. Along the γ interface the field Π is continuous, hence

$$\overset{1}{\beta}_\tau=\overset{2}{\beta}_\tau\,, \qquad \overset{\sigma}{\beta}_\tau=\overset{\sigma}{\beta}_\alpha\,\tau^\alpha\,, \tag{5.6.44}$$

where $\boldsymbol{\tau} = (\tau^\alpha)$ is defined as in Sec. 3.8. Thus

$$\overset{\sigma}{\boldsymbol{\beta}}= \beta_\tau\boldsymbol{\tau}+ \overset{\sigma}{\beta}_n\,\boldsymbol{n}\,, \qquad \beta_\tau=\overset{1}{\beta}_\tau=\overset{2}{\beta}_\tau \tag{5.6.45}$$

and hence

$$\overset{1}{\boldsymbol{\beta}} - \overset{2}{\boldsymbol{\beta}}= k\boldsymbol{n}\,, \qquad k =\overset{1}{\beta}_n - \overset{2}{\beta}_n\,. \tag{5.6.46}$$

The following equations are at our disposal:

$$\overset{\sigma}{Q}{}^\alpha = H_\sigma^{\alpha\gamma}\,\overset{\sigma}{\beta}_\gamma\,, \qquad \beta^h = \theta_1\,\overset{1}{\beta} +\theta_2\,\overset{2}{\beta}\,,$$
$$Q_h = \theta_1\,\overset{1}{Q} +\theta_2\,\overset{2}{Q}\,, \qquad \overset{1}{Q}{}^\alpha n_\alpha =\overset{2}{Q}{}^\alpha n_\alpha\,. \tag{5.6.47}$$

We proceed as in Sec. 3.8 and find subsequently

$$\beta_\lambda^h =\overset{1}{\beta}_\lambda -\rho\theta_2 n_\lambda\,, \tag{5.6.48}$$

$$(H_2^{\alpha\lambda} - H_H^{\alpha\lambda})\beta_\lambda^h = \theta_1(H_2^{\alpha\lambda} - H_1^{\alpha\lambda})\,\overset{1}{\beta}_\lambda\,, \tag{5.6.49}$$

$$\rho = \frac{\mathcal{M}^\gamma n_\gamma}{n_\alpha H_2^{\alpha\beta} n_\beta}\,, \qquad \mathcal{M}^\alpha = (H_2^{\alpha\lambda} - H_1^{\alpha\lambda})\,\overset{1}{\beta}_\lambda\,. \tag{5.6.50}$$

Hence

$$(H_2^{\alpha\lambda} - H_H^{\alpha\lambda})\beta_\lambda^h = \theta_1\mathcal{M}^\alpha\,, \tag{5.6.51}$$

and

$$\theta_1(\boldsymbol{H}_2 - \boldsymbol{H}_H)_{\alpha\lambda}^{-1}\mathcal{M}^\alpha = (\boldsymbol{H}_2 - \boldsymbol{H}_1)_{\alpha\lambda}^{-1}\mathcal{M}^\alpha - \frac{\theta_2}{n_\alpha H_2^{\alpha\beta} n_\beta}n_\gamma n_\lambda\mathcal{M}^\gamma\,, \tag{5.6.52}$$

which ends the derivation.

If $\boldsymbol{H}_2$ is isotropic: $H_2^{\alpha\beta} = H_2\delta^{\alpha\beta}$, we find $n_\alpha H_2^{\alpha\beta} n_\beta = H_2$ and Eq. (5.6.42) assumes the form suitable for subsequent layering process, as described in Sec. 3.8.

5.7. *Other linear and nonlinear models of plates with moderate thickness. Homogenization study*

In the case of transversely nonhomogeneous plates both the Kirchhoff and Reissner-Hencky displacement assumptions can lead to unacceptable inaccuracies, see Christensen (1979). In this section a possible improvement of these displacement assumptions is analyzed and the homogenized formulae for this refined plate model are derived. A justification of the displacement-based hypothesis (5.7.23) needs introduction of the Reissner averaged generalized displacements. They are derived in Sec. 5.7.1 that precedes the section in which the refined, geometrically nonlinear plate model is put forward. Homogenization is dealt with in Sec. 5.7.3.

5.7.1. Reissner's model

There are two reasons why the Reissner-Hencky modelling (Sec. 5.1) is unsatisfactory. First, it introduces the assumption of the in-plane stress state in an arbitrary manner, thus introducing the contradiction between the kinematic assumptions and the constitutive relations. Moreover, it leads to overestimation of the transverse shear stiffnesses. The original model of Reissner (1945) is free of these drawbacks. This model is recalled here to clear up the choice of the generalized displacements of a refined plate theory considered in Sec. 5.7.2.

Assume that a plate occupies a domain $B = \Omega \times (-c, c)$, where $2c$ represents the plate thickness and the mid-plane Ω is parametrized by Cartesian coordinates (x_α); $x = (x_\alpha, x_3) \in B$. The x_3 axis is perpendicular to Ω. Thus the plate is of constant thickness.

Let us assume that the plate material is transversely homogeneous with moduli $C^{ijkl}(x)$ and flexibilities $c_{ijkl}(x)$; $c = C^{-1}$. The planes $x_3 = $ constant are planes of material symmetry, hence (2.4.1) holds. On a part $\Upsilon_0 = \Gamma_0 \times (-c, c)$, $\Gamma_0 \subset \partial\Omega$, of the lateral surface the displacements are prescribed

$$w_i(s, x_3) = w_i^0(s, x_3), \qquad s \in \Gamma_0, \qquad x_3 \in [-c, c]. \tag{5.7.1}$$

The other part $\Upsilon_1 = \Gamma_1 \times (-c, c)$ is subject to the tractions $T^i(s, x_3)$; $\overline{\Gamma}_1 \cup \overline{\Gamma}_0 = \Gamma = \partial\Omega$. The faces $x_3 = \pm c$ are subject to the distributed transverse loading of intensity $\frac{1}{2}q(x)$. Thus the stress type boundary conditions are given by

$$\sigma^{\alpha 3}(x, \pm c) = 0, \quad \sigma^{33}(x, \pm c) = \pm\frac{1}{2}q(x), \quad x \in \Omega$$

$$\sigma^{ij}(s, x_3)n_j(s) = T^i(s, x_3), \quad s \in \Gamma_1, \quad x_3 \in [-c, c]. \tag{5.7.2}$$

Let us confine our consideration to the case of the tractions being of the following form:

$$T^\alpha(s, x_3) = \frac{1}{2c}\bar{N}^\alpha(s) + \frac{3}{2}\frac{x_3}{c^3}\bar{M}^\alpha(s),$$

$$T^3(s, x_3) = \frac{3}{4c}\left(1 - \left(\frac{x_3}{c}\right)^2\right)\bar{Q}(s), \tag{5.7.3}$$

where $\bar{N}^\alpha(s)$, $\bar{M}^\alpha(s)$, $\bar{Q}(s)$ are given functions of $s \in \Gamma_1$. The body forces are omitted. A trial stress field (τ^{ij}) is said to be statically admissible if its components $\tau^{ij} = \tau^{ji}$ satisfy the equilibrium equations

$$\tau^{\alpha\beta}_{,\beta} + \tau^{\alpha 3}_{,3} = 0, \qquad \tau^{\alpha 3}_{,\alpha} + \tau^{33}_{,3} = 0, \tag{5.7.4}$$

within B and the boundary conditions (5.7.2). According to the Castigliano theorem the unknown stress field minimizes the functional:

$$J(\tau) = \frac{1}{2}\int_B (\tau^{\alpha\beta}c_{\alpha\beta\lambda\mu}\tau^{\lambda\mu} + 2\tau^{\alpha\beta}c_{\alpha\beta 33}\tau^{33}$$

$$+ \tau^{33}c_{3333}\tau^{33} + 4\tau^{\alpha 3}c_{\alpha 3\lambda 3}\tau^{\lambda 3})dx - \int_{\Upsilon_0} \tau^{ij}\nu_j \mathring{w}_i \, ds dx_3. \tag{5.7.5}$$

Thus $J(\boldsymbol{\sigma}) = \min\{J(\boldsymbol{\tau}) | \boldsymbol{\tau}$ - statically admissible$\}$. Consequently, the functional

$$\mathcal{L}(\boldsymbol{\tau}, \boldsymbol{\lambda}, \boldsymbol{\eta}) = J(\boldsymbol{\tau}) + \int_B [\lambda_\alpha(\boldsymbol{x})(\tau^{\alpha\beta}_{,\beta} + \tau^{\alpha 3}_{,3}) + \lambda_3(\boldsymbol{x})(\tau^{\alpha 3}_{,\alpha} + \tau^{33}_{,3})]d\boldsymbol{x}$$

$$+ \int_{\Upsilon_1} [\eta_\alpha(s, x_3)(\tau^{\alpha\beta}\nu_\beta - T^\alpha) + \eta_3(s, x_3)(\tau^{\alpha 3}\nu_\alpha - T^3)]dsdx_3 , \qquad (5.7.6)$$

assumes a stationary value at the solution $(\boldsymbol{\sigma}, \widetilde{\boldsymbol{\lambda}}, \widetilde{\boldsymbol{\eta}})$. Here λ_i, η_i are Lagrange multipliers. We observe that a rigorous proof of the stationarity of the functional $\mathcal{L}$ can be performed by using Brezzi's theorem, cf. Sec. 1.2.2. The same concerns the functional $\mathcal{L}_1$ below.

Let us assume that the stresses within B are of the form

$$\sigma^{\alpha\beta}(\boldsymbol{x}) = \frac{1}{2c}N^{\alpha\beta}(x) + \frac{3}{2}\frac{x_3}{c^3}M^{\alpha\beta}(x) ,$$

$$\sigma^{\alpha 3}(\boldsymbol{x}) = \frac{3}{4c}\left(1 - \left(\frac{x_3}{c}\right)^2\right)Q^\alpha(x) , \qquad (5.7.7)$$

$$\sigma^{33}(\boldsymbol{x}) = \frac{3}{4c}x_3\left[1 - \frac{1}{3}\left(\frac{x_3}{c}\right)^2\right]q(x) .$$

Note that the boundary conditions $(5.7.2)_1$ are identically satisfied. The fields $N^{\alpha\beta}, M^{\alpha\beta}, Q^\alpha$ are unknown fields defined on Ω. The stress fields (5.7.7) satisfy the equilibrium equations (5.7.4) in B if the fields $\boldsymbol{N}, \boldsymbol{M}, \boldsymbol{Q}$ and q are linked by

$$N^{\alpha\beta}_{,\beta} = 0 , \qquad M^{\alpha\beta}_{,\beta} + Q^\alpha = 0 , \qquad Q^\alpha_{,\alpha} + q = 0 , \qquad \text{in } \Omega . \qquad (5.7.8)$$

The trial stress fields are assumed in a form similar to (5.7.7):

$$\tau^{\alpha\beta}(\boldsymbol{x}) = \frac{1}{2c}\mathcal{N}^{\alpha\beta}(x) + \frac{3}{2}\frac{x_3}{c^3}\mathcal{M}^{\alpha\beta}(x) ,$$

$$\tau^{\alpha 3}(\boldsymbol{x}) = \frac{3}{4c}\left(1 - \left(\frac{x_3}{c}\right)^2\right)\mathcal{Q}^\alpha(x) , \qquad (5.7.9)$$

$$\tau^{33}(\boldsymbol{x}) = \frac{3}{4c}x_3\left[1 - \frac{1}{3}\left(\frac{x_3}{c}\right)^2\right]q(x) .$$

The idea of Reissner's (1945) modelling consists in requiring that $\mathcal{L}$ assumes a stationary value for the trial stress fields of the form (5.7.9). Thus the set of all statically admissible stress fields is confined to the set of the stress fields which are statically admissible within B but outside a boundary layer along Υ. In fact, the boundary conditions $(5.7.2)_2$ will not be satisfied identically.

Substituting (5.7.9) into (5.7.6) yields

$$\mathcal{L}_1(\mathcal{N}, \mathcal{M}, \mathcal{Q} \ ; \ v, \psi, \upsilon \ ; \ z, \phi, z) = J_1(\mathcal{N}, \mathcal{M}, \mathcal{Q})$$

$$+ \int_{\Omega} [v_\alpha(\mathcal{N}^{\alpha\beta}_{,\beta}) + \psi_\alpha(\mathcal{M}^{\alpha\beta}_{,\beta} + \mathcal{Q}^\alpha) + v(\mathcal{Q}^\alpha_{,\alpha} + q)]dx \tag{5.7.10}$$

$$+ \int_{\Gamma_1} [z_\alpha(\mathcal{N}^{\alpha\beta}n_\beta - \bar{N}^\alpha) + \phi_\alpha(\mathcal{M}^{\alpha\beta}_{,\beta} - \bar{M}^\alpha) + z(\mathcal{Q}^\alpha n_\alpha - \bar{Q})]ds ,$$

where

$$J_1 = \frac{1}{2} \int_{\Omega} (\mathcal{N}^{\alpha\beta} a_{\alpha\beta\lambda\mu} \mathcal{N}^{\lambda\mu} + \mathcal{M}^{\alpha\beta} d_{\alpha\beta\lambda\mu} \mathcal{M}^{\lambda\mu} + \mathcal{Q}^\alpha h_{\alpha\beta} \mathcal{Q}^\beta$$

$$+ b_{\alpha\beta} \mathcal{M}^{\alpha\beta} q + A q^2)dx - \int_{\Gamma_0} (\overset{\circ}{u}_\alpha \mathcal{N}^{\alpha\beta} n_\beta + \overset{\circ}{\varphi}_\alpha \mathcal{M}^{\alpha\beta} n_\beta + \overset{\circ}{w} \mathcal{Q}^\alpha n_\alpha)ds , \tag{5.7.11}$$

$$a_{\alpha\beta\lambda\mu} = \frac{1}{2c} c_{\alpha\beta\lambda\mu} , \quad d_{\alpha\beta\lambda\mu} = \frac{3}{2c^3} c_{\alpha\beta\lambda\mu} , \quad h_{\alpha\beta} = \frac{12}{5c} c_{\alpha3\beta3} , \quad b_{\alpha\beta} = \frac{6}{5c} c_{\alpha\beta33} , \tag{5.7.12}$$

and the constant A does not play any role below. The Lagrangian multipliers involved in (5.7.10) and (5.7.11) are weighted averages of the multipliers of the original three-dimensional problem:

$$v_\alpha(x) = \frac{1}{2c} \int_{-c}^{c} \lambda_\alpha(x, x_3) dx_3 ,$$

$$\psi_\alpha(x) = \frac{3}{2c^3} \int_{-c}^{c} x_3 \lambda_\alpha(x, x_3) dx_3 , \tag{5.7.13}$$

$$v(x) = \frac{3}{4c} \int_{-c}^{c} \left[1 - \left(\frac{x_3}{c}\right)^2 \right] \lambda(x, x_3) dx_3 .$$

The fields $(z_\alpha, \phi_\alpha, z)$ and $(u^0_\alpha, \varphi^0_\alpha, w^0)$ depend on (η_α, η_3) and (w^0_α, w^0_3) in a similar manner. At the stationary point we have:

$$\mathcal{N}^{\alpha\beta} = N^{\alpha\beta} , \quad \mathcal{M}^{\alpha\beta} = M^{\alpha\beta} , \quad \mathcal{Q}^\alpha = Q^\alpha ,$$

$$v_\alpha = u_\alpha , \quad \psi_\alpha = \varphi_\alpha , \quad v = w . \tag{5.7.14}$$

The stationary conditions of $\mathcal{L}_1$ with respect to v_α, ψ_α and v give (5.7.8). These equations are equilibrium equations of the plate model derived. The stationary conditions with respect to z_α, ϕ_α and z yield the boundary conditions on Γ_1:

$$N^{\alpha\beta} n_\beta = \overline{N}^\alpha , \quad M^{\alpha\beta} n_\beta = \overline{M}^\alpha , \quad Q^\alpha n_\alpha = \overline{Q} . \tag{5.7.15}$$

The stationary conditions with respect to $\mathcal{N}^{\alpha\beta}, \mathcal{M}^{\alpha\beta}$ and $\mathcal{Q}^\alpha$ yield the constitutive relations in the inverted form:

$$u_{(\alpha,\beta)} = a_{\alpha\beta\lambda\mu} N^{\lambda\mu} , \quad \varphi_{(\alpha,\beta)} = d_{\alpha\beta\lambda\mu} M^{\lambda\mu} + b_{\alpha\beta} q , \quad \gamma_\alpha = h_{\alpha\beta} Q^\beta , \tag{5.7.16}$$

where $\gamma_\alpha = w,_\alpha + \varphi_\alpha$. Moreover, the displacement boundary conditions on Γ_0 are satisfied:

$$u_\alpha = \overset{\circ}{u}_\alpha\,, \qquad \varphi_\alpha = \overset{\circ}{\varphi}_\alpha\,, \qquad w = \overset{\circ}{w} \qquad \text{on} \quad \Gamma_0\,. \tag{5.7.17}$$

The last conditions provide a physical interpretation of the Lagrangian multipliers u, φ, w. Along Γ_0 they are weighted averages of given displacements $\overset{\circ}{w}_\alpha$, $\overset{\circ}{w}_3$. The weighted functions of x_3 that occur in (5.7.13) follow from the structure of the equilibrium equations (5.7.4).

The variational equilibrium equation encompasses Eqs. (5.7.8) and (5.7.15). It is given by

$$\int_\Omega [N^{\alpha\beta} e_{\alpha\beta}(\boldsymbol{v} - \boldsymbol{u}) + M^{\alpha\beta} e_{\alpha\beta}(\boldsymbol{\psi} - \boldsymbol{\varphi}) + Q^\alpha(\psi_\alpha - \varphi_\alpha + (v - w),_\alpha)]dx$$

$$= \int_\Omega q(v - w)dx + \int_{\Gamma_1} [\overline{N}^\alpha(v_\alpha - u_\alpha) + \overline{M}^\alpha(\psi_\alpha - \varphi_\alpha) + \overline{Q}(v - w)]ds\,, \tag{5.7.18}$$

for every $(\boldsymbol{v}, \boldsymbol{\psi}, v)$ satisfying:

$$v_\alpha = \overset{\circ}{u}_\alpha\,, \quad \psi_\alpha = \overset{\circ}{\varphi}_\alpha\,, \quad v = \overset{\circ}{w} \quad \text{on} \quad \Gamma_1\,. \tag{5.7.19}$$

By inverting (5.7.16) we find the primal form of the constitutive relations:

$$N^{\alpha\beta} = A^{\alpha\beta\lambda\mu} e_{\lambda\mu}(\boldsymbol{u})\,, \quad M^{\alpha\beta} = D^{\alpha\beta\lambda\mu} \varrho_{\lambda\mu}(\boldsymbol{\varphi}) - B^{\alpha\beta}q\,, \quad Q^\alpha = H^{\alpha\lambda}\gamma_\lambda(\boldsymbol{\varphi}, w)\,, \tag{5.7.20}$$

where $\gamma_\lambda(\boldsymbol{\varphi}, w) = w,_\lambda + \varphi_\lambda$ and $\varrho_{\lambda\mu}(\boldsymbol{\varphi}) = e_{\lambda\mu}(\boldsymbol{\varphi})$. Moreover

$$A^{\alpha\beta\lambda\mu} = 2c\widetilde{C}^{\alpha\beta\lambda\mu}\,, \quad D^{\alpha\beta\lambda\mu} = \frac{2}{3}c^3\widetilde{C}^{\alpha\beta\lambda\mu}\,,$$

$$H^{\alpha\beta} = \frac{5c}{12}\widehat{c}^{\alpha\beta}\,, \quad B^{\alpha\beta} = D^{\alpha\beta\lambda\mu}b_{\lambda\mu}\,, \tag{5.7.21}$$

where $\widetilde{C} = [c_{\alpha\beta\lambda\mu}]^{-1}$, $\widehat{c} = [c_{\alpha3\beta3}]^{-1}$.

Formally, there are two differences in comparison with the Reissner-Hencky model of Sec. 5.1. Now the expression for M contains a free term depending on q and the stiffnesses $H^{\alpha\beta}$ correspond to the case of $k = \dfrac{5}{6} \neq 1$, see Eqs. (5.1.5), (5.1.6) in the case of orthotropy.

Remark 5.7.1. The Reissner (1945) model derived here cannot be easily generalized to the case of a plate being transversely nonhomogeneous. Then the distribution of stresses should be carefully predicted taking into account the material distribution across the thickness. For the three-layer plates loaded in plane such a modelling is discussed in Sec. 11.

Remark 5.7.2. From the three-dimensional elasticity theory it is known that the Lagrangian multipliers $\lambda_\alpha, \lambda_3$ involved in (5.7.6) can be identified with the displacements w_α, w_3 of the body. Thus the formulae (5.7.13) define the two-dimensional generalized displacements in terms of the displacements within the plate body B.

Let us check the result of substitution of the Hencky assumptions (5.1.1) into (5.7.13) for $\lambda_\alpha = w_\alpha$, $\lambda = w_3$. We find the following identities:

$$v_\alpha = u_\alpha , \qquad \psi_\alpha = \varphi_\alpha , \qquad v = w \tag{5.7.22}$$

which confirm that Hencky's and Reissner's generalized displacements have a similar meaning.

5.7.2. A refined theory of moderately thick plates undergoing moderately large deflections

In the present section the von Kármán theory of thin plates undergoing moderately large deflections (Sec. 4.1) is refined to take into account the effect of transverse shear deformation. Having at our disposal the kinematic modelling, such as that of Sec. 5, or stress-based approach, such as that of Sec. 5.7.1, we shall use the former. Such a choice is motivated by difficulties arising in constructing the statically admissible stress fields in the case of transversely nonhomogeneous plates.

Let us consider the plate problem of Sec. 5.7.1. The model construction will be based on the kinematic assumptions of the form:

$$w_\alpha(x, x_3) = u_\alpha(x) - x_3 w_{,\alpha}(x) + f(x_3)[\varphi_\alpha(x) + w_{,\alpha}(x)] ,$$
$$w_3(x, x_3) = w(x) , \tag{5.7.23}$$

where

$$f(x_3) = \frac{5}{4}x_3 \left[1 - \frac{1}{3}\left(\frac{x_3}{c}\right)^2\right] \tag{5.7.24}$$

and $2c = $ const represents the plate thickness, see Sec. 5.7.1. Note that

$$\int_{-c}^{c} (x_3)^2 dx_3 = \int_{-c}^{c} x_3 f(x_3) dx_3 = \frac{2}{3}c^3 . \tag{5.7.25}$$

Thus substitution of (5.7.23) into (5.7.13), with $\lambda_\alpha = w_\alpha$, $\lambda = w_3$, again gives the identities (5.7.22). This justifies using the same notation: u_α, φ_α and w, as in Secs. 5.1 and 5.7.1, and clears up physical meaning of these unknowns.

The infinitesimal transverse shear deformations associated with (5.7.23) are given by

$$e_{\alpha 3}(\boldsymbol{w}) = \frac{5}{8}\left[1 - \left(\frac{x_3}{c}\right)^2\right] (\varphi_\alpha + w_{,\alpha}) . \tag{5.7.26}$$

Hence the boundary conditions $(5.7.2)_1$ concerning the stresses $\sigma^{\alpha 3}$ are identically satisfied, which links the present approach with that of Reissner (1945). Here, however, the boundary condition concerning $\sigma^{33}(x, \pm c)$ is violated. Thus we expect that the plate theory based on (5.7.23) should be more accurate than Kirchhoff's and Reissner-Hencky's theories, but whether it is better than the Reissner (1945) theory is an open problem. This model will, however, be used mostly in the analysis of transversely nonhomogeneous plates, like laminated plates, for which the Reissner approach cannot be applied.

Let us proceed now to deriving a nonlinear plate model based on the assumptions (5.7.23) and on the stress assumption of the generalized plane stress, cf. (5.1.2).

Within the nonlinear elasticity theory the components of the deformation tensor are given by

$$E_{ij}(\boldsymbol{w}) = e_{ij}(\boldsymbol{w}) + \frac{1}{2}\delta^{kl}w_{i,k}w_{j,l} \, . \tag{5.7.27}$$

We shall adopt here the von Kármán approximation according to which only the nonlinear terms depending on the deflection w are retained. Hence we assume, cf. (5.7.23)

$$E_{\alpha\beta}(\boldsymbol{w}) = e_{\alpha\beta}(\boldsymbol{w}) + \frac{1}{2}w_{,\alpha}w_{,\beta} \, , \qquad E_{i3}(\boldsymbol{w}) = e_{i3}(\boldsymbol{w}) \, . \tag{5.7.28}$$

The components of the linear strain measures have the form:

$$e_{\alpha\beta}(\boldsymbol{w}) = e_{\alpha\beta}(\boldsymbol{u}) + x_3\kappa_{\alpha\beta}(\boldsymbol{w}) + f(x_3)[\rho_{\alpha\beta}(\boldsymbol{\varphi}) - \kappa_{\alpha\beta}(\boldsymbol{w})] \, , \quad e_{33}(\boldsymbol{w}) = 0 \, , \tag{5.7.29}$$

and $e_{\alpha3}(\boldsymbol{w})$ is given by (5.7.26). Here

$$\rho_{\alpha\beta}(\boldsymbol{\varphi}) = e_{\alpha\beta}(\boldsymbol{\varphi}) \, , \qquad \kappa_{\alpha\beta}(\boldsymbol{w}) = -w_{,\alpha\beta} \, . \tag{5.7.30}$$

The energy density $\sigma^{ij}E_{ij}$ reduces to $\sigma^{ij}E_{ij} = \sigma^{\alpha\beta}E_{\alpha\beta} + 2\sigma^{\alpha3}E_{\alpha3}$. Hence

$$\sigma^{ij}E_{ij} = \sigma^{\alpha\beta}[e_{\alpha\beta}(\boldsymbol{u}) + \frac{1}{2}w_{,\alpha}w_{,\beta}] + x_3\sigma^{\alpha\beta}\kappa_{\alpha\beta}(\boldsymbol{w})$$
$$+ f(x_3)\sigma^{\alpha\beta}[\rho_{\alpha\beta}(\boldsymbol{\varphi}) - \kappa_{\alpha\beta}(\boldsymbol{w})] + \frac{5}{4}\left[1 - \left(\frac{x_3}{c}\right)^2\right]\sigma^{\alpha3}\gamma_\alpha(\boldsymbol{\varphi}, w) \, , \tag{5.7.31}$$

where $\gamma_\alpha(\boldsymbol{\varphi}, w) = \varphi_\alpha + w_{,\alpha}$. Thus the energy of the plate

$$\mathcal{E}(\boldsymbol{u}, \boldsymbol{\varphi}, w) = \frac{1}{2}\int_\Omega \int_{-c}^{c} \sigma^{ij}E_{ij}\,dx_3dx \, , \tag{5.7.32}$$

assumes the form

$$\mathcal{E}(\boldsymbol{u}, \boldsymbol{\varphi}, w) = \frac{1}{2}\int_\Omega [N^{\alpha\beta}(e_{\alpha\beta}(\boldsymbol{u}) + \frac{1}{2}w_{,\alpha}w_{,\beta}) + M^{\alpha\beta}\kappa_{\alpha\beta}(\boldsymbol{w})$$
$$+ \mathbf{M}^{\alpha\beta}(\rho_{\alpha\beta}(\boldsymbol{\varphi}) - \kappa_{\alpha\beta}(\boldsymbol{w})) + Q^\alpha\gamma_\alpha(\boldsymbol{\varphi}, w)]dx \, , \tag{5.7.33}$$

where the stress resultants are:

$$\begin{bmatrix} N^{\alpha\beta} \\ M^{\alpha\beta} \\ \mathbf{M}^{\alpha\beta} \end{bmatrix} = \int\limits_{-c}^{c} \begin{bmatrix} 1 \\ x_3 \\ f(x_3) \end{bmatrix} \sigma^{\alpha\beta} dx_3$$

$$Q^{\alpha} = \frac{5}{4} \int\limits_{-c}^{c} \left[1 - \left(\frac{x_3}{c}\right)^2 \right] \sigma^{\alpha 3} dx_3 \ . \tag{5.7.34}$$

In the case of transverse homogeneity, the substitution of the constitutive relation:

$$\sigma^{\alpha\beta} = \widetilde{C}^{\alpha\beta\lambda\mu} E_{\lambda\mu}(\boldsymbol{w}) \ , \qquad \sigma^{\alpha 3} = 2C^{\alpha 3\lambda 3} E_{\lambda 3}(\boldsymbol{w}) \ , \tag{5.7.35}$$

into (5.7.34) yields:

$$N^{\alpha\beta} = A^{\alpha\beta\lambda\mu}[e_{\lambda\mu}(\boldsymbol{u}) + \frac{1}{2} w_{,\alpha} w_{,\beta}] \ ,$$

$$M^{\alpha\beta} = D^{\alpha\beta\lambda\mu} \rho_{\lambda\mu}(\boldsymbol{\varphi}) \ ,$$

$$\mathbf{M}^{\alpha\beta} = \frac{1}{84} D^{\alpha\beta\lambda\mu}[85\rho_{\lambda\mu}(\boldsymbol{\varphi}) - \kappa_{\lambda\mu}(w)] \ , \tag{5.7.36}$$

$$Q^{\alpha} = H^{\alpha\beta} \gamma_{\beta}(\boldsymbol{\varphi}, w) \ ,$$

where the tensors A, D and H are given by (5.7.19). In the general case the relations (5.7.36) should be appropriately modified.

The expression (5.7.33) will be simplified by introducing

$$S^{\alpha\beta} = M^{\alpha\beta} - \mathbf{M}^{\alpha\beta} \ , \tag{5.7.37}$$

or

$$S^{\alpha\beta} = \frac{1}{84} D^{\alpha\beta\lambda\mu}[\kappa_{\lambda\mu}(w) - \rho_{\lambda\mu}(\boldsymbol{\varphi})] \ , \tag{5.7.38}$$

in the transversely homogeneous case. Since

$$(M^{\alpha\beta} - \mathbf{M}^{\alpha\beta})\kappa_{\alpha\beta} + \mathbf{M}^{\alpha\beta}\rho_{\alpha\beta} - S^{\alpha\beta}(\kappa_{\alpha\beta} - \rho_{\alpha\beta}) + M^{\alpha\beta}\rho_{\alpha\beta} \ ,$$

therefore $\mathcal{E}$ assumes the form

$$\mathcal{E}(\boldsymbol{u}, \boldsymbol{\varphi}, w) = \frac{1}{2} \int\limits_{\Omega} \{ A^{\alpha\beta\lambda\mu}[e_{\lambda\mu}(\boldsymbol{u}) + \frac{1}{2} w_{,\lambda} w_{,\mu}][e_{\alpha\beta}(\boldsymbol{u}) + \frac{1}{2} w_{,\alpha} w_{,\beta}]$$

$$+ \ D^{\alpha\beta\lambda\mu} \rho_{\alpha\beta}(\boldsymbol{\varphi}) \rho_{\lambda\mu}(\boldsymbol{\varphi})$$

$$+ \ \widetilde{D}^{\alpha\beta\lambda\mu}[\kappa_{\alpha\beta}(w) - \rho_{\alpha\beta}(\boldsymbol{\varphi})][\kappa_{\lambda\mu}(w) - \rho_{\lambda\mu}(\boldsymbol{\varphi})]$$

$$+ \ H^{\alpha\beta} \gamma_{\alpha}(\boldsymbol{\varphi}, w) \gamma_{\beta}(\boldsymbol{\varphi}, w) \} dx \ , \tag{5.7.39}$$

where $\widetilde{D} = \dfrac{1}{84} D$ in the case of transverse homogeneity.

Remark 5.7.3. If the stresses $\sigma^{\alpha\beta}$ are linear across the thickness, then $M^{\alpha\beta} = \mathbf{M}^{\alpha\beta}$ and $S^{\alpha\beta} = 0$, cf. Eq. (5.7.37). Thus the term underlined in (5.7.39) represents the difference between the total bending energy and the energy due to linear distribution of the stresses $\sigma^{\alpha\beta}$.
□

Remark 5.7.4. If the displacements satisfy the Kirchhoff constraints (3.1.1) then $\varphi_\alpha = -w_{,\alpha}$, hence

$$\gamma_\alpha(\varphi, w) = 0 \,, \quad \rho_{\alpha\beta}(\varphi) = \kappa_{\alpha\beta}(w) \,. \tag{5.7.40}$$

The expression (5.7.39) reduces to the von Kármán form:

$$\mathcal{E}_1(\boldsymbol{u}, w) = \frac{1}{2}\int_\Omega \{A^{\alpha\beta\lambda\mu}[e_{\lambda\mu}(\boldsymbol{u}) + \frac{1}{2}w_{,\lambda}w_{,\mu}][e_{\alpha\beta}(\boldsymbol{u}) + \frac{1}{2}w_{,\alpha}w_{,\beta}]$$

$$+ D^{\alpha\beta\lambda\mu}\kappa_{\alpha\beta}(w)\kappa_{\lambda\mu}(w)\}dx \,. \tag{5.7.41}$$

□

Assume that the tractions T^i on Υ_1 are distributed according to (5.7.3). Then

$$\int_{\Upsilon_1} T^i w_i dS = \int_{\Gamma_1} (\overline{N}^\alpha u_\alpha + \overline{M}^\alpha \varphi_\alpha + \overline{Q}w)ds \,. \tag{5.7.42}$$

Thus the functional of the total potential energy of the plate has the following form:

$$J(\boldsymbol{u}, \varphi, w) = \mathcal{E}_1(\boldsymbol{u}, \varphi, w) - \int_\Omega qw dx - \int_{\Gamma_1} (\overline{N}^\alpha u_\alpha + \overline{M}^\alpha \varphi_\alpha + \overline{Q}w)ds \,. \tag{5.7.43}$$

5.7.3. Homogenization study

The aim of this section is to perform homogenization of the nonlinear plate model with the constitutive equations given by (5.7.36)$_1$, (5.7.36)$_{3,4}$ and (5.7.38). The loading functional $L(\boldsymbol{u}, \varphi, w)$ is assumed to play the role of a perturbation functional in the homogenization procedure. For instance, it may be given by the two integrals appearing on the r.h.s. of (5.7.43).

As previously, let Y be a two-dimensional basic cell. Below, we shall be concerned with the non-uniform homogenization. Therefore we assume that the elastic moduli are now given by: $A_\varepsilon^{\alpha\beta\lambda\mu}(x) = A^{\alpha\beta\lambda\mu}(x, \frac{x}{\varepsilon})$, $D_\varepsilon^{\alpha\beta\lambda\mu}(x) = D_\varepsilon^{\alpha\beta\lambda\mu}(x, \frac{x}{\varepsilon})$, $H_\varepsilon^{\alpha\beta}(x) = H^{\alpha\beta}(x, \frac{x}{\varepsilon})$, $x \in \Omega$, and the functions $A^{\alpha\beta\lambda\mu}(x, y)$, $D^{\alpha\beta\lambda\mu}(x, y)$, $H^{\alpha\beta}(x, y)$ are Y-periodic in the second variable. It is reasonable to assume that $A^{\alpha\beta\lambda\mu}$, $D^{\alpha\beta\lambda\mu}$, $H^{\alpha\beta}$ are functions belonging to $L^\infty(\Omega \times Y)$. These moduli satisfy the usual symmetry and ellipticity conditions:

(a) $A^{\alpha\beta\lambda\mu} = A^{\lambda\mu\alpha\beta} = A^{\beta\alpha\lambda\mu}$, $D^{\alpha\beta\lambda\mu} = D^{\lambda\mu\alpha\beta} = D^{\beta\alpha\lambda\mu}$, $H^{\alpha\beta} = H^{\beta\alpha}$.

(b) There exist constants $c_1 \geq c_0 > 0$ such that

$$\forall \, \boldsymbol{\epsilon} \in \mathbb{E}_s^2 \qquad c_0|\boldsymbol{\epsilon}|^2 \leq A^{\alpha\beta\lambda\mu}(x,y) \leq c_1|\boldsymbol{\epsilon}|^2 \,,$$
$$\forall \, \boldsymbol{\epsilon} \in \mathbb{E}_s^2 \qquad c_0|\boldsymbol{\epsilon}|^2 \leq D^{\alpha\beta\lambda\mu}(x,y) \leq c_1|\boldsymbol{\epsilon}|^2 \,,$$
$$\forall \, \boldsymbol{\eta} \in \mathbb{R}^2 \qquad c_0|\boldsymbol{\eta}|^2 \leq H^{\alpha\beta}(x,y) \leq c_1|\boldsymbol{\eta}|^2 \,,$$

for each $x \in \Omega$, $y \in Y$.

The functional $J_\varepsilon(\boldsymbol{u}, \boldsymbol{\varphi}, w)$ of the total potential energy now has the following form:

$$J_\varepsilon(\boldsymbol{u}, \boldsymbol{\varphi}, w) = \mathcal{E}_\varepsilon(\boldsymbol{u}, \boldsymbol{\varphi}, w) - L(\boldsymbol{u}, \boldsymbol{\varphi}, w) \,, \qquad (5.7.44)$$

where

$$\mathcal{E}_\varepsilon(\boldsymbol{u}, \boldsymbol{\varphi}, w) = \int_\Omega \mathcal{W}[x, \frac{x}{\varepsilon}, \boldsymbol{E}(\boldsymbol{u}, w), \boldsymbol{\rho}(\boldsymbol{\varphi}), \boldsymbol{\kappa}(w), \boldsymbol{\gamma}(\boldsymbol{\varphi}, w)]dx \,, \qquad (5.7.45)$$

$$\mathcal{W}\,[x, y, \boldsymbol{E}(\boldsymbol{u}, w), \boldsymbol{\rho}(\boldsymbol{\varphi}), \boldsymbol{\kappa}(w), \boldsymbol{\gamma}(\boldsymbol{\varphi}, w)] = \frac{1}{2}A^{\alpha\beta\lambda\mu}(x,y)E_{\alpha\beta}(\boldsymbol{u}, w)E_{\lambda\mu}(\boldsymbol{u}, w)$$

$$+ \frac{1}{2}D^{\alpha\beta\lambda\mu}(x,y)[\rho_{\alpha\beta}(\boldsymbol{\varphi})\rho_{\lambda\mu}(\boldsymbol{\varphi}) + \frac{1}{84}(\kappa_{\alpha\beta}(w) - \rho_{\alpha\beta}(\boldsymbol{\varphi}))(\kappa_{\lambda\mu}(w) - \rho_{\lambda\mu}(\boldsymbol{\varphi}))]$$

$$+ \frac{1}{2}H^{\alpha\beta}(x,y)\gamma_\alpha(\boldsymbol{\varphi}, w)\gamma_\beta(\boldsymbol{\varphi}, w) \,. \qquad (5.7.46)$$

For each fixed $\varepsilon > 0$ the minimization problem:

$$(P_\varepsilon) \qquad J_\varepsilon(\boldsymbol{u}^\varepsilon, \boldsymbol{\varphi}^\varepsilon, w^\varepsilon) = \inf\{J_\varepsilon(\boldsymbol{u}, \boldsymbol{\varphi}, w)|(\boldsymbol{u}, \boldsymbol{\varphi}, w) \in V(\Omega)\}$$

possesses a solution. Here

$$V(\Omega) = \{(\boldsymbol{u}, \boldsymbol{\varphi}, w) \in H^1(\Omega)^2 \times H^1(\Omega)^2 \times H_0^2(\Omega)|\boldsymbol{u} = \boldsymbol{0}, \boldsymbol{\varphi} = \boldsymbol{0}, \text{ on } \Gamma_0\} \,, \qquad (5.7.47)$$

and Γ_0 is the part of the boundary Γ, meas $\Gamma_0 > 0$, $\Gamma = \overline{\Gamma}_0 \cup \overline{\Gamma}_1$. Different boundary conditions are also possible. Detailed study of the existence of solution for this type of nonlinear plates was performed by Bielski and Telega (1996). Under the assumption of "small loading", a uniqueness theorem was established.

To pass with ε to zero we shall use the Γ-convergence method. We note that the method of two-scale asymptotic expansions requires more regular elastic moduli; then $A^{\alpha\beta\lambda\mu}$ and $H^{\alpha\beta}$ are to be elements of $W^{1,\infty}(\Omega \times Y)$ while $D^{\alpha\beta\lambda\mu} \in W^{2,\infty}(\Omega \times Y)$. According to Theorem 1.3.28 one can significantly weaken the assumptions specified by the condition (b).

Γ-convergence of a class of functionals

Prior to finding the Γ-limit of the sequence $\{J_\varepsilon\}_{\varepsilon>0}$ specified by (5.7.44) we shall extend Theorem 1.3.28. As we shall see, this extension implies much weaker assumptions on the stored energy function $\mathcal{W}$ than those sufficient for the existence of solutions.

Consider the following integrand

$$f = f(x, y, \mathbf{u}, \mathbf{w}, \mathbf{f}, \mathbf{p}, \mathbf{q}, \mathbf{r}, \mathbf{s}) \,,$$

where $f(\cdot,\cdot,\cdot,\cdot,\cdot,\cdot,\cdot,\cdot,\cdot) : \Omega \times Y \times \mathbb{R}^2 \times \mathbb{R} \times \mathbb{R}^2 \times \mathbb{E}_s^2 \times \mathbb{E}_s^2 \times \mathbb{E}_s^2 \times \mathbb{R}^2 \mapsto \mathbb{R}^+$. We make the following assumptions:

(i) f is a measurable function, Y-periodic in y, continuous in $(\mathbf{u}, \mathbf{w}, \mathbf{f})$ and convex with respect to $(\mathbf{p}, \mathbf{q}, \mathbf{r}, \mathbf{s})$,

(ii) $0 \le f(x, y, \mathbf{u}, \mathbf{w}, \mathbf{f}, \mathbf{p}, \mathbf{q}, \mathbf{r}, \mathbf{s}) \le a(x)[A(y) + |\mathbf{u}|^2 + |\mathbf{w}|^2 + |\mathbf{f}|^2 + |\mathbf{p}|^2 + |\mathbf{q}|^2 + |\mathbf{r}|^2$
$\qquad\qquad . + |\mathbf{s}|^2]$,

(iii) $|f(x, y, \mathbf{u}, \mathbf{w}, \mathbf{f}, \mathbf{p}, \mathbf{q}, \mathbf{r}, \mathbf{s}) - f(x', y, \mathbf{u}', \mathbf{w}', \mathbf{f}', \mathbf{p}, \mathbf{q}, \mathbf{r}, \mathbf{s})| \le g(|x' - x| + |\mathbf{u}' - \mathbf{u}|$
$\qquad\qquad + |\mathbf{w}' - \mathbf{w}| + |\mathbf{f}' - \mathbf{f}|)(A(y) + f(x, y, \mathbf{u}, \mathbf{w}, \mathbf{f}, \mathbf{p}, \mathbf{q}, \mathbf{r}, \mathbf{s}))$,

for every $x, x', y \in \mathbb{R}^2$, and every $\mathbf{u}', \mathbf{u} \in \mathbb{R}^2, \mathbf{w}, \mathbf{w}' \in \mathbb{R}$, and $\mathbf{r}, \mathbf{s}, \mathbf{p}, \mathbf{q} \in \mathbb{E}_s^2$. Here $A \in L^1_{loc}(\mathbb{R}^2)$ is a Y-periodic function while $g : \mathbb{R}^+ \to \mathbb{R}^+$ is an increasing function continuous at 0, and such that $g(0) = 0$; $a : \mathbb{R}^2 \to \mathbb{R}^+$ is a non-negative continuous function.

For a fixed $\varepsilon > 0$, we set

$$G_\varepsilon(\mathbf{u}, \boldsymbol{\varphi}, w) = \int_\Omega f[x, \frac{x}{\varepsilon}, \mathbf{u}(x), \boldsymbol{\varphi}(x), w(x), E(\mathbf{u}(x), w(x)),$$

$$\boldsymbol{\rho}(\boldsymbol{\varphi}(x)), \boldsymbol{\kappa}(w(x)), \boldsymbol{\gamma}(\boldsymbol{\varphi}(x), w(x))]dx , \qquad (5.7.48)$$

where $\mathbf{u}, \boldsymbol{\varphi} \in H^1(\Omega)^2$ and $w \in H_0^2(\Omega)$. We set

$$W_{per}(Y) = H^1_{per}(Y)^2 \times H^1_{per}(Y)^2 \times H^2_{per}(Y) , \qquad H^1_{per}(Y)^2 = [H^1_{per}(Y)]^2 ,$$

$$\widetilde{W}_{per}(Y) = \{(\mathbf{u}, \boldsymbol{\varphi}, w) \in W_{per}(Y)| \int_Y \mathbf{u}(y)dy = 0, \int_Y \boldsymbol{\varphi}(y)dy = 0, \int_Y w(y)dy = 0\} .$$

Let τ denote the strong topology of $L^2(\Omega)^2 \times L^2(\Omega)^2 \times H^1(\Omega)$. Under the assumptions (i), (ii) and (iii) the Γ-limit of the sequence of functionals (5.7.48) is given by the formula

$$G(\mathbf{u}, \boldsymbol{\varphi}, w) = \Gamma(\tau) - \lim_{\varepsilon \to 0} G_\varepsilon(\mathbf{u}, \boldsymbol{\varphi}, w)$$

$$= \int_\Omega f_h[x, \mathbf{u}, \boldsymbol{\varphi}, w, E(\mathbf{u}, w), \boldsymbol{\rho}(\boldsymbol{\varphi}), \boldsymbol{\kappa}(w), \boldsymbol{\gamma}(\boldsymbol{\varphi}, w)]dx , \qquad (5.7.49)$$

where

$$f_h(x, \mathbf{u}, \boldsymbol{\varphi}, w, E^h, \boldsymbol{\rho}^h, \boldsymbol{\kappa}^h, \boldsymbol{\gamma}^h) = \inf\{\frac{1}{|Y|}\int_Y f(x, \mathbf{u}, \boldsymbol{\varphi}, w, E^h$$

$$+ e^y(\mathbf{u}), \boldsymbol{\kappa}^h + \kappa^y(w), \boldsymbol{\rho}^h + e^y(\boldsymbol{\varphi}), \boldsymbol{\gamma}^h)dy|(\mathbf{u}, \boldsymbol{\varphi}, w) \in W_{per}(Y)\} . \qquad (5.7.50)$$

Moreover, $E^h, \boldsymbol{\rho}^h, \boldsymbol{\kappa}^h \in \mathbb{E}_s^2$ and $\boldsymbol{\gamma}^h \in \mathbb{R}^2$ are macroscopic strain measures. The introduced natural strong topology τ implies that the strain measure $\boldsymbol{\gamma}$ plays the role of a parameter in the determination of f_h.

In the case of the sequence of functionals $\{J_\varepsilon\}_{\varepsilon>0}$, the loading functional L has been assumed to play the role of the perturbation functional. Therefore it suffices to find the $\Gamma(\tau)$-limit of $\{\mathcal{E}_\varepsilon\}_{\varepsilon>0}$. Obviously, this limit is a special case of (5.7.49).

Homogenized elastic potential

Let us proceed to the derivation of formulae for the effective moduli. Now the integrand f appearing in (5.7.48) is just the elastic potential $\mathcal{W}$. For the plate model considered, we have

$$f(x,y,\boldsymbol{u},\boldsymbol{\varphi},w,\boldsymbol{E},\boldsymbol{\rho},\boldsymbol{\kappa},\boldsymbol{\gamma}) = \mathcal{W}(x,y,\boldsymbol{E},\boldsymbol{\rho},\boldsymbol{\kappa},\boldsymbol{\gamma})\,. \tag{5.7.51}$$

Then the formula (5.7.50) yields two independent *convex* minimization problems

$$\begin{aligned}
&\mathcal{W}_h\,(x,\boldsymbol{E}^h,\boldsymbol{\rho}^h,\boldsymbol{\kappa}^h,\boldsymbol{\gamma}^h)\\
&= \inf\{\frac{1}{|Y|}\int_Y \frac{1}{2}A^{\alpha\beta\lambda\mu}(x,y)(E^h_{\alpha\beta}+e^y_{\alpha\beta}(\boldsymbol{v}))(E^h_{\lambda\mu}+e^y_{\lambda\mu}(\boldsymbol{v}))dy|\boldsymbol{v}\in\widetilde{H}^1_{per}(Y)^2\}\\
&+ \inf\{\frac{1}{|Y|}\int_Y \frac{1}{2}D^{\alpha\beta\lambda\mu}(x,y)[(\rho^h_{\alpha\beta}+e^y_{\alpha\beta}(\boldsymbol{\eta}))(\rho^h_{\lambda\mu}+e^y_{\lambda\mu}(\boldsymbol{\eta}))\\
&+ \frac{1}{84}(\rho^h_{\alpha\beta}-\kappa^h_{\alpha\beta}+e^y_{\alpha\beta}(\boldsymbol{\eta})-\kappa^h_{\alpha\beta}(\omega))(\rho^h_{\lambda\mu}-\kappa^h_{\lambda\mu}\\
&+ e^y_{\lambda\mu}(\boldsymbol{\eta})-\kappa^y_{\lambda\mu}(\omega))]dy|\boldsymbol{\eta}\in\widetilde{H}^1_{per}(Y)^2,\omega\in\widetilde{H}^2_{per}(Y)\}\\
&+ \frac{1}{2|Y|}\int_Y H^{\alpha\beta}(x,y)\gamma^h_\alpha\gamma^h_\beta dy\,.
\end{aligned} \tag{5.7.52}$$

The existence of minimizer for functionals of the type occurring on the r.h.s. of (5.7.52) follows from the properties of the elastic moduli $\boldsymbol{A}$, $\boldsymbol{D}$, and $\boldsymbol{H}$, and the fact that $e^y(\boldsymbol{v})$ is a linear strain measure. Let $(\widetilde{\boldsymbol{v}},\widetilde{\boldsymbol{\eta}},\widetilde{\omega})\in\widetilde{W}_{per}(Y)$ be a minimizer of the minimized functional in (5.7.52). Obviously it is determined uniquely. The minimization problem (5.7.52) can also be solved in a larger space $W_{per}(Y)$. Then $\widetilde{\omega}$ is unique up to a constant while $\widetilde{\boldsymbol{v}}$ and $\widetilde{\boldsymbol{\eta}}$ are unique up to constant vectors. It is worth noting that for prescribed $\boldsymbol{E}^h$, $\boldsymbol{\rho}^h$, $\boldsymbol{\kappa}^h$, and $\boldsymbol{\gamma}^h$, the local minimization problems on the r.h.s. of (5.7.52) are similar to the case of linear plates. The nonlinearity of the strain measure has been included in $\boldsymbol{E}^h$.

We can write

$$\begin{aligned}
&\mathcal{W}_h\,(x,\boldsymbol{E}^h,\boldsymbol{\rho}^h,\boldsymbol{\kappa}^h,\boldsymbol{\gamma}^h)\\
&= \inf\{\frac{1}{|Y|}\int_Y \frac{1}{2}A^{\alpha\beta\lambda\mu}(x,y)(E^h_{\alpha\beta}+e^y_{\alpha\beta}(\boldsymbol{v}))(E^h_{\lambda\mu}+e^y_{\lambda\mu}(\boldsymbol{v}))dy|\boldsymbol{v}\in\widetilde{H}^1_{per}(Y)^2\}\\
&+ \inf\{\frac{1}{|Y|}\int_Y \frac{1}{2}D^{\alpha\beta\lambda\mu}(x,y)[(\rho^h_{\alpha\beta}+e^y_{\alpha\beta}(\boldsymbol{\eta}))(\rho^h_{\lambda\mu}+e^y_{\lambda\mu}(\boldsymbol{\eta}))\\
&+ \frac{1}{84}(\rho^h_{\alpha\beta}-\kappa^h_{\alpha\beta}+e^y_{\alpha\beta}(\boldsymbol{\eta})-\kappa^y_{\alpha\beta}(\omega))
\end{aligned}$$

$$
\times \ (\rho_{\lambda\mu}^h - \kappa_{\lambda\mu}^h + e_{\lambda\mu}^y(\boldsymbol{\eta}) - \kappa_{\lambda\mu}^y(\omega))]dy \,|\, (w,\boldsymbol{\eta}) \in \widetilde{H}_{per}^2(Y) \times \widetilde{H}_{per}^1(Y)^2 \}
$$

$$
+ \ \frac{1}{|Y|} \int_Y \frac{1}{2} H^{\alpha\beta}(x,y)\gamma_\alpha^h\gamma_\beta^h dy
$$

$$
= \ \frac{1}{|Y|} \int_Y \{ \frac{1}{2} A^{\alpha\beta\lambda\mu}(x,y)(E_{\alpha\beta}^h + e_{\alpha\beta}^y(\widetilde{v}))(E_{\lambda\mu}^h + e_{\lambda\mu}^y(\widetilde{v}))
$$

$$
+ \ \frac{1}{2} D^{\alpha\beta\lambda\mu}(x,y)[(\rho_{\alpha\beta}^h + e_{\alpha\beta}^y(\widetilde{\boldsymbol{\eta}}))(\rho_{\lambda\mu}^h + e_{\lambda\mu}^y(\widetilde{\boldsymbol{\eta}}))
$$

$$
+ \ \frac{1}{84}(\rho_{\alpha\beta}^h - \kappa_{\alpha\beta}^h + e_{\alpha\beta}^y(\widetilde{\boldsymbol{\eta}}) - \kappa_{\alpha\beta}^y(\widetilde{\omega}))(\rho_{\lambda\mu}^h - \kappa_{\lambda\mu}^h + e_{\lambda\mu}^y(\widetilde{\boldsymbol{\eta}}) - \kappa_{\lambda\mu}^y(\widetilde{\omega}))]
$$

$$
+ \ \frac{1}{2} H^{\alpha\beta}(x,y)\gamma_\alpha^h\gamma_\beta^h \}dy \ . \tag{5.7.53}
$$

We pass to a brief discussion of the properties of the effective elastic stored energy function $\mathcal{W}_h$. The homogenized coefficients $A_h^{\alpha\beta}$, $D_h^{\alpha\beta\lambda\mu}$, $E_h^{\alpha\beta\lambda\mu}$, $F_h^{\alpha\beta\lambda\mu}$ and $H_h^{\alpha\beta\lambda\mu}$, appearing in the expression for $\mathcal{W}_h$ are derived below.

Properties of the elastic potential $\mathcal{W}_h$
(1)

$$
\mathcal{W}_h\ (x, \boldsymbol{E}^h, \boldsymbol{\rho}^h, \boldsymbol{\kappa}^h, \boldsymbol{\gamma}^h) = \frac{1}{2} A_h^{\alpha\beta\lambda\mu} E_{\alpha\beta}^h E_{\lambda\mu}^h + \frac{1}{2} D_h^{\alpha\beta\lambda\mu} \rho_{\alpha\beta}^h \rho_{\lambda\mu}^h
$$

$$
+ \ \frac{1}{2} E_h^{\alpha\beta\lambda\mu} \kappa_{\alpha\beta}^y \kappa_{\lambda\mu}^h + F_h^{\alpha\beta\lambda\mu} \rho_{\alpha\beta}^h \kappa_{\lambda\mu}^h + \frac{1}{2} H_h^{\alpha\beta} \gamma_\alpha^h \gamma_\beta^h \ ,
$$

where $\boldsymbol{A}_h$, $\boldsymbol{D}_h$, $\boldsymbol{E}_h$, $\boldsymbol{F}_h$ and $\boldsymbol{H}_h$ are given by (5.7.67) – (5.7.71) below; moreover

$$
A_h^{\alpha\beta\lambda\mu}(x) = \frac{\partial^2 \mathcal{W}_h}{\partial E_{\alpha\beta}^h \partial E_{\lambda\mu}^h} \ , \qquad D_h^{\alpha\beta\lambda\mu}(x) = \frac{\partial^2 \mathcal{W}_h}{\partial \rho_{\alpha\beta}^h \partial \rho_{\lambda\mu}^h} \ ,
$$

$$
E_h^{\alpha\beta\lambda\mu}(x) = \frac{\partial^2 \mathcal{W}_h}{\partial \kappa_{\alpha\beta}^h \partial \kappa_{\lambda\mu}^h} \ , \qquad F_h^{\alpha\beta\lambda\mu}(x) = \frac{\partial^2 \mathcal{W}_h}{\partial \rho_{\alpha\beta}^h \partial \kappa_{\lambda\mu}^h} \ , \qquad H_h^{\alpha\beta}(x) = \frac{\partial^2 \mathcal{W}_h}{\partial \gamma_\alpha^h \partial \gamma_\beta^h} \ .
$$

(2)

$$
\mathcal{W}_h(x, \boldsymbol{E}^h, \boldsymbol{\rho}^h, \boldsymbol{\kappa}^h, \boldsymbol{\gamma}^h) \le \frac{c_1}{2}(|\boldsymbol{E}^h|^2 + |\boldsymbol{\rho}^h|^2 + \frac{1}{84}|\boldsymbol{\rho}^h - \boldsymbol{\kappa}^h|^2 + |\boldsymbol{\gamma}^h|^2) \ ,
$$

(3)

$$
\mathcal{W}_h(x, \boldsymbol{E}^h, \boldsymbol{\rho}^h, \boldsymbol{\kappa}^h, \boldsymbol{\gamma}^h) \ge \frac{c_0}{2}(|\boldsymbol{E}^h|^2 + |\boldsymbol{\rho}^h|^2 + \frac{1}{84}|\boldsymbol{\rho}^h - \boldsymbol{\kappa}^h|^2 + |\boldsymbol{\gamma}^h|^2) \ ,
$$

for $x \in \Omega$. To prove these properties we use (5.7.53) and (b). We observe that $\int_Y \mathrm{tr}\ [\boldsymbol{E}^h e^y(\boldsymbol{v}(y))]dy = 0$, $\int_Y \mathrm{tr}\ [\boldsymbol{\rho}^h e^y(\boldsymbol{\eta}(y))]dy = 0$ and $\int_Y \mathrm{tr}\ [\boldsymbol{\kappa}^h \kappa^y(\omega(y))]dy = 0$, by periodicity of $\boldsymbol{v}, \boldsymbol{\eta}$, and ω. We recall that $|\boldsymbol{\rho}^h - \boldsymbol{\kappa}^h|^2 \le 2(|\boldsymbol{\rho}^h|^2 + |\boldsymbol{\kappa}^h|^2)$.

Local problems

The minimization problem on the r.h.s. of (5.7.52) is equivalent to:

$$\int_Y A^{\alpha\beta\lambda\mu}(x,y)(E^h_{\alpha\beta} + e^y_{\alpha\beta}(\widetilde{v}))e^y_{\lambda\mu}(\nu)dy$$

$$+ \int_Y D^{\alpha\beta\lambda\mu}(x,y)(\rho^h_{\alpha\beta} + e^y_{\alpha\beta}(\widetilde{\varphi}))e^y_{\lambda\mu}(\phi)dy$$

$$+ \int_Y \frac{1}{84}D^{\alpha\beta\lambda\mu}(x,y)(\rho^h_{\alpha\beta} - \kappa^h_{\alpha\beta} + e^y_{\alpha\beta}(\widetilde{\eta}) - \kappa^y_{\lambda\mu}(\widetilde{\tau}))(e^y_{\lambda\mu}(\phi) - \kappa^y_{\lambda\mu}(\tau))dy = 0\,,$$

$$\forall\,\nu,\phi \in \widetilde{H}^1_{per}(Y)^2\,,\quad \tau \in \widetilde{H}^2_{per}(Y)\,. \tag{5.7.54}$$

Consequently one has

$$\int_Y A^{\alpha\beta\lambda\mu}(x,y)(E^h_{\alpha\beta} + e^y_{\alpha\beta}(\widetilde{v}))e^y_{\lambda\mu}(\nu)dy = 0 \quad \forall\,\nu \in \widetilde{H}^1_{per}(Y)^2\,, \tag{5.7.55}$$

$$\int_Y D^{\alpha\beta\lambda\mu}(x,y)(\rho^h_{\alpha\beta} + e^y_{\alpha\beta}(\widetilde{\eta}))e^y_{\lambda\mu}(\phi)dy + \int_Y \frac{1}{84}D^{\alpha\beta\lambda\mu}(x,y)[\rho^h_{\alpha\beta} - \kappa^h_{\alpha\beta} + e^y_{\alpha\beta}(\widetilde{\eta})$$

$$- \kappa^y_{\alpha\beta}(\widetilde{\omega})]e^y_{\lambda\mu}(\phi)dy = 0\,,\qquad \forall\,\phi \in \widetilde{H}^1_{per}(Y)^2\,, \tag{5.7.56}$$

$$\int_Y D^{\alpha\beta\lambda\mu}(x,y)[\rho^h_{\alpha\beta} - \kappa^h_{\alpha\beta} + e^y_{\alpha\beta}(\widetilde{\eta}) - \kappa^y_{\alpha\beta}(\widetilde{\omega})]\kappa^y_{\lambda\mu}(\tau)dy = 0, \forall\tau \in \widetilde{H}^2_{per}(Y). \tag{5.7.57}$$

Due to the linearity of the local problem (5.7.55), the solution $\widetilde{v}$ has the following form

$$\widetilde{v}_\alpha(y) = E^h_{\sigma\tau}\chi^{(\sigma\tau)}_\alpha(y)\,, \tag{5.7.58}$$

where $\chi^{(\sigma\tau)}$ are Y- periodic functions. Substituting (5.7.58) into (5.7.55) we obtain

$$\int_Y A^{\alpha\beta\lambda\mu}(x,y)(E^h_{\lambda\mu} + E^h_{\sigma\tau}(\chi^{(\sigma\tau)}))e^y_{\alpha\beta}(\nu)dy = 0\,,\quad \forall\,\nu \in \widetilde{H}^1_{per}(Y)^2\,.$$

After simple calculations the last relation transforms to

$$E^h_{\sigma\tau}\int_Y A^{\alpha\beta\lambda\mu}(x,y)(\delta_{\lambda\sigma}\delta_{\mu\tau} + e^y_{\lambda\mu}(\chi^{(\sigma\tau)}))e^y_{\alpha\beta}(\nu)dy = 0,\quad \forall\,\nu \in \widetilde{H}^1_{per}(Y)^2. \tag{5.7.59}$$

Now we can formulate the local problems.
Problem P^1_{loc}. Find $\chi^{(\sigma\tau)} \in \widetilde{H}^1_{per}(Y)^2$ such that

$$\int_Y A^{\alpha\beta\lambda\mu}(x,y)(\delta_{\lambda\sigma}\delta_{\mu\tau} + e^y_{\lambda\mu}(\chi^{(\sigma\tau)}))e^y_{\alpha\beta}(\nu)dy = 0\,,\quad \forall\,\nu \in \widetilde{H}^1_{per}(Y)^2\,. \tag{5.7.60}$$

On account of the linearity of the local problems (5.7.56), (5.7.57) we can express $\widetilde{\eta}$ and $\widetilde{\omega}$ in the following form

$$\widetilde{\eta}_\alpha = \rho^h_{\sigma\tau}\Phi^{(\sigma\tau)}_\alpha + \kappa^h_{\sigma\tau}\Xi^{(\sigma\tau)}_\alpha \,, \tag{5.7.61}$$

$$\widetilde{\omega} = \rho^h_{\sigma\tau}\Psi^{(\sigma\tau)} + \kappa^h_{\sigma\tau}\Theta^{(\sigma\tau)} \,, \tag{5.7.62}$$

where $\Phi^{(\sigma\tau)}$, $\Xi^{(\sigma\tau)}$, $\Theta^{(\sigma\tau)}$ and $\Psi^{(\sigma\tau)}$ are Y-periodic functions. Substituting (5.7.61) and (5.7.62) into (5.7.56) and (5.7.57) we obtain the following problems

$$\rho^h_{\sigma\tau}\int_Y D^{\alpha\beta\lambda\mu}(x,y)[\delta_{\sigma\lambda}\delta_{\tau\mu} + e^y_{\lambda\mu}(\Phi^{(\sigma\tau)})$$

$$+ \frac{1}{84}\delta_{\sigma\lambda}\delta_{\tau\mu} + \frac{1}{84}e^y_{\lambda\mu}(\Phi^{(\sigma\tau)}) - \frac{1}{84}\kappa^h_{\lambda\mu}(\Phi^{(\sigma\tau)})]e^y_{\alpha\beta}(\phi)dy$$

$$+ \kappa^h_{\sigma\tau}\int_Y D^{\alpha\beta\lambda\mu}(x,y)[e^y_{\lambda\mu}(\Xi^{(\sigma\tau)}) - \frac{1}{84}\delta_{\sigma\lambda}\delta_{\tau\mu} + \frac{1}{84}e^y_{\lambda\mu}(\Xi^{(\sigma\tau)})$$

$$- \frac{1}{84}\kappa^h_{\lambda\mu}(\Theta^{(\sigma\tau)})]e^y_{\alpha\beta}(\phi)dy = 0 \,, \quad \forall\,\phi \in \widetilde{H}^1_{per}(Y)^2 \,, \tag{5.7.63}$$

$$\rho^h_{\sigma\mu}\int_Y D^{\alpha\beta\lambda\mu}(x,y)[\delta_{\sigma\lambda}\delta_{\tau\mu} + e^y_{\lambda\mu}(\Phi^{(\sigma\tau)}) - \kappa^y_{\lambda\mu}(\Psi^{(\sigma\tau)})]\kappa^y_{\alpha\beta}(\tau)dy$$

$$+ \kappa^h_{\sigma\tau}\int_Y D^{\alpha\beta\lambda\mu}[-\delta_{\sigma\lambda}\delta_{\tau\mu} + e^y_{\lambda\mu}(\Xi^{(\sigma\tau)})$$

$$- \kappa^y_{\lambda\mu}(\Theta^{(\sigma\tau)})]\kappa^y_{\alpha\beta}(\tau)dy = 0 \,, \quad \forall\,\tau \in \widetilde{H}^2_{per}(Y) \,. \tag{5.7.64}$$

Since the matrices ρ^h and κ^h are arbitrary, we have to solve the following independent local problems resulting from (5.7.63) and (5.7.64), respectively.

Problem P^2_{loc}. Find $\Phi^{(\sigma\tau)} \in \widetilde{H}^1_{per}(Y)^2$ and $\Psi^{(\sigma\tau)} \in \widetilde{H}^2_{per}(Y)$ such that

$$\int_Y D^{\alpha\beta\lambda\mu}(x,y)[\frac{85}{84}(\delta_{\sigma\lambda}\delta_{\tau\mu} + e^y_{\lambda\mu}(\Phi^{(\sigma\tau)})) - \frac{1}{84}\kappa^y_{\lambda\mu}(\Psi^{(\sigma\tau)})]e^y_{\alpha\beta}(\phi)dy = 0 \;\; \forall\phi \in \widetilde{H}^1_{per}(Y)^2 \,,$$

$$\int_Y D^{\alpha\beta\lambda\mu}(x,y)[\delta_{\sigma\lambda}\delta_{\tau\mu} + e^y_{\lambda\mu}(\Phi^{(\sigma\tau)}) - \kappa^y_{\lambda\mu}(\Psi^{(\sigma\tau)})]\kappa^y_{\alpha\beta}(\tau)dy = 0 \;\; \forall\tau \in \widetilde{H}^2_{per}(Y) \,.$$

Problem P^3_{loc}. Find $\Xi^{(\sigma\tau)} \in \widetilde{H}^1_{per}(Y)^2$ and $\Theta^{(\sigma\tau)} \in \widetilde{H}^2_{per}(Y)$ such that

$$\int_Y D^{\alpha\beta\lambda\mu}(x,y)[\frac{85}{84}e^y_{\tau\mu}(\Xi^{(\sigma\lambda)}) - \frac{1}{84}(\delta_{\sigma\lambda}\delta_{\tau\mu} + \kappa^y_{\lambda\mu}(\Theta^{(\sigma\tau)}))]e^y_{\alpha\beta}(\phi)dy = 0 \;\; \forall\,\phi \in \widetilde{H}^1_{per}(Y)^2 \,,$$

$$\int_Y D^{\alpha\beta\lambda\mu}(x,y)[-\delta_{\sigma\lambda}\delta_{\tau\mu} - \kappa^y_{\lambda\mu}(\Theta^{(\sigma\tau)}) + e^y_{\lambda\mu}(\Xi^{(\sigma\tau)})]\kappa^y_{\alpha\beta}(\tau)dy = 0 \;\; \forall\tau \in \widetilde{H}^2_{per}(Y) \,.$$

Observe that the local problems P^i_{loc} $(i = 1, 2, 3)$ are equivalent to the convex minimization problem appearing in (5.7.53).

Effective moduli

The density of the stored energy function of the effective plate can be expressed in an alternative form. By virtue of (5.7.53) we have

$$\mathcal{W}_h\left(x, \boldsymbol{E}^h, \boldsymbol{\rho}^h, \boldsymbol{\kappa}^h, \boldsymbol{\gamma}^h\right) = \frac{1}{|Y|}\int_Y \{\frac{1}{2}A^{\alpha\beta\lambda\mu}(x,y)(E_{\alpha\beta}^h + e_{\alpha\beta}^y(\widetilde{\boldsymbol{v}}))(E_{\lambda\mu}^h + e_{\lambda\mu}^y(\widetilde{\boldsymbol{v}}))$$

$$+ \ \frac{1}{2}D^{\alpha\beta\lambda\mu}(x,y)[(\rho_{\alpha\beta}^h + e_{\alpha\beta}^y(\widetilde{\boldsymbol{\eta}}))(\rho_{\lambda\mu}^h + e_{\lambda\mu}^y(\widetilde{\boldsymbol{\eta}}))$$

$$+ \ \frac{1}{84}(\rho_{\alpha\beta}^h - \kappa_{\alpha\beta}^h + e_{\alpha\beta}^y(\widetilde{\boldsymbol{\eta}}) - \kappa_{\alpha\beta}^y(\widetilde{w}))(\rho_{\lambda\mu}^h - \kappa_{\lambda\mu}^h + e_{\lambda\mu}^y(\widetilde{\boldsymbol{\eta}}) - \kappa_{\lambda\mu}^y(\widetilde{\omega}))]$$

$$+ \ \frac{1}{2}H^{\alpha\beta}(x,y)\gamma_\alpha^h\gamma_\beta^h\}dy \ . \tag{5.7.65}$$

Taking into account (5.7.55) – (5.7.57) and setting $\boldsymbol{\nu} = \widetilde{\boldsymbol{v}}$, $\omega = \widetilde{\omega}$ and $\boldsymbol{\phi} = \widetilde{\boldsymbol{\eta}}$, we conclude that

$$\mathcal{W}_h\left(x, \boldsymbol{E}^h, \boldsymbol{\rho}^h, \boldsymbol{\kappa}^h, \boldsymbol{\gamma}^h\right) = \frac{1}{|Y|}\int_Y \{\frac{1}{2}A^{\alpha\beta\lambda\mu}(x,y)(E_{\alpha\beta}^h + e_{\alpha\beta}^h(\widetilde{\boldsymbol{v}}))E_{\lambda\mu}^h$$

$$+ \ \frac{1}{2}D^{\alpha\beta\lambda\mu}(x,y)[(\rho_{\alpha\beta}^h + e_{\alpha\beta}^y(\widetilde{\boldsymbol{\eta}}))\rho_{\lambda\mu}^h + \frac{1}{84}(\rho_{\alpha\beta}^h$$

$$- \ \kappa_{\alpha\beta}^h + e_{\alpha\beta}^y(\widetilde{\boldsymbol{\eta}}) - \kappa_{\alpha\beta}^y(\widetilde{\omega}))(\rho_{\lambda\mu}^h - \kappa_{\lambda\mu}^h)] + \frac{1}{2}H^{\alpha\beta}(x,y)\gamma_\alpha^h\gamma_\beta^h\}dy \ . \tag{5.7.66}$$

Using local problems (P_{loc}^1), (P_{loc}^2), and (P_{loc}^3) we obtain the following formulae for the homogenized coefficients

$$A_h^{\alpha\beta\lambda\mu}(x) = \frac{\partial^2 \mathcal{W}_h}{\partial E_{\alpha\beta}^h \partial E_{\lambda\mu}^h} = \frac{1}{|Y|}\int_Y A^{\alpha\beta\sigma\tau}(x,y)(\delta_{\sigma\lambda}\delta_{\tau\mu} + e_{\sigma\tau}^y(\chi^{(\lambda\mu)}))dy \ , \tag{5.7.67}$$

$$D_h^{\alpha\beta\lambda\mu}(x) = \frac{\partial^2 \mathcal{W}_h}{\partial \rho_{\alpha\beta}^h \partial \rho_{\lambda\mu}^h} = \frac{1}{|Y|}\int_Y D^{\alpha\beta\sigma\tau}(x,y)[\frac{85}{84}(\delta_{\sigma\lambda}\delta_{\tau\mu} + e_{\lambda\mu}^y(\boldsymbol{\Phi}^{(\sigma\tau)}))$$

$$-\frac{1}{84}\kappa_{\lambda\mu}^y(\boldsymbol{\Psi}^{(\sigma\tau)})]dy \ , \tag{5.7.68}$$

$$E_h^{\alpha\beta\lambda\mu}(x) = \frac{\partial^2 \mathcal{W}_h}{\partial \kappa_{\alpha\beta}^h \partial \kappa_{\lambda\mu}^h} = \frac{1}{|Y|}\int_Y \frac{1}{84}D^{\alpha\beta\sigma\tau}(x,y)[\delta_{\sigma\lambda}\delta_{\tau\mu} + \kappa_{\lambda\mu}^y(\boldsymbol{\Theta}^{(\sigma\tau)}))$$

$$-e_{\lambda\mu}^y(\boldsymbol{\Xi}^{(\sigma\tau)})]dy \ , \tag{5.7.69}$$

$$F_h^{\sigma\tau\lambda\mu}(x) = \frac{\partial^2 \mathcal{W}_h}{\partial \rho_{\sigma\tau}^h \partial \kappa_{\lambda\mu}^h} = \frac{1}{|Y|}\int_Y \frac{1}{84}D^{\alpha\beta\lambda\mu}(x,y)[-\delta_{\alpha\sigma}\delta_{\beta\tau} - e_{\lambda\mu}^y(\boldsymbol{\Phi}^{(\sigma\tau)})$$

$$+\kappa_{\alpha\beta}^y(\boldsymbol{\Psi}^{(\sigma\tau)})]dy \ , \tag{5.7.70}$$

$$H_h^{\alpha\beta}(x) = \frac{\partial^2 \mathcal{W}_h}{\partial d_\alpha^h \partial d_\beta^h} = \frac{1}{|Y|}\int_Y H^{\alpha\beta}(x,y)dy \ . \tag{5.7.71}$$

Hence the homogenized potential can be written in the form

$$\mathcal{W}_h\left(x, \boldsymbol{E}^h, \boldsymbol{\rho}^h, \boldsymbol{\kappa}^h, \boldsymbol{\gamma}^h\right) = \frac{1}{2}A_h^{\alpha\beta\lambda\mu}(x)E_{\alpha\beta}^h E_{\lambda\mu}^h + \frac{1}{2}D_h^{\alpha\beta\lambda\mu}(x)\rho_{\alpha\beta}^h\rho_{\lambda\mu}^h$$

$$+\ \frac{1}{2}E_h^{\alpha\beta\lambda\mu}(x)\kappa_{\alpha\beta}^h\kappa_{\lambda\mu}^h + F_h^{\alpha\beta\lambda\mu}(x)\rho_{\alpha\beta}^h\kappa_{\lambda\mu}^h + \frac{1}{2}H_h^{\alpha\beta}(x)\gamma_\alpha^h\gamma_\beta^h\ . \tag{5.7.72}$$

Note that the homogenized coefficients $H_h^{\alpha\beta}(x)$ are calculated as the mean value of $H^{\alpha\beta}(x, y)$ with respect to the microscopic variable y.

Homogenized constitutive equations

In view of (5.7.72), the effective (homogenized) constitutive relationships are given by

$$\begin{aligned}
N_h^{\alpha\beta} &= A_h^{\alpha\beta\lambda\mu}(x)E_{\lambda\mu}^h\ ,\\
S_h^{\alpha\beta} &= D_h^{\alpha\beta\lambda\mu}(x)\rho_{\lambda\mu}^h + F_h^{\alpha\beta\lambda\mu}(x)\kappa_{\lambda\mu}^h\ ,\\
\mathbf{M}_h^{\alpha\beta} &= E_h^{\alpha\beta\lambda\mu}(x)\kappa_{\lambda\mu}^h + F_h^{\alpha\beta\lambda\mu}(x)\rho_{\lambda\mu}^h\ ,\\
Q_h^\alpha &= H_h^{\alpha\beta}(x)\gamma_\beta^h\ .
\end{aligned} \tag{5.7.73}$$

On account of the presence of elastic moduli $F_h^{\alpha\beta\lambda\mu}$, Eqs. $(5.7.73)_2$ and $(5.7.73)_3$ are coupled, in contrast to the primal Eqs. (5.7.38) and $(5.7.36)_3$.

Remark 5.7.5. Deleting the nonlinear term in the strain measure $\boldsymbol{E}(\boldsymbol{u}, w)$ we recover the *linear* homogenized model derived in Sec. 5.2.

As we already know from Remark 5.7.4, for $\boldsymbol{\varphi} = -\nabla w$, the nonlinear model just studied reduces to the von Kármán plate model. The homogenized model reduces then to the one derived in Sec. 4.2.

Moderately thick perforated plates and moderately thick plates stiffened with rigid inclusions can be investigated similarly to thin plates, cf. Secs. 3.12, 3.13 and 4.3.

6.   Sandwich plates with soft core

6.1.   Hoff's theory

The subject of consideration is an elastic sandwich plate of constant thickness $2h$, composed of a core layer of thickness $2c$ and two external layers (face-plates) of thickness d, see Fig. 6.1.1.

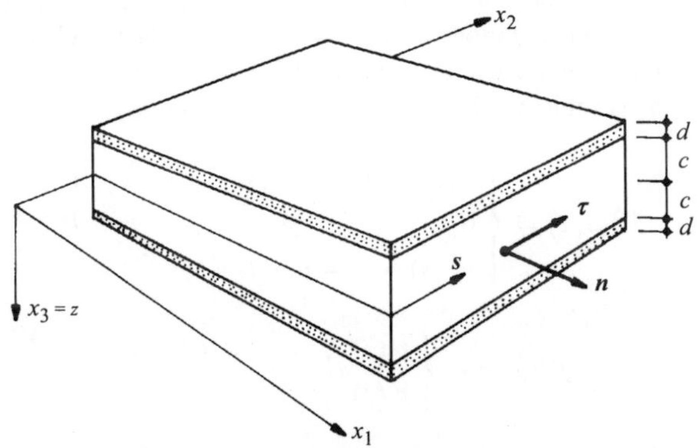

Fig. 6.1.1. Sandwich plate

The plate is symmetric with respect to its middle plane $\Omega(x_3 \equiv z = 0)$. The face-plates are assumed to be made of the same linearly elastic material of moduli C_f^{ijkl} satisfying (2.4.1). The core is assumed to be made of a soft material with moduli:

$$C_c^{\alpha\beta\lambda\mu} = 0, \quad C_c^{\alpha\beta33} = 0 , \quad C_c^{\alpha333} = C_c^{3\beta\gamma\delta} = 0 ,$$
$$C_c^{\alpha3\beta3} = \mu_c\delta^{\alpha\beta} , \qquad C_c^{3333} = E_c . \tag{6.1.1}$$

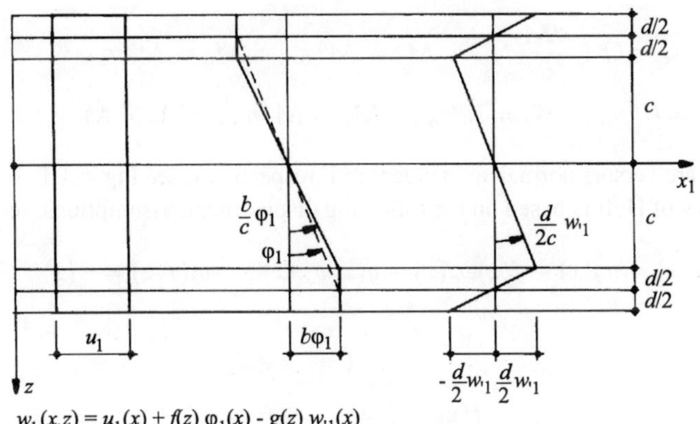

Fig. 6.1.2. Kinematic assumption

Simplifications (6.1.1) can be justified if the only role of the core is to keep a constant distance between the face-plates. Typical cores are made from aluminium honeycombs manufactured by sticking thin aluminium sheets and then stretched. Such a honeycomb has negligible in-plane stiffnesses: it can be easily squeezed in hands. However, its transverse shearing and normal stiffnesses cannot be neglected. The hexagonal geometry justifies the equalities: $C_c^{1313} = C_c^{2323}$, $C_c^{1323} = 0$.

Assume that the upper and lower faces $z = \pm h$ are subjected to transverse loadings $p^{\pm}$. We shall confine our considerations to the case when the plate is clamped on the lateral surface $\Upsilon_u = \Gamma_u \times (-h, h)$, where Γ_u is a part of $\Gamma = \partial\Omega$; $\Gamma\backslash\Gamma_u = \Gamma_\sigma$. The lateral surface $\Upsilon_\sigma = \Gamma_\sigma \times (-h, h)$ is subject to tractions T^i. They are assumed in the form

$$T^\alpha(s, z) = \begin{cases} p_\alpha^-(s) + (z + b)r_\alpha^-(s) \;, & z \in I_1 \;, \\ 0 & , \quad z \in I_2 \;, \\ p_\alpha^+(s) + (z - b)r_\alpha^+(s) \;, & z \in I_3 \;, \end{cases} \tag{6.1.2}$$

$$T^3(s, z) = \begin{cases} t_3(s) \;, & z \in I_1 \cup I_3 \;, \\ t_3^c(s) \;, & z \in I_2 \end{cases} \tag{6.1.3}$$

where

$$b = c + \frac{d}{2} \;, \quad h = c + d \;, \qquad I_1 = [-c - d, -c] \;, \quad I_2 = [-c, c] \;, \quad I_3 = [c, c + d] \;.$$

Let us define the following stress resultants on Γ_σ

$$\tilde{N}^\alpha = d(p_\alpha^+ + p_\alpha^-) \;, \qquad Q = 2(dt_3 + ct_3^c) \;,$$

$$\mathcal{\tilde{M}}^\alpha = \frac{d^3}{12}(r_\alpha^+ + r_\alpha^-) \;, \qquad \mathbf{\tilde{M}}^\alpha = db(p_\alpha^+ - p_\alpha^-) \;,$$

$$\tilde{Q} = Q + \frac{\partial \mathcal{\tilde{M}}_\tau}{\partial s} \;, \qquad \mathcal{\tilde{M}}_\tau = \mathcal{\tilde{M}}^\alpha \tau_\alpha \;, \qquad \mathcal{\tilde{M}}_n = \mathcal{\tilde{M}}^\alpha n_\alpha \;,$$

$$\tilde{N}_n = \tilde{N}^\alpha n_\alpha \;, \quad \tilde{N}_\tau = \tilde{N}^\alpha \tau_\alpha \;, \quad \mathbf{\tilde{M}}_n = \mathbf{\tilde{M}}^\alpha n_\alpha \;, \quad \mathbf{\tilde{M}}_\tau = \mathbf{\tilde{M}}^\alpha \tau_\alpha \;,$$

$$\tag{6.1.4}$$

where n, τ are versors normal and tangent to Γ respectively, see Fig. 6.1.1.

The theory of Hoff is based on the following displacement assumptions, see Fig. 6.1.2

$$w_\alpha(x, z) = u_\alpha(x) + f(z)\varphi_\alpha(x) - g(z)w_{,\alpha}(x) \;, \quad w_3(x, z) = w(x) \;, \tag{6.1.5a}$$

where

$$f(z) = \begin{cases} -b \;, & z \in I_1 \\ b\dfrac{z}{c} \;, & z \in I_2 \\ b \;, & z \in I_3 \;; \end{cases} \tag{6.1.5b}$$

$$g(z) = \begin{cases} z + b, & z \in I_1 \\ -\dfrac{d}{2c}z, & z \in I_2 \\ z - b, & z \in I_3 . \end{cases} \tag{6.1.5c}$$

The functions u_α describe in-plane displacements of the middle plane, φ_α represent angles of rotation of cross-sections (cf. the angle φ_1 in Fig. 6.1.2). The last term in $(6.1.5a)_1$ is proportional to the quantity d and describes bending states of the face-plates. For sandwich plates with thin faces $(d \ll c)$ its role is negligible.

Moreover, the plane-stress state in the face-plates is assumed. Thus

$$\sigma^{\alpha\beta}(x, z) = \begin{cases} \tilde{C}_f^{\alpha\beta\lambda\mu} e_{\lambda\mu}(\boldsymbol{w}), & z \in I_1 \cup I_3 \\ 0 & z \in I_2 ; \end{cases}$$

$$\sigma^{\alpha 3}(x, z) = \begin{cases} 2C_f^{\alpha 3\lambda 3} e_{\lambda 3}(\boldsymbol{w}), & z \in I_1 \cup I_3 \\ 2\mu_c\, e_{\alpha 3}(\boldsymbol{w}) & z \in I; \end{cases} \tag{6.1.6}$$

$$\sigma^{33}(x, z) = \begin{cases} C_f^{33\lambda\mu} e_{\lambda\mu}(\boldsymbol{w}) + C_f^{3333} e_{33}(\boldsymbol{w}), & z \in I_1 \cup I_3 \\ E_c\, e_{33}(\boldsymbol{w}), & z \in I_2 , \end{cases}$$

with $\tilde{C}_f$ defined according to the rule (2.7.6). The strains associated with (6.1.5) are

$$e^{\alpha\beta}(\boldsymbol{w}) = \epsilon_{\alpha\beta} + f(z)\rho_{\alpha\beta} + g(z)k_{\alpha\beta} ,$$

$$e_{\alpha 3}(\boldsymbol{w}) = \begin{cases} 0, & z \in I_1 \cup I_3 \\ \dfrac{b}{c}\gamma_\alpha, & z \in I_2 , \end{cases} \tag{6.1.7}$$

$$e_{33}(\boldsymbol{w}) = 0 ,$$

where

$$\epsilon_{\alpha\beta} = e_{\alpha\beta}(\boldsymbol{u}) , \quad \rho_{\alpha\beta} = e_{\alpha\beta}(\boldsymbol{\varphi}) , \quad \gamma_\alpha(\boldsymbol{\varphi}, w) = w_{,\alpha} + \varphi_\alpha , \quad k_{\alpha\beta} = \kappa_{\alpha\beta}(w) . \tag{6.1.8}$$

Note that:
(i) displacements w_α given by (6.1.5) are continuous along the thickness, in particular – at points $z = \pm c$,
(ii) the faces do not undergo transverse shearing, cf. (6.1.7).

The same kinematic assumptions are imposed on the trial displacement field $v = \tilde{v} = (\tilde{v}_i)$ involved in the variational equation of equilibrium (2.1.5). By using assumptions (6.1.5), (6.1.7) and integrating over z one finds the variational equilibrium equation for the theory of Hoff

$$\int_{\Omega} [N^{\alpha\beta} e_{\alpha\beta}(\boldsymbol{v}) + \mathcal{M}^{\alpha\beta}\kappa_{\alpha\beta}(v) + \mathbf{M}^{\alpha\beta} e_{\alpha\beta}(\boldsymbol{\psi}) + Q^{\alpha}\gamma_{\alpha}(\boldsymbol{\psi}, v)]dx \tag{6.1.9}$$

$$= \int_{\Omega} pv dx + \int_{\Gamma_{\sigma}} (\tilde{N}_n v_n + \tilde{N}_\tau v_\tau + \tilde{Q}v - \tilde{\mathcal{M}}_n\frac{\partial v}{\partial n} + \tilde{\mathbf{M}}_n\psi_n + \tilde{\mathbf{M}}_\tau\psi_\tau)ds \ ,$$

that holds for all $(\boldsymbol{v}, v, \boldsymbol{\psi})$ kinematically admissible; $v_n = v_\alpha n^\alpha$, $v_\tau = v_\alpha \tau^\alpha$, $\psi_n = \psi_\alpha n^\alpha$, $\psi_\tau = \psi_\alpha \tau^\alpha$. The stress and couple resultants are defined as follows

$$N^{\alpha\beta} = \int_{-(c+d)}^{c+d} \sigma^{\alpha\beta}dz \ , \qquad \mathbf{M}^{\alpha\beta} = \int_{-(c+d)}^{c+d} f(z)\sigma^{\alpha\beta}dz \ ,$$

$$\mathcal{M}^{\alpha\beta} = \int_{-(c+d)}^{c+d} g(z)\sigma^{\alpha\beta}dz \ , \qquad Q^{\alpha} = \frac{b}{c}\int_{-c}^{c} \sigma^{\alpha 3}dz \ . \tag{6.1.10}$$

Substitution of (6.1.6) into (6.1.10) yields

$$N^{\alpha\beta} = A^{\alpha\beta\lambda\mu}\epsilon_{\lambda\mu} \ , \qquad \mathbf{M}^{\alpha\beta} = D^{\alpha\beta\lambda\mu}\rho_{\lambda\mu} \ ,$$

$$\mathcal{M}^{\alpha\beta} = D_f^{\alpha\beta\lambda\mu}k_{\lambda\mu} \ , \qquad Q^{\alpha} = H^{\alpha\lambda}\gamma_{\lambda} \tag{6.1.11}$$

with

$$A^{\alpha\beta\lambda\mu} = 2d\tilde{C}^{\alpha\beta\lambda\mu} \ , \qquad D^{\alpha\beta\lambda\mu} = 2db^2\tilde{C}_f^{\alpha\beta\lambda\mu} \ ,$$

$$D_f^{\alpha\beta\lambda\mu} = 2 \cdot \frac{d^3}{12}\tilde{C}_f^{\alpha\beta\lambda\mu} \ , \qquad H^{\alpha\lambda} = \frac{2}{c}b^2\mu_c\delta^{\alpha\lambda} \ . \tag{6.1.12}$$

Thus the constitutive relations (6.1.11) are decoupled. $\boldsymbol{D}$ represents the resulting bending stiffness tensor of the whole plate, while $\boldsymbol{D}_f$ stands for the bending stiffness tensor of the face-plates. Note that

$$D_f^{\alpha\beta\lambda\mu}/D^{\alpha\beta\lambda\mu} = \frac{1}{12}\left(\frac{d}{b}\right)^2 \tag{6.1.13}$$

and for sandwich plates of thin face-plates

$$D_f^{\alpha\beta\lambda\mu} \ll D^{\alpha\beta\lambda\mu} \ . \tag{6.1.14}$$

Substitution of relations (6.1.11) into (6.1.9) leads to the variational form of the equilibrium problem for the plate problem studied. It involves a symmetric and coercive bilinear form. By using the Lax-Milgram theorem one can easily prove that the boundary value problem obtained in this way is well-posed. Note that this problem splits up into the membrane problem for u and bending problem for (φ, w). In the sequel only this latter problem will be discussed.

6.2. *Effective stiffnesses in the periodic case*

Stiffnesses A, D_f, D, and H of Hoff's sandwich plates depend on: the thicknesses of face-plates d, the thickness of the core $2c$, the elastic moduli of faces for the plane-stress case $(\tilde{C}^{\alpha\beta\lambda\mu})$ and transverse shear modulus of the core μ_c. These quantities, in particular some of them, can vary periodically in x_α directions. For instance, this is the case if the core is made of aluminium honeycombs, as commonly used in airspace technology.

To find effective stiffnesses of Hoff's plate with periodic characteristics one can apply:
(i) the in-plane scaling homogenization,
(ii) the refined scaling method.

Let us focus our attention on the method (i). We proceed as in Sec. 5.2 and consider only the bending problem. After introducing the parameter ε as in Sec. 5.2 one considers the εY-periodic problem for a clamped plate

$$(P^\varepsilon_{Hoff}) \quad \left| \begin{array}{l} \text{find } (\varphi^\varepsilon, w^\varepsilon) \in V^0_K(\Omega) = H^1_0(\Omega)^2 \times H^2_0(\Omega) \text{ such that} \\[2mm] \displaystyle\int_\Omega [\mathcal{M}^{\alpha\beta} \kappa_{\alpha\beta}(v^\varepsilon) + \mathbf{M}^{\alpha\beta}_\varepsilon e_{\alpha\beta}(\psi^\varepsilon) + Q^\alpha_\varepsilon (v^\varepsilon_{,\alpha} + \psi^\varepsilon_\alpha)]dx \\[2mm] \displaystyle = \int_\Omega pv^\varepsilon dx \qquad \forall\, (\psi^\varepsilon, v^\varepsilon) \in V^0_K(\Omega)\,. \end{array} \right. \tag{6.2.1}$$

Here we assume that p is independent of ε; moreover,

$$\mathcal{M}^{\alpha\beta}_\varepsilon = D^{\alpha\beta\lambda\mu}_f \left(\frac{x}{\varepsilon}\right) \kappa_{\lambda\mu}(w^\varepsilon)\,,$$

$$\mathbf{M}^{\alpha\beta}_\varepsilon = D^{\alpha\beta\lambda\mu} \left(\frac{x}{\varepsilon}\right) e_{\lambda\mu}(\varphi^\varepsilon)\,, \qquad Q^\alpha_\varepsilon = H^{\alpha\lambda} \left(\frac{x}{\varepsilon}\right) \gamma_\lambda(\varphi^\varepsilon, w^\varepsilon)\,. \tag{6.2.2}$$

The functions $D_f(\cdot)$, $D(\cdot)$ and $H(\cdot)$ are Y-periodic; $Y = (0, l_1) \times (0, l_2)$.

The deflection function w^ε is represented in the form $(3.2.4)_2$ and the function φ^ε in the form of $(3.2.4)_1$. The main terms determining $\mathcal{M}^{\alpha\beta}_\varepsilon$, $\mathbf{M}^{\alpha\beta}_\varepsilon$ and Q^α_ε are

$$\mathcal{M}^{\alpha\beta}_0 = D^{\alpha\beta\lambda\mu}_f \left(\frac{x}{\varepsilon}\right) [\kappa^h_{\lambda\mu} + \kappa^y_{\lambda\mu}(w^{(2)})]\,,$$

$$\mathbf{M}^{\alpha\beta}_0 = D^{\alpha\beta\lambda\mu} \left(\frac{x}{\varepsilon}\right) [\rho^h_{\lambda\mu} + e^y_{\lambda\mu}(\varphi^{(1)})]\,, \qquad Q^\alpha_0 = H^{\alpha\lambda} \left(\frac{x}{\varepsilon}\right) \gamma^h_\lambda\,, \tag{6.2.3}$$

where

$$\kappa^h_{\lambda\mu} = \kappa_{\lambda\mu}(w^{(0)})\,, \qquad \rho^h_{\lambda\mu} = e_{\lambda\mu}(\varphi^{(0)})\,, \qquad \gamma^h_\lambda = w^{(0)}_{,\lambda} + \varphi^{(0)}_\lambda\,. \tag{6.2.4}$$

The local strains $e^y_{\lambda\mu}(\cdot)$, $\kappa^y_{\lambda\mu}(\cdot)$ have been defined in Sec. 2.8. The effective stress and couple resultants are defined by

$$\mathcal{M}_h = \langle \mathcal{M}_0 \rangle\,, \qquad \mathbf{M}_h = \langle \mathbf{M}_0 \rangle\,, \qquad Q_h = \langle Q_0 \rangle\,. \tag{6.2.5}$$

Similarly to Secs. 3 and 5 one proves that

$$w^{(2)}(x) = \chi^{(\lambda\mu)}\left(\frac{x}{\varepsilon}\right)\kappa_{\lambda\mu}^h(x)\,,\qquad \varphi^{(1)}(x) = \Upsilon^{(\lambda\mu)}\left(\frac{x}{\varepsilon}\right)\rho_{\lambda\mu}^h(x)\,,\qquad (6.2.6)$$

up to additive functions in x.

The functions $\chi^{(\lambda\mu)}$ are solutions to the local problem $(P^2_{KS,Y})$ of Sec. 3.2 in which stiffnesses $D^{\alpha\beta\lambda\mu}$ should be replaced with $D_f^{\alpha\beta\lambda\mu}$. The functions $\Upsilon^{(\lambda\mu)}$ are solutions to local problems $(P^1_{H,Y})$ of Sec. 5.2 with stiffnesses $D^{\alpha\beta\lambda\mu}$ given by $(6.1.5)_2$. Thus the formulae (6.2.3) can be written in the form

$$\mathcal{M}_0^{\alpha\beta} = [D_f^{\alpha\beta\lambda\mu}\left(\frac{x}{\varepsilon}\right) + D_f^{\alpha\beta\gamma\delta}\left(\frac{x}{\varepsilon}\right)\kappa_{\gamma\delta}^y(\chi^{(\lambda\mu)})|_{y=x/\varepsilon}]\kappa_{\lambda\mu}^h\,,$$

$$\mathbf{M}_0^{\alpha\beta} = [D^{\alpha\beta\lambda\mu}\left(\frac{x}{\varepsilon}\right) + D^{\alpha\beta\gamma\delta}\left(\frac{x}{\varepsilon}\right)e_{\gamma\delta}^y(\Upsilon^{(\lambda\mu)})|_{y=x/\varepsilon}]\rho_{\lambda\mu}^h\,,\qquad (6.2.7)$$

$$Q_0^{\alpha} = H^{\alpha\lambda}\left(\frac{x}{\varepsilon}\right)\gamma_\lambda^h\,.$$

After averaging and using (6.2.5) one finds

$$\mathcal{M}_h^{\alpha\beta} = \mathcal{D}_h^{\alpha\beta\lambda\mu}\kappa_{\lambda\mu}^h\,,\qquad \mathbf{M}_h^{\alpha\beta} = \mathbf{D}_h^{\alpha\beta\lambda\mu}\rho_{\lambda\mu}^h\,,\qquad Q_h^\alpha = \mathcal{H}_h^{\alpha\beta}\gamma_\beta^h\,,\qquad (6.2.8)$$

where

$$\mathcal{D}_h^{\alpha\beta\lambda\mu} = \langle D_f^{\alpha\beta\gamma\delta}X_{\gamma\delta}^{\lambda\mu}\rangle\,,\qquad X_{\gamma\delta}^{\lambda\mu} = \delta_\gamma^\lambda\delta_\delta^\mu + \kappa_{\gamma\delta}^y(\chi^{(\lambda\mu)})\,;\qquad (6.2.9)$$

$$\mathbf{D}_h^{\alpha\beta\lambda\mu} = \langle D_f^{\alpha\beta\gamma\delta}Z_{\gamma\delta}^{\lambda\mu}\rangle\,,\qquad Z_{\gamma\delta}^{\lambda\mu} = \delta_\gamma^\lambda\delta_\delta^\mu + e_{\gamma\delta}^y(\Upsilon^{(\lambda\mu)})\,;\qquad (6.2.10)$$

$$\mathcal{H}_h^{\alpha\beta} = \langle H^{\alpha\beta}\rangle\,.\qquad (6.2.11)$$

Note that the main terms determining the strain measures κ and ρ are

$$\overset{o}{\kappa}_{\gamma\delta} = X_{\gamma\delta}^{\lambda\mu}\kappa_{\lambda\mu}^h\,,\qquad \overset{o}{\rho}_{\gamma\delta} = Z_{\gamma\delta}^{\lambda\mu}\rho_{\lambda\mu}^h\,.\qquad (6.2.12)$$

The effective elastic potential

$$\mathcal{W}_{Hf} = \frac{1}{2}(\mathcal{M}_h^{\alpha\beta}\kappa_{\alpha\beta}^h + \mathbf{M}_h^{\alpha\beta}\rho_{\alpha\beta}^h + Q_h^\alpha\gamma_\alpha^h)\qquad (6.2.13)$$

can be written in the form

$$\mathcal{W}_{Hf} = \frac{1}{2}\langle\overset{o}{\kappa}_{\alpha\beta}\,D_f^{\alpha\beta\lambda\mu}\,\overset{o}{\kappa}_{\lambda\mu} + \overset{o}{\rho}_{\alpha\beta}\,D^{\alpha\beta\lambda\mu}\,\overset{o}{\rho}_{\lambda\mu}\rangle + \frac{1}{2}\gamma_\alpha^h\langle H^{\alpha\beta}\rangle\gamma_\beta^h\,.\qquad (6.2.14)$$

Usual symmetry conditions of tensors (6.2.9) – (6.2.11) and positive definiteness of potential $\mathcal{W}_{Hf}$ can be proved in a standard way, similarly to Sec. 2.8. The homogenized plate is still a Hoff's plate, the homogenized problem being formulated as follows

$$(P_h^{Hf})\quad\left|\begin{array}{l}\text{find }(\varphi^{(0)}, w^{(0)}) \in V_K^0(\Omega)\text{ such that}\\[2mm]\displaystyle\int_\Omega [\mathcal{M}_h^{\alpha\beta}\kappa_{\alpha\beta}(v) + \mathbf{M}_h^{\alpha\beta}e_{\alpha\beta}(\psi) + Q_h^\alpha\gamma_\alpha(v,\psi)]dx\\[4mm]\displaystyle = \int_\Omega pv\,dx\qquad \forall\,(\psi, v) \in V_K^0(\Omega)\,,\end{array}\right.\qquad (6.2.15)$$

where M_h, $\mathbf{M}_h$, Q_h are given by (6.2.8), (6.2.4).

If one applies the method (ii) with scaling all dimensions of the periodicity cell: l_1, l_2, c, d one imposes in this way Kirchhoff's constraints: $\varphi = -\nabla w$ on the homogenized solution. The homogenized plate becomes a thin plate with effective stiffnesses determined by a solution to a local problem. This local problem has the mathematical form of the Hoff's problem. Thus the refined scaling method transmits mathematical structure of the original problem to the local level and simplifies the homogenized problem to Kirchhoff's form thus suppressing transverse deformations at the macroscale.

6.3. *Reissner's approximation and relevant homogenization formulae*

If thicknesses of the face-plates are much smaller than the whole plate thickness, then stiffnesses $D_f^{\alpha\beta\lambda\mu}$ can be neglected in comparison with $D^{\alpha\beta\lambda\mu}$, cf. (6.1.14). Consequently the external moments $\tilde{\mathcal{M}}^\alpha$ cannot be applied. Such model of sandwich plates has been proposed by E. Reissner. Then the variational equilibrium equation (6.1.9) is reduced to the form

$$\int_\Omega [N^{\alpha\beta} e_{\alpha\beta}(\boldsymbol{v}) + \mathbf{M}^{\alpha\beta} e_{\alpha\beta}(\boldsymbol{\psi}) + Q^\alpha \gamma_\alpha(\boldsymbol{\psi}, v)]dx$$

$$= \int_\Omega pv\,dx + \int_{\Gamma_\sigma} (\tilde{N}_n v_n + \tilde{N}_\tau v_\tau + Qv + \tilde{\mathbf{M}}_n \psi_n + \tilde{\mathbf{M}}_\tau \psi_\tau)ds , \qquad (6.3.1)$$

the constitutive relations (6.1.11) being unchanged. Thus the Reissner model for sandwich plates coincides with the Reissner-Hencky model of transversely homogeneous plates. Only the stiffnesses involved there have different meaning. Due to this analogy we do not have to derive the formulae for homogenized stiffnesses, they look the same as those that follow from the in-plane scaling (Sec. 5.2) or a refined scaling (Sec. 5.4).

7. Comments and bibliographical notes

One can indicate two methods of formulation of the two-dimensional equations of homogeneous plates of constant thickness:

a) the unknown displacement, strain or stress fields are expanded into series (in particular, into power series) with respect to the x_3 variable. In the most popular approaches the linear expansions of displacement fields are assumed;

b) the ratio of the plate thickness to the in-plane dimensions of the plate is considered as a small parameter ε. The plate domain is then transformed to a domain of dimensions independent of ε and all the variables involved are scaled appropriately. By an asymptotic procedure one finds limits of the ε-dependent displacements and stress fields as ε tends to zero.

Let us repeat after the review paper by Jemielita (1991) that the origin of the method (a) can be found in the papers by Cauchy (1828) and Poisson (1829). To arrive at the two-dimensional plate equations one should use appropriate variational techniques, cf. Reissner (1985).

The method (b) was proposed by Friedrichs and Dressler (1961) and by Goldenveizer (1962) and Goldenveizer and Kolos (1965). This method assumed its rigorous formulation in the paper by Ciarlet and Destuynder (1979) and was further developed in the books by Ciarlet (1997) and Destuynder (1986).

In the first order approximation the method (b) provides a justification of the thin plate equations, along with its boundary conditions. These boundary conditions were first derived by Kirchhoff (1850). The methods of (a) type lead usually to the thin plate equations, if one neglects the transverse shear and transverse normal deformations. This might not necessary be the rule. Let us recall that Vekua (1982) found a formula for the bending stiffness of thin isotropic plates which differs from that commonly accepted! Thus one should be careful, since not all thin plate models derived by the method (a) must comply with the model of Kirchhoff. This is the merit of the method (b) which proves that the Kirchhoff model must inevitably emerge. Moreover, Miara (1994) proved that only one scaling exists (up to a multiplicative power of ε) which leads to a thin plate model. Other scalings result in a rigid plate or lead to a plate of vanishing stiffnesses, see Ciarlet (1997, Sec. 1.10).

The other disadvantages of the method (a) is disregarding of the boundary layer phenomena. Just in the boundary layers the errors of the thin plate models go to extremes. Recently Dauge and Gruais (1996, 1998a, 1998b) found the estimates of the error of the asymptotic method with taking into account the terms describing the boundary layer deformations, cf. Ciarlet (1997, Sec. 1.12).

The asymptotic method can be applied to the nonlinearly elastic plate models. Appropriate plate models are derived in Chapter 4 of the book by Ciarlet (1997).

The von Kármán plate model requires caution. The von Kármán equations can be derived by imposing appropriate constraints on displacements, cf. Fung (1965) and Landau and Lifschitz (1967), but some simplifications are difficult to justify. Let us recall that in the original paper of von Kármán these equations were not derived, but postulated, see Ciarlet

(1997). One of the advantages of the asymptotic method is that the von Kármán equations can be rigorously derived, as shown in the last chapter of the book of Ciarlet (1997); cf. also Blanchard and Xiang (1992), Ciarlet and Rabier (1980), and Sławianowska and Telega (1999).

The problem of rigorous derivation of two-dimensional models of plates with strong inhomogeneities, in particular, of periodically varying elastic characteristics, is not so well analyzed as for homogeneous plates. In this problem two simultaneous modelling procedures must be applied. The reduction of the transverse dimension must be indissolubly bonded with smearing out the nonhomogeneities in the longitudinal directions.

The method of averaging stiffnesses of periodic plates described in Secs. 2.2, 2.3 and 2.10.4 was proposed by Caillerie (1982, 1984, 1987) – there it is called the $e \approx \varepsilon$ model – and by Kohn and Vogelius (1984, 1985, 1986a) – model $a = 1$. The paper of Caillerie was concerned with plates of constant thickness. The papers of Kohn and Vogelius dealt with the case of plates of varying thickness. From this point of view the papers of Caillerie and Kohn and Vogelius are complementary. However, the methods used in these papers differ considerably.

The common feature of models $e \approx \varepsilon$ and $a = 1$ is that the local problems are posed on a three-dimensional cell of periodicity. Solutions to these local problems determine the effective stiffnesses of the effective plate which is viewed as thin. Thus the homogenized equilibrium problem assumes the form known from the Kirchhoff theory of anisotropic thin plates.

It turns out that the effective stiffnesses determined in this manner are just proper effective stiffnesses. They cannot be corrected. Other formulae, whether derived in this book or reported in other papers, can only be better or worse approximations of them, see Lewiński (1986a, 1991a – 1991d). The justifications given in Sec. 2.10 seem to be original and different from those available in the literature. Also, we considered dual problems. They are confined to plates of constant thickness. It may be conjectured that Γ-convergence theorems formulated and proved in Sec. 2.10 can be extended to plates with variable thickness. A possibility which comes to mind is similar to the one used in the proof of Theorem 3.12.1. In this case it would mean that a plate of constant thickness is constructed by appending to the original plate a soft material. An alternative justification of the asymptotic analysis of thin isotropic, homogeneous plates with constant thickness was devised by Bourguin et al. (1992). By analogy with the asymptotic analysis Anzellotti et al. (1994) developed the Γ-convergence method for functionals with varying domains. The general approach was then applied to the justification of well known rod and plate models made of an isotropic linear elastic and homogeneous material. Bouchitté et al. (1997b) proposed a general and rather abstract approach to modelling structures, which consists in describing the initial, say 3D structure by means of a measure. The low dimensional elastic model is obtained by a suitable relaxation of the original functional. The examples given are confined to linear isotropic strings and membranes.

A numerical algorithm of solving the three-dimensional local problems of Sec. 2.3 has recently been reported in Lefik (1995), Urbański (1998) and Bourgeois et al. (1997). Moreover, in this last paper a careful discussion of the accuracy of the homogenization methods discussed in Secs. 2, 3 and 5 has led the authors to the conclusion that for a wide class of sandwich periodic plates the two-dimensional formulae of Secs. 2.7, 2.9, 5.2 and 5.3 yield reasonable results and hence the complicated method of Sec. 2.3 can be omitted. The authors confirm the observation of Lewiński (1992) that the formula for D_h^{1212} of Duvaut and Metellus can drastically overestimate the correct value of this stiffness, see Table 3, case 6 of the quoted paper.

In the derivation of Sec. 2.2 the initial plate problem is reformulated to a form in which the unknown displacement fields are functions of two independent variables: $x \in \Omega$ and $y \in \mathcal{Y}$. Such a formulation simplifies the averaging process, see Lewiński (1991a). This approach has much in common with the concept of Nguetseng (1989), developed by Allaire (1992), of the two-scale convergence.

In the case of some specific shapes of the basic periodicity cell the formulae of Sec. 2.3 can be approximated by analytical methods. If the cell has the shape of a thin plate, then the Kirchhoff-type modelling can be applied, where some results of the paper by Lewiński (1991c, Sec. 2) are reported, see Sec. 2.8. A weaker assumption of the cell being moderately thick justifies the Hencky-type modelling, cf. Sec. 2.7 and Lewiński (1991c, Sec. 5). In the case of cells composed of thin-walled plates and slabs Kalamkarov (1992) derived closed analytical formulae, convenient for further analysis and optimization, cf. Kolpakov (1997). Effective stiffnesses of reinforced plates are the subject of the paper by Destuynder and Theodory (1986). In this paper the homogenization process precedes the reduction of the transverse dimension. The first results on homogenization of reinforced plates are due to Artola and Duvaut (1977, 1978), cf. also Loboda (1981).

The modelling of Sec. 2.10.2 has been originally proposed by Caillerie (1984). It leads to the homogenization results of Duvaut and Metellus (1976) and Duvaut (1976). They are rederived in Sec. 3.2 by the direct homogenization of the Kirchhoff plate equations. The formulae of Duvaut apply to the case when the basic cell has a shape of a thin plate, cf. also Lewiński (1986a).

The opposite case of the cells being transversely slender is considered in Secs. 2.9 and 2.10.3, the latter derivation being similar to that of Caillerie (1984).

In all cases the reference plane of the plate is assumed to lie within the plate body. If the plate is wrinkled such that the middle surface is a wavy surface of reference, then different homogenization technique should be applied, see Aganović et al. (1996), Aganović et al. (1998). This problem is not considered in the book.

A generalization of the analysis of Kohn and Vogelius (1984, 1985, 1986a) to the case of thin plates undergoing moderately large deflections can be found in Quintela-Estevez (1989) and Alvarez-Vázquez and Quintela-Estevez (1992), cf. also Pruchnicki (1998).

Direct homogenization of transversely symmetric periodic plates in bending or stretching was first performed by Duvaut and Metellus (1976) and Duvaut (1976). In the case of plates with circular openings or inclusions these results lead to the formulae for effective stiffnesses derived by Grigoliuk and Filshtinskii (1970) by means of the complex potentials method. Since the date of publishing the papers of Caillerie (1982, 1984, 1987) and Kohn and Vogelius (1984, 1985, 1986a) the homogenized model of Duvaut and Metellus has become a part of a more general asymptotic modelling, see Secs. 2.8 and 2.10.2, cf. also Panasenko and Reztsov (1987) and Reztsov (1990).

In Sec. 3.4 the variational form of the equations defining effective stiffnesses is derived. The compatibility equations that enter these formulae concern arbitrary domains, even multiconnected. The compatibility equations of such form are due to Moreau (1976). The results on correctors, reported in Theorems 3.5.1 and 3.5.3 are original.

In the case of one-dimensional periodicity the homogenization formulae of Duvaut and Metellus assume closed forms, see Duvaut (1976), where the isotropic case was considered, and Lewiński and Telega (1988a, Sec. 5) for the general case. In 1986 Francfort and Murat proved that these formulae can be put in a specific invariant form (Sec. 3.8.1). The derivation of Francfort and Murat (1986) concerned the elasticity problem. The counterpart of this formula for thin plates in bending was derived by Lipton (1994a). Another invariant formula for one-dimensional periodicity case was found by Lurie, Cherkaev and Fedorov (1984), see also Lurie and Cherkaev (1986). This formula has been derived in Sec. 3.8.3.

The formulae (3.9.38) have been discovered by Norris (1989) for $X = A$.

The concept of non-uniform homogenization recalled in Sec. 3.10 (concerning plates periodic with respect to a curvilinear parametrization) has been introduced in Bensoussan et al. (1978), cf. also Braides (1983).

The analysis of Sec. 3.11 concerning plates with quadratic inclusions is due to Olszewski (1990). This analysis is similar to that of Bourgat (1978) and Bourgat and Dervieux (1978) concerning the heat conduction problem.

The Γ-convergence approach, presented in Sec. 3.12, to the study of periodically perforated Kirchhoff plates is new. Duvaut (1997b) solved the same problem by using the energy method (H-convergence). To use the last method it is necessary to formulate appropriate extension theorems for functions defined on Ω^ε to functions defined on Ω, cf. also Secs. 8-10, 11.5 and 11.7. The proof of Theorem 3.12.1 also furnishes the homogenization formula for a Kirchhoff plate with soft inclusions. The problem of homogenization of three-dimensional elastic solids weakened with periodically distributed voids was examined by Lené (1978, 1984) and Oleinik et al. (1992). Homogenization problems in the case of perforated domains were studied by many authors, cf. Acerbi et al. (1992), Allaire and Murat (1993), Attouch (1984), Cioranescu and Saint Jean Paulin (1979), Jikov et al. (1994), Oleinik et al. (1992) and the references cited therein. The first two papers just mentioned, though dealing only with second order problems, are of special interest from the point of view of regularity of perforated domains. Particularly, Allaire and Murat (1993) studied the n-dimensional problem and assumed that the holes may meet the boundary, contrary to the paper by Cioranescu and Saint Jean Paulin (1979). Acerbi et al. (1992) considered also the n-dimensional problem and assumed that $\Omega^\varepsilon = \Omega \cap E_\varepsilon$, where $E_\varepsilon = \varepsilon E$ and E is an arbitrary periodic open subset of $\mathbb{R}^n$ with Lipschitz boundary. The extension result was proved under the only assumption that E is connected. It would be interesting to extend these weaker assumptions concerning the holes in the n-dimensional bodies to perforated plates.

El Otmani et al. (1995) derived the effective plate model assuming as a starting point the dynamic equations of a thin perforated linear elastic body characterized by three small parameters: the thickness e, the period dimension ε and the thickness of the material $\varepsilon\delta$ (in the plane of the plate). The body has the form of a thin perforated parallelepiped with a lattice structure, cf. Figs. 1 and 2 in the paper by El Otmani et al. (1995). The inertial forces were assumed in the form $e^r \dfrac{\partial^2 \boldsymbol{u}^{e\varepsilon\delta}}{\partial t^2}$, where $\boldsymbol{u}^{e\varepsilon\delta}$ stands for the displacement vector of the thin body. For $r > 2$, the time derivatives vanish in the limit when e tends to zero. Therefore special attention was devoted to the critical case $r = 2$, which preserves time derivatives. The asymptotic analysis was performed first when $e \to 0$, next when $\varepsilon \to 0$ (homogenization) and finally when $\delta \to 0$. The first passage to the limit yields dynamic equations of an anisotropic plate (the initial lattice structure was assumed to be made of an anisotropic material). To pass with δ to zero the methods used primarily by Cioranescu and Saint Jean Paulin (1986, 1988) to the study of reinforced and honeycomb structures were exploited, cf. also Bakhvalov and Panasenko (1984, Chap. 8). Assuming that the material is isotropic it was shown that in the limit when $\delta \to 0$ the Poisson coefficient is negative. The same problem was also examined in the theses by Sac-Épée (1994) and El Otmani (1994). Sac-Épée (1994) investigated also the case $r > 0$ (the main results were summarized in the paper by El Otmani et al. (1995)). By using similar methods one can derive effective models for different initial geometry of thin lattice structures. The following problems seem to be left open: (i) derivation of the effective model which takes into account the rotational inertia term, i.e., a term proportional to $\Delta \ddot{w}$, where w stands for the transverse deflection of the plate and $\ddot{w} = \dfrac{\partial^2 w}{\partial t^2}$. According to Raoult (1985) such a model can be derived by taking into account the first corrector. (ii) The asymptotic analysis similar to the one carried out in Sec. 2. Specifically of interest would be the model obtained by simultaneous passage to zero with these three parameters.

The results of Sec. 3.13 on homogenization of plates with rigid inclusions seem to be new. It is not difficult to perform homogenization of other models of plates with voids or rigid inclusions.

The first paper on the homogenized properties of von Kármán plates (or plates undergoing moderately large deflections) was written by Duvaut (1977a) by using the energy method (H-convergence). To apply this method it is necessary to assume, that the loading is "small" thus ensuring, for each $\varepsilon > 0$, the existence of a unique solution. Our approach to homogenization of von Kármán plates, presented in Sec. 4.2, exploits Theorem 1.3.28. Then it suffices to assume, that the loading functional is a continuous perturbation functional. Dual approach to homogenization of von Kármán plates remains an open problem. Since the primal problem is a nonconvex one, the dual formulation is nontrivial even in the standard case, cf. Telega (1989), Bielski and Telega (1996). Section 4.3 is based on the paper by Mignot et al. (1981), cf. also Mignot and Puel (1978), Mignot et al. (1980), Suquet (1981). The relevant results on bifurcation points for potential operators can be found in the book by Krasnosel'skii (1964, Chap. VI). Various eigenvalue problems in elasticity were discussed by Bendsøe (1995), Oleinik (1987) and Oleinik et al. (1992), cf. also Santosa and Vogelius (1993).

The plate model used in Sec. 5.1 is due to Hencky (1947). A direct homogenization of the Hencky plate has been first performed in Bourgeat and Tapiéro (1983, 1985), cf. Tadlaoui and Tapiéro (1988). In Sec. 5.2 this approach is called the in-plane scaling based homogenization. The refined scaling approach (Sec. 5.3) has been proposed by Lewiński (1992) and justified by Telega (1992), cf. also Lewiński (1997).

In the case of ribbed plates (unidirectional periodicity) the three homogenization methods: of Bourgeat and Tapiéro (1983), Duvaut and Metellus (1976) and Lewiński (1992) give the same formulae for the stiffnesses of indices: 1111, 2222, 1122, see Eq. (5.6.16). The drastic difference between the predictions of the torsional stiffness D^{1212} by the methods of Duvaut and Metellus and Bourgeat and Tapiéro was first noted in Lewiński and Telega (1988a) and then discussed in Lewiński (1991c, 1992). The refined formula (5.6.29) for the effective torsional stiffness was derived in Lewiński (1992).

The method of Sec. 5.3 is equivalent to the method of Sec. 2.7. This method can be used to find closed formulae for the effective stiffnesses of transversely asymmetric plates periodic in one direction, especially to plates with asymmetric ribs, see Lewiński (1995).

The Francfort and Murat (1986) formula for the layered media applies directly to assessing the effective bending stiffnesses of moderately thick plates, see Eq. (5.6.40). The formula (5.6.42) for the effective transverse shear stiffness tensor of ribbed plates is similar to the formula of Tartar (1985) for the effective conductivity moduli of the layered media. Just this paper of Tartar had been an inspiration for Francfort and Murat for finding their formula for effective moduli of a layered elastic media.

Section 5.7 concerns the problem of improving the Reissner-Hencky plate theory by imposing more accurate displacement assumptions. Sec. 5.7.1 is of preliminary character and recalls the stress-based modelling of Reissner (1944, 1945, 1950). The proof of the dual formulation or Castigliano's theorem reported in Sec. 5.7.1 can be found in Duvaut and Lions (1972) and Nečas and Hlaváček (1981), cf. also Ekeland and Temam (1976). The definitions (5.7.13) of generalized displacements are due to Reissner (1945).

The history of the displacement assumption (5.7.23) is described in the review paper of Jemielita (1991). Usually this assumption has been proposed in a form similar, but not identical to (5.7.23). According to Jemielita (1991) the assumption similar to (5.7.23) was first proposed by Vlasov (1957). It was then rediscovered by Kączkowski (1968), Levinson (1980) and others. The choice of unknowns is crucial here. According to the Reissner (1945) definitions the fields $(\boldsymbol{u}, \varphi, w)$ are correct generalized displacements. Instead of this choice some authors use $\boldsymbol{\theta}$ instead of φ, where $\boldsymbol{\theta} = a\nabla w + b\varphi$, a, b being differently chosen. In 1987 Hutchinson showed that only the choice $a = 0$ and $b = 1$ leads to a correct approximation of the boundary conditions, which gave one more argument for the Reissner definitions (5.7.13) and the choice $(\boldsymbol{u}, \varphi, w)$ assumed in Sec. 5.7.2.

A theory being energy consistent with the Vlasov assumption was first derived by Reddy (1984). Its reformulation involving the Reissner unknowns $(\boldsymbol{u}, \varphi, w)$ is presented in Sec.

5.7.2 after the papers of Lewiński (1986b, 1991c), see also Lewiński (1987).

The problem of existence of solutions for the model presented in Sec. 5.7.2 was solved by Bielski and Telega (1996), cf. also Bielski and Telega (1998). Other, energy inconsistent plate theories based on the Vlasov assumption are due to Vlasov (1957), Kączkowski (1968) and Levinson (1980). They introduce improvements in comparison to the Hencky (1947) plate theory, see Rychter's (1987) proof based on the hypercircle method of Synge and Prager, cf. Synge (1957).

The results of Sec. 5.7.3 concerning homogenization of nonlinear moderately thick plates are reported after Bielski and Telega (1997).

The sandwich plates are the subject of the monographs of Plantema (1966) and Stamm and Witte (1974). A variational derivation of the Hoff theory can be found in Lewiński (1991e). Its simplified version is given in Sec. 6.1. The approximation considered in Sec. 6.3 follows the paper of Reissner (1947).

Many problems are not comprised in Chapter II. In particular, we have not considered the aperiodic homogenization. Some qualitative results for plates are due to Damlamian and Vogelius (1985). Two-dimensional non-uniform homogenization for a large class of nonlinear problems was presented in Sec. 1.3.5. Moreover, the boundary layer phenomena lie outside the scope of the book. In the case of homogeneous plates they were the subject of the thesis by Pecastaings (1985), and the papers by Gregory and Wan (1984), Coutris and Monavon (1986), Li et al. (1997), Schwab and Wright (1995). In the last two papers hierarchic plate models were examined. The development of the $h - p$ version of the finite element method allows one to generate a natural hierarchy of plate models based on polynomial approximations through the thickness, cf. Babuška and Li (1992), Schwab (1996).

Let us also mention recent results related to error estimation and boundary layers in thin plates. Destuynder and Gruais (1995) considered the asymptotic expansion of the solution to variational problems in linear elasticity posed over a three-dimensional plate whose thickness e tends to zero. It was established that the unknown displacement field converges with an estimated error of $e^{1/2}$ for the plane component and $e^{3/2}$ for the vertical component. Dauge and Gruais (1996, 1998b) developed higher order asymptotics and studied the boundary layer for an anisotropic and inhomogeneous plate made of a monoclinic material and clamped along its lateral face, cf. also Dauge and Gruais (1998a), Dauge et al. (1998). Dauge et al. (1997a, 1997b) investigated the limit behavior and the boundary layers of three-dimensional displacements in isotropic and homogeneous plates as the thickness tends to zero, in each of the eight main types of boundary conditions on their edges. Pruchnicki (1998) combined the formal asymptotic procedure with homogenization and the analysis of the boundary layer. This author considered geometrically nonlinear elastic isotropic composite plates with a periodic structure. Both constituents of the plate: steel and elastomer were assumed to obey a hyperelastic constitutive equation.

Boundary layers in an elastic stratified material with a periodic structure were considered by Dumontet (1985a, 1985b, 1986) and Sanchez-Palencia (1987) and extended to piezo-electric composites by Gambin and Gałka (1995). Moreover, boundary layers in laminated plates were investigated by Davet and Destuynder (1985, 1986) and Davet et al. (1985). The mixed periodic boundary conditions were considered by Constanda (1995).

The effective stiffnesses of laminated plates composed of laminae having planes of material symmetry parallel to one plane are given by Eqs. (2.11.4). These formulae can be derived directly by imposing Kirchhoff's constraints on the distribution of displacements across the laminate thickness, cf. Christensen (1979). If some laminae are stiff and some are soft the thin plate model should not be used, since the transverse shear deformations cannot be neglected. More complicated laminate models, capable of including these effects are discussed in Noor and Burton (1989) and Matysiak and Nagórko (1989), cf. also Kubik (1993).

The perforated plates (see Sec. 3.12) become gridworks if the walls between openings are thin. The equilibrium problem of such structures can be easily analyzed within the framework of skeletal structures, which makes the problem algebraic. If such structures are regular (or periodic) the algebraic equations of equilibrium assume the form of recurrent equations of constant coefficients, which enables one to find their explicit analytical solutions, see Bleich and Melan (1927) and Gutkowski (1973), and develop discrete analogues of the differential geometry methods, see Frąckiewicz (1970). In the theory of skeletal structures the joints displace and rotate. Consequently, the continuum descriptions of dense regular gridworks involve rotations as independent variables, see Woźniak (1970) and the review paper by Noor (1988).

Passing to zero with a characteristic distance between joints makes it possible to eliminate the rotational degrees of freedom and form an effective plate model, see (Lewiński, 1984b, Sec. 8). For a given structure this model is uniquely determined and the formulae for effective moduli are similar to those found by Cioranescu and Saint Jean Paulin (1986) and Bakhvalov and Panasenko (1984), see also Shi and Tong (1995). On the other hand, the micropolar effective moduli of dense periodic gridworks cannot be uniquely determined, cf. Lewiński (1984a, 1984b), which restricts applications of this approach to some qualitative analyses.

The asymptotic analysis of thermoelastic and piezoelastic plates and shells is also not covered by the book, cf. Blanchard and Francfort (1987), Rogacheva (1994), Taghite and Lanchon-Ducauquis (1993), Taghite et al. (1997) and the references cited therein. Kolpakov (1992) performed formal asymptotic homogenization of stationary equations of thermo-elasticity in a thin domain exhibiting a periodic structure.

Chapter III

ELASTIC PLATES WITH CRACKS

Introduction

There are two ways of finding the elastic effective characteristics of cracked solids. The first concept is to assume periodic distribution of cracks and find the "exact" effective characteristics. They are usually implicit or given by complicated formulae. To find approximate closed formulae one should apply special estimation techniques, like the translation method (Chap. VI). The first attempt to homogenize a cracked solid is due to Sanchez-Palencia (1980). In this approach the cracks are treated as unilateral cracks that open or close. Their unilateral behavior is modelled by the internal Signorini conditions on the crack faces. Sanchez-Palencia proved that the effective solid has non-linear hyperelastic properties. Then this work of Sanchez-Palencia has been justified by Attouch and Murat (1985). The homogenization approach decomposes the initial problem into the nonlinear problem for the solid with smeared-out cracks and the nonlinear problem posed on the periodicity cell. If the non-overlapping condition is neglected, both problems become linear and independent.

The second approach in the theory of cracked solids has been initiated by Budiansky and O'Connell (1976) and then developed by Laws and Brockenbrough (1987). It refers to the self-consistent method of Hill (1963), cf. also Nemat-Nasser and Hori (1993). The resulting effective solid is a priori treated as isotropic. Its two effective constants are expressed by closed formulae, depending on the crack geometry. These formulae could probably be derived from the homogenization ones, but this passage is until now unknown.

To assess the effective properties of cracked plates we apply the first method to various plate models. In Sec. 8 we consider all cracking modes admissible within the most popular model of Kirchhoff and perform the asymptotic homogenization. The effective plate turns out to be hyperelastic. More accurate analysis can be performed by using the Hencky plate theory. In Sec. 9 we consider three types of cracks due to bending, tension and shearing. In all cases the homogenization method leads to specific hyperelastic plate models with smeared-out cracks. Using Kirchhoff and Hencky models to assess deformation of plates with cracks can be critized, since the kinematic assumptions on which these models are based cease to apply in the vicinity of cracks. In Sec. 10 we construct a special two-layer plate model. Its kinematic assumption is suitable for describing the flexural cracks. The cracks introduce the membrane-bending coupling, because the presence of the crack introduces transverse asymmetry. Within this model the homogenization is performed and the effective hyperelastic potential is found.

Different problems are treated in Sec. 11 devoted to assessing stiffness loss in the $[0_m^o/90_n^o]_s$ laminates, cracked in the 90° – layer. To treat this problem in a possibly accurate manner we start with forming a new laminate model capable of describing the transverse cracks in the internal layer. Then we assume that the cracks form a periodic layout and next perform homogenization. This method was used to assess the loss of the Young moduli, Poisson ratios and the Kirchhoff modulus of laminates with aligned cracks in the 90° – layer. The final results turn out to be similar to the theoretical predictions of Hashin (1985) and McCartney (1992) and describe fairly well the experimental results available in the literature.

8. Unilateral cracks in thin plates

Two dimesional plate models admit cracking modes corresponding to the kinematic assumptions on which a given plate theory is based. We are interesting in cracking modes intrinsic to a given plate theory. If one concludes that admissible cracking modes are too simple to describe the phenomena observed it is suggested rather to change the plate model and consider a larger class of cracking modes, but still inherent in a given plate model. In this section we consider cracking modes admissible within Kirchhoff's plate modelling. Although the admissible cracking modes can be viewed as too simple to describe complicated through – the thickness material phenomena of cracked plates, it seems reasonable to analyse all modelling possibilities of the simplest and most popular plate model.

8.1. Cracking modes

Consider a thin transversely symmetric elastic plate weakened by a crack whose projection on the middle plane Ω forms an arc F. The crack is understood here as a certain discontinuity line, which will be explained later.

Kirchhoff's theory admits the following sets of kinematically admissible deflection for the case of a clamped plate with a unilateral crack

$$K^1 = \{v \in H_1^2(\Omega \backslash F) \mid [\![v]\!] \leq 0 \text{ on } F\}\,,$$

$$K^2 = \{v \in H_1^2(\Omega \backslash F) \mid [\![v]\!] \leq 0\,, \quad [\![\partial v/\partial \boldsymbol{n}]\!] = 0 \text{ on } F\}\,,$$

$$K^3 = \{v \in H_1^2(\Omega \backslash F) \mid [\![v]\!] = 0\,, \quad [\![\partial v/\partial \boldsymbol{n}]\!] \leq 0 \text{ on } F\}\,, \qquad (8.1.1)$$

$$K^4 = \{v \in H_1^2(\Omega \backslash F) \mid [\![\partial v/\partial \boldsymbol{n}]\!] \leq 0 \text{ on } F\}\,,$$

$$K^5 = \{v \in H_1^2(\Omega \backslash F) \mid [\![v]\!] \leq 0\,, \quad [\![\partial v/\partial \boldsymbol{n}]\!] \leq 0 \text{ on } F\}\,,$$

where

$$H_1^2(\Omega \backslash F) = \{v \in H^2(\Omega \backslash F) \mid v = 0 \text{ and } \partial v/\partial \boldsymbol{n} = 0 \text{ on } \Gamma = \partial \Omega$$

$$\text{in the sense of traces}\}\,, \qquad (8.1.2)$$

and $\boldsymbol{n}$ represents a vector normal to F and Γ. It is assumed that F is closed as a set, $\bar{F} = F$, and of class C^1.

Because of a linear distribution of stresses $\sigma^{\alpha\beta}$ across the plate thickness a cracking due to bending precedes a transverse shearing cracking that can result in a discontinuity of deflection. From this point of view the cracking mode (3) (set K^3) is a primal one.

Assume that the plate is subject to transverse loading q. Then the virtual work of the external loading is represented by the linear continuous form

$$f(v) = \int_{\Omega\backslash F} qv dx = \int_{\Omega} qv dx , \quad v \in H^2_1(\Omega\backslash F) . \tag{8.1.3}$$

We assume that $q \in L^2(\Omega)$. Let us define the bilinear form

$$a_F(w, v) = \int_{\Omega\backslash F} D^{\alpha\beta\lambda\mu}\kappa_{\alpha\beta}(w)\kappa_{\lambda\mu}(v)dx , \tag{8.1.4}$$

where $D = (D^{\alpha\beta\lambda\mu})$ represents the bending stiffness tensor which may depend on $x \in \Omega\backslash F$ and satisfies the usual conditions; $\kappa_{\alpha\beta}(v) = -v_{,\alpha\beta}$.
The variational formulation of the equilibrium problem of a fissured (cracked) plate reads:

$$(P_j) \quad \left| \begin{array}{l} \text{find } w \in K^j \text{ such that} \\[2mm] a_F(w, v - w) \geq f(v - w) \quad \forall\ v \in K^j . \end{array} \right. \tag{8.1.5}$$

The bilinear form $a_F(\cdot,\cdot)$ is coercive on $H^2_1(\Omega\backslash F)$ and K^j is convex and closed in this space, hence each of problems $(P_j)(j = 1, 2, \dots, 5)$ is uniquely solvable. In fact the coerciveness can readily be proved by dividing Ω into two subdomains, say Ω_1 and Ω_2, along F. Denote by Σ the line separating those subdomains. On $\Sigma\backslash F$ both v and $\partial v/\partial n$ are continuous. We may write

$$a_F(v, v) = \int_{\Omega_1\cup\Omega_2\cup(\Sigma\backslash F)} D^{\alpha\beta\lambda\mu}\kappa_{\alpha\beta}(v)\kappa_{\lambda\mu}(v)dx$$

$$\geq c_0(||v||^2_{H^2(\Omega_1)} + ||v||^2_{H^2(\Omega_2)}) = c_0||v||^2_{H^2(\Omega\backslash F)} , \tag{8.1.6}$$

where c_0 is a positive constant and v is an arbitrary function from $H^2_1(\Omega\backslash F)$.

8.2.   *Periodic layout of cracks. Homogenization*

Suppose that the plate is weakened by microcracks forming an εY-periodic layout; $Y = (0, l_1) \times (0, l_2)$. Here $\varepsilon > 0$ is a small parameter. All cracks are of the same unilateral mode. Every small rectangle εY, homothetic to the so-called basic cell Y, is damaged by a microcrack εF. Thus Y is weakened by F, cf. Fig. 8.2.1. Note that F may be a sum of disjoint cracks yet F is a closed set and $F \subset Y$. Moreover $YF = Y\backslash F$ is assumed to be connected. The local fields defined on YF belong to the space:

$$H^2_{per}(YF) = \{v \in H^2(YF)\big| v \text{ and } \frac{\partial v}{\partial y_\alpha} \text{ are } Y\text{-periodic}\} .$$

Microdeflections defined on YF will belong to the sets K^j_{YF}, being counterparts of the sets K^j with the space $H^2_1(\Omega\setminus F)$ replaced by $H^2_{per}(YF)$ and normal $\boldsymbol{n}$ – with normal $\boldsymbol{N}$, see Fig. 8.2.1. For instance we have

$$K^3_{YF} = \{v \in H^2_{per}(YF) \mid [\![v]\!] = 0, \quad [\![\partial v/\partial \boldsymbol{N}]\!] \le 0 \ \text{on} \ F\}. \tag{8.2.1}$$

The remaining sets K^j_{YF} are defined similarly.

Let F^ε denote the sum of all microcracks εF and $\Omega^\varepsilon = \Omega\setminus F^\varepsilon$. Depending on the type of the cracking mode the deflection w^ε will be sought in the sets K^j_ε; these sets are formed similarly as the sets K^j, (8.1.1), by replacing F with F^ε. In the same manner the bilinear form a^ε is defined – on putting F^ε instead of F in (8.1.4).

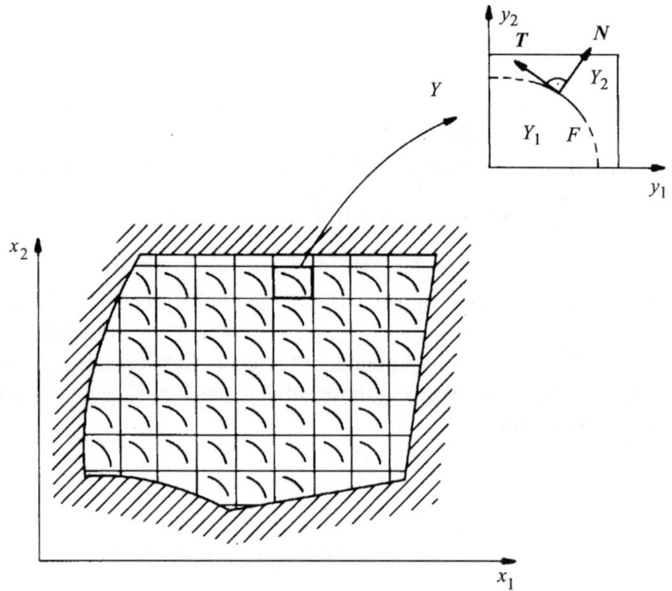

Fig. 8.2.1. Plate with cracks of periodic layout. Geometry of the rescaled cell of periodicity

The problem for w^ε reads:

$$(P^\varepsilon_j) \quad \left|\begin{array}{l} \text{find } w^\varepsilon \in K^j_\varepsilon \text{ such that} \\[2mm] a^\varepsilon(w^\varepsilon, v - w^\varepsilon) \ge f(v - w^\varepsilon) \qquad \forall \ v \in K^j_\varepsilon . \end{array}\right. \tag{8.2.2}$$

Due to coerciveness of the bilinear form a^ε the function w^ε is unique. The coerciveness is proved in the next subsection.

Homogenization means a passage to zero with the parameter ε. Similar to Sec. 3 we postulate the formal two-scale expansion

$$w^\varepsilon(x) = w^{(0)}(x) + \varepsilon^2\chi(x,y) + \varepsilon^3 w^{(3)}(x,y) + \ldots, \quad y = x/\varepsilon, \tag{8.2.3}$$

where functions χ, w are defined on $\Omega \times YF$. We assume that $\chi(x, \cdot) \in K_{YF}^j$ and $w^{(0)} \in H_0^2(\Omega)$. A trial field v is assumed in a similar form

$$v(x) = v^{(0)}(x) + \varepsilon^2 v^{(2)}(x, y) + \varepsilon^3 v^{(3)}(x, y) + \ldots, \; y = x/\varepsilon , \tag{8.2.4}$$

where $v^{(0)} \in H_0^2(\Omega)$, $v^{(k)}(x, \cdot) \in K_{YF}^j$.

Substituting (8.2.3), (8.2.4) into (8.2.2) and next passing to zero with ε we obtain

$$
\begin{aligned}
a(w^{(0)}, v^{(0)} - w^{(0)}) &+ \int_\Omega D^{\alpha\beta\lambda\mu} \kappa_{\alpha\beta}(w^{(0)}) \langle \kappa_{\lambda\mu}^y(v^{(2)} - \chi) \rangle_{YF} dx \\
&+ \int_\Omega D^{\alpha\beta\lambda\mu} \langle \kappa_{\alpha\beta}^y(\chi) \rangle_{YF} \kappa_{\lambda\mu}(v^{(0)} - w^{(0)}) dx \\
&+ \int_\Omega D^{\alpha\beta\lambda\mu} \langle \kappa_{\alpha\beta}^y(\chi) \kappa_{\lambda\mu}^y(v^{(2)} - \chi) \rangle_{YF} dx \geq \int_\Omega q(v^{(0)} - w^{(0)}) dx ,
\end{aligned}
\tag{8.2.5}
$$

where $\langle \cdot \rangle_{YF} = (l_1 l_2)^{-1} \int_{YF} (\cdot) dy$; moreover

$$a(v, w) = \int_\Omega D^{\alpha\beta\lambda\mu} \kappa_{\alpha\beta}(v) \kappa_{\lambda\mu}(w) dx .$$

Taking in (8.2.5) $v^{(2)} = \chi$ and knowing that $\pm(v^{(0)} - w^{(0)}) \in H_0^2(\Omega)$ we arrive at the variational equation

$$\int_\Omega D^{\alpha\beta\lambda\mu} \langle \kappa_{\alpha\beta}^h + \kappa_{\alpha\beta}^y(\chi) \rangle_{YF} \kappa_{\lambda\mu}(v^{(0)} - w^{(0)}) dx = \int_\Omega q(v^{(0)} - w^{(0)}) dx , \tag{8.2.6}$$

where $\kappa_{\alpha\beta}^h = \kappa_{\alpha\beta}(w^{(0)})$. This equation implies

$$M_{h,\alpha\beta}^{\alpha\beta} \mid q = 0 , \tag{8.2.7}$$

$$M_h^{\alpha\beta} = \langle M_0^{\alpha\beta} \rangle_{YF} , \qquad M_0^{\alpha\beta} = D^{\alpha\beta\lambda\mu}[\kappa_{\lambda\mu}^y(\chi) + \kappa_{\lambda\mu}^h] . \tag{8.2.8}$$

Combining (8.2.5) with (8.2.6) we get

$$\int_\Omega D^{\alpha\beta\lambda\mu} \kappa_{\alpha\beta}^h \langle \kappa_{\lambda\mu}^y(v^{(2)} - \chi) \rangle_{YF} dx + \int_\Omega D^{\alpha\beta\lambda\mu} \langle \kappa_{\alpha\beta}^y(\chi) \kappa_{\lambda\mu}^y(v^{(2)} - \chi) \rangle_{YF} dx \geq 0$$

$$\forall \, v^{(2)} \in K_{YF}^j . \tag{8.2.9}$$

Let $u \in K_{YF}^j$ and $v^{(2)} = \chi + \varphi(u - \chi)$, $\varphi \in \mathbf{D}(\Omega)$, $0 \leq \varphi \leq 1$. Then $v^{(2)} \in K_{YF}^j$ and the variational inequality (8.2.9) becomes

$$\int_\Omega \varphi \langle M_0^{\alpha\beta} \kappa_{\alpha\beta}^y(u-\chi)\rangle_{YF}\, dx \geq 0 \quad \forall\, u \in K_{YF}^j \quad \forall\, \varphi \in \mathbf{D}(\Omega)^+ . \qquad (8.2.10)$$

Thus we arrive at the local problem posed on the basic cell:

$$(P_Y^j) \quad \left|\; \begin{aligned} &\text{find } \chi(x,\cdot) \in K_{YF}^j \text{ such that} \\ &\quad d(\chi, u-\chi) \geq L^j(u-\chi) \qquad \forall\, u \in K_{YF}^j , \end{aligned}\right. \qquad (8.2.11)$$

provided that $w^{(0)}$ is given. Here

$$\begin{aligned} d(u,v) &= \langle D^{\alpha\beta\lambda\mu}\kappa_{\alpha\beta}^y(u)\kappa_{\lambda\mu}^y(v)\rangle_{YF} , \\ L^j(v) &= -\langle D^{\alpha\beta\lambda\mu}\kappa_{\alpha\beta}^h\kappa_{\lambda\mu}^y(v)\rangle_{YF} , \end{aligned} \qquad (8.2.12)$$

for $u, v \in H^2(YF)$; j indicates that $\kappa_{\alpha\beta}^h$ depends on $w^{(0)}$ relevant to the j-th cracking mode. The bilinear form d is obviously coercive on the space $H^2(YF)\backslash\mathcal{P}_1$, where $\mathcal{P}_1$ is the space of polynomials of the first order. Hence we conclude that the homogenized constitutive equation

$$M_h^{\alpha\beta} = \langle D^{\alpha\beta\lambda\mu}(\kappa_{\lambda\mu}^h + \kappa_{\lambda\mu}^y(\chi))\rangle_{YF} , \qquad (8.2.13)$$

is well defined.

In this manner we eventually arrive at the *nonlinear* bending problem of the homogenized plate:

$$(P_h^j) \quad \left|\; \begin{aligned} &\text{find } w^{(0)} \in H_0^2(\Omega) \text{ such that equilibrium equation (8.2.7) is satisfied and} \\ &\text{macroscopic moments } M_h^{\alpha\beta} \text{ are determined by (8.2.13) provided that } \chi \\ &\text{is a solution of the local problem } (P_Y^j) . \end{aligned}\right.$$

It can easily be shown that problem (P_Y^j) is equivalent to a convex minimization problem over K_{YF}^j while the homogenized (effective) elastic potential is given by, cf. also Secs 9-11

$$\mathcal{W}(\boldsymbol{\kappa}^h) = \frac{1}{2}\left\langle D^{\alpha\beta\lambda\mu}[\kappa_{\alpha\beta}^h + \kappa_{\alpha\beta}^y(\chi)][\kappa_{\lambda\mu}^h + \kappa_{\lambda\mu}^y(\chi)]\right\rangle_{YF} . \qquad (8.2.14)$$

We recall that χ depends on $\boldsymbol{\kappa}^h$. The elastic potential $\mathcal{W}$ has the following properties:
(i) $\mathcal{W}$ is of class C^1, positive and strictly convex
(ii)

$$M_h^{\alpha\beta} = \frac{\partial \mathcal{W}}{\partial \kappa_{\alpha\beta}^h} . \qquad (8.2.15)$$

(iii) $\mathcal{W}(\boldsymbol{\kappa}^h)$ is positively homogeneous of order 2

$$\mathcal{W}(\lambda\boldsymbol{\kappa}^h) = \lambda^2 \mathcal{W}(\boldsymbol{\kappa}^h) , \quad \lambda \in \mathbb{R}^+ .$$

Now this dependence is in general nonlinear, because $K_{YF}^j \neq H_{per}^2(YF)$.

(iv) $\mathcal{W}$ and its gradients M_h satisfy

$$\mathcal{W} = \frac{1}{2} M_h^{\alpha\beta} \kappa_{\alpha\beta}^h \, .$$

(v) There exist positive constants c_α such that

$$c_1 |\kappa^h|^2 \le \mathcal{W}(\kappa^h) \le c_2 |\kappa^h|^2 \, .$$

(vi) There exists a positive constant c such that

$$|M_h(\kappa_1) - M_h(\kappa_2)| \le c|\kappa_1 - \kappa_2| \, , \quad \kappa_\alpha \in \mathbb{E}_s^2 \, .$$

(vii) The operator $M_h = M_h(\kappa^h)$ is strictly monotone

$$(M_h(\kappa_1) - M_h(\kappa_2)) : (\kappa_1 - \kappa_2) > 0 \text{ for } \kappa_1 \ne \kappa_2 \, .$$

Moreover one can easily prove

Theorem 8.2.1. Let $q \in H^{-2}(\Omega)$ and let us define an operator $A : H_0^2(\Omega) \to H^{-2}(\Omega)$ as follows

$$A(w) = \frac{\partial^2 M_h^{\alpha\beta}(w)}{\partial x_\alpha \partial x_\beta} \, .$$

Then a weak or variational solution of the equilibrium problem

$$A(w^{(0)}) + q = 0$$

for the homogenized plate exists and is unique. $\qquad\qquad\qquad\qquad\square$

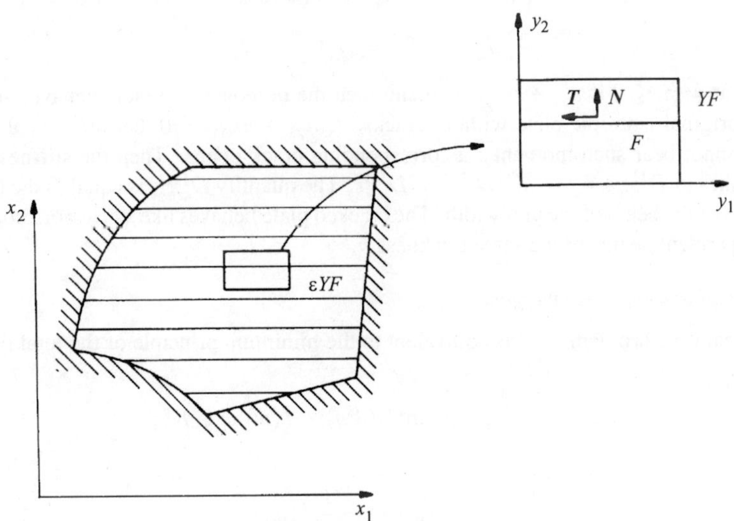

Fig. 8.2.2. Plate with aligned cracks

To illustrate the method let us consider a simple case of aligned cracks of the third type $(j = 3)$. Geometry of the basic cell is shown in Fig. 8.2.2.

Assume that the plate is isotropic of constant thickness h, hence

$$D^{\alpha\alpha\alpha\alpha} = D = \frac{Eh^3}{12(1-\nu^2)}\,, \qquad D^{1122} = \nu D\,, \qquad D^{1212} = \frac{1-\nu}{2}D\,.$$

The local problem (P_Y^3) can be analytically solved. The elastic potential of the homogenized plate then has the following form

$$\mathcal{W}(\kappa^h) = \begin{cases} D[(\kappa_{11}^h)^2 + 2\nu\kappa_{11}^h\kappa_{22}^h + (\kappa_{22}^h)^2 \\ +(1-\nu)((\kappa_{12}^h)^2 + (\kappa_{21}^h)^2)]/2 & \text{if } \kappa_{22}^h + \nu\kappa_{11}^h \le 0\,, \\ D[(1-\nu^2)(\kappa_{11}^h)^2 + (1-\nu)((\kappa_{12}^h)^2 + (\kappa_{21}^h)^2)]/2 & \text{otherwise.} \end{cases}$$

The elastic potential $\mathcal{W}$ is convex, non-negative and of class C^1, but not strictly convex, cf. Fig. 8.2.3. Consequently, $\mathcal{W}$ is not strictly monotone. These violations are "admissible" since crack F intersects ∂Y and consequently the very assumption of connectedness of YF is not satisfied here.

The constitutive relation (8.2.15) assumes now the form

$$M_h^{11} = \begin{cases} D(\kappa_{11}^h + \nu\kappa_{22}^h) & \text{if } \kappa_{22}^h + \nu\kappa_{11}^h \le 0\,, \\ D(1-\nu^2)\kappa_{11}^h & \text{otherwise;} \end{cases}$$

$$M_h^{22} = \begin{cases} D(\kappa_{22}^h + \nu\kappa_{11}^h) & \text{if } \kappa_{22}^h + \nu\kappa_{11}^h \le 0\,, \\ 0 & \text{otherwise;} \end{cases}$$

$$M_h^{12} = M_h^{21} = D(1-\nu)\kappa_{12}^h\,.$$

Note that $M_h^{22} \le 0$ if $\kappa_{22}^h + \nu\kappa_{11}^h \le 0$ and then the homogenized plate behaves similarly to the original isotropic plate without cracks. If $\kappa_{22}^h + \nu\kappa_{11}^h > 0$ then $M_h^{22} = 0$ and the plate cannot bear such moments; a corresponding crack opens. Then the stiffness D^{1111} diminishes to $D_{crack}^{1111} = (1-\nu^2)D^{1111} < D^{1111}$. The quantity D_{crack}^{1111} is equal to the bending stiffness of the beam of the unit width. The cracked plate behaves like a gridwork composed of independent beams of the same thickness h.

8.3. *Justification: Γ-convergence*

The variational problem (P_j^ε) is equivalent to the minimum principle of the total potential energy:

$$J_\varepsilon(w^\varepsilon) - f(w^\varepsilon) = \inf\{J_\varepsilon(w) - f(w)|w \in K_\varepsilon^j\}\,, \tag{8.3.1}$$

where

$$J_\varepsilon(w) = \frac{1}{2}a^\varepsilon(w,w)\,. \tag{8.3.2}$$

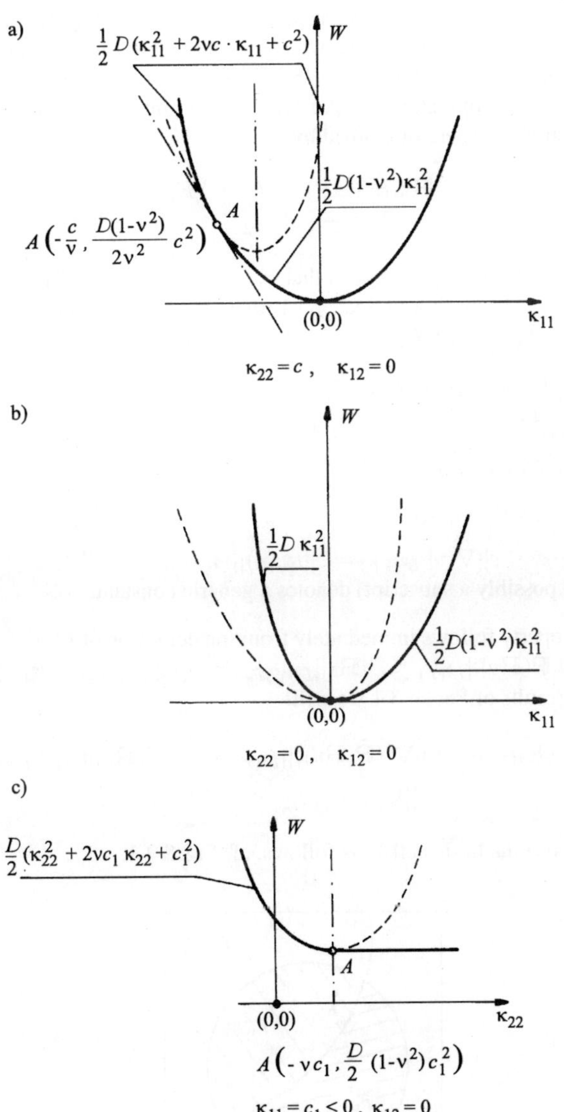

Fig. 8.2.3. Representation of the potential $\mathcal{W}(\kappa)$ for three specific sections:
$\kappa_{22}=0, \kappa_{22}=c, \kappa_{11}=c_1; c>0, c_1<0$

The functional f plays the role of a perturbation functional. The study of Γ-convergence of the sequence $\{J_\varepsilon\}_{\varepsilon>0}$ is similar to the corresponding and more complicated considerations carried out in Secs. 9 – 11. Therefore we shall focus only on those points which are connected with extensions operators from $H^2(YF)$ to $H^2(Y)$ and from $H^2(\Omega^\varepsilon)$ to $H^2(\Omega)$. To construct an extension operator, say $\mathcal{Q}$, from $H^2(YF)$ into $H^2(Y)$ we set $\mathcal{Q} := \tilde{\mathbb{P}} \circ \mathbb{R}$ where $\mathbb{R}$ is the restriction operator defined by

$$\mathbb{R} : H^2(YF) \to H^2(Y\backslash F_\eta)$$
$$v \longmapsto v\big|_{Y\backslash F_\eta} \, . \tag{8.3.3}$$

Here F_η is a sufficiently smooth hole such that $F_\eta \subset Y$ and $F \subset F_\eta$. The parameter $\eta > 0$ is kept fixed, $\eta = \eta_0$. According to Sec. 4.3.1, there exists a continuous linear extension operator $\tilde{\mathbb{P}} : H^2(Y\backslash F_\eta) \to H^2(Y)$.

Properties of the extension operator $\mathcal{Q}$
(i) $\mathcal{Q}v = v \qquad$ in $\quad F_\eta$,

(ii) $||\mathcal{Q}v||_{L^2(Y)} \leq c||v||_{L^2(Y\backslash F_\eta)}$,

(iii) $||\nabla^2 \mathcal{Q}v||_{L^2(Y)} \leq c||\nabla^2 v||_{L^2(YF)}$,

(iv) $||\mathcal{Q}v - v||_{H^2(YF)} \leq c||\nabla^2 v||_{L^2(YF)} = c||\kappa_y(v)||_{L^2(YF)}$,
where $c > 0$ (with possibly a subscript) denotes a generic constant.

Proof. The first property follows immediately from the definition of $\mathcal{Q}$.
(ii) $||\mathcal{Q}v||_{L^2(Y)} = ||\tilde{\mathbb{P}}(\mathbb{R}v)||_{L^2(Y)} \leq c||\mathbb{R}v||_{L^2(Y\backslash F_\eta)} = c||v||_{L^2(Y\backslash F_\eta)} \leq c||v||_{L^2(Y\backslash F_\eta)}$.
(iii) Applying the results of Sec. 4.3.1 we write

$$||\nabla^2 \mathcal{Q}v||_{L^2(Y)} = ||\nabla^2(\tilde{\mathbb{P}}(\mathbb{R}v))||_{L^2(Y)} \leq c||\nabla^2(\mathbb{R}v)||_{L^2(Y\backslash F_\eta)}$$
$$= c||\nabla^2 v||_{L^2(Y\backslash F_\eta)} \leq c||\nabla^2 v||_{L^2(YF)} \, .$$

(iv) Let us decompose the basic cell Y as follows, cf. Fig. 8.3.1.

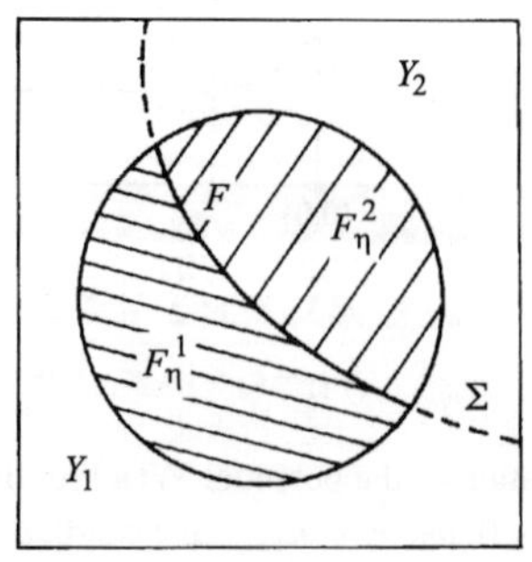

Fig. 8.3.1.

$$Y = Y_1 \cup Y_2 \cup \Sigma \quad \text{with } Y_1 \cap Y_2 = \emptyset .$$

Next we set

$$F_\eta = F_\eta^1 \cup F_\eta^2 \cup (\Sigma \cap F_\eta) , \quad F_\eta^\alpha = F_\eta \cap Y_\alpha , \quad \alpha = 1, 2.$$

We observe that $\mathcal{Q}v - v = 0$ and $\nabla(\mathcal{Q}v - v) = 0$ on $Y \backslash F_\eta$. Consequently, $\mathcal{Q}v - v = 0$ and $\dfrac{\partial}{\partial N}(\mathcal{Q}v - v) = 0$ on a part of the boundary of F_η^1 (resp. F_η^2) with strictly positive measure, where N is the outer unit vector normal to ∂F_η. Thus we have

$$||\mathcal{Q}v - v||_{H^2(F_\eta^\alpha)} \leq c_1 ||\nabla^2(\mathcal{Q}v - v)||_{L^2(F_\eta^\alpha)} , \quad \alpha = 1, 2 .$$

Hence

$$\begin{aligned}
||\mathcal{Q}v - v||_{H^2(YF)} &\leq c_1 \sum_{\alpha=1}^{2} ||\nabla^2(\mathcal{Q}v - v)||_{L^2(F_\eta^\alpha)} \\
&\leq c_1 \sum_{\alpha=1}^{2} (||\nabla^2(\mathcal{Q}v)||_{L^2(Y_\alpha)} + ||\nabla^2 v||_{L^2(Y_\alpha)}) \\
&\leq c_1 (||\nabla^2(\mathcal{Q}v)||_{L^2(Y)} + ||\nabla^2 v||_{L^2(YF)}) .
\end{aligned}$$

Taking into account the property (iii) we finally obtain

$$||\mathcal{Q}v - v||_{H^2(YF)} \leq c_1(1 + c)||\kappa_y(v)||_{L^2(YF)} . \qquad \square$$

After rescaling $y \rightsquigarrow x/\varepsilon$, we have

$$\frac{\partial}{\partial y_\alpha} = \varepsilon \frac{\partial}{\partial x_\alpha} , \quad \frac{\partial^2}{\partial y_\alpha \partial y_\beta} = \varepsilon^2 \frac{\partial^2}{\partial x_\alpha \partial x_\beta} .$$

Hence the property (iv) yields

$$\begin{aligned}
||\mathcal{Q}^\varepsilon v - v||_{L^2(\varepsilon Y)} + \varepsilon ||\nabla(\mathcal{Q}^\varepsilon v - v)||_{L^2(\varepsilon(YF))} &+ \varepsilon^2 ||\nabla^2(\mathcal{Q}^\varepsilon v - v)||_{L^2(\varepsilon(YF))} \\
&\leq \varepsilon^2 c ||\kappa(v)||_{L^2(\varepsilon(YF))} .
\end{aligned} \tag{8.3.4}$$

The global extension operator $\mathcal{Q}^\varepsilon : H^2(\Omega^\varepsilon) \to H^2(\Omega)$ is constructed similarly to the corresponding operator in Sec. 9.3. We assume that near the boundary $\partial\Omega$ there are no cracks (one can always choose a suitable subsequence of the sequence $\{\varepsilon \to 0\}$). Anyway we have

$$v \in H^2(\Omega^\varepsilon) , \quad v = 0 \quad \text{and} \quad \frac{\partial v}{\partial n} = 0 \text{ on } \partial\Omega \quad \Rightarrow \quad \mathcal{Q}^\varepsilon v \in H_0^2(\Omega) .$$

Summing over all cells $\varepsilon(YF)$ contained in Ω we finally get

$$\begin{aligned}
||\mathcal{Q}^\varepsilon v - v||_{L^2(\Omega)} + \varepsilon ||\nabla(\mathcal{Q}^\varepsilon v - v)||_{L^2(\Omega^\varepsilon)} &+ \varepsilon^2 ||\nabla^2(\mathcal{Q}^\varepsilon v - v)||_{L^2(\Omega^\varepsilon)} \\
&\leq \varepsilon^2 c ||\kappa(v)||_{L^2(\Omega^\varepsilon)} .
\end{aligned} \tag{8.3.5}$$

The last inequality implies

$$\|\mathcal{Q}^\varepsilon v - v\|_{L^2(\Omega)} \leq \varepsilon^2 c \|\boldsymbol{\kappa}(v)\|_{L^2(\Omega)} ,$$
$$\|\nabla^2(\mathcal{Q}^\varepsilon v - v)\|_{L^2(\Omega^\varepsilon)} \leq \varepsilon c \|\boldsymbol{\kappa}(v)\|_{L^2(\Omega^\varepsilon)} , \qquad (8.3.6)$$
$$\|\nabla^2(\mathcal{Q}^\varepsilon v - v)\|_{L^2(\Omega^\varepsilon)} \leq c \|\boldsymbol{\kappa}(v)\|_{L^2(\Omega)} .$$

We recall that $\|\nabla^2 v\|_{L^2(\Omega^\varepsilon)} = \|\boldsymbol{\kappa}(v)\|_{L^2(\Omega^\varepsilon)}$. For $\mathcal{Q}^\varepsilon v \in H_0^2(\Omega)$ we have

$$\|\mathcal{Q}^\varepsilon v\|_{H^2(\Omega)} \leq c \|\boldsymbol{\kappa}(\mathcal{Q}^\varepsilon v)\|_{L^2(\Omega)} \leq c_1 \|\boldsymbol{\kappa}(v)\|_{L^2(\Omega^\varepsilon)} , \qquad (8.3.7)$$

because the property (iii) yields

$$\|\boldsymbol{\kappa}(\mathcal{Q}^\varepsilon v)\|_{L^2(\Omega)} \leq c \|\boldsymbol{\kappa}(v)\|_{L^2(\Omega^\varepsilon)} .$$

We can formulate an important inequality.

Lemma 8.3.1. For any $v \in H_1^2(\Omega^\varepsilon)$ the following inequality is satisfied

$$\|v\|_{H^2(\Omega^\varepsilon)} \leq c \|\boldsymbol{\kappa}(v)\|_{L^2(\Omega^\varepsilon)} . \qquad (8.3.8)$$

Proof. The triangle inequality furnishes

$$\|v\|_{H^2(\Omega^\varepsilon)} \leq \|v - \mathcal{Q}^\varepsilon v\|_{H^2(\Omega^\varepsilon)} + \|\mathcal{Q}^\varepsilon v\|_{H^2(\Omega)} .$$

Taking into account (8.3.6) and (8.3.7) we write

$$\|v\|_{H^2(\Omega^\varepsilon)} \leq (\varepsilon^2 c_1 + \varepsilon c_1 + c_2) \|\boldsymbol{\kappa}(v)\|_{L^2(\Omega^\varepsilon)} .$$

For $0 < \varepsilon < \varepsilon_0$ with ε_0 held fixed we obtain the required inequality. $\qquad \square$

Remark 8.3.2. In Sec. 13.1 it will be shown that a function $v \in HB(Y)$ is continuous. Also, there exists a continuous embedding from $H^1(YF)$ into $BV(Y)$. On the other hand a function $v \in H^2(YF)$ is in general discontinuous and consequently such function cannot belong to the space $HB(Y)$. However, under the assumption that $[\![v]\!] = 0$ on F there exists a continuous embedding of $H^2(YF)$ into $HB(Y)$. $\qquad \square$

The formal homogenization procedure performed in Sec. 8.2 is justified by the following result.

Theorem 8.3.3. The sequence of functionals $\{J_\varepsilon\}_{\varepsilon>0}$ given by (8.3.2) is Γ-convergent in the strong topology of $H^1(\Omega)$ to

$$J_h(w) = \int_\Omega \mathcal{W}[\boldsymbol{\kappa}(w(x))]dx , \quad w \in H^2(\Omega) . \qquad (8.3.9)$$

Proof. It is similar to that of Theorem 10.3.1, which describes physically more elaborate model. Now, however, the inequality (8.3.8) has to be used and the density of C^1 functions with piecewise constant second gradient in $H^2(\Omega)$ plays an essential role, cf. Prop. 1.4.16. $\qquad \square$

Remark 8.3.4. Equivalently, one can work directly with the variational inequality (8.2.2), similarly to Sec. 9.3. Dual homogenization is left to the reader as an exercise.

9. Unilateral cracks in plates with transverse shear deformation

The theory of bending of thin plates (or Kirchhoff's theory) applies to the transversely symmetric plates. Only under this assumption the bending effects are not associated with membrane deformations. Therefore the analysis of cracked plates put forward in Sec. 8 concerns the case of cracks that weaken the plate material over and below the middle plane in the same way. A generalization of this approach to the case of other cracking modes is not possible within the framework of the bending theory – one should augment the model with membrane stress resultants and associated with them in-plane displacement fields.

Another shortcoming of the Kirchhoff theory is the underlying assumption of constraint angles of rotations of the plate cross-sections; the deformation state within the plate is fully determined by the deformation of the plate middle plane. In the case when the cracks are present this assumption has a limited applicability, since densely distributed cracks weaken the transverse shear plate stiffness and hence make the rotations of the cross-sections independent of the middle plane deflection.

The aim of the present section is to put forward an analysis of reduction of stiffnesses of cracked plates by taking into account: the transverse shear deformation and transverse asymmetry of the cracking modes. This is feasible if we base the analysis on the Reissner-Hencky plate model.

9.1. Admissible cracking modes

Consider a clamped homogeneous plate of constant thickness h subject to resulting surface transverse loading q, see Sec. 5.1. Through-the-thickness distribution of displacements is taken of the form

$$w_\alpha(x, x_3) = u_\alpha(x) + x_3\varphi_\alpha(x) , \quad w_3(x, x_3) = w(x) , \qquad (9.1.1)$$

cf. (5.1.1). The variational equilibrium equation within the theory of Hencky's plates reads

$$\int_\Omega [N^{\alpha\beta} e_{\alpha\beta}(\widetilde{v}) + M^{\alpha\beta} e_{\alpha\beta}(\widetilde{\psi}) + Q^\alpha(\widetilde{\psi}_\alpha + \widetilde{v}_{,\alpha})]dx = \int_\Omega q\widetilde{v}dx , \qquad (9.1.2)$$

for each $(\widetilde{v}, \widetilde{v}, \widetilde{\psi})$ vanishing on $\Gamma = \partial\Omega$. The constitutive relations have the decoupled form (5.1.5); index "Z" should be neglected. To take into account bending cracking modes with arbitrary penetration zones we shift the reference domain from $x_3 = 0$ to $x_3 = -e$. Instead of (9.1.1) we assume

$$w_\alpha(x, \widetilde{x}_3) = v_\alpha(x) + \widetilde{x}_3\varphi_\alpha(x) , \qquad w_3(x, \widetilde{x}_3) = w(x) ,$$
$$\widetilde{x}_3 = x_3 + e , \qquad v_\alpha = u_\alpha - e\varphi_\alpha . \qquad (9.1.3)$$

The variational equilibrium equation (9.1.2) preserves its form, but the constitutive relations now become coupled

$$N^{\alpha\beta} = A^{\alpha\beta\lambda\mu} e_{\lambda\mu}(v) + E^{\alpha\beta\lambda\mu} e_{\lambda\mu}(\varphi) ,$$

$$M^{\alpha\beta} = E^{\alpha\beta\lambda\mu}e_{\lambda\mu}(\boldsymbol{v}) + G^{\alpha\beta\lambda\mu}e_{\lambda\mu}(\boldsymbol{\varphi})\,, \qquad (9.1.4)$$
$$Q^{\alpha} = H^{\alpha\lambda}\gamma_{\lambda}\,, \qquad \gamma_{\lambda} = w_{,\lambda} + \varphi_{\lambda}\,,$$

and

$$E^{\alpha\beta\lambda\mu} = eA^{\alpha\beta\lambda\mu}\,, \qquad G^{\alpha\beta\lambda\mu} = e^2 A^{\alpha\beta\lambda\mu} + D^{\alpha\beta\lambda\mu}\,. \qquad (9.1.5)$$

Assume that the plate is weakened by a crack the projection of which on Ω forms an arc F ; F is closed as a set, $F = \bar{F}$, of class C^1 and strictly contained in Ω. Let us define

$$H_1^1(\Omega\backslash F) = \{v|v \in L^2(\Omega\backslash F)\,,\ v_{,\alpha} \in L^2(\Omega\backslash F)\,,\ v = 0 \text{ on } \Gamma\}\,. \qquad (9.1.6)$$

Kinematically admissible in-plane displacements and bending slopes will be elements of appropriate closed and convex sets. Those sets are defined as follows

$$\begin{aligned}
C_{dc} &= \{\boldsymbol{v} = (v_{\alpha}) \in H_1^1(\Omega\backslash F)^2 \mid [\![v_n]\!] \geq 0\,,\ [\![v_{\tau}]\!] = 0 \text{ on } F\}\,, \\
C_{cd} &= \{\boldsymbol{v} \in H_1^1(\Omega\backslash F)^2 \mid [\![v_n]\!] = 0\,,\ [\![v_{\tau}]\!] \geq 0 \text{ on } F\}\,, \\
C_{dd} &= \{\boldsymbol{v} \in H_1^1(\Omega\backslash F)^2 \mid [\![v_n]\!] \geq 0\,,\ [\![v_{\tau}]\!] \geq 0 \text{ on } F\}\,, \\
C_{di} &= \{\boldsymbol{v} \in H_1^1(\Omega\backslash F)^2 \mid [\![v_n]\!] \geq 0 \text{ on } F\}\,, \\
C_{id} &= \{\boldsymbol{v} \in H_1^1(\Omega\backslash F)^2 \mid [\![v_{\tau}]\!] \geq 0 \text{ on } F\}\,, \\
C_{ci} &= \{\boldsymbol{v} \in H_1^1(\Omega\backslash F)^2 \mid [\![v_n]\!] = 0 \text{ on } F\}\,, \\
C_{ic} &= \{\boldsymbol{v} \in H_1^1(\Omega\backslash F)^2 \mid [\![v_{\tau}]\!] = 0 \text{ on } F\}\,, \\
C_{cc} &= H_0^1(\Omega)^2\,, \qquad C_{ii} = H_1^1(\Omega\backslash F)^2\,,
\end{aligned} \qquad (9.1.7)$$

(c – continuous, d – discontinuous, i – indeterminate though not necessarily continuous). Here $v_n = \boldsymbol{v}\cdot\boldsymbol{n}$, $v_{\tau} = \boldsymbol{v}\cdot\boldsymbol{\tau}$, where $\boldsymbol{n}, \boldsymbol{\tau}$, are versors normal and tangent to F, respectively.

Kinematically admissible deflections will be elements of one of the following closed and convex sets

$$C_d = \{v \in H_1^1(\Omega\backslash F) \mid [\![v]\!] \geq 0 \text{ on } F\}\,, \quad C_c = H_0^1(\Omega)\,, \quad C_i = H_1^1(\Omega\backslash F)\,. \qquad (9.1.8)$$

By

$$\mathcal{K}_{\mathbf{abcdh}} = C_{\mathbf{ab}} \times C_{\mathbf{c}} \times C_{\mathbf{dh}}\,; \quad \mathbf{a,b,c,d,h} \in \{d,c,i\} \qquad (9.1.9)$$

we denote the set of kinematically admissible fields (z, u, ψ). Of particular interest is the bending cracking mode for which, see Fig. 9.1.1

$$(\boldsymbol{z}, u, \boldsymbol{\psi}) \in K^{ben} = \mathcal{K}_{cccdc} = C_{cc} \times C_c \times C_{dc}\,. \qquad (9.1.10)$$

If the plate undergoes a transverse cracking, then

$$(\boldsymbol{z}, v, \boldsymbol{\psi}) \in K^{shear} = C_{cc} \times C_d \times C_{cc}\,,$$

see Fig. 9.1.2.

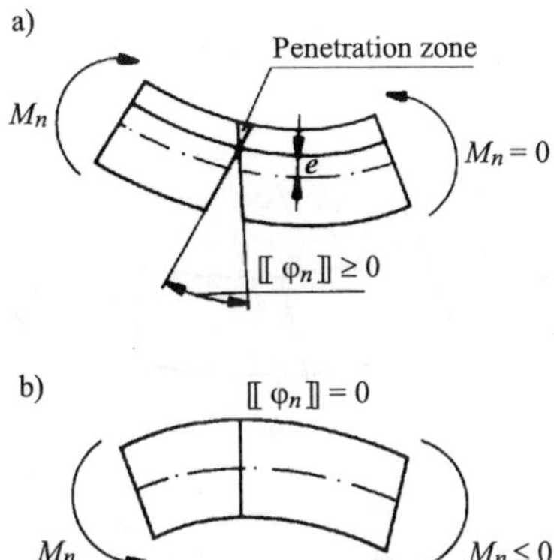

Fig. 9.1.1. The bending cracking mode;
a) the crack is open: $M_n = M^{\alpha\beta} n_\alpha n_\beta = 0$, $[\![\varphi_n]\!] \geq 0$;
b) the crack is closed, $M_n \leq 0$, $[\![\varphi_n]\!] = 0$.

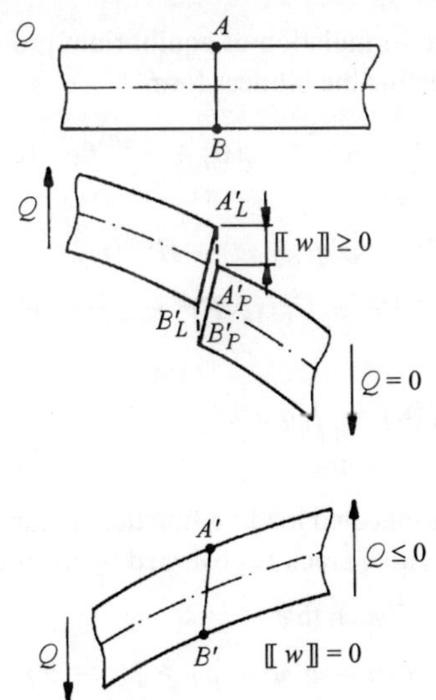

Fig. 9.1.2. The transverse shearing cracking mode;
a) the crack is open: $Q = Q^\alpha n_\alpha = 0$, $[\![w]\!] \geq 0$;
b) the crack is closed, $Q \leq 0$, $[\![w]\!] = 0$.

For the tension cracking mode

$$K^{ten} = C_{dc} \times C_c \times C_{cc} \,,$$

cf. Fig. 9.1.3.

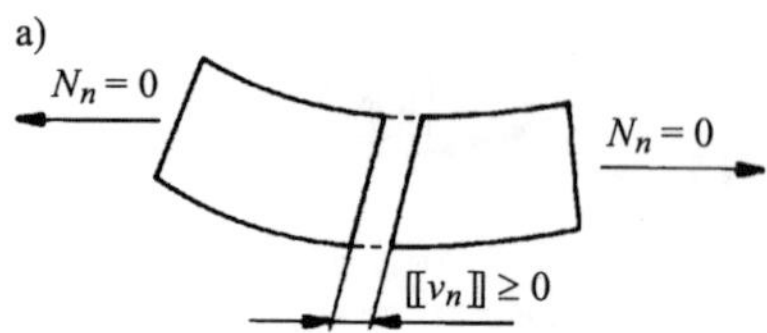

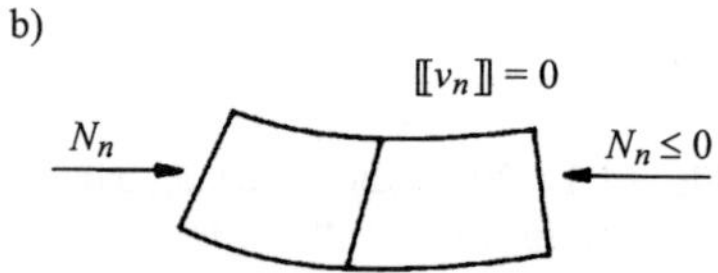

Fig. 9.1.3. The tension cracking mode;
a) the crack is open: $N_n = N^{\alpha\beta}n_\alpha n_\beta = 0$, $[\![v_n]\!] \geq 0$;
b) the crack is closed, $N_n \leq 0$, $[\![v_n]\!] = 0$.

We proceed now with the formulation of equilibrium problems for the plates with unilateral cracks. First, let us define the bilinear form

$$a_F(\boldsymbol{v}, w, \boldsymbol{\varphi}; \, \boldsymbol{z}, u, \boldsymbol{\psi}) = \int\limits_{\Omega\backslash F} [(A^{\alpha\beta\lambda\mu}e_{\lambda\mu}(\boldsymbol{v}) + E^{\alpha\beta\lambda\mu}e_{\lambda\mu}(\boldsymbol{\varphi}))e_{\alpha\beta}(\boldsymbol{z})$$

$$+(E^{\alpha\beta\lambda\mu}e_{\lambda\mu}(\boldsymbol{v}) + G^{\alpha\beta\lambda\mu}e_{\lambda\mu}(\boldsymbol{\varphi}))e_{\alpha\beta}(\boldsymbol{\psi}) + H^{\alpha\beta}(w_{,\alpha} + \varphi_\alpha)(u_{,\beta} + \psi_\beta)]dx \,, \qquad (9.1.11)$$

for all $(\boldsymbol{v}, w, \boldsymbol{\varphi}), (\boldsymbol{z}, u, \boldsymbol{\psi}) \in V_F = H_0^1(\Omega\backslash F)^2 \times H_0^1(\Omega\backslash F) \times H_0^1(\Omega\backslash F)^2$, and the linear form

$$f(u) = \int\limits_{\Omega\backslash F} qudx \,, \quad u \in H_0^1(\Omega\backslash F) \,, \qquad (9.1.12)$$

where $q \in L^2(\Omega)$. More complicated loading functionals can be considered similarly.

Equilibrium of the plate with a crack is governed by the problem

$$(P_F) \quad \left|\begin{array}{l} \text{find } (\boldsymbol{v}, w, \boldsymbol{\varphi}) \in K \text{ such that} \\[2mm] a_F(\boldsymbol{v}, w, \boldsymbol{\varphi}; \, \boldsymbol{z} - \boldsymbol{v}, u - w, \boldsymbol{\psi} - \boldsymbol{\varphi}) \geq f(u - w) \quad \forall \, (\boldsymbol{z}, u, \boldsymbol{\psi}) \in K, \end{array}\right. \qquad (9.1.13)$$

where K represents K^{ben}, K^{shear}, K^{ten} or any other possible set of type $\mathcal{K}_{\mathbf{abcdh}}$. It can easily be shown that the bilinear form a_F is coercive provided that $e < h$. Each of the sets of type $\mathcal{K}_{\mathbf{abcdh}}$ is convex and closed. Consequently, the variational inequality (9.1.13) is uniquely solvable.

9.2. *Periodic layout of cracks. Homogenization*

Consider a plate with cracks εF that form an εY-periodic layout, as in Fig. 8.2.1. Similarly to Sec. 8.2 one can define the convex sets $K_{YF}^{\mathbf{c}}, K_{YF}^{\mathbf{ab}}$ that correspond to the sets $C_{\mathbf{c}}, C_{\mathbf{ab}}$. Moreover, we define the convex sets K_ε by replacing F with F^ε representing the family of cracks. Thus we can formulate problem (P_ε) for finding $(v^\varepsilon, w^\varepsilon, \varphi^\varepsilon) \in K_\varepsilon$ such that

$$(P_\varepsilon) \quad \left| \begin{array}{l} a^\varepsilon(v^\varepsilon, w^\varepsilon, \varphi^\varepsilon \; ; \; z - v^\varepsilon, u - w^\varepsilon, \psi - \varphi^\varepsilon) \geq f(u - w^\varepsilon) \\ \forall \; (z, u, \psi) \in K_\varepsilon \,. \end{array} \right. \qquad (9.2.1)$$

In the next section we shall derive the Poincaré and Korn inequalities applicable to the highly irregular domain Ω^ε, cf. (9.3.21) and (9.3.39). By using them, the reader can easily prove that $(v^\varepsilon, w^\varepsilon, \varphi^\varepsilon) \in K_\varepsilon$ solving (P_ε) exists and is unique. To find the effective behavior of the plate when $\varepsilon \to 0$ we apply the asymptotic homogenization method. The unknowns as well as trial fields are expanded in the form:

$$\eta^\varepsilon(x) = \eta^{(0)}(x) + \varepsilon \eta^{(1)}(x, y) + \varepsilon^2 \eta^{(2)}(x, y) + \ldots, \quad y = x/\varepsilon \qquad (9.2.2)$$

for $\eta \in \{v_\alpha, w, \varphi_\alpha, z_\alpha, u, \psi_\alpha\}$ with $\eta^{(0)} \in H_0^1(\Omega)$ and

$$w^{(1)}(x, \cdot), u^{(1)}(x, \cdot) \in K_{YF}^{\mathbf{c}} , \qquad v^{(1)}(x, \cdot), z^{(1)}(x, \cdot) \in K_{YF}^{\mathbf{ab}} ,$$
$$\varphi^{(1)}(x, \cdot), \psi^{(1)}(x, \cdot) \in K_{YF}^{\mathbf{dh}} , \qquad (9.2.3)$$

where $\mathbf{a}, \mathbf{b}, \mathbf{c}, \mathbf{d}, \mathbf{h} \in \{d, c, i\}$.

Now we substitute (9.2.2) into the variational inequality (9.2.1). Performing the passage to the limit $(\varepsilon \to 0)$ in the standard manner, see Sec. 8.2, we obtain the variational formulation of the equilibrium problem of the homogenized Reissner-like plate:

$$(P_h) \quad \left| \begin{array}{l} \text{find } (v^{(0)}, w^{(0)}, \varphi^{(0)}) \in V = H_0^1(\Omega)^2 \times H_0^1(\Omega) \times H_0^1(\Omega)^2 \text{ such that} \\[2mm] a_h(v^{(0)}, w^{(0)}, \varphi^{(0)}; \; z^{(0)}, u^{(0)}, \psi^{(0)}) = \displaystyle\int_\Omega q u^{(0)} dx \quad \forall \, (z^{(0)}, u^{(0)}, \psi^{(0)}) \in V , \end{array} \right. \qquad (9.2.4)$$

where

$$a_h(v^{(0)}, w^{(0)}, \varphi^{(0)}; z^{(0)}, u^{(0)}, \varphi^{(0)}) = \int_\Omega \{\langle N_0^{\alpha\beta}\rangle_{YF} e_{\alpha\beta}(z^{(0)}) \qquad (9.2.5)$$
$$+ \langle M_0^{\alpha\beta}\rangle_{YF} e_{\alpha\beta}(\psi^{(0)}) + \langle Q_0^\alpha\rangle_{YF} (u_{,\alpha}^{(0)} + \psi_\alpha^{(0)})\} dx ,$$

and

$$N_0^{\alpha\beta} = A^{\alpha\beta\lambda\mu}[e_{\lambda\mu}(v^{(0)}) + e_{\lambda\mu}^y(v^{(1)})] + E^{\alpha\beta\lambda\mu}[e_{\lambda\mu}(\varphi^{(0)}) + e_{\lambda\mu}^y(\varphi^{(1)})] ,$$
$$M_0^{\alpha\beta} = E^{\alpha\beta\lambda\mu}[e_{\lambda\mu}(v^{(0)}) + e_{\lambda\mu}^y(v^{(1)})] + G^{\alpha\beta\lambda\mu}[e_{\lambda\mu}(\varphi^{(0)}) + e_{\lambda\mu}^y(\varphi^{(1)})] , \qquad (9.2.6)$$
$$Q_0^\alpha = H^{\alpha\beta}\left(\varphi_\beta^{(0)} + w_{,\beta}^{(0)} + \frac{\partial w^{(1)}}{\partial y_\beta} \right) .$$

The averaging operation $\langle \cdot \rangle_{YF}$ is defined as in the previous section; for the definition of $e^y_{\lambda\mu}(\cdot)$ see Sec. 2.8.

The local fields $v^{(1)}$, $w^{(1)}$, $\varphi^{(1)}$ are solutions to the local problems posed on the basic cell Y, where the independent variable x is treated as a parameter. These problems assume the form:

$$(P^1_{YF}) \quad \left| \begin{array}{l} \text{find } v^{(1)} \in K^{\mathbf{ab}}_{YF} \text{ and } \varphi^{(1)} \in K^{\mathbf{dh}}_{YF} \text{ such that} \\[2mm] b(v^{(1)}, z - v^{(1)}) + e(\varphi^{(1)}, z - v^{(1)}) \geq L_1(z - v^{(1)}) \quad \forall z \in K^{\mathbf{ab}}_{YF}, \quad (9.2.7) \\[2mm] e(v^{(1)}, \psi - \varphi^{(1)}) + g(\varphi^{(1)}, \psi - \varphi^{(1)}) \geq L_2(\psi - \varphi^{(1)}) \quad \forall \psi \in K^{\mathbf{dh}}_{YF} \quad (9.2.8) \end{array} \right.$$

and

$$(P^2_{YF}) \quad \left| \begin{array}{l} \text{find } w^{(1)} \in K^{\mathbf{c}}_{YF} \text{ such that} \\[2mm] h(w^{(1)}, u - w^{(1)}) \geq L_3(u - w^{(1)}) \qquad \forall u \in K^{\mathbf{c}}_{YF}, \quad (9.2.9) \end{array} \right.$$

where the bilinear forms $b(\cdot, \cdot)$, $e(\cdot, \cdot)$, $g(\cdot, \cdot)$, $h(\cdot, \cdot)$ are defined by

$$b(v, z) = \langle A^{\alpha\beta\lambda\mu} e^y_{\lambda\mu}(v) e^y_{\alpha\beta}(z) \rangle_{YF}, \qquad e(v, \psi) = \langle E^{\alpha\beta\lambda\mu} e^y_{\lambda\mu}(v) e^y_{\alpha\beta}(\psi) \rangle_{YF},$$

$$g(\varphi, \psi) = \langle G^{\alpha\beta\lambda\mu} e^y_{\lambda\mu}(\varphi) e^y_{\alpha\beta}(\psi) \rangle_{YF}, \qquad h(u, v) = \left\langle H^{\alpha\beta} \frac{\partial u}{\partial y_\alpha} \frac{\partial v}{\partial y_\beta} \right\rangle_{YF}. \tag{9.2.10}$$

The linear forms L_i $(i = 1, 2, 3)$ are defined as follows

$$\begin{array}{l} L_1(z) = -\langle (E^{\alpha\beta\lambda\mu} e_{\alpha\beta}(\varphi^{(0)}) + A^{\alpha\beta\lambda\mu} e_{\lambda\mu}(v^{(0)})) e^y_{\alpha\beta}(z) \rangle_{YF}, \\[2mm] L_2(z) = -\langle (E^{\alpha\beta\lambda\mu} e_{\lambda\mu}(v^{(0)}) + G^{\alpha\beta\lambda\mu} e_{\lambda\mu}(\varphi^{(0)})) e^y_{\alpha\beta}(z) \rangle_{YF}, \\[2mm] L_3(u) = -\left\langle H^{\alpha\beta}(\varphi^{(0)}_\alpha + w^{(0)}_{,\alpha}) \dfrac{\partial u}{\partial y_\beta} \right\rangle_{YF}. \end{array} \tag{9.2.11}$$

The macroscopic quantities $\kappa^h_{\alpha\beta} = e_{\alpha\beta}(\varphi^{(0)})$, $\epsilon^h_{\alpha\beta} = e_{\alpha\beta}(v^{(0)})$, $\gamma^h_\alpha = \varphi^{(0)}_\alpha + w^{(0)}_{,\alpha}$ are assumed as given when the local problems are considered.

Note that the variational inequalities (9.2.7), (9.2.8) are coupled unless $e = 0$. Some specific cases will be studied later on.

One can easily prove that the fields $v^{(1)}$, $\varphi^{(1)}$ are determined up to constant vectors while $w^{(1)}$ up to a constant. Moreover, the fields $v^{(1)}$, $\varphi^{(1)}$ depend on κ^h, ϵ^h whereas $w^{(1)}$ depends on γ^h.

The homogenized stress and couple resultants are determined by

$$N^{\alpha\beta}_h = \langle N^{\alpha\beta}_0 \rangle_{YF}, \qquad M^{\alpha\beta}_h = \langle M^{\alpha\beta}_0 \rangle_{YF}, \qquad Q^\alpha_h = \langle Q^\alpha_0 \rangle_{YF}. \tag{9.2.12}$$

By localizing the variational equality (9.2.4) one finds the equilibrium equations of the homogenized plate

$$M^{\alpha\beta}_{h,\beta} - Q^\alpha_h = 0, \qquad N^{\alpha\beta}_{h,\beta} = 0, \qquad Q^\alpha_{h,\alpha} + q = 0. \tag{9.2.13}$$

Various aspects of localization are studied in the next section.

The elastic potential is inferred from the local problems (P^α_{YF}) $(\alpha = 1, 2)$ and is given by, cf. also Secs. 2.10, 11.4 − 11.7,

$$
\begin{aligned}
\mathcal{W}(\epsilon^h, \boldsymbol{\kappa}^h, \boldsymbol{\beta}^h) = \frac{1}{2}\langle A^{\alpha\beta\lambda\mu}(\epsilon^h_{\lambda\mu} + \epsilon^y_{\lambda\mu}(\boldsymbol{v}^{(1)})) \cdot (\epsilon^h_{\alpha\beta} + \epsilon^y_{\alpha\beta}(\boldsymbol{v}^{(1)})) \\
+ 2E^{\alpha\beta\lambda\mu}(\kappa^h_{\lambda\mu} + \epsilon^h_{\lambda\mu}(\boldsymbol{\varphi}^{(1)}))(\epsilon^h_{\alpha\beta} + \epsilon^y_{\alpha\beta}(\boldsymbol{v}^{(1)})) \\
+ G^{\alpha\beta\lambda\mu}(\kappa^y_{\lambda\mu} + \epsilon^y_{\lambda\mu}(\boldsymbol{\varphi}^{(1)}))(\kappa^h_{\alpha\beta} + \epsilon^y_{\alpha\beta}(\boldsymbol{\varphi}^{(1)})) \\
+ H^{\alpha\lambda}\left(\gamma^h_\alpha + \frac{\partial w^{(1)}}{\partial y_\alpha}\right)\left(\gamma^h_\lambda + \frac{\partial w^{(1)}}{\partial y_\lambda}\right)\rangle_{YF} .
\end{aligned}
\tag{9.2.14}
$$

This potential is of class C^1, positive and strictly convex. The homogenized constitutive relations are:

$$
N^{\alpha\beta}_h = \frac{\partial \mathcal{W}}{\partial \epsilon^h_{\alpha\beta}} , \qquad M^{\alpha\beta}_h = \frac{\partial \mathcal{W}}{\partial \kappa^h_{\alpha\beta}} , \qquad Q^\alpha_h = \frac{\partial \mathcal{W}}{\partial \gamma^h_\alpha} .
\tag{9.2.15}
$$

Thus the homogenized plate is hyperelastic. Moreover, we conclude that
(i) $\mathcal{W}$ is positively homogeneous of order 2

$$
\mathcal{W}(\lambda\epsilon^h, \lambda\boldsymbol{\kappa}^h, \lambda\boldsymbol{\gamma}^h) = \lambda^2 \mathcal{W}(\epsilon^h, \boldsymbol{\kappa}^h, \boldsymbol{\gamma}^h) , \qquad \lambda \geq 0 .
\tag{9.2.16}
$$

(ii) There exist positive constants c_1, c_2 such that

$$
c_1(|\epsilon^h|^2 + |\boldsymbol{\kappa}^h|^2 + |\boldsymbol{\gamma}^h|^2) \leq \mathcal{W}(\epsilon^h, \boldsymbol{\kappa}^h, \boldsymbol{\gamma}^h) \leq c_2(|\epsilon^h|^2 + |\boldsymbol{\kappa}^h|^2 + |\boldsymbol{\gamma}^h|^2) ,
\tag{9.2.17}
$$

for each $\epsilon^h, \boldsymbol{\kappa}^h, \boldsymbol{\gamma}^h \in \mathbb{E}^2_s$ and $\boldsymbol{\gamma}^h \in \mathbb{R}^2$. Due to the above properties of $\mathcal{W}$, the problem (P_h) is uniquely solvable. In the next section we shall prove that

$$
\begin{aligned}
v^\epsilon_\alpha \to v^{(0)}_\alpha , \; w^\epsilon \to w^{(0)}, \; \varphi^\epsilon_\alpha \to \varphi^{(0)}_\alpha & \qquad \text{strongly in } L^2(\Omega) , \\
N^{\alpha\beta}_\epsilon \rightharpoonup N^{\alpha\beta}_h, \; M^{\alpha\beta}_\epsilon \rightharpoonup M^{\alpha\beta}_h, \; Q^\alpha_\epsilon \rightharpoonup Q^\alpha_h & \qquad \text{weakly in} L^2(\Omega) .
\end{aligned}
\tag{9.2.18}
$$

Let us consider now particular modes of cracking.

Bending cracking mode
The condition $(z, u, \psi) \in K^{ben}$ means that $v^{(1)}, z \in H^1_{per}(Y)^2$; $w^{(1)}, u \in H^1_{per}(Y)$ and $\varphi^{(1)}, \psi \in K^{dc}_{YF}$. Analysis of (P^2_{YF}) yields $w^{(1)} = w^{(1)}(x)$ and consequently

$$
Q^\alpha_h = H^{\alpha\lambda}\gamma^h_\lambda .
\tag{9.2.19}
$$

Next, an analysis of (P^1_{YF}) yields $v^{(1)} = v^{(1)}(x)$ and the local problem reduces to

$$
(P^{ben}_Y) \quad \left|
\begin{aligned}
&\text{find } \varphi^{(1)} \in K^{dc}_{YF} \text{ such that} \\
&g(\varphi^{(1)}, \psi - \varphi^{(1)}) \geq L_2(\psi - \varphi^{(1)}) \qquad \forall\, \psi \in K^{dc}_{YF} .
\end{aligned}
\right.
\tag{9.2.20}
$$

The homogenized constitutive relations then assume the form

$$N_h^{\alpha\beta} = A^{\alpha\beta\lambda\mu}\epsilon_{\lambda\mu}^h + \langle E^{\alpha\beta\lambda\mu}[\kappa_{\lambda\mu}^h + e_{\lambda\mu}^y(\varphi^{(1)}(\epsilon^h,\kappa^h))]\rangle_{YF} \,,$$

$$M_h^{\alpha\beta} = E^{\alpha\beta\lambda\mu}\epsilon_{\lambda\mu}^h + \langle G^{\alpha\beta\lambda\mu}[\kappa_{\lambda\mu}^h + e_{\lambda\mu}^y(\varphi^{(1)}(\epsilon^h,\kappa^h))]\rangle_{YF} \,. \qquad (9.2.21)$$

Let us check whether these constitutive equations are really coupled. To this end let us return to the mid-plane displacement $u^{(0)} = v^{(0)} + e\varphi^{(0)}$. The bending moments $\widetilde{M}_h^{\alpha\beta}$ referred to the mid-plane are

$$\widetilde{M}_h^{\alpha\beta} = M_h^{\alpha\beta} - eN_h^{\alpha\beta} \,.$$

Since

$$A^{\alpha\beta\lambda\mu}\epsilon_{\lambda\mu}^h + E^{\alpha\beta\lambda\mu}\kappa_{\lambda\mu}^h = A^{\alpha\beta\lambda\mu}e_{\lambda\mu}(u^{(0)}) \,,$$

we can rearrange the constitutive relations (9.2.21) to the form

$$N_h^{\alpha\beta} = A^{\alpha\beta\lambda\mu}[e_{\lambda\mu}(u^{(0)}) + e\langle e_{\lambda\mu}^y(\varphi^{(1)}(\epsilon^h,\kappa^h))\rangle_{YF}] \,,$$

$$\widetilde{M}_h^{\alpha\beta} = D^{\alpha\beta\lambda\mu}[\kappa_{\lambda\mu}^h + \langle e_{\lambda\mu}^y(\varphi^{(1)}(\epsilon^h,\kappa^h))\rangle_{YF}] \,. \qquad (9.2.22)$$

Thus coupling is absent only if $e = 0$. If $e \neq 0$ the curvature tensor influences the membrane forces and vice versa, the moment tensor is influenced by the in-plane strain tensor ϵ^h. Moreover, note that the quantity $\langle e_{\lambda\mu}^y(\varphi^{(1)})\rangle_{YF}$ is completely determined by the jumps $[\![\varphi_\alpha^{(1)}]\!]$ along F:

$$\langle e_{\alpha\beta}^y(\varphi^{(1)})\rangle_{YF} = -\frac{1}{2|Y|}\int_F ([\![\varphi_\alpha^{(1)}]\!]\check{N}_\beta + [\![\varphi_\beta^{(1)}]\!]\check{N}_\alpha)ds \,, \qquad (9.2.23)$$

where $\check{N} = (\check{N}_\alpha)$ is a versor normal to F.

Transverse shear cracking mode

The condition $(z, u, \psi) \in K^{shear}$ means that $z, v^{(1)}, \psi, \varphi^{(1)} \in H_{per}^1(Y)^2$; $u, w^{(1)} \in K_{YF}^d$. Thus $v^{(1)} = v^{(1)}(x)$, $\varphi^{(1)} = \varphi^{(1)}(x)$ and $w^{(1)} \in K_{YF}^d$ satisfies

$$(P_{YF}^{shear}) \qquad \Big|\, h(w, u - w^{(1)}) \geq L_3(u - w^{(1)}) \ \forall\, u \in K_{YF}^d \,.$$

The homogenized constitutive equations assume the form

$$N_h^{\alpha\beta} = A^{\alpha\beta\lambda\mu}e_{\lambda\mu}(u^{(0)}) \,, \qquad \widetilde{M}_h^{\alpha\beta} = D^{\alpha\beta\lambda\mu}\kappa_{\lambda\mu}^h \,,$$

$$Q_h^\alpha = \left\langle H^{\alpha\lambda}\left(\gamma_\lambda^h + \frac{\partial w^{(1)}(\gamma^h)}{\partial y_\lambda}\right)\right\rangle_{YF} \,.$$

Hence the stretching and bending effects are uncoupled and the homogenized properties do not depend upon e. The shear force Q_h^α depends on the distribution of $[\![w^{(1)}]\!]$ along F.

Tension cracking mode

In this case $(z, u, \psi) \in K^{ten}$. Consequently, $w^{(1)} = w^{(1)}(x)$ and $\varphi^{(1)} = \varphi^{(1)}(x)$. The problem is unaffected by e. Putting $e = 0$ we identify $u^{(1)} = v^{(1)}$ that solves the problem

$$(P_{YF}^{ten}) \quad \left| \begin{array}{l} \text{find } u^{(1)} \in K_{YF}^{dc} \text{ such that} \\[2mm] b(u^{(1)}, z - u^{(1)}) \geq L_1(z - u^{(1)}) \qquad \forall\, z \in K_{YF}^{dc}\,, \end{array} \right.$$

where

$$L_1(z) = -\langle A^{\alpha\beta\lambda\mu} \epsilon_{\lambda\mu}^h e_{\alpha\beta}^y(z)\rangle_{YF}\,.$$

The homogenized constitutive relations

$$N_h^{\alpha\beta} = \langle A^{\alpha\beta\lambda\mu}[\epsilon_{\lambda\mu}^h + \epsilon_{\lambda\mu}^y(u^{(1)}(\epsilon^h))]\rangle_{YF}\,, \qquad M_h^{\alpha\beta} = D^{\alpha\beta\lambda\mu}\kappa_{\lambda\mu}^h\,, \quad Q_h^\alpha = H^{\alpha\lambda}\gamma_\lambda^h$$

are decoupled. Thus the homogenized bending / transverse shearing problem is conventional, that is linear within the framework of moderately thick plates. On the other hand, the homogenized membrane problem is nonlinear.

9.3.   *Justification: variational inequality (9.2.1) and the energy method*

In the previous section we have performed homogenization of elastic Reissner-like plates weakened by periodically distributed fissures. The method of two-scale asymptotic expansions has been used. Our aim now is to justify those results by passing rigorously to the limit ($\varepsilon \to 0$) in the variational inequality (9.2.1). To this end Murat-Tartar's energy method will be applied. We are going to study the case where

$$K_\varepsilon = C_{di}^\varepsilon \times C_d^\varepsilon \times C_{di}^\varepsilon\,, \tag{9.3.1}$$

and

$$C_{di}^\varepsilon = \{v \in H_1^1(\Omega^\varepsilon)^2|\; [\![v_n]\!] \geq 0 \text{ on } F^\varepsilon\}\,, \quad C_d^\varepsilon = \{v \in H_1^1(\Omega^\varepsilon)|\; [\![v]\!] \geq 0 \text{ on } F^\varepsilon\}\,. \tag{9.3.2}$$

The remaining cases can be justified similarly.

Convex and closed sets of kinematically admissible local fields are now defined by:

$$\begin{aligned} K_{YF}^{di} &= \{v \in H_{per}^1(YF)^2|\; [\![v_N]\!] \geq 0 \text{ on } F\}\,, \\ K_{YF}^d &= \{v \in H_{per}^1(YF)|\; [\![v]\!] \geq 0 \text{ on } F\}\,, \end{aligned} \tag{9.3.3}$$

where $v_N = v_\alpha \check{N}^\alpha$.

Prior to the study of convergence, we have to construct extension operators, cf. Sec. 8.3.

Extension operator $\mathcal{Q}_1^\varepsilon : H^1(\Omega^\varepsilon) \to H^1(\Omega)$

This is a model case and we are going to study it carefully. A closed set $F_\eta \subset Y$ with smooth boundary has the same meaning as in Sec. 8.3. There exists an extension operator $\mathbb{P}_1 : H^1(Y \backslash F_\eta) \to H^1(Y)$ satisfying the properties

$$\|\mathbb{P}_1 v\|_{H^1(Y)} \leq c\|v\|_{H^1(Y \backslash F_\eta)}\,, \tag{9.3.4}$$

$$\|\mathbb{P}_1 v\|_{L^2(Y)} \leq c\|v\|_{L^2(Y \backslash F_\eta)}\,, \tag{9.3.5}$$

where $c > 0$ (with possibly a subscript) is a generic constant. We observe that c depends on η and tends to $+\infty$ when $\eta \to 0$; however, we argue with $\eta = \eta_0$ held fixed.

Next we define

$$\widetilde{\mathbb{P}}_1 : H^1(Y\backslash F_\eta) \to H^1(Y), \qquad v \mapsto \widetilde{\mathbb{P}}_1(v) = \mathbb{P}_1(v - \langle v\rangle_{Y\backslash F_\eta}) + \langle v\rangle_{Y\backslash F_\eta}, \qquad (9.3.6)$$

where

$$\langle v\rangle_{Y\backslash F_\eta} = \frac{1}{|Y\backslash F_\eta|} \int_{Y\backslash F_\eta} v(y)dy.$$

Lemma 9.3.1. For each $v \in H^1(Y\backslash F_\eta)$ the following inequalities hold

$$\|\widetilde{\mathbb{P}}_1 v\|_{L^2(Y)} \leq c\|v\|_{L^2(Y\backslash F_\eta)}, \qquad (9.3.7)$$

$$\|\nabla(\widetilde{\mathbb{P}}_1 v)\|_{L^2(Y)} \leq c\|\nabla v\|_{L^2(Y\backslash F_\eta)}. \qquad (9.3.8)$$

Proof. To prove the first inequality we calculate, cf. (9.3.6)

$$\|\widetilde{\mathbb{P}}_1 v\|_{L^2(Y)} \leq \|\mathbb{P}_1(v - \langle v\rangle_{Y\backslash F_\eta})\|_{L^2(Y)} + \|\langle v\rangle_{Y\backslash F_\eta}\|_{L^2(Y)} \leq c\|v - \langle v\rangle_{Y\backslash F_\eta}\|_{L^2(Y\backslash F_\eta)}$$

$$+ |Y|\,|\langle v\rangle_{Y\backslash F_\eta}| \leq c\|v\|_{L^2(Y\backslash F_\eta)} + (c|Y\backslash F_\eta| + |Y|)|\langle v\rangle_{Y\backslash F_\eta}|$$

$$\leq \left(c + \frac{(c|Y\backslash F_\eta| + |Y|)}{|Y\backslash F_\eta|^{1/2}}\right)\|v\|_{L^2(Y\backslash F_\eta)}.$$

We pass now to proving the second inequality. We have

$$\|\nabla(\widetilde{\mathbb{P}}_1 v)\|_{L^2(Y)} = \|\nabla\mathbb{P}_1(v - \langle v\rangle_{Y\backslash F_\eta}\|_{L^2(Y)}$$

$$\leq c\{\|v - \langle v\rangle_{Y\backslash F_\eta}\|_{L^2(Y)} + \|\nabla(v - \langle v\rangle_{Y\backslash F_\eta})\|_{L^2(Y\backslash F_\eta)}\}.$$

By applying Poincaré-Wirtinger's inequality to the function $v - \langle v\rangle_{Y\backslash F_\eta}$ we infer that there exists a constant $c_1 > 0$ such that

$$\|v - \langle v\rangle_{Y\backslash F_\eta}\|_{L^2(Y\backslash F_\eta)} \leq c_1\|\nabla v\|_{L^2(Y\backslash F_\eta)}.$$

From the last two inequalities we finally obtain

$$\|\nabla(\widetilde{\mathbb{P}}_1 v)\|_{L^2(Y)} \leq c(1 + c_1)\|\nabla v\|_{L^2(Y\backslash F_\eta)}. \qquad \square$$

We are now in a position to construct the extension operator $\mathcal{Q}_1 : H^1(Y\backslash F) \to H^1(Y)$.

Definition 9.3.2. The extension operator $\mathcal{Q}_1 : H^1(Y\backslash F) \to H^1(Y)$ is equal to $\mathcal{Q}_1 = \widetilde{\mathbb{P}}_1 \circ \mathbb{R}_1$, where

$$\mathbb{R}_1 : H^1(Y\backslash F) \to H^1(Y\backslash F_\eta), \qquad v \mapsto v|_{Y\backslash F_\eta}$$

is the restriction operator.

Properties of $\mathcal{Q}_1$

For each $v \in H^1(YF)$ the following properties are satisfied

$$\mathcal{Q}_1 v = v \quad \text{in} \quad Y \backslash F_\eta . \tag{9.3.9}$$

$$\|\mathcal{Q}_1 v\|_{L^2(Y)} \leq c\|v\|_{L^2(Y \backslash F_\eta)} . \tag{9.3.10}$$

$$\|\nabla_y (\mathcal{Q}_1 v)\|_{L^2(Y)} \leq c\|\nabla_y v\|_{L^2(Y \backslash F_\eta)} . \tag{9.3.11}$$

$$\|\mathcal{Q}_1 v - v\|_{L^2(Y)} \leq c\|\nabla_y v\|_{L^2(Y \backslash F_\eta)} \leq c\|\nabla_y v\|_{L^2(Y \backslash F)} . \tag{9.3.12}$$

Proof. By using (9.3.7) we have

(a) $\|\mathcal{Q}_1 v\|_{L^2(Y)} = \|\widetilde{\mathbf{P}}_1(\mathbf{R}_1 v)\|_{L^2(Y)} \leq c\|\mathbf{R}_1 v\|_{L^2(Y \backslash F_\eta)} = c\|v\|_{L^2(Y \backslash F_\eta)} .$

(b) In order to prove (9.3.11) we take into account (9.3.8):

$$\|\nabla(\mathcal{Q}_1 v)\|_{L^2(Y)} = \|\nabla(\widetilde{\mathbf{P}}_1(\mathbf{R}_1 v))\|_{L^2(Y)} \leq c\|\nabla_y(\mathbf{R}_1 v)\|_{L^2(Y \backslash F_\eta)}$$
$$= c\|\nabla_y v\|_{L^2(Y \backslash F_\eta)} \leq c\|\nabla_y v\|_{L^2(Y \backslash F)} .$$

(c) By dividing the basic cell Y in the same manner as in Sec. 8.3 (Fig. 8.3.1) we have

$$\|\mathcal{Q}_1 v - v\|_{L^2(F_\eta^\alpha)} \leq c_1\|\nabla_y(\mathcal{Q}_1 v - v)\|_{L^2(F_\eta^\alpha)} , \quad \alpha = 1, 2 .$$

Hence

$$\|\mathcal{Q}_1 v - v\|_{L^2(Y)} \leq c_1 \sum_{\alpha=1}^{2} \|\nabla_y(\mathcal{Q}_1 v - v)\|_{L^2(F_\eta^\alpha)} \leq c_1 \sum_{\alpha=1}^{2} (\|\nabla_y(\mathcal{Q}_1 v)\|_{L^2(Y_\alpha)}$$
$$+ \|\nabla_y v\|_{L^2(Y_\alpha)}) \leq c_1(\|\nabla_y \mathcal{Q}_1 v\|_{L^2(Y)} + \|\nabla_y v\|_{L^2(Y \backslash F)}) ,$$

because $\mathcal{Q}_1 v - v = 0$ on a part of the boundary of F_η^1 (resp. F_η^2) of strictly positive measure. Obviously c with possibly a subscript denotes a positive constant. Taking into account (9.3.11) we finally obtain

$$\|\mathcal{Q}_1 v - v\|_{L^2(Y)} \leq c_1(1 + c)\|\nabla_y v\|_{L^2(Y \backslash F)} . \qquad \square$$

We pass now to constructing the extension operator $\mathcal{Q}_1^\varepsilon \colon H^1(\Omega^\varepsilon) \to H^1(\Omega)$. To this end, by

$$F^\varepsilon = \bigcup_{i \in \mathcal{I}(\varepsilon)} F_{\varepsilon,i} , \tag{9.3.13}$$

we denote the union of all the microcracks which are periodically distributed and of size ε with $F_{\varepsilon,i} \subset Y_{\varepsilon,i}$ for every $i \in \mathcal{I}(\varepsilon)$; moreover

$$Y_{\varepsilon,i} = \varepsilon Y + \boldsymbol{\xi}_\varepsilon^i , \qquad \boldsymbol{\xi}_\varepsilon^i \in \mathbb{R}^2 . \tag{9.3.14}$$

The operator $\mathcal{Q}_{1,i}^\varepsilon$ is obtained from $\mathcal{Q}_1$ by changing scale, the global operator $\mathcal{Q}_1^\varepsilon$ being obtained by stitching all the $\mathcal{Q}_{1,i}^\varepsilon$ operators. More precisely, the operator $\mathcal{Q}_1^\varepsilon$ is equal to the following chain of mappings:

$$u \in H^1(\Omega^\varepsilon)$$

$\downarrow$ restriction

$$u_1 \in H^1(Y_{\varepsilon,i} \backslash F_{\varepsilon,i}) = H^1(\varepsilon(Y \backslash F) + \boldsymbol{\xi}_\varepsilon^i)$$

$\downarrow$ translation + change of scale

$$u_2 \in H^1(Y \backslash F) \,, \ \ u_2(y) = u_1(\varepsilon y + \boldsymbol{\xi}_\varepsilon^i)$$

$\downarrow$ restriction + smooth extension

$$u_3 = \mathcal{Q}_1 u_2 \in H^1(Y)$$

$\downarrow$ translation + change of scale

$$u_4 \in H^1(\varepsilon Y + \boldsymbol{\xi}_\varepsilon^i) \,, \ \ u_4(x) = u_3\left(\frac{x - \boldsymbol{\xi}_\varepsilon^i}{\varepsilon}\right)$$

$\downarrow$ stitching with respect to $i \in \mathcal{I}(\varepsilon)$

$$u_5 = \mathcal{Q}_1^\varepsilon u \,. \tag{9.3.15}$$

We observe that by these operations u is not affected on $\bigcup\limits_{i \in \mathcal{I}(\varepsilon)} (Y_{\varepsilon,i} \backslash F_\eta^{\varepsilon,i})$, where $F_\eta^{\varepsilon,i}$ is the ε-homothetic of the neighborhood F_η of F which is included in $Y_{\varepsilon,i}$. Thus u remains unchanged near the boundary of the periodicity cell $Y_{\varepsilon,i}$. Consequently, $\mathcal{Q}_1^\varepsilon u$ belongs to $H^1(\Omega)$. As in Sec. 8.3 we assume that near the boundary of Ω there are no cracks. Thus $\mathcal{Q}_1^\varepsilon$ may be assumed to be equal to the identity near $\partial\Omega$ so that

$$u \in H^1(\Omega^\varepsilon) \,, \ \ \ u = 0 \text{ on } \partial\Omega \ \Rightarrow \ \mathcal{Q}_1^\varepsilon u \in H_0^1(\Omega) \,. \tag{9.3.16}$$

Being a composition of linear continuous operators $\mathcal{Q}_1^\varepsilon$ acts as a continuous linear operator from $H^1(\Omega^\varepsilon)$ to $H^1(\Omega)$.

Properties of $\mathcal{Q}_1^\varepsilon$

For every $v \in H^1(\Omega^\varepsilon)$ the following properties hold:

$$\mathcal{Q}_1^\varepsilon v = v \ \ \text{ in } \ \ \Omega \backslash F_\eta^\varepsilon \,, \tag{9.3.17}$$

$$||\mathcal{Q}_1^\varepsilon v||_{L^2(\Omega)} \leq c||v||_{L^2(\Omega^\varepsilon)} \,, \tag{9.3.18}$$

$$||\nabla(\mathcal{Q}_1^\varepsilon v)||_{L^2(\Omega)} \leq c||\nabla v||_{L^2(\Omega^\varepsilon)} \,, \tag{9.3.19}$$

$$||\mathcal{Q}_1^\varepsilon v - v||_{L^2(\Omega)} \leq c\varepsilon||\nabla v||_{L^2(\Omega^\varepsilon)} \,. \tag{9.3.20}$$

Here $F_\eta^\varepsilon = \bigcup\limits_{i \in \mathcal{I}(\varepsilon)} F_\eta^{\varepsilon,i}$ and $c > 0$ is independent of ε.

Proof. The first relation follows directly from the construction of $\mathcal{Q}_1^\varepsilon$. The second and the third properties are obvious since the change of scale affects similarly both the left and the right members of (9.3.10) and (9.3.11).

It remains to prove the last property. To this end we argue on a cell, say εY. After the change of scale $y = x/\varepsilon$ we have

$$\int_{\varepsilon Y} |\mathcal{Q}_1^\varepsilon v(x) - v(x)|^2 dx = \int_Y |\mathcal{Q}_1(v(\varepsilon y)) - v(\varepsilon y)|^2 \varepsilon^2 dy .$$

The inequality (9.3.12) yields

$$\int_{\varepsilon Y} |\mathcal{Q}_1^\varepsilon v(x) - v(x)|^2 dx \le c \int_{Y\backslash F} |\nabla_y(v(\varepsilon y))|^2 \varepsilon^2 dy = c \int_{Y\backslash F} |(\nabla_x v)(\varepsilon y)|^2 \varepsilon^2 \varepsilon^2 dy .$$

Changing scale again, $x = \varepsilon y$, we write

$$\int_{\varepsilon Y} |\mathcal{Q}_1^\varepsilon v(x) - v(x)|^2 dx \le c\varepsilon^2 \int_{\varepsilon(YF)} |\nabla_x(v(x))|^2 dx .$$

The last inequality is valid on each cell $Y_{\varepsilon,i}$. Adding all these inequalities we obtain

$$\int_\Omega |\mathcal{Q}_1^\varepsilon v - v|^2 dx \le c\varepsilon^2 \int_{\Omega^\varepsilon} |\nabla v(x)|^2 dx ,$$

and (9.3.20) follows. $\qquad\square$

Now we are in a position to formulate the Poincaré inequality for the highly irregular domain Ω^ε.

Proposition 9.3.3. There exists $c > 0$, independent of ε, such that for any $v \in H^1(\Omega^\varepsilon)$, which satisfies $v = 0$ on $\partial\Omega$, for any $\varepsilon > 0$ the following inequality holds

$$||v||_{L^2(\Omega)} \le c||\nabla v||_{L^2(\Omega^\varepsilon)} . \tag{9.3.21}$$

Proof. Applying Poincaré's inequality to $\mathcal{Q}_1^\varepsilon v$ on Ω we have

$$||\mathcal{Q}_1^\varepsilon v||_{L^2(\Omega)} \le c_1 ||\nabla(\mathcal{Q}_1^\varepsilon v)||_{L^2(\Omega)} .$$

On account of (9.3.19) we obtain

$$||\mathcal{Q}_1^\varepsilon v||_{L^2(\Omega)} \le c_1 c_2 ||\nabla v||_{L^2(\Omega^\varepsilon)} . \tag{9.3.22}$$

The triangle inequality yields

$$||v||_{L^2(\Omega)} \le ||\mathcal{Q}_1^\varepsilon v - v||_{L^2(\Omega)} + ||\mathcal{Q}_1^\varepsilon v||_{L^2(\Omega)} .$$

From (9.3.20) and (9.3.22) we obtain

$$||v||_{L^2(\Omega)} \le c_2(\varepsilon + c_1)||\nabla v||_{L^2(\Omega^\varepsilon)} .$$

Taking now $0 < \varepsilon < \varepsilon_0$ with ε_0 fixed we conclude the proof. $\qquad\square$

Extension operator $Q_2^\varepsilon : H^1(\Omega^\varepsilon)^2 \to H^1(\Omega)^2$

In order to formulate Korn's inequality for the same domain Ω^ε we shall proceed as previously by constructing an extension operator $Q_2^\varepsilon\colon H^1(\Omega^\varepsilon)^2 \to H^1(\Omega)^2$.

Lemma 9.3.4. There exists an extension operator $\mathbb{P}_2\colon H^1(Y\backslash F_\eta)^2 \to H^1(Y)^2$ such that

$$\|\mathbb{P}_2 v\|_{0,Y} \le c\|v\|_{0,Y\backslash F_\eta} . \tag{9.3.23}$$

$$\sum_{\alpha,\beta=1}^{2} \|e_{\alpha\beta}(\mathbb{P}_2 v)\|_{0,Y} \le c \sum_{\alpha,\beta=1}^{2} \|e_{\alpha\beta}(v)\|_{0,Y\backslash F_\eta} \le c\|e(v)\|_{0,YF} , \tag{9.3.24}$$

where $\|\cdot\|_{0,Y}$, etc. denotes L^2-norm.

Proof. Such an operator can be constructed in the following way. Let $\mathcal{R}$ denote the space of rigid displacements. Then each $v \in H^1(Y\backslash F_\eta)^2$ can be decomposed according to

$$v = v_1 + r , \tag{9.3.25}$$

where $r \in \mathcal{R}$ and $v_1 \bot \mathcal{R}$ in $L^2(Y\backslash F_\eta)^2$. To extend v to Y we extend v_1 continuously, which can be done since the boundary ∂F_η is assumed to be sufficiently smooth. This linear and continuous operator is denoted by $\mathbb{P}_2$. Thus

$$\mathbb{P}_2 v = \mathbb{P}_2 v_1 + r . \tag{9.3.26}$$

Now let us prove (9.3.23). We have

$$\|\mathbb{P}_2 v\|_{0,Y} = \|\mathbb{P}_2 v_1 + r\|_{0,Y} = \|\mathbb{P}_2(v_1 + r)\|_{0,Y}$$
$$\le c\|v_1 + r\|_{0,Y\backslash F_\eta} = c\|v\|_{0,Y\backslash F_\eta} .$$

Applying Korn's inequality to v_1 we have

$$\|v_1\|_{1,Y\backslash F_\eta}^2 \le c\sum_{\alpha,\beta}\|e_{\alpha\beta}(v_1)\|_{0,Y\backslash F_\eta}^2 , \tag{9.3.27}$$

where $\|\cdot\|_{1,Y\backslash F_\eta}$ denotes here $H^1(Y\backslash F_\eta)^2$-norm. Since

$$\sum_{\alpha,\beta}\|e_{\alpha\beta}(\mathbb{P}_2 v)\|_{0,Y}^2 = \sum_{\alpha,\beta}\|e_{\alpha\beta}(\mathbb{P}_2 v_1)\|_{0,Y} = \sum_{\alpha,\beta}\|e_{\alpha\beta}(\mathbb{P}_2 v_1)\|_{0,F_\eta}^2$$

$$+ \sum_{\alpha,\beta}\|e_{\alpha\beta}(\mathbb{P}_2 v_1)\|_{0,Y\backslash F_\eta}^2 \le c_1\|v_1\|_{1,Y\backslash F_\eta}^2 + \sum_{\alpha,\beta}\|e_{\alpha\beta}(v_1)\|_{0,Y\backslash F_\eta}^2 .$$

Combining the last inequality with (9.3.27) one readily obtains (9.3.24). $\qquad\square$

The extension operator Q_2 is defined in much the same way as the operator Q_1.

Definition 9.3.5. The extension operator

$$Q_2 : H^1(YF)^2 \to H^1(Y)^2$$

is equal to

$$Q_2 = \mathbb{P}_2 \circ \mathbb{R}_2 , \tag{9.3.28}$$

where $\mathbb{R}_2 : H^1(YF)^2 \to H^1(Y\backslash F_\eta)^2$ is the restriction operator.

Properties of the operator Q_2

$$Q_2 v = v \quad \text{on} \quad Y \backslash F_\eta , \tag{9.3.29}$$

$$\|Q_2 v\|_{0,Y} \leq c\|v\|_{0,Y\backslash F_\eta} \leq c\|v\|_{0,YF} = c\|v\|_{0,Y} , \tag{9.3.30}$$

$$\|e(Q_2 v)\|_{0,Y} \leq c\|e(v)\|_{0,Y\backslash F_\eta} \leq c\|e(v)\|_{0,YF} , \tag{9.3.31}$$

$$\|Q_2 v - v\|_{1,YF} \leq c\|e(v)\|_{0,YF} . \tag{9.3.32}$$

Proof. It is similar to the earlier given proof for the operator Q_1. Now, however, Korn's inequality should be used instead of the Poincaré inequality. $\square$

To construct the global operator Q_2^ε we proceed analogously to the case of Q_1^ε, cf. the scheme (9.3.15). As previously, the operator Q_2^ε may be set equal to the identity near the boundary $\partial\Omega$, see the property (9.3.33) below. Then

$$u \in H_1^1(\Omega^\varepsilon)^2 \implies Q_2^\varepsilon u \in H_0^1(\Omega)^2 .$$

Properties of Q_2^ε

$$Q_2^\varepsilon u = u \quad \text{on} \quad \Omega \backslash F_\eta^\varepsilon , \tag{9.3.33}$$

$$\|Q_2^\varepsilon u\|_{0,\Omega} \leq c\|u\|_{0,\Omega} , \tag{9.3.34}$$

$$\|e(Q_2^\varepsilon u)\|_{0,\Omega} \leq c\|e(u)\|_{0,\Omega^\varepsilon} , \tag{9.3.35}$$

$$\|Q_2^\varepsilon u - u\|_{0,\Omega} \leq c\varepsilon\|e(u)\|_{0,\Omega^\varepsilon} , \tag{9.3.36}$$

$$\|\nabla(Q_2^\varepsilon u - u)\|_{0,\Omega^\varepsilon} \leq c\|e(u)\|_{0,\Omega^\varepsilon} . \tag{9.3.37}$$

Those properties can be proved by applying arguments similar to the proof of the properties (9.3.17)-(9.3.20) of the operator Q_1^ε.

By noting that (9.3.36) and (9.3.37) yield

$$\|Q_2^\varepsilon u - u\|_{1,\Omega^\varepsilon} \leq (c_1\varepsilon + c)\|e(u)\|_{0,\Omega^\varepsilon} , \tag{9.3.38}$$

we can formulate a Korn-type inequality.

Proposition 9.3.6 (*Korn's inequality for Ω^ε*). For each $u \in H_1^1(\Omega^\varepsilon)^2$ the following inequality is satisfied:

$$\|u\|_{1,\Omega^\varepsilon} \leq (c\varepsilon + c_1)\|e(u)\|_{0,\Omega^\varepsilon} . \tag{9.3.39}$$

Here $0 < \varepsilon < \varepsilon_0$ and ε_0 is held fixed. The positive constants c and c_1 are independent of ε.
Proof. It is similar to the proof of Proposition 9.3.3 and details are left to the reader. $\square$

Corollary 9.3.7.
(i) For any sequence $\{w^\varepsilon\}_{\varepsilon\to 0}$ satisfying $\sup_{\varepsilon>0}\|w^\varepsilon\| < \infty$ the sequence $\{Q_1^\varepsilon w^\varepsilon\}_{\varepsilon\to 0}$ is bounded in $H^1(\Omega)$ and

$$\|Q_1^\varepsilon w^\varepsilon - w^\varepsilon\|_{0,\Omega} \to 0 \quad \text{as} \quad \varepsilon \to 0 .$$

(ii) For any sequence $\{v^\varepsilon\}_{\varepsilon\to 0}$ satisfying $\sup_{\varepsilon>0} ||v^\varepsilon||_{1,\Omega^\varepsilon} < \infty$ the sequence $\{\mathcal{Q}_2^\varepsilon v^\varepsilon\}_{\varepsilon\to 0}$ is bounded in $H^1(\Omega)^2$ and

$$||\mathcal{Q}_2^\varepsilon v^\varepsilon - v^\varepsilon||_{0,\Omega} \to 0 \text{ as } \varepsilon \to 0 .$$

Proof. It follows immediately from (9.3.20) for (i), and from (9.3.36) for (ii). $\qquad\square$

After these lengthy, though indispensable preparations we pass to the justification of the effective model derived formally in Sec. 9.2. The justification procedure is divided into three major steps.

Step 1: boundedness

The variational problem (P_ε) is equivalent to

$$\min\left\{ \frac{1}{2}a^\varepsilon(v,w,\varphi;v,w,\varphi) - f(w)|(v,w,\varphi) \in K_\varepsilon \right\} . \tag{9.3.40}$$

Hence

$$\frac{1}{2}a^\varepsilon(v^\varepsilon,w^\varepsilon,\varphi^\varepsilon;v^\varepsilon,w^\varepsilon,\varphi^\varepsilon) - f(w^\varepsilon)$$

$$\leq \frac{1}{2}a^\varepsilon(v,w,\varphi;v,w,\varphi) - f(w) , \qquad \forall\, (v,w,\varphi) \in K_\varepsilon . \tag{9.3.41}$$

Since $(v,w,\varphi) = (0,0,0) \in K_\varepsilon$, from the last inequality we obtain

$$c\int_{\Omega^\varepsilon} (|e(v^\varepsilon)|^2 + |e(\varphi^\varepsilon)|^2 + |\nabla w^\varepsilon + \varphi^\varepsilon|^2)dx \leq (\int_\Omega q^2 dx)^{1/2}(\int_\Omega (w^\varepsilon)^2 dx)^{1/2} = c_1||w^\varepsilon||_{0,\Omega} .$$

Hence, by taking into account (9.3.21) and (9.3.39) we readily obtain

$$\sup_{\varepsilon>0}(||e(v^\varepsilon)||^2_{0,\Omega^\varepsilon} + ||e(\varphi^\varepsilon)||^2_{0,\Omega^\varepsilon} + ||\nabla w^\varepsilon||^2_{0,\Omega^\varepsilon}) \leq \text{ const } < \infty , \tag{9.3.42}$$

and

$$\sup_{\varepsilon>0}(||v^\varepsilon||_{1,\Omega^\varepsilon} + ||\varphi^\varepsilon||^2_{1,\Omega^\varepsilon} + ||w^\varepsilon||^2_{1,\Omega^\varepsilon}) \leq \text{ const } < \infty . \tag{9.3.43}$$

Let us denote by N_ε, M_ε and Q_ε the generalized stresses which are given by (9.1.4) with v, w, φ being replaced by v^ε, w^ε, φ^ε respectively. The estimate (9.3.42) implies that the sequences $\{N_\varepsilon^{\alpha\beta}\}_{\varepsilon\to 0}$, $\{M_\varepsilon^{\alpha\beta}\}_{\varepsilon\to 0}$ and $\{Q_\varepsilon^\alpha\}_{\varepsilon\to 0}$ are bounded in L^2-norm.

Next, by using (9.3.42) and (9.3.43) and the properties of the extension operators $\mathcal{Q}_1^\varepsilon$, $\mathcal{Q}_2^\varepsilon$ combined with Korn's inequality applied to the domain Ω we deduce that the sequences

$$\{\mathcal{Q}_1^\varepsilon w^\varepsilon\}_{\varepsilon\to 0} , \quad \{\mathcal{Q}_2^\varepsilon v^\varepsilon\}_{\varepsilon\to 0} , \quad \text{and} \quad \{\mathcal{Q}_2^\varepsilon \varphi^\varepsilon\}_{\varepsilon\to 0}$$

are bounded in the norm $||\cdot||_{1,\Omega}$. Consequently, there exist subsequences, still indexed with ε, such that

$$Q_1^\varepsilon w^\varepsilon \to \bar{w}\,, \quad \text{strongly in} \quad L^2(\Omega)\,, \tag{9.3.44}$$

$$Q_2^\varepsilon v^\varepsilon \to \bar{v}\,, \quad Q_2^\varepsilon \varphi^\varepsilon \to \bar{\varphi} \text{ strongly in } L^2(\Omega)^2\,. \tag{9.3.45}$$

Hence, by using anew the properties of the extension operators Q_1^ε and Q_2^ε we have

$$w^\varepsilon \to \bar{w}\,, \; v_\alpha^\varepsilon \to \bar{v}_\alpha\,, \; \varphi_\alpha^\varepsilon \to \bar{\varphi}_\alpha \text{ strongly in} \quad L^2(\Omega)\,, \tag{9.3.46}$$

$$M_\varepsilon^{\alpha\beta} \rightharpoonup \bar{M}^{\alpha\beta}\,, \; N_\varepsilon^{\alpha\beta} \rightharpoonup \bar{N}^{\alpha\beta}\,, \; Q_\varepsilon^\alpha \rightharpoonup \bar{Q}^\alpha \text{ weakly in } L^2(\Omega)\,, \tag{9.3.47}$$

since, for instance

$$||Q_2^\varepsilon v^\varepsilon - v^\varepsilon||_{0,\Omega} = ||Q_2^\varepsilon v^\varepsilon - \bar{v} - (v^\varepsilon - \bar{v})||_{0,\Omega} \geq \left| ||Q_2^\varepsilon v^\varepsilon - \bar{v}||_{0,\Omega} - ||v^\varepsilon - \bar{v}||_{0,\Omega} \right|\,.$$

Step 2: localization

We shall now derive the local relations resulting from the variational inequality (9.2.1), which may also be written as follows

$$\int_{\Omega^\varepsilon} \{N_\varepsilon^{\alpha\beta} e_{\alpha\beta}(v - v^\varepsilon) + M_\varepsilon^{\alpha\beta} e_{\alpha\beta}(\varphi - \varphi^\varepsilon) + Q_\varepsilon^\alpha[(w - w^\varepsilon)_{,\alpha} + \varphi_\alpha - \varphi_\alpha^\varepsilon]\}dx$$

$$\geq f(w - w^\varepsilon) \qquad \forall\,(v, w, \varphi) \in K_\varepsilon\,. \tag{9.3.48}$$

Let us take $v_\alpha = v_\alpha^\varepsilon \pm \theta_\alpha$, $\varphi_\alpha = \varphi_\alpha^\varepsilon \pm \eta$, $w = w^\varepsilon \pm \xi$ where $\xi, \theta_\alpha, \eta_\alpha \in \mathbf{D}(\Omega)$. Noting that $(v, w, \varphi) \in K_\varepsilon$ and applying the Green formula we obtain

$$\int_{\Omega^\varepsilon} (-N_\varepsilon^{\alpha\beta}{}_{,\beta}\theta_\alpha - M_\varepsilon^{\alpha\beta}{}_{,\beta}\eta_\alpha - Q_{\varepsilon,\alpha}^\alpha \xi + Q_\varepsilon^\alpha \eta_\alpha)dx$$

$$+ \int_{F^\varepsilon} ([N_\varepsilon^{\alpha\beta}]n_\beta\theta_\alpha + [M_\varepsilon^{\alpha\beta}]n_\beta\eta_\alpha + [Q_\varepsilon^\alpha]n_\alpha\xi)ds = \int_\Omega q\xi dx\,, \tag{9.3.49}$$

for each $\xi, \theta_\alpha, n_\alpha \in \mathbf{D}(\Omega)$. Hence

$$M_\varepsilon^{\alpha\beta}{}_{,\beta} - Q_\varepsilon^\alpha = 0 \qquad \text{in} \qquad \Omega^\varepsilon\,,$$

$$N_\varepsilon^{\alpha\beta}{}_{,\beta} = 0\,, \quad Q_{\varepsilon,\alpha}^\alpha + q = 0 \quad \text{in} \quad \Omega^\varepsilon\,, \tag{9.3.50}$$

and

$$[M_\varepsilon^{\alpha\beta}]n_\beta = 0\,, \quad [N_\varepsilon^{\alpha\beta}]n_\beta = 0\,, \quad [Q_\varepsilon^\alpha]n_\alpha = 0 \text{ on } F^\varepsilon\,. \tag{9.3.51}$$

The equilibrium equations (9.3.50) are to be understood in the sense of distributions. In (9.3.51) we recognize the action and reaction principle for the generalized stresses.

Performing now integration by parts in (9.3.48) and taking into account (9.3.50), (9.3.51) we get

$$\int_{F^\varepsilon} \{-N_\varepsilon^{\alpha\beta} n_\beta [v_\alpha - v_\alpha^\varepsilon] - M_\varepsilon^{\alpha\beta} n_\beta [\varphi_\alpha - \varphi_\alpha^\varepsilon] - Q^\varepsilon [w - w^\varepsilon]\} ds \geq 0 \,,$$

for every $(\boldsymbol{v}, w, \boldsymbol{\varphi}) \in K_\varepsilon$, where

$$N_\varepsilon^{\alpha\beta} n_\beta = \overset{1}{N}{}_\varepsilon^{\alpha\beta} n_\beta = \overset{2}{N}{}_\varepsilon^{\alpha\beta} n_\beta \,, \text{ etc.}$$

We recall that $(\overset{1}{N}{}_\varepsilon^{\alpha\beta} n_\beta)$ and $(\overset{2}{N}{}_\varepsilon^{\alpha\beta} n_\beta)$ denote stress vectors at the same point of F^ε calculated from both sides of the surface (line).

The localization of the last inequality is carried out in the following way. First, we observe that it can be written in the equivalent form of three inequalities

$$-\int_{F^\varepsilon} N_\varepsilon^{\alpha\beta} n_\beta [v_\alpha - v_\alpha^\varepsilon] ds = -\int_{F^\varepsilon} (N_n^\varepsilon [v_n - v_n^\varepsilon] + N_\tau^\varepsilon [v_\tau - v_\tau^\varepsilon]) ds \geq 0$$

$$\forall \, \boldsymbol{v} \in C_{di}^\varepsilon \quad (9.3.52)$$

$$-\int_{F^\varepsilon} M_\varepsilon^{\alpha\beta} n_\beta [\varphi_\alpha - \varphi_\alpha^\varepsilon] ds = -\int_{F^\varepsilon} (M_n^\varepsilon [\varphi_n - \varphi_n^\varepsilon] + M_\tau^\varepsilon [\varphi_\tau - \varphi_\tau^\varepsilon]) ds \geq 0$$

$$\forall \, \boldsymbol{\varphi} \in C_{di}^\varepsilon \,, \quad (9.3.53)$$

$$-\int_{F^\varepsilon} Q^\varepsilon [w - w^\varepsilon] ds \geq 0 \qquad \forall \, \boldsymbol{\varphi} \in C_d^\varepsilon \,, \tag{9.3.54}$$

where the subscripts n and τ denote the normal and tangential components of a considered quantity respectively; $Q^\varepsilon = Q_2^\alpha n_\alpha = \overset{1}{Q}{}_\varepsilon^\alpha n_\alpha = \overset{2}{Q}{}_\varepsilon^\alpha n_\alpha$. In the case considered no constraints are imposed on v_τ and φ_τ, therefore by taking $v_n = v_n^\varepsilon$, $\varphi_n = \varphi_n^\varepsilon$ we conclude that $N_\tau^\varepsilon = 0$ and $M_\tau^\varepsilon = 0$. Then (9.3.52) and (9.3.53) reduce to

$$-\int_{F^\varepsilon} N_n^\varepsilon [v_n - v_n^\varepsilon] ds \geq 0 \qquad \forall \, \boldsymbol{v} \in C_{di}^\varepsilon \,, \tag{9.3.55}$$

$$-\int_{F^\varepsilon} M_n^\varepsilon [\varphi_n - \varphi_n^\varepsilon] ds \geq 0 \qquad \forall \, \boldsymbol{\varphi} \in C_{di}^\varepsilon \,, \tag{9.3.56}$$

respectively. It is thus sufficient to show how to localize one of the inequalities (9.3.54) – (9.3.56). For instance, let us investigate the second one. To this end we take $v_n = (1 - \theta)v_n^\varepsilon + \theta\eta$, where $\theta \in \mathbf{D}(\Omega)$, $0 \leq \theta \leq 1$ while $[\eta] \geq 0$ on F^ε. Noting that the aforementioned inequalities are positively homogeneous we obtain

$$\int_{F^\varepsilon} \theta N_n^\varepsilon [\eta - v_n^\varepsilon] ds \leq 0 \qquad \forall \, \theta \in \mathbf{D}^+(\Omega), \qquad \forall \, \eta \,, \, [\eta]_{F^\varepsilon} \geq 0 \,, \tag{9.3.57}$$

where

$$\mathbf{D}^+(\Omega) = \{\theta \in \mathbf{D}(\Omega) \mid \theta(x) \geq 0,\ x \in \Omega\}\ .$$

Now we take $\eta = 0$ and next $\eta = 2v_n^\varepsilon$, then we get

$$N_n^\varepsilon[\![v_n^\varepsilon]\!] = 0 \quad \text{on}\quad F^\varepsilon\ . \tag{9.3.58}$$

By taking $\eta = v_n^\varepsilon + \zeta$ with $[\![\zeta]\!]_{F^\varepsilon} = 1$, from (9.3.57) we obtain

$$N_n^\varepsilon \leq 0\ , \tag{9.3.59}$$

since $\theta \in \mathbf{D}^+(\Omega)$.

All in all, the unilateral conditions satisfied on F^ε are of *the Signorini - type* and are given by

$$
\begin{aligned}
[\![v_n^\varepsilon]\!] \geq 0\ , &\qquad N_n^\varepsilon \leq 0\ , &\qquad N_\tau^\varepsilon = 0\ , &\qquad N_n^\varepsilon[\![v_n^\varepsilon]\!] = 0\ , \\
[\![\varphi_n^\varepsilon]\!] \geq 0\ , &\qquad M_n^\varepsilon \leq 0\ , &\qquad M_\tau^\varepsilon = 0\ , &\qquad M_n^\varepsilon[\![\varphi_n^\varepsilon]\!] = 0\ , \\
[\![w^\varepsilon]\!] \geq 0\ , &\qquad Q^\varepsilon \leq 0\ , &\qquad Q^\varepsilon[\![w^\varepsilon]\!] = 0\ .
\end{aligned}
\tag{9.3.60}
$$

We now pass to the investigation of the local variational inequalities. First, consider the variational inequality (9.2.7) which can be written as follows

$$\int_{YF} [A^{\alpha\beta\lambda\mu}(\epsilon_{\lambda\mu}^h + e_{\lambda\mu}^y(\boldsymbol{v}^{(1)})) + E^{\alpha\beta\lambda\mu}(\kappa_{\lambda\mu}^h + \kappa_{\lambda\mu}^y(\boldsymbol{\varphi}^{(1)}))]e_{\alpha\beta}^y(\boldsymbol{v} - \boldsymbol{v}^{(1)})dy \geq 0\ , \tag{9.3.61}$$

and is valid for each $\boldsymbol{v} \in K_{YF}^{di}$. Let us take $\boldsymbol{v} = \boldsymbol{v}^{(1)} + \boldsymbol{\theta},\, \boldsymbol{\theta} \in K_{YF}^{di}$. Then

$$\int_{YF} [A^{\alpha\beta\lambda\mu}(\epsilon_{\lambda\mu}^h + e_{\lambda\mu}^y(\boldsymbol{v}^{(1)})) + E^{\alpha\beta\lambda\mu}(\kappa_{\lambda\mu}^h + \kappa_{\lambda\mu}^y(\boldsymbol{\varphi}^{(1)}))]e_{\alpha\beta}^y(\boldsymbol{\theta})dy \geq 0\ . \tag{9.3.62}$$

Let us now take $\theta_\alpha \in \mathbf{D}(YF)$, that is θ_α vanishes in a neighborhood of $\partial(YF) = \partial Y \cup F$. Then we readily get

$$-\frac{\partial N_0^{\alpha\beta}}{\partial y_\beta} = 0 \quad \text{in} \quad \mathbf{D}'(YF)\ , \tag{9.3.63}$$

where Eq. $(9.2.6)_1$ has been taken into account. Similarly, from (9.2.8) and $(9.2.6)_2$ we obtain

$$-\frac{\partial M_0^{\alpha\beta}}{\partial y_\beta} = 0 \quad \text{in} \quad \mathbf{D}'(YF)\ . \tag{9.3.64}$$

Finally, the variational inequality (9.2.9) combined with $(9.2.6)_3$ give

$$-\frac{\partial Q_0^\alpha}{\partial y_\alpha} = 0 \quad \text{in} \quad \mathbf{D}'(YF)\ . \tag{9.3.65}$$

Let us return to (9.3.62) and take $\theta \in K_{YF}^{di}$ vanishing in a neighborhood of F. Integrating by parts we conclude that

$$N_0^{\alpha\beta}\mu_\beta \quad \text{is} \quad Y\text{-antiperiodic,} \tag{9.3.66}$$

where (9.3.63) has been taken into account while μ stands for the outward unit normal vector to ∂Y. In a quite similar manner we get

$$M_0^{\alpha\beta}\mu_\beta \quad \text{and} \quad Q_0^\alpha\mu_\alpha \quad \text{are} \quad Y\text{-antiperiodic,} \tag{9.3.67}$$

i.e. they assume opposite values on the opposite sides of Y.

Let us examine the variational inequality (9.3.61). Taking v such that $v = v^{(1)}$ in a neighborhood of ∂Y and performing the integration by parts we obtain

$$\int_F [\![N_0^{\alpha\beta}(v_\alpha - v_\alpha^{(1)})]\!]\check{N}_\beta ds \geq 0 , \tag{9.3.68}$$

for any v with $[\![v_N]\!] \geq 0$ on F, since (9.3.63) is satisfied. Now we have the following decomposition: $N_0^{\alpha\beta}\check{N}_\beta = N_N^0\check{N}^\alpha + N_T^0 T^\alpha$, where

$$\check{N}^\alpha = \check{N}_\alpha , \quad T^\alpha = T_\alpha , \quad N_N^0 = N_0^{\alpha\beta}\check{N}_\alpha\check{N}_\beta , \quad N_T^0 = N_0^{\alpha\beta}\check{N}_\alpha T_\beta ,$$
$$N_0^{\alpha\beta}\check{N}_\beta v_\alpha = N_N^0 v_N + N_T^0 v_T .$$

Hence

$$\int_F \{[\![N_N^0(v_N - v_N^{(1)})]\!] + [\![N_T^0(v_T - v_T^{(1)})]\!]\}ds \geq 0 , \tag{9.3.69}$$

for any v with $[\![v_N]\!] \geq 0$ on F. We recall that $v_N = v_\alpha\check{N}^\alpha$.

By a reasoning similar to that which resulted in (9.3.57) we arrive at

$$\int_F [\![N_N^0(v_N - v_N^{(1)})]\!]ds \geq 0 , \tag{9.3.70}$$

for any v with $[\![v_N]\!] \geq 0$ on F.

Next, the variational inequality (9.2.8) gives

$$\int_F [\![M_N^0(\varphi_N - \varphi_N^{(1)})]\!]ds \geq 0 , \tag{9.3.71}$$

for any φ such that $[\![\varphi_N]\!] \geq 0$ on F.

From (9.2.9) we obtain

$$\int_F [\![Q^0(w - w^{(1)})]\!]ds \geq 0 , \tag{9.3.72}$$

for any w with $[\![w]\!] \geq 0$ on F, where $Q^0 = Q_0^\alpha N_\alpha$.

Let us set

$$\widetilde{w}(y) = \langle \boldsymbol{\gamma}^h, y \rangle + w^{(1)}(y) = \sum_\alpha \gamma_\alpha^h y_\alpha + w^{(1)}(y)$$

$$\widetilde{v}_\alpha(y) = \sum_\alpha \epsilon_{\alpha\beta}^h y_\beta + v_\alpha^{(1)}(y) \,, \qquad \widetilde{\varphi}_\alpha(y) = \sum_\alpha \kappa_{\alpha\beta}^h y_\beta + \varphi_\alpha^{(1)}(y) \,. \tag{9.3.73}$$

By using the localization technique similar to that which resulted in (9.3.57) and replacing v in (9.370) by $v - \epsilon^h y$ we arrive at the following inequality

$$\int_F \theta [\![(A^{\alpha\beta\lambda\mu}(e_{\lambda\mu}^y(\widetilde{v}) + E^{\alpha\beta\lambda\mu}\kappa_{\lambda\mu}^y(\widetilde{\varphi}))(v_\alpha - \widetilde{v}_\alpha)]\!] \check{N}_\beta ds \geq 0 \,, \tag{9.3.74}$$

for any v with $[\![v_N]\!] \geq 0$ on F and $\theta \in \mathbf{D}^+(Y)$.

In an analogous manner we derive

$$\int_F \theta [\![(E^{\alpha\beta\lambda\mu}e_{\lambda\mu}^y(\widetilde{v}) + G^{\alpha\beta\lambda\mu}e_{\lambda\mu}^y(\widetilde{\varphi}))(\varphi_\alpha - \widetilde{\varphi}_\alpha)]\!] \check{N}_\beta ds \geq 0 \,, \tag{9.3.75}$$

and

$$\int_F \theta [\![H^{\alpha\beta} \frac{\partial \widetilde{w}}{\partial y_\beta}(w - \widetilde{w})]\!] \check{N}_\alpha ds \geq 0 \,, \tag{9.3.76}$$

for each φ, w with $[\![\varphi_N]\!] \geq 0$ and $[\![w]\!] \geq 0$ on F.

We now change the scale: $y = x/\varepsilon$ and put

$$\widetilde{w}^\varepsilon(x) = \varepsilon \widetilde{w}(x/\varepsilon) = \langle \boldsymbol{\gamma}^h, x \rangle + \varepsilon w^{(1)}(x/\varepsilon) \,,$$

$$\widetilde{v}_\alpha^\varepsilon(x) = \varepsilon \widetilde{v}_\alpha(x/\varepsilon) = \sum_\beta \epsilon_{\alpha\beta}^h x_\beta + \varepsilon v_\alpha^{(1)}(x/\varepsilon) \,, \tag{9.3.77}$$

$$\widetilde{\varphi}_\alpha^\varepsilon(x) = \varepsilon \widetilde{\varphi}_\alpha(x/\varepsilon) = \sum_\beta \kappa_{\alpha\beta}^h x_\beta + \varphi_\alpha^{(1)}(x/\varepsilon) \,.$$

We recall that $\boldsymbol{\gamma}^h \in \mathbb{R}^2$ and $\epsilon^h, \kappa^h \in \mathbb{E}_s^2$. It is evident that on F^ε we have $[\![\widetilde{w}^\varepsilon]\!] \geq 0$, $[\![\widetilde{v}_n^\varepsilon]\!] \geq 0$ and $[\![\widetilde{\varphi}_n^\varepsilon]\!] \geq 0$.

After the rescaling, the local equilibrium equations (9.3.63) – (9.3.65) can be written in the following form

$$-[A^{\alpha\beta\lambda\mu}e_{\lambda\mu}(\widetilde{v}^\varepsilon) + E^{\alpha\beta\lambda\mu}e_{\lambda\mu}(\widetilde{\varphi}^\varepsilon)]_{,\beta} = 0 \qquad \text{in } \mathbf{D}'(\Omega^\varepsilon) \,,$$

$$-[E^{\alpha\beta\lambda\mu}e_{\lambda\mu}(\widetilde{v}^\varepsilon) + G^{\alpha\beta\lambda\mu}e_{\lambda\mu}(\widetilde{\varphi}^\varepsilon)]_{,\beta} = 0 \qquad \text{in } \mathbf{D}'(\Omega^\varepsilon) \,, \tag{9.3.78}$$

$$-(H^{\alpha\beta}\widetilde{w}_{,\beta}^\varepsilon)_{,\alpha} = 0 \quad \text{in} \quad \mathbf{D}'(\Omega^\varepsilon) \,.$$

We recall that the rescaling $y = x/\varepsilon$ means that $\partial/\partial y = \varepsilon \partial/\partial x$.

Then the inequalities (9.3.74) – (9.3.76) transform into

$$\int_{F^\varepsilon} \theta[\![(A^{\alpha\beta\lambda\mu}e_{\lambda\mu}(\widetilde{\boldsymbol{v}}^\varepsilon) + E^{\alpha\beta\lambda\mu}e_{\lambda\mu}(\widetilde{\boldsymbol{\varphi}}^\varepsilon))(v_\alpha - \widetilde{v}_\alpha^\varepsilon)]\!]n_\beta ds \geq 0 \,,$$

$$\int_{F^\varepsilon} \theta[\![(E^{\alpha\beta\lambda\mu}e_{\lambda\mu}(\widetilde{\boldsymbol{v}}^\varepsilon) + G^{\alpha\beta\lambda\mu}e_{\lambda\mu}(\widetilde{\boldsymbol{\varphi}}^\varepsilon))(\varphi_\alpha - \widetilde{\varphi}_\alpha^\varepsilon)]\!]n_\beta ds \geq 0 \,, \qquad (9.3.79)$$

$$\int_{F^\varepsilon} \theta[\![H^{\alpha\beta}\widetilde{w}_{,\beta}^\varepsilon(w - \widetilde{w}^\varepsilon)]\!]n_\alpha ds \geq 0 \,,$$

respectively, for each $\boldsymbol{v}, \boldsymbol{\varphi} \in C_{di}^\varepsilon$, $w \in C_d^\varepsilon$ and every $\theta \in \mathbf{D}^+(\Omega)$.

Step 3: Identification of $\bar{v}, \bar{w}$ and $\bar{\varphi}$

The final step consists in proving that $\bar{\boldsymbol{v}} = \boldsymbol{v}^{(0)}$, $\bar{w} = w^{(0)}$ and $\bar{\boldsymbol{\varphi}} = \boldsymbol{\varphi}^{(0)}$, cf. (9.2.4) and (9.3.46).

The stored energy function of the virgin material:

$$\mathcal{W}(\boldsymbol{\gamma}, \boldsymbol{\rho}, \boldsymbol{d}) := (A^{\alpha\beta\lambda\mu}\gamma_{\alpha\beta}\gamma_{\lambda\mu} + 2E^{\alpha\beta\lambda\mu}\gamma_{\alpha\beta}\rho_{\lambda\mu}$$
$$+ G^{\alpha\beta\lambda\mu}\rho_{\alpha\beta}\rho_{\lambda\mu} + H^{\alpha\beta}d_\alpha d_\beta)/2 \,, \qquad (9.3.80)$$

is obviously convex and differentiable, where $\boldsymbol{\gamma}, \boldsymbol{\rho} \in \mathbb{E}_s^2$ and $\boldsymbol{d} \in \mathbb{R}^2$. Consequently, it is subdifferentiable and its subdifferential is a maximal monotone operator. The last property gives

$$J_1^\varepsilon := \int_{\Omega^\varepsilon} \theta(x)\{A^{\alpha\beta\lambda\mu}e_{\lambda\mu}(\boldsymbol{v}^\varepsilon) + E^{\alpha\beta\lambda\mu}e_{\lambda\mu}(\boldsymbol{\varphi}^\varepsilon)$$
$$- [A^{\alpha\beta\lambda\mu}e_{\lambda\mu}(\widetilde{\boldsymbol{v}}^\varepsilon) + E^{\alpha\beta\lambda\mu}e_{\lambda\mu}(\widetilde{\boldsymbol{\varphi}}^\varepsilon)]\}e_{\alpha\beta}(\boldsymbol{v}^\varepsilon - \widetilde{\boldsymbol{v}}^\varepsilon)dx \geq 0 \,, \qquad (9.3.81)$$

$$J_2^\varepsilon := \int_{\Omega^\varepsilon} \theta(x)\{E^{\alpha\beta\lambda\mu}e_{\lambda\mu}(\boldsymbol{v}^\varepsilon) + G^{\alpha\beta\lambda\mu}e_{\lambda\mu}(\boldsymbol{\varphi}^\varepsilon)$$
$$- [E^{\alpha\beta\lambda\mu}e_{\lambda\mu}(\widetilde{\boldsymbol{v}}^\varepsilon) + G^{\alpha\beta\lambda\mu}e_{\lambda\mu}(\widetilde{\boldsymbol{\varphi}}^\varepsilon)]\}e_{\alpha\beta}(\boldsymbol{\varphi}^\varepsilon - \widetilde{\boldsymbol{\varphi}}^\varepsilon)dx \geq 0 \,, \qquad (9.3.82)$$

$$J_3^\varepsilon := \int_{\Omega^\varepsilon} \theta(x)[H^{\alpha\beta}(w_{,\beta}^\varepsilon + \varphi_\beta^\varepsilon) - H^{\alpha\beta}\widetilde{w}_{,\beta}^\varepsilon][(w_{,\alpha}^\varepsilon + \varphi_\alpha^\varepsilon) - \widetilde{w}_{,\alpha}^\varepsilon]dx \geq 0 \,, \qquad (9.3.83)$$

where $\theta \in \mathbf{D}^+(\Omega)$. Let us now pass to the limit in (9.3.81). Integration by parts combined with $(9.1.4)_1$ written for Ω^ε, $(9.3.50)_2$, $(9.3.78)_1$ and $(9.3.79)_1$ yield

$$- \int_{\Omega^\varepsilon} \{N_\varepsilon^{\alpha\beta} - [A^{\alpha\beta\lambda\mu}e_{\lambda\mu}(\widetilde{\boldsymbol{v}}^\varepsilon) + E^{\alpha\beta\lambda\mu}e_{\lambda\mu}(\widetilde{\boldsymbol{\varphi}}^\varepsilon)]\}\theta_{,\beta}(v_\alpha^\varepsilon - \widetilde{v}_\alpha^\varepsilon)dx \geq 0 \,.$$

Passing now with ε to zero we obtain

$$- \int_\Omega \left(\bar{N}^{\alpha\beta} - \frac{\partial \mathcal{W}}{\partial \epsilon_{\alpha\beta}^h}\right)(\bar{v}_\alpha - \sum_\beta \epsilon_{\alpha\beta}^h x_\beta)\theta_{,\beta}dx \geq 0 \,, \qquad (9.3.84)$$

where $(9.3.47)_2$ has been taken into account and the fact that

$$v_\alpha^\varepsilon - \widetilde{v}_\alpha^\varepsilon \to \widetilde{v}_\alpha - \sum_\beta \epsilon_{\alpha\beta}^h x_\beta \quad \text{strongly in } L^2(\Omega)^2 \text{ as } \varepsilon \to 0 \,,$$

$$A^{\alpha\beta\lambda\mu} e_{\lambda\mu}(\widetilde{v}^\varepsilon) + E^{\alpha\beta\lambda\mu} e_{\lambda\mu}(\widetilde{\varphi}^\varepsilon) \rightharpoonup \frac{1}{|Y|} \int_{YF} [A^{\alpha\beta\lambda\mu} e_{\lambda\mu}^y(\widetilde{v}) + E^{\alpha\beta\lambda\mu} e_{\lambda\mu}^y(\widetilde{\varphi})] dy$$

$$= \frac{\partial \mathcal{W}}{\partial \epsilon_{\alpha\beta}^h} \quad \text{weakly in } L^2(\Omega^2) \text{ as } \varepsilon \to 0 \,.$$

Integrating by parts in (9.3.84) and recalling that $\epsilon^h \in \mathbb{E}_s^2$ we get

$$-\int_\Omega \theta (\bar{N}^{\alpha\beta}{}_{,\beta}(\bar{v}_\alpha - \sum_\beta \epsilon_{\alpha\beta}^h x_\beta)) dx + \int_\Omega \theta \left(\bar{N}^{\alpha\beta} - \frac{\partial \mathcal{W}}{\partial \epsilon_{\alpha\beta}^h} \right) (e_{\alpha\beta}(\bar{v}) - \epsilon_{\alpha\beta}^h) dx \geq 0 \,. \quad (9.3.85)$$

From $(9.3.47)_2$ and $(9.3.50)_2$ we conclude that

$$\bar{N}^{\alpha\beta}{}_{,\beta} = 0 \quad \text{in} \quad \mathbf{D}'(\Omega) \,. \quad (9.3.86)$$

Substitution of (9.3.86) into (9.3.85) yields

$$\int_\Omega \theta \left(\bar{N}^{\alpha\beta} - \frac{\partial \mathcal{W}}{\partial \epsilon_{\alpha\beta}^h} \right) (e_{\alpha\beta}(\bar{v}) - \epsilon_{\alpha\beta}^h) dx \geq 0 \,,$$

for each $\theta \in \mathbf{D}^+(\Omega)$. Hence

$$\left(\bar{N}^{\alpha\beta}(x) - \frac{\partial \mathcal{W}}{\partial \epsilon_{\alpha\beta}^h} \right) (e_{\alpha\beta}(\bar{v}(x)) - \epsilon_{\alpha\beta}^h) \geq 0 \,, \quad (9.3.87)$$

for every $\epsilon^h \in \mathbb{E}_s^2$ and a.e. $x \in \Omega$.

The functionals J_2^ε and J_3^ε, defined by (9.3.82) and (9.3.83) respectively, can be examined in a similar manner. Then we conclude that

$$\bar{M}^{\alpha\beta}{}_{,\beta} - \bar{Q}^\alpha = 0 \qquad \text{in } \mathbf{D}'(\Omega) \,, \quad (9.3.88)$$
$$\bar{Q}^\alpha{}_{,\alpha} + q = 0 \qquad \text{in } \mathbf{D}'(\Omega) \,, \quad (9.3.89)$$

and

$$\left(\bar{M}^{\alpha\beta}(x) - \frac{\partial \mathcal{W}}{\partial \kappa_{\alpha\beta}^h} \right) (e_{\alpha\beta}(\bar{\varphi}(x)) - \kappa_{\alpha\beta}^h) \geq 0 \qquad \forall \, \kappa^h \in \mathbb{E}_s^2 \,, \quad a.e. \; x \in \Omega \,, \quad (9.3.90)$$

$$\left(\bar{Q}^\alpha(x) - \frac{\partial \mathcal{W}}{\partial \gamma_\alpha^h} \right) (\bar{\varphi}_\alpha(x) + \bar{w}_{,\alpha}(x) - \gamma_\alpha^h) \geq 0 \qquad \forall \, \gamma^h \in \mathbb{R}^2 \,, \quad a.e. \; x \in \Omega \,. \quad (9.3.91)$$

Besides the properties (9.2.16) and (9.2.17) of $\mathcal{W}$ its subdifferential $\partial\mathcal{W}$ enjoys the property of maximal monotonicity. The last property follows immediately from the convexity and finiteness of $\mathcal{W}$, cf. Sec. 1.2.1. In our case $\partial\mathcal{W}$ is a maximal monotone mapping from $\mathbb{E}_s^2 \times \mathbb{E}_s^2 \times \mathbb{R}^2$ to $\mathbb{E}_2^s \times \mathbb{E}_2^s \times \mathbb{R}^2$. Consequently, (9.3.87), (9.3.90) and (9.3.91) imply

$$\bar{N}^{\alpha\beta}(x) = \partial\mathcal{W}/\partial e_{\alpha\beta}(\bar{v}(x)) , \qquad \bar{M}^{\alpha\beta}(x) = \partial\mathcal{W}/\partial e_{\alpha\beta}(\bar{\varphi}(x)) ,$$
$$\bar{Q}^{\alpha}(x) = \partial\mathcal{W}/\partial\gamma_{\alpha}(\bar{\varphi}(x), \bar{w}(x)) ,$$

for a.e. $x \in \Omega$, where $\mathcal{W} = \mathcal{W}[e(\bar{v}(x)), e(\bar{\varphi}(x)), \gamma(\bar{\varphi}(x)), \bar{w}(x)]$.

By virtue of (9.3.92), from (9.3.86), (9.3.88) and (9.3.89) we recover the equilibrium equations (9.2.13) where $M_h = \bar{M}$, $N_h = \bar{N}$, $Q_h = \bar{Q}$ and $v^{(0)} = \bar{v}$, $\varphi^{(0)} = \bar{\varphi}$, $w^{(0)} = \bar{w}$. Thus the proof of convergence is complete. $\qquad\square$

Remark 9.3.8. The method of Γ-convergence additionally ensures the convergence of the total potential energy of the cracked plate to the total potential energy of the homogenized plate, i.e.:

$$\frac{1}{2}a^{\varepsilon}(v^{\varepsilon}, w^{\varepsilon}, \varphi^{\varepsilon} ; v^{\varepsilon}, w^{\varepsilon}, \varphi^{\varepsilon}) - f(w^{\varepsilon}) \rightarrow \int_{\Omega} \mathcal{W}[e(v^{(0)}), e(\varphi^{(0)}) , \gamma(\varphi^{(0)}, w^{(0)})]dx - f(w^{(0)}) .$$

10. Part-through the thickness cracks

10.1. Two-layer description

This section is aimed at formulating the boundary value problem of an elastic plate transversely cut with a crack of constant depth smaller than the whole plate thickness h. The crack divides the plate into two layers of constant thicknesses a and b. The upper face $x_3 = 0$ will be viewed as a reference plane and will be denoted by Ω. This domain is parametrized by Cartesian coordinates x_α, $\alpha = 1, 2$. Thus the plate occupies a domain $B = \Omega \times (0, h)$, $h = a + b$; $\boldsymbol{x} \in B$ and $\boldsymbol{x} = (x, x_3)$, $x = (x_1, x_2) \in \Omega$, $x_3 \in [0, h]$. The elastic moduli $C^{ijkl} \in L^\infty(B)$ can vary in longitudinal and transverse directions; the planes $x_3 =$const are planes of material symmetry, cf. (2.4.1). The elastic moduli satisfy the usual symmetry conditions, while the matrix C is positive definite. The plate is subject to transverse loading of intensity $p = p(x)$, applied to the upper face $x_3 = 0$.

The two-layer two-dimensional model of the plate is based upon two assumptions:

(i) the in-plane displacements w_α are assumed to be piece-wise linear through the thickness, while transverse displacements w_3 are assumed to be constant across the thickness:

$$w_\alpha(x, x_3) = \begin{cases} r_\alpha(x) + (x_3 - a)\varphi_\alpha(x) & \text{if} \quad 0 \leq x_3 \leq a \,, \\ r_\alpha(x) + (x_3 - a)\psi_\alpha(x) & \text{if} \quad a \leq x_3 \leq h \,, \end{cases} \tag{10.1.1}$$

$$w_3(x, x_3) = w(x) \,, \qquad \text{if } 0 \leq x_3 \leq h \,.$$

The fields r_α represent the in-plane displacements of the interface plane; the fields $\varphi_\alpha, \psi_\alpha$ represent the angles of rotation of the transverse sections of both layers, cf. Fig. 10.1.1.

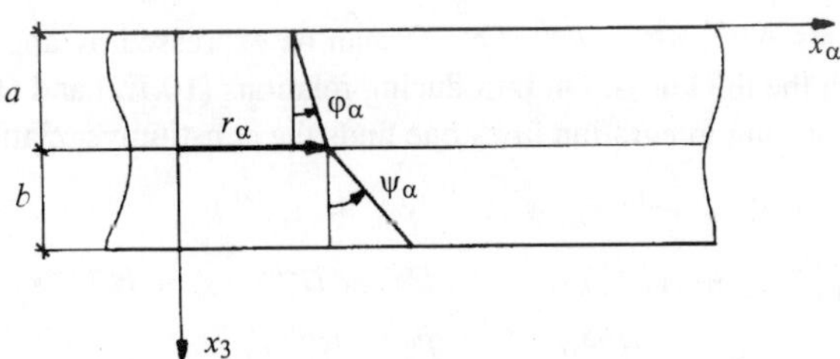

Fig. 10.1.1. Kinematic assumption

(ii) The generalized plane stress state holds within the plate, hence the constitutive relations have the form

$$\sigma^{\alpha\beta} = \tilde{C}^{\alpha\beta\lambda\mu}e_{\lambda\mu}(\boldsymbol{w}) \,, \quad \sigma^{\alpha3} = 2C^{\alpha3\lambda3}e_{\lambda3}(\boldsymbol{w}) \,, \quad \sigma^{33} = 0 \,, \tag{10.1.2}$$

where $\tilde{C}$ is defined according to the rule (3.1.4).

The deformations associated with kinematic assumptions (10.1.1) are given by

$$e_{\alpha\beta}(\boldsymbol{w}) = \begin{cases} \epsilon_{\alpha\beta}(\boldsymbol{r}) + (x_3 - a)\rho_{\alpha\beta}(\boldsymbol{\varphi}) & \text{for} \quad 0 \le x_3 \le a\,, \\ \epsilon_{\alpha\beta}(\boldsymbol{r}) + (x_3 - a)k_{\alpha\beta}(\boldsymbol{\psi}) & \text{for} \quad a \le x_3 \le h\,; \end{cases}$$

$$2e_{\alpha 3}(\boldsymbol{w}) = \begin{cases} g_\alpha(w, \boldsymbol{\psi}) & \text{for} \quad 0 \le x_3 \le a\,, \\ d_\alpha(w, \boldsymbol{\psi}) & \text{for} \quad a \le x_3 \le h\,, \end{cases}$$

$$e_{33}(\boldsymbol{w}) = 0 \qquad \text{for} \qquad 0 \le x_3 \le h\,,$$

$$(10.1.3)$$

where

$$\epsilon_{\alpha\beta}(\boldsymbol{r}) = e_{\alpha\beta}(\boldsymbol{r})\,, \qquad \rho_{\alpha\beta}(\boldsymbol{\varphi}) = e_{\alpha\beta}(\boldsymbol{\varphi})\,,$$

$$k_{\alpha\beta}(\boldsymbol{\psi}) = e_{\alpha\beta}(\boldsymbol{\psi})\,, \qquad e_{\alpha\beta}(\boldsymbol{f}) = \frac{1}{2}(f_{\alpha,\beta} + f_{\beta,\alpha})\,, \qquad (10.1.4)$$

$$g_\alpha(w, \boldsymbol{\varphi}) = w_{,\alpha} + \varphi_\alpha\,, \qquad d_\alpha(w, \boldsymbol{\psi}) = w_{,\alpha} + \psi_\alpha\,.$$

Thus the transverse shear deformations are piece-wise constant across the thickness. Consequently the shearing stiffnesses will be overestimated and introduction of shear-correction factor (k_s) will be necessary.

The two-layer plate model is constructed as follows. The point of departure is the variational equilibrium equation (2.5.1) for the three-dimensional problem of elasticity. The trial displacement field $\boldsymbol{v}(x, x_3)$ is represented according to assumptions (10.1.1); the trial fields for $\boldsymbol{r}$, $\boldsymbol{\varphi}$, $\boldsymbol{\psi}$, w are denoted by $\boldsymbol{s}$, $\boldsymbol{\eta}$, $\boldsymbol{\chi}$ and v respectively. After integration through the thickness one finds

$$\int_\Omega [N^{\alpha\beta}\epsilon_{\alpha\beta}(\boldsymbol{s}) + M^{\alpha\beta}\rho_{\alpha\beta}(\boldsymbol{\eta}) + L^{\alpha\beta}k_{\alpha\beta}(\boldsymbol{\chi}) + Q^\alpha g_\alpha(v, \boldsymbol{\eta}) + T^\alpha d_\alpha(v, \boldsymbol{\chi})]dx = \int_\Omega pv dx\,.$$

$$(10.1.5)$$

The stress resultants $N^{\alpha\beta}$, $M^{\alpha\beta}$, $L^{\alpha\beta}$, Q^α, T^α can be expressed by appropriate integrals of stresses through the thickness. On introducing relations (10.1.2) and (10.1.3) into these formulae and performing integration in x_3 one finds the constitutive relations

$$N^{\alpha\beta} = A^{\alpha\beta\lambda\mu}\epsilon_{\lambda\mu} + A_1^{\alpha\beta\lambda\mu}\rho_{\lambda\mu} + B_1^{\alpha\beta\lambda\mu}k_{\lambda\mu}\,,$$

$$M^{\alpha\beta} = A_1^{\alpha\beta\lambda\mu}\epsilon_{\lambda\mu} + A_2^{\alpha\beta\lambda\mu}\rho_{\lambda\mu}\,, \qquad L^{\alpha\beta} = B_1^{\alpha\beta\lambda\mu}\epsilon_{\lambda\mu} + B_2^{\alpha\beta\lambda\mu}k_{\lambda\mu}\,, \qquad (10.1.6)$$

$$Q^\alpha = H_a^{\alpha\beta}g_\beta\,, \qquad T^\alpha = H_b^{\alpha\beta}d_\beta\,,$$

with stiffnesses defined by

$$A^{\alpha\beta\lambda\mu} = \int_0^h \tilde{C}^{\alpha\beta\lambda\mu}dx_3\,, \qquad A_\sigma^{\alpha\beta\lambda\mu} = \int_0^a (x_3 - a)^\sigma \tilde{C}^{\alpha\beta\lambda\mu}dx_3\,,$$

$$B_\sigma^{\alpha\beta\lambda\mu} = \int_a^h (x_3 - a)^\sigma \tilde{C}^{\alpha\beta\lambda\mu}dx_3\,, \qquad \sigma = 1, 2\,; \qquad (10.1.7)$$

$$H_a^{\alpha\beta} = k_s \int_0^a C^{\alpha 3 \beta 3} dx_3 \,, \qquad H_b^{\alpha\beta} = k_s \int_a^h C^{\alpha 3 \beta 3} dx_3 \,.$$

In the case of a plate made of a homogeneous material the stiffnesses assume the form

$$\boldsymbol{A} = h\tilde{\boldsymbol{C}} \,, \qquad \boldsymbol{A}_1 = -\frac{a^2}{2}\tilde{\boldsymbol{C}} \,, \qquad \boldsymbol{B}_1 = -\frac{b^2}{2}\tilde{\boldsymbol{C}} \,,$$

$$\boldsymbol{A}_2 = \frac{a^3}{3}\tilde{\boldsymbol{C}} \,, \qquad \boldsymbol{B}_2 = \frac{b^3}{3}\tilde{\boldsymbol{C}} \,, \tag{10.1.8}$$

$$\boldsymbol{H}_a = k_s a\hat{\boldsymbol{C}} \,, \qquad \boldsymbol{H}_b = k_s b\hat{\boldsymbol{C}} \,, \qquad \hat{\boldsymbol{C}} = [C^{\alpha 3 \beta 3}] \,.$$

On substituting equations (10.1.6) into (10.1.5) one observes that the left-hand side of Eq. (10.1.5) defines a bilinear form $b_\Omega(\cdot,\cdot)$ with respect to $(\boldsymbol{r},\boldsymbol{\varphi},\boldsymbol{\psi},w)$ and $(\boldsymbol{s},\boldsymbol{\eta},\boldsymbol{\chi},v)$ and the equation itself can be written as follows

$$b_\Omega(\boldsymbol{r},\boldsymbol{\varphi},\boldsymbol{\psi},w \,;\, \boldsymbol{s},\boldsymbol{\eta},\boldsymbol{\chi},v) = f(v) \,, \tag{10.1.9}$$

where

$$f(v) = \int_\Omega pv\,dx \,. \tag{10.1.9a}$$

For the sake of simplicity the problem of a clamped plated will be discussed. Then $w_i(x,x_3) = 0$ for $x \in \partial\Omega$ and $x_3 \in [0,h]$. By (10.1.1) this condition is satisfied if the fields $\boldsymbol{r}$, $\boldsymbol{\varphi}$, $\boldsymbol{\psi}$ and w vanish on $\partial\Omega$. The space of kinematically admissible displacement has the form

$$V_{2L}^0(\Omega) = H_0^1(\Omega)^2 \times H_0^1(\Omega)^2 \times H_0^1(\Omega)^2 \times H_0^1(\Omega) \,,$$

($2L$ = two-layer model).

The equilibrium problem of the two-layered plate reads as follows:

$$(P_{2L}) \quad \left| \begin{array}{l} \text{find the kinematic fields } (\boldsymbol{r},\boldsymbol{\varphi},\boldsymbol{\psi},w) \in V_{2L}^0(\Omega) \text{ such that the variational} \\ \text{equation (10.1.9) holds for each } (\boldsymbol{s},\boldsymbol{\eta},\boldsymbol{\chi},v) \in V_{2L}^0(\Omega) \,. \end{array} \right.$$

One can prove that the problem above is well-posed, in particular the bilinear form $b_\Omega(\cdot,\cdot)$ is $V_{2L}^0(\Omega)$-elliptic. In fact, the matrices $\boldsymbol{H}_a$ and $\boldsymbol{H}_b$ are positive definite. Moreover, it is not difficult to verify that

$$\boldsymbol{A} = \begin{bmatrix} \boldsymbol{A} & \boldsymbol{A}_1 & \boldsymbol{B}_1 \\ \boldsymbol{A}_1 & \boldsymbol{A}_2 & \boldsymbol{0} \\ \boldsymbol{B}_1 & \boldsymbol{0} & \boldsymbol{B}_2 \end{bmatrix}$$

is also a positive definite matrix. The proof is similar to that of Sec. 11.1, where unique solvability of a two-dimensional model of cross-ply laminates is shown.

Since fields (s, η, χ, v) are arbitrary, the variational equilibrium equation (10.1.4) implies the local equilibrium equations

$$N^{\alpha\beta}_{,\beta} = 0 , \qquad -M^{\alpha\beta}_{,\beta} + Q^\alpha = 0 , \qquad -L^{\alpha\beta}_{,\beta} + T^\alpha = 0 ,$$

$$-(Q^\alpha + T^\alpha)_{,\alpha} = p , \tag{10.1.10}$$

satisfied in Ω.

Apart from the clamped edge conditions the model admits the boundary conditions concerning

$$
\begin{array}{llllll}
N_n & \text{or} & r_n , & N_\tau & \text{or} & r_\tau , \\
M_n & \text{or} & \varphi_n , & M_\tau & \text{or} & \varphi_\tau , \\
L_n & \text{or} & \psi_n , & L_\tau & \text{or} & \psi_\tau , \\
Q + T & \text{or} & & w ,
\end{array}
\tag{10.1.11}
$$

where

$$F_n = F^{\alpha\beta} n_\alpha n_\beta , \qquad F_\tau = F^{\alpha\beta} n_\alpha \tau_\beta ; \qquad F \in \{N, M, L\} ,$$

$$Q = Q^\alpha n_\alpha , \qquad\qquad\qquad T = T^\alpha n_\alpha ; \tag{10.1.12}$$

$$g_n = g_\alpha n^\alpha , \qquad\qquad g_\tau = g_\alpha \tau^\alpha ; \qquad g \in \{r, \varphi, \psi\} ,$$

while n, τ are versors outward normal and tangent to $\partial\Omega$, respectively. In particular, along the clamped edge

$$r_n = 0 , \quad r_\tau = 0 , \quad \varphi_n = 0 , \quad \varphi_\tau = 0 , \quad \psi_n = 0 , \quad \psi_\tau = 0 , \quad w = 0$$

or these fields assume given values. If the edge is supported and sliding, then

$$N_n = 0 , \quad N_\tau = 0 , \quad M_n = 0 , \quad \varphi_\tau = 0 , \quad L_n = 0 , \quad \psi_\tau = 0 , \quad w = 0 ,$$

or these fields are given. Along the free, unloaded edge we have

$$N_n = 0 , \ N_\tau = 0 , \qquad M_n = 0 , \ M_\tau = 0 , \qquad L_n = 0 , \ L_\tau = 0 , \qquad Q + T = 0 .$$

The two-layer plate model enables one to describe deformation of a plate with a transverse crack in one layer. In the present section we shall deal with a crack in the layer: $a \le x_3 \le h$, the crack surface being perpendicular to Ω. The projection of the crack on Ω forms an arc F. F can be treated as a part of a curve Σ dividing the domain Ω into two subdomains $\Omega_1, \Omega_2 (\Omega = \Omega_1 \cup \Omega_2 \cup \Sigma)$ which do not overlap, cf. Fig. 10.1.2. The versors normal and tangent to F are denoted once again by n, τ, respectively.

According to (10.1.1) the distribution of the in-plane displacement $w_n = w_\alpha n^\alpha$, normal to F, is given by

$$w_n(x, x_3) = \begin{cases} r_n(x) + (x_3 - a)\varphi_n(x) , & 0 \le x_3 \le a \\ r_n(x) + (x_3 - a)\psi_n(x) , & a \le x_3 \le h . \end{cases} \tag{10.1.13}$$

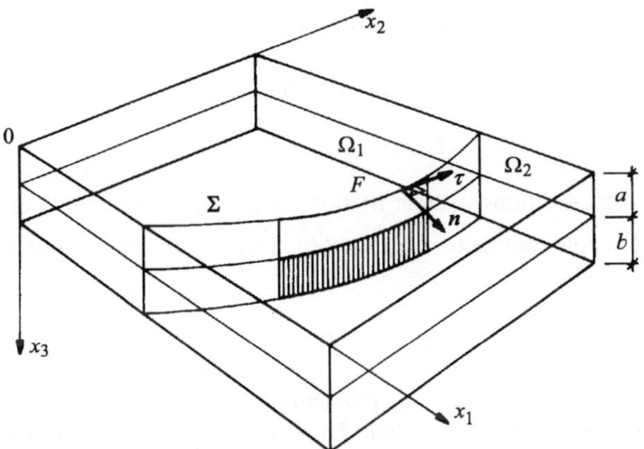

Fig. 10.1.2. Plate with a transverse part-through-the thickness crack

The opening of the crack is characterized by a jump $[\![\psi_n]\!] = \psi_{n|2} - \psi_{n|1}$ along F; here $\psi_{n|\alpha}$ represents the value of ψ_n at the αth side of F; $\alpha = 1, 2$. If this jump is positive: $[\![\psi_n]\!] > 0$, then the crack is open. The crack is closed if $[\![\psi_n]\!] = 0$. We shall consider the cracks such that

$$[\![\psi_n]\!] \geq 0 , \qquad [\![\psi_\tau]\!] - \text{arbitrary} , \qquad [\![r]\!] = 0 , \qquad [\![\varphi]\!] = 0 , \qquad [\![w]\!] = 0 . \quad (10.1.14)$$

Consequently, the upper layer deforms continuously while the lower layer $(a \leq x_3 \leq h)$ undergoes cracking, see Fig. 10.1.3. Such a cracking mode can be figured out with the help of the two-layer plate model presented here and could not be described by one-layer plate theories, like those due to Kirchhoff or Reissner-Hencky in which the penetration zones cannot be avoided, see Secs. 8 and 9.

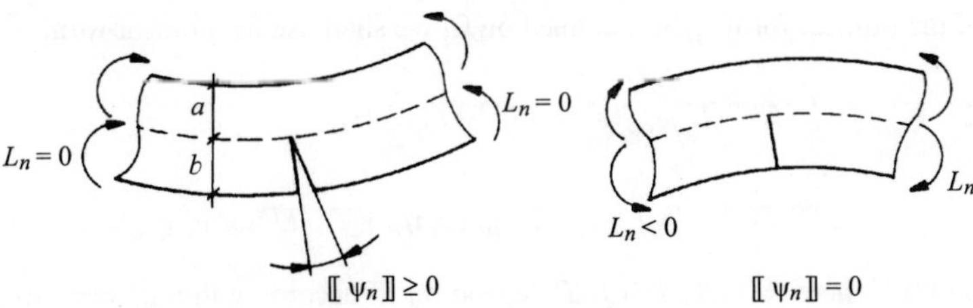

Fig. 10.1.3. Opening and closure of the crack. Signorini conditions

The stress couples $\overset{\alpha}{L}_n = L^{\lambda\mu}_{|\alpha} n_\lambda n_\mu$ are equal on both sides of F:

$$L_n = \overset{1}{L}_n = \overset{2}{L}_n , \qquad\qquad (10.1.15)$$

which expresses the principle of action and reaction. The moment L_n is always non-positive and becomes zero if the crack opens. Thus the relations

$$L_n \leq 0 \,, \qquad L_n[\![\psi_n]\!] = 0 \,, \qquad [\![\psi_n]\!] \geq 0 \,,$$

$$L_\tau = 0 \,, \qquad L_\tau = \overset{1}{L}_\tau = \overset{2}{L}_\tau \,, \qquad \overset{\alpha}{L}_\tau = L^{\lambda\mu}{}_{|\alpha} n_\lambda \tau_\mu \,, \tag{10.1.16}$$

are Signorini-type conditions without friction for the considered mode of cracking. The other stress and couple resultants: $\overset{\alpha}{N}_n, \overset{\alpha}{N}_\tau, \overset{\alpha}{M}_n, \overset{\alpha}{M}_\tau, \overset{\alpha}{Q}_n, \overset{\alpha}{T}_n$ are equal at both sides of the crack:

$$\overset{1}{R}_n = \overset{2}{R}_n \,, \qquad \overset{1}{R}_\tau = \overset{2}{R}_\tau \,, \qquad \boldsymbol{R} \in \{\boldsymbol{N}, \boldsymbol{M}, \boldsymbol{Q} + \boldsymbol{T}\} \,. \tag{10.1.17}$$

Now one can formulate the equilibrium problem for a plate with the crack described above. This problem amounts to finding the field $\boldsymbol{U} = (\boldsymbol{r}, \boldsymbol{\varphi}, \boldsymbol{\psi}, w)$ on Ω, vanishing on $\partial\Omega$, such that the equilibrium equations (10.1.10) are satisfied in $\Omega \backslash F$, the constitutive relations (10.1.6) hold and along F the Signorini-type conditions (10.1.16), (10.1.17) are fulfilled. This formulation will be called strong.

Let us pass to the variational formulation. First let us define the set

$$K = \{\boldsymbol{v} \in H^1_1(\Omega \backslash F) | \; [\![v_n]\!] \geq 0 \text{ on } F\} \,, \tag{10.1.18}$$

where

$$H^1_1(\Omega \backslash F) = \{\boldsymbol{v} \in H^1(\Omega \backslash F)^2 \mid \boldsymbol{v} = \boldsymbol{0} \text{ on } F\} \,.$$

Note that K is a convex set closed in $H^1(\Omega \backslash F)$. The set of kinematically admissible fields $\boldsymbol{v}$ is given by

$$\boldsymbol{U} \in \mathbb{K} = H^1_0(\Omega)^2 \times H^1_0(\Omega)^2 \times K \times H^1_0(\Omega) \,. \tag{10.1.19}$$

Instead of the bilinear form $b_\Omega(\cdot, \cdot)$ defined on Ω, we shall use the bilinear form:

$$a(\boldsymbol{U}, \boldsymbol{V}) = \int\limits_{\Omega \backslash F} [N^{\alpha\beta}(\boldsymbol{U}) e_{\alpha\beta}(\boldsymbol{s}) + M^{\alpha\beta}(\boldsymbol{U}) \rho_{\alpha\beta}(\boldsymbol{\eta})$$

$$+ L^{\alpha\beta}(\boldsymbol{U}) k_{\alpha\beta}(\boldsymbol{\chi}) + Q^\alpha(\boldsymbol{U}) g_\alpha(v, \boldsymbol{\eta}) + T^\alpha(\boldsymbol{U}) d_\alpha(v, \boldsymbol{\chi})] dx \,, \tag{10.1.20}$$

defined on $\Omega \backslash F$, here $\boldsymbol{N}, \boldsymbol{M}, \boldsymbol{L}, \boldsymbol{Q}, \boldsymbol{T}$ depend on $\boldsymbol{U}$ according to relations (10.1.6) and (10.1.4).

The variational formulation of the boundary value problem for the considered clamped plate transversely loaded and weakened by the crack F reads:

$$(P_{2LF}) \quad \left| \begin{array}{l} \text{find } \boldsymbol{U} \in \mathbb{K} \text{ such that} \\[4pt] a(\boldsymbol{U}, \boldsymbol{V} - \boldsymbol{v}) \geq f(v - w) \qquad \forall \; \boldsymbol{V} \in \mathbb{K} \end{array} \right. \tag{10.1.21}$$

This variational formulation can be derived from the strong formulation given before by using a standard technique known from the theory of three-dimensional bodies with Signorini's cracks. On the other hand, under stronger assumptions the variational formulation implies all conditions involved in the strong formulation. Moreover, it is not difficult to prove that problem (P_{2LF}) is well posed. It is sufficient to prove that the form $a(\cdot,\cdot)$ is coercive. This is a consequence of the Korn inequality being true for the domain $\Omega\backslash F$, cf. Secs. 8 and 9.

10.2. *In-plane scaling and effective model*

This section is aimed at discussing effective properties of transversely homogeneous plates weakened by periodically distributed cracks of the type described in Sec. 10.1. Thus one can choose a rectangular basic cell of periodicity the repetition of which forms a given periodic layout. Assume that dimensions of this cell are proportional to a small parameter ε. Thus the domain $\Omega\backslash F$ is now replaced with $\Omega^\varepsilon = \Omega\backslash F^\varepsilon$, F^ε being a sum of all cracks $F_{\varepsilon,i}$, $i \in \mathcal{I}(\varepsilon)$; $\mathcal{I}(\varepsilon)$ represents the number of cells which lie within Ω. Periodicity cells are εY, $Y = (0, l_1) \times (0, l_2)$ being a rescaled cell of periodicity. Instead of the sets K and $\mathbb{K}$ we shall handle the ε-dependent sets

$$K_\varepsilon = \{v \in H_1^1(\Omega^\varepsilon) \mid [\![v_n]\!] \geq 0 \text{ on } F^\varepsilon\}, \qquad (10.2.1)$$

$$\mathbb{K}_\varepsilon = H_0^1(\Omega)^2 \times H_0^1(\Omega)^2 \times K_\varepsilon \times H_0^1(\Omega)^2. \qquad (10.2.2)$$

The bilinear form $a^\varepsilon(\cdot,\cdot)$ is defined by (10.1.20) with $\Omega\backslash F$ replaced by Ω^ε.

The variational formulation of the εY-periodic problem reads:

$$(P_{2LF}^\varepsilon) \quad \left| \begin{array}{l} \text{find } U^\varepsilon \in \mathbb{K}_\varepsilon \text{ such that} \\ a^\varepsilon(U^\varepsilon, V - U^\varepsilon) \geq f(v - w^\varepsilon) \qquad \forall\ V \in \mathbb{K}_\varepsilon. \end{array} \right. \qquad (10.2.3)$$

From Secs. 8 and 9 we know that the Poincaré and Korn inequalities can be extended to the domain Ω^ε. Thus it is not difficult to prove that problem (P_{2LF}^ε) is uniquely solvable. The parameter ε scales the in-plane dimensions of the periodicity cell, which means that for $\varepsilon \to 0$ the three-dimensional cells of periodicity become infinitely slender in the transverse dimension. This takes place since these transverse dimensions are present in the formulation (P_{2LF}^ε) : they are concealed in the definition of stiffnesses (10.1.8). Thus the effective plate model to be derived will be applicable for plates with densely distributed cracks. This means that crack spacing should be much smaller than the plate thickness.

Prior to performing the asymptotic analysis let us define the set K_{YF} and the space $H_{per}^1(YF)$ as follows

$$K_{YF} = \{v \in H_{per}^1(YF)^2 \mid [\![v_N]\!] \geq 0 \text{ on } F\},$$
$$H_{per}^1(YF) = \{v \in H^1(YF) \mid v \text{ is } Y\text{-periodic}\}. \qquad (10.2.4)$$

Here $\check{\boldsymbol{N}} = (\check{N}^\alpha)$ represents the versor outward normal to F and $v_N = v_\alpha \check{N}^\alpha$. The set K_{YF} is closed and convex in $H_{per}^1(YF)^2$. Now let us represent the solution U^ε to the problem

(P^ε_{2LF}) in the form of a two-scale expansion:

$$f^\varepsilon_\alpha(x) = f^{(0)}_\alpha(x) + \varepsilon f^{(1)}_\alpha(x,y) + \varepsilon^2 f^{(2)}_\alpha(x,y) + \ldots \qquad (10.2.5)$$

where $y = x/\varepsilon$; here f stands for r, φ, ψ and w. All functions $f^{(0)}_\alpha \in H^1_0(\Omega)$ and $f^{(1)}_\alpha(x,\cdot)$ are elements of $H^1_{per}(Y)$, if f stands for (r, φ, w) and $\psi^{(1)}(x,\cdot) \in K_{YF}$.

Similarly we expand the trial functions $s^\varepsilon_\alpha, \eta^\varepsilon_\alpha, \chi^\varepsilon_\alpha, v^\varepsilon$. The functions $s^{(0)}_\alpha, \eta^{(0)}_\alpha, \chi^{(0)}_\alpha$ and $v^{(0)}$ are of class $H^1_0(\Omega)$; $s^{(1)}_\alpha(x,\cdot), \eta^{(1)}_\alpha(x,\cdot), v^{(1)}(x,\cdot)$ belong to $H^1_{per}(Y)$, while $\chi^{(1)}(x,\cdot) \in K_{YF}$.

To find the homogenized problem we substitute representations (10.2.5) into (10.2.3), taking the first terms of (10.2.5) for the trial fields. On passing to zero with ε one finds

$$(P^{2L}_h) \quad \left| \begin{array}{l} \text{find } \boldsymbol{U}^\varepsilon = (\boldsymbol{r}^{(0)}, \boldsymbol{\varphi}^{(0)}, \boldsymbol{\psi}^{(0)}, w^{(0)}) \in V^0_{2L}(\Omega) \text{ such that} \\[2mm] b_h(\boldsymbol{U}^{(0)}, \boldsymbol{V}^{(0)}) = f(v^{(0)}) \qquad \forall\, \boldsymbol{V}^{(0)} = (s^{(0)}, \boldsymbol{\eta}^{(0)}, \boldsymbol{\chi}^{(0)}, v^{(0)}) \in V^0_{2L}(\Omega) \end{array} \right. \qquad (10.2.6)$$

where the bilinear form $b_h(\cdot,\cdot)$ has the form

$$b_h(\boldsymbol{U}^{(0)}, \boldsymbol{V}^{(0)}) = \int\limits_\Omega [N^{\alpha\beta}_h \epsilon_{\alpha\beta}(s^{(0)}) + M^{\alpha\beta}_h \rho_{\alpha\beta}(\boldsymbol{\eta}^{(0)}) + L^{\alpha\beta}_h k_{\alpha\beta}(\boldsymbol{\chi}^{(0)})$$
$$+ Q^\alpha_h g_\alpha(v^{(0)}, \boldsymbol{\eta}^{(0)}) + T^\alpha_h d_\alpha(v^{(0)}, \boldsymbol{\chi}^{(0)})]dx \ . \qquad (10.2.7)$$

The homogenized constitutive relations are given by

$$\boldsymbol{N}_h = \langle \boldsymbol{N}_0 \rangle_{YF} \ , \quad \boldsymbol{M}_h = \langle \boldsymbol{M}_0 \rangle_{YF} \ , \quad \boldsymbol{L}_h = \langle \boldsymbol{L}_0 \rangle_{YF} \ ,$$
$$\boldsymbol{Q}_h = \langle \boldsymbol{Q}_0 \rangle_{YF} \ , \quad \boldsymbol{T}_h = \langle \boldsymbol{T}_0 \rangle_{YF} \ , \qquad (10.2.8)$$

where the brackets $\langle \cdot \rangle_{YF}$ mean averaging over YF (see Eq. (8.2.5)), and

$$N^{\alpha\beta}_0 = A^{\alpha\beta\lambda\mu} \epsilon^h_{\lambda\mu} + A^{\alpha\beta\lambda\mu}_1 \rho^h_{\lambda\mu} + B^{\alpha\beta\lambda\mu}_1 [k^h_{\lambda\mu} + e^y_{\lambda\mu}(\boldsymbol{\psi}^{(1)})] \ ,$$
$$M^{\alpha\beta}_0 = A^{\alpha\beta\lambda\mu}_1 \epsilon^h_{\lambda\mu} + A^{\alpha\beta\lambda\mu}_2 \rho^h_{\lambda\mu} \ ,$$
$$L^{\alpha\beta}_0 = B^{\alpha\beta\lambda\mu}_1 \epsilon^h_{\lambda\mu} + B^{\alpha\beta\lambda\mu}_2 [k^h_{\lambda\mu} + e^y_{\lambda\mu}(\boldsymbol{\psi}^{(1)})] \ , \qquad (10.2.9)$$
$$Q^\alpha_0 = H^{\alpha\beta}_a g^h_\beta \ , \qquad T^\alpha_0 = H^{\alpha\beta}_b d^h_\beta \ .$$

The homogenized generalized strains are given by

$$\epsilon^h_{\alpha\beta} = \epsilon_{\alpha\beta}(\boldsymbol{r}^{(0)}) \ , \qquad \rho^h_{\alpha\beta} = \rho_{\alpha\beta}(\boldsymbol{\varphi}^{(0)}) \ , \qquad k^h_{\alpha\beta} = k_{\alpha\beta}(\boldsymbol{\psi}^{(0)}) \ ,$$
$$g^h_\alpha = w^{(0)}_{,\alpha} + \varphi^{(0)}_\alpha \ , \qquad d^h_\alpha = w^{(0)}_{,\alpha} + \psi^{(0)}_\alpha \ . \qquad (10.2.10)$$

We recall that $e^y_{\alpha\beta}(\boldsymbol{v}) = \left(\dfrac{\partial v_\alpha}{\partial y_\beta} + \dfrac{\partial v_\beta}{\partial y_\alpha} \right)/2$.

The field $\psi^{(1)}$ involved in Eqs. (10.2.9) is a solution to the following local problem:

$$(P_{2L,Y}) \quad \left| \begin{array}{l} \text{find } \psi^{(1)} \in K_{YF} \text{ such that} \\ C^{\alpha\beta\lambda\mu} \langle e^y_{\alpha\beta}(\psi^{(1)}) e^y_{\lambda\mu}(t - \psi^{(1)}) \rangle_{YF} \geq l(t - \psi^{(1)}) \qquad \forall\, t \in K_{YF} \,, \end{array} \right. \quad (10.2.11)$$

where

$$l(t) = -C^{\alpha\beta\lambda\mu} \left(\frac{3}{2b} \epsilon^h_{\alpha\beta} + k^h_{\alpha\beta} \right) \langle e^y_{\alpha\beta}(t) \rangle_{YF} \,. \tag{10.2.12}$$

We have used here the property of transverse homogeneity of the plate, see relations (10.1.8). It turns out that fields $r^{(1)}$, $\varphi^{(1)}$ and $w^{(1)}$ do not depend on y and hence do not affect the effective constitutive relations (10.2.8). Note, moreover, that the non-linear terms of these constitutive relations depend on the function $[\![\psi^{(1)}]\!](\epsilon^h, k^h)$ defining a jump of $\psi^{(1)}$ across F. Indeed, by (10.2.8) – (10.2.12) one finds the constitutive relations

$$N^{\alpha\beta}_h = \tilde{C}^{\alpha\beta\lambda\mu}\{h\epsilon^h_{\lambda\mu} - \frac{a^2}{2}\rho^h_{\lambda\mu} + \frac{b^2}{2}[k^h_{\lambda\mu} - k^F_{\lambda\mu}(\epsilon^h, k^h)]\} \,,$$

$$M^{\alpha\beta}_h = \tilde{C}^{\alpha\beta\lambda\mu}[-\frac{a^2}{2}\epsilon^h_{\lambda\mu} + \frac{a^3}{3}\rho^h_{\lambda\mu}] \,,$$

$$L^{\alpha\beta}_h = \tilde{C}^{\alpha\beta\lambda\mu}\{\frac{b^2}{2}\epsilon^h_{\lambda\mu} + \frac{b^3}{3}[k^h_{\lambda\mu} - k^F_{\lambda\mu}(\epsilon^h, k^h)]\} \,,$$

$$Q^\alpha_h = k_s a C^{\alpha 3\beta 3} g^h_\beta \,, \qquad T^\alpha_h = k_s b C^{\alpha 3\beta 3} d^h_\beta \,,$$

$$\tag{10.2.13}$$

involving the crack deformation measures

$$k^F_{\alpha\beta}(\epsilon^h, k^h) = -\langle e^y_{\alpha\beta}(\psi^{(1)}(\epsilon^h, k^h)) \rangle_{YF} \,, \tag{10.2.14}$$

which can equivalently be put in the form

$$k^F_{\alpha\beta} = \frac{1}{2l_1 l_2} \int_F ([\![\psi^{(1)}_\alpha]\!]\check{N}_\beta + [\![\psi^{(1)}_\beta]\!]\check{N}_\alpha)\,ds \,. \tag{10.2.15}$$

This proves that the jump $[\![\psi^{(1)}]\!]$ determines the nonlinear part of the homogenized constitutive relationships. The tensor k^F describes the opening of the crack F. Note, moreover, that opening of the crack affects only the constitutive relations for N_h and L_h. Coupling between those relations follows from transverse asymmetry of the problem.

One can show that the effective plate deforms like a nonlinear and hyperelastic structure. Its hyperelastic potential is given by

$$\mathcal{W} = \frac{1}{2}\langle N^{\alpha\beta}_0 \epsilon^h_{\alpha\beta} + M^{\alpha\beta}_0 \rho^h_{\alpha\beta} + L^{\alpha\beta}_0 (k^h_{\alpha\beta} + e^y_{\alpha\beta}(\psi^{(1)})) + Q^\alpha_0 g^h_\alpha + T^\alpha_0 d^h_\alpha \rangle \,, \tag{10.2.16}$$

or, using (10.2.13) – (10.2.15), one finds

$$
\mathcal{W} = \frac{1}{2}\tilde{C}^{\alpha\beta\lambda\mu}\left[h\epsilon^h_{\alpha\beta}\epsilon^h_{\lambda\mu} + \frac{a^3}{3}\rho^h_{\alpha\beta}\rho^h_{\lambda\mu} + \frac{b^3}{3}\langle(k^h_{\alpha\beta} + e^y_{\alpha\beta}(\psi^{(1)}))(k^h_{\lambda\mu} + e^y_{\lambda\mu}(\psi^{(1)}))\rangle_{YF} \right.
$$

$$
\left. + \; e^h_{\alpha\beta}[b^2(k^h_{\lambda\mu} - k^F_{\lambda\mu}) - a^2\rho^h_{\lambda\mu}]\right] + \frac{1}{2}k_s C^{\alpha3\beta3}(ag^h_\alpha g^h_\beta + bd^h_\alpha d^h_\beta)\,. \tag{10.2.17}
$$

The potential $\mathcal{W}$ satisfies the following condition: there exist positive constants c_0, c_1 such that

$$
c_0(|e|^2 + |\rho|^2 + |k|^2 + |g|^2 + |d|^2)
$$
$$
\leq \mathcal{W}(e,\rho,k,g,d) \leq c_1(|e|^2 + |\rho|^2 + |k|^2 + |g|^2 + |d|^2) \tag{10.2.18}
$$

for each $e,\rho,k \in \mathbb{E}^2_s$ and for each $g,d \in \mathbb{R}^2$. Moreover, $\mathcal{W}$ is of class C^1 and

$$
N^{\alpha\beta}_h = \frac{\partial\mathcal{W}}{\partial\epsilon^h_{\alpha\beta}}\,, \qquad M^{\alpha\beta}_h = \frac{\partial\mathcal{W}}{\partial\rho^h_{\alpha\beta}}\,, \qquad L^{\alpha\beta}_h = \frac{\partial\mathcal{W}}{\partial k^h_{\alpha\beta}}\,,
$$
$$
Q^\alpha_h = \frac{\partial\mathcal{W}}{\partial g^h_\alpha}\,, \qquad T^\alpha_h = \frac{\partial\mathcal{W}}{\partial d^h_\alpha}\,, \tag{10.2.19}
$$

which proves that the effective plate is hyperelastic. For transversely homogeneous material the potential $\mathcal{W}$ can equivalently be represented in the form

$$
\mathcal{W}(\epsilon,\rho,k,g,d) = \frac{1}{2}\tilde{C}^{\alpha\beta\lambda\mu}(h\epsilon_{\alpha\beta}\epsilon_{\lambda\mu} + \frac{a^3}{3}\rho_{\alpha\beta}\rho_{\lambda\mu} - a^2\epsilon_{\alpha\beta}\rho_{\lambda\mu})
$$
$$
+ \frac{1}{2}k_s C^{\alpha3\beta3}(ag_\alpha g_\beta + bd_\alpha d_\beta) + \mathcal{W}_1(\epsilon,k)\,, \tag{10.2.20}
$$

where

$$
\mathcal{W}_1(\epsilon,k) \doteq \min_{v\in K_{YF}} \langle j_1(\epsilon, e^y(v) + k)\rangle_{YF}\,, \tag{10.2.21}
$$

and

$$
j_1(\epsilon,k) = \frac{1}{2}C^{\alpha\beta\lambda\mu}\left(\frac{b^3}{3}k_{\alpha\beta}k_{\lambda\mu} + b^2\epsilon_{\alpha\beta}k_{\lambda\mu}\right)\,. \tag{10.2.22}
$$

We observe that the convex minimization problem in (10.2.21) is equivalent to solving the variational inequality (10.2.11) of the local problem $(P_{2L,Y})$. Representation (10.2.20) is a convenient starting point for proving the estimates (10.2.18).

By virtue of introducing the potential $\mathcal{W}$ one can represent the effective problem (P^{2L}_h), see Eq. (10.2.6), as a minimization problem

$$
(\tilde{P}^{2L}_h) \quad\left|\;\begin{aligned}&\inf\{\int_\Omega \mathcal{W}(\epsilon(s),\rho(\eta),k(\chi),g(v,\eta),d(v,\chi))dx\\[4pt]&\qquad\qquad -f(v)\,|\,(s,\eta,\chi,v) \in V^0_{2L}(\Omega)\}\,.\end{aligned}\right. \tag{10.2.23}
$$

Taking account of the properties of the macroscopic potential $\mathcal{W}$ given above we conclude that problems (P^{2L}_h) and $(\tilde{P}^{2L}_h)$ are equivalent and uniquely solvable; $U^{(0)} = (r^{(0)}, \varphi^{(0)}, \psi^{(0)}, w^{(0)}) \in V^0_{2L}(\Omega)$ exists and is unique.

10.3. The study of convergence

For a fixed $\varepsilon > 0$ the functional J_ε of the total potential energy may be written in the form

$$J_\varepsilon(\boldsymbol{r}, \boldsymbol{\varphi}, \boldsymbol{\psi}, w) = G_1^\varepsilon(\boldsymbol{r}, \boldsymbol{\varphi}, \boldsymbol{\psi}) + \Phi(\boldsymbol{\varphi}, \boldsymbol{\psi}, w) - f(w) \,, \tag{10.3.1}$$

where

$$\begin{aligned}
G_1^\varepsilon(\boldsymbol{r}, \boldsymbol{\varphi}, \boldsymbol{\psi}) = \frac{1}{2} \int_\Omega &[A^{\alpha\beta\lambda\mu} e_{\alpha\beta}(\boldsymbol{r}) e_{\lambda\mu}(\boldsymbol{r}) + A_1^{\alpha\beta\lambda\mu} e_{\alpha\beta}(\boldsymbol{r}) \rho_{\lambda\mu}(\boldsymbol{\varphi}) \\
&+ M^{\alpha\beta}(\boldsymbol{r}, \boldsymbol{\varphi}) \rho_{\alpha\beta}(\boldsymbol{\varphi})] dx + \frac{1}{2} \int_{\Omega^\varepsilon} [B_2^{\alpha\beta\lambda\mu} k_{\alpha\beta}(\boldsymbol{\psi}) k_{\lambda\mu}(\boldsymbol{\psi}) \\
&+ 2 B_1^{\alpha\beta\lambda\mu} e_{\alpha\beta}(\boldsymbol{r}) k_{\lambda\mu}(\boldsymbol{\psi})] dx \,,
\end{aligned} \tag{10.3.2}$$

$$\Phi(\boldsymbol{\varphi}, \boldsymbol{\psi}, w) = \frac{1}{2} \int_\Omega [Q^\alpha(w, \boldsymbol{\varphi}) g_\alpha(w, \boldsymbol{\varphi}) + T^\alpha(w, \boldsymbol{\psi}) d_\alpha(w, \boldsymbol{\psi})] dx \,, \tag{10.3.3}$$

and f is given by (10.1.9a).

We observe that the functional $\Phi - f$, being continuous in the strong topology of $L^2(\Omega)^2$ is a perturbation functional. Consequently we have to solve the problem of the Γ-convergence of the sequence of functionals $\{J_\varepsilon(\boldsymbol{r}, \boldsymbol{\varphi}, \cdot, w)\}_{\varepsilon>0}$. Thus in the present case this convergence will concern only the integral over Ω^ε.

We set

$$j(\boldsymbol{\epsilon}, \boldsymbol{\rho}, \boldsymbol{k}) = \frac{1}{2} [\boldsymbol{\epsilon}, \boldsymbol{\rho}, \boldsymbol{k}] \mathbf{A} [\boldsymbol{\epsilon}, \boldsymbol{\rho}, \boldsymbol{k}]^T = j_0(\boldsymbol{\epsilon}, \boldsymbol{\rho}) + j_1(\boldsymbol{\epsilon}, \boldsymbol{k}) \,, \tag{10.3.4}$$

where

$$j_0(\boldsymbol{\epsilon}, \boldsymbol{\rho}) = \frac{1}{2} (A^{\alpha\beta\lambda\mu} \epsilon_{\alpha\beta} \epsilon_{\lambda\mu} + A_1^{\alpha\beta\lambda\mu} \epsilon_{\alpha\beta} \rho_{\lambda\mu} + M^{\alpha\beta} \rho_{\alpha\beta}) \,, \tag{10.3.5}$$

$$j_1(\boldsymbol{\epsilon}, \boldsymbol{k}) = \frac{1}{2} (B_2^{\alpha\beta\lambda\mu} k_{\alpha\beta} k_{\lambda\mu} + 2 B_1^{\alpha\beta\lambda\mu} \epsilon_{\alpha\beta} k_{\lambda\mu}) \,. \tag{10.3.6}$$

Here $(\boldsymbol{\epsilon}, \boldsymbol{\rho}, \boldsymbol{k}) \in \mathbb{E}_s^2$ and $M^{\alpha\beta}$ is given by $(10.1.6)_2$. For a plate made of an inhomogeneous material j_0 and j_1 depend additionally on $x \in \Omega$ because $\mathbf{A}$ does. In the case of transversely homogeneous plates j_1 is given by (10.2.22). The partial stored energy function j is non-negative: $j \geq 0$. We observe also that if $\boldsymbol{\epsilon}$ is prescribed then j_1 is a sum of a positive definite quadratic function and of a linear function.

Let us set

$$G^\varepsilon(\boldsymbol{r}, \boldsymbol{\varphi}, \boldsymbol{\psi}) = \begin{cases} G_1^\varepsilon(\boldsymbol{r}, \boldsymbol{\varphi}, \boldsymbol{\psi}) & \text{if } \boldsymbol{r}, \boldsymbol{\varphi} \in H^1(\Omega)^2 \,, \ \boldsymbol{\psi} \in K_\varepsilon \\ +\infty & \text{otherwise.} \end{cases} \tag{10.3.7}$$

The main result of this section is formulated in the form of the following theorem, where $\mathcal{W}_1$ is given by (10.2.21). In the general case j_1 is defined by (10.3.6).

Theorem 10.3.1. The sequence of functionals $\{G^\varepsilon(r, \varphi, \cdot)\}_{\varepsilon > 0}$ is Γ-convergent in the strong topology of $L^2(\Omega)^2$ to

$$G(r, \varphi, \psi) = \begin{cases} \displaystyle\int_\Omega j_0[e(r), \rho(\varphi)]dx + \int_\Omega \mathcal{W}_1[e(r), k(\psi)]dx \\ \qquad\qquad\qquad\qquad \text{if } r, \varphi, \psi \in H^1(\Omega)^2 \, ; \\ +\infty \qquad\qquad\qquad\quad \text{otherwise.} \end{cases} \qquad (10.3.8)$$

Proof. It will be divided into four steps.

In steps 1 and 2 we shall prove that for any $\psi \in H^1(\Omega)^2$ there exists a sequence $\{\psi^\varepsilon\}_{\varepsilon > 0} \subset K_\varepsilon$ strongly convergent to ψ in the topology of $L^2(\Omega)^2$ such that

$$G(r, \varphi, \psi) \geq \lim_{\varepsilon \to 0} \sup G^\varepsilon(r, \varphi, \psi^\varepsilon) \, . \qquad (10.3.9)$$

Step 1. Let $\{\Omega_K\}_{K \in \mathcal{K}}$ be a finite partition of Ω by polygonal sets. Such a partition enables us to exploit the local character of the functionals $G^\varepsilon(r, \varphi, \cdot)$. In fact the Lebesgue measure of $\Omega \backslash \sum_{K \in \mathcal{K}} \Omega_K$ tends to zero when $\varepsilon \to 0$. We set

$$\Omega_K^\delta = \{x \in \Omega_K| \text{ dist } (x, \partial\Omega_K) > \delta\} \, , \quad \delta > 0 \, ,$$

and let $\varphi_K^\delta \in \mathbf{D}(\Omega_K)$ be such that $0 \leq \varphi_K^\delta \leq 1, \varphi_{K|_{\Omega_K^\delta}}^\delta = 1$.

Let $\psi \in H^1(\Omega)^2$ be a piecewise affine continuous function:

$$\psi_\alpha(x) = k_{\alpha\beta}^K x_\beta + a_\alpha^K \, , \quad k^K \in \mathbb{E}_s^2 \, , \quad a^K \in \mathbb{R}^2 \, , \quad \forall\, x \in \Omega_K \, . \qquad (10.3.10)$$

Hence we obviously have

$$k(\psi(x)) = e(\psi(x)) = k^K \, , \quad x \in \Omega_K \, , \quad K \in \mathcal{K} \, . \qquad (10.3.11)$$

With every family of functions $\{v^K\}_{K \in \mathcal{K}} \subset K_{YF}$ we associate the sequence

$$\psi^{\varepsilon,\delta}(x) = \psi(x) + \varepsilon\varphi_K^\delta(x)v^K\left(\frac{x}{\varepsilon}\right) \, , \qquad \varepsilon > 0 \, , \qquad (10.3.12)$$

where the summation convention obviously applies to K. Since

$$\int_\Omega \left|v^K\left(\frac{x}{\varepsilon}\right)\right|^2 dx \leq n_0^2 \int_Y \left|v^K(y)\right|^2 dy \, ,$$

where $n_0(\varepsilon)$ is sufficiently large, therefore

$$\psi^{\varepsilon,\delta} \to \psi \qquad \text{in} \qquad L^2(\Omega)^2 \text{ strongly.}$$

From Eq. (10.3.12) we conclude that $\psi^{\varepsilon,\delta} \in K_\varepsilon$ because

$$[\psi_n^{\varepsilon,\delta}]_{F^\varepsilon} = \varepsilon\varphi_K^\delta[v_n^K]_{F^\varepsilon} \geq 0 \, . \qquad (10.3.13)$$

Let $t < 1$ (intended to go to 1) and set $\Omega_K^\varepsilon = \Omega^\varepsilon \cap \Omega_K$. Recalling that $t\varphi_K^\delta + t(1 - \varphi_K^\delta) + (1 - t) = 1$ and using the convexity of $j_1(e(r), \cdot)$ we obtain

$$G^\varepsilon(r, \varphi, t\psi^{\varepsilon,\delta}) = \sum_{K \in \mathcal{K}} \int_{\Omega_K^\varepsilon} j[e(r(x)), \rho(\varphi(x)), t\varphi_K^\delta(x)\left(k^K + k(v^K)\left(\frac{x}{\varepsilon}\right)\right)$$

$$+ t(1 - \varphi_K^\delta(x))k^K + (1 - t)\frac{\varepsilon t}{1 - t}\mathrm{sym}\left(v^K\left(\frac{x}{\varepsilon}\right) \otimes \nabla\varphi_K^\delta(x))\right]dx$$

$$\leq \sum_{K \in \mathcal{K}} \int_{\Omega_K^\varepsilon} j\left[e(r(x)), \rho(\psi(x)), k^K + k(v^K)\left(\frac{x}{\varepsilon}\right)\right] dx$$

$$+ \int_{\Omega_K^\varepsilon} (1 - \varphi_K^\delta(x))j[e(r(x)), \rho(\varphi(x)), k^K]dx$$

$$+ (1 - t)\int_{\Omega_K^\varepsilon} j[e(r(x)), \rho(\varphi(x)), \frac{\varepsilon t}{1 - t}\mathrm{sym}\left(v^K\left(\frac{x}{\varepsilon}\right) \otimes \nabla\varphi_K^\delta(x))]dx$$

$$\leq \sum_{K \in \mathcal{K}} \{\int_{\Omega_K^\varepsilon} j[e(r(x)), \rho(\varphi(x)), k^K + k(v^K)\left(\frac{x}{\varepsilon}\right)]dx$$

$$+ c_1 \int_{\Omega_K^\varepsilon} (1 - \varphi_K^\delta(x))[|e(r(x))|^2 + |\rho(\varphi(x))|^2 + |k^K|^2]dx + c_1(1 - t)\int_{\omega_K^\varepsilon} [|e(r(x))|^2$$

$$+ |\rho(\varphi(x))|^2 + |\frac{\varepsilon t}{1 - t}\mathrm{sym}\left(v^K\left(\frac{x}{\varepsilon}\right) \otimes \nabla\varphi_K^\delta(x))|^2]dx\}.$$

Here we have used the fact that $j \geq 0$ and the following notation

$$[\mathrm{sym}(v^K \otimes \nabla\varphi_K^\delta)] = \frac{1}{2}\left(v_\alpha^K\frac{\partial\varphi_K^\delta}{\partial x_\beta} + v_\beta^K\frac{\partial\varphi_K^\delta}{\partial x_\alpha}\right) \quad \text{(no summation on } K\text{).} \qquad (10.3.14)$$

The function $j[e(r), \rho(\varphi), k^K + k(v^K)\left(\frac{\cdot}{\varepsilon}\right)]$ is εY-periodic, hence by applying Theorem 1.1.5 we arrive at the following inequality

$$\limsup_{\substack{\delta \to 0 \\ t \to 1^-}} \limsup_{\varepsilon \to 0} G^\varepsilon(r, \varphi, t\psi^{\varepsilon,\delta})$$

$$\leq \sum_{K \in \mathcal{K}} \int_{\Omega_K} \frac{1}{|Y|}\int_{YF} j[e(r(x)), \rho(\varphi(x)), k^K + e^y(v^K(y))]dy\,dx . \qquad (10.3.15)$$

By applying Lemma 1.3.27 we construct a mapping $\varepsilon \to (t(\varepsilon), \delta(\varepsilon))$ with $(t(\varepsilon), \delta(\varepsilon)) \to (1^-, 0)$ as $\varepsilon \to 0$. Now we set

$$\psi^\varepsilon = t(\varepsilon)\psi^{\varepsilon,\delta(\varepsilon)} .$$

Thus $\psi^\varepsilon \to \psi$ strongly in $L^2(\Omega)^2$ when $\varepsilon \to 0$. Consequently (10.3.15) yields

$$\limsup_{\varepsilon \to 0} G^\varepsilon(\boldsymbol{r}, \boldsymbol{\varphi}, \psi^\varepsilon) \leq \limsup_{\substack{\delta \to 0 \\ t \to 1^-}} \limsup_{\varepsilon \to 0} G^\varepsilon(\boldsymbol{r}, \boldsymbol{\varphi}, t\psi^{\varepsilon,\delta})$$

$$\leq \sum_{K \in \mathcal{K}} \int_{\Omega_K} \frac{1}{|Y|} \int_{YF} j[e(\boldsymbol{r}(x)), \boldsymbol{\rho}(\boldsymbol{\varphi}(x)), \boldsymbol{k}^K + e^y(\boldsymbol{v}^K(y))] dy\, dx \ .$$

Taking now the infimum on the r.h.s. of the last inequality when $\boldsymbol{v}^K$ runs over the set K_{YF} we arrive at the relation

$$G^s(\boldsymbol{r}, \boldsymbol{\varphi}, \psi) \leq \limsup_{\varepsilon \to 0} G^\varepsilon(\boldsymbol{r}, \boldsymbol{\varphi}, \psi) \tag{10.3.16}$$

$$\leq \int_\Omega j_0[(e(\boldsymbol{r}(x)), \boldsymbol{\rho}(\boldsymbol{\varphi}(x))) + \mathcal{W}_1(e(\boldsymbol{r}(x)), \boldsymbol{k}(\psi(x)))] dx \ .$$

Step 2. For each $\boldsymbol{\epsilon}, \boldsymbol{\theta}, \boldsymbol{\chi} \in \mathbb{E}_s^2$ we have $j_0(\boldsymbol{\epsilon}, \boldsymbol{\theta}) + \mathcal{W}_1(\boldsymbol{\epsilon}, \boldsymbol{\chi}) \leq c_1(|\boldsymbol{\epsilon}|^2 + |\boldsymbol{\theta}|^2 + |\boldsymbol{\chi}|^2)$, where c_1 is a positive constant. Hence we obtain

$$G^s(\boldsymbol{r}, \boldsymbol{\varphi}, \psi) \leq c_1 \int_\Omega (|e(\boldsymbol{r}(x))|^2 + |\boldsymbol{\rho}(\boldsymbol{\varphi}(x))|^2 + |\boldsymbol{k}(\psi(x))|^2) dx \ ,$$

for each $\psi \in H^1(\Omega)^2$. Being convex and finite the functional $G^s(\boldsymbol{r}, \boldsymbol{\varphi}, \cdot)$ is continuous on $H^1(\Omega)^2$. Consequently, by virtue of density of piecewise affine continuous functions in $H^1(\Omega)$, the inequality $G^s(\boldsymbol{r}, \boldsymbol{\varphi}, \psi) \leq G(\boldsymbol{r}, \boldsymbol{\varphi}, \psi)$ can be extended to the whole space $H^1(\Omega)^2$.

We now pass to proving that for any sequence $\{\psi^\varepsilon\}_{\varepsilon > 0} \subset K_\varepsilon$ such that $\psi^\varepsilon \xrightarrow[\varepsilon \to 0]{} \psi$ in $L^2(\Omega)^2$ strongly, the following inequality is satisfied

$$G(\boldsymbol{r}, \boldsymbol{\varphi}, \psi^\varepsilon) \leq \liminf_{\varepsilon \to 0} G^\varepsilon(\boldsymbol{r}, \boldsymbol{\varphi}, \psi^\varepsilon) \ . \tag{10.3.17}$$

Step 3. First, let us show that if ψ is an affine function:

$$\psi(x) = \boldsymbol{\chi}x + \boldsymbol{a} \ , \quad \boldsymbol{\chi} \in \mathbb{E}_s^2, \quad \boldsymbol{a} \in \mathbb{R}^2 \tag{10.3.18}$$

then

$$\lim_{\varepsilon \to 0} G^\varepsilon(\boldsymbol{r}, \boldsymbol{\varphi}, \psi^\varepsilon) = G(\boldsymbol{r}, \boldsymbol{\varphi}, \psi) \ , \tag{10.3.19}$$

where

$$\psi^\varepsilon(x) = \psi(x) + \varepsilon\psi_{(\boldsymbol{\epsilon}, \boldsymbol{\chi})}\left(\frac{x}{\varepsilon}\right) \ . \tag{10.3.20}$$

Here $\psi_{(\boldsymbol{\epsilon}, \boldsymbol{\chi})}$ is a solution to the following local problem:

$$\inf\{\frac{1}{|Y|} \int_{YF} j_1[\boldsymbol{\epsilon}, \boldsymbol{\chi} + e^y(\boldsymbol{v})] dy \mid \boldsymbol{v} \in K_{YF}\} \ . \tag{10.3.21}$$

We immediately obtain

$$\lim_{\varepsilon \to 0} G^{\varepsilon}(\boldsymbol{r}, \boldsymbol{\varphi}, \boldsymbol{\psi}^{\varepsilon}) = \int_{\Omega} [j_0(\boldsymbol{e}(\boldsymbol{r}(x)), \boldsymbol{\rho}(\boldsymbol{\varphi}(x))) + \mathcal{W}_1(\boldsymbol{e}(\boldsymbol{r}(x)), \boldsymbol{\chi})]dx .$$

Equation (10.3.18) gives $\boldsymbol{k}(\boldsymbol{\psi}) = \boldsymbol{\chi}$, and thus

$$\lim_{\varepsilon \to 0} G^{\varepsilon}(\boldsymbol{r}, \boldsymbol{\varphi}, \boldsymbol{\psi}^{\varepsilon}) = \int_{\Omega} [j_0(\boldsymbol{e}(\boldsymbol{r}(x)), \boldsymbol{\rho}(\boldsymbol{\varphi}(x))) + \mathcal{W}_1(\boldsymbol{e}(\boldsymbol{r}(x)), \boldsymbol{k}(\boldsymbol{\psi}(x)))]dx = G(\boldsymbol{r}, \boldsymbol{\varphi}, \boldsymbol{\psi}) .$$

Step 4. Let $\{\boldsymbol{\psi}^{\varepsilon}\}_{\varepsilon>0} \subset K_{\varepsilon}$ be a sequence of functions strongly convergent in $L^2(\Omega)^2$ to a certain $\boldsymbol{\psi}$. We take $\boldsymbol{q}$, a continuous affine function:

$$\boldsymbol{q}(x) = \boldsymbol{\chi}^K x + \boldsymbol{a} , \qquad \boldsymbol{\chi}^K \in \mathbb{E}_s^2 , \quad \boldsymbol{a} \in \mathbb{R}^2 , \ x \in \Omega_K . \qquad (10.3.22)$$

Let us denote by $\tilde{\boldsymbol{v}}^K (K \in \mathcal{K})$ a solution to the problem (10.3.21) with $\boldsymbol{\chi}$ being replaced by $\boldsymbol{\chi}^K$. Obviously, $\tilde{\boldsymbol{v}}^K$ depends on ϵ and $\boldsymbol{\chi}^K$. We set

$$\boldsymbol{q}^{\varepsilon,K}(x) = \boldsymbol{q}(x) + \varepsilon \tilde{\boldsymbol{v}}^K \left(\frac{x}{\varepsilon}\right) , \qquad K \in \mathcal{K} . \qquad (10.3.23)$$

Clearly, for each $K \in \mathcal{K}$ we have: $\boldsymbol{q}^{\varepsilon,K} \to \boldsymbol{q}$ strongly in $L^2(\Omega)^2$ as $\varepsilon \to 0$.

Next, let us introduce a function $\varphi_K \in \mathbf{D}(\Omega_K)$ such that $0 \leq \varphi_K(x) \leq 1$ for $x \in \Omega_K$. Since $j \geq 0$ therefore we may write

$$\int_{\Omega} j_0(\boldsymbol{e}(\boldsymbol{r}), \boldsymbol{\rho}(\boldsymbol{\varphi}))dx + \int_{\Omega^{\varepsilon}} j_1(\boldsymbol{e}(\boldsymbol{r}), \boldsymbol{k}(\boldsymbol{\psi}^{\varepsilon}))dx$$

$$\geq \sum_{K \in \mathcal{K}} \int_{\Omega_K} \varphi_K(x) j_0(\boldsymbol{e}(\boldsymbol{r}), \boldsymbol{\rho}(\boldsymbol{\varphi}))dx + \sum_{K \in \mathcal{K}} \int_{\Omega_K^{\varepsilon}} \varphi_K(x) j_1(\boldsymbol{e}(\boldsymbol{r}), \boldsymbol{k}(\boldsymbol{\psi}^{\varepsilon}))dx . \quad (10.3.24)$$

Applying the subdifferential inequality to the function $j(\boldsymbol{e}(\boldsymbol{r}), \boldsymbol{\rho}(\boldsymbol{\varphi}), \cdot)$ we obtain

$$\sum_{K \in \mathcal{K}} \int_{\Omega_K} \varphi_K(x) j_0(\boldsymbol{e}(\boldsymbol{r}), \boldsymbol{\rho}(\boldsymbol{\varphi}))dx + \sum_{K \in \mathcal{K}} \int_{\Omega_K^{\varepsilon}} \varphi_K(x) j_1(\boldsymbol{e}(\boldsymbol{r}), \boldsymbol{k}(\boldsymbol{\psi}^{\varepsilon}))dx$$

$$- \sum_{K \in \mathcal{K}} \int_{\Omega_K} \varphi_K(x) j_0(\boldsymbol{e}(\boldsymbol{r}), \boldsymbol{\rho}(\boldsymbol{\varphi}))dx - \sum_{K \in \mathcal{K}} \int_{\Omega_K^{\varepsilon}} \varphi_K(x) j_1(\boldsymbol{e}(\boldsymbol{r}), \boldsymbol{k}(\boldsymbol{q}^{\varepsilon,K}))dx$$

$$\geq \sum_{K \in \mathcal{K}} \int_{\Omega_K} \varphi_K(x) D^{\alpha\beta}[\boldsymbol{e}(\boldsymbol{r}), \boldsymbol{k}(\boldsymbol{q}^{\varepsilon,K})] k_{\alpha\beta}(\boldsymbol{\psi}^{\varepsilon} - \boldsymbol{q}^{\varepsilon,K})dx , \qquad (10.3.25)$$

where

$$D^{\alpha\beta}[\boldsymbol{e}(\boldsymbol{r}), \boldsymbol{k}(\boldsymbol{q}^{\varepsilon,K})] = [D_2 j_1(\boldsymbol{e}(\boldsymbol{r}), \boldsymbol{k}(\boldsymbol{q}^{\varepsilon,K}))]^{\alpha\beta} .$$

Here $D_2 j_1(\boldsymbol{e}(\boldsymbol{r}), \cdot)$ stands for the gradient of the function $j_1(\boldsymbol{e}(\boldsymbol{r}), \cdot)$.

By using (10.3.19) we arrive at

$$\lim_{\varepsilon \to 0}[\int_{\Omega_K} \varphi_K j_0(e(r), \rho(\varphi))dx + \int_{\Omega_K^\varepsilon} \varphi_K j_1(e(r), k(q^{\varepsilon,K}))dx]$$

$$= \int_{\Omega_K} \varphi_K [j_0(e(r), \rho(\varphi)) + \mathcal{W}_1(e(r), k(q))]dx , \quad K \in \mathcal{K} . \tag{10.3.26}$$

Performing integration by part on the r.h.s. of (10.3.25) we readily obtain (no summation over K!)

$$\int_{\Omega_K^\varepsilon} \varphi_K D^{\alpha\beta}(e(r), k(q^{\varepsilon,K}))k_{\alpha\beta}(\psi^\varepsilon - q^{\varepsilon,K})dx = A_K^\varepsilon + B_K^\varepsilon + C_K^\varepsilon , \tag{10.3.27}$$

where

$$A_K^\varepsilon = -\int_{\Omega_K^\varepsilon} \varphi_K [D^{\alpha\beta}(e(r), k(q^{\varepsilon,K}))]_{,\beta}(\psi_\alpha^\varepsilon - q_\alpha^{\varepsilon,K})dx , \tag{10.3.28}$$

$$B_K^\varepsilon = -\int_{\Omega_K^\varepsilon} D^{\alpha\beta}(e(r), k(q^{\varepsilon,K}))\varphi_{K,\beta}(\psi_\alpha^\varepsilon - q_\alpha^{\varepsilon,K})dx , \tag{10.3.29}$$

$$C_K^\varepsilon = \int_{F_K^\varepsilon} \varphi_K [\![(\psi_\alpha^\varepsilon - q_\alpha^{\varepsilon,K})D^{\alpha\beta}(e(r), k(q^{\varepsilon,K}))n_\beta]\!]ds . \tag{10.3.30}$$

Here $F_K^\varepsilon = F^\varepsilon \cap \Omega_K$.

Before proceeding further, some localization considerations are indispensable. The local minimization problem appearing in (10.2.21) with j_1 given by (10.3.6) is equivalent to:

$$\left|\begin{array}{l} \text{find } \tilde{v}^K \in K_{YF} \text{ such that} \\[2mm] \int_{YF} D^{\alpha\beta}[e(r), \chi^K + e^y(v^K)]e_{\alpha\beta}^y(v - \tilde{v}^K)dy \geq 0 \quad \forall\, v \in K_{YF} . \end{array}\right. \tag{10.3.31}$$

Taking $v = \tilde{v}^K + \varphi$, $\varphi \in K_{YF}$, from the last inequality we obtain

$$\int_{YF} D^{\alpha\beta}[e(r), \chi^K + e^y(\tilde{v}^K)]e_{\alpha\beta}^y(\varphi)dy \geq 0 \quad \forall\, \varphi \in K_{YF} . \tag{10.3.32}$$

Particularly, one can take $\varphi \in \mathbf{D}(YF)^2$, that is φ vanishes in a vicinity of $\partial(YF) = \partial Y \cup F$. In such a case the inequality (10.3.32) yields

$$-\frac{\partial}{\partial y_\beta}D^{\alpha\beta}(e(r), \chi^K + e^y(\tilde{v}^K)) = 0 \quad \text{in} \quad \mathbf{D}'(YF)^2 . \tag{10.3.33}$$

Let us consider (10.3.32) once again and take $\varphi \in K_{YF}$ such that $\varphi = (0,0)$ in a neighborhood of F. Integrating by parts and taking (10.3.33) into account we conclude that

$$(D^{\alpha\beta}(e(r), \chi^K + e^y(v^K))\check{N}_\beta) \qquad \text{is } Y\text{-antiperiodic.} \tag{10.3.34}$$

Consequently (10.3.33) can be extended to $\mathbb{R}^2 \setminus \cup [F + (m_1 y_1, m_2 y_2)]$, where $m_1, m_2 \in \mathbb{Z}$, and $\mathbb{Z}$ stands for the set of integers.

Let us return to the variational inequality (10.3.31). Integrating by parts and taking (10.3.33) and (10.3.34) into account we get

$$\int_F [\![(v_\alpha - \tilde{v}_\alpha^K)D^{\alpha\beta}(e(r), e^y(\tilde{v}^K) + \chi^K)\check{N}_\beta]\!]ds \geq 0 \quad \forall v \in K_{YF}. \tag{10.3.35}$$

Let us take $v = (1 - \varphi)(\tilde{v}^K + \chi^K y) + \varphi\xi - \chi^K y$, where $\xi \in K_{YF}$ and $\varphi \in \mathbf{D}(Y)$ with $0 \leq \varphi \leq 1$. Obviously $[\![v_N]\!] \geq 0$ on F and thus $v \in K_{YF}$. In such a case (10.3.35) yields

$$\int_F \varphi[\![(\xi_\alpha - (\tilde{v}_\alpha^K + \chi_{\alpha\gamma}^K y_\gamma))D^{\alpha\beta}(e(r), e^y(\tilde{v}^K) + \chi^K)]\!]\check{N}_\beta ds \geq 0, \tag{10.3.36}$$

for each $\xi \in K_{YF}$, $\varphi \in \mathbf{D}^+(Y)$.

By changing scale $(y \rightsquigarrow \dfrac{x}{\varepsilon})$, from (10.3.33) and (10.3.34) we get

$$\frac{\partial}{\partial y_\beta} D^{\alpha\beta}[e(r(x)), k(q^{\varepsilon,K}(x))] = 0 \qquad \text{in } \mathbf{D}'(\Omega^\varepsilon). \tag{10.3.37}$$

Next, (10.3.36) yields

$$\int_{F_K^\varepsilon} \varphi_K [\![(\psi_\alpha^\varepsilon - q_\alpha^{\varepsilon,K})D^{\alpha\beta}(e(r), k(q^{\varepsilon,K}))]\!]\check{N}_\beta ds \geq 0, \tag{10.3.38}$$

where $\varphi_K \in \mathbf{D}^+(\Omega_K^\varepsilon)$.

Taking (10.3.37) and (10.3.38) into account from (10.3.24), (10.3.25) and (10.3.27) – (10.3.30) we derive the following relation

$$G^\varepsilon(r, \varphi, \psi^\varepsilon) = \int_\Omega j_0(e(r), \rho(\varphi))dx + \int_{\Omega^\varepsilon} j_1(e(r), k(\psi^\varepsilon))dx$$

$$\geq \sum_{K \in \mathcal{K}} \int_{\Omega_K} \varphi_K j_0(e(r), \rho(\varphi))dx + \sum_{K \in \mathcal{K}} \int_{\Omega_K^\varepsilon} j_1(e(r), k(q^{\varepsilon,K}))dx$$

$$- \sum_{K \in \mathcal{K}} \int_{\Omega_K^\varepsilon} D^{\alpha\beta}(e(r), k(q^{\varepsilon,K}))\varphi_{K,\beta}(\psi_\alpha^\varepsilon - q_\alpha^{\varepsilon,K})dx. \tag{10.3.39}$$

Prior to passing to the limit ($\varepsilon \to 0$) in the last inequality we shall demonstrate the following property

$$D_2 \mathcal{W}_1(\boldsymbol{\epsilon}, \boldsymbol{\chi}) = \frac{1}{|Y|} \int\limits_{YF} D_2 j_1(\boldsymbol{\epsilon}, e^y(\tilde{\boldsymbol{v}}) + \boldsymbol{\chi}) dy \; . \tag{10.3.40}$$

Since the function $j_1(\boldsymbol{\epsilon}, \cdot)$ is convex and finite, therefore we have

$$D_2 j_1(\boldsymbol{\epsilon}, \cdot) = \partial_2 j_1(\boldsymbol{\epsilon}, \cdot) \; . \tag{10.3.41}$$

Here $\partial_2 j$ is the subdifferential of the function $j_1(\boldsymbol{\epsilon}, \cdot)$. Similarly we can write

$$D_2 \mathcal{W}_1(\boldsymbol{\epsilon}, \boldsymbol{\chi}) = \partial_2 \mathcal{W}_1(\boldsymbol{\epsilon}, \boldsymbol{\chi}) \; . \tag{10.3.42}$$

Let us denote by $m^{\alpha\beta}(y)(\alpha, \beta = 1, 2; y \in YF)$ the microscopic bending moment corresponding to $\tilde{\boldsymbol{v}}(y)$, that is

$$m^{\alpha\beta}(y) = D^{\alpha\beta}(\boldsymbol{e}(\boldsymbol{r}), e^y(\tilde{\boldsymbol{v}}) + \boldsymbol{\chi}) \tag{10.3.43}$$

and

$$\int\limits_{YF} m^{\alpha\beta}(y) e^y_{\alpha\beta}(\boldsymbol{v}(y) - \tilde{\boldsymbol{v}}(y)) dy \geq 0 \qquad \forall \, \boldsymbol{v} \in K_{YF} \; . \tag{10.3.44}$$

By virtue of (10.3.41) and (10.3.42) we write

$$\mathcal{W}_1\left(\boldsymbol{e}(\boldsymbol{r}), \boldsymbol{\chi}^{(1)}\right) - \mathcal{W}_1(\boldsymbol{e}(\boldsymbol{r}), \boldsymbol{\chi}) = \frac{1}{|Y|} \int\limits_{YF} [D_2 j_1(\boldsymbol{e}(\boldsymbol{r}), e^y(\tilde{\boldsymbol{v}}^{(1)}) + \boldsymbol{\chi}^{(1)})$$

$$- \; D_2 j_1(\boldsymbol{e}(\boldsymbol{r}), e^y(\tilde{\boldsymbol{v}}) + \boldsymbol{\chi})] dy \geq \frac{1}{|Y|} \int\limits_{YF} m^{\alpha\beta}(y) [e^y_{\alpha\beta}(\tilde{\boldsymbol{v}}^{(1)}) + \chi^{(1)}_{\alpha\beta} - (e^y_{\alpha\beta}(\tilde{\boldsymbol{v}}) + \chi_{\alpha\beta})] dy \; .$$

Here $\tilde{\boldsymbol{v}}^{(1)}$ corresponds to $\boldsymbol{\chi}^{(1)}$ and $\tilde{\boldsymbol{v}}$ to $\boldsymbol{\chi}$.

Hence, taking into account (10.3.44) we obtain

$$\mathcal{W}_1(\boldsymbol{e}(\boldsymbol{r}), \boldsymbol{\chi}^{(1)}) - \mathcal{W}_1(\boldsymbol{e}(\boldsymbol{r}), \boldsymbol{\chi}) \geq \left(\frac{1}{|Y|} \int\limits_{YF} m^{\alpha\beta}(y) dy \right) (\chi^{(1)}_{\alpha\beta} - \chi_{\alpha\beta}) = L_h^{\alpha\beta}(\chi^{(1)}_{\alpha\beta} - \chi_{\alpha\beta}) \; ,$$

$$\tag{10.3.45}$$

where

$$L_h^{\alpha\beta} = \frac{1}{|Y|} \int\limits_{YF} m^{\alpha\beta}(y) dy \; . \tag{10.3.46}$$

The last formula coincides with Eq. (10.2.13)$_3$ in the case of transversely homogeneous plates.

Now we are in a position to pass to the limit in (10.3.39). First, we observe that $\boldsymbol{\psi}^\varepsilon - \boldsymbol{q}^{\varepsilon,K} \to \boldsymbol{\psi} - \boldsymbol{q}$ strongly in $L^2(\Omega)^2$ as $\varepsilon \to 0$ and

$$D^{\alpha\beta}(\boldsymbol{e}(\boldsymbol{r}), \boldsymbol{k}(\boldsymbol{q}^{\varepsilon,K}))$$
$$= D^{\alpha\beta}\left[\boldsymbol{e}(\boldsymbol{r}), \boldsymbol{\chi}^K + \boldsymbol{k}(\tilde{\boldsymbol{v}}^K)\left(\frac{x}{\varepsilon}\right)\right] \to \frac{1}{|Y|}\int_{YF} D^{\alpha\beta}[\boldsymbol{e}(\boldsymbol{r}), \boldsymbol{\chi}^K + \boldsymbol{e}^y(\tilde{\boldsymbol{v}}^K(y))]dy ,$$

$$\tag{10.3.47}$$

weakly in $L^2(\Omega)$ as $\varepsilon \to 0$. Due to (10.3.40) the limit in (10.3.47) is equal just to $D_2 \mathcal{W}_1(\boldsymbol{e}(\boldsymbol{r}), \boldsymbol{\chi}^K)$. Thus we arrive at

$$\liminf_{\varepsilon \to 0} G^\varepsilon(\boldsymbol{r}, \boldsymbol{\varphi}, \boldsymbol{\psi}^\varepsilon) \geq \sum_{K \in \mathcal{K}} \int_{\Omega_K} \varphi_K(x)[j_0(\boldsymbol{e}(\boldsymbol{r}), \boldsymbol{\rho}(\boldsymbol{\varphi})) + \mathcal{W}_1(\boldsymbol{e}(\boldsymbol{r}), \boldsymbol{k}(\boldsymbol{q}))]dx$$
$$- \sum_{K \in \mathcal{K}} \int_{\Omega_K} [D_2\mathcal{W}_1(\boldsymbol{e}(\boldsymbol{r}), \boldsymbol{\chi}^K)]^{\alpha\beta} \varphi_{K,\beta}(\psi_\alpha - q_\alpha)dx . \tag{10.3.48}$$

According to Sec. 9 there exists an extension operator $\mathcal{Q}_2^\varepsilon$ such that the sequence $\{\mathcal{Q}_2^\varepsilon \boldsymbol{\psi}^\varepsilon\}_{\varepsilon \to 0}$ is bounded in $H^1(\Omega)^2$ and $\|\mathcal{Q}^\varepsilon \boldsymbol{\psi}^\varepsilon - \boldsymbol{\psi}\|_{L^2(\Omega)^2} \to 0$ as $\varepsilon \to 0$. By using this result we conclude that $\boldsymbol{\psi} \in H^1(\Omega)^2$. Integrating by parts the last term of (10.3.48) and next passing with φ_K to 1 for each $K \in \mathcal{K}$, we obtain

$$\liminf_{\varepsilon \to 0} G^\varepsilon(\boldsymbol{r}, \boldsymbol{\varphi}, \boldsymbol{\psi}^\varepsilon) \geq \int_\Omega [j_0(\boldsymbol{e}(\boldsymbol{r}), \boldsymbol{\rho}(\boldsymbol{\varphi})) + \mathcal{W}_1(\boldsymbol{e}(\boldsymbol{r}), \boldsymbol{k}(\boldsymbol{q}))]dx$$
$$+ \int_\Omega (\psi_\alpha - q_\alpha)\frac{\partial}{\partial x_\beta}[D_2\mathcal{W}_1(\boldsymbol{e}(\boldsymbol{r}), \boldsymbol{k}(\boldsymbol{q}))]^{\alpha\beta}dx$$
$$+ \int_\Omega [D_2\mathcal{W}_1(\boldsymbol{e}(\boldsymbol{r}), \boldsymbol{k}(\boldsymbol{q}))]^{\alpha\beta} k_{\alpha\beta}(\boldsymbol{\psi} - \boldsymbol{q})dx , \tag{10.3.49}$$

since $\boldsymbol{k}(\boldsymbol{q}) = \boldsymbol{\chi}^K$ on Ω_K. In (10.3.49) $\boldsymbol{q}$ is a piecewise affine continuous function. By using the density argument we extend the inequality (10.3.49) to the whole space $H^1(\Omega)^2$. Setting now $\boldsymbol{\psi} = \boldsymbol{q}$ we finally arrive at (10.3.17) and the proof is complete. $\qquad\square$

Remark 10.3.2. Let us denote by J_h the functional of the total potential energy of the homogenized plate, which is given by

$$J_h(\boldsymbol{r}, \boldsymbol{\varphi}, \boldsymbol{\psi}, w) = G(\boldsymbol{r}, \boldsymbol{\varphi}, \boldsymbol{\psi}) + \Phi(\boldsymbol{\varphi}, \boldsymbol{\psi}, w) - f(w). \tag{10.3.50}$$

Now we formulate the minimum principles of the total potential energy for fissured and homogenized plate:

$$(P_\varepsilon) \quad \left| \begin{array}{l} J_\varepsilon(\boldsymbol{r}^\varepsilon, \boldsymbol{\varphi}^\varepsilon, \boldsymbol{\psi}^\varepsilon, w^\varepsilon) \\[6pt] = \inf\{J_\varepsilon(\boldsymbol{r}, \boldsymbol{\varphi}, \boldsymbol{\psi}, w)|\, \boldsymbol{r}, \boldsymbol{\varphi} \in H_0^1(\Omega)^2, \boldsymbol{\psi} \in K_\varepsilon, w \in H_0^1(\Omega)^2\} , \end{array} \right.$$

$$(P_h) \quad \left| \begin{array}{l} J_h(\boldsymbol{r}^{(0)}, \boldsymbol{\varphi}^{(0)}, \boldsymbol{\psi}^{(0)}, w^{(0)}) \\[6pt] = \inf\{J_h(\boldsymbol{r}, \boldsymbol{\varphi}, \boldsymbol{\psi}, w)|\, \boldsymbol{r}, \boldsymbol{\varphi}, \boldsymbol{\psi} \in H_0^1(\Omega)^2, w \in H_0^1(\Omega)^2\} . \end{array} \right.$$

We observe that (P_ε) and (P_h) are equivalent to (P^ε_{2LF}) and (P^{2L}_h) respectively.

We claim that

$$J_\varepsilon(r^\varepsilon, \varphi^\varepsilon, \psi^\varepsilon, w^\varepsilon) \longrightarrow J_h(r^{(0)}, \varphi^{(0)}, \psi^{(0)}, w^{(0)}) \quad \text{when} \quad \varepsilon \to 0 \,.$$

Indeed, according to Theorem 10.3.1, for $(r^{(0)}, \varphi^{(0)}, \psi^{(0)}, w^{(0)})$ as above, there exists a sequence $\{(\widetilde{r}^\varepsilon, \widetilde{\varphi}^\varepsilon, \widetilde{\psi}^\varepsilon, \widetilde{w}^\varepsilon)\}_{\varepsilon>0} \subset H_0^1(\Omega)^2 \times H_0^1(\Omega)^2 \times K_\varepsilon \times H_0^1(\Omega)$ strongly convergent in $L^2(\Omega)^2 \times L^2(\Omega)^2 \times L^2(\Omega)^2 \times L^2(\Omega)$ and such that

$$J_h(r^{(0)}, \varphi^{(0)}, \psi^{(0)}, w^{(0)}) \geq \limsup_{\varepsilon \to 0} J_\varepsilon(\widetilde{r}^\varepsilon, \widetilde{\varphi}^\varepsilon, \widetilde{\psi}^\varepsilon, \widetilde{w}^\varepsilon) \,. \tag{10.3.51}$$

We also have

$$\liminf_{\varepsilon \to 0} J_\varepsilon(r^\varepsilon, \varphi^\varepsilon, \psi^\varepsilon, w^\varepsilon) \geq J_h(r^{(0)}, \varphi^{(0)}, \psi^{(0)}, w^{(0)}) \,, \tag{10.3.52}$$

$$J_\varepsilon(r^\varepsilon, \varphi^\varepsilon, \psi^\varepsilon, w^\varepsilon) \leq J_\varepsilon(\widetilde{r}^\varepsilon, \widetilde{\varphi}^\varepsilon, \widetilde{\psi}^\varepsilon, \widetilde{w}^\varepsilon) \,. \tag{10.3.53}$$

The relations (10.3.51) – (10.3.53) corroborate our statement, i.e.:

$$\inf P_\varepsilon \longrightarrow \inf P_h \quad \text{when} \quad \varepsilon \to 0 \,.$$

10.4. Dual homogenization

This section is concerned with the dual homogenization involving generalized stresses. As usual, to derive the dual problem (P^*_ε) we apply the theory of duality presented in Sec. 1.2.5. In the present case the operator Λ maps the space $H_0^1(\Omega)^2 \times H_0^1(\Omega)^2 \times H_1^1(\Omega^\varepsilon)^2 \times H_0^1(\Omega)^2$ into $L^2(\Omega, \mathbb{E}_s^2) \times L^2(\Omega, \mathbb{E}_s^2) \times L^2(\Omega^\varepsilon, \mathbb{E}_s^2) \times L^2(\Omega)^2 \times L^2(\Omega)^2$ and is defined as follows

$$\Lambda(r, \varphi, \psi, w) = (\Lambda_1 r, \Lambda_2 \varphi, \Lambda_3 \psi, \Lambda_4(w, \varphi), \Lambda_5(w, \varphi))$$

$$= (e(r), \rho(\varphi), k(\psi), g(w, \varphi), d(w, \varphi)) \,.$$

To find the conjugate operator Λ^* we have to provide explicit expressions for the operators $\Lambda_1^*, \dots, \Lambda_5^*$. Simple calculation yields

$$\langle N, \Lambda_1 r \rangle_{L^2 \times L^2} = \int_\Omega N^{\alpha\beta} e_{\alpha\beta}(r) dx = -\int_\Omega N^{\alpha\beta}{}_{,\beta} r_\alpha dx = \langle \Lambda_1^* N, r \rangle_{H^{-1} \times H_0^1} \,, \tag{10.4.1}$$

where $N \in L^2(\Omega, \mathbb{E}_2^s)$. In a similar manner we obtain

$$\Lambda_2^* M = -(M^{\alpha\beta}{}_{,\beta}) \quad \text{in} \quad \Omega \,; \tag{10.4.2}$$

$$\Lambda_3^* L = \begin{cases} (-L^{\alpha\beta}{}_{,\beta}) & \text{in} \quad \Omega^\varepsilon \,, \\ -L_n & \text{on} \quad F^\varepsilon \,; \end{cases} \tag{10.4.3}$$

$$\Lambda_4^* Q = \begin{cases} -Q^\alpha{}_{,\alpha} & \text{in} \quad \Omega \,, (w) \,, \\ -Q & \text{in} \quad \Omega \,, (\varphi) \,; \end{cases} \tag{10.4.4}$$

$$\Lambda_5^* T = \begin{cases} -T^\alpha{}_{,\alpha} & \text{in} \quad \Omega \,, (w) \,, \\ -T & \text{in} \quad \Omega \,, (\psi) \,. \end{cases} \tag{10.4.5}$$

Further we set

$$\mathcal{L}_\varepsilon(v,\psi) = -f(v) + I_{K_\varepsilon}(\psi) , \qquad (10.4.6)$$

where I_{K_ε} stands for the indicator function of the set K_ε.

Standard calculation yields

$$\mathcal{L}_\varepsilon^*(-\Lambda^*(\boldsymbol{N},\boldsymbol{M},\boldsymbol{L},\boldsymbol{Q},\boldsymbol{T})) = \sup\left\{ \int_\Omega (N^{\alpha\beta}{}_{,\beta}r_\alpha + M^{\alpha\beta}{}_{,\beta}\varphi_\alpha \right.$$

$$+Q^\alpha{}_{,\alpha}w - Q^\alpha\varphi_\alpha + T^\alpha{}_{,\alpha}w - T^\alpha\psi_\alpha + pw)dx + \int_{\Omega^\varepsilon} L^{\alpha\beta}{}_{,\beta}\psi_\alpha dx$$

$$\left. +\int_{F^\varepsilon} L_n[\![\psi_n]\!]ds - I_{K_\varepsilon}(\psi)\,\middle|\, \boldsymbol{r},\boldsymbol{\varphi}\in H_0^1(\Omega)^2, \boldsymbol{\psi}\in H_1^1(\Omega^\varepsilon),\ w\in H_0^1(\Omega) \right\}$$

$$= \begin{cases} 0 & \text{if } \mathrm{div}\boldsymbol{N}=0,\ -\mathrm{div}\boldsymbol{M}+\boldsymbol{Q}=0\ \text{in } \Omega, \\ & \quad \mathrm{div}(\boldsymbol{Q}+\boldsymbol{T})+p=0\ \text{in } \Omega, \\ & \quad -\mathrm{div}\boldsymbol{L}+\boldsymbol{T}=0\ \text{in } \Omega^\varepsilon,\quad L_n\le 0\ \text{on } F^\varepsilon; \\ +\infty & \text{otherwise.} \end{cases} \qquad (10.4.7)$$

We introduce the functional of the total elastic energy of the fissured plate

$$G_2^\varepsilon(\Lambda(\boldsymbol{r},\boldsymbol{\varphi},\boldsymbol{\psi},w)) = \int_{\Omega^\varepsilon} j[\boldsymbol{e}(\boldsymbol{r}),\boldsymbol{\rho}(\boldsymbol{\varphi}),\boldsymbol{k}(\boldsymbol{\psi})]dx + \int_\Omega j_2[\boldsymbol{g}(w,\boldsymbol{\varphi}),\boldsymbol{d}(w,\boldsymbol{\psi})]dx , \quad (10.4.8)$$

where

$$j_2(\boldsymbol{g},\boldsymbol{d}) = \frac{1}{2}H_a^{\alpha\beta}g_\alpha g_\beta + \frac{1}{2}H_b^{\alpha\beta}d_\alpha d_\beta . \qquad (10.4.9)$$

To find the conjugate functional $(G_2^\varepsilon)^*$ of G_2^ε we calculate

$$j^*(\boldsymbol{N},\boldsymbol{M},\boldsymbol{L}) = \sup\{N^{\alpha\beta}e_{\alpha\beta} + M^{\alpha\beta}\rho_{\alpha\beta} + L^{\alpha\beta}k_{\alpha\beta} - j(\boldsymbol{e},\boldsymbol{\rho},\boldsymbol{k})\,|\,\boldsymbol{e},\boldsymbol{\rho},\boldsymbol{k}\in\mathbb{E}_s^2\}$$

$$= \frac{1}{2}[\boldsymbol{N},\boldsymbol{M},\boldsymbol{L}]\mathbf{A}^{-1}[\boldsymbol{N},\boldsymbol{M},\boldsymbol{T}]^T , \qquad (10.4.10)$$

$$j_2^*(\boldsymbol{Q},\boldsymbol{T}) = \sup\{Q^\alpha g_\alpha + T^\alpha d_\alpha - j_2(\boldsymbol{g},\boldsymbol{d})\,|\,\boldsymbol{g},\boldsymbol{d}\in\mathbb{R}^2\}$$

$$= \frac{1}{2}h_{\alpha\beta}^a Q^\alpha Q^\beta + \frac{1}{2}h_{\alpha\beta}^b T^\alpha T^\beta , \qquad (10.4.11)$$

where $h^a = H_a^{-1}$, $h^b = H_1^b$.

The set of statically admissible generalized stresses is defined by, cf. (10.4.7)

$$S_s^\varepsilon := \{\boldsymbol{N}\in L^2(\Omega,\mathbb{E}_2^s), \boldsymbol{M}\in L^2(\Omega,\mathbb{E}_2^s), \boldsymbol{L}\in L^2(\Omega^\varepsilon,\mathbb{E}_2^s),$$

$$\boldsymbol{Q},\boldsymbol{T}\in L^2(\Omega)^2|\ \mathrm{div}\boldsymbol{N}=0,\ -\mathrm{div}\boldsymbol{M}+\boldsymbol{Q}=0,\ \text{in }\Omega$$

$$\mathrm{div}(\boldsymbol{Q}+\boldsymbol{T})+p=0\ \text{in }\Omega;\ -\mathrm{div}\boldsymbol{L}+\boldsymbol{T}=0\ \text{in }\Omega^\varepsilon;\ L_n\le 0\ \text{on }F^\varepsilon\} . \qquad (10.4.12)$$

Now we are in a position to formulate

Problem (P_ε^*)

$$
\begin{vmatrix}
\text{Find} \\[4pt]
\sup\{ -\int_\Omega [j^*(\boldsymbol{N}(x), \boldsymbol{M}(x), \boldsymbol{L}(x)) + j_2^*(\boldsymbol{Q}(x), \boldsymbol{T}(x))]dx \\[4pt]
\quad -I_{\boldsymbol{S}_s^\varepsilon}(\boldsymbol{N}, \boldsymbol{M}, \boldsymbol{L}, \boldsymbol{Q}, \boldsymbol{T})|\ (\boldsymbol{N}, \boldsymbol{M}, \boldsymbol{L}, \boldsymbol{Q}, \boldsymbol{T}) \in \mathbf{H}\} \,,
\end{vmatrix}
$$

where

$$
\mathbf{H} = L^2(\Omega, \mathbb{E}_2^s) \times L^2(\Omega, \mathbb{E}_2^s) \times L^2(\Omega^\varepsilon, \mathbb{E}_2^s) \times L^2(\Omega)^2 \times L^2(\Omega)^2 \,. \qquad (10.4.13)
$$

Problem (P_ε^*) is obviously uniquely solvable and

$$
\inf P_\varepsilon = \sup P_\varepsilon^* \,. \qquad (10.4.14)
$$

Let us set

$$
\mathcal{G}_\varepsilon(\boldsymbol{N}, \boldsymbol{M}, \boldsymbol{L}, \boldsymbol{Q}, \boldsymbol{T}) = \int_\Omega [j^*(\boldsymbol{N}(x), \boldsymbol{M}(x), \boldsymbol{L}(x)) + j_2^*(\boldsymbol{Q}(x), \boldsymbol{T}(x))]dx
$$
$$
+I_{\boldsymbol{S}_s^\varepsilon}(\boldsymbol{N}, \boldsymbol{M}, \boldsymbol{L}, \boldsymbol{Q}, \boldsymbol{T}) \,. \qquad (10.4.15)
$$

In order to study the Γ-convergence of the sequence $\{\mathcal{G}_\varepsilon\}_{\varepsilon>0}$ it is indispensable to derive the dual macroscopic potential $\mathcal{W}^*$, where $\mathcal{W}$ is given by (10.2.16). We write

$$
\begin{aligned}
\mathcal{W}^*(\boldsymbol{N}, \boldsymbol{M}, \boldsymbol{L}, \boldsymbol{Q}, \boldsymbol{T}) &= \sup\{N^{\alpha\beta}\epsilon_{\alpha\beta} + M^{\alpha\beta}\theta_{\alpha\beta} + L^{\alpha\beta}\chi_{\alpha\beta} + Q^\alpha g_\alpha \\
&\quad + T^\alpha d_\alpha - \mathcal{W}(\boldsymbol{\epsilon}, \boldsymbol{\theta}, \boldsymbol{\chi}, \boldsymbol{g}, \boldsymbol{d})|\ \boldsymbol{\epsilon}, \boldsymbol{\theta}, \boldsymbol{\chi} \in \mathbb{E}_s^2 \,;\ \boldsymbol{g}, \boldsymbol{d} \in \mathbb{R}^2\} \\
&= (j_0 + \mathcal{W}_1)^*(\boldsymbol{N}, \boldsymbol{M}, \boldsymbol{L}) + j_2^*(\boldsymbol{Q}, \boldsymbol{T}) \,, \qquad (10.4.16)
\end{aligned}
$$

where $\boldsymbol{N}, \boldsymbol{M}, \boldsymbol{L} \in \mathbb{E}_2^s$ and $\boldsymbol{Q}, \boldsymbol{T} \in \mathbb{R}^2$; moreover

$$
(j_0 + \mathcal{W}_1)^*(\boldsymbol{N}, \boldsymbol{M}, \boldsymbol{L}) = \sup_{\boldsymbol{\epsilon},\boldsymbol{\theta},\boldsymbol{\chi}\in\mathbb{E}_s^2} \{N^{\alpha\beta}\epsilon_{\alpha\beta} + M^{\alpha\beta}\theta_{\alpha\beta} + L^{\alpha\beta}\chi_{\alpha\beta} \qquad (10.4.17)
$$
$$
- j_0(\boldsymbol{\epsilon}, \boldsymbol{\theta}) - \frac{1}{|Y|} \inf_{\boldsymbol{v}\in K_{YF}} \int_{YF} j_1(\boldsymbol{\epsilon}, e^y(\boldsymbol{v}) + \boldsymbol{\chi})dy\} \,.
$$

To be more specific, consider the case of transversely homogeneous plates. To this end we set $c = \tilde{C}^{-1}$. Proceeding similarly to Sec. 11.5 in the next chapter it can be shown that

$$
(j_0 + \mathcal{W}_1)^*(\boldsymbol{N}, \boldsymbol{M}, \boldsymbol{L}) = c_{\alpha\beta\lambda\mu}\frac{2}{h}N^{\alpha\beta}N^{\lambda\mu} + \left(\frac{6}{ah}N^{\alpha\beta}M^{\lambda\mu}\right.
$$
$$
\left. -\frac{6}{bh}N^{\alpha\beta}L^{\lambda\mu} + \frac{3(4a+b)}{2a^3h}M^{\alpha\beta}M^{\lambda\mu} - \frac{9}{abh}M^{\alpha\beta}L^{\lambda\mu} + \frac{9}{2b^2h}L^{\alpha\beta}L^{\lambda\mu}\right)
$$
$$
+ \inf\left\{\frac{1}{|Y|}\int_{YF} \frac{3}{2b^3}c_{\alpha\beta\lambda\mu}(L^{\alpha\beta} - m^{\alpha\beta})(L^{\lambda\mu} - m^{\lambda\mu})dy|\ \boldsymbol{m} \in \boldsymbol{S}_{per}\right\} \,, \qquad (10.4.18)
$$

where

$$\mathbf{S}_{per} = (\mathbf{L} - \mathbf{S}^c_{per}) \cap (\mathbb{E}^2_s)^{\perp} \,, \qquad (10.4.19)$$

and

$$\mathbf{S}^c_{per} = \{\mathbf{m} \in L^2(YF, \mathbb{E}^s_2)|\ \mathrm{div}_y \mathbf{m} = 0 \ \text{in}\ YF\,;\, \mathbf{m}_T = 0\,,\, m_N \le 0 \ \text{on}\ F\,;$$
$$\mathbf{m}\boldsymbol{\mu}\ \text{takes opposite values on the opposite sides of}\ Y\}\,,$$
$$(\mathbb{E}^2_s)^{\perp} = \{\mathbf{m} \in L^2(YF, \mathbb{E}^s_2)|\ \int\limits_{YF} \mathbf{m}(y)dy = 0\}\,.$$

Here $\boldsymbol{\mu}$ represents a unit vector outward normal to ∂Y.

We observe that in order to find $\mathcal{W}^*$ one has to solve only the following local problem:

$$\left|\ \begin{array}{l} \text{for a given}\ \mathbf{L} \in \mathbb{E}^s_2\ \text{find} \\[4pt] \inf\{\dfrac{1}{|Y|} \int\limits_{YF} \dfrac{3}{2b^3}C_{\alpha\beta\lambda\mu}(L^{\alpha\beta} - m^{\alpha\beta}(y))(L^{\lambda\mu} - m^{\lambda\mu}(y))dy|\ \mathbf{m} \in \mathbf{S}_{per}\}\,. \end{array}\right.$$

This problem can be written in an equivalent form. Namely for an element $\mathbf{m} \in \mathbf{S}_{per}$ we have, cf. (10.4.19)

$$\mathbf{m} = \mathbf{L} - \mathbf{m}_1\,, \qquad \langle \mathbf{m} \rangle = \frac{1}{|Y|} \int\limits_{YF} \mathbf{m}(y)dy = 0\,, \qquad \mathbf{m}_1 \in \mathbf{S}^c_{per}$$

and thus $\langle \mathbf{m}_1 \rangle = \mathbf{L}$. Hence the last local problem means evaluating

$$\inf\{\frac{1}{|Y|} \int\limits_{YF} \frac{3}{2b^3}C_{\alpha\beta\lambda\mu}m^{\alpha\beta}m^{\lambda\mu}dy|\ \mathbf{m} \in \mathbf{S}^c_{per}, \langle \mathbf{m} \rangle = \mathbf{L}\}\,.$$

We are now in a position to solve the problem of Γ-convergence of the sequence $\{\mathcal{G}_\varepsilon\}_{\varepsilon>0}$ defined by (10.4.15). To this end we set

$$J^p_\varepsilon(\mathbf{r}, \boldsymbol{\varphi}, \boldsymbol{\psi}, w; \mathbf{p}_i) = \int\limits_{\Omega^\varepsilon} \{j[e(\mathbf{r}) + \mathbf{p}_1, \rho(\boldsymbol{\varphi}) + \mathbf{p}_2, k(\boldsymbol{\psi} + \mathbf{p}_3)]$$
$$+ j_2[g(w, \boldsymbol{\varphi}) + \mathbf{p}_4, d(w, \boldsymbol{\psi}) + \mathbf{p}_5]\}dx\,.$$

Here $i = 1, 2, 3, 4, 5$ and $\mathbf{p}_\alpha \in L^2(\Omega, \mathbb{E}^2_s)$, $\mathbf{p}_3 \in L^2(\Omega^\varepsilon, \mathbb{E}^2_s)$, $\mathbf{p}_4, \mathbf{p}_5 \in L^2(\Omega)^2$; obviously $\alpha = 1, 2$. The Γ-convergence of the sequence $\{J_\varepsilon\}_{\varepsilon>0}$ to J_h implies the Γ-convergence of $\{J^p_\varepsilon\}_{\varepsilon>0}$, in the strong topology of $L^2(\Omega)^2 \times L^2(\Omega)^2 \times L^2(\Omega)^2 \times L^2(\Omega) \times \mathbf{H}$ to the following functional

$$J^p_h(\mathbf{r}, \boldsymbol{\varphi}, \boldsymbol{\psi}, w; \mathbf{p}_i) = \int\limits_{\Omega} \{\mathcal{W}[e(\mathbf{r}) + \mathbf{p}_1, \rho(\boldsymbol{\varphi}) + \mathbf{p}_2, k(\boldsymbol{\psi})$$
$$+ \mathbf{p}_3, g(w, \boldsymbol{\varphi}) + \mathbf{p}_4, d(w, \boldsymbol{\psi}) + \mathbf{p}_5]\}dx\,. \qquad (10.4.20)$$

More precisely, one has to choose approximating sequence for $\mathbf{p}_i$ $(i = 1, 2, 3, 4, 5)$. To this end we first consider the case where $\mathbf{p}_i$ are constant functions on the sets Ω_K introduced in the previous section. Then it is not difficult to prove the following inequalities, cf. Sec. 5.4

$$J_h^p(\boldsymbol{r}, \boldsymbol{\varphi}, \boldsymbol{\psi}, w \ \boldsymbol{p}_i) \leq \liminf_{\varepsilon \to 0} J_\varepsilon^p(\boldsymbol{r}^\varepsilon, \boldsymbol{\varphi}^\varepsilon, \boldsymbol{\psi}^\varepsilon, w^\varepsilon; \boldsymbol{p}_i) , \qquad (10.4.21)$$

$$J_h^p(\boldsymbol{r}, \boldsymbol{\varphi}, \boldsymbol{\psi}, w \ \boldsymbol{p}_i) \geq \limsup_{\varepsilon \to 0} J_\varepsilon^p(\boldsymbol{r}^\varepsilon, \boldsymbol{\varphi}^\varepsilon, \boldsymbol{\psi}^\varepsilon, w^\varepsilon; \boldsymbol{p}_i) . \qquad (10.4.22)$$

Here $\{\boldsymbol{r}^\varepsilon, \boldsymbol{\varphi}^\varepsilon, \boldsymbol{\psi}^\varepsilon, w^\varepsilon\}_{\varepsilon>0} \subset H_0^1(\Omega)^2 \times H_0^1(\Omega)^2 \times K_\varepsilon \times H_0^1(\Omega)$ is the sequence involved in the property (iv) of the Γ-limit, cf. Sec. 1.3.4. Next, arbitrary elements $\boldsymbol{p}_i$, $(i = 1, 2, 3, 4, 5)$ from the corresponding L^2-space are approximated by piecewise constant functions and inequalities like (10.4.21), (10.4.22) are proved for such a general case.

Now we have to verify the coercivity condition (1.3.75). For this purpose we set

$$\mathcal{J}_\varepsilon^p(\boldsymbol{r}, \boldsymbol{\varphi}, \boldsymbol{\psi}, w; \boldsymbol{p}_i) = J_\varepsilon^p(\boldsymbol{r}, \boldsymbol{\varphi}, \boldsymbol{\psi}, w; \boldsymbol{p}_i) - f(w) . \qquad (10.4.23)$$

It is easily seen that

$$\mathcal{J}_\varepsilon^p(\boldsymbol{r}, \boldsymbol{\varphi}, \boldsymbol{\psi}, w; \ \boldsymbol{0}, \boldsymbol{0}, \boldsymbol{0}, \boldsymbol{0}, \boldsymbol{0})$$
$$\geq c(||\boldsymbol{r}||_{1,\Omega}^2 + ||\boldsymbol{\varphi}||_{1,\Omega}^2 + ||\boldsymbol{\psi}||_{1,\Omega^\varepsilon}^2 + ||w||_{1,\Omega}^2 - ||p||_{0,\Omega} ||w||_{1,\Omega}) ,$$

where $c > 0$ is a constant independent of ε. The assumptions of Theorem 1.3.38 being satisfied, we conclude that

$$\mathcal{G}_h = (w - \mathbf{H}) - \lim_\varepsilon \mathcal{G}_\varepsilon , \qquad (10.4.24)$$

$$\mathcal{G}_h(\boldsymbol{N}, \boldsymbol{M}, \boldsymbol{L}, \boldsymbol{Q}, \boldsymbol{T}) = \int_\Omega \mathcal{W}^*[\boldsymbol{N}(x), \boldsymbol{M}(x), \boldsymbol{L}(x), \boldsymbol{Q}(x), \boldsymbol{T}(x)] dx$$
$$+ I_{\mathbf{S}_s}(\boldsymbol{N}, \boldsymbol{M}, \boldsymbol{L}, \boldsymbol{Q}, \boldsymbol{T}) , \qquad (10.4.25)$$

and

$$\mathbf{S}_s := \{\boldsymbol{N}, \boldsymbol{M}, \boldsymbol{L} \in L^2(\Omega, \mathbb{E}_s^2); \quad \boldsymbol{Q}, \boldsymbol{T} \in L^2(\Omega)^2 | \ \mathrm{div}\boldsymbol{N} = 0,$$
$$-\mathrm{div}\boldsymbol{M} + \boldsymbol{Q} = 0, \ -\mathrm{div}\boldsymbol{L} + \boldsymbol{T} = 0 , \ \mathrm{div}(\boldsymbol{Q} + \boldsymbol{T}) + p = 0 \text{ in } \Omega\}. \qquad (10.4.26)$$

We recall that $w - \mathbf{H}$ stands for the weak topology of the space $\mathbf{H}$.

We observe that (10.4.24) still applies to transversely inhomogeneous plates since the explicit form of $\mathcal{W}^*$ has not been used and the sets $\mathbf{S}_s^\varepsilon$, $\mathbf{S}_s$ are independent of material properties.

Further consequences resulting from application of dual homogenization Theorem 1.3.36 to the plate model studied are left to the reader.

Remark 10.4.1. For technical reasons it is convenient to assume that divergences of all fields appearing in the present section belong to the corresponding L^2-spaces.

10.5. *Passage to classical models of cracked plates*

The homogenization process based on the in-plane scaling (10.2.5) has preserved the initial mathematical structure: models of an εY-periodic plate (P^ε_{2LF}) and the homogenized model (P^{2L}_h) are both models of a two-layer plate. In the next section an alternative homogenized model will be derived. By using a so-called refined scaling we shall arrive at a homogenized model of a physically non-linear Kirchhoff-type plate. In this case the homogenization process goes beyond the limits of the original framework of the two-layer plate model and leads to the thin plate model of Kirchhoff type. To bridge a gap between these two approaches a passage from the results of the previous sections to the Hencky-Reissner-and then to the Kirchhoff plate description will be presented.

Let us impose the constraints

$$\varphi^{(0)} = \psi^{(0)} \tag{10.5.1}$$

on problem $(\tilde{P}^{2L}_h)$. The potential $\mathcal{W}$ is replaced by

$$\mathcal{W}_H = \frac{1}{2}\tilde{C}^{\alpha\beta\lambda\mu}[h\epsilon^h_{\alpha\beta}\epsilon^h_{\lambda\mu} + \frac{a^3}{3}k^h_{\alpha\beta}k^h_{\lambda\mu} + \frac{b^3}{3}\langle(k^h_{\alpha\beta} + e^y_{\alpha\beta}(\psi^{(1)}))(k^h_{\lambda\mu} + e^y_{\lambda\mu}(\psi^{(1)})))\rangle_{YF}$$
$$+ [(b^2 - a^2)k^h_{\lambda\mu} - b^2 k^F_{\lambda\mu}]e^h_{\alpha\beta}] + \frac{1}{2}k_s h C^{\alpha3\beta3}\beta^h_\alpha\beta^h_\beta \,, \tag{10.5.2}$$

where $\beta^h_\alpha = w^{(0)}_{,\alpha} + \varphi^{(0)}_\alpha$. Specifying the constitutive relations

$$N^{\alpha\beta}_h = \frac{\partial \mathcal{W}_H}{\partial \epsilon^h_{\alpha\beta}} \,, \qquad \mathbf{M}^{\alpha\beta}_h = \frac{\partial \mathcal{W}_H}{\partial k^h_{\alpha\beta}} \,, \qquad \mathbf{Q}^\alpha_h = \frac{\partial \mathcal{W}_H}{\partial \beta^h_\alpha} \,, \tag{10.5.3}$$

we obtain

$$N^{\alpha\beta}_h = \tilde{C}^{\alpha\beta\lambda\mu}[h\epsilon^h_{\lambda\mu} + \frac{h}{2}(b - a)k^h_{\lambda\mu} - \frac{b^2}{2}k^F_{\lambda\mu}] \,,$$
$$\tag{10.5.4}$$
$$\mathbf{M}^{\alpha\beta}_h = \tilde{C}^{\alpha\beta\lambda\mu}[\frac{h}{2}(b - a)\epsilon^h_{\lambda\mu} + \frac{1}{3}(a^3 + b^3)k^h_{\lambda\mu} - \frac{b^3}{3}k^F_{\lambda\mu}] \,;$$
$$\mathbf{Q}^\alpha_h = k_s h C^{\alpha3\beta3}\beta^h_\beta \,. \tag{10.5.5}$$

Note that $\mathbf{M}_h = M_h + L_h$, $\mathbf{Q}_h = Q_h + T_h$. The equilibrium equations assume the form

$$N^{\alpha\beta}_{h,\beta} = 0 \,, \quad -\mathbf{M}^{\alpha\beta}_{h,\beta} + \mathbf{Q}^\alpha_h = 0 \,, \quad -\mathbf{Q}^\alpha_{h,\alpha} = p \,, \tag{10.5.6}$$

well known from the theory of Hencky-Reissner plates.

The properties of the potential $\mathcal{W}$ mentioned above imply that the reduced potential $\mathcal{W}_H$ is still smooth, strictly convex and enjoys similar properties, cf. Sec. 10.2.

To arrive at the Kirchhoff-type plate model one should, in addition, impose $\beta^h = 0$ or

$$\varphi^{(0)}_\alpha = \psi^{(0)}_\alpha = -w^{(0)}_{,\alpha} \,.$$

Then $k^h_{\alpha\beta} = \kappa_{\alpha\beta}(w^{(0)})$. The equilibrium equations reduce to the form

$$N^{\alpha\beta}_{h,\beta} = 0 , \qquad -\mathbf{M}^{\alpha\beta}_{h,\alpha\beta} = p . \tag{10.5.7}$$

The constitutive relations linking $\mathbf{N}_h, \mathbf{M}_h$ with $\epsilon^h, \mathbf{k}^h$ are still coupled, because the presence of the cracks makes the plate transversely non-homogeneous (unbalanced).

10.6. Refined scaling and effective model

Consider the problem of homogenization of effective properties of the plate discussed in Sec. 10.2. The plate is weakened with a family of cracks that form an εY-periodic layout. As it was mentioned in Sec. 10.2 the effective models found in Sec. 10.2 and studied in Secs. 10.3 – 10.5, concern cracks which are densely distributed: cracks spacing should be much smaller than the plate thickness. To consider a larger class of cracks layouts it is necessary to introduce a new scaling, called here refined, that preserves the three-dimensional shape of the cell of periodicity. Thus instead of the in-plane scaling: $Y \rightsquigarrow \varepsilon Y$ we replace:

$$Y \rightsquigarrow \varepsilon Y, \quad a \rightsquigarrow \varepsilon a, \quad b \rightsquigarrow \varepsilon b, \quad h \rightsquigarrow \varepsilon h . \tag{10.6.1}$$

To compensate the decrease of stiffnesses as $\varepsilon \to 0$ we scale the loading

$$p \rightsquigarrow \varepsilon^3 p . \tag{10.6.2}$$

Below, our study is restricted to transversely homogeneous plates.

The bilinear form $a^\varepsilon(\cdot,\cdot)$ on the left-hand side of Eq. (10.2.3) is now replaced by $\varepsilon^3 \tilde{a}^\varepsilon(\boldsymbol{U},\boldsymbol{V})$ and the bilinear form $\tilde{a}^\varepsilon(\cdot,\cdot)$ reads

$$
\begin{aligned}
\tilde{a}^\varepsilon(\boldsymbol{U},\boldsymbol{V}) = \int_{\Omega^\varepsilon} \Big\{ &\tilde{C}^{\alpha\beta\lambda\mu} \Big[\frac{h^3}{\varepsilon^2} e_{\alpha\beta}(\boldsymbol{r})e_{\lambda\mu}(\boldsymbol{s}) + \frac{a^3}{3}\rho_{\alpha\beta}(\boldsymbol{\varphi})\rho_{\lambda\mu}(\boldsymbol{\eta}) \\
&+ \frac{b^3}{3}k_{\alpha\beta}(\boldsymbol{\psi})k_{\lambda\mu}(\boldsymbol{\chi}) + \frac{1}{2\varepsilon}e_{\alpha\beta}(\boldsymbol{s})[b^2 k_{\lambda\mu}(\boldsymbol{\psi}) - a^2\rho_{\lambda\mu}(\boldsymbol{\varphi})] \\
&+ \frac{1}{2\varepsilon}e_{\alpha\beta}(\boldsymbol{r})[b^2 k_{\lambda\mu}(\boldsymbol{\chi}) - a^2\rho_{\lambda\mu}(\boldsymbol{\eta})] \Big] \\
&+ \frac{k_s}{\varepsilon^2}C^{\alpha 3\beta 3}[a(w_{,\alpha} + \varphi_\alpha)(v_{,\beta} + \eta_\beta) + b(w_{,\alpha} + \psi_\alpha)(v_{,\beta} + \chi_\beta)] \Big\} dx ,
\end{aligned} \tag{10.6.3}
$$

where $\boldsymbol{U} = (\boldsymbol{r}, \boldsymbol{\varphi}, \boldsymbol{\psi}, w); \boldsymbol{V} = (\boldsymbol{s}, \boldsymbol{\eta}, \boldsymbol{\chi}, v)$. According to the scaling (10.6.2) the linear form $F(v)$ is replaced with $\varepsilon^3 f(v)$. Thus the problem (P^ε_{2LF}) is rearranged as follows

$$(\tilde{P}^\varepsilon_{2LF}) \quad \left| \begin{array}{l} \text{find } \boldsymbol{U}^\varepsilon \in \mathbb{K}_\varepsilon \text{ such that} \\[4pt] \tilde{a}^\varepsilon(\boldsymbol{U}^\varepsilon; \boldsymbol{V} - \boldsymbol{U}^\varepsilon) \geq f(v - w^\varepsilon) \quad \forall \ \boldsymbol{V} \in \mathbb{K}_\varepsilon . \end{array} \right. \tag{10.6.4}$$

The solution U^ε will be represented in the form compatible with singularities inherent in the definition of $\tilde{a}^\varepsilon(\cdot,\cdot)$:

$$r_\alpha^\varepsilon(x) = \varepsilon r_\alpha^{(0)}(x) + \varepsilon^2 r_\alpha^{(1)}\left(x,\frac{x}{\varepsilon}\right) + \varepsilon^3 r_\alpha^{(2)}\left(x,\frac{x}{\varepsilon}\right) + \dots ,$$

$$\varphi_\alpha^\varepsilon(x) = \varphi_\alpha^{(0)}(x) + \varepsilon\varphi_\alpha^{(1)}\left(x,\frac{x}{\varepsilon}\right) + \varepsilon^2\varphi_\alpha^{(2)}\left(x,\frac{x}{\varepsilon}\right) + \dots ,$$

$$\psi_\alpha^\varepsilon(x) = \psi_\alpha^{(0)}(x) + \varepsilon\psi_\alpha^{(1)}\left(x,\frac{x}{\varepsilon}\right) + \varepsilon^2\psi_\alpha^{(2)}\left(x,\frac{x}{\varepsilon}\right) + \dots ,$$

$$w^\varepsilon(x) = w^{(0)}(x) + \varepsilon w^{(1)}\left(x,\frac{x}{\varepsilon}\right) + \varepsilon^2 w^{(2)}\left(x,\frac{x}{\varepsilon}\right) + \dots$$

$$(10.6.5)$$

and the trial fields s_α, η_α, χ_α, v are represented similarly. The functions $r_\alpha^{(k)}(x,\cdot)$, $s_\alpha^{(k)}(x,\cdot)$, $\varphi_\alpha^{(k)}(x,\cdot)$, $\eta_\alpha^{(k)}(x,\cdot)$, $w^{(k)}(x,\cdot)$, $v^{(k)}(x,\cdot)$, $k \geq 1$, are elements of the corresponding spaces $H_{per}^m(Y)$. More precisely, we assume that $r_\alpha^{(1)}(x,\cdot)$, $s_\alpha^{(1)}(x,\cdot)$, $\varphi_\alpha^{(1)}(x,\cdot)$, $\eta_\alpha^{(1)}(x,\cdot)$, $w^{(1)}(x,\cdot)$, $v^{(1)}(x,\cdot) \in H_{per}^1(Y)$ and $\psi^{(1)}(x,\cdot)$, $\chi^{(1)}(x,\cdot) \in K_{YF}$.

Moreover, to suppress singularities of the integral of (10.6.3), we assume that

$$\varphi_\alpha^{(0)} = -w_{,\alpha}^{(0)} , \qquad w^{(1)} = w^{(1)}(x), , \qquad \eta_\alpha^{(0)} = -v_{,\alpha}^{(0)} , \qquad v^{(1)} = v^{(1)}(x) ,$$

$$\psi_\alpha^{(0)} = \varphi_\alpha^{(0)} , \qquad \chi_\alpha^{(0)} = \eta_\alpha^{(0)} .$$

$$(10.6.6)$$

Let us define

$$\tilde{\varphi}_\alpha = \varphi_\alpha^{(1)} + w_{,\alpha}^{(1)} , \qquad \tilde{\psi}_\alpha = \psi_\alpha^{(1)} + w_{,\alpha}^{(1)} ,$$

$$\tilde{\eta}_\alpha = \eta_\alpha^{(1)} + v_{,\alpha}^{(1)} , \qquad \tilde{\chi}_\alpha = \chi_\alpha^{(1)} + v_{,\alpha}^{(1)} .$$

$$(10.6.7)$$

Substitution of representations (10.6.5) – (10.6.7) into (10.6.4) gives the variational inequality

$$\int_\Omega \{N_0^{\alpha\beta}[e_{\alpha\beta}(s^{(0)} - r^{(0)}) + e_{\alpha\beta}^y(s^{(1)} - r^{(1)})] + M_0^{\alpha\beta}[\kappa_{\alpha\beta}(v^{(0)} - w^{(0)}) + e_{\alpha\beta}^y(\eta^{(1)} - \varphi^{(1)})]$$

$$+L_0^{\alpha\beta}[\kappa_{\alpha\beta}(v^{(0)} - w^{(0)}) + e_{\alpha\beta}^y(\chi^{(1)} - \psi^{(1)})] + Q_0^\beta[(\tilde{\eta}_\beta - \tilde{\varphi}_\beta) + \frac{\partial}{\partial y_\beta}(v^{(2)} - w^{(2)})]$$

$$+T_0^\beta[(\tilde{\chi}_\beta - \tilde{\psi}_\beta) + \frac{\partial}{\partial y_\beta}(v^{(2)} - w^{(2)})]\}dx \geq \int_\Omega p(v^{(0)} - w^{(0)})dx + 0(\varepsilon) , \qquad (10.6.8)$$

where $\kappa_{\alpha\beta}(v^{(0)}) = -v_{,\alpha\beta}^{(0)}$ and

$$N_0^{\alpha\beta} = \tilde{C}^{\alpha\beta\lambda\mu}[h\epsilon_{\lambda\mu}^0 - \frac{a^2}{2}\rho_{\lambda\mu}^0 + \frac{b^2}{2}k_{\lambda\mu}^0] ,$$

$$M_0^{\alpha\beta} = \tilde{C}^{\alpha\beta\lambda\mu}[-\frac{a^2}{2}\epsilon_{\lambda\mu}^0 + \frac{a^3}{3}\rho_{\lambda\mu}^0] , \qquad L_0^{\alpha\beta} = \tilde{C}^{\alpha\beta\lambda\mu}[\frac{b^2}{2}\epsilon_{\lambda\mu}^0 + \frac{b^3}{3}k_{\lambda\mu}^0] , \qquad (10.6.9)$$

$$Q_0^\alpha = k_s a C^{\alpha 3\beta 3}(\tilde{\varphi}_\beta + \frac{\partial w^{(2)}}{\partial y_\beta}) , \qquad T_0^\alpha = k_s b C^{\alpha 3\beta 3}(\tilde{\psi}_\beta + \frac{\partial w^{(2)}}{\partial y_\beta}) ,$$

with the following definitions of strain measures:

$$\epsilon^0_{\alpha\beta} = \epsilon^h_{\alpha\beta} + e^y_{\alpha\beta}(\boldsymbol{r}^{(1)}) , \qquad \epsilon^h_{\alpha\beta} = e_{\alpha\beta}(\boldsymbol{r}^{(0)}) ,$$

$$\rho^0_{\alpha\beta} = \kappa^h_{\alpha\beta} + e^y_{\alpha\beta}(\boldsymbol{\varphi}^{(1)}) , \tag{10.6.10}$$

$$k^0_{\alpha\beta} = \kappa^h_{\alpha\beta} + e^y_{\alpha\beta}(\boldsymbol{\psi}^{(1)}) , \qquad \kappa^h_{\alpha\beta} = \kappa_{\alpha\beta}(w^{(0)}) .$$

Now we put

$$\boldsymbol{s}^{(0)}(x) = \boldsymbol{r}^{(0)}(x) + \bar{\boldsymbol{s}}^{(0)}(x) , \qquad \boldsymbol{\eta}^{(0)}(x) = \boldsymbol{\varphi}^{(0)}(x) + \bar{\eta}(x) ,$$

$$\boldsymbol{\chi}^{(0)}(x) = \boldsymbol{\psi}^{(0)}(x) + \bar{\chi}(x) , \qquad v^{(0)}(x) = w^{(0)}(x) + \bar{v}(x) ,$$

$$\boldsymbol{s}^{(1)} = \boldsymbol{r}^{(1)} , \quad \boldsymbol{\eta}^{(1)} = \boldsymbol{\varphi}^{(1)} , \qquad \boldsymbol{\chi}^{(1)} = \boldsymbol{\psi}^{(1)} , v^{(1)} = w^{(1)} , v^{(2)} = w^{(2)} ,$$

into (10.6.8). Since the virtual fields $\bar{s}$, $\bar{\eta}$, $\bar{\chi}$, $\bar{v}$ are sufficiently regular, otherwise arbitrary, the inequality (10.6.8) assume the form of the variational equality

$$\int_\Omega N^{\alpha\beta}_0 [e_{\alpha\beta}(\bar{s}) + \mathbf{m}^{\alpha\beta}_0 \kappa_{\alpha\beta}(\bar{v})]dx = \int_\Omega p\bar{v}dx + 0(\varepsilon) , \qquad \forall \, (\bar{s}, \bar{v}) \in V^0_K(\Omega) \quad (10.6.11)$$

where

$$\mathbf{m}^{\alpha\beta}_0 = M^{\alpha\beta}_0 + L^{\alpha\beta}_0 \tag{10.6.12}$$

and $V^0_K(\Omega) = H^1_0(\Omega)^2 \times H^2_0(\Omega)$.

Letting $\varepsilon \to 0$ we reduce the equality (10.6.11) to the form of the variational equation of equilibrium of the homogenized plate:

$$\int_\Omega [N^{\alpha\beta}_h e_{\alpha\beta}(\boldsymbol{s}) + \mathbf{m}^{\alpha\beta}_h \kappa_{\alpha\beta}(v)]dx = \int_\Omega pvdx , \qquad \forall \, (\boldsymbol{s}, v) \in V^0_K(\Omega) , \tag{10.6.13}$$

with

$$\boldsymbol{N}_h = \langle \boldsymbol{N}_0 \rangle_{YF} , \qquad \mathbf{m}_h = \langle \mathbf{m}_0 \rangle_{YF} . \tag{10.6.14}$$

Substitution of expressions (10.6.9) into (10.6.14) gives the homogenized constitutive relations in the following form

$$N^{\alpha\beta}_h = \tilde{C}^{\alpha\beta\lambda\mu}[h\epsilon^h_{\lambda\mu} + \frac{b^2 - a^2}{2}\kappa^h_{\lambda\mu} - \frac{b^2}{2}\kappa^F_{\lambda\mu}] ,$$

$$\tag{10.6.15}$$

$$\mathbf{m}^{\alpha\beta}_h = \tilde{C}^{\alpha\beta\lambda\mu}[\frac{b^2 - a^2}{2}\epsilon^h_{\lambda\mu} + \frac{a^3 + b^3}{3}\kappa^h_{\lambda\mu} - \frac{b^3}{3}\kappa^F_{\lambda\mu}] ,$$

involving the crack opening measures

$$\kappa^F_{\alpha\beta} = -\langle e^y_{\alpha\beta}(\boldsymbol{\psi}^{(1)}) \rangle_{YF} , \quad \text{or} \quad \kappa^F_{\alpha\beta} = -\langle e^y_{\alpha\beta}(\tilde{\boldsymbol{\psi}}) \rangle_{YF} , \tag{10.6.16}$$

or equivalently

$$\kappa^F_{\alpha\beta} = \frac{1}{2l_1 l_2} \int_F ([\tilde{\psi}_\alpha]\check{N}_\beta + [\tilde{\psi}_\beta]\check{N}_\alpha)ds . \tag{10.6.17}$$

Let us take (10.6.11) into account in the inequality (10.6.8). We thus arrive at a new inequality which after passing to zero with ε assumes the form

$$\int_\Omega \langle N_0^{\alpha\beta} e_{\alpha\beta}^y(\boldsymbol{s}^{(1)} - \boldsymbol{r}^{(1)}) + M_0^{\alpha\beta} e_{\alpha\beta}^y(\tilde{\boldsymbol{\eta}} - \tilde{\boldsymbol{\varphi}}) + L_0^{\alpha\beta} e_{\alpha\beta}^y(\tilde{\boldsymbol{\chi}} - \tilde{\boldsymbol{\psi}})$$

$$+ Q_0^\beta[\tilde{\eta}_\beta - \tilde{\varphi}_\beta + \frac{\partial}{\partial y_\beta}(v^{(2)} - w^{(2)})] + T_0^\beta[\tilde{\chi}_\beta - \tilde{\psi}_\beta + \frac{\partial}{\partial y_\beta}(v^{(2)} - w^{(2)})]\rangle_{YF}\, dx \geq 0 \,.$$

$$(10.6.18)$$

Let us take

$$\boldsymbol{s}^{(1)} - \boldsymbol{r}^{(1)} = \alpha\bar{\boldsymbol{s}} \,, \qquad \bar{s} \in H_{per}^1(Y)^2 \qquad (x - \text{fixed})$$

$$\tilde{\boldsymbol{\eta}} - \tilde{\boldsymbol{\varphi}} = \beta\bar{\boldsymbol{\eta}} \,, \qquad \bar{\eta} \in H_{per}^1(Y)^2 \qquad (x - \text{fixed})$$

$$\tilde{\boldsymbol{\chi}} = \tilde{\boldsymbol{\psi}} + \gamma(\bar{\boldsymbol{\chi}} - \tilde{\boldsymbol{\psi}}) \,, \qquad \bar{\chi} \in K_{YF} \qquad (x - \text{fixed})$$

$$v^{(2)} - w^{(2)} = \lambda\bar{v} \,, \qquad \bar{v} \in H_{per}^1(Y) \,,$$

$$(10.6.19)$$

where α, β, γ, λ are functions from the space $\mathbf{D}(\Omega)$ that assume values from the interval $[0,1]$. Note that $\tilde{\boldsymbol{\psi}} \in K_{YF}$ since $\tilde{\boldsymbol{\psi}} = \boldsymbol{\psi}^{(1)} + \nabla w^{(1)}$ and $\nabla w^{(1)}$ does not depend on y. Thus $\boldsymbol{\psi}^{(1)} \in K_{YF}$ implies $\tilde{\boldsymbol{\psi}} \in K_{YF}$. Hence $[\![\tilde{\chi}_N]\!] = (1 - \gamma)[\![\tilde{\psi}_N]\!] + \gamma[\![\bar{\chi}_N]\!] \geq 0$, which means that $\tilde{\chi} \in K_{YF}$. Let us recall that $f_N = f_\alpha \check{N}^\alpha$. On substituting representations (10.6.19) into (10.6.18) and making use of fields $\bar{s}$, $\bar{\eta}$, $\bar{\chi}$, $\bar{v}$ and functions α, β, γ, λ being independent, one arrives at the local problem:

$$(\bar{P}_{2LY}) \quad \left| \begin{array}{l} \text{find } (\boldsymbol{r}^{(1)}, \tilde{\boldsymbol{\varphi}}, \tilde{\boldsymbol{\psi}}, w^{(2)}) \in H_{2L,per}(YF) \\[4pt] = H_{per}^1(Y)^2 \times H_{per}^1(Y)^2 \times K_{YF} \times H_{per}^1(Y) \text{ such that} \\[4pt] \langle N_0^{\alpha\beta} e_{\alpha\beta}^y(\bar{s})\rangle_{YF} = 0 \,, \\[4pt] \langle M_0^{\alpha\beta} e_{\alpha\beta}^y(\bar{\eta}) + Q_0^\beta \bar{\eta}_\beta \rangle_{YF} = 0 \,, \\[4pt] \langle L_0^{\alpha\beta} e_{\alpha\beta}^y(\bar{\chi} - \tilde{\psi}) + T_0^\beta(\bar{\chi}_\beta - \tilde{\psi}_\beta)\rangle_{YF} \geq 0 \,, \\[4pt] \langle W_0^{\prime\beta} \bar{v}_{|\beta}\rangle_{YF} = 0 \qquad \forall\, (\bar{s}, \bar{\eta}, \bar{\chi}, \bar{v}) \in H_{2L,per}(YF) \end{array} \right.$$

$$(10.6.20)$$

where

$$N_0^{\alpha\beta} = \tilde{C}^{\alpha\beta\lambda\mu}[h e_{\lambda\mu}^y(\boldsymbol{r}^{(1)}) - \frac{a^2}{2} e_{\lambda\mu}^y(\tilde{\boldsymbol{\varphi}}) + \frac{b^2}{2} e_{\lambda\mu}^y(\tilde{\boldsymbol{\psi}})] + n_0^{\alpha\beta} \,,$$

$$M_0^{\alpha\beta} = \tilde{C}^{\alpha\beta\lambda\mu}[-\frac{a^2}{2} e_{\lambda\mu}^y(\boldsymbol{r}^{(1)}) + \frac{a^3}{3} e_{\lambda\mu}^y(\tilde{\boldsymbol{\varphi}})] + m_0^{\alpha\beta} \,,$$

$$(10.6.21)$$

$$L_0^{\alpha\beta} = \tilde{C}^{\alpha\beta\lambda\mu}[\frac{b^2}{2} e_{\lambda\mu}^y(\boldsymbol{r}^{(1)}) + \frac{b^3}{3} e_{\lambda\mu}^y(\tilde{\boldsymbol{\psi}})] + l_0^{\alpha\beta} \,,$$

$$W_0^\alpha = k_s C^{\alpha3\beta3}(h w_{|\beta}^{(2)} + a\tilde{\varphi}_\beta + b\tilde{\psi}_\beta) \,,$$

and Q_0^α, T_0^α are given by $(10.6.9)_{4,5}$. Moreover

$$n_0^{\alpha\beta} = \tilde{C}^{\alpha\beta\lambda\mu}\left(h\epsilon_{\lambda\mu}^h + \frac{b^2 - a^2}{2}\kappa_{\lambda\mu}^h \right),$$

$$m_0^{\alpha\beta} = \tilde{C}^{\alpha\beta\lambda\mu}\left(-\frac{a^2}{2}\epsilon_{\lambda\mu}^h + \frac{a^3}{3}\kappa_{\lambda\mu}^h \right), \qquad l_o^{\alpha\beta} = \tilde{C}^{\alpha\beta\lambda\mu}\left(\frac{b^2}{2}\epsilon_{\lambda\mu}^h + \frac{b^3}{3}\kappa_{\lambda\mu}^h \right). \tag{10.6.22}$$

Thus solution $(r^{(1)}, \tilde{\varphi}, \tilde{\psi}, w^{(2)})$ depends on (ϵ^h, κ^h). In particular, κ^F given by (10.6.17) is a function of (ϵ^h, κ^h), which determines the constitutive relations: $N_h = N_h(\epsilon^h, \kappa^h)$, $m_h = m_h(\epsilon^h, \kappa^h)$, cf. (10.6.15). The mathematical structure of the local problem $(\bar{P}_{2LY})$ is similar to the mathematical structure of the original problem (P_{2LF}) of Sec. 10.1. Thus both problems are simultaneously well posed. Note only that the fields $r_\alpha^{(1)}(\alpha = 1, 2)$ and $w^{(2)}$ are determined up to additive constants. These constants, however, do not affect the final form of the constitutive relations, since these relations depend only on the jumps $[\![\tilde{\psi}_\alpha(\epsilon^h, \kappa^h)]\!]$ on F, see Eq. (10.6.17).

The homogenized constitutive relations (10.6.15) can be put in the form characteristic for a hyperelastic medium:

$$N_h^{\alpha\beta} = \frac{\partial W}{\partial \epsilon_{\alpha\beta}^h}, \qquad m_h^{\alpha\beta} = \frac{\partial W}{\partial \kappa_{\alpha\beta}^h}. \tag{10.6.23}$$

The potential $W(\epsilon^h, \kappa^h)$ is given by

$$W = \frac{1}{2}\langle N_0^{\alpha\beta}\epsilon_{\alpha\beta}^0 + M_0^{\alpha\beta}\rho_{\alpha\beta}^0 + L_0^{\alpha\beta}k_{\alpha\beta}^0 + Q_0^\beta(\tilde{\varphi}_\beta + \frac{\partial w^{(2)}}{\partial y_\beta}) + T_0^\beta(\tilde{\psi}_\beta + \frac{\partial w^{(2)}}{\partial y_\beta})\rangle_{YF}. \tag{10.6.24}$$

Although differentiable, the potential W is only of class C^1. Moreover, it can easily be proved that there exist positive constants c_0, c_1 such that

$$c_0(|\epsilon|^2 + |\kappa|^2) \le W(\epsilon, \kappa) \le c_1(|\epsilon|^2 + |\kappa|^2)$$

for each $\epsilon, \kappa \in \mathbb{E}_s^2$.

The homogenized problem:

$$(\bar{P}_h^{2L}) \quad \left| \begin{array}{l} \text{find } (r^{(0)}, w^{(0)}) \in V_K^0(\Omega) \text{ such that the variational equilibrium} \\ \text{equation (10.6.3) holds with } N_h \text{ and } M_h \text{ determined by (10.6.23)} \end{array} \right.$$

can be rearranged to the minimization problem

$$(\check{P}_h^{2L}) \quad \inf\{\int_\Omega W(\epsilon(s), \kappa(v)) - f(v) | \, (s, v) \in V_K^0(\Omega)\}.$$

In both formulations this problem is uniquely solvable.

Remark 10.6.1. The homogenized model derived in the present section can be justified by the method of Γ-convergence. The procedure is similar to that already used in Sec. 5.4 and to that which will be presented in Sec. 11.5.

10.7. Plates with aligned cracks

The present section is aimed at evaluating the stiffness loss of an orthotropic plate weakened by aligned part-through the thickness cracks going along the orthotropy axis. The transverse dimensions of the plate are εa, εb, $\varepsilon h = \varepsilon(a + b)$, since the refined scaling method of Sec. 10.6 will be used here. The crack distance is εl, cf. Fig. 10.7.1, where the cross-section $x_2 = \mathrm{const}$ is shown.

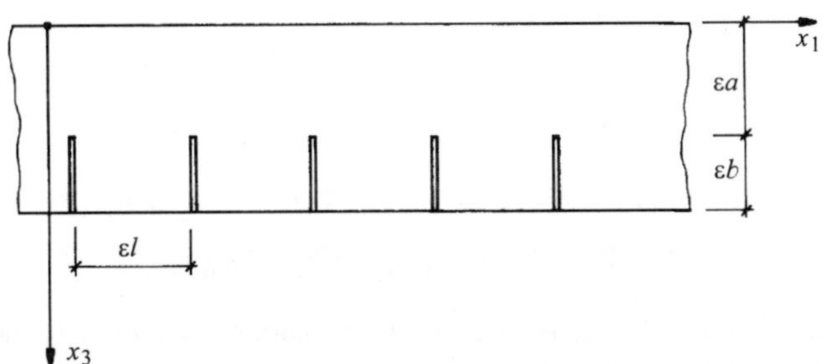

Fig. 10.7.1. Plate with aligned cracks. Section $x_2 =$const

The rescaled periodicity cell Y degenerates to the interval $y_1 \in (0, l)$, cf. Fig. 10.7.2.

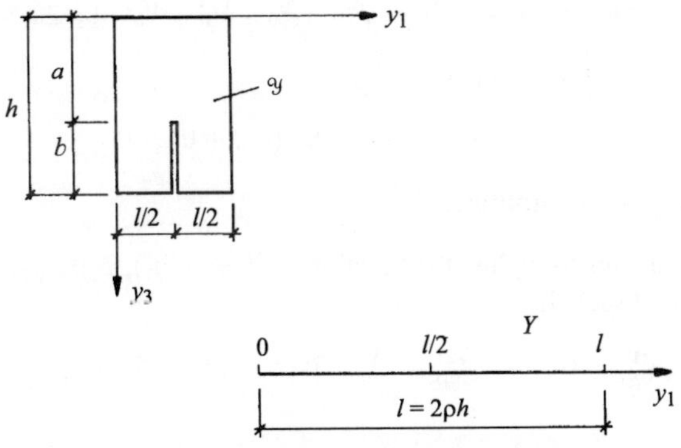

Fig. 10.7.2. Basic cells $\mathcal{Y}$ and Y

Position of the crack within Y can be arbitrary. Thus we choose the simplest case when the crack is located in the middle, see Fig. 10.7.2.

Let us consider first the case when the crack is open. Due to orthotropy the problem $(\bar{P}_{2L,Y})$ splits up into two local problems: a) bending problem (P_a), b) torsional problem (P_b). Both are one-dimensional.

a) Problem (P_a) amounts to finding the functions $r_1^{(1)} = r_a(y_1)$, $\tilde{\varphi}_1 = \varphi_a(y_1)$, $\tilde{\psi}_1 = \psi_a(y_1)$, $w^{(2)} = w(y_1)$ defined for $y_1 \in [0, l]$ such that

$$N_{0|1}^{11} = 0 , \quad -M_{0|1}^{11} + Q_0^1 = 0 , \quad -L_{0|1}^{11} + T_0^1 = 0 , \quad W_{0|1}^1 = 0 , \qquad (10.7.1)$$

where $(\cdot)_{|1} = d(\cdot)/dy_1$ and

$$N_0^{11} = \tilde{C}^{1111} \left(h r_{a|1} - \frac{a^2}{2} \varphi_{a|1} + \frac{b^2}{2} \psi_{a|1} \right) + n_0^{11} ,$$

$$M_0^{11} = \tilde{C}^{1111} \left(-\frac{a^2}{2} r_{a|1} + \frac{a^3}{3} \varphi_{a|1} \right) + m_0^{11} ,$$

$$L_0^{11} = \tilde{C}^{1111} \left(\frac{b^2}{2} r_{a|1} + \frac{b^3}{3} \psi_{a|1} \right) + l_0^{11} , \qquad (10.7.2)$$

$$Q_0^1 = a C^{1313} (\varphi_a + w_{|1}) , \quad T_0^1 = b C^{1313} (\psi_a + w_{|1}) ,$$

$$W_0^1 = k_s C^{1313} (a \varphi_a + b \psi_a + h w_{|1}) .$$

The shear correction factor k_s is taken as 1. Quantities n_0^{11}, m_0^{11}, l_0^{11} are defined by (10.6.22). The boundary conditions are:

$$f(0) = f(l) , \qquad f \in \{N_0^{11}, M_0^{11}, L_0^{11}, W_0^1 ; r_a, \varphi_a, \psi_a, w\} .$$

The switching conditions at $y_1 = l/2$ are given by

$$f(l/2 - 0) = f(l/2 + 0) , \quad f \in \{N_0^{11}, M_0^{11}, W_0^1 ; r_a, \varphi_a, w\} .$$

Since the crack is considered as open, we have

$$L_0^{11}(l/2 - 0) = 0 , \quad L_0^{11}(l/2 + 0) = 0 .$$

The jump $[\![\psi_a(l/2)]\!]$ is here arbitrary. $\qquad\qquad\qquad\qquad\qquad\qquad\qquad$ $\square$

b) Problem (P_b) amounts to finding the functions $r_2^{(1)} = r_b(y_1)$, $\tilde{\varphi}_2 = \varphi_b(y_1)$, $\tilde{\psi}_2 = \psi_b(y_1)$ defined for $y_1 \in [0, l]$ such that

$$N_{0|1}^{12} = 0 , \quad -M_{0|1}^{21} + Q_0^2 = 0 , \quad -L_{0|1}^{21} + T_0^2 = 0 , \qquad (10.7.3)$$

where

$$N_0^{21} = \tilde{C}^{2121} \left(h r_{b|1} - \frac{a^2}{2} \varphi_{b|1} + \frac{b^2}{2} \psi_{b|1} \right) + n_0^{21} ,$$

$$M_0^{21} = \tilde{C}^{2121} \left(-\frac{a^2}{2} r_{b|1} + \frac{a^3}{3} \varphi_{b|1} \right) + m_0^{21} ,$$

$$L_0^{21} = \tilde{C}^{2121} \left(\frac{b^2}{2} r_{b|1} + \frac{b^3}{3} \psi_{b|1} \right) + l_0^{21} , \qquad (10.7.4)$$

$$Q_0^2 = a C^{2323} \varphi_b , \quad T_0^2 = b C^{2323} \psi_b ,$$

where

$$n_0^{21} = 2\tilde{C}^{2121}\left(h\epsilon_{12}^h + \frac{1}{2}(b^2 - a^2)\kappa_{12}^h\right),$$

$$m_0^{21} = 2\tilde{C}^{2121}\left(-\frac{1}{2}a^2\epsilon_{12}^h + \frac{1}{3}a^3\kappa_{12}^h\right), \qquad l_0^{21} = 2\tilde{C}^{2121}\left(\frac{b^2}{2}\epsilon_{12}^h + \frac{1}{3}b^3\kappa_{12}^h\right).$$

$$\text{(10.7.5)}$$

The boundary conditions are:

$$f(0) = f(l), \qquad f \in \{N_0^{21}, M_0^{21}, L_0^{21} \, ; \, r_b, \varphi_b, \psi_b\} \,.$$

The switching conditions at $y_1 = l/2$ are given by

$$f(l/2 - 0) = f(l/2 + 0), \quad f \in \{r_b, \varphi_b\} \,.$$

The jump $[\![\psi_b(l/2)]\!]$ is here arbitrary. $\qquad\qquad\qquad\qquad\qquad\qquad\qquad\quad\square$

The homogenized equations (10.6.15) reduce to the form

$$N_h^{\alpha\alpha} = \tilde{C}^{\alpha\alpha 11}\left(h\epsilon_{11}^h + \frac{b^2 - a^2}{2}\kappa_{11}^h - \frac{b^2}{2}\kappa_{11}^F\right)$$
$$+ \tilde{C}^{\alpha\alpha 22}\left(h\epsilon_{22}^h + \frac{b^2 - a^2}{2}\kappa_{22}^h\right),$$
$$N_h^{12} = \tilde{C}^{1212}[2h\epsilon_{12}^h + (b^2 - a^2)\kappa_{12}^h - b^2\kappa_{12}^F] \,;$$

$$m_h^{\alpha\alpha} = \tilde{C}^{\alpha\alpha 11}\left(\frac{b^2 - a^2}{2}\epsilon_{11}^h + \frac{a^3 + b^3}{3}\kappa_{11}^h - \frac{b^3}{3}\kappa_{11}^F\right)$$
$$+ \tilde{C}^{\alpha\alpha 22}\left(\frac{b^2 - a^2}{2}\epsilon_{22}^h + \frac{a^3 + b^3}{3}\kappa_{22}^h\right),$$
$$m_h^{12} = \tilde{C}^{1212}\left[(b^2 - a^2)\epsilon_{12}^h + \frac{2}{3}(a^3 + b^3)\kappa_{12}^h - \frac{2}{3}b^3\kappa_{12}^F\right],$$

$$\text{(10.7.6)}$$

where, according to (10.6.17):

$$\kappa_{11}^F = \frac{1}{l}[\![\psi_a]\!], \qquad \kappa_{12}^F = \frac{1}{2l}[\![\psi_b]\!] \,. \qquad\qquad\qquad\text{(10.7.7)}$$

Here $[\![f]\!] = f(l/2 + 0) - f(l/2 - 0)$. Note that

$$\kappa_{11}^F = \kappa_{11}^F(\epsilon_{11}^h, \epsilon_{22}^h, \kappa_{11}^h, \kappa_{22}^h), \qquad \kappa_{12}^F = \kappa_{12}^F(\epsilon_{12}^h, \kappa_{12}^h) \,,$$

which confirm that problem (P_a) and (P_b) are independent. Both problems can be analytically solved. Let us sketch how to solve the problem (P_a) in the case of $a = b = h/2$.

Substitution of (10.7.2) into (10.7.1) gives the homogeneous system of ordinary differential equations:

$$\frac{d^2 r_a}{dy_1^2} = \frac{6\sigma}{h}(\varphi_a - \psi_a) \,,$$

$$\frac{d^2 \varphi_a}{dy_1^2} = \frac{3\sigma}{ah}\left(\frac{h}{a}\frac{dw}{dy_1} + \frac{3a+h}{a}\varphi_a - \psi_a\right) \,,$$

$$\frac{d^2 \psi_a}{dy_1^2} = \frac{3\sigma}{bh}\left(\frac{h}{b}\frac{dw}{dy_1} + \frac{3b+h}{b}\psi_a - 3\varphi_a\right) \,,$$

$$\frac{d}{dy_1}\left(\frac{dw}{dy_1} + \frac{a}{h}\varphi_a + \frac{b}{h}\psi_a\right) = 0 \,,$$

where $\sigma = C^{1313}/\tilde{C}^{1111}$.

Let us substitute $y_1 = h\xi$, $a = b = h/2$ and eliminate the unknowns r_a and w. One finds

$$\frac{d^2 \varphi_a}{d\xi^2} = 24\sigma(\varphi_a - \psi_a) + 12\sigma K \,, \qquad \frac{d^2 \psi_a}{d\xi^2} = -24\sigma(\varphi_a - \psi_a) + 12\sigma K \,,$$

where K is a constant. The system above is easy to solve. Within the whole interval $(0, 2\rho)$, $2\rho = l/h$, the solution can be represented as follows

$$\varphi_a = \begin{cases} C_p e^{-\alpha_1 \xi} + C_k e^{-\alpha_1(\rho-\xi)} + 6\sigma K\xi^2 + C_3\xi + C_4 \,, & \text{if } \xi \in (0, \rho) \\ D_p e^{-\alpha_1(\xi-\rho)} + D_k e^{-\alpha_1(2\rho-\xi)} + 6\sigma L\xi^2 + D_3\xi + D_4 \,, & \text{if } \xi \in (\rho, 2\rho) \,, \end{cases}$$

$$\psi_a = \begin{cases} -C_p e^{-\alpha_1 \xi} - C_k e^{-\alpha_1(\rho-\xi)} + 6\sigma K\xi^2 + C_3\xi + C_4 \,, & \text{if } \xi \in (0, \rho) \\ -D_p e^{-\alpha_1(\xi-\rho)} - D_k e^{-\alpha_1(2\rho-\xi)} + 6\sigma L\xi^2 + D_3\xi + D_4 \,, & \text{if } \xi \in (\rho, 2\rho) \,, \end{cases}$$

$$r_a = \begin{cases} \dfrac{h}{4}[C_p e^{-\alpha_1 \xi} + C_k e^{-\alpha_1(\rho-\xi)} + C_5\xi + C_6] \,, & \text{if } \xi \in (0, \rho) \\[2mm] \dfrac{h}{4}[D_p e^{-\alpha_1(\xi-\rho)} + D_k e^{-\alpha_1(2\rho-\xi)} + D_5\xi + D_6] \,, & \text{if } \xi \in (\rho, 2\rho) \,, \end{cases}$$

$$w = \begin{cases} h[-2\sigma K\xi^3 - \dfrac{1}{2}C_3\xi^2 + (K - C_4)\xi + C_7] \,, & \text{if } \xi \in (0, \rho) \\[2mm] h[-2\sigma K\xi^3 - \dfrac{1}{2}D_3\xi^2 + (L - D_4)\xi + D_7] \,, & \text{if } \xi \in (0, 2\rho) \,, \end{cases}$$

where $\alpha_1 = (48\sigma)^{1/2}$, $\rho = l/2h$ and $C_p, C_k, C_3, C_4, D_3, D_4, K, L, C_5, C_6, D_5, D_6, C_7, D_7$ are integration constants. By using the boundary conditions, the conditions on the crack and the switching conditions one finds that

$$C_3 = D_3 = C_5 = D_5 \,, \qquad K = L = 0 \,, \qquad C_4 = 0 \,.$$

Further analysis gives

$$[\![\psi_a]\!]/l = \frac{24 l_0^{11}}{h^3 \tilde{C}^{1111}} \cdot \frac{8}{\alpha_1 \rho \coth(\alpha_1 \rho) + 7}$$

if $l_0^{11} \geq 0$ and this result turns out to be sufficient to find all formulae for the reduced effective stiffnesses with indices: 1111, 1122, 2222.

Finding the solution to the problem (P_b) is easier and will be omitted.

The final results for the case of $a = b = h/2$ have the following form. The crack deformation measures read

$$\kappa_{11}^{F} = [(\kappa_{11}^{h} + \nu_{12}\kappa_{22}^{h}) + \frac{1}{R}(\epsilon_{11}^{h} + \nu_{12}\epsilon_{22}^{h})]g_1(\alpha_1\rho) \,,$$

$$\kappa_{12}^{F} = (\kappa_{12}^{h} + \frac{1}{R}\epsilon_{12}^{h})g_2(\alpha_2\rho) \tag{10.7.8}$$

where $R = 2a/3$,

$$\alpha_1 = (48 \cdot C^{1313}/\tilde{C}^{1111})^{1/2} \,, \qquad \nu_{12} = \tilde{C}^{1122}/\tilde{C}^{1111} \,,$$

$$\alpha_2 = (12 \cdot C^{2323}/\tilde{C}^{1212})^{1/2} \,, \qquad \rho = \frac{l}{2h} \tag{10.7.9}$$

and functions $g_1(x), g_2(x)$ are given by

$$g_1(x) = \frac{8}{x \cdot \coth x + 7} \,, \tag{10.7.10}$$

$$g_2(x) = \frac{8\sinh 2x}{3\sinh 2x + 2x(2 + 3\cosh 2x)} \,. \tag{10.7.11}$$

Note that

$$\lim_{x \to 0} g_\alpha(x) = 1 \,, \qquad \lim_{x \to \infty} g_\alpha(x) = 0 \,, \qquad \alpha = 1, 2 \,. \tag{10.7.12}$$

For isotropic case we have

$$\alpha_1 = [24(1 - \nu)]^{1/2} \,, \qquad \alpha_2 = (12)^{1/2} \,, \qquad \nu_{12} = \nu \,, \tag{10.7.13}$$

where ν is the Poisson ratio.

The results (10.7.8) concern the case of the crack being open. One can prove that the crack is open if $l_0^{11} \geq 0$, which means here that

$$\eta := \frac{1}{R}(\epsilon_{11}^{h} + \nu_{12}\epsilon_{22}^{h}) + (\kappa_{11}^{h} + \nu_{12}\kappa_{22}^{h}) \geq 0 \,. \tag{10.7.14}$$

Thus

$$\kappa_{11}^{F} = \begin{cases} 0 & \text{if } \eta \leq 0 \ \text{(the crack is closed)} \\ \eta g_1(\alpha_1\rho) & \text{if } \eta > 0 \ \text{(the crack is open)} \,, \end{cases} \tag{10.7.15}$$

while formula (10.7.8)$_2$ for κ_{12}^{F} holds good in both cases. Having found the rule (10.7.15) we can write down the final form of the homogenized constitutive relations

$$N_h^{\alpha\alpha} = h(A^{\alpha\alpha 11}\epsilon_{11}^{h} + A^{\alpha\alpha 22}\epsilon_{22}^{h}) + h^2(E^{\alpha\alpha 11}\kappa_{11}^{h} + E^{\alpha\alpha 22}\kappa_{22}^{h}) \,,$$

$$N_h^{12} = 2h(A^{1212}\epsilon_{12}^{h} + hE^{1212}\kappa_{12}^{h}) \,; \tag{10.7.16}$$

$$\mathbf{m}_h^{\alpha\alpha} = h^2(F^{\alpha\alpha 11}\epsilon_{11}^{h} + F^{\alpha\alpha 22}\epsilon_{22}^{h}) + h^3(D^{\alpha\alpha 11}\kappa_{11}^{h} + D^{\alpha\alpha 22}\kappa_{22}^{h}) \,,$$

$$\mathbf{m}_h^{12} = 2h^2(F^{1212}\epsilon_{12}^{h} + hD^{1212}\kappa_{12}^{h}) \,. \tag{10.7.17}$$

The explicit formulae for the reduced stiffnesses will be given for the case $a = b = h/2$.

The stiffnesses

$$A^{\alpha\alpha\beta\beta} = A^{\beta\beta\alpha\alpha} \, , \quad E^{\alpha\alpha\beta\beta} = F^{\beta\beta\alpha\alpha} \, , \quad D^{\alpha\alpha\beta\beta} = D^{\beta\beta\alpha\alpha} \, , \tag{10.7.18}$$

are reduced or not:

i) if $\eta \le 0$ the crack is closed and then

$$A^{\alpha\alpha\beta\beta} = \tilde{C}^{\alpha\alpha\beta\beta} \, , \quad E^{\alpha\alpha\beta\beta} = F^{\beta\beta\alpha\alpha} = 0 \, , \quad D^{\alpha\alpha\beta\beta} = \tilde{C}^{\alpha\alpha\beta\beta}/12 \, . \tag{10.7.19}$$

ii) If $\eta > 0$ the crack is open and then the moduli are reduced as follows

$$A^{1111}/\tilde{C}^{1111} = 1 - \frac{3}{8}g_1(\alpha_1\rho) \, ,$$

$$A^{1122}/\tilde{C}^{1111} = \nu_{12}\left(1 - \frac{3}{8}g_1(\alpha_1\rho)\right) \, , \quad A^{2222}/\tilde{C}^{2222} = 1 - \frac{3}{8}\nu_{12}\nu_{21}g_1(\alpha_1\rho) \, , \tag{10.7.20}$$

$$E^{1111}/\tilde{C}^{1111} = -\frac{1}{8}g_1(\alpha_1\rho) \, ,$$

$$E^{1122}/\tilde{C}^{1111} = -\frac{1}{8}\nu_{12}g_1(\alpha_1\rho) \, , \quad E^{2222}/\tilde{C}^{2222} = -\frac{1}{8}\nu_{12}\nu_{21}g_1(\alpha_1\rho) \, ; \tag{10.7.21}$$

$$F^{\alpha\alpha\beta\beta} = E^{\beta\beta\alpha\alpha} \, ; \tag{10.7.22}$$

$$D^{1111}/\tilde{D}^{1111} = 1 - \frac{1}{2}g_1(\alpha_1\rho) \, ,$$

$$D^{1122}/\tilde{D}^{1111} = \nu_{12}\left[1 - \frac{1}{2}g_1(\alpha_1\rho)\right] \, , \quad D^{2222}/\tilde{D}^{2222} = 1 - \frac{1}{2}\nu_{12}\nu_{21}g_1(\alpha_1\rho) \, , \tag{10.7.23}$$

where

$$\tilde{D}^{\alpha\beta\lambda\mu} = \frac{1}{12}\tilde{C}^{\alpha\beta\lambda\mu} \, , \quad \nu_{12} = \tilde{C}^{1122}/\tilde{C}^{1111} \, , \quad \nu_{21} = \tilde{C}^{1122}/\tilde{C}^{2222} \, . \tag{10.7.24}$$

Reduction of stiffnesses:

$$K^{1212} = K^{2121} = K^{2112} = K^{1221} \, , \quad K^{\alpha\beta\alpha\beta} \in \{A^{\alpha\beta\alpha\beta}, E^{\alpha\beta\alpha\beta}, F^{\alpha\beta\alpha\beta}, D^{\alpha\beta\alpha\beta}\} \, , \quad (\alpha \ne \beta) \, ,$$

is independent of η:

$$A^{1212}/C^{1212} = 1 - \frac{3}{8}g_2(\alpha_2\rho) \, , \qquad E^{1212}/C^{1212} = -\frac{1}{8}g_2(\alpha_2\rho) \, ,$$

$$D^{1212}/\tilde{D}^{1212} = 1 - \frac{1}{2}g_2(\alpha_2\rho) \, , \qquad F^{1212} = E^{1212} \, . \tag{10.7.25}$$

The opening of the crack is determined by the crack deformation measures κ_{11}^F, κ_{12}^F, see (10.7.7), (10.7.8). They depend on the crack density $c_d = h/l$, since $\rho = \frac{1}{2}(c_d)^{-1}$. According to (10.7.12)

$$\lim_{c_d \to 0} \kappa_{12}^F = 0 \, , \quad \lim_{c_d \to \infty} = \eta(\epsilon^h, \kappa^h) \, , \qquad \lim_{c_d \to \infty} \kappa_{12}^F = \vartheta(\epsilon_{12}^h, \kappa_{12}^h) \, , \tag{10.7.26}$$

where η is defined by (10.7.14) and $\vartheta = \kappa_{12}^h + \epsilon_{12}^h/R$. Shapes of functions $\bar{\kappa}_{11}^F(c_d) = \kappa_{11}^F/\eta$,

$\bar{\kappa}^F_{12}(c_d) = \kappa^F_{12}/\vartheta$ are shown in Fig. 10.7.3. This figure concerns an isotropic plate of $\nu = 0.3$.

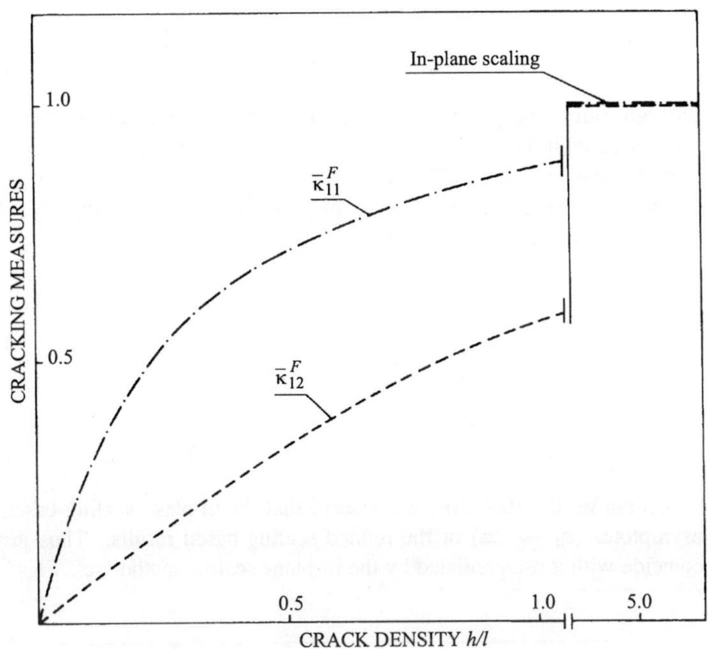

Fig. 10.7.3. Cracking measures $\bar{\kappa}^F_{1\alpha}$ versus crack density h/l

One can see that the cracking measures $\bar{\kappa}^F_{1\alpha}$ tend asymptotically to 1, as the crack density increases.

Note that

$$A^{1111}/\tilde{C}^{1111} = A^{1122}/\tilde{C}^{1122} , \qquad D^{1111}/\tilde{D}^{1111} = D^{1122}/\tilde{D}^{1122} . \tag{10.7.27}$$

The stiffnesses $A^{\alpha\beta\lambda\mu}$ and $D^{\alpha\beta\lambda\mu}$ decrease when c_d increases and tend asymptotically to the following values, see Figs. 10.7.4 – 10.7.6.

$$A^{1111}(c_d = \infty)/\tilde{C}^{1111} = A^{1122}(c_d = \infty)/\tilde{C}^{1122} = A^{1212}(c_d = \infty)/\tilde{C}^{1212} = \frac{5}{8} ,$$

$$A^{2222}(c_d = \infty)/\tilde{C}^{2222} = 1 - \frac{3}{8}\nu^2 ,$$

$$\tag{10.7.28}$$

$$D^{1111}(c_d = \infty)/\tilde{D}^{1111} = D^{1122}(c_d = \infty)/\tilde{D}^{1122} = D^{1212}(c_d = \infty)/\tilde{D}^{1212} = \frac{1}{2} ,$$

$$D^{2222}(c_d = \infty)/\tilde{D}^{2222} = 1 - \frac{\nu^2}{2} .$$

The reciprocal stiffnesses tend asymptotically to

$$
E^{1111}(c_d = \infty)/\tilde{C}^{1111} = E^{1122}(c_d = \infty)/\tilde{C}^{1122} = E^{1212}(c_d = \infty)/\tilde{C}^{1212} = -\frac{1}{8} ,
$$

$$
E^{2222}(c_d = \infty)/\tilde{C}^{2222} = -\frac{\nu^2}{8} , \tag{10.7.29}
$$

thus their absolute values increase with c_d. This means that the membrane – bending coupling becomes essential if the crack density increases, see Fig. 10.7.6.

The same problem of evaluating effective stiffnesses of the plate with aligned cracks (see Fig. 10.7.1) could be solved by means of the in-plane scaling method of Sec. 10.5. Problem $(P_{2L,Y})$ can be solved exactly, thus making it possible to find relations $k^F_{\lambda\mu}(\epsilon^h, k^h)$. It turns out that $k^F_{22} = 0$ and

$$
k^F_{11} = \begin{cases} 0 & \text{if} & \eta' \le 0 \\ \eta' & \text{if} & \eta' > 0 , \end{cases} \qquad k^F_{12} = k^h_{12} + \frac{1}{R}\epsilon^h_{12} \tag{10.7.30}
$$

where

$$
\eta' = \frac{1}{R}(\epsilon^h_{11} + \nu_{12}\epsilon^h_{22}) + (k^h_{11} + \nu_{12}k^h_{22}) . \tag{10.7.31}
$$

Taking into account result (10.7.26)$_2$ we observe that the in-plane scaling-based method produces asymptotes $(c_d \to \infty)$ of the refined scaling-based results. Thus just results (10.7.28) coincide with those predicted by the in-plane scaling method, cf. Figs. 10.7.4 – 10.7.6.

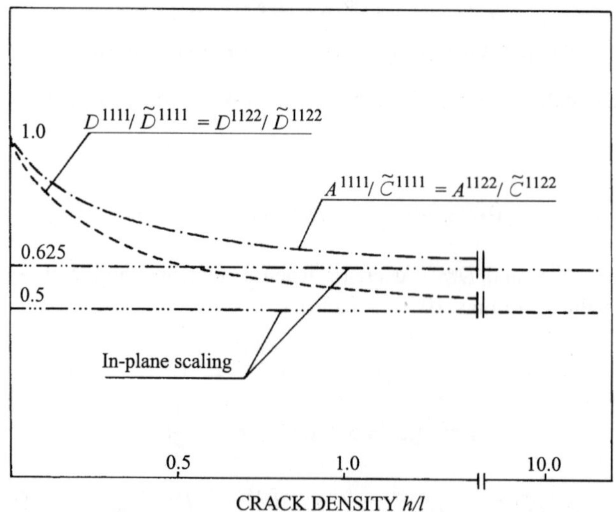

Fig. 10.7.4. Reduction of the stiffnesses $A^{\alpha\alpha 11}$, $D^{\alpha\alpha 11}$ versus crack density

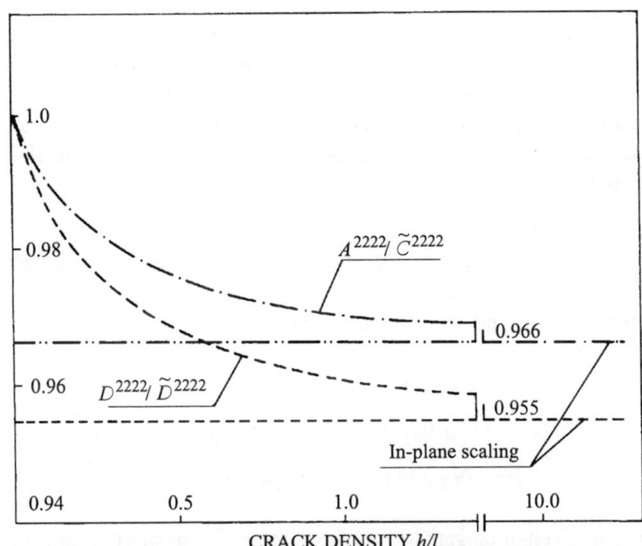

Fig. 10.7.5. Reduction of the stiffnesses A^{2222}, D^{2222}

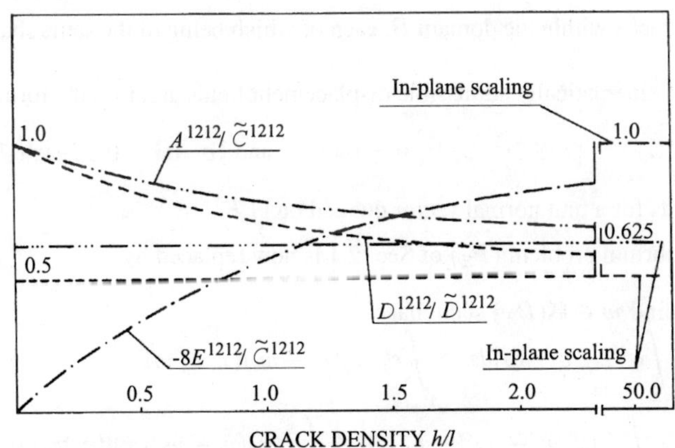

Fig. 10.7.6. Reduction of the stiffnesses A^{1212}, E^{1212}, D^{1212}

This confirms that the in-plane scaling assumptions refer to the case of plates with high density of crack distribution.

10.8.   *Cracks of arbitrary position. Three-dimensional local analysis*

In Sec. 2.3 the effective characteristics of a thin $\mathcal{Z}$-periodic plate have been determined. Two types of inhomogeneities: material and geometric have been considered. The elastic moduli and the bounding surfaces of the plate were Z-periodic functions. In the present section it is additionally assumed that each cell $\mathcal{Z}$ is weakened by an identical crack $\mathcal{F}$. The aim is still the same: to smear-out the cracks and find the effective potential of a hypothetic homogenized plate.

10.8.1.   Asymptotic analysis

The surface of the crack divides the cell into two parts, denoted by $\mathcal{Z}_\alpha$. Assume that at each point $x \in \mathcal{F}$ a versor $\check{N}$ is directed outward to $\mathcal{Z}_1$. The friction on the crack $\mathcal{F}$ is neglected and the non-penetration condition is assumed. Thus

$$\overset{1}{\sigma}{}^{ij}\check{N}_i\check{N}_j = \overset{2}{\sigma}{}^{ij}\check{N}_i\check{N}_j = \sigma_N\,, \qquad \sigma_N \leq 0$$

$$\overset{1}{\sigma}{}^{ij}\check{N}_iT_j = \overset{2}{\sigma}{}^{ij}\check{N}_iT_j = 0\,, \qquad\qquad (10.8.1)$$

$$[\![\boldsymbol{w}\cdot\check{\boldsymbol{N}}]\!]_{\mathcal{F}} \geq 0\,, \qquad \sigma_N[\![\boldsymbol{w}\cdot\check{\boldsymbol{N}}]\!]_{\mathcal{F}} = 0\,.$$

Here $\boldsymbol{T}$ represents a versor tangent to $\mathcal{F}$ at the point $\boldsymbol{x}$, $\overset{\alpha}{\sigma}{}^{ij}$ means the value of the stress σ^{ij} taken from the α-th side of $\mathcal{F}(\alpha = 1, 2)$ and $[\![\ ]\!]_{\mathcal{F}}$ represents the jump on $\mathcal{F}$: $[\![f]\!]_{\mathcal{F}} := \overset{2}{f} - \overset{1}{f}$. The subscript $\mathcal{F}$ will further be omitted.

Let $B_{\mathcal{F}}$ represent the cracked three-dimensional domain of the plate or $B_{\mathcal{F}} = B\backslash\underset{j}{\cup}\mathcal{F}_j$ and $\mathcal{F}_j$ are cracks within the domain B, each of which being of the same shape as the crack $\mathcal{F}$.

The set of kinematically admissible displacement fields assumes the form

$$\mathbb{K}(B_{\mathcal{F}}) = \{\boldsymbol{v} \in H^1(B_{\mathcal{F}})^3|\ \boldsymbol{v} = 0 \text{ on } \Gamma_0 \text{ and } [\![\boldsymbol{v}\cdot\boldsymbol{n}]\!] \geq 0 \text{ on each } \mathcal{F}\}\,.$$

Here $\boldsymbol{n}$ stands for a unit normal vector defined on $\underset{j}{\cup}\mathcal{F}_j$.

The equilibrium problem $(P_{\mathcal{Z}})$ of Sec. 2.1 is now replaced by

$$(P_{\mathcal{Z}})\quad\left|\begin{array}{l} \text{find } \boldsymbol{w} \in \mathbb{K}(B_{\mathcal{F}}) \text{ such that}\\[4pt] \displaystyle\int_{B_{\mathcal{F}}} \sigma^{ij}e_{ij}(\boldsymbol{v} - \boldsymbol{w})dx \geq \int_{\Gamma_+} r^i_+(x)(v_i - w_i)(x, x_3^+)d\Gamma\\[14pt] \displaystyle + \int_{\Gamma_-} r^i_-(x)(v_i - w_i)(x, x_3^-)d\Gamma + \int_B b^i(x)(v_i - w_i)(\boldsymbol{x})dx\ \ \forall\ \boldsymbol{v} \in \mathbb{K}(B_{\mathcal{F}})\,. \end{array}\right.$$

$$(10.8.2)$$

Since our aim is to find effective characteristics of the cracked plate we introduce a small parameter ε according to the rules $(2.2.1) - (2.2.3)$. The cracks $\mathcal{F}$ are replaced by $\varepsilon\mathcal{F}$. Instead of $\mathcal{Y}$ we shall deal here with $\mathcal{Y}\mathcal{F} = \mathcal{Y}\backslash\mathcal{F}$, $\mathcal{Y}$ being defined as in Sec. 2.2. The surface of the crack is still denoted by $\mathcal{F}$, see Fig. 10.8.1. The domain $B_{\mathcal{F}}$ is replaced with

$B_\varepsilon^\varepsilon = B_\varepsilon \backslash \mathcal{F}_e^\varepsilon$, where the domain B_ε has been introduced in Sec. 2.2 and $\mathcal{F}_\varepsilon^\varepsilon = \bigcup_j \varepsilon \mathcal{F}_j$. The set $\mathbb{K}(B_\mathcal{F})$ is replaced with $\mathbb{K}(B_\varepsilon^\varepsilon)$ defined by

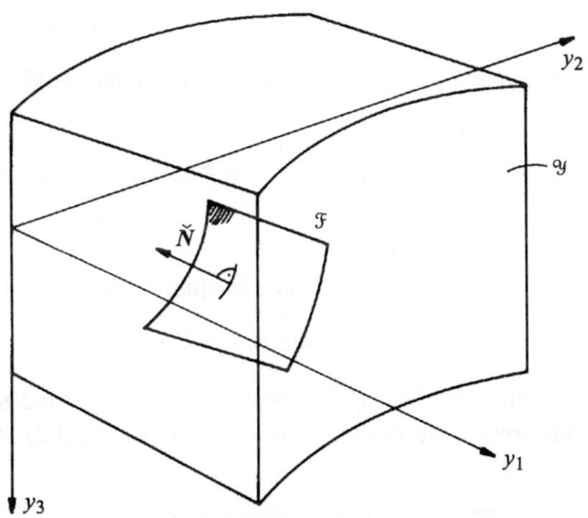

Fig. 10.8.1. Periodicity cell with a crack

$$\mathbb{K}(B_\varepsilon^\varepsilon) = \{v \in H^1(B_\varepsilon^\varepsilon)^3 | \ v = 0 \text{ on } \Gamma_0^\varepsilon \text{ and } [\![v \cdot n]\!] \geq 0 \text{ on } \mathcal{F}_\varepsilon^\varepsilon\} \ .$$

Obviously, the unit normal vector n is now defined on $\mathcal{F}^\varepsilon$.

The unknown displacement field w^ε solves the problem:

$$(P_\varepsilon) \quad \left|
\begin{aligned}
&\text{find } w^\varepsilon \in \mathbb{K}(B_\varepsilon^\varepsilon) \text{ such that for each } v \in \mathbb{K}(B_\varepsilon^\varepsilon) \\
&\int_{B_\varepsilon^\varepsilon} C^{ijkl}\left(\frac{x}{\varepsilon}\right) e_{kl}(w^\varepsilon)e_{ij}(v - w^\varepsilon)dx \\
&\geq \int_{\Gamma_+^\varepsilon} [\varepsilon^2 p_+^\alpha(v_\alpha - w_\alpha^\varepsilon)(x, \varepsilon c^+) + \varepsilon^3 q_+(v_3 - w_3^\varepsilon)(x, \varepsilon c^+)]d\Gamma \\
&+ \int_{\Gamma_-^\varepsilon} [\varepsilon^2 p_-^\alpha(v_\alpha - w_\alpha^\varepsilon)(x, \varepsilon c^-) + \varepsilon^3 q_-(v_3 - w_3^\varepsilon)(x, \varepsilon c^-)]d\Gamma \\
&+ \int_{B_\varepsilon} [\varepsilon b^\alpha\left(\frac{x}{\varepsilon}\right)(v_\alpha - w_\alpha)(x) + \varepsilon^2 b^3\left(\frac{x}{\varepsilon}\right)(v_3 - w_3^\varepsilon)(x)]dx \ .
\end{aligned}
\right. \tag{10.8.3}$$

Now we introduce the assumptions (2.2.9), set $y = x/\varepsilon$ as an independent variable and integrate both sides of (10.8.3) over $\mathcal{Y}$. Problem (P_ε) is rearranged to the form:

$$(\widetilde{P}_\varepsilon) \quad \left| \begin{array}{l} \text{find } \widetilde{w}^\varepsilon \in \mathbb{K}(\Omega \times \mathcal{Y}\mathcal{F}) \text{ such that} \\[2mm] A^\varepsilon(\widetilde{w}^\varepsilon, v - \widetilde{w}^\varepsilon) \geq F^\varepsilon(v - \widetilde{w}^\varepsilon) \quad \forall\, v \in \mathbb{K}(\Omega \times \mathcal{Y}\mathcal{F}), \end{array} \right. \qquad (10.8.4)$$

where A^ε and F^ε are defined by (2.2.15) and (2.2.16) while the set $\mathbb{K}(\Omega \times \mathcal{Y}\mathcal{F})$ is defined as follows

$$\mathbb{K}(\Omega \times \mathcal{Y}\mathcal{F}) = \{ v = (v_i(x, y)) |\ v(x, \cdot) \in \mathbb{K}(\mathcal{Y}\mathcal{F}), v(\cdot, y) \in H_0^1(\Omega)^3 \},$$

where

$$\mathbb{K}(\mathcal{Y}\mathcal{F}) = \{ v \in H^1(\mathcal{Y}\mathcal{F})^3 |\ v \text{ assumes equal values on the opposite}$$

$$\text{lateral faces of } \mathcal{Y}\mathcal{F} \text{ and } [\![v \cdot \check{N}]\!] \geq 0 \text{ on } \mathcal{F} \}.$$

The solution $\widetilde{w}^\varepsilon$ is sought in the form (2.3.1) with $u^{(0)} \in H_0^1(\Omega)^3$ and $u^{(p)} \in K(\Omega \times \mathcal{Y}\mathcal{F})$ for $p = 1, 2, \ldots.$. The stresses are expanded according to Eq. (2.3.2). Moreover, any trial field v is expanded as follows

$$v = v^{(0)}(x) + \varepsilon v^{(1)}(x, y) + \varepsilon^2 v^{(2)}(x, y) + \ldots, \qquad (10.8.5)$$

where $v^{(0)} \in H_0^1(\Omega)^3$, $v^{(p)} \in \mathbb{K}(\Omega \times \mathcal{Y}\mathcal{F})$, $p \geq 1$.

Let us substitute the expansions (2.3.2), (2.3.1) and (10.8.5) into the l.h.s. of (10.8.4). We find

$$A^\varepsilon(\widetilde{w}^\varepsilon, v - \widetilde{w}^\varepsilon) = \varepsilon h_0 \int_\Omega \prec (\sigma_0^{i\alpha} + \varepsilon \sigma_1^{i\alpha} + \varepsilon^2 \sigma_2^{i\alpha} + \ldots)[v_{i,\alpha}^{(0)} - u_{i,\alpha}^{(0)}$$

$$+ \varepsilon(v_{i,\alpha}^{(1)} - u_{i,\alpha}^{(1)}) + \ldots] + \frac{1}{\varepsilon}(\sigma_0^{ij} + \varepsilon \sigma_1^{ij}$$

$$+ \varepsilon^2 \sigma_2^{ij} + \ldots)[\varepsilon(v_{i|j}^{(1)} - u_{i|j}^{(1)}) + \varepsilon^2(v_{i|j}^{(2)} - u_{i|j}^{(2)}) + \ldots] \succ dx, \qquad (10.8.6)$$

and the inequality (10.8.4) gives

$$\varepsilon h_0 \int_\Omega \prec \sigma_0^{i\alpha}(v_{i,\alpha}^{(0)} - u_{i,\alpha}^{(0)}) + \sigma_0^{ij}(v_{i|j}^{(1)} - u_{i|j}^{(1)}) \succ dx + 0(\varepsilon^2) \geq 0(\varepsilon^2). \qquad (10.8.7)$$

Hence

$$\int_\Omega [\prec \sigma_0^{i\alpha} \succ (v_i^{(0)} - u_i^{(0)})_{,\alpha} + \prec \sigma_0^{ij}(v_{i|j}^{(1)} - u_{i|j}^{(1)}) \succ] dx \geq 0. \qquad (10.8.8)$$

We take $v^{(0)} = u^{(0)}$ and

$$v^{(1)} = u^{(1)} + \theta(u - u^{(1)}), \qquad 0 \leq \theta \leq 1, \qquad (10.8.9)$$

where $\theta \in \mathbf{D}(\Omega)$, $\boldsymbol{u} \in \mathbb{K}(\mathcal{YF})$. Note that

$$[\![\boldsymbol{v}^{(1)} \cdot \check{\boldsymbol{N}}]\!] = (1 - \theta)[\![\boldsymbol{u}^{(1)} \cdot \check{\boldsymbol{N}}]\!] + \theta[\![\boldsymbol{u} \cdot \check{\boldsymbol{N}}]\!] \geq 0 \,,$$

hence $\boldsymbol{v}^{(1)} \in \mathbb{K}(\Omega \times \mathcal{YF})$. Then the inequality (10.8.8) reduces to

$$\int_\Omega \theta(x) \prec \sigma_0^{ij}(u_{i|j} - u_{i|j}^{(1)}) \succ dx \geq 0 \qquad \forall\, \theta \in \mathbf{D}^+(\Omega) \,, \quad \forall\, \boldsymbol{u} \in \mathbb{K}(\mathcal{YF}) \,. \qquad (10.8.10)$$

By (2.3.3) the last inequality yields

$$\prec [C^{ijkl}u_{k|l}^{(1)} + C^{ijk\beta}u_{k,\beta}^{(0)}](u_{i|j} - u_{i|j}^{(1)}) \succ \geq 0 \,. \qquad (10.8.11)$$

Step 2 of Sec. 2.3 implies that $u_\alpha^{(0)} = 0$. One can prove that the presence of cracks does not change this conclusion. We thus have $u_\alpha^{(0)} = 0$. Moreover, by the standard theorem concerning well-posedness of the variational inequalities one can show that $\boldsymbol{u}^{(1)}$ is determined up to an additive constant vector (depending on x, x being treated here as a parameter). On the other hand, one notes easily that

$$\widehat{u}_\alpha^{(1)} = -u_{3,\alpha}^{(0)}\widehat{y}_3 \,, \qquad \widehat{u}_3^{(1)} = 0$$

satisfy the condition:

$$C^{ijkl}\widehat{u}_{k|l}^{(1)} + C^{ij3\beta}u_{3,\beta}^{(0)} = 0$$

and $\prec \widehat{\boldsymbol{u}}^{(1)} \succ = 0$. Thus the fields

$$u_\alpha^{(1)} = u_\alpha(x) - u_{3,\alpha}^{(0)}\widehat{y}_3 \,, \qquad u_3^{(1)} = u_3(x) \qquad (10.8.12)$$

satisfy (10.8.11) and $\prec \boldsymbol{u}^{(1)} \succ = \prec \boldsymbol{u} \succ$, $\boldsymbol{u}$ being an arbitrary field defined on Ω. Consequently, $\sigma_0^{ij} = 0$. From now on we shall write: $u_3^{(0)} = w$.

Let us return to the inequality (10.8.4). By using the expansion (10.8.6) and taking $\boldsymbol{v}^{(0)} = \boldsymbol{u}^{(0)}$, $\boldsymbol{v}^{(1)} = \boldsymbol{u}^{(1)}$ one has

$$\varepsilon^3 h_0 \int_\Omega \prec \sigma_1^{ij}(v_{i|j}^{(2)} - u_{i|j}^{(2)}) \succ dx \geq 0(\varepsilon^4) \,. \qquad (10.8.13)$$

Substitution

$$\boldsymbol{v}^{(2)} = \boldsymbol{u}^{(2)} + \theta(x)(\boldsymbol{\varphi} - \boldsymbol{u}^{(2)}) \,, \qquad \boldsymbol{\varphi} \in \mathbb{K}(\mathcal{YF}) \,,$$

where $\theta \in \mathbf{D}(\Omega)$, $0 \leq \theta \leq 1$, gives

$$\int_\Omega \theta(x) \prec \sigma_1^{ij}(\varphi_{i|j} - u_{i|j}^{(2)}) \succ dx \geq 0 \qquad \forall\, \theta \in \mathbf{D}^+(\Omega). \qquad (10.8.14)$$

Hence we find the local problem in the following form

$$(P_{\mathcal{YF}}) \quad \left| \begin{array}{l} \text{find } \boldsymbol{u}^{(2)}(x, \cdot) \in \mathbb{K}(\mathcal{YF}) \text{ such that} \\ \prec \sigma_1^{ij}(\varphi_{i|j} - u_{i|j}^{(2)}) \succ \geq 0 \qquad \forall\, \varphi \in \mathbb{K}(\mathcal{YF}), \end{array} \right. \tag{10.8.15}$$

where

$$\sigma_1^{ij} = C^{ijkl}(u_{k|l}^{(2)} + u_{k,l}^{(1)}) \,. \tag{10.8.16}$$

Equivalently, by (10.8.12) we obtain

$$\sigma_1^{ij} = C^{ijkl}u_{k|l}^{(2)} + C^{ij3\beta}u_{3,\beta} + C^{ij\alpha\beta}(e_{\alpha\beta}(\boldsymbol{u}) + \widehat{y}_3\kappa_{\alpha\beta}(w)) \,, \tag{10.8.17}$$

with

$$e_{\alpha\beta}(\boldsymbol{u}) = \frac{1}{2}(u_{\alpha,\beta} + u_{\beta,\alpha}) \,, \qquad \kappa_{\alpha\beta}(w) = -w_{,\alpha\beta} \,. \tag{10.8.18}$$

The field $\boldsymbol{u}^{(1)}$ is assumed as given. Let us note that the component $u_3^{(1)} = u_3(x)$ does not affect the relation: $\boldsymbol{\sigma}_1 = \boldsymbol{\sigma}_1(\boldsymbol{u}^{(1)})$. Indeed let us take $\boldsymbol{u}^{(1)} = (0, u_3(x))$. Then the solution of (10.8.7) assumes the form

$$\boldsymbol{u}^{(2)} = (-\widehat{y}_3 u_{3,1}, \; -\widehat{y}_3 u_{3,2}, \; 0) \,, \tag{10.8.19}$$

which implies $\sigma_1^{ij} = 0$. Thus $\boldsymbol{\sigma}_1$ depends solely on $u_\alpha^{(1)}$ or on the deformation fields $e_{\alpha\beta}(\boldsymbol{u})$ and $\kappa_{\alpha\beta}(w)$.

Let us complete $\boldsymbol{\epsilon} = e(\boldsymbol{u})$ and $\boldsymbol{\rho} = \kappa(w)$ with zero components: $\epsilon_{i3} = 0$, $\rho_{i3} = 0$. Then the local problem $(P_{\mathcal{YF}})$ can be rearranged to the form:

$$(P'_{\mathcal{YF}}) \quad \left| \begin{array}{l} \text{find } \boldsymbol{u}^{(2)}(x, \cdot) = \chi \in \mathbb{K}(\mathcal{YF}) \text{ such that} \\ \prec C^{klij}(e_{ij}^y(\chi) + \epsilon_{ij} + \widehat{y}_3\rho_{ij})e_{kl}^y(\boldsymbol{v} - \chi) \succ \geq 0 \,, \quad \forall\, \boldsymbol{v} \in \mathbb{K}(\mathcal{YF}) \,. \end{array} \right. \tag{10.8.20}$$

The local stresses σ_1^{ij} are expressed by

$$\sigma_1^{ij} = C^{ijkl}(e_{kl}^y(\chi) + \epsilon_{kl} + \widehat{y}_3\rho_{kl}) \,. \tag{10.8.21}$$

Given $\boldsymbol{\epsilon}, \boldsymbol{\rho} \in \mathbb{E}_s^3$ with $\epsilon_{i3} = 0$, $\rho_{i3} = 0$ the solution χ of the problem $(P'_{\mathcal{YF}})$ is determined up to an additive constant vector while the stress field $\boldsymbol{\sigma}_1$ is determined uniquely.

Let us define the rescaled stress and couple resultants, see (2.3.41)

$$\mathcal{N}^{\alpha\beta} = h_0 \prec \sigma_1^{\alpha\beta} \succ \,, \qquad \mathcal{M}^{\alpha\beta} = h_0 \prec \widehat{y}_3\sigma_1^{\alpha\beta} \succ \,. \tag{10.8.22}$$

Note that local problem $(P'_{\mathcal{YF}})$ defines a nonlinear operator $A : \mathbb{E}_s^2 \times \mathbb{E}_s^2 \to \mathbb{E}_2^s \times \mathbb{E}_2^s$ such that

$$A(\boldsymbol{\epsilon}, \boldsymbol{\rho}) = (\mathcal{N}, \mathcal{M}) \,. \tag{10.8.23}$$

Proceeding similarly to Secs. 8 and 9 one can show that

$$\mathcal{N} = \frac{\partial \mathcal{W}}{\partial \epsilon}, \qquad \mathcal{M} = \frac{\partial \mathcal{W}}{\partial \rho}, \qquad (10.8.24)$$

where the potential $\mathcal{W}$ is expressed by, cf. Eq. (2.10.109)

$$\mathcal{W} = \frac{1}{2} h_0 \prec C^{ijkl}[\epsilon_{ij} + \widehat{y}_3 \rho_{ij} + e^y_{ij}(\chi)][\epsilon_{kl} + \widehat{y}_3 \rho_{kl} + e^y_{kl}(\chi)] \succ . \qquad (10.8.25)$$

The function χ depends on ϵ, ρ according to $(P'_{y\mathcal{F}})$. The potential $\mathcal{W}$ is convex, of class C^1 and positive definite: $\mathcal{W} \geq 0$ and the equality holds only for $\epsilon = 0$, $\rho = 0$. Moreover, the following conditions are satisfied:
(i) $\mathcal{W}$ is positively homogeneous of degree 2, i.e.

$$\mathcal{W} = (\lambda\epsilon, \lambda\rho) = \lambda^2 \mathcal{W}(\epsilon, \rho), \qquad \lambda \geq 0, \ \epsilon \in \mathbb{E}^2_s, \quad \rho \in \mathbb{E}^2_s,$$

and

$$\mathcal{W}(\epsilon, \rho) = \frac{1}{2}(\mathcal{N}^{\alpha\beta}\epsilon_{\alpha\beta} + \mathcal{M}^{\alpha\beta}\rho_{\alpha\beta}) . \qquad (10.8.26)$$

(ii) There exist two positive constants m_0 and m_1 such that

$$m_0 \sum_{\alpha,\beta=1}^{2} [(\epsilon_{\alpha\beta})^2 + (\rho_{\alpha\beta})^2] \leq \mathcal{W}(\epsilon, \rho) \leq m_1 \sum_{\alpha,\beta=1}^{2} [(\epsilon_{\alpha\beta})^2 + (\rho_{\alpha\beta})^2] . \qquad (10.8.27)$$

(iii) $\mathcal{W}$ is strictly convex.
(iv) $\mathcal{W}$ is strictly monotone, i.e.

$$(\mathcal{N}_2^{\alpha\beta} - \mathcal{N}_1^{\alpha\beta})(\epsilon^2_{\alpha\beta} - \epsilon^1_{\alpha\beta}) + (\mathcal{M}_2^{\alpha\beta} - \mathcal{M}_1^{\alpha\beta})(\rho^2_{\alpha\beta} - \rho^1_{\alpha\beta}) \geq 0 , \qquad (10.8.28)$$

and equality holds iff $\epsilon^{(2)} = \epsilon^{(1)}$ and $\rho^{(2)} = \rho^{(1)}$.

Let us proceed now to finding the homogenized equilibrium equation of the plate with smeared-out cracks. The first terms of the asymptotic expansion for displacements have the form

$$\widetilde{w}^\varepsilon_\alpha = \varepsilon(u_\alpha(x) - \widehat{y}_3 w(x)_{,\alpha}) + 0(\varepsilon^2) , \qquad \widetilde{w}^\varepsilon_3 = w(x) + \varepsilon u_3(x) + 0(\varepsilon^2) . \qquad (10.8.29)$$

According to this representation we take the trial field v of (10.8.4):

$$v_\alpha = \widetilde{w}^\varepsilon_\alpha + \varepsilon(v_\alpha(x) - \widehat{y}_3 v(x)_{,\alpha}) , \qquad v_3 = \widetilde{w}^\varepsilon_3 + v(x) + \varepsilon v_3(x) . \qquad (10.8.30)$$

Note that $v \in \mathbb{K}(\Omega \times \mathcal{YF})$ and hence the field v can be substituted into (10.8.4). We compute

$$A^\varepsilon(\widetilde{w}^\varepsilon, v - \widetilde{w}^\varepsilon) = \varepsilon^3 \int_\Omega (\mathcal{N}^{\alpha\beta} e_{\alpha\beta}(v) + \mathcal{M}^{\alpha\beta} \kappa_{\alpha\beta}(v))dx + 0(\varepsilon^4) , \qquad (10.8.31)$$

and

$$F^\varepsilon(\boldsymbol{v} - \widetilde{\boldsymbol{w}}^\varepsilon) = \varepsilon^3 \int\limits_\Omega (\widehat{p}^\alpha v_\alpha - \widehat{m}^\alpha v_{,\alpha} + \widehat{q}v)dx + 0(\varepsilon^4)\,, \tag{10.8.32}$$

where $\widehat{p}^\alpha$, $\widehat{m}^\alpha$ and $\widehat{q}$ are given by Eqs. (2.3.46) and (2.3.48). Since the signs of $\boldsymbol{v}$ and v are arbitrary the variational inequality (10.8.4) assumes the form of the variational equality:

$$\int\limits_\Omega (\mathcal{N}^{\alpha\beta}e_{\alpha\beta}(\boldsymbol{v}) + \mathcal{M}^{\alpha\beta}\kappa_{\alpha\beta}(v))dx = \int\limits_\Omega (\widehat{p}^\alpha v_\alpha - \widehat{m}^\alpha v_{,\alpha} + \widehat{q}v)dx\,, \tag{10.8.33}$$

that holds for each $(\boldsymbol{v}, v) \in V_K^0(\Omega) = (H_0^1(\Omega))^2 \times H_0^2(\Omega)$.

Now we are in a position to formulate the homogenized problem

(P_h) | find $(\boldsymbol{u}, w) \in V_K^0(\Omega)$such that the variational equation (10.8.33) holds, while $\mathcal{N}$ and $\mathcal{M}$ are determined by (10.8.24).

The last problem is equivalent to the minimization problem:

(P_h') $\min\{J(\boldsymbol{v}, v)|(\boldsymbol{v}, v) \in V_K^0(\Omega)\}\,,$

where J, representing the total potential energy, is defined by

$$J(\boldsymbol{v}, v) = \int\limits_\Omega \mathcal{W}(e(\boldsymbol{v}), \boldsymbol{\kappa}(v))dx - f(\boldsymbol{v}, v)\,, \tag{10.8.34}$$

with

$$f(\boldsymbol{v}, v) = \int\limits_\Omega (\widehat{p}^\alpha v_\alpha - \widehat{m}^\alpha v_{,\alpha} + \widehat{q}v)dx\,. \tag{10.8.35}$$

The equivalency of (P_h) and (P_h') follows from the following property of J: J is strictly convex, finite and coercive on $V_K^0(\Omega)$. These properties can easily be inferred from the properties of $\mathcal{W}$ mentioned above. Consequently a solution $(\boldsymbol{u}, w)$ of the problems (P_h) or (P_h') exists and is unique.

The state of deformation of the original plate can be recovered by the formulae (2.3.60), (2.3.61).

10.8.2.   Justification by Γ-convergence

The homogenized plate model derived in the previous section will now be rigorously justified.

For the sake of simplicity we assume that $c^+ = -c^- = c$. Dividing then both sides of the inequality (10.8.3) by ε^3 and taking into account (2.10.1) we conclude that this inequality is equivalent to the minimization problem, which means evaluating

$$\inf\{\widetilde{J}_{\varepsilon\varepsilon}(\widetilde{\boldsymbol{v}})|\widetilde{\boldsymbol{v}} \in \mathbb{K}(B_\varepsilon^\varepsilon)\}\,. \tag{10.8.36}$$

Here $\widetilde{J}_{\varepsilon\varepsilon}$ is given by (2.10.9) with $e = \varepsilon$ while B_ε is to be replaced by $B_\varepsilon^\varepsilon$. The body forces are equal to $b(x_\alpha, x_3/\varepsilon)$ or to $b\left(\dfrac{x}{\varepsilon}\right)$.

After the rescaling (2.10.2) the functional $\widetilde{J}_{\varepsilon\varepsilon}$ becomes, cf. Eq. (2.10.14)

$$J_{\varepsilon\varepsilon}(v) = \frac{1}{2}\int_{B_\varepsilon} C_\varepsilon^{ijkl}(z)(\mathbf{Q}^\varepsilon e^z(v))_{ij}(z)(\mathbf{Q}^\varepsilon e^z(v))_{kl}dz - L_\varepsilon(v) , \qquad (10.8.37)$$

where L_ε is defined by (2.10.16) with $e = \varepsilon$ whilst $B^\varepsilon = B\backslash\mathcal{F}^\varepsilon$. The set $\mathcal{F}^\varepsilon$ is obtained from the set of fissures $\mathcal{F}_\varepsilon^\varepsilon$ by the rescaling (2.10.2). In (10.8.37) v is a function from $H^1(B^\varepsilon)^3$. We observe that B^ε is a domain with constant thickness equal to $2c$.

The proof of Theorem 2.10.13 is based on Lemmas 2.10.1, 2.10.12 and exploits also the dual effective potential given by Eq. (2.10.126). We shall now extend those results to the case of the fissured domain B^ε. We introduce the space, cf. (2.10.12),

$$V_0(B^\varepsilon) = \{v \in H^1(B^\varepsilon)^3|\ v = 0 \text{ on } \Upsilon_0\} .$$

Lemma 10.8.1. Let $\{v^\varepsilon\}_{\varepsilon>0} \subset V_0(B^\varepsilon)$ be a sequence strongly convergent to $v \in L^2(B)^3$ in $L^2(B)^3$ as $\varepsilon \to 0$. Suppose that there exists a constant $K > 0$, which is independent of ε and such that

$$\int_{B^\varepsilon} j[\frac{z_\alpha}{\varepsilon}, z_\varepsilon, \mathbf{Q}^\varepsilon e^z(v^\varepsilon)]dz \leq K .$$

Then $e_{i3}^z(v) = 0$ and $v \in V_K^0(B)$, where $V_K^0(B)$ is defined by (2.10.25).

Proof. Similarly to the proof of Lemma 2.10.1 we obtain

$$m||\mathbf{Q}^\varepsilon e^z(v^\varepsilon)||_{0,B^\varepsilon} \leq K .$$

In Sec. 9.3 the extension operator $\mathcal{Q}_2^\varepsilon$ was constructed for a two-dimensional domain. The same extension procedure can be applied to the domain B^ε. In this manner we construct an extension operator $\mathcal{Q}_3^\varepsilon : H_0^1(B^\varepsilon)^3 \to H_0^1(B)^3$ with the properties similar to the estimates (9.3.33) – (9.3.37). Thus we get

$$||\mathbf{Q}^\varepsilon e^z(\mathcal{Q}_3^\varepsilon v^\varepsilon)||_{0,B} \leq K_1 .$$

Hence

$$\frac{1}{\varepsilon}||e_{\alpha3}^z(\mathcal{Q}_3^\varepsilon v^\varepsilon)||_{0,B} \leq K_1 , \qquad \frac{1}{\varepsilon^2}||e_{33}^z(\mathcal{Q}_3^\varepsilon v^\varepsilon)||_{0,B} \leq K_1 .$$

Moreover we have, cf. (9.3.36)

$$\mathcal{Q}_3^\varepsilon v^\varepsilon \to v \qquad \text{in } L^2(B)^3 \text{ strongly.}$$

The properties of the extension operator $\mathcal{Q}_3^\varepsilon$ imply that $v \in V_0(B)$. Thus $e_{i3}^z(v) = 0$ and consequently v belongs to $V_K^0(B)$. $\qquad\qquad\square$

In the case considered $\widehat{y}_3 = y_3$ and the effective elastic potential is given by, cf. (10.8.25)

$$\mathcal{W}(\epsilon, \rho) = \inf\{\frac{h_0}{2} \prec C^{ijkl}(\boldsymbol{y})[e_{ij}^y(\boldsymbol{v}) + \epsilon_{ij} + y_3\rho_{ij}][e_{kl}^y(\boldsymbol{v})$$
$$+ \epsilon_{kl} + y_3\rho_{kl}] \succ |\boldsymbol{v} \in \mathbb{K}(\mathcal{YF})\}\,, \qquad (10.8.38)$$

where $\epsilon, \rho \in \mathbb{E}_s^3$ with $\epsilon_{i3} = 0$ and $\rho_{i3} = 0$.

To obtain the dual potential $\mathcal{W}^*$ we can proceed similarly to the derivation of the formula (2.10.125). We eventually arrive at, cf. also Sec. 10.4,

$$\mathcal{W}^*(\epsilon^*, \rho^*) = \inf\{h_0 \prec j^*(\boldsymbol{y}, \frac{1}{h_0}\epsilon^* - \boldsymbol{p}^*(\boldsymbol{y})) \succ |\boldsymbol{p}^* \in S(\epsilon^*)\} \qquad (10.8.39)$$

where

$$S(\epsilon^*) = \left(\frac{1}{h_0}\epsilon^* - S_{per}(\mathcal{YF})\right) \cap \{\boldsymbol{q}^* \in L^2(\mathcal{YF}, \mathbb{E}_3^s)| \prec \boldsymbol{p}^* \succ = 0\,, h_0 \prec y_3\boldsymbol{q}^* \succ = \rho^*\}\,,$$
$$(10.8.40)$$

and

$$S_{per}(\mathcal{YF}) = \{\boldsymbol{p}^* \in L^2(\mathcal{YF}, \mathbb{E}_3^s)| \operatorname{div}_y\boldsymbol{p}^* = 0 \text{ in } \mathcal{YF},\ p_N^* \leq 0 \text{ and } \boldsymbol{p}_T = 0 \text{ on } \mathcal{F},$$
$$\boldsymbol{p}^*\boldsymbol{\nu} \text{ assumes opposite values on the opposite faces}$$
$$\text{of } \partial Y \times (-c, c),\quad \boldsymbol{p}^*\boldsymbol{\nu} = 0 \text{ on } Y \times \{\pm c\}\}\,. \qquad (10.8.41)$$

Here $\boldsymbol{\nu}$ stands for the outward unit normal vector to $\partial \mathcal{Y}$. We recall that in (10.8.39) ϵ^* and ρ^* are arbitrary elements from the space $\mathbb{E}_3^s$ with $\epsilon^{*i3} = 0$ and $\rho^{*i3} = 0$, $i = 1, 2, 3$. Note that if $\boldsymbol{n} \in S(\epsilon^*)$ then $\boldsymbol{n} = \frac{1}{h_0}\epsilon^* - \boldsymbol{p}^*$, where $\boldsymbol{p}^* \in S_{per}(\mathcal{YF})$. Hence we conclude that $\boldsymbol{n}\boldsymbol{\mu} = 0$ on $Y \times \{\pm c\}$ because $\epsilon^{*i3} = 0$.

A counterpart of Lemma 2.10.12 is formulated in the form of the following result.

Lemma 10.8.2. Assume that $\boldsymbol{n} \in S(\epsilon^*)$ and a sequence $\{\boldsymbol{v}^\varepsilon\}_{\varepsilon>0} \subset H^1(B^\varepsilon)^3$ is such that $\boldsymbol{v}^\varepsilon \to \boldsymbol{v}$ strongly in $L^2(B)^3$ as $\varepsilon \to 0$ and $\{\mathbb{Q}^\varepsilon e^z(\boldsymbol{v}^\varepsilon)\}_{\varepsilon>0}$ is bounded in $L^2(B^\varepsilon, \mathbb{E}_s^3)$. Then $\boldsymbol{v} \in V_K(B)$ and

$$\lim_{\varepsilon \to 0} \int_{B^\varepsilon} \psi(z_\alpha)\boldsymbol{n}^{ij}\left(\frac{z_\beta}{\varepsilon}, z_3\right)(\mathbb{Q}^\varepsilon e^z(\boldsymbol{v}^\varepsilon))_{ij}dz = \int_\Omega \psi\rho^* : \kappa^z(w)dz\,, \qquad (10.8.42)$$

where $\psi \in \mathbf{D}(\Omega)$.

Proof. We can proceed analogously to the proof of Lemma 2.10.12. After integration by parts, however, an integral over $\mathcal{F}^\varepsilon$ will appear. It is not difficult to show that this integral tends to zero when $\varepsilon \to 0$. $\qquad \square$

The sequence of loading functionals $\{L_\varepsilon\}_{\varepsilon>0}$ is continuously convergent, cf. Sec. 2.10. Consequently it suffices to study the Γ-convergence of the following sequence of functionals

$$
J_\varepsilon^1(\boldsymbol{v}) = \begin{cases} \dfrac{1}{2} \displaystyle\int_{B^\varepsilon} C_\varepsilon^{ijkl}(\mathbf{Q}^\varepsilon e^z(\boldsymbol{v}))_{ij}(\mathbf{Q}^\varepsilon e^z(\boldsymbol{v}))_{kl}dz & \text{if } \boldsymbol{v} \in \mathbb{K}(B^\varepsilon), \\ +\infty & \text{otherwise,} \end{cases}
\tag{10.8.43}
$$

where

$$
\mathbb{K}(B^\varepsilon) = \{\boldsymbol{v} \in H^1(B^\varepsilon)^3 \mid \boldsymbol{v} = 0 \text{ on } \Upsilon_0 = \Gamma \times (-c,c)\} .
\tag{10.8.44}
$$

Theorem 10.8.3. The sequence of functionals $\{J_\varepsilon^1\}_{\varepsilon>0}$ defined by (10.8.43) is Γ-convergent in the strong topology of $L^2(B)^3$ to the functional

$$
J_h^1(\boldsymbol{v}) = \int_\Omega \mathcal{W}(e(\boldsymbol{u}), \kappa(w))dx .
\tag{10.8.45}
$$

Here $\boldsymbol{v} \in V_K^0(B)$, $v_\alpha(x, z_3) = u_\alpha(x) - z_3\dfrac{\partial w}{\partial x_\alpha}$, $v_3 = w(x) = w(x_\alpha)$ and $u_\alpha \in H_0^1(\Omega)$, $w \in H_0^2(\Omega)$.

Proof. We can proceed analogously to the proof of Theorem 2.10.13, where instead of Lemmas 2.10.1 and 2.10.12 we exploit Lemmas 10.6.1 and 10.6.2. Details are left to the reader. $\qquad\square$

Remark 10.8.4. Having at our disposal the above theorem and the dual potential $\mathcal{W}^*$, given by (10.8.39), one can perform dual homogenization. The assumption that $\mathcal{F}$ does not intersect the boundary of $\mathcal{Y}$ can be weakened. It suffices to treat $\mathcal{Y}$ as a torus by identification of the opposite sides. However, this torus must still be a connected set.

11. Stiffness loss of cracked laminates

11.1. *Two-dimensional model of transversely symmetric laminates in stretching and in-plane shearing*

This section is aimed at forming a new mathematical model of the in-plane deformations of three-layer, balanced laminates subject to membrane boundary loadings. The layers of the laminate are composed of fiber-reinforced plies. Since only the cross-ply laminates will be considered, the layers will be viewed as orthotropic. When stretched, the laminates crack under relatively low in-plane loadings. This is due to discrepancy between the values of thermal expansion coefficients and elastic moduli of fibers and the matrix. The appearing intralaminar cracks go usually across the whole thickness of some layers and are almost equally spaced, cf. Fig. 11.1.1.

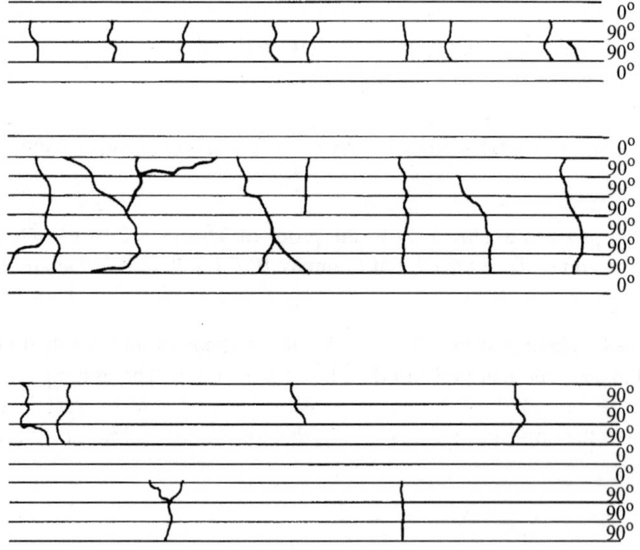

Fig. 11.1.1. Images of intralaminar cracks in the $[0^0_m/90^0_n]_s$ laminates subjected to tensile forces in the 0^0 direction; after Highsmith and Reifsnider (1982) (published by courtesy of ASTM STP)

At a certain level of loading the crack patterns attain a saturation state in which the layout of cracks is nearly uniform, called CDS-characteristic damage state.

In the next sections the problem of stiffness loss caused by transverse cracking in the internal layer of the three-layered balanced laminates will be discussed. Having this in mind we develop below a special two-dimensional model of the three-layered laminates that, by distinguishing between in-plane displacements of the layers, will be capable of describing opening and closing of the transverse cracks in the internal layer. The modelling proposed is of Reissner type: we put forward stress assumptions that satisfy the equilibrium equations pointwise outside a boundary zone. To avoid artificial recovering of displacements through Lagrangian multipliers we augment the stress assumptions with displacement assumptions and use the saddle-point variational principle of Reissner. The model thus derived involves five displacement components. Unfortunately, the model cannot be simpler. Let us note, however, that after performing homogenization with appropriate scaling, the homogenized model is reduced to the conventional model of an in-plane loaded plate. Only its potential is determined by solving more complicated equations of the three-layer plate model.

Consider a transversely symmetric (balanced) laminate composed of the face of thickness d and the internal layer of thickness $2c$. The middle plane Ω of the internal layer is parametrized by Cartesian coordinates x_α; $x = (x_\alpha) \in \Omega$. The whole laminate occupies a cylindrical domain $B = \Omega \times (-h, h)$, $h = c + d$. We write $\boldsymbol{x} \in B$, $\boldsymbol{x} = (x, x_3)$, x_3 being an axis perpendicular to Ω.

The lower and upper faces $x_3 = \pm h$ are free of loads while the lateral edge $S = \partial\Omega \times (-h, h)$ is subjected to the tractions $p^i(s, x_3)$ on its part S_σ; $S_\sigma = \Gamma_\sigma \times (-h, h)$, $\Gamma_\sigma \subset \partial\Omega$. The remaining part of S, $S_w = \Gamma_w \times (-h, h)$ is clamped; $\bar{\Gamma}_w \cup \bar{\Gamma}_\sigma = \partial\Omega$, $\bar{S}_w \cup \bar{S}_\sigma = S$; $s \in \partial\Omega$. For the scale of simplicity the loading p^i is assumed to have the following through-the-thickness distribution

$$
p^\alpha(s, x_3) = \begin{cases} \dfrac{1}{2c}\bar{L}^\alpha(s) & \text{if } x_3 \in I_2 , \\[2ex] \dfrac{1}{2d}[\bar{N}^\alpha(s) - \bar{L}^\alpha(s)] & \text{if } x_3 \in I_1 \cup I_3 ; \end{cases}
$$

$$
p^3(s, x_3) = \begin{cases} \dfrac{1}{2d}(x_3 + h)\bar{Q}(s) & \text{if } x_3 \in I_1 , \\[2ex] -\dfrac{x_3}{2c}\bar{Q}(s), & \text{if } x_3 \in I_2 , \\[2ex] \dfrac{1}{2d}(x_3 - h)\bar{Q}(s) & \text{if } x_3 \in I_3 , \end{cases}
\tag{11.1.1}
$$

where

$$
I_1 = (-h, -c) , \quad I_2 = (-c, c) , \quad I_3 = (c, h) .
\tag{11.1.2}
$$

The body forces are omitted.

The through-the thickness distribution of elastic compliances is layer-wise constant

$$
c_{ijkl}(\boldsymbol{x}) = \begin{cases} c_{ijkl}^{(m)}(x) & \text{if } x_3 \in I_2 \\[1.5ex] c_{ijkl}^{(f)}(x) & \text{otherwise.} \end{cases}
\tag{11.1.3}
$$

The planes $x_3 = \text{const}$ are assumed as planes of material symmetry, hence

$$
c_{3\alpha\beta\gamma}^{(n)} = c_{333\alpha}^{(n)} = 0 , \qquad n \in \{m, f\} .
\tag{11.1.4}
$$

The tensor (c_{ijkl}) satisfies usual symmetry conditions and is positive definite, cf. (2.1.6), (2.1.7).

Within the three-dimensional framework, the problem of equilibrium of the laminate considered amounts to finding the stress field $(\tilde{\sigma}^{ij})$ and the displacement field $(\tilde{w}_i)$ for which the two-field Reissner functional

$$I(\boldsymbol{w}, \boldsymbol{\sigma}) = \int_B \left[\frac{1}{2}(w_{\alpha,\beta} + w_{\beta,\alpha})\sigma^{\alpha\beta} + (w_{\alpha,3} + w_{3,\alpha})\sigma^{\alpha 3} \right.$$
$$\left. + w_{33}\sigma^{33} - \frac{1}{2}c_{ijkl}\sigma^{ij}\sigma^{kl} \right] d\boldsymbol{x} - \int_{S_\sigma} p^i w_i ds dx_3 , \qquad (11.1.5)$$

assumes its stationary value at a saddle point $(\tilde{w}, \tilde{\sigma})$. In linear elasticity, such a boundary value problem is standard and can be solved by using Brezzi's theorem, see Sec. 1.2.2.

The two-dimensional modelling is based on the following stress assumptions, cf. Fig. 11.1.2.

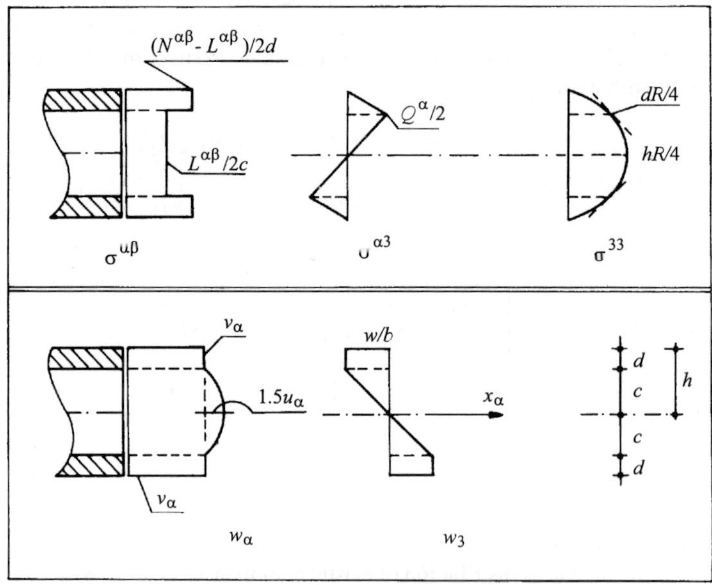

Fig. 11.1.2. Stress and displacement assumptions

$$\sigma^{\alpha\beta}(x) = \begin{cases} L^{\alpha\beta}(x)/(2c) & \text{if } x_3 \in I_2 , \\ \left[N^{\alpha\beta}(x) - L^{\alpha\beta}(x)\right]/(2d) & \text{otherwise} ; \end{cases} \qquad (11.1.6)$$

$$\sigma^{\alpha 3}(x) = \begin{cases} (x_3 + h)Q^\alpha(x)/(2d) & \text{if } x_3 \in I_1 , \\ -Q^\alpha(x)/(2c) & \text{if } x_3 \in I_2 , \\ (x_3 - h)Q^\alpha(x)/(2d) & \text{if } x_3 \in I_3 ; \end{cases} \qquad (11.1.7)$$

$$\sigma^{33}(x) = \begin{cases} \dfrac{1}{4d}(x_3 + h)^2 R(x) & \text{if } x_3 \in I_1 \,, \\[2mm] \dfrac{1}{4c}(-(x_3)^2 + ch)R(x) & \text{if } x_3 \in I_2 \,, \\[2mm] \dfrac{1}{4d}(x_3 - h)^2 R(x) & \text{if } x_3 \in I_3 \,. \end{cases} \qquad (11.1.8)$$

Here $L^{\alpha\beta}, N^{\alpha\beta}, Q^\alpha, R$ are unknown fields defined on Ω. On the other hand the following distribution of displacements is assumed, cf. Fig. 11.1.2

$$w_\alpha(x) = \begin{cases} v_\alpha(x) + \dfrac{3}{2c^2}[c^2 - (x_3)^2]u_\alpha(x) & \text{if } x_3 \in I_2 \,, \\[2mm] v_\alpha(x), & \text{otherwise} \,; \end{cases} \qquad (11.1.9)$$

$$w_3(x) = \begin{cases} -w(x)/b & \text{if } x_3 \in I_1 \,, \\[2mm] \dfrac{x_3}{c \cdot b}w(x) & \text{if } x_3 \in I_2 \,, \\[2mm] w(x)/b & \text{if } x_3 \in I_3 \,; \end{cases} \qquad (11.1.10)$$

where $b = \dfrac{d}{2} + \dfrac{c}{3}$.

Note that the assumptions (11.1.7), (11.1.8) satisfy the boundary condition $\sigma^{k3}(x, \pm h) = 0$ and continuity conditions on the interfaces $x_3 = \pm c$. We shall see later that also the equilibrium equations $\sigma^{ij}_{,j} = 0$ are satisfied for $x \in \Omega$, $x_3 \in (-h, h)$, which makes the fields σ^{ij} statically admissible in the interior of B and on the faces $x_3 = \pm h$.

Substitution of (11.1.6) − (11.1.8) and (11.1.9) − (11.1.10) into Reissner functional (11.1.5) and integration over x_3 yields

$$I(w, \sigma) = J(v, u, w; N, L, Q, R) \,, \qquad (11.1.11)$$

where

$$J = \int_\Omega [v_{\alpha,\beta}N^{\alpha\beta} + u_{\alpha,\beta}L^{\alpha\beta} + (u_\alpha - w_{,\alpha})Q^\alpha + wR \quad W^*(x, N, L, Q, R)]dx$$

$$- \int_{\Gamma_\sigma} (\bar{N}^\alpha r_\alpha + \bar{L}^\alpha u_\alpha - \bar{Q}w)ds \,. \qquad (11.1.12)$$

The existence of a unique saddle point $(\tilde{v}, \tilde{u}, \tilde{w}; \tilde{N}, \tilde{L}, \tilde{Q}, \tilde{R})$ of the functional J can be proved by using Brezzi's theorem, see Sec. 1.2.2.

The density of complementary energy has the form

$$W^* = (D^N_{\alpha\beta\lambda\mu} N^{\alpha\beta} N^{\lambda\mu} + D^L_{\alpha\beta\lambda\mu} L^{\alpha\beta} L^{\lambda\mu} + 2D^{NL}_{\alpha\beta\lambda\mu} N^{\alpha\beta} L^{\lambda\mu}$$

$$+ D^Q_{\alpha\beta} Q^\alpha Q^\beta + 2D^{RL}_{\alpha\beta} RL^{\alpha\beta} + 2D^{RN}_{\alpha\beta} RN^{\alpha\beta} + D^R R^2)/2 \,, \qquad (11.1.13)$$

where

$$D^N_{\alpha\beta\lambda\mu} = \frac{1}{2d}c^{(f)}_{\alpha\beta\lambda\mu}\,, \qquad D^{LN}_{\alpha\beta\lambda\mu} = -D^N_{\alpha\beta\lambda\mu}\,,$$

$$D^L_{\alpha\beta\lambda\mu} = D^N_{\alpha\beta\lambda\mu} + \frac{1}{2d}c^{(m)}_{\alpha\beta\lambda\mu}\,, \qquad D^Q_{\alpha\beta} = \frac{2}{3}(dc^{(f)}_{\alpha3\beta3} + c\,c^{(m)}_{\alpha3\beta3})\,,$$

$$D^{RL}_{\alpha\beta} = -\frac{d}{12}c^{(f)}_{\alpha\beta33} + \frac{1}{4}\left(h - \frac{c}{3}\right)c^{(m)}_{\alpha\beta33}\,, \qquad D^{RN}_{\alpha\beta} = \frac{d}{12}c^{(f)}_{\alpha\beta33}\,,$$

$$D^R = \frac{d^3}{40}c^{(f)}_{3333} + \frac{c}{8}\left(h^2 - \frac{2}{3}ch + \frac{c^2}{5}\right)c^{(m)}_{3333}\,. \tag{11.1.14}$$

The functional J attains its stationary value if the following conditions are fulfilled:
(i) the two-dimensional equilibrium equations

$$-N^{\alpha\beta}_{,\beta} = 0\,, \qquad -L^{\alpha\beta}_{,\beta} + Q^\alpha = 0\,, \qquad Q^\alpha_{,\alpha} + R = 0 \tag{11.1.15}$$

for $x \in \Omega$,
(ii) the constitutive relations

$$\varepsilon_{\alpha\beta} = D^N_{\alpha\beta\lambda\mu}N^{\lambda\mu} + D^{NL}_{\alpha\beta\lambda\mu}L^{\lambda\mu} + D^{RN}_{\alpha\beta}R\,,$$

$$\gamma_{\alpha\beta} = D^{NL}_{\alpha\beta\lambda\mu}N^{\lambda\mu} + D^L_{\alpha\beta\lambda\mu}L^{\lambda\mu} + D^{RL}_{\alpha\beta}R\,,$$

$$w = D^{RN}_{\alpha\beta}N^{\alpha\beta} + D^{RL}_{\alpha\beta}L^{\alpha\beta} + D^R R\,,$$

$$\kappa_\alpha = D^Q_{\alpha\beta}Q^\beta\,, \tag{11.1.16}$$

where the deformation measures are defined as follows

$$\varepsilon_{\alpha\beta} = \varepsilon_{\alpha\beta}(\boldsymbol{v}) = e_{\alpha\beta}(\boldsymbol{v})\,, \qquad \gamma_{\alpha\beta} = \gamma_{\alpha\beta}(\boldsymbol{u}) = e_{\alpha\beta}(\boldsymbol{u})\,,$$

$$\kappa_\alpha = \kappa_\alpha(\boldsymbol{u}, w) = u_\alpha - w_{,\alpha}\,. \tag{11.1.17}$$

(iii) The stress-type boundary conditions along Γ_σ

$$N_n = \bar{N}_n\,, \qquad N_\tau = \bar{N}_\tau\,, \qquad L_n = \bar{L}_n\,, \qquad L_\tau = \bar{L}_\tau\,, \qquad Q = \bar{Q}\,, \tag{11.1.18}$$

where

$$N_n = N^{\alpha\beta}n_\alpha n_\beta\,, \qquad N_\tau = N^{\alpha\beta}n_\beta\tau_\alpha$$

$$L_n = L^{\alpha\beta}n_\alpha n_\beta\,, \qquad L_\tau = L^{\alpha\beta}n_\beta\tau_\alpha\,, \qquad Q = Q^\alpha n_\alpha\,,$$

$$\bar{N}_n = \bar{N}^\alpha n_\alpha\,, \qquad \bar{N}_\tau = \bar{N}^\alpha_\tau \tau_\alpha\,. \tag{11.1.19}$$

Here $\boldsymbol{n}, \boldsymbol{\tau}$ are unit vectors: outward normal and tangent to Γ_σ respectively.

Now we can clear up the motivation of using assumptions (11.1.6) – (11.1.10). Note that the equilibrium equations

$$\sigma^{\alpha\beta}_{,\beta} + \sigma^{\alpha3}_{,3} = 0\,, \qquad \sigma^{\alpha3}_{,\alpha} + \sigma^{33}_{,3} = 0 \tag{11.1.20}$$

are satisfied identically for all $x \in \Omega$ and $x_3 \in I_1 \cup I_2 \cup I_3$ if we take into account Eqs. (11.1.15). Moreover $\sigma^{k3}(x, \pm h) = 0$ and σ^{ij} are continuous on the interfaces $x_3 = \pm c$. Obviously, the stress field $\boldsymbol{\sigma}$ cannot satisfy exactly the boundary conditions along the lateral surface S. Nevertheless one can say that σ is statically admissible except for a boundary zone around S.

Let us invert the constitutive relations (11.1.16):

$$N^{\lambda\mu} = A_v^{\lambda\mu\alpha\beta}\varepsilon_{\alpha\beta} + A_{vu}^{\lambda\mu\alpha\beta}\gamma_{\alpha\beta} + A_{vw}^{\lambda\mu}w\ ,$$
$$L^{\lambda\mu} = A_{vu}^{\lambda\mu\alpha\beta}\varepsilon_{\alpha\beta} + A_u^{\lambda\mu\alpha\beta}\gamma_{\alpha\beta} + A_{uw}^{\lambda\mu}w\ ,$$
$$R = A_{vw}^{\alpha\beta}\varepsilon_{\alpha\beta} + A_{uw}^{\alpha\beta}\gamma_{\alpha\beta} + A_w w\ ,$$
$$Q^\alpha = H^{\alpha\beta}\kappa_\beta\ .$$

$$(11.1.21)$$

The matrices A_v, A_{vu}, A_{vw}, A_u, A_{uw}, A_w, H are stiffnesses of the composite. Having found the primal form of the constitutive relations given by (11.1.21), one can proceed to the primal formulation of the equilibrium problem. The space of kinematically admissible fields has the form

$$V = \{(v, u, w)|\ v \in H^1(\Omega)^2\ ,\ u \in H^1(\Omega)^2\ ,\ w \in H^1(\Omega)\ ;$$
$$v = 0\ ,\ u = 0\ ,\ w = 0\quad \text{on}\quad \Gamma_w\}\ .$$

For $(v, u, w) \in V$, $(v', u', w') \in V$ let us introduce the bilinear form

$$a_\Omega(v, u, w\ ; v', u', w') = \int_\Omega [N^{\alpha\beta}(v, u, w)\varepsilon_{\alpha\beta}(v')$$

$$+L^{\alpha\beta}(v, u, w)\gamma_{\alpha\beta}(u') + Q^\alpha(u, w)\kappa_\alpha(u', w') + R(v, u, w)w']dx\ ,\quad (11.1.22)$$

where N, L, Q, R depend on (v, u, w) according to (11.1.21) and (11.1.17), and the linear form

$$f(v', u', w') = \int_{\Gamma_\sigma} (\bar{N}^\alpha v'_\alpha + \bar{L}^\alpha u'_\alpha - \bar{Q}w')ds\ .$$

$$(11.1.23)$$

The equilibrium problem in its primal variational formulation reads:

$$(P_\Omega)\quad \left|\begin{array}{l} \text{find } (v, u, w) \in V \text{ such that} \\ a_\Omega(v, u, w\ ;\ v', u', w') = f(v', u', w') \qquad \forall\ (v', u', w') \in V\ . \end{array}\right.\quad (11.1.24)$$

We observe that the elastic potential

$$j(x, \varepsilon, \gamma, \kappa, r) = \frac{1}{2}EAE^T + \frac{1}{2}H^{\alpha\beta}\kappa_\alpha\kappa_\beta\ ,$$

$$(11.1.25)$$

where $E = (\varepsilon, \gamma, r) \in \mathbb{E}_s^2 \times \mathbb{E}_s^2 \times \mathbb{R}$, $\kappa \in \mathbb{R}^2$ and

$$A = \begin{bmatrix} A_v, & A_{vu}, & A_{vw} \\ A_{vu}, & A_u, & A_{uw} \\ A_{vw}, & A_{uw}, & A_w \end{bmatrix}$$

$$(11.1.26)$$

constitute a positive-definite form:

$$\exists c_1 > 0\ ,\qquad EA(x)E^T \geq c_1(|\varepsilon|^2 + |\gamma|^2 + r^2)\ ,$$
$$H^{\alpha\beta}(x)a_\alpha a_\beta \geq c_1|a|^2\ ,$$

$$(11.1.27)$$

for a.e. $x \in \Omega$ and each $\varepsilon, \gamma \in \mathbb{E}^2_s$, $r \in \mathbb{R}$, $a \in \mathbb{R}^2$. Consequently the bilinear form $a_\Omega(\cdot, \cdot)$ is V-elliptic and continuous. If

$$\bar{N}^\alpha, \bar{L}^\alpha, \bar{Q} \in L^2(\Gamma_\sigma) , \tag{11.1.28}$$

then the bilinear form f is continuous and we conclude that the problem (P_Ω) is well-posed. We recall that each element of the matrices $\boldsymbol{A}$, $\boldsymbol{H}$ is an element of the space $L^\infty(\Omega)$.

Remark 11.1.1. It is instructive to prove (11.1.27). Let us consider $(11.1.27)_2$. Since $\boldsymbol{H}(x) = (\boldsymbol{D}^Q)^{-1}$, therefore it is sufficient to prove that there exists a constant $c_2 > 0$ such that

$$D^Q_{\alpha\beta}(x)b^\alpha b^\beta \geq c_2|\boldsymbol{b}|^2 , \tag{11.1.29}$$

for a.e. $x \in \Omega$ and for each $\boldsymbol{b} \in \mathbb{R}^2$. To corroborate this statement we write

$$\frac{1}{2}c_{ijkl}\sigma^{ij}\sigma^{kl} = \frac{1}{2}c_{\alpha\beta\lambda\mu}\sigma^{\alpha\beta}\sigma^{\lambda\mu} + c_{\alpha\beta 33}\sigma^{\alpha\beta}\sigma^{33} + 2c_{\alpha 3\beta 3}\sigma^{\alpha 3}\sigma^{\beta 3} + \frac{1}{2}c_{3333}(\sigma^{33})^2 .$$

Taking $\sigma^{\alpha\beta} = \sigma^{33} = 0$, $\sigma^{\alpha 3} \neq 0$ we have

$$c_{\alpha 3\beta 3}\sigma^{\alpha 3}\sigma^{\beta 3} > 0 ,$$

because the matrix $[c_{ijkl}]$ is positive definite. Treating $(\sigma^{\alpha 3}) \neq \boldsymbol{0}$ and $\boldsymbol{Q} \neq \boldsymbol{0}$ in (11.1.7) as arbitrary we obtain

$$\int_{-h}^{h} c_{\alpha 3\beta 3}\sigma^{\alpha 3}\sigma^{\beta 3} dx_3 = [\int_{-h}^{h} g(x_3)c_{\alpha 3\beta 3}(\boldsymbol{x})dx_3]Q^\alpha(x)Q^\beta(x)$$

$$= D^Q_{\alpha\beta}(x)Q^\alpha(x)Q^\beta(x) > 0 , \quad \text{a.e. } x \in \Omega ,$$

where $g(x_3)$ is deduced from (11.1.7). Hence (11.1.29) follows, since c_2 may be taken as the minimum eigenvalue of the symmetric matrix $\boldsymbol{D}^Q$. The proof of $(11.1.27)_1$ is similar to the above.

11.2. *Modelling the unilateral crack within the internal layer*

The cross-ply laminates subjected to in-plane shearing and tension incur intralaminar cracks or cracks which go across the lamina of a given orientation of fibers. The two-dimensional model of Sec. 11.1 has been formed with the intention of considering the cross cracks in the internal layer. The aim of this section is to put forward this construction.

Consider the laminate with a crack in the internal layer or for $x_3 \in I_2 = (-c, c)$. The crack surface $S_F = F \times I_2$ is perpendicular to the domain Ω; F represents its projection on Ω, see Fig. 11.2.1. Thus the crack is assumed as present at the stress-free state of the laminate. Its geometry is given a priori, independently of the applied loading.

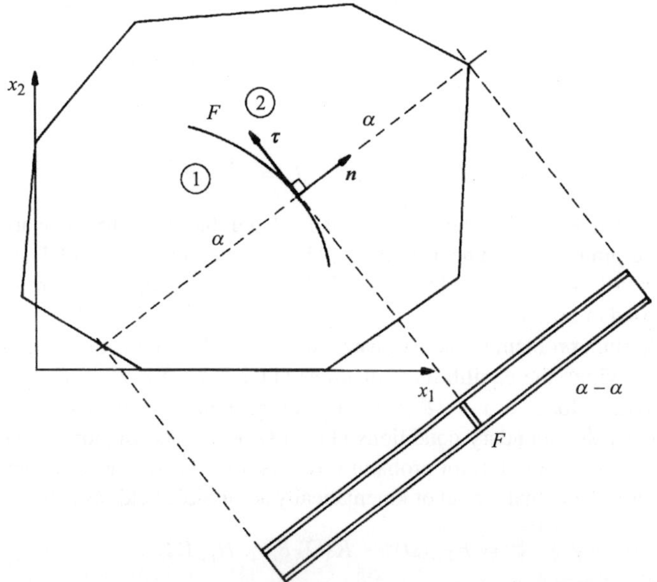

Fig. 11.2.1. Laminate with an intralaminar crack in the internal layer

Let us extend the arc F so that it divides the domain Ω into two subdomains Ω_α, see Fig. 11.2.1. Since the surface S_F is cylindrical the versor $n = (n_\alpha)$ normal to F at point $(x, 0)$ is parallel to all versors normal to S_F at points (x, x_3), $x_3 \in (-h, h)$. The versors tangent to S_F are represented by the versor τ tangent to F at point $(x, 0)$. A jump on S_F of a field g defined on $\Omega \times I_2$ is denoted by

$$[\![g]\!]_{SF} = g_{|2} - g_{|1} \; ;$$

here $g_{|\alpha}$ represents the value of g at the α-th side of S_F. Now we can express the unilateral contact conditions on the crack surface:

$$\sigma_n = \overset{1}{\sigma}_n = \overset{2}{\sigma}_n \leq 0 \,, \quad [\![w_n]\!]_{SF} \geq 0 \,, \quad \sigma_n [\![w_n]\!]_{SF} = 0 \,,$$
$$\overset{1}{\sigma}_{n\tau} = \overset{2}{\sigma}_{n\tau} = 0 \,, \quad \overset{1}{\sigma}_{nx3} = \overset{2}{\sigma}_{nx3} = 0 \,, \tag{11.2.1}$$

where

$$\overset{\delta}{\sigma}_n = \sigma^{\alpha\beta}{}_{|\delta} n_\alpha n_\beta \,, \quad \overset{\delta}{\sigma}_{n\tau} = \sigma^{\alpha\beta}{}_{|\delta} n_\alpha \tau_\beta \,, \quad \overset{\delta}{\sigma}_{nx3} = \sigma^{\alpha3}{}_{|\delta} n_\alpha \,, \quad w_n = w_\alpha n^\alpha \,. \tag{11.2.2}$$

Since the friction between crack lips is neglected, the jump $[\![w_\tau]\!]_{SF}$ assumes arbitrary values. On account of the stress and displacement assumptions (11.1.6)-(11.1.10), the contact conditions $(11.2.1)_1$ are approximated by the Signorini conditions expressed in terms of the stress resultants:

$$L_n = \overset{1}{L_n} = \overset{2}{L_n} \leq 0 \,, \quad [\![u_n]\!] \geq 0 \,, \quad L_n[\![u_n]\!] = 0 \,, \quad \overset{1}{L_\tau} = \overset{2}{L_\tau} = 0 \,, \quad \text{on } F \,. \quad (11.2.3)$$

Now $[\![\cdot]\!]$ denotes a jump on F and

$$\overset{\delta}{L_n} = L^{\alpha\beta}{}_{|_\delta} n_\alpha n_\beta \,, \qquad \overset{\delta}{L_\tau} = L^{\alpha\beta}{}_{|_\delta} n_\alpha \tau_\beta \,; \qquad\qquad (11.2.4)$$

$f_{|_\delta}$ means the values of f at the δ-th side of F. Note that the fields v, w are continuous on F and the jump $[\![u_\tau]\!]$ is unconstrained. The stress assumptions (11.1.7), (11.1.8) do not allow for expressing the conditions $(11.2.1)_2$ in terms of the two-dimensional laminate model proposed in Sec. 11.1.

The equilibrium problem of the laminate with the crack F amounts to finding the fields (v, u, w) satisfying: the equilibrium equations (11.1.15) for $x \in \Omega \backslash F$, the constitutive relations (11.1.21) for $x \in x \in \Omega \backslash F$, the boundary conditions involved in the definition of the space V, the boundary conditions (11.1.18) on Γ_σ and Signorini-type conditions (11.2.3) on F. This equilibrium problem posed above in a strong form can be put in a variational form. Note that the set of kinematically admissible fields assumes now the form

$$\mathbb{K} = H_{\Gamma_w}(\Omega)^2 \times K(\Omega \backslash F) \times H_{\Gamma_w}(\Omega) \,, \qquad\qquad (11.2.5)$$

where

$$H_{\Gamma_w}(\Omega) = \{ v \in H^1(\Omega) | \; v = 0 \quad \text{on } \Gamma_w \} \,, \qquad\qquad (11.2.6)$$
$$K(\Omega \backslash F) = \{ u \in H^1(\Omega \backslash F)^2 | \; u = 0 \quad \text{on } \Gamma_w \text{ and } [\![u_n]\!] \geq 0 \quad \text{on } F \} \,. \quad (11.2.7)$$

One can easily prove that the equilibrium problem considered is formally equivalent to solving the following variational inequality (V.I.):

$$(P_{\Omega \backslash F}) \left| \begin{array}{l} \text{find } (v, u, w) \in \mathbb{K} \text{ such that} \\[4pt] a_{\Omega \backslash F}(v, u, w \,; v', u' - u, w') \geq f(v', u' - u, w') \;\; \forall \, (v', u', w') \in \mathbb{K} \,. \end{array} \right. \quad (11.2.8)$$

The bilinear form $a_{\Omega \backslash F}(\cdot, \cdot)$ is defined by (11.1.22) with integration over $a_{\Omega \backslash F}$.

Variational inequality (11.2.8) is uniquely solvable. In fact, from Sec. 8 we already know that Korn's inequality remains valid for functions from the space $H_{\Gamma_w}(\Omega)^2$. Consequently, the bilinear form involved in V.I. (11.2.8) is coercive on $\mathbb{K}$.

11.3. Regular crack systems

In the case of fiber reinforced laminates the crack patterns are fairly regular. This observation justifies the assumption of a regular crack layout that will be considered in the sequel. Nevertheless, even if it is not the case, the analysis of a regular (or periodic) cracking layout is a first stage of more general analyses. For instance, the determination of effective properties of materials with stochastically distributed cracks is significantly facilitated once the periodic case has been solved.

Consider the laminate of Sec. 11.2 weakened by a family $\mathcal{F}$ of cracks F_j of the type described previously. Then one can divide the domain Ω (except for a boundary zone) into homothetic rectangular cells Z_j, each of which is weakened by a crack F_j. Assume that F_j do not intersect the boundaries ∂Z_j and the boundary $\partial\Omega$, see Fig. 11.3.1.

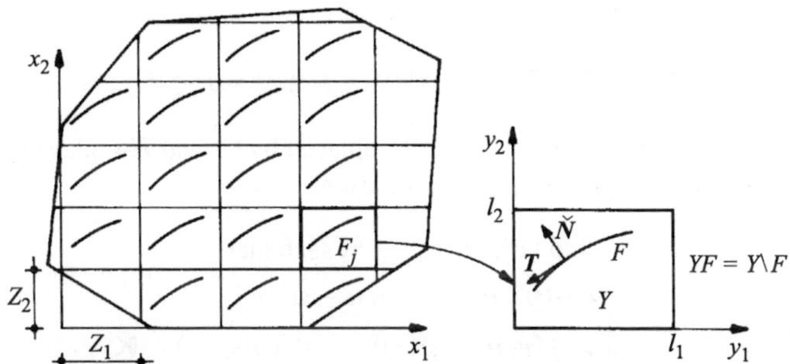

Fig. 11.3.1 Z-periodic layout of cracks. Geometry of the basic cell $YF = Y\backslash F$

Denote by $Z = (l_1^z, l_2^z)$ a master cell to which all Z_j are homothetic. The cell Z includes the crack F; F is homothetic to all F_j.

Mathematical description of the previous section applies here. A single crack F is here replaced with $\mathcal{F}$. Both strong and variational formulations of the previous section remain formally unchanged.

Although posed correctly, the formulation of the equilibrium problem of the laminate with a great number of cracks is impractical, since no results can be wrung from it, even if numerical methods are employed. To have these results in hand we shall resort to the two-scale asymptotic expansion method. This is reasonable especially in the case if we are interested in this part of the solution which does not change if the number of cracks increases. Just this part of the solution provides us with the crucial information about the effective or overall properties of the laminate.

The two-scale asymptotic expansion method is not uniquely defined, since the small parameter ε can enter into it in various manners. Two versions of the asymptotic expansion method corresponding to two methods of introducing ε into the original problem will be discussed in the succeeding sections.

11.4. *Moderately thick laminate weakened by transverse cracks of high density.* *Model* (h, l_0)

The aim of this section is to derive the effective model of cracked, moderately thick laminate with crack spacings much smaller than the thickness $2h$. The derivation will be based on the two-scale asymptotic expansions method. The starting point is the introduction of a small parameter ε. Since crack spacings are much smaller than the laminate thickness we assume that the dimensions of the periodicity cell are proportional to ε and keep h as

ε-independent. Thus we replace

$$Z \rightsquigarrow \varepsilon Y \,,\ Y = (l_1, l_2) \,;\quad h \rightsquigarrow h \,,\ c \rightsquigarrow c \,,\ d \rightsquigarrow d \,,$$
$$F_j \rightsquigarrow \varepsilon F_j \,;\quad \mathcal{F} \rightsquigarrow F^\varepsilon \,;\quad \Omega^\varepsilon = \Omega \backslash F^\varepsilon \,, \tag{11.4.1}$$

and F^ε represents the sum of all cracks εF_j. The set of kinematically admissible fields assumes the form

$$\mathbb{K}_\varepsilon := \mathbb{K}(\Omega^\varepsilon) = H_{\Gamma_w}(\Omega)^2 \times K(\Omega^\varepsilon) \times H_{\Gamma_w}(\Omega) \,, \tag{11.4.2}$$

where $K(\Omega^\varepsilon)$ is given by (11.2.7) with $\Omega \backslash F$ replaced by Ω^ε and F-replaced with F^ε.

For a fixed $\varepsilon > 0$ the equilibrium problem reads:

$$(P^1_{\Omega^\varepsilon}) \quad \left|\begin{array}{l} \text{find } (\boldsymbol{v}^\varepsilon, \boldsymbol{u}^\varepsilon, w^\varepsilon) \in \mathbb{K} \text{ such that} \\[4pt] a_{\Omega^\varepsilon}(\boldsymbol{v}^\varepsilon, \boldsymbol{u}^\varepsilon, w^\varepsilon \,;\, \boldsymbol{v}, \boldsymbol{u} - \boldsymbol{u}^\varepsilon, w) \\[4pt] \geq f(\boldsymbol{v}, \boldsymbol{u} - \boldsymbol{u}^\varepsilon, w) \qquad \forall\ (\boldsymbol{v}, \boldsymbol{u}, w) \in \mathbb{K}_\varepsilon \,. \end{array}\right. \tag{11.4.3}$$

The bilinear form $a_{\Omega^\varepsilon}(\cdot\,;\cdot)$ is defined by (11.1.22) with Ω replaced by Ω^ε.

Remark 11.4.1. Applying Korn's inequality derived in Sec. 9, we conclude that the bilinear form a_{Ω^ε} is coercive on the space $V(\Omega^\varepsilon) = H_{\Gamma_w}(\Omega)^2 \times H_{\Gamma_w}(\Omega^\varepsilon)^2 \times H_{\Gamma_w}(\Omega)$, provided that $F \subset Y$. It can easily be proved that $\mathbb{K}_\varepsilon$ is a convex and closed set in $V(\Omega^\varepsilon)$. Under the assumption (11.1.28), the linear form (11.1.23) is continuous and consequently problem $(P^1_{\Omega^\varepsilon})$ is uniquely solvable. $\square$

The scaling (11.4.1) used here will be called *in-plane scaling*, since the transverse dimensions are kept ε-independent.

We shall now employ the method of a two-scale expansions to the problem $(P^1_{\Omega^\varepsilon})$. First, let us define the sets:

$$K_{YF} = \{\boldsymbol{u} \in H^1_{per}(YF)^2 |\ [\![u_N]\!] \geq 0 \ \text{ on } F\} \,,$$
$$\mathbb{K}_{YF} = H^1_{per}(Y)^2 \times K_{YF} \times H^1_{per}(Y) \,. \tag{11.4.4}$$

Here $YF = Y \backslash F$, see Fig. 11.3.1, the space $H^1_{per}(YF)$ has been defined in Sec. 10.1; $u_N = u_\alpha \check{N}^\alpha$, $\check{\boldsymbol{N}} = (\check{N}^\alpha)$ stands for the versor normal to F; the brackets $[\![\ ]\!]$ denote a jump on F.

The solution to the problem $(P^1_{\Omega^\varepsilon})$ as well as the trial fields are assumed in the form

$$\eta^\varepsilon = \eta^{(0)}(x) + \varepsilon\eta^{(1)}(x, y) + \varepsilon^2\eta^{(2)}(x, y) + \ldots \tag{11.4.5}$$

where $y = x/\varepsilon$; $\eta^{(0)} \in H_{\Gamma_w}(\Omega)$ for $\eta \in \{v_\alpha, u_\alpha, w\}$ and $\eta^{(1)}(x, \cdot) \in H^1_{per}(Y)$ for $\eta \in \{v_\alpha, w\}$. Moreover, we assume that $\boldsymbol{u}^{(1)}(x, \cdot) \in K_{YF}$. The functions $\eta^{(k)}(x, \cdot)$, $k \geq 2$ are Y-periodic and sufficiently regular. The functions $\eta^{(k)}(\cdot, y)$ are defined on Ω and are sufficiently regular; $y \in YF$. Similar assumptions hold for trial fields.

The stress resultants associated with the kinematic fields in the form (11.4.5) assume the form

$$\mathcal{S}_\varepsilon = S_0 + 0(\varepsilon) , \qquad \mathcal{S} \in \{N_\varepsilon^{\alpha\beta}, L_\varepsilon^{\alpha\beta}, R_\varepsilon, Q_\varepsilon^\alpha\} , \tag{11.4.6}$$

where

$$N_0^{\lambda\mu} = A_v^{\lambda\mu\alpha\beta}\varepsilon_{\alpha\beta}^{(0)} + A_{vu}^{\lambda\mu\alpha\beta}\gamma_{\alpha\beta}^{(0)} + A_{vw}^{\lambda\mu}w^{(0)} ,$$
$$L_0^{\lambda\mu} = A_{vu}^{\lambda\mu\alpha\beta}\varepsilon_{\alpha\beta}^{(0)} + A_u^{\lambda\mu\alpha\beta}\gamma_{\alpha\beta}^{(0)} + A_{uw}^{\lambda\mu}w^{(0)} ,$$
$$R_0 = A_{vw}^{\alpha\beta}\varepsilon_{\alpha\beta}^{(0)} + A_{uw}^{\alpha\beta}\gamma_{\alpha\beta}^{(0)} + A_w w^{(0)} , \tag{11.4.7}$$
$$Q_0^\alpha = H^{\alpha\beta}\left(\kappa_\beta^h - \frac{\partial w^{(1)}}{\partial y_\beta}\right)_{|y=x/\varepsilon} .$$

The deformation measures are defined by

$$\varepsilon_{\alpha\beta}^{(0)} = \varepsilon_{\alpha\beta}^h + \varepsilon_{\alpha\beta}^y(\boldsymbol{v}^{(1)})_{|y=x/\varepsilon} , \qquad \gamma_{\alpha\beta}^{(0)} = \gamma_{\alpha\beta}^h + \gamma_{\alpha\beta}^y(\boldsymbol{u}^{(1)})_{|y=x/\varepsilon} , \tag{11.4.8}$$

where

$$\varepsilon_{\alpha\beta}^h = \varepsilon_{\alpha\beta}(\boldsymbol{v}^{(0)}) , \qquad \gamma_{\alpha\beta}^h = \gamma_{\alpha\beta}(\boldsymbol{u}^{(0)}) , \qquad \kappa_\beta^h = u_\beta^{(0)} - \frac{\partial w^{(0)}}{\partial x_\beta} , \tag{11.4.9}$$
$$\varepsilon_{\alpha\beta}^y(\boldsymbol{v}^{(1)}) = e_{\alpha\beta}^y(\boldsymbol{v}^{(1)}) , \qquad \gamma_{\alpha\beta}^y(\boldsymbol{u}^{(1)}) = e_{\alpha\beta}^y(\boldsymbol{u}^{(1)}) . \tag{11.4.10}$$

The homogenized stress resultants are given by

$$N_h^{\alpha\beta} = \langle N_0^{\alpha\beta}\rangle , \quad L_h^{\alpha\beta} = \langle L_0^{\alpha\beta}\rangle , \quad R_h = \langle R_0\rangle , \quad Q_h^\alpha = \langle Q_0^\alpha\rangle , \tag{11.4.11}$$

where

$$\langle\cdot\rangle = \frac{1}{|Y|}\int_{YF}(\cdot)dy, \quad |Y| = l_1 l_2 . \tag{11.4.12}$$

To find the effective model of the laminate one should substitute expansions (11.4.5) into the variational inequality (11.4.3) and let ε tend to zero. Following the lines of the derivations used in the previous sections one eventually arrives at the homogenized problem and at the local problems.

The homogenized problem assumes the form:

$$(P_h^1) \quad \left|\begin{array}{l} \text{find } (\boldsymbol{v}^{(0)}, \boldsymbol{u}^{(0)}, w^{(0)}) \in V \text{ such that} \\ a_h(\boldsymbol{v}^{(0)}, \boldsymbol{u}^{(0)}, w^{(0)} ; \boldsymbol{v}, \boldsymbol{u}, w) = f(\boldsymbol{v}, \boldsymbol{u}, w) \qquad \forall\,(\boldsymbol{v}, \boldsymbol{u}, w) \in V , \end{array}\right. \tag{11.4.13}$$

where

$$a_h(\boldsymbol{v}^{(0)}, \boldsymbol{u}^{(0)}, w^{(0)} ; \boldsymbol{v}, \boldsymbol{u}, w)$$
$$= \int_\Omega [N_h^{\alpha\beta}(\boldsymbol{v}^{(0)}, \boldsymbol{u}^{(0)}, w^{(0)})\varepsilon_{\alpha\beta}(\boldsymbol{v}) + L_h^{\alpha\beta}(\boldsymbol{v}^{(0)}, \boldsymbol{u}^{(0)}, w^{(0)})\gamma_{\alpha\beta}(\boldsymbol{u})$$
$$+ R_h(\boldsymbol{v}^{(0)}, \boldsymbol{u}^{(0)}, w^{(0)})w + Q_h^\alpha(\boldsymbol{u}^{(0)}, w^{(0)})\kappa_\alpha(\boldsymbol{u}, w)]dx . \tag{11.4.14}$$

The homogenized stress resultants are expressed by

$$
\begin{aligned}
N_h^{\lambda\mu} &= A_v^{\lambda\mu\alpha\beta}\varepsilon_{\alpha\beta}^h + A_{vu}^{\lambda\mu\alpha\beta}[\gamma_{\alpha\beta}^h + \langle\gamma_{\alpha\beta}^y(\boldsymbol{u}^{(1)})\rangle] + A_{vw}^{\lambda\mu}w^h \, , \\
L_h^{\lambda\mu} &= A_{vu}^{\lambda\mu\alpha\beta}\varepsilon_{\alpha\beta}^h + A_u^{\lambda\mu\alpha\beta}[\gamma_{\alpha\beta}^h + \langle\gamma_{\alpha\beta}^y(\boldsymbol{u}^{(1)})\rangle] + A_{uw}^{\lambda\mu}w^h \, , \\
R_h &= A_{vw}^{\alpha\beta}\varepsilon_{\alpha\beta}^h + A_{uw}^{\alpha\beta}[\gamma_{\alpha\beta}^h + \langle\gamma_{\alpha\beta}^y(\boldsymbol{u}^{(1)})\rangle] + A_w w^h \, , \\
Q_h^\alpha &= H^{\alpha\beta}\kappa_\beta^h \, .
\end{aligned}
\tag{11.4.15}
$$

The notation $w^h \equiv w^{(0)}$ indicates that this field plays the role of a deformation field. The field $\boldsymbol{u}^{(1)}$ depends upon the homogenized deformations (11.4.9). This dependence is governed by the local problem:

$$
(P_{loc}^1) \quad \left|
\begin{aligned}
&\text{find } (\boldsymbol{v}^{(1)}, \boldsymbol{u}^{(1)}, w^{(1)}) \in \mathbb{K}_{YF} \text{ such that} \\
&\langle N_0^{\alpha\beta}\varepsilon_{\alpha\beta}^y(\boldsymbol{v})\rangle = 0 \, , \\
&\langle L_0^{\alpha\beta}\gamma_{\alpha\beta}^y(\boldsymbol{u} - \boldsymbol{u}^{(1)})\rangle \geq 0 \, , \\
&\langle Q_0^\alpha \frac{\partial w}{\partial y_\alpha}\rangle = 0 \, , \qquad \forall \, (\boldsymbol{v}, \boldsymbol{u}, w) \in \mathbb{K}_{YF} \, .
\end{aligned}
\right.
\tag{11.4.16}
$$

In this problem the x-dependent fields $\varepsilon_{\alpha\beta}^h, \gamma_{\alpha\beta}^h, w^h$ are treated as given and x is viewed as a parameter.

The elastic potential of the homogenized laminate has the following form

$$
\begin{aligned}
U_h = \frac{1}{2}\langle N_0^{\alpha\beta}[\varepsilon_{\alpha\beta}^h + \varepsilon_{\alpha\beta}^y(\boldsymbol{v}^{(1)})] + L_0^{\alpha\beta}[\gamma_{\alpha\beta}^h + \gamma_{\alpha\beta}^y(\boldsymbol{u}^{(1)})] \\
+ R_0 w^h + Q_0^\alpha(\kappa_\alpha^h - \frac{\partial w^1}{\partial y_\alpha})\rangle \, ,
\end{aligned}
\tag{11.4.17}
$$

or, equivalently

$$
U_h = \frac{1}{2}(N_h^{\alpha\beta}\varepsilon_{\alpha\beta}^h + L_h^{\alpha\beta}\gamma_{\alpha\beta}^h + R_h w^h + Q_h^\alpha\kappa_\alpha^h) = U_1 + U_2 \, ,
\tag{11.4.18}
$$

where

$$
\begin{aligned}
&U_1(x, \boldsymbol{\varepsilon}^h, \boldsymbol{\gamma}^h, w^h) \\
&= \inf\{\frac{1}{|Y|}\int_{YF} j_1(x, \boldsymbol{\varepsilon}^h + \boldsymbol{\varepsilon}^y(\boldsymbol{v}), \boldsymbol{\gamma}^h + \boldsymbol{\gamma}^y(\boldsymbol{u}), w^h)dy| \, (\boldsymbol{v}, \boldsymbol{u}) \in H_{per}^1(Y)^2 \times K_{YF}\} \\
&= \frac{1}{|Y|}\int_{YF} j_1(x, \boldsymbol{\varepsilon}^h + \boldsymbol{\varepsilon}^y(\boldsymbol{v}^{(1)}), \boldsymbol{\gamma}^h + \boldsymbol{\gamma}^y(\boldsymbol{u}^{(1)}), w^h)dy \, , \\
&U_2(x, \boldsymbol{\kappa}^h) = \frac{1}{2}H^{\alpha\beta}(x)\kappa_\alpha^h\kappa_\beta^h \, .
\end{aligned}
$$

The homogenized constitutive relations (11.4.15) can be written as follows

$$
N_h^{\alpha\beta} = \frac{\partial U_h}{\partial\varepsilon_{\alpha\beta}^h} \, , \quad L_h^{\alpha\beta} = \frac{\partial U_h}{\partial\gamma_{\alpha\beta}^h} \, , \quad R_h = \frac{\partial U_h}{\partial w^h} \, , \quad Q_h^\alpha = \frac{\partial U_h}{\partial\kappa_\alpha^h} \, .
\tag{11.4.19}
$$

The potential $U_h(x, \varepsilon^h, \gamma^h, \kappa^h, w^h)$ has the following properties:

(i) $U_h(x, \cdot, \cdot, \cdot, \cdot)$ is a strictly convex function provided that F does not separate Y into disjoint subdomains. For other types of cracks this function is convex.

(ii) $U_h(x, \cdot, \cdot, \cdot, \cdot)$ is of class C^1.

(iii) $\exists\, c_1 > 0$ such that

$$U_h(x, \varepsilon^h, \gamma^h, \kappa^h, w^h) \le c_1(|\varepsilon^h|^2 + |\gamma^h|^2 + |\kappa^h|^2 + |w^h|^2)$$

for a.e. $x \in \Omega$ and all $(\varepsilon^h, \gamma^h, \kappa^h, w^h) \in \mathbb{E}_s^2 \times \mathbb{E}_s^2 \times \mathbb{R}^2 \times \mathbb{R}$.

(iv) $\exists\, c_0 > 0$ such that

$$U_h(x, \varepsilon^h, \gamma^h, \kappa^h, w^h) \ge c_0(|\varepsilon^h|^2 + |\gamma^h|^2 + |\kappa^h|^2 + |w^h|^2)$$

for a.e. $x \in \Omega$ and all $(\varepsilon^h, \gamma^h, \kappa^h, w^h) \in \mathbb{E}_s^2 \times \mathbb{E}_s^2 \times \mathbb{R}^2 \times \mathbb{R}$, provided that F does not separate Ω into disjoint subdomains.

The proof is left to the reader as an exercise.

By virtue of the properties (i) – (iv) the problem (P_h^1) is uniquely solvable, provided that the length of Γ_w is positive.

Remark 11.4.2. The solution to the problem (P_{loc}^1) is not sensitive to the ratios $\rho_\alpha = l_\alpha/2h$ (crack spacing / laminate thickness), hence it is not sensitive to the transverse shape of the periodicity cell. Consequently, the effective potential U_h and formulae for the effective stiffnesses do not depend upon ρ_α. This means that they apply to the case when ρ_α are very small, which is a consequence of keeping h as ε-independent in the in-plane scaling (11.4.1).

The formulation of the problem (P_{loc}^1) can be simplified. Equation $(11.4.16)_3$ implies $w^1 = w^1(x)$. Thus the problem (P_{loc}^1) reduces to

$$(\bar{P}_{loc}^1) \quad \left|
\begin{array}{ll}
\text{find } v^{(1)} \in H_{per}^1(Y)^2 \text{ and } u^{(1)} \in K_{YF} \text{ such that} & \\[2mm]
\langle N_0^{\alpha\beta} \dfrac{\partial v_\alpha}{\partial y_\beta} \rangle = 0\,, & \forall\, v \in H_{per}^1(Y)^2 \\[4mm]
\langle T_0^{\alpha\beta} \dfrac{\partial}{\partial y_\beta}(u_\alpha - u_\alpha^{(1)}) \rangle \ge 0\,, & \forall\, u \in K_{YF}\,.
\end{array}
\right.$$

11.5. Thin laminate with transverse cracks of high density. Model (h_0, l_0)

In the conventional engineering computations the longitudinal displacements of the laminate are usually treated as uniform through the thickness. Moreover, the effect of stresses σ^{33} is neglected. Such an approach, justifiable for thin laminates, corresponds to the following assumptions

$$u_\alpha^{(0)} = 0\,, \qquad R_0 = 0\,. \tag{11.5.1}$$

Note that only at the macrolevel the graph $w_\alpha(z)$ is assumed to be uniform; in general $u^{(1)} \ne 0$.

Let us substitute $\boldsymbol{u} = \boldsymbol{0}$ and $w = 0$ into (P_h^1). One finds

$$\int_\Omega N_h^{\alpha\beta} \varepsilon_{\alpha\beta}(\boldsymbol{v}) dx = \int_{\Gamma_\sigma} \bar{N}^\alpha v_\alpha ds \qquad \forall \ \boldsymbol{v} \in H_{\Gamma_w}(\Omega)^2 . \tag{11.5.2}$$

The field w^h appearing in Eq. $(11.4.15)_1$ can be eliminated by means of the equation $R_h = 0$. Then we find so-called Kachanov's form of the constitutive relation:

$$N_h^{\lambda\mu} = A_1^{\lambda\mu\alpha\beta} \varepsilon_{\alpha\beta}^h - A_2^{\lambda\mu\alpha\beta} \varepsilon_{\alpha\beta}^F , \tag{11.5.3}$$

where

$$A_1^{\alpha\beta\lambda\mu} = A_v^{\alpha\beta\lambda\mu} - A_{vw}^{\alpha\beta} A_{vw}^{\lambda\mu}(A_w)^{-1} , \qquad A_2^{\alpha\beta\lambda\mu} = A_{vu}^{\alpha\beta\lambda\mu} - A_{vw}^{\alpha\beta} A_{uw}^{\lambda\mu}(A_w)^{-1} , \tag{11.5.4}$$

and

$$\varepsilon_{\alpha\beta}^F = -\langle \gamma_{\alpha\beta}^y(\boldsymbol{u}^{(1)}) \rangle . \tag{11.5.5}$$

Since

$$\varepsilon_{\alpha\beta}^F = \frac{1}{2l_1 l_2} \int_F ([\![u_\alpha^{(1)}]\!]\check{N}_\beta + [\![u_\beta^{(1)}]\!]\check{N}_\beta) ds , \tag{11.5.6}$$

therefore the tensor $\boldsymbol{\varepsilon}^F$ describes the crack opening. Decomposing $\boldsymbol{u}^{(1)}$ into normal and tangential components we obtain

$$\varepsilon_{\alpha\beta}^F = \frac{1}{2l_1 l_2} \int_F \left[[\![u_N^{(1)}]\!]\check{N}_\alpha \check{N}_\beta + \frac{1}{2}[\![u_T^{(1)}]\!](T_\alpha \check{N}_\beta + T_\beta \check{N}_\alpha) \right] ds , \tag{11.5.7}$$

where

$$u_\alpha^{(1)} = u_N^{(1)} \check{N}_\alpha + u_T^{(1)} T_\alpha , \tag{11.5.8}$$

and $\check{\boldsymbol{N}}, \boldsymbol{T}$ are versors normal and tangent to F, see Fig. 11.3.1.

To be consistent, we reformulate the local problem (P_{loc}^1). By $(11.4.7)_3$, the equality $R_0 = 0$ implies

$$A_{vw}^{\alpha\beta}[\varepsilon_{\alpha\beta}^h + \varepsilon_{\alpha\beta}^y(\boldsymbol{v}^{(1)})] + A_{uw}^{\alpha\beta} \gamma_{\alpha\beta}^y(\boldsymbol{u}^{(1)}) + A_w w^h = 0 , \tag{11.5.9}$$

since $\gamma_{\alpha\beta}^h = 0$ by $(11.5.1)_1$. Consequently, Eqs $(11.4.7)_{1,2}$ assume the form

$$\begin{aligned}
N_0^{\lambda\mu} &= A_1^{\lambda\mu\alpha\beta}[\varepsilon_{\alpha\beta}^h + \varepsilon_{\alpha\beta}^y(\boldsymbol{v}^{(1)})] + A_2^{\lambda\mu\alpha\beta} \gamma_{\alpha\beta}^y(\boldsymbol{u}^{(1)}) , \\
L_0^{\lambda\mu} &= A_3^{\lambda\mu\alpha\beta}[\varepsilon_{\alpha\beta}^h + \varepsilon_{\alpha\beta}^y(\boldsymbol{v}^{(1)})] + A_4^{\lambda\mu\alpha\beta} \gamma_{\alpha\beta}^y(\boldsymbol{u}^{(1)}) ,
\end{aligned} \tag{11.5.10}$$

where

$$A_3^{\lambda\mu\alpha\beta} = A_2^{\alpha\beta\lambda\mu} , \qquad A_4^{\alpha\beta\lambda\mu} = A_u^{\alpha\beta\lambda\mu} - A_{uw}^{\alpha\beta} A_{uw}^{\lambda\mu}(A_w)^{-1} . \tag{11.5.11}$$

The local problem assumes the form $(\bar{P}_{loc}^1)$ in which

$$N_0^{\alpha\beta} = n^{\alpha\beta} + n_0^{\alpha\beta} , \qquad L_0^{\alpha\beta} = l^{\alpha\beta} + l_0^{\alpha\beta} , \tag{11.5.12}$$

where

$$n^{\lambda\mu} = A_1^{\lambda\mu\alpha\beta} \varepsilon^y_{\alpha\beta}(\boldsymbol{v}^{(1)}) + A_2^{\lambda\mu\alpha\beta} \gamma^y_{\alpha\beta}(\boldsymbol{u}^{(1)}) \,,$$
$$l^{\lambda\mu} = A_3^{\lambda\mu\alpha\beta} \varepsilon^y_{\alpha\beta}(\boldsymbol{v}^{(1)}) + A_4^{\lambda\mu\alpha\beta} \gamma^y_{\alpha\beta}(\boldsymbol{u}^{(1)}) \,,$$

$$(11.5.13)$$

and

$$n_0^{\lambda\mu} = A_1^{\lambda\mu\alpha\beta} \varepsilon^h_{\alpha\beta} \,, \qquad l_0^{\lambda\mu} = A_3^{\lambda\mu\alpha\beta} \varepsilon^h_{\alpha\beta} \,. \qquad (11.5.14)$$

The new local problem thus obtained will be referred to as (P^0_{loc}). If the crack F does not separate Y into two disjoint subdomains, then the problem (P^0_{loc}) is well-posed; $(\boldsymbol{v}^{(1)}, \boldsymbol{u}^{(1)})$ are determined up to additive constants. Let us define the potential

$$\mathcal{V}_h = \frac{1}{2}\langle N_0^{\alpha\beta}[\varepsilon^h_{\alpha\beta} + \varepsilon^y_{\alpha\beta}(\boldsymbol{v}^{(1)})] + L_0^{\alpha\beta}\gamma^y_{\alpha\beta}(\boldsymbol{u}^{(1)})\rangle \qquad (11.5.15a)$$

or

$$\mathcal{V}_h = \frac{1}{2}N_h^{\alpha\beta}(\boldsymbol{\varepsilon}^h)\varepsilon^h_{\alpha\beta} \,. \qquad (11.5.15b)$$

Then

$$N_h^{\alpha\beta} = \frac{\partial \mathcal{V}_h}{\partial \varepsilon^h_{\alpha\beta}} \,. \qquad (11.5.16)$$

The potential $\mathcal{V}_h(x, \boldsymbol{\varepsilon}^h)$ has the following properties:
(i) $\mathcal{V}_h(x, \cdot)$ is a strictly convex function of class C^1 in the space $\mathbb{E}^2_s$
(ii) $\exists\, c_1 > c_0 > 0$ such that for a.e. $x \in \Omega$

$$c_0|e|^2 \leq \mathcal{V}_h(x, e) \leq c_1|e|^2 \quad \forall e \in \mathbb{E}^2_s \,.$$

These properties are also satisfied when F divides Y into two disjoint subdomains. The macroscopic constitutive relation (11.5.16) is a by-product of (i). The variational equation (11.5.2) along with (11.5.16) constitutes a new homogenized problem (P^0_h). Due to the property (ii) this problem admits a unique solution.

Remark 11.4.2 applies here. The effective constitutive relations (11.5.16) are not sensitive to the quantities $\rho_\alpha = l_\alpha/2h$ representing the crack spacing to the laminate thickness ratios. Thus the results refer to the case of ρ_α being very small or to the case of a dense distribution of the cracks.

11.6.   Thin laminate weakened by transverse cracks of arbitrary density. Model (h_0, l)

This section is aimed at deriving a new effective model of a thin three-layer laminate with transverse cracks in the internal layer. No limitations concerning the crack spacing-to-the thickness ratio are imposed. The notation (h_0, l) means that $h \ll \text{diam}\,(\Omega)$.

The following quantities

$$c, \; d, \; h, \; b, \; l_1^z, \; l_2^z \,, \qquad (11.6.1)$$

are internal length scales of the model of Sec. 11.3. In this section we shall assume that all these parameters are proportional to the one small parameter ε. Thus we replace:

$$c \rightsquigarrow \varepsilon c \,, \ d \rightsquigarrow \varepsilon d \,, \ h \rightsquigarrow \varepsilon h \,, \ b \rightsquigarrow \varepsilon b \,, \ l_\alpha^z \rightsquigarrow \varepsilon l_\alpha \,, \quad Z \rightsquigarrow \varepsilon Y \,, \ F \rightsquigarrow F^\varepsilon \,. \tag{11.6.2}$$

If ε tends to zero, the thickness of the laminate also diminishes to zero. To compensate for this degeneracy we scale the loading:

$$\bar{N}^\alpha \rightsquigarrow \varepsilon \bar{N}^\alpha \,, \ \bar{L}^\alpha \rightsquigarrow \varepsilon \bar{L}^\alpha \,, \ \bar{Q} \rightsquigarrow \bar{Q} \,. \tag{11.6.3}$$

The scaling (11.6.2) implies the following scaling of the stiffnesses involved in $(P^1_{\Omega^\varepsilon})$

$$(A_v^{\alpha\beta\lambda\mu}, A_{vu}^{\alpha\beta\lambda\mu}, A_u^{\alpha\beta\lambda\mu}) \rightsquigarrow (\varepsilon A_v^{\alpha\beta\lambda\mu}, \varepsilon A_{vu}^{\alpha\beta\lambda\mu}, \varepsilon A_u^{\alpha\beta\lambda\mu}) \,,$$

$$(A_{vw}^{\alpha\beta}, A_{uw}^{\alpha\beta}) \rightsquigarrow \left(\frac{1}{\varepsilon} A_{vw}^{\alpha\beta}, \frac{1}{\varepsilon} A_{uw}^{\alpha\beta}\right) \,, \qquad H^{\alpha\beta} \rightsquigarrow \frac{1}{\varepsilon} H^{\alpha\beta} \,, \qquad A_w \rightsquigarrow \frac{1}{\varepsilon^3} A_w \,. \tag{11.6.4}$$

Note that the scaling (11.6.2) preserves homothety between three-dimensional periodicity cells $\varepsilon \mathcal{Y}$, $\mathcal{Y} = YF \times (-h, h)$ for all values of ε. This scaling will be called refined. According to this scaling the constitutive relations (11.1.21) read as follows

$$N_\varepsilon^{\lambda\mu} = \varepsilon A_v^{\lambda\mu\alpha\beta} \varepsilon_{\alpha\beta}(\boldsymbol{v}^\varepsilon) + \varepsilon A_{vu}^{\lambda\mu\alpha\beta} \gamma_{\alpha\beta}(\boldsymbol{u}^\varepsilon) + \frac{1}{\varepsilon} A_{vw}^{\lambda\mu} w^\varepsilon \,,$$

$$L_\varepsilon^{\lambda\mu} = \varepsilon A_{vu}^{\lambda\mu\alpha\beta} \varepsilon_{\alpha\beta}(\boldsymbol{v}^\varepsilon) + \varepsilon A_u^{\lambda\mu\alpha\beta} \gamma_{\alpha\beta}(\boldsymbol{u}^\varepsilon) + \frac{1}{\varepsilon} A_{uw}^{\lambda\mu} w^\varepsilon \,,$$

$$R_\varepsilon = \frac{1}{\varepsilon} A_{vw}^{\lambda\mu} \varepsilon^{\lambda\mu}(\boldsymbol{u}^\varepsilon) + \frac{1}{\varepsilon} A_{uw}^{\lambda\mu} \gamma_{\lambda\mu}(\boldsymbol{u}^\varepsilon) + \frac{1}{\varepsilon^3} A_w w^\varepsilon \,, \tag{11.6.5}$$

$$Q_\varepsilon^\alpha = \frac{1}{\varepsilon} H^{\alpha\beta} \kappa_\beta(\boldsymbol{u}^\varepsilon, w^\varepsilon) \,.$$

These relations determine the counterpart of the bilinear form a_{Ω^ε} involved in $(P^1_{\Omega^\varepsilon})$:

$$b_{\Omega^\varepsilon}(\boldsymbol{v}^\varepsilon, \boldsymbol{u}^\varepsilon, w^\varepsilon ; \ \boldsymbol{v}, \boldsymbol{u}, w) = \int_{\Omega^\varepsilon} [N_\varepsilon^{\alpha\beta}(\boldsymbol{v}^\varepsilon, \boldsymbol{u}^\varepsilon, w^\varepsilon) \varepsilon_{\alpha\beta}(\boldsymbol{v})$$

$$+ L_\varepsilon^{\alpha\beta}(\boldsymbol{v}^\varepsilon, \boldsymbol{u}^\varepsilon, w^\varepsilon) \gamma_{\alpha\beta}(\boldsymbol{u}) + Q_\varepsilon^\alpha(\boldsymbol{u}^\varepsilon, w^\varepsilon) \kappa_\alpha(\boldsymbol{u}, w) + R_\varepsilon(\boldsymbol{v}^\varepsilon, \boldsymbol{u}^\varepsilon, w^\varepsilon) w] dx \,. \tag{11.6.6}$$

The scaling (11.6.3) defines a new linear form representing the work of the boundary loading:

$$g^\varepsilon(\boldsymbol{v}, \boldsymbol{u}, w) = \int_{\Gamma_\sigma} (\varepsilon \bar{N}^\alpha v_\alpha + \varepsilon \bar{L}^\alpha u_\alpha - \bar{Q} w) ds \,. \tag{11.6.7}$$

Thus the equilibrium problem reads:

$$(P^2_{\Omega^\varepsilon}) \qquad \left| \begin{array}{l} \text{find } (\boldsymbol{v}^\varepsilon, \boldsymbol{u}^\varepsilon, w^\varepsilon) \in \mathbb{K}(\Omega^\varepsilon) \text{ such that} \\[4pt] b_{\Omega^\varepsilon}(\boldsymbol{v}^\varepsilon, \boldsymbol{u}^\varepsilon, w^\varepsilon ; \ \boldsymbol{v}, \boldsymbol{u} - \boldsymbol{u}^\varepsilon, w) \geq g^\varepsilon(\boldsymbol{v}, \boldsymbol{u} - \boldsymbol{u}^\varepsilon, w) \\[4pt] \hspace{4cm} \forall \ (\boldsymbol{v}, \boldsymbol{u}, w) \in \mathbb{K}(\Omega^\varepsilon) \,. \end{array} \right. \tag{11.6.8}$$

This problem will be studied in the next section.

Let us pass now to constructing an asymptotic solution of $(P^2_{\Omega\varepsilon})$. We use the following representations

$$v^\varepsilon_\alpha = v^{(0)}_\alpha(x) + \varepsilon v^{(1)}_\alpha(x,y) + \varepsilon^2 v^{(2)}_\alpha(x,y) + \ldots,$$
$$u^\varepsilon_\alpha = \varepsilon u^{(1)}_\alpha(x,y) + \varepsilon^2 u^{(2)}_\alpha(x,y) + \ldots,$$
$$w^\varepsilon = \varepsilon^2 w^{(2)}(x,y) + \varepsilon^3 w^{(3)}(x,y) + \ldots, \quad y = x/\varepsilon.$$

(11.6.9)

The trial fields are expanded similarly

$$v_\alpha = v'^{(0)}_\alpha(x) + \varepsilon v'^{(1)}_\alpha(x,y) + \varepsilon^2 v'^{(2)}_\alpha(x,y) + \ldots,$$
$$u_\alpha = \varepsilon u'^{(1)}_\alpha(x,y) + \varepsilon^2 u'^{(2)}_\alpha(x,y) + \ldots,$$
$$w = \varepsilon^2 w'^{(2)}(x,y) + \varepsilon^3 w'^{(3)}(x,y) + \ldots, \quad y = x/\varepsilon.$$

(11.6.10)

It is assumed that

$$v^{(0)}_\alpha, v'^{(0)}_\alpha \in H_{\Gamma_w}(\Omega),$$
$$\eta \in H^1_{per}(Y), \ \eta \in \{v^{(1)}_\alpha(x,\cdot), v'^{(1)}_\alpha(x,\cdot), w^{(2)}(x,\cdot), w'^{(2)}(x,\cdot)\},$$
$$\boldsymbol{u}^{(1)}(x,\cdot), \boldsymbol{u}'^{(1)}(x,\cdot) \in K_{YF}.$$

(11.6.11)

The deformation measures associated with the kinematic fields (11.6.9) are

$$\varepsilon_{\alpha\beta}(\boldsymbol{v}^\varepsilon) = \varepsilon_{\alpha\beta}(\boldsymbol{v}^{(0)}) + \varepsilon^y_{\alpha\beta}(\boldsymbol{v}^{(1)})_{|y=x/\varepsilon} + 0(\varepsilon),$$
$$\gamma_{\alpha\beta}(\boldsymbol{u}^\varepsilon) = \gamma^y_{\alpha\beta}(\boldsymbol{u}^{(1)})_{|y=x/\varepsilon} + 0(\varepsilon),$$
$$\kappa_\alpha(\boldsymbol{u}^\varepsilon, w^\varepsilon) = \varepsilon \kappa^y_\alpha(\boldsymbol{u}^{(1)}, w^{(2)})_{|y=x/\varepsilon} + 0(\varepsilon^2),$$

(11.6.12)

where

$$\kappa^y_\alpha(\boldsymbol{u}^{(1)}, w^{(2)}) = u^{(1)}_\alpha - \frac{\partial w^{(2)}}{\partial y_\alpha}.$$

(11.6.13)

The stress resultants (11.6.5) associated with the kinematics (11.6.9) are

$$N^{\alpha\beta}_\varepsilon = \varepsilon N^{\alpha\beta}_0 + 0(\varepsilon^2), \qquad L^{\alpha\beta}_\varepsilon = \varepsilon L^{\alpha\beta}_0 + 0(\varepsilon^2),$$
$$Q^\alpha_\varepsilon = Q^\alpha_0 + 0(\varepsilon), \qquad R_\varepsilon = \frac{1}{\varepsilon} R_0 + 0(1)$$

(11.6.14)

where

$$N^{\lambda\mu}_0 = A^{\lambda\mu\alpha\beta}_v[\varepsilon^h_{\alpha\beta} + \varepsilon^y_{\alpha\beta}(\boldsymbol{v}^{(1)})] + A^{\lambda\mu\alpha\beta}_{vu}\gamma^y_{\alpha\beta}(\boldsymbol{u}^{(1)}) + A^{\lambda\mu}_{vw}w^{(2)},$$
$$L^{\lambda\mu}_0 = A^{\lambda\mu\alpha\beta}_{vu}[\varepsilon^h_{\alpha\beta} + \varepsilon^y_{\alpha\beta}(\boldsymbol{v}^{(1)})] + A^{\lambda\mu\alpha\beta}_u\gamma^y_{\alpha\beta}(\boldsymbol{u}^{(1)}) + A^{\lambda\mu}_{uw}w^{(2)},$$
$$R_0 = A^{\alpha\beta}_{vw}[\varepsilon^h_{\alpha\beta} + \varepsilon^y_{\alpha\beta}(\boldsymbol{v}^{(1)})] + A^{\alpha\beta}_{uw}\gamma^y_{\alpha\beta}(\boldsymbol{u}^{(1)}) + A_w w^{(2)},$$
$$Q^\alpha_0 = H^{\alpha\beta}\kappa^y_\beta(\boldsymbol{u}^{(1)}, w^{(2)}),$$

(11.6.15)

and $\varepsilon_{\alpha\beta}^h = \varepsilon_{\alpha\beta}(\boldsymbol{v}^{(0)})$. By using the representations (11.6.9) – (11.6.15) one can express the bilinear form (11.6.6) as follows

$$b_{\Omega^\varepsilon}(\boldsymbol{v}^\varepsilon, \boldsymbol{u}^\varepsilon, w^\varepsilon \,;\, \boldsymbol{v}, \boldsymbol{u}, w) = \varepsilon \int\limits_{\Omega^\varepsilon} \{ N_0^{\alpha\beta} [\varepsilon_{\alpha\beta}(\boldsymbol{v}'^{(0)}) + \varepsilon_{\alpha\beta}^y(\boldsymbol{v}'^{(1)})]$$

$$+ L_0^{\alpha\beta} \gamma_{\alpha\beta}^y(\boldsymbol{u}'^{(1)}) + Q_0^\alpha \kappa_\alpha^y(\boldsymbol{u}'^{(1)}, w'^{(2)}) + R_0 w'^{(2)} \} dx + 0(\varepsilon^2) \,. \qquad (11.6.16)$$

On the other hand the linear form (11.6.7) is represented by

$$g^\varepsilon(\boldsymbol{v}, \boldsymbol{u}, w) = \varepsilon \int\limits_{\Gamma_\sigma} \bar{N}^\alpha v_\alpha'^{(0)} ds + 0(\varepsilon^2) \,. \qquad (11.6.17)$$

To find the main terms of the solution $(v_\alpha^\varepsilon, u_\alpha^\varepsilon, w^\varepsilon)$ we take first

$$\boldsymbol{v} = \pm \boldsymbol{v}'^{(0)}(x) \,, \qquad \boldsymbol{u} = 0 \,, \qquad w = 0 \,, \qquad (11.6.18)$$

in the inequality (11.6.8), with the left- and right-hand sides represented by (11.6.16) and (11.6.17), respectively and next divide both sides of the inequalities (for $\pm \boldsymbol{v}'^{(0)}$) by ε. Letting ε tend to zero we find the variational equation

$$\int\limits_\Omega N_h^{\alpha\beta} \varepsilon_{\alpha\beta}(\boldsymbol{v}'^{(0)}) dx = \int\limits_{\Gamma_\sigma} \bar{N}^\alpha v_\alpha'^{(0)} ds \,, \qquad (11.6.19)$$

with

$$N_h^{\alpha\beta} = \langle N_0^{\alpha\beta} \rangle \,. \qquad (11.6.20)$$

Let us return now to the inequality (11.6.8) with the left- and right-hand sides given by (11.6.16) and (11.6.17) respectively. Let us divide it by ε, pass to zero with ε and take into account the equality (11.6.19). Then the inequality reduces to

$$\int\limits_\Omega \langle N_0^{\alpha\beta} \varepsilon_{\alpha\beta}(\boldsymbol{v}'^{(1)}) + L_0^{\alpha\beta} \gamma_{\alpha\beta}^y(\boldsymbol{u}'^{(1)} - \boldsymbol{u}^{(1)}) + Q_0^\alpha \left(u_\alpha'^{(1)} - u_\alpha^{(1)} - \frac{\partial w'^{(2)}}{\partial y_\alpha} \right)$$

$$+ R_0 w'^{(2)} \rangle dx \geq 0 \,, \qquad \forall \, (\boldsymbol{v}'^{(1)}, \boldsymbol{u}'^{(1)}, w'^{(2)}) \in \mathbb{K}(\Omega^\varepsilon) \,. \qquad (11.6.21)$$

To derive local problem we put

$$\boldsymbol{v}'^{(1)}(x, y) = \pm \vartheta(x) \boldsymbol{v}'(y) \,, \qquad w'^{(2)}(x, y) = \pm \psi(x) w'(y) \,,$$

$$\boldsymbol{u}'^{(1)}(x, y) = \boldsymbol{u}^{(1)}(x, y) + \chi(x)[\boldsymbol{u}'(y) - \boldsymbol{u}^{(1)}(x, y)] \,, \qquad (11.6.22)$$

$$\vartheta, \psi, \chi \in \mathbf{D}(\Omega) \,, \qquad 0 \leq \chi \leq 1 \,, \qquad (\boldsymbol{v}', \boldsymbol{u}', w') \in \mathbb{K}_{YF} \,.$$

Substituting $(11.6.22)_{1,2}$ into $(11.6.21)$ and passing with ε to zero we find the local problem in the form:

$$(P^2_{loc}) \quad \left| \begin{array}{l} \text{find } (\boldsymbol{v}^{(1)}, \boldsymbol{u}^{(1)}, w^{(1)}) \in \mathbb{K}_{YF} \text{ such that} \\[4pt] \langle N_0^{\alpha\beta} \varepsilon^y_{\alpha\beta}(\boldsymbol{v}') \rangle = 0 \,, \\[4pt] \langle L_0^{\alpha\beta} \gamma^y_{\alpha\beta}(\boldsymbol{u}' - \boldsymbol{u}^{(1)}) + Q_0^\alpha (u'_\alpha - u^{(1)}_\alpha) \rangle \geq 0 \,, \\[4pt] \langle R_0 - Q_0^\alpha \dfrac{\partial w'}{\partial y_\alpha} \rangle = 0 \,, \qquad \forall \ (\boldsymbol{v}', \boldsymbol{u}', w') \in \mathbb{K}_{YF} \,, \end{array} \right. \tag{11.6.23}$$

where N_0, L_0, Q_0, R_0, are given by $(11.6.15)$. The strain $\varepsilon^h(x)$ is treated here as given and x is viewed as a parameter.

The problem (P^2_{loc}) is equivalent to the following minimization problem:

$$(\tilde{P}^2_{loc}) \quad \text{find} \quad \inf\{\langle j(x, \varepsilon^h + \varepsilon^y(\boldsymbol{v}), \gamma^y(\boldsymbol{v}), \kappa^y(\boldsymbol{u}, w), w) \rangle \,|\, (\boldsymbol{v}, \boldsymbol{u}, w) \in \mathbb{K}_{YF}\} \,, \tag{11.6.24}$$

where

$$j(x, \varepsilon, \gamma, \kappa, r) = j_1(x, \varepsilon, \gamma, r) + \frac{1}{2} H^{\alpha\beta}(x) \kappa_\alpha \kappa_\beta \,, \tag{11.6.25}$$

and for $\boldsymbol{E} = (\varepsilon, \gamma, r) \in \mathbb{E}^2_s \times \mathbb{E}^2_s \times \mathbb{R}$, we define

$$j_1(x, \boldsymbol{E}) = \frac{1}{2} \boldsymbol{E} \mathbf{A}(x) \boldsymbol{E}^T \,, \tag{11.6.26}$$

with $\mathbf{A}$ given by $(11.1.26)$. The property $(11.1.27)$ implies that a solution $(\boldsymbol{v}^{(1)}, \boldsymbol{u}^{(1)}, w^{(2)}) \in \mathbb{K}_{YF}$ to the problem $(\tilde{P}^2_{loc})$ and hence to (P^2_{loc}), exists. The fields $v^{(1)}_\alpha$, $u^{(1)}_\alpha$, $w^{(2)}$ are determined up to additive constants.

Now we are ready to formulate the homogenized problem:

$$(P^2_h) \quad \left| \begin{array}{l} \text{find } \boldsymbol{v}^{(0)} \in H_{\Gamma_w}(\Omega)^2 \text{ such that the variational equation } (11.6.19) \text{ is satisfied, where} \\ N_h^{\alpha\beta} \text{ are given by } (11.6.20) \text{ and } (11.6.15)_1, \text{ and the fields } (\boldsymbol{v}^{(1)}, \boldsymbol{u}^{(1)}, w^{(2)}) \text{ appear-} \\ \text{ing in the constitutive relation } (11.6.20) \text{ depend on the tensor } \varepsilon^h_{\alpha\beta} = \varepsilon_{\alpha\beta}(\boldsymbol{v}^{(0)}) \\ \text{according to the implicit relation determined by the problem } (P^2_{loc}). \end{array} \right.$$

Let us define the homogenized potential

$$\mathcal{W}_h(x, \varepsilon^h) = \langle j(x, \varepsilon^h + \varepsilon^y(\boldsymbol{v}^{(1)}), \gamma^y(\boldsymbol{u}^{(1)}), \kappa^y(\boldsymbol{u}^{(1)}, w^{(2)}), w^{(2)}) \rangle$$

$$= \frac{1}{|Y|} \inf\{ \int_{YF} j_1[x, \varepsilon^h + \varepsilon^y(\boldsymbol{v}), \gamma^y(\boldsymbol{u}), \kappa^y(\boldsymbol{u}, w), w] dy \,|\, (\boldsymbol{v}, \boldsymbol{u}, w) \in \mathbb{K}_{YF}\} \,, \tag{11.6.27}$$

or

$$\mathcal{W}_h(x, \varepsilon^h) = \frac{1}{2} \langle N_0^{\alpha\beta} [\varepsilon^h_{\alpha\beta} + \varepsilon^y_{\alpha\beta}(\boldsymbol{v}^{(1)})] + L_0^{\alpha\beta} \gamma^y_{\alpha\beta}(\boldsymbol{u}^{(1)}) $$

$$+ Q_0^\alpha \kappa^y_\alpha(\boldsymbol{u}^{(1)}, w^{(2)}) + R_0 w^{(2)} \rangle \,. \tag{11.6.28}$$

The last formula can be simplified as follows. Let us substitute $v' = v^{(1)}$, $w' = w^{(2)}$, $u'^{(1)} = 2u^{(1)}$ and then $u'^{(1)} = 0$ into (11.6.23). After adding the equalities obtained in this manner one finds

$$\langle N_0^{\alpha\beta}\varepsilon_{\alpha\beta}^y(v^{(1)}) + L_0^{\alpha\beta}\gamma_{\alpha\beta}^y(u^{(1)}) + Q_0^\alpha\kappa_\alpha(u^{(1)}, w^{(2)}) + R_0 w^{(2)}\rangle = 0 \ . \quad (11.6.29)$$

Consequently the expression (11.6.28) reduces to

$$\mathcal{W}_h(x, \varepsilon^h) = \frac{1}{2}N_0^{\alpha\beta}(\varepsilon^h)\varepsilon_{\alpha\beta}^h \ . \qquad (11.6.30)$$

The potential $\mathcal{W}_h$ has the following properties:
(i)

$$N_h^{\alpha\beta} = \frac{\partial \mathcal{W}_h}{\partial \varepsilon_{\alpha\beta}^h} \ . \qquad (11.6.31)$$

(ii) $\mathcal{W}_h(x, \cdot)$ is strictly convex and of class C^1.
(iii) $\exists c_1 > c_0 > 0$ such that for a.e. $x \in \Omega$

$$c_0|e|^2 \le \mathcal{W}_h(x, e) \le c_1|e|^2 \qquad \forall e \in \mathbb{E}_s^2 \ . \qquad (11.6.32)$$

We observe that formula (11.6.31) and the properties (ii), (iii) are preserved if F divides the basic cell Y into two disjoint subdomains. The property (iii) ensures unique solvability of the homogenized problem (P_h^2), provided that meas $(\Gamma_w) > 0$.

It turns out that the homogenized constitutive relations (11.6.20) can be rearranged to Kachanov's form (11.5.3). To corroborate this statement we recall that $v^{(1)} \in H_{per}^1(Y)^2$, then $\langle\varepsilon_{\alpha\beta}^y(v^{(1)})\rangle = 0$. Thus the expression (11.6.20) assumes the form

$$N_h^{\lambda\mu} = A_v^{\lambda\mu\alpha\beta}\varepsilon_{\alpha\beta}^h + A_{vu}^{\lambda\mu\alpha\beta}\langle\gamma_{\lambda\mu}^y(u^{(1)})\rangle + A_{vw}^{\lambda\mu}\langle w^{(2)}\rangle \ . \qquad (11.6.33)$$

Let us substitute $w' =$const into Eq. (11.6.23)$_3$. Hence $\langle R_0\rangle = 0$, which makes it possible to express $\langle w^{(2)}\rangle$ in terms of $\langle\gamma_{\lambda\mu}^y(u^{(1)})\rangle$.

Elimination of $\langle w^{(2)}\rangle$ transforms Eq. (11.6.33) to the form

$$N_h^{\lambda\mu} = A_1^{\lambda\mu\alpha\beta}\varepsilon_{\alpha\beta}^h - A_2^{\lambda\mu\alpha\beta}\tilde{\varepsilon}_{\alpha\beta}^F \ , \qquad (11.6.34)$$

with $\tilde{\varepsilon}_{\alpha\beta}^F = -\langle\gamma_{\alpha\beta}^y(u^{(1)})\rangle$, cf. Eqs. (11.5.6) and (11.5.7). The tilde over $\varepsilon_{\alpha\beta}^F$ indicates that these quantities are evaluated according to the (h_0, l) method. The tensors $(A_\sigma^{\lambda\mu\alpha\beta})$ coincide with those defined by (11.5.4). Only now the field $u^{(1)}$ is a solution to a different local problem. Note that the knowledge of the functions $[u_N^{(1)}(\varepsilon^h)]$, $[u_T^{(1)}(\varepsilon^h)]$ along F suffices for the determination of the homogenized constitutive relationship (11.6.34). The quantities $\tilde{\varepsilon}_{\alpha\beta}^F$ will be referred to as crack deformation measures.

The constitutive relations of Kachanov type (11.6.34) are usually postulated in the damage mechanics. In contrast to Kachanov's phenomenological approach, the formula

(11.6.34) has not been proposed but rigorously derived. The internal state variables $(\tilde{\varepsilon}^F_{\alpha\beta})$ are directly connected with macrodeformation fields $(\varepsilon^h_{\alpha\beta})$ through the local problem. The tensor A_2 may be called a "damage moduli tensor".

Unlike the results based on the in-plane scaling, the problem (P^2_{loc}) is sensitive to the change of the ratios: $\rho_\alpha = l_\alpha/2h$. The refined scaling (11.6.2) preserves the relations: $l_\alpha/d, l_\alpha/c$ for each ε. Consequently, the homogenized constitutive relation $N^{\alpha\beta}_h(\varepsilon^h)$ depends upon the crack spacing measured with respect to the laminate thickness.

Remark 11.6.1. The model (h_0, l_0) derived in Sec. 11.5 turns out to be a limit of the model (h_0, l) as $\rho_\alpha = l_\alpha/2h$ tend to zero. Thus

$$\mathcal{V}_h = \lim_{\rho_1 \to 0, \rho_2 \to 0} \mathcal{W}_h(\rho_1, \rho_2) \qquad (11.6.35)$$

for l_1/l_2 fixed. Two ways of deriving the model (h_0, l_0) are outlined in Diagram 11.6.1.

Diagram 11.6.1

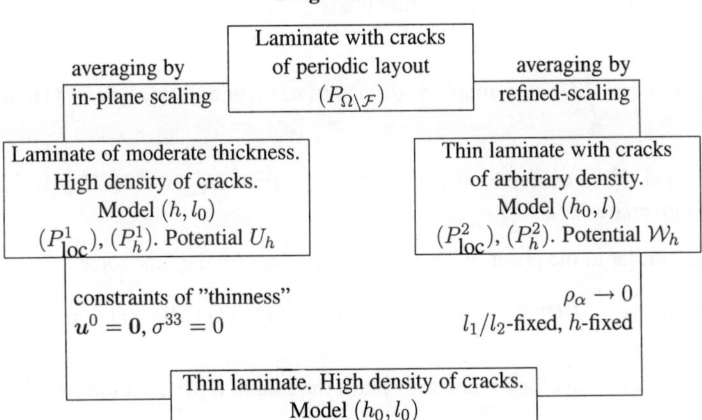

11.7. Justification by Γ-convergence and duality

The physical models derived in Secs. 11.4, 11.6 will now be justified rigorously by applying the method of Γ-convergence. From the mathematical point of view, the in-plane scaling results in a model which can be justified similarly to the one describing partially penetrating fissures, cf. Sec. 10.4. Therefore in Sec. 11.7.1 we shall only formulate a convergence theorem and provide indispensable comments.

On account of the appearance of singular terms in the functional of the total potential energy a rigorous justification of the refined scaling is more complicated. Previously, in Sec. 5.4, we have already dealt with a much simpler case of the refined scaling of an uncracked plate.

11.7.1.　Moderately thick laminate

The model itself has been developed in Sec. 11.4. The notations used there are still preserved. Problem $(P^1_{\Omega^\varepsilon})$, being convex is equivalent to the following minimization problem:

$$J^\varepsilon_1(\boldsymbol{v}^\varepsilon, \boldsymbol{u}^\varepsilon, w^\varepsilon) = \inf\{J^\varepsilon_1(\boldsymbol{v}, \boldsymbol{u}, w) \mid (\boldsymbol{v}, \boldsymbol{u}, w) \in \mathbb{K}(\Omega^\varepsilon)\}\,, \tag{11.7.1}$$

where the functional of the total potential energy J^ε_1 is given by

$$J^\varepsilon_1(\boldsymbol{v}, \boldsymbol{u}, w) = \frac{1}{2}a_{\Omega^\varepsilon}(\boldsymbol{v}, \boldsymbol{u}, w;\ \boldsymbol{v}, \boldsymbol{u}, w) - f(\boldsymbol{v}, \boldsymbol{u}, w)$$

$$= \int_{\Omega^\varepsilon} [j_1(x, \boldsymbol{\varepsilon}(\boldsymbol{v}), \boldsymbol{\gamma}(\boldsymbol{u}), w) + \frac{1}{2}H^{\alpha\beta}(x)\kappa_\alpha(\boldsymbol{u}, w)\kappa_\beta(\boldsymbol{u}, w)]dx - f(\boldsymbol{v}, \boldsymbol{u}, w)\,. \tag{11.7.2}$$

We recall that j_1 is defined by (11.6.26). From Remark 11.4.1 it follows that the functional J^ε_1 is coercive on the space $V(\Omega^\varepsilon) = H_{\Gamma_w}(\Omega)^2 \times H_{\Gamma_w}(\Omega^\varepsilon)^2 \times H_{\Gamma_w}(\Omega)$ and $(\boldsymbol{v}^\varepsilon, \boldsymbol{u}^\varepsilon, w^\varepsilon) \in \mathbb{K}_\varepsilon = \mathbb{K}(\Omega^\varepsilon)$, solving the minimization problem (11.7.1), is unique.

Now we are in a position to formulate the Γ-convergence theorem.

Theorem 11.7.1. The sequence of functionals $\{J^\varepsilon_1\}_{\varepsilon>0}$ is Γ-convergent in the topology $\tau = (w - H^1(\Omega)^2) \times (s - L^2(\Omega)^2) \times (w - H^1(\Omega))$ to

$$J^h_1(\boldsymbol{v}, \boldsymbol{u}, w) = \int_\Omega U_h[x, \boldsymbol{\varepsilon}(\boldsymbol{v}(x)), \boldsymbol{\gamma}(\boldsymbol{u}(x)),\ \boldsymbol{\kappa}(\boldsymbol{u}(x), w(x)), w(x)]dx - f(\boldsymbol{v}, \boldsymbol{u}, w)\,.$$

$$\tag{11.7.3}$$

The functional J^h_1 is coercive on the space $H_{\Gamma_w}(\Omega)^2 \times H_{\Gamma_w}(\Omega)^2 \times H_{\Gamma_w}(\Omega)$ and $U_h = U_1 + U_2$ is given by (11.4.18).

Proof. It is similar to the proof of Theorem 10.3.1. Indeed, the functional

$$\Phi(\boldsymbol{v}, \boldsymbol{u}, w) := \frac{1}{2}\int_\Omega H^{\alpha\beta}(x)\kappa_\alpha(\boldsymbol{u}(x), w(x))\kappa_\beta(\boldsymbol{u}(x), w(x))dx - f(\boldsymbol{v}, \boldsymbol{u}, w) \tag{11.7.4}$$

is continuous in the topology τ and thus plays the role of a perturbation functional. Next, we set

$$G^\varepsilon(\boldsymbol{v}, \boldsymbol{u}, w) = \begin{cases} \displaystyle\int_{\Omega^\varepsilon} j_1[x, \boldsymbol{\varepsilon}(\boldsymbol{v}(x)), \boldsymbol{\gamma}(\boldsymbol{u}(x)), w(x)]dx, \\ \qquad \text{if } \boldsymbol{v} \in H^1(\Omega)^2, w \in H^1(\Omega) \text{ and } \boldsymbol{u} \in K^\varepsilon\,; \\ +\infty \qquad \text{otherwise,} \end{cases} \tag{11.7.5}$$

where

$$K^\varepsilon = \{\boldsymbol{u} \in H^1(\Omega^\varepsilon)^2 \mid [\![u_n]\!] \geq 0 \text{ on } F^\varepsilon\}\,, \tag{11.7.6}$$

while j_1 is given by (11.1.30). Consequently the proof reduces to showing that the sequence of functionals $\{G^\varepsilon(\boldsymbol{v}, \cdot, w)\}_{\varepsilon>0}$ Γ-converges in the strong topology of $L^2(\Omega)^2$ to

$$G(\boldsymbol{v}, \boldsymbol{u}, w) = \begin{cases} \displaystyle\int_\Omega U_1[x, \boldsymbol{\varepsilon}(\boldsymbol{v}(x)), \boldsymbol{\gamma}(\boldsymbol{u}(x)), w(x)]dx, & \text{if } \boldsymbol{v},\ \boldsymbol{u} \in H^1(\Omega)^2,\ w \in H^1(\Omega), \\ +\infty & \text{otherwise.} \end{cases} \tag{11.7.7}$$

Now the reader will be able to perform both direct and dual homogenization by following the consideration of Secs. 10.3 and 10.4. $\square$

11.7.2.   Refined scaling and Γ-convergence

The model (h_0, l), proposed in Sec. 11.6, will now be justified rigorously by using the method of Γ-convergence. The scaling leading to this model is also called by us "refined s-caling". Preserving the notation introduced in Sec. 11.6, the functional of the total potential energy has the following form

$$J_{p\varepsilon}(\boldsymbol{v}, \boldsymbol{u}, w) = \frac{1}{2} b_{\Omega^\varepsilon}(\boldsymbol{v}, \boldsymbol{u}, w \; ; \; \boldsymbol{v}, \boldsymbol{u}, w) - g^\varepsilon(\boldsymbol{v}, \boldsymbol{u}, w)$$

$$= \varepsilon \int_{\Omega^\varepsilon} j_\varepsilon[x, \boldsymbol{\varepsilon}(\boldsymbol{v}(x)), \boldsymbol{\gamma}(\boldsymbol{u}(x)), \; \boldsymbol{\kappa}(\boldsymbol{u}(x), w(x)), w(x)]dx - \varepsilon f_\varepsilon(\boldsymbol{v}, \boldsymbol{u}, w), \qquad (11.7.8)$$

where

$$j_\varepsilon(x, \boldsymbol{\epsilon}_1, \boldsymbol{\epsilon}_2, \boldsymbol{a}, r) = \frac{1}{2}[\boldsymbol{\epsilon}_1, \boldsymbol{\epsilon}_2, r]\mathbf{A}_\varepsilon(x)[\boldsymbol{\epsilon}_1, \boldsymbol{\epsilon}_2, r]^T + \frac{1}{2\varepsilon^2}H^{\alpha\beta}(x)a_\alpha a_\beta , \qquad (11.7.9)$$

$$f_\varepsilon(\boldsymbol{v}, \boldsymbol{u}, w) = \int_{\Gamma_\sigma}(\bar{N}^\alpha v_\alpha + \bar{L}^\alpha u_\alpha - \frac{1}{\varepsilon}\bar{Q}w)ds . \qquad (11.7.10)$$

Here $\epsilon_\alpha \in \mathbb{E}_s^2 \; (\alpha = 1, 2)$, $\boldsymbol{a} \in \mathbb{R}^2$, $r \in \mathbb{R}$ and

$$\mathbf{A}_\varepsilon(x) = \begin{bmatrix} \boldsymbol{A}_v & \boldsymbol{A}_{vu} & \dfrac{1}{\varepsilon^2}\boldsymbol{A}_{vw} \\[2mm] \boldsymbol{A}_{vu} & \boldsymbol{A}_u & \dfrac{1}{\varepsilon^2}\boldsymbol{A}_{uw} \\[2mm] \dfrac{1}{\varepsilon^2}\boldsymbol{A}_{vw} & \dfrac{1}{\varepsilon^2}\boldsymbol{A}_{uw} & \dfrac{1}{\varepsilon^4}\boldsymbol{A}_w \end{bmatrix} . \qquad (11.7.11)$$

As we know, all the elements constituting $\boldsymbol{A}_\varepsilon$ may depend on x, cf. Sec. 11.1. The variational inequality (11.6.8) is then equivalent to the convex minimization problem:

$$J_{p\varepsilon}(\tilde{\boldsymbol{v}}^\varepsilon, \tilde{\boldsymbol{u}}^\varepsilon, \tilde{w}^\varepsilon) = \inf\{J_{p\varepsilon}(\boldsymbol{v}, \boldsymbol{u}, w) \mid (\boldsymbol{v}, \boldsymbol{u}, w) \in \mathbb{K}_\varepsilon\} . \qquad (11.7.12)$$

We set

$$J_\varepsilon = \frac{1}{\varepsilon}J_{p\varepsilon} . \qquad (11.7.13)$$

From (11.7.8) and (11.7.13) we conclude that

$$J_\varepsilon(\boldsymbol{v}, \boldsymbol{u}, w) = \int_{\Omega^\varepsilon} j_\varepsilon[x, \boldsymbol{\varepsilon}(\boldsymbol{v}(x)), \boldsymbol{\gamma}(\boldsymbol{u}(x)), \; \boldsymbol{\kappa}(\boldsymbol{u}(x), w(x)), w(x)]dx - f_\varepsilon(\boldsymbol{v}, \boldsymbol{u}, w) .$$

$$(11.7.14)$$

We observe that the functional $J_{p\varepsilon}$ models the physical problem. The mathematical situation is described by the functional J_ε. Of our main interest will be the Γ-convergence of the sequence $\{J_\varepsilon\}_{\varepsilon>0}$. Primarily, however, we shall provide some auxiliary results. First, we note that for a fixed $\varepsilon > 0$ the minimization problem (11.7.12) is equivalent to

$$J_\varepsilon(\bar{v}^\varepsilon, \bar{u}^\varepsilon, \bar{w}^\varepsilon) = \inf\{J_\varepsilon(v, u, w) \mid (v, u, w) \in \mathbb{K}_\varepsilon\}. \tag{11.7.15}$$

The existence and uniqueness of a $(\bar{v}^\varepsilon, \bar{u}^\varepsilon, \bar{w}^\varepsilon) \in \mathbb{K}_\varepsilon$ will now be proved.

Theorem 11.7.2. Assume that $0 < \varepsilon < 1$ and

$$\bar{N}^\alpha \in L^2(\Gamma_\sigma), \quad \bar{Q} \in L^2(\Gamma_\sigma), \quad \bar{L}^\alpha \in L^\infty(\Gamma_\sigma), \quad \alpha = 1, 2. \tag{11.7.16}$$

Then the minimization problem appearing on the r.h.s. of (11.7.15) is uniquely solvable.

Proof. The set $K(\Omega^\varepsilon)$ defined by (11.2.7) with $\Omega\backslash F$ replaced by Ω^ε is convex and weakly closed in $H^1(\Omega^\varepsilon)^2$. It is thus sufficient to show that J_ε is coercive. Obviously, this functional is strictly convex on $H^1(\Omega)^2 \times H^1(\Omega^\varepsilon)^2 \times H^1(\Omega)$. Let us rewrite the matrix $\mathbf{A}_\varepsilon$ in the following form

$$\mathbf{A}_\varepsilon = \mathbf{A} + \begin{bmatrix} 0 & 0 & \dfrac{1}{\varepsilon^2}\mathbf{A}_{vw} - \mathbf{A}_{vw} \\[2ex] 0 & 0 & \dfrac{1}{\varepsilon^2}\mathbf{A}_{uw} - \mathbf{A}_{uw} \\[2ex] \dfrac{1}{\varepsilon^2}\mathbf{A}_{vw} - \mathbf{A}_{vw} & \dfrac{1}{\varepsilon^2}\mathbf{A}_{uw} - \mathbf{A}_{uw} & \dfrac{1}{\varepsilon^4}\mathbf{A}_w - \mathbf{A}_w \end{bmatrix}. \tag{11.7.17}$$

For each $u \in H_{\Gamma_w}(\Omega^\varepsilon)^2$ (meas $\Gamma_w > 0$) Korn's inequality is given by, cf. Prop. 9.3.6.

$$\|u\|_{1,\Omega^\varepsilon} \le (\tilde{c}\varepsilon + c_1)\|\gamma(u)\|_{0,\Omega^\varepsilon}, \tag{11.7.18}$$

where $\tilde{c}$ and c_1 are positive constants independent of ε. Recalling that the matrices $\mathbf{A}$ and $(H^{\alpha\beta})$ are positive definite, by using (11.7.17) and (11.7.18) for $(v, u, w) \in V(\Omega^\varepsilon)$ and $0 < \varepsilon < 1$ we get

$$J_\varepsilon(v, u, w) \ge c_2(\|\varepsilon(v)\|_{0,\Omega}^2 + \|\gamma(u)\|_{0,\Omega^\varepsilon}^2 + \|w\|_{0,\Omega}^2) + \frac{c_3}{\varepsilon^2}\|u - \nabla(w)\|_{0,\Omega^\varepsilon}^2$$

$$+ \frac{1 - \varepsilon^2}{\varepsilon^2}\int_\Omega A_{vw}^{\alpha\beta}\varepsilon_{\alpha\beta}(v)w\,dx + \frac{1 - \varepsilon^2}{\varepsilon^2}\int_\Omega A_{uw}^{\alpha\beta}\gamma_{\alpha\beta}(u)w\,dx$$

$$+ \frac{c_4(1 - \varepsilon^4)}{\varepsilon^4}\|w\|_{0,\Omega}^2 - c_5(\|v\|_{0,\Gamma_\sigma} + \|u\|_{L^1,\Gamma_\sigma} + \frac{1}{\varepsilon}\|w\|_{0,\Gamma_\sigma}), \tag{11.7.19}$$

where $c_2, \ldots, c_5$ are positive constants, which do not depend on ε. A trace theorem for the space $H^1(\Omega^\varepsilon)^2$ seems not to be available. However, in Sec. 14.1 it will be shown that the injection $H^1(\Omega^\varepsilon)^2 \subset BD(\Omega)$ is continuous, where $BD(\Omega)$ is the space of functions with bounded deformation. It is well known that the trace of a function $u \in BD(\Omega)$ is in $L^1(\Gamma)^2$. Anyway we have

$$\|u\|_{L^1(\Gamma_\sigma)^2} \le c_6\|u\|_{BD(\Omega)} \le c_7\|u\|_{1,\Omega^\varepsilon}, \tag{11.7.20}$$

and the constants c_6 and c_7 do not depend on ε.

For any $f_1, f_2 \in L^2(\Omega)$ and each $\rho > 0$ the following elementary inequality holds true:

$$||f_1 + f_2||_{0,\Omega}^2 \geq \frac{\rho}{1+\rho}||f_1||_{0,\Omega}^2 - \rho||f_2||_{0,\Omega}^2 \, . \qquad (11.7.21)$$

Taking into account (11.7.20) and (11.7.21) in (11.7.19), after simple calculations we finally obtain

$$J_\varepsilon(\boldsymbol{v}, \boldsymbol{u}, w) \geq c_8(||\boldsymbol{\varepsilon}(\boldsymbol{v})||_{0,\Omega}^2 + ||\boldsymbol{\gamma}(\boldsymbol{u})||_{0,\Omega^\varepsilon}^2 + ||w||_{0,\Omega}^2)$$

$$+ \frac{c_9\rho}{\varepsilon^2(1+\rho)}||\nabla w||_{0,\Omega}^2 - c_{10}(||\boldsymbol{v}||_{1,\Omega} + ||\boldsymbol{u}||_{1,\Omega^\varepsilon} + \frac{1}{\varepsilon}||w||_{1,\Omega}) \, .$$

As previously, c_8, c_9 and c_{10} are positive constants independent of ε and ρ has to be suitably chosen. For an $\varepsilon > 0$ and fixed the coercivity of the functional J_ε follows, which concludes the proof. $\qquad \square$

We pass now to the formulation of a counterpart of Lemma 5.4.4, cf. also Lemma 2.10.1.

Lemma 11.7.3. Let $\{\boldsymbol{v}^\varepsilon, \boldsymbol{u}^\varepsilon, w^\varepsilon\}_{\varepsilon>0}$ be a sequence in $H^1(\Omega)^2 \times H^1(\Omega^\varepsilon)^2 \times H^1(\Omega)$ such that

$$\sup_{\varepsilon>0}\{||\boldsymbol{v}^\varepsilon||_{1,\Omega} + ||\boldsymbol{u}^\varepsilon||_{1,\Omega^\varepsilon} + ||w^\varepsilon||_{1,\Omega}\} < \infty \, , \qquad (11.7.22)$$

$$J_\varepsilon(\boldsymbol{v}^\varepsilon, \boldsymbol{u}^\varepsilon, w^\varepsilon) \leq c_1 \, , \qquad (11.7.23)$$

where c_1 is a non-negative constant, independent of ε and (11.7.16) holds true. Denote by $(\boldsymbol{v}, \boldsymbol{u}, w)$ a limit of a convergent subsequence in the topology $(w - H^1(\Omega)^2) \times (s - L^2(\Omega)^2) \times (w - H^1(\Omega))$. Then $\boldsymbol{u} = \nabla w = 0$ and $w = 0$.

Proof. Taking into account (11.7.17) and recalling that $\mathbf{A}$ and $(H^{\alpha\beta})$ are positive definite matrices we have

$$J_\varepsilon(\boldsymbol{v}^\varepsilon, \boldsymbol{u}^\varepsilon, w^\varepsilon) \geq c_2(||\boldsymbol{\varepsilon}(\boldsymbol{v}^\varepsilon)||_{0,\Omega}^2 + ||\boldsymbol{\gamma}(\boldsymbol{u}^\varepsilon)||_{0,\Omega^\varepsilon}^2 + ||w^\varepsilon||_{0,\Omega}^2)$$

$$+ \frac{1}{2\varepsilon^2}\int_{\Omega^\varepsilon} H^{\alpha\beta}\kappa_\alpha(\boldsymbol{u}^\varepsilon, w^\varepsilon)\kappa_\beta(\boldsymbol{u}^\varepsilon, w^\varepsilon)dx$$

$$+ \int_\Omega \left(\frac{1}{\varepsilon^2}A_{vw}^{\alpha\beta}\varepsilon_{\alpha\beta}(\boldsymbol{v}^\varepsilon)w^\varepsilon - A_{vw}^{\alpha\beta}\varepsilon_{\alpha\beta}(\boldsymbol{v}^\varepsilon)w^\varepsilon \right) dx$$

$$+ \int_{\Omega^\varepsilon} \left(\frac{1}{\varepsilon^2}A_{uw}^{\alpha\beta}\gamma_{\alpha\beta}(\boldsymbol{u}^\varepsilon)w^\varepsilon - A_{uw}^{\alpha\beta}\gamma_{\alpha\beta}(\boldsymbol{u}^\varepsilon)w^\varepsilon \right) dx$$

$$+ \int_\Omega \left(\frac{1}{2\varepsilon^4}A_w(w^\varepsilon)^2 - \frac{1}{2}A_w(w^\varepsilon)^2 \right) dx$$

$$- \int_{\Gamma_\sigma} \left(\bar{N}^\alpha v_\alpha^\varepsilon + \bar{L}^\alpha u_\alpha^\varepsilon - \frac{1}{\varepsilon}\bar{Q}w^\varepsilon \right) ds \, , \qquad (11.7.24)$$

where c_2 is a positive constant appearing in (11.7.19). Elementary inequalities hold true:

$$\frac{1}{\varepsilon^2}A^{\alpha\beta}_{vw}\varepsilon_{\alpha\beta}(\boldsymbol{v}^\varepsilon)w^\varepsilon = \frac{1}{\sqrt{\varepsilon}}A^{\alpha\beta}_{vw}\varepsilon_{\alpha\beta}(\boldsymbol{v}^\varepsilon)\frac{w^\varepsilon}{\varepsilon^{3/2}} \geq -\frac{1}{2\varepsilon}(A^{\alpha\beta}_{vw}\varepsilon_{\alpha\beta}(\boldsymbol{v}^\varepsilon))^2 - \frac{(w^\varepsilon)^2}{2\varepsilon^3}\,,$$

$$\frac{1}{\varepsilon^2}A^{\alpha\beta}_{uw}\gamma_{\alpha\beta}(\boldsymbol{u}^\varepsilon)w^\varepsilon \geq -\frac{1}{2\varepsilon}(A^{\alpha\beta}_{uw}\gamma_{\alpha\beta}(\boldsymbol{u}^\varepsilon))^2 - \frac{(w^\varepsilon)^2}{2\varepsilon^3}\,,$$

$$-A^{\alpha\beta}_{vw}\varepsilon_{\alpha\beta}(v^\varepsilon)w^\varepsilon \geq -\frac{1}{2}(A^{\alpha\beta}_{vw}\varepsilon_{\alpha\beta}(\boldsymbol{v}))^2 - \frac{(w^\varepsilon)^2}{2}\,,$$ (11.7.25)

$$-A^{\alpha\beta}_{uw}\gamma_{\alpha\beta}(\boldsymbol{u}^\varepsilon)w^\varepsilon \geq -\frac{1}{2}(A^{\alpha\beta}_{uw}\gamma_{\alpha\beta}(\boldsymbol{u}^\varepsilon))^2 - \frac{(w^\varepsilon)^2}{2}\,.$$

Taking into account (11.7.10), (11.7.20), (11.7.22), (11.7.23) and (11.7.25) in (11.7.24), we get

$$M + \frac{M_1}{\varepsilon}(\|\boldsymbol{\varepsilon}(\boldsymbol{v}^\varepsilon)\|^2_{0,\Omega} + \|\boldsymbol{\gamma}(\boldsymbol{u}^\varepsilon)\|^2_{0,\Omega^\varepsilon} + \|w^\varepsilon\|_{1,\Omega})$$

$$\geq \frac{1}{2\varepsilon^2}\int_{\Omega^\varepsilon} H^{\alpha\beta}\kappa_\alpha(\boldsymbol{u}^\varepsilon,w^\varepsilon)\kappa_\beta(\boldsymbol{u}^\varepsilon,w^\varepsilon)dx + \int_{\Omega^\varepsilon}\left(\frac{A_w}{2\varepsilon^4} - \frac{1}{\varepsilon^3}\right)(w^\varepsilon)^2dx\,,$$

where M and M_1 are positive constants which do not depend on ε. Hence we obtain

$$M\varepsilon^2 + M_2\varepsilon \geq \frac{1}{2}\int_{\Omega^\varepsilon} H^{\alpha\beta}\kappa_\alpha(\boldsymbol{u}^\varepsilon,w^\varepsilon)\kappa_\beta(\boldsymbol{u}^\varepsilon,w^\varepsilon)dx + \int_{\Omega^\varepsilon}\left(\frac{M_3}{2\varepsilon^2} - \frac{1}{\varepsilon}\right)(w^\varepsilon)^2dx\,,$$

since $A_w(x) \geq M_3 > 0$ for a.e. $x \in \Omega$. Consequently, for a subsequence ε' of ε we have

$$\int_{\Omega^{\varepsilon'}} H^{\alpha\beta}\kappa_\alpha(\boldsymbol{u}^{\varepsilon'},w^{\varepsilon'})\kappa_\beta(\boldsymbol{u}^{\varepsilon'},w^{\varepsilon'})dx \to \int_{\Omega} H^{\alpha\beta}\kappa_\alpha(\boldsymbol{u},w)\kappa_\beta(\boldsymbol{u},w)dx\,,$$

$$\int_{\Omega^{\varepsilon'}}\left(\frac{M_3}{2(\varepsilon')^2} - \frac{1}{\varepsilon'}\right)(w^{\varepsilon'})^2dx \to 0\,,$$

as $\varepsilon' \to 0$. We observe that for ε' sufficiently small $\left(\dfrac{M_3}{2(\varepsilon')^2} - \dfrac{1}{\varepsilon'}\right)$ is positive and tends to infinity. Thus $w = 0$ and $\boldsymbol{\kappa}(\boldsymbol{u},w) = 0$ or $\boldsymbol{u} = \nabla w$. $\square$

A basic property of the sequence of minimizers $\{\bar{\boldsymbol{v}}^\varepsilon, \bar{\boldsymbol{u}}^\varepsilon, \bar{w}^\varepsilon\}_{\varepsilon>0}$ (see (11.7.15)) is specified by the following result.

Lemma 11.7.4. There exists a positive constant $c_1 > 0$, which is independent of ε and such that $(0 < \varepsilon < 1)$

$$\sup_{\varepsilon>0}(\|\bar{\boldsymbol{v}}^\varepsilon\|_{1,\Omega} + \|\bar{\boldsymbol{u}}^\varepsilon\|_{1,\Omega^\varepsilon} + \frac{1}{\varepsilon}\|\bar{w}^\varepsilon\|_{1,\Omega}) \leq c_1\,.$$

Proof. For any $(v, u, w) \in \mathbb{K}_\varepsilon$ we have

$$J_\varepsilon(\bar{v}^\varepsilon, \bar{u}^\varepsilon, \bar{w}^\varepsilon) \le J_\varepsilon(v, u, w) \,.$$

Particularly, by taking $v = 0$, $u = 0$, $w = 0$, since $(0, 0, 0) \in \mathbb{K}_\varepsilon$ we get

$$J_\varepsilon(\bar{v}^\varepsilon, \bar{u}^\varepsilon, \bar{w}^\varepsilon) \le 0 \,.$$

Hence

$$\int_{\Omega^\varepsilon} j_\varepsilon[x, \varepsilon(\bar{v}^\varepsilon), \gamma(\bar{u}^\varepsilon), \kappa(\bar{u}^\varepsilon, \bar{w}^\varepsilon), \bar{w}^\varepsilon] dx \le c_2(||\bar{v}^\varepsilon||_{0,\Gamma_\sigma} + ||\bar{u}^\varepsilon||_{L^1(\Gamma_\sigma)^2} + \frac{1}{\varepsilon}||\bar{w}^\varepsilon||_{0,\Gamma_\sigma}) \,.$$

By using (11.7.17), (11.7.18), (11.7.20) and (11.7.21), after simple calculations we arrive at $(0 < \varepsilon < 1)$

$$(||\varepsilon(\bar{v}^\varepsilon)||_{0,\Omega}^2 + ||\gamma(\bar{u}^\varepsilon)||_{0,\Omega^\varepsilon}^2 + \frac{1}{\varepsilon}||\bar{w}^\varepsilon||_{0,\Omega}^2) + \frac{1}{\varepsilon^2}||\bar{u}^\varepsilon - \nabla\bar{w}^\varepsilon||_{0,\Omega^\varepsilon}^2$$

$$\le c_3(||\varepsilon(\bar{v}^\varepsilon)||_{0,\Omega}^2 + ||\gamma(\bar{u}^\varepsilon)||_{0,\Omega^\varepsilon}^2 + \frac{1}{\varepsilon}||\bar{w}^\varepsilon||_{0,\Omega}^2)^{\frac{1}{2}} \,, \qquad (11.7.26)$$

where c_3 is a positive constant independent of ε, which concludes the proof. $\qquad \square$

Remark 11.7.5. Since $0 < \varepsilon < 1$, therefore by applying the Korn and Poincaré inequalities we can write

$$\sup_{\varepsilon > 0}(||\bar{v}^\varepsilon||_{1,\Omega} + ||\bar{u}^\varepsilon||_{1,\Omega^\varepsilon} + ||\bar{w}^\varepsilon||_{1,\Omega}) \le c_1 \,. \qquad (11.7.27)$$

Recall that for the domain Ω^ε Korn's inequality is given by (11.7.18).

From Sec. 9.3 we know that there exists an extension operator $\mathcal{Q}_2^\varepsilon \colon H^1(\Omega^\varepsilon)^2 \to H^1(\Omega)^2$ such that

$$||\mathcal{Q}_2^\varepsilon \bar{u}^\varepsilon - \bar{u}^\varepsilon||_{0,\Omega} \to 0 \,, \qquad (11.7.28)$$

provided that $\sup_{\varepsilon > 0} ||\bar{u}^\varepsilon||_{1,\Omega^\varepsilon} < \infty$. The estimate (11.7.27) combined with (11.7.28) imply the existence of a subsequence ε' of ε such that

$$v^{\varepsilon'} \rightharpoonup v \quad \text{weakly in} \quad H^1(\Omega)^2 \,,$$
$$\bar{u}^{\varepsilon'} \to u \quad \text{strongly in} \quad L^2(\Omega)^2 \,,$$
$$\bar{w}^{\varepsilon'} \rightharpoonup w \quad \text{weakly in} \quad H^1(\Omega) \,,$$

as $\varepsilon' \to 0$. Furthermore, (11.7.26) yields

$$||\bar{w}^{\varepsilon'}||_{0,\Omega} \to 0 \,, \quad \text{when} \quad \varepsilon' \to 0 \,.$$

Hence $w(x) = 0$ for $a.e.$ $x \in \Omega$. Similarly we get

$$||\bar{u}^{\varepsilon'} - \nabla\bar{w}^{\varepsilon'}||_{0,\Omega} \to 0 \,, \quad \text{when} \quad \varepsilon' \to 0 \,,$$

and thus $u(x) = 0$ for $a.e.$ $x \in \Omega$. $\qquad \square$

Dual effective potential

Later on, to prove that the sequence $\{J_\varepsilon\}_{\varepsilon>0}$ is Γ-convergent we shall exploit in an essential manner some duality arguments. Therefore we have to derive first the complementary (dual) effective stored energy functions $\mathcal{W}_h^*$. The procedure is similar to the one applied in Secs. 5.4 and 10.4. It is thus sufficient to provide main points of the derivation.

The complementary potential is given by the Fenchel conjugate of $\mathcal{W}_h$:

$$\mathcal{W}_h^*(x,\epsilon^*) := \sup\{\epsilon^* : \epsilon - \mathcal{W}_h(x,\epsilon) \mid \epsilon \in \mathbb{E}_s^2\}, \quad \epsilon^* \in \mathbb{E}_2^s. \tag{11.7.29}$$

By using (11.6.24), we write

$$\mathcal{W}_h^*(x,\epsilon^*)$$

$$= \sup_{\epsilon\in\mathbb{E}_s^2}\left\{\epsilon^* : \epsilon - \inf_{(u,v,w)\in\mathbb{K}_{YF}} \frac{1}{|Y|}\int_{YF} j[x,\epsilon+\varepsilon^y(v),\gamma^y(u),\kappa^y(u,w),w]dy\right\}. \tag{11.7.30}$$

We observe that

$$\inf\left\{\int_{YF} j[x,\epsilon+\varepsilon^y(v),\gamma^y(u),\kappa(u,w),w]dy \mid (v,u,w)\in\mathbb{K}_{YF}\right\}$$

$$= \inf\left\{\int_{YF} j[x,\epsilon+\varepsilon^y(v),\gamma^y(u),\kappa^y(u,w),w]dy\right.$$

$$\left.+I_{K_{YF}}(u) \mid v\in H_{per}^1(Y)^2, u\in H_{per}^1(YF)^2, w\in H_{per}^1(Y)\right\}, \tag{11.7.31}$$

where $I_{K_{YF}}$ is the indicator function of the set K_{YF}, the last being defined by $(11.4.4)_1$. The minimization problem (11.7.31) is here called (P_ϵ)-problem. To apply Rockafellar's theory of duality we introduce

$$G_\epsilon(x,\Lambda(v,u,w)) = \int_{YF} j_\epsilon[x,\varepsilon^y(v),\gamma^y(u),\kappa^y(u,w),w]dy$$

$$= \int_{YF} j[x,\epsilon+\varepsilon^y(v),\gamma^y(u),\kappa^y(u,w),w]dy, \tag{11.7.32}$$

$$\mathbf{F}(x,u,w) = I_{K_{YF}}(u). \tag{11.7.33}$$

Here $\epsilon \in \mathbb{E}_s^2$ is held fixed whilst $\Lambda = (\Lambda_1,\Lambda_2,\Lambda_3,\Lambda_4)$ and $\Lambda_1 v = \varepsilon^y(v)$, $\Lambda_2 u = \gamma^y(u)$, $\Lambda_3(u,w) = \kappa^y(u,w)$, $\Lambda_4 w = w$. Obviously, the operator Λ maps $H^1(Y)^2 \times H^1(YF)^2 \times H^1(Y)$ into $L^2(Y,\mathbb{E}_s^2) \times L^2(YF,\mathbb{E}_s^2) \times L^2(Y)^2 \times L^2(Y)$. Standard calculations yield, cf. Secs. 5.4 and 10.4

$$\Lambda_1^*\mathbf{n} = \begin{cases} -div_y\mathbf{n} & \text{in } Y, \\ \mathbf{n}\mu & \text{on } \partial Y; \end{cases} \tag{11.7.34}$$

$$\Lambda_2^*\mathbf{l} = \begin{cases} -div_y\mathbf{l} & \text{in } YF, \\ \mathbf{l}\mu & \text{on } \partial Y, \\ -\mathbf{l}_N & \text{on } F, \\ -\mathbf{l}_T & \text{on } F; \end{cases} \tag{11.7.35}$$

$$\Lambda_3^* q = \begin{cases} q & \text{in } Y, & (u) \\ div_y q & \text{in } Y, & (w) \\ -q \cdot \mu & \text{on } \partial Y & (w) \, ; \end{cases} \tag{11.7.36}$$

$$\Lambda_4^* r = r \quad \text{in } Y, \tag{11.7.37}$$

where $n \in L^2(Y, \mathbb{E}_2^s)$, $l \in L^2(YF, \mathbb{E}_2^s)$, $q \in L^2(Y)^2$, $r \in L^2(Y)$, while μ stands for the outward unit normal to ∂Y. We recall that $n\mu = (n^{\alpha\beta}\mu_\beta)$, $q \cdot \mu = q^\alpha \mu_\alpha$, $l_N = l^{\alpha\beta} \rfloor_1 \check{N}_\alpha \check{N}_\beta$ and $l_T = (l^{\alpha\beta} \rfloor_1 \check{N}_\beta) - l_N \check{N}$. For instance, let us derive (11.7.35). We have

$$\langle l, \Lambda_2 u \rangle_{L^2(YF, \mathbb{E}_2^s) \times L^2(YF, \mathbb{E}_s^2)} = \int_{YF} l : \gamma^y(u) dy \tag{11.7.38}$$

$$= -\int_{YF} l^{\alpha\beta}{}_{|\beta} u_\alpha dy + \int_{\partial Y} l^{\alpha\beta} \mu_\beta u_\alpha ds - \int_F l_N [\![u_N]\!] ds - \int_F l_T \cdot [\![u_T]\!] ds = \langle \Lambda_2^* l, u \rangle \, ,$$

where $u \in H^1(YF)^2$, $l \in L^2(YF, \mathbb{E}_2^s)$. The duality pairing $\langle \Lambda_2^* l, u \rangle$ in the last relation is to be understood in the following sense:

$$\langle \Lambda_2^* l, u \rangle = -\langle div_y l, u \rangle_{[H^1(YF)^2]^* \times H^1(YF)^2} + \langle l\mu, u \rangle_{H^{-1/2}(\partial Y)^2 \times H^{1/2}(\partial Y)^2}$$

$$- \langle l_N, [\![u_N]\!] \rangle_{H^{-1/2}(F) \times H^{1/2}(F)} - \langle l_T, [\![u_T]\!] \rangle_{[H^{-1/2}(F)^2]^2 \times H^{1/2}(F)^2} \, . \tag{11.7.39}$$

Next we calculate

$$\mathbf{F}^*(-\Lambda^*(n, l, q, r)) = \sup\{ \int_Y (\text{div}_y n) \cdot v dy + \int_{YF} (\text{div}_y l) \cdot u dy$$

$$- \int_Y (div_y q) w dy - \int_Y q \cdot u dy - \int_Y r w dy - \int_{\partial Y} n^{\alpha\beta} \mu_\alpha v_\alpha ds - \int_{\partial Y} l^{\alpha\beta} \mu_\beta u_\alpha ds$$

$$+ \int_{\partial Y} q^\alpha \mu_\alpha w ds + \int_F l_N [\![u_N]\!] ds + \int_F l_T \cdot [\![u_T]\!] ds - I_{K_{YF}}(u) \mid v \in H^1_{per}(Y)^2 \, ,$$

$$u \in H^1_{per}(YF)^2 \, , \; w \in H^1_{per}(Y) \} = \begin{cases} 0 & \text{if } (n, l, q, r) \in S^c_{per}(YF) \, , \\ +\infty & \text{otherwise} \, , \end{cases} \tag{11.7.40}$$

where

$$S^c_{per}(YF) = \{(n, l, q, r) \in L^2(Y, \mathbb{E}_2^s) \times L^2(YF, \mathbb{E}_2^s) \times L^2(Y)^2 \times L^2(Y) \mid$$
$$div_y n = 0 \text{ in } Y \, ; \; div_y l - q = 0 \text{ in } YF \, ; \; div_y q + r = 0 \text{ in } Y \, ;$$
$$n\mu \, , \; l\mu \text{ and } q \cdot \mu \text{ assume opposite values on the opposite}$$
$$\text{sides of } Y \, ; \; l_N \leq 0 \text{ and } l_T = 0 \text{ on } F \} \, . \tag{11.7.41}$$

We observe that various integrals appearing in (11.7.40) are always taken in the sense of duality pairings when an integral is not in L^1.

For a fixed $\epsilon \in \mathbb{E}_s^2$, the dual problem means evaluating

$$(P_\epsilon^*) \qquad \sup\{-G_\epsilon^*(x, \mathbf{n}, \mathbf{l}, \boldsymbol{q}, \mathbf{r}) - \mathbf{F}^*(-\Lambda^*(\mathbf{n}, \mathbf{l}, \boldsymbol{q}, \mathbf{r})) \mid (\mathbf{n}, \mathbf{l}, \boldsymbol{q}, \mathbf{r}) \in S_{per}^c(YF)\} \ .$$

To find the functional G_ϵ^* we calculate

$$G_\epsilon^*(x, \mathbf{n}, \mathbf{l}, \boldsymbol{q}, \mathbf{r}) = \int\limits_{YF} j_\epsilon^*(x, \mathbf{n}(y), \mathbf{l}(y), \boldsymbol{q}(y), \mathbf{r}(y))dy \ , \tag{11.7.42}$$

where

$$\begin{aligned}
j_\epsilon^*(x, \boldsymbol{a}^*, \boldsymbol{b}^*, \boldsymbol{c}^*, d^*) &= \sup\{\boldsymbol{a}^* : \boldsymbol{a} + \boldsymbol{b}^* : \boldsymbol{b} + \boldsymbol{c}^* \cdot \boldsymbol{c} + d^* d \\
&\qquad - j_\epsilon(x, \boldsymbol{a}, \boldsymbol{b}, \boldsymbol{c}, d) \mid \boldsymbol{a}, \boldsymbol{b} \in \mathbb{E}_s^2, \boldsymbol{c} \in \mathbb{R}^2, d \in \mathbb{R}\} \\
&= \sup\{\boldsymbol{a}^* : \boldsymbol{a} + \boldsymbol{b}^* : \boldsymbol{b} + \boldsymbol{c}^* \cdot \boldsymbol{c} + d^* d \\
&\qquad - j(x, \epsilon + \boldsymbol{a}, \boldsymbol{b}, \boldsymbol{c}, d) \mid \boldsymbol{a}, \boldsymbol{b} \in \mathbb{E}_s^2, \boldsymbol{c} \in \mathbb{R}^2, d \in \mathbb{R}\}
\end{aligned}$$
$$= -\boldsymbol{a}^* : \epsilon + j^*(x, \boldsymbol{a}^*, \boldsymbol{b}^*, \boldsymbol{c}^*, d^*) = -\boldsymbol{a}^* : \epsilon + \mathcal{W}^*(x, \boldsymbol{a}^*, \boldsymbol{b}^*, \boldsymbol{c}^*, d^*) \ , \tag{11.7.43}$$

since $j^* = \mathcal{W}^*$, see Eq. (11.1.13); moreover $\boldsymbol{a}^*, \boldsymbol{b}^* \in \mathbb{E}_2^s$, $\boldsymbol{c}^* \in \mathbb{R}^2$ and $d^* \in \mathbb{R}$.

As we already know, the problem (P_ϵ) is solvable and we may write

$$\inf P_\epsilon = \sup P_\epsilon^* \ . \tag{11.7.44}$$

Combining (11.7.42) – (11.7.44) and (11.7.29) we get

$$\begin{aligned}
\mathcal{W}_h^*(x, \epsilon^*) &= \sup_{\epsilon \in \mathbb{E}_s^2} \frac{1}{|Y|}\{\int\limits_Y \epsilon^* : \epsilon dy + \inf_{(\mathbf{n}, \mathbf{l}, \boldsymbol{q}, \mathbf{r}) \in S_{per}^c(YF)} \int\limits_{YF} j_\epsilon^*(x, \mathbf{n}, \mathbf{l}, \boldsymbol{q}, \mathbf{r})dy\} \\
&= \sup_{\epsilon \in \mathbb{E}_s^2} \inf_{(\mathbf{n}, \mathbf{l}, \boldsymbol{q}, \mathbf{r})} \frac{1}{|Y|}\{\int\limits_{YF} \mathcal{W}^*[x, \mathbf{n}(y), \mathbf{l}(y), \boldsymbol{q}(y), \mathbf{r}(y)]dy + \int\limits_{YF} (\epsilon^* - \mathbf{n}(y)) : \epsilon dy\}.
\end{aligned}$$

The supremum over $\epsilon \in \mathbb{E}_s^2$ is finite provided that

$$\frac{1}{|Y|}\int\limits_{YF} \mathbf{n}(y)dy = \epsilon^* \ . \tag{11.7.45}$$

The last two relations yield the final form of the dual effective potential:

$$\begin{aligned}
&W_h^*(x, \epsilon^*) \\
&= \inf \frac{1}{|Y|}\{\int\limits_{YF} \mathcal{W}^*[x, \epsilon^* + \mathbf{n}(y), \mathbf{l}(y), \boldsymbol{q}(y), \mathbf{r}(y)]dy \mid (\mathbf{n}, \mathbf{l}, \boldsymbol{q}, \mathbf{r}) \in S_{per}(YF)\}, \tag{11.7.46}
\end{aligned}$$

where

$$S_{per}(YF) = \{(\mathbf{n}, \mathbf{l}, \boldsymbol{q}, \mathbf{r}) \in S_{per}^c(YF) \mid \int\limits_Y \mathbf{n}(y)dy = 0\} \ . \tag{11.7.47}$$

Physically, the dual variable ϵ^* is identified with the effective generalized stresses $(N_h^{\alpha\beta})$. We also observe that the condition

$$\int\limits_{YF} \mathbf{n}(y)\,dy = \int\limits_{Y} \mathbf{n}(y)\,dy = 0\,, \tag{11.7.48}$$

equivalent to (11.7.45), can be obtained by using the inf-convolution operation, cf. Secs. 2.10 and 5.4. It suffices to substitute $\mathbf{n}(y) = \tilde{n}(y) - |Y|\epsilon^*$ in (11.7.48), where $\mathbf{n} \in S_{per}(YF), \tilde{n} \in S^c_{per}(YF)$. Relation (11.7.45) is then recovered.

Remark 11.7.6. It can easily be verified that

$$S_{per}(YF) = [S^c_{per}(YF)] \cap (\mathbb{E}^2_s)^\perp\,, \tag{11.7.49}$$

where $(\mathbb{E}^2_s)^\perp$ denotes the orthogonal complement of $\mathbb{E}^2_s$ in L^2.

Remark 11.7.7. The local equilibrium equations appearing in the definition of $S_{per}(YF)$ via (11.7.41) are obviously to be understood in the sense of distributions. Since $q \in L^2(Y)^2$ and $r \in l^2(Y)$, therefore $div_y l \in L^2(YF)^2$ and $div_y q \in L^2(Y)$. Additionally, it is reasonable to assume that $div_y \mathbf{n} \in L^2(Y)^2$. The minimization problem appearing on the r.h.s. of (11.7.46) is then uniquely solvable as a strictly convex (dual) problem.

Remark 11.7.8.
(i) The local equilibrium equation

$$div_y q + \mathbf{r} = 0 \quad \text{in} \quad Y\,, \tag{11.7.50}$$

yields

$$\int\limits_{Y} \mathbf{r}(y)\,dy = -\int\limits_{Y} div_y q(y)\,dy = -\int\limits_{\partial Y} q^\alpha \mu_\alpha\,ds = 0\,,$$

because $q \cdot \mu$ is Y-antiperiodic.
(ii) From the local equilibrium equation

$$div_y l - q = 0 \quad \text{in} \quad YF\,, \tag{11.7.51}$$

we obtain

$$\int\limits_{Y} q^\alpha(y)\,dy = \int\limits_{YF} \frac{\partial l^{\alpha\beta}}{\partial y_\beta}\,dy = \int\limits_{\partial Y} l^{\alpha\beta}\mu_\beta\,ds + \int\limits_{F} [\![l^{\alpha\beta}\check{N}_\beta]\!]\,ds = 0\,,$$

since $l\check{N}$ is antiperiodic and $[\![l^{\alpha\beta}\check{N}_\beta]\!] = 0$ (on F), according to the action and reaction principle.
(iii) Equations (11.7.50) and (11.7.51) imply

$$div_y div_y l = div_y q = -\mathbf{r}\,, \tag{11.7.52}$$

and thus $div_y div_y l \in L^2(YF)$. Consequently the following equation makes sense:

$$\int\limits_{YF} [div_y(div_y l - q)]w\,dy = 0\,,$$

for each $w \in H^1_{per}(Y)$. Integrating by parts we conclude that $(div_y l) \cdot \mu$ is Y-antiperiodic.
(iv) Considering Eq. (11.7.50) once again, by formal differentiation and integration we get

$$\sum_{\alpha=1}^{2} \int_Y v_\alpha \frac{\partial}{\partial y_\alpha}(div_y q + r) dy = 0 \,,$$

for each $v \in H^1_{per}(Y)^2$. Hence

$$\int_{\partial Y} (div_y q + r) v \cdot \mu ds = 0 \,.$$

The periodicity of a function v implies that $v \cdot \mu$ is Y-antiperiodic and thus $div_y q + r$ is
Y-periodic. Similarly we deduce that $div_y div_y l - div_y q = div_y div_y l + r$ is also Y-periodic.
   The above periodicity and antiperiodicity properties of the local fields
$(n, l, q, r) \in S_{per}(Y)$ enable us to extend them to the whole plane $\mathbb{R}^2$ or to $\mathbb{R}^2 \backslash \underset{i \in \mathbb{Z}^2}{\cup} F_i$
whenever a field is discontinuous across F; here $\mathbb{Z}$ is the set of integers and F_i denotes
F translated by a vector $i = (i_1, i_2) \in \mathbb{Z}^2$. Consequently, after rescaling $y \rightsquigarrow x/\varepsilon$, and
recalling that $\dfrac{\partial}{\partial y_\alpha} = \varepsilon \dfrac{\partial}{\partial x_\alpha}$, from the local equilibrium equations one readily gets

$$\varepsilon div n_\varepsilon = 0 \quad \text{in} \quad \Omega \,, \tag{11.7.53}$$

$$\varepsilon div l_\varepsilon - q_\varepsilon = 0 \quad \text{in} \quad \Omega^\varepsilon \,, \tag{11.7.54}$$

$$\varepsilon div q_\varepsilon + r_\varepsilon = 0 \quad \text{in} \quad \Omega \,, \tag{11.7.55}$$

$$\varepsilon^2 div\, div l_\varepsilon - \varepsilon div q_\varepsilon = \varepsilon^2 div\, div l_\varepsilon + r_\varepsilon = 0 \quad \text{in} \quad \Omega^\varepsilon \,, \tag{11.7.56}$$

where $n_\varepsilon(x) = n\left(\frac{x}{\varepsilon}\right)$, etc.
   We shall now formulate a counterpart of Lemmas 2.10.12 and 5.4.3, which will be used
in the proof of Γ-convergence of the sequence of functionals (11.7.14).

Lemma 11.7.9. Let $(n, l, q, r) \in S_{per}(YF)$ and $\varphi \in D(\Omega)$. Suppose that $\{v^\varepsilon, u^\varepsilon,$
$w^\varepsilon\}_{\varepsilon>0} \subset H^1(\Omega)^2 \times H^1(\Omega^\varepsilon)^2 \times H^1(\Omega)$ is a sequence strongly convergent to $(v, 0, w)$
in $L^2(\Omega)^2 \times L^2(\Omega)^2 \times L^2(\Omega)$. Then we have

$$a = \lim_{\varepsilon \to 0} \int_{\Omega^\varepsilon} \varphi(x)[n_\varepsilon^{\alpha\beta}\varepsilon_{\alpha\beta}(v^\varepsilon) + l_\varepsilon^{\alpha\beta}\gamma_{\alpha\beta}(u^\varepsilon) + \frac{1}{\varepsilon}q_\varepsilon^\alpha \kappa_\alpha(u^\varepsilon, w^\varepsilon) + \frac{1}{\varepsilon^2}r_\varepsilon w^\varepsilon] dx = 0$$

where $n_\varepsilon(x) = n\left(\frac{x}{\varepsilon}\right)$, etc.

Proof. We set $a = \lim_{\varepsilon \to 0} a^\varepsilon$, where

$$a^\varepsilon = \int_{\Omega^\varepsilon} \varphi(x)[n_\varepsilon^{\alpha\beta}\varepsilon_{\alpha\beta}(v^\varepsilon) + l_\varepsilon^{\alpha\beta}\gamma_{\alpha\beta}(u^\varepsilon) + \frac{1}{\varepsilon}q_\varepsilon^\alpha \kappa_\alpha(u^\varepsilon, w^\varepsilon) + \frac{1}{\varepsilon^2}r_\varepsilon w^\varepsilon] dx \,.$$

Integration by parts yields

$$a = \lim_{\varepsilon \to 0} a^\varepsilon = \lim_{\varepsilon \to 0} \int_{\Omega^\varepsilon} [- \left(\varphi(x)\mathbf{n}_\varepsilon^{\alpha\beta}\right)_{,\beta} v_\alpha^\varepsilon(x) - (\varphi(x)\mathbf{l}_\varepsilon^{\alpha\beta})_{,\beta} u_\alpha^\varepsilon(x) + \frac{1}{\varepsilon}\varphi(x)q_\varepsilon^\alpha u_\alpha^\varepsilon(x)$$

$$+ \frac{1}{\varepsilon}\varphi(x)q_\varepsilon^\alpha w^\varepsilon(x) + \frac{1}{\varepsilon^2}\varphi(x)\mathbf{r}_\varepsilon w^\varepsilon]dx - \lim_{\varepsilon \to 0} \int_{F^\varepsilon} \varphi(x)\mathbf{l}_N\left(\frac{x}{\varepsilon}\right)[\![u_N^\varepsilon(x)]\!]ds \ , \quad (11.7.57)$$

where $\varphi_{,\alpha} = \dfrac{\partial\varphi}{\partial x_\alpha}$, etc.

Since $\mathbf{l}\left(\dfrac{\cdot}{\varepsilon}\right)$ is εY-periodic therefore

$$\sup_{\varepsilon > 0}\left\| \mathbf{l}_N\left(\frac{\cdot}{\varepsilon}\right) \right\|_{H^{-1/2}(F^\varepsilon)} < +\infty \ .$$

Recalling that u^ε tends to 0 we have

$$\lim_{\varepsilon \to 0} \int_{F^\varepsilon} \varphi(x)\mathbf{l}_N\left(\frac{x}{\varepsilon}\right)[\![u_N^\varepsilon(x)]\!]ds = 0 \ . \tag{11.7.58}$$

From (11.7.57) and (11.7.58) we get

$$a = \lim_{\varepsilon \to 0} \int_{\Omega^\varepsilon} \left\{ \varphi(x)(-div\mathbf{n}_\varepsilon) \cdot \mathbf{v}^\varepsilon(x) + \varphi(x)\left[-(div\mathbf{l}_\varepsilon) + \frac{1}{\varepsilon}\mathbf{q}\left(\frac{x}{\varepsilon}\right)\right] \cdot \mathbf{u}(x) \right.$$

$$+ \varphi(x)\left[\frac{1}{\varepsilon}div\mathbf{q}_\varepsilon + \frac{1}{\varepsilon^2}\mathbf{r}\left(\frac{x}{\varepsilon}\right)\right] w^\varepsilon(x) - \varphi_{,\beta}(x)\mathbf{n}_\varepsilon^{\alpha\beta}v_\alpha^\varepsilon(x)$$

$$\left. - \varphi_{,\beta}(x)\mathbf{l}_\varepsilon^{\alpha\beta}u_\alpha^\varepsilon(x) + \frac{1}{\varepsilon}\varphi_{,\alpha}(x)q^\alpha\left(\frac{x}{\varepsilon}\right) w^\varepsilon(x) \right\} dx \ . \tag{11.7.59}$$

By using Theorem 1.1.5 and Remark 11.7.7 we write

$$\lim_{\varepsilon \to 0} \int_{\Omega} \varphi_{,\beta}(x)\mathbf{n}^{\alpha\beta}\left(\frac{x}{\varepsilon}\right) v_\alpha^\varepsilon(x)dx = \int_{\Omega} \varphi_{,\beta}(x)v_\alpha(x)dx\frac{1}{|Y|}\int_Y \mathbf{n}^{\alpha\beta}(y)dy = 0 \ ,$$

$$\lim_{\varepsilon \to 0} \int_{\Omega} \varphi_{,\alpha}(x)\frac{1}{\varepsilon}q^\alpha\left(\frac{x}{\varepsilon}\right) w^\varepsilon(x)dx = 0 \ . \tag{11.7.60}$$

Taking into account (11.7.53) – (11.7.55) and (11.7.60) in (11.7.59) we conclude the proof.

$$\square$$

Γ-convergence of the sequence (11.7.14)

As we already know, the sequence of functionals $\{J_\varepsilon\}_{\varepsilon > 0}$, defined by (11.7.14), characterizes the class studied of cross-ply laminates in the case of the refined scaling specified by (11.6.2) and (11.6.3). After indispensable preparations we are now in a position to formulate and prove the basic result concerning the Γ-convergence of this sequence.

Theorem 11.7.10. Under the assumption (11.7.16) the sequence of functionals (11.7.14) Γ-converges in the topology $\tau = s - [L^2(\Omega)^2 \times L^2(\Omega)^2 \times L^2(\Omega)]$ to the functional

$$J_h(\boldsymbol{v}) = \int_\Omega W_h[x, \boldsymbol{\varepsilon}(\boldsymbol{v}(x))]dx - L(\varepsilon), \quad \boldsymbol{v} \in H_{\Gamma_w}(\Omega)^2 , \tag{11.7.61}$$

where

$$L(\boldsymbol{v}) = \int_{\Gamma_\sigma} \bar{N}^\alpha v_\alpha d\Gamma , \tag{11.7.62}$$

and $\mathcal{W}_h$ is given by (11.6.27).

Proof. As usual, it is divided into two parts.

I. For any $\boldsymbol{v} \in H_{\Gamma_w}(\Omega)^2$ we have to find a sequence $\{v^\varepsilon, u^\varepsilon, w^\varepsilon\}_{\varepsilon>0} \subset \mathbb{K}_\varepsilon$ such that

$$(v^\varepsilon, u^\varepsilon, w^\varepsilon) \xrightarrow{\tau} (\boldsymbol{v}, 0, 0) \quad \text{when } \varepsilon \to 0 , \tag{11.7.63}$$

and

$$J_h(\boldsymbol{v}) \geq \lim_{\varepsilon \to 0} \sup J_\varepsilon(v^\varepsilon, u^\varepsilon, w^\varepsilon) . \tag{11.7.64}$$

Step 1. Let $\{\Omega_K\}_{K \in \mathcal{K}}$ be a finite partition of Ω formed by polygonal sets. Such a partition enables one to exploit the local character of the functionals J_ε. One can single out those elements, denote them by $\{\Omega_K^b\}_{K \in \mathcal{K}_b}$, with the following property

$$\Gamma_\sigma = \bigcup_{K \in \mathcal{K}_b} \bar{\Omega}_K^b \cap \Gamma_\sigma , \tag{11.7.65}$$

where $\bar{\Omega}_K^b$ stands for the closure of Ω_K^b. The case of a curvilinear part Γ_σ of the boundary, though excluded at this stage, can be examined similarly by assuming that $\Omega_K^b(K \in \mathcal{K}_b)$ are curvilinear finite elements. For all sets Ω_K, different from Ω_K^b, we introduce

$$\Omega_K^\delta := \{x \in \Omega_K \mid dist(x, \partial\Omega_K) > \delta\} , \quad \delta > 0 \tag{11.7.66}$$

where $K \in \mathcal{K}_1$ and $\mathcal{K} = \mathcal{K}_1 \cup \mathcal{K}_b$. Obviously the sets of indices $\mathcal{K}_1$ and $\mathcal{K}_b$ are disjoint. Furthermore, for $\Gamma_\sigma^{bK} = \Gamma_\sigma \cap \bar{\Omega}_K^b(K \in \mathcal{K}_b)$ we set

$$\Omega_K^{b\delta} = \{x \in \Omega_K^b \mid dist(x, (\partial\Omega_K^b)\backslash\Gamma_\sigma^{bK}) > \delta\} , \quad \delta > 0 . \tag{11.7.67}$$

Now for the subfamily (11.7.66) we take $\varphi_K^\delta \in \mathbf{D}(\Omega_K)$, such that $0 \leq \varphi_K^\delta \leq 1, \varphi_K^\delta|_{\Omega_K^\delta} = 1$. Similarly, for the subfamily (11.7.67) we take $\varphi_K^\delta \in \mathbf{D}(\bar{\Omega}_K^b), 0 \leq \varphi_K^\delta \leq 1, \varphi_K^\delta|_{\Omega_K^{b\delta} \cup \Gamma_\sigma^{bK}} = 1$. Let $\boldsymbol{v} \in H_{\Gamma_w}^1(\Omega)^2$ be a piecewise affine and continuous function, that is

$$v_\alpha(x) = \sum_\beta \epsilon_{\alpha\beta}^K x_\beta + a_\alpha^K , \quad \epsilon^K \in \mathbb{E}_s^2 , \quad a^K \in \mathbb{R}^2 , \tag{11.7.68}$$

where $K \in \mathcal{K}$. Hence we have

$$\boldsymbol{\varepsilon}(\boldsymbol{v}(x)) = \boldsymbol{\epsilon}^K , \quad x \in \Omega_K , \; K \in \mathcal{K} . \qquad (11.7.69)$$

With every family of functions $\{\boldsymbol{v}^K, \boldsymbol{u}^K, w^K\}_{K \in \mathcal{K}} \subset \mathbb{K}_{YF}$ we associate the following sequences

$$\boldsymbol{v}^{\varepsilon,\delta}(x) = \boldsymbol{v}(x) + \varepsilon \sum_{K \in \mathcal{K}} \varphi_K^\delta(x) \boldsymbol{v}^K \left(\frac{x}{\varepsilon}\right) ,$$

$$\boldsymbol{u}^{\varepsilon,\delta}(x) = \varepsilon \sum_{K \in \mathcal{K}} \varphi_K^\delta \boldsymbol{u}^K \left(\frac{x}{\varepsilon}\right) , \qquad (11.7.70)$$

$$w^{\varepsilon,\delta}(x) = \varepsilon^2 \sum_{K \in \mathcal{K}} \varphi_K^\delta(x) w^K \left(\frac{x}{\varepsilon}\right) .$$

Hence

$$\begin{aligned}
\boldsymbol{v}^{\varepsilon,\delta} &\to \boldsymbol{v} && \text{strongly in} \quad L^2(\Omega)^2 , \\
\boldsymbol{u}^{\varepsilon,\delta} &\to 0 && \text{strongly in} \quad L^2(\Omega)^2 , \qquad (11.7.71) \\
w^{\varepsilon,\delta} &\to 0 && \text{strongly in} \quad L^2(\Omega) ,
\end{aligned}$$

when $\varepsilon \to 0$. Moreover, $(11.7.70)_2$ implies

$$[\![u_n^{\varepsilon,\delta}]\!] = \varepsilon \sum_{K \in \mathcal{K}} \varphi_K^\delta(x) [\![u_n^K]\!] \geq 0 \quad \text{on } F^\varepsilon .$$

We put $\omega_K^\varepsilon = \Omega^\varepsilon \cap \Omega_K$, $K \in \mathcal{K}$. Let $t > 1$ (intended to tend to 1). Noting that $t\varphi_K^\delta + t(1 - \varphi_K^\delta) + (1 - t) = 1$ and exploiting the convexity of the function $j_\varepsilon(x, \cdot, \cdot, \cdot, \cdot)$ we obtain

$$J_\varepsilon\left(t\boldsymbol{v}^{\varepsilon,\delta}, t\boldsymbol{u}^{\varepsilon,\delta}, tw^{\varepsilon,\delta}\right) = \sum_{K \in \mathcal{K}} \int_{\omega_K^\varepsilon} j_\varepsilon[x, \boldsymbol{\varepsilon}(t\boldsymbol{v}^{\varepsilon,\delta}), \boldsymbol{\gamma}(t\boldsymbol{u}^{\varepsilon,\delta}), \boldsymbol{\kappa}(t\boldsymbol{u}^{\varepsilon,\delta}, tw^{\varepsilon,\delta}), tw^{\varepsilon,\delta}] dx$$

$$- \sum_{K \in \mathcal{K}_b} \int_{\Gamma_\sigma^{bK}} (t\bar{N}^\alpha v_\alpha^{\varepsilon,\delta} + t\bar{L}^\alpha u_\alpha^{\varepsilon,\delta} - \frac{1}{\varepsilon} t\bar{Q} w^{\varepsilon,\delta}) d\Gamma$$

$$= \sum_{K \in \mathcal{K}} \int_{\omega_K^\varepsilon} j_\varepsilon \Big[x, t\varphi_K^\delta(x) \Big(\boldsymbol{\epsilon}^K + \boldsymbol{\varepsilon}(\boldsymbol{v}^K) \Big(\frac{x}{\varepsilon}\Big) \Big) + t(1 - \varphi_K^\delta(x))\boldsymbol{\epsilon}^K$$

$$+ (1 - t)\frac{\varepsilon t}{1 - t} \Big(\boldsymbol{v}^K \Big(\frac{x}{\varepsilon}\Big) \otimes \nabla \varphi_K^\delta(x) \Big)_s , \; t\varphi_K^\delta(x)\boldsymbol{\gamma}(\boldsymbol{u}^K) \Big(\frac{x}{\varepsilon}\Big) + t(1 - \varphi_K^\delta(x))\boldsymbol{0}$$

$$+ (1 - t)\frac{\varepsilon t}{1 - t} \Big(\boldsymbol{u}^K \Big(\frac{x}{\varepsilon}\Big) \otimes \nabla \varphi_K^\delta(x) \Big)_s ,$$

$$t\varphi_K^\delta(x) \Big(\varepsilon \boldsymbol{u}^K \Big(\frac{x}{\varepsilon}\Big) - \varepsilon(\nabla w^K) \Big(\frac{x}{\varepsilon}\Big) \Big) + t(1 - \varphi_K^\delta(x))\boldsymbol{0}$$

$$- (1 - t)\frac{\varepsilon^2 t}{1 - t} w^K(x)\nabla \varphi_K^\delta(x), \; t\varphi_K^\delta(x)\varepsilon^2 w^K \Big(\frac{x}{\varepsilon}\Big) + t(1 - \varphi_K^\delta(x))0 + (1 - t)0 \Big] dx$$

$$- \sum_{K \in \mathcal{K}_b} t \int_{\Gamma_\sigma^{bK}} [\bar{N}^\alpha(v_\alpha + \varepsilon\varphi_K^\delta v_\alpha^K) + \varepsilon\varphi_K^\delta \bar{L}^\alpha u_\alpha^K - \varepsilon\varphi_K^\delta \bar{Q} w^K] d\Gamma$$

$$\leq \sum_{K\in\mathcal{K}} \int_{\omega_K^\varepsilon} j\left[x, \boldsymbol{\epsilon}^K + \boldsymbol{\epsilon}(\boldsymbol{v}^K)\left(\frac{x}{\varepsilon}\right), \boldsymbol{\gamma}(\boldsymbol{u}^K)\left(\frac{x}{\varepsilon}\right), \boldsymbol{u}^K\left(\frac{x}{\varepsilon}\right) - (\nabla w^K)\left(\frac{x}{\varepsilon}\right), w^K\left(\frac{x}{\varepsilon}\right)\right] dx$$

$$+ M_1 |\boldsymbol{\epsilon}^K|^2 \int_{\omega_K^\varepsilon} (1 - \varphi_K^\delta(x)) dx + M_2(1-t) \int_{\omega_K^\varepsilon} \left[\frac{\varepsilon t}{1-t}\left|\left(\boldsymbol{v}^K\left(\frac{x}{\varepsilon}\right)\otimes\nabla\varphi_K^\delta(x)\right)_s\right|^2\right.$$

$$\left. + \frac{\varepsilon t}{1-t}\left|\left(\boldsymbol{u}^K\left(\frac{x}{\varepsilon}\right)\otimes\nabla\varphi_K^\delta(x)\right)_s\right|^2 + \frac{\varepsilon^2 t}{1-t}\left|w^K\left(\frac{x}{\varepsilon}\right)\nabla\varphi_K^\delta(x)\right|^2\right] dx$$

$$- \sum_{K\in\mathcal{K}_b} t \int_{\Gamma_\sigma^{bK}} [\bar{N}^\alpha(v_\alpha + \varepsilon\varphi_K^\delta v_\alpha^K) + \varepsilon\varphi_K^\delta \bar{L}^\alpha u_\alpha^K - \varepsilon\varphi_K^\delta \bar{Q}w^K] ds, \qquad (11.7.72)$$

since $j_\varepsilon \geq 0$ (and likewise $j \geq 0$); here M_1 and M_2 are positive constants, independent of ε and t. We recall that

$$[(\boldsymbol{v}^K \otimes \varphi_K^\delta)_s]_{\alpha\beta} = \frac{1}{2}\left(v_\alpha^K \frac{\partial\varphi_K^\delta}{\partial x_\beta} + v_\beta^K \frac{\partial\varphi_K^\delta}{\partial x_\alpha}\right).$$

The function $j[x, \boldsymbol{\epsilon}^K + \boldsymbol{\epsilon}(\boldsymbol{v}^K)\left(\frac{\cdot}{\varepsilon}\right), \boldsymbol{\gamma}(\boldsymbol{u}^K)\left(\frac{\cdot}{\varepsilon}\right), \boldsymbol{u}^K\left(\frac{\cdot}{\varepsilon}\right) - (\nabla w^K)\left(\frac{\cdot}{\varepsilon}\right), w^K\left(\frac{\cdot}{\varepsilon}\right)]$ is εY-periodic, hence by applying Theorem 1.1.5 we get

$$\limsup_{\varepsilon\to 0} J_\varepsilon(t\boldsymbol{v}^{\varepsilon\delta}, t\boldsymbol{u}^{\varepsilon,\delta}, tw^{\varepsilon,\delta})$$

$$\leq \sum_{K\in\mathcal{K}} \int_{\Omega_K} \frac{1}{|Y|} \int_{YF} j[x, \boldsymbol{\epsilon}^K + \boldsymbol{\epsilon}^y(\boldsymbol{v}^K(y)), \boldsymbol{\gamma}^y(\boldsymbol{u}^K(y)), \boldsymbol{\kappa}^y(\boldsymbol{u}^K(y), w^K(y)), w^K(y)] dy\, dx$$

$$+ M_1 \sum_{K\in\mathcal{K}} |\boldsymbol{\epsilon}^K|^2 \int_{\Omega_K} (1 - \varphi_K^\delta(x)) dx - t \int_{\Gamma_\sigma} \bar{N}^\alpha v_\alpha ds.$$

Now let $t \to 1^-$ and $\delta \to 0$, then

$$\limsup_{\substack{t\to 1^- \\ \delta\to 0}} \limsup_{\varepsilon\to 0} J_\varepsilon(t\boldsymbol{v}^{\varepsilon\delta}, t\boldsymbol{u}^{\varepsilon,\delta}, tw^{\varepsilon,\delta}) \leq \sum_{K\in\mathcal{K}} \int_{\Omega_K} \frac{1}{|Y|} \int_{YF} j[x, \boldsymbol{\epsilon}^K$$

$$+ \boldsymbol{\epsilon}^y(\boldsymbol{v}^K(y)), \boldsymbol{\gamma}^y(\boldsymbol{u}^K(y)), \boldsymbol{\kappa}^y(\boldsymbol{u}^K(y), w^K(y))] dy\, dx - \int_{\Gamma_\sigma} \bar{N}^\alpha v_\alpha ds.$$

By applying Lemma 1.3.27 one can construct a mapping $\varepsilon \to (t(\varepsilon), \delta(\varepsilon))$ with $(t(\varepsilon), \delta(\varepsilon)) \to (1^-, 0)$ as $\varepsilon \to 0$ such that setting

$$\boldsymbol{v}^\varepsilon(x) = t(\varepsilon)\boldsymbol{v}^{\varepsilon,\delta(\varepsilon)}, \qquad \boldsymbol{u}^\varepsilon(x) = t(\varepsilon)\boldsymbol{u}^{\varepsilon,\delta(\varepsilon)}, \qquad w^\varepsilon(x) = t(\varepsilon)w^{\varepsilon,\delta(\varepsilon)}, \qquad (11.7.73)$$

we conclude that

$$\limsup_{\varepsilon\to 0} J_\varepsilon(\boldsymbol{v}^\varepsilon, \boldsymbol{u}^\varepsilon, w^\varepsilon) \leq \sum_{K\in\mathcal{K}} \int_{\Omega_K} \frac{1}{|Y|} \int_{YF} j[x, \boldsymbol{\epsilon}^K$$

$$+\varepsilon^y(\boldsymbol{v}^K(y)),\boldsymbol{\gamma}^y(\boldsymbol{u}^K(y)),\boldsymbol{\kappa}^y(\boldsymbol{u}^K(y),w^K(y)),w^K(y)]dy\,dx - \int_{\Gamma_\sigma}\bar{N}^\alpha v_\alpha ds \ . \qquad (11.7.74)$$

By taking the infimum on the r.h.s. of the last inequality when $(\boldsymbol{v}^K,\boldsymbol{u}^K,w^K)$ run over the set $\mathbb{K}_{YF}$ we obtain

$$J_h^s(\boldsymbol{v},0,0) \le \limsup_{\varepsilon\to 0} J_\varepsilon(\boldsymbol{v}^\varepsilon,\boldsymbol{u}^\varepsilon,w^\varepsilon)$$

$$\le \sum_{K\in\mathcal{K}} \int_{\Omega_K} \mathcal{W}_h(x,\epsilon^K)dx - L(\boldsymbol{v}) = \int_\Omega \mathcal{W}_h[x,\varepsilon(\boldsymbol{v}(x))]dx - L(\boldsymbol{v}) = J_h(\boldsymbol{v}) \ .$$

Here (11.7.71) has been taken into account and J_h^s denotes the Γ-limit superior, cf. Sec. 1.1.
Step 2. The convexity of J_ε is preserved by the Γ-limit superior J_h^s, cf. Sec. 1.3.4. By virtue of the quadratic growth of $\mathcal{W}_h(x,\cdot)$ (see (11.6.27)), we have

$$J_h^s(\boldsymbol{v},0,0) \le c_1 \int_\Omega |\varepsilon(\boldsymbol{v}(x))|^2 dx - L(\boldsymbol{v}) \ , \ \boldsymbol{v}\in H_{\Gamma_w}(\Omega)^2 \ . \qquad (11.7.75)$$

Being convex and finite, the functional J_h^s is continuous on $H_{\Gamma_w}(\Omega)^2$; in fact those two properties hold true on the space $H^1(\Omega)^2$. Basic properties of the effective potential $\mathcal{W}_h$ and linearity of the functional L imply that J_h is also a convex and continuous functional. Due to the density of piecewise affine continuous functions in $H^1(\Omega)^2$, and consequently also in the subspace $H_{\Gamma_w}(\Omega)^2$, the inequality

$$J_h^s(\boldsymbol{v},0,0) \le J_h(\boldsymbol{v}) \ ,$$

can be extended to this subspace. In fact, for $\boldsymbol{v}\in H_{\Gamma_w}(\Omega)^2$ there exists a sequence $\{\boldsymbol{v}^k\}_{k\in\mathbb{N}}$ of piecewise affine continuous functions vanishing on Γ_w and convergent to $\boldsymbol{v}$ in the strong topology of $H^1(\Omega)^2$. In view of the continuity of the functional J_h we have

$$J_h(\boldsymbol{v}^k) \to J_h(\boldsymbol{v}) \quad \text{as} \quad k\to\infty \ . \qquad (11.7.76)$$

Following the previous step, for each $k\in\mathbb{N}$ there exists a sequence $\{\boldsymbol{v}^{k,\varepsilon},\boldsymbol{u}^{k,\varepsilon},w^{k,\varepsilon}\}_{\varepsilon>0}\subset \mathbb{K}_\varepsilon$ such that

$$\boldsymbol{v}^{k,\varepsilon}\to\boldsymbol{v}^k \quad \text{strongly in} \quad L^2(\Omega)^2 \ ,$$
$$\boldsymbol{u}^{k,\varepsilon}\to\boldsymbol{u}^k \quad \text{strongly in} \quad L^2(\Omega)^2 \ ,$$
$$w^{k,\varepsilon}\to w^k \quad \text{strongly in} \quad L^2(\Omega) \ ,$$

when $\varepsilon\to 0$; moreover

$$\limsup_{\varepsilon\to 0} J_\varepsilon(\boldsymbol{v}^{k,\varepsilon},\boldsymbol{u}^{k,\varepsilon},w^{k,\varepsilon}) \le J_h(\boldsymbol{v}^k) \ . \qquad (11.7.77)$$

Combining (11.7.76) and (11.7.77) we deduce that

$$\limsup_{k\to\infty}\limsup_{\varepsilon\to 0} J_\varepsilon(\boldsymbol{v}^{k,\varepsilon},\boldsymbol{u}^{k,\varepsilon},w^{k,\varepsilon}) \le J_h(\boldsymbol{v}) \ , \qquad \limsup_{k\to\infty}\limsup_{\varepsilon\to 0} ||\boldsymbol{v}^{k,\varepsilon}-\boldsymbol{v}||_{0,\Omega} = 0 \ ,$$

$$\limsup_{k\to\infty}\limsup_{\varepsilon\to 0} ||\boldsymbol{u}^{k,\varepsilon}||_{0,\Omega} = 0 \ , \qquad \limsup_{k\to\infty}\limsup_{\varepsilon\to 0} ||w^{k,\varepsilon}||_{0,\Omega} = 0 \ .$$

By applying now Lemma 1.3.27 we infer existence of a mapping $\varepsilon \to k(\varepsilon)$ with $k(\varepsilon) \to \infty$ such that setting $\boldsymbol{v}^\varepsilon = \boldsymbol{v}^{k(\varepsilon),\varepsilon}$, $\boldsymbol{u}^\varepsilon = \boldsymbol{u}^{k(\varepsilon),\varepsilon}$, $w^\varepsilon = w^{k(\varepsilon),\varepsilon}$, one finally obtains

$$J_h^s(\boldsymbol{v},\boldsymbol{0},0) \leq \lim_{\varepsilon \to 0} \sup J_\varepsilon(\boldsymbol{v}^v, \boldsymbol{u}^\varepsilon, w^\varepsilon) \leq J_h(\boldsymbol{v}) \,,$$

for each $\boldsymbol{v} \in H_{\Gamma_w}(\Omega)^2$, which completes the proof of (11.7.64).

II. The second part of the proof of Γ-convergence of the sequence $\{J_\varepsilon\}_{\varepsilon>0}$ consists in showing that for any sequence $\{\boldsymbol{v}^\varepsilon, \boldsymbol{u}^\varepsilon, w^\varepsilon\}_{\varepsilon>0} \subset \mathbb{K}_\varepsilon$ and convergent to $(\boldsymbol{v},\boldsymbol{u},w)$ in the topology τ, the following inequality is satisfied

$$J_h(\boldsymbol{v}) \leq \lim_{\varepsilon \to 0} \inf J_\varepsilon(\boldsymbol{v}^\varepsilon, \boldsymbol{u}^\varepsilon, w^\varepsilon) \,. \tag{11.7.78}$$

We recall that if $\boldsymbol{u} \neq \boldsymbol{0}$ and/or $w \neq 0$ (see Lemma 11.7.3), then obviously

$$\lim_{\varepsilon \to 0} \inf J_\varepsilon(\boldsymbol{v}^\varepsilon, \boldsymbol{u}^\varepsilon, w^\varepsilon) = +\infty \,,$$

and (11.7.78) is trivially satisfied. Of interest is thus the case where $\boldsymbol{u} = \boldsymbol{0}$, $w = 0$. Similarly to the proofs of Theorems 2.10.13 and 5.4.6, some results from the theory of duality play an important role. More precisely, it will be shown that

$$\lim_{\varepsilon \to 0} \inf J_\varepsilon(\boldsymbol{v}^\varepsilon, \boldsymbol{u}^\varepsilon, w^\varepsilon) \geq J_h(\boldsymbol{v}) = \int_\Omega \mathcal{W}_h[x, \boldsymbol{\varepsilon}(\boldsymbol{v}(x))]dx - L(\boldsymbol{v})$$

$$= \sup\{\int_\Omega [N^{\alpha\beta}(x)\varepsilon_{\alpha\beta}(\boldsymbol{v}(x)) - \mathcal{W}_h^*(x,\boldsymbol{N}(x))]dx \mid \boldsymbol{N} \in L^2(\Omega, \mathbb{E}_2^s)\} - L(\boldsymbol{v}) \,. \tag{11.7.79}$$

Step 3. Let $\{\Omega_K\}_{K \in \mathcal{K}}$ be a finite family of open disjoint sets such that $\bar{\Omega} = \bigcup_{K \in \mathcal{K}} \bar{\Omega}_K$. Note that such a partition may be completely different from the one introduced in Step 1. We take

$$\boldsymbol{N}(x) = \sum_{K \in \mathcal{K}} \chi^K(x)\boldsymbol{N}_K \,, \quad \boldsymbol{N}_K \in \mathbb{E}_2^s \,, \tag{11.7.80}$$

where χ^K is the characteristic function of Ω_K.

Let $(\boldsymbol{n}_K, \boldsymbol{l}_K, \boldsymbol{q}_K, \boldsymbol{r}_K) \in \boldsymbol{S}_{per}(YF)$, $\omega_K^\varepsilon = \Omega_K \cap \Omega^\varepsilon$ and take $\varphi_K^\delta \in \boldsymbol{D}(\Omega_K), 0 \leq \varphi_K^\delta \leq 1$, $\varphi_K^\delta(x) = 1$ for $x \in \Omega_K^\delta$, where

$$\Omega_K^\delta = \{x \in \Omega_K \mid dist(x,\partial\Omega_K) > \delta\} \,, \quad \delta > 0 \,.$$

Applying Lemma 11.7.9 and recalling that $j_\varepsilon \geq 0$ we obtain

$$\lim_{\varepsilon \to 0} \inf J_\varepsilon(\boldsymbol{v}^\varepsilon, \boldsymbol{u}^\varepsilon, w^\varepsilon)$$

$$\geq \lim_{\varepsilon \to 0} \inf\{\sum_{K \in \mathcal{K}} \int_{\omega_K^\varepsilon} j_\varepsilon[x, \boldsymbol{\varepsilon}(\boldsymbol{v}^\varepsilon), \boldsymbol{\gamma}(\boldsymbol{u}^\varepsilon), \boldsymbol{\kappa}(\boldsymbol{u}^\varepsilon, w^\varepsilon), w^\varepsilon]dx - \int_{\Gamma_\sigma^{bK}} (\bar{N}^\alpha v_\alpha^\varepsilon + \bar{L}^\alpha u_\alpha^\varepsilon - \frac{1}{\varepsilon}\bar{Q}w^\varepsilon)ds\}$$

$$\geq \liminf_{\varepsilon \to 0} \sum_{K \in \mathcal{K}} \int_{\omega_K^\varepsilon} \varphi_K^\delta(x) j_\varepsilon[x, \boldsymbol{\varepsilon}(\boldsymbol{v}^\varepsilon), \boldsymbol{\gamma}(\boldsymbol{u}^\varepsilon), \boldsymbol{\kappa}(\boldsymbol{u}^\varepsilon, w^\varepsilon), w^\varepsilon] dx - L(\boldsymbol{v})$$

$$= \liminf_{\varepsilon \to 0} \sum_{K \in \mathcal{K}} \int_{\omega_K^\varepsilon} \varphi_K^\delta(x) \{ j_\varepsilon[x, \boldsymbol{\varepsilon}(\boldsymbol{v}^\varepsilon), \boldsymbol{\gamma}(\boldsymbol{u}^\varepsilon), \boldsymbol{\kappa}(\boldsymbol{u}^\varepsilon, w^\varepsilon), w^\varepsilon] dx - [\mathbf{n}_K \left(\frac{x}{\varepsilon} \right) : \boldsymbol{\varepsilon}(\boldsymbol{v}^\varepsilon)$$

$$+ \mathbf{l}_K \left(\frac{x}{\varepsilon} \right) : \boldsymbol{\gamma}(\boldsymbol{u}^\varepsilon) + \frac{1}{\varepsilon} \boldsymbol{q} \left(\frac{x}{\varepsilon} \right) \cdot \boldsymbol{\kappa}(\boldsymbol{u}^\varepsilon, w^\varepsilon) + \frac{1}{\varepsilon^2} \mathbf{r}_K \left(\frac{x}{\varepsilon} \right) w^\varepsilon] \} dx - L(\boldsymbol{v})$$

$$= \liminf_{\varepsilon \to 0} \sum_{K \in \mathcal{K}} \int_{\omega_K^\varepsilon} \varphi_K^\delta(x) j_{\varepsilon, (\mathbf{n}_K, \mathbf{l}_K, \boldsymbol{q}_K, \mathbf{r}_K)}[x, \boldsymbol{\varepsilon}(\boldsymbol{v}^\varepsilon), \boldsymbol{\gamma}(\boldsymbol{u}^\varepsilon), \boldsymbol{\kappa}(\boldsymbol{u}^\varepsilon, w^\varepsilon), w^\varepsilon] - L(\boldsymbol{v}) ,$$

where

$$j_{\varepsilon, (\mathbf{n}_K, \mathbf{l}_K, \boldsymbol{q}_K, \mathbf{r}_K)}(x, \boldsymbol{a}, \boldsymbol{b}, \boldsymbol{c}, d)$$
$$= j_\varepsilon(x, \boldsymbol{a}, \boldsymbol{b}, \boldsymbol{c}, d) - \mathbf{n}_K : \boldsymbol{a} - \mathbf{l}_K : \boldsymbol{b} - \frac{1}{\varepsilon} \boldsymbol{q}_K \cdot \boldsymbol{c} - \frac{1}{\varepsilon^2} \mathbf{r}_K \, d . \qquad (11.7.81)$$

Fenchel's inequality applied to the function $j_{\varepsilon, (\mathbf{n}_K, \mathbf{l}_K, \boldsymbol{q}_K, \mathbf{r}_K)}(x, \cdot, \cdot, \cdot, \cdot)$ at

$$[(\boldsymbol{\varepsilon}(\boldsymbol{v}^\varepsilon(x)), \boldsymbol{N}_K); (\boldsymbol{\gamma}(\boldsymbol{u}^\varepsilon), \boldsymbol{0}); (\boldsymbol{\kappa}(\boldsymbol{u}^\varepsilon(x), w^\varepsilon(x)), \boldsymbol{0}); (w^\varepsilon(x), 0)]$$

gives

$$j^*_{\varepsilon, (\mathbf{n}_K, \mathbf{l}_K, \boldsymbol{q}_K, \mathbf{r}_K)}(x, \boldsymbol{N}_K, 0, 0, 0)$$
$$\geq \boldsymbol{N}_K : \boldsymbol{\varepsilon}(\boldsymbol{v}^\varepsilon(x)) - j_{\varepsilon, (\mathbf{n}_K, \mathbf{l}_K, \boldsymbol{q}_K, \mathbf{r}_K)}[x, \boldsymbol{\varepsilon}(\boldsymbol{v}^\varepsilon), \boldsymbol{\gamma}(\boldsymbol{u}^\varepsilon), \boldsymbol{\kappa}(\boldsymbol{v}^\varepsilon, w^\varepsilon), w^\varepsilon] .$$

Hence

$$\liminf_{\varepsilon \to 0} J_\varepsilon(\boldsymbol{v}^\varepsilon, \boldsymbol{u}^\varepsilon, w^\varepsilon)$$

$$\geq \liminf_{\varepsilon \to 0} \sum_{K \in \mathcal{K}} \int_{\omega_K^\varepsilon} \varphi_K^\delta(x) [\boldsymbol{N}_K : \boldsymbol{\varepsilon}(\boldsymbol{v}^\varepsilon(x)) - j^*_{\varepsilon, (\mathbf{n}_K, \mathbf{l}_K, \boldsymbol{q}_K, \mathbf{r}_K)}(x, \boldsymbol{N}_K, 0, 0, 0)] dx - L(\boldsymbol{v})$$

$$= \liminf_{\varepsilon \to 0} \sum_{K \in \mathcal{K}} \int_{\Omega_K} \varphi_K^\delta(x) \left[\boldsymbol{N}_K : \boldsymbol{\varepsilon}(\boldsymbol{v}^\varepsilon(x)) - j^*_{\varepsilon, (\mathbf{n}_K, \mathbf{l}_K, \boldsymbol{q}_K, \mathbf{r}_K)}(x, \boldsymbol{N}_K, 0, 0, 0) \right] dx - L(\boldsymbol{v}) ,$$

$$(11.7.82)$$

since $\boldsymbol{v}^\varepsilon \in H^1(\Omega)^2$.

The sequence $\{\boldsymbol{v}^\varepsilon\}_{\varepsilon > 0}$ is a bounded sequence in $H^1(\Omega)^2$, therefore for a subsequence still denoted in the same manner, we have

$$\boldsymbol{\varepsilon}(\boldsymbol{v}^\varepsilon) \rightharpoonup \boldsymbol{\varepsilon}(\boldsymbol{v}) \quad \text{weakly in} \quad L^2(\Omega, \mathbb{E}_s^2) .$$

Thus

$$\lim_{\varepsilon \to 0} \int_{\Omega} \varphi_K^\delta(x) \boldsymbol{N}_K : \boldsymbol{\varepsilon}(\boldsymbol{v}^\varepsilon(x)) dx = \int_{\Omega} \varphi_K^\delta(x) \boldsymbol{N}_K : \boldsymbol{\varepsilon}(\boldsymbol{v}(x)) dx .$$

The explicit form of $j^*_{\varepsilon,(\mathbf{n}_K,\mathbf{l}_K,\boldsymbol{q}_K,\mathbf{r}_K)}(\cdot,0,0,0)$ is found by performing standard calculation:

$$j^*_{\varepsilon,(\mathbf{n}_K,\mathbf{l}_K,\boldsymbol{q}_K,\mathbf{r}_K)}(x,\boldsymbol{\epsilon}^*,0,0,0)$$

$$= \sup\{\boldsymbol{\epsilon}^* : \boldsymbol{\epsilon} - j_{\varepsilon,(\mathbf{n}_K,\mathbf{l}_K,\boldsymbol{q}_K,\mathbf{r}_K)}(x,\boldsymbol{\epsilon},\boldsymbol{b},\boldsymbol{c},d) \,|\,\boldsymbol{\epsilon},\boldsymbol{b} \in \mathbb{E}^2_s,\ \boldsymbol{c} \in \mathbb{R}^2,\ d \in \mathbb{R}\}$$

$$= \sup\{(\boldsymbol{\epsilon}^* + \mathbf{n}_K) : \boldsymbol{\epsilon} + \mathbf{l}_K : \boldsymbol{b} + \frac{1}{\varepsilon}\boldsymbol{q}_K\cdot\boldsymbol{c} + \frac{1}{\varepsilon^2}\mathbf{r}_K d$$

$$-j_\varepsilon(x,\boldsymbol{\epsilon},\boldsymbol{b},\boldsymbol{c},d) \,|\, \boldsymbol{\epsilon},\boldsymbol{b} \in \mathbb{E}^2_s,\ \boldsymbol{c} \in \mathbb{R}^2,\ d \in \mathbb{R}\}$$

$$= \frac{1}{2}[D^N_{\alpha\beta\lambda\mu}(x)(\epsilon^{*\alpha\beta} + \mathbf{n}^{\alpha\beta}_K)(\epsilon^{*\lambda\mu} + \mathbf{n}^{\lambda\mu}_K) + D^L_{\alpha\beta\lambda\mu}(x)\mathbf{l}^{\alpha\beta}_K \mathbf{l}^{\lambda\mu}_K$$

$$+2D^{NL}_{\alpha\beta\lambda\mu}(x)(\epsilon^{*\alpha\beta} + \mathbf{n}^{\alpha\beta}_K)\mathbf{l}^{\lambda\mu}_K + D^Q_{\alpha\beta}(x)q^\alpha_K q^\beta_K$$

$$+2D^{RL}_{\alpha\beta}(x)\mathbf{l}^{\alpha\beta}_K \mathbf{r}_K + D^{RN}_{\alpha\beta}(x)\mathbf{n}^{\alpha\beta}_K \mathbf{r}_K + D^R \mathbf{r}^2_K]\,, \quad \text{(no sumation on } K!)$$

where $\boldsymbol{\epsilon}^* \in \mathbb{E}^s_2$.

The sequence of periodic functions $j^*_{\varepsilon,(\mathbf{n}_K(\frac{\cdot}{\varepsilon}),\mathbf{l}_K(\frac{\cdot}{\varepsilon}),\boldsymbol{q}_K(\frac{\cdot}{\varepsilon}),\mathbf{r}_K(\frac{\cdot}{\varepsilon}))}(x,\boldsymbol{N}_K,0,0,0)$

is bounded in $L^1(\Omega)$ and converges weakly (in $L^1(\Omega)$) to

$$\frac{1}{|Y|}\int_{YF} j^*[x,\boldsymbol{N}_K + \mathbf{n}_K(y),\mathbf{l}_K(y),\boldsymbol{q}_K(y),\mathbf{r}_K(y)]dy\,,$$

where $j^* = \mathcal{W}^*$, cf. (11.1.13).

Thus inequality (11.7.82) becomes

$$\liminf_{\varepsilon\to 0} J_\varepsilon(\boldsymbol{v}^\varepsilon,\boldsymbol{u}^\varepsilon,w^\varepsilon) \geq \sum_{K\in\mathcal{K}}\{\int_{\Omega_K} \varphi^\delta_K(x)\boldsymbol{N}_K : \boldsymbol{\varepsilon}(\boldsymbol{v}(x))$$

$$-\frac{1}{|Y|}\int_Y \mathcal{W}^*[x,\boldsymbol{N}_K + \mathbf{n}_K(y),\mathbf{l}_K(y),\boldsymbol{q}_K(y),\mathbf{r}_K(y)dy]dx\} - L(\boldsymbol{v})\,. \quad (11.7.83)$$

Next, taking the supremum on the r.h.s. of (11.7.83) when $(\mathbf{n}_K,\mathbf{l}_K,\boldsymbol{q}_K,\mathbf{r}_K)$ runs over the set $\boldsymbol{S}_{per}(YF)$ of statically admissible (local) generalized stresses one obtains

$$\liminf_{\varepsilon\to 0} J_\varepsilon(\boldsymbol{v}^\varepsilon,\boldsymbol{u}^\varepsilon,w^\varepsilon) \geq \sum_{K\in\mathcal{K}}\int_{\Omega_K} \varphi^\delta_K(x)[\boldsymbol{N}_K : \boldsymbol{\varepsilon}(\boldsymbol{v}(x)) - \mathcal{W}^*_h(x,\boldsymbol{N}_K)]dx - L(\boldsymbol{v})\,.$$

We recall that $\mathcal{W}^*_h$ is given by (11.7.46) and $\sup(-\int) = -\inf\int$.

Since $\varphi^\delta_K \geq 0$ and $\boldsymbol{N}$ is a piecewise constant function (see (11.7.80)), therefore

$$\liminf_{\varepsilon\to 0} J_\varepsilon(\boldsymbol{v}^\varepsilon,\boldsymbol{u}^\varepsilon,w^\varepsilon) \geq \sum_{K\in\mathcal{K}}\int_{\Omega_K} \varphi^\delta_K(x)\boldsymbol{N}(x) : \boldsymbol{\varepsilon}(\boldsymbol{v}(x))dx$$

$$-\sum_{K\in\mathcal{K}}\int_{\Omega_K} \varphi^\delta_K(x)W^*_h(x,\boldsymbol{N}(x))dx - L(\boldsymbol{v})\,.$$

The inequality $0 \leq \sum\limits_{K \in \mathcal{K}} \varphi_K^\delta(x) \leq 1$, $x \in \Omega$, implies

$$0 \leq \sum_{K \in \mathcal{K}} \varphi_K^\delta(x) W_h^*(x, \boldsymbol{N}(x)) \leq W_h^*(x, \boldsymbol{N}(x)) \,,$$

because $\mathcal{W}_h^* \geq 0$. Thus

$$\lim_{\varepsilon \to 0} \inf J_\varepsilon(\boldsymbol{v}^\varepsilon, \boldsymbol{u}^\varepsilon, w^\varepsilon) \geq \sum_{K \in \mathcal{K}} \int_\Omega \varphi_K^\delta(x) \boldsymbol{N}(x) : \boldsymbol{\varepsilon}(\boldsymbol{v}(x)) dx - \int_\Omega W_h^*(x, \boldsymbol{N}(x)) dx - L(\boldsymbol{v}) \,.$$

Now let δ tend to zero, then $\sum\limits_{K \in \mathcal{K}} \varphi_K^\delta(x)$ tends to 1 for almost every $x \in \Omega$ and consequently

$$\lim_{\varepsilon \to 0} \inf J_\varepsilon(\boldsymbol{v}^\varepsilon, \boldsymbol{u}^\varepsilon, w^\varepsilon) \geq \int_\Omega [\boldsymbol{N}(x) : \boldsymbol{\varepsilon}(\boldsymbol{v}(x)) - \mathcal{W}_h^*(x, \boldsymbol{N}(x))] dx - L(\boldsymbol{v}) \,.$$

Step 4. For each $\boldsymbol{N} \in L^2(\Omega, \mathbb{E}_2^s)$ there exists a sequence $\{\boldsymbol{N}_k\}_{k \in \mathbb{N}}$ of simple functions such that

$$\boldsymbol{N}_k \to \boldsymbol{N} \quad \text{strongly in} \quad L^2(\Omega, \mathbb{E}_2^s) \text{ when } k \to \infty \,.$$

Here

$$\boldsymbol{N}_k(x) = \sum_{K(k)} \chi_{K(k)}^{\delta_k}(x) \boldsymbol{N}^{K(k)} \,,$$

where

$$\chi_{K(k)}^{\delta_k}(x) = \begin{cases} 1 & \text{if } x \in \Omega_{K(k)} \,, \\ 0 & \text{otherwise,} \end{cases}$$

and $\delta_k = 1/k$, $\mathrm{diam}\Omega_{K(k)} \leq \delta_k$ and $\bar{\Omega} = \bigcup\limits_{K(k)} \bar{\Omega}_{K(k)}$. By virtue of the previous step we have

$$\lim_{\varepsilon \to 0} \inf J_\varepsilon(\boldsymbol{v}^\varepsilon, \boldsymbol{u}^\varepsilon, w^\varepsilon) \geq \int_\Omega [\boldsymbol{N}_k(x) : \boldsymbol{\varepsilon}(\boldsymbol{v}(x)) - \mathcal{W}_h^*(x, \boldsymbol{N}_k(x))] dx - L(\boldsymbol{v}) \,.$$

Passing now to the limit when k tends to infinity we finally obtain

$$\lim_{\varepsilon \to 0} \inf J_\varepsilon(\boldsymbol{v}^\varepsilon, \boldsymbol{u}^\varepsilon, w^\varepsilon) \geq \int_\Omega [\boldsymbol{N}(x) : \boldsymbol{\varepsilon}(\boldsymbol{v}(x)) - \mathcal{W}_h^*(x, \boldsymbol{N}(x))] dx - L(\boldsymbol{v}) \,.$$

This inequality completes the proof. $\square$

Remark 11.7.11. Rigorously, we should write $J_h(\boldsymbol{v}, 0, 0)$ instead of $J_h(\boldsymbol{v})$. Also, we observe that in the second part of the proof $\boldsymbol{N} \in L^2(\Omega, \mathbb{E}_2^s)$ is not necessarily a statically admissible field.

Corollary 11.7.12. According to Step 3 of our proof, the dual of $j_\varepsilon(x, \cdot, \cdot, \cdot, \cdot)$ has the following form

$$j_\varepsilon^*(x, \boldsymbol{N}, \boldsymbol{L}, \boldsymbol{Q}, R) = \frac{1}{2}\{D_{\alpha\beta\lambda\mu}^N(x)N^{\alpha\beta}N^{\lambda\mu} + D_{\alpha\beta\lambda\mu}^L(x)L^{\alpha\beta}L^{\lambda\mu} + 2D_{\alpha\beta\lambda\mu}^{NL}(x)N^{\alpha\beta}L^{\lambda\mu}$$
$$+2\varepsilon^2[D_{\alpha\beta}^{RN}(x)N^{\alpha\beta}R + D_{\alpha\beta}^{RL}(x)L^{\alpha\beta}R] + \varepsilon^4 D^R R^2\}$$
$$+\frac{1}{2}\varepsilon^2 D_{\alpha\beta}^Q(x)Q^\alpha Q^\beta . \tag{11.7.84}$$

Dual homogenization

By comparing the complementary stored energy function (11.1.13) with (11.7.84) it is clear that now the refined scaling involves the following replacement

$$\boldsymbol{Q} \rightsquigarrow \varepsilon\boldsymbol{Q} , \qquad R \rightsquigarrow \varepsilon^2 R . \tag{11.7.85}$$

To formulate a problem dual to the one appearing on the r.h.s. of (11.7.15) it remains to find the set of statically admissible generalized stresses. To this end we introduce the functional

$$f_{\varepsilon 1}(\boldsymbol{v}, \boldsymbol{u}, w) = -f_\varepsilon(\boldsymbol{v}, \boldsymbol{u}, w) + I_{\mathbf{K}_\varepsilon}(\boldsymbol{v}, \boldsymbol{u}, w) , \tag{11.7.86}$$

where f_ε is defined by (11.7.10), and

$$\Lambda = (\Lambda_1, \Lambda_2, \Lambda_3, \Lambda_4) , \quad \Lambda(\boldsymbol{v}, \boldsymbol{u}, w) = (\boldsymbol{\varepsilon}(\boldsymbol{v}), \boldsymbol{\gamma}(\boldsymbol{u}), \boldsymbol{\kappa}(\boldsymbol{u}, w), w) ,$$
$$\Lambda : H_{\Gamma_w}(\Omega)^2 \times H_{\Gamma_w}(\Omega^\varepsilon)^2 \times H_{\Gamma_w}(\Omega) \to L^2(\Omega, \mathbb{E}_2^s) \times L^2(\Omega^\varepsilon, \mathbb{E}_2^s) \times L^2(\Omega)^2 \times L^2(\Omega) .$$

It can now easily be found that, cf. Secs. 2.10, 5.4, 10.4

$$f_{\varepsilon 1}^*(-\Lambda^*(\boldsymbol{N}, \boldsymbol{L}, \boldsymbol{Q}, R)) = \begin{cases} 0 & \text{if } (\boldsymbol{N}, \boldsymbol{L}, \boldsymbol{Q}, R) \in S(\Omega^\varepsilon) , \\ \infty & \text{otherwise} , \end{cases} \tag{11.7.87}$$

where

$$S(\Omega^\varepsilon) = \{(\boldsymbol{N}, \boldsymbol{L}, \boldsymbol{Q}, R) \mid \boldsymbol{N} \in L^2(\Omega, \mathbb{E}_2^s), \operatorname{div}\boldsymbol{N} \in L^2(\Omega)^2 ,$$
$$\boldsymbol{L} \in L^2(\Omega^\varepsilon, \mathbb{E}_2^s) , \operatorname{div}\boldsymbol{L} \in L^2(\Omega^\varepsilon)^2 , \boldsymbol{Q} \in L^2(\Omega)^2 , \operatorname{div}\boldsymbol{Q} \in L^2(\Omega) ,$$
$$\operatorname{div}\boldsymbol{N} = 0 \text{ in } \Omega, \operatorname{div}\boldsymbol{L} - \boldsymbol{Q} = 0 \text{ in } \Omega^\varepsilon , \operatorname{div}\boldsymbol{Q} + R = 0 \text{ in } \Omega ;$$
$$N^{\alpha\beta}n_\beta = \bar{N}^\alpha , L^{\alpha\beta}n_\beta = \bar{L}^\alpha , Q^\alpha n_\alpha = \frac{1}{\varepsilon}\bar{Q} \text{ on } \Gamma_\sigma ; L_n \leq 0 ,$$
$$L_\tau = 0 \quad \text{on} \quad F^\varepsilon\} . \tag{11.7.88}$$

Here L_n and L_τ denote normal and tangential generalized contact stresses on F^ε. For a fixed $\varepsilon > 0$ we can now formulate the dual problem, which means evaluating

$$G_\varepsilon(\bar{\boldsymbol{N}}_\varepsilon, \bar{\boldsymbol{L}}_\varepsilon, \bar{\boldsymbol{Q}}_\varepsilon, \bar{R}_\varepsilon) = \sup\{G_\varepsilon(\boldsymbol{N}, \boldsymbol{L}, \boldsymbol{Q}, R) \mid \boldsymbol{N} \in L^2(\Omega, \mathbb{E}_2^s),$$
$$\boldsymbol{L} \in L^2(\Omega^\varepsilon, \mathbb{E}_2^s) , \boldsymbol{Q} \in L^2(\Omega)^2 , R \in L^2(\Omega)\} , \tag{11.7.89}$$

where

$$G_\varepsilon(\boldsymbol{N}, \boldsymbol{L}, \boldsymbol{Q}, R) = -\int_{\Omega^\varepsilon} j_\varepsilon^*[x, \boldsymbol{N}(x), \boldsymbol{L}(x), \boldsymbol{Q}(x), R(x)]dx$$
$$- I_{\boldsymbol{S}(\Omega^\varepsilon)}(\boldsymbol{N}, \boldsymbol{L}, \boldsymbol{Q}, R) . \tag{11.7.90}$$

By using Proposition 1.2.47, 1.2.48 and 1.2.49, we conclude that

$$J_\varepsilon(\bar{\boldsymbol{v}}^\varepsilon, \bar{\boldsymbol{u}}^\varepsilon, \bar{w}^\varepsilon) = G_\varepsilon(\bar{\boldsymbol{N}}_\varepsilon, \bar{\boldsymbol{L}}_\varepsilon, \bar{\boldsymbol{Q}}_\varepsilon, \bar{R}_\varepsilon) , \tag{11.7.91}$$

and $(\bar{\boldsymbol{N}}_\varepsilon, \bar{\boldsymbol{L}}_\varepsilon, \bar{\boldsymbol{Q}}_\varepsilon, \bar{R}_\varepsilon) \in \boldsymbol{S}(\Omega^\varepsilon)$ is uniquely determined. We are now in a position to formulate a theorem on Γ-convergence of the sequence $\{G_\varepsilon\}_{\varepsilon>0}$.

Theorem 11.7.13.
(i) The sequence of functionals (11.7.90) is Γ-convergent in the weak topology of the space $L^2(\Omega, \mathbb{E}_2^s) \times L^2(\Omega, \mathbb{E}_2^s) \times L^2(\Omega)^2 \times L^2(\Omega)$ to

$$G_h(\boldsymbol{N}) = -\int_\Omega \mathcal{W}_h^*(x, \boldsymbol{N}(x))dx - I_{\boldsymbol{S}_h(\Omega)}(\boldsymbol{N}) , \tag{11.7.92}$$

where

$$\boldsymbol{S}_h(\Omega) = \{\boldsymbol{N} \in L^2(\Omega, \mathbb{E}_2^s) \mid \mathrm{div}\boldsymbol{N} \in L^2(\Omega)^2 , \ \mathrm{div}\boldsymbol{N} = \boldsymbol{0} \text{ in } \Omega ,$$
$$N^{\alpha\beta}n_\beta = \bar{N}^\alpha \text{ on } \Gamma_\sigma\} . \tag{11.7.93}$$

(ii) If $\{\boldsymbol{v}^{\varepsilon'}, \boldsymbol{u}^{\varepsilon'}, w^{\varepsilon'}\}_{\varepsilon'>0}$ is a convergent subsequence of $\{\bar{\boldsymbol{v}}^\varepsilon, \bar{\boldsymbol{u}}^\varepsilon, \bar{w}^\varepsilon\}_{\varepsilon>0}$, i.e.

$$(\boldsymbol{v}^{\varepsilon'}, \boldsymbol{u}^{\varepsilon'}, w^{\varepsilon'}) \xrightarrow[\varepsilon'\to 0]{} (\bar{\boldsymbol{v}}, \boldsymbol{0}, 0) \text{ in } L^2(\Omega)^2 \times L^2(\Omega)^2 \times L^2(\Omega) \text{ strongly } ,$$

then

$$J_h(\bar{\boldsymbol{v}}) = \inf\{J_h(\boldsymbol{v}) \mid \boldsymbol{v} \in H_{\Gamma_w}(\Omega)^2\}$$

and

$$\inf J_{\varepsilon'} \to \inf J_h \quad \text{when } \varepsilon' \to 0 .$$

(iii) If $(\bar{\boldsymbol{N}}_\varepsilon, \bar{\boldsymbol{L}}_\varepsilon, \bar{\boldsymbol{Q}}_\varepsilon, \bar{R}_\varepsilon)$ is a solution of the dual problem (11.7.89) and $\{\boldsymbol{N}_{\varepsilon'}, \boldsymbol{L}_{\varepsilon'}, \boldsymbol{Q}_{\varepsilon'}, R_{\varepsilon'}\}_{\varepsilon'>0}$ is a convergent subsequence

$$(\boldsymbol{N}_{\varepsilon'}, \boldsymbol{L}_{\varepsilon'}, \boldsymbol{Q}_{\varepsilon'}, R_{\varepsilon'}) \xrightarrow[\varepsilon'\to 0]{} (\bar{\boldsymbol{N}}, \boldsymbol{0}, \boldsymbol{0}, 0)$$
$$\text{in } L^2(\Omega, \mathbb{E}_2^s) \times L^2(\Omega, \mathbb{E}_2^s) \times L^2(\Omega)^2 \times L(\Omega) \text{ weakly } ,$$

then

$$G_h(\bar{\boldsymbol{N}}) = \sup\{-\int_\Omega \mathcal{W}_h^*(x, \boldsymbol{N}(x))dx \mid x \in \boldsymbol{S}_h(\Omega)\} ,$$

and

$$\sup G_{\varepsilon'} \to \sup G_h , \qquad \inf J_h = \sup G_h ,$$

when $\varepsilon' \to 0$.

Proof. One has to verify the assumptions of Theorems 1.3.36, 1.3.38, cf. Sec. 5.5.   □

Remark 11.7.14. To be precise, we should write $\mathcal{W}_h^*(x, \boldsymbol{N}, 0, 0, 0)$ and $G_h(\boldsymbol{N}, 0, 0, 0)$. Then the above Γ-convergence of G_ε to G_h, from the notational point of view is quite clear.

11.8. *Application of the augmented Lagrangian method to solving local problems with unilateral constraints*

To find effective potentials describing macroscopic behavior of plates and cross-ply laminates one has to solve variational inequalities posed on a basic cell. Equivalently, one has to find minimizers of the corresponding convex optimization problems.

In the present section we shall apply the augmented Lagrangian algorithm, expounded in Sec. 1.5, to solving the minimization problem appearing in the definition of the effective elastic potential $\mathcal{W}_h$ given by (11.6.27). The remaining cases can be examined in the same manner. This problem can be written as follows:

$$\min\left\{\frac{1}{2}a_{YF}(x, \boldsymbol{U}, \boldsymbol{U}) - l(x, \boldsymbol{U}) \mid \boldsymbol{U} \in \tilde{H}_{per},\ g(\boldsymbol{u}) \le 0 \text{ on } F\right\} + M(x) . \quad (11.8.1)$$

Here $\boldsymbol{U} = (\boldsymbol{v}, \boldsymbol{u}, w)$, $g(\boldsymbol{u}) = -[\![(\gamma_0 \boldsymbol{u})_N]\!]$ on F; γ_0 is the zeroth-order trace operator (on F). For the sake of simplicity we write $g(\boldsymbol{u}) = -[\![u_N]\!]$. Moreover we have

$$\frac{1}{2}a_{YF}(x, \boldsymbol{U}, \boldsymbol{U}) = \int\limits_{YF} j[x, \boldsymbol{\varepsilon}^y(\boldsymbol{v}), \boldsymbol{\gamma}^y(\boldsymbol{u}), \boldsymbol{\kappa}^y(\boldsymbol{u}, w), w]dy ,$$

$$l(x, \boldsymbol{U}) = \int\limits_{YF} (A_v^{\alpha\beta\lambda\mu}(x)\varepsilon_{\alpha\beta}^y(\boldsymbol{v})\epsilon_{\lambda\mu} + A_{vu}^{\alpha\beta\lambda\mu}(x)\gamma_{\alpha\beta}^y(\boldsymbol{u})\epsilon_{\lambda\mu} + A_{vw}^{\alpha\beta}\epsilon_{\alpha\beta}w)dy ,$$

$$M(x) = \frac{|Y|}{2}A_v^{\alpha\beta\lambda\mu}(x)\epsilon_{\alpha\beta}\epsilon_{\lambda\mu} ,$$

where $\boldsymbol{U} = (\boldsymbol{v}, \boldsymbol{u}, w) \in H^1(Y)^2 \times H^1(YF)^2 \times H^1(Y)$, $\epsilon \in \mathbb{E}_s^2$. The space $\tilde{H}_{per}$ is the subspace of $H_{per}^1(Y)^2 \times H_{per}^1(YF)^2 \times H_{per}^1(Y)$ and consists of functions with zero mean value over Y. Thus we write $\tilde{H}_{per} = \tilde{H}_{per}^1(Y)^2 \times \tilde{H}_{per}^1(YF)^2 \times \tilde{H}_{per}^1(Y)$. The minimization problem (11.8.1) is then uniquely solvable, its solution is denoted by $\tilde{\boldsymbol{U}}$. We now define

$$V = H^1(Y)^2 \times H^1(YF)^2 \times H^1(Y) , \quad V_1 = \tilde{H}_{per} , \quad H = H^{1/2}(F) , \quad H^* = H^{-1/2}(F) .$$

The coerciveness constant is still denoted by c_0. The inequality $g(\boldsymbol{u}) \le 0$ is satisfied almost everywhere on F. In fact, $g(\boldsymbol{u}) \in H^{1/2}(F)$ and hence $g(\boldsymbol{u}) \in L^2(F)$. Normal contact stress l_N is an element of the space $H^{-1/2}(F)$. Therefore it is necessary to introduce the Riesz operator $\mathcal{R} : H^{1/2}(F) \to H^{-1/2}(F)$. We recall that if $f \in H^{-1/2}(F)$, there exists one and only one $v_f \in H^{1/2}(F)$ such that $f(v) = \langle f, v \rangle_{H^{-1/2}(F) \times H^{1/2}(F)} = (v_f, v)_{H^{1/2}(F)}$, $\mathcal{R}(v_f) = f$ and $\|f\|_{H^{-1/2}(F)} = \|v_f\|_{H^{1/2}(F)}$, cf. Yosida (1978), Kikuchi and Oden (1988,

Chap. 3). Here $(\cdot,\cdot)_{H^{1/2}(F)}$ stands for an inner product on $H^{1/2}(F)$. Then we obviously have $\mathcal{R}^{-1}(l_N) \in H^{1/2}(F)$.

Theorem 5.1. of Ekeland and Temam (1976, Chap. III) implies existence of $\tilde{\lambda} \in H^* = H^{-1/2}(F)$, such that $\langle \tilde{\lambda}, g(\tilde{u}) \rangle_{H^* \times H} = 0$. Equivalently, by Riesz' representation theorem we have $(\mathcal{R}^{-1}\tilde{\lambda}, g(\tilde{u}))_{H \times H} = 0$. We recall that the condition $g(u) \leq 0$ defines a cone in $H^{1/2}(F)$. Now g is a linear and continuous operator from $H^1(YF)^2$ onto $H^{1/2}(F)$. The assumption (A) of Sec. 1.5 is obviously satisfied and takes now the following form:

(a) $\langle \tilde{\lambda}, g(\tilde{u}) \rangle_{H^* \times H} = (\mathcal{R}^{-1}\tilde{\lambda}, g(\tilde{u}))_{H \times H} = 0$,

(b) $a_{YF}(x, \tilde{U}, U) - l(x, U) + \langle \tilde{\lambda}, g(\tilde{u}) \rangle_{H^* \times H} = 0 \quad \forall U = (v, u, w) \in V_1$.

We observe that $(\tilde{U}, \tilde{\lambda}) \in V_1 \times H^{-1/2}(F)$ is a unique saddle point of the Lagrangian defined by

$$L(x, \tilde{U}, U) = \frac{1}{2}a_{YF}(x, U, U) - l(x, U) + \langle \lambda, g(u) \rangle_{H^* \times H} . \qquad (11.8.2)$$

Now we are in a position to apply the augmented Lagrangian algorithm of Sec. 1.5 to solve the local problem (11.8.1). This algorithm involves now the Riesz mapping $\mathcal{R}$. To this end we define a family of augmented Lagrangian problems by

$$(P)_{\eta,\lambda} \quad \min\{L_\eta(x, U, \lambda) | U \in V_1\} ,$$

where

$$L_\eta(x, U, \lambda) = \frac{1}{2}a_{YF}(x, U, U) - l(x, U) + \frac{1}{2\eta}(\| \sup(0, \mathcal{R}^{-1}\lambda + \eta g(u)) \|_H^2 - \|\mathcal{R}^{-1}\lambda\|_H)$$

$$= \frac{1}{2}a_{YF}(x, U, U) - l(x, U) + \frac{1}{2\eta}(\| \sup(0, \mathcal{R}^{-1}\lambda + \eta g(u)) \|_H^2 - \|\lambda\|_{H^*}^2) ,$$

and $\lambda \in H^*, \mathcal{R}^{-1}\lambda \in H, \|\lambda\|_{H^*} = \|\mathcal{R}^{-1}\lambda\|_H, \eta > 0, \eta \in \mathbb{R}$.

The Algorithm

(1) Choose $\lambda_1 \in H^*, \lambda_1 \geq 0$, and $\eta > 0$.

(2) Put $n = 1$.

(3) Solve $(P)_{\eta,\lambda_n}$ for U_n.

(4) Put $\mathcal{R}^{-1}\lambda_{n+1} = \mathcal{R}^{-1}\lambda_n + \eta\hat{g}(U_n, \lambda_n, \eta) = \sup(0, \mathcal{R}^{-1}\lambda + \eta g(u_n))$.

(5) Put $n = n + 1$ and return to (3).

Here $\hat{g} : V_1 \times H \times \mathbb{R}^+ \to H$ is defined by $\hat{g}(U, \mathcal{R}^{-1}\lambda, \eta) = \sup(g(u), -\frac{1}{\eta}\mathcal{R}^{-1}\lambda), \lambda \in H^* = H^{-1/2}(F)$. We observe that $\lambda \in H^*, \lambda \geq 0$ satisfies the inequality $\langle \lambda, v \rangle_{H^* \times H} \geq 0$ for each $v \in H, v \geq 0$; moreover we have $\langle \lambda, v \rangle_{H^* \times H} = (\mathcal{R}^{-1}\lambda, v)_H \geq 0$ for all $v \in H$, $v \geq 0$.

From Theorem 1.5.3 we conclude the following estimate

$$c_0\|\tilde{U}, U_n\|_V^2 + \frac{1}{2\eta}\|\mathcal{R}^{-1}\lambda_{n+1} - \mathcal{R}^{-1}\tilde{\lambda}\|_H^2 \leq \frac{1}{2\eta}\|\mathcal{R}^{-1}\lambda_n - \mathcal{R}^{-1}\tilde{\lambda}\|_H^2 , \qquad (11.8.3)$$

for all $n = 1, 2, \ldots$ Here

$$\|\boldsymbol{U}\|_V^2 = \|\boldsymbol{v}\|_{H^1(Y)^2}^2 + \|\boldsymbol{u}\|_{H^1(YF)^2}^2 + \|w\|_{H^1(Y)}^2 . \tag{11.8.4}$$

By using the properties of the Riesz mapping $\mathcal{R}$ we obtain the estimate equivalent to the previous one

$$c_0 \|\tilde{\boldsymbol{U}}, \boldsymbol{U}_n\|_V^2 + \frac{1}{2\eta} \|\lambda_{n+1} - \tilde{\lambda}\|_{H^*}^2 \leq \frac{1}{2\eta} \|\lambda_n - \tilde{\lambda}\|_{H^*}^2 . \tag{11.8.5}$$

The augmented Lagrangian algorithm can obviously be replaced by the variable stepsize algorithm that was explained in Sec. 1.5.

11.9. Case of aligned parallel intralaminar cracks. Effective characteristics according to the (h_0, l) approach

The aim of this section is to assess the loss of effective elastic characteristics of the $[0_m^\circ/90_n^\circ]_s$ laminates with the straight matrix cracks going transversely through the whole thickness of the 90°-plies. The in-plane boundary loading will be considered. Let us start with considering simplifications implied by the assumptions of the plies being orthotropic and the cracks being aligned. Assume that the axes x_1 and x_2 are axes of orthotropy. The cracks are parallel to the axis x_2 for a laminate with short aligned cracks, see Fig. 11.9.1. In view of the orthotropy assumption the only non-zero components of the stiffness tensors involved in (11.1.21) are

$$A_v^{\alpha\alpha\beta\beta}, A_v^{\alpha\beta\alpha\beta}; A_{vw}^{\alpha\alpha}; A_{vu}^{\alpha\alpha\beta\beta}, A_{vu}^{\alpha\beta\alpha\beta}, \qquad A_{uw}^{\alpha\alpha}, A_u^{\alpha\alpha\beta\beta}, A_u^{\alpha\beta\alpha\beta}; A_w . \tag{11.9.1}$$

Consequently, the only non-zero components of the tensors $\boldsymbol{A}_\sigma$ are given by, cf. (11.5.4)

$$A_1^{\alpha\alpha\beta\beta}, A_2^{\alpha\alpha\beta\beta}, A_1^{\alpha\beta\alpha\beta} = A_v^{\alpha\beta\alpha\beta}, A_2^{\alpha\beta\alpha\beta} = A_{vu}^{\alpha\beta\alpha\beta} . \tag{11.9.2}$$

Note that $A_1^{1122} = A_1^{2211}$ but $A_2^{1122} \neq A_2^{2211}$. The components of the vectors $\check{\boldsymbol{N}}, \boldsymbol{T}$ are $(0, 1)$ and $(1, 0)$, respectively; the vectors $\check{\boldsymbol{N}}$ and $\boldsymbol{T}$ are normal and tangent to the crack F, see Fig. 11.3.1. According to Eq. (11.5.6) we find expressions for the components of the matrix $(\varepsilon_{\alpha\beta}^F)$

$$\varepsilon_{11}^F = \frac{1}{l_1 l_2} \int_{s_1}^{s_2} [\![u_1^{(1)}]\!]\, dy_2 , \qquad \varepsilon_{22}^F = 0 , \qquad \varepsilon_{12}^F = \frac{1}{2l_1 l_2} \int_{s_1}^{s_2} [\![u_2^{(1)}]\!]\, dy_2 . \tag{11.9.3}$$

Thus, the homogenized constitutive relations (11.5.3) assume the form

$$\begin{aligned}
N_h^{11} &= A_v^{1111} \left(\alpha_{11}\varepsilon_{11}^h + \alpha_{12}\varepsilon_{22}^h - \beta_{11}\varepsilon_{11}^F\right) , \\
N_h^{22} &= A_v^{1111} \left(\alpha_{12}\varepsilon_{11}^h + \alpha_{22}\varepsilon_{22}^h - \beta_{21}\varepsilon_{11}^F\right) , \\
N_h^{12} &= 2A_v^{1212} \left(\varepsilon_{12}^h - \widehat{\alpha}\varepsilon_{12}^F\right) ,
\end{aligned} \tag{11.9.4}$$

where $\varepsilon_{\alpha\beta}^F = \varepsilon_{\alpha\beta}^F(\varepsilon_{11}^h, \varepsilon_{22}^h, \varepsilon_{12}^h)$. We cannot expect, in general, that $\varepsilon_{\alpha\alpha}^F = \varepsilon_{\alpha\alpha}^F(\varepsilon_{11}^h, \varepsilon_{22}^h)$ and $\varepsilon_{12}^F = \varepsilon_{12}^F(\varepsilon_{12}^h)$, since the cracks considered are of a unilateral type. The coefficients

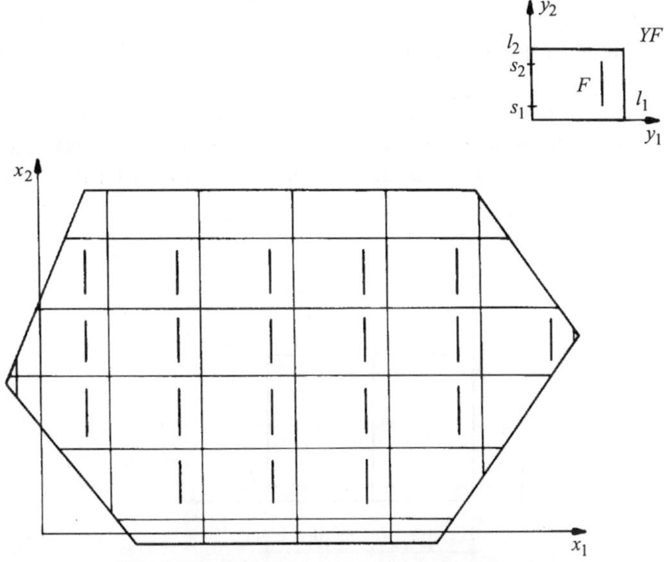

Fig. 11.9.1. Case of parallel cracks

involved in Eqs. (11.9.4) are defined as follows

$$\alpha_{\lambda\mu} = A^{\lambda\lambda\mu\mu}1/A_v^{1111} \,, \quad \beta_{\lambda\mu} = A_2^{\lambda\lambda\mu\mu}/A_v^{1111} \,, \quad \hat{\alpha} = A_{vu}^{2121}/A_v^{2121} \,. \tag{11.9.5}$$

Note that the formulae (11.9.3) – (11.9.4) hold true within the framework of the (h_0, l) model introduced in Sec. 11.6. Only the meaning of the vector $u^{(1)}$ is different.

Let us consider now the case of straight-line cracks lying at equal distances l and going along the whole domain Ω, thus intersecting the opposite edges. Since these cracks cut only

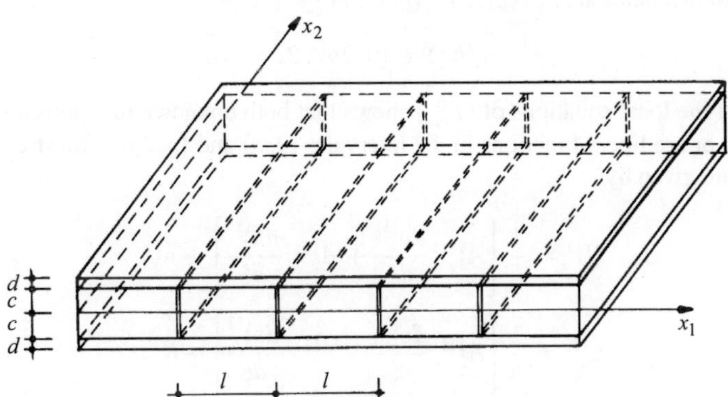

Fig. 11.9.2. Laminate with straight cracks going along the whole domain Ω

the internal layer, the laminate remains capable of carrying the in-plane boundary loading. The crack lines coincide with the $x_1 = nl$ lines, $n = 1, 2 \ldots$, see Fig. 11.9.2.

The cracks weaken the overall stiffnesses of the laminate. This degradation will be assessed by using model (h_0, l_0) of Sec. 11.5. Let us recall that the predictions will be insensitive to the l/h ratio, see Remark 11.4.2 and the comments at the end of Sec. 11.5.

In order to find the formulae for effective stiffnesses one should start by solving the local problem (P_{loc}^0) of Sec. 11.5. Geometry of the periodicity cell is here simple, see Fig. 11.9.3.

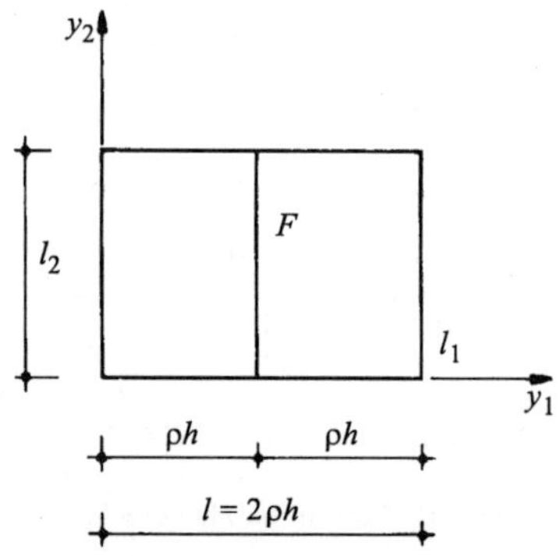

Fig. 11.9.3. Rescaled cell of periodicity

For the sake of simplicity, the crack F is located at the centre. The solutions to the local problem will be independent of the variable y_1. The unknowns $v_1^{(1)}(y_1)$, $u_1^{(1)}(y_1)$ and $v_2^{(1)}(y_1)$, $u_2^{(1)}(y_1)$ will be found from the independent: stretching and shearing problems.

Solution of the local stretching problem. Stiffnesses $A_c^{\alpha\alpha\beta\beta}$
The unknown fields are $v_1^{(1)}(y_1)$, $u_1^{(1)}(y_1)$. Let

$$\xi = y_1/h \; ; \xi \in (0, 2\varrho) \; ; 2\varrho = l/h \, . \tag{11.9.6}$$

Analysis of the local equations of (P_{loc}^0) shows that both unknown functions are piecewise linear in ξ, or are linear functions in the intervals: $(0, \varrho)$ and $(\varrho, 2\varrho)$. The stress resultants (11.5.12) are given by

$$N_0^{11} = \frac{1}{h} \left[A_1^{1111} \frac{dv_1^{(1)}}{d\xi} + A_2^{1111} \frac{du_1^{(1)}}{d\xi} \right] + n_0^{11},$$

$$L_0^{11} = \frac{1}{h} \left[A_2^{1111} \frac{dv_1^{(1)}}{d\xi} + A_4^{1111} \frac{du_1^{(1)}}{d\xi} \right] + l_0^{11}, \tag{11.9.7}$$

where n_0^{11}, l_0^{11} are defined by (11.5.14). The unknown functions satisfy:

- periodicity conditions

$$v_1^{(1)}(0) = v_1^{(1)}(2\varrho), \qquad u_1^{(1)}(0) = u_1^{(1)}(2\varrho) ,$$
$$N_0^{11}(0) = N_0^{11}(2\varrho), \qquad L_0^{11}(0) = L_0^{11}(2\varrho) ; \tag{11.9.8}$$

- switching and contact conditions

$$v_1^{(1)}(\varrho - 0) = v_1^{(1)}(\varrho + 0), \qquad N_0^{11}(\varrho - 0) = N_0^{11}(\varrho + 0),$$
$$L_0^{11}(\varrho - 0) = L_0^{11}(\varrho + 0) \leq 0, \qquad L_0^{11}(\varrho - 0)[\![u_1^{(1)}]\!] = 0, \tag{11.9.9}$$
$$[\![u_1^{(1)}]\!] = u_1^{(1)}(\varrho + 0) - u_1^{(1)}(\varrho - 0) \geq 0 .$$

Knowing that functions $v_1^{(1)}$, $u_1^{(1)}$ are piece-wise linear one can choose appropriate constants such that the conditions (11.9.8) – (11.9.9) are fulfilled. In this way we find

$$\varepsilon_{11}^y(\mathbf{v}^{(1)}) = 0, \qquad \langle \gamma_{11}^y(\mathbf{u}^{(1)}) \rangle = \begin{cases} 0 & \text{for } l_0^{11} \leq 0, \\ -\dfrac{l_0^{11}}{A_4^{1111}} & \text{otherwise.} \end{cases}$$

Here $\langle \cdot \rangle = \dfrac{1}{l} \displaystyle\int_0^l (\cdot) dy_1$. Due to periodicity we have $\langle \gamma_{11}^y(\boldsymbol{u}^{(1)}) \rangle = -[\![u_1^{(1)}]\!]/l$. Hence we find a non-zero component of the crack deformation measure $\varepsilon_{\alpha\beta}^F$, see Eq. (11.9.3)

$$\varepsilon_{11}^F = \frac{1}{l}[\![u_1^{(1)}]\!] = \begin{cases} 0, & \text{if } E_h \leq 0 \text{ (the crack is closed)}, \\ F_{11}^0 E_h, & \text{if } E_h > 0 \text{ (the crack is open)}, \end{cases} \tag{11.9.10}$$

where

$$E_h = l_0^{11}/A_v^{1111} \qquad F_{11}^0 = A_v^{1111}/A_4^{1111}, \qquad l_0^{11} = A_3^{1111}\varepsilon_{11}^h + A_3^{1122}\varepsilon_{22}^h , \tag{11.9.11}$$

or

$$E_h = \beta_{11}\varepsilon_{11}^h + \beta_{21}\varepsilon_{22}^h , \qquad F_{11}^0 = [\gamma - (\lambda_1)^2/\mu]^{-1} , \tag{11.9.12}$$

with

$$(\gamma, \lambda_1, \mu) = (A_u^{1111}, h^2 A_{uw}^{11}, h^4 A_w)/A_v^{1111} . \tag{11.9.13}$$

On account of Eqs. (11.9.10) and (11.9.4) we find the homogenized constitutive relations for the axial stress resultants

$$N_h^{\alpha\alpha} = \begin{cases} A_1^{\alpha\alpha 11}\varepsilon_{11}^h + A_1^{\alpha\alpha 22}\varepsilon_{22}^h , & \text{for } E_h \leq 0, \\ A_c^{\alpha\alpha 11}\varepsilon_{11}^h + A_c^{\alpha\alpha 22}\varepsilon_{22}^h , & \text{for } E_h > 0, \end{cases} \tag{11.9.14}$$

where

$$A_c^{\alpha\alpha\beta\beta} = A_1^{\alpha\alpha\beta\beta} - A_2^{\alpha\alpha 11} A_3^{11\beta\beta}(A_4^{1111})^{-1} . \tag{11.9.15}$$

It is not difficult to verify that $N_h^{\alpha\alpha}$ given by (11.9.14) is continuous along the line $E_h = 0$.

Note, moreover, that

$$A_c^{\alpha\alpha\beta\beta} = A_c^{\beta\beta\alpha\alpha} \, . \tag{11.9.16}$$

These effective (reduced due to cracking) stiffnesses can be put in the form

$$A_c^{\lambda\lambda\mu\mu}/A_v^{1111} = \alpha_{\lambda\mu} - \beta_{\lambda 1}\beta_{\mu 1}F_{11}^0 \, . \tag{11.9.17}$$

In the "technical" notation the constitutive relations read as follows:
- in the case of closed cracks $(E_h \leq 0)$

$$\begin{bmatrix} N_h^{11} \\ N_h^{22} \end{bmatrix} = \frac{2h}{1 - \nu_{12}\nu_{21}} \begin{bmatrix} E_1 & \nu_{12}E_1 \\ \nu_{21}E_2 & E_2 \end{bmatrix} \begin{bmatrix} \varepsilon_{11}^h \\ \varepsilon_{22}^h \end{bmatrix} , \tag{11.9.18}$$

- in the case of open cracks $(E_h > 0)$

$$\begin{bmatrix} N_h^{11} \\ N_h^{22} \end{bmatrix} = \frac{2h}{1 - \nu_{12}^c\nu_{21}^c} \begin{bmatrix} E_1^c & \nu_{12}^c E_1^c \\ \nu_{21}^c E_2^c & E_2^c \end{bmatrix} \begin{bmatrix} \varepsilon_{11}^h \\ \varepsilon_{22}^h \end{bmatrix} \tag{11.9.19}$$

with

$$\nu_{12} = \frac{A_1^{1122}}{A_1^{1111}} \, , \qquad \nu_{21} = \frac{A_1^{1122}}{A_1^{2222}} \, , \qquad E_\alpha = (1 - \nu_{12}\nu_{21})\frac{A_1^{\alpha\alpha\alpha\alpha}}{2h} \, . \tag{11.9.20}$$

The components ν_{12}^c, ν_{21}^c and E_α^c are defined in terms of $A_c^{\alpha\alpha\beta\beta}$ in a similar manner.

Solution of the shearing local problem. Stiffness A_c^{1212}

Since the unknown functions $v_2^{(1)}$, $u_2^{(1)}$ are piecewise linear, the stress resultants

$$N_0^{21} = \frac{1}{h}\left[A_1^{2121}\frac{dv_2^{(1)}}{dy_1} + A_2^{2121}\frac{du_2^{(1)}}{dy_1} \right] + n_0^{21} \, ,$$

$$L_0^{21} = \frac{1}{h}\left[A_2^{2121}\frac{dv_2^{(1)}}{dy_1} + A_4^{2121}\frac{du_2^{(1)}}{dy_1} \right] + l_0^{21}, \tag{11.9.21}$$

are piecewise constant. The periodicity conditions are given by

$$\begin{aligned} v_2^{(1)}(0) &= v_2^{(1)}(2\varrho), & u_2^{(1)}(0) &= u_2^{(1)}(2\varrho) \, , \\ N_0^{21}(0) &= N_0^{21}(2\varrho), & L_0^{21}(0) &= L_0^{21}(2\varrho) \, , \end{aligned} \tag{11.9.22}$$

while the switching conditions are

$$\begin{aligned} v_2^{(1)}(\varrho - 0) &= v_2^{(1)}(\varrho + 0), \\ N_0^{21}(\varrho - 0) &= N_0^{21}(\varrho + 0), & L_0^{21}(\varrho - 0) &= L_0^{21}(\varrho + 0) \, . \end{aligned} \tag{11.9.23}$$

Using (11.9.22)–(11.9.23), the equality $A_4^{2121} = A_2^{2121}$ which follows from the orthotropy, and the definition of $l_0^{21} = 2A_2^{2121}\varepsilon_{12}^h$ we obtain

$$\varepsilon_{12}^F = \varepsilon_{12}^h \, . \tag{11.9.24}$$

The homogenized constitutive relation $(11.9.4)_3$ assumes the form

$$N_h^{12} = 2A_c^{1212}\varepsilon_{12}^h\,, \qquad A_c^{1212} = (1 - \widehat{\alpha})A_v^{1212}\,, \tag{11.9.25}$$

with $\widehat{\alpha}$ defined by $(11.9.5)_3$. One can prove that $1 - \widehat{\alpha} = d/h$. Thus the effective Kirchhoff moduli of the uncracked and cracked laminate are

$$G_{12} = A_v^{1212}/2h, \qquad G_{12}^c = A_c^{1212}/2h\,. \tag{11.9.26}$$

The homogenized potential

Having found relationships (11.9.14) and (11.9.25), one can express the homogenized constitutive relations in the hyperelastic form (11.5.16) with the effective potential $\mathcal{V}_h$ now given by

$$\mathcal{V}_h = \begin{cases} \mathcal{V}_h^0 & \text{for} \quad E_h \leq 0, \\ \mathcal{V}_h^c & \text{for} \quad E_h \geq 0, \end{cases} \tag{11.9.27}$$

where

$$\mathcal{V}_h^0 = \left(\sum_{\alpha,\beta} A_1^{\alpha\alpha\beta\beta}\varepsilon_{\alpha\alpha}^h\varepsilon_{\beta\beta}^h + 2A_c^{1212}\left[(\varepsilon_{12}^h)^2 + (\varepsilon_{21}^h)^2\right] \right) \Big/2$$

$$\mathcal{V}_h^c = \left(\sum_{\alpha,\beta} A_c^{\alpha\alpha\beta\beta}\varepsilon_{\alpha\alpha}^h\varepsilon_{\beta\beta}^h + 2A_c^{1212}\left[(\varepsilon_{12}^h)^2 + (\varepsilon_{21}^h)^2\right] \right) \Big/2\,. \tag{11.9.28}$$

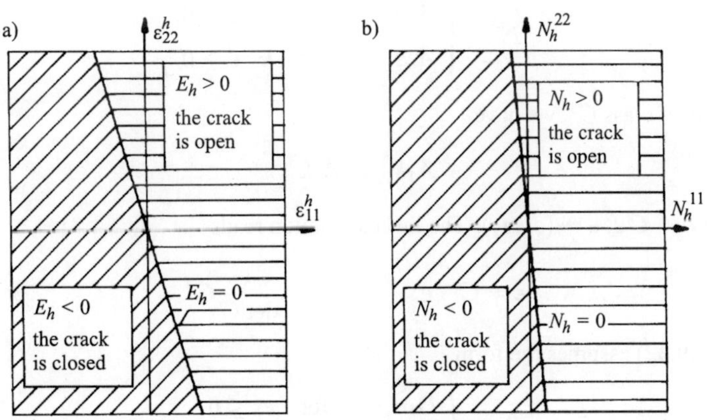

Fig. 11.9.4.

a) $(\varepsilon_{11}^h, \varepsilon_{22}^h)$-plane; the condition $E_h > 0$ for the crack opening in the glass/epoxy $[0°, 90_3°]_s$ laminate tested by Highsmith and Reifsnider [1982]; $E_h = 0.4298\varepsilon_{11}^h + 0.1286\varepsilon_{22}^h$.

b) (N_h^{11}, N_h^{22})-plane. The condition $N_h > 0$ for the crack opening; the same laminate, $N_h = 0.632N_h^{11} + 0.0408N_h^{22}$.

Note that

$$\mathcal{V}_h^c = \mathcal{V}_h^0 - \frac{1}{2}A_v^{1111}F_{11}^0(E_h)^2 ,$$ (11.9.29)

which shows that $\mathcal{V}_h$ is of class C^1 (not C^2), the result already known from Sec. 11.5. Moreover, we see that the line $E_h = 0$, or

$$\beta_{11}\varepsilon_{11}^h + \beta_{21}\varepsilon_{22}^h = 0 ,$$ (11.9.30)

is a line of discontinuity of the second order derivatives of the potential $\mathcal{V}_h$, cf. Fig. 11.9.4a. This figure characterizes the $[0°/90°_3]_s$ glass/epoxy laminate examined in detail in Sec. 11.11.

Inverted form of the homogenized constitutive relations

In order to invert the relations (11.9.4) we introduce a matrix notation:

$$k = [\beta_{11}, \beta_{21}], \qquad k_\perp = [-\beta_{21}, \beta_{11}], \qquad C = \begin{bmatrix} k \\ k_\perp \end{bmatrix} ,$$

$$A_m = \begin{bmatrix} A_m^{1111} & A_m^{1122} \\ A_m^{1122} & A_m^{2222} \end{bmatrix} , \qquad m = 1 \quad \text{or} \quad c,$$ (11.9.31)

$$\varepsilon = \left[\varepsilon_{11}^h, \varepsilon_{22}^h\right]^T , \qquad N = \left[N_h^{11}, N_h^{22}\right]^T .$$

The constitutive relations (11.9.14) can be written as follows

$$N = \begin{cases} A_1\varepsilon, & \text{for} \quad k \cdot \varepsilon \le 0, \\ A_c\varepsilon, & \text{for} \quad k \cdot \varepsilon \ge 0. \end{cases}$$ (11.9.32)

Since $\mathcal{V}_h$ is of class C^1, we have

$$A_1 k_\perp^\top = A_c k_\perp^\top .$$ (11.9.33)

Let us set $e = C\varepsilon$, $e = (e_1, e_2)$. The inverse relation reads

$$\varepsilon = C^{-1}e, \qquad C^{-1} = \frac{1}{\det C}C^T .$$ (11.9.34)

Hence (11.9.32) assumes the form

$$N = \begin{cases} B_1 e, & \text{for} \quad e_1 \le 0 , \\ B_c e, & \text{for} \quad e_1 \ge 0 , \end{cases}$$ (11.9.35)

with $B_1 = A_1 C^{-1}$, $B_c = A_c C^{-1}$. Consequently

$$e = \begin{cases} D_1 N, & \text{for} \quad e_1 \le 0 , \\ D_c N, & \text{for} \quad e_1 \ge 0 , \end{cases}$$ (11.9.36)

with $D_1 = CA_1^{-1}$, $D_c = CA_c^{-1}$. Our aim now is to express the conditions $e_1 < 0$ or $e_1 > 0$ in terms of N. To this end let us introduce a new vector $\mathcal{E} = D_1 N$. Hence

$$e = \begin{cases} \mathcal{E}, & \text{for} \quad e_1 \le 0 , \\ P\mathcal{E}, & \text{for} \quad e_1 \ge 0 , \end{cases} \tag{11.9.37}$$

where

$$P = D_c D_1^{-1} \qquad \text{or} \qquad P = \frac{1}{\det C} CA_c^{-1} A_1 C^T . \tag{11.9.38}$$

One can prove that (11.9.33) implies

$$P_{11} = \frac{\det A_1}{\det A_c} \qquad \text{and} \qquad P_{12} = 0 . \tag{11.9.39}$$

Note that $P_{11} > 0$. Thus (11.9.37) yields

$$e_1 = \begin{cases} \mathcal{E}_1, & \text{for} \quad e_1 \le 0 , \\ P_{11}\mathcal{E}_1, & \text{for} \quad e_1 \ge 0 , \end{cases} \tag{11.9.40}$$

hence sign $e_1 = $ sign $\mathcal{E}_1$, which makes it possible to rewrite (11.9.37) in the form

$$C\varepsilon = \begin{cases} D_1 N, & \text{for} \quad \mathcal{E}_1 \le 0 , \\ D_c N, & \text{for} \quad \mathcal{E}_1 \ge 0 , \end{cases} \tag{11.9.41}$$

and, finally, to find

$$\varepsilon = \begin{cases} A_1^{-1} N, & \text{for} \quad \mathcal{E}_1 \le 0 , \\ A_c^{-1} N, & \text{for} \quad \mathcal{E}_1 \ge 0 . \end{cases} \tag{11.9.42}$$

The condition $\mathcal{E}_1 \le 0$ can be expressed as $N_h \le 0$, with

$$N_h = (\beta_{11}\alpha_{22} - \beta_{21}\alpha_{21})N_h^{11} + (\alpha_{11}\beta_{21} - \beta_{11}\alpha_{12})N_h^{22} , \tag{11.9.43}$$

and sign $N_h = $ sign E_h. The inverted form of the homogenized relations (11.9.14) reads

$$\varepsilon_{\alpha\alpha}^h = \begin{cases} \dfrac{1}{2hE_\alpha}(N_h^{\alpha\alpha} - \nu_{\beta\alpha}N_h^{\beta\beta}), & \text{for} \quad N_h \le 0 , \\[2ex] \dfrac{1}{2hE_\alpha^c}(N_h^{\alpha\alpha} - \nu_{\beta\alpha}^c N_h^{\beta\beta}), & \text{for} \quad N_h \ge 0 , \end{cases} \tag{11.9.44}$$

and $\beta = 3 - \alpha$; do not sum over α and β. Since the relation $(11.9.25)_1$ can be easily inverted one can express now the potential $\mathcal{V}_h$ in terms of $N_h^{\alpha\beta}$. The line of discontinuity of its second order derivatives is $N_h = 0$, see Fig. 11.9.4b. The relations (11.9.44) could be found by using the Fenchel conjugate of $\mathcal{V}_h$, i.e.

$$\mathcal{V}_h^*(E^*) = \sup\left\{ E^{*\alpha\beta} E_{\alpha\beta} - \mathcal{V}_h(E) \mid E \in \mathbb{E}_s^2 \right\} , \qquad E^* \in \mathbb{E}_2^s . \tag{11.9.45}$$

The complementary potential $\mathcal{V}_h^*$ is strictly convex, of class C^1 and

$$\varepsilon^h = \frac{\partial \mathcal{V}_h^*}{\partial N^h} , \qquad \varepsilon^h \in \mathbb{E}_s^2 , \qquad N^h \in \mathbb{E}_s^2 . \tag{11.9.46}$$

11.10. Degradation of effective stiffnesses of laminates with aligned parallel cracks. The refined scaling approach – model (h_0, l)

The aim of this section is to evaluate the loss of the overall stiffnesses of the cracked laminates considered in Sec. 11.9 by using the (h_0, l) model developed in Sec. 11.6. The results of this section *will be sensitive* to the l/h ratio and hence will apply to arbitrary values of this ratio. Similarly to the in-plane scaling method, the local problem (P_{loc}^2) of Sec. 11.6 splits up into two: stretching (P_a) and shearing (P_s) problems. The unknown functions depend solely on $y_1 = h\xi$. To distinguish between the predictions of the models: (h_0, l_0) and (h_0, l), the last predictions will be denoted with a tilde $(\sim)$.

Solution of the stretching local problem (P_a)

The unknown functions are $v_1^{(1)}$, $u_1^{(1)}$, $w^{(2)}$. Let us introduce the non-dimensional functions:

$$v = v_1^{(1)}/h , \qquad u = u_1^{(1)}/h , \qquad w = w^{(2)}/h^2 + w_0 , \tag{11.10.1}$$

where

$$w_0 = \frac{1}{\mu}\left(\beta\varepsilon_{11}^h + \beta_2\varepsilon_{22}^h\right) , \tag{11.10.2}$$

and

$$\beta_2 = h^2 A_{vw}^{22}/A_v^{1111} , \qquad \beta = h^2 A_{vw}^{11}/A_v^{1111} ; \tag{11.10.3}$$

μ has been defined by (11.9.13). The non-vanishing stress resultants are given by (11.6.15); they assume the form

$$
\begin{aligned}
N_0^{11} &= A_v^{1111}(v' + \alpha u' + \beta w) + n_0^{11}, \\
N_0^{22} &= A_v^{1111}(\beta_1 v' + \beta_4 u' + \beta_2 w) + n_0^{22}, \\
L_0^{11} &= A_v^{1111}(\alpha v' + \gamma u' + \lambda_1 w) + l_0^{11}, \\
L_0^{22} &= A_v^{1111}(\beta_4 v' + \gamma_3 u' + \beta_3 w) + l_0^{22}, \\
R_0 &= \tfrac{1}{h^2} A_v^{1111}(\beta v' + \lambda_1 u' + \mu w), \\
Q_0^{(1)} &= \tfrac{\delta}{h} A_v^{1111}(u - w'), \qquad (\cdot)' = d(\cdot)/d\xi ,
\end{aligned}
\tag{11.10.4}
$$

with

$$\alpha = A_{vu}^{1111}/A_v^{1111} , \qquad \delta = h^2 H^{11}/A_v^{1111} ;$$
$$(\beta_1, \beta_3, \beta_4, \gamma_3) = (A_v^{1122}, h^2 A_{uw}^{22}, A_{vu}^{1122}, A_u^{1122})/A_v^{1111} ; \tag{11.10.5}$$

the coefficients γ, λ_1 being defined by (11.9.13). The equilibrium equations reduce to the form

$$\frac{dN_0^{11}}{d\xi} = 0, \qquad \frac{dL_0^{11}}{d\xi} = hQ_0^{(1)} , \qquad \frac{dQ_0^1}{d\xi} = -hR_0 . \tag{11.10.6}$$

Substitution of (11.10.4) into Eqs. (11.10.6) gives

$$v'' + \alpha u'' + \beta w' = 0,$$
$$\alpha v'' + (\gamma u'' - \delta u) + \lambda w' = 0,$$
$$-\beta v' - \lambda u' + (\delta w'' - \mu w) = 0.$$
(11.10.7)

The strong formulation of the local problem (P_{loc}^2) reduces to finding the fields (v, u, w) defined on the interval $[0, 2\varrho]$ which satisfy:
- the equations (11.10.7) for $\xi \in (0, \varrho) \cup (\varrho, 2\varrho)$,
- the periodicity conditions

$$f(0) = f(2\varrho), \qquad f \in \{v, u, w, N_0^{11}, L_0^{11}, Q_0^1\},$$
(11.10.8)

- the switching conditions at $\xi = \varrho$

$$f(\varrho - 0) = f(\varrho + 0), \qquad f \in \{v, w, N_0^{11}, L_0^{11}, Q_0^1\},$$
(11.10.9)

and

$$L = L_0^{11}(\varrho - 0) = L_0^{11}(\varrho + 0) \le 0, \qquad L[\![u]\!] = 0,$$
$$[\![u]\!] = u(\varrho + 0) - u(\varrho - 0) \ge 0.$$
(11.10.10)

We note by (11.6.34) that the only unknown which is really needed for assessing the loss of stiffnesses is the jump $[\![u]\!]$; we have

$$\tilde{\varepsilon}_{11}^F = \frac{[\![u]\!]}{2\varrho}, \qquad \tilde{\varepsilon}_{22}^F = 0.$$
(11.10.11)

The analysis of the system (11.10.7) implies that the function u satisfies the following fourth order ordinary differential equation

$$a_1 \frac{d^4 u}{d\xi^4} + a_2 \frac{d^2 u}{d\xi^2} + a_3 = 0.$$
(11.10.12)

To make this derivation complete we report the formulae for the coefficients a_i:

$$a_1 = \mu_{12}\mu_{21} - \mu_{11}\mu_{23}, \quad a_3 = \mu_{13}\mu_{22}, \quad a_2 = \mu_{22}\mu_{12} + \mu_{13}\mu_{21} - \mu_{24}\mu_{11}, \quad (11.10.13)$$

where

$$\mu_{11} = 1 - \alpha\beta/\lambda, \quad \mu_{12} = \alpha - \beta\gamma/\lambda, \quad \mu_{13} = \delta\beta/\lambda,$$
$$\mu_{21} = \alpha\mu_{13}, \quad \mu_{22} = \beta^2 - \mu, \quad \mu_{23} = \gamma\mu_{13},$$
$$\mu_{24} = \beta\lambda - \delta\mu_{13} - \mu\alpha, \quad \mu_{44} = \mu_{13}/\mu_{11}.$$
(11.10.14)

Let $\pm\sigma$, $\pm\omega$ be the roots of the characteristic equation

$$a_1 x^4 + a_2 x^2 + a_3 = 0.$$
(11.10.15)

If the roots are complex, then $\sigma = p - iq$, $\omega = p + iq$, $p, q \in \mathbb{R}$. In general one cannot say, whether the roots are real or complex. The derivation of the formula that interrelates $[\![u]\!]$ with $\varepsilon^h_{\alpha\alpha}$ is left to the reader. The final result has the form

$$\widetilde{\varepsilon}^F_{11} = \begin{cases} 0, & \text{for} \quad E_h \leq 0 , \\ F_{11}(\varrho)E_h, & \text{for} \quad E_h \geq 0 , \end{cases} \tag{11.10.16}$$

with E_h defined by $(11.9.12)_1$ and

$$F_{11}(\varrho) = f_{11}\left[\frac{\beta_{11}}{\sigma^2\omega^2}g_1(\sigma,\omega) + g_2(\sigma,\omega)F(\varrho;\omega,\sigma)\right]^{-1} ,$$

$$g_\alpha(\sigma,\omega) = \gamma_{\alpha 1} + \gamma_{\alpha 2}\frac{a_3}{a_1} - \gamma_{\alpha 3}\frac{a_2}{a_1} , \tag{11.10.17}$$

$$F(\varrho;\omega,\sigma) = \frac{\varrho}{\sigma^2 - \omega^2}\left(\frac{\coth\omega\varrho}{\omega} - \frac{\coth\sigma\varrho}{\sigma}\right) .$$

The coefficients $\gamma_{\alpha i}$ and f_{11} are given by

$$\begin{aligned}
f_{11} &= \alpha - \mu_{12}/\mu_{11} , & \gamma_{11} &= \mu_{44}(\beta - \mu_{44}) , \\
\gamma_{12} &= \mu_{12}f_{11}/\mu_{11} , & \gamma_{22} &= \frac{\delta}{\beta}(f_{11})^2 , \\
\gamma_{13} &= \mu_{44}f_{11} , & \gamma_{21} = \frac{\delta}{\beta}(\beta - \mu_{44})^2 , & \gamma_{23} &= \frac{\delta}{\beta}f_{11}(\beta - \mu_{44}) .
\end{aligned} \tag{11.10.18}$$

Note that

$$\sigma^2 + \omega^2 = -a_2/a_1 , \qquad \sigma^2\omega^2 = a_3/a_1 . \tag{11.10.19}$$

If $\sigma = p - iq$, $\omega = p + iq$ $(p, q \in \mathbb{R})$, the function $F(\varrho;\omega,\sigma)$ assumes the form $F_0(\varrho;p,q)$, namely

$$F_0(\varrho;p,q) = \frac{f(p\varrho,q\varrho)}{2pq(p^2 + q^2)} , \tag{11.10.20}$$

with

$$f(x,y) = \frac{y\sinh 2x + x\sin 2y}{\cosh 2x - \cos 2y} . \tag{11.10.21}$$

Assessing the loss of the $\widetilde{A}^{\alpha\alpha\beta\beta}_c$ stiffnesses
  Substitution of (11.10.16) into (11.9.4) gives

$$N^{\alpha\alpha}_h = \begin{cases} A^{\alpha\alpha 11}_1\varepsilon^h_{11} + A^{\alpha\alpha 22}_1\varepsilon^h_{22}, & \text{for} \quad E_h \leq 0, \\ \widetilde{A}^{\alpha\alpha 11}_c\varepsilon^h_{11} + \widetilde{A}^{\alpha\alpha 22}_c\varepsilon^h_{22}, & \text{for} \quad E_h \geq 0, \end{cases} \tag{11.10.22}$$

where

$$\widetilde{A}_c^{\lambda\lambda\mu\mu}/A_v^{1111} = \alpha_{\lambda\mu} - \beta_{\lambda1}\beta_{\mu1}F_{11}(\varrho) \; . \tag{11.10.23}$$

The tildes indicate that the model (h_0, l) is used. The relations (11.10.22) are continuous along the line (11.9.30). The constitutive relations (11.10.22) can be expressed in terms of technical notation usually used for orthotropic materials, cf. (11.9.18) – (11.9.20). For the case of closed cracks $(E_h \leq 0)$ these relations have the form (11.9.18). For the case of open cracks $(E_h > 0)$ they read as follows

$$\begin{bmatrix} N_h^{11} \\ N_h^{22} \end{bmatrix} = \frac{2h}{1 - \widetilde{\nu}_{12}^c\widetilde{\nu}_{21}^c} \begin{bmatrix} \widetilde{E}_1^c & \widetilde{\nu}_{12}^c\widetilde{E}_1^c \\ \widetilde{\nu}_{21}^c\widetilde{E}_2^c & \widetilde{E}_2^c \end{bmatrix} \begin{bmatrix} \varepsilon_{11}^h \\ \varepsilon_{22}^h \end{bmatrix}, \tag{11.10.24}$$

where

$$\widetilde{\nu}_{12}^c = \frac{\widetilde{A}_c^{1122}}{\widetilde{A}_c^{1111}} \; , \qquad \widetilde{\nu}_{21}^c = \frac{\widetilde{A}_c^{1122}}{\widetilde{A}_c^{2222}} \; , \qquad \widetilde{E}_\alpha^c = (1 - \widetilde{\nu}_{12}^c\widetilde{\nu}_{21}^c)\frac{\widetilde{A}_c^{\alpha\alpha\alpha\alpha}}{2h} \; . \tag{11.10.25a}$$

The last formula can be rearranged to the form

$$\widetilde{E}_\alpha^c = E_\alpha + \frac{(\nu_{\beta\alpha} - \widetilde{\nu}_{\beta\alpha}^c)^2}{(1 - \nu_{12}\nu_{21})}E_\beta - (\beta_{\alpha1} - \widetilde{\nu}_{\beta\alpha}^c\beta_{\beta1})^2 F_{11}(\varrho)E_v \; , \tag{11.10.25b}$$

where $\beta = 3 - \alpha$; $E_v = A_v^{1111}/2h$ (do not sum over β). The constitutive relations (11.10.22) can be inverted to the form similar to (11.9.43), where instead of E_α^c, $\nu_{\alpha\beta}^c$ one should put $\widetilde{E}_\alpha^c$, $\widetilde{\nu}_{\alpha\beta}^c$.

Solution of the shearing local problem (P_s)
  The non-dimensional fields

$$\widehat{u} = u_2^{(1)}/h \; , \qquad \widehat{v} = v_2^{(1)}/h \; , \tag{11.10.26}$$

are the unknowns of the problem. They determine the stress resultants

$$N_0^{21} = A_v^{2121}\left(\frac{d\widehat{v}}{d\xi} + \widehat{\alpha}\frac{d\widehat{u}}{d\xi} + 2\varepsilon_{21}^h\right) \; ,$$

$$L_0^{21} = \widehat{\alpha}A_v^{2121}\left(\frac{d\widehat{v}}{d\xi} + \frac{d\widehat{u}}{d\xi} + 2\varepsilon_{21}^h\right) \; , \qquad Q_0^2 = \frac{\widehat{\delta}}{h}A_v^{2121}\widehat{u} \; . \tag{11.10.27}$$

Here

$$\widehat{\delta} = hH^{22}/A_v^{2121} \; . \tag{11.10.28}$$

Substitution of these expressions into the equilibrium equations

$$\frac{dN_0^{21}}{d\xi} = 0, \qquad \frac{dL_0^{21}}{d\xi} = hQ_0^2 \; , \tag{11.10.29}$$

gives the set of ordinary differential equations

$$\widehat{v}'' + \widehat{\alpha}\widehat{u}'' = 0 \,, \qquad \widehat{\alpha}\widehat{v}'' + (\widehat{\alpha}\widehat{u}'' - \widehat{\delta}\widehat{u}) = 0 \,, \tag{11.10.30}$$

where $(\cdot)'' = d^2(\cdot)/d\xi^2$. Our goal is to find the crack deformation measure

$$\widetilde{\varepsilon}_{12}^{F} = \frac{[\![\widehat{u}]\!]}{4\varrho} \tag{11.10.31}$$

that depends on the field $\widehat{u}$. To find $\widehat{u}$ one should solve the set (11.10.30) with
• the conditions of periodicity

$$f(0) = f(2\varrho) \,, \qquad f \in \{\widehat{v}, \widehat{u}, N_0^{21}, L_0^{21}\} \,, \tag{11.10.32}$$

• the switching conditions at $\xi = \varrho$

$$f(\varrho - 0) = f(\varrho + 0) \,, \quad f \in \{\widehat{v}, N_0^{21}\} \,, \quad L_0^{21}(\varrho - 0) = L_0^{21}(\varrho + 0) = 0 \,. \tag{11.10.33}$$

Both fields $\widehat{v}, \widehat{u}$ satisfy

$$\frac{d^4 f}{d\xi^4} - (\widehat{\lambda})^2 \frac{d^2 f}{d\xi^2} = 0 \,, \quad f \in \{\widehat{v}, \widehat{u}\} \,, \tag{11.10.34}$$

where $\widehat{\lambda} = h(\widehat{\delta}/(cd))^{1/2}$. Solving the problem above is much easier than solving (11.10.7)–(11.10.10). The final result has the form

$$\widetilde{\varepsilon}_{12}^{F} = \frac{h}{c} F_{12}(\widehat{\lambda}\varrho)\varepsilon_{12}^{h} \,, \tag{11.10.35}$$

$$F_{12}(x) = \left(1 + \frac{d}{c}x \coth x\right)^{-1} \,. \tag{11.10.36}$$

Assessing the loss of the Kirchhoff modulus
   Substitution of (11.10.35) into (11.9.4)$_3$ gives

$$N_h^{12} = 2\widetilde{A}_c^{1212}\varepsilon_{12}^{h} \,, \tag{11.10.37}$$

$$\widetilde{A}_c^{1212}/A_v^{1212} = 1 - F_{12}(\widehat{\lambda}\varrho) \,. \tag{11.10.38}$$

The reduced Kirchhoff modulus of the laminate is defined by

$$\widetilde{G}_{12}^{c} = \widetilde{A}_c^{1212}/2h \,. \tag{11.10.39}$$

Homogenized potential
   The homogenized constitutive relationships (11.10.22) and (11.10.37) obey the hyperelastic rule (11.6.31) with the hyperelastic potential $\mathcal{W}_h$ given by

$$\mathcal{W}_h = \begin{cases} \mathcal{W}_h^0 \,, & \text{for} \quad E_h \leq 0 \,, \\ \mathcal{W}_h^c \,, & \text{for} \quad E_h \geq 0 \,, \end{cases} \tag{11.10.40}$$

where

$$\mathcal{W}_h^0 = \left(\sum_{\alpha,\beta} A_1^{\alpha\alpha\beta\beta} \varepsilon_{\alpha\alpha}^h \varepsilon_{\beta\beta}^h + 2\widetilde{A}_c^{1212} \left[(\varepsilon_{12}^h)^2 + (\varepsilon_{21}^h)^2 \right] \right) / 2 \,,$$

$$\mathcal{W}_h^c = \left(\sum_{\alpha,\beta} \widetilde{A}_c^{\alpha\alpha\beta\beta} \varepsilon_{\alpha\alpha}^h \varepsilon_{\beta\beta}^h + 2\widetilde{A}_c^{1212} \left[(\varepsilon_{12}^h)^2 + (\varepsilon_{21}^h)^2 \right] \right) / 2 \,. \tag{11.10.41}$$

The last equation can be expressed as follows

$$\mathcal{W}_h^c = \mathcal{W}_h^0 - \frac{1}{2} A_v^{1111} F_{11}(\varrho)(E_h)^2. \tag{11.10.42}$$

Hence we readily see that $\mathcal{W}_h$ is of class C^1, $E_h = 0$ (see Eq. (11.9.30)) being its line of non-smoothness of its first order derivatives. The complementary effective potential $\mathcal{W}_h^*(N_h^{\alpha\beta})$ is defined by Fenchel's transformation, cf. Eq. (11.9.44). The potential $\mathcal{W}_h^*$, defined on the space $\mathbb{E}_2^s$ is smooth, the equation $N_h = 0$ (see (11.9.42)) being the line of non-smoothness of the first derivative of $\mathcal{W}_h^*$.

Comparison of (h_0, l_0) and (h_0, l) predictions
    One can easily prove that

$$\lim_{\varrho \to 0} F_{11}(\varrho) = F_{11}^0 \,, \tag{11.10.43}$$

where $F_{11}(\varrho)$ and F_{11}^0 are defined as in $(11.10.17)_1$ and $(11.9.11)_2$, respectively. Let us recall that $\varrho = l/2h$, l being the crack spacing. The relation (11.10.43) implies

$$\lim_{\varrho \to 0} \widetilde{\varepsilon}_{11}^F = \varepsilon_{11}^F \,, \tag{11.10.44}$$

and hence

$$\lim_{\varrho \to 0} \widetilde{A}_c^{\alpha\alpha\beta\beta} = A_c^{\alpha\alpha\beta\beta} \,. \tag{11.10.45}$$

On the other hand, one can show that

$$\lim_{\varrho \to 0} F_{12}(\widehat{\lambda}\varrho) = \frac{c}{h} \,, \tag{11.10.46}$$

which implies

$$\lim_{\varrho \to 0} \widetilde{\varepsilon}_{12}^F = \varepsilon_{12}^F \,, \tag{11.10.47}$$

and

$$\lim_{\varrho \to 0} \widetilde{A}_c^{1212} = A_c^{1212} \,. \tag{11.10.48}$$

Thus, if the crack density $c_d := \varrho^{-1}$ tends to infinity, the (h_0, l) predictions tend to (h_0, l_0) predictions. The (h_0, l_0) model provides asymptotes for the curves $A_c^{\alpha\beta\lambda\mu}(c_d)$ predicted by the (h_0, l) model:

$$A_c^{\alpha\beta\lambda\mu} = \lim_{c_d \to \infty} \widetilde{A}_c^{\alpha\beta\lambda\mu}(c_d) , \qquad (11.10.49)$$

hence

$$(E_\alpha^c, \nu_{\alpha\beta}^c, G_{12}^c) = \lim_{c_d \to \infty} (\widetilde{E}_\alpha^c(c_d), \widetilde{\nu}_{\alpha\beta}^c(c_d), \widetilde{G}_{12}^c(c_d)) , \quad V_h = \lim_{c_d \to \infty} W_h(c_d) . \qquad (11.10.50)$$

Moreover, one can prove that

$$\lim_{\varrho \to \infty} \widetilde{A}_c^{\alpha\alpha\beta\beta} = A_1^{\alpha\alpha\beta\beta} , \qquad \lim_{\varrho \to \infty} \widetilde{A}_c^{1212} = A_v^{1212} . \qquad (11.10.51)$$

Hence

$$(E_\alpha, \nu_{\alpha\beta}, G_{12}) = \lim_{c_d \to \infty} (\widetilde{E}_\alpha^c(c_d), \widetilde{\nu}_{\alpha\beta}^c(c_d), \widetilde{G}_{12}^c(c_d)) , \qquad (11.10.52)$$

Thus the stiffnesses of the uncracked laminate are recovered if the crack density c_d tends to zero.

11.11. Degradation of effective stiffnesses of laminates $[0_n^\circ/90_m^\circ]_s$. Comparison with experimental results and with other analytical predictions

11.11.1. Scope of the section

We consider the laminate of Fig. 11.9.2. The material of the faces of the laminate is reinforced with fibers lying along the x_1 axis, hence the faces are called 0° layers. The internal layer of thickness $2c$ is reinforced with fibers lying along the x_2 axis, hence this layer is called 90°. The fibrous microstructure introduces orthotropy properties characterized by the orthotropic moduli: E_A, G_A, ν_A and E_T, G_T, ν_T; "A" indicates the fibrous direction; T means the direction that is transverse to the direction of the fibers. If referred to the coordinate system of Fig. 11.9.2, the elastic compliances (11.1.3) of both layers are expressed in terms of the orthotropic moduli as follows

$$c_{1111}^f = c_{2222}^m = (E_A)^{-1} , \qquad c_{1111}^m = c_{2222}^f = c_{3333}^f = c_{3333}^m = (E_T)^{-1} ,$$
$$c_{1122}^f = c_{1122}^m = c_{1133}^f = c_{2233}^m = -\nu_A/E_A , \quad c_{1133}^m = c_{2233}^f = -\nu_T/E_T , \qquad (11.11.1)$$
$$c_{1212}^f = c_{1212}^m = c_{2323}^m = c_{1313}^f = (4G_A)^{-1} , \quad c_{1313}^m = c_{2323}^f = (4G_T)^{-1} .$$

Consequently the compliances (11.1.14) and stiffnesses involved in (11.1.21) are fully determined by the orthotropic moduli and by the transverse dimensions c and d of the laminate. The 0°-layers are composed of n 0° plies of thickness t, while the 90° layer is formed by $2m$ 90° plies of the same thickness t. Hence the laminate manufactured in this manner is called a $[0_n^\circ/90_m^\circ]_s$ laminate. The subscript "s" indicates that the laminate is transversely symmetric with respect to the $x_3 = 0$ plane. Such laminates are sometimes called "balanced" laminates.

The first damage mode observed in the in-plane loaded, three-layer, balanced crossply laminates is usually transverse cracking along the fibers of the outer or inner layers. When stretched along the fibers of the outer layers or sheared in its plane, samples of the balanced $[0^\circ_m/90^\circ_n]_s$ laminates undergo transverse cracking in the 90° layer, with values of crack density c_d determined by magnitude of the in-plane loads applied. The crack density c_d will be further defined either as the number of cracks per 1 mm (then $c_d = 1mm/l$, l represents the crack spacing) or as the ratio: $2c/l$, see Fig. 11.9.2.

The intralaminar cracks lead to degradation of the effective elastic characteristics of the laminate. The aim of this section is to assess the loss of the effective Young moduli, Poisson ratios and the effective Kirchhoff moduli of the selected glass/epoxy, graphite/epoxy and carbon/epoxy laminates for which the elastic characteristics are available; they are collected in Table 11.11.1. All results will be predicted by the formulae derived in Sec. 11.10. Let us emphasize here that the (h_0, l) model provides a unified algorithm for finding decay of all components of the stiffness matrix, including the off-diagonal components, like the Poisson coefficients. Our theoretical predictions are compared with available experimental data of Highsmith and Reifsnider (1982), Groves (1986), Ogin et al. (1985) and Smith and Wood (1990).

11.11.2. $[0^\circ/\dot{9}0^\circ_3]_s$ glass/epoxy laminate tested by Highsmith and Reifsnider (1982)

The complete characteristics of the plies of this laminate can be found in Hashin (1985); they are given in Table 11.11.1.

Laminate	Tested in	t [mm]	E_A [GPa]	E_T [GPa]	ν_A	ν_T	G_A [GPa]	G_T [GPa]
$[0^0/90^0_3]_s$ glass/epoxy	Highsmith and Reifsnider [1982]	0.203	41.7	13.0	0.30	0.42	3.40	4.58
$[0^0/90^0]_s$ glass/epoxy	Ogin et al. [1985]	0.125	40.0	11.0	0.3	0.42	5.0	3.87
$[0^0_m/90^0_n]_s gl$ graphite/epoxy	Groves [1986]	0.127	144.8	9.6	0.31	0.461	4.8	3.285
$[0^0/90^0]_s$ glass/epoxy	Smith and Wood [1990]	0.155	40.0	10.0	0.31	0.42	5.0	3.52
$[0^0/90^0]_s$ carbon/epoxy	Smith and Wood [1990]	0.125	145.0	9.5	0.31	0.42	5.6	3.345

Table 11.11.1. Mechanical properties of the $[0^0_m/90^0_n]_s$ laminates referred to in the text. Thicknesses of the layers are $d = mt$, $c = nt$, G_T is computed by: $G_T = E_T/(2(1 + \nu_T))$

Here $c = 3d$, $d = t = 0.203\,\mathrm{mm}$. Knowing the orthotropic moduli of the plies (see the first row in Table 11.11.1) one can compute the compliances c^f_{ijkl}, c^m_{ijkl} by Eq. (11.11.1). Having c and d one can compute now the compliances of the uncracked laminate by Eqs. (11.1.14). By inverting the constitutive equations (11.1.16) we find the stiffnesses of the uncracked laminate, involved in the primal constitutive relations (11.1.21). The conventional orthotropic moduli of the uncracked laminate are computed by Eqs. (11.9.20) and (11.9.26). We find

$$E_1 = 20.30\,\text{GPa}\,, \qquad E_2 = 34.75\,\text{GPa}\,, \qquad G_{12} = 3.40\,\text{GPa}\,,$$
$$\nu_{12} = 0.193\,, \qquad \nu_{21} = 0.113\,. \tag{11.11.2}$$

According to the in-plane scaling (h_0, l_0) (see Sec. 11.9) the effective moduli of the cracked laminate are crack density independent. Using formulae (11.9.17) one finds

$$E_1^c = 10.70\,\text{GPa}\,, \qquad E_2^c = 34.53\,\text{GPa}\,, \qquad G_{12}^c = 0.85\,\text{GPa}\,,$$
$$\nu_{12}^c = 0.0943\,, \qquad \nu_{21}^c = 0.0292\,. \tag{11.11.3}$$

According to the experimental data of Highsmith and Reifsnider (1982) the minimum value of E_1^c is achieved for $c_d = 0.75$ cracks/mm and equals $11.0\,\text{GPa}$. In this case the agreement is very good.

To describe the relations $E_\alpha^c(c_d)$, $\nu_{\alpha\beta}^c(c_d)$, $G_{12}^c(c_d)$ one should resort to using the (h_0, l) model. In the case considered the roots of the equation (11.10.15) are complex. They are $(p, \pm q)$, with $p = 1.98025$ and $q = 0.8934$. Thus the function $F_{11}(\varrho)$ defined by (11.10.17) is determined by $F = F_0(\varrho; p, q)$, F_0 being defined by (11.10.20). The decay of stiffnesses is given by (11.10.23) – (11.10.25a) and by (11.10.36) – (11.10.39).

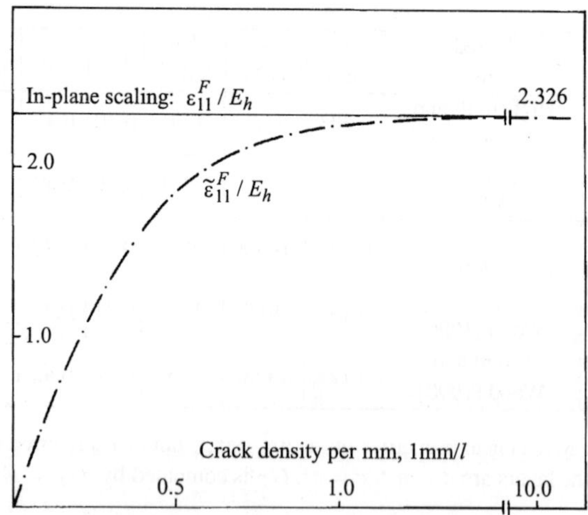

Fig. 11.11.1. $[0^0/90_3^0]_s$ glass/epoxy laminate tested by Highsmith and Reifsnider (1982). Longitudinal crack deformation versus crack density. The in-plane scale prediction $\varepsilon_{11}^F \approx \varepsilon_{11}^h + 0.299\varepsilon_{22}^h$ is an asymptote for the (h_0, l) or refined-scale prediction: $\widetilde{\varepsilon}_{11}^F$

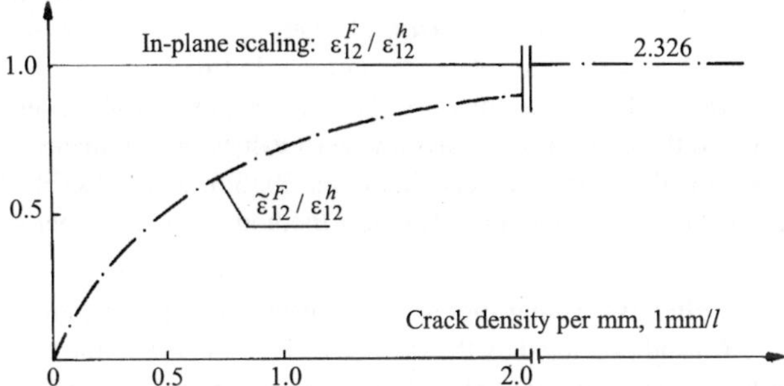

Fig. 11.11.2. The same laminate. Shear crack deformation versus crack density. The in-plane scaling prediction $\varepsilon^F_{12} = \varepsilon^h_{12}$ is an asymptote for the (h_0, l) prediction: $\widetilde{\varepsilon}^F_{12}$

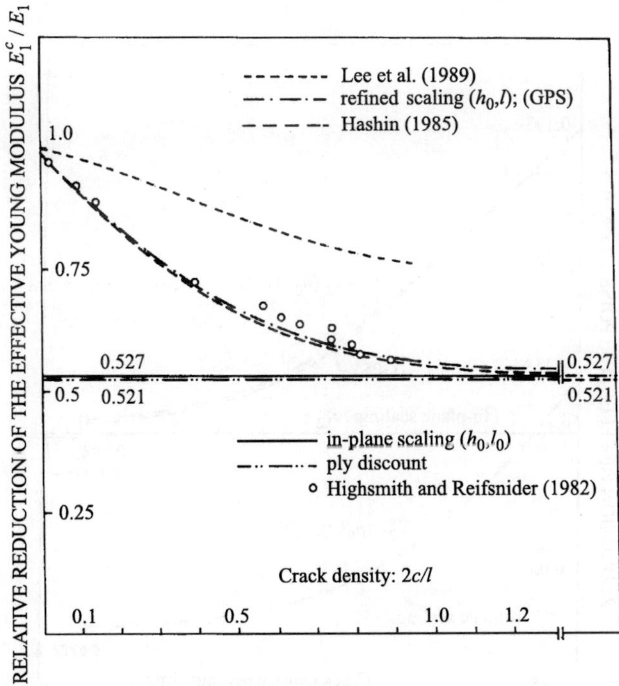

Fig. 11.11.3. The same laminate. Assessing the loss of the effective Young modulus E^c_1 versus crack density. Experimental results of Highsmith and Reifsnider (1982) are denoted by circles.

Prior to finding the curves describing a decay of stiffnesses, let us consider the relations $\widetilde{\varepsilon}^F_{1\alpha}(c_d)$ representing the crack deformation measures. Let us recall that $\varrho = l/2h$, hence $c_d = 2c/l = (c/h)\varrho^{-1}$; if $c_d = 1mm/l$, then $c_d = (1mm/2h)\varrho^{-1}$. Thus relations (11.10.16) and (11.10.35) interrelate $\widetilde{\varepsilon}^F_{1\alpha}$ with c_d. Let us recall that according to the (h_0, l_0) model the crack deformation measures $\varepsilon^F_{1\alpha}$ are independent of c_d, see (11.9.10) and (11.9.24). For $c_d = 0$ the quantities $\widetilde{\varepsilon}^F_{1\alpha}$ assume zero values, which means that the (h_0, l) approach encompasses the case of a very dilute distribution of cracks, cf. Fig. 11.11.1, 11.11.2, contrary to the (h_0, l_0) method. This last method gives the upper asymptotes for $\widetilde{\varepsilon}^F_{1\alpha}$.

The decay of the effective Young modulus $\widetilde{E}^c_1$ observed in experiments by Highsmith and Reifsnider (1982) and predicted by the method of Hashin (1985), the GPS (generalized plane strain) method of McCartney (1992), the method of Lee et al. (1989) and Allen et al. (1987) and by the (h_0, l) method is presented in Fig. 11.11.3.

The Hashin's curve has not been repeated after Fig. 3 in Hashin (1985) but has been independently plotted by the present authors. The experimental data are placed according

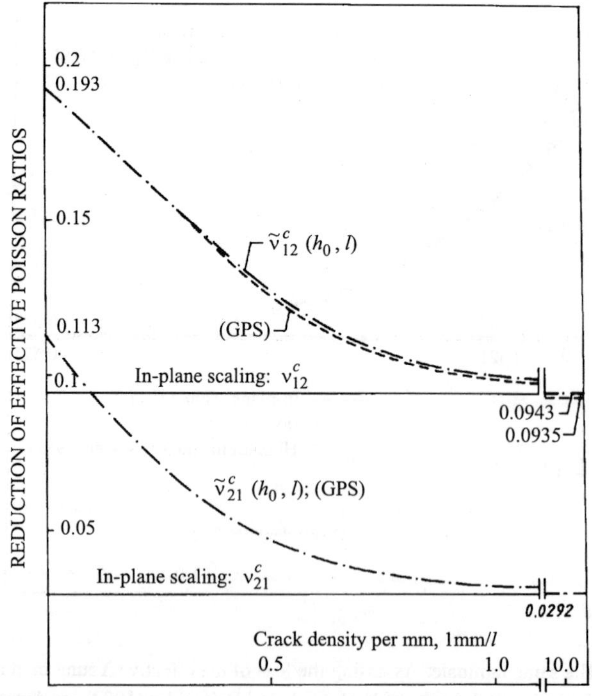

Fig. 11.11.4. The same laminate. Effective Poisson ratios as functions of the crack density

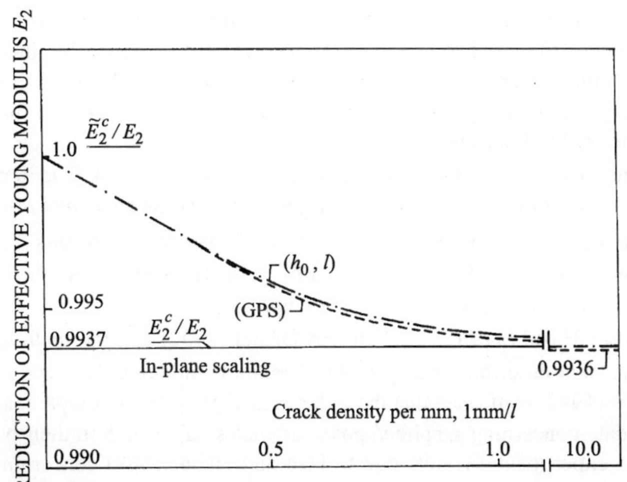

Fig. 11.11.5. The same laminate. Decay of the effective Young modulus E_1^c

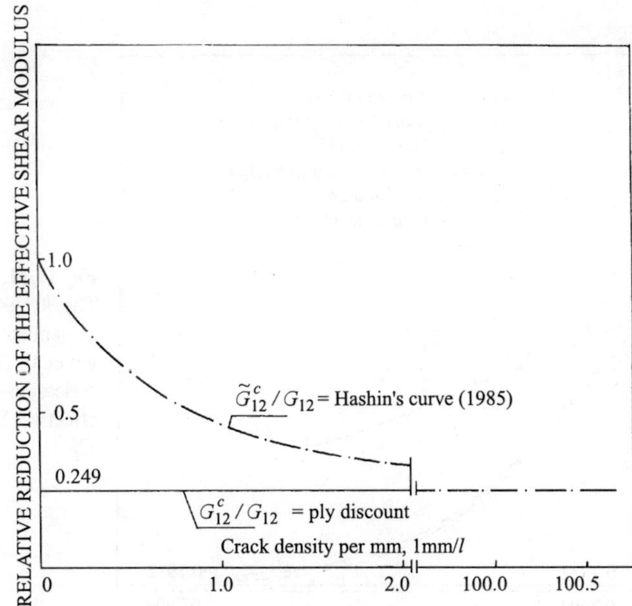

Fig. 11.11.6. The same laminate. Decay of the effective Kirchhoff modulus. Hashin's (1985) and (h_0 , l) predictions coincide. Predictions based on the in-plane scaling coincide with ply-discount results

to Fig. 14 of Highsmith and Reifsnider (1982) and Fig. 1a in Lee et al. (1989). The Hashin's curve lies slightly below the curves predicted by the GPS model and by the (h_0, l) models, the juxtaposition of the last two curves being too close to be noticeable in Fig. 11.11.3. The in-plane scaling method (h_0, l_0) gives a horizontal asymptote for the $\widetilde{E}_1^c / E_1$ curve; the conventional ply-discount assessment (see Garrett and Bailey (1977)) lies a little below and is an asymptote for Hashin's curve.

Decay of the effective Poisson ratios $\widetilde{\nu}_{\alpha\beta}^c$ is shown in Fig. 11.11.4. The in-plane scaling method (h_0, l_0) produces the asymptotes for more realistic refined-scaling results.

A very small decay of $\widetilde{E}_2^c$ is shown in Fig. 11.11.5. The results of the GPS method are slightly smaller. The decay of $\widetilde{E}_2^c$ and $\widetilde{\nu}_{\alpha\beta}^c$ could not be described within the framework of Hashin's (1985) approach, hence the lack of comparisons.

The method of Hashin (1985), of Tsai and Daniel (1992) and the method (h_0, l) lead to the same formula describing decay of the Kirchhoff modulus, $\widetilde{G}_{12}^c$, see Fig. 11.11.6. Tsai and Daniel (1992) confirmed that these theoretical predictions compare favorably with experimental data concerning graphite/epoxy laminates, cf. Fig. 5 in their paper. On the other hand the experimental results due to Han and Hahn (1989) concerning the GFRP $[0_2^\circ, 90_2^\circ]$ laminates lie far away from these theoretical predictions. Experimental data for $\widetilde{G}_{12}^c$ of the laminate considered there were not available to the present authors.

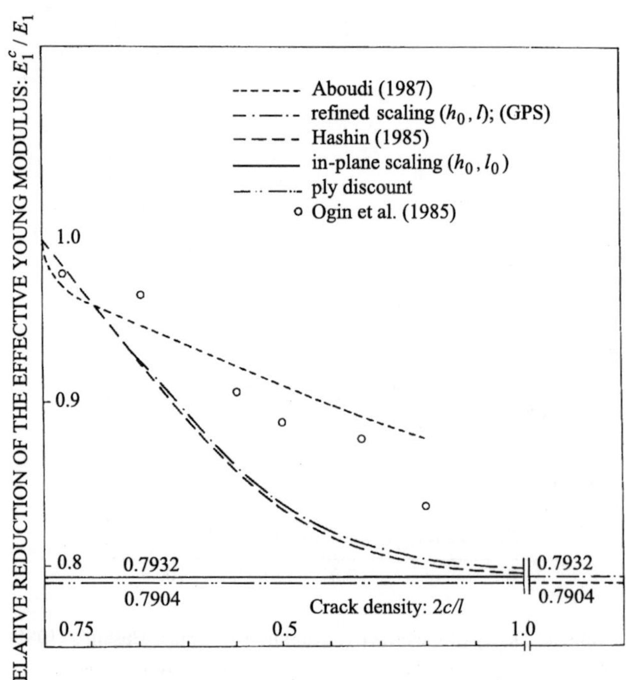

Fig. 11.11.7.
The $[0^0/90^0]_s$ glass/epoxy laminate tested by Ogin et al. (1985) (circles). Assessing the loss of the effective Young modulus E_1^c

11.11.3. $[0°/90°]_s$ glass/epoxy laminate tested by Ogin et al. (1985)

The physical characteristics of the plies of this laminate are given in Table 11.11.1, the second row. The values of t, E_A, E_T, G_A, G_T are repeated after Aboudi (1987); the values ν_A, ν_T are assumed by the present authors. The Hashin curve $E_1^c(c_d = 2c/l)$ as well as the almost coinciding curve provided by the GPS model and by the model (h_0, l) yield lower bounds for the experimental results of Ogin et al. (1985), cf. Fig. 11.11.7. The accuracy, however, is not so satisfactory as for the laminate considered in Sec. 11.11.2. Better results are provided by the displacement-based method of Aboudi (1987).

11.11.4. $[0°_m/90°_n]_s$ graphite/epoxy laminates tested by Groves (1986)

The elastic characteristics of the plies of these laminates are collected in the 3rd row of Table 11.11.1. In the case of the $[0°/90°_2]$ laminate the experimental results for $\widetilde{E}_1^c$ of Groves (1986) lie between the curve of Lee et al. (1989) and the curve of Hashin (1985). The almost coinciding curves provided by the GPS model and the model (h_0, l) lie slightly over the Hashin curve; the differences could not be shown in Fig. 11.11.8.

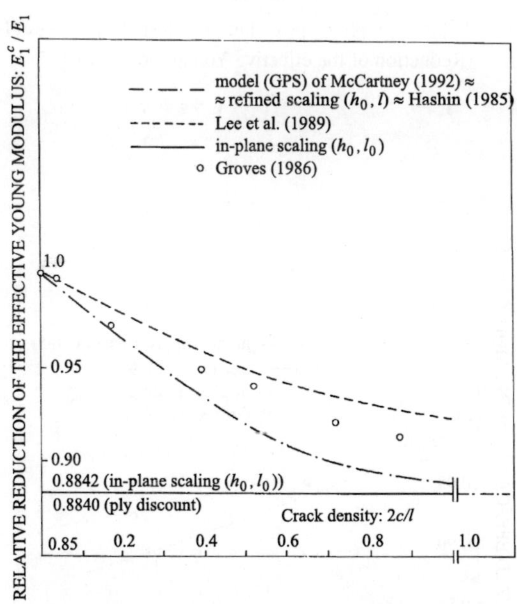

Fig. 11.11.8. The $[0°/90°_2]_s$ graphite/epoxy laminate tested by Groves (1986) (circles). Assessing the loss of the effective Young modulus E_1^c

Similarly, for the $[0°/90°]_s$, $[0°/90°_3]_s$ and $[0°_2/90°_2]_s$ laminates the experimental values of E_1^c lie between theoretical predictions of Lee et al. (1989) and the (h_0, l) results, see Figs. 11.11.9 – 11.11.11.

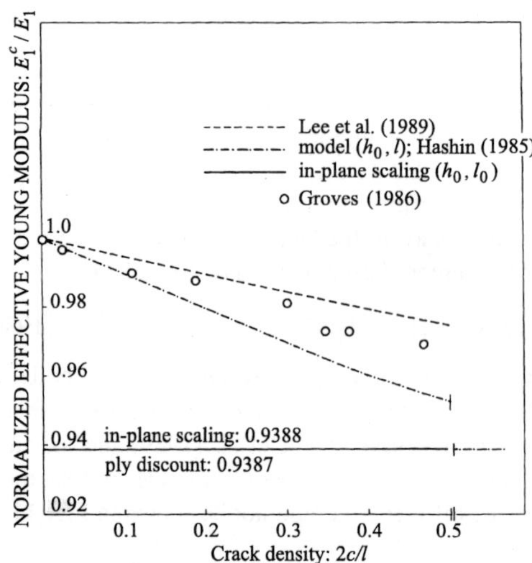

Fig. 11.11.9. The $[0^0/90^0]_s$ graphite/epoxy laminate tested by Groves (1986) (circles).
Reduction of the effective Young modulus E_1^c

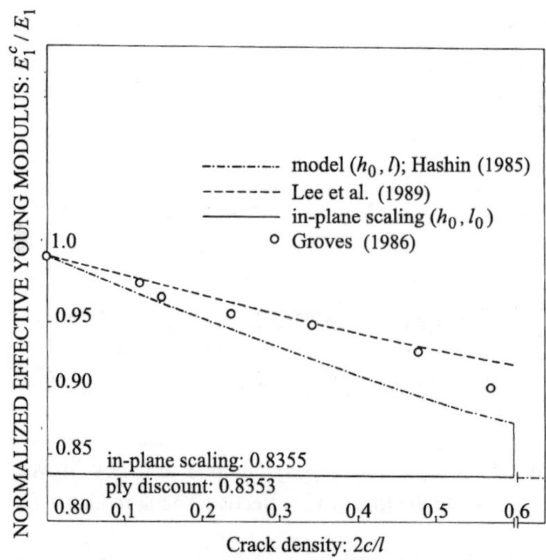

Fig. 11.11.10. The $[0^0/90_3^0]_s$ graphite/epoxy laminate tested by Groves (1986) (circles). Reduction
of the effective Young modulus E_1^c

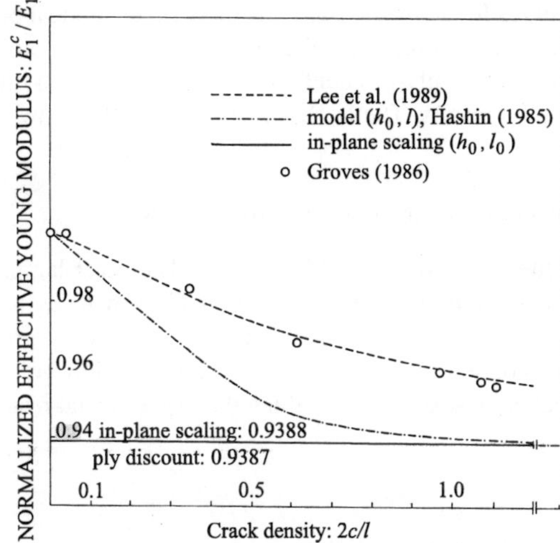

Crack density: $2c/l$

Fig. 11.11.11. The $[0_2^0/90_2^0]_s$ graphite/epoxy laminate tested by Groves (1986) (circles). Reduction of the effective Young modulus E_1^c

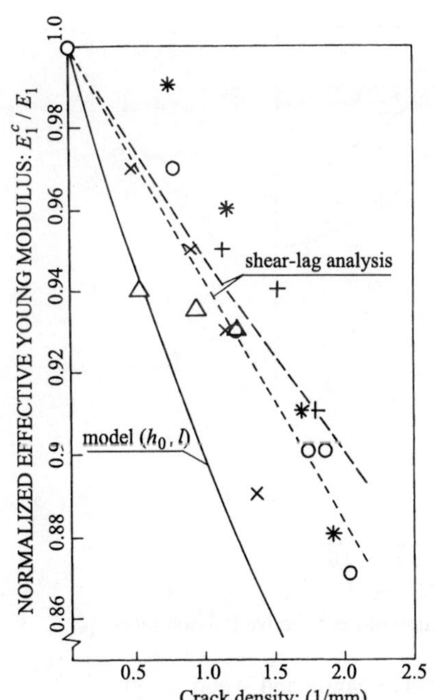

Crack density: (1/mm)

Fig. 11.11.12. The $[0^0/90^0]_s$ glass/epoxy laminate tested by Smith and Wood (1990). Reduction of the effective Young modulus E_1^c. The signs $*$, $\circ$, $\triangle$, $\times$, $+$ refer to the different samples used in experiments. The shear-lag predictions are taken from the same paper

The horizontal asymptotes of the curves found by the (h_0, l) method are theoretical predictions of the (h_0, l_0) method (the in-plane scaling). On the other hand, the asymptotes to the Hashin curves can be interpreted as the "ply-discount" predictions. They lie a little below the in-plane scaling – based (h_0, l_0) results. The Hashin curves lie a little below the (h_0, l) predictions.

11.11.5. $[0°/90°]_s$ glass/epoxy laminates tested by Smith and Wood (1990)

The moduli of the plies are given in the fourth row of Table 11.11.1. The shear-lag predictions of Smith and Wood (1990) concerning the effective modulus E_1^c provide us with better curves than the (h_0, l) model, see Fig. 11.11.12. The model (h_0, l), however, gives better information about decay of the Poisson ratio ν_{12}^c, see Fig. 11.11.13. For greater values of crack density, also this model is incapable of describing the experimental results correctly.

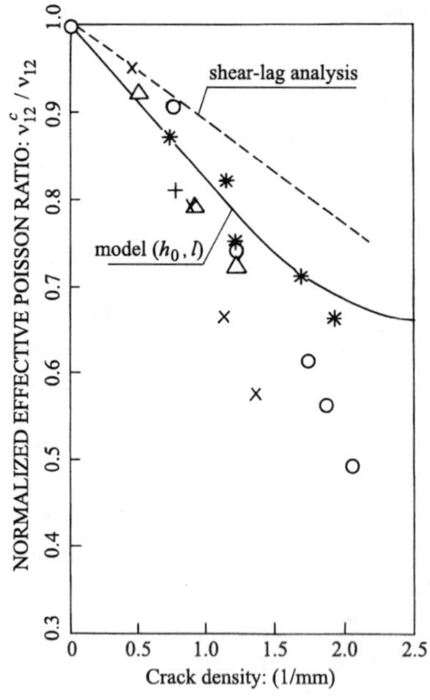

Fig. 11.11.13. The same laminate. Reduction of the effective Poisson ratio ν_{12}^c

11.11.6. $[0°/90°]_s$ carbon/epoxy laminate tested by Smith and Wood (1990)

The plies characteristics are set up in the last row of Table 11.11.1. The shear-lag curve of Smith and Wood (1990) overestimates the experimental values of ν_{12}^c, while the (h_0, l) description looks fairly well, cf. Fig. 11.11.14.

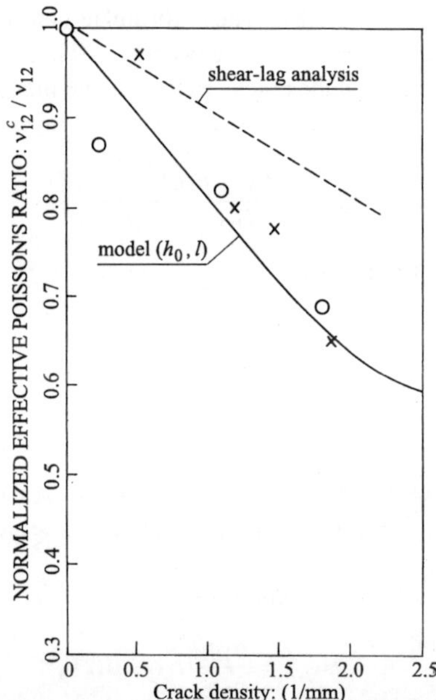

Fig. 11.11.14. The $[0^0/90^0]_s$ carbon/epoxy laminate tested by Smith and Wood (1990) (the signs $\circ$, $\times$ refer to various samples). Reduction of the effective Poisson ratio ν_{12}^c. The shear – lag predictions are taken from the same paper

11.12.   Stress distribution around crack tips

The subject of our consideration is the same as in Secs. 11.9 – 11.11; we consider the laminate with straight cracks in the internal layer, see Fig. 11.9.2. The analysis of Sec. 11.10 was confined to assessing stiffness loss due to cracking. The aim of the present section is to recover the stress distribution between the cracks and around an isolated crack by the (h_0, l) method. Such a recovery is feasible due to the stress assumptions (11.1.6) – (11.1.8) a priori imposed. Moreover, the formulae for stress distribution are closed, since solutions to the local problems of the (h_0, l) model were found in closed forms, see Sec. 11.10.

We start from the formulae for the in-plane stresses $\sigma^{1\alpha}$ that arise in the laminate before cracking starts. By stress assumptions (11.1.6) and by formulae (11.10.4), (11.10.27) we find

$$\sigma_{m.o}^{11} = (A_v^{1111}/2c)E_h , \qquad \sigma_{m.o}^{12} = 2G_A \varepsilon_{12}^h ,$$
$$\sigma_{f.o}^{11} = (A_v^{1111}/2d)(\alpha_{11}\varepsilon_{11}^h + \alpha_{21}\varepsilon_{22}^h - E_h) , \qquad \sigma_{f.o}^{12} = \sigma_{m.o}^{12} , \tag{11.12.1}$$

where E_h is defined by $(11.9.12)_1$; subscript m labels the central layer, while f refers to the faces. The subscript o indicates that the stresses refer to the uncracked state. Having found the solutions to local problems (P_a) and (P_s) of Sec. 11.10 one can determine distribution of stresses by Eqs. (11.1.6) – (11.1.8). These formulae can be cast in closed forms in the intervals $(0, \varrho)$ and $(\varrho, 2\varrho)$, $\varrho = l/2h$. The point $\xi = 0$ lies in the middle between two cracks while the point $\xi = \varrho$ lies on the crack. Here only the final form of same expressions for stresses will be provided. The axial stress in the central layer is given by

$$\sigma_m^{11}/\sigma_{m.o}^{11} = \frac{F_0(\varrho; p, q) - S(\varrho, \xi)}{B_{11} + F_0(\varrho; p, q)} , \qquad (11.12.2)$$

with F_0 defined by Eq. (11.10.20) and

$$S(\varrho, \xi) = \frac{f_1(\varrho, \xi) - f_2(\varrho, \xi)}{pq(p^2 + q^2)(\cosh 2p\varrho - \cos 2q\varrho)} , \qquad (11.12.3)$$

where

$$f_1(\varrho, \xi) = q\varrho(\cosh p\xi \cos q\xi \sinh p\varrho \cos q\varrho + \sinh p\xi \sin q\xi \cosh p\varrho \sin q\varrho) ,$$
$$f_2(\varrho, \xi) = p\varrho(\sinh p\xi \sin q\xi \sinh p\varrho \cos q\varrho - \cosh p\xi \cos q\xi \cosh p\varrho \sin q\varrho) , \quad (11.12.4)$$

and the quantity B_{11} follows from the expressions

$$(A_{\lambda\mu}, B_{\lambda\mu}) = \frac{a_1 g_1}{a_3 g_2}(\alpha_{\lambda\mu}, \beta_{\lambda\mu}) , \qquad (11.12.5)$$

where a_i, g_i have been defined by (11.10.13) and $(11.10.17)_2$, respectively. The axial stress in the external layers is calculated from

$$\sigma_f^{11} - \sigma_{f.o}^{11} = \frac{(B_{11} - A_{11}) + S(\varrho, \xi)}{B_{11}} + F_0(\varrho; p, q) \cdot \frac{c}{d}\sigma_{m.o}^{11} . \qquad (11.12.6)$$

The transverse shear stress on the interface $x_3 = -c$ is given by

$$\sigma^{13}(x_3 = -c)/\sigma_{m.o}^{11} = \frac{c}{h}\frac{S_{13}(\varrho, \xi)}{B_{11} + F_0(\varrho; p, q)} , \qquad (11.12.7)$$

where

$$S_{13}(\varrho, \xi) = \frac{\varrho(\cosh p\xi \sin q\xi \sinh p\varrho \cos q\varrho - \sinh p\xi \cos q\xi \cosh p\varrho \sin q\varrho)}{pq(\cosh 2p\varrho - \cos 2q\varrho)} . \qquad (11.12.8)$$

The transverse normal stress on the middle plane $x_3 = 0$ has the form

$$\sigma^{33}(x_3 = 0)/\sigma_{m.o}^{11} = \frac{c}{2h}\frac{S_{33}(\varrho, \xi)}{B_{11} + F_0(\varrho; p, q)} , \qquad (11.12.9)$$

where

$$S_{33}(\varrho, \xi) = -\frac{f_1(\varrho, \xi) + f_2(\varrho, \xi)}{pq(\cosh 2p\varrho - \cos 2q\varrho)} . \qquad (11.12.10)$$

The shearing in-plane stress in the central layer is expressed by

$$\sigma_m^{12}/\sigma_{m.o}^{12} = \frac{x(\cosh(\widehat{\lambda}\varrho) - \cosh(\widehat{\lambda}\xi))}{\frac{c}{d}\cdot\sinh(\widehat{\lambda}\varrho) + \widehat{\lambda}\varrho\cosh(\widehat{\lambda}\varrho)} \, . \tag{11.12.11}$$

Other formulae can be found similarly. Distribution of stresses σ_m^{11}, σ_m^{22}, σ^{33} ($x_3 = 0$), σ_m^{12}, σ_f^{12} are symmetric, and stresses σ^{13} ($x_3 = -c$), σ^{23} ($x_3 = c$) – antisymmetric with respect to the central line between two cracks, see Figs. 11.12.1, 11.12.2 concerning the laminate of Sec. 11.11.2 ($\varrho = 0.75$). The graph of σ_m^{11} attains its maximum at $\xi = 0$. As the crack density increases the stress $\sigma_m^{11}(\xi = 0)$ decays to zero, while $\max|\sigma^{13}(x_3 = \pm c)|$ and $\max|\sigma^{33}|$ increase.

If the crack density decreases the crack interaction effects cease to be essential. Thus it is interesting to consider a limiting case when this interaction is not present, hence the case of $\varrho = \infty$. Let us introduce a new ordinate $y_1^0 = 0.5l - y_1$, $\xi_0 = y_1^0/h$.

The variable ξ_0 equals zero at the crack tip. We have, in particular

$$\sigma_m^{11}(\rho = \infty)/\sigma_{m.o}^{11} = 1 - \exp(-p\xi_0)(\cos q\xi_0 + \frac{p}{q}\sin q\xi_0) \, ,$$

$$\sigma^{13}(x_3 = c, \varrho = \infty)/\sigma_{m.o}^{11} = -\frac{c}{h}\frac{p^2 + q^2}{q}\exp(-p\xi_0)\sin q\xi_0 \, , \tag{11.12.12}$$

$$\sigma_m^{12}(\varrho = \infty)/\sigma_{m.o}^{12} = 1 - \exp(-\widehat{\lambda}\xi_0) \, .$$

The stress σ_m^{11} tends to $\sigma_{m.o}^{11}$ if $\xi_0 \to \infty$, while the stresses $\sigma^{13}(x_3 = c)$, $\sigma^{33}(x_3 = 0)$ tend to zero, cf. Fig. 11.12.3.

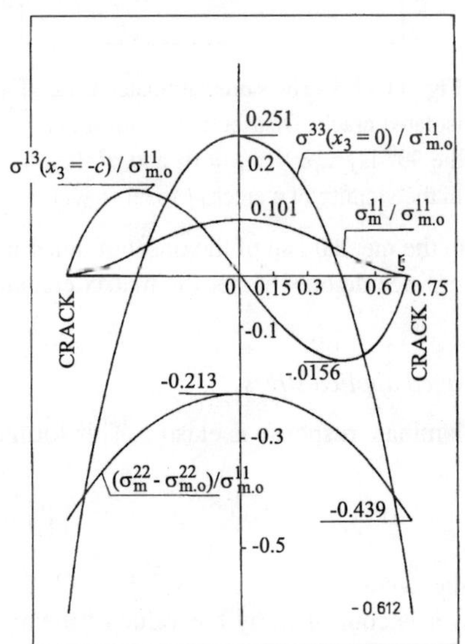

Fig. 11.12.1. The $[0/90_3^0]_s$ glass/epoxy laminate tested by Highsmith and Reifsnider (1982). Crack spacing $l = 3c$; $\rho = 0.75$. Distribution of stresses σ^{11}, σ^{22} in the 90^0 layer and stress $\sigma^{33}(x_3 = 0)$ between two interacting cracks. Model (h_0, l)

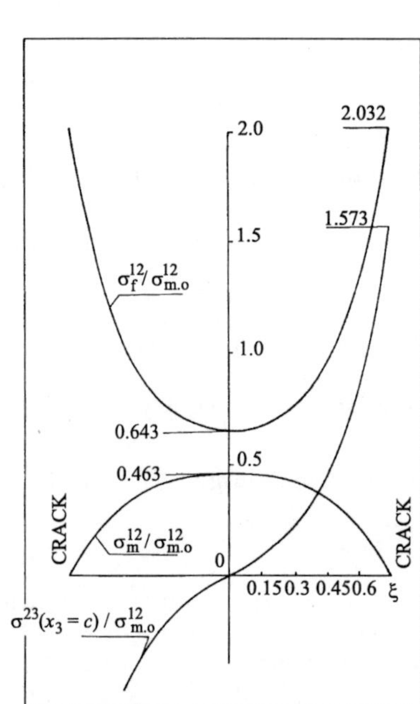

Fig. 11.12.2. The same laminate. Distribution of stresses σ^{12} in the 0^0 and 90^0 layers and of $\sigma^{23}(x_3 = c)$ between two interacting cracks. Model (h_0, l)

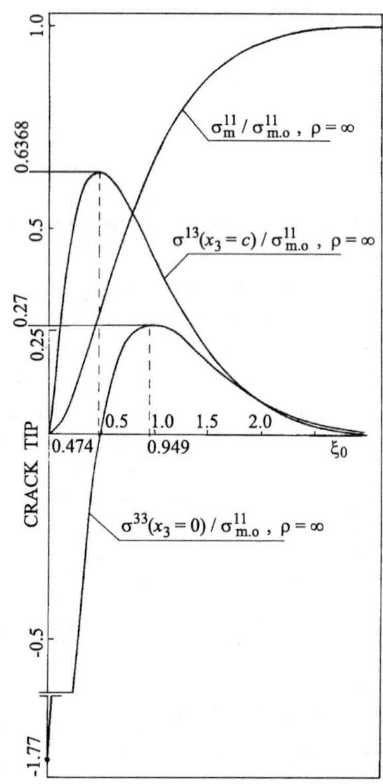

Fig. 11.12.3. The same laminate. Case of an isolated crack. Distribution of stresses σ^{11} in the 90^0-layer, $\sigma^{33}(x_3 = 0)$ and $\sigma^{13}(x_3 = c)$ in the vicinity of a crack. Model (h_0, l)

The variaton of the stress $\sigma^{13}(x_3 = c)$ clears up the mechanism of the onset of delamination. On the other hand, the values of the stress σ^{11}_m decide on whether the matrix cracking will precede delamination or not.

11.13. Crack spacing as a function of the averaged applied stress

Relation (11.12.2) for σ^{11}_m is valid as far as the laminate response is elastic. This formula may be useful in deriving a relation

$$l = l(\sigma_h) , \qquad \sigma_h = N^{11}_h/2h \tag{11.13.1}$$

expressing the crack spacing l as a function of the averaged stress σ_h (at $N^{22}_h = 0$) if one assumes that the cracking process in the 90°-layer is controlled by the value of the axial stress $\sigma^{11}_m(\xi = 0)$ in the middle point between two cracks. At this stage of the analysis a delamination due to σ^{13}, σ^{33} is neglected. Thus we put

$$\sigma_m^{11}(\xi = 0) = \sigma_{\text{crit}} \, . \tag{11.13.2}$$

Taking into account Eq. (11.12.2) we find a relation of the type (11.13.1). For the $[0°/90°_3]_s$ glass/epoxy laminate discussed in Sec. 11.11.2 this relation is presented in Fig. 11.13.1. The quantity σ_{crit} is taken as 36.3 MPa (see Fig. 9 in Hashin (1985)).

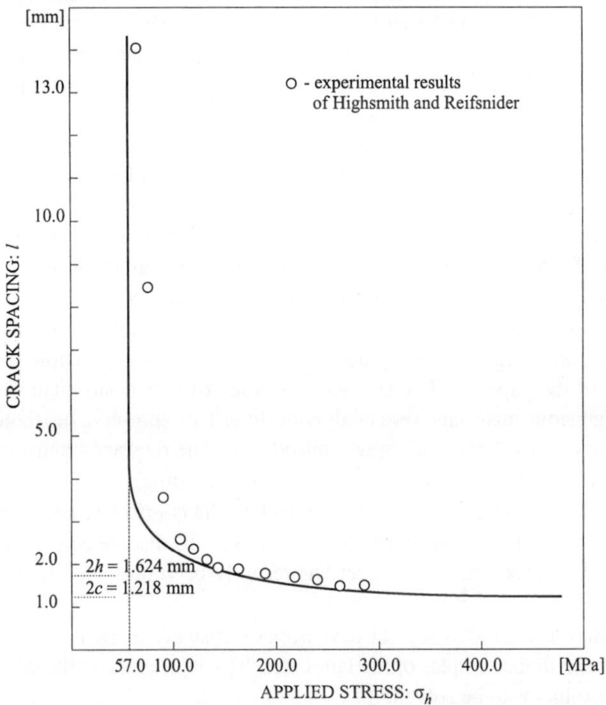

Fig. 11.13.1. The same laminate. Crack spacing l versus the applied stress $\sigma_h = N_h^{11}/2h$ in case of $N_h^{22} = 0$. Experimental results are reported after Highsmith and Reifsnider (1982, Fig. 6) and Hashin (1985, Fig. 9) versus theoretical predictions of the (h_0, l) model

The forces $N_h = 2h(\sigma_h)_{\text{crit}}$ applied to the samples of the laminate trigger the first cracks. It turns out that $(\sigma_h)_{\text{crit}} = 57.0$ MPa, see Fig. 11.13.1. With an increase of the stress σ_h over this critical value a great number of new cracks appear. The crack spacing decreases rapidly and for some level of the applied stress the crack spacing l attains an almost constant value, cf. Fig. 11.13.1. Such a state of saturation has been observed by Garrett and Bailey (Fig. 4 in their paper of 1977) and by Highsmith and Reifsnider (1982), whose experimental data are depicted in Fig. 11.13.1. Thus a very simple damage criterion (11.13.1) turns out to be sufficient to predict relation (11.13.1)$_1$ fairly well, see also Hashin (1996).

12.   Comments and bibliographical notes

The cracks of surfaces perpendicular to the plate middle plane (or through cracks) can run through its whole thickness or can have a depth smaller than the plate thickness. A three-dimensional stress analysis around such cracks can be found in Iyengar et al. (1988), Folias and Reuter (1990), see also Sih (1971) and Movchan and Movchan (1995). The formulae for effective stiffnesses of plates with periodically distributed cracks, otherwise arbitrary, were found by Chacha and Sanchez-Palencia (1992), where the local problems are three-dimensional, see Sec. 10.8. Similar problem was studied by Kolpakov (1991).

A two-dimensional analysis of stress resultants around the cracks in thin and moderately thick plates is the subject of the monograph by Berezhnitskii et al. (1979) and of the papers by Joseph and Erdogan (1987), cf. also Haryadi et al. (1998). The effective properties of thin plates with periodically distributed cracks were found in Lewiński and Telega (1985, 1988b); these results are outlined in Sec. 8. The effective stiffnesses of moderately thick plates with periodically distributed cracks were the subject of the papers by Lewiński and Telega (1988c, 1988d) and Telega and Lewiński (1988), where the Reissner-Hencky plate model was used. These results are briefly put forward in Sec. 9.

The problem of flexural cracks was examined in more detail in Lewiński and Telega (1989, 1991a) and Telega (1993) by using an in-plane scaling. Section 10.3 generalizes consideration of the paper by Telega (1993). In the last paper only plates made of transversely homogeneous materials were dealt with. In order to preserve the three-dimensional shape of periodicity a refined scaling was introduced. The relevant results are summarized in Sec. 10.6 and are published in this section for the first time.

The effect of crack closure is usually neglected. This effect is taken into account in Secs. 8-10 by the method of Sanchez-Palencia (1980), cf. also Telega (1990). A different methodology of incorporating this effect has recently been presented by Prat and Bažant (1997).

The problems considered in Sec. 11 have been extensively studied since late '70s. When stretched or sheared, the samples of the laminates $[0^\circ_m/90^\circ_n]_s$ crack in the 90° – layer. There are two main problems to be solved:

(i) assess the loss of effective stiffnesses as functions of the crack density,

(ii) assess the crack density as a function of the applied boundary forces.

The simplest method for solving problem (i) is the shear-lag approach. At least two versions of this method are worth recalling, one is due to Garrett and Bailey (1977) and the second-due to Highsmith and Reifsnider (1982), cf. also Groves et al. (1987). Both methods need assuming some material constants that interrelate the shear stresses at the interfaces with the values of jumps of displacements. Upon fixing these modelling constants one can arrive at fairly good predictions of experimental data, see Chou (1992).

To solve the problem (ii) one should put forward a failure criterion. The simplest failure criterion used by Garrett and Bailey (1977) says that a new crack in the 90° layer occurs if the maximal tensile stress in this layer attains a certain critical value. Just this criterion is used in Sec. 11.3. However, the critical value of the tensile stress turns out to be dependent on the whole transverse geometry of the laminate, which means that it is not a material constant. Parvizi et al. (1978) were the first to replace this failure criterion by a new one based on the notion of the energy release rate. This approach has found its continuation in the papers of Nairn (1989) and Nairn and Hu (1994).

The shear-lag approach of Garrett and Bailey (1977) was incapable of predicting decay of the effective Poisson ratios. Smith and Wood (1990) have shown how to improve this approach to describe the Poisson ratios reduction. Despite the apparent successes of the shear-lag model one should mention that this approach is characterized by the following drawbacks:

– it is not able to distinguish between laminates $[(s)/90_n^\circ]_s$ and $[90_n^\circ/(s)]_s$, where (s) represents any non-90° ply orthotropic sublaminate;
– the transverse normal stresses are neglected;
– an undetermined constant is involved.

These drawbacks are removed in the Hashin (1985, 1987, 1995) variational approach. Hashin used Reissner's idea (as in Reissner (1945)) of constructing a statically admissible stress field and using the Castigliano's principle. The construction concerns a three-layer plate loaded in plane. The crucial assumption is that the in-plane stresses are piecewise constant across the thickness. The distribution of other stress components are determined by the requirement of static admissibility. This approach has made it possible to find the reduction of the Young (E_1) and Kirchhoff (G) moduli of the cracked laminate. In an indirect way one can also assess the decay of the Poisson ratio ν_{12}. Hashin's (1985) approach was generalized by Nairn (1989) to comprise the thermal strains.

The modelling presented in Sec. 11 refers to the Hashin (1985, 1987) stress assumptions. However, the stress assumptions (11.1.6) – (11.1.8) do not coincide with those of Hashin, where the analogues of the $N^{\alpha\beta}$ forces are treated as boundary forces. Moreover, the whole averaging procedure is completely different.

The effective properties of the cracked laminate are found by the homogenization method. Thus the analysis of Sec. 11 is performed at two levels: at macro level, where the laminate behavior turns out to be nonlinear and hyperelastic and at the microlevel, where the analysis concerns the periodicity cell. If the cracks are treated as unilateral, the basic cell problem involves a variational inequality, which makes the local problem nonlinear. In most papers on laminates with cracks the unilateral effect is neglected. Here this effect is taken into account by the method of Sanchez-Palencia (1980).

The homogenization method used draws upon the methodology developed by Sanchez-Palencia (1980), Leguillon and Sanchez-Palencia (1982), Attouch and Murat (1985) and Chacha and Sanchez-Palencia (1992). However, a special scaling is adopted which makes it possible to interrelate the effective stiffnesses with the crack density. A similar scaling was successfully used in the analysis of periodic plates, see Sec. 5.

In contrast to the Hashin (1985) approach, where the Castigliano principle is used, the modelling of Sec. 11 is based on the Reissner (1950) stress-displacement variational principle. In this way the generalized displacements are introduced directly, i.e., without any need of introducing Lagrangian multipliers. The kinematic assumptions (11.1.9), (11.1.10) are helpful, but do not determine the accuracy of the model. The constitutive relations (11.1.16) are found in an inverted form, which is a characteristic feature of the Reissner modelling.

The results of Sec. 11 were announced in Lewiński and Telega (1991b, 1992, 1993), Telega and Lewiński (1993) and published *in extenso* in the papers by Lewiński and Telega (1996a, 1996b, 1996c, 1996d, 1997b, 1998) and Telega and Lewiński (1994).

In the relevant literature one can also find the displacement-based refinements of the shear-lag model. Han et al. (1988) proposed a parabolic distribution of displacements across the laminate thickness. Han and Hahn (1989) extended the analysis to comprise shear loading and Poisson effects. More complicated equations were found by Tsai et al. (1990) and Tsai and Daniel (1992). This model, called the *interlaminar shear analysis method*, is capable of predicting a decay of the Kirchhoff modulus. The result coincides with that of Hashin (1985) and with the formula (11.10.38) of the (h_0, l) method of Sec. 11.10. The method of Tsai applies also to the case of cracks in the external layers, but then the equations are too complicated to find closed formulae.

The aim of the McCartney (1992) contribution was to generalize Hashin's (1985) results to the case of finite width specimens. McCartney recovered displacement fields compatible with Hashin's stress assumptions.

Accuracy of the Hashin (1985) approach was examined by Lee and Hong (1993). These authors expressed the displacement fields in terms of power series with respect to the x_3 variable. The method leads to a second order matrix differential equation whose solution can only be found numerically.

Similar in spirit is the approach of Gamby and Rebiere (1993). These authors solved the basic cell problem by using the Fourier series method.

The approaches outlined above are restricted to the cross-ply laminates and their possible generalizations to angle-ply laminates are unknown. For such more complicated cases one can apply the finite strip method of Li et al. (1994). Alternatively, one can resort to the methods of continuum damage mechanics. Talreja (1985, 1986) indicated that the constitutive relations of this approach should involve new vector unknowns representing possible damage modes. The scalar damage fields are insufficient here. The Talreja model was later improved by Allen et al. (1987), Allen (1994) and by Li et al. (1997).

The hybrid methods were employed in the papers of Gudmundson and Östlund (1992a, 1992b) and Gudmundson and Zang (1993), where both analytical elasticity solutions as well as finite element solutions are utilized for predicting effective characteristics of laminates with transverse cracks.

If the laminates are used in the environment where high temperatures prevail, their viscoelastic properties cannot be neglected. Zocher et al. (1997) have shown recently how to generalize the Lee et al. (1989) model to the viscoelastic case.

The transverse cracking in $90°$ plies trigger the delamination at the $0°/90°$ interfaces. The passage between these two failure modes are analytically described in the paper by Yang and Boehler (1992). The fiber breaks are discussed in the insightful papers by Bayerlein et al. (1996) and Beyerlein and Phoenix (1997). Unresolved seems to be the problem of constructing two-dimensional plate models with randomly distributed microcracks, cf. Telega and Gambin (1996) and Telega and Lewiński (1993). More precisely, of interest would be stochastic homogenization combined with asymptotic methods of constructing two-dimensional plate models.

Chapter IV

ELASTIC-PERFECTLY PLASTIC PLATES

Introduction

Until now we have dealt with plates and laminates made of elastic materials. The present chapter is concerned with geometrically linear plates made of an elastic-perfectly plastic Hencky-Nadai-Iliushin material, or Hencky material for short. Such plasticity, also called deformational theory of plasticity can be viewed as elasticity with nonlinear constraints imposed by the yield condition. Suppose that a solid made of a Hencky material is subject to a monotone loading characterized by a load multiplier λ. For certain $\overline{\lambda}$ the so called limit state is attained and for $\lambda < \overline{\lambda}$ the *safe load hypothesis* is valid. Consequently, it is indispensable to determine $\overline{\lambda}$ and associated stress and velocity fields. In other words one has to study also the *limit analysis* problems.

13. Mathematical complements, homogenization of functionals with linear growth

In this section we shall introduce mathematical tools of vital importance for the study of plate models made of a Hencky material and exhibiting a periodic structure. In contrast to the elastic case, the functional involved in the kinematical formulation has only linear growth. It means that one has to work with nonreflexive Banach spaces, which are briefly introduced in Sec. 13.1. Physically this necessity is caused by perfectly-plastic behavior. Next, strain fields (or strain rate fields in the case of limit analysis) are no longer integrable functions but measures, which are due to inherent discontinuities of displacement (velocity) fields. For instance, plastic hinges at a clamped part of the boundary of a Kirchhoff plate provide an example of discontinuity of the flexion angle.

To study asymptotic and homogenization problems for functionals depending on measures, in Sec. 13.2 we shall introduce and concisely discuss convex functionals of a measure. Moreover, the convergence in Kuratowski's sense of a particular sequence of sets will also be considered. The last case will prove useful in the study of Γ-convergence for plastic plates loaded by bending moments along a part of the boundary, cf. Sec. 14.1.

Section 13.3 is concerned with the formulation of homogenization theorems for functionals with linear growth depending on the second gradient (hessian). Such functionals depending on the first gradient of a scalar function or on the symmetric gradient (strain) of a vector field are also briefly examined.

The last section is concerned with the Γ-convergence of the same sequence of functionals with linear growth provided that Dirichlet type boundary conditions are taken into account. In contrast to the elastic homogenization, the homogenized functional contains now the term defined on the boundary and provides an example of relaxation of boundary conditions.

The spaces $BV(\Omega)$, $BD(\Omega)$ and $HB(\Omega)$ are convenient in mathematical studies of functionals with linear growth. For instance, the space $W^{2,1}(\Omega)$ is too small to ensure existence of solutions for perfectly plastic plates. A common feature of functionals with linear growth is that from a sequence bounded in L^1 one can extract a subsequence weakly-$*$ convergent in $\mathbb{M}^1$. Hence, the need for "large" spaces such as $BV(\Omega)$, $BD(\Omega)$ or $HB(\Omega)$. We observe that Chacon's Biting Lemma offers another possibility of characterization of bounded sequences in L^1, see Ball and Zhang (1990).

13.1. Functional setting: spaces $W^{1,1}$, $W^{2,1}$, LD, BV, BD, and HB

In this section Ω is a sufficiently regular domain in $\mathbb{R}^n$. For three-dimensional problems $n = 3$; if Ω stands for the mid-plane or a reference plane of a plate, then $n = 2$. The regularity of the boundary has been discussed in Sec. 1.1.2.

The spaces $W^{1,1}(\Omega)$ and $W^{2,1}(\Omega)$

General properties of these spaces have been discussed in Sec. 1.1.2. Here we are going to specify mainly the trace spaces.

The space $W^{1,1}(\Omega)$ is defined by

$$W^{1,1}(\Omega) = \{u \in L^1(\Omega) | u_{,i} \in L^1(\Omega) , \ i = 1, \ldots, n\} , \qquad (13.1.1)$$

where $u_{,i} = \dfrac{\partial u}{\partial x_i}$. This space is equipped with the natural norm following from the formula (1.1.6) for $m = p = 1$.

The trace mapping $tr = \gamma_0$ is a surjective mapping from $W^{1,1}(\Omega)$ onto $L^1(\Gamma)$, where $\Gamma = \partial\Omega$. Moreover, $\gamma_0 u = u$ for all u in $C^1(\overline{\Omega})$.

The next Sobolev space is defined by

$$W^{2,1}(\Omega) = \{u \in W^{1,1}(\Omega) | u_{,ij} \in L^1(\Omega)\} , \qquad (13.1.2)$$

where $u_{,ij} = \dfrac{\partial^2 u}{\partial x_i \partial x_j}$. For $n = 2$ it is convenient to define this space as follows

$$W^{2,1}(\Omega) = \{w \in W^{1,1}(\Omega) | \kappa_{\alpha\beta}(w) \in L^1(\Omega)\} , \qquad (13.1.3)$$

where $\kappa_{\alpha\beta}(w) = -w_{,\alpha\beta}$, $\alpha, \beta = 1, 2$.

There exist linear continuous mappings γ_0, γ_1 of $W^{2,1}(\Omega)$ into $L^1(\Gamma)$ such that $\gamma_0 u = u$ and $\gamma_1 u = \dfrac{\partial u}{\partial \boldsymbol{n}}$ for all u in $C^2(\overline{\Omega})$, where $\boldsymbol{n}$ denotes the unit normal on Γ, exterior to Ω. The trace space $\gamma_0(W^{2,1}(\Omega))$ has yet been identified. It is only known that this space is continuously imbedded into $W^{1,1}(\Gamma)$. Demengel (1984) proved that the trace mapping γ_1 :

$W^{2,1}(\Omega) \to L^1(\Gamma)$ is surjective provided that Ω is uniformly C^2-regular. More precisely, the surjectivity means here that for g in $L^1(\Gamma)$ there exists u in $W^{2,1}(\Omega)$, such that $\gamma_0 u = 0$, $\gamma_1 u = g$.

The space $LD(\Omega)$

This space is defined by

$$LD(\Omega) = \{u \in L^1(\Omega)^n | e_{ij}(u) \in L^1(\Omega) \, , \, i,j = 1,\ldots,n\} \, . \tag{13.1.4}$$

We recall that $e_{ij}(u) = u_{(i,j)} = \left(\dfrac{\partial u_i}{\partial x_j} + \dfrac{\partial u_j}{\partial x_i}\right)/2$. This space is a Banach space for the natural norm

$$||u||_{LD(\Omega)} = ||u||_{L^1(\Omega)^n} + \sum_{i,j=1}^{n} ||e_{ij}(u)||_{L^1(\Omega)} \, . \tag{13.1.5}$$

The space $C^\infty(\overline{\Omega})^n$ is dense in $LD(\Omega)$. Since Korn's inequality does not hold in L^1, therefore $W^{1,1}(\Omega)^n \subset LD(\Omega)$.

There exists a surjective continuous linear operator γ_0 from $LD(\Omega)$ onto $L^1(\Gamma)^n$ such that

$$\gamma_0 u = u_{|\Gamma} \, , \quad \text{for all} \quad u \in LD(\Omega) \cap C(\Omega)^n \, .$$

The imbedding of $LD(\Omega)$ into $L^p(\Omega)$, $1 \leq p \leq n/(n-1)$, is continuous; it is compact for all p such that $1 \leq p < n/(n-1)$.

The space $BV(\Omega)$

A function $u \in L^1(\Omega)$ is said to be of *bounded variation* if

$$\int_\Omega |\nabla u| = \sup\{\int_\Omega u(x)\operatorname{div}\varphi(x)dx | \varphi \in C_0^\infty(\Omega)^n \, , \, \sum_{i=0}^n \varphi_i^2 \leq 1\} < +\infty \, . \tag{13.1.6}$$

Equipped with the norm

$$||u||_{BV(\Omega)} = \int_\Omega |u(x)|dx + \int_\Omega |\nabla u| \, , \tag{13.1.7}$$

$BV(\Omega)$ is a separable nonreflexive Banach space and $W^{1,1}(\Omega) \subset BV(\Omega)$, the inclusion being strict. The space $W^{1,1}(\Omega)$ is obviously a closed subspace of $BV(\Omega)$. Particularly, if Ω_1 is a Borel subset of Ω with sufficiently regular boundary, its characteristic function χ_{Ω_1} is in $BV(\Omega)\backslash W^{1,1}(\Omega)$ and

$$\int_\Omega |\nabla \chi_{\Omega_1}| = H_{n-1}(\partial\Omega_1 \cap \Omega) \, ,$$

where H_{n-1} is $(n-1)$-dimensional Hausdorff measure, cf. Eq. (1.1.27).

Equivalently, $BV(\Omega)$ can be introduced in terms of measures. Let $\mathbb{M}^1(\Omega)$ denote the space of bounded measures on Ω. This space coincides with the space of distributions μ on

Ω such that

$$\|\mu\|_{\mathbf{M}^1(\Omega)} = \sup\{\langle \mu, \varphi \rangle \,|\, \varphi \in C_0^\infty(\Omega), |\varphi(x)| \leq 1\} < +\infty . \tag{13.1.8}$$

The norm is also denoted as follows

$$\|\mu\|_{\mathbf{M}^1(\Omega)} = |\mu|(\Omega) = \int_\Omega |\mu| , \qquad \mu \in \mathbf{M}^1(\Omega) ,$$

$|\mu| = \mu^+ + \mu^-$ being the total variation measure associated to μ and $|\mu|(\Omega)$ is the total variation of μ in Ω. It is known that $\mathbf{M}^1(\Omega)$ is the dual space of $C_0(\Omega)$, the space of continuous functions vanishing on the boundary of Ω and

$$\int_\Omega |\mu| = \sup\{ \int_\Omega \varphi(x) d\mu(x) \,|\, \varphi \in C_0(\Omega) , \|\varphi\|_{C(\Omega)} \leq 1 \} .$$

Thus $BV(\Omega)$ can equivalently be expressed as the space of functions $u \in L^1(\Omega)$ for which $u_{,i} \in \mathbf{M}^1(\Omega)$, for every $1 \leq i \leq 1$, with the partial derivatives $u_{,i}$ understood in the distributional sense. If the vector valued measure space $\mathbf{M}^1(\Omega)^n = \mathbf{M}^1(\Omega, \mathbb{R}^n)$ is endowed with the l^1-norm, in the sense that

$$\int_\Omega |\boldsymbol{\mu}| = \sum_{i=1}^n \int_\Omega |\mu_i| , \qquad \boldsymbol{\mu} \in \mathbf{M}^1(\Omega)^n ,$$

then

$$\|u\|_{BV(\Omega)} = \|u\|_{L^1(\Omega)} + \int_\Omega |\nabla u| = \|u\|_{L^1(\Omega)} + \sum_{i=1}^n \int_\Omega |u_{,i}| , \qquad u \in BV(\Omega) .$$

Since the space $\mathbf{M}^1(\Omega)$ is the dual of a normed space, it is *weakly-$*$ sequentially compact*, which means that for any bounded sequence $\{\mu_j\}_{j \in \mathbb{N}}$ of $\mathbf{M}^1(\Omega)$, there exists a $\mu \in \mathbf{M}^1(\Omega)$ and a subsequence μ_{j_k} converging weakly-$*$ (or vaguely) to μ, i.e.:

$$\forall \varphi \in C_0(\Omega) \quad \langle \mu_{j_k}, \varphi \rangle \to \langle \mu, \varphi \rangle \quad \text{as} \quad j_k \to \infty .$$

The weak-$*$ topology on $\mathbf{M}^1(\Omega)$ is denoted by $\sigma(\mathbf{M}^1(\Omega), C_0(\Omega))$. We recall that here $\langle \cdot, \cdot \rangle$ stands for the duality pairing between $\mathbf{M}^1(\Omega)$ and $C_0(\Omega)$.

We shall also need the space of bounded measures on $\overline{\Omega}$, the closure of Ω. This space is the dual of $C(\overline{\Omega})$, i.e.:

$$\mathbf{M}(\Omega) = [C(\overline{\Omega})]^* . \tag{13.1.9}$$

Let us summarize basic properties of the space $BV(\Omega)$.

Proposition 13.1.1.
(a) If $\{u_m\}_{m \in \mathbb{N}} \subset BV(\Omega)$ and $\lim_{m \to \infty} u_m = n$ in $L^1(\Omega)$ then

$$\int_\Omega |\nabla u| \leq \liminf_{m \to \infty} \int_\Omega |\nabla u| .$$

(b) The imbedding $BV(\Omega) \subset L^p(\Omega)$ is continuous for $1 \le p \le n/(n-1)$, and compact for $1 \le p < n/(n-1)$.
(c) (Poincaré type inequality) There exists a constant $K = K(n) > 0$ such that, cf. Eq. (1.49)

$$\left(\int_\Omega |u - \langle u \rangle|^p dx \right)^{\frac{1}{p}} \le K \int_\Omega |\nabla u|, \quad \text{for all } u \in BV(\Omega),$$

where $p = n/(n-1)$ and $\langle u \rangle = \dfrac{1}{|\Omega|} \int_\Omega u dx$. If $n = 1$, then the $L^{n/(n-1)}$-norm is understood to be the L^∞-norm.
(d) There exists a linear, continuous and surjective trace operator $\gamma_0 : BV(\Omega) \to L^1(\Gamma)$ such that $\gamma_0 u = u_{|\Gamma}$ for each $u \in BV(\Omega) \cap C(\overline{\Omega})$. $\qquad\square$

Corollary 13.1.2. From a bounded sequence $\{u_m\}_{m \in \mathbb{N}} \subset BV(\Omega)$ one can extract a subsequence u_{m_k} weakly convergent to u in the following sense:

$$u_{m_k} \to u \quad \text{in } L^1(\Omega) \quad \text{strongly},$$
$$\nabla u_{m_k} \rightharpoonup \nabla u \quad \text{in } \mathbb{M}^1(\Omega) \quad \text{weak-*}.$$
$\qquad\square$

Corollary 13.1.3. (Chavent and Kunisch, 1997) If $n = 1$ or 2 then $BV(\Omega)$ and $X = L^2(\Omega) \cap BV(\Omega)$ are equivalent Banach spaces. $\qquad\square$

Remark 13.1.4. If $\Omega_1 \subset\subset \Omega$ is an open set of class C^1, by u^+ and u^- we denote the trace of $u_{|\Omega \setminus \Omega_1}$ and $u_{|\Omega_1}$ on $\partial\Omega_1$. Then

$$\int_{\partial\Omega_1} |\nabla u| = \int_{\partial\Omega_1} |u^+ - u^-| dH_{n-1}.$$

We recall that $\Omega_1 \subset\subset \Omega$ means that $\overline{\Omega}_1 \subset \Omega$. $\qquad\square$

The lemma which follows is due to Anzellotti (1986). This result and its easy to conceive variants are useful in the study of homogenization problems for bodies or plates with loaded boundary, see Bouchitté and Suquet (1991) and Sec. 14.1.

Lemma 13.1.5. Let $\Omega \subset \mathbb{R}^n (n \ge 2)$ be a bounded domain with Lipschitz boundary and let n stands for the outward unit normal to $\partial\Omega$. If $u \in L^\infty(\Omega)^n$ and $\operatorname{div} u \in L^n(\Omega)$ then at every Lebesgue point of $u \cdot n$ one has

$$(u \cdot n)(x) = \lim_{\rho \to 0^+} \lim_{r \to 0^+} \frac{1}{|Q_{r,\rho}(x, n(x))|} \int_{Q_{r,\rho}} u(\xi) \cdot n(x) d\xi,$$

where

$$Q_{r,\rho}(x, n(x)) = \{\xi - t n(x) \,|\, |\xi - x| < \rho, \quad 0 < t < r\}. \qquad\square$$

The space $BD(\Omega)$

This space is defined by

$$BD(\Omega) = \{u \in L^1(\Omega)^n \,|\, e_{ij}(u) \in \mathbb{M}^1(\Omega)\,,\ i,j = 1,\ldots,n\}\,. \qquad (13.1.10)$$

Endowed with a norm

$$||u||_{BD(\Omega)} = ||u||_{L^1(\Omega)^n} + \sum_{i,j=1}^{n} ||e_{ij}(u)||_{\mathbb{M}^1(\Omega)}\,, \qquad (13.1.11)$$

the space $BD(\Omega)$ becomes a Banach space. It is known that

$$BV(\Omega)^n \subset BD(\Omega)\,, \quad BV(\Omega)^n \neq BD(\Omega)$$

and $LD(\Omega)$ is a closed subspace of $BD(\Omega)$, $LD(\Omega) \neq BD(\Omega)$.

It is also known that $BD(\Omega)$ is the dual of a normed space. Consequently one can define a weak-$*$ topology on this space and the closed balls of $BD(\Omega)$ are then compact and sequentially compact for this topology. This topology is called the *weak topology* of $BD(\Omega)$. Particularly, $\{u^m\}_{m\in\mathbb{N}} \subset BD(\Omega)$ converges to u in this topology if, for $m \to \infty$

$$
\begin{aligned}
u^m &\to u \quad \text{in} \quad L^1(\Omega)^n\,, \\
e(u) &\rightharpoonup e(u) \quad \text{in} \quad \mathbb{M}^1(\Omega, \mathbb{E}_s^n) \ \text{weak-}*\,.
\end{aligned}
\qquad (13.1.12)
$$

Another natural topology on $BD(\Omega)$ is the *intermediate topology* defined by the distance

$$||u - v||_{L^1(\Omega)^n} + |\int_\Omega |e(u)| - \int_\Omega |e(v)||\,. \qquad (13.1.13)$$

In fact, this topology is intermediate between the weak topology and the topology of the norm on $BD(\Omega)$. The intermediate topology is not uniquely defined.

Let us summarize main properties of the space $BD(\Omega)$.

Proposition 13.1.6.
(i) If Ω is an open set of class C^1, there exists a continuous surjective linear operator γ_0 from $BD(\Omega)$ into $L^1(\Gamma)^n$ such that

$$\gamma_0 u = u_{|\Gamma} \quad \text{for all} \quad u \in BD(\Omega) \cap C(\overline{\Omega})^n\,.$$

The trace operator is a continuous operator from $BD(\Omega)$ into $L^1(\Gamma)^n$, when $BD(\Omega)$ is endowed with the intermediate topology.
(ii) The imbedding $BD(\Omega) \subset L^p(\Omega)^n$ is continuous for $1 \leq p \leq n/(n-1)$ and compact for $1 \leq p < n/(n-1)$.
(iii) $C^\infty(\overline{\Omega})$ is dense in the space $BD(\Omega)$ endowed with the intermediate topology.

(iv) For all $i, j = 1, \ldots, n$ and all $\varphi \in C^1(\overline{\Omega})$ one has the generalized Green formula

$$\frac{1}{2}\int_\Omega (u_i\varphi_{,j} + u_j\varphi_{,i})dx + \int_\Omega \varphi e_{ij}(\boldsymbol{u})dx = \int_\Gamma \varphi \mathsf{T}_{ij}(\gamma_0\boldsymbol{u})d\Gamma ,\qquad (13.1.14)$$

where

$$\mathsf{T}_{ij}(\boldsymbol{p}) = \frac{1}{2}(p_in_j + p_jn_i) .\qquad\qquad \square$$

Remark 13.1.7. (see Temam, 1985, Chap. II, Sec. 7) An appropriate stress space is

$$\mathcal{S}(\Omega) = \{\boldsymbol{\sigma} \in L^\infty(\Omega, \mathbb{E}_n^s) \,|\, \mathrm{div}\boldsymbol{\sigma} \in L^{n+1}(\Omega)^n\} .\qquad (13.1.15)$$

For materials incompressible in the plastic range one can take

$$\mathcal{S}_c(\Omega) = \{\boldsymbol{\sigma} \in L^2(\Omega, \mathbb{E}_n^s) \,|\, \mathrm{div}\boldsymbol{\sigma} \in L^{n+1}(\Omega)^n \,,\, \boldsymbol{\sigma}_D \in L^\infty(\Omega, \mathbb{E}_n^s)\} ,\quad (13.1.16)$$

where $\boldsymbol{\sigma}_D$ stands for the stress deviator, i.e.:

$$\sigma_D^{ij} = \sigma^{ij} - \frac{1}{3}(tr\boldsymbol{\sigma})\delta_{ij} .$$

More precisely, the condition $\mathrm{div}\boldsymbol{\sigma} \in L^{n+1}(\Omega)^n$ can be replaced by the condition $\mathrm{div}\boldsymbol{\sigma} \in L^{n+\alpha}(\Omega)^n$, with $\alpha > 0$ arbitrarily small; one can even allow $\alpha = 0$.

One can easily define $\sigma^{ij}e_{ij}(\boldsymbol{u})$ as a distribution on Ω. Moreover, it can be shown that $\sigma^{ij}e_{ij}(\boldsymbol{u})$ and $\sigma_D^{ij}e_{ij}^D(\boldsymbol{u})$ are *bounded measures* on Ω. Here $e_{ij}^D(\boldsymbol{u})$ are components of the strain deviator. $\qquad\qquad \square$

Remark 13.1.8. For the study of bodies made of plastically incompressible materials it is convenient to introduce a subspace of the space $BD(\Omega)$:

$$U(\Omega) = \{\boldsymbol{u} \in BD(\Omega) \,|\, \mathrm{div}\boldsymbol{u} \in L^2(\Omega)\} ,\qquad (13.1.17)$$

which is a Banach space for the natural norm

$$||\boldsymbol{u}||_{U(\Omega)} = ||\boldsymbol{u}||_{BD(\Omega)} + ||\mathrm{div}\boldsymbol{u}||_{L^2(\Omega)} .\qquad (13.1.18)$$

If $\{\boldsymbol{u}^m\}_{m\in\mathbb{N}} \subset U(\Omega)$ is a bounded sequence then there exists a subsequence $\{\boldsymbol{u}^{m_k}\}$ which converges weakly to $\boldsymbol{u}$ in $U(\Omega)$, i.e.:

$$\boldsymbol{u}^{m_k} \to \boldsymbol{u} \quad \text{in} \quad L^1(\Omega)^n \quad \text{strongly,}$$

$$\mathrm{div}\boldsymbol{u}^{m_k} \rightharpoonup \mathrm{div}\boldsymbol{u} \quad \text{in} \quad L^2(\Omega) \quad \text{weakly,}$$

$$e_{ij}(\boldsymbol{u}^{m_k}) \rightharpoonup e_{ij}(\boldsymbol{u}) \quad \text{in} \quad \mathbb{M}^1(\Omega) \quad \text{weakly-}* ;$$

for $i, j = 1, \ldots, n$.

For more details the reader is referred to Temam (1985, Chap. II, Secs. 3 and 5).

The space $HB(\Omega)$

This space is defined by

$$HB(\Omega) = \{v \in W^{1,1}(\Omega) \,|\, \kappa(v) \in \mathbf{M}^1(\Omega, \mathbb{E}_s^2)\} \,. \tag{13.1.19}$$

Having in mind applications to mathematical studies of plates, here $n = 2$. Only such case is of interest to us. Endowed with a natural norm

$$\|v\|_{HB(\Omega)} = \|v\|_{W^{1,1}(\Omega)} + \|\kappa(v)\|_{\mathbf{M}^1(\Omega, \mathbb{E}_s^2)} \,, \tag{13.1.20}$$

the space $HB(\Omega)$ becomes a Banach space.

A sequence $\{v^m\}_{m \in \mathbf{N}} \subset HB(\Omega)$ converges to v *weakly* if

$$
\begin{aligned}
v^m &\to v \quad \text{in} \quad W^{1,1}(\Omega) \quad \text{strongly,} \\
\kappa(v^m) &\rightharpoonup \kappa(v) \quad \text{in} \quad \mathbf{M}^1(\Omega, \mathbb{E}_s^2) \quad \text{weak-}* \,.
\end{aligned}
\tag{13.1.21}
$$

An intermediate topology can be defined by the distance

$$d(v, w) = \|v - w\|_{W^{1,1}(\Omega)} + \left| \int_\Omega |\kappa(v)| - \int_\Omega |\kappa(w)| \right| \,. \tag{13.1.22}$$

Let us specify some of the basic properties of the space $HB(\Omega)$.

Proposition 13.1.9.

(i) The imbedding $HB(\Omega) \subset W^{1,p}(\Omega)$ is continuous with $1 \le p \le 2$ and compact if $1 \le p < 2$.

(ii) Let Ω be a bounded domain with C^2-regularity property, except a finite number of points. Then the imbedding

$$HB(\Omega) \subset C(\overline{\Omega})$$

is continuous.

(iii) The trace mapping

$$HB(\Omega) \to \gamma_0(HB(\Omega)) \times L^1(\Gamma) = \gamma_0(W^{2,1}(\Omega)) \times L^1(\Gamma) \,,$$

is linear, continuous and surjective. The trace mapping is such that

$$\gamma_0 v = v_{|\Gamma} \,, \qquad \gamma_1 v = \frac{\partial v}{\partial \boldsymbol{n}}_{|\Gamma} \,, \qquad \forall\, v \in C^2(\overline{\Omega}) \,.$$

(iv) Let Ω_1 be a relatively compact subset of Ω ($\Omega_1 \subset\subset \Omega$) whose boundary $\partial\Omega_1$ is of class C^2 and let $\Omega_2 = \Omega \backslash \overline{\Omega}_1$. Let v be a function of $L^1(\Omega)$ whose restriction v_α to Ω_α is in $HB(\Omega_\alpha)$, $\alpha = 1, 2$. Then $v \in HB(\Omega)$ if and only if the traces of v_1 and v_2 on $\partial\Omega_1$ are equal: $\gamma_0 v_1 = \gamma_0 v_2$.

In this case one has

$$\kappa(v) = \chi_{\Omega_1}\kappa(v_1) + \chi_{\Omega_2}\kappa(v_2) + \mathbf{F}\left(\frac{\partial}{\partial\boldsymbol{n}}(v_2 - v_1)\right)\delta_{\partial\Omega_1} \,,$$

where χ_{Ω_α} is the characteristic function of Ω_α and $\dfrac{\partial v_\alpha}{\partial n}$ is the value on $\partial\Omega_1$ of $\gamma_1 v_\alpha$, n is the unit normal on $\partial\Omega_1$ in the direction from Ω_1 to Ω_2, and $\delta_{\partial\Omega_1}$ is the Dirac distribution on $\partial\Omega_1$. $\qquad\qquad\square$

Let us introduce the space of moments

$$S = \{M \in L^\infty(\Omega, \mathbb{E}_2^s)|\ \mathrm{div}\ \mathrm{div}\, M \in \mathbb{M}^1(\Omega)\}\,, \qquad (13.1.23)$$

where $\mathrm{div}\ \mathrm{div}\, M = M^{\alpha\beta}{}_{,\beta\alpha}$ in the sense of distributions.

For v in $W^{2,1}(\Omega)$ and M in S the *generalized Green formula* holds true

$$\langle \mathrm{div}\ \mathrm{div}\, M, v\rangle + \int_\Omega M^{\alpha\beta}\kappa_{\alpha\beta}(v)dx = \langle b_0(M), v\rangle - \langle b_1(M), \frac{\partial v}{\partial n}\rangle\,. \qquad (13.1.24)$$

If $M \in C^2(\Omega, \mathbb{E}_2^s)$, then $b_0(M)$ and $b_1(M)$ coincide with the known expressions for the effective shear force and bending moment on Γ, respectively. Obviously, on the r.h.s. of the last formula $v = \gamma_0 v$, $\dfrac{\partial v}{\partial n} = \gamma_1 v$.

When $M \in S$ and $v \in HB(\Omega)$, we can define a distribution $M^{\alpha\beta}\kappa_{\alpha\beta}(v)$ by putting

$$\langle M^{\alpha\beta}\kappa_{\alpha\beta}(v), \varphi\rangle = \int_\Omega v M^{\alpha\beta}\varphi_{,\alpha\beta}dx + 2\int_\Omega M^{\alpha\beta}v_{,\beta}\varphi_{,\alpha}dx - \int_\Omega v\varphi\, \mathrm{div}\ \mathrm{div}\, M\,, \qquad (13.1.25)$$

for all $\varphi \in C_0^\infty(\Omega)$. Let us specify the properties of the measure just introduced.

Proposition 13.1.10.
(i) For $M \in S$ and $v \in HB(\Omega)$, the distribution $M^{\alpha\beta}\kappa_{\alpha\beta}(v)$ is a bounded measure on Ω, absolutely continuous with respect to $|\kappa(v)|$ and

$$\left|\int_\Omega \varphi M^{\alpha\beta}\kappa_{\alpha\beta}(v)\right| \leq \|M\|_{L^\infty(\Omega, \mathbb{E}_2^s)}\int_\Omega |\varphi|\,|\kappa(v)|\,, \qquad (13.1.26)$$

for all $\varphi \in C_0(\Omega)$.
(ii) If $\{v^m\}_{m\in\mathbb{N}} \subset HB(\Omega)$ is a sequence of functions converging to v under the metric (13.1.22) then $M^{\alpha\beta}\kappa_{\alpha\beta}(v^m)$ converges to $M^{\alpha\beta}\kappa_{\alpha\beta}(v)$ in the following sense

$$\int_\Omega \varphi M^{\alpha\beta}\kappa_{\alpha\beta}(v^m) \to \int_\Omega \varphi M^{\alpha\beta}\kappa_{\alpha\beta}(v)\,, \quad \forall\, \varphi \in C(\overline{\Omega})\,. \qquad\square$$

Let us pass to a generalization of the Green formula (13.1.24). In fact, there exist two continuous linear surjective mappings b_0, b_1 from S onto $[\gamma_0(W^{2,1}(\Omega)]^* \times L^\infty(\Gamma)$, such that for $M \in C^2(\overline{\Omega}, \mathbb{E}_2^s)$, $b_0(M)$ and $b_1(M)$ have the same meaning as in (13.1.24). Thus $b_0(M)$ and $b_1(M)$ are generalized effective shear force and bending moment on Γ, respectively. For $M \in S$, $v \in HB(\Omega)$ and any $\varphi \in H^2(\Omega)$ the following *generalized Green* *formula* holds true

$$\int_\Omega (M^{\alpha\beta}\kappa_{\alpha\beta}(v))\varphi = \int_\Omega v M^{\alpha\beta}\varphi_{,\alpha\beta} - \int_\Omega v\varphi\,\mathrm{div}\,\mathrm{div}\,\boldsymbol{M} \tag{13.1.27}$$

$$+ 2\int_\Omega M^{\alpha\beta}v_{,\beta}\varphi_{,\alpha} + \langle b_0(\boldsymbol{M}), v\varphi\rangle - \langle b_1(\boldsymbol{M}), \frac{\partial v}{\partial \boldsymbol{n}}\varphi + v\frac{\partial\varphi}{\partial \boldsymbol{n}}\rangle .$$

For $\varphi = 1$, we obtain a more familiar form of this formula.

Remark 13.1.11. In Secs. 8 – 10 plates weakened by periodically fissures have been studied. Preserving the notations introduced there, Attouch and Murat (1985) proved that: (a) there exists a continuous imbedding from $H^1(Y\backslash F)$ into $BV(Y)$; n is not necessarily equal to two. (b) There exists a constant K independent of ε such that for any $v \in H^1(\Omega^\varepsilon)$, for any $\varepsilon > 0$, the following inequality holds

$$||v||_{BV(\Omega)} \leq K||v||_{H^1(\Omega^\varepsilon)} .$$

Telega (1990c) showed that the following imbeddings are continuous:

$$LD(Y\backslash F) \subset BD(Y) ,$$
$$LD(\Omega^\varepsilon) \subset BD(\Omega) . \tag{13.1.28}$$

Similar imbeddings obviously hold for $H^1(Y\backslash F)^n$ and $H^1(\Omega^\varepsilon)^n$.

13.2.   *Convex functions and functionals of a measure*

If $v \in BV(\Omega), u \in BD(\Omega)$ and $w \in HB(\Omega)$, then $\nabla v, e(u)$ and $\kappa(w)$ are bounded measures. Suppose now that $j(x,\mu), \mu = \nabla v, e(u)$ or $\kappa(w)$, where j is an appropriate potential. A natural question arises: what is the precise mathematical meaning of a function of a measure? How to interpret the integral $\int_\Omega j(x,\mu)$? Such problems are discussed in the present section.

Let $j : (x,\epsilon) \in \Omega \times \mathbb{E}_s^n \to \mathbb{R}\cup\{+\infty\}$ be a convex integrand, see Sec. 1.2.3. We make further assumptions:

(H_1) there exist a function $\varphi_0 \in L^1(\Omega, \mathbb{E}_n^s)$ and a function $a \in L^1(\Omega)$ such that dx-almost everywhere in x, for each $\epsilon, j(x,\epsilon) \geq \varphi_0^{ij}\epsilon_{ij} - a(x)$ (equivalently, there exists $\varphi_0 \in C_0(\Omega, \mathbb{E}_n^s)$ such that $J_{j*}(\varphi_0) < \infty$).

(H_2) There exists $v_0 \in L^1(\Omega, \mathbb{E}_s^n)$ such that $J_j(v_0) < \infty$ (equivalently, there exist $v_0 \in L^1(\Omega, \mathbb{E}_s^n)$ and $a_1 \in L^1(\Omega)$ such that dx-almost everywhere, for each $\rho^*, j^*(x,\rho^*) \geq \rho^{*ij}v_{0ij}(x) - a_1(x)$).

We recall that

$$J_j(v_0) = \int_\Omega j(x, v_0(x))dx .$$

We set

$$\tilde{J}(\boldsymbol{\mu}) = \begin{cases} \displaystyle\int_\Omega j(x,\boldsymbol{h}(x))dx\,, & \text{if } \boldsymbol{\mu} << dx \text{ and } \boldsymbol{h} \in L^1(\Omega,\mathbb{E}^n_s)\,, \\[4mm] +\infty & \text{otherwise.} \end{cases} \tag{13.2.1}$$

We recall that $\boldsymbol{\mu} \ll dx$ means that $\boldsymbol{\mu}$ is absolutely continuous with respect to the Lebesgue measure $dx = dx_1 dx_2 \ldots dx_n$, cf. Yosida (1978, Preliminaries).

Under the assumptions (H$_1$), (H$_2$) the functional

$$J(\boldsymbol{\mu}) = \sup\{\langle \boldsymbol{\mu},\boldsymbol{\varphi}\rangle - \int_\Omega j^*(x,\boldsymbol{\varphi}(x))dx \,|\, \boldsymbol{\varphi} \in C_0(\Omega,\mathbb{E}^s_n)\}\,, \tag{13.2.2}$$

represents the lower semicontinuous regularization of the functional $\tilde{J}$ in the weak-$*$ topology of $\mathbb{M}^1(\Omega,\mathbb{E}^n_s)$, i.e.:

$$J = \sup\{G \,|\, G \leq \tilde{J}, G - \sigma(\mathbb{M}^1(\Omega,\mathbb{E}^n_s),C_0(\Omega,\mathbb{E}^s_n)) \text{ lower semicontinuous}\}\,.$$

Now the problem is to give an integral representation of J. To this end we introduce the notion of the *principal part (recession function)* of j defined on $\Omega \times \mathbb{E}^n_s$

$$j_\infty(x,\boldsymbol{\epsilon}) = \lim_{t \to \infty} \frac{1}{t} j(x,t\boldsymbol{\epsilon}) = \lim_{\rho \to 0} \rho j(x,\frac{\boldsymbol{\epsilon}}{\rho}) = \sup\{\epsilon^{*ij}\epsilon_{ij}\,|\, \boldsymbol{\epsilon}^* \in \mathit{dom}\, j^*(x,\cdot)\}\,.$$

The mechanical meaning of j_∞ is given in Secs. 13.3, 13.4 and 14.

Under the assumptions (H$_1$), (H$_2$) and

(H$_3$) $\forall\, \eta > 0$, $\exists\, \delta > 0$, $|x' - x''| < \delta \Rightarrow \forall\, \boldsymbol{\epsilon}$, $|j_\infty(x',\boldsymbol{\epsilon}) - j_\infty(x'',\boldsymbol{\epsilon})| \leq \eta(1 + |\boldsymbol{\epsilon}|)$,

the functional J given by (13.2.2) is $\sigma(\mathbb{M}^1(\Omega,\mathbb{E}^n_s),C_0(\Omega,\mathbb{E}^s_n))$ lower semicontinuous and

$$J(\boldsymbol{\mu}) = \int_\Omega j(x,\frac{d\boldsymbol{\mu}_a}{dx})dx + \int_\Omega j_\infty(x,\frac{d\boldsymbol{\mu}_s}{d|\boldsymbol{\mu}_s|})d|\boldsymbol{\mu}_s|\,, \tag{13.2.3}$$

where $\boldsymbol{\mu} \in \mathbb{M}^1(\Omega,\mathbb{E}^n_s)$. Here

$$\boldsymbol{\mu} = \boldsymbol{\mu}_a(x) + \boldsymbol{\mu}_s\,, \tag{13.2.4}$$

is the Lebesgue decomposition of $\boldsymbol{\mu}$ into absolutely continuous (regular) part and singular part with respect to dx. Obviously, $|\boldsymbol{\mu}_s|$ stands for the total variation measure associated to $\boldsymbol{\mu}_s$ and $d\boldsymbol{\mu}_s/d|\boldsymbol{\mu}_s|$ is the density of $\boldsymbol{\mu}_s$ with respect to $|\boldsymbol{\mu}_s|$. The singular term in (13.2.3) is often written as $\int_\Omega j_\infty(x,d\boldsymbol{\mu}_s)$ or even just $\int_\Omega j_\infty(x,\boldsymbol{\mu}_s)$.

Remark 13.2.1. Having in mind applications to plasticity, let $\mathcal{C}$ be a closed convex-valued and lower semicontinuous function such that for each $x, 0 \in \mathcal{C}(x)$. Moreover, we assume that

$$\forall\, \boldsymbol{\varphi} \in C_0(\Omega,\mathbb{E}^s_n), \boldsymbol{\varphi}(x) \in \mathcal{C}(x) \text{ a.e.} \Rightarrow \boldsymbol{\varphi}(x) \in \mathcal{C}(x) \text{ everywhere}\,.$$

For materials incompressible in the plastic range $\mathcal{C}(x) = \mathcal{C}_D(x) + \mathbb{R}I$, where $\mathcal{C}_D(x)$ is a convex compact subset of the deviatoric space and $0 \in \mathcal{C}_D(x)$. In the subsequent sections of this chapter the integrand of the following type will be used

$$f(x, \epsilon) = [\mathcal{W}_e^*(x, \cdot) + I_{C(x)}(\cdot)]^*(\epsilon) \ , \quad \epsilon \in \mathbb{E}_s^n \ . \qquad (13.2.5)$$

Here $\mathcal{W}_e^*$ is a convex normal integrand on $\Omega \times \mathbb{E}_n^s$ such that $0 \leq \mathcal{W}_e^*(x, \cdot) \leq g(\cdot)$, where $g : \mathbb{E}_n^s \to \mathbb{R}$ is a continuous function. When the behavior of the material is linear in the elastic range then, for instance

$$\mathcal{W}_e^*(x, \boldsymbol{\sigma}) = \frac{1}{2} c_{ijkl}(x) \sigma^{ij} \sigma^{kl} \ , \qquad \boldsymbol{\sigma} \in \mathbb{E}_n^s \ .$$

Let $h(x, \epsilon) = I_{C(x)}^*(\epsilon)$. Then

$$\forall \, \boldsymbol{\mu} \in \mathbb{M}^1(\Omega, \mathbb{E}_s^n) \ , \qquad J(\boldsymbol{\mu}) = \int_\Omega f(x, \frac{d\boldsymbol{\mu}_a}{dx}) dx + \int_\Omega h(x, \boldsymbol{\mu}_s) \ .$$

Obviously, $h(x, \cdot) = f_\infty(x, \cdot)$ provided that $dom \, f^*(x, \cdot) = C(x)$.

In the case when $C(x) = C_D(x) + \mathbb{R}I$ one has

$$h(x, \epsilon) = \begin{cases} I_{C_D(x)}^*(\epsilon) & \text{if } \epsilon \in \mathbb{E}^D \ , \\ +\infty & \text{otherwise,} \end{cases}$$

where $\mathbb{E}^D$ stands for the deviatoric space. $\qquad\qquad\qquad\qquad\qquad\qquad \Box$

The two lemmas which follow are useful in proving homogenization theorems. Let f be a convex function on $\mathbb{E}_s^n$ satisfying the following property

$$\exists \, \Lambda_0 \geq \lambda_0 > 0 \ , \ \exists \, k_0 > 0 \ , \ \forall \, \epsilon \in \mathbb{E}_s^n \ , \qquad \lambda_0 |\epsilon| - k_0 \leq f(\epsilon) \leq \Lambda_0(1 + |\epsilon|) \ . \quad (13.2.6)$$

Lemma 13.2.2. If $f : \mathbb{E}_s^n \to \mathbb{R}$ is convex and satisfies (13.2.6) then

$$\forall \, \boldsymbol{\mu}_1, \boldsymbol{\mu}_2 \in \mathbb{M}^1(\Omega, \mathbb{E}_s^n), \qquad \int_\Omega f(\boldsymbol{\mu}_1 + \boldsymbol{\mu}_2) \leq \int_\Omega f(\boldsymbol{\mu}_1) + \int_\Omega f_\infty(\boldsymbol{\mu}_2) \ .$$

Lemma 13.2.3. Let $\boldsymbol{\mu} \in \mathbb{M}^1(\Omega, \mathbb{E}_s^n)$ and denote by $A_{\boldsymbol{\mu}}$ the space of functions $v : \Omega \to \mathbb{E}_n^s$ defined by

$$v(x) = \sum_{i \in \mathcal{I}} \chi_{\Omega_i}(x) \epsilon_i^* \ ,$$

where $\epsilon_i^* \in \mathbb{E}_n^s$ and $\{\Omega_i\}_{i \in \mathcal{I}}$ is a family of open disjoint sets such that $\overline{\Omega}_i \subset \Omega$ and

$$meas \, (\partial\Omega_i) = \int_{\partial\Omega_i} |\boldsymbol{\mu}| = |\boldsymbol{\mu}|(\partial\Omega_i) = 0 \ .$$

If f is a convex function satisfying (13.2.6) then

$$\int_\Omega f(\boldsymbol{\mu}) = \sup_{v \in A_{\boldsymbol{\mu}}} \{ \int_\Omega v^{\alpha\beta} \mu_{\alpha\beta} - \int_\Omega f^*(v(x)) dx \} \ .$$

Remark 13.2.4. The last two lemmas remain valid for $f(x, \epsilon)$ provided that (13.2.6) holds for every $x \in \Omega$. Obviously, f is a measurable function convex in ϵ.

Remark 13.2.5. Positive convex functions of a measure enable us to define intermediate topologies on $BV(\Omega)$, $BD(\Omega)$ and $HB(\Omega)$ and to state approximation results. For instance, on $HB(\Omega)$ we can define the intermediate topology corresponding to the distance

$$d_f(v, w) = ||v - w||_{W^{1,1}(\Omega)} + |\int_\Omega |\kappa(v)| - \int_\Omega |\kappa(w)|| + |\int_\Omega f(x, \kappa(v)) - \int_\Omega f(x, \kappa(w))| \,.$$

Then for all $v \in HB(\Omega)$, there exists a sequence $\{v^m\}_{m \in \mathbb{N}} \subset C^\infty(\Omega) \cap W^{2,1}(\Omega)$ with the following properties:
(i) $\gamma_0 v^m = \gamma_0 v$, $\gamma_1 v^m = \gamma_1 v$, $\forall\, m$.
(ii) As $m \to \infty$, v^m converges to v for the distance d_f.

Let us formulate a useful result on multifunctions.

Lemma 13.2.6. Let C be a multifunction on Ω to the convex subsets of $\mathbb{R}^d$. Then C is lower semicontinuous on Ω if and only if $(x, \epsilon) \to I^*_{C(x)}(\epsilon)$ is l.s.c. on $\Omega \times \mathbb{R}^d$. $\qquad\square$

Commutativity of integral and inf revisited

Rockafellar's Theorem 1.2.33 involves decomposable spaces and precludes practically important classes of spaces. A different approach was proposed by Bouchitté and Valadier (1988), cf. also Bouchitté and Valdier (1989).

Let $L^0(\Omega)^d$ denote the vector space of real measurable functions.

Definition 13.2.7. A subset C of $L^0(\Omega)^d$ is said to be PCU-stable (or C_0-stable) if for any continuous partition of unity $\alpha_0, \alpha_1, \ldots, \alpha_K \in C(\Omega; [0, 1])$ such that $\alpha_1, \ldots, \alpha_K$ have compact supports, for every $u_0, \ldots, u_K$ in C, the function $\sum_{k=0}^{K} \alpha_k u_k$ belongs to C. $\qquad\square$

Possible variants:
(i) $\alpha_0, \ldots, \alpha_K \in \mathbf{D}(\Omega)$,, $\alpha_0 \in C^\infty(\Omega)$. Then C is said to be C_c^∞-stable.
(ii) Lip-stability: $\alpha_0, \ldots, \alpha_K$ are Lipschitz functions.

Remark 13.2.8. The notion and existence of the partition of unity is discussed, for instance, in Stein (1970) and Yosida (1978). $\qquad\square$

For any subset C_1 of $L^0(\Omega)^d$ there exists a smallest closed-valued measurable multivalued function F such that $\forall\, u \in C_1$, $u(x) \in F(x)$ almost everywhere. We write $F = ess\ sup\{u(\cdot)|\ u \in C_1\}$ and say that F is the *essential supremum* of the multifunctions $x \to \{u(x)\}$ $(u \in C_1)$. If C_1 is *convex*, then F is (a.e.) *convex-valued*. This remains true if C_1 is PCU-stable.

We set

$$J(u) = \int_\Omega j(x, u(x))dx \,, \qquad u \in L^0(\Omega)^d \,.$$

The functional J assumes its values in $\overline{\mathbb{R}}$, and as usual in convex analysis, $\int_{\Omega} j(x, \boldsymbol{u}(x))dx = +\infty$ as soon as $\int_{\Omega} j(x, \boldsymbol{u}(x))^+ dx = +\infty$, where a^+ stands for the positive part of a.

Theorem 13.2.9. Let $\mathcal{C}$ be a PCU-stable subset of $L^0(\Omega)^d$ (variants: Lip-stable, C_c^∞-stable). Suppose there exists $\boldsymbol{u}^0 \in \mathcal{C}$ with $J(\boldsymbol{u}^0) \in \mathbb{R}$. Then $F = ess \sup\{\boldsymbol{u}(\cdot)|\ \boldsymbol{u} \in \mathcal{C} \cap dom\ J\}$ is convex-valued,

$$\inf\{J(\boldsymbol{u}) \mid \boldsymbol{u} \in \mathcal{C}\} = \int_{\Omega} [\inf_{\boldsymbol{z} \in F(x)} j(x, \boldsymbol{z})]dx ,$$

and

$$\inf\{j(x, \boldsymbol{z}) \mid \boldsymbol{z} \in F(x)\} = ess \sup\{j(x, \boldsymbol{u}(x)) \mid \boldsymbol{u} \in \mathcal{C} \cap dom\ J\} . \qquad \square$$

Specific case of convergence of a sequence of sets

Let us consider now a particular case of convergence in Kuratowski's sense, cf. Sec. 1.3.7. By $\mathbf{O}$ we denote a compact or locally compact set of $\mathbb{R}^m$. The statements which follow are due to Bouchitté (1987).

Theorem 13.2.10. Let μ be a positive bounded Radon measure, $\mu \in \mathbb{M}_+^1(\mathbf{O})$, and let $F_\varepsilon : \mathbf{O} \to \mathbb{R}^d(\varepsilon \in \mathcal{E})$ be a multivalued, μ-measurable mapping with values in the closed and convex sets of $\mathbb{R}^d$ and such that $0 \in F_\varepsilon(x)$ μ-a.e.

Then the sequence of convex sets

$$\mathcal{C}_\varepsilon = \{\varphi \in C_0(\mathbf{O})^d \mid \varphi(x) \in F_\varepsilon(x)\ \mu\text{-a.e.}\}$$

admits, in $C_0(\mathbf{O})^d$, a strongly convergent subsequence in Kuratowski's sense. The limit set $\mathcal{C}$ of such a subsequence has the following form

$$\mathcal{C}_\varepsilon = \{\varphi \in C_0(\mathbf{O})^d \mid \varphi(x) \in F_\varepsilon(x)\ \forall\, x \in \mathbf{O}\} ,$$

where F is a lower semicontinuous multivalued map with values in the convex and closed sets containing the zero element of $\mathbb{R}^d$. $\qquad \square$

To be more specific, let us investigate the following case:
(1) $\mu \in \mathbb{M}_+^1(\mathbf{O})$ and supp $\mu = \mathbf{O}$,
(2) $\{A_\varepsilon\}_{\varepsilon>0}$ - a family of Borel subset of $\mathbf{O}$ such that $\mu(\partial A_\varepsilon) = 0$,
(3) F_1, F_2 - continuous multivalued maps with values in convex and closed subsets of $\mathbb{R}^d$ containing 0 as the interior point.
(4) $F_\varepsilon(x) = \begin{cases} F_1(x) & \text{if } x \in A_\varepsilon , \\ F_2(x) & \text{if } x \in \mathbf{O} \backslash A_\varepsilon ; \end{cases}$
(5) $\mathcal{C}_\varepsilon = \{\varphi \in C_0(\mathbf{O})^d \mid \varphi(x) \in F_\varepsilon(x)\ \mu\text{ - a.e. on } \mathbf{O}\}$.

For instance, μ can be a surface measure. We have the following convergence result.

Theorem 13.2.11. If $\mathrm{int}A_\varepsilon \to A$ and $\mathrm{int}(\mathbf{O}\backslash A_\varepsilon) \to B$ as $\varepsilon \to 0$, then A and B are closed sets such that $A \cup B = \mathbf{O}$. Moreover, the sequence $\{\mathcal{C}_\varepsilon\}_{\varepsilon>0}$ is strongly convergent in Kuratowski's sense to the convex set $\mathcal{C}$ defined by

$$\mathcal{C} = \{\varphi \in C_0(\mathbf{O})^d \mid \varphi(x) \in F(x), \ \forall\, x \in \mathbf{O}\},$$

where F is the following l.s.c. multivalued mapping

$$F(x) = \begin{cases} F_1(x) \cap F_2(x) & \text{if } x \in A \cap B, \\ F_1(x) & \text{if } x \in A\backslash B, \\ F_2(x) & \text{if } x \in B\backslash A. \end{cases} \qquad \square$$

We recall a practical criterion of continuity: F_1 is continuous on $\mathbf{O}$ if and only if F_1 is lower semicontinuous with closed graph in $\mathbf{O}\times\mathbb{R}^d$. Also, the assumption $0 \in int(F_1(x)\cap F_2(x))$, $\forall\, x \in \mathbf{O}$, implies that the multivalued mapping $x \to F_1(x) \cap F_2(x)$ is l.s.c.

A simple criterion ensures that $A = B = \mathbf{O}$. It means that the media occupying A_ε and $\mathbf{O}\backslash A_\varepsilon$ undergo perfect mixing when $\varepsilon \to 0$.

Proposition 13.2.12. If the sequence of characteristic function $\{\chi_{A_\varepsilon}\}_{\varepsilon>0}$ converges in $\sigma(L_\mu^\infty, L_\mu^1)$ to a function $\theta(x)$ such that $0 < \|\theta\|_{L_\mu^\infty(\mathbf{O})} < 1$ then the sets $\mathrm{int}A_\varepsilon$ and $\mathrm{int}(\mathbf{O}\backslash A_\varepsilon)$ converge to $\mathbf{O}$ and one has:

$$A = B = \mathbf{O}, \qquad F(x) = F_1(x) \cap F_2(x) \qquad \forall\, x \in \mathbf{O}. \qquad \square$$

13.3. General homogenization theorems for functionals with linear growth

Prior to the presentation in Sec. 14 of homogenization problems related to thin elastic-perfectly plastic plates we shall first consider a simpler case, which does not involve any boundary condition and loading.

Suppose that a plate made of a Hencky material exhibits a microperiodic structure characterized by the complementary elasto-plastic potential

$$j_\varepsilon^*(x, \boldsymbol{\rho}^*) = j\left(\frac{x}{\varepsilon}, \boldsymbol{\rho}^*\right) = \frac{1}{2}\widehat{d}_{\alpha\beta\lambda\mu}\left(\frac{x}{\varepsilon}\right)\rho^{*\alpha\beta}\rho^{*\lambda\mu} + I_{\mathcal{C}_\varepsilon(x)}(\boldsymbol{\rho}^*), \qquad (13.3.1)$$

where $\rho^* \in \mathbb{E}_2^s$, $x \in \Omega$ and $\mathcal{C}_\varepsilon(x) = \mathcal{C}\left(\dfrac{x}{\varepsilon}\right)$ stands for the periodic elasticity convex. The elasticity convex $\mathcal{C}\left(\dfrac{x}{\varepsilon}\right)$ is the set of plastically admissible moments at $x \in \Omega$. Obviously, $\widehat{D} = \widehat{d}^{-1}$ possesses usual properties, cf. $(5.4.6)_1$. The variable ρ^* is just the moment tensor or $\rho^* = M$. For instance, in the case of the Huber-Mises yield condition the elasticity convex $\mathcal{C}(y)$ has the following form

$$\mathcal{C}(y) = \{M \mid (M^{11})^2 - M^{11}M^{22} + (M^{22})^2 + 3(M^{12})^2 \le M_0^2(y)\},$$

where $y = x/\varepsilon$ and $M_0\left(\dfrac{x}{\varepsilon}\right)$ denotes the limit bending moment at a point $x \in \Omega$. The function $M_0(y)$ is obviously assumed to be Y-periodic. Equation (13.3.1) implies $\mathcal{C}_\varepsilon(x) = \mathrm{dom}\, j_\varepsilon^*(x, \cdot)$.

In general, $\mathcal{C}$ is a closed convex-valued l.s.c. multifunction such that for each $y \in Y$, $0 \in \text{int}\, \mathcal{C}(y)$. Moreover, we assume that there exist constants k_0 and k_1, $0 < k_0 \le k_1 < +\infty$ such that

$$\{M \in \mathbb{E}_2^s : |M| \le k_0\} \subset \mathcal{C}_\varepsilon(x) \subset \{M \in \mathbb{E}_2^s : |M| \le k_1\}\,, \tag{13.3.2}$$

for every $x \in \Omega$, where $|M|^2 = \displaystyle\sum_{\alpha,\beta=1}^{2} M^{\alpha\beta} M^{\alpha\beta}$.

The elasto-plastic potential is calculated by using Fenchel's transform

$$j\left(\frac{x}{\varepsilon}, \rho\right) = \sup\{\rho^* : \rho - j^*\left(\frac{x}{\varepsilon}, \rho^*\right) | \rho^* \in \mathbb{E}_2^s\}\,, \tag{13.3.3}$$

where $\rho \in \mathbb{E}_s^2$. This function is a convex normal integrand and has the following property

$$\exists\, k_3 \ge k_2 > 0\,, \quad k_2(|\rho| - 1) \le j(y, \rho) \le k_3(1 + |\rho|)\,, \tag{13.3.4}$$

for each $(y, \rho) \in Y \times \mathbb{E}_s^2$; moreover $j(y, 0) = 0$ and $j(y, \rho) \ge 0$ for each $\rho \in \mathbb{E}_s^2$.

The recession function $j_\infty\left(\frac{x}{\varepsilon}, \cdot\right)$ of $j\left(\frac{x}{\varepsilon}, \cdot\right)$ or the support function of $\mathcal{C}\left(\frac{x}{\varepsilon}\right)$ is given by

$$j_\infty\left(\frac{x}{\varepsilon}, \rho\right) = \lim_{\eta \to 0^+} \eta j\left(\frac{x}{\varepsilon}, \frac{\rho}{\eta}\right) = \sup\{\rho^* : \rho | \rho^* \in \mathcal{C}\left(\frac{x}{\varepsilon}\right)\}\,, \tag{13.3.5}$$

for each $\rho \in \mathbb{E}_s^2$. This function is also a convex normal integrand, positively homogeneous with respect to $\rho \in \mathbb{E}_s^2$ and satisfies the condition

$$\exists\, k_5 \ge k_4 > 0\,, \quad k_4|\rho| \le j_\infty(y, \rho) \le k_5(1 + |\rho|)\,,$$

for each $(y, \rho) \in Y \times \mathbb{E}_s^2$. From the physical point of view, in the case of limit analysis the function j_∞ represents the density of plastic dissipation. For the elasto-plastic Hencky plate j_∞ stands for the density of plastic work.

Example 13.3.1. Let the elasticity convex be defined by

$$\mathcal{C}\left(\frac{x}{\varepsilon}\right) = \{M \in \mathbb{E}_2^s | a(M) \le \Phi\left(\frac{x}{\varepsilon}\right)\}\,,$$

where Φ is εY-periodic and $a(M)$ a convex function, not necessarily isotropic. Particularly, for

$$a(M) = \frac{1}{2} a_{\alpha\beta\lambda\mu} M^{\alpha\beta} M^{\lambda\mu}\,,$$

one has

$$j_\infty\left(\frac{x}{\varepsilon}, \rho\right) = \left[2\widetilde{A}^{\alpha\beta\lambda\mu} \rho_{\alpha\beta}\rho_{\lambda\mu} \Phi\left(\frac{x}{\varepsilon}\right)\right]^{1/2}\,, \qquad \widetilde{A} = a^{-1}\,,$$

provided that a is positive definite; here $x \in \Omega$ and $\rho \in \mathbb{E}_s^2$. The case of a being only positive semidefinite is left to the reader. $\qquad\square$

Γ-convergence in $W^{1,1}(\Omega)$ of functionals with linear growth

Let $Y \subset \mathbb{R}^2$ stand for a basic cell. Of primal importance for two-dimensional homogenization of thin elasto-plastic plates is the following theorem.

Theorem 13.3.2. Let $\Omega \subset \mathbb{R}^2$ be a bounded domain with the uniform C^2-regularity property except possibly at a finite number of points. Let $j : (y, \rho) \in \mathbb{R}^2 \times \mathbb{E}_s^2 \to \mathbb{R}$ be measurable, convex in ρ, Y-periodic in y, and such that

$$\exists\, \Lambda_0 \geq \lambda_0 > 0 \,, \qquad \exists\, k_0 > 0 \,, \qquad \lambda_0|\rho| - k_0 \leq j(y,\rho) \leq \Lambda_0(1 + |\rho|) \,, \tag{13.3.6}$$

is satisfied for each (y, ρ).

For every $\varepsilon > 0$ we define the functional J_ε on $W^{1,1}(\Omega)$ by

$$J_\varepsilon(w) = \begin{cases} \displaystyle\int_\Omega j\left(\frac{x}{\varepsilon}, \kappa(w(x))\right) dx & \text{if } w \in W^{2,1}(\Omega) \,, \\[4mm] +\infty & \text{if } w \in W^{1,1}(\Omega)\backslash W^{2,1}(\Omega) \,. \end{cases} \tag{13.3.7}$$

Then

$$\Gamma(W^{1,1}(\Omega)) - \lim_{\varepsilon \to 0} J_\varepsilon = J_h \,, \tag{13.3.8}$$

where

$$J_h(w) = \begin{cases} \displaystyle\int_\Omega j_h(\kappa(w)) & \text{if } w \in HB(\Omega) \,, \\[4mm] +\infty & \text{if } w \in W^{1,1}(\Omega)\backslash HB(\Omega) \,, \end{cases} \tag{13.3.9}$$

and

$$j_h(\rho) = \inf\{\langle j(y, \kappa^y(v) + \rho)\rangle|\, v \in W_{per}^{2,1}(Y)\} \,, \quad \rho \in \mathbb{E}_s^2 \,, \tag{13.3.10}$$

and

$$W_{per}^{2,1}(Y) = \{v \in W^{2,1}(Y)|\, v \text{ and } \frac{\partial v}{\partial y_\alpha} \text{ are } Y\text{-periodic}\} \,. \tag{13.3.11}$$

$\square$

We recall that $\kappa_{\alpha\beta}(w) = -w_{,\alpha\beta} = -\dfrac{\partial^2 w}{\partial x_\alpha \partial x_\beta}$, $\kappa_{\alpha\beta}^y(v) = -v_{|\alpha\beta} = -\dfrac{\partial^2 v}{\partial y_\alpha \partial y_\beta}$ while the averaging in (13.3.10) is over Y. The Y-periodicity involved in (13.3.10) means that the traces (values) of $v \in W_{per}^{2,1}(Y)$ are equal on the opposite sides of Y. The same pertains to $\partial v/\partial y_\alpha$.

Since the growth of the integrand j is linear, the infimum in (13.3.10) will, in general, be attained not in $W_{per}^{2,1}(Y)$ but in $HB(Y)$. The assumptions concerning j ensure that for a function $v \in HB(Y)$, $\langle j(y, \kappa^y(v) + \rho)$ is a convex functional of a measure, cf. Sec. 13.2.

Throughout this section k with a subscript denotes a positive constant.

Properties of j_h

(i) The function j_h is convex and such that

$$\exists\, \Lambda_0' \geq \lambda_0' > 0\,, \qquad \exists\, k_0' > 0\,, \qquad \lambda_0'|\rho| - k_0' \leq j_h(\rho) \leq \Lambda_0'(1 + |\rho|)\,,$$

for each $\rho \in \mathbb{E}_s^2$.

The proof results directly from (13.3.6) and (13.3.10).

(ii) The polar function j_h^* of j_h is given by

$$j_h^*(\rho^*) = \inf\{\langle j^*(y, m(y) + \rho^*)\rangle | m \in S_{per}(Y)\}\,, \quad \rho^* \in \mathbb{E}_2^s \qquad (13.3.12)$$

where

$$S_{per}(Y) = \{m \in L^\infty(Y, \mathbb{E}_2^s)|\, \text{div}_y\, \text{div}_y m = 0 \quad \text{in } Y, \int_Y m(y)dy = 0\,, \qquad (13.3.13)$$

$$m_\nu \text{ assume equal and } q \text{ opposite values on the opposite sides of } Y\}\,,$$

and

$$m_\nu = m^{\alpha\beta}\nu_\alpha\nu_\beta\,, \qquad (13.3.14)$$

$$q = \nu_\alpha \frac{\partial m^{\alpha\beta}}{\partial y_\beta} + \frac{\partial m_\tau}{\partial s}\,, \qquad m_t = m^{\alpha\beta}\nu_\alpha\tau_\beta\,. \qquad (13.3.15)$$

In Chapter IV $\nu = (\nu_\alpha)$ denotes the exterior unit normal to ∂Y. The meaning of s and $\tau = (\tau_\alpha)$ is obvious. More precisely, m_ν and q are to be understood in the sense of traces.

Proof. The Fenchel conjugate of j_h has the form

$$j_h^*(\rho^*) = \sup\{\rho^* : \rho - j_h(\rho)|\, \rho \in \mathbb{E}_s^2\}$$

$$= \sup \frac{1}{|Y|}\{\int_Y [\rho^* : (\kappa^y(v) + \rho) - j(y, \kappa^y(v) + \rho)]dy|\, \rho \in \mathbb{E}_s^2, v \in W_{per}^{2,1}(Y)\}\,,$$

where $\rho^* \in \mathbb{E}_2^s$.

If we set

$$W = \kappa^y(W_{per}^{2,1}(Y)) \oplus \mathbb{E}_s^2\,, \qquad J(m) = \int_Y j(y, m(y))dy\,,$$

then we have

$$j_h^*(\rho^*) = \frac{1}{|Y|}(J + I_W)^*(\rho^*) = (J^*\square I_{S_{per}(Y)})(\rho^*)\,. \qquad (13.3.16)$$

Here $\square$ denotes the inf-convolution, cf. Sec. 1.2.1. Moreover, ρ^* is to be identified with an element of $L^\infty(Y, \mathbb{E}_2^s)$ and

$$J^*(m) = \int_Y j^*(y, m(y))dy\,,$$

$$S_{per}(Y) := W^\perp = [\kappa^y(W_{per}^{2,1}(Y))]^\perp \cap (\mathbb{E}_s^2)^\perp\,. \qquad (13.3.17)$$

We calculate

$$(\mathbb{E}_s^2)^\perp = \{ m \in L^\infty(Y, \mathbb{E}_2^s) | \langle m, \rho \rangle_{L^\infty(Y,\mathbb{E}_2^s) \times L^1(Y,\mathbb{E}_s^2)} = 0, \quad \forall \rho \in \mathbb{E}_s^2 \}$$

$$= \{ m \in L^\infty(Y, \mathbb{E}_2^s) | \int_Y m(y) dy = 0 \} . \tag{13.3.18}$$

To find $[\kappa^y(W_{per}^{2,1}(Y))]^\perp$ we use the integration by parts, at least formally

$$0 = \langle m, \kappa^y(v) \rangle_{L^\infty(Y,\mathbb{E}_2^s) \times L^1(Y,\mathbb{E}_s^2)} = \int_Y m^{\alpha\beta}(y) \kappa_{\alpha\beta}^y(v(y)) dy$$

$$= -\int_Y m^{\alpha\beta}{}_{|\beta\alpha} v\, dy + \int_{\partial Y} q\gamma_0 v\, ds - \int_{\partial Y} m_\nu \gamma_1 v\, ds , \quad \forall v \in W_{per}^{2,1}(Y) , \tag{13.3.19}$$

where q (the local effective shear force) and m_ν (the local bending moment on ∂Y) are given by (13.3.15) and (13.3.14), respectively.

Since

$$[\kappa^y(W_{per}^{2,1}(Y))]^\perp$$
$$= \{ m \in L^\infty(Y, \mathbb{E}_2^s) | \langle m, \kappa^y(v) \rangle_{L^\infty(Y,\mathbb{E}_2^s) \times L^1(Y,\mathbb{E}_s^2)} = 0, \forall v \in W_{per}^{2,1}(Y) \} , \tag{13.3.20}$$

from (13.3.17) – (13.3.20) we readily obtain (13.3.13). We recall that due to periodicity, the last term in (13.3.19) vanishes.

Finally, (13.3.16) takes the form

$$j_h^*(\rho^*) = \frac{1}{|Y|} \inf\{ J^*(m_1) + I_{S_{per}(Y)}(m_2) | \rho^* = m_1 + m_2, m_\alpha \in S_{per}(Y) \}$$

$$= \frac{1}{|Y|} \inf\{ J^*(\rho^* - m_2) | m_2 \in S_{per}(Y) \}$$

$$= \inf\{ \langle j^*(y, m(y) + \rho^*) \rangle | m \in S_{per}(Y) \} ,$$

because $S_{per}(Y)$ is a linear space. This completes the proof of (13.3.12).

(iii) The function $j_{h\infty}$ defined by

$$j_{h\infty}(\rho) = \lim_{\eta \to 0^+} \eta j_h(\rho/\eta)$$

coincides with the function

$$j_{\infty h}(\rho) = \inf\{ \langle j_\infty(y, \kappa^y(v) + \rho) \rangle | v \in W_{per}^{2,1}(Y) \} .$$

Proof. Calculating $(j_{\infty h})^*$ similarly as j_h^* we conclude that

$$\text{dom}\, (j_{\infty h})^* = \text{dom}\, j_h^* .$$

Thus we have

$$j_{\infty h}(\rho) = \sup\{ \rho^* : \rho^* | \rho^* \in \text{dom}\, (j_{\infty h})^* \}$$
$$= \sup\{ \rho : \rho^* | \rho^* \in \text{dom}\, j_h^* \} = j_{h\infty}(\rho) . \qquad \square$$

Physically, that set $C^h := \mathrm{dom}\,(j_{\infty h})^* = \mathrm{dom}\,j_h^*$ represents the elasticity convex or the set of plastically admissible macroscopic moments. We observe that convexity and boundedness of C^h are inferred from the properties of the recession function $j_{\infty h}$.

The lemma which follows is a counterpart of Lemmas 2.10.12, 11.7.8.

Lemma 13.3.3. Let $m \in \mathbf{S}_{per}(Y)$ and $\varphi \in \mathbf{D}(\Omega)$. Let $\{w^\varepsilon\}_{\varepsilon>0} \subset W^{2,1}(\Omega)$ be a sequence strongly convergent in $W^{1,1}(\Omega)$. Then we have

$$\lim_{\varepsilon \to 0} \int_\Omega \varphi(x) m^{\alpha\beta}\left(\frac{x}{\varepsilon}\right) w^\varepsilon_{,\alpha\beta}dx = 0 \ . \tag{13.3.21}$$

Proof. We set

$$r_\varepsilon = \int_\Omega \varphi(x) m^{\alpha\beta}\left(\frac{x}{\varepsilon}\right) w^\varepsilon_{,\alpha\beta}(x)dx \ .$$

Applying twice integration by parts, recalling that $w^\varepsilon \in C(\bar\Omega)$ and knowing that

$$\mathrm{div}\,\mathrm{div}\,\boldsymbol{m}\left(\frac{\cdot}{\varepsilon}\right) = 0 \quad \text{in} \quad \Omega \ ,$$

we get

$$r_\varepsilon = -\int_\Omega \left[2m^{\alpha\beta}\left(\frac{x}{\varepsilon}\right) \varphi_{,\alpha}(x)w^\varepsilon_{,\beta}(x) + m^{\alpha\beta}\left(\frac{x}{\varepsilon}\right) \varphi_{,\alpha\beta}(x)w^\varepsilon(x)\right]dx \ .$$

The sequence $\boldsymbol{m}\left(\frac{\cdot}{\varepsilon}\right)$, being bounded in $L^\infty(\Omega, \mathbb{E}_2^s)$, is convergent in the topology $\sigma(L^\infty, L^1)$ to the mean value $\langle m(y)\rangle = 0$, whereas the sequences $\{\varphi_{,\alpha}w^\varepsilon_{,\beta}\}_{\varepsilon>0}$, $\{\varphi_{,\alpha\beta}w^\varepsilon\}_{\varepsilon>0}$ are strongly convergent in L^1 and (13.3.21) is proved. $\qquad\square$

Proof of Theorem 13.3.2. We may assume that $j(y, \boldsymbol{\rho}) \geq 0$, which is the physical case, cf. (13.3.3). Otherwise it is sufficient to consider the function $j_1(y, \boldsymbol{\rho}) = j(y, \boldsymbol{\rho}) + j^*(y, 0)$.
I. We are going to prove that

$$\Gamma(W^{1,1}(\Omega)) - \liminf_{\varepsilon \to 0} J_\varepsilon \geq J_h \ . \tag{13.3.22}$$

Let $\{w^\varepsilon\}_{\varepsilon>0} \subset W^{2,1}(\Omega)$ be strongly convergent to w in $W^{1,1}(\Omega)$ as $\varepsilon \to 0$ and let $\liminf_\varepsilon J_\varepsilon(w^\varepsilon) < +\infty$. Otherwise, when $\liminf_\varepsilon J_\varepsilon(w^\varepsilon) = +\infty$, the inequality (13.3.22) is trivially satisfied.

From (13.3.6) we conclude that $\kappa(w^\varepsilon)$ is bounded in $L^1(\Omega, \mathbb{E}_s^2)$ and consequently converges weakly in $\mathbb{M}^1(\Omega, \mathbb{E}_s^2)$. Hence w belongs to $HB(\Omega)$. By Lemma 13.2.3, it suffices to prove that for each $M \in A_\mu$, $\mu = \kappa(w)$, the following inequality holds true

$$\liminf_{\varepsilon \to 0} J_\varepsilon(w^\varepsilon) \geq \int_\Omega M^{\alpha\beta}\kappa_{\alpha\beta}(w) - \int_\Omega j_h^*(\boldsymbol{M}(x))dx \ . \tag{13.3.23}$$

If $M \in A_\mu$ then we write

$$M(x) = \sum_{i \in \mathcal{I}} \chi_{\Omega_i}(x) M_i \, .$$

Lemma 13.3.3 implies that for every family of functions $\{m_i\}_{i \in \mathcal{I}} \subset S_{per}(Y)$ we have

$$\lim_{\varepsilon \to 0} \int_\Omega [\sum_{i \in \mathcal{I}} \varphi_i^\delta(x) m_i \left(\frac{x}{\varepsilon}\right)] : \kappa(w^\varepsilon(x)) dx = 0 \, . \tag{13.3.24}$$

The properties of the functions φ_i^δ are specified in the proof of Theorem 2.10.13, cf. also Th. 11.7.9. Recalling that $j \geq 0$ and taking into account the last relation we obtain

$$\lim_{\varepsilon \to 0} \inf J_\varepsilon(w^\varepsilon) = \lim_{\varepsilon \to 0} \inf \int_\Omega j \left(\frac{x}{\varepsilon}, \kappa(w^\varepsilon(x))\right) dx \tag{13.3.25}$$

$$\geq \sum_{i \in \mathcal{I}} \lim_{\varepsilon \to 0} \inf \int_{\Omega_i} \varphi_i^\delta(x)[M_i : \kappa(w^\varepsilon(x)) - j_{m_i}^*(\frac{x}{\varepsilon}, M_i)] dx \, ,$$

where $j_{m_i}(y, \rho) = j(y, \rho) - m_i(y) : \rho$.

For $\int_Y j_{m_i}^*(y, M_i) dy < +\infty$, the function $j_{m_i}^*(\cdot, M_i)$ is bounded from below by $(-\Lambda_0)$.

Indeed, if G_1, G_2 are two convex functions such that $G_1 \leq G_2$, then $G_2^* \leq G_1^*$. Consequently, the sequence of εY-periodic functions $j_{m_i}^*$ is bounded in $L^1(\Omega)$ and converges weakly to

$$\langle j_{m_i}^*(y, M_i) \rangle = \langle j^*(y, m_i(y) + M_i) \rangle \, ,$$

because

$$j_{m_i}^*(y, M_i) = \sup\{M_i : \rho - j(y, \rho) + \rho : m_i(y) | \rho \in \mathbb{E}_s^2\} = j^*(y, m_i(y) + M_i) \, .$$

It follows that

$$\lim_{\varepsilon \to 0} \inf J_\varepsilon(w^\varepsilon) \geq \sum_{i \in \mathcal{I}} [\int_{\Omega_i} \varphi_i^\delta(x) M_i : \kappa(w) - \langle j^*(y, m_i(y) + M_i) \rangle \int_{\Omega_i} \varphi_i^\delta dx] \, .$$

Taking now the supremum on the r.h.s. of the last inequality with respect to m_i running over $S_{per}(Y)$ and taking into account (13.3.12) we obtain

$$\lim_{\varepsilon \to 0} \inf J_\varepsilon(w^\varepsilon) \geq \int_\Omega (\sum_{i \in \mathcal{I}} \varphi_i^\delta(x)) M(x) : \kappa(w) - \int_{\Omega_i} (\sum_{i \in \mathcal{I}} \varphi_i^\delta(x)) j_h^*(M(x)) dx \, . \tag{13.3.26}$$

We recall that the functions φ_i^δ are, ex definitione, such that $0 \leq \sum_i \varphi_i^\delta(x) \leq 1, x \in \Omega$.

Hence

$$0 \leq \sum_i \varphi_i^\delta(x)[j_h^*(\boldsymbol{M}(x)) + j_h(\boldsymbol{0})] \leq j_h^*(\boldsymbol{M}(x)) + j_h(\boldsymbol{0}) \ ,$$

which implies

$$\sum_i \varphi_i^\delta(x) j_h^*(\boldsymbol{M}(x)) \leq j_h^*(\boldsymbol{M}(x)) + j_h(\boldsymbol{0})[1 - \sum_i \varphi_i^\delta(x)] \ , \qquad (13.3.27)$$

since $[j_h^*(\boldsymbol{M}(x)) + j_h(\boldsymbol{0})]$ is positive.

Combining (13.3.26) with (13.3.27) we get

$$\liminf_{\varepsilon \to 0} J_\varepsilon(w^\varepsilon) \geq \int_\Omega \sum_i \varphi_i^\delta(x) \boldsymbol{M}(x) : \boldsymbol{\kappa}(w)$$
$$- \{ \int_\Omega j_h^*(\boldsymbol{M}(x)) dx + j_h(\boldsymbol{0}) \int_\Omega [1 - \sum_i \varphi_i^\delta(x)] dx \} \ .$$

The sequence $\{\sum_i \varphi_i^\delta\}_{\delta > 0}$ converges to 1 for a.e. $x \in \Omega$ when $\delta \to 0$, and simultaneously a.e. with respect to the measure $|\boldsymbol{\kappa}(w)|$, since meas $(\partial\Omega_i) = |\boldsymbol{\kappa}(w)|(\partial\Omega_i) = 0$. Applying now the dominated convergence theorem in $L^1_{|\boldsymbol{\kappa}(w)|}(\Omega)$ to the integral $\int_\Omega \varphi_i^\delta(x) \boldsymbol{M}(x) : \boldsymbol{\kappa}(w)$, in the limit we obtain (13.3.23).

II. The proof of the inequality

$$J^s = \Gamma(W^{1,1}(\Omega)) - \limsup_{\varepsilon \to 0} J_\varepsilon \leq J_h$$

is subdivided into four steps.

Step 1. Let $\{\Omega_i\}$ be a finite partition of Ω formed by polygonal sets. Let $w \in W^{2,1}(\Omega)$ be a function of class $C^1(\Omega)$ such that

$$\boldsymbol{\kappa}(w(x)) = \boldsymbol{\rho}_i \ , \qquad x \in \Omega_i \ , \qquad \boldsymbol{\rho}_i \in \mathbb{E}_s^2 \ . \qquad (13.3.28)$$

We set

$$\Omega_i^\delta = \{x \in \Omega_i | \ \mathrm{dist}\,(x, \partial\Omega_i) > \delta\} \ , \qquad \delta > 0 \ ,$$

and take a family of functions $\{w_i\}_{i \in \mathcal{I}} \subset W^{2,1}_{per}(Y)$. The sequence of functions

$$w_\varepsilon^\delta(x) = w(x) + \varepsilon^2 \sum_{i \in \mathcal{I}} \varphi_i^\delta(x) w_i \left(\frac{x}{\varepsilon}\right) \ , \qquad \varepsilon > 0$$

is convergent to w in $W^{1,1}(\Omega)$ as $\varepsilon \to 0$.

Proceeding similarly to Secs. 2.10 and 11.7 it is not difficult to show that

$$\limsup_{\substack{\delta \to 0 \\ t \to 1^-}} \limsup_{\varepsilon \to 0} J_\varepsilon(t w_\varepsilon^\delta) \leq \sum_{i \in \mathcal{I}} |\Omega_i| \langle j[y, \boldsymbol{\kappa}^y(w_i)(y) + \boldsymbol{\rho}_i] \rangle \ .$$

Employing now Lemma 1.3.27 we infer that there exists a mapping $\varepsilon \to (\delta(\varepsilon), t(\varepsilon))$ such that

$$\lim_{\varepsilon \to 0} \delta(\varepsilon) = 0 \,, \qquad \lim_{\varepsilon \to 0} t(\varepsilon) = 1^- \,.$$

Consequently, we can construct a sequence $w^\varepsilon = t(\varepsilon) w_\varepsilon^{\delta(\varepsilon)}$ such that $w^\varepsilon \to w$ strongly in $W^{1,1}(\Omega)$ when $\varepsilon \to 0$ and satisfying

$$\limsup_{\varepsilon \to 0} J_\varepsilon(w^\varepsilon) \leq \limsup_{\substack{\delta \to 0 \\ t \to 1^-}} \limsup_{\varepsilon \to 0} J_\varepsilon(tw^\delta) \leq \sum_{i \in \mathcal{I}} |\Omega_i| \langle j(y, \kappa^y(w_i) + \rho_i) \rangle \,.$$

Taking the infimum with respect to w_i running over $W_{per}^{2,1}(Y)$ one readily finds

$$\Gamma(W^{1,1}(\Omega)) - \limsup_{\varepsilon \to 0} J_\varepsilon(w^\varepsilon) \leq \sum_{i \in \mathcal{I}} |\Omega_i| j_h(\rho_i) = \int_\Omega j_h(\kappa^y(w)) dx \,.$$

Step 2. From (13.3.6) and (13.3.10) we have

$$j_h(\rho) \leq \langle j(y, \rho) \rangle \leq \Lambda_0(1 + |\rho|) \,.$$

Hence

$$J^s(v) \leq \Lambda_0 \int_\Omega (1 + |\kappa(v(x))|) dx \,, \tag{13.3.29}$$

for each $v \in W^{2,1}(\Omega)$.

As we known, the Γ-limit superior preserves convexity, cf. Sec. 1.3.4. From (13.3.29) we conclude that J^s is a convex and finitely valued functional on $W^{2,1}(\Omega)$. Consequently, this functional is continuous on $W^{2,1}(\Omega)$, as already J_h is.

Step 3. Applying Proposition 1.4.16 we prove that $J^s(w) \leq J_h(w)$ for every $w \in W^{2,1}(\Omega)$. Indeed, for $w \in W^{2,1}(\Omega)$ let $\{w_k\}_{k \in \mathbf{N}}$ be a sequence of functions of class $C^1(\Omega)$ with piecewise constant second derivatives and such that $w_k \to w$ in $W^{2,1}(\Omega)$ as $k \to \infty$. The continuity of J_h on $W^{2,1}(\Omega)$ yields

$$J_h(w_k) \to J_h(w) \quad \text{when} \quad k \to \infty \,.$$

On account of step 1, for each $k \in \mathbf{N}$, there exists a sequence $\{w_{k,\varepsilon}\}_{\varepsilon > 0}$ such that $w_{k,\varepsilon} \to w_k$ in $W^{1,1}(\Omega)$ as $\varepsilon \to 0$ and

$$J_h(w_k) \geq \limsup_{\varepsilon \to 0} J_\varepsilon(w_{k,\varepsilon}) \,.$$

Hence

$$\limsup_{k \to \infty} \limsup_{\varepsilon \to 0} J_\varepsilon(w_{k,\varepsilon}) \leq J_h(w) \,.$$

Applying Lemma 1.3.27 once again we conclude that there exists a strictly increasing mapping $\varepsilon \to k(\varepsilon)$, such that setting $w^\varepsilon = w_{k(\varepsilon),\varepsilon}$, the sequence $\{w^\varepsilon\}_{\varepsilon > 0}$ tends to w in the norm of $W^{1,1}(\Omega)$ and

$$\limsup_{\varepsilon \to 0} J_\varepsilon(w^\varepsilon) \leq J_h(w) \,.$$

Step 4. Let us set

$$\widetilde{J}(w) = \begin{cases} J_h(w) & \text{if } w \in W^{2,1}(\Omega) \,, \\ +\infty & \text{if } w \in W^{1,1}(\Omega)/W^{2,1}(\Omega) \,. \end{cases}$$

Hence $J^s(w) \leq \widetilde{J}(w)$ for each $w \in W^{1,1}(\Omega)$. We perform now the lower semicontinuous regularization of J^s and $\widetilde{J}$ on $W^{1,1}(\Omega)$, cf. Sec. 13.2. Consequently $J^s(w) \leq J_h(w)$ for each $w \in HB(\Omega)$. This proves the theorem. $\qquad\qquad\qquad\square$

Comments on other cases

Consider now the following specific complementary elasto-plastic potential

$$j^*(y, \boldsymbol{\epsilon}^*, \boldsymbol{\Delta}^*) = \frac{1}{2}(\widehat{d}_{\alpha\beta\lambda\mu}(y)\epsilon^{*\alpha\beta}\epsilon^{*\lambda\mu} + \widehat{h}_{\alpha\beta}\Delta^{*\alpha}\Delta^{*\beta}) + I_{\mathcal{C}(y)}(\boldsymbol{\epsilon}^*, \boldsymbol{\gamma}^*) \,, \qquad (13.3.30)$$

where $\boldsymbol{\epsilon}^* \in \mathbb{E}_2^s$, $\boldsymbol{\Delta}^* \in \mathbb{R}^2$ while $\widehat{\boldsymbol{D}} = \widehat{\boldsymbol{d}}^{-1}$ and $\widehat{\boldsymbol{H}} = \widehat{\boldsymbol{h}}^{-1}$ satisfy (5.4.6). The elasticity convex is now defined in the space $\mathbb{E}_2^s \times \mathbb{R}^2$. The reader can easily formulate the assumption similar to (13.3.2).

The Fenchel conjugate of $j^*(y, \cdot, \cdot)$ is given by

$$j(y, \boldsymbol{\epsilon}, \boldsymbol{\Delta}) = \sup\{\boldsymbol{\epsilon}^* : \boldsymbol{\epsilon} + \boldsymbol{\Delta}^* \cdot \boldsymbol{\Delta} - j(y, \boldsymbol{\epsilon}, \boldsymbol{\Delta})|\, \boldsymbol{\epsilon} \in \mathbb{E}_s^2 \,, \boldsymbol{\Delta} \in \mathbb{R}^2\} \,, \quad (13.3.31)$$

and represents the specific elasto-plastic potential of a *Reissner plate* made of a Hencky material. The properties of the function $j(y, \boldsymbol{\epsilon}, \boldsymbol{\Delta})$ are similar to those of the function $j(y, \boldsymbol{\rho})$ given by (13.3.3). One can also introduce the recession function

$$\begin{aligned} j_\infty(y, \boldsymbol{\epsilon}, \boldsymbol{\Delta}) &= \lim_{\eta \to 0^+} \eta j(y, \frac{\boldsymbol{\epsilon}}{\eta}, \frac{\boldsymbol{\Delta}}{\eta}) \\ &= \sup\{\boldsymbol{\epsilon}^* : \boldsymbol{\epsilon} + \boldsymbol{\Delta}^* \cdot \boldsymbol{\Delta}|\, (\boldsymbol{\epsilon}, \boldsymbol{\Delta}^*) \in \mathcal{C}(y)\} \,. \end{aligned} \qquad (13.3.32)$$

Now we are in a position to formulate the second homogenization theorem for functionals with linear growth.

Theorem 13.3.4. Let $\Omega \subset \mathbb{R}^2$ be a bounded domain with Lipschitzian boundary. Let $j : (y, \boldsymbol{\epsilon}, \boldsymbol{\Delta}) \in \mathbb{R}^2 \times \mathbb{E}_s^2 \times \mathbb{R}^2 \to \mathbb{R}$ be measurable, convex in $(\boldsymbol{\epsilon}, \boldsymbol{\Delta})$, Y-periodic in y, and such that

$$\exists\, \Lambda_0 \geq \lambda_0 > 0 \,, \quad \exists\, k_0 > 0 \,,$$
$$\lambda_0(|\boldsymbol{\epsilon}| + |\boldsymbol{\Delta}|) - k_0 \leq j(y, \boldsymbol{\epsilon}, \boldsymbol{\Delta}) \leq \Lambda_0(1 + |\boldsymbol{\epsilon}| + |\boldsymbol{\Delta}|) \,, \qquad (13.3.33)$$

is satisfied for each $(y, \boldsymbol{\epsilon}, \boldsymbol{\Delta})$.

For every $\varepsilon > 0$, we define the functional J_ε on $L^1(\Omega)^2 \times L^1(\Omega)$ by

$$J_\varepsilon(\boldsymbol{\varphi}, w) = \begin{cases} \displaystyle\int_\Omega j\left[\frac{x}{\varepsilon}, e(\boldsymbol{\varphi}), \boldsymbol{\gamma}(\boldsymbol{\varphi}, w)\right] dx & \text{if } \boldsymbol{\varphi} \in LD(\Omega) \,, w \in W^{1,1}(\Omega) \,; \\ +\infty & \text{if } \boldsymbol{\varphi} \in L^1(\Omega)^2 \backslash LD(\Omega), \, w \in L^1(\Omega)\backslash W^{1,1}(\Omega) \,. \end{cases} \qquad (13.3.34)$$

Then

$$\Gamma(L^1(\Omega)^2 \times L^1(\Omega)) - \lim_{\varepsilon \to 0} J_\varepsilon = J_h \,, \tag{13.3.35}$$

where

$$J_h(\varphi, w) = \begin{cases} \displaystyle\int_\Omega j_h[e(\varphi), \gamma(\varphi, w)] & \text{if } (\varphi, w) \in BD(\Omega) \times BV(\Omega) \,, \\ +\infty & \text{if } \varphi \in L^1(\Omega)^2 \backslash BD(\Omega) \,, \, w \in L^1(\Omega) \backslash BV(\Omega) \,, \end{cases} \tag{13.3.36}$$

and, for $\epsilon \in \mathbb{E}_s^2, \Delta \in \mathbb{R}^2$

$$j_h(\epsilon, \Delta) = \inf\{\langle j(y, e^y(\xi) + \epsilon, \nabla_y v + \Delta)\rangle | \xi \in LD_{per}(Y), v \in W_{per}^{1,1}(Y)\} \,, \tag{13.3.37}$$

$$LD_{per}(Y) = \{\xi \in LD(Y)| \, \xi \quad \text{is} \quad Y\text{-periodic}\} \,, \tag{13.3.38}$$

$$W_{per}^{1,1}(Y) = \{v \in W^{1,1}(Y)| \, v \quad \text{is} \quad Y\text{-periodic}\} \,. \tag{13.3.39}$$

Proof. It is similar to the proof of Theorem 13.3.2 and is left to the reader. Obviously, the proof involves now the approximation result formulated as Proposition 1.4.2 and the dual potential j_h^* given by

$$j_h^*(\epsilon^*, \Delta^*) = \inf\{\langle j[y, m(y) + \epsilon^*, q(y) + \Delta^*]\rangle | \, (m, q) \in S_{per}^R(Y)\} \,, \tag{13.3.40}$$

where $\epsilon^* \in \mathbb{E}_2^s, \Delta^* \in \mathbb{R}^2$ and

$$S_{per}^R(Y) = \{(m, q) \in L^\infty(Y, \mathbb{E}_2^s \times \mathbb{R}^2)| \, \mathrm{div}_y m - q = 0 \text{ in } Y \,, \mathrm{div}_y q = 0 \text{ in } Y \,,$$
$$\langle m(y)\rangle = 0 \,, \quad \langle q(y)\rangle = 0, \, m\nu \text{ and } q \cdot \nu = q^\alpha \nu_\alpha \tag{13.3.41}$$
$$\text{take opposite values on the opposite sides of } Y\} \,. \qquad \square$$

The physical meaning of ϵ^* and Δ^* is obvious: $\epsilon^* = M, \Delta^* = Q$, cf. Sec. 5. To provide an example of the elasticity convex, one may consider the following

$$\mathcal{C}(y) = \{(M, Q) \in \mathbb{E}_2^s \times \mathbb{R}^2 | M^2/M_0^2(y) + Q^2/Q_0^2(y) \leq 1\} \,, \tag{13.3.42}$$

where $M^2 = (M^{11})^2 - M^{11}M^{22} + (M^{22})^2 + 3(M^{12})^2, Q^2 = (Q^1)^2 + (Q^2)^2$.

Remark 13.3.5. It is now not difficult to formulate the homogenization theorem for a sequence of functionals with linear growth on $BD(\Omega) \times HB(\Omega)$. Such a situation is typical for plates in which the interaction between bending and membrane forces becomes important. For thin plates this effect is due to geometrical nonlinearities. If the strain measures are given by Eq. (4.1.17) and $\kappa_{\alpha\beta}(w) = -w_{,\alpha\beta}$ then the local problem has the following form

$$j_h(\epsilon, \rho) = \inf\{\langle j(y, e^y(\xi) + \tilde{\epsilon} + \frac{1}{2}a \otimes a, \kappa^y(v) + \rho)\rangle$$
$$| \, \xi \in LD_{per}(Y) \,, v \in W_{per}^{2,1}(Y)\} \,, \tag{13.3.43}$$

where $\epsilon = \tilde{\epsilon} + \frac{1}{2}a \otimes a, \tilde{\epsilon} \in \mathbb{E}_s^2, a \in \mathbb{R}^2$ and $\rho \in \mathbb{E}_s^2$.

The function $j_h(\cdot, \cdot)$ is still convex independent of whether the strain measures are linear or not. It means that for the von Kármán plate made of an elastic perfectly-plastic Hencky material, the notion of the convex functional of a measure is well defined.

13.4. Γ-convergence and Dirichlet boundary conditions, relaxation

The Γ-convergence problems for functionals with linear growth have left out Dirichlet boundary conditions. In contrast to the elastic case, now the Γ-limit functionals will involve terms (integrals) defined on $\Gamma = \partial\Omega$. Physically, these terms are due to possible discontinuities (hinges), appearing on Γ.

13.4.1.   Thin Kirchhoff plates made of Hencky material

In the previous section we have studied the problem of $\Gamma(W^{1,1}(\Omega))$ convergence of a sequence of functionals with linear growth without imposing boundary conditions. Now we assume the following boundary conditions

$$\gamma_0 w = w_0 , \quad \gamma_1 w = w_1 \quad \text{on } \Gamma , \tag{13.4.1}$$

where $w \in W^{2,1}(\Omega)$, $w_0 \in \gamma_0(W^{2,1}(\Omega))$ and $w_1 \in L^1(\Gamma)$, cf. Sec. 13.1.

Let us denote by $\widetilde{w}_0$ a function defined on $\widetilde{\Omega} \supset \overline{\Omega}$ and such that $\gamma_0\widetilde{w}_0 = w_0$, $\gamma_1\widetilde{w}_0 = w_1$ on Γ; moreover the support of the function $\widetilde{w}_0$ is contained in $\widetilde{\Omega}$. Under the assumption that Ω has the uniform C^2-regularity property we can take $\widetilde{w}_0 \in W^{2,1}(\widetilde{\Omega})$. Further, we set

$$\mathcal{K}(w_0, w_1) = \{w \in HB(\Omega)|\, \gamma_0 w = w_0 , \quad \gamma_1 w = w_1 \quad \text{on } \Gamma\} , \tag{13.4.2}$$

$$\mathbf{F}_{\alpha\beta}(p) = p n_\alpha n_\beta , \qquad p \in L^1(\Gamma) , \tag{13.4.3}$$

where $\boldsymbol{n} = (n_\alpha)$ stands for the outer unit normal vector to Γ. Denoting by $I_{\mathcal{K}(w_0,w_1)}$ the indicator function of the affine space $\mathcal{K}(w_0, w_1)$ we formulate the first result.

Theorem 13.4.1. Let a sequence of functionals $\{J_\varepsilon\}_{\varepsilon>0}$ be defined by (13.3.7). Then the sequence $\{J_\varepsilon + I_{\mathcal{K}(w_0,w_1)}\}_{\varepsilon>0}$ is Γ-convergent in the strong topology of the space $W^{1,1}(\Omega)$ to the functional

$$\Phi(w) = J_h(w) + \int_\Gamma j_{h\infty}(\mathbf{F}(w_1 - \gamma_1 w))d\Gamma , \tag{13.4.4}$$

where $w \in HB(\Omega)$, $\gamma_0 w = w_0$ and J_h is given by (13.3.9).

Proof.
I. Let us first prove that

$$\Gamma(W^{1,1}(\Omega)) - \lim_{\varepsilon\to 0} \inf(J_\varepsilon + I_{\mathcal{K}(w_0,w_1)}) \geq \Phi . \tag{13.4.5}$$

With any function $w \in HB(\Omega)$, $\gamma_0 w = w_0 = \gamma_0\widetilde{w}_0$, we associate the function $\widetilde{w} \in HB(\mathbb{R}^2)$ defined by $\widetilde{w} = w\chi_\Omega + \widetilde{w}_0\chi_{\Omega^c}$. Here χ_Ω stands for the characteristic function of Ω and Ω^c is the complement of Ω. We have, cf. Sec. 13.1,

$$\boldsymbol{\kappa}(\widetilde{w}) = \chi_\Omega\boldsymbol{\kappa}(w) + \chi_{\Omega^c}\boldsymbol{\kappa}(\widetilde{w}_0) + \mathbf{F}\left(\frac{\partial}{\partial\boldsymbol{n}}(\widetilde{w}_0 - w)\right)\delta_\Gamma , \tag{13.4.6}$$

where δ_Γ denotes the Dirac distribution on Γ, $\widetilde{\Omega} = \Omega \cup \Omega^c \cup \Gamma$.

For $w \in HB(\Omega)$, let $\{w^\varepsilon\}_{\varepsilon>0} \subset W^{2,1}(\Omega)$ be such that w^ε satisfies (13.4.1), $w^\varepsilon \to w$ in $W^{1,1}(\Omega)$ strongly as $\varepsilon \to 0$ and

$$\lim_\varepsilon \inf(J_\varepsilon + I_{\mathcal{K}(w_o,w_1)})(w^\varepsilon) < +\infty \,.$$

Hence for every bounded and regular domain $\widetilde{\Omega} \supset \overline{\Omega}$ we conclude that $\widetilde{w}^\varepsilon \to \widetilde{w}$ in $W^{1,1}(\widetilde{\Omega})$ strongly when $\varepsilon \to 0$. We observe that $\{\widetilde{w}^\varepsilon\}_{\varepsilon>0} \subset W^{2,1}(\widetilde{\Omega})$.

For the domain $\widetilde{\Omega}$, Theorem 13.3.2 gives

$$\lim_{\varepsilon \to 0} \inf \int_{\widetilde{\Omega}} j\left(\frac{x}{\varepsilon}, \boldsymbol{\kappa}(\widetilde{w}^\varepsilon)\right) dx \geq \int_{\widetilde{\Omega}} j_h(\boldsymbol{\kappa}(\widetilde{w})) \,. \tag{13.4.7}$$

The convex functional of the measure on the r.h.s. is decomposed as follows, cf. Sec. 13.2,

$$\int_{\widetilde{\Omega}} j_h(\boldsymbol{\kappa}(\widetilde{w})) = \int_{\Omega} j_h(\boldsymbol{\kappa}(w)) + \int_{\Gamma} j_{h\infty}(\mathbf{F}(w_1 - \gamma_1 w)) d\Gamma + \int_{\widetilde{\Omega}\backslash\overline{\Omega}} j_h(\boldsymbol{\kappa}(\widetilde{w}_0)) dx \,. \tag{13.4.8}$$

We recall that $\widetilde{w}_0 \in W^{2,1}(\widetilde{\Omega})$. By using (13.3.6), we readily get

$$J_\varepsilon(w^\varepsilon) = \int_{\Omega} j\left(\frac{x}{\varepsilon}, \boldsymbol{\kappa}(w^\varepsilon)\right) dx = \int_{\widetilde{\Omega}} j\left(\frac{x}{\varepsilon}, \boldsymbol{\kappa}(\widetilde{w}^\varepsilon)\right) dx$$

$$- \int_{\widetilde{\Omega}\backslash\Omega} j\left(\frac{x}{\varepsilon}, \boldsymbol{\kappa}(\widetilde{w}_0)\right) dx \geq \int_{\widetilde{\Omega}} j\left(\frac{x}{\varepsilon}, \boldsymbol{\kappa}(\widetilde{w}^\varepsilon)\right) dx - \Lambda_0 \int_{\widetilde{\Omega}\backslash\Omega} (1 + |\boldsymbol{\kappa}(\widetilde{w}_0)|) dx \,. \tag{13.4.9}$$

Taking into account (13.4.7) – (13.4.9) we obtain

$$\lim_{\varepsilon \to 0} \inf \, (J_\varepsilon + I_{\mathcal{K}(w_o,w_1)})(w^\varepsilon) \geq \lim_{\varepsilon \to 0} \inf \int_{\widetilde{\Omega}} j\left(\frac{x}{\varepsilon}, \boldsymbol{\kappa}(\widetilde{w}^\varepsilon)\right) dx$$

$$- \Lambda_0 \int_{\widetilde{\Omega}\backslash\Omega} (1 + \boldsymbol{\kappa}(\widetilde{w}_0)) dx \geq \Phi(w) + \int_{\widetilde{\Omega}\backslash\overline{\Omega}} [j_h(\boldsymbol{\kappa}(\widetilde{w}_0)) - \Lambda_0(1 + |\boldsymbol{\kappa}(\widetilde{w}^\varepsilon)|)] dx \,,$$

and (13.4.5) follows, since for $|\widetilde{\Omega}\backslash\overline{\Omega}| \to 0$ the last integral over $\widetilde{\Omega}\backslash\overline{\Omega}$ tends to zero.

II. To prove that

$$J^s = \Gamma(W^{1,1}(\Omega)) - \lim_{\varepsilon \to 0} \sup(J_\varepsilon + I_{\mathcal{K}(w_0,w_1)}) \leq \Phi \,,$$

one can first show that for each $w \in \mathcal{K}(w_0, w_1) \cap C^\infty(\Omega)$ the following inequality is satisfied

$$J^s(w) \leq J_h(w) \,.$$

Next, it is sufficient to verify that the functional Φ represents the lower semicontinuous regularization in $W^{1,1}(\Omega)$ of the functional $J_h + I_{\mathcal{K}(w_0,w_1)\cap C^\infty(\Omega)}$. The proof is then complete.

□

Remark 13.4.2. The limit (homogenized) functional Φ involves one boundary integral over Γ. Just this term represents a relaxation of one of the Dirichlet boundary conditions.

Remark 13.4.3. If the boundary conditions (13.4.1) are imposed on a part Γ_0 of Γ then in Eq. (13.4.4) the integral over Γ is to be replaced by the integral over Γ_0. Similarly, (13.4.2) involves Γ_0.

13.4.2. Moderately thick plates, refined scaling

Let us consider now a moderately thick plate made of an elastic perfectly plastic Hencky material and subject to the following boundary conditions

$$\varphi = \varphi_0, \qquad w = w_0 \quad \text{on } \Gamma. \tag{13.4.10}$$

Obviously, the functions $\varphi_0 \in L^1(\Gamma)^2$, $w_0 \in L^1(\Gamma)$ are prescribed and (13.4.10) is to be understood in the sense of traces.

We set

$$\mathcal{K}(\varphi_0, w_0) = \{(\varphi, w) \in LD(\Omega) \times W^{1,1}(\Omega)| \ \varphi = \varphi_0, \ w = w_0, \ \text{on } \Gamma\}, \tag{13.4.11}$$

$$\mathsf{T}_{\alpha\beta} = \frac{1}{2}(p_\alpha n_\beta + p_\beta n_\alpha). \tag{13.4.12}$$

We are now in a position to formulate the homogenization theorem.

Theorem 13.4.4. Under the assumptions of Theorem 13.3.4 and (13.4.10), the sequence of functionals $\{J_\varepsilon + I_{\mathcal{K}(\varphi_0, w_0)}\}_{\varepsilon > 0}$ is Γ-convergent in the strong topology of the space $L^1(\Omega)^2 \times L^1(\Omega)$ to the functional

$$\Phi(\varphi, w) = J_h(\varphi, w) + \int_\Gamma j_{h\infty}[\mathsf{T}(\varphi_0 - \varphi), (w_0 - w)n]d\Gamma, \tag{13.4.13}$$

where J_h is given by (13.3.36) and $\varphi \in BD(\Omega)$, $w \in BV(\Omega)$.

Proof. It is similar to the proof of Theorem 13.4.1 and is therefore omitted. $\qquad\square$

Remark 13.4.5. In Secs. 5.3 – 5.5 the so-called refined scaling of a moderately thick elastic plate has been studied. Let us discuss now the problem of the refined scaling of plates made of the Hencky material. To obtain the refined model we replace the strain measure $\gamma(\varphi, w)$ by $\frac{1}{\varepsilon}\gamma(\varphi, w)$ or, equivalently the shear force vector Q by εQ. Then in the condition (13.3.33) one should also take $\frac{1}{\varepsilon}\Delta$ instead of Δ ($\varepsilon > 0$ and fixed). It is not difficult to formulate counterparts of Lemmas 5.4.3, 5.4.4. The homogenized potential is now given by, cf. (5.4.10), (13.3.37)

$$j_h(\rho) = \inf\{\langle j(y, e^y(\xi) + \rho, \gamma^y(\xi, v))\rangle| \ \xi \in LD_{per}(Y), v \in W^{1,1}_{per}(Y)\}, \tag{13.4.14}$$

where $\rho \in \mathbb{E}_s^2$. We observe that (13.3.37) involves the gradient ∇_y while in (13.4.14) we have γ^y. The limit functional has now the following form

$$\Phi(w) = \int_\Omega j_h(\kappa(w)) + \int_\Gamma j_{h\infty}[\mathbf{F}(-\varphi_0 \cdot \mathbf{n} - \gamma_1 w)]d\Gamma , \qquad (13.4.15)$$

where $w \in HB(\Omega)$ and $w = w_0$. Obviously this functional describes a thin Kirchhoff plate made of the elastic perfectly plastic Hencky material.

13.4.3.   On von Kármán plates made of Hencky material

For an elastic perfectly plastic von Kármán plate with a periodic microstructure, the macroscopic potential has been defined by (13.3.43). On the other hand, the sequence of functionals

$$J_\varepsilon(\mathbf{u}, w) = \begin{cases} \displaystyle\int_\Omega j\left[\frac{x}{\varepsilon}, \mathbf{E}(\mathbf{u}, w), \kappa(w)\right] dx & \text{if } \mathbf{u} \in LD(\Omega),\ w \in W^{2,1}(\Omega) , \\ +\infty & \text{if } \mathbf{u} \in L^1(\Omega)^2 \backslash LD(\Omega),\ W^{1,1}(\Omega)\backslash W^{2,1}(\Omega) , \end{cases} \qquad (13.4.16)$$

is a sequence of functionals nonconvex in w, where the strain measure $\mathbf{E}(\mathbf{u}, w)$ is given by (4.1.17). Fortunately, the local (microscopic) potential $j(y, \cdot, \cdot)$ is still a convex function on $\mathbb{E}_s^2 \times \mathbb{E}_s^2$ for each $y \in Y$.

The Dirichlet boundary conditions are given by (13.4.1) and

$$\mathbf{u} = \mathbf{u}_0 \qquad \text{on } \Gamma , \qquad (13.4.17)$$

where $\mathbf{u}_0 \in L^1(\Gamma)^2$. Similarly to (13.4.10), (13.4.17) is to be understood in the sense of the trace of $\mathbf{u}$ on Γ. The homogenized potential is given by (13.3.43) and the limit functional has now the following form

$$\Phi(\mathbf{u}, w) = J_h(\mathbf{u}, w) + \int_\Gamma j_{h\infty}[\mathbf{T}(\mathbf{u}_0 - \mathbf{u}), \mathbf{F}(w_1 - \frac{\partial w}{\partial \mathbf{n}})]d\Gamma , \qquad (13.4.18)$$

where $\mathbf{u} \in BD(\Omega)$, $w \in HB(\Omega)$, $w = w_0$ on Γ and

$$J_h(\mathbf{u}, w) = \int_\Omega j_h(\mathbf{E}(\mathbf{u}, w), \kappa(w)) . \qquad (13.4.19)$$

14. Homogenization of plates loaded by forces and moments

This section is concerned with elastic perfectly plastic plates provided that loading is distributed over Ω and a part Γ_1 of the boundary. We assume that $\Gamma = \overline{\Gamma}_0 \cup \overline{\Gamma}_1$, $\Gamma_0 \cap \Gamma_1 = \emptyset$. As we know from the previous section, the Γ-limit comprises an integral over Γ_0 where the Dirichlet boundary conditions are prescribed. In the case of plastic plates made of Hencky material and loaded along Γ_0 the limit (homogenized) functional may also contain an integral defined over Γ_1. For instance, if a plastic Kirchhoff plate is loaded by boundary bending moments then the strength of the boundary of the homogenized plate may be weaker than that strength which follows from the bulk properties of this plate.

The present section deals mainly with plastic Kirchhoff plates, though moderately thick and von Kármán plates will also be briefly discussed. Moreover, in Sec. 14.2 three models of thin plastic plates will be investigated, similarly to the approach of Sec. 2.10.

14.1. *Thin Kirchhoff plates loaded by boundary forces and moments*

As previously, we assume that $\Gamma = \partial\Omega = \overline{\Gamma}_0 \cup \overline{\Gamma}_1$, $\Gamma_0 \cap \Gamma_1 = \emptyset$. The length of Γ_0 is positive. On Γ_0 the plate is clamped. The loading functional is assumed in the form

$$L(w) = L_1(w) + \int_{\Gamma_1} Q^0 w d\Gamma - \int_{\Gamma_1} M_n^0 \frac{\partial w}{\partial \boldsymbol{n}} d\Gamma = L_2(w) - \int_{\Gamma_1} M_n^0 \frac{\partial w}{\partial \boldsymbol{n}} d\Gamma , \qquad (14.1.1)$$

where

$$L_1(w) = \langle b, w \rangle_{\mathbb{M}(\Omega) \times C(\overline{\Omega})} , \qquad (14.1.2)$$

$$L_2(w) = L_1(w) + \int_{\Gamma_1} Q^0 w d\Gamma . \qquad (14.1.3)$$

Here $Q^0 \in L^\infty(\Gamma_1)$, $M_n^0 \in C_0(\Gamma_1)$ and $b \in \mathbb{M}(\Omega) = [C(\overline{\Omega})]^*$. Since $b \in \mathbb{M}(\Omega)$, therefore concentrated transverse forces are not excluded. The continuity of the imbedding $HB(\Omega) \subset C(\overline{\Omega})$ implies that the linear form (14.1.2) is continuous on $HB(\Omega)$ and consequently also weak-$*$ continuous.

The functional L_2, given by (14.1.3), is weakly-$*$ continuous. The trace operator mapping $HB(\Omega)$ (or $W^{2,1}(\Omega)$) onto $L^1(\Gamma)$ is continuous in the strong topologies, but not in the weak-$*$ topology of $HB(\Omega)$. Consequently, the functional L is not weakly-$*$ continuous.

Assume now that the plate, being subject to the loading (14.1.1), is a plastic plate with an εY-periodic structure.

Let us pass to the formulation of dual extremum principles. To this end we introduce the set $\mathcal{K}_0$ of kinematically admissible fields

$$\mathcal{K}_0 = \{ w \in W^{2,1}(\Omega) | \; w = 0 \, , \; \frac{\partial w}{\partial \boldsymbol{n}} = 0 \, , \quad \text{on } \Gamma_0 \} . \qquad (14.1.4)$$

For a given load multiplier λ, to be precised below in Lemma 14.1.1, the kinematical principle is formulated as follows:

Problem (P_λ^ε)

> Find
> $$\inf\{\int_\Omega j\left[\frac{x}{\varepsilon}, \kappa(w(x))\right] dx - \lambda L(w)| \ w \in \mathcal{K}_0\} \ .$$

As usual, the dual problem can be derived by using Rockafellar's theory of duality.
Problem $(P_\lambda^\varepsilon)^*$

> Find
> $$\sup\{-\int_\Omega j^*\left(\frac{x}{\varepsilon}, M(x)\right) dx| \ M \in \mathcal{S}_\lambda^\varepsilon\} \ .$$

The set $\mathcal{S}_\lambda^\varepsilon$ of statically admissible fields lying within the yield locus $\mathcal{C}\left(\frac{x}{\varepsilon}\right)$ is given by

$$\mathcal{S}_\lambda^\varepsilon = \{M \in L^\infty(\Omega, \mathbb{E}_2^s)| \ M^{\alpha\beta}{}_{,\beta\alpha} + \lambda b = 0 \ , \quad \text{in } \Omega; \quad Q = \lambda Q^0 \ ,$$
$$M_n = \lambda M_n^0 \ , \quad \text{on } \Gamma_1; \quad M(x) \in \mathcal{C}\left(\frac{x}{\varepsilon}\right) \ , \quad x \in \Omega\} \ . \tag{14.1.5}$$

Here Q is the effective shear force. More precisely, instead of Q and M_n we should use trace operators, say b_0 and b_1, cf. Sec. 13.1. In general, we have $b_0(M) \in [\gamma_0(W^{2,1}(\Omega))]^*$, $b_1(M) \in L^\infty(\Gamma_1)$, provided that $M \in \{M \in L^\infty(\Omega, \mathbb{E}_2^s)| \ M^{\alpha\beta}{}_{,\beta\alpha} \in L^1(\Omega)\}$. Particularly, if $M \in C^2(\bar{\Omega}, \mathbb{E}_2^s)$ then

$$Q = M^{\alpha\beta}{}_{,\beta}n_\alpha + \frac{\partial}{\partial s}(M^{\alpha\beta}n_\beta t_\alpha) \ , \quad \text{on } \Gamma_1 \ ,$$
$$M_n = M^{\alpha\beta}n_\alpha n_\beta \ , \quad \text{on } \Gamma_1 \ . \tag{14.1.6}$$

Here t is the tangent unit vector and s denotes a curvi-linear abscissa on Γ measured positively in the direction t.

We now pass to limit analysis problems.
Problem P_{LA}^ε

> Find
> $$\inf\{\int_\Omega j_\infty\left(\frac{x}{\varepsilon}, \kappa(v(x))\right) dx| \ v \in \mathcal{K}_0, \ L(v) = 1\} \ .$$

Problem $(P_{LA}^\varepsilon)^*$

> Find
> $$\sup\{\lambda(M)| \ M \in \mathcal{S}_\lambda^\varepsilon\} \ .$$

We recall that v occurring in the problem (P_{LA}^ε) stands for a velocity field.
As in the case of homogeneous plates we have the following result.

Lemma 14.1.1. The following conditions are equivalent:
(i) $\inf P_\lambda^\varepsilon = \sup(P_\lambda^\varepsilon)^* > -\infty.$

(ii) $\mathcal{S}_\lambda^\varepsilon \neq \emptyset.$

(iii) $\lambda^\varepsilon = \inf P_{LA}^\varepsilon = \sup(P_{LA}^\varepsilon)^* \geq \lambda.$ $\square$

Let us set

$$\widehat{L}(w,\mu) = L_2(w) - \int_{\Gamma_1} M_n^0 d\mu \,, \tag{14.1.7}$$

where $w \in HB(\Omega)$, $\mu \in \mathbf{M}^1(\Gamma_1)$.

We can now formulate relaxed problems.

Problem RP_λ^ε

$\left|\begin{array}{l}\text{Find} \\[4pt] \inf\{\displaystyle\int_\Omega j\left(\frac{x}{\varepsilon},\boldsymbol{\kappa}(w)\right) + \int_{\Gamma_0} j_\infty\left(\frac{x}{\varepsilon},\mathbf{F}\left(-\frac{\partial w}{\partial \boldsymbol{n}}\right)\right) d\Gamma + \int_{\Gamma_1} j_\infty\left(\frac{x}{\varepsilon},\mathbf{F}\left(\mu - \frac{\partial w}{\partial \boldsymbol{n}}d\Gamma\right)\right) \\[10pt] \qquad -\lambda\widehat{L}(w,\mu)|\, w \in HB(\Omega),\, \mu \in \mathbf{M}^1(\Gamma_1)\,,\, w = 0 \text{ on } \Gamma_0\}\,.\end{array}\right.$

Problem (RP_{LA}^ε)

$\left|\begin{array}{l}\text{Find} \\[4pt] \inf\{\displaystyle\int_\Omega j_\infty\left(\frac{x}{\varepsilon},\boldsymbol{\kappa}(v)\right) + \int_{\Gamma_0} j_\infty\left(\frac{x}{\varepsilon},\mathbf{F}\left(-\frac{\partial v}{\partial \boldsymbol{n}}\right)\right) d\Gamma \\[10pt] \qquad + \int_{\Gamma_1} j_\infty\left(\frac{x}{\varepsilon},\mathbf{F}\left(\mu - \frac{\partial v}{\partial \boldsymbol{n}}d\Gamma\right)\right)|\, \widehat{L}(v,\mu) = 1,\, v \in HB(\Omega),\, \mu \in \mathbf{M}^1(\Gamma_1)\,, \\[10pt] \hspace{9cm} v = 0 \text{ on } \Gamma_0\}\,.\end{array}\right.$

It can be shown that

$$(P_\lambda^\varepsilon)^* = (RP_\lambda^\varepsilon)^*\,, \qquad (P_{LA}^\varepsilon)^* = (RP_{LA}^\varepsilon)^*\,,$$
$$\inf P_\lambda^\varepsilon = \sup(P_\lambda^\varepsilon)^* = \sup(RP_\lambda^\varepsilon)^* = \min RP_\lambda^\varepsilon\,,$$

and similarly for the limit analysis problems. Consequently, we conclude that the problems RP_λ^ε and RP_{LA}^ε attain their infima in the space $HB(\Omega) \times \mathbf{M}^1(\Gamma_1)$. Thus we have

$$\lambda^\varepsilon = \inf P_{LA}^\varepsilon = \min RP_{LA}^\varepsilon\,. \tag{14.1.8}$$

The proof follows by the same method as in the case of homogeneous plates.

Homogenization

We are now in a position to find the effective plate model. We recall that $\dfrac{\partial w}{\partial \boldsymbol{n}}$ or $\dfrac{\partial v}{\partial \boldsymbol{n}}$ are elements of the space $L^1(\Gamma_1)$. Moreover, bounded sequences in $L^1(\Gamma_1)$ contain weakly-$*$ convergent subsequences with limits in $\mathbf{M}^1(\Gamma_1)$, in general.

Consider a sequence of functionals defined by

$$\Phi_\varepsilon(w,\mu) = \begin{cases} \displaystyle\int_\Omega \left(\frac{x}{\varepsilon}, \kappa(w(x))\right) dx\,, & \text{if } w \in W^{2,1}(\Omega) \text{ and } w = 0 \text{ on } \Gamma_0, \\[2mm] & \dfrac{\partial w}{\partial \boldsymbol{n}} ds = \mu \quad \text{on } \Gamma_1\,; \\[4mm] +\infty & \text{otherwise.} \end{cases} \qquad (14.1.9)$$

The functional $\widehat{L}$ defined on $HB(\Omega) \times \mathbb{M}^1(\Gamma)$ by (14.1.7) is continuous in the weak-$*$ topology of this space. Similarly, the set $\{(v,\mu) \in HB(\Omega) \times \mathbb{M}^1(\Gamma_1)\mid \widehat{L}(v,\mu) = 1, v = 0$ on $\Gamma_0\}$ is sequentially weakly-$*$ closed in $HB(\Omega) \times \mathbb{M}^1(\Gamma)$. Since the embedding of $HB(\Omega)$ into $W^{1,1}(\Omega)$ is compact therefore it suffices to find the Γ-limit of the sequence $\{\Phi_\varepsilon\}_{\varepsilon>0}$ in the topology $\tau = (s - W^{1,1}(\Omega)) \times (\text{ weak-}* - \mathbb{M}^1(\Gamma_1))$. Obviously, the functional $\widehat{L}$ then plays the role of a continuous perturbation functional.

We set

$$\mathcal{S}(\Omega) = \{\boldsymbol{M} \in L^\infty(\Omega, \mathbb{E}_2^s)\mid \text{ div div}\boldsymbol{M} \in L^1(\Omega)\,, \; M_n \in C_0(\Gamma_1)\}\,,$$
$$\mathcal{S}_\lambda(b, Q^0, M_n^0) = \{\boldsymbol{M} \in \mathcal{S}(\Omega)\mid \text{ div div}\boldsymbol{M} + \lambda b = 0 \quad \text{in} \quad \Omega\,;$$
$$Q = \lambda Q^0\,, \quad M_n = \lambda M_n^0 \text{ on } \Gamma_1\}\,,$$

where $b \in L^1(\Omega)$, $Q^0 \in L^\infty(\Gamma_1)$, $M_n^0 \in C_0(\Gamma_1)$. The space $\mathcal{S}(\Omega)$ is equipped with the following topology denoted by τ_S

$$\boldsymbol{M}_\varepsilon \overset{\tau_S}{\to} \boldsymbol{M} \text{ iff } \begin{cases} \boldsymbol{M}_\varepsilon \rightharpoonup \boldsymbol{M} & \text{in} \quad L^\infty(\Omega, \mathbb{E}_2^s) \quad \text{weak-}*\,, \\ \text{div div}\boldsymbol{M}_\varepsilon \rightharpoonup \text{div div}\boldsymbol{M} & \text{in} \quad L^1(\Omega) \text{ weakly}\,, \\ (M_\varepsilon^{\alpha\beta} n_\alpha n_\beta)|_{\Gamma_1} \to M_{n|\Gamma_1} & \text{in} \quad C_0(\Gamma_1) \text{ strongly}\,. \end{cases}$$

If the loading $b \in \mathbb{M}(\Omega)$ is admissible then we assume that div div$\boldsymbol{M}_\varepsilon \rightharpoonup$ div div$\boldsymbol{M}$ in $\mathbb{M}(\Omega)$ weak-$*$.

Let us define the set

$$\widehat{\mathcal{S}}^\varepsilon = \{\boldsymbol{M} \in \mathcal{S}(\Omega)\mid \boldsymbol{M}(x) \in \mathcal{C}_\varepsilon(x)\,, \text{ a.e. } x \in \Omega\}\,,$$

and make the following assumption:
(A) $\widehat{\mathcal{S}}^\varepsilon$ convergences to a closed set $\widehat{\mathcal{S}}$ in Kuratowski's sense for the topology τ_S on $\mathcal{S}(\Omega)$. $\qquad \square$

For every $s \in \Gamma_1$ we define

$$\mathcal{C}_b(s) = \text{ closure } \{M^{\alpha\beta} n_\alpha(s) n_\beta(s)\mid \boldsymbol{M} \in \widehat{\mathcal{S}}\}\,.$$

Obviously, $\mathcal{C}_b$ is a lower semicontinuous multivalued mapping with convex and closed values.

The main result of the present section is now formulated as the following theorem.

Theorem 14.1.2. The Γ-limit of $\{\Phi_\varepsilon\}_{\varepsilon>0}$ is given by

$$\Phi_h(w,\mu) = \int_\Omega j_h(\kappa(w)) + \int_{\Gamma_0} j_{h\infty}\left(\mathbf{F}\left(-\frac{\partial w}{\partial \boldsymbol{n}}\right)\right) ds + \int_{\Gamma_1} d_b\left(s, \mu - \frac{\partial w}{\partial \boldsymbol{n}} ds\right)\,, \qquad (14.1.10)$$

where $w \in HB(\Omega)$, $\mu \in \mathbb{M}^1(\Gamma_1)$ and d_b is the l.s.c., convex and positively homogeneous of degree-one function defined as follows

$$d_b(s,a) = I^*_{C_b(s)}(-a) \; . \qquad \qquad \Box$$

Prior to passing to the proof we shall provide some auxiliary results.

Corollary 14.1.3. Under the assumptions of Theorem 14.1.2 we have

$$\lim_{\varepsilon \to 0} \lambda^\varepsilon = \lambda^h \; , \tag{14.1.11}$$

where λ^ε is defined by (14.1.8) and

$$\lambda^h = \min\{\Phi_{h\infty}(v,\mu) \mid v \in HB(\Omega), \gamma_0 w = 0 \quad \text{on } \Gamma_0 \, , \mu \in \mathbb{M}^1(\Gamma_1) \, , \widehat{L}(v,\mu) = 1\} \; .$$

Here $\Phi_{h\infty}$ has the form (14.1.10) with j_h being replaced by $j_{h\infty}$.

In fact, (14.1.11) is a straightforward consequence of Theorem 14.1.2, the aforementioned continuity of $\widehat{L}$ and of the continuity of the trace operator γ_0. $\qquad \Box$

We are going to show that

$$\lambda^h = \min\left(\lambda^h_\Omega, \lambda^h_b\right) \; , \tag{14.1.12}$$

where

$$\lambda^h_\Omega = \inf\{\int_\Omega j_{h\infty}(\boldsymbol{\kappa}(v)) + \int_{\Gamma_0} j_{h\infty}\left(\mathbf{F}\left(-\frac{\partial v}{\partial \boldsymbol{n}}\right)\right) ds \mid v \in W^{2,1}(\Omega) \, ,$$

$$v = 0 \text{ on } \Gamma_0 \, , \; L(v) = 1\} \; , \tag{14.1.13}$$

$$\lambda^h_b = \inf\{\int_{\Gamma_1} d_b(s,\mu) \mid \int_{\Gamma_1} M^0_n d\mu = 1 \, , \; \mu \in \mathbb{M}^1(\Gamma_1)\}$$

$$= \sup\{\lambda \mid \lambda M^0_n(s) \in C_b(s) \text{ for every } s \in \Gamma_1\} \; . \tag{14.1.14}$$

Indeed, we may write

$$\lambda^h = \min_{\substack{v \in HB(\Omega) \\ v_{|\Gamma_0}=0}} \{\min \Phi_{h\infty}(v,\mu) \mid \mu \in \mathbb{M}^1(\Gamma_1), \int_{\Gamma_1} M^0_n d\mu = -1 + \int_\Omega bv dx + \int_{\Gamma_1} Q^0 v ds\}$$

$$= \min_{\substack{v \in HB(\Omega) \\ v_{|\Gamma_0}=0}} \{\int_\Omega j_{h\infty}(\boldsymbol{\kappa}(v)) + \int_{\Gamma_0} j_{h\infty}\left(\mathbf{F}\left(-\frac{\partial v}{\partial \boldsymbol{n}}\right)\right) ds + \min_{\mu \in \mathbb{M}^1(\Gamma_1)} \int_{\Gamma_1} d_b(s, \mu - \frac{\partial v}{\partial \boldsymbol{n}} ds)\} \; .$$

Let us set

$$\int_\Omega bv dx + \int_{\Gamma_1} Q^0 v ds - 1 = \mathcal{L}(b, Q^0, v) \; .$$

By using the definition of $d_b(\cdot, \cdot)$ we obtain

$$\inf\{\int_{\Gamma_1} d_b(x, \mu - \frac{\partial v}{\partial \boldsymbol{n}} ds) | \ \mu \in \mathbb{M}^1(\Gamma_1), \ \int_{\Gamma_1} M_n^0 d\mu = \mathcal{L}(b, Q^0, v)\} = \lambda_b^h |1 - L(v)| \ ,$$

where

$$\lambda_b^h = \sup\{\lambda | \ \lambda M_n^0(s) \in \mathcal{C}_b(s) \text{ for each } s \in \Gamma_1\} \ .$$

Here $L(v)$ is given by (14.1.1). Consequently

$$\lambda^h = \inf\{\int_\Omega j_{h\infty}(\boldsymbol{\kappa}(v)) + \int_{\Gamma_0} j_{h\infty}\left(\mathbf{F}\left(-\frac{\partial v}{\partial \boldsymbol{n}}\right)\right) ds$$
$$+ \ \lambda_b^h |1 - L(v)| \ \big| \ 0 \le L(v) \le 1, \ v \in HB(\Omega) \ , \ v = 0 \text{ on } \Gamma_0\}$$
$$= \min\{\lambda_b^h, \lambda_\Omega^h\} \ ,$$

because the infimum in the expression defining λ^h is attained for $L(v) = 0$ or $L(v) = 1$.
   We observe that if $L(v) \ne 0$, the function

$$t \in \mathbb{R} \to \int_\Omega j_{h\infty}(\boldsymbol{\kappa}(tv)) + \int_{\Gamma_0} j_{h\infty}\left(\mathbf{F}\left(-\frac{\partial (tv)}{\partial \boldsymbol{n}}\right)\right) ds + \lambda_b^h |1 - L(tv)|$$

is piecewise affine, decreasing for $t < 0$ and increasing for $t > 1/L(v)$.
   We pass to the second auxiliary result.

Lemma 14.1.4. Let a sequence $\{M_\varepsilon\}_{\varepsilon>0} \subset \mathcal{S}(\Omega)$ be convergent to $M \in \mathcal{S}(\Omega)$ in the topology τ_S as $\varepsilon \to 0$. Then
(i)

$$\liminf_{\varepsilon \to 0} \int_\Omega j^*\left(\frac{x}{\varepsilon}, M_\varepsilon(x)\right) dx \ge \int_\Omega j_h^*(M(x)) dx \ . \tag{14.1.15}$$

(ii) If M_ε belongs to $\widehat{\mathcal{S}}^\varepsilon$ then M belongs to

$$\widehat{\mathcal{S}}^h = \{M \in \mathcal{S}(\Omega) | \ M(x) \in \mathcal{C}^h \text{ a.e. } x \in \Omega\} \ . \tag{14.1.16}$$

Proof. Let $\{\Omega_i\}_{i \in \mathcal{I}}$ be a finite family of disjoint, sufficiently regular open subsets of Ω such that meas $(\Omega \backslash \bigcup_{i \in \mathcal{I}} \Omega_i) = 0$. For instance, Ω_i may be a domain with the uniform C^2-regularity property, except possibly at a finite number of points of $\partial \Omega_i$.
   We set

$$\boldsymbol{\rho}(x) = \sum_{i \in \mathcal{I}} \chi_{\Omega_i}(x) \boldsymbol{\rho}^i \ , \qquad \boldsymbol{\rho}^i \in \mathbb{E}_s^2 \ . \tag{14.1.17}$$

Applying Theorem 13.3.2 to each open set Ω_i (the local character of Γ-limit), we conclude the existence of a sequence $\{w_\varepsilon^i\}_{\varepsilon>0} \subset HB(\Omega_i)$ such that:

$$w_\varepsilon^i = 0 \quad \text{and} \quad \frac{\partial w_\varepsilon^i}{\partial n} = 0 \quad \text{on } \partial\Omega_i , \qquad \lim_{\varepsilon\to 0} w_\varepsilon^i = 0 \quad \text{in} \quad W^{1,1}(\Omega_i) ,$$

$$\lim_{\varepsilon\to 0} \int_{\Omega_i} j\left(\frac{x}{\varepsilon}, \boldsymbol{\rho}^i + \boldsymbol{\kappa}(w_\varepsilon^i)\right) dx = \int_{\Omega_i} j_h(\boldsymbol{\rho}^i) dx .$$

Hence, for $w^\varepsilon = \sum_{i\in\mathcal{I}} \chi_{\Omega_i}(x) w_\varepsilon^i$ we have

$$w^\varepsilon = 0 \quad \text{and} \quad \frac{\partial w^\varepsilon}{\partial n} = 0 \quad \text{on } \partial\Omega , \qquad \lim_{\varepsilon\to 0} w^\varepsilon = 0 \quad \text{in } W^{1,1}(\Omega_i) ,$$

$$\lim_{\varepsilon\to 0} \int_{\Omega} j\left(\frac{x}{\varepsilon}, \boldsymbol{\rho}(x) + \boldsymbol{\kappa}(w^\varepsilon)\right) dx = \int_{\Omega} j_h(\boldsymbol{\rho}(x)) dx .$$

For a sequence $\{\boldsymbol{M}_\varepsilon\}_{\varepsilon>0} \subset \mathcal{S}(\Omega)$ τ_S-converging to $\boldsymbol{M}$ Fenchel's inequality gives:

$$\int_{\Omega} j^*\left(\frac{x}{\varepsilon}, \boldsymbol{M}_\varepsilon(x)\right) dx \geq \int_{\Omega} \boldsymbol{M}_\varepsilon(x) : [\boldsymbol{\rho}(x) + \boldsymbol{\kappa}(w^\varepsilon(x))] dx$$

$$- \int_{\Omega} j\left(\frac{x}{\varepsilon}, \boldsymbol{\rho}(x) + \boldsymbol{\kappa}(w^\varepsilon(x))\right) dx .$$

Hence

$$\lim_{\varepsilon\to 0} \inf \int_{\Omega} j^*\left(\frac{x}{\varepsilon}, \boldsymbol{M}_\varepsilon(x)\right) dx \geq \int_{\Omega} [\boldsymbol{M}(x) : \boldsymbol{\rho}(x) - j_h(\boldsymbol{\rho}(x))] dx , \qquad (14.1.18)$$

since

$$\lim_{\varepsilon\to 0} \int_{\Omega} \boldsymbol{M}_\varepsilon(x) : \boldsymbol{\kappa}(w^\varepsilon(x)) dx = -\lim_{\varepsilon\to 0} \int_{\Omega} (\operatorname{div} \operatorname{div} \boldsymbol{M}_\varepsilon) w^\varepsilon(x) dx = 0 .$$

We recall that $w^\varepsilon \in C(\overline{\Omega})$.

Piecewise constant functions of the form (14.1.17) are dense in $L^1(\Omega, \mathbb{E}_s^2)$. Taking the supremum on the r.h.s. of (14.1.18) and using Rockafellar's Theorem 1.2.33 we get

$$\lim_{\varepsilon\to 0} \inf \int_{\Omega} j^*\left(\frac{x}{\varepsilon}, \boldsymbol{M}_\varepsilon(x)\right) dx$$

$$\geq \sup\left\{ \int_{\Omega} [\boldsymbol{M}(x) : \boldsymbol{\rho}(x) - j_h(\boldsymbol{\rho}(x))] dx \mid \boldsymbol{\rho} \in L^1(\Omega, \mathbb{E}_s^2) \right\}$$

$$= \int_{\Omega} \sup_{\boldsymbol{\rho}\in\mathbb{E}_s^2} [\boldsymbol{M}(x) : \boldsymbol{\rho} - j_h(\boldsymbol{\rho})] dx = \int_{\Omega} j_h^*(\boldsymbol{M}(x)) dx ,$$

and thus (i) is proved.

To prove (ii) we observe that if $M_\varepsilon \in \widehat{\mathcal{S}}^\varepsilon$ then the left hand side of (4.1.15) is finite and consequently $M(x) \in C^h$ for a.e. $x \in \Omega$. We recall that $C^h = \operatorname{dom} j_h^*$. The proof of the lemma is complete. $\qquad\qquad\square$

Proof of Theorem 14.1.2. It will be divided into two parts.
I. First it will be shown that

$$\Gamma(\tau) - \lim_{\varepsilon \to 0} \inf \Phi_\varepsilon \geq \Phi_h \ .$$

Obviously, it suffices to prove that

$$\lim_{\varepsilon \to 0} \inf \Phi_\varepsilon \geq \Phi_h \ .$$

Since the role of Theorem 14.1.2 is to reveal the influence of Γ_1 therefore, to simplify the proof we assume that $\Gamma_0 = \emptyset$ or $\Gamma_1 = \partial\Omega$.

Take a sequence $\{w^\varepsilon, \mu^\varepsilon\}_{\varepsilon > 0} \subset W^{1,1}(\Omega) \times \mathbb{M}^1(\partial\Omega)$ converging to (w, μ) in the topology τ, and such that

$$\Phi_\varepsilon(w^\varepsilon, \mu^\varepsilon) \leq \text{const.}$$

Hence $w^\varepsilon \in W^{2,1}(\Omega)$ and $\mu^\varepsilon = \dfrac{\partial w^\varepsilon}{\partial \boldsymbol{n}} ds$. Since the sequence $\{w^\varepsilon\}_{\varepsilon > 0}$ is bounded in $W^{2,1}(\Omega)$ and the imbedding of $W^{1,1}(\Omega)$ into $W^{2,1}(\Omega)$ is compact, therefore

$$\lim_{\varepsilon \to 0} w^\varepsilon = w \quad \text{in} \quad W^{1,1}(\Omega) \text{ strongly} \ ,$$

$$\lim_{\varepsilon \to 0} \frac{\partial w^\varepsilon}{\partial \boldsymbol{n}} ds = \mu \quad \text{in} \quad \mathbb{M}^1(\partial\Omega) \text{ weak-}* \ .$$

Now let M be an element of $\widehat{S}$ and $\{M_\varepsilon\}_{\varepsilon > 0} \subset \widehat{\mathcal{S}}^\varepsilon$ a sequence τ_S - converging to M. From Sec. 13.2 we know that $\xi^\varepsilon = M_\varepsilon : \kappa(w^\varepsilon)$ is a measure on Ω defined as follows

$$\langle \xi^\varepsilon, \varphi \rangle = -\int_\Omega \varphi w^\varepsilon \operatorname{div} \operatorname{div} M_\varepsilon + 2 \int_\Omega M_\varepsilon : (\nabla w^\varepsilon \otimes \nabla \varphi) dx + \int_\Omega w^\varepsilon M_\varepsilon : (\nabla\nabla\varphi) dx \ ,$$

$$(14.1.19)$$

for all $\varphi \in \mathbf{D}(\Omega)$. We recall that $\kappa_{\alpha\beta}(w) = -w_{,\alpha\beta}$. Hence

$$\lim_{\varepsilon \to 0} \langle \xi^\varepsilon, \varphi \rangle = \langle \xi, \varphi \rangle = -\int_\Omega \varphi w \operatorname{div} \operatorname{div} M + 2 \int_\Omega M : (\nabla w \otimes \nabla \varphi) dx$$

$$+ \int_\Omega w M : (\nabla\nabla\varphi) dx \ . \qquad\qquad (14.1.20)$$

The r.h.s. of (14.1.20) defines a measure $\xi = M : \kappa(w)$ and $\xi^\varepsilon \rightharpoonup \xi$ in $\mathbb{M}^1(\Omega)$ weak-$*$ when $\varepsilon \to 0$.

We set

$$\Omega^\delta = \{x \in \Omega \,|\, \operatorname{dist}(x, \partial\Omega) > \delta\} \ , \qquad \delta > 0 \ .$$

Then

$$\lim_{\varepsilon\to 0}\xi^{\varepsilon}(\Omega^{\delta}) = \xi(\Omega^{\delta}) , \tag{14.1.21}$$

for δ outside at most a countable set (on $\partial(\Omega^{\delta})$ the measures may have atoms, the set of such δ - is at most countable).

Let us take such a δ. We may write

$$\int_{\Omega} j\left(\frac{x}{\varepsilon},\boldsymbol{\kappa}(w^{\varepsilon})\right) dx = \int_{\Omega^{\delta}} j\left(\frac{x}{\varepsilon},\boldsymbol{\kappa}(w^{\varepsilon})\right) dx + \int_{\Omega\setminus\Omega^{\delta}} j\left(\frac{x}{\varepsilon},\boldsymbol{\kappa}(w^{\varepsilon})\right) dx . \tag{14.1.22}$$

On account of Theorem 13.3.2 we have

$$\lim_{\varepsilon\to 0}\int_{\Omega^{\delta}} j\left(\frac{x}{\varepsilon},\boldsymbol{\kappa}(w^{\varepsilon})\right) dx \geq \int_{\Omega^{\delta}} j_h(\boldsymbol{\kappa}(w)) . \tag{14.1.23}$$

Moreover

$$\int_{\Omega\setminus\Omega^{\delta}} j\left(\frac{x}{\varepsilon},\boldsymbol{\kappa}(w^{\varepsilon})\right) dx \geq \int_{\Omega\setminus\Omega^{\delta}} j_{\infty}\left(\frac{x}{\varepsilon},\boldsymbol{\kappa}(w^{\varepsilon})\right) dx - K|\Omega\setminus\Omega^{\delta}|$$

$$\geq \int_{\Omega\setminus\Omega^{\delta}} \boldsymbol{M}_{\varepsilon} : \boldsymbol{\kappa}(w^{\varepsilon})dx - K|\Omega\setminus\Omega^{\delta}| , \tag{14.1.24}$$

where K is a strictly positive constant such that

$$j^*(y,\boldsymbol{\rho}^*) \leq K \quad \text{for every } \boldsymbol{\rho}^* \text{ in } \mathcal{C}(y) . \tag{14.1.25}$$

In (14.1.24) the following estimate of the recession function j_{∞} is used:

$$j_{\infty}(y,\boldsymbol{\rho}) \leq j(y,\boldsymbol{\rho}) + K , \tag{14.1.26}$$

which results from (14.1.25). Indeed, we may write

$$j^*(y,\boldsymbol{\rho}^*) - K \leq 0 \leq I_{\mathcal{C}(y)}(\boldsymbol{\rho}^*) ,$$

for each $\boldsymbol{\rho}^* \in \mathbb{E}_2^s$. Hence

$$I^*_{\mathcal{C}(y)}(\boldsymbol{\rho}) = j_{\infty}(y,\boldsymbol{\rho}) \leq (j^*(y,\cdot) - K)^*(\boldsymbol{\rho}) = j(y,\boldsymbol{\rho}) + K ,$$

for ever $\boldsymbol{\rho} \in \mathbb{E}_s^2$.

We can now return to (14.1.24). Generalized Green's formula yields, cf. Sec. 13.1

$$\int_{\Omega} \boldsymbol{M}_{\varepsilon} : \boldsymbol{\kappa}(w^{\varepsilon})dx = -\int_{\Omega} w^{\varepsilon}\,\text{div}\;\text{div}\boldsymbol{M}_{\varepsilon}dx$$

$$+ \int_{\partial\Omega} b_0(\boldsymbol{M}_{\varepsilon})\gamma_0 w^{\varepsilon}ds - \int_{\partial\Omega} b_1(\boldsymbol{M}_{\varepsilon})\gamma_1 w^{\varepsilon}ds , \tag{14.1.27}$$

We recall that for $M \in C^2(\overline{\Omega}, \mathbb{E}_2^s)$:

$$b_0(M) = Q = M^{\alpha\beta}{}_{,\beta} n_\alpha + \frac{\partial}{\partial s}(M^{\alpha\beta} n_\beta \tau_\alpha) \,, \qquad b_1(M) = M_n = M^{\alpha\beta} n_\alpha n_\beta \,.$$

To simplify the notation we shall also write Q, M_n in the case studied, where $M \in \mathcal{S}(\Omega)$.

Passing to the limit in (14.1.27) we obtain

$$\lim_{\varepsilon \to 0} \int_\Omega M_\varepsilon : \kappa(w^\varepsilon)dx = -\int_\Omega w\mathrm{div}\ \mathrm{div}M\,dx + \int_{\partial\Omega} Qw\,ds - \int_{\partial\Omega} M_n d\mu \,. \quad (14.1.28)$$

Taking into account (14.1.21) we get

$$\lim_{\varepsilon \to 0} \int_{\Omega^\delta} M_\varepsilon : \kappa(w^\varepsilon)dx = \int_{\Omega^\delta} M : \kappa(w) \,. \quad (14.1.29)$$

By virtue of (14.1.28) and (14.1.29) we obtain

$$\lim_{\varepsilon \to 0} \int_{\Omega\backslash\Omega^\delta} M_\varepsilon : \kappa(w^\varepsilon)dx = -\int_\Omega w\mathrm{div}\,\mathrm{div}M\,dx + \int_{\partial\Omega} Qw\,ds - \int_{\partial\Omega} M_n d\mu - \int_{\Omega^\delta} M : \kappa(w) \,.$$

Now (14.1.22) – (14.1.24) and the last relation yield

$$\liminf_{\varepsilon \to 0} \Phi_\varepsilon(w^\varepsilon, \mu^\varepsilon) \geq \int_{\Omega^\delta} j_h(\kappa(w)) - \int_\Omega w\mathrm{div}\,\mathrm{div}M\,dx + \int_{\partial\Omega} Qw\,ds$$

$$- \int_{\partial\Omega} M_n d\mu - \int_{\Omega^\delta} M : \kappa(w) - K|\Omega\backslash\Omega^\delta| \,.$$

Let now δ tend to zero. Applying the formula (13.1.27) with $\varphi = 1$ we arrive at

$$\liminf_{\varepsilon \to 0} \Phi_\varepsilon(w^\varepsilon, \mu^\varepsilon) \geq \int_\Omega j_h(\kappa(w)) + \int_{\partial\Omega} M_n(-(d\mu - \frac{\partial w}{\partial n}ds)) \,.$$

Taking the supremum on the r.h.s. with respect to $M \in \widehat{\mathcal{S}}$ and assuming for the moment that

$$\sup\{\int_{\partial\Omega} M_n(-(d\mu - \frac{\partial w}{\partial n}ds))|\ M \in \widehat{\mathcal{S}}\} = \int_{\partial\Omega} \sup_{a\in\mathcal{C}_b(s)}[a(-(d\mu - \frac{\partial w}{\partial n}ds))]$$

$$= \int_{\partial\Omega} d_b(s, \mu - \frac{\partial w}{\partial n}ds) \,, \quad (14.1.30)$$

we complete the first part of the proof. Thus it remains to show Eq. (14.1.30). To this end, it suffices to prove that the set

$$\widehat{\mathcal{S}}_b = \{b_1(M) \mid M \in \widehat{\mathcal{S}}\} \,, \quad (14.1.31)$$

is C_c^∞-stable, cf. Sec. 13.2. It means that for a partition of unity $(\alpha_0, \ldots, \alpha_n)$, $\alpha_i \in C^\infty(\partial\Omega, [0,1])$, $\sum_{i=0}^{n} \alpha_i(s) = 1$ for each $s \in \partial\Omega$, and for each finite family of functions $M_i \in \widehat{\mathcal{S}}_b$ $(i = 0, \ldots, n)$, the function

$$M = \sum_{i=0}^{n} \alpha_i M_i \,, \tag{14.1.32}$$

is also an element of $\widehat{\mathcal{S}}_b$.

Indeed, by definition of $\widehat{\mathcal{S}}_b$ we conclude that there exists a family of functions M_i $(i = 0, \ldots, n)$ in $\widehat{\mathcal{S}}$ such that $\varphi_i = M_{in} = b_1(M_i) = M_i^{\alpha\beta} n_\alpha n_\beta$ on $\partial\Omega$. The C^2-uniform regularity property of Ω, which we assume, implies the existence of open sets $\mathbf{O}_i$ covering $\partial\Omega$ for $i \geq 1$ and of a partition of unity β_i subordinate to $\mathbf{O}_i$ $(i = 0, \ldots, n)$, $\beta_i \in \mathbf{D}(\mathbf{O}_i)$, $\sum_{i=1}^{n} \beta_i = 1$ with $\beta_0 \in \mathbf{D}(\Omega)$. We choose α_i $(i = 1, \ldots, n)$ as a restriction of β_i :

$$\alpha_i = \beta_{i|\mathbf{O}_i \cap \partial\Omega} \,, \qquad i = 1, \ldots, n \,.$$

The function α_0 is a restriction of β_0 extended by zero to $\overline{\Omega}$, i.e. $\alpha_0 \equiv 0$ on $\partial\Omega$. We have thus constructed a partition of unity $\overline{\beta}_i$ $(i = 0, \ldots, n)$ on $\overline{\Omega}$, where $\overline{\beta}_i = \beta_i$ for $i = 1, \ldots, n$, and $\overline{\beta}_0$ is the aforementioned extension of β_0.

It is clear that the function

$$\boldsymbol{M} = \sum_{i=0}^{n} \overline{\beta}_i \boldsymbol{M}_i \,,$$

satisfies the condition: $M = M_n = b_1(\boldsymbol{M}) = M^{\alpha\beta} n_\alpha n_\beta$ on $\partial\Omega$. It remains to prove that $\boldsymbol{M}$ is an element of $\widehat{\mathcal{S}}$. According to the definition of $\widehat{\mathcal{S}}$, each $\boldsymbol{M}_i$ can be approximated in the topology τ_S by a sequence $\boldsymbol{M}_i^\varepsilon$ of elements of $\widehat{\mathcal{S}}^\varepsilon$. Let

$$\boldsymbol{M}^\varepsilon = \sum_{i=0}^{n} \overline{\beta}_i \boldsymbol{M}_i^\varepsilon \,.$$

Having in mind the regularity of $\overline{\beta}_i$ and convexity of $\mathcal{C}\left(\dfrac{x}{\varepsilon}\right)$ we conclude that $\boldsymbol{M}^\varepsilon \in \widehat{\mathcal{S}}^\varepsilon$ and that $\boldsymbol{M}^\varepsilon$ converges to $\boldsymbol{M}$ in the topology τ_S . Thus $\boldsymbol{M}$ belongs to $\widehat{\mathcal{S}}$ and the set $\widehat{\mathcal{S}}_b$ defined by (14.1.31) is C_c^∞-stable.

Remark 14.1.5. An alternative form of Theorem 14.1.2 will be more suitable in proving its second part. Let us set

$$\widehat{L}(w, \vartheta, \mu) = L_1(w) + \int_{\Gamma_1} Q^0 \vartheta ds - \int_{\Gamma_1} M_n^0 d\mu \,, \tag{14.1.33}$$

$$\Psi_\varepsilon(w, \vartheta, \mu) = \begin{cases} \displaystyle\int_\Omega j\left(\frac{x}{\varepsilon}, \kappa(w(x))\right) dx & \text{if } w \in W^{2,1}(\Omega) \, , \; w = 0 \text{ on } \Gamma_0 \, , \\[2ex] & w = \vartheta \text{ and } \dfrac{\partial w}{\partial n} = \mu \quad \text{on} \quad \Gamma_1 \, ; \\[2ex] +\infty & \text{otherwise} \, . \end{cases}$$

Next, we introduce the following topology:

$$\tau_\vartheta = (s - W^{1,1}(\Omega)) \times (s - L^1(\Gamma_1)) \times (\text{weak-*} - \mathbb{M}^1(\Gamma_1))$$

and the topology τ_Q on the space $\mathcal{S}(\Omega)$:

$$\boldsymbol{M}_\varepsilon \overset{\tau_Q}{\to} \boldsymbol{M} \text{ iff } \begin{cases} \boldsymbol{M}_\varepsilon \rightharpoonup \boldsymbol{M} \quad \text{in } L^\infty(\Omega, \mathbb{E}_2^s) & \text{weak-*} \, , \\ \operatorname{div} \operatorname{div} \boldsymbol{M}_\varepsilon \rightharpoonup \operatorname{div} \operatorname{div} \boldsymbol{M} & \text{in } L^1(\Omega) \text{ weakly} \, , \\ Q^\varepsilon \rightharpoonup Q \text{ in } L^\infty(\Gamma_1) & \text{weak-*} \\ M_\varepsilon^{\alpha\beta} n_\alpha n_\beta \to M_n \quad \text{in } C_0(\Gamma_1) & \text{strongly} \, , \end{cases}$$

where $Q^\varepsilon = b_0(\boldsymbol{M}_\varepsilon), Q = b_0(\boldsymbol{M})$.

Now the sets $\widehat{\mathcal{S}}$ and $\mathcal{C}_b(s)$ are to be understood in the sense of the topology τ_Q. Let us formulate the relevant homogenization theorem.

Theorem 14.1.6. The sequence of functionals $\{\Psi_\varepsilon\}_{\varepsilon>0}$ is $\Gamma(\tau_\vartheta)$ - convergent to

$$\Psi_h(w, \vartheta, \mu) = \begin{cases} \Phi_h(w, \mu) & \text{if } w \in HB(\Omega) \, , \; w = 0 \text{ on } \Gamma_0 \, , \\ & w = \vartheta \text{ on } \Gamma_1 \, , \; \mu \in \mathbb{M}^1(\Gamma_1) \, ; \\ +\infty & \text{otherwise} \, , \end{cases} \qquad (14.1.34)$$

where the limit functional Φ_h is given by (14.1.10). $\qquad\qquad \square$

The proof of the first part runs as before. Therefore we now turn to the second part of the proof. It is worth noting that for $w = \vartheta$ on Γ_1, the functionals Ψ_h and Φ_h coincide. Also, we recall that $HB(\Omega)$ is continuously embedded into $L^\infty(\Omega)$. Therefore we introduce the following topology

$$\tau_\infty = (\text{wcak-*} - L^\infty(\Omega)) \times (s - L^1(\Gamma_1)) \times (\text{weak-*} - \mathbb{M}^1(\Gamma_1)) \, .$$

II. We pass to proving that

$$\Gamma(\tau_\vartheta) - \lim_{\varepsilon \to 0} \sup \Psi_\varepsilon \le \Psi_h \, .$$

Primarily, however, the conjugate functionals Ψ_ε^* and Ψ_h^* will be derived.

Lemma 14.1.7. In the duality between $L^\infty(\Omega) \times L^1(\Gamma_1) \times \mathbb{M}^1(\Gamma_1)$ and $L^1(\Omega) \times L^\infty(\Gamma_1) \times C_0(\Gamma_1)$ we have

$$\Psi_\varepsilon^*(b, Q^0, M_n^0) = \inf\{\int_\Omega j^*\left(\frac{x}{\varepsilon}, \boldsymbol{M}(x)\right) dx - \int_{\Gamma_0} Q w_0 \, ds$$

$$+ \int_{\Gamma_0} M_n w_1 ds \mid \boldsymbol{M} \in \mathcal{S}(b, Q^0, M_n^0)\} \ , \tag{14.1.35}$$

$$\Psi_h^*(b, Q^0, M_n^0) = \begin{cases} \inf\{\displaystyle\int_\Omega j_h^*(\boldsymbol{M}(x))dx - \int_{\Gamma_0} Q w_0 ds \\ \qquad + \displaystyle\int_{\Gamma_0} M_n w_1 ds \mid \boldsymbol{M} \in \mathcal{S}(b, Q^0, M_n^0)\} \\ \qquad\quad \text{if } (-M_n(s)) \in \mathcal{C}_b(s) \text{ for each } s \in \Gamma_1 \ ; \\ +\infty \qquad \text{otherwise.} \end{cases} \tag{14.1.36}$$

Proof. Let us take (b, Q^0, M_n^0) in $L^1(\Omega) \times L^\infty(\Gamma_1) \times C_0(\Gamma_1)$. We find

$$\Psi_\varepsilon^*(b, Q^0, M_n^0) = \sup\{ \int_\Omega bw dx + \int_{\Gamma_1} Q^0 \vartheta ds + \int_{\Gamma_1} M_n^0 d\mu - \Psi_\varepsilon(w, \vartheta, \mu)$$

$$\mid (w, \vartheta, \mu) \in L^\infty(\Omega) \times L^1(\Gamma_1) \times \mathbf{M}^1(\Gamma_1)\}$$

$$= -\inf\{ \int_\Omega j\left(\frac{x}{\varepsilon}, \boldsymbol{\kappa}(w)\right) dx - \int_{\Gamma_1} Q^0 \vartheta ds$$

$$- \int_{\Gamma_1} M_n^0 d\mu \mid (w, \vartheta, \mu) \in L^\infty(\Omega) \times L^1(\Gamma_1) \times \mathbf{M}^1(\Gamma_1)\} \ .$$

According to the theory of duality outlined in Sec. 1.2.5 we set

$$G(\Lambda w) = \int_\Omega j\left(\frac{x}{\varepsilon}, \Lambda w\right) dx \ , \quad F(w, \vartheta, \mu) = -\widehat{\mathcal{L}}(w, \vartheta, \mu) \ ,$$

$$\Lambda w = \boldsymbol{\kappa}(w) \ , \quad \Lambda : W^{2,1}(\Omega) \to L^1(\Omega, \mathbf{E}_s^2) \ .$$

Here

$$\widehat{\mathcal{L}}(v, \vartheta, \mu) = \int_\Omega bw dx + \int_{\Gamma_1} Q^0 \vartheta ds - \int_{\Gamma_1} M_n^0 d\mu \ .$$

Hence

$$\langle \boldsymbol{M}, \Lambda w \rangle_{L^\infty(\Omega, \mathbf{E}_2^s) \times L^1(\Omega, \mathbf{E}_s^2)} = \int_\Omega \boldsymbol{M} : \boldsymbol{\kappa}(w) dx = - \int_\Omega w \operatorname{div} \operatorname{div} \boldsymbol{M} dx$$

$$+ \int_{\partial\Omega} Q w ds - \int_{\partial\Omega} M_n \frac{\partial w}{\partial \boldsymbol{n}} ds = \langle \Lambda^* \boldsymbol{M}, w \rangle \ ,$$

and consequently

$$\Lambda^* \boldsymbol{M} = \begin{cases} -\operatorname{div} \operatorname{div} \boldsymbol{M} & \text{in } \Omega \ , \\ Q & \text{on } \partial\Omega \ , \ (w) \ , \\ -M_n & \text{on } \partial\Omega \ , \ \left(\dfrac{\partial w}{\partial \boldsymbol{n}}\right) \ . \end{cases}$$

Next we find

$$(-\widehat{\mathcal{L}})^*(-\Lambda^* M) = \sup\{\langle -\Lambda^* M,\, w\rangle + \int_\Omega bw\,dx + \int_{\Gamma_1} Q^0 \vartheta\, ds - \int_{\Gamma_1} M_n^0 d\mu$$

$$| \; w \in W^{2,1}(\Omega)\,,\; w = w_0 \quad \text{and} \quad \frac{\partial w}{\partial n} = w_1,\; \text{on } \Gamma_0\}$$

$$= \begin{cases} -\displaystyle\int_{\Gamma_0} Q w_0\, ds + \int_{\Gamma_0} M_n w_1\, ds\,, & \text{if div div} M = 0 \text{ in } \Omega\,,\; Q = Q^0 \text{ and } M_n = M_n^0 \text{ on } \Gamma_1\,, \\[2mm] & \qquad w = \vartheta \text{ and } \dfrac{\partial w}{\partial n} = \mu \text{ on } \Gamma_1\,; \\[2mm] +\infty & \text{otherwise}\,. \end{cases}$$

$$(14.1.37)$$

Further we have, cf. the formula (1.2.31)

$$G^*(M) = \int_\Omega j^*\left(\frac{x}{\varepsilon},\, M(x)\right) dx\,, \qquad M \in L^\infty(\Omega, \mathbb{E}_2^s)\,. \tag{14.1.38}$$

Finally, from (14.1.37) and (14.1.38) we obtain (14.1.35).

We proceed now to proving (14.1.36). Since the functional Ψ_h is defined on the nonreflexive space $HB(\Omega) \times L^1(\Gamma_1) \times \mathbb{M}^1(\Gamma_1)$, therefore to avoid difficulties in applying the theory of duality, we introduce the following functional:

$$\Theta(w, \vartheta, \mu) = \begin{cases} \Psi_h(w, \vartheta, \mu) & \text{if } w \in W^{2,1}(\Omega)\,,\; w = \vartheta \text{ on } \Gamma_1\,,\; \mu \in L^1(\Gamma_1), \\ +\infty & \text{otherwise.} \end{cases} \tag{14.1.39}$$

We assert that Θ and Ψ_h have the same dual functions. More precisely, it will be shown that their l.s.c. regularized functionals in $L^\infty(\Omega) \times L^1(\Gamma_1) \times \mathbb{M}^1(\Gamma_1)$ coincide.

First, suppose that $\Theta \geq \Psi_h$. According to the definition of the lower semicontinuous regularization of a functional, cf. Sec. 1.2.1, we have $\Theta \geq \Psi_h \geq \overline{\Psi}_h$ and consequently $\overline{\Theta} \geq \overline{\Psi}_h$. Here $\overline{\Theta}$ denotes the l.s.c. regularization of Θ.

We shall prove now the reverse inequality: $\overline{\Theta} \leq \overline{\Psi}_h$. To this end, for any $(w, \vartheta, \mu) \in HB(\Omega) \times L^1(\Gamma_1) \times \mathbb{M}^1(\Gamma_1)$ with $w = \vartheta$ on Γ_1 we consider a sequence $\{w^\varepsilon, \vartheta^\varepsilon, \mu^\varepsilon\}_{\varepsilon>0} \subset W^{2,1}(\Omega) \times L^1(\Gamma_1) \times L^1(\Gamma_1)$ such that:

(a)

$$w^\varepsilon \to w \quad \text{in } W^{1,1}(\Omega) \quad \text{as } \varepsilon \to 0\,,$$

$$w^\varepsilon = w \quad \text{on } \Gamma_0\,,\; w^\varepsilon = \vartheta^\varepsilon \quad \text{on } \Gamma_1\,,\; \frac{\partial w^\varepsilon}{\partial n} = \frac{\partial w}{\partial n} \quad \text{on } \partial\Omega\,,$$

$$\vartheta^\varepsilon \to \vartheta \quad \text{in } L^1(\Gamma_1) \quad \text{as } \varepsilon \to 0\,, \qquad \lim_{\varepsilon \to 0} \int_\Omega j_h(\kappa(w^\varepsilon))\,dx = \int_\Omega j_h(\kappa(w))\,.$$

Particularly, we may take $w^\varepsilon = \vartheta^\varepsilon = w$ also on Γ_1.

(b)

$$\mu^\varepsilon \rightharpoonup \mu \quad \text{in } \mathbb{M}^1(\Gamma_1) \quad \text{weak-}* \quad \text{as } \varepsilon \to 0\,,$$

$$\lim_{\varepsilon \to 0} \int_{\Gamma_1} d_b(s, \mu^\varepsilon - \frac{\partial w}{\partial \boldsymbol{n}}) ds = \int_{\Gamma_1} d_b(s, \mu - \frac{\partial w}{\partial \boldsymbol{n}} ds) \ .$$

The existence of sequences $\{w^\varepsilon\}_{\varepsilon > 0}$, $\{\mu^\varepsilon\}_{\varepsilon > 0}$ follows from Sec. 13.1. For this particular sequence $\{w^\varepsilon, \vartheta^\varepsilon, \mu^\varepsilon\}_{\varepsilon > 0}$ we have:

$$\lim_{\varepsilon \to 0} \Theta(w^\varepsilon, \vartheta^\varepsilon, \mu^\varepsilon) = \Psi_h(w, \vartheta, \mu) \ ,$$

and consequently $\overline{\Theta} \leq \overline{\Psi}_h$.

We will now compute Θ^*. For (b, Q^0, M_n^0) in $L^1(\Omega) \times L^\infty(\Gamma_1) \times C_0(\Gamma_1)$ we find:

$$\Theta^*(b, Q^0, M_n^0) = \sup\{ \int_\Omega (bw - j_h(\boldsymbol{\kappa}(w)))dx + \int_{\Gamma_1} Q^0 \vartheta ds + \int_{\Gamma_1} [M_n^0 \theta$$
$$- d_b(s, \theta - \frac{\partial w}{\partial \boldsymbol{n}})]ds| \ (w, \vartheta, \theta) \in W^{2,1}(\Omega) \times L^1(\Gamma_1) \times C_0(\Gamma_1) \ ,$$
$$w = w_0 \text{ on } \Gamma_0 \ , \ w = \vartheta \text{ on } \Gamma_1\} \ . \tag{14.1.40}$$

Let us fix w and ϑ and calculate the supremum in θ. By using Rockafellar's Theorem 1.2.33 we obtain

$$\sup\{ \int_{\Gamma_1} [M_n^0 \theta - d_b(s, \theta - \frac{\partial w}{\partial \boldsymbol{n}})]ds| \ \theta \in L^1(\Gamma_1)\}$$
$$= \int_{\Gamma_1} \sup_{a \in \mathbb{R}} [M_n^0(x)a - d_b(s, a - \frac{\partial w}{\partial \boldsymbol{n}})]ds = \int_{\Gamma_1} \left(M_n^0 \frac{\partial w}{\partial \boldsymbol{n}} + I_{C_b(s)}(-M_n^0) \right) ds$$
$$= \begin{cases} \int_{\Gamma_1} M_n^0 \frac{\partial w}{\partial \boldsymbol{n}} ds & \text{if } (-M_n^0(s)) \in C_b(s), \text{ for every } s \in \Gamma_1 \ , \\ +\infty & \text{otherwise.} \end{cases} \tag{14.1.41}$$

The proof of (14.1.36) follows now immediately from (14.1.40) and (14.1.41) by applying Rockafellar's theory of duality. $\qquad\square$

To complete the proof of Theorem 14.1.2 we have to show that

$$\Gamma(\tau_\vartheta) - \lim_{\varepsilon \to 0} \sup \Psi_\varepsilon \leq \Psi_h \ . \tag{14.1.42}$$

To show that (14.1.42) really holds it suffices to establish the following inequality:

$$\Gamma(\tau_\infty^*) - \lim_{\varepsilon \to 0} \inf \Psi_\varepsilon^* \geq \Psi_h^* \ , \tag{14.1.43}$$

where $\tau_\infty^* = (w - L^1(\Omega)) \times (w^* - L^\infty(\Gamma_1)) \times (s - C_0(\Gamma_1))$. Indeed, let $(w, \vartheta, \mu) \in HB(\Omega) \times L^1(\Gamma_1) \times \mathbb{M}^1(\Gamma_1)$, $w = \vartheta$ on Γ_1 and assume that a sequence $\{w^\varepsilon, \vartheta^\varepsilon, \theta^\varepsilon\}_{\varepsilon > 0} \subset$

$W^{2,1}(\Omega) \times L^1(\Gamma_1) \times L^1(\Gamma_1)$, $w^\varepsilon = \vartheta^\varepsilon$ on Γ_1, converges to (w, ϑ, μ) in the topology τ_ϑ and is such that

$$\limsup_{\varepsilon \to 0} \Psi_\varepsilon(w^\varepsilon, \vartheta^\varepsilon, \theta^\varepsilon) \leq \Psi_h(w, \vartheta, \mu) . \tag{14.1.44}$$

Take now a sequence $\{b^\varepsilon, Q^\varepsilon, M^\varepsilon\}_{\varepsilon>0}$ in $L^1(\Omega) \times L^\infty(\Gamma_1) \times C_0(\Gamma_1)$ convergent to (b, Q^0, M_n^0) in τ_∞^*. Fenchel's transformation yields:

$$\Psi_\varepsilon^*(b^\varepsilon, Q^\varepsilon, M^\varepsilon) \geq \int_\Omega b^\varepsilon w^\varepsilon dx + \int_{\Gamma_1} (Q^\varepsilon \vartheta^\varepsilon + M^\varepsilon \theta^\varepsilon) ds - \Psi_\varepsilon(w^\varepsilon, \vartheta^\varepsilon, \theta^\varepsilon) .$$

Hence

$$\liminf_{\varepsilon \to 0} \Psi_\varepsilon^*(b^\varepsilon, Q^\varepsilon, M^\varepsilon) \geq \liminf_{\varepsilon \to 0} \{ \int_\Omega b^\varepsilon w^\varepsilon dx + \int_{\Gamma_1} (Q^\varepsilon \vartheta^\varepsilon + M^\varepsilon \theta^\varepsilon) ds \}$$
$$- \limsup_{\varepsilon \to 0} \Psi_\varepsilon(w^\varepsilon, \vartheta^\varepsilon, \theta^\varepsilon) .$$

By using (14.1.44) we obtain

$$\liminf_{\varepsilon \to 0} \Psi_\varepsilon^*(b^\varepsilon, Q^\varepsilon, M^\varepsilon) \geq \int_\Omega bw dx + \int_{\Gamma_1} (Q^0 \vartheta + M_n^0) ds - \Psi_h(w, \vartheta, \mu) .$$

The last inequality being valid for each (w, ϑ, μ) with $w = \vartheta$ on Γ_1, we arrive at the desired conclusion or (14.1.43).

Let $\{b^\varepsilon, Q^\varepsilon, M^\varepsilon\}_{\varepsilon>0} \subset L^1(\Omega) \times L^\infty(\Gamma_1) \times C_0(\Gamma_1)$ be a sequence convergent to (b, Q^0, M_n^0) in the topology τ_∞^*, and such that

$$\liminf_{\varepsilon \to 0} \Psi_\varepsilon^*(b^\varepsilon, Q^\varepsilon, M^\varepsilon) < +\infty .$$

By Lemma 14.1.7, there exists a sequence $\{M_\varepsilon\}_{\varepsilon>0}$ in $\mathcal{S}(\Omega)$ such that:
(i) $M_\varepsilon \in \mathcal{S}(b^\varepsilon, Q^\varepsilon, M^\varepsilon)$, thus

$$M_\varepsilon \in \widehat{\mathcal{S}}^\varepsilon ,$$

(ii)

$$\Psi_\varepsilon^*(b^\varepsilon, Q^\varepsilon, M^\varepsilon) \geq \int_\Omega j^* \left(\frac{x}{\varepsilon}, M_\varepsilon(x) \right) dx - \int_{\Gamma_0} Q^\varepsilon w_0 ds + \int_{\Gamma_0} M^\varepsilon w_1 ds - \varepsilon . \tag{14.1.45}$$

The condition (13.3.4) implies

$$\begin{aligned} j^*(y, \boldsymbol{\rho}^*) &\leq k_2 && \text{if } \boldsymbol{\rho}^* \in B(0, k_2) , \\ -k_3 &\leq j^*(y, \boldsymbol{\rho}^*) && \text{if } \boldsymbol{\rho}^* \in B(0, k_3) , \end{aligned} \tag{14.1.46}$$

where $B(0, k) \subset \mathbb{E}_2^s$ is a ball with a radius k. By virtue of (14.1.45) and (14.1.46), the sequence $\{M_\varepsilon\}_{\varepsilon>0}$ is bounded in $L^\infty(\Omega, \mathbb{E}_2^s)$; more precisely it belongs to $\mathcal{S}(b^\varepsilon, Q^\varepsilon, M^\varepsilon) \cap$

$\widehat{\mathcal{S}}^\varepsilon$. Consequently, it contains a τ_Q - convergent subsequence, say $\{M_{\varepsilon'}\}_{\varepsilon'>0}$. Its limit is denoted by M. Obviously M is an element of $\mathcal{S}(b, Q^0, M^0) \cap \widehat{\mathcal{S}}$. Then we find

$$\lim_{\varepsilon \to 0} \int_{\Gamma_0} Q^\varepsilon w_0 ds = \int_{\Gamma_0} Q w_0 ds , \qquad \lim_{\varepsilon \to 0} \int_{\Gamma_0} M^\varepsilon w_1 ds = \int_{\Gamma_0} M_n w_1 ds . \qquad (14.1.47)$$

Applying Lemma 14.1.4, combining it with (14.1.45) and (14.1.47) we get

$$\liminf_{\varepsilon \to 0} \Psi_\varepsilon^*(b^\varepsilon, Q^\varepsilon, M^\varepsilon) \geq \Psi_h^*(b, Q^0, M_n^0) , \qquad (14.1.48)$$

and thus the inequality (14.1.42) is established. The proof of Theorem 14.1.2 is complete.

$\square$

The assumption (A): some comments

The reader is certainly aware that the determination of the set $C_b(x)$ is not a simple problem. Therefore one may ask whether it is possible to find an estimate of this set. In some cases at least the answer is positive. Indeed, the limit in Kuratowski's sense in the strong topology of $C_0(\Gamma_1)$ of the sets

$$\widehat{\mathcal{S}}_1^\varepsilon = \{M^{\alpha\beta} n_\alpha n_{\beta|\Gamma_1}| \, M \in \widehat{\mathcal{S}}^\varepsilon\} ,$$

can sometimes be either determined or estimated. By using Theorem 13.2.11 with $\mu_n = ds$, $\mathbf{O} = \Gamma_1, p = 1$, we conclude that if $\widehat{\mathcal{S}}_1^{\varepsilon'}$ strongly converges in Kuratowski's sense in $C_0(\Gamma_1)$ to $\widehat{\mathcal{S}}_1$, then

$$\widehat{\mathcal{S}}_1 = \{\varphi \in C_0(\Gamma_1) \,,| \, \varphi(x) \in \widehat{C}(x) \quad \forall \, x \in \Gamma_1\} ,$$

where $\widehat{C}(x)$ is a l.s.c. multivalued mapping with closed convex values.

An explicit formula for $\widehat{C}(x)$ follows from Proposition 13.2.12 provided that the plate is made of two materials with strength domains C_1 and C_2. More precisely, let us define:

$$\Omega_1^\varepsilon = \{x \in \Omega \,|\, C\left(\frac{x}{\varepsilon}\right) = C_1\} , \qquad A_\varepsilon = \overline{\Omega}_1^\varepsilon \cap \Gamma_1 .$$

According to the aforementioned proposition we assume that

$$|\partial A_\varepsilon| = 0 , \qquad \text{int } A_\varepsilon \to A , \qquad \text{int } (\Gamma_1 \backslash A_\varepsilon) \to B .$$

Then A and B are closed and

$$\widehat{C}(x) = \begin{cases} C_{2b}(x) & \text{if } x \in A \backslash B , \\ C_{1b}(x) & \text{if } x \in B \backslash A , \\ C_{2b}(x) \cap C_{1b}(x) & \text{if } x \in A \cap B , \end{cases}$$

where $C_{\alpha b}(x) = (C_\alpha n(x)) \cdot n(x), \alpha = 1, 2$. Next, we introduce

$$C_b^h(x) = \text{closure } \{M^{\alpha\beta} n_\alpha(x) n_\beta(x) \,,| \, M \in \widehat{\mathcal{S}}^h\} ,$$

where $\widehat{\mathcal{S}}^h$ is defined by (14.1.16).

The following lemma confirms an intuitive fact.

Lemma 14.1.8. Let $\Omega \subset \mathbb{R}^2$ be bounded and of class C^2. Assume that div div M is in $L^2(\Omega)$. Then

$$\mathcal{C}_b^h(x) = (\mathcal{C}^h \boldsymbol{n}(x)) \cdot \boldsymbol{n}(x) \quad \text{for every } x \in \Gamma_1 .$$

Proof. The inclusion

$$(\mathcal{C}^h \boldsymbol{n}(x)) \cdot \boldsymbol{n}(x) \subset \mathcal{C}_b^h(x)$$

follows immediately. Indeed, if M is in $(\mathcal{C}^h \boldsymbol{n}(x)) \cdot \boldsymbol{n}(x)$ for one $x \in \Gamma_1$, there exists ρ^* in $\mathcal{C}^h$ such that $M = (\rho^* \boldsymbol{n}(x)) \cdot \boldsymbol{n}(x)$. The constant field $\boldsymbol{M}(x) = \rho^*$ belongs to $\widehat{S}^h$ and consequently $(\rho^* \boldsymbol{n}(x)) \cdot \boldsymbol{n}(x)$ is in $\mathcal{C}_b^h(x)$. To prove the reverse inclusion we extend Lemma 13.1.5. In our case we have the following characterization of $(\boldsymbol{M}\boldsymbol{n}) \cdot \boldsymbol{n}$ on $\partial\Omega$ at every Lebesgue point of $(\boldsymbol{M}\boldsymbol{n}) \cdot \boldsymbol{n}$:

$$(\boldsymbol{M}\boldsymbol{n}(x)) \cdot \boldsymbol{n}(x) = \lim_{\rho \to 0^+} \lim_{r \to 0^+} \frac{1}{|Q_{r,\rho}(x, \boldsymbol{n}(x))|} \int_{Q_{r,\rho}(x,\mathbf{n}(x))} (\boldsymbol{M}(\xi)\boldsymbol{n}(x)) \cdot \boldsymbol{n}(x)d\xi ,$$

where

$$Q_{r,\rho}(x, \boldsymbol{n}(x)) = \{\xi - t\boldsymbol{n}(x)| \; |\xi - x| < \rho \, , \, 0 < t < r\} .$$

Consequently $(\boldsymbol{M}\boldsymbol{n}(x)) \cdot \boldsymbol{n}(x)$ is in $(\mathcal{C}^h \boldsymbol{n}(x)) \cdot \boldsymbol{n}(x)$ for a.e. x in $\partial\Omega$ provided that $\boldsymbol{M}(\xi)$ belongs to $\mathcal{C}^h$ for a.e. ξ in Ω. Moreover, for M being an element of $\widehat{S}^h$, the continuity of both $(\boldsymbol{M}\boldsymbol{n}(x))\cdot\boldsymbol{n}(x)$ and of the multivalued mapping $(\mathcal{C}^h\boldsymbol{n}(x))\cdot\boldsymbol{n}(x)$ imply that $(\boldsymbol{M}\boldsymbol{n}(x))\cdot \boldsymbol{n}(x)$ is in $(\mathcal{C}^h\boldsymbol{n}(x)) \cdot \boldsymbol{n}(x)$ for every $x \in \Gamma_1$. Thus $(\mathcal{C}^h\boldsymbol{n}(x)) \cdot \boldsymbol{n}(x)$ contains $\{(\boldsymbol{M}\boldsymbol{n}(x)) \cdot \boldsymbol{n}(x)| \; M \in \widehat{S}^h\}$. The set $(\mathcal{C}^h\boldsymbol{n}(x)) \cdot \boldsymbol{n}(x)$ is closed and therefore $\mathcal{C}_b^h(x)$ is contained in it and the proof is complete. $\qquad\square$

The sets $\mathcal{C}_b(x), \mathcal{C}_b^h(x)$ and $\widehat{\mathcal{C}}(x)$ are interrelated according to the following proposition.

Proposition 14.1.9. Under the above assumptions:

(i) $$[(\mathcal{C}_1 \cap \mathcal{C}_2)\boldsymbol{n}(x)] \cdot \boldsymbol{n}(x) \subset \mathcal{C}_b(x) \subset \widehat{\mathcal{C}}(x) \cap \mathcal{C}_b^h(x) .$$

(ii) $$d_b(x, a) \leq j_{h\infty}(-a\boldsymbol{n}(x) \otimes \boldsymbol{n}(x)) .$$

Proof. The first inclusion in (i) is evident.

To prove the second inclusion we take $M \in \mathcal{C}_b(x)$ and $M' = \eta M$ with $0 < \eta < 1$. One can find φ in $C_0(\Gamma_1)$ such that $\varphi(x) = M'$, and $\boldsymbol{M}$ in $\widehat{S}$ such $M^{\alpha\beta}n_\alpha n_\beta = \varphi$ on Γ_1. Let $\{\boldsymbol{M}_\varepsilon\}_{\varepsilon>0}$ be a sequence of the elements of $\widehat{S}^\varepsilon$ converging to $\boldsymbol{M}$. By virtue of Lemma 14.1.4 we then have

$$\boldsymbol{M}(x) \in \mathcal{C}^h \quad \text{for} \quad \text{a.e. } x \in \Omega .$$

Therefore $\varphi(x) = (\boldsymbol{M}\boldsymbol{n}(x)) \cdot \boldsymbol{n}(x) \in (\mathcal{C}^h\boldsymbol{n}(x)) \cdot \boldsymbol{n}(x)$ for every x in Γ_1; moreover

$$((\boldsymbol{M}_\varepsilon\boldsymbol{n}) \cdot \boldsymbol{n})_{|\Gamma_1} \to ((\boldsymbol{M}\boldsymbol{n}) \cdot \boldsymbol{n})_{|\Gamma_1} \quad \text{in } C_0(\Gamma_1) \text{ strongly} .$$

Thus $\varphi(x) = (\boldsymbol{M}\boldsymbol{n}(x)) \cdot \boldsymbol{n}(x) \in \widehat{\mathcal{C}}(x)$ for every x in Γ_1 and M' belongs to $\widehat{\mathcal{C}}(x)$. Let now $\eta \to 1^-$. In this manner the conclusion is extended to M and (i) follows. Now (ii) is a direct consequence of the second inclusion in (i) and the definition of $d_b(x, \cdot)$. This completes the proof. $\qquad\qquad\qquad\qquad\qquad\qquad\qquad\qquad\qquad\qquad\qquad\qquad\qquad\qquad\qquad\quad$ $\square$

Remark 14.1.10. In the specific case when $C_1 \subset C_2$ we obtain: $C_b(x) = (C_1 n(x)) \cdot n(x)$ for every x in B. On B the strength of the boundary is governed by the strength of the weakest material.

14.2.   *Three models of thin, transversely inhomogeneous and anisotropic plates with constant thickness*

In Sec. 2.10 three models of thin linear elastic plates were justified by the Γ-convergence method. The present section is devoted to similar problems provided that the plate is made of Hencky material. As we already know from the Sec. 14.1, Hencky's plasticity leads to the study of functionals with linear growth and homogenization procedure involves relaxation of at least some of boundary conditions. Moreover, the loading functional influences the choice of functional spaces.

14.2.1.   Basic relations

A thin three-dimensional domain B_e and its boundary have been introduced in Sec. 2.10.1. In this section we set $c = 1$. Now only the loading is rescaled, but not the elastic moduli. The complementary elasto-plastic potential is assumed in the following form

$$j_{e\varepsilon}^*(x, \sigma) = j^*\left(\frac{x_\alpha}{\varepsilon}, \frac{x_3}{e}, \sigma\right) = \frac{1}{2} c_{ijkl}\left(\frac{x_\alpha}{\varepsilon}, \frac{x_3}{e}\right) \sigma^{ij}\sigma^{kl} + I_{C\left(\frac{x_\alpha}{\varepsilon}, \frac{x_3}{e}\right)}(\sigma), \qquad (14.2.1)$$

where σ is the three-dimensional stress tensor and $x \in B_e$. From now on we omit the superscripts x and z, because it will be evident to which independent variables the quantities like strains are referred to.

    Within the framework of Hencky plasticity the constitutive relationship is assumed in the subdifferential form

$$e_{ij} = e_{ij}^{el} + e_{ij}^{pl} = c_{ijkl}\sigma^{kl} + e_{ij}^{pl}, \qquad (14.2.2)$$

where

$$e^{pl} \in \partial I_{C\left(\frac{x_\alpha}{\varepsilon}, \frac{x_3}{e}\right)}(\sigma). \qquad (14.2.3)$$

The elasticity convex $C\left(\dfrac{x_\alpha}{\varepsilon}, \dfrac{x_3}{e}\right)$ is a closed and convex set in $\mathbb{E}_3^s$. We assume that there exist constants k_0, k_1, such that $0 < k_0 \leq k_1 < +\infty$ and satisfying

$$\{\sigma \in \mathbb{E}_3^s : |\sigma| \leq k_0\} \subset C\left(\frac{x_\alpha}{\varepsilon}, \frac{x_3}{e}\right) \subset \{\sigma \in \mathbb{E}_3^s : |\sigma| \leq k_1\}, \qquad (14.2.4)$$

for every $x \in B_e$.

    Let $\sigma = \sigma_D + \dfrac{1}{3}, tr\,\sigma$, where $tr\,\sigma = \sum_{i=1}^{3}\sigma^{ii} = \delta_{ij}\sigma^{ij}$, and σ_D is the stress deviator. For plastically incompressible isotropic materials the yield condition does not depend on $tr\,\sigma$.

Particularly, in the case of the Huber-Mises-Hencky yield condition we have

$$C\left(\frac{x_\alpha}{\varepsilon},\frac{x_3}{e}\right) = \left\{\sigma \in \mathbb{E}_3^s : |\sigma_D| \le \tau_0\left(\frac{x_\alpha}{\varepsilon},\frac{x_3}{e}\right)\right\}, \quad x \in B_e . \tag{14.2.5}$$

For isotropic materials we have

$$c\sigma = a_1(tr\,\sigma)I + a_2\sigma_D , \tag{14.2.6}$$

where $a_1 = (1-2\nu)/3E$, $a_2 = (1+\nu)/E$; here ν denotes Poisson's ratio and, as usual, E is Young's modulus. The yield limit in pure shear $\tau_0\left(\frac{x_\alpha}{\varepsilon},\frac{x_3}{e}\right)$ is now εY-periodic, where the basic cell Y is two-dimensional.

The elasto-plastic potential $j\left(\frac{x_\alpha}{\varepsilon},\frac{x_3}{e},e\right)$ is found to be given by, cf. Eq. (13.3.3)

$$j\left(\frac{x_\alpha}{\varepsilon},\frac{x_3}{e},e\right) = \sup\{\sigma : e - j_{e\varepsilon}^*(x,\sigma)|\,\sigma \in \mathbb{E}_3^s\} , \tag{14.2.7}$$

where $e \in \mathbb{E}_s^3$. This function is a convex normal integrand and possesses the following property

$$\exists\, k_3 \ge k_2 > 0 , \quad k_2(|e|-1) \le j_{e\varepsilon}(x,e) \le k_3(1+|e|) , \tag{14.2.8}$$

for each $(x,e) \in B_e \times \mathbb{E}_s^3$; moreover $j_{e\varepsilon}(x,0) = 0$.

To prove (14.2.8) it is sufficient to use (14.2.4). For plastically incompressible materials the analogous condition takes the form:

$$\exists\, k_2, k_3, k_4, k_5 > 0 , \quad \text{such that}$$
$$k_4(tr\,e)^2 + k_2(|e^D|-1) \le j_{e\varepsilon}(x,e) \le k_5(tr\,e)^2 + k_3(|e^D|+1) , \tag{14.2.9}$$

where e^D is the deviator of $e \in \mathbb{E}_s^3$.

The recession function $j_\infty\left(\frac{x}{\varepsilon},\frac{x_3}{e},\cdot\right)$ of $j\left(\frac{x}{\varepsilon},\frac{x_3}{e},\cdot\right)$ is now defined by, cf. Eq. (13.3.5),

$$j_\infty\left(\frac{x}{\varepsilon},\frac{x_3}{e},e\right) = \sup\left\{\sigma : e|\,\sigma \in C\left(\frac{x}{\varepsilon},\frac{x_3}{e}\right)\right\} , \tag{14.2.10}$$

where, as previously $x = (x_\alpha)$. It satisfies the condition:

$$\exists\, k_7 \ge k_6 > 0 , \quad k_6|e| \le j_{e\varepsilon}(x,e) \le k_7(1+|e|) , \tag{14.2.11}$$

for each $(x,e) \in B_e \times \mathbb{E}_s^3$. The plate is subject to body forces $\widetilde{b} = (\widetilde{b}^i) = (b_\alpha^e, eb_3^e)$, where $b^e(x) = b(x_\alpha, x_3/e)$, $b \in L^\infty(B)^3$ and to tractions $\widetilde{g}_\pm^e = (eg_\pm^\alpha, e^2 g_\pm^3)$, where $g \in L^\infty(\Omega)^3$.

The minimum principle describing the equilibrium of the body B_e means evaluating

$$(\widetilde{P}_{e\varepsilon}) \quad \inf\left\{\int_{B_e} j\left(\frac{x}{\varepsilon},\frac{x_3}{e},e(\widetilde{u})\right)dx - L_e(u)|\,u \in H^1(B_e)^3, u = 0 \text{ on } \Upsilon^e\right\} ,$$

where

$$L_e(\boldsymbol{u}) = \int_{B_e} \widetilde{\boldsymbol{b}}(x) \cdot \widetilde{\boldsymbol{u}}(x)dx + \int_{\Gamma_\pm^e} \widetilde{\boldsymbol{g}}_\pm^e \cdot \widetilde{\boldsymbol{u}}d\Gamma \ . \tag{14.2.12}$$

We observe that the scaling of loading is now different from the one performed in Sec. 2.10.

The stress problem means evaluating

$$(\widetilde{P}_{e\varepsilon}^*) \qquad\qquad \sup\{-\frac{1}{2}\int_{B_e} c_{ijkl}\left(\frac{x}{\varepsilon},\frac{x_3}{e}\right)\widetilde{\sigma}^{ij}\widetilde{\sigma}^{kl}dx|\ \widetilde{\boldsymbol{\sigma}} \in \mathcal{S}^\varepsilon(B_e)\}$$

where

$$\mathcal{S}^\varepsilon(B_e) = \{\widetilde{\boldsymbol{\sigma}} \in L^2(B_e,\mathbb{E}_3^s)|\ \widetilde{\boldsymbol{\sigma}}(x) \in \mathcal{C}\left(\frac{x}{\varepsilon},\frac{x_3}{e}\right)\ \text{a.e. } \boldsymbol{x} \in B_{e^.},$$
$$\operatorname{div}\widetilde{\boldsymbol{\sigma}} + \widetilde{\boldsymbol{b}} = 0 \text{ in } B_e\ ,\ \widetilde{\boldsymbol{\sigma}}\boldsymbol{n} = \widetilde{\boldsymbol{g}}_\pm^e \text{ on } \Gamma_\pm^e\} \ . \tag{14.2.13}$$

In general, the problem $\widetilde{P}_{e\varepsilon}$ has no solution. From the mechanical point of view, plastic hinges can appear. Consequently, the space $H^1(B_e)^3$ is "too small" to allow for a solution. Also, the loading cannot be arbitrary.

Let us formulate the limit analysis problems. Their variational forms are formulated as follows:

$$(\widetilde{P}_{e\varepsilon LA})\quad \overline{\lambda} := \inf\{\int_{B_e} j_\infty\left(\frac{x}{\varepsilon},\frac{x_3}{e},e(\widetilde{\boldsymbol{v}})\right)dx|\ L(\widetilde{\boldsymbol{v}}) = 1,\ \widetilde{\boldsymbol{v}} \in H^1(B_e)^3, \widetilde{\boldsymbol{v}} = 0 \text{ on } \Upsilon^e\} \ ,$$

$$(\widetilde{P}_{e\varepsilon LA}^*) \qquad\qquad \sup\{\lambda|\ \widetilde{\boldsymbol{\sigma}} \in \mathcal{S}_\lambda^\varepsilon(B_e)\} \ ,$$

where

$$\mathcal{S}_\lambda^\varepsilon(B_e) = \{\widetilde{\boldsymbol{\sigma}} \in L^2(B_e,\mathbb{E}_3^s)|\ \widetilde{\boldsymbol{\sigma}} \in \mathcal{C}\left(\frac{x}{\varepsilon},\frac{x_3}{e}\right)\quad \text{a.e. } \boldsymbol{x} \in B_e\ ,$$
$$\operatorname{div}\widetilde{\boldsymbol{\sigma}} + \lambda\widetilde{\boldsymbol{b}} = 0 \text{ in } B_e\ ,\ \widetilde{\boldsymbol{\sigma}}\boldsymbol{n} = \lambda\widetilde{\boldsymbol{g}}_\pm^e \text{ on } \Gamma_\pm^e\} \ . \tag{14.2.14}$$

Here $\lambda \geq 0$ is the load multiplier and $\boldsymbol{n}$ denotes the outward unit normal vector to the boundary. In the problem $(\widetilde{P}_{e\varepsilon LA})$ $\widetilde{\boldsymbol{v}}$ denotes a *velocity* field. For plastically incompressible materials $\operatorname{div}\widetilde{\boldsymbol{v}} = 0$ and consequently $e(\widetilde{\boldsymbol{v}}) = e^D(\widetilde{\boldsymbol{v}})$. Replacing $L_e(\widetilde{\boldsymbol{u}})$ by $\lambda L_e(\widetilde{\boldsymbol{u}})$ in $(\widetilde{P}_{e\varepsilon})$, the corresponding extremum problem is denoted by $(\widetilde{P}_{e\varepsilon}^\lambda)$. We have the following characterization.

Lemma 14.2.1. The following conditions are equivalent:

(i) $\inf \widetilde{P}_{e\varepsilon}^\lambda = \sup(\widetilde{P}_{e\varepsilon}^\lambda)^* > -\infty.$

(ii) $\mathcal{S}_\lambda^\varepsilon(B_e) \neq \emptyset \quad (\emptyset - \text{empty set}).$

(iii) $\overline{\lambda} = \inf \widetilde{P}_{e\varepsilon LA} \geq \lambda.$ $\qquad\qquad\qquad\qquad\qquad\qquad\qquad\square$

We observe that if $\lambda = 1$, then $\widetilde{P}_{e\varepsilon}^{\lambda} = \widetilde{P}_{e\varepsilon}$ and the condition (iii) becomes

$$\inf P_{e\varepsilon LA} \geq 1 \,. \tag{14.2.15}$$

Our first aim is to pass with e to zero provided that the second small parameter or ε is fixed. We assume the following hypothesis:

there exists an $\alpha > \lambda$ such that $\inf \widetilde{P}_{e\varepsilon LA} \geq \alpha$ for all $e \in \{1/n,\ n \in \mathbb{N}\,,\ n \geq 2\}$.
$$\tag{14.2.16}$$

The problem $\widetilde{P}_{e\varepsilon}$ is relaxed to:

$$(R\widetilde{P}_{e\varepsilon}) \quad \inf\{\int_{B_e} j_{e\varepsilon}(\boldsymbol{x}, e(\boldsymbol{u})) + \int_{\Upsilon^e} j_\infty\left(\frac{x_\alpha}{\varepsilon}, \frac{x_3}{e}, -\mathbf{T}(\widetilde{\boldsymbol{u}})\right) dS - L_e(\widetilde{\boldsymbol{u}})|\ \widetilde{\boldsymbol{u}} \in BD(B_e)\}$$

where $\mathbf{T}_{ij}(\widetilde{\boldsymbol{u}}) = (\widetilde{\boldsymbol{u}} \otimes \boldsymbol{n})_s$. For plastically incompressible materials the infimum is to be taken over $\widetilde{\boldsymbol{u}} \in U(B_e)$ with $\widetilde{\boldsymbol{u}} \cdot \boldsymbol{n} = 0$ on $\Upsilon = \Gamma \times (-e, e)$, where

$$U(B_e) = \{\widetilde{\boldsymbol{u}} \in BD(B_e)|\ \mathrm{div}\ \widetilde{\boldsymbol{u}} \in L^2(B_e)\} \,. \tag{14.2.17}$$

Moreover, $\mathbf{T}(\widetilde{\boldsymbol{u}})$ is to be replaced by $\mathbf{T}^D(\widetilde{\boldsymbol{u}})$. The following theorem holds.

Theorem 14.2.2.

(i)　　$\inf \widetilde{P}_{e\varepsilon} = \inf R\widetilde{P}_{e\varepsilon}$.

(ii)　　$\widetilde{P}_{e\varepsilon}$ and $R\widetilde{P}_{e\varepsilon}$ have the same dual problem $\widetilde{P}_{e\varepsilon}^*$.

(iii)　Under the hypothesis (14.2.16), each problem $R\widetilde{P}_{e\varepsilon}$ possesses a solution $\widetilde{\boldsymbol{u}}^\varepsilon$ and $\widetilde{P}_{e\varepsilon}^*$ has a unique solution $\widetilde{\boldsymbol{\sigma}}_\varepsilon$. Moreover, we have $\|\widetilde{\boldsymbol{u}}^\varepsilon\|_{BD(B_e)} \leq$ constant $(\|\widetilde{\boldsymbol{u}}^\varepsilon\|_{U(B_e)} \leq$ constant for plastically incompressible materials) and $\|\widetilde{\boldsymbol{\sigma}}_\varepsilon\|_{L^2(B_e, \mathbb{E}_3^s)} \leq$ constant.　□

Similar to Sec. 2.10 we now pass to the change of variables so that the e-dependence is changed from the integration region to the integrand. To this end we introduce a mapping $\boldsymbol{F}^e : \mathbb{R}^3 \to \mathbb{R}^3$ defined by, cf. (2.10.2)

$$\boldsymbol{F}^e(z_\alpha, z_3) = (z_\alpha, ez_3) = (x_\alpha, x_3), \qquad \forall\, e > 0 \,. \tag{14.2.18}$$

To a function $h(\boldsymbol{z})$ defined on B corresponds the function $\widetilde{h}$ on B_e :

$$\widetilde{h}(\boldsymbol{x}) = h(\boldsymbol{F}^{-e}(\boldsymbol{x})) = h(x_\alpha, x_3/e) \,, \tag{14.2.19}$$

where $\boldsymbol{x} \in B_e$ and $\boldsymbol{F}^{-e} = (\boldsymbol{F}^e)^{-1}$.

Then, for stresses $\widetilde{\boldsymbol{\sigma}} = (\widetilde{\sigma}^{ij}) \in L^2(B_e, \mathbb{E}_3^s)$ we set

$$\boldsymbol{\sigma} = \mathbf{Q}^e \circ \widetilde{\boldsymbol{\sigma}} \circ \boldsymbol{F}^e \,, \tag{14.2.20}$$

where $\mathbf{Q}^e$ is given by (2.10.15). Thus $\boldsymbol{\sigma} \in L^2(B, \mathbb{E}_3^s)$. We recall that $\mathbf{Q}^{-e}\big|_{e=0^+}$ is also a continuous mapping from $\mathbb{E}_3^s$ to $\mathbb{E}_3^s$, though $\mathbf{Q}^0$ does not exist; here $(\mathbf{Q}^e)^{-1} = \mathbf{Q}^{-e}$.

For a function $\widetilde{u} \in H^1(B_e)^3$ (resp. $BD(B_e)$ or $U(B_e)$) we define $\widetilde{u} \to u \in H^1(B)^3$ (resp. $BD(B)$ or $U^e(B)$) by

$$u = F^e \circ \widetilde{u} \circ F^e \; . \tag{14.2.21}$$

Here the function space $U^e(B)$ is defined by

$$U^e(B) = \{u \in L^1(B)^3 | \; e(u) \in \mathbb{M}^1(B, \mathbb{E}_s^3), \; \mathrm{tr}\, \mathbb{Q}^e e(u) \in L^2(B)\} \; .$$

Obviously, it is a subspace of $BD(B)$. In a more explicit form, the relations (14.2.20) and (14.2.21) are written as

$$\sigma(z_\alpha, z_3) = \mathbb{Q}^e \widetilde{\sigma}(z_\alpha, e z_3) \; , \quad u(z_\alpha, z_3) = (\widetilde{u}_\alpha(z_\beta, e z_3), e\widetilde{u}_3(z_\beta, e z_3)) \; . \tag{14.2.22}$$

Having performed the change of variables we transform the variational problems $(\widetilde{P}_{e\varepsilon}^\lambda)$, $(\widetilde{P}_{e\varepsilon}^\lambda)^*$ and $(R\widetilde{P}_{e\varepsilon}^\lambda)$ to new problems posed on the domain B:

$$(P_{e\varepsilon}) \quad \inf\{\int_B j\left(\frac{z}{\varepsilon}, z_3, \mathbb{Q}^e e(u)\right) dx - \lambda L(u)| \; u \in H^1(B)^3, u = 0 \text{ on } \Upsilon\}, \tag{14.2.23}$$

$$(P_{e\varepsilon}^\lambda)^* \quad \sup\{-\frac{1}{2}\int_B c_{ijkl}\left(\frac{z}{\varepsilon}, z_3\right)(\mathbb{Q}^{-e}\sigma)^{ij}(\mathbb{Q}^{-e}\sigma)^{kl} dx| \; \sigma \in \mathcal{S}_\lambda^\varepsilon(B)\} \; , \tag{14.2.24}$$

$$(RP_{e\varepsilon}^\lambda) \quad \inf\{\int_B j\left(\frac{z}{\varepsilon}, z_3, \mathbb{Q}^e e(u)\right)$$

$$+ \int_{\Upsilon_0} j_\infty\left(\frac{z}{\varepsilon}, z_3; -\mathbb{Q}^e T(u)\right) dS - \lambda L(u)| \; u \in BD(B)\} \; , \tag{14.2.25}$$

with obvious modifications in the case of materials incompressible in the plastic range. We recall that $z = (z_\alpha)$, $\Upsilon = \Upsilon^1$, while the space $\mathcal{S}_\lambda^\varepsilon(B)$ is defined by

$$\mathcal{S}_\lambda^\varepsilon(B) = \{\sigma \in L^2(B, \mathbb{E}_3^s)| \; \sigma \in \mathcal{C}\left(\frac{z}{\varepsilon}, z_3\right) \text{ a.e. } z \in B, \; \mathrm{div}\, \sigma + \lambda b = 0 \text{ in } B,$$

$$\sigma n = \lambda g_\pm \text{ on } \Gamma_\pm\} \; . \tag{14.2.26}$$

The loading functional $L(u)$ is now given by

$$L(u) = \int_B b(z) \cdot u(z) dz + \int_{\Gamma_\pm} g_\pm \cdot u d\Gamma \; . \tag{14.2.27}$$

The new variational problems posed on B readily follow by observing that

$$\int_{B_e} j\left(\frac{x}{\varepsilon}, \frac{x_3}{e}; e(\widetilde{u})\right) dx - L_e(\widetilde{u}) = e[\int_B j\left(\frac{z}{\varepsilon}, z_3; \mathbb{Q}^e e(u)\right) dz - L(u)] \; .$$

The following theorem interrelates the problems posed on B_e with the corresponding ones defined on B in an evident manner.

Theorem 14.2.3. $(\widetilde{u}^{e\varepsilon}, \widetilde{\sigma}_{e\varepsilon})$ is a solution to the problem $(R\widetilde{P}_{e\varepsilon}, \widetilde{P}^*_{e\varepsilon})$ if and only if $(u^{e\varepsilon}, \sigma_{e\varepsilon}) = (F^e \circ \widetilde{u}^{e\varepsilon} \circ F^e, Q^e \circ \widetilde{\sigma}_{e\varepsilon} \circ F^e)$ is a solution to the problem $(RP_{e\varepsilon}, P^*_{e\varepsilon})$. $\square$

The hypothesis (14.2.16) is changed to:

$$\text{there exists an } \alpha > \lambda \text{ such that } \inf P_{e\varepsilon LA} > \alpha \text{ for all } e , \tag{14.2.28}$$

where

$$(P_{e\varepsilon LA}) \quad \inf\{\int_B j_\infty \left(\frac{z}{\varepsilon}, z_3, Q^e e(v)\right) dz \mid v = H^1(B)^3, \; v = 0 \text{ on } \Upsilon, L(v) = 1\} .$$

For materials incompressible in the plastic range we add the condition tr $Q^e e(v) = 0$. Then in the dissipation density j_∞ appears $(Q^e e(v))^D$ instead of $Q^e e(v)$. We have the following limit analysis theorem.

Theorem 14.2.4. $\inf \widetilde{P}_{e\varepsilon LA} = \inf P_{e\varepsilon LA}$.

Proof. It readily follows by noting that $j_\infty \left(\frac{z}{\varepsilon}, z_3; \cdot\right)$ is positively homogeneous. More precisely, the limit load multiplier $\overline{\lambda}$ can equivalently be calculated as follows:

$$\overline{\lambda} = \inf \{\frac{\int_{B_e} j_\infty(x/\varepsilon, x_3/e, e(\widetilde{v}))dx}{L(\widetilde{v})} \mid \widetilde{v} \in H^1(B_e)^3, \widetilde{v} = 0 \quad \text{on } \Upsilon^e\}$$

$$= \inf\{\frac{e\int_B j_\infty(z/\varepsilon, z_3, Q^e e(v))dz}{eL(v)} \mid v \in H^1(B)^3, v = 0 \text{ on } \Upsilon\} . \qquad \square$$

14.2.2.   Derivation of the effective plate model by passing to zero: $e \to 0$ and next $\varepsilon \to 0$

The passage to the two-dimensional plate model is now obviously more complicated than in the elastic case. We shall point out the main steps of the derivation of the effective two-dimensional model mainly for plates made of Hencky material incompressible in the plastic range.

Theorem 14.2.5. Under the hypothesis (14.2.16) and assuming that $g_\pm = (0, 0, g^3_\pm)$ with $g^3_\pm \in C^1_0(\Omega)$ we have

$$\lim_{e \to 0} (\inf P_{e\varepsilon}) = \inf P_\varepsilon , \tag{14.2.29}$$

where

$$(P_\varepsilon) \inf\{\int_B J[z/\varepsilon, z_3, w_{(\alpha;\beta)} + z_3\kappa_{\alpha\beta}(w)]dz - L(u)\mid u \in H^1(B)^3, \; u = 0 \text{ on } \Upsilon,$$

$$u = (w_\alpha - z_3 w_{;3}, w), \; (w_\alpha) \in BD(\Omega), w \in HB(\Omega)\} . \tag{14.2.30}$$

The relaxed problem of (14.2.30) is

$$(RP_\varepsilon) \ \inf\Big\{ \int_B \mathrm{J}(z/\varepsilon, z_3, w_{(\alpha;\beta)} + z_3\kappa_{\alpha\beta}(w)) + \int_\Upsilon \mathrm{J}_{\varepsilon\infty}(z_\alpha, -\mathbf{T}_1(w_\alpha) + z_3\mathbf{F}(w))dS$$

$$-L(\boldsymbol{u})\Big|\ \boldsymbol{u} \in \overset{\approx}{\tilde{U}}(B)\ ,\ \boldsymbol{u} = (w_\alpha - z_3 w_{;3}, w),\ w = 0 \text{ on } \Upsilon\Big\}\ . \qquad (14.2.31)$$

Here

$$(\mathbf{T}_1(w_\alpha))_{\beta\gamma} = \frac{1}{2}(w_\beta n_\gamma + w_\gamma n_\beta)\ , \qquad (14.2.32)$$

$$\mathbf{F}_{\alpha\beta}(w) = \frac{\partial w}{\partial \boldsymbol{n}} n_\alpha n_\beta\ , \qquad (14.2.33)$$

$$\overset{\approx}{\tilde{U}}(B) = \{\boldsymbol{u} \in BD(B)|\ e_{i3}(\boldsymbol{u}) = 0\}\ , \qquad (14.2.34)$$

and

$$\mathrm{J}(z/\varepsilon, z_3, \boldsymbol{\epsilon}) = \mathrm{J}(z_\alpha/\varepsilon, z_3, \boldsymbol{\epsilon}) = \sup\{\boldsymbol{\sigma} : \boldsymbol{\epsilon} - j^*(z/\varepsilon, z_3, \boldsymbol{\sigma})|\ \boldsymbol{\sigma} \in \mathbb{E}_3^s,\ \sigma^{i3} = 0\}\ , \quad (14.2.35)$$

for each $\boldsymbol{\epsilon} \in \mathbb{E}^s$ with $\epsilon_{i3} = 0$. Particularly, for materials obeying the Huber-Mises yield criterion we have

$$\mathrm{J}(z/\varepsilon, z_3, \boldsymbol{\epsilon}) = \sup\Big\{\boldsymbol{\sigma} : \boldsymbol{\epsilon} - \frac{1}{2}c_{ijkl}(z/\varepsilon, z_3, \boldsymbol{\sigma})|\ \boldsymbol{\sigma} \in \mathbb{E}_3^s,$$

$$\sigma^{i3} = 0,\ |\boldsymbol{\sigma}_D| \leq \tau_0(z/\varepsilon, z_3)\Big\}\ , \qquad (14.2.36)$$

where

$$|\boldsymbol{\sigma}_D| = \Big|\boldsymbol{\sigma} - \frac{1}{3}\begin{bmatrix} \mathrm{tr}\,\boldsymbol{\sigma} & 0 \\ 0 & \mathrm{tr}\,\boldsymbol{\sigma} \end{bmatrix}\Big| \leq \tau_0(z/\varepsilon, z_3)\ .$$

Moreover, if $\boldsymbol{u}^{e\varepsilon}$ is a solution of $(RP_{e\varepsilon})$, then $\{\boldsymbol{u}^{e\varepsilon}\}_{e>0}$ is a bounded sequence in $BD(B)$ and any of its subsequences which converges in $L^1(B)^3$ is such that its limit is a solution to (RP_ε).

Proof. The proof will be divided into several steps, cf. Sec. 2.10.2.

Step 1. The space $BD(B)$ equipped with the topology of $L^1(B)^3$ satisfies the first axiom of countability. Thus the following limits exist

$$\mathcal{J}_\varepsilon(\boldsymbol{u}) = \inf_{\substack{\{u^e\} \subset H^1(B)^3 \\ u^e \to u \in L^1(B)^3}} \liminf_{e \to 0} \int_B j\left(\frac{z}{\varepsilon}, z_3, \mathbf{Q}^e e(\boldsymbol{u})\right) dz$$

$$= \inf_{\substack{\{u^e\} \subset H^1(B)^3 \\ u^e \to u \in L^1(B)^3}} \limsup_{e \to 0} \int_B j\left(\frac{z}{\varepsilon}, z_3, \mathbf{Q}^e e(\boldsymbol{u})\right) dz\ .$$

Next one proves that

$$\mathcal{J}_\varepsilon(\boldsymbol{u}) = \inf_{\substack{\{\boldsymbol{u}^e\} \subset BD(B) \\ \boldsymbol{u}^e \to \boldsymbol{u} \in L^1(B)^3}} \liminf_{e \to 0} \int_B j\left(\frac{z}{\varepsilon}, z_3, \mathbf{Q}^e \boldsymbol{e}(\boldsymbol{u})\right)$$

$$= \inf_{\substack{\{\boldsymbol{u}^e\} \subset BD(B) \\ \boldsymbol{u}^e \to \boldsymbol{u} \in L^1(B)^3}} \limsup_{e \to 0} \int_B j\left(\frac{z}{\varepsilon}, z_3, \mathbf{Q}^e \boldsymbol{e}(\boldsymbol{u})\right) .$$

For plastically incompressible materials the sequences $\{\boldsymbol{u}^e\}_{e>0}$ belong to $U(B)$.

Step 2. One shows that for any $\boldsymbol{u} \in BD(B) \backslash \{\boldsymbol{u} \in BD(B)|\ e_{i3}(\boldsymbol{u}) = 0\}$

$$\mathcal{J}_\varepsilon(\boldsymbol{u}) = +\infty . \tag{14.2.37}$$

Moreover, for any $\boldsymbol{u} = \overset{\approx}{U}(B)$, if $w = (w_\alpha, w) \in \overset{\approx}{W}(\Omega)$ is such that $\boldsymbol{u} = (w_\alpha - z_3 w_{;\alpha}, w)$, then

$$c\int_\Omega (|\boldsymbol{e}(w_\alpha)| + |\boldsymbol{\kappa}(w)|) + \widetilde{c}(B) \leq \mathcal{J}_\varepsilon(\boldsymbol{u}) \leq c'\int_\Omega (|\boldsymbol{e}(w_\alpha)| + |\boldsymbol{\kappa}(w)|) + \widetilde{c}'(B), \tag{14.2.38}$$

for some constants $c, c' > 0$ and $\widetilde{c}, \widetilde{c}' \in \mathbb{R}$. Here

$$\overset{\approx}{W}(\Omega) = \{(w_\alpha, w)|\ (w_\alpha) \in BD(\Omega), w \in HB(\Omega)\} . \tag{14.2.39}$$

The proof of (14.2.37) is similar to the proof of (2.10.31). It is then not difficult to show that, cf. Lemma 2.10.1,

$$\{\boldsymbol{u} \in BD(B)|\ e_{i3}(\boldsymbol{u}) = 0, i = 1, 2, 3\} = \{\boldsymbol{u}|\ \exists (w_\alpha, w) \in \overset{\approx}{W}(\Omega) \tag{14.2.40}$$

$$\text{such that } \boldsymbol{u} = (w_\alpha - z_3 w_{;\alpha}, w)\} .$$

We observe that on $\overset{\approx}{W}(\Omega)$ one can define an intermediate topology by, cf. Sec. 13.2,

$$(w_\alpha^n, w^n) \to w \quad \text{in} \quad L^1(\Omega)^3 , \quad w^n \to w \quad \text{in} \quad W^{1,1}(\Omega) ,$$

$$\int_\Omega (|\boldsymbol{e}(w_\alpha^n)| + |\boldsymbol{\kappa}(w^n)|) \to \int_\Omega (|\boldsymbol{e}(w_\alpha)| + |\boldsymbol{\kappa}(w)|) \tag{14.2.41}$$

when $n \to \infty$. Then $C^\infty(\overline{\Omega})^3$ is dense in $\overset{\approx}{W}(\Omega)$.

To prove the left-hand side of inequality appearing in (14.2.38) it suffices to take a sequence $\{\boldsymbol{u}^e\}_{e>0} \subset H^1(B)^3$ such that $\boldsymbol{u}^e \to \boldsymbol{u}$ in $L^1(B)$ and

$$\mathcal{J}_\varepsilon(\boldsymbol{u}) = \lim_{e \to 0} \int_B j\left(\frac{z}{\varepsilon}, z_3, \mathbf{Q}^e \boldsymbol{e}(\boldsymbol{u}^e)\right) dz .$$

By using (14.2.8) (or (14.2.9) for plastically incompressible materials) we establish the desired inequality.

To show the right-hand side of (14.2.38), for each $u \in \{u \in BD(B)|\ e_{i3}(u) = 0\} \cap C^\infty(B)^3$, $(w_\alpha, w) \in \overset{\approx}{W}(\Omega) \cap C^\infty(\overline{\Omega})$ such that $u = (w_\alpha - z_3 w_{;\alpha}, w)$, we define

$$u^e = \left(w_\alpha - z_3 w_{;\alpha},\ w - e^2\left(-\frac{1}{2}z_3^2 \Delta w + z_3 w_{\alpha;\alpha}\right)\right) .$$

Then $u^e \to u$ in $L^1(B)$ and simple calculation yields

$$\mathcal{J}_\varepsilon(u) \le \lim_{e \to 0} \inf \int_B j\left(\frac{z}{\varepsilon}, z_3, \mathbf{Q}^e e(u^e)\right) dz \ . \le k_3 \int_B |e(u)| + k_3 \int_B |z_3 \Delta w|$$

$$+ \sum_\alpha w_{\alpha;\alpha}| + k_3\ \text{meas}\ (B) \le c' \int_\Omega (|e(w_\alpha)| + |\kappa(w)|) + \widetilde{c}'(B) , \quad (14.2.42)$$

where k_3 is the constant appearing in (14.2.8). To extend (14.2.42) to $\overset{\approx}{U}(B)$, we exploit the intermediate topology (14.2.41) and lower semicontinuity of J_ε.

Step 3. To show that

$$\mathrm{J}_\varepsilon(u) = \sup\{\int_B (e(u) : \boldsymbol{\sigma} - \frac{1}{2}c_{ijkl}\left(\frac{z}{\varepsilon}, z_3\right)\sigma^{ij}\sigma^{kl})dz|\ \boldsymbol{\sigma} \in C_0^\infty(B, \mathbb{E}_3^s),$$

$$\boldsymbol{\sigma}(z) \in \mathcal{C}\left(\frac{z}{\varepsilon}, z_3\right),\ \sigma_{i3} = 0\} , \quad (14.2.43)$$

we proceed similarly to step (iii) of the proof of Theorem 2.10.2. Now one has to consider separately the case of compressible and plastically incompressible materials.

Step 4. For $g_\pm$ and $\{u^{e\varepsilon}\}_{e>0}$ as in theorem which is being proved and assuming that $u^{e\varepsilon} \to u^\varepsilon$ when $e \to 0$, we have

$$\int_{\Gamma_\pm} g_\pm \cdot u^{e\varepsilon} d\Gamma \to \int_{\Gamma_\pm} g_\pm \cdot u^\varepsilon d\Gamma . \quad (14.2.44)$$

Indeed, since $u^{e\varepsilon} \rightharpoonup u^\varepsilon$ in $BD(B)$ weak-$*$, therefore $u^{e\varepsilon}, u^\varepsilon \in BD(B)$. By the trace theorem we conclude that $u^{e\varepsilon}_{|\partial B}, u^\varepsilon_{|\partial B} \in L^1(\partial B)^3$. Thus for any $\varphi \in C^1(\overline{B}), \varphi = 0$ on Υ, we have

$$\lim_{e \to 0} \left(\int_{\Gamma_+} g_+^3 u_3^{e\varepsilon} d\Gamma + \int_{\Gamma_-} g_-^3 u_3^{e\varepsilon} d\Gamma\right) = \int_B \partial_3 \varphi u_3^\varepsilon dz = \int_{\Gamma_+} g_+^3 u_3^\varepsilon d\Gamma + \int_{\Gamma_-} g_-^3 u_3^\varepsilon d\Gamma ,$$

provided that $\varphi_{|\Gamma_+} = g_+^3, -\varphi_{|\Gamma_-} = g_-^3$ and noting that $|\partial_3 u_3^{e\varepsilon}|_{\mathbf{M}^1(B)} = O(e^2)$.

Step 5.
(i) To show that

$$\lim_{e \to 0} (\inf P_{e\varepsilon}) \le \inf P_\varepsilon , \quad (14.2.45)$$

we take $u^\delta \in H^1(B)^3 \cap \tilde{\tilde{U}}(B)$ such that $u^\delta = 0$ on Υ, $u^\delta = (w_\alpha - z_3 w_{;\alpha}, w)$, $(w_\alpha, w) \in \tilde{\tilde{W}}(\Omega)$. If

$$\int_B \textsf{J}[z/\varepsilon, z_3, e(w_\alpha) + z_3 \kappa(w)]dz - L(u^\delta) \leq \inf P_\varepsilon + \delta \,,$$

then there exists a sequence $\{u^e\}_{e>0} \subset H^1(B)$ satisfying

$$u^e = 0 \quad \text{on} \quad \Upsilon, \quad u^e \to u^\delta \quad \text{in} \quad L^1(B)^3 \quad \text{as} \quad e \to 0 \,,$$

and

$$\lim_{e \to 0} \int_B \textsf{J}[z/\varepsilon, z_3, \mathbf{Q}^e e(u^e)]dz = \int_B \textsf{J}[z/\varepsilon, z_3, e(w_\alpha) + z_3 \kappa(w)] \,.$$

Moreover, we have

$$\lim_{e \to 0} L(u^e) = L(u^\delta) \,.$$

(ii) To show the inverse inequality of (14.2.45) suppose that $\{u^e\}_{e>0} \subset H^1(B)^3$ is such that $u^e = 0$ on Υ and

$$\int_B \textsf{J}[z/\varepsilon, z_3, \mathbf{Q}^e e(u^e)]dz - L(u^e) \leq \inf P + e \,. \tag{14.2.46}$$

Next we introduce

$$\Omega_\delta = \{(z_\alpha) \in \Omega | \text{ dist}\,((z_\alpha), \Omega) < \delta\} \,, \quad \delta > 0 \,,$$

$$B_\delta = \Omega_\delta \times (-1, 1) \,, \quad \overline{u}^e = \begin{cases} u^e & \text{if} \quad z \in B \,, \\ 0 & \text{if} \quad z \in B_\delta \backslash B \,. \end{cases}$$

We readily conclude that $\|\overline{u}^e\|_{BD(B_\delta)} \leq$ constant (or $\|\overline{u}^e\|_{U^e(B_\delta)} \leq$ constant for plastically incompressible materials). If $u^e \to u^0$ in $L^1(B)^3$, then $u^e \to u^0$ in $L^1(B_\delta)^3$, where

$$\overline{u}^0 = \begin{cases} u^0 & \text{if} \quad z \in B \,, \\ 0 & \text{if} \quad z \in B_\delta \backslash \overline{B} \end{cases} \quad \text{and} \quad \overline{u}^0 \in \tilde{\tilde{U}}(B_\delta) \,.$$

Hence $u_3^0 = 0$ on Υ. We observe that u^0 may depend on ε which is held fixed. Applying now (14.2.43) to B_δ we get

$$\lim_{e \to 0} \int_{B_\delta} \textsf{J}[z/\varepsilon, z_3, \mathbf{Q}^e e(\overline{u}^e)] \geq \int_{B_\delta} \textsf{J}[z/\varepsilon, z_3, e(\overline{w}_\alpha^0) + z_3 \kappa(\overline{w}^0)] \,,$$

where $(\overline{w}_\alpha^0, \overline{w}^0) \in \widetilde{\widetilde{W}}\,(\Omega_\delta)$ is such that $\overline{u}^0 = (\overline{w}_\alpha^0 - z_3\overline{w}_{;\alpha}^0, \overline{w}^0)$. More precisely, the last inequality yields

$$\lim_{e \to 0} \int_B \jmath[z/\varepsilon, z_3, \mathbf{Q}^e e(u^e)] \geq \int_B \jmath[z/\varepsilon, z_3, e(w_\alpha) + z_3\boldsymbol{\kappa}(w)]$$

$$+ \int_\Upsilon \jmath\infty[z_\alpha, z_3, -\mathbf{T}_1(w_\alpha^0) + z_3\mathbf{F}(w^0)]dS\,.$$

Combining the last inequality with (14.2.44) we conclude that

$$\lim_{e \to 0} (\inf P_{e\varepsilon}) \geq \inf P_\varepsilon\,.$$

(iii) To complete the proof one has to show the convergence of solutions $\{u^{e\varepsilon}\}_{e>0}$. To this end one can proceed similarly to (ii). $\square$

Remark 14.2.6. By performing a relaxation of the loading functional on $\Gamma_\pm$ it seems possible to consider more general loading g_+. Indeed, by replacing the loading functional L with

$$\widehat{L}(u, \boldsymbol{\mu}_+, \boldsymbol{\mu}_-) = \int_B b \cdot u dz + \int_{\Gamma_+} g_+ \cdot d\boldsymbol{\mu}_+ + \int_{\Gamma_-} g_- \cdot d\boldsymbol{\mu}_-\,, \tag{14.2.47}$$

where now $g_\pm \in C_0(\Omega)^3$ and $\boldsymbol{\mu}_\pm \in \mathbf{M}^1(\Omega)^3$, we conclude that $\widehat{L}$ is continuous in the weak-$*$ topology of $BD(B) \times \mathbf{M}^1(\Omega)^3 \times \mathbf{M}^1(\Omega)^3$. Consequently, the functional $\widehat{L}$ plays the role of a perturbation functional when $e \to 0$ and $\varepsilon \to 0$ in their three combinations. Additionally, one has to add relaxed terms over Γ_+ and Γ_- to the functionals appearing in the problem $RP_{e\varepsilon}$, cf. Suquet (1988). $\square$

Convergence of $\{\lambda^{e\varepsilon}\}_{e>0}$

    The limit load multiplier $\lambda^{e\varepsilon}$ has been defined in Sec. 14.2.1. Assume that $b \in L^\infty(B)^3$ and $g_\pm = (0, 0, g_\pm^3)$ with $g_\pm^3 \in C_0^1(\Omega)$. Then it can be shown that

$$\lim_{e \to 0} \lambda^{e\varepsilon} = \lambda^\varepsilon\,, \tag{14.2.48}$$

where

$$(P_{\varepsilon LA}) \quad \lambda^\varepsilon = \inf \{ \int_B \jmath\infty[z/\varepsilon, z_3, e(w_\alpha) + z_3\boldsymbol{\kappa}(w)]dz | \; L(v) = 1,$$

$$v \in H^1(B)^3 \cap \widetilde{\widetilde{U}}\,(B)\,, v = 0 \quad \text{on} \quad \Upsilon, \; (w_\alpha, w) \in \widetilde{\widetilde{W}}\,(\Omega),$$

$$v = (w_\alpha - z_3 w_{;\alpha}, w)\}\,. \tag{14.2.49}$$

The hypothesis (14.2.16) or (14.2.28) can now be replaced by

$$\lambda^\varepsilon > \lambda\,, \tag{14.2.50}$$

where λ is the load multiplier associated with the loading functional L, cf. (14.2.23) - (14.2.25). Obviously, if $\lambda = 1$ then the safe load hypothesis (14.2.50) is given by $\lambda^\varepsilon > 1$.

We observe that the relaxed problem $RP_{\varepsilon LA}$ of $P_{\varepsilon LA}$ means evaluating

$$(RP_{\varepsilon LA}) \quad \inf \{ \int_B \mathrm{J}_\infty[z/\varepsilon, z_3, e(w_\alpha) + z_3 \kappa(w)]$$

$$+ \int_\Upsilon \mathrm{J}_\infty[z_\alpha, z_3, -\mathbf{T}_1(w_\alpha) + z_3 \mathbf{F}(w^0)] dS | \, L(v_\alpha, v) = 1,$$

$$(v_\alpha, v) \in \widetilde{\widetilde{U}} \, (B) , v_3 = 0 \quad \text{on} \quad \Upsilon, \, (w_\alpha, w) \in \widetilde{\widetilde{W}} \, (\Omega), \, \boldsymbol{v} = (w_\alpha - z_3 w_{;\alpha}, w) \}.$$

Obviously we have

$$\lambda^\varepsilon = \inf P_{\varepsilon LA} = \inf RP_{\varepsilon LA} = \min RP_{\varepsilon LA} . \tag{14.2.51}$$

Homogenization: $\varepsilon \to 0$

Having constructed the two-dimensional plate model with a microperiodic structure we are in a position to perform the homogenization. We assume that the loading functional L plays the role of a perturbation functional.

Theorem 14.2.7. The sequence of functionals defined by

$$\mathcal{J}_\varepsilon(w_\alpha, w) = \begin{cases} \int_\Omega \int_{-1}^1 \mathrm{J}[z/\varepsilon, z_3, e(w_\alpha) + z_3 \kappa(w)] dz & \text{if } (w_\alpha) \in LD(\Omega), \\ \qquad w \in W^{2,1}(\Omega), \quad w_\alpha = 0 , \, w = 0 , \, \partial w / \partial \boldsymbol{n} = 0 \quad \text{on } \Gamma ; \\ +\infty & \text{otherwise,} \end{cases} \tag{14.2.52}$$

is Γ-convergent in the topology $\tau = s - [L^p(\Omega)^2 \times W^{1,1}(\Omega)], 1 \le p < 2$, to

$$\mathcal{J}_h(w_\alpha, w) = \begin{cases} \int_\Omega \mathrm{J}_h[e(w_\alpha) \, \kappa(w)] + \int_\Gamma \mathrm{J}_{h\infty}[-\mathbf{T}_1(w_\alpha), \mathbf{F}(w)] d\Gamma \\ \qquad \text{if} (w_\alpha) \in BD(\Omega), \, w \in HB(\Omega), \, w = 0 \text{ on } \Gamma ; \\ +\infty \qquad \text{otherwise.} \end{cases} \tag{14.2.53}$$

Here

$$\mathrm{J}_h(\boldsymbol{\epsilon}, \boldsymbol{\rho}) = \inf \{ \frac{1}{|Y|} \int_Y \int_{-1}^1 \mathrm{J}[y, z_3, e^y(\boldsymbol{\xi}) + \boldsymbol{\epsilon} + z_3(\kappa^y(\eta) + \boldsymbol{\rho})] dy dz_3$$

$$| \, \boldsymbol{\xi} \in LD_{per}(Y), \, \eta \in W^{2,1}_{per}(Y) \} , \tag{14.2.54}$$

where $\boldsymbol{\epsilon}, \boldsymbol{\rho} \in \mathbb{E}^2_s, y = (y_\alpha)$.

Proof. It is a straightforward combination of the proof of Theorem 13.4.1 and the results due to Bouchitté and Suquet (1991). $\square$

Remark 14.2.8. Since the function $\jmath(y, z_3, \cdot)$ has only linear growth therefore the infimum on the right-hand side of (14.2.54) is attained, in general, in the space $BD_{per}(Y) \times HB_{per}(Y)$, where

$$
\begin{aligned}
BD_{per}(Y) &= \{u \in BD_{loc}(\mathbb{R}^2) |\ u(z + i) \\
&= u(z) \text{ for a.e. } z \text{ in } \mathbb{R}^2 \text{ and } i \text{ in } \mathbb{Z}^2\}, \quad (14.2.55) \\
HB_{per}(Y) &= \{w \in HB_{loc}(\mathbb{R}^2) |\ w(z + i) = w(z), \\
&\qquad \nabla w(z + i) = \nabla w \text{ for a.e. } z \text{ in } \mathbb{R}^2 \text{ and } i \in \mathbb{Z}^2\}, \quad (14.2.56)
\end{aligned}
$$

where $\mathbb{Z}$ is the set of integers.

We recall that functions of $BD_{per}(Y)$ are allowed to have discontinuities on ∂Y, although the Y-periodicity implies that the external trace on some face of Y must be equal to the internal trace on the opposite face. The same pertains to the gradient of a function of $HB_{per}(Y)$, the function itself being continuous. $\qquad\square$

Remark 14.2.9. The elasto-plastic thin plate model with the potential $\jmath_h$ couples membrane and bending effects. Once they can be decoupled the resulting bending problem is still more general than the one studied in Sec. 14.1, provided that $\Gamma_1 = \emptyset$, since the present model admits transverse inhomogeneities. For instance, the effective plate model (14.2.54) includes layered plates with possibly a two-dimensional microperiodic structure in the plane of layers. $\qquad\square$

Let us pass to the formulation of the limit analysis problem for the plate made of Hencky material characterized by $\jmath_h$.

$$
(P_{hLA}) \quad \lambda^h = \inf\{\int_\Omega \jmath_{h\infty}[e(v), \kappa(v)]dz|\ L(v, v) = 1,\ v \in H_0^1(\Omega)^2,\ v \in H_0^2(\Omega)\}.
$$

The relaxed problem can also be easily formulated. Under the assumption of continuity of L in the topology τ, as a consequence of Theorem 14.2.7 we get:

$$
\lim_{\varepsilon \to 0} \lambda^\varepsilon = \lambda^h . \quad (14.2.57)
$$

The equilibrium problem for the homogenized or effective elasto-plastic plate means evaluating

$$
(P_h^\lambda) \quad \inf\{\int_\Omega \jmath_h[e(u), \kappa(w)]dz - \lambda L(u, w)|\ u \in H_0^1(\Omega)^2,\ w \in H_0^2(\Omega)\},
$$

where $\lambda < \lambda^h$ (the safe load hypothesis). The relaxed problem has the following form:

$$
(RP_h^\lambda) \quad \inf\{\jmath_h(u, (w) - \lambda L(u, w)|\ u \in BD(\Omega),\ w \in HB(\Omega),\ w = 0 \text{ on } \Gamma\} .
$$

14.2.3.   Derivation of the second effective plate model: $\varepsilon \to 0$ and next $e \to 0$

Both in the present and in the next section the loading functional is assumed to play the role of a perturbation functional. We set

$$J_{e\varepsilon}(\boldsymbol{u}) = \begin{cases} \displaystyle\int_B j[z/\varepsilon, z_3, \mathbf{Q}^e e(\boldsymbol{u})]dz & \text{if } \boldsymbol{u} \in LD(B)\,,\ \boldsymbol{u} = 0 \text{ on } \Upsilon\,, \\ +\infty & \text{otherwise.} \end{cases} \tag{14.2.58}$$

For a fixed $e > 0$, the homogenization process $(\varepsilon \to 0)$ is standard and yields the following $\Gamma(L^p(B)^3)$ -limit, $1 \le p < 3/2$,

$$J_e^0(\boldsymbol{u}) = \int_B j_0[z_3, \mathbf{Q}^e e(\boldsymbol{u})] + \int_{\Upsilon_0} j_{0\infty}[z_3, -\mathbf{Q}^e \mathbf{T}(\boldsymbol{u})]dS\,,\ \boldsymbol{u} \in BD(B)\,, \quad (14.2.59)$$

where

$$j_0(z_3, \epsilon) = \inf\{\langle j(y_\alpha, z_3, e^y(\boldsymbol{\xi}) + \epsilon)\rangle|\ \boldsymbol{\xi} \in LD_{per}^3(Y)\}\,. \tag{14.2.60}$$

Here $\epsilon \in \mathbb{E}_s^3$ and

$$LD_{per}^3(Y) = \{\boldsymbol{\xi} \in L^1(Y)^3|\ e_{ij}^y(\boldsymbol{\xi}) \in L^1(Y)\,,\xi_i \quad \text{is } Y\text{-periodic; } i,j = 1,2,3\}\,. \tag{14.2.61}$$

We recall that for $\boldsymbol{\xi} \in LD_{per}^3(Y)$ we have

$$e_{\alpha 3}^y(\boldsymbol{\xi}) = \frac{1}{2}\frac{\partial \xi_3}{\partial y_\alpha}\,, \quad e_{33}^y(\boldsymbol{\xi}) = 0\,. \tag{14.2.62}$$

For plastically incompressible materials, $\boldsymbol{u}$ in (14.2.59) belongs to $U^e(B)$ and $\boldsymbol{u} \cdot \boldsymbol{n} = 0$ on Υ. Formulation of problems P_e^λ and RP_e^λ is left to the reader. The load multiplier λ satisfies the safe load hypothesis, now given by $\lambda < \lambda^e$, where

$$(P_{eLA}) \qquad \lambda^e = \inf\{\int_B j_{0\infty}[z_3, \mathbf{Q}^e e(\boldsymbol{v})]dz|\ L(\boldsymbol{v}) = 1,\ \boldsymbol{v} \in H^1(B)^3,\ \boldsymbol{v} = 0 \text{ on } \Upsilon\}\,.$$

We have

$$\lim_{\varepsilon \to 0} \lambda^{e\varepsilon} = \lambda^e\,. \tag{14.2.63}$$

By using the results of Sec. 14.2.2 we pass to zero with e. We conclude that the sequence of functionals $\{J_e^0\}_{e>0}$ is $\Gamma(L^p(B)^3)$ -convergent to

$$J_{\mathbf{h}}^0(w_\alpha, w) = \int_B j_0[z_3, e(w_\alpha) + z_3\boldsymbol{\kappa}(w)] + \int_{\Upsilon_0} j_{0\infty}[z_3, -\mathbf{T}_1(w_\alpha) + z_3\mathbf{F}(w)]dS\,, \quad (14.2.64)$$

where $(w_\alpha) \in BD(\Omega)$, $w \in HB(\Omega)$ with $w = 0$ on Υ; $1 \le p < 3/2$.

The relaxed problem RP_h^λ means evaluating

$$(RP_h^\lambda) \quad \inf\{J_h^0(w_\alpha, w) - \lambda L(\boldsymbol{u})|\ \boldsymbol{u} \in \overset{\approx}{U}(B),\ u_3 = 0 \quad \text{on } \Upsilon,$$

$$(w_\alpha, w) \in \overset{\approx}{W}(\Omega) \text{ with } \boldsymbol{u} = (w_\alpha - z_3 w_{;\alpha}, w)\},$$

where $\lambda < \lambda^h$ and

$$(RP_{hLA}) \quad \lambda^h = \inf\ \{\int_B j_{0\infty}[z_3, \boldsymbol{e}(w_\alpha) + z_3\boldsymbol{\kappa}(w)]$$

$$+ \int_{Y_0} j_{0\infty}[z_3, -\mathbf{T}_1(w_\alpha) + z_3\mathbf{F}(w)]dS|\ L(\boldsymbol{v}) = 1,\ \boldsymbol{v} \in \overset{\approx}{U}(B),\ v_3 = 0 \text{ on } \Upsilon,$$

$$(w_\alpha, w) \in \overset{\approx}{W}(\Omega) \text{ with } \boldsymbol{v} = (w_\alpha - z_3 w_{;\alpha}, w)\}.$$

We recall that in (RP_h^λ) a field (w_α, w) is a displacement field while in (RP_{hLA}) such a field is a velocity field.

We also have

$$\lim_{\varepsilon \to 0} \lambda^e = \lambda^h. \tag{14.2.65}$$

Remark 14.2.10. Outside of discontinuities the effective potential of the homogenized Hencky plate is now given by

$$j_{oh}(\boldsymbol{\epsilon}, \boldsymbol{\rho}) = \int_{-1}^{1} j_0(z_3, \boldsymbol{\epsilon} + z_3\boldsymbol{\rho})dz_3, \tag{14.2.66}$$

where $\boldsymbol{\epsilon}, \boldsymbol{\rho} \in \mathbb{E}_s^2$.

14.2.4. Derivation of the third effective plate model: $e \to 0$ and $\varepsilon \to 0$ simultaneously

In Sec. 10.2.4 we have justified a thin plate model made of a linear elastic material with a periodic microstructure provided that $e \to 0$ and $\varepsilon \to 0$ simultaneously. We proceed now to extension of such a practically important model to thin plates made of Hencky material. As one can expect, we have to find the Γ-limit of the sequence of functionals involved in $P_{e\varepsilon}^\lambda$ when $e \to 0$ and $\varepsilon \to 0$ *simultaneously.*

Once the elastoplastic potential $j(z/\varepsilon, z_3, \mathbb{Q}^e e)$ is known, one can generalize the results of Sec. 10.2.4. First, let us introduce the effective elasto-plastic potential

$$j_H(\boldsymbol{\epsilon}, \boldsymbol{\rho}) = \inf\{\prec j(y_\alpha, y_3, \boldsymbol{e}^y(\boldsymbol{\eta}) + \boldsymbol{\epsilon} + y_3\boldsymbol{\rho}) \succ |\ \boldsymbol{\eta} \in LD_{per}^Y(\mathcal{Y})\}, \tag{14.2.67}$$

where $\boldsymbol{\epsilon}, \boldsymbol{\rho} \in \mathbb{E}_s^3$ with $\epsilon_{i3} = 0$ and $\rho_{i3} = 0$. We recall that $z_3 = y_3$, $\mathcal{Y} = Y \times (-1, 1)$ and now $c = 1$.

The space $LD_{per}^Y(\mathcal{Y})$ is defined by

$$LD_{per}^Y(\mathcal{Y}) = \{\boldsymbol{\eta} \in LD_{per}^Y(\mathcal{Y})|\ \boldsymbol{\eta}(y_\alpha, y_3) \text{ is } Y\text{-periodic in } y_\alpha,\ \alpha = 1, 2\}. \tag{14.2.68}$$

The properties of the potential j_H as well as of J_h and j_{0h} follow from those of j, the last being given by (14.2.7). Such a simple study is left to the reader.

The dual potential j_H^* is derived according to the formula (2.10.115), where $\widetilde{W}(\mathcal{Y})$ is to be replaced by $LD_{per}^Y(\mathcal{Y})$. The final formula for this potential is given by (2.10.125), where $h_0 = 2$ and the space $L^2(\mathcal{Y}, \mathbb{E}_3^s)$ appearing in (2.10.126) is to be replaced by $L^\infty(\mathcal{Y}, \mathbb{E}_3^s)$. Then, one can formulate a lemma similar to Lemma 2.10.12. More precisely, a sequence $\{v^\varepsilon\}_{\varepsilon>0} \subset LD(B)$ converges strongly in $L^1(B)^3$ to $v \in \widetilde{\widetilde{U}}(B)$. Now we are in a position to formulate a counterpart of Theorem 2.10.13, which holds for periodic plates with the thickness comparable with the period. More precisely, e and ε jointly tend to zero, staying proportional. It is sufficient to consider the case $\varepsilon = e$, since otherwise one can always perform a scale change in y_1 and y_2 in the cell Y.

Theorem 14.2.11.

(i) The sequence of functionals $\{J_{\varepsilon\varepsilon}\}_{\varepsilon>0}$ given by (14.2.58) is $\Gamma(L^p(B^3)$ -convergent to

$$J_H(u, w) = \int_\Omega j_H(e(u), \kappa(w)) + \int_\Gamma j_{H\infty}(-\mathsf{T}_1(u), \mathsf{F}(w))d\Gamma , \qquad (14.2.69)$$

where $u \in BD(\Omega)$, $w \in HB(\Omega)$, $w = 0$ on Γ and $1 \le p < 3/2$.

(ii) The limit load multiplier λ^H is determined by

$$\lambda^H = \inf\{\int_\Omega j_{H\infty}(e(u), \kappa(w)) + \int_\Gamma j_{H\infty}(-\mathsf{T}_1(u), \mathsf{F}(w))d\Gamma|\, L(u, w) = 1 ,$$

$$u \in BD(\Omega), w \in HB(\Omega), \ w = 0 \text{ on } \Gamma\} . \qquad (14.2.70)$$

Moreover, we have

$$\lim_{\varepsilon \to 0} \lambda^{\varepsilon\varepsilon} = \lambda^H . \qquad (14.2.71)$$

Proof. To show (14.2.69), one can exploit the proof of Theorem 2.10.13 combined with the relaxation of boundary conditions on Γ. The relation (14.2.71) is then a straightforward consequence of (i) provided that L is a continuous perturbation. Details are left to the reader. $\qquad\qquad\square$

15. Comments and bibliographical notes

The space $LD(\Omega)$ is systematically studied in Temam (1985, Chap. II). For more details concerning the space $BV(\Omega)$ the reader is referred to the book by Giusti (1984), cf. also Anzellotti and Giaquinta (1978), Demengel (1989). Anzellotti and Giaquinta (1978) solved the problem of approximation of functions in $BV(\Omega)$ by C^∞-functions in the intermediate topology. The corresponding convergence is called T-convergence. Under rather weak assumptions it was also proved that the trace operator:

$$\gamma_0 : \; BV(\Omega) \to L^1(\partial^*\Omega)$$

is continuous in the intermediate topology; moreover this operator is surjective. Here $\partial^*\Omega$ denotes the *reduced boundary* of Ω, see Giusti (1984). Roughly speaking, $\partial^*\Omega$ is the regular part of $\Gamma = \partial\Omega$, thus allowing for domains with corners.

The book by Temam (1985) offers a systematic presentation of the properties of the space $BD(\Omega)$, cf. also Demengel (1989). The space $HB(\Omega)$ is studied in Demengel (1984) and Temam (1985), cf. also Demengel (1982, 1989).

The integral representation formula is due to Bouchitté and Valadier (1988); for related studies on convex functionals of a measure, see Hadhri (1985a), Temam (1985).

The deformational theory of plasticity or Hencky plasticity and its relation to the flow theory of plasticity is presented in the books by Kachanov (1971) and Olszak et al. (1965). For more details on yield conditions for plastic plates the reader is referred to Janas et al. (1972), Save and Massonet (1972), Olszak et al. (1965), see also Telega and Wojnar (1996). The influence of shear forces on plastic behavior of plates was studied by Sawczuk and Duszek (1963) and Papadopoulos and Taylor (1990).

Bouchitté (1986-1987) found the Γ-limit on $BV(\Omega)$ of sequence of functionals with linear growth. This author also solved the problem of homogenization in the presence of a Dirichlet boundary condition. Amar (1998) extended the notion of two-scale convergence to the case of sequences of bounded measures. It is thus possible to carry out homogenization in $BV(\Omega)$ in an alternative manner. However, boundary conditions were not taken into account. Braides and Chiadò Piat (1995) solved the problem of homogenization in $BV(\Omega)$ when the integrand j is possibly a discontinuous function in the space variables. The homogenization potential j_h is then characterized by a minimum value problem on sets of finite perimeter, see Giusti (1984). The paper by Demengel and Tang Qi (1990) offers a study of periodic homogenization in the space $U(\Omega)$ defined by (13.1.17).

Homogenization theorems in $HB(\Omega)$ presented in Secs. 13.3, 13.4 are due to Telega (1995). The results concerning homogenization of Reissner and von Kármán plates made of Henecky materials are original. Though only plates clamped along their boundaries were examined, one can extend those results to plates subject to boundary loading. In this case at least formal homogenization can be performed for von Kármán plates by extending the results of Secs. 13.4.3, 14.1 and combining them with those due to Hadhri (1985b). To prove Lemma 14.1.1, one can proceed similarly to the case of homogeneous plates studied in Demengel (1982) and Temam (1985), cf. also Christiansen (1986) for an alternative approach to duality.

Telega (1991) performed homogenization of Kirchhoff plate loaded by boundary forces and moments. Section 14.1 provides refinements and proofs, which were not given in Telega (1991). We follow the approach developed by Bouchitté and Suquet (1991) for three-dimensional Hencky solids loaded on the boundary.

Percivale (1990) and Tang Qi (1990) used the theory of Γ-convergence to a mathematical justification of two-dimensional plate models made of isotropic and homogeneous Hencky material. The study performed by Tang Qi (1990) is more complete. The results presented in Sec. 14.2 are original and extend those presented in Sec. 2.10. Particularly, Sec. 14.2.1 extends the results presented in the book by Temam (1985) valid only for homogeneous materials incompressible in the plastic range. In fact, from the mathematical point of view it is easier to handle with plastically compressible materials. For inhomogeneous materials, the elasto-plastic potential depends on $x \in B_e$ (or $z \in B$). One has to use then the notion of convex functionals of a measure introduced in Sec. 13.2.

To accomplish the passage with e to zero we assumed, after Tang Qi (1990), a specific form of the loading functional. It seems that one can also consider the case of loading functional with nonvanishing components $g^e_{\pm\alpha}$. Then a relaxation term on the faces of B_e (or B) would appear.

The load multiplier λ usually affects both the body forces and boundary loading. Often physically more realistic is the case where this multiplier is associated to the boundary loading only, cf. Telega (1985, 1988, 1990).

Chapter V

ELASTIC AND PLASTIC SHELLS

Introduction

Even in the case of homogeneous shells their statical analysis is complex and usually requires using computational tools, like the finite element packages. These difficulties grow essentially if the shell is nonhomogeneous, especially if the properties of the shell vary rapidly. The homogenization methods make it possible to split up the analysis into an analysis of a slowly nonhomogeneous shell and a local analysis within the cells of periodicity. The aim of this chapter is to put forward the details of the homogenization method within the context of the selected thin shell models. It is worth emphasizing that the homogenized model remains nonhomogeneous, because the periodicity assumption concerns the distribution of elastic properties in the plane reference domain.

To put the homogenization analysis in a broader perspective, this chapter begins with a short discussion of some linear and nonlinear models of thin shells. In the case of balanced elastic shells the first order constitutive relations are decoupled: the membrane forces are not affected by the changes of curvature and the tangent deformation does not influence the moments. Such approximation is called the first order approximation of Love. It seems interesting that the homogenization process makes these constitutive relations coupled. Such coupling occurs for non-shallow shells only. This property is confirmed in Sec. 17 by using the Γ-convergence method of homogenization.

Except elastic shells, in Sec. 19 we study perfectly plastic shells, thus extending partially the analysis performed in the previous chapter.

16. Linear and nonlinear models of elastic shells

A natural starting point for deriving a thin shell theory is the Naghdi theory of elastic shells with transverse shear deformations, cf. Naghdi (1963). We recall this theory and then proceed to the justification of the two most important theories of thin shells subject to the Kirchhoff-Love constraints. The last two sections concern some simple models of shells undergoing moderately large rotations.

16.1. *Theory of shells with transverse shear deformation*

Before defining the shell itself let us start with geometry of its mid-surface. Assume that $\Omega \subset \mathbb{R}^2$ is a bounded sufficiently regular domain parametrized by a Cartesian coordinate system (ξ^α), $\alpha = 1, 2$. Let $\mathbf{\Phi} : \Omega \to S \subset \mathbb{R}^3$ be a mapping of class $C^3(\overline{\Omega})$. Here S denotes a surface in $\mathbb{R}^3$ determined by: $x^i = x^i(\xi^1, \xi^2)$, $i = 1, 2, 3$, $\xi = (\xi^1, \xi^2) \in \Omega$, or $\mathbf{\Phi}(\xi) = x^i(\xi)e_i$, where e_i are the versors of the orthogonal Cartesian coordinate system $\{x^i\}$ with the origin at the point $\mathbf{0}$. The vectors tangent to the coordinate lines ξ^α on S are, cf. Fig. 16.1.1

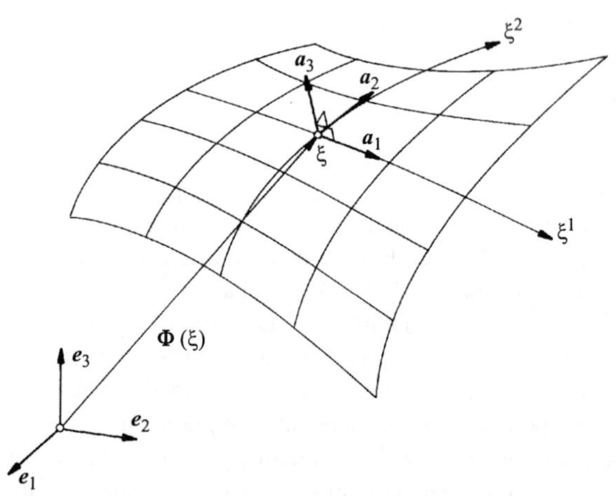

Fig. 16.1.1. Parametrization of the shell middle surface

$$a_\alpha = \frac{\partial \mathbf{\Phi}}{\partial \xi^\alpha} = \mathbf{\Phi},_\alpha \, . \tag{16.1.1}$$

Let a_3 be the unit vector normal to S at a point ξ. This vector is defined by

$$a_3 = \frac{a_1 \times a_2}{|a_1 \times a_2|} \, . \tag{16.1.2}$$

The vector a_3 determines a straight line parametrized by ξ^3. The coordinate ξ^3 is equal to zero at S and measures the distance to S. Let us define the domain

$$B = \{ x^i(\xi^j) \mid \xi = (\xi^1, \xi^2) \in \Omega, \qquad |\xi^3| < h/2 \} \, .$$

The three dimensional body that occupies this domain will be called a shell of constant thickness h.

The symmetric covariant metric tensor of S is given by the scalar product

$$a_{\alpha\beta} = a_\alpha \cdot a_\beta \, . \tag{16.1.3}$$

The covariant components of the curvature tensor of S are

$$b_{\alpha\beta} = -a_\alpha \cdot a_{3,\beta} = a_3 \cdot a_{\alpha,\beta} \, . \tag{16.1.4}$$

The contravariant metric tensor $(a^{\alpha\beta})$ satisfying the relation: $a^{\alpha\beta}a_{\beta\gamma} = \delta^{\alpha}_{\gamma}$ is used to raise the indices. The Christoffel symbols of S are given by

$$\Gamma^{\alpha}_{\beta\gamma} = a^{\alpha\lambda}\Gamma_{\lambda\beta\gamma}\,, \qquad \Gamma_{\alpha\beta\gamma} = \frac{1}{2}(a_{\alpha\beta,\gamma} + a_{\alpha\gamma,\beta} - a_{\beta\gamma,\alpha})\,. \qquad (16.1.5)$$

The covariant derivative is defined as follows

$$u_{\alpha||\beta} = u_{\alpha,\beta} - \Gamma^{\lambda}_{\alpha\beta}u_{\lambda}\,, \qquad (16.1.6)$$

for an arbitrary covariant vector (u_{α}).

Note that

$$\boldsymbol{a}_{\alpha} \times \boldsymbol{a}_{\beta} = \varepsilon_{\alpha\beta}\boldsymbol{a}_3\,,$$

where

$$\varepsilon_{\alpha\beta} = \sqrt{a}\,\bar{\varepsilon}_{\alpha\beta}\,, \qquad \bar{\varepsilon}_{11} = \bar{\varepsilon}_{22} = 0\,, \qquad \bar{\varepsilon}_{12} = -\bar{\varepsilon}_{21} = 1\,. \qquad (16.1.7)$$

Here $(\varepsilon_{\alpha\beta})$ is the Ricci pseudotensor and $\bar{\varepsilon}_{\alpha\beta}$ is the permutation symbol; $a = \det(a_{\alpha\beta})$. At regular points $a \neq 0$. The elementary area of S is given by

$$dS = \sqrt{a}d\xi^1 d\xi^2\,. \qquad (16.1.8)$$

The third fundamental tensor of S is defined by

$$c_{\alpha\beta} = b^{\lambda}_{\alpha}b_{\lambda\beta}\,. \qquad (16.1.9)$$

The Gauss formula assumes the form

$$\boldsymbol{a}_{\alpha,\beta} = \Gamma^{\gamma}_{\alpha\beta}\boldsymbol{a}_{\gamma} + b_{\alpha\beta}\boldsymbol{a}_3\,. \qquad (16.1.10)$$

The Weingarten formula reads

$$\boldsymbol{a}_{3,\alpha} = -b^{\lambda}_{\alpha}\boldsymbol{a}_{\lambda}\,. \qquad (16.1.11)$$

The Mainardi-Codazzi equations assume the form

$$b^{\gamma}_{[\alpha||\beta]} = 0\,, \qquad (16.1.12)$$

where $[\cdot]$ denotes the skew symmetry.

The domain B is composed of the surfaces $\xi^3 = $ const. Let P be a point (ξ^i); its position vector is given by

$$\widehat{\boldsymbol{\Phi}}(\xi^i) = \boldsymbol{\Phi}(\xi) + \xi^3 \boldsymbol{a}_3(\xi)\,. \qquad (16.1.13)$$

The vectors tangent to the coordinate lines ξ^{α} at P are given by

$$\boldsymbol{g}_{\alpha} = \widehat{\boldsymbol{\Phi}}_{,\alpha} = \mu^{\sigma}_{\alpha}\boldsymbol{a}_{\sigma}\,, \qquad (16.1.14)$$

where

$$\mu^{\sigma}_{\alpha} = \delta^{\sigma}_{\alpha} - \xi^3 b^{\sigma}_{\alpha}\,, \qquad (16.1.15)$$

is the so-called shifter; moreover, $g_3 = a_3$. The metric tensor of the surface $\xi^3 = $ const has the following components

$$g_{\alpha\beta} = \boldsymbol{g}_\alpha \cdot \boldsymbol{g}_\beta \, , \qquad g_{\alpha 3} = 0 \, , \qquad g_{33} = 1 \, , \tag{16.1.16}$$

where, by (16.1.3) and (16.1.9)

$$g_{\alpha\beta} = a_{\alpha\beta} - 2\xi^3 b_{\alpha\beta} + (\xi^3)^2 c_{\alpha\beta} \, . \tag{16.1.17}$$

Along the lines ξ^3 the coordinate system $\{\xi^i\}$ is orthogonal, see (16.1.16). Such a parametrization of B is usually called normal.

The elementary area of the surface $\xi^3 = const$ named $S(\xi^3)$ is given by

$$dS(\xi^3) = \sqrt{\det(g_{\alpha\beta})}d\xi \tag{16.1.18}$$

or

$$dS(\xi^3) = \mu\sqrt{a}d\xi = \mu dS \, , \tag{16.1.19}$$

where $\mu = \det(\mu_\alpha^\gamma)$.

A vector w can be referred to both bases $\{a_i\}$ and $\{g_i\}$, which will be written as follows

$$\boldsymbol{w} = \overline{w}^\alpha \boldsymbol{a}_\alpha + \overline{w}^3 \boldsymbol{a}_3 \, , \qquad \boldsymbol{w} = w^\alpha \boldsymbol{g}_\alpha + w^3 \boldsymbol{g}_3 \, . \tag{16.1.20}$$

Note, that not the type of indices but a bar distinguishes between both representations. By (16.1.14) we have

$$\overline{w}^\alpha = \mu_\gamma^\alpha w^\gamma \, , \qquad \overline{w}^3 = w^3 \, . \tag{16.1.21}$$

Let $\{a^i\}$ and $\{g^i\}$ be co-bases, such that $a^i \cdot a_j = \delta_j^i$, $g^i \cdot g_j = \delta_j^i$. Then

$$w_\alpha = \mu_\alpha{}^\gamma \overline{w}_\gamma \, , \qquad w_3 = \overline{w}_3 \, . \tag{16.1.22}$$

The simplest theory of shells is based on the following assumption:

$$\overline{w}_\alpha(\xi, \xi^3) = u_\alpha(\xi) + \xi^3 \varphi_\alpha(\xi) \, , \qquad \overline{w}_3(\xi, \xi^3) = w(\xi) \, , \tag{16.1.23}$$

where $\overline{w}_i$ are components of the displacement vector referred to the basis $\{a^i\}$ and $u_\alpha, \varphi_\alpha, w$ are unknown fields defined on the middle surface S. The components w_i referred to the basis $\{a^i\}$ are given by (16.1.22). Having found these components one can determine the components of the strain tensor e_{ij} referred to the basis $g^i \otimes g^j$. They read as follows, see Eqs. (4.30) – (4.31) of Naghdi (1963),

$$2e_{\alpha\beta}(\boldsymbol{w}) = \mu_\alpha^\gamma[\theta_{\gamma\beta}(\boldsymbol{u}, w) + \xi^3 h_{\gamma\beta}(\boldsymbol{\varphi})] + \mu_\beta^\gamma[\theta_{\gamma\alpha}(\boldsymbol{u}, w) + \xi^3 h_{\gamma\alpha}(\boldsymbol{\varphi})] \, ,$$
$$2e_{\alpha 3}(\boldsymbol{w}) = \varphi_\alpha + w_{,\alpha} + b_\alpha^\beta u_\beta \, , \qquad e_{33} = 0 \, , \tag{16.1.24}$$

where

$$\theta_{\gamma\beta}(\boldsymbol{u}, w) = u_{\gamma||\beta} - b_{\gamma\beta}w \, , \qquad h_{\alpha\beta}(\boldsymbol{\varphi}) = \varphi_{\alpha||\beta} \, . \tag{16.1.25}$$

Assume that the shell is clamped at its lateral cylindrical surface. Kinematically admissible fields (η_i) obey the assumption (16.1.23). Hence

$$\bar{\eta}_\alpha(\xi, \xi^3) = v_\alpha(\xi) + \xi^3 \psi_\alpha(\xi) , \qquad \bar{\eta}_3(\xi, \xi^3) = v(\xi) , \tag{16.1.26}$$

where v_α, ψ_α, v are certain fields defined on S. Let n, t be unit vectors: outward normal and tangent to ∂S and lying in the tangent planes of ∂S, respectively. The condition of clamping implies

$$v_n = 0 , \quad v_t = 0 , \quad \psi_n = 0 , \quad \psi_t = 0 , \quad v = 0 , \quad \text{on } \partial S . \tag{16.1.27}$$

Here $f_n = f_\alpha n^\alpha$, $f_t = f_\alpha t^\alpha$ and n^α, t^α are referred to the basis $\{a_\alpha\}$; $f \in \{v, \psi\}$. The unknown fields satisfy similar conditions:

$$u_n = 0 , \quad u_t = 0 , \quad \varphi_n = 0 , \quad \varphi_t = 0 , \quad w = 0 . \tag{16.1.28}$$

Assume that the shell is subject to body forces of density $p(\xi, \xi^3)$. The variational equilibrium equation:

$$\int_B \sigma^{ij} e_{ij}(\eta) d\xi = \int_B p^i \eta_i d\xi , \tag{16.1.29}$$

should hold for all η satisfying (16.1.27). Here the stress components σ^{ij} are referred to the basis $g_i \otimes g_j$ at a point ξ. We compute

$$\sigma^{ij} e_{ij}(\eta) = \sigma^{\alpha\beta} e_{\alpha\beta}(\eta) + 2\sigma^{\alpha 3} e_{\alpha 3}(\eta) , \tag{16.1.30}$$

substitute (16.1.24) for $w = \eta$ and next integrate from $-h/2$ to $+h/2$. Thus we find the two-dimensional form of the variational equilibrium equation:

$$\int_S [N^{\beta\gamma}\theta_{\gamma\beta}(v, v) + M^{\beta\gamma}h_{\gamma\beta}(\psi) + Q^\alpha(\psi_\alpha + v_{,\alpha} + b^\beta_\alpha v_\beta)]dS = \int_S (q^\alpha v_\alpha + qv)dS , \tag{16.1.31}$$

for each (v, ψ, v) satisfying (16.1.27). Here

$$\mathbf{N}^{\beta\gamma} = \int_{-h/2}^{h/2} \mu\sigma^{\beta\lambda}\mu^\gamma_\lambda d\xi^3 , \qquad \mathbf{M}^{\beta\gamma} = \int_{-h/2}^{h/2} \mu\sigma^{\beta\lambda}\mu^\gamma_\lambda \xi^3 d\xi^3 ,$$

$$Q^\alpha = \int_{-h/2}^{h/2} \mu\sigma^{\alpha 3} d\xi^3 , \tag{16.1.32}$$

and

$$q^\alpha = \int_{-h/2}^{h/2} \mu^\alpha_\beta p^\beta \mu d\xi^3 , \qquad q = \int_{-h/2}^{h/2} \mu p^3 d\xi^3 . \tag{16.1.33}$$

The work of surface moments on the angles ψ_α has been neglected.

The stress and couple resultants $\mathbf{N}^{\beta\gamma}, \mathbf{M}^{\beta\gamma}$ are not independent. The condition of symmetry: $\sigma^{12} = \sigma^{21}$ implies the algebraic equation

$$\varepsilon_{\lambda\mu}(\mathbf{N}^{\mu\lambda} - b^{\mu}_{\beta}\mathbf{M}^{\beta\lambda}) = 0 \ . \tag{16.1.34}$$

The variational equation (16.1.31) implies the following five local equations of equilibrium:

$$\mathbf{N}^{\beta\alpha}{}_{\|\beta} - b^{\alpha}_{\gamma}Q^{\gamma} + q^{\alpha} = 0 \ ,$$
$$Q^{\alpha}{}_{\|\alpha} + b_{\alpha\gamma}\mathbf{N}^{\gamma\alpha} + q = 0 \ , \tag{16.1.35}$$
$$\mathbf{M}^{\beta\alpha}{}_{\|\beta} - Q^{\alpha} = 0 \ .$$

Equation (16.1.34) is called the sixth equation of equilibrium.

The constitutive relations are proposed by assuming that $|\sigma^{33}| \ll |\sigma^{\alpha\beta}|$, which means accepting the constitutive relations in the form

$$\sigma^{\alpha\beta} = \widetilde{C}^{\alpha\beta\lambda\mu}e_{\lambda\mu} \ , \qquad \sigma^{\alpha 3} = 2C^{\alpha 3\lambda 3}e_{\lambda 3} \ , \qquad \sigma^{33} = 0 \ , \tag{16.1.36}$$

where, cf. (5.12)

$$\widetilde{C}^{\alpha\beta\lambda\mu} = C^{\alpha\beta\lambda\mu} - C^{\alpha\beta 33}C^{33\lambda\mu}/C^{3333} \ . \tag{16.1.37}$$

Substitution of (16.1.36) into (16.1.32) gives

$$\mathbf{N}^{\beta\alpha} = {}_{0}B^{\alpha\beta\lambda\mu}\theta_{\lambda\mu}(\boldsymbol{u}, w) + {}_{1}B^{\alpha\beta\lambda\mu}h_{\lambda\mu}(\boldsymbol{\varphi}) \ ,$$
$$\mathbf{M}^{\beta\alpha} = {}_{1}B^{\alpha\beta\lambda\mu}\theta_{\lambda\mu}(\boldsymbol{u}, w) + {}_{2}B^{\alpha\beta\lambda\mu}h_{\lambda\mu}(\boldsymbol{\varphi}) \ , \tag{16.1.38}$$
$$Q^{\alpha} = H^{\alpha\lambda}\beta_{\lambda}(\boldsymbol{u}, w, \boldsymbol{\varphi}) \ ,$$

where

$$_{n}B^{\gamma\beta\delta\sigma} = \int\limits_{-h/2}^{h/2} (\xi^{3})^{n}\mu\mu^{\gamma}_{\lambda}\mu^{\delta}_{\alpha}\widetilde{C}^{\beta\lambda\sigma\alpha}d\xi^{3} \ , \quad n = 0, 1, 2 \ ,$$

$$H^{\alpha\lambda} = \int\limits_{-h/2}^{h/2} \mu C^{\alpha 3\lambda 3}d\xi^{3} \ . \tag{16.1.39}$$

The quantities

$$\beta_{\lambda}(\boldsymbol{u}, w, \boldsymbol{\varphi}) = \varphi_{\lambda} + w_{,\lambda} + b^{\beta}_{\lambda}u_{\beta} \tag{16.1.40}$$

represent transverse shear deformations.

Upon substitution of the constitutive relations (16.1.38) into (16.1.31) one finds the variational equation of the form:

$$\mathbf{a}(\boldsymbol{u}, w, \boldsymbol{\varphi}; \boldsymbol{v}, v, \boldsymbol{\psi}) = \int\limits_{\Omega} (q^{\alpha}v_{\alpha} + qv)\sqrt{a}d\xi \tag{16.1.41}$$

with

$$\mathbf{a}(\boldsymbol{u}, w, \boldsymbol{\varphi}; \boldsymbol{v}, v, \boldsymbol{\psi}) = \int_{\Omega} [{}_0 B^{\alpha\beta\lambda\mu} \theta_{\lambda\mu}(\boldsymbol{u}, w)\theta_{\alpha\beta}(\boldsymbol{v}, v)$$

$$+ {}_1 B^{\alpha\beta\lambda\mu}(\theta_{\lambda\mu}(\boldsymbol{u}, w)h_{\alpha\beta}(\boldsymbol{\psi}) + \theta_{\alpha\beta}(\boldsymbol{v}, v)h_{\lambda\mu}(\boldsymbol{\varphi})) \qquad (16.1.42)$$

$$+ {}_2 B^{\alpha\beta\lambda\mu}h_{\lambda\mu}(\boldsymbol{\varphi})h_{\alpha\beta}(\boldsymbol{\psi}) + H^{\lambda\mu}\beta_\mu(\boldsymbol{u}, w, \boldsymbol{\varphi})\beta_\lambda(\boldsymbol{v}, v, \boldsymbol{\psi})]\sqrt{a}d\xi .$$

The symmetry conditions

$$\,_n B^{\alpha\beta\lambda\mu} = \,_n B^{\lambda\mu\alpha\beta} \qquad (16.1.43)$$

imply

$$\mathbf{a}(\boldsymbol{u}, w, \boldsymbol{\varphi}; \boldsymbol{v}, v, \boldsymbol{\psi}) = \mathbf{a}(\boldsymbol{v}, v, \boldsymbol{\psi}; \boldsymbol{u}, w, \boldsymbol{\varphi}) . \qquad (16.1.44)$$

In the case of clamped shells the set of kinematically admissible fields V is obviously given by

$$(\boldsymbol{v}, v, \boldsymbol{\psi}) \in V , \quad V = H_0^1(\Omega)^2 \times H_0^1(\Omega) \times H_0^1(\Omega)^2 .$$

Within the assumption of a generalized plane-stress the constitutive relations (16.1.38) are exact. Thus the positive definiteness property of the bilinear form

$$\int_S [\widetilde{C}^{\alpha\beta\lambda\mu} e_{\alpha\beta}(\boldsymbol{w})e_{\lambda\mu}(\boldsymbol{v}) + 4C^{\alpha3\lambda3}e_{\alpha3}(\boldsymbol{w})e_{\lambda3}(\boldsymbol{v})]dS$$

implies V-ellipticity of the form (16.1.42). Consequently the following problem:

$$(P) \quad \left| \begin{array}{l} \text{find } (\boldsymbol{u}, w, \boldsymbol{\varphi}) \in V \text{ such that} \\ \text{Eq. (16.1.41) holds for every } (\boldsymbol{v}, v, \boldsymbol{\psi}) \in V \end{array} \right.$$

is uniquely solvable.

16.2.   Koiter's version of thin shell model

If a shell is thin with respect to its characteristic dimension L then the transverse shear deformations β_λ can be neglected. The angles of rotations φ_α are then determined by distortion of the shell mid-surface

$$\varphi_\lambda = -w_{,\lambda} - b_\lambda^\beta u_\beta . \qquad (16.2.1)$$

The quantity L is usually defined as $\min\{R_1, R_2, d\}$, where R_α represent the principal radii of curvature of the surface S and d represents the global dimension of the body B of the shell. Some authors associate L with a characteristic wavelength of the deformation pattern, thus linking the notion of thinness with the loading applied.

Substitution of (16.2.1) into $(16.1.25)_2$ or the definition of $(h_{\alpha\beta})$ transforms this tensor to the form

$$k_{\alpha\beta}(\boldsymbol{u}, w) = -w_{\|\alpha\beta} - b_{\alpha\|\beta}^\gamma u_\gamma - b_\alpha^\gamma u_{\gamma\|\beta} . \qquad (16.2.2)$$

The deformation measures $(\theta_{\alpha\beta})$, $(k_{\alpha\beta})$ are interrelated by the compatibility equations

$$\varepsilon^{\nu\rho}(b^{\alpha}_{\gamma}\varepsilon^{\gamma\mu}\theta_{\mu\gamma\|\rho} + \varepsilon^{\alpha\mu}k_{\mu\nu\|\rho}) = 0 ,$$

$$\varepsilon^{\mu\alpha}\varepsilon^{\nu\rho}(\theta_{\mu\nu\|\rho\alpha} - b_{\alpha\rho}k_{\mu\nu}) = 0 , \tag{16.2.3}$$

$$\varepsilon^{\alpha\beta}(k_{\alpha\beta} + b^{\nu}_{\alpha}\theta_{\nu\beta}) = 0 .$$

They can be expressed by the following variational equation:

$$\int_{\Omega} \varepsilon^{\alpha\mu}\varepsilon^{\nu\gamma}[k_{\mu\nu}\theta_{\alpha\gamma}(\boldsymbol{v},v) - \theta_{\mu\nu}k_{\alpha\gamma}(\boldsymbol{v},v)]\sqrt{a}d\xi = 0 \qquad \forall\, (\boldsymbol{v},v) \in V^0(\Omega) ,$$

with $V^0(\Omega) = H^1_0(\Omega)^2 \times H^2_0(\Omega)$. If we compare this equation with the homogeneous variational equation of equilibrium

$$\int_{\Omega} [\mathbf{N}^{\gamma\alpha}\theta_{\alpha\gamma}(\boldsymbol{v},v) + \mathbf{M}^{\gamma\alpha}k_{\alpha\gamma}(\boldsymbol{v},v)]\sqrt{a}d\xi = 0 \quad \forall\; (\boldsymbol{v},v) \in V^0(\Omega) ,$$

we note that the quantities

$$\mathbf{N}^{\gamma\alpha} \quad \text{and} \quad \varepsilon^{\alpha\mu}\varepsilon^{\nu\gamma}k_{\mu\nu} , \qquad \mathbf{M}^{\gamma\alpha} \quad \text{and} \quad -\varepsilon^{\alpha\mu}\varepsilon^{\nu\gamma}\theta_{\mu\nu} \tag{16.2.4}$$

are conjugate in the sense of the analogy between the equilibrium equations and compatibility equations. This analogy tells us that the following representation of the stress and couple resultants:

$$\mathbf{N}^{\gamma\alpha} = \varepsilon^{\alpha\mu}\varepsilon^{\nu\gamma}k_{\mu\nu}(\boldsymbol{\psi},\psi) , \qquad \mathbf{M}^{\gamma\alpha} = -\varepsilon^{\alpha\mu}\varepsilon^{\nu\gamma}\theta_{\mu\nu}(\boldsymbol{\psi},\psi)$$

where $\boldsymbol{\psi} = (\psi_1, \psi_2)$, satisfy the homogeneous equilibrium equations identically, irrespective of the choice of stress functions ψ_1, ψ_2, ψ.

Let us assume now that the shell is transversely homogeneous. A formal substitution of (16.2.1) into the formulae of Sec. 16.1 leads to a correct shell model. However, if we take advantage of the shell being thin, we arrive at some contradictions. For thin shells one may assume $\mu^{\gamma}_{\lambda} \approx \delta^{\gamma}_{\lambda}$ and $\mu = 1$ in (16.1.39), which leads to

$$_0B^{\gamma\beta\delta\sigma} = A^{\gamma\beta\delta\sigma} , \qquad _1B^{\gamma\beta\delta\sigma} = 0, \qquad _2B^{\gamma\beta\delta\sigma} = D^{\gamma\beta\delta\sigma} , \tag{16.2.5}$$

where $\boldsymbol{A} = h\widetilde{\boldsymbol{C}}$, $\boldsymbol{D} = (h^2/12)\boldsymbol{A}$. This approximation, however, results in $\mathbf{N}^{\alpha\beta} = \mathbf{N}^{\beta\alpha}$, $\mathbf{M}^{\alpha\beta} = \mathbf{M}^{\beta\alpha}$, which violates the sixth equilibrium equation (16.1.34). Thus a passage to a thin shell model requires special caution. Let us define the deformation measures

$$\epsilon_{\alpha\beta} = \frac{1}{2}(\theta_{\alpha\beta} + \theta_{\beta\alpha}) , \qquad \rho_{\alpha\beta} = k_{\alpha\beta} - b^{\lambda}_{\beta}\theta_{\lambda\alpha} . \tag{16.2.6}$$

On expressing them in terms of displacements one finds

$$\epsilon_{\alpha\beta}(\boldsymbol{u},w) = \frac{1}{2}(u_{\alpha\|\beta} + u_{\beta\|\alpha}) - b_{\alpha\beta}w ,$$

$$\rho_{\alpha\beta}(\boldsymbol{u},w) = -w_{\|\alpha\beta} - b^{\gamma}_{\alpha\|\beta}u_{\gamma} - b^{\gamma}_{\alpha}u_{\gamma\|\beta} - b^{\gamma}_{\beta}u_{\gamma\|\alpha} + c_{\alpha\beta}w , \tag{16.2.7}$$

and by the Mainardi-Codazzi relation (16.1.12) we conclude that the tensor ρ is symmetric. This property follows also directly from the compatibility equation $(16.2.3)_3$.

Let us rearrange the variational equation of equilibrium (16.1.31) by substituting: $k_{\gamma\beta} = \rho_{\gamma\beta} + b^\lambda_\beta \theta_{\lambda\gamma}$ and $\beta_\alpha(\boldsymbol{v}, v, \boldsymbol{\psi}) = 0$. We find

$$\int_S [\mathbf{N}^{\beta\alpha}\theta_{\alpha\beta}(\boldsymbol{v}, v) + \mathbf{M}^{\beta\alpha}(\rho_{\alpha\beta}(\boldsymbol{v}, v) + b^\lambda_\beta \theta_{\lambda\alpha}(\boldsymbol{v}, v))]dS = \int_S (q^\alpha v_\alpha + qv)dS \ .$$

Hence

$$\int_S [N^{\alpha\beta}\epsilon_{\alpha\beta}(\boldsymbol{v}, v) + M^{\alpha\beta}\rho_{\alpha\beta}(\boldsymbol{v}, v)]dS = \int_S (q^\alpha v_\alpha + qv)dS \ , \qquad (16.2.8)$$

where

$$N^{\alpha\beta} = \mathbf{N}^{\alpha\beta} + b^\beta_\gamma \mathbf{M}^{\gamma\alpha} \ , \qquad M^{\alpha\beta} = \mathbf{M}^{(\alpha\beta)} \ , \qquad (16.2.9)$$

or

$$N^{\alpha\beta} = \mathbf{N}^{(\alpha\beta)} + \frac{1}{2}(b^\beta_\gamma \mathbf{M}^{\gamma\alpha} + b^\alpha_\gamma \mathbf{M}^{\gamma\beta}) \ , \qquad (16.2.10)$$

since by (16.1.34) the antisymmetric part of $(N^{\alpha\beta})$ vanishes. Thus the variational equilibrium equation (16.2.8) involves the symmetric stress and couple resultants as well as conjugate symmetric deformation measures.

It turns out that the deformation measures ϵ and ρ determine the change of the tensors $(a_{\alpha\beta})$ and $b_{\alpha\beta}$ describing the internal and external geometry of the surface S. To show the role of these tensors let us assume that the deformed configuration of the shell is still parametrized by the coordinates $\{\xi^i\}$ and this parametrization is still normal, which reflects the assumption of the transverse deformations $e_{\alpha 3}$ being zero. Consequently the relation (16.1.17) holds for deformed configuration

$$\breve{g}_{\alpha\beta} = \breve{a}_{\alpha\beta} - 2\xi^3 \breve{b}_{\alpha\beta} + (\xi^3)^2 \breve{c}_{\alpha\beta} \ , \qquad (16.2.11)$$

where the symbol "$\cup$" indicates that the geometrical object is referred to the deformed shell configuration. Within the nonlinear theory of elasticity the deformations of a thin shell are defined by

$$\gamma_{\alpha\beta} = \frac{1}{2}(\breve{g}_{\alpha\beta} - g_{\alpha\beta}) \ , \qquad \gamma_{\alpha 3} = 0 \ , \qquad \gamma_{33} = 0 \ . \qquad (16.2.12)$$

Hence

$$\gamma_{\alpha\beta} = \overset{o}{\gamma}_{\alpha\beta} + \xi^3 \widetilde{\rho}_{\alpha\beta} + \frac{1}{2}(\xi^3)^2(\widetilde{c}_{\alpha\beta} - c_{\alpha\beta}) \ ,$$

$$\overset{o}{\gamma}_{\alpha\beta} = \frac{1}{2}(\breve{a}_{\alpha\beta} - a_{\alpha\beta}) \ , \qquad \widetilde{\rho}_{\alpha\beta} = b_{\alpha\beta} - \breve{b}_{\alpha\beta} \ . \qquad (16.2.13)$$

Upon expressing the tensors $(\overset{o}{\gamma}_{\alpha\beta})$ and $(\widetilde{\rho}_{\alpha\beta})$ in terms of displacements $(\boldsymbol{u}, w)$ and performing the linearization one finds

$$\overset{o}{\gamma}_{\alpha\beta} = \epsilon_{\alpha\beta} + \text{nonlinear terms,}$$

$$\widetilde{\rho}_{\alpha\beta} = \rho_{\alpha\beta} + \text{nonlinear terms.} \qquad (16.2.14)$$

According to the theorem of Bonnet, the tensors $(\breve{a}_{\alpha\beta})$ and $(\breve{b}_{\alpha\beta})$ determine the deformed configuration $\breve{S}$ up to a position in space. This means that tensors $(\epsilon_{\alpha\beta})$ and $(\rho_{\alpha\beta})$ vanish if the shell is subject to translations and infinitely small rotations. This justifies calling them strain tensors. Let us emphasize that the deformation measures $\theta_{\alpha\beta}$ and $k_{\alpha\beta}$ do not vanish under rigid body motions and that is why these measures cannot be viewed as strain measures.

Since the tensor $(\breve{c}_{\alpha\beta})$ is determined by $(\breve{a}_{\alpha\beta})$ and $(\breve{b}_{\alpha\beta})$, the strain measures $(\overset{o}{\gamma}_{\alpha\beta})$ and $(\widetilde{\rho}_{\alpha\beta})$ fully determine the state of deformation (γ_{ij}) of the shell body. The linear part of $(\gamma_{\alpha\beta})$ assumes the form $(e_{\alpha\beta})$ with $k_{\alpha\beta}$ given by (16.2.2). Thus in the linear theory the state of deformation of the shell body is fully determined by the tensors $(\epsilon_{\alpha\beta})$ and $(\rho_{\alpha\beta})$. Let us stress once again that the only approximations that rearrange the variational equilibrium equation (16.1.29) of the three-dimensional shell body to its two-dimensional counterpart (16.2.8) are the kinematic assumptions (16.1.26) and replacing the tensor $C^{\alpha\beta\lambda\mu}$ with $\widetilde{C}^{\alpha\beta\lambda\mu}$, cf. (16.1.42).

Now it is clear that the strains $(e_{\alpha\beta})$ of the shell can be expressed in terms of the tensors $(\epsilon_{\alpha\beta})$ and $(\rho_{\alpha\beta})$. Thus substitution of (16.1.36) into (16.1.32) makes it possible to express the stress and couple resultants in terms of tensors $(\epsilon_{\alpha\beta})$ and $(\rho_{\alpha\beta})$. If we apply the assumptions of thinness (the terms of the order of $O(h/R)$ may be neglected) we arrive at the simplest constitutive relations

$$N^{\alpha\beta} = A^{\alpha\beta\lambda\mu}\epsilon_{\lambda\mu} , \qquad M^{\alpha\beta} = D^{\alpha\beta\lambda\mu}\rho_{\lambda\mu} . \qquad (16.2.15)$$

Now the shell modelling is completed.

Let us define the bilinear form

$$a(\boldsymbol{u}, w; \boldsymbol{v}, v) \qquad\qquad\qquad\qquad\qquad\qquad\qquad (16.2.16)$$
$$= \int_{\Omega} [A^{\alpha\beta\lambda\mu}\epsilon_{\alpha\beta}(\boldsymbol{u}, w)\epsilon_{\lambda\mu}(\boldsymbol{v}, v) + D^{\alpha\beta\lambda\mu}\rho_{\alpha\beta}(\boldsymbol{u}, w)\rho_{\lambda\mu}(\boldsymbol{v}, v)]\sqrt{a}\, d\xi ,$$

and the linear form

$$f(\boldsymbol{v}, v) = \int_{\Omega} (q^{\alpha}v_{\alpha} + qv)\sqrt{a}d\xi . \qquad\qquad (16.2.17)$$

The equilibrium problem of a clamped shell means finding $(\boldsymbol{u}, w) \in V^0(\Omega) = H_0^1(\Omega)^2 \times H_0^2(\Omega)$ such that

$$(P) \qquad a(\boldsymbol{u}, w; \boldsymbol{v}, v) = f(\boldsymbol{v}, v) \qquad \forall\, (\boldsymbol{v}, v) \in V^0(\Omega) .$$

Both the forms (16.2.16) and (16.2.17) are continuous. The bilinear form $a(;)$ is symmetric. Since $\boldsymbol{A}$ and $\boldsymbol{D}$ are positive definite we always have $a(\boldsymbol{v}, v; \boldsymbol{v}, v) \geq 0$ and the equality holds if and only if $\epsilon_{\alpha\beta}(\boldsymbol{v}, v) = 0$ and $\rho_{\alpha\beta}(\boldsymbol{v}, v) = 0$. We already know that these conditions are satisfied if the fields $(\boldsymbol{v}, v)$ describe a rigid body motion of the shell. Such a motion is not possible due to clamping. Thus $a(\boldsymbol{v}, v; \boldsymbol{v}, v) > 0$ if $(\boldsymbol{v}, v) \in V^0(\Omega)$ and $(\boldsymbol{v}, v) \neq 0$. In fact one can prove more stringent condition of $V^0(\Omega)$ – ellipticity, a detailed proof can be found in Bernadou and Ciarlet (1976), cf. also Bernadou (1996). Thus the problem (P) is uniquely solvable.

The shell model composed of equations (16.2.7), (16.2.8) and (16.2.15) is called Koiter's model. Although derived from the previous shell model, for which the static-geometric analogy (16.2.4) holds, it is not characterized by this remarkable property. It turns out that to preserve this analogy one should perform the "symmetrization" procedure in a different manner. This will be the subject of the next section.

16.3. Budiansky-Sanders-Koiter version of a thin shell model

In this section we show that the following stress and couple resultants

$$\widehat{N}^{\alpha\beta} = \mathbf{N}^{(\alpha\beta)} - d^{(\alpha\beta)}\mathcal{M} , \qquad \widehat{M}^{\alpha\beta} = \mathbf{M}^{(\alpha\beta)} , \tag{16.3.1}$$

and the strain measures

$$\epsilon_{\alpha\beta} , \qquad \widehat{\rho}_{\alpha\beta} = k_{(\alpha\beta)} + d_{(\alpha\beta)}\Phi , \tag{16.3.2}$$

are energy conjugated. The following notation is introduced:

$$d_{\alpha\gamma} = \varepsilon_{\alpha\beta}b^{\beta}_{\gamma} , \qquad \mathcal{M} = \frac{1}{2}\varepsilon_{\alpha\beta}\mathbf{M}^{\alpha\beta} , \qquad \Phi = \frac{1}{2}\varepsilon^{\alpha\beta}\theta_{\alpha\beta} . \tag{16.3.3}$$

Thus we shall prove that

$$U = \mathbf{N}^{\beta\alpha}\theta_{\alpha\beta} + \mathbf{M}^{\beta\alpha}k_{\alpha\beta} = \widehat{N}^{\alpha\beta}\epsilon_{\alpha\beta} + \widehat{M}^{\alpha\beta}\widehat{\rho}_{\alpha\beta} . \tag{16.3.4}$$

It will turn out that such a choice results in a thin shell theory in which a static-geometric analogy holds good.

Let us introduce decompositions

$$\mathbf{N}^{\alpha\beta} = \mathbf{N}^{(\alpha\beta)} + \varepsilon^{\alpha\beta}T , \quad \mathbf{M}^{\alpha\beta} = \mathbf{M}^{(\alpha\beta)} + \varepsilon^{\alpha\beta}\mathcal{M} ,$$
$$\theta_{\alpha\beta} = \epsilon_{\alpha\beta} + \varepsilon_{\alpha\beta}\Phi , \quad k_{\alpha\beta} = k_{(\alpha\beta)} + \varepsilon_{\alpha\beta}\tau , \tag{16.3.5}$$

where $\mathcal{M}$ and Φ are defined by (16.3.4) and

$$T = \frac{1}{2}\varepsilon_{\alpha\beta}\mathbf{N}^{\alpha\beta} , \qquad \tau = \frac{1}{2}\varepsilon^{\alpha\beta}k_{\alpha\beta} . \tag{16.3.6}$$

The sixth equilibrium equation (16.1.34) and the fourth compatibility equation (16.2.3)$_3$ can be written in the form

$$2T + d_{\alpha\gamma}(\mathbf{M}^{(\gamma\alpha)} + \varepsilon^{\gamma\alpha}\mathcal{M}) = 0 ,$$
$$2\tau - d^{\beta\sigma}(\epsilon_{\sigma\beta} + \varepsilon_{\sigma\beta}\Phi) = 0 . \tag{16.3.7}$$

Taking into account (16.3.7) we rearrange U to the form

$$U = (\mathbf{N}^{(\alpha\beta)} + \varepsilon^{\beta\alpha}T)(\epsilon_{\alpha\beta} + \varepsilon_{\alpha\beta}\Phi) + (\mathbf{M}^{(\alpha\beta)} + \varepsilon^{\beta\alpha}\mathcal{M})(k_{(\alpha\beta)} + \varepsilon_{\alpha\beta}\tau) \tag{16.3.8}$$
$$+ \lambda_1(2T + d_{(\alpha\gamma)}\mathbf{M}^{(\gamma\alpha)} + d_{\alpha\gamma}\varepsilon^{\gamma\alpha}\mathcal{M}) + \lambda_2(2\tau - d^{(\beta\sigma)}\epsilon_{\sigma\beta} - d^{\beta\sigma}\varepsilon_{\sigma\beta}\Phi) ,$$

where λ_1 and λ_2 are Lagrangian multipliers. Let us choose $\lambda_1 = \Phi$ and $\lambda_2 = \mathcal{M}$. Then the expression (16.3.8) rearranges into (16.3.4)$_2$.

Let us prove that the tensor $(\widehat{\rho}_{\alpha\beta})$ vanishes at a rigid body motion. To this end we calculate the symmetric part of $(k_{\alpha\beta})$. By (16.2.6) we have

$$k_{(\alpha\beta)} = \rho_{\alpha\beta} + \frac{1}{2}(b_\beta^\lambda \theta_{\lambda\alpha} + b_\alpha^\lambda \theta_{\lambda\beta}) \, . \tag{16.3.9}$$

Substitution of $(16.3.5)_3$ gives

$$k_{(\alpha\beta)} = \rho_{\alpha\beta} + \frac{1}{2}b_\beta^\lambda \epsilon_{\lambda\alpha} + \frac{1}{2}b_\alpha^\lambda \epsilon_{\lambda\beta} - d_{(\alpha\beta)}\Phi \, . \tag{16.3.10}$$

By (16.3.2) we find

$$\widehat{\rho}_{\alpha\beta} = \rho_{\alpha\beta} + \frac{1}{2}(b_\alpha^\lambda \epsilon_{\lambda\beta} + b_\beta^\lambda \epsilon_{\lambda\alpha}) \, . \tag{16.3.11}$$

If the shell is subject to translations or infinitesimal rotations then the tensors $(\epsilon_{\alpha\beta})$ and $(\rho_{\alpha\beta})$ vanish. Consequently the tensor $(\widehat{\rho}_{\alpha\beta})$ behaves in the same way. This tensor depends on $(\boldsymbol{u}, w)$ as follows

$$\widehat{\rho}_{\alpha\beta}(\boldsymbol{u}, w) = -w_{\|\alpha\beta} - b_{\alpha\|\beta}^\gamma u_\gamma + \frac{1}{4}b_\alpha^\gamma(u_{\beta\|\gamma} - 3u_{\gamma\|\beta}) + \frac{1}{4}b_\beta^\gamma(u_{\alpha\|\gamma} - 3u_{\gamma\|\alpha}) \, . \tag{16.3.12}$$

The assumption of thinness justifies taking the constitutive equations in the form

$$\widehat{N}_{\alpha\beta} = A^{\alpha\beta\lambda\mu}\epsilon_{\lambda\mu} \, , \qquad \widehat{M}_{\alpha\beta} = A^{\alpha\beta\lambda\mu}\widehat{\rho}_{\lambda\mu} \, , \tag{16.3.13}$$

provided that the shell is transversely homogeneous.

The variational equilibrium equation has the following form

$$\int_S [\widehat{N}^{\alpha\beta}\epsilon_{\alpha\beta}(\boldsymbol{v}, v) + \widehat{M}^{\alpha\beta}\widehat{\rho}_{\alpha\beta}(\boldsymbol{v}, v)]dS = f(\boldsymbol{v}, v) \, , \qquad \forall \, (\boldsymbol{v}, v) \in V^0(\Omega) \, . \tag{16.3.14}$$

Further steps are similar to those concerning Koiter's theory. The bilinear form involved in the equilibrium problem is similar to (16.2.16) with ρ replaced by $\widehat{\rho}$. The bilinear form is $V^0(\Omega)$ - elliptic and the equilibrium problem is uniquely solvable.

The compatibility equations interrelating the measures $(\epsilon_{\alpha\beta})$ and $(\widehat{\rho}_{\alpha\beta})$ can be put in the form of a variational equation:

$$\int_\Omega \varepsilon^{\alpha\mu}\varepsilon^{\nu\gamma}[\widehat{\rho}_{\mu\nu}\epsilon_{\alpha\gamma}(\boldsymbol{v}, v) - \epsilon_{\mu\nu}\widehat{\rho}_{\alpha\gamma}(\boldsymbol{v}, v)]\sqrt{a}d\xi = 0 \, , \qquad \forall \, (\boldsymbol{v}, v) \in V^0(\Omega) \, . \tag{16.3.15}$$

By comparing (16.3.14) with (16.3.15) we conclude that the representations

$$\widehat{N}^{\gamma\alpha} = \varepsilon^{\alpha\mu}\varepsilon^{\nu\gamma}\widehat{\rho}_{\mu\nu}(\boldsymbol{\psi}, \psi) \, , \qquad \widehat{M}^{\gamma\alpha} = -\varepsilon^{\alpha\mu}\varepsilon^{\nu\gamma}\epsilon_{\mu\nu}(\boldsymbol{\psi}, \psi) \tag{16.3.16}$$

satisfy the homogeneous equilibrium equations identically; $\boldsymbol{\psi} = (\psi_\alpha)$ and ψ_α are sufficiently regular functions defined on Ω.

16.4. Thin shell model with moderately large rotations around tangents

The aim of this section is to put forward a nonlinear generalization of the linear Koiter theory of thin shells. Within this model the kinematic assumptions (16.1.23) hold true, hence the nonlinear formula (16.2.13)$_1$ remains valid. Let us consider once again the tensor $(\overset{o}{\gamma}_{\alpha\beta})$ defined by (16.2.13)$_2$. We have

$$\breve{a}_\alpha = a_\alpha + \theta^\gamma_{.\alpha} a_\gamma - \varphi_\alpha a_3 \,, \tag{16.4.1}$$

where φ_α are given by (16.2.1). Hence

$$\breve{a}_{\alpha\beta} = (\delta^\gamma_\alpha + \theta^\gamma_{.\alpha})(\delta^\sigma_\beta + \theta^\sigma_{.\beta}) a_{\gamma\sigma} + \varphi_\alpha\varphi_\beta$$

and

$$2\,\overset{o}{\gamma}_{\alpha\beta} = 2\epsilon_{\alpha\beta} + \underline{\theta_{\gamma\alpha}\theta_{\sigma\beta} a^{\gamma\sigma}} + \varphi_\alpha\varphi_\beta \,, \tag{16.4.2}$$

where $(\epsilon_{\alpha\beta})$ is defined by (16.2.7)$_1$. If the shell is subject to deformations associated with moderately large rotations (φ_α) one can neglect the terms underlined in (16.4.2). The elastic potential is taken in the form

$$\mathcal{W}(\gamma, \rho) = \frac{1}{2} A^{\alpha\beta\lambda\mu} \gamma_{\alpha\beta}\gamma_{\lambda\mu} + \frac{1}{2} D^{\alpha\beta\lambda\mu} \rho_{\alpha\beta}\rho_{\lambda\mu} \,, \tag{16.4.3}$$

where

$$\gamma_{\alpha\beta} = \epsilon_{\alpha\beta} + \frac{1}{2}\varphi_\alpha\varphi_\beta \,, \tag{16.4.4}$$

and the tensor $(\rho_{\alpha\beta})$ is defined as in the Koiter theory, cf. Eq. (16.2.7)$_2$. The equilibrium problem of the clamped plate means evaluating

$$(P) \qquad \inf\{J(v,v)|(v,v) \in V^0(\Omega)\} \,,$$

where

$$J(v,v) = \int\limits_\Omega [\mathcal{W}(\gamma(v,v), \rho(v,v)) - (q^\alpha v_\alpha + qv)]\sqrt{a}\,d\zeta \,. \tag{16.4.5}$$

We observe that the functional J is non-convex.

Theorem 6.4.1 Let Φ be of class $C^3(\overline{\Omega})$ and assume that

$$A^{\alpha\beta\lambda\mu} \in L^\infty(\Omega) \,, \qquad D^{\alpha\beta\lambda\mu} \in L^\infty(\Omega) \,, \qquad q^\alpha \in L^2(\Omega) \,, \qquad q \in L^2(\Omega) \,,$$

$$A^{\alpha\beta\lambda\mu}\rho_{\alpha\beta}\rho_{\lambda\mu} \geq c \sum_{\alpha,\beta=1}^{2} (\rho_{\alpha\beta})^2 \,, \qquad D^{\alpha\beta\lambda\mu}\rho_{\alpha\beta}\rho_{\lambda\mu} \geq c \sum_{\alpha,\beta=1}^{2} (\rho_{\alpha\beta})^2 \,, \tag{16.4.6}$$

for all $\rho \in \mathbb{E}^2_s$; c is a positive constant. Moreover, the usual symmetry conditions hold. Then the functional J attains its minimum on $V^0(\Omega)$.

Proof. Let

$$\mathbf{K} = \{(\boldsymbol{v}, v) \in V^0(\Omega) \mid J(\boldsymbol{v}, v) < \infty\} \,.$$

To show that

$$\inf(P) = \inf\{J(\boldsymbol{v}, v) \mid (\boldsymbol{v}, v) \in \mathbf{K}\}$$

one can use the direct methods of the calculus of variations. Let us denote by $\{\boldsymbol{v}^n, v^n\} \subset \mathbf{K}$ a minimizing sequence. By definition, the following inequalities hold:

$$\forall\, n \in \mathbb{N}, \quad J(\boldsymbol{v}^{n+1}, v^{n+1}) \leq J(\boldsymbol{v}^n, v^n) \leq \ldots \leq J(\boldsymbol{v}^0, v^0) \,,$$

and

$$\lim_{n \to \infty} J(\boldsymbol{v}^n, v^n) = \inf\{J(\boldsymbol{v}, v) \mid (\boldsymbol{v}, v) \in \mathbf{K}\}$$

Destuynder (1983) proved that the sequence $\{\boldsymbol{v}^n, v^n\}_{n \in \mathbb{N}}$ is bounded. Consequently there exists a subsequence $\{\boldsymbol{v}^{n_k}, v^{n_k}\}_{k \in \mathbb{N}}$ weakly convergent to $(\boldsymbol{u}, w) \in V^0(\Omega)$. Since the functional J is weakly lower semicontinuous, therefore we finally get:

$$\lim_{n_k \to \infty} \inf J(\boldsymbol{v}^{n_k}, v^{n_k}) \geq J(\boldsymbol{u}, w) = \inf(P) \,. \qquad \square$$

Remark 6.4.2. Destuynder's (1983) considerations are confined to isotropic shells. Obviously, an extension to anisotropic shells is straightforward.

An alternative approach to solving the same problem was proposed by Bielski and Telega (1988) with details given in Telega and Bielski (1987). This approach is based on the results due to Ball, Currie and Olver (1981).

We observe that to construct the minimizing sequences one can exploit Ekeland's ε-variational principle (Ekeland, 1974, 1979, 1990; Ekeland and Temam, 1976). $\qquad \square$

16.5. *The models of Mushtari-Donnell-Vlasov and Mushtari-Marguerre*

Essential difficulties connected with solving the equations of the Koiter or Budiansky-Sanders-Koiter shell models justify less general approaches that lead to equations of lesser complexity.

The idea underlying further simplifications is that there exist domains in the shell in which the bending deformations prevail. The case of concentrated loads serves as an example: in the vicinity of a concentrated force the shell behaves like a curved plate whose deflections w are much greater than the tangential displacements (u_α). In such cases the tensor of change of curvature may be determined as in the theory of plate bending:

$$\kappa_{\alpha\beta}(w) = -w_{\|\alpha\beta} \,. \tag{16.5.1}$$

Assume that the starting point is the Budiansky-Sanders-Koiter theory. Thus the expression (16.5.1) can be viewed as derived from (16.3.12) by neglecting the terms involving the tangential displacements (u_α). The bilinear form of this model is

$$a_1(\boldsymbol{u}, w; \boldsymbol{v}, v) = \int_\Omega \{A^{\alpha\beta\lambda\mu}\epsilon_{\lambda\mu}(\boldsymbol{u}, w)\epsilon_{\alpha\beta}(\boldsymbol{v}, v) + D^{\alpha\beta\lambda\mu}\kappa_{\lambda\mu}(w)\kappa_{\alpha\beta}(v)\}\sqrt{a}d\xi \ . \quad (16.5.2)$$

The equilibrium problem of a clamped shell reads:

$$(P_1) \quad \left| \begin{array}{l} \text{find } (\boldsymbol{u}, w) \in V^0(\Omega) \text{ such that} \\ a_1(\boldsymbol{u}, w; \boldsymbol{v}, v) = f(\boldsymbol{v}, v) \quad \forall\, (\boldsymbol{v}, v) \in V^0(\Omega) \ . \end{array} \right.$$

There is one more assumption on which the Mushtari-Donnell-Vlasov modelling is based. One should additionally *assume* that the compatibility equation (16.3.15) preserves its form, namely

$$\int_\Omega \varepsilon^{\alpha\mu}\varepsilon^{\nu\gamma}[\kappa_{\mu\nu}\epsilon_{\alpha\gamma}(\boldsymbol{v}, v) - \epsilon_{\mu\nu}\kappa_{\alpha\gamma}(v)]\sqrt{a}d\xi = 0 \quad \forall\, (\boldsymbol{v}, v) \in V^0(\Omega) \ . \quad (16.5.3)$$

Obviously, the differential equations that follow from the variational equation (16.5.3) will not be satisfied exactly, but the errors introduced in this manner are consistent with the assumption (16.5.1).

Note that the condition: $a_1(\boldsymbol{v}, v; \boldsymbol{v}, v) = 0$ does not imply that $(\boldsymbol{v}, v)$ are associated with a rigid body motion. The problem (P_1) is not uniquely solvable unconditionally, as in the previous shell models. It turns out that the problem is well-posed provided that the quantities $|b^\alpha_\beta|$ are *sufficiently small*, see Bernadou (1996, Sec. 7 of Part I). This confirms that the model applies for sufficiently shallow shells.

The nonlinear generalization of the shell model presented above is based on the following choice of deformation measures:

$$\gamma_{\alpha\beta}(\boldsymbol{u}, w) = \frac{1}{2}(u_{\alpha||\beta} + u_{\beta||\alpha}) - b_{\alpha\beta}w + \frac{1}{2}w_{,\alpha}w_{,\beta} \ , \qquad (16.5.4)$$

$$\kappa_{\alpha\beta}(w) = -w_{||\alpha\beta} \ ,$$

cf. (16.4.4). The bilinear form is given by:

$$a_2(\boldsymbol{u}, w; \boldsymbol{v}, v) = \int_\Omega [N^{\alpha\beta}\eta_{\alpha\beta}(w, \boldsymbol{v}, v) + M^{\alpha\beta}\kappa_{\alpha\beta}(v)]\sqrt{a}d\xi \ , \qquad (16.5.5)$$

where

$$N^{\alpha\beta} = A^{\alpha\beta\lambda\mu}\gamma_{\lambda\mu}(\boldsymbol{u}, w) \ , \qquad M^{\alpha\beta} = D^{\alpha\beta\lambda\mu}\kappa_{\lambda\mu}(w) \ , \qquad (16.5.6)$$

and

$$\eta_{\alpha\beta}(w, \boldsymbol{v}, v) = v_{(\beta||\alpha)} - b_{\alpha\beta}v + \frac{1}{2}(w_{,\alpha}v_{,\beta} + w_{,\beta}v_{,\alpha}) \ . \qquad (16.5.7)$$

The equilibrium problem amounts to finding $(\boldsymbol{u}, w) \in V^0(\Omega)$ such that

$$(P_2) \quad a_2(\boldsymbol{u}, w; \boldsymbol{v}, v) = f(\boldsymbol{v}, v) \quad \forall\, (\boldsymbol{v}, v) \in V^0(\Omega) \ . \qquad (16.5.8)$$

Equations (16.5.4) – (16.5.8) form the model called the Mushtari-Marguerre shell theory. Mathematical aspects of this model were studied by Bernadou and Oden (1981) and Vorovich (1989). More precisely, in the paper by the first two authors the problem of existence of a solution was solved by applying the theory of pseudomonotone operators. This solution is unique provided that the normal load is sufficiently small. The book by Vorovich (1989) provides a systematic account of not only variational methods, including Bubnov-Galerkin and Ritz methods, but topological methods are also presented.

17.   Homogenization of stiffnesses of thin periodic elastic shells. Linear approach

17.1.   Koiter's shell. Asymptotic analysis and the convergence theorem

Let us consider the plate of a periodic structure analyzed in Sec. 2. Let us imagine a large deformation of this plate, admissible by the Kirchhoff kinematic constraints, according to which normals to the middle plane remain normal to the plane after deformation. This deformed plane is a surface that can be viewed as a middle surface of a shell of oscillating geometric and material characteristics. Assume now that just the shell of this shape is given in the stress-free state. The present section is aimed at analyzing the effective moduli of shells of such geometry and elastic characteristics.

A fully correct starting point for assessing effective characteristics of such a shell is the three-dimensional setting, as it has been chosen in Sec. 2 regarding the plate problem. More precisely, the process of averaging should be performed simultaneously with the process of reducing the third dimension. Such a simultaneous process has been applied by Kalamkarov, Kudryavtsev and Parton (1987), cf. also Kalamkarov (1992). Since up till now this method has not been rigorously justified we shall confine our consideration to the averaging method which is based on a two-dimensional shell model. The reduction of the transverse dimension will precede homogenization of stiffnesses thus making the final results applicable only for shells, the periodicity cells of which have shapes of thin shells. However, the same formulae are of crucial importance in constructing the relaxed formulation of the problem of optimal layout for which the total compliance of the shell is the functional to be minimized, cf. Sec. 28.

The homogenization formulae for the Koiter shell model are substantiated by the Γ-convergence theorem.

17.1.1.   Asymptotic analysis

The objective is a statical analysis of a thin shell of a structure periodic with respect to the fixed curvilinear coordinates $\xi = (\xi^\alpha)$, see Fig. 17.1.1.

The Koiter description will be applied, cf. Sec. 16.2. We assume that the reduced moduli $\widetilde{C}^{\alpha\beta\lambda\mu}(\xi, y_0)$ and/or the shell thickness $h(\xi, y_0)$ are slowly varying with respect to ξ and are $\varepsilon_0 Y$-periodic with respect to the second variable $y_0 = \xi/\varepsilon_0$. Here $Y = (0, l_1) \times (0, l_2)$ is a rectangle and $\varepsilon_0 > 0$. The rectangle $\varepsilon_0 Y$ transforms into $\Phi(\varepsilon_0 Y)$ – a curvilinear rectangle, cf. Fig 17.1.1; the definition of Φ is given in Sec. 16.1.

To consider the averaged characteristics of the $\varepsilon_0 Y$-periodic shell we replace ε_0 with ε and consider ε as a small parameter, thus constructing a family of shell problems. The stiffnesses $A^{\alpha\beta\lambda\mu}$ and $D^{\alpha\beta\lambda\mu}$ are replaced by

$$A_\varepsilon^{\alpha\beta\lambda\mu}(\xi) = A^{\alpha\beta\lambda\mu}(\xi, y) , \quad D_\varepsilon^{\alpha\beta\lambda\mu}(\xi) = D^{\alpha\beta\lambda\mu}(\xi, y) , \quad y = \xi/\varepsilon , \qquad (17.1.1)$$

where $A^{\alpha\beta\lambda\mu}(\xi, \cdot)$ and $D^{\alpha\beta\lambda\mu}(\xi, \cdot)$ are Y-periodic and of class $L_{loc}^\infty(\mathbb{R}^2)$. The shell will be assumed as transversely homogeneous, hence

$$A^{\alpha\beta\lambda\mu}(\xi, y) = h(\xi, y)\widetilde{C}^{\alpha\beta\lambda\mu}(\xi, y) , \quad D^{\alpha\beta\lambda\mu}(\xi, y) = \frac{1}{12}h^3(\xi, y)\widetilde{C}^{\alpha\beta\lambda\mu}(\xi, y) . \qquad (17.1.2)$$

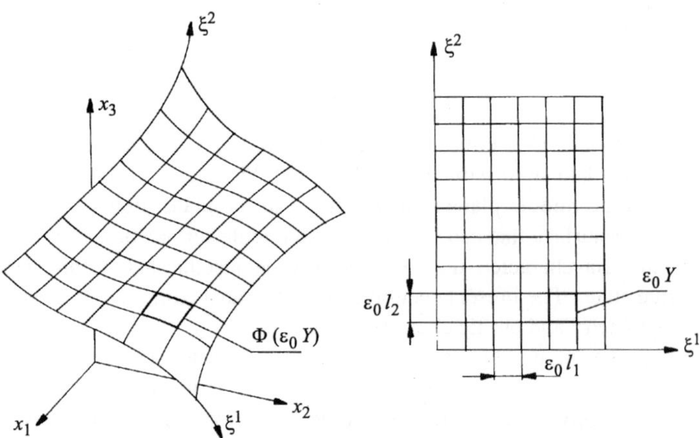

Fig. 17.1.1. Middle plane of the $\varepsilon_0 Y$-periodic thin shell

The functions $\widetilde{C}^{\alpha\beta\lambda\mu}(\cdot, y)$ describe ε-independent variations of the metric of the shell middle surface. The constitutive relations (16.2.15) assume now the form

$$N_\varepsilon^{\alpha\beta} = A_\varepsilon^{\alpha\beta\lambda\mu}\epsilon_{\lambda\mu}(\boldsymbol{u}^\varepsilon, w^\varepsilon), \quad M_\varepsilon^{\alpha\beta} = D_\varepsilon^{\alpha\beta\lambda\mu}\rho_{\lambda\mu}(\boldsymbol{u}^\varepsilon, w^\varepsilon), \tag{17.1.3}$$

where

$$\epsilon_{\alpha\beta}(\boldsymbol{u}^\varepsilon, w^\varepsilon) = \frac{1}{2}(u_{\alpha||\beta}^\varepsilon + u_{\beta||\alpha}^\varepsilon) - b_{\alpha\beta}w^\varepsilon,$$

$$\rho_{\alpha\beta}(\boldsymbol{u}^\varepsilon, w^\varepsilon) = -w_{||\alpha\beta}^\varepsilon + c_{\alpha\beta}w^\varepsilon - b_\alpha^\sigma u_{\sigma||\beta}^\varepsilon - b_\beta^\sigma u_{\sigma||\alpha}^\varepsilon - b_{\alpha||\beta}^\sigma u_\sigma^\varepsilon. \tag{17.1.4}$$

We note that the tensor $\boldsymbol{b}$ is ε-independent. The covariant derivative is defined by (16.1.6).

The shell is assumed to be clamped and subject to surface loadings of density q^α and q, independent of the parameter ε. The bilinear form (16.2.16) reads now:

$$a^\varepsilon(\boldsymbol{u}^\varepsilon, w^\varepsilon; \boldsymbol{v}, v) = \int_\Omega [N_\varepsilon^{\alpha\beta}\epsilon_{\alpha\beta}(\boldsymbol{v}, v) + M_\varepsilon^{\alpha\beta}\rho_{\alpha\beta}(\boldsymbol{v}, v)]\sqrt{a}\, d\xi. \tag{17.1.5}$$

Problem (P) of Sec. 16.2 assumes the form

$$(P_\varepsilon) \quad \left| \begin{array}{l} \text{find } (\boldsymbol{u}^\varepsilon, w^\varepsilon) \in V^0(\Omega) \text{ such that} \\ a^\varepsilon(\boldsymbol{u}^\varepsilon, w^\varepsilon; \boldsymbol{v}, v) = f(\boldsymbol{v}, v), \qquad \forall\, (\boldsymbol{v}, v) \in V^0(\Omega). \end{array} \right. \tag{17.1.6}$$

Similar to the problem (P), the last problem is uniquely solvable for each $\varepsilon > 0$.

Similar to the case of plates periodic with respect to a curvilinear parametrization (cf. Sec. 3.10), we postulate that the solution $(\boldsymbol{u}^\varepsilon, w^\varepsilon)$ can be expanded in the form

$$u_\alpha^\varepsilon = u_\alpha^{(0)}(\xi) + \varepsilon u_\alpha^{(1)}(\xi, y) + \ldots, \quad y = \xi/\varepsilon,$$

$$w^\varepsilon = w^{(0)}(\xi) + \varepsilon^2 w^{(2)}(\xi, y) + \varepsilon^3 w^{(3)}(\xi, y) + \ldots. \tag{17.1.7}$$

In a similar manner we expand the trial functions. We assume that $(\boldsymbol{u}^{(0)}, w^{(0)}) \in V^0(\Omega)$, $u_\alpha^{(1)}(\xi, \cdot), v_\alpha^{(1)}(\xi, \cdot) \in H^1_{per}(Y)$, $w^{(2)}(\xi, \cdot), v^{(2)}(\xi, \cdot) \in H^2_{per}(Y)$.

According to the kinematic assumptions (17.1.7) we find

$$N_\varepsilon^{\alpha\beta} = N_0^{\alpha\beta} + O(\varepsilon)\,, \qquad N_0^{\alpha\beta} = A^{\alpha\beta\lambda\mu}(\xi, y)[\epsilon_{\lambda\mu}^h + u_{\lambda|\mu}^{(1)}]\,,$$
$$M_\varepsilon^{\alpha\beta} = M_0^{\alpha\beta} + O(\varepsilon)\,, \qquad M_0^{\alpha\beta} = D^{\alpha\beta\lambda\mu}(\xi, y)[\rho_{\lambda\mu}^h - w_{|\lambda\mu}^{(2)} - 2b_\lambda^\sigma u_{\sigma|\mu}^{(1)}]\,, \tag{17.1.8}$$

where

$$\epsilon_{\lambda\mu}^h = \epsilon_{\lambda\mu}(\boldsymbol{u}^{(0)}, w^{(0)})\,, \qquad \rho_{\lambda\mu}^h = \rho_{\lambda\mu}(\boldsymbol{u}^{(0)}, w^{(0)}) \tag{17.1.9}$$

and $u_{\lambda|\mu}^{(1)} = \dfrac{\partial u_\lambda^{(1)}}{\partial y_\mu}$, etc., see Sec. 2. The averaged stress and couple resultants are defined by

$$N_h^{\alpha\beta}(\xi) = \langle N_0^{\alpha\beta}(\xi, y)\rangle\,, \qquad M_h^{\alpha\beta}(\xi) = \langle M_0^{\alpha\beta}(\xi, y)\rangle\,, \tag{17.1.10}$$

where the parentheses $\langle\cdot\rangle$ imply averaging over Y :

$$\langle g\rangle = \frac{1}{l_1 l_2} \int\limits_Y g\, dy_1 dy_2\,. \tag{17.1.11}$$

Let us put $v_\alpha = v_\alpha^{(0)}$ and $v = v^{(0)}$ into Eq. (17.1.6) and let ε tend to zero. By applying (1.1.1) we arrive at the macroscopic variational equation of equilibrium

$$\int\limits_\Omega [N_h^{\alpha\beta}(v_{\alpha||\beta}^{(0)} - b_{\alpha\beta}v^{(0)}) \tag{17.1.12}$$
$$+ M_h^{\alpha\beta}(-v_{||\alpha\beta}^{(0)} + c_{\alpha\beta}v^{(0)} - 2b_\alpha^\gamma v_{\gamma||\beta}^{(0)} - b_{\alpha||\beta}^\gamma v_\gamma^{(0)})]\sqrt{a}\, d\xi = f(\boldsymbol{v}^{(0)}, v^{(0)})\,,$$

with $f(\cdot, \cdot)$ defined as before, cf. (16.2.17). Now let us put

$$v_\alpha^\varepsilon = v_\alpha^{(0)}(\xi) + \varepsilon v_\alpha^{(1)}(\xi, y) + \ldots\,, \qquad v^\varepsilon = v^{(0)}(\xi) + \varepsilon^2 v^{(2)}(\xi, y) + \ldots\,, \tag{17.1.13}$$

into Eq. (17.1.6), then pass to zero with ε. By combining this equation with (17.1.12) we get

$$\int\limits_\Omega [\langle N_0^{\alpha\beta} v_{\alpha|\beta}^{(1)}\rangle - \langle M_0^{\alpha\beta}(v_{|\alpha\beta}^{(2)} + 2b_\alpha^\gamma v_{\gamma|\beta}^{(1)})\rangle]\sqrt{a}\, d\xi = 0 \tag{17.1.14}$$
$$\forall\, (\boldsymbol{v}^{(1)}(\xi, \cdot), v^{(2)}(\xi, \cdot)) \in H_{K,per}(Y)\,,$$

where the space $H_{K,per}(Y)$ has been defined by (3.2.14). Let

$$v_\alpha^{(1)} = v_\alpha(y)\varphi(\xi)\,, \qquad v_\alpha \in H^1_{per}(Y)\,, \qquad \varphi \in \mathbf{D}(\Omega)\,,$$
$$v^{(1)} = v(y)\psi(\xi)\,, \qquad v \in H^2_{per}(Y)\,, \qquad \psi \in \mathbf{D}(\Omega)\,. \tag{17.1.15}$$

Substitution of (17.1.15) into (17.1.14) gives

$$\int\limits_\Omega \varphi\langle(N_0^{\gamma\beta} - 2b_\alpha^\gamma M_0^{\alpha\beta})v_{\gamma|\beta}\rangle\sqrt{a}\, d\xi = 0\,, \qquad \forall\, v_\alpha \in H^1_{per}(Y)\quad \forall\, \varphi \in \mathbf{D}(\Omega), \tag{17.1.16}$$

and

$$\int_{\Omega} \psi \langle M_0^{\alpha\beta} v_{|\alpha\beta} \rangle \sqrt{a}\, d\xi = 0 \, , \qquad \forall\, v \in H^2_{per}(Y) \, , \quad \psi \in \mathbf{D}(\Omega) \, . \tag{17.1.17}$$

Thus we find the local problem:

$$(P_Y) \quad \left|\; \begin{aligned} &\text{find } (\boldsymbol{u}^{(1)}(\xi,\cdot), w^{(2)}(\xi,\cdot)) \in H_{K,per}(Y) \text{ such that} \\ &a_L(\boldsymbol{u}^{(1)}, w^{(2)}; \boldsymbol{v}, v) = \widetilde{f}(\boldsymbol{v}, v) \quad \forall(\boldsymbol{v}, v) \in H_{K,per}(Y) \, , \end{aligned} \right.$$

where

$$\begin{aligned} a_L(\boldsymbol{u}, w; \boldsymbol{v}, v) = \langle B^{\alpha\beta\lambda\mu} u_{\alpha|\beta} v_{\lambda|\mu} &- E^{\alpha\beta\lambda\mu}(u_{\alpha|\beta} v_{|\lambda\mu} + v_{\alpha|\beta} w_{|\lambda\mu}) \\ &+ D^{\alpha\beta\lambda\mu} w_{|\alpha\beta} v_{|\lambda\mu} \rangle \, , \end{aligned} \tag{17.1.18}$$

$$\widetilde{f}(\boldsymbol{v}, v) = -\langle A^{\alpha\beta\lambda\mu}(\delta^\gamma_\alpha \epsilon^h_{\lambda\mu} - \frac{h^2}{6} b^\gamma_\alpha \rho^h_{\lambda\mu}) v_{\gamma|\beta} - D^{\alpha\beta\lambda\mu} \rho^h_{\lambda\mu} v_{|\alpha\beta} \rangle \tag{17.1.19}$$

with

$$u_{\alpha|\beta} = \partial u_\alpha / \partial y_\beta \, . \tag{17.1.20}$$

The tensors $\boldsymbol{B}, \boldsymbol{E}$ are defined by

$$B^{\gamma\beta\sigma\mu}(\xi, y) = [\delta^\gamma_\alpha \delta^\sigma_\lambda + \frac{1}{3} h^2(\xi, y) b^\gamma_\alpha(\xi) b^\sigma_\lambda(\xi)] A^{\alpha\beta\lambda\mu}(\xi, y) \, , \tag{17.1.21}$$

$$E^{\gamma\beta\lambda\mu}(\xi, y) = -\frac{1}{6} h^2(\xi, y) b^\gamma_\alpha(\xi) A^{\alpha\beta\lambda\mu}(\xi, y) \, . \tag{17.1.22}$$

We note that both the bilinear form a_L and the linear form $\widetilde{f}$ depend on $\xi \in \Omega$. The local problem can now be rearranged to the form

$$(P'_Y) \quad \left|\; \begin{aligned} &\text{find } (\boldsymbol{u}^{(1)}(\xi,\cdot), w^{(2)}(\xi,\cdot)) \in H_{K,per}(Y) \text{ such that} \\ &b(\boldsymbol{u}^{(1)}, \boldsymbol{v}) + e(\boldsymbol{v}, w^{(2)}) + f_1(\boldsymbol{v}) = 0 \quad \forall\, \boldsymbol{v} \in H^1_{per}(Y)^2 \, , \\ &e(\boldsymbol{u}^{(1)}, v) + d(w^{(2)}, v) + g(v) = 0 \quad \forall\, v \in H^2_{per}(Y) \, , \end{aligned} \right.$$

where

$$\begin{aligned} b(\boldsymbol{u}, \boldsymbol{v}) &= \langle B^{\gamma\beta\sigma\mu}(\xi, y) u_{\sigma|\mu} v_{\gamma|\beta} \rangle \, , \\ d(w, v) &= \langle D^{\alpha\beta\lambda\mu}(\xi, y) w_{|\lambda\mu} v_{|\alpha\beta} \rangle \, , \\ e(\boldsymbol{u}, w) &= -\langle E^{\gamma\beta\lambda\mu}(\xi, y) w_{|\lambda\mu} u_{\gamma|\beta} \rangle \, , \\ f_1(\boldsymbol{v}) &= \langle A^{\alpha\beta\lambda\mu}(\xi, y)(\delta^\gamma_\alpha \epsilon^h_{\lambda\mu} - \frac{h^2}{6} b^\gamma_\alpha \rho^h_{\lambda\mu}) v_{\gamma|\beta} \rangle \, , \\ g(v) &= -\langle D^{\alpha\beta\lambda\mu}(\xi, y) \rho^h_{\lambda\mu} v_{|\alpha\beta} \rangle \, . \end{aligned} \tag{17.1.23}$$

Note that the solutions $u^{(1)}$, $w^{(2)}$ are linear with respect to ϵ^h and ρ^h. Hence there exist functions $\Psi^{(\alpha\beta)}(\xi, y)$, $\Phi^{(\alpha\beta)}(\xi, y)$, $\Xi^{(\alpha\beta)}(\xi, y)$, $\chi^{(\alpha\beta)}(\xi, y)$ such that

$$u^{(1)} = \Psi^{(\alpha\beta)}(\xi, y)\epsilon^h_{\alpha\beta}(\xi) + \Phi^{(\alpha\beta)}(\xi, y)\rho^h_{\alpha\beta}(\xi) \,,$$
$$w^{(2)} = \Xi^{(\alpha\beta)}(\xi, y)\epsilon^h_{\alpha\beta}(\xi) + \chi^{(\alpha\beta)}(\xi, y)\rho^h_{\alpha\beta}(\xi) \,. \tag{17.1.24}$$

The problem (P'_Y) is equivalent to

$$(P^1_Y) \quad \left| \begin{array}{l} \text{find } \Psi^{(\alpha\beta)}(\xi, \cdot), \Xi^{(\alpha\beta)}(\xi, \cdot) \in H_{K,per}(Y) \text{ such that} \\[2mm] b(\Psi^{(\lambda\mu)}, v) + e(v, \Xi^{(\lambda\mu)}) + \langle A^{\alpha\beta\lambda\mu}v_{\alpha|\beta}\rangle = 0 \,, \\[2mm] e(\Psi^{(\lambda\mu)}, v) + d(\Xi^{(\lambda\mu)}, v) = 0 \\[1mm] \forall(v, v) \in H_{K,per}(Y) \end{array} \right.$$

and

$$(P^2_Y) \quad \left| \begin{array}{l} \text{find } \Phi^{(\lambda\mu)}(\xi, \cdot), \chi^{(\lambda\mu)}(\xi, \cdot) \in H_{K,per}(Y) \text{ such that} \\[2mm] b(\Phi^{(\lambda\mu)}, v) + e(v, \chi^{(\lambda\mu)}) + \langle -\dfrac{h^2}{6} A^{\alpha\beta\lambda\mu}b^\gamma_\alpha v_{\gamma|\beta}\rangle = 0 \,, \\[2mm] e(\Phi^{(\lambda\mu)}, v) + d(\chi^{(\lambda\mu)}, v) - \langle D^{\alpha\beta\lambda\mu}v_{|\alpha\beta}\rangle = 0 \\[1mm] \forall(v, v) \in H_{K,per}(Y) \,. \end{array} \right.$$

We observe that the problems (P^α_Y) are well-posed. Indeed, they are equivalent to the problem (P_Y), which is equivalent to the minimization problem appearing on the r.h.s. of (17.1.50) below. The last problem is uniquely solvable in the space $H_{K,per}(Y)$.

Now we pass to the study of the homogenization constitutive equations. Substitution of the representations (17.1.24) into (17.1.10) gives

$$N^{\alpha\beta}_h = A^{\alpha\beta\lambda\mu}_h \epsilon^h_{\lambda\mu} + E^{\alpha\beta\lambda\mu}_h \rho^h_{\lambda\mu} \,, \qquad M^{\alpha\beta}_h = F^{\alpha\beta\lambda\mu}_h \epsilon^h_{\lambda\mu} + D^{\alpha\beta\lambda\mu}_h \rho^h_{\lambda\mu} \,, \tag{17.1.25}$$

where

$$A^{\alpha\beta\delta\gamma}_h = \langle A^{\alpha\beta\lambda\mu}(\delta^\gamma_\lambda \delta^\delta_\mu + \Psi^{(\delta\gamma)}_{\lambda|\mu})\rangle \,, \qquad E^{\alpha\beta\delta\gamma}_h = \langle A^{\alpha\beta\lambda\mu}\Phi^{(\delta\gamma)}_{\lambda|\mu}\rangle \,,$$
$$F^{\alpha\beta\delta\gamma}_h = -\langle \frac{h^2}{12} A^{\alpha\beta\lambda\mu}(2b^\sigma_\lambda \Psi^{(\delta\gamma)}_{\sigma|\mu} + \Xi^{(\delta\gamma)}_{|\lambda\mu})\rangle \,, \tag{17.1.26}$$
$$D^{\alpha\beta\delta\gamma}_h = \langle \frac{h^2}{12} A^{\alpha\beta\lambda\mu}(\delta^\delta_\lambda \delta^\gamma_\mu - 2b^\sigma_\lambda \Phi^{(\delta\gamma)}_{\sigma|\mu} - \chi^{(\delta\gamma)}_{|\lambda\mu})\rangle \,.$$

We shall prove that the following symmetries hold:

$$A^{\alpha\beta\delta\gamma}_h = A^{\delta\gamma\alpha\beta}_h \,, \qquad D^{\alpha\beta\delta\gamma}_h = D^{\delta\gamma\alpha\beta}_h \,, \qquad F^{\delta\gamma\alpha\beta}_h = E^{\alpha\beta\delta\gamma}_h \,. \tag{17.1.27}$$

Let us take $v = \Phi^{(\delta\gamma)}$ and $v = \chi^{(\delta\gamma)}$ in (P^2_Y) and $v = \Psi^{(\lambda\mu)}$ and $v = \Xi^{(\lambda\mu)}$ in (P^1_Y). Then we obtain the following identities

$$b(\Psi^{(\lambda\mu)}, \Phi^{(\delta\gamma)}) + e(\Phi^{(\delta\gamma)}, \Xi^{(\lambda\mu)}) + \langle A^{\alpha\beta\lambda\mu}\Phi^{(\delta\gamma)}_{\alpha|\beta}\rangle = 0 \,, \tag{17.1.28}$$

$$e(\mathbf{\Psi}^{(\lambda\mu)}, \chi^{(\delta\gamma)}) = -d(\Xi^{(\lambda\mu)}), \chi^{(\delta\gamma)}) \,, \tag{17.1.29}$$

$$b(\mathbf{\Phi}^{(\delta\gamma)}, \mathbf{\Psi}^{(\lambda\mu)}) + e(\mathbf{\Psi}^{(\lambda\mu)}, \chi^{(\delta\gamma)}) + \langle -\frac{h^2}{6} A^{\alpha\beta\delta\gamma} b_\alpha^\sigma \Psi_{\sigma|\beta}^{(\lambda\mu)} \rangle = 0 \,, \tag{17.1.30}$$

$$e(\mathbf{\Phi}^{(\delta\gamma)}, \Xi^{(\lambda\mu)}) + d(\chi^{(\delta\gamma)}, \Xi^{(\lambda\mu)}) - \langle D^{\alpha\beta\delta\gamma} \Xi_{|\alpha\beta}^{(\lambda\mu)} \rangle = 0 \,. \tag{17.1.31}$$

Equation (17.1.28) implies

$$E_h^{\lambda\mu\delta\gamma} = -b(\mathbf{\Psi}^{(\lambda\mu)}, \mathbf{\Phi}^{(\delta\gamma)}) - e(\mathbf{\Phi}^{(\delta\gamma)}, \Xi^{(\lambda\mu)}) \,. \tag{17.1.32}$$

By adding equations (17.1.30) and (17.1.31) and next taking into account (17.1.26)$_3$ one obtains

$$\begin{aligned}
F_h^{\delta\gamma\lambda\mu} &+ b(\mathbf{\Phi}^{(\delta\gamma)}, \mathbf{\Psi}^{(\lambda\mu)}) + e(\mathbf{\Psi}^{(\lambda\mu)}, \chi^{(\delta\gamma)}) \\
&+ e(\mathbf{\Phi}^{(\delta\gamma)}, \Xi^{(\lambda\mu)}) + d(\chi^{(\delta\gamma)}, \Xi^{(\lambda\mu)}) = 0 \,.
\end{aligned} \tag{17.1.33}$$

Taking into account (17.1.29) one finds

$$F^{\delta\gamma\lambda\mu} = -b(\mathbf{\Phi}^{(\delta\gamma)}, \mathbf{\Psi}^{(\lambda\mu)}) - e(\mathbf{\Psi}^{(\delta\gamma)}, \Xi^{(\lambda\mu)}) \,. \tag{17.1.34}$$

Let us compare this formula with Eq. (17.1.32). Symmetry of the form $b(\cdot, \cdot)$ implies the relations (17.1.27)$_3$ between the tensors $\mathbf{F}_h$ and $\mathbf{E}_h$.

To prove the symmetry relations (17.1.27)$_{1,2}$ let us define new symmetric tensors

$$\begin{aligned}
\widehat{A}_h^{\alpha\beta\delta\gamma} &= \langle A^{\nu\rho\lambda\mu}(\delta_\lambda^\gamma \delta_\mu^\delta + \Psi_{\lambda|\mu}^{(\delta\gamma)})(\delta_\nu^\alpha \delta_\rho^\beta + \Psi_{\nu|\rho}^{(\alpha\beta)}) \rangle \,, \\
\widehat{D}_h^{\alpha\beta\delta\gamma} &= \langle D^{\nu\rho\lambda\mu}(\delta_\lambda^\gamma \delta_\mu^\delta - \chi_{|\lambda\mu}^{(\delta\gamma)})(\delta_\nu^\alpha \delta_\rho^\beta - \chi_{|\nu\rho}^{(\alpha\beta)}) \rangle \,.
\end{aligned} \tag{17.1.35}$$

Let us take $v = \mathbf{\Psi}^{(\alpha\beta)}$ in the first equation of (P_Y^1) and $v = \Xi^{(\alpha\beta)}$ in the second equation of this problem. On combining these identities with (17.1.26)$_1$ and (17.1.35)$_1$ one finds

$$A_h^{\alpha\beta\delta\gamma} = \widehat{A}_h^{\alpha\beta\delta\gamma} - d(\Xi^{(\alpha\beta)}, \Xi^{(\delta\gamma)}) + g(\mathbf{\Psi}^{(\alpha\beta)}, \mathbf{\Psi}^{(\delta\gamma)}) \,, \tag{17.1.36}$$

where the bilinear form $g(\cdot, \cdot)$ is defined by

$$g(\mathbf{u}, \mathbf{v}) = \langle \frac{h^2}{3} b_\alpha^\gamma b_\lambda^\sigma A^{\alpha\beta\lambda\mu} u_{\sigma|\mu} v_{\gamma|\beta} \rangle \,. \tag{17.1.37}$$

Symmetry of the forms $d(\cdot, \cdot)$ and $g(\cdot, \cdot)$ imply the symmetry condition (17.1.27)$_1$.

Similarly one can derive the formula

$$D_h^{\alpha\beta\delta\gamma} = \widehat{D}_h^{\alpha\beta\delta\gamma} - d(\mathbf{\Phi}^{(\alpha\beta)}, \mathbf{\Phi}^{(\delta\gamma)}) \,, \tag{17.1.38}$$

which proves the symmetry condition (17.1.27)$_2$ due to the symmetry of the form $b(\cdot, \cdot)$.

Let us substitute now the representations (17.1.24) into the relations defining N_0 and M_0, cf. (17.1.8). We find

$$N_0^{\alpha\beta} = A^{\alpha\beta\lambda\mu}\epsilon_{\lambda\mu}^0 , \qquad M_0^{\alpha\beta} = D^{\alpha\beta\lambda\mu}\rho_{\lambda\mu}^0 , \tag{17.1.39}$$

with

$$\epsilon_{\lambda\mu}^0 = \epsilon_{\lambda\mu}^h + \Psi_{\lambda|\mu}^{(\delta\gamma)}\epsilon_{\delta\gamma}^h + \Phi_{\lambda|\mu}^{(\delta\gamma)}\rho_{\delta\gamma}^h ,$$
$$\rho_{\lambda\mu}^0 = \rho_{\lambda\mu}^h - \Xi_{|\lambda\mu}^{(\delta\gamma)}\epsilon_{\delta\gamma}^h - \chi_{|\lambda\mu}^{(\delta\gamma)}\rho_{\delta\gamma}^h - 2b_\lambda^\sigma\Psi_{\sigma|\mu}^{(\delta\gamma)}\epsilon_{\delta\gamma}^h - 2b_\lambda^\sigma\Phi_{\sigma|\mu}^{(\delta\gamma)}\rho_{\delta\gamma}^h . \tag{17.1.40}$$

Drawing upon all previous results one can prove the Hill-type identity

$$\langle N_0^{\alpha\beta}\epsilon_{\alpha\beta}^0 + M_0^{\alpha\beta}\rho_{\alpha\beta}^0\rangle = \langle N_0^{\alpha\beta}\rangle\langle\epsilon_{\alpha\beta}^0\rangle + \langle M_0^{\alpha\beta}\rangle\langle\rho_{\alpha\beta}^0\rangle . \tag{17.1.41}$$

The elastic potential of the homogenized shell

$$\mathcal{W}_h(\xi, \epsilon^h, \rho^h) = \frac{1}{2}(N_h^{\alpha\beta}\epsilon_{\alpha\beta}^h + M_h^{\alpha\beta}\rho_{\alpha\beta}^h) \tag{17.1.42}$$

can be put in the form, cf. (17.1.44)

$$\mathcal{W}_h(\xi, \epsilon^h, \rho^h) = \frac{1}{2}\langle A^{\alpha\beta\lambda\mu}\epsilon_{\alpha\beta}^0\epsilon_{\lambda\mu}^0 + D^{\alpha\beta\lambda\mu}\rho_{\alpha\beta}^0\rho_{\lambda\mu}^0\rangle . \tag{17.1.43}$$

The constitutive relations (17.1.25) can be inferred from the formulae

$$N_h^{\alpha\beta} = \frac{\partial\mathcal{W}_h}{\partial\epsilon_{\alpha\beta}^h} , \qquad M_h^{\alpha\beta} = \frac{\partial\mathcal{W}_h}{\partial\rho_{\alpha\beta}^h} . \tag{17.1.44}$$

Remark 17.1.1. The whole asymptotic homogenization can be repeated taking the Budiansky-Sanders-Koiter model as a departure point, cf. Sec. 16.3. The derivation will differ in details only and that is why it is left to the reader as an exercise.

17.1.2.   Justification by the Γ-convergence method

The variational problem (P_ε) given by (17.1.6) is equivalent to the following minimization problem

$$(Q_\varepsilon) \quad J_\varepsilon(u^\varepsilon, w^\varepsilon) - f(u^\varepsilon, w^\varepsilon) = \inf\{J_\varepsilon(v, v) - f(v, v)|(v, v) \in V^0(\Omega)\} , \tag{17.1.45}$$

where

$$J_\varepsilon(v, v) = \frac{1}{2}a^\varepsilon(v, v; v, v) = \int_\Omega \mathcal{W}(\xi, \xi/\varepsilon, \epsilon(v, v), \rho(v, v))\sqrt{a}\, d\xi ,$$

for each $(v, v) \in H^1(\Omega)^2 \times H^2(\Omega)$. Here

$$\mathcal{W}(\xi, y, \epsilon, \rho) = \frac{1}{2}A^{\alpha\beta\lambda\mu}(\xi, y)\epsilon_{\alpha\beta}\epsilon_{\lambda\mu} + \frac{1}{2}D^{\alpha\beta\lambda\mu}(\xi, y)\rho_{\alpha\beta}\rho_{\lambda\mu} . \tag{17.1.46}$$

We make the following assumptions:

(H_1) The function $\mathcal{W} : \Omega \times \mathbb{R}^2 \times \mathbb{E}_s^2 \times \mathbb{E}_s^2 \to [0, +\infty)$ is measurable, Y-periodic in y and continuous in ξ.

(H_2) There exist Y-periodic function $a \in L_{loc}^1(\mathbb{R}^2)$, increasing function $\omega : \mathbb{R}^+ \to \mathbb{R}^+$, continuous at zero and such that $\omega(0) = 0$, and continuous non-negative function $B : \Omega \to \mathbb{R}$, for which the following inequalities are satisfied:

$$|\mathcal{W}(\xi, y, \epsilon, \rho) - \mathcal{W}(\xi', y, \epsilon, \rho)| \leq \omega(|\xi - \xi'|)(a(y) + \mathcal{W}(\xi, y, \epsilon, \rho)) , \quad (17.1.47)$$

$$c(|\epsilon|^2 + |\rho|^2) \leq \mathcal{W}(\xi, y, \epsilon, \rho) \leq B(\xi)(a(y) + |\epsilon|^2 + |\rho|^2) , \quad (17.1.48)$$

for each ξ, $\xi' \in \Omega$, $y \in \mathbb{R}^2$, ϵ, $\rho \in \mathbb{E}_s^2$; here c is a positive constant. Under the assumptions $(H_1), (H_2)$ a solution $(u^\varepsilon, w^\varepsilon) \in V^0(\Omega)$ of the minimization problem appearing in (17.1.45) exists and is unique.

We observe that in order to perform the homogenization itself ($\varepsilon \to 0$) it suffices to take $c = 0$ in (17.1.48), cf. Sec. 1.3.5. Now we are in a position to formulate the homogenization theorem.

Theorem 17.1.2. Let q^α, $q \in L^2(\Omega)$. Under the assumptions (H_1), (H_2) the sequence of functionals $\{J_\varepsilon - f\}_{\varepsilon > 0}$ is $\Gamma(L^2(\Omega)^2 \times H^1(\Omega))$ – convergent to the functional $J_h - f$, where

$$J_h(u, w) = \int_\Omega \mathcal{W}_h[\xi, \epsilon(u, w), \rho(u, w)]\sqrt{a}\, d\xi , \quad (u, w) \in V^0(\Omega) , \quad (17.1.49)$$

$$\begin{aligned}
\mathcal{W}_h(\xi, \epsilon, \rho) = \frac{1}{2} \inf\{ &\langle A^{\alpha\beta\lambda\mu}(\xi, y)[e_{\alpha\beta}^y(v) + \epsilon_{\alpha\beta}][e_{\lambda\mu}^y(v) + \epsilon_{\lambda\mu}] \\
&+ D^{\alpha\beta\lambda\mu}(\xi, y)[\kappa_{\alpha\beta}^y(v) - 2b_\alpha^\sigma(\xi)v_{\sigma|\beta} + \rho_{\alpha\beta}][\kappa_{\lambda\mu}(v) \\
&- 2b_\lambda^\delta(\xi)v_{\delta|\mu} + \rho_{\lambda\mu}]\rangle|(v, v) \in H_{K,per}(Y) \} , \quad (17.1.50)
\end{aligned}$$

for each $\epsilon, \rho \in \mathbb{E}_s^2$. We recall that $e_{\alpha\beta}^y(v) = \left(\dfrac{\partial v_\alpha}{\partial y_\beta} + \dfrac{\partial v_\beta}{\partial y_\alpha}\right)/2$ and $\kappa_{\alpha\beta}^y(v) = -v_{|\alpha\beta} = -\dfrac{\partial^2 v}{\partial y_\alpha \partial y_\beta}$.

Proof. This theorem is a straightforward consequence of Theorem 1.3.28. $\square$

To better grasp the term $2b_\alpha^\sigma(\xi)v_{\sigma|\beta}$ we write the deformation measure $\rho(u, w)$ in the following way

$$\rho_{\alpha\beta}(u, w) = -w_{\|\alpha\beta} + c_{\alpha\beta}w - 2e_{\alpha\beta}^\|(bu) , \quad (17.1.51)$$

where

$$e_{\alpha\beta}^\|(bu) = \frac{1}{2}[(b_\alpha^\sigma u_\sigma)_{\|\beta} + (b_\beta^\sigma u_\sigma)_{\|\alpha}] . \quad (17.1.52)$$

We observe that the solution $(\overline{u}, \overline{w})$ of the homogenized problem

$$(Q_h) \quad J_h(\overline{u}, \overline{w}) - f(\overline{u}, \overline{w}) = \inf\{J_h(u, w) - f(u, w)|(u, w) \in V^0(\Omega)\}, \quad (17.1.53)$$

coincides with the leading term $(u^{(0)}, w^{(0)})$ of the asymptotic expansion (17.1.7); moreover, $\epsilon = \epsilon(u^{(0)}, w^{(0)}), \rho = \rho(u^{(0)}, w^{(0)})$.

The solvability of the convex minimization problem appearing on the r.h.s. of (17.1.50) is readily inferred by applying Korn's inequality.

Remark 17.1.3. Until now all mathematical developments concerning shells have been posed on the region Ω. One can also work directly with the shell middle surface S. Then, however, one has to introduce the spaces like $L^2(S)$, $H^1(S)$, $H^2(S)$, etc. Let us introduce these spaces. First, we set

$$L^2(S) = \{u|\, \widehat{u} = u \circ \Phi \in L^2(\Omega)\}\,,$$

with the natural norm

$$\|u\|^2_{L^2(S)} = \int_\Omega |\,\widehat{u}|^2 \sqrt{a}\, d\xi = \int_S |u|^2 dS\,.$$

Next we introduce

$$\mathbf{H}_t(S) = \{\boldsymbol{v} = v^\alpha \boldsymbol{a}_\alpha|\, v^\alpha \in L^2(S)\}\,, \tag{17.1.54}$$

with the norm defined by

$$\|\boldsymbol{v}\|^2_{\mathbf{H}_t(S)} = \int_S \boldsymbol{v}^T \cdot \boldsymbol{v}\, dS = \int_S v_\alpha v^\alpha dS\,.$$

The last norm is equivalent to

$$\left\{ \sum_{\alpha=1,2} \|v^\alpha\|^2_{L^2(S)} \right\}^{1/2}\,.$$

We recall that $\boldsymbol{v}^T$ stands for the transpose of $\boldsymbol{v}$.

Now we are in a position to define

$$H^1(S) = \left\{ u \in L^2(S)|\, \left(\frac{\partial u}{\partial \boldsymbol{x}}\right)^T \in \mathbf{H}_t(S) \right\}\,, \tag{17.1.55}$$

which is equipped with its natural norm:

$$\|u\|_{H^1(S)} = \|u\|_{L^2(S)} + \left\|\left(\frac{\partial u}{\partial \boldsymbol{x}}\right)^T\right\|_{\mathbf{H}_t(S)}\,.$$

Here $\boldsymbol{x} = \Phi(\xi^\alpha)$ is any point of S. Finally we set

$$H^2(S) = \{w|\, \widehat{w} = w \circ \Phi \in H^2(\Omega)\}\,, \tag{17.1.56}$$

with the norm

$$\|w\|_{H^2(S)} = \|\widehat{w}\|_{H^2(\Omega)}\,.$$

In particular we have

$$H^2_0(S) = \{w|\, \widehat{w} = w \circ \Phi \in H^2_0(\Omega)\}\,.$$

17.2. *Dual homogenization*

Let us first formulate the dual problem (Q^*_ϵ). As usual, we apply Rockafellar's theory of duality, cf. Sec. 1.2.5. We set

$$\Lambda(\boldsymbol{u}, w) = (\Lambda_1(\boldsymbol{u}, w), \Lambda_2(\boldsymbol{u}, w)) = (\boldsymbol{\epsilon}(\boldsymbol{u}, w), \boldsymbol{\rho}(\boldsymbol{u}, w)),$$

where $\Lambda : H^1_0(S)^2 \times H^2_0(S) \to L^2(S, \mathbb{E}^2_s) \times L^2(S, \mathbb{E}^2_s)$. Standard calculation yields

$$\langle \boldsymbol{N}, \Lambda_1(\boldsymbol{u}, w) \rangle_{L^2 \times L^2} = \int_S N^{\alpha\beta} u_{\alpha||\beta} dS - \int_S N^{\alpha\beta} b_{\alpha\beta} w dS$$

$$= -\int_S N^{\alpha\beta}{}_{||\beta} u_\alpha dS - \int_S b_{\alpha\beta} N^{\alpha\beta} w dS = \langle \Lambda^*_1 \boldsymbol{N}, (\boldsymbol{u}, w) \rangle_{(H^{-1} \times L^2) \times (H^1_0 \times H^2_0)},$$

where $\boldsymbol{N} \in L^2(S, \mathbb{E}^s_2)$ and $\boldsymbol{u} \in H^1_0(S)^2$, $w \in H^2_0(S)$. Hence we have

$$\Lambda^*_1 \boldsymbol{N} = \begin{cases} (-N^{\alpha\beta}{}_{||\beta}) & \text{in } S \quad (\boldsymbol{u}), \\ -b_{\alpha\beta} N^{\alpha\beta} & \text{in } S \quad (w). \end{cases} \tag{17.2.1}$$

Similarly we find

$$\Lambda^*_2 \boldsymbol{M} = \begin{cases} (2 b^\sigma_\alpha M^{\alpha\beta}{}_{||\beta} + b^\sigma_{\alpha||\beta} M^{\alpha\beta}) & \text{in } S \quad (\boldsymbol{u}), \\ -M^{\alpha\beta}{}_{||\alpha\beta} + c_{\alpha\beta} M^{\alpha\beta} & \text{in } S \quad (w), \end{cases} \tag{17.2.2}$$

where $\boldsymbol{M} \in L^2(S, \mathbb{E}^s_2)$.

    Next we calculate

$$(-f)^*(-\Lambda^*(\boldsymbol{N}, \boldsymbol{M})) = \sup\{\langle -\Lambda^*_1 \boldsymbol{N}, (\boldsymbol{u}, w) \rangle + \langle -\Lambda^*_2 \boldsymbol{M}, (\boldsymbol{u}, w) \rangle$$
$$+ f(\boldsymbol{u}, w) | (\boldsymbol{u}, w) \in H^1_0(S)^2 \times H^2_0(S)\}$$

$$= \sup\{\int_S (N^{\alpha\beta}{}_{||\beta} - 2 b^\alpha_\sigma M^{\sigma\beta}{}_{||\beta} - b^\alpha_{\sigma||\beta} M^{\sigma\beta} + q^\alpha) u_\alpha dS | \boldsymbol{u} \in H^1_0(S)^2\}$$

$$+ \sup\{\int_S (b_{\alpha\beta} N^{\alpha\beta} + M^{\alpha\beta}{}_{||\alpha\beta} - c_{\alpha\beta} M^{\alpha\beta} + q) w dS | w \in H^2_0(S)\}$$

$$= \begin{cases} 0 & \text{if } N^{\alpha\beta}{}_{||\beta} - 2 b^\alpha_\sigma M^{\sigma\beta}{}_{||\beta} - b^\alpha_{\sigma||\beta} M^{\sigma\beta} + q^\alpha = 0 \quad \text{in } S, \\ & \text{and } b_{\alpha\beta} N^{\alpha\beta} + M^{\alpha\beta}{}_{||\alpha\beta} - c_{\alpha\beta} M^{\alpha\beta} + q = 0 \text{ in } S; \\ +\infty & \text{otherwise.} \end{cases} \tag{17.2.3}$$

The equilibrium equations appearing in (17.2.3) coincide with those derived in a different manner by Niordson (1985, p.99) provided that $(M^{\alpha\beta})$ is replaced with $(-M^{\alpha\beta})$.

    The space of statically admissible generalized stresses is defined by

$$\mathcal{S}(S) = \{(\boldsymbol{N}, \boldsymbol{M}) \in L^2(S, \mathbb{E}^s_2) \times L^2(S, \mathbb{E}^s_2) | N^{\alpha\beta}{}_{||\beta} \in L^2(S),$$
$$M^{\alpha\beta}{}_{||\alpha\beta} \in L^2(S) \text{ and the equilibrium equations are satisfied}\}. \tag{17.2.4}$$

Let us set

$$\boldsymbol{a}^\varepsilon = \boldsymbol{A}_\varepsilon^{-1}\,, \quad \boldsymbol{d}^\varepsilon = \boldsymbol{D}_\varepsilon^{-1}\,. \tag{17.2.5}$$

The density of the complementary energy is then expressed by

$$\mathcal{W}_\varepsilon^*(\xi, \boldsymbol{N}, \boldsymbol{M}) = W^*(\xi, \xi/\varepsilon, \boldsymbol{N}, \boldsymbol{M}) = \frac{1}{2}a_{\alpha\beta\lambda\mu}^\varepsilon N^{\alpha\beta}N^{\lambda\mu} + \frac{1}{2}d_{\alpha\beta\lambda\mu}^\varepsilon M^{\alpha\beta}M^{\lambda\mu}, \tag{17.2.6}$$

where $\boldsymbol{N}, \boldsymbol{M} \in \mathbb{E}_2^s$. Now we are in a position to formulate the dual problem:

$$(Q_\varepsilon^*) \quad \left| \begin{array}{l} \text{find} \\ -G_\varepsilon(\boldsymbol{N}^\varepsilon, \boldsymbol{M}^\varepsilon) = \sup\{-G_\varepsilon(\boldsymbol{N}, \boldsymbol{M})|(\boldsymbol{N}, \boldsymbol{M}) \in \mathcal{S}(S)\} \end{array} \right.$$

where

$$G_\varepsilon(\boldsymbol{N}, \boldsymbol{M}) = \int_S \mathcal{W}_\varepsilon^*(\xi, \boldsymbol{N}(x), \boldsymbol{M}(x))dS\,, \tag{17.2.7}$$

and $\boldsymbol{x} = \Phi(\xi)$. The duality Proposition (1.2.50) implies

$$\inf Q_\varepsilon = \min Q_\varepsilon = \sup Q_\varepsilon^* = \max Q_\varepsilon^*\,.$$

Since the complementary potential $\mathcal{W}_\varepsilon^*(\xi, \cdot, \cdot)$ is strictly convex, therefore $(\boldsymbol{N}^\varepsilon, \boldsymbol{M}^\varepsilon) \in \mathcal{S}(S)$ solving the dual problem (Q_ε^*) is unique.

Dual effective potential $\mathcal{W}_h^*$

Prior to the formulation of the dual homogenization theorem pertaining to the Γ-convergence of the sequence of functionals $\{G_\varepsilon\}_{\varepsilon>0}$ given by (17.2.7), it is indispensable to derive the dual effective potential $\mathcal{W}_h^*$. We recall that for $\epsilon, \rho \in \mathbb{E}_s^2$ the primal homogenized potential is given by (17.1.50). For $\boldsymbol{N}_h, \boldsymbol{M}_h \in \mathbb{E}_2^s$ the dual potential is calculated as the Fenchel conjugate:

$$\mathcal{W}_h^*(\xi, \boldsymbol{N}_h, \boldsymbol{M}_h) = \sup\{N_h^{\alpha\beta}\epsilon_{\alpha\beta} + M_h^{\alpha\beta}\rho_{\alpha\beta} - \mathcal{W}_h(\xi, \epsilon, \rho)|\,\epsilon, \rho \in \mathbb{E}_s^2\}$$

$$= \sup_{\epsilon, \rho \in \mathbb{E}_s^2} \{N_h^{\alpha\beta}\epsilon_{\alpha\beta} + M_h^{\alpha\beta}\rho_{\alpha\beta}$$

$$- \frac{1}{2|Y|} \inf_{(\boldsymbol{v},v)\in H_{K,per}(Y)} \int_Y [A^{\alpha\beta\lambda\mu}(\xi, y)(e_{\alpha\beta}^y(\boldsymbol{v}) + \epsilon_{\alpha\beta})(e_{\lambda\mu}^y(\boldsymbol{v}) + \epsilon_{\lambda\mu})$$

$$+ D^{\alpha\beta\lambda\mu}(\xi, y)(\kappa_{\alpha\beta}^y(v) - 2b_\alpha^\sigma(\xi)\frac{\partial v_\sigma}{\partial y_\beta} + \rho_{\alpha\beta})(\kappa_{\lambda\mu}^y(v) - 2b_\lambda^\delta(\xi)\frac{\partial v_\delta}{\partial y_\mu} + \rho_{\lambda\mu})]dy$$

$$= \frac{1}{|Y|} \sup_{\epsilon, \rho \in \mathbb{E}_s^2} \{\int_Y [N_h^{\alpha\beta}(e_{\alpha\beta}^y(\boldsymbol{v}) + \epsilon_{\alpha\beta}) + M_h^{\alpha\beta}(\kappa_{\alpha\beta}^y(v) - 2b_\alpha^\sigma(\xi)\frac{\partial v_\sigma}{\partial y_\beta} + \rho_{\alpha\beta})]dy$$

$$- \inf_{(\boldsymbol{v},v)\in H_{K,per}(Y)} \frac{1}{2} \int_Y [A^{\alpha\beta\lambda\mu}(\xi, y)(e_{\alpha\beta}^y(\boldsymbol{v}) + \epsilon_{\alpha\beta})((e_{\lambda\mu}^y(\boldsymbol{v}) + \epsilon_{\lambda\mu})$$

$$+ D^{\alpha\beta\lambda\mu}(\xi, y)(\kappa_{\alpha\beta}^y(v) - 2b_\alpha^\sigma(\xi)\frac{\partial v_\sigma}{\partial y_\beta} + \rho_{\alpha\beta})(\kappa_{\lambda\mu}^y(v) - 2b_\lambda^\delta(\xi)\frac{\partial v_\delta}{\partial y_\mu} + \rho_{\lambda\mu})]dy\}\,,$$

because

$$\int_Y N_h^{\alpha\beta} e_{\alpha\beta}^y(\boldsymbol{v}) dy = 0 , \qquad \boldsymbol{v} \in H_{per}^1(Y)^2 ,$$

$$\int_Y M_h^{\alpha\beta} \kappa_{\alpha\beta}^y(v) dy = 0 , \qquad v \in H_{per}^2(Y) ,$$

$$\int_Y M_h^{\alpha\beta} b_\alpha^\sigma(\xi) \frac{\partial v_\sigma}{\partial y_\beta} dy = 0 , \qquad \boldsymbol{v} \in H_{per}^1(Y)^2 .$$

Thus we may write

$$\mathcal{W}_h^*(\xi, \boldsymbol{N}_h, \boldsymbol{M}_h) = |Y|^{-1} \sup \{ \int_Y \{ N_h^{\alpha\beta}(e_{\alpha\beta}^y(\boldsymbol{v}) + \epsilon_{\alpha\beta})$$

$$+ M_h^{\alpha\beta}(\kappa_{\alpha\beta}^y(v) - 2b_\alpha^\sigma(\xi) \frac{\partial v_\sigma}{\partial y_\beta} + \rho_{\alpha\beta})$$

$$- \frac{1}{2} [A^{\alpha\beta\lambda\mu}(\xi, y)(e_{\alpha\beta}^y(\boldsymbol{v}) + \epsilon_{\alpha\beta})(e_{\lambda\mu}^y(\boldsymbol{v}) + \epsilon_{\lambda\mu})$$

$$+ D^{\alpha\beta\lambda\mu}(\xi, y)(\kappa_{\alpha\beta}^y(v) - 2b_\alpha^\sigma(\xi) \frac{\partial v_\sigma}{\partial y_\beta} + \rho_{\alpha\beta})(\kappa_{\lambda\mu}^y(v)$$

$$- 2b_\lambda^\delta(\xi) \frac{\partial v_\delta}{\partial y_\mu} + \rho_{\lambda\mu})] \} dy | (\boldsymbol{v}, v) \in H_{K,per}(Y), \epsilon, \rho \in \mathbb{E}_s^2 \} .$$

Now we introduce the following space

$$X := [e^y(H_{per}^1(Y)^2) \oplus \mathbb{E}_s^2] \times \{ [\kappa^y(H_{per}^2(Y)) - 2b\nabla_y(H_{per}^1(Y)^2)] \} \oplus \mathbb{E}_s^2 ,$$

and set

$$\mathbf{J}(\xi, \boldsymbol{\Delta}, \boldsymbol{\psi}) = \int_Y \mathcal{W}[\xi, y, \boldsymbol{\Delta}(y), \boldsymbol{\psi}(y)] dy .$$

We observe that the space X depends on ξ, since $\boldsymbol{b}$ does. The notation $\boldsymbol{b}\nabla_y(H_{per}^1(Y)^2)$ has the obvious meaning: for $\boldsymbol{v} \in H_{per}^1(Y)^2$, $\boldsymbol{b}(\xi)\nabla_y\boldsymbol{v} = (b_\lambda^\delta(\xi)(\partial v_\delta/\partial y_\mu))$. Thus we can write

$$\mathcal{W}_h^*(\xi, \boldsymbol{N}_h, \boldsymbol{M}_h) = |Y|^{-1}(\mathbf{J} + I_X)^*(\boldsymbol{N}_h, \boldsymbol{M}_h) ,$$

where I_X stands for the indicator function of the space X. The conjugate or dual functional $(\mathbf{J} + I_X)^*$ takes the form, cf. Sec. 1.2.5

$$(\mathbf{J} + I_X)^*(\boldsymbol{N}_h, \boldsymbol{M}_h) = (\mathbf{J}^* \square I_{\mathcal{S}_{per}(Y)})(\boldsymbol{N}_h, \boldsymbol{M}_h) , \tag{17.2.8}$$

where

$$\mathcal{S}_{per}(Y) := X^\perp , \tag{17.2.9}$$

while $\square$ denotes the operation of inf-convolution. We calculate

$$X^\perp = \{(\boldsymbol{n}, \boldsymbol{m}) \in L^2(Y, \mathbb{E}_2^s) \times L^2(Y, \mathbb{E}_2^s)|$$

$$\int_Y [n^{\alpha\beta}(y)(e^y_{\alpha\beta}(\boldsymbol{v}) + \epsilon_{\alpha\beta}) + m^{\alpha\beta}(y)(\kappa^y_{\alpha\beta}(v) - 2b^\sigma_\alpha(\xi)\frac{\partial v_\sigma}{\partial y_\beta} + \rho_{\alpha\beta})]dy = 0$$

$$\forall\, (\boldsymbol{v}, v) \in H_{K,per}(Y)\} \, . \tag{17.2.10}$$

To determine $X^\perp$ we successively find:

(i) For $\boldsymbol{\rho} = \boldsymbol{0}$ and $(\boldsymbol{v}, v) = (\boldsymbol{0}, 0)$ we obtain $\displaystyle\int_Y n^{\alpha\beta}(y)\epsilon_{\alpha\beta}dy = 0$, and hence

$$\int_Y n^{\alpha\beta}(y)dy = 0 \, . \tag{17.2.11}$$

Here $\epsilon_{\alpha\beta}$ is treated as constant, thus square integrable, element of $L^2(Y)$.

(ii) For $\boldsymbol{\epsilon} = \boldsymbol{0}$ and $(\boldsymbol{v}, v) = (\boldsymbol{0}, 0)$ one similarly gets

$$\int_Y \boldsymbol{m}(y)dy = \boldsymbol{0} \, . \tag{17.2.12}$$

(iii) Taking into account (17.2.11) and (17.2.12) in (17.2.10) we obtain

$$X^\perp = \{(\boldsymbol{n}, \boldsymbol{m}) \in L^2(Y, \mathbb{E}_2^s) \times L^2(Y, \mathbb{E}_2^s)| \int_Y \boldsymbol{n}(y)dy = \boldsymbol{0} \, , \ \int_Y \boldsymbol{m}(y)dy = \boldsymbol{0} \, ,$$

$$\int_Y [n^{\alpha\beta}(y)e^y_{\alpha\beta}(\boldsymbol{v}) + m^{\alpha\beta}(y)(\kappa^y_{\alpha\beta}(v) - 2b^\sigma_\alpha(\xi)\frac{\partial v_\sigma}{\partial y_\beta})]dy = 0$$

$$\forall\, (\boldsymbol{v}, v) \in H_{K,per}(Y)\} \, . \tag{17.2.13}$$

For $(\boldsymbol{v}, 0) \in H_{K,per}(Y)$ we get

$$\int_Y [n^{\alpha\beta}(y)e^y_{\alpha\beta}(\boldsymbol{v}) - m^{\alpha\beta}(y)2b^\sigma_\alpha(\xi)\frac{\partial v_\sigma}{\partial y_\beta}]dy = 0 \quad \forall\, \boldsymbol{v} \in H^1_{per}(Y)^2 \, .$$

Integration by part yields

$$\int_Y \left(-\frac{\partial n^{\alpha\beta}}{\partial y_\beta} + 2b^\alpha_\sigma(\xi)\frac{\partial m^{\sigma\beta}}{\partial y_\beta}\right) v_\alpha dy + \int_{\partial Y} (n^{\alpha\beta} - 2b^\alpha_\lambda(\xi)m^{\lambda\beta})\nu_\beta v_\alpha ds = 0$$

$$\forall v \in H^1_{per}(Y)^2 \, ,$$

where $\boldsymbol{\nu}$ stands for the outer unit normal vector to ∂Y. Hence we conclude that

$$\frac{\partial n^{\alpha\beta}}{\partial y_\beta} - 2b^\alpha_\lambda(\xi)\frac{\partial m^{\lambda\beta}}{\partial y_\beta} = 0 \quad \text{in } Y \, ,$$

$$(n^{\alpha\beta} - 2b^\alpha_\lambda(\xi)m^{\lambda\beta})\nu_\beta \text{ takes opposite values on the opposite sides of } Y. \tag{17.2.14}$$

It remains to study the following variational equation

$$\int_Y m^{\alpha\beta}(y)\kappa^y_{\alpha\beta}(v)\,dy = 0 \quad \forall\, v \in H^2_{per}(Y)\,.$$

Integration by parts gives

$$-\int_Y m^{\alpha\beta}(y)\kappa^y_{\alpha\beta}(v)\,dy = \int_Y v\frac{\partial^2 m^{\alpha\beta}}{\partial y_\alpha \partial y_\beta}\,dy - \int_{\partial Y} \mathbf{q}v\,ds$$

$$+ \int_{\partial Y} m_\nu \frac{\partial v}{\partial \boldsymbol{\nu}}\,ds + \sum_{i=1}^{4} R_i v(O_i) = 0 \quad \forall\, v \in H^2_{per}(Y)\,, \tag{17.2.15}$$

where O_i $(i = 1, 2, 3, 4)$ denote the corners of Y and

$$m_\nu = m^{\alpha\beta}\nu_\alpha\nu_\beta\,, \quad R_i = 2(-1)^{i+1}m^{12}(O_i)\,,$$

$$\mathbf{q} = \nu_\alpha \frac{\partial m^{\alpha\beta}}{\partial y_\beta} + \frac{\partial m_t}{\partial s}\,, \quad m_t = m^{\alpha\beta}\nu_\alpha\tau_\beta\,. \tag{17.2.16}$$

Here $\boldsymbol{\tau}$ is the unit vector tangent to ∂Y. From Eq. (17.2.15) we conclude that:

$$\mathrm{div}_y\,\mathrm{div}_y \boldsymbol{m} = \frac{\partial^2 m^{\alpha\beta}}{\partial y_\alpha \partial y_\beta} = 0 \quad \text{in } Y\,,$$

m_ν takes equal values and $\mathbf{q}$ – opposite values on the opposite sides of Y,   (17.2.17)

the condition $\sum_{i=1}^{4} R_i = 0$ is identically satisfied.

On account of (17.2.10), (17.2.13), (17.2.14) and (17.2.17) the space $\mathcal{S}_{per}(Y)$ of local generalized stresses $(\boldsymbol{n}, \boldsymbol{m})$ is finally given by

$$\mathcal{S}_{per}(Y) = \left\{(\boldsymbol{n}, \boldsymbol{m}) \in L^2(Y, \mathbb{E}^s_2) \times L^2(Y, \mathbb{E}^s_2)\bigg| \int_Y \boldsymbol{n}(y)\,dy = 0\,, \int_Y \boldsymbol{m}(y)\,dy = 0\,,\right.$$

$$\frac{\partial n^{\alpha\beta}}{\partial y_\beta} - 2b^\alpha_\lambda(\xi)\frac{\partial m^{\lambda\beta}}{\partial y_\beta} = 0 \quad \text{in } Y\,, \mathrm{div}_y\,\mathrm{div}_y\boldsymbol{m} = 0 \quad \text{in } Y\,,$$

$(n^{\alpha\beta} - 2b^\alpha_\lambda(\xi)m^{\lambda\beta})\nu_\beta$ take opposite values on the opposite sides of Y,

$$\left. m_\nu \text{ takes equal and } -\mathbf{q} \text{ opposite values on the opposite sides of } Y\right\}\,. \tag{17.2.18}$$

In accordance with (17.2.8) and the definition of inf-convolution we have

$$\mathcal{W}^*_h(\xi, \boldsymbol{N}_h, \boldsymbol{M}_h) = |Y|^{-1}\inf\{\mathbf{J}^*(\xi, \boldsymbol{n}_1, \boldsymbol{m}_1) + I_{\mathcal{S}_{per}(Y)}(\boldsymbol{n}_2, \boldsymbol{m}_2)|\,\boldsymbol{N}_h = \boldsymbol{n}_1 + \boldsymbol{n}_2,$$

$$\boldsymbol{M}_h = \boldsymbol{m}_1 + \boldsymbol{m}_2, (\boldsymbol{n}_1, \boldsymbol{m}_1), (\boldsymbol{n}_2, \boldsymbol{m}_2) \in \mathcal{S}_{per}(Y)\}$$

$$= |Y|^{-1}\inf\{\mathbf{J}^*(\xi, \boldsymbol{N}_h - \boldsymbol{n}_2, \boldsymbol{M}_h - \boldsymbol{m}_2)|(\boldsymbol{n}_2, \boldsymbol{m}_2) \in \mathcal{S}_{per}(Y)\}\,,$$

where

$$\mathbf{J}^*(\xi, \boldsymbol{n}, \boldsymbol{m}) = \int_Y \mathcal{W}^*(\xi, y, \boldsymbol{n}(y), \boldsymbol{m}(y)) dy \,,$$

and

$$\mathcal{W}^*(\xi, y, \boldsymbol{n}, \boldsymbol{m}) = \frac{1}{2} a_{\alpha\beta\lambda\mu}(\xi, y) n^{\alpha\beta} n^{\lambda\mu} + \frac{1}{2} d_{\alpha\beta\lambda\mu}(\xi, y) m^{\alpha\beta} m^{\lambda\mu} \,.$$

Since $\mathcal{S}_{per}(Y)$ is a linear space therefore

$$\mathcal{W}_h^*(\xi, \boldsymbol{N}_h, \boldsymbol{M}_h)$$
$$= |Y|^{-1} \inf\{ \int_Y \mathcal{W}^*(\xi, y, \boldsymbol{N}_h + \boldsymbol{n}(y), \boldsymbol{M}_h + \boldsymbol{m}(y)) dy | (\boldsymbol{n}, \boldsymbol{m}) \in \mathcal{S}_{per}(Y) \}$$
$$= \frac{1}{2} |Y|^{-1} \inf\{ \int_Y [a_{\alpha\beta\lambda\mu}(\xi, y)(N_h^{\alpha\beta} + n^{\alpha\beta}(y))(N_h^{\lambda\mu} + n^{\lambda\mu}(y))$$
$$+ d_{\alpha\beta\lambda\mu}(\xi, y)(M_h^{\alpha\beta} + m^{\alpha\beta}(y))(M_h^{\lambda\mu} + m^{\lambda\mu}(y))] dy | (\boldsymbol{n}, \boldsymbol{m}) \in \mathcal{S}_{per}(Y) \} \,, \qquad (17.2.19)$$

where $\boldsymbol{N}_h \in \mathbb{E}_2^s, \boldsymbol{M}_h \in \mathbb{E}_2^s$.

Suppose that $(\widetilde{\boldsymbol{n}}, \widetilde{\boldsymbol{m}}) \in \mathcal{S}_{per}(Y)$ solves the minimization problem occurring in (17.2.19). Such a solution obviously exists and is unique. Due to linearity of the problem studied, there exist functions $(\mathbf{N}_{(\alpha\beta)}, \mathbf{O}_{(\alpha\beta)}) \in \mathcal{S}_{per}(Y)$, $(\mathbf{M}_{(\alpha\beta)}, \mathbf{R}_{(\alpha\beta)}) \in \mathcal{S}_{per}(Y)$ such that

$$\widetilde{\boldsymbol{n}} = N_h^{\alpha\beta} \mathbf{N}_{(\alpha\beta)} + M_h^{\alpha\beta} \mathbf{M}_{(\alpha\beta)} \,, \qquad \widetilde{\boldsymbol{m}} = N_h^{\alpha\beta} \mathbf{O}_{(\alpha\beta)} + M_h^{\alpha\beta} \mathbf{R}_{(\alpha\beta)} \,.$$

These functions can be found by solving two minimization problems. Let

$$(\boldsymbol{I}_{(\gamma\delta)}, \mathbf{0}) \in \mathbb{E}_2^s \times \mathbb{E}_2^s \quad (\gamma, \delta\text{-fixed}) \,,$$

where $I_{(\gamma\delta)}^{\alpha\beta}$ are components of the identity tensor. Then

$$\mathcal{W}_h^*(\xi, \boldsymbol{I}_{(\gamma\delta)}, \mathbf{0}) = \frac{1}{2|Y|} \inf\{ \int_Y [a_{\alpha\beta\lambda\mu}(\xi, y)(I_{(\gamma\delta)}^{\alpha\beta} + n^{\alpha\beta}(y))(I_{(\gamma\delta)}^{\lambda\mu} + n^{\lambda\mu}(y))$$
$$+ d_{\alpha\beta\lambda\mu}(\xi, y) m^{\alpha\beta}(y) m^{\lambda\mu}(y)] dy | (\boldsymbol{n}, \boldsymbol{m}) \in \mathcal{S}_{per}(Y) \} \,. \qquad (17.2.20)$$

The infimum on the r.h.s. is obviously attained at $(\mathbf{N}_{(\gamma\delta)}, \mathbf{O}_{(\gamma\delta)}) \in \mathcal{S}_{per}(Y)$.

To formulate the second minimization problem we take $(\mathbf{0}, \boldsymbol{I}_{(\gamma\delta)}) \in \mathbb{E}_2^s \times \mathbb{E}_2^s, \gamma, \delta\text{-fixed}$. Next we find

$$\mathcal{W}_h^*(\xi, \mathbf{0}, \boldsymbol{I}_{(\gamma\delta)}) = \frac{1}{2|Y|} \inf\{ \int_Y [a_{\alpha\beta\lambda\mu}(\xi, y) n^{\alpha\beta}(y) n^{\lambda\mu}(y)$$
$$+ d_{\alpha\beta\lambda\mu}(\xi, y)(I_{(\gamma\delta)}^{\alpha\beta} + m^{\alpha\beta}(y))(I_{(\gamma\delta)}^{\lambda\mu} + m^{\lambda\mu}(y))] dy | (\boldsymbol{n}, \boldsymbol{m}) \in \mathcal{S}_{per}(Y) \} \,. \qquad (17.2.21)$$

Now the infimum is attained at $(\mathbf{M}_{(\gamma\delta)}, \mathbf{R}_{(\gamma\delta)}) \in \mathcal{S}_{per}(Y)$.

Remark 17.2.1. The form (17.2.18) of $\mathcal{S}_{per}(Y)$ implies that both in (17.2.20) and (17.2.21) the infima over n and m are, in general, coupled. If $b = (b_{\alpha\beta})$ vanishes then the minimization over n and m can be performed independently.

Dual homogenization theorem: Γ-convergence of the sequence $\{G_\varepsilon\}_{\varepsilon>0}$
   We are in a position to formulate the dual homogenization theorem.

Theorem 17.2.2. The sequence of functionals $G_\varepsilon + I_{S(S)}$, where G_ε is defined by (17.2.7), is Γ-convergent in the weak topology of $\mathbf{H} = L^2(S, \mathbb{E}_2^s) \times L^2(S, \mathbb{E}_2^s)$ to $G_h + I_{S(S)}$. The functional G_h has the form

$$G_h(\boldsymbol{N}, \boldsymbol{M}) = \int_S \mathcal{W}_h^*(\xi, \boldsymbol{N}(x), \boldsymbol{M}(x))dS \quad \text{if} \quad (\boldsymbol{N}, \boldsymbol{M}) \in \mathbf{H}, \qquad (17.2.22)$$

where $\mathcal{W}_h^*$ is given by (17.2.19) and $x = \Phi(\xi)$.
Sketch of the proof.
(i) We set

$$J_\varepsilon^p(\boldsymbol{u}, w; \boldsymbol{p}, \boldsymbol{q}) = \int_S \mathcal{W}_\varepsilon[\xi, \boldsymbol{\epsilon}(\boldsymbol{u}, w) + \boldsymbol{p}, \boldsymbol{\rho}(\boldsymbol{u}, w) + \boldsymbol{q}]dS,$$

$$J_h^p(\boldsymbol{u}, w; \boldsymbol{p}, \boldsymbol{q}) = \int_S \mathcal{W}_h[\xi, \boldsymbol{\epsilon}(\boldsymbol{u}, w) + \boldsymbol{p}, \boldsymbol{\rho}(\boldsymbol{u}, w) + \boldsymbol{q}]dS,$$

where $\boldsymbol{p}, \boldsymbol{q} \in L^2(S, \mathbb{E}_s^2)$. By using Theorem 17.1.2 we then prove that

$$J_h^p = \Gamma(\tau \times s_{\mathbf{H}'}) - \lim_{\varepsilon \to 0} J_\varepsilon^p,$$

where τ denotes the strong topology of the space $L^2(S)^2 \times H^1(S)$ while $s_{\mathbf{H}'}$ stands for the strong topology of the space $\mathbf{H}' = L^2(S, \mathbb{E}_s^2) \times L^2(S, \mathbb{E}_s^2)$.
(ii) According to Azé's Theorem 1.3.36 we then have

$$G_h + I_{S(S)} = \Gamma(w_{\mathbf{H}}) - \lim_{\varepsilon \to 0} (G_\varepsilon + I_{S(S)}).$$

Further, let $(\boldsymbol{u}^\varepsilon, w^\varepsilon) \in H_0^1(S)^2 \times H_0^2(S)$ be a minimizer of (Q_ε) and $\{\boldsymbol{u}^{\varepsilon'}, w^{\varepsilon'}\}_{\varepsilon'>o}$ a convergent subsequence: $(\boldsymbol{u}^{\varepsilon'}, w^{\varepsilon'}) \xrightarrow{\tau} (\overline{\boldsymbol{u}}, \overline{w})$ when $\varepsilon' \to 0$. Then

$$\inf Q_{\varepsilon'} \to \inf Q_h \quad \text{when} \quad \varepsilon' \to 0.$$

Similarly, if $(\boldsymbol{N}_\varepsilon, \boldsymbol{M}_\varepsilon)$ solves the problem (Q_ε^*) and $\{\boldsymbol{N}_{\varepsilon''}, \boldsymbol{M}_{\varepsilon''}\}$ is a convergent subsequence:

$$(\boldsymbol{N}_{\varepsilon''}, \boldsymbol{M}_{\varepsilon''}) \xrightarrow{w_{\mathbf{H}}} (\overline{\boldsymbol{N}}, \overline{\boldsymbol{M}}) \quad \text{when} \quad \varepsilon'' \to 0$$

then

$$(Q_h^*) \quad -G_h(\overline{\boldsymbol{N}}, \overline{\boldsymbol{M}}) = \sup\{-\int_S \mathcal{W}_h^*[\xi, \boldsymbol{N}(x), \boldsymbol{M}(x)]dS| \, (\boldsymbol{N}, \boldsymbol{M}) \in \mathcal{S}(S)\},$$

and

$$\sup Q^*_{\varepsilon''} \to \sup Q^*_h , \qquad \inf Q_h = \sup Q^*_h .$$

Obviously, both the infimum and supremum are attained. $\qquad\qquad\square$

17.3. Effective stiffnesses of ribbed orthotropic cylindrical shells

The homogenization will be performed within the framework of the Koiter shell model studied in Secs. 17.1 and 17.2. Two cases will be dealt with. In the first one the shell is stiffened circumferentially. The second case concerns the stiffening coinciding with the generating lines. The final results turn out to differ considerably.

17.3.1.   Geometrical and material characteristics of a cylindrical shell

Consider a closed circular orthotropic cylindrical shell, the mid-surface of which is given by the mapping

$$\boldsymbol{\Phi}(\xi) = [R\cos\theta, R\sin\theta, \xi^2] , \tag{17.3.1}$$

where $\xi^1 = \theta, \xi^2 = x^3$, see Fig. 17.3.1 below. The fundamental tensors of the mid-plane are

$$a_{11} = R^2, \; a_{12} = 0, \; a_{22} = 1, \; b_{11} = -R, \; b_{12} = b_{22} = 0 . \tag{17.3.2}$$

Due to orthotropy the following stiffnesses vanish

$$K^{1211} = K^{2111} = K^{2221} = K^{1222} , \qquad \boldsymbol{K} \in \{\boldsymbol{A}, \boldsymbol{D}\} . \tag{17.3.3}$$

According to (17.1.21), (17.1.22) the non-zero components of the tensors $\boldsymbol{B}$ and $\boldsymbol{E}$ are

$$B^{1111} = \left(1 + \frac{h^2}{3R^2}\right) A^{1111} , \qquad B^{1122} = B^{2211} = A^{1122} ,$$

$$B^{1212} = \left(1 + \frac{h^2}{3R^2}\right) A^{1212} , \tag{17.3.4}$$

$$B^{2121} = B^{1221} = B^{2112} = A^{1212} , \qquad B^{2222} = A^{2222} ;$$

$$E^{1111} = \frac{h^2}{6R} A^{1111} , \qquad E^{1122} = \frac{h^2}{6R} A^{1122} ,$$

$$E^{1212} = \frac{h^2}{6R} A^{1212} . \tag{17.3.5}$$

We recall that $D^{\alpha\beta\lambda\mu} = (h^2/12) A^{\alpha\beta\lambda\mu}$. The following identities will be used

$$D^{1111} B^{1111} - (E^{1111})^2 = A^{1111} D^{1111} ,$$

$$E^{1111} (A^{1111} D^{1111})^{-1} = 2(R A^{1111})^{-1} , \tag{17.3.6}$$

$$B^{1111} (A^{1111} D^{1111})^{-1} = \left(\frac{4}{R^2} + \frac{12}{h^2}\right) (A^{1111})^{-1} .$$

17.3.2. Strong formulations of the local problems (P_Y^α)

Let us define the auxiliary local fields

$$
\overset{(\lambda\mu)}{N_1}{}^{\alpha\beta} = B^{\alpha\beta\sigma\rho}\Psi^{(\lambda\mu)}_{\sigma|\rho} - E^{\alpha\beta\delta\rho}\Xi^{(\lambda\mu)}_{|\delta\rho} + A^{\alpha\beta\lambda\mu}\,,
$$

$$
\overset{(\lambda\mu)}{M_1}{}^{\alpha\beta} = E^{\gamma\rho\alpha\beta}\Psi^{(\lambda\mu)}_{\gamma|\rho} - D^{\alpha\beta\delta\rho}\Xi^{(\lambda\mu)}_{|\delta\rho}\,;
$$

(17.3.7)

$$
\overset{(\lambda\mu)}{N_2}{}^{\alpha\beta} = B^{\alpha\beta\sigma\rho}\Phi^{(\lambda\mu)}_{\sigma|\rho} + E^{\alpha\beta\delta\rho}[-\chi^{(\lambda\mu)}_{|\delta\rho} + \delta^\lambda_\delta\delta^\mu_\rho]\,,
$$

$$
\overset{(\lambda\mu)}{M_2}{}^{\alpha\beta} = E^{\gamma\rho\alpha\beta}\Phi^{(\lambda\mu)}_{\gamma|\rho} + D^{\gamma\rho\alpha\beta}[-\chi^{(\lambda\mu)}_{|\gamma\rho} + \delta^\lambda_\gamma\delta^\mu_\rho]\,,
$$

(17.3.8)

where $(\ )_{|\alpha} = \partial/\partial y_\alpha$ and

$$
\overset{1}{\boldsymbol{U}}{}^{(\lambda\mu)} = (\boldsymbol{\Psi}^{(\lambda\mu)}(\xi,\cdot), \Xi^{(\lambda\mu)}(\xi,\cdot))\,, \qquad \overset{2}{\boldsymbol{U}}{}^{(\lambda\mu)} = (\boldsymbol{\Phi}^{(\lambda\mu)}(\xi,\cdot), \chi^{(\lambda\mu)}(\xi,\cdot))\,.
$$

(17.3.9)

Then the problems (P_Y^σ) can be rearranged to the form:

$$
(\widetilde{P}_Y^\sigma) \quad \left|
\begin{array}{l}
\text{find } \overset{\sigma}{\boldsymbol{U}}{}^{(\lambda\mu)} \in H_{K,per}(Y) \text{ such that} \\[2mm]
\langle \overset{(\lambda\mu)}{N_\sigma}{}^{\alpha\beta} v_{\alpha|\beta}\rangle = 0\,,\ \langle \overset{(\lambda\mu)}{M_\sigma}{}^{\alpha\beta} v_{|\alpha\beta}\rangle = 0 \quad \forall(\boldsymbol{v},v) \in H_{K,per}(Y)\,.
\end{array}
\right.
$$

(17.3.10)

The variational equations (17.3.10) imply:
(i) the differential equations

$$
\overset{(\lambda\mu)}{N_\sigma}{}^{\alpha\beta}{}_{|\beta} = 0\,, \qquad \overset{(\lambda\mu)}{M_\sigma}{}^{\alpha\beta}{}_{|\alpha\beta} = 0\,, \quad \text{in } Y\,.
$$

(17.3.11)

(ii) The periodicity conditions: the quantities

$$
\overset{(\lambda\mu)}{N_\sigma}{}^{\alpha\beta}\nu_\alpha\nu_\beta\,, \qquad \overset{(\lambda\mu)}{N_\sigma}{}^{\alpha\beta}\nu_\beta\tau_\alpha\,, \qquad \overset{(\lambda\mu)}{M_\sigma}{}^{\alpha\beta}\nu_\alpha\nu_\beta
$$

(17.3.12)

assume the same values on the opposite sides of Y, and

$$
\overset{(\lambda\mu)}{M}{}^{\alpha\beta}{}_{|\alpha}\nu_\beta + \frac{\partial}{\partial s}(\overset{(\lambda\mu)}{M_\sigma}{}^{\alpha\beta}\nu_\beta\tau_\alpha)
$$

(17.3.13)

assumes opposite values on the opposite sides of Y.

17.3.3. Case of stiffeners along the generating lines

The functions $A^{\alpha\beta\lambda\mu}(\xi,y)$, $h(\xi,y)$ do not depend on y_2 and are $l_1 = \beta$-periodic with respect to $y_1 = \xi^1/\varepsilon$; ε is a small parameter, cf. Fig. 17.3.1. The solutions $\overset{\sigma}{\boldsymbol{U}}{}^{(\lambda\mu)}$ will be functions of one variable y_1.

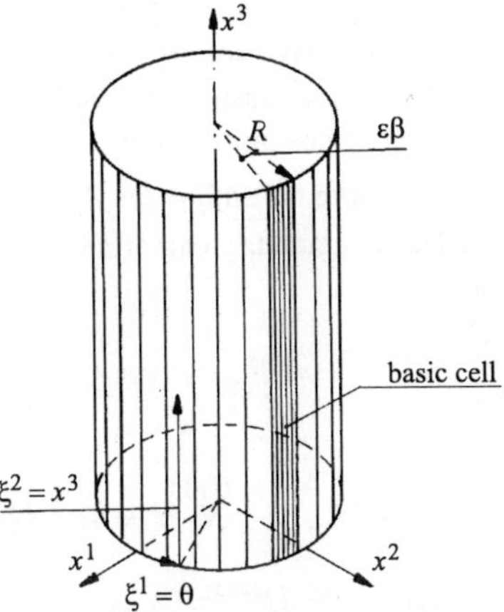

Fig. 17.3.1. Cylindrical shell with stiffeners along the generating lines

A detailed analysis of the solution of the problems $(\widetilde{P}^{\sigma}_{Y})$ shows that the final formulae for the effective stiffnesses are unaffected by the curvature of the shell and coincide with the formulae for periodic plates. Thus the membrane stiffnesses $A_h^{\alpha\beta\lambda\mu}$ are given by Eqs. (3.7.14), the bending stiffnesses $D_h^{\alpha\beta\lambda\mu}$ are expressed by Eqs. (3.7.7) and the reciprocal stiffnesses $E_h^{\alpha\beta\lambda\mu}$ and $F_h^{\alpha\beta\lambda\mu}$ vanish.

17.3.4. Case of circumferential stiffening

The geometric and material characteristics of the shell are $\varepsilon l_1 = \varepsilon a$-periodic with respect to the coordinate $\xi^2 = x^3$, cf. Fig. 17.3.2.

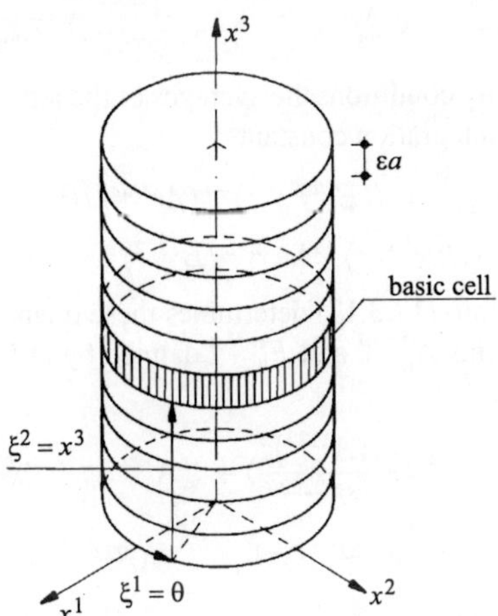

Fig. 17.3.2. Cylindrical shell with circumferential stiffening

Thus we have $A^{\alpha\beta\lambda\mu} = A^{\alpha\beta\lambda\mu}(\xi, y_2)$, $h = h(\xi, y_2)$, $y_2 \in (0, a)$ and these functions are a-periodic with respect to y_2. For ξ fixed the solutions to the $(\widetilde{P}_Y^\sigma)$ problems are functions of one variable y_2. Contrary to the previous case of stiffening the formulae for $\boldsymbol{A}_h, \boldsymbol{E}_h, \boldsymbol{D}_h$ do not coincide with those for plates. Thus more detailed analysis seems necessary.

a) *Solution to the $(\widetilde{P}_Y^1)$ problem. Finding the stiffnesses $A_h^{\alpha\beta\lambda\mu}$ and $F_h^{\alpha\beta\lambda\mu}$*

The variable ξ will be viewed as a parameter. Our aim is to find the functions $\Psi_1^{(\lambda\mu)}(y_2)$, $\Psi_2^{(\lambda\mu)}(y_2)$, $\Xi^{(\lambda\mu)}(y_2)$ such that

$$
\overset{(\lambda\mu)}{N_1}{}^{\sigma 2}{}_{|2} = 0 \,, \qquad \overset{(\lambda\mu)}{M_1}{}^{22}{}_{|22} = 0 \,, \qquad \sigma = 1, 2 \tag{17.3.14}
$$

where

$$
\overset{(\lambda\mu)}{N_1}{}^{12} = B^{1212}\Psi_1^{(\lambda\mu)}{}_{|2} + A^{12\lambda\mu} \,, \qquad \overset{(\lambda\mu)}{N_1}{}^{22} = B^{2222}\Psi_2^{(\lambda\mu)}{}_{|2} + A^{22\lambda\mu} \,,
$$
$$
\overset{(\lambda\mu)}{M_1}{}^{22} = -D^{2222}\Xi^{(\lambda\mu)}{}_{|22} \,; \tag{17.3.15}
$$

we recall that $\varphi_{|2} = \partial\varphi/\partial y_2$. The periodicity conditions have the form

$$
g(0) = g(a) \,, \tag{17.3.16}
$$

where g represents the following functions: $\Psi_\alpha^{(\lambda\mu)}$, $\Xi^{(\lambda\mu)}$, $\Xi^{(\lambda\mu)}{}_{|2}$, $\overset{(\lambda\mu)}{N_1}{}^{\sigma 2}$, $\overset{(\lambda\mu)}{M_1}{}^{22}$, $\overset{(\lambda\mu)}{M_1}{}^{22}{}_{|2}$. Integration of (17.3.14) gives

$$
\overset{(\lambda\mu)}{N_1}{}^{12} = C_2^{(\lambda\mu)} \,, \qquad \overset{(\lambda\mu)}{N_1}{}^{22} = C_1^{(\lambda\mu)} \,, \qquad \overset{(\lambda\mu)}{M_1}{}^{22} = C_3^{(\lambda\mu)} + y_2 C_4^{(\lambda\mu)} \,.
$$

One of the periodicity conditions gives $C_4^{(\lambda\mu)} = 0$. We can rewrite Eqs. (17.3.15) in the form

$$
\Psi_{1|2}^{(\lambda\mu)} = \frac{C_2^{(\lambda\mu)}}{B^{1212}} - \frac{A^{12\lambda\mu}}{B^{1212}} \,, \qquad \Psi_{2|2}^{(\lambda\mu)} = \frac{C_1^{(\lambda\mu)}}{B^{2222}} - \frac{A^{22\lambda\mu}}{B^{2222}} \,, \qquad \Xi_{|22}^{(\lambda\mu)} = -\frac{C_3^{(\lambda\mu)}}{D^{2222}} \,. \tag{17.3.17}
$$

According to the periodicity conditions the averages of the left-hand sides of Eqs. (17.3.17) vanish. Thus we find the integration constants:

$$
C_1^{(\lambda\mu)} = \langle (B^{2222})^{-1}\rangle^{-1}\langle A^{22\lambda\mu}/B^{2222}\rangle \,,
$$
$$
C_2^{(\lambda\mu)} = \langle (B^{1212})^{-1}\rangle^{-1}\langle A^{12\lambda\mu}/B^{1212}\rangle \,, \qquad C_3^{(\lambda\mu)} = 0 \,. \tag{17.3.18}
$$

Substitution of (17.3.18) into (17.3.17) determines these quantities. Then one can find the formulae for the components $A_h^{\alpha\beta\lambda\mu}$ and $F_h^{\alpha\beta\lambda\mu}$ defined by (17.1.26). Simple computation gives

$$
A_h^{1111} = \langle A^{1111} - \frac{(A^{1122})^2}{A^{2222}}\rangle + \langle (A^{2222})^{-1}\rangle^{-1}\langle \frac{A^{1122}}{A^{2222}}\rangle^2 \,,
$$

$$
A_h^{1122} = \langle \frac{A^{1122}}{A^{2222}}\rangle\langle (A^{2222})^{-1}\rangle^{-1} \,, \qquad A_h^{2222} = \langle (A^{2222})^{-1}\rangle^{-1} \,, \tag{17.3.19}
$$

$$
A_h^{1212} = \langle \frac{\sigma}{1+\sigma}A^{1212}\rangle + \langle (1+\sigma)^{-1}\rangle^2\langle \frac{1}{(1+\sigma)A^{1212}}\rangle^{-1} \,,
$$

and

$$F_h^{1212} = -\frac{R}{2}\left[\langle\frac{\sigma}{1+\sigma}A^{1212}\rangle - \langle\frac{\sigma}{1+\sigma}\rangle\langle(1+\sigma)^{-1}\rangle\langle\frac{1}{(1+\sigma)A^{1212}}\rangle^{-1}\right],\qquad (17.3.20)$$

where $\sigma = \dfrac{h^2}{3R^2}$. The remaining stiffnesses are given by

$$A_h^{2211} = A_h^{1122}, \quad A_h^{1221} = A_h^{2121} = A_h^{2112} = A_h^{1212},$$
$$F_h^{2112} = F_h^{2121} = F_h^{1221} = F_h^{1212}, \quad F_h^{\alpha\alpha\beta\beta} = 0. \qquad (17.3.21)$$

In the case of h being constant the formulae for A_h^{1212} and F_h^{1212} simplify to

$$A_h^{1212} = \langle(A^{1212})^{-1}\rangle^{-1} + \frac{\sigma}{1+\sigma}(\langle A^{1212}\rangle - \langle(A^{1212})^{-1}\rangle^{-1}),$$
$$F_h^{1212} = -\frac{\sigma R}{2(1+\sigma)}(\langle A^{1212}\rangle - \langle(A^{1212})^{-1}\rangle^{-1}). \qquad (17.3.22)$$

Note that the formulae for $A_h^{\alpha\alpha\beta\beta}$ coincide with those for periodic plates, found in Sec. 3.7. The membrane – bending coupling does not occur, but we note *the membrane shearing – torsion coupling*. This coupling tends to zero when $h/R \to 0$.

b) *Solution to the* $(\widetilde{P}_Y^2)$*-problem. Finding the stiffnesses* $E_h^{\alpha\beta\lambda\mu}$ *and* $D_h^{\alpha\beta\lambda\mu}$
The functions $\Phi_\sigma^{(\lambda\mu)}(y_2)$ and $\chi^{(\lambda\mu)}(y_2)$ that satisfy the following differential equations are unknown:

$$\overset{(\lambda\mu)}{N_2}{}^{\gamma 2}{}_{|2} = 0, \qquad \overset{(\lambda\mu)}{M_2}{}^{22}{}_{|22} = 0, \qquad \gamma = 1, 2, \qquad (17.3.23)$$

where

$$\overset{(\lambda\mu)}{N_2}{}^{12} = B^{1212}\overset{(\lambda\mu)}{\Phi_1}{}_{|2} + E^{12\lambda\mu}, \qquad \overset{(\lambda\mu)}{N_2}{}^{22} = B^{2222}\overset{(\lambda\mu)}{\Phi_2}{}_{|2}, \qquad (17.3.24)$$

$$\overset{(\lambda\mu)}{M_2}{}^{22} = -D^{2222}\chi^{(\lambda\mu)}{}_{|22} + D^{22\lambda\mu}.$$

The periodicity condition (17.3.16) holds for g representing: $\Phi_\alpha^{(\lambda\mu)}$, $\chi^{(\lambda\mu)}$, $\chi^{(\lambda\mu)}{}_{|2}$, $\overset{(\lambda\mu)}{N_2}{}^{\sigma 2}$, $\overset{(\lambda\mu)}{M_2}{}^{22}$, and $\overset{(\lambda\mu)}{M_2}{}^{22}{}_{|2}$. Proceeding similarly to the previous case we find the gradients of the unknown functions

$$\Phi_{1|2}^{(\lambda\mu)} = \frac{1}{B^{1212}}\left[\langle(B^{1212})^{-1}\rangle^{-1}\langle\frac{E^{12\lambda\mu}}{B^{1212}}\rangle - E^{12\lambda\mu}\right],$$

$$\Phi_{2|2}^{(\lambda\mu)} = 0, \qquad (17.3.25)$$

$$\chi_{|22}^{(\lambda\mu)} = -\frac{1}{D^{2222}}\left[\langle(D^{2222})^{-1}\rangle^{-1}\langle\frac{D^{22\lambda\mu}}{D^{2222}}\rangle - D^{22\lambda\mu}\right].$$

Substitution into (17.1.29) gives

$$E_h^{\alpha\beta\lambda\mu} = F_h^{\lambda\mu\alpha\beta} \, ,$$

$$D_h^{1111} = \langle D^{1111} - \frac{(D^{1122})^2}{D^{2222}}\rangle + \langle \frac{D^{1122}}{D^{2222}}\rangle^2 \langle (D^{2222})^{-1}\rangle^{-1} \, ,$$

$$D_h^{1122} = \langle (D^{2222})^{-1}\rangle^{-1} \langle \frac{D^{1122}}{D^{2222}}\rangle \, ,$$

$$D_h^{2211} = D_h^{1122} \, , \quad D_h^{2222} = \langle (D^{2222})^{-1}\rangle^{-1} \, ,$$

$$D_h^{1212} = \langle \frac{1}{1+\sigma} D^{1212}\rangle + \langle \frac{\sigma}{1+\sigma}\rangle^2 \langle \frac{\sigma}{1+\sigma}\frac{1}{D^{1212}}\rangle^{-1} \, ,$$

$$D_h^{1221} = D_h^{2112} = D_h^{2121} = D_h^{1212} \, .$$

$$(17.3.26)$$

If $h = \text{const}$, then

$$D_h^{1212} = \langle D^{1212}\rangle - \frac{\sigma}{(1+\sigma)}(\langle D^{1212}\rangle - \langle (D^{1212})^{-1}\rangle^{-1}) \, . \qquad (17.3.27)$$

Thus in the case of $h =$const we have

$$A_h^{1212} \geq \langle (A^{1212})^{-1}\rangle^{-1} \, , \qquad D_h^{1212} \leq \langle D^{1212}\rangle \, . \qquad (17.3.28)$$

This means that the curving of the basic cell increases the effective shearing stiffness and decreases the effective torsional stiffness. If $h/R \to 0$, then the effective stiffnesses of the cylindrical shell tend to the values of the effective stiffnesses of the plate, see Sec. 3.7. Note that only the effective stiffnesses of indices (1212) depend upon the ratio h/R.

17.4. Shallow shells of periodic structure: effective properties

Only in rare cases the whole shell structure can be viewed as a shell obeying the Mushtari-Donnell-Vlasov shallow shell approximation, see Sec. 16.5. However, this approximation can be applied to some regions in which bending prevails. A possible periodic structure of such a shell region may be smeared-out with using less complicated formulae. Their derivation is the aim of the present section.

Assume that the membrane and bending stiffnesses are given by Eq. (17.1.1), where $A^{\alpha\beta\lambda\mu}(\xi, \cdot)$ and $D^{\alpha\beta\lambda\mu}(\xi, \cdot)$ are Y-periodic and of class $L^\infty(Y)$. According to (16.5.1) and (16.5.2) the constitutive relations assume the form

$$N_\varepsilon^{\alpha\beta} = A^{\alpha\beta\lambda\mu}(\xi, \xi/\varepsilon)\epsilon_{\lambda\mu}(\boldsymbol{u}^\varepsilon, w^\varepsilon) \, , \qquad M_\varepsilon^{\alpha\beta} = D^{\alpha\beta\lambda\mu}(\xi, \xi/\varepsilon)\kappa_{\lambda\mu}(w^\varepsilon) \, . \qquad (17.4.1)$$

Note that now the measures of change of curvature do not depend on the field $\boldsymbol{u}^\varepsilon$.

Assume that the shell is clamped along the boundary and that the loads q^α, q are ε-independent. The bilinear form (16.5.2) depends now on ε and has the following form

$$a_1^\varepsilon(\boldsymbol{u}^\varepsilon, w^\varepsilon; \boldsymbol{v}, v) = \int_\Omega [N_\varepsilon^{\alpha\beta}\epsilon_{\alpha\beta}(\boldsymbol{v}, v) + M_\varepsilon^{\alpha\beta}\kappa_{\alpha\beta}(v)]\sqrt{a}d\xi \, , \qquad (17.4.2)$$

and the equilibrium problem has the form (P_ε) (see Eq. (17.1.6)) with the form a^ε replaced by a_1^ε. This problem will be called $(P_{1\varepsilon})$. The solution to this problem is represented by (17.1.7). The homogenization results follow from those of Sec. 17.1 by crossing out appropriate terms depending on u^ε.

The homogenized constitutive relations turn out to be decoupled

$$N_h^{\alpha\beta} = A_h^{\alpha\beta\lambda\mu}\epsilon_{\lambda\mu}^h \,, \qquad M_h^{\alpha\beta} = D_h^{\alpha\beta\lambda\mu}\kappa_{\lambda\mu}^h \,, \tag{17.4.3}$$

where $\epsilon_{\alpha\beta}^h = \epsilon_{\alpha\beta}(u^0, w^0)$, $\kappa_{\alpha\beta}^h = \kappa_{\lambda\mu}(w^0)$.

The effective stiffnesses are given by

$$A_h^{\alpha\beta\delta\gamma} = \langle A^{\alpha\beta\lambda\mu}(\delta_\lambda^\gamma\delta_\mu^\delta + \psi_{\lambda|\mu}^{(\delta\gamma)})\rangle \,, \qquad D_h^{\alpha\beta\delta\gamma} = \langle D^{\alpha\beta\lambda\mu}(\delta_\lambda^\gamma\delta_\mu^\delta - \chi_{|\lambda\mu}^{(\gamma\delta)})\rangle \,, \tag{17.4.4}$$

where the auxiliary functions $\psi^{(\delta\gamma)}, \chi^{(\delta\gamma)}$ are solutions to the basic cell problems:

$$(P_{loc}^1) \quad \left|\begin{array}{l} \text{find } \psi^{(\gamma\delta)}(\xi, \cdot) \in H_{per}^1(Y)^2 \text{ such that} \\[2mm] \displaystyle\int_Y A^{\alpha\beta\sigma\mu}(\xi, y)(\delta_\sigma^\gamma\delta_\mu^\delta + \psi_{\sigma|\mu}^{(\delta\gamma)})v_{\alpha|\beta}dy = 0 \qquad \forall\, v \in H_{per}^1(Y)^2 \,; \end{array}\right.$$

$$(P_{loc}^2) \quad \left|\begin{array}{l} \text{find } \chi^{(\gamma\delta)}(\xi, \cdot) \in H_{per}^2(Y) \text{ such that} \\[2mm] \displaystyle\int_Y D^{\alpha\beta\lambda\mu}(\xi, y)(\delta_\lambda^\gamma\delta_\mu^\delta - \chi_{|\lambda\mu}^{\gamma\delta})v_{|\alpha\beta}dy = 0 \qquad \forall\, v \in H_{per}^2(Y) \,. \end{array}\right.$$

The homogenized equilibrium problem amounts to finding $(u^0, w^0) \in V^0(\Omega)$ such that

$$(P_h) \quad \left|\int_\Omega [N_h^{\alpha\beta}\epsilon_{\alpha\beta}(v, v) + M_h^{\alpha\beta}\kappa_{\alpha\beta}(v)]\sqrt{a}d\xi = f(v, v) \qquad \forall (v, v) \in V^0(\Omega)\right.$$

where $f(\cdot, \cdot)$ is given by (16.2.17); the constitutive relations have the form (17.4.3).

The homogenized potential

$$\mathcal{W}_h = \frac{1}{2}(N_h^{\alpha\beta}\epsilon_{\alpha\beta}^h + M_h^{\alpha\beta}\kappa_{\alpha\beta}^h) \,, \tag{17.4.5}$$

can be put in the form similar to (17.1.50):

$$\mathcal{W}_h\,(\xi, \epsilon^h, \kappa^h) = \frac{1}{2}\inf\{\langle A^{\alpha\beta\lambda\mu}(\xi, y)[e_{\alpha\beta}^y(v) + \epsilon_{\alpha\beta}^h][e_{\lambda\mu}^y(v) + \epsilon_{\lambda\mu}^h]$$
$$+\; D^{\alpha\beta\lambda\mu}(\xi, y)[\kappa_{\alpha\beta}^y(v) + \kappa_{\alpha\beta}^h][\kappa_{\lambda\mu}^y(v) + \kappa_{\lambda\mu}^h]\rangle \mid (v, v) \in H_{K,per}(Y)\} \,, \tag{17.4.6}$$

for $\epsilon^h, \kappa^h \in \mathbb{E}_s^2$.

Thus the homogenization formulae are similar to those derived for periodic plates, see Sec. 3.2. The curvatures of the shell do not affect the formulae for the tensors A_h and D_h. The reciprocal stiffnesses do not occur.

18. Homogenized properties of thin periodic elastic shells undergoing moderately large rotations around tangents

As we already know, this model is characterized by the strain measures $(16.2.7)_2$ and $(16.4.4)$, where $\epsilon_{\alpha\beta}(u, w)$ is defined by $(16.2.7)_1$. The homogenization problem is formulated similarly to the case of Koiter's shell model. Thus for a fixed $\varepsilon > 0$ the functional of the total potential energy J_ε is given by

$$J_\varepsilon(u, w) = \int_\Omega \mathcal{W}(\xi, \xi/\varepsilon, \gamma(u, w), \rho(u, w))\sqrt{a}\,d\xi - f(u, w) , \qquad (18.1.1)$$

where $(u, w) \in V^0(\Omega)$ and the loading functional f is defined by $(16.2.17)$.

The stored energy function $\mathcal{W}$ satisfies the assumptions specified in Secs. 17.1.1 and 17.1.2.

Applying Theorem 1.3.28 we conclude that the $\Gamma(\tau)$-limit of the sequence of functionals specified by (18.1) has the following form

$$J_h(u, w) = \int_\Omega \mathcal{W}_h(\xi, \gamma(u, w), \rho(u, w))\sqrt{a}\,d\xi - f(u, w) . \qquad (18.1.2)$$

Here $\tau = s - [L^2(\Omega)^2 \times H^1(\Omega)]$ and the homogenized potential is given by

$$\mathcal{W}_h(\xi, \gamma^h, \rho^h) = \frac{1}{2|Y|} \inf\{\int_Y [A^{\alpha\beta\lambda\mu}(\xi, y)(e^y_{\alpha\beta}(v) + \gamma^h_{\alpha\beta})(e^y_{\lambda\mu}(v) + \gamma^h_{\lambda\mu})$$

$$+ D^{\alpha\beta\lambda\mu}(\xi, y)(\kappa^y_{\alpha\beta}(v) - 2b^\rho_\alpha(\xi)\frac{\partial v_\rho}{\partial y_\beta} + \rho^h_{\alpha\beta})(\kappa^y_{\lambda\mu}(v) - 2b^\delta_\lambda \frac{\partial v_\delta}{\partial y_\mu} + \rho^h_{\lambda\mu})]dy$$

$$|\,(v, v) \in H^1_{per}(Y)^2 \times H^2_{per}(Y)\} , \qquad (18.1.3)$$

where $\gamma^h, \rho^h \in \mathbb{E}^2_s$.

The homogenized constitutive equations are

$$N^{\alpha\beta}_h = \partial\mathcal{W}_h/\partial\gamma^h_{\alpha\beta} , \qquad M^{\alpha\beta}_h = \partial\mathcal{W}_h/\partial\rho^h_{\alpha\beta} . \qquad (18.1.4)$$

It is clear that they assume the form similar to $(17.1.25)$ with ϵ^h replaced by γ^h. These equations are in general coupled.

Uncoupling occurs for the Mushtari-Marguerre shell model. In the last case the effective tensors A_h and D_h are determined similarly to the case of thin plate homogenization, cf. Eqs. $(3.2.32)$ and the local problems $(P^\sigma_{KS,Y})$, $\sigma = 1, 2$, as well as Sec. 17.4.

19. Perfectly plastic shells

The goal of this section is to examine Koiter's shells made of a Hencky material. The strain measures are given by (16.2.7) and the periodic structure has been described in Sec. 17. Let $\mathcal{C}(\xi, y) \subset \mathbb{E}_2^s \times \mathbb{E}_2^s$ be a convex bounded and closed set such that $(0,0) \in int\, \mathcal{C}(\xi, y)$. We recall that $x^i = \Phi^i(\xi^\alpha)$, $i = 1,2,3$; $\alpha = 1,2$, and $y = \xi/\varepsilon$. Similarly to the case of plastic plates the set $\mathcal{C}(\xi, y)$ is generated by the yield condition. The density of the complementary energy of the elastic-perfectly plastic shell made of the Hencky material is assumed in the following form

$$j^*(\xi, y, \boldsymbol{N}, \boldsymbol{M}) = \mathcal{W}^*(\xi, y, \boldsymbol{N}, \boldsymbol{M}) + I_{\mathcal{C}(\xi,y)}(\boldsymbol{N}, \boldsymbol{M}) \,, \tag{19.1}$$

where $\mathcal{W}^*$ stands for the density of the complementary elastic energy, cf. Sec. 17.2.

The constitutive equation is conveniently written in the subdifferential form

$$(\boldsymbol{\epsilon}, \boldsymbol{\rho}) \in \partial j(\xi, y, \boldsymbol{N}, \boldsymbol{M}) \,. \tag{19.2}$$

Here ∂ denotes the subdifferential of the function $j^*(\xi, y, \cdot, \cdot)$.

The density of the elasto-plastic potential is given by, cf. Sec. 13.1

$$\begin{aligned} j^*(\xi, y, \boldsymbol{\epsilon}, \boldsymbol{\rho}) &= j^{**}(\xi, y, \boldsymbol{\epsilon}, \boldsymbol{\rho}) \\ &= \sup\{N^{\alpha\beta}\epsilon_{\alpha\beta} + M^{\alpha\beta}\rho_{\alpha\beta} - j^*(\xi, y, \boldsymbol{N}, \boldsymbol{M}) \mid (\boldsymbol{N}, \boldsymbol{M}) \in \mathbb{E}_2^s \times \mathbb{E}_2^s\} \,, \end{aligned} \tag{19.3}$$

for each $\boldsymbol{\epsilon}, \boldsymbol{\rho} \in \mathbb{E}_s^2$.

The inverse constitutive relationship is written as follows

$$(\boldsymbol{N}, \boldsymbol{M}) \in \partial j(\xi, y, \boldsymbol{\epsilon}, \boldsymbol{\rho}) \,. \tag{19.4}$$

The recession function $j_\infty(\xi, y, \cdot, \cdot)$ of $\mathcal{C}(\xi, y)$ is calculated as follows

$$j_\infty(\xi, y, \boldsymbol{\epsilon}, \boldsymbol{\rho}) = \sup\{N^{\alpha\beta}\epsilon_{\alpha\beta} + M^{\alpha\beta}\rho_{\alpha\beta} \mid (\boldsymbol{N}, \boldsymbol{M}) \in \mathcal{C}(\xi, y)\} \,. \tag{19.5}$$

Properties of j and j_∞

There exist $R > r > 0$ and $k > 0$ such that

(i) $r(|\boldsymbol{\epsilon}| + |\boldsymbol{\rho}|) - k \leq j(\xi, y, \boldsymbol{\epsilon}, \boldsymbol{\rho}) \leq R(|\boldsymbol{\epsilon}| + |\boldsymbol{\rho}|)$,

(ii) $r(|\boldsymbol{\epsilon}| + |\boldsymbol{\rho}|) \leq j_\infty(\xi, y, \boldsymbol{\epsilon}, \boldsymbol{\rho}) \leq R(|\boldsymbol{\epsilon}| + |\boldsymbol{\rho}|)$,

(iii) $j_\infty^*(\xi, y, \boldsymbol{N}, \boldsymbol{M}) = I_{\mathcal{C}(\xi,y)}(\boldsymbol{N}, \boldsymbol{M})$,

for $\boldsymbol{\epsilon}, \boldsymbol{\rho} \in \mathbb{E}_s^2$, $\xi \in \Omega$ and $y \in Y$.

Let us assume that the shell is clamped on $\partial S = \boldsymbol{\Phi}(\Gamma)$, $\Gamma = \partial\Omega$. The loading functional is given by (16.2.17). We assume that in the homogenization procedure it plays the role of a continuous perturbation. The kinematical or displacement formulation of the equilibrium of the shell considered is formulated as a minimization problem.

Problem P_λ^ε

    Find

$$\inf\{\int_\Omega j[\xi, \xi/\varepsilon, \boldsymbol{\epsilon}(\boldsymbol{u}, w), \boldsymbol{\rho}(\boldsymbol{u}, w)]\sqrt{a}d\xi - \lambda f(\boldsymbol{u}, w) \mid (\boldsymbol{u}, w) \in LD(\Omega) \times W^{2,1}(\Omega),$$

$$\boldsymbol{u} = 0, \ w = 0, \ \frac{\partial w}{\partial \boldsymbol{n}} = 0 \quad \text{on} \quad \Gamma\}.$$

Here $\lambda \geq 0$ is the load multiplier. The limit analysis problem is formulated in the usual way.

Problem P_{LA}^ε

    Find

$$\lambda^\varepsilon = \inf\{\int_\Omega j_\infty[\xi, \xi/\varepsilon, \boldsymbol{\epsilon}(\dot{\boldsymbol{u}}, \dot{w}), \boldsymbol{\rho}(\dot{\boldsymbol{u}}, \dot{w})]\sqrt{a}d\xi \mid (\dot{\boldsymbol{u}}, \dot{w}) \in LD(\Omega) \times W^{2,1}(\Omega),$$

$$f(\dot{\boldsymbol{u}}, \dot{w}) = 1; \ \dot{\boldsymbol{u}} = 0, \ \dot{w} = 0, \ \frac{\partial \dot{w}}{\partial \boldsymbol{n}} = 0 \quad \text{on} \quad \Gamma\}.$$

The existence of solutions to problems (P_λ^ε) and (P_{LA}^ε) is discussed in Bojarski and Telega (1994). As in the case of plastic plates, these problems are solvable in the space $BD(\Omega) \times HB(\Omega)$. More precisely, the infima of the corresponding relaxed functionals are attained just in $BD(\Omega) \times HB(\Omega)$. Obviously, the function $j[\xi, \xi/\varepsilon, \boldsymbol{\epsilon}(\boldsymbol{u}, w), \boldsymbol{\rho}(\boldsymbol{u}, w)]$, $\boldsymbol{u} \in BD(\Omega), w \in HB(\Omega)$, is to be treated as a function of measures and $\lambda < \lambda^\varepsilon$.

We observe that in the general case if $\boldsymbol{u} \in BD(\Omega)$, $w \in HB(\Omega)$, then $\boldsymbol{\rho}(\boldsymbol{u}, w)$ is not necessarily a bounded measure. There two simple cases ensuring that ρ is a bounded measure: (i) $\boldsymbol{\rho}(\boldsymbol{u}, w) = -w_{\|\alpha\beta}$, (ii) $b_2^1(\xi) = b_1^2(\xi) = 0$ and $b_1^1(\xi) = b_2^2(\xi)$ for each $\xi \in \Omega$. In the first case the shell is shallow whilst in the second case the shell is a part of a sphere. Another way of ensuring that $\boldsymbol{\rho}(\boldsymbol{u}, w)$ would be a bounded measure is to assume that for $\boldsymbol{u} \in BD(\Omega)$, $(b_\alpha^\lambda u_\lambda) \in BD(\Omega)$ also. This problem deserves deeper examination. Anyway, in the homogenization study which follows, $\boldsymbol{\rho}(\boldsymbol{u}, w)$ is assumed to be a bounded measure, when needed.

Let us pass now with ε to zero. We find the $\Gamma(L^p(\Omega)^2 \times W^{1,1}(\Omega))$-limit problems, where $1 \leq p < 2$.

Problem P_λ^h

    Find

$$\inf\{\int_\Omega j_h[\xi, \boldsymbol{\epsilon}(\boldsymbol{u}, w), \boldsymbol{\rho}(\boldsymbol{u}, w)] + \int_\Gamma j_{h\infty}[\xi, \mathbf{T}(-\boldsymbol{u}), \mathbf{F}\left(-\frac{\partial w}{\partial \boldsymbol{n}}\right)]a_1 d\Gamma$$

$$-\lambda f(\boldsymbol{u}, w) \mid (\boldsymbol{u}, w), \in BD(\Omega) \times HB(\Omega), w = 0 \quad \text{on} \quad \Gamma\}.$$

Problem P_{LA}^h

Find

$$\lambda^h = \inf\{\int_\Omega j_{h\infty}[\xi, \epsilon(\dot{\boldsymbol{u}}, \dot{w}), \boldsymbol{\rho}(\dot{\boldsymbol{u}}, \dot{w})] + \int_\Gamma j_{h\infty}[\xi, \mathbf{T}(-\dot{\boldsymbol{u}}), \mathbf{F}\left(-\frac{\partial \dot{w}}{\partial \boldsymbol{n}}\right)]a_1 d\Gamma$$

$$| f(\dot{\boldsymbol{u}}, \dot{w}) = 1, \ (\dot{\boldsymbol{u}}, \dot{w}) \in BD(\Omega) \times HB(\Omega), \dot{w} = 0 \quad \text{on} \quad \Gamma\}\,.$$

Here

$$a_1 = \sqrt{\left(\frac{dx^1}{ds}\right)^2 + \left(\frac{dx^2}{ds}\right)^2 + \left(\frac{dx^3}{ds}\right)^2}\,,$$

and s represents a parametrization of Γ. The homogenized potential is given by

$$j_h(\xi, \epsilon^h, \boldsymbol{\rho}^h) = \inf\{\frac{1}{|Y|}\int_Y j[\xi, y, e_{\alpha\beta}^y(\boldsymbol{v}) + \epsilon_{\alpha\beta}^h, \kappa_{\alpha\beta}^y(v) - b_\alpha^\lambda\frac{\partial v_\lambda}{\partial y_\beta} - b_\beta^\lambda\frac{\partial v_\lambda}{\partial y_\alpha} + \rho_{\alpha\beta}^h]dy$$

$$|(\boldsymbol{v}, v) \in LD_{per}(Y) \times W_{per}^{2,1}(Y)\}\,, \tag{19.6}$$

where $\epsilon^h, \boldsymbol{\rho}^h \in \mathbb{E}_s^2$ and $\xi \in \Omega$. Obviously we have

$$\lambda^h = \lim_{\varepsilon \to} \lambda^\varepsilon\,.$$

Remark 19.1. One can also consider dual formulations and shells loaded on the boundary, cf. Bojarski and Telega (1994) and Sec. 14.1.

                                          Elastic and plastic shells

20. Comments and bibliographical notes

The first complete set of equations of linear vibrations of thin elastic shells were derived by Love (1889). However, these equations were not mathematically correct. Novozhilov was the first who proposed a linear theory free of any inherent errors, but restricted to the special parametrization determined by the lines of principal curvatures, see Novozhilov (1951), cf. also Leissa (1969, 1993, 1998). In 1960 Koiter proved that there is no unique version of the first order linear shell theory. He showed how to construct energy equivalent shell models that satisfy all requirements of mechanical and mathematical correctness. The next step was done by Budiansky and Sanders (1963) who restricted this class to those shell models in which the static-geometric analogy (discovered by Goldenveizer) takes place. They noticed that among all such shell models only one version uses the strain measures which for shells of revolution reduce to those forms which are widely accepted. Just this shell model was called "the best"; it was derived in Sec. 16.3. For theoretical consideration some authors prefer an energy equivalent shell model, called in Sec. 16 the Koiter shell model. In this model the formulae for strain measures are shorter than in the "best" version, but the static-geometric analogy does not hold. A rigorous mathematical proof of correctness of the Koiter model was given by Bernadou and Ciarlet (1976), cf. also Bernadou et al. (1994), Bernadou et al. (1997), and Ciarlet and Sanchez-Palencia (1996).

The contemporary literature on shell modelling is ample but, in our opinion, the article by Naghdi (1963) is conspicuous due to its rigor, lucidity and elegancy of presentation, cf. also Destuynder (1985) and Lewiński (1980). This article starts with the shell theory with transverse shear deformation, which was reported briefly in Sec. 16.1. Further shell models can be derived by restricting the class of admissible deformations. The nonlinear shell models of Secs. 16.4 and 16.5 are reported after Pietraszkiewicz (1979).

Periodicity of thin shells can be viewed as periodicity with respect to the curvilinear parametrization of its middle surface, cf. also Lutoborski and Telega (1982, 1984). This concept was used by Lewiński and Telega (1988a), where the effective shell model was derived by using the two-scale asymptotic technique and justified in Lewiński and Telega (1991c) and Telega and Lewiński (1998a) by the Γ-convergence method. This effective shell model (see Sec. 17.1) involves displacements as independent (primal) variables. The effective problem can be rearranged to the dual form involving stress and couple resultants as unknowns. The dual formulae of Sec. 17.2 were derived in Telega and Lewiński (1998b).

The effective properties of ribbed cylindrical shells given in Sec. 17.3 were found by Lewiński (1991f).

The homogenization results of Sec. 18 concerning shells undergoing moderately large deflections were published for the first time in Lewiński and Telega (1991c).

In the homogenization method used in Sec. 17 the two-dimensional thin shell model of Koiter is treated as a departure point. Assuming that the curvature tensor $b = (b_{\alpha\beta})$ vanishes we come back to the model of Duvaut and Metellus (Sec. 3) for thin plates. As explained in Sec. 5 this model applies to the case when the periodicity cells have a shape of a thin plate. By analogy, the homogenized model of Sec. 17 applies to the case when periodicity cells are thin shells themselves. If we apply the homogenization technique to the solutions of moderately thick shells (Sec. 16.1) we find the homogenization formulae similar to those derived in Sec. 5.2 for plates. This derivation is feasible but is not included in this volume.

The most general averaging should start from the three-dimensional setting. Such averaging was performed for the first time by Kalamkarov et al. (1987) by the two-scale expansion method, see also Kalamkarov (1992, Chap. 5). This derivation has been based on the assumption of the radii of curvature being much greater than the shell thickness, which has led to the homogenized and linearized shell model of Mushtari-Donnell-Vlasov type (see Sec. 16.5). Consequently, the transverse shear forces do not enter the two first equilibrium equations (see Eqs. (15.89) in Kalamkarov (1992)) and the tensor of changes of curvature does not depend on the tangent displacements (see Eqs. (15.89) and (15.49) in Kalamkarov (1992)). Thus one can conjecture that the formulae of Kalamkarov apply to periodic shells of slowly varying curvatures.

For some specific geometries of the basic cell of periodicity the algorithm of Kalamkarov leads to the closed approximate formulae for effective stiffnesses of composite or reinforced shells. Since these formulae depend explicitly on the geometrical characteristics of the basic cells they turn out to be useful in solving the optimum design problems for such shells, see Kalamkarov and Kolpakov (1996, 1997).

A derivation of the homogenization formulae of larger range of application is up till now unknown. The reason is that such derivation would combine difficulties of homogenization and reduction of the transverse dimension of a shell. In the case of plates one knows how to derive the thin plate equations by a systematic and variationally consistent method, see Ciarlet and Destuynder (1979). In the case of shells one should not expect such unique asymptotic results since any complete shell theory involves two stiffness tensors of various units, which reflects the fact that such a theory is spanned between the membrane theory and the pure flexure theory and, as such, cannot be justified by one asymptotic process, see Ciarlet and Lods (1996a, 1996b, 1996c), Busse et al. (1997), Genevey (1996a, 1996b), Miara and Sanchez-Palencia (1998). Consequently one should not expect that homogenization of three-dimensional shell equations can lead us smoothly just to the Koiter shell equations with uniquely determined effective stiffnesses. We observe, however, that Genevey (1996a, 1998) applied Γ-convergence theory to a justification of both the membrane and flexural theories of linear isotropic elastic shells. The same method was used by Le Dret and Raoult (1996) and Genevey (1997) to justify of a geometrically nonlinear membrane model, cf. also Collard and Miara (1997), Lods and Miara (1998), Miara (1998). As was already mentioned the problem of combination of rigorous justification with homogenization remains open.

Section 19 presents first results on two-dimensional homogenization of perfectly plastic shells made of a Hencky material. The relaxation of the functionals involved follows the paper by Bojarski and Telega (1994).

Chapter VI

APPLICATION OF HOMOGENIZATION METHODS IN OPTIMUM DESIGN OF PLATES AND SHELLS

Introduction

One should clearly distinguish between the optimization problems of finite number of design variables and those in which design variables are functions. The former problems arise upon discretization. Having a truss or a frame we deal with a finite number of unknowns and design variables. By the nonlinear programming methods such problems can be solved, provided that the number of unknowns is not too great; nowadays it should be less than one thousand. In practice, even the case of one hundred unknowns can be almost unsolvable, if the chosen merit function has a lot of local extrema.

On the other hand, the interest of researchers has been focused on shape optimization problems. One of the most interesting problem is how to place some given materials into a fixed domain to optimize a global behavior of the body. If augmented with an isoperimetric condition like that of a given volume, such problems sound reasonably. However, it turns out that partitioning of the materials usually makes the design better. And there is no barrier for such partitioning. To achieve the best design we are compelled to admit an infinitely dense partitioning, which means admitting composite domains in which the original materials enter at the microlevel.

The elastic characteristics of these composite materials are new design variables. Instead of looking for the optimum shape of the constituents we shall look for an optimum layout of microstructural parameters. The overall elastic characteristics of the composite domains are determined by the formulae of homogenization. In this manner the homogenization theory enters into the realm of the layout optimization problems. One of the aims of the present chapter is to show this application of the homogenization method by the example of the minimum compliance problem of thin elastic plates, see Sec. 26. To understand this topic one should start with studying the optimal lower and upper bounds of the plate energy. They are derived by the translation method. The translation method is also a tool which makes it possible to derive the Hashin-Shtrikman bounds for two-phase plates. The derivation is nevertheless far from being automatic. To derive the optimal bounds for the Kirchhoff modulus one should handle two strain or stress fields and estimate the sum of energies, see Secs. 22.3, 22.4 and 23.2. The bounds found in this manner are attainable, which can be proved by checking that the isotropic plates constructed by three subsequent "layerings" have the moduli that coincide with these bounds, see Secs. 22.5 and 23.3.

If the merit function is the weight of the structure then optimal layouts consist of infinite number of infinitely thin bars in tension or compression, see Fig. 29.2. The optimal features of such unusual structures – which are neither trusses nor plates – were discovered in 1904 by Michell. Their theory is yet not developed. An intuitive introduction to the theory of these structures is given in Sec. 29, the introduction being given in the spirit of the arguments of Strang and Kohn (1983).

The difficulties arise if in the layout problem the properties of the phases can vary within certain limits, like in the problem of a plate of varying thickness. To overcome mathematical difficulties arising due to this arbitrariness one resorts to the techniques of the Young measures, cf. Secs. 21.6 and 27.

The layout theory of shells is less developed. Even in the case of the minimum compliance problem we do not know the microstructures that realize the optimum. Nevertheless we can formulate the relaxed problem in terms of the dual variables and then rearrange it to the primal formulation involving the displacement fields, see Sec. 28.

21. Mathematical complements

Optimization methods require using some peculiar tools of description and analysis. First, it turns out that instead of the standard bases $e_\alpha \otimes e_\beta$, $e_\alpha \otimes e_\beta \otimes e_\lambda \otimes e_\mu$ for second and fourth order tensors it is frequently helpful to introduce orthonormal vectorial bases. This formal change of the reference is considered in Sec. 21.1.

While estimating energy by the so-called translation method we come across the matrix inequality

$$D_0 \geq \langle (D - T)^{-1} \rangle^{-1} + T \,.$$

In Sec. 21.2 we show that this inequality is equivalent to

$$Y(D_0) + T \geq 0 \,,$$

which defines Y-transformation of D_0. Its properties are explained in the same section.

Some useful integral inequalities follow from the quasiaffine and quasiconvex properties of particular functions. They are reported in Sec. 21.4. To clear up some proofs we precede this section with Sec. 21.3 in which basic informations on the Fourier representations are summarized. Section 21.5 is devoted to estimating the effective energy, stored in a periodicity cell, from below. A proof is given that the harmonic mean plays the role of this estimate. Section 21.6 introduces the reader to Young measures, often called parametrized measures. Our presentation will be oriented towards application to optimal design of structures.

21.1. Alternative representation of second and fourth order tensors

Two-dimensional second order tensors are usually referred to the basis $e_\alpha \otimes e_\beta$, where (e_1, e_2) represent versors of the Cartesian orthogonal coordinate system. Thus for any

$\kappa \in \mathbb{E}^2$ we have

$$\kappa = \sum_{\alpha,\beta=1}^{2} \kappa_{\alpha\beta} e_\alpha \otimes e_\beta . \tag{21.1.1}$$

Let us introduce the tensors

$$\mathbf{a}_1 = \frac{1}{\sqrt{2}}(e_1 \otimes e_1 + e_2 \otimes e_2) , \qquad \mathbf{a}_2 = \frac{1}{\sqrt{2}}(e_1 \otimes e_1 - e_2 \otimes e_2) ,$$

$$\mathbf{a}_3 = \frac{1}{\sqrt{2}}(e_1 \otimes e_2 + e_2 \otimes e_1) , \qquad \mathbf{a}_4 = \frac{1}{\sqrt{2}}(e_1 \otimes e_2 - e_2 \otimes e_1) . \tag{21.1.2}$$

The scalar product of two tensors

$$\mathbf{p} = p^{\alpha\beta} e_\alpha \otimes e_\beta , \qquad \mathbf{q} = q^{\alpha\beta} e_\alpha \otimes e_\beta \tag{21.1.3}$$

is defined by

$$\mathbf{p} : \mathbf{q} = p^{\alpha\beta} q_{\alpha\beta} . \tag{21.1.4}$$

One can easily prove that

$$\mathbf{a}_I \cdot \mathbf{a}_J = \delta_{IJ} , \qquad I, J \in \{1,2,3,4\}$$

and we say that the basis $\mathbf{a}_I$ is orthonormal. Each tensor $\kappa \in \mathbb{E}^2$ can be decomposed in the basis $(\mathbf{a}_I)$ as follows

$$\kappa = \sum_{J=1}^{4} \kappa^J \mathbf{a}_J . \tag{21.1.5}$$

If κ is symmetric, then $\kappa^4 = 0$ and we shall write $\kappa \in \mathbb{E}_s^2$. We then have

$$\kappa = \sum_{i=1}^{3} \kappa^i \mathbf{a}_i , \tag{21.1.6}$$

where

$$\kappa^1 = \frac{1}{\sqrt{2}}(\kappa_{11} + \kappa_{22}) , \qquad \kappa^2 = \frac{1}{\sqrt{2}}(\kappa_{11} - \kappa_{22}) ,$$

$$\kappa^3 = \frac{1}{\sqrt{2}}(\kappa_{12} + \kappa_{21}) = \sqrt{2}\kappa_{12} . \tag{21.1.7}$$

Consequently

$$\kappa_{11} = \frac{1}{\sqrt{2}}(\kappa^1 + \kappa^2) , \qquad \kappa_{22} = \frac{1}{\sqrt{2}}(\kappa^1 - \kappa^2) ,$$

$$\kappa_{12} = \frac{1}{\sqrt{2}}\kappa^3 . \tag{21.1.8}$$

Let us introduce the convention that the indices i, j, k run over 1, 2, 3. The summation sign will be omitted, if the summation concerns indices at different levels.

Consider two orthonormal bases: (e_α) and $(\widetilde{e}_\alpha)$. According to Fig. 21.1.1 we have

$$e_1 = \cos\alpha\,\widetilde{e}_1 + \sin\alpha\,\widetilde{e}_2\,, \qquad e_2 = -\sin\alpha\,\widetilde{e}_1 + \cos\alpha\,\widetilde{e}_2\,. \qquad (21.1.9)$$

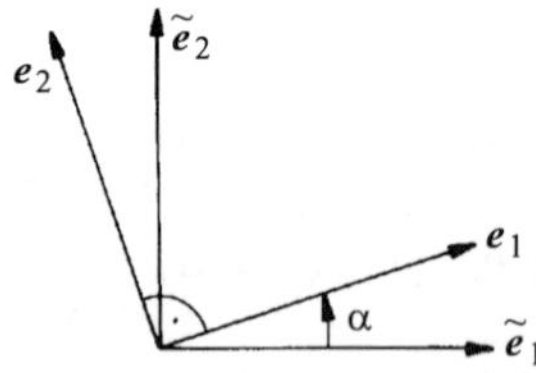

Fig. 21.1.1. Two orthogonal bases (e_α) and $(\widetilde{e}_\alpha)$

Let us associate the basis $(\widetilde{a}_i)$ with the basis $(\widetilde{e}_\alpha)$ by the rules (21.1.2), that is

$$\widetilde{a}_1 = \frac{1}{\sqrt{2}}(\widetilde{e}_1 \otimes \widetilde{e}_1 + \widetilde{e}_2 \otimes \widetilde{e}_2)\,, \qquad (21.1.10)$$

and the tensors $\widetilde{a}_I$, $I = 2, 3, 4$, are defined similarly. The bases (a_i) and $(\widetilde{a}_i)$ are linked by the transformation relation:

$$\begin{bmatrix} a_1 \\ a_2 \\ a_3 \end{bmatrix} = \begin{bmatrix} 1 & 0 & 0 \\ 0 & \cos 2\alpha & \sin 2\alpha \\ 0 & -\sin 2\alpha & \cos 2\alpha \end{bmatrix} \begin{bmatrix} \widetilde{a}_1 \\ \widetilde{a}_2 \\ \widetilde{a}_3 \end{bmatrix}\,. \qquad (21.1.11)$$

Let us decompose $\kappa \in \mathbb{E}_s^2$ in both bases:

$$\kappa = \kappa^i a_i\,, \qquad \kappa = \widetilde{\kappa}^i \widetilde{a}_i\,. \qquad (21.1.12)$$

Hence

$$\widetilde{\kappa}^1 = \kappa^1\,, \qquad \widetilde{\kappa}^2 = \cos 2\alpha\,\kappa^2 - \sin 2\alpha\,\kappa^3\,, \qquad \widetilde{\kappa}^3 = \sin 2\alpha\,\kappa^2 + \cos 2\alpha\,\kappa^3\,. \qquad (21.1.13)$$

Thus the tensor $\kappa \in \mathbb{E}_s^2$ can be viewed as a vector whose components are observed in the basis $(\widetilde{a}_i)$, see Fig. 21.1.2. The projection of κ on $\widetilde{a}_1$ gives the first invariant of κ

$$I(\kappa) = \kappa^1 = \frac{1}{\sqrt{2}}(\kappa_{11} + \kappa_{22}) = \frac{1}{\sqrt{2}} tr\kappa\,. \qquad (21.1.14)$$

The second invariant represents the length of the projection $\Pi\kappa$ of κ on the $(\widetilde{a}_2, \widetilde{a}_3)$ plane:

$$II(\kappa) = [(\widetilde{\kappa}^2)^2 + (\widetilde{\kappa}^3)^2]^{1/2}\,. \qquad (21.1.15)$$

Note that

$$II(\kappa) = \left\{ \frac{1}{2}[(\kappa_{11} - \kappa_{22})^2 + 4(\kappa_{12})^2] \right\}^{1/2} \qquad (21.1.16)$$

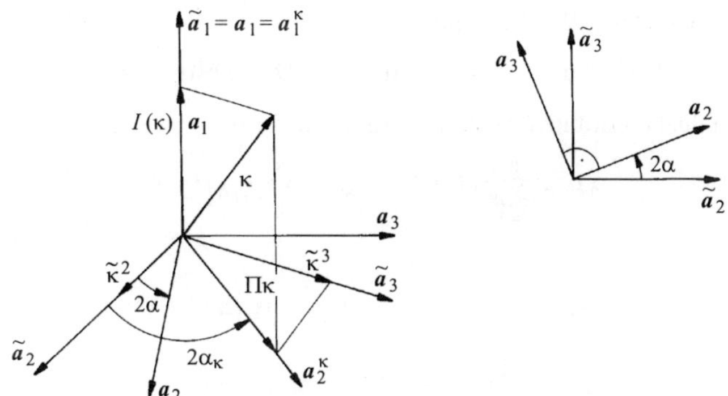

Fig. 21.1.2. Vectorial representation of a symmetric second order tensor κ. Note that
$$\Pi\kappa = II(\kappa)\mathbf{a}_2^\kappa$$

or

$$II(\kappa) = \frac{1}{\sqrt{2}}[(tr\kappa)^2 - 4\det\kappa]^{1/2} . \tag{21.1.17}$$

The determinant of κ : $\det\kappa = \kappa_{11}\kappa_{22} - (\kappa_{12})^2$ can be represented in the form

$$\det\kappa = -\frac{1}{2}\kappa^T \overset{0}{\boldsymbol{T}} \kappa , \tag{21.1.18}$$

where $\kappa = [\kappa^1, \kappa^2, \kappa^3]^T$ and $\overset{0}{\boldsymbol{T}}=\overset{0}{T}{}^{ij}\mathbf{a}_i \otimes \mathbf{a}_j$. The components $\overset{0}{T}{}^{ij}$ form the diagonal matrix:

$$[\overset{0}{T}{}^{ij}] = \text{diag}[-1, 1, 1] . \tag{21.1.19}$$

Note that $\overset{0}{\boldsymbol{T}}= -\boldsymbol{I}_1 + \boldsymbol{I}_2$, where the tensors $\boldsymbol{I}_\alpha$ are defined by (3.8.29).

In Eq. (21.1.1) κ represents the tensor and, at the right-hand side, the same notation is used for the column of numbers $(\kappa^1, \kappa^2, \kappa^3)$. This ambiguity in notation should not lead to misunderstandings.

The constitutive relations $\boldsymbol{M} = \boldsymbol{D}\kappa$ for an orthotropic thin plate have the form

$$\begin{aligned}
M^{11} &= D^{1111}\kappa_{11} + D^{1122}\kappa_{22} , \\
M^{22} &= D^{2211}\kappa_{11} + D^{2222}\kappa_{22} , \\
M^{12} &= 2D^{1212}\kappa_{12} ,
\end{aligned} \tag{21.1.20}$$

where

$$\boldsymbol{M} = M^{\alpha\beta}\boldsymbol{e}_\alpha \otimes \boldsymbol{e}_\beta , \qquad \kappa = \sum_{\alpha,\beta}\kappa_{\alpha\beta}\boldsymbol{e}_\alpha \otimes \boldsymbol{e}_\beta ,$$
$$\boldsymbol{D} = D^{\alpha\beta\lambda\mu}\boldsymbol{e}_\alpha \otimes \boldsymbol{e}_\beta \otimes \boldsymbol{e}_\lambda \otimes \boldsymbol{e}_\mu . \tag{21.1.21}$$

The versors e_α coincide with the orthotropy directions. Let us transform the relations (21.1.20) to the basis $(\mathbf{a}_i)$. We decompose

$$M = M^i \mathbf{a}_i , \qquad \boldsymbol{\kappa} = \kappa^i \mathbf{a}_i , \qquad D = D^{ij} \mathbf{a}_i \otimes \mathbf{a}_j . \qquad (21.1.22)$$

The primal and dual constitutive relations assume the form

$$M^i = \sum_{j=1}^{3} D^{ij} \kappa^j , \qquad \kappa^i = \sum_{j=1}^{3} d_{ij} M^j , \qquad (21.1.23)$$

where

$$D^{11} = \frac{1}{2}(D^{1111} + 2D^{1122} + D^{2222}) ,$$

$$D^{22} = \frac{1}{2}(D^{1111} - 2D^{1122} + D^{2222}) ,$$

$$D^{12} = \frac{1}{2}(D^{1111} - D^{2222}) , \quad D^{33} = 2D^{1212} ,$$

$$D^{13} = 0 , \quad D^{23} = 0 . \qquad (21.1.24)$$

The relations between d_{ij} and $d_{\alpha\beta\lambda\mu}$ are similar. If we refer to the basis $(\widetilde{e}_\alpha)$ the constitutive relations assume the form $\widetilde{M} = \widetilde{D}\widetilde{\kappa}$ with

$$\widetilde{D}^{11} = D^{11} ,$$

$$\widetilde{D}^{12} = \cos 2\alpha \, D^{12} , \qquad \widetilde{D}^{13} = \sin 2\alpha \, D^{12} ,$$

$$\widetilde{D}^{22} = \cos^2 2\alpha \, D^{22} + \sin^2 2\alpha \, D^{33} , \qquad (21.1.25)$$

$$\widetilde{D}^{23} = \sin 2\alpha \cos 2\alpha (D^{22} - D^{33}) ,$$

$$\widetilde{D}^{33} = \sin^2 2\alpha \, D^{22} + \cos^2 2\alpha \, D^{33} .$$

Consider now the case of isotropy. The representations of the tensors D and $d = D^{-1}$ have the form

$$D = 2k\boldsymbol{I}_1 + 2\mu\boldsymbol{I}_2 , \qquad d = \frac{1}{2}\mathcal{K}\boldsymbol{I}_1 + \frac{1}{2}\mathcal{L}\boldsymbol{I}_2 , \qquad (21.1.26)$$

where

$$\mathcal{K} = k^{-1} , \qquad \mathcal{L} = \mu^{-1} . \qquad (21.1.27)$$

The moduli k and μ are called Kelvin's and Kirchhoff's moduli, respectively. The tensors $\boldsymbol{I}_\sigma$ are defined by

$$\boldsymbol{I}_1 = \mathbf{a}_1 \otimes \mathbf{a}_2 , \qquad \boldsymbol{I}_2 = \mathbf{a}_2 \otimes \mathbf{a}_2 + \mathbf{a}_3 \otimes \mathbf{a}_3 . \qquad (21.1.28)$$

They can also be represented as

$$\boldsymbol{I}_\sigma = I_\sigma^{\alpha\beta\lambda\mu} e_\alpha \otimes e_\beta \otimes e_\lambda \otimes e_\mu ; \qquad (21.1.29)$$

the components $I_\sigma^{\alpha\beta\lambda\mu}$ have been reported in Sec. 3.8.2, see Eqs. (3.8.29). Let us recall that

$$D^{1111} = D^{2222} = k + \mu , \quad D^{1122} = k - \mu , \quad D^{1212} = \mu ,$$
$$D^{2211} = D^{1122} , \quad D^{2121} = D^{2112} = D^{1221} = D^{1212} \tag{21.1.30}$$

and

$$d_{1111} = \frac{1}{4}(\mathcal{K} + \mathcal{L}) , \quad d_{1122} = \frac{1}{4}(\mathcal{K} - \mathcal{L}) , \quad d_{1212} = \frac{1}{4}\mathcal{L} ,$$
$$d_{2211} = d_{1122} , \quad d_{2121} = d_{2112} = d_{1221} = d_{1212} . \tag{21.1.31}$$

Moreover, referring to the basis $\mathbf{a}_i \otimes \mathbf{a}_j$ one finds

$$D^{11} = 2k , \quad D^{22} = D^{33} = 2\mu , \quad D^{12} = 0 , \quad D^{\alpha 3} = 0 ,$$
$$d_{11} = \frac{1}{2}\mathcal{K} , \quad d_{22} = d_{33} = \frac{1}{2}\mathcal{L} , \quad d_{12} = 0 , \quad d_{\alpha 3} = 0 , \tag{21.1.32}$$

hence the matrices $[D^{ij}]$ and $[d_{ij}]$ are diagonal

$$[D^{ij}] = \text{diag}\,[2k, 2\mu, 2\mu] , \qquad [d_{ij}] = \text{diag}\,\left[\frac{1}{2}\mathcal{K}, \frac{1}{2}\mathcal{L}, \frac{1}{2}\mathcal{L}\right] . \tag{21.1.33}$$

The density of elastic energy is given by

$$\frac{1}{2}\mathbf{M} : \boldsymbol{\kappa} = \frac{1}{2}M^{\alpha\beta}\kappa_{\alpha\beta} = \frac{1}{2}\sum_{i=1}^{3}M^i\kappa^i , \tag{21.1.34}$$

and

$$\mathcal{W}(\boldsymbol{\kappa}) = \frac{1}{2}\sum_{i,j=1}^{3} D^{ij}\kappa^i\kappa^j , \tag{21.1.35a}$$

or

$$\mathcal{W}^*(\mathbf{M}) = \frac{1}{2}d_{ij}M^i M^j . \tag{21.1.35b}$$

In the isotropic case we have

$$\mathcal{W} = k(\kappa^1)^2 + \mu[(\kappa^2)^2 + (\kappa^3)^2] , \tag{21.1.36}$$

or

$$\mathcal{W} = k[I(\boldsymbol{\kappa})]^2 + \mu[II(\boldsymbol{\kappa})]^2 . \tag{21.1.37}$$

Alternatively we can express the elastic energy in terms of invariants of $\mathbf{M}$

$$\mathcal{W}^* = \frac{1}{4}\mathcal{K}[I(\mathbf{M})]^2 + \frac{1}{4}\mathcal{L}[II(\mathbf{M})]^2 . \tag{21.1.38}$$

Let us consider now the problem of principal values of strains and moments. To this end we choose the angle α such that the vector $\mathbf{a}_2 = \mathbf{a}_2^\kappa$ is collinear with $\Pi\kappa$, cf. Fig. 21.1.2. Then $\kappa^3 = 0$. We denote $\alpha = \alpha_\kappa$, then we have

$$\cos 2\alpha_\kappa = \frac{\widetilde{\kappa}^2}{II(\kappa)} \,, \qquad \Pi\kappa = II(\kappa)\mathbf{a}_2^\kappa \,. \tag{21.1.39}$$

Consequently

$$\widetilde{\kappa}^1 = I(\kappa) \,, \quad \widetilde{\kappa}^2 = \cos 2\alpha_\kappa II(\kappa) \,, \quad \widetilde{\kappa}^3 = \sin 2\alpha_\kappa II(\kappa) \,. \tag{21.1.40}$$

Note that $\mathbf{a}_1^\kappa = \mathbf{a}_1$, by (21.1.11). We represent κ in the basis $(\mathbf{a}_i^\kappa)$

$$\kappa = \kappa_\kappa^1 \mathbf{a}_1^\kappa + \kappa_\kappa^2 \mathbf{a}_2^\kappa \,, \tag{21.1.41}$$

or

$$\kappa = I(\kappa)\mathbf{a}_1 + II(\kappa)\mathbf{a}_2^\kappa \,, \tag{21.1.42}$$

since

$$\kappa_\kappa^1 = I(\kappa) \,, \qquad \kappa_\kappa^2 = II(\kappa) \,, \qquad \mathbf{a}_2^\kappa = \cos 2\alpha_\kappa \, \widetilde{\mathbf{a}}_2 + \sin 2\alpha_\kappa \, \widetilde{\mathbf{a}}_3 \,. \tag{21.1.43}$$

The principal strains κ_I, κ_{II} are determined according to (21.1.8)

$$\kappa_I = \frac{1}{\sqrt{2}}(I(\kappa) + II(\kappa)) \,, \qquad \kappa_{II} = \frac{1}{\sqrt{2}}(I(\kappa) - II(\kappa)) \,. \tag{21.1.44}$$

By (21.1.44) we have $\kappa_I \geq \kappa_{II}$, since $II(\kappa) \geq 0$.

The decomposition (21.1.41), (21.1.42) applied to the tensor of moments reads:

$$M = M_M^1 \mathbf{a}_1^M + M_M^2 \mathbf{a}_2^M \,, \tag{21.1.45}$$

or

$$M = I(M)\mathbf{a}_1 + II(M)\mathbf{a}_2^M \,, \tag{21.1.46}$$

since

$$M_M^1 = I(M) \,, \quad M_M^2 = II(M) \,. \tag{21.1.47}$$

We see one more reason for introducing the basis (21.1.2): the decomposition of M in the eigenbasis $\mathbf{a}_i^M$ involves invariants. The density of the complementary energy (21.1.35) can be expressed in the form

$$\mathcal{W}^* = \frac{1}{2} \sum_{\alpha,\beta=1}^{3} d_{\alpha\beta}^M M_M^\alpha M_M^\beta \,, \tag{21.1.48}$$

where $d_{\alpha\beta}^M$ are referred to the basis $\mathbf{a}_\alpha^M$.

21.2. *Y-transformation*

While estimating the effective conductivity of a two-component isotropic composite, Milton (1991) introduced a fractional linear transformation by means of which the Hashin-Shtrikman bounds assumed a surprisingly simple form, seemingly independent of the volume fraction. According to Milton (1991) the same transformation had been independently discovered by Cherkaev and Gibiansky in the same year. Moreover, Milton (1991) explained the geometrical meaning of this transformation and named it $Y(\cdot)$.

In this section we define the Y-transformation and report its main properties. Its role will become clear in Sec. 22, where the translation method is used to bound the elastic moduli of the two-phase isotropic plates.

Let us start with

Lemma 21.2.1. Let S_1, S_2 be positive definite quadratic matrices such that

$$S_1 S_2 = S_2 S_1 . \tag{21.2.1}$$

Let us define

$$\langle S \rangle_m = m_1 S_1 + m_2 S_2 ,$$
$$\langle S^{-1} \rangle_m^{-1} = (m_1 S_1^{-1} + m_2 S_2^{-1})^{-1} , \qquad [S]_m = m_1 S_2 + m_2 S_1 , \tag{21.2.2}$$

for $m_1, m_2 \in [0, 1]$ such that $m_1 + m_2 = 1$. Then

$$\langle S \rangle_m - \langle S^{-1} \rangle_m^{-1} = m_1 m_2 (S_1 - S_2)[S]_m^{-1}(S_1 - S_2) . \tag{21.2.3}$$

Proof. By positive definiteness of S_α the matrix $[S]_m$ is invertible. Let us define the matrix

$$R = m_1 S_1 + m_2 S_2 - m_1 m_2 (S_1 - S_2)(m_1 S_2 + m_2 S_1)^{-1}(S_1 - S_2) .$$

Let us rearrange the above expression as follows

$$R = [(m_1 S_1 + m_2 S_2)(m_1 S_2 + m_2 S_1)$$
$$- m_1 m_2 (S_1 - S_2)(m_1 S_2 + m_2 S_1)^{-1}(S_1 - S_2)(m_1 S_2 + m_2 S_1)](m_1 S_2 + m_2 S_1)^{-1} .$$

In view of (21.2.1) we have

$$(S_1 - S_2)(m_1 S_2 + m_2 S_1) = (m_1 S_2 + m_2 S_1)(S_1 - S_2) .$$

Thus

$$R = [(m_1 S_1 + m_2 S_2)(m_1 S_2 + m_2 S_1)$$
$$- m_1 m_2 (S_1 - S_2)(S_1 - S_2)](m_1 S_2 + m_2 S_1)^{-1} .$$

Using (21.2.1) one finds

$$R = (m_1 + m_2)^2 S_1 S_2 (m_1 S_2 + m_2 S_1)^{-1}$$

or

$$R = [(m_1 S_2 + m_2 S_1)(S_1 S_2)^{-1}]^{-1} .$$

Applying (21.2.1) once again we find (21.2.3). $\qquad\square$

Lemma 21.2.2. The relation (21.2.3) remains valid without the condition (21.2.1).

Proof. Let A_1, A_2 be positive definite quadratic matrices of the same dimension as S_α. Let $S_1 = A_1 A_2^{-1}$ and $S_2 = I$, I being the unit matrix. Thus equality (21.2.1) is fulfilled and by (21.2.3) the following identity holds

$$(m_1 A_1 A_2^{-1} + m_2 I) - (m_1 A_2 A_1^{-1} + m_2 I)^{-1}$$
$$= m_1 m_2 (A_1 A_2^{-1} - I)(m_1 I + m_2 A_1 A_2^{-1})^{-1}(A_1 A_2^{-1} - I) \,.$$

By multiplying this identity by A_2 we find

$$(m_1 A_1 + m_2 A_2) - [A_2^{-1}(m_1 A_2 A_1^{-1} + m_2 I)]^{-1}$$
$$= m_1 m_2 (A_1 - A_2) A_2^{-1}(m_1 I + m_2 A_1 A_2^{-1})^{-1}(A_1 - A_2) \,.$$

Hence

$$\langle A \rangle_m - \langle A^{-1} \rangle_m^{-1} = m_1 m_2 (A_1 - A_2)[(m_1 I + m_2 A_1 A_2^{-1}) A_2]^{-1}(A_1 - A_2) \,,$$

which gives (21.2.3) for $S_\alpha = A_\alpha$. The assumption (21.2.1) has turned out to be redundant and this is precisely the assertion of the lemma. $\qquad\qquad\square$

Lemma 21.2.3. Let D_1, D_2 be $n \times n$ positive definite matrices such that the matrix $D_1 - D_2$ is invertible. Let T be an $n \times n$ matrix such that the matrices $D_\alpha - T$ are positive definite. Let D_0 be an $n \times n$ matrix such that the matrix

$$D_0 - (\langle (D - T)^{-1} \rangle_m^{-1} + T)$$

be semi positive definite, which will be symbolically written as

$$D_0 \geq \langle (D - T)^{-1} \rangle_m^{-1} + T \,. \tag{21.2.4}$$

The operation $\langle \cdot \rangle_m$ is defined by (21.2.2)$_1$, with $m_\alpha \in [0, 1], m_1 + m_2 = 1$. Then the condition (21.2.4) can be rearranged to the form

$$Y(D_0) + T \geq 0 \,, \tag{21.2.5}$$

where[*]

$$Y(D_0) = -m_2 D_1 - m_1 D_2$$
$$- m_1 m_2 (D_1 - D_2)(D_0 - m_1 D_1 - m_2 D_2)^{-1}(D_1 - D_2) \,, \tag{21.2.6}$$

provided that the matrix $D_0 - m_1 D_1 - m_2 D_2$ is invertible.

Proof. Let us substitute $S_1 = D_1 - T$ and $S_2 = D_2 - T$ into the identity (21.2.3). By Lemma 21.2.2 we know that the condition $S_1 S_2 = S_2 S_1$ does not need to be satisfied. The conditions $S_\alpha \geq 0$ are fulfilled according to the assumptions. Thus we find

$$\langle (D - T)^{-1} \rangle_m^{-1} = \langle D \rangle_m - T$$
$$- m_1 m_2 (D_1 - D_2)(m_1 D_2 + m_2 D_1 - T)^{-1}(D_1 - D_2) \,, \tag{21.2.7}$$

---

[*] The operator $Y(\cdot)$ has nothing to do with the cell of periodicity Y. This ambiguity in notation should not lead to misunderstandings.

and the condition (21.2.4) can be rearranged to the form

$$m_1 \boldsymbol{D}_1 + m_2 \boldsymbol{D}_2 - \boldsymbol{D}_0$$
$$\leq m_1 m_2 (\boldsymbol{D}_1 - \boldsymbol{D}_2)(m_1 \boldsymbol{D}_2 + m_2 \boldsymbol{D}_1 - \boldsymbol{T})^{-1}(\boldsymbol{D}_1 - \boldsymbol{D}_2) . \qquad (21.2.8)$$

Let us rewrite (21.2.6) in the form

$$Y(\boldsymbol{D}_0) + m_2 \boldsymbol{D}_1 + m_1 \boldsymbol{D}_2$$
$$= m_1 m_2 (\boldsymbol{D}_1 - \boldsymbol{D}_2)(m_1 \boldsymbol{D}_1 + m_2 \boldsymbol{D}_2 - \boldsymbol{D}_0)^{-1}(\boldsymbol{D}_1 - \boldsymbol{D}_2) .$$

Hence

$$(Y(\boldsymbol{D}_0) + m_2 \boldsymbol{D}_1 + m_1 \boldsymbol{D}_2)^{-1} m_1 m_2$$
$$= (\boldsymbol{D}_1 - \boldsymbol{D}_2)^{-1}(m_1 \boldsymbol{D}_1 + m_2 \boldsymbol{D}_2 - \boldsymbol{D}_0)(\boldsymbol{D}_1 - \boldsymbol{D}_2)^{-1} .$$

Consequently we have

$$m_1 m_2 (\boldsymbol{D}_1 - \boldsymbol{D}_2)[Y(\boldsymbol{D}_0) + m_1 \boldsymbol{D}_2 + m_2 \boldsymbol{D}_1]^{-1}(\boldsymbol{D}_1 - \boldsymbol{D}_2)$$
$$= m_1 \boldsymbol{D}_1 + m_2 \boldsymbol{D}_2 - \boldsymbol{D}_0 . \qquad (21.2.9)$$

By (21.2.8) and (21.2.9) we find the estimate

$$(\boldsymbol{D}_1 - \boldsymbol{D}_2)(Y(\boldsymbol{D}_0) + [\boldsymbol{D}]_m)(\boldsymbol{D}_1 - \boldsymbol{D}_2) \leq (\boldsymbol{D}_1 - \boldsymbol{D}_2)([\boldsymbol{D}]_m - \boldsymbol{T})^{-1}(\boldsymbol{D}_1 - \boldsymbol{D}_2) .$$

Since $\boldsymbol{D}_1 - \boldsymbol{D}_2$ is nonsingular, we conclude that

$$(Y(\boldsymbol{D}_0) + [\boldsymbol{D}]_m)^{-1} \leq ([\boldsymbol{D}]_m - \boldsymbol{T})^{-1} .$$

Hence we conclude the formula (21.2.5).

Remark 21.2.1. Consider the case when $\boldsymbol{D}_\alpha$ and $\boldsymbol{D}_0$ are isotropic. Then the Y-transformation (21.2.6) transmits to the level of the components. Indeed, assume that

$$[\boldsymbol{D}_\alpha^{ij}] = \operatorname{diag}(d_{\alpha 1}, d_{\alpha 2}, \ldots, d_{\alpha n}) , \qquad [\boldsymbol{D}_0^{ij}] = \operatorname{diag}(d_{01}, d_{02}, \ldots, d_{0n}) . \qquad (21.2.10)$$

Then

$$[Y(\boldsymbol{D}_0)^{ij}] = \operatorname{diag}(z_1, z_2, \ldots, z_n) \qquad (21.2.11)$$

and

$$z_p = -m_2 d_{1p} - m_1 d_{2p} - m_1 m_2 (d_{1p} - d_{2p})^2 (d_{0p} - m_1 d_{1p} - m_2 d_{2p})^{-1} , \qquad (21.2.12)$$

or

$$z_p = y(d_{0p}, d_{1p}, d_{2p}, m_1, m_2) , \qquad (21.2.13)$$

where $p \in \{1, \ldots, n\}$. The function $y^{(*)}$ is defined by

$$y(x, a, b, m_1, m_2) = -m_2 a - m_1 b - m_1 m_2 (a - b)^2 (x - m_1 a - m_2 b)^{-1} ,$$

(*) Function y has nothing to do with points $y = (y_1, y_2)$ of the periodicity cell. This notation should not be misleading.

or, equivalently

$$y(x, a, b, m_1, m_2) = \frac{ab - x(m_1 b + m_2 a)}{x - (m_1 a + m_2 b)} \; . \tag{21.2.14}$$

The function y has the following properties

(i)

$$y(a, a, b, m_1, m_2) = -a \; , \tag{21.2.15}$$

(ii)

$$y(b, a, b, m_1, m_2) = -b \; , \tag{21.2.16}$$

(iii)

$$y\left(\frac{1}{x}, \frac{1}{a}, \frac{1}{b}, m_1, m_2\right) = \frac{1}{y(x, a, b, m_1, m_2)} \; , \tag{21.2.17}$$

(iv)

$$y(\alpha x, \alpha a, \alpha b, m_1, m_2) = \alpha y(x, a, b, m_1, m_2) \quad \text{for } \alpha \in \mathbb{R} \; ; \tag{21.2.18}$$

(v) the equality

$$y(x, a, b, m_1, m_2) = \lambda \tag{21.2.19}$$

is equivalent to

(a)

$$x = y(\lambda, -b, -a, m_1, m_2) \; , \tag{21.2.20}$$

(b)

$$x = m_1 a + m_2 b - \frac{m_1 m_2 (a - b)^2}{m_1 b + m_2 a + \lambda} \; , \tag{21.2.21}$$

if $m_1 b + m_2 a + \lambda \neq 0$,

(c)

$$\frac{m_2}{x - a} = \frac{1}{b - a} + \frac{m_1}{a + \lambda} \; , \tag{21.2.22}$$

if $x \neq a, a \neq b, a + \lambda \neq 0,$

(d)

$$\frac{m_1}{x - b} = \frac{1}{a - b} + \frac{m_2}{b + \lambda} \; , \tag{21.2.23}$$

if $x \neq b, a \neq b, b + \lambda \neq 0,$

(e)

$$\frac{m_1}{a + \lambda} + \frac{m_2}{b + \lambda} = \frac{1}{x + \lambda} \; , \tag{21.2.24}$$

if $a + \lambda \neq 0, b + \lambda \neq 0, x + \lambda \neq 0,$

(f)

$$\frac{m_1}{x-b} + \frac{m_2}{x-a} = \frac{1}{x+\lambda} \, , \tag{21.2.25}$$

if $x \neq b, x \neq a, x + \lambda \neq 0$.

21.3. Fourier representation of Y-periodic functions

The notions of quasiconvexity and quasiaffinity were introduced in Sec. 1.2.4. Now these properties will be exploited by expanding the periodic functions in the Fourier series. The aim of this section is to study briefly this problem.

Let f be Y-periodic; $Y = (0, l_1) \times (0, l_2)$. By periodicity, such a function can be extended to the whole space $\mathbb{R}^2$. The periodic extension to $\mathbb{R}^2$ is still denoted by f. Y is parametrized by $y = (y_1, y_2)$. Let us introduce

$$\xi_1 = y_1/l_1 \, , \quad \xi_2 = y_2/l_2 \, , \quad k_\alpha = \frac{2\pi i n_\alpha}{l_\alpha} \, , \quad k = (k_1, k_2) \, , \tag{21.3.1}$$

where $i = \sqrt{-1}$ and $n_\alpha (\alpha = 1, 2)$ are integers. Assume that f is sufficiently regular to be represented by the Fourier series

$$f(y) = \sum_{n_1, n_2 = -\infty}^{\infty} f_{n_1, n_2} e^{2\pi i (n_1 \xi_1 + n_2 \xi_2)} \, . \tag{21.3.2}$$

This formula will be written as follows

$$f(y) = \sum_k \widehat{f}(k) e^{k_1 y_1 + k_2 y_2} \, , \tag{21.3.3}$$

where

$$\widehat{f}(k) = \langle e^{-k_1 y_1 - k_2 y_2} f(y) \rangle \, . \tag{21.3.4}$$

Here $\langle \cdot \rangle$ means averaging over Y. In particular

$$\widehat{f}(0) = \langle f(y) \rangle \, . \tag{21.3.5}$$

The Fourier representation of the first and second derivatives are

$$f_{|\alpha} = \sum_k \widehat{f}_{|\alpha}(k) e^{k_1 y_1 + k_2 y_2} \, , \qquad f_{|\alpha\beta} = \sum_k \widehat{f}_{|\alpha\beta}(k) e^{k_1 y_1 + k_2 y_2} \, , \tag{21.3.6}$$

where $f_{|\alpha} = \partial f / \partial y_\alpha$ and

$$\widehat{f}_{|\alpha}(k) = k_\alpha \widehat{f}(k) \, , \quad \widehat{f}_{|\alpha\beta}(k) = k_\alpha k_\beta \widehat{f}(k) \, . \tag{21.3.7}$$

Consider two functions: f and g, both of them Y-periodic. Their scalar product is represented by

$$\langle f\bar{g} \rangle = \sum_k \widehat{f}(k)\overline{\widehat{g}(k)} \tag{21.3.8}$$

or

$$\langle f\bar{g}\rangle = \langle f\rangle\langle\bar{g}\rangle + \sum_{\mathbf{k}\neq 0}\widehat{f}(\mathbf{k})\overline{\widehat{g}(\mathbf{k})}\,. \qquad (21.3.9)$$

Here $\overline{(\cdot)}$ means complex conjugate. If both functions assume real values, then

$$\langle fg\rangle = \langle f\rangle\langle g\rangle + \frac{1}{2}\sum_{\mathbf{k}\neq 0}(\widehat{f}(\mathbf{k})\overline{\widehat{g}(\mathbf{k})} + \overline{\widehat{f}(\mathbf{k})}\widehat{g}(\mathbf{k}))\,. \qquad (21.3.10)$$

21.4.   *Examples of quasiconvex and quasiaffine functions*

The aim of this section is to report and prove some integral inequalities concerning functions whose arguments are kinematically admissible local strains or statically admissible local stresses. These inequalities enter the so-called translation method used in Sec. 22 for finding optimal estimates of the effective moduli of the two-phase plates.

The inequalities to be considered below follow from the theory of quasiconvexity, see Dacorogna (1982, 1989). Their applications are outlined in Gibiansky (1993).

Let us denote by $S(Y)$ a set of vector functions defined on $Y \subset \mathbb{R}^2$, assuming equal values on the opposite sides of Y and satisfying some additional differentiability conditions.

We recall that a function $F : S(Y) \to \mathbb{R}$ is called *quasiconvex* on $S(Y)$ if

$$\langle F(\boldsymbol{\sigma})\rangle \geq F(\langle\boldsymbol{\sigma}\rangle) \quad \forall\boldsymbol{\sigma} \in S(Y)\,. \qquad (21.4.1)$$

A function $F : S(Y) \to \mathbb{R}$ is called *quasiaffine* on $S(Y)$ if

$$\langle F(\boldsymbol{\sigma})\rangle = F(\langle\boldsymbol{\sigma}\rangle) \quad \forall\boldsymbol{\sigma} \in S(Y)\,. \qquad (21.4.2)$$

In this section we shall consider the cases when $S(Y)$ coincides with $\mathcal{K}_{\boldsymbol{\kappa}}^{per}(Y)$, $\mathcal{S}_1^{per}(Y)$, $\mathcal{S}_2^{per}(Y)$ or their Cartesian products. Those sets are defined in Secs. 3.4 and 3.6.

21.4.1.   A quasiaffine function of the strain tensor $\boldsymbol{\kappa}$

Consider the function $F : \mathcal{K}_{\boldsymbol{\kappa}}^{per}(Y) \to \mathbb{R}$ defined by

$$F(\boldsymbol{\kappa}) = \det\boldsymbol{\kappa}\,, \qquad \boldsymbol{\kappa} \in \mathcal{K}_{\boldsymbol{\kappa}}^{per}(Y)\,. \qquad (21.4.3)$$

We shall prove that

$$\langle\det\boldsymbol{\kappa}\rangle = \det\langle\boldsymbol{\kappa}\rangle \qquad \forall\boldsymbol{\kappa} \in \mathcal{K}_{\boldsymbol{\kappa}}^{per}(Y)\,, \qquad (21.4.4)$$

which means that $F(\boldsymbol{\kappa}) = \det\boldsymbol{\kappa}$ is quasiaffine on $\mathcal{K}_{\boldsymbol{\kappa}}^{per}(Y)$.

In fact, according to Sec. 3.4 $\boldsymbol{\kappa} \in \mathcal{K}_{\boldsymbol{\kappa}}^{per}(Y)$ can be represented as

$$\kappa_{\alpha\beta} = \kappa_{\alpha\beta}^h - v_{|\alpha\beta}\,, \qquad v \in H_{per}^2(Y)\,,$$

where $\kappa^h \in \mathbb{E}_s^2$.

Let us assume that $v \in C^{\infty}_{per}(Y)$ and compute

$$\langle \det \boldsymbol{\kappa} \rangle = \langle (\kappa^h_{11} - v_{|11})(\kappa^h_{22} - v_{|22}) - (\kappa^h_{12} - v_{|12})^2 \rangle$$
$$= \det \boldsymbol{\kappa}^h - \kappa^h_{11}\langle v_{|22}\rangle - \kappa^h_{22}\langle v_{|11}\rangle + 2\kappa^h_{12}\langle v_{|12}\rangle + \langle (v_{|1}v_{|22})_{|1} - (v_{|1}v_{|12})_{|2}\rangle .$$

By periodicity of v we arrive at (21.4.4). Finally, by density arguments the same equality is extended to $v \in H^2_{per}(Y)$. This establishes the formula (21.4.4).

It is instructive to repeat this proof by using Fourier representations. By (21.3.9) we calculate

$$\langle \det \boldsymbol{\kappa} \rangle = \langle \kappa_{11}\bar\kappa_{22} - \kappa_{12}\bar\kappa_{12} \rangle = \langle \kappa_{11}\rangle\langle\bar\kappa_{22}\rangle - \langle\kappa_{12}\rangle\langle\bar\kappa_{12}\rangle + R ,$$

where

$$R = \sum_{k \neq 0} [\widehat{\kappa}_{11}(\boldsymbol{k})\overline{\widehat{\kappa}_{22}(\boldsymbol{k})} - \widehat{\kappa}_{12}(\boldsymbol{k})\overline{\widehat{\kappa}_{12}(\boldsymbol{k})}] .$$

Since $\kappa_{\alpha\beta}(w) = -w_{|\alpha\beta}$, we have $\widehat{\kappa}_{\alpha\beta} = -k_\alpha k_\beta \widehat{w}$. Hence

$$R = \sum_{k \neq 0} [(k_1)^2(\overline{k_2})^2 - k_1 k_2(\overline{k_1 k_2})]\widehat{w} = 0 ,$$

because $\overline{k_\alpha} = -k_\alpha$.

21.4.2. A quasiaffine function of two strain tensors

Consider the functions $F_\alpha : \mathcal{K}^{per}_{\boldsymbol{\kappa}}(Y) \times \mathcal{K}^{per}_{\boldsymbol{\kappa}}(Y) \to \mathbb{R}$ defined by

$$F_1(\boldsymbol{\kappa}, \boldsymbol{\rho}) = \kappa_{12}\rho_{11} - \rho_{12}\kappa_{11} , \qquad F_2(\boldsymbol{\kappa}, \boldsymbol{\rho}) = \kappa_{22}\rho_{12} - \rho_{22}\kappa_{12} , \qquad (21.4.5)$$

where $\boldsymbol{\kappa}, \boldsymbol{\rho} \in \mathcal{K}^{per}_{\boldsymbol{\kappa}}(Y)$. Then

$$\langle \kappa_{12}\rho_{11} - \rho_{12}\kappa_{11} \rangle = \langle \kappa_{12}\rangle\langle\rho_{11}\rangle - \langle\rho_{12}\rangle\langle\kappa_{11}\rangle ,$$
$$\langle \rho_{12}\kappa_{22} - \rho_{22}\kappa_{12} \rangle = \langle \rho_{12}\rangle\langle\kappa_{22}\rangle - \langle\rho_{22}\rangle\langle\kappa_{12}\rangle , \qquad (21.4.6)$$

i.e. the functions F_α are quasiaffine on $\mathcal{K}^{per}_{\boldsymbol{\kappa}}(Y) \times \mathcal{K}^{per}_{\boldsymbol{\kappa}}(Y)$. Obviously, it is sufficient to prove the first equality. There exist $u, v \in H^2_{per}(Y)$ such that

$$\kappa_{\alpha\beta} = \kappa^h_{\alpha\beta} - v_{|\alpha\beta} , \qquad \rho_{\alpha\beta} = \rho^h_{\alpha\beta} - u_{|\alpha\beta} ,$$

where $\boldsymbol{\kappa}^h, \boldsymbol{\rho}^h \in \mathbb{E}^2_s$. Assume that $u, v \in C^{\infty}_{per}(Y)$ and compute

$$\langle \rho_{11}\kappa_{12} - \rho_{12}\kappa_{11} \rangle = \langle (\rho^h_{11} - u_{|11})(\kappa^h_{12} - v_{|12}) - (\rho^h_{12} - u_{|12})(\kappa^h_{11} - u_{|11})\rangle$$
$$= (\rho^h_{11}\kappa^h_{12} - \rho^h_{12}\kappa^h_{11})$$
$$- \kappa^h_{12}\langle u_{|11}\rangle - \rho^h_{11}\langle v_{|21}\rangle + \rho^h_{12}\langle v_{|11}\rangle$$
$$+ \kappa^h_{11}\langle u_{|12}\rangle + \langle (u_{|1}v_{|12})_{|1} - (u_{|1}v_{|11})_{|2}\rangle .$$

By periodicity we find $(21.4.6)_1$. By density arguments this equality is next extended to the space $H^2_{per}(Y)$. $\qquad\qquad\square$

21.4.3.   An aggregate form of the previous results

Let us express the relations (21.4.4) and (21.4.6) in terms of the components κ^i, ρ^i referred to the basis $(\boldsymbol{a}_i)$, cf. Sec. 21.1. For arbitrary $\boldsymbol{\kappa}, \boldsymbol{\rho} \in \mathcal{K}_{\boldsymbol{\kappa}}^{per}(Y)$ we have

$$\langle \boldsymbol{\kappa}^T \overset{o}{\boldsymbol{T}} \boldsymbol{\kappa} \rangle = \langle \boldsymbol{\kappa} \rangle^T \overset{o}{\boldsymbol{T}} \langle \boldsymbol{\kappa} \rangle , \qquad \langle \boldsymbol{\rho}^T \boldsymbol{E} \boldsymbol{\kappa} \rangle = \langle \boldsymbol{\rho} \rangle^T \boldsymbol{E} \langle \boldsymbol{\kappa} \rangle , \tag{21.4.7}$$

where $\overset{o}{\boldsymbol{T}}$ is defined by (21.1.19) and

$$[E_{ij}] = \begin{bmatrix} 0 & 0 & 0 \\ 0 & 0 & 1 \\ 0 & -1 & 0 \end{bmatrix} . \tag{21.4.8}$$

Equivalence of (21.4.6) and (21.4.7)$_2$ follows from the identity

$$\begin{aligned}
\langle \boldsymbol{\rho}^T \boldsymbol{E} \boldsymbol{\kappa} \rangle &= \langle \rho^2 \kappa^3 - \rho^3 \kappa^2 \rangle \\
&= \frac{1}{2} \langle (\rho_{11} - \rho_{22}) \kappa_{12} - \rho_{12}(\kappa_{11} - \kappa_{22}) \rangle \\
&= \frac{1}{2} [(\langle \rho_{11} \rangle - \langle \rho_{22} \rangle) \langle \kappa_{12} \rangle - \langle \rho_{12} \rangle (\langle \kappa_{11} \rangle - \langle \kappa_{22} \rangle)] \\
&= (\langle \rho^2 \rangle \langle \kappa^3 \rangle - \langle \rho^3 \rangle \langle \kappa^2 \rangle) = \langle \boldsymbol{\rho} \rangle^T \boldsymbol{E} \langle \boldsymbol{\kappa} \rangle .
\end{aligned} \tag{21.4.9}$$

Let us define the (6×6) - matrix

$$\boldsymbol{\mathcal{T}} = \begin{bmatrix} t_1 \overset{0}{\boldsymbol{T}} & t_2 \boldsymbol{E} \\ t_2 \boldsymbol{E}^T & t_1 \overset{0}{\boldsymbol{T}} \end{bmatrix} , \tag{21.4.10}$$

where $t_1, t_2 \in \mathbb{R}$. Let

$$\boldsymbol{\varepsilon} = [\kappa^1, \kappa^2, \kappa^3, \rho^1, \rho^2, \rho^3]^T . \tag{21.4.11}$$

By (21.4.7) we have

$$\langle \boldsymbol{\varepsilon}^T \boldsymbol{\mathcal{T}} \boldsymbol{\varepsilon} \rangle = \langle \boldsymbol{\varepsilon} \rangle^T \boldsymbol{\mathcal{T}} \langle \boldsymbol{\varepsilon} \rangle , \tag{21.4.12}$$

for each $t_\alpha \in \mathbb{R}$ and any $\boldsymbol{\kappa}, \boldsymbol{\rho} \in \mathcal{K}_{\boldsymbol{\kappa}}^{per}(Y)$.

21.4.4.   A quasiaffine function of the stress tensor $\boldsymbol{m}$

Let us define the function $F : \mathcal{S}_2^{per}(Y) \to \mathbb{R}$ given by $F(\boldsymbol{m}) = -\det \boldsymbol{m}$, $\boldsymbol{m} \in \mathcal{S}_2^{per}(Y)$. We are going to prove that this function is quasiconvex on $\mathcal{S}_2^{per}(Y)$, i.e.:

$$\langle -\det \boldsymbol{m} \rangle \geq -\det(\langle \boldsymbol{m} \rangle) , \quad \boldsymbol{m} \in \mathcal{S}_2^{per}(Y) . \tag{21.4.13}$$

Indeed, by definition of the set $\mathcal{S}_2^{per}(Y)$ the tensor $\boldsymbol{m} = (m^{\alpha\beta})$ satisfies the differential equation $m_{|\alpha\beta}^{\alpha\beta} = 0$. Hence

$$(k_1)^2 \widehat{m}^{11} + 2 k_1 k_2 \widehat{m}^{12} + (k_2)^2 \widehat{m}^{22} = 0 ,$$

or

$$\widehat{m}^{12} = -\frac{1}{2} \left(\frac{k_1}{k_2} \widehat{m}^{11} + \frac{k_2}{k_1} \widehat{m}^{22} \right) , \tag{21.4.14}$$

if $k \neq 0$. Let us compute now the quantity $\langle -F(\boldsymbol{m}) \rangle$ by applying the formula (21.3.10)

$$\langle \det \boldsymbol{m} \rangle = \langle \frac{1}{2} m^{11} \overline{m^{22}} + \frac{1}{2} \overline{m^{11}} m^{22} - m^{12} \overline{m^{12}} \rangle$$

$$= \frac{1}{2} \langle m^{11} \rangle \langle \overline{m^{22}} \rangle + \frac{1}{2} \langle \overline{m^{11}} m^{22} \rangle - \langle m^{12} \rangle \langle \overline{m^{12}} \rangle + R \, ,$$

where

$$R = \sum_{k \neq 0} \left[\frac{1}{2} \widehat{m}^{11}(\boldsymbol{k}) \overline{\widehat{m}^{22}}(\boldsymbol{k}) + \frac{1}{2} \overline{\widehat{m}^{11}}(\boldsymbol{k}) \widehat{m}^{22}(\boldsymbol{k}) - \widehat{m}^{12}(\boldsymbol{k}) \overline{\widehat{m}^{12}}(\boldsymbol{k}) \right] \, .$$

Substitution of (21.4.14) into the definition of R gives

$$R = -\frac{1}{4} \sum_{k \neq 0} \left| \frac{k_1}{k_2} \widehat{m}^{11} - \frac{k_2}{k_1} \widehat{m}^{22} \right|^2 \, ,$$

since $\overline{(k_1/k_2)} = k_1/k_2$. The inequality $R \leq 0$ implies (21.4.13).

By using (21.1.18) the inequality (21.4.13) can be written in the form

$$\langle \boldsymbol{m}^T \overset{0}{\boldsymbol{T}} \boldsymbol{m} \rangle \geq \langle \boldsymbol{m} \rangle^T \overset{0}{\boldsymbol{T}} \langle \boldsymbol{m} \rangle \quad \forall \boldsymbol{m} \in \mathcal{S}_2^{per}(Y) \, , \tag{21.4.15}$$

where $\boldsymbol{m} = (m^1, m^2, m^3)^T$.

21.4.5.   A quasiaffine function of the stress tensor $\boldsymbol{n}$

Consider the function $F : \mathcal{S}_1^{per}(Y) \to \mathbb{R}$ defined by

$$F(\boldsymbol{n}) = -\det \boldsymbol{n} \, , \quad \boldsymbol{n} \in \mathcal{S}_1^{per}(Y) \, . \tag{21.4.16}$$

This function is quasiaffine. In view of (21.1.18) this condition assumes the form:

$$\langle \boldsymbol{n}^T \overset{0}{\boldsymbol{T}} \boldsymbol{n} \rangle = \langle \boldsymbol{n} \rangle^T \overset{0}{\boldsymbol{T}} \langle \boldsymbol{n} \rangle \quad \forall \boldsymbol{n} \in \mathcal{S}_1^{per}(Y) \, , \tag{21.4.17}$$

where $\boldsymbol{n} = (n^1, n^2, n^3)^T$ and $n^i = \boldsymbol{n} \cdot \boldsymbol{a}_i$.

In fact, the condition $\boldsymbol{n} \in \mathcal{S}_1^{per}(Y)$, see Sec. 3.6, implies that $n^{\alpha\beta}_{|\beta} = 0$ or

$$k_1 \widehat{n}^{11}(\boldsymbol{k}) + k_2 \widehat{n}^{12}(\boldsymbol{k}) = 0 \, , \qquad k_1 \widehat{n}^{12}(\boldsymbol{k}) + k_2 \widehat{n}^{22}(\boldsymbol{k}) = 0 \, . \tag{21.4.18}$$

The solution of the above system is nontrivial. Thus

$$\widehat{n}^{11}(\boldsymbol{k}) \overline{\widehat{n}^{22}(\boldsymbol{k})} - \widehat{n}^{12}(\boldsymbol{k}) \overline{\widehat{n}^{12}(\boldsymbol{k})} = 0 \, , \tag{21.4.19}$$

since $k_1 \overline{k}_2 = \overline{k}_1 k_2$. Let us compute

$$-\frac{1}{2} \langle \boldsymbol{n}^T \overset{0}{\boldsymbol{T}} \boldsymbol{n} \rangle = \langle \det \boldsymbol{n} \rangle = \det \langle \boldsymbol{n} \rangle + R \, ,$$

where

$$R = \sum_{k \neq 0} \left[\widehat{n}_{11}(\boldsymbol{k}) \overline{\widehat{n}_{22}(\boldsymbol{k})} - \widehat{n}^{12}(\boldsymbol{k}) \overline{\widehat{n}^{12}(\boldsymbol{k})} \right] \, .$$

By (21.4.19) we note that $R = 0$, which completes the proof.

21.4.6. A quasiaffine function of two stress tensors

Consider the function $F : \mathcal{S}_1^{per}(Y) \times \mathcal{S}_1^{per}(Y) \to \mathbb{R}$ defined by

$$F(\boldsymbol{\tau}, \boldsymbol{n}) = \boldsymbol{\tau}^T \boldsymbol{E} \boldsymbol{n} \,, \quad \boldsymbol{\tau}, \boldsymbol{n} \in \mathcal{S}_1^{per}(Y) \,, \tag{21.4.20}$$

where the tensor $\boldsymbol{E}$ has representation (21.4.8) in the basis $\boldsymbol{a}_i \otimes \boldsymbol{a}_j$. We shall prove that the function F is quasiaffine on $\mathcal{S}_1^{per}(Y) \times \mathcal{S}_1^{per}(Y)$. Indeed, by applying the first two equalities of (21.4.9) we find

$$2\langle \boldsymbol{\tau}^T \boldsymbol{E} \boldsymbol{n} \rangle = \langle \tau^{11} n^{12} - \tau^{12} n^{11} \rangle + \langle \tau^{12} n^{22} - \tau^{22} n^{12} \rangle \,. \tag{21.4.21}$$

Application of the formula (21.3.9) gives

$$2\langle \boldsymbol{\tau}^T \boldsymbol{E} \boldsymbol{n} \rangle = \langle \tau^{11} \rangle \langle n^{12} \rangle - \langle \tau^{12} \rangle \langle n^{11} \rangle + \langle \tau^{12} \rangle \langle n^{22} \rangle - \langle \tau^{22} \rangle \langle n^{12} \rangle + R_1 + R_2 \,,$$

where

$$R_1 = \sum_{k \neq 0} \left[\overline{\hat{\tau}^{11}(\boldsymbol{k})} \hat{n}^{12}(\boldsymbol{k}) - \overline{\hat{\tau}^{12}(\boldsymbol{k})} \hat{n}^{11}(\boldsymbol{k}) \right] \,, \quad R_2 = \sum_{k \neq 0} \left[\overline{\hat{\tau}^{12}(\boldsymbol{k})} \hat{n}^{22}(\boldsymbol{k}) - \overline{\hat{\tau}^{22}(\boldsymbol{k})} \hat{n}^{12}(\boldsymbol{k}) \right] \,.$$

Let $k_\alpha \neq 0$. Let $\boldsymbol{\tau}, \boldsymbol{n} \in \mathcal{S}_1^{per}(Y)$. Then, by (21.4.18) we have

$$\hat{n}^{12}(\boldsymbol{k}) = -\frac{k_1}{k_2} \hat{n}^{11}(\boldsymbol{k}) \,, \qquad \overline{\hat{\tau}^{12}}(\boldsymbol{k}) = -\frac{k_1}{k_2} \overline{\hat{\tau}^{11}}(\boldsymbol{k}) \,,$$

$$\hat{n}^{12}(\boldsymbol{k}) = -\frac{k_2}{k_1} \hat{n}^{22}(\boldsymbol{k}) \,, \qquad \overline{\hat{\tau}^{12}}(\boldsymbol{k}) = -\frac{k_2}{k_1} \overline{\hat{\tau}^{22}}(\boldsymbol{k}) \,,$$

since $\overline{(k_1/k_2)} = k_1/k_2$. Hence $R_\alpha = 0$, which completes the proof.

21.4.7. An aggregate form of the two previous results

Consider the function $F : \mathcal{S}_1^{per}(Y) \times \mathcal{S}_1^{per}(Y) \to \mathbb{R}$ defined by

$$F(\boldsymbol{\tau}, \boldsymbol{n}) = \boldsymbol{s}^T \boldsymbol{\mathcal{T}} \boldsymbol{s} \,, \tag{21.4.22}$$

where $\boldsymbol{s} = (\tau^1, \tau^2, \tau^3, n^1, n^2, n^3)$ and $\boldsymbol{\mathcal{T}}$ is defined by (21.4.10) with $t_\alpha \in \mathbb{R}$. By quasi-affine properties of the functions defined by (21.4.16) and (21.4.20) we conclude that $\boldsymbol{\mathcal{T}}$ given by (21.4.22) is also quasiaffine for each $t_\alpha \in \mathbb{R}$, or

$$\langle \boldsymbol{s}^T \boldsymbol{\mathcal{T}} \boldsymbol{s} \rangle = \langle \boldsymbol{s} \rangle^T \mathcal{T} \langle \boldsymbol{s} \rangle \tag{21.4.23}$$

for $\boldsymbol{s} \in \mathcal{S}_1^{per}(Y) \times \mathcal{S}_1^{per}(Y)$.

21.5. *Harmonic mean as a lower bound for effective energy*

Let $b^h \in \mathbb{E}_s^2$, $b \in L^2(Y, \mathbb{E}_s^2)$ and $A = A^{ij} a_i \otimes a_j$ with $A^{ij} \in L^\infty(Y)$. By introducing the decomposition $b = b^i a_i$, we can write $b \cdot (A \cdot b) = b^T A b$, where b^T is identified with (b^1, b^2, b^3). Let us define the function $F : \mathbb{E}_s^2 \to \mathbb{R}$ by

$$F(b^h) = \min\{\langle b^T A b \rangle \mid b \in L^2(Y, \mathbb{E}_s^2) \,, \ \langle b \rangle = b^h \} \,. \tag{21.5.1}$$

Assume that A is positive definite, hence nonsingular. Let us define

$$b_0 = A^{-1} \langle A^{-1} \rangle^{-1} b^h \tag{21.5.2}$$

and note that $\langle b_0 \rangle = b^h$. We are going to prove that

$$F(b^h) = \langle b_0^T A b_0 \rangle \,, \tag{21.5.3}$$

or

$$\langle b^T A b \rangle \geq (b^h)^T \langle A^{-1} \rangle^{-1} b^h \,, \tag{21.5.4}$$

for $b \in L^2(Y, \mathbb{E}_s^2)$ and $\langle b \rangle = b^h$. To this end we define the Lagrangian

$$F_\lambda(b) = \langle b^T A b + 2\lambda^T (b - b^h) \rangle \,,$$

where $\lambda \in \mathbb{E}_s^2$. The necessary condition of stationarity implies

$$2 A b_0 + 2\lambda = 0$$

or $b_0 = -A^{-1}\lambda$. Hence $\langle b_0 \rangle = -\langle A^{-1} \rangle \lambda$. Consequently, $\lambda = -\langle A^{-1} \rangle^{-1} b^h$ and we find Eq. (21.5.2). Since A is positive definite, the function $\langle b^T A b \rangle$ assumes minimum value at $b = b_0$. Thus for any $b \in L^2(Y, \mathbb{E}_s^2)$

$$\begin{aligned}
\langle b^T A b \rangle &\geq \langle b_0^T A b_0 \rangle = \langle ((b^h)^T \langle A^{-1} \rangle^{-1} A^{-1} A A^{-1} \langle A^{-1} \rangle^{-1} b^h \rangle \\
&= (b^h)^T \langle A^{-1} \rangle^{-1} \langle A^{-1} \rangle \langle A^{-1} \rangle^{-1} b^h = (b^h)^T \langle A^{-1} \rangle^{-1} b^h \,,
\end{aligned}$$

which completes the proof.

21.6. *Elements of the theory of Young measures*

The Young measure was introduced by Young (1937, 1969) as a means of treating problems of the calculus of variations for which there does not exist a minimizer in a classical sense. Afterwards, the Young measure has become a tool for the study of nonlinear partial differential equations following the work of Tartar (1979) and DiPerna (1983), cf. also the references cited in Pedregal (1997) and in Valadier (1994). The use we make of it here is the one for which it was originally introduced by Young (1937, 1969), cf. also Ball (1989), Ball and Knowles (1990), Pedregal (1997, 1999), Roubiček (1997).

The aim of this section is to introduce the notion of the Young measure in a rather simple manner. We follow Ball (1989), Ball and Knowles (1990) and Pedregal (1997). For a more sophisticated approach exploiting the notion of measure theory like *disintegration*, the reader is referred to Valadier (1994).

Let $\{u^m\}_{m\in\mathbb{N}} \subset L^\infty(\Omega)^p$ be a sequence convergent to u in the weak-$*$ topology. If $f : \mathbb{R}^p \to \mathbb{R}$ is a continuous function, the sequence $\{f(u^m)\}_{m\in\mathbb{N}}$ is bounded in $L^\infty(\Omega)$. Consequently, up to a subsequence one has

$$f(u^m) \rightharpoonup g \quad \text{in} \quad L^\infty(\Omega) \text{ weak-}* \text{ when} \quad m \to \infty .$$

Naturally, a question which arises is: what is the relation between g and $f(u)$? *Parametrized measures* are a device to answer this basic question.

In general, both $\{u^m\}_{m\in\mathbb{N}}$ and $\{f(u^m)\}_{m\in\mathbb{N}}$ are oscillating sequences. The answer is straightforward in two simple cases:

(i) if f is affine then $g = f(u)$,

(ii) if f is convex and continuous then $g \geq f(u)$.

We recall that f is affine means that both f and $(-f)$ are convex functions.

The idea of Young consists in associating with $u^m(x)$ a measure on $\Omega \times \mathbb{R}^p$ by putting for each continuous and bounded function φ on $\Omega \times \mathbb{R}^p$:

$$\langle \mu_m, \varphi \rangle = \int_\Omega \varphi(x, u^m(x))dx . \tag{21.6.1}$$

Particularly, one may take $\varphi \in C_0(\Omega \times \mathbb{R}^p)$. We observe that for each Borel set A from $\Omega \times \mathbb{R}^p$

$$\mu_m(A) = |\{x \in \Omega \mid (x, u^m(x)) \in A\}| .$$

Similarly, for every Borel set $B \subset \Omega$ one has

$$\mu_m(B \times \mathbb{R}^p) = |B| ,$$

where bars $| \cdot |$ denote the Lebesgue measure. Thus if Ω is bounded then, cf. Sec. 13.1

$$||\mu_m|| = \sup_{|\varphi|\leq 1} |\langle \mu_m, \varphi \rangle| = |\Omega| < +\infty .$$

In such case $\{\mu^m\}_{m\in\mathbb{N}}$ is sequentially relatively compact in the weak-$*$ topology of $\mathbb{M}^1(\Omega \times \mathbb{R}^p)$. We can now formulate the first result on Young measures.

Theorem 21.6.1. Let $\Omega \subset \mathbb{R}^n$ be a bounded open set and $\mathcal{C} \subset \mathbb{R}^p$ a compact set. Then for each sequence $\{u^m(x)\}_{m\in\mathbb{N}}$ such that $u^m(x) \in \mathcal{C}$ for a.e. $x \in \Omega$ and for every $m \in \mathbb{N}$ there exists a subsequence m_j and a measure $\mu_\infty \in \mathbb{M}^1(\Omega \times \mathbb{R}^p)$ such that

(a) $\forall \varphi \in C(\overline{\Omega} \times \mathbb{R}^p), \quad \lim_{j\to\infty} \int_\Omega \varphi(x, u^{m_j}(x))dx = \langle \mu_\infty, \varphi \rangle.$

Moreover

$$\mu_\infty \geq 0 , \qquad\qquad \operatorname{supp} \mu_\infty \subset \Omega \times \mathcal{C} ,$$

$$\forall \, \psi \in C(\overline{\Omega}) , \qquad \langle \mu_\infty, \psi(\boldsymbol{x}) \otimes 1 \rangle = \int_\Omega \psi(\boldsymbol{x}) d\boldsymbol{x} ,$$

$$\forall \, \boldsymbol{\phi} \in C(\overline{\Omega})^p , \qquad \langle \mu_\infty, \boldsymbol{\phi}(\boldsymbol{x}) \cdot \boldsymbol{z} \rangle = \int_\Omega \boldsymbol{\phi}(\boldsymbol{x}) \cdot \boldsymbol{u} d\boldsymbol{x} ,$$

where $\boldsymbol{u}$ is the weak-$*$ limit of $\{\boldsymbol{u}^m\}_{m\in\mathbf{N}}$.

(b) There exists a parametrized family of probability measures $\{\nu_{\boldsymbol{x}}\}_{\boldsymbol{x}\in\Omega}$ on $\mathbf{R}^p$ such that $\operatorname{supp} \nu_{\boldsymbol{x}} \subset \mathcal{C}$ and

$$\langle \mu_\infty, \varphi \rangle = \int_\Omega [\int_{\mathbf{R}^p} \varphi(\boldsymbol{x}, \boldsymbol{z}) \nu_{\boldsymbol{x}}(d\boldsymbol{z})] d\boldsymbol{x} ,$$

$$\varphi(\cdot, \boldsymbol{u}^{m_j}(\cdot)) \rightharpoonup g(\cdot) = \int_{\mathbf{R}^p} \varphi(\cdot, \boldsymbol{z}) \nu_{\boldsymbol{x}}(d\boldsymbol{z}) \quad \text{in } L^\infty(\Omega) \quad \text{weak-}* ,$$

for each $\varphi \in C(\overline{\Omega} \times \mathbf{R}^p)$. In particular, for $\varphi(\boldsymbol{x}, \boldsymbol{z}) = \boldsymbol{z}$ we get

$$\boldsymbol{u}(\boldsymbol{x}) = \int_{\mathbf{R}^p} \boldsymbol{z} \nu_{\boldsymbol{x}}(d\boldsymbol{z}) . \tag{21.6.2}$$

Remark 21.6.2. The Young measure $\nu = \{\nu_{\boldsymbol{x}}\}_{\boldsymbol{x}\in\Omega}$ can intuitively be thought of as giving the limiting probability distribution of the values of $\boldsymbol{u}^{m_j}$ near $\boldsymbol{x}$ as $j \to \infty$. To be more precise, suppose that $\boldsymbol{x} \in \Omega$ and denote by $B(\boldsymbol{x}, \eta)$ the open ball with center $\boldsymbol{x}$ and radius $\eta > 0$. Keeping $\boldsymbol{x}, j$ and η fixed, let $\nu^j_{\boldsymbol{x},\eta} = \nu^{m_j}_{\boldsymbol{x},\eta}$ be the probability distribution of the values of $\boldsymbol{u}^{m_j}(\widetilde{\boldsymbol{x}})$ as $\widetilde{\boldsymbol{x}}$ is chosen uniformly at random from $B(\boldsymbol{x}, \eta)$. Then

$$\nu_{\boldsymbol{x}} = \lim_{\delta \to 0} \lim_{j \to \infty} \nu^j_{\boldsymbol{x},\eta} . \qquad\qquad \square$$

Let us consider now a more general case where $\mathcal{C} \subset \mathbf{R}^p$ is only closed. We shall formulate now a theorem similar to the previous one.

Theorem 21.6.3. Let $\Omega \subset \mathbf{R}^n$ be open and bounded and let $\mathcal{C} \subset \mathbf{R}^p$ be closed. Let $\boldsymbol{u}^m: \Omega \to \mathbf{R}^p$, $m = 1, 2, \ldots$, be a sequence of Lebesgue measurable functions satisfying $\boldsymbol{u}^m(\cdot) \to \mathcal{C}$ in measure as $m \to \infty$, i.e. given any open neighborhood $\mathcal{N}$ of $\mathcal{C}$ in $\mathbf{R}^p$

$$\lim_{m \to \infty} \operatorname{meas} \{\boldsymbol{x} \in \Omega \mid \boldsymbol{u}^{(m)}(\boldsymbol{x}) \notin \mathcal{N}\} = 0 .$$

Then there exists a subsequence $\{\boldsymbol{u}^{m_j}\}$ of $\{\boldsymbol{u}^m\}$ and a family $\{\nu_{\boldsymbol{x}}\}_{\boldsymbol{x}\in\Omega}$ of positive measures on $\mathbf{R}^p$, depending measurably on $\boldsymbol{x}$, such that

(i)

$$\|\nu_{\boldsymbol{x}}\|_{\mathbf{M}^1} = \int_{\mathbf{R}^p} d\nu_{\boldsymbol{x}} \leq 1 \text{ for } a.e.\ \boldsymbol{x} \in \Omega ,$$

(ii)

$$\operatorname{supp} \nu_{\boldsymbol{x}} \subset C \quad \text{for } a.e. \ \boldsymbol{x} \in \Omega ,$$

(iii)

$$f(\boldsymbol{u}^{m_j}) \rightharpoonup \langle \nu_{\boldsymbol{x}}, f \rangle = \int_{\mathbb{R}^p} f(\boldsymbol{z}) \nu_{\boldsymbol{x}}(d\boldsymbol{z}) \text{ in } L^\infty(\Omega) \text{ weak-} * , \qquad (21.6.3)$$

for each continuous function $f : \mathbb{R}^p \to \mathbb{R}$ satisfying $\lim_{|\boldsymbol{z}| \to \infty} f(\boldsymbol{z}) = 0$.

Suppose further that $\{\boldsymbol{u}^{m_j}\}$ satisfies the boundedness condition:

$$\lim_{k \to \infty} \sup_j \text{ meas } \{\boldsymbol{x} \in \Omega \cap B(0, R) : \ |\boldsymbol{u}^{m_j}(\boldsymbol{x})| \geq 0\} = 0 ,$$

for every $R > 0$. Then $\|\nu_{\boldsymbol{x}}\|_{\mathbf{M}^1} = 1$ for a.e. $\boldsymbol{x} \in \Omega$ (i.e. $\nu_{\boldsymbol{x}}$ is a probability measure), and given any measurable subset Ω_1 of Ω

$$f(\boldsymbol{u}^{m_j}) \rightharpoonup \langle \nu_{\boldsymbol{x}}, f \rangle \text{ in } L^1(\Omega_1) \text{ when } j \to \infty ,$$

for any continuous function $f : \mathbb{R}^p \to \mathbb{R}$ such that $\{f(\boldsymbol{u}^{m_j})\}$ is sequentially weakly relatively compact in $L^1(\Omega_1)$. $\qquad\qquad\square$

Measurable dependence of $\nu_{\boldsymbol{x}}$ on $\boldsymbol{x}$ means that $\langle \nu_{\boldsymbol{x}}, f \rangle$ is measurable.

Example 21.6.4. Let $\Omega = (0, 1) \times (0, 1)$ and consider the problem of minimizing

$$J(u) = \int_\Omega [(u_{,x_1} - 1)^2 + u_{,x_2}^2] dx_1 dx_2 , \qquad (21.6.4)$$

among scalar functions $u = u(x_\alpha)$ satisfying the boundary condition

$$u|_{x_2=0} = 0 . \qquad (21.6.5)$$

In (21.6.4), $u_{,x_1}$, $u_{,x_2}$ denote the weak partial derivatives of u. We claim that the infimum of J subject to (21.6.5) is zero, but that it is not attained. To prove the former statement, define $\overline{u} : \mathbb{R} \times (0, \infty) \to \mathbb{R}$ by

$$\overline{u}(x_\alpha) = \begin{cases} x_1 \varphi(x_2) & \text{if } \ 0 \leq x_1 \leq \dfrac{1}{2} , \\[2mm] (1 - x_1)\varphi(x_2) & \text{if } \ \dfrac{1}{2} \leq x_1 \leq 1 , \end{cases} \qquad (21.6.6)$$

where $\varphi(x_2) = x_2$ if $0 \leq x_2 \leq 1$, and $\varphi(x_2) = 1$ if $x_2 \leq 1$, extended as a 1-periodic function of x_1 to the whole of $\mathbb{R} \times (0, \infty)$. Then define $u^m(x_\alpha) = m^{-1}\overline{u}(mx_1, mx_2)$. Now $\nabla u^m(x_\alpha) = (\overline{u}_{,x_1}, \overline{u}_{,x_2})(mx_1, mx_2)$ is uniformly bounded and so

$$\lim_{m \to \infty} J(u^m) = \lim_{m \to \infty} \int_{\Omega \cap \{x_2 \leq m^{-1}\}} [((u^m_{,x_1})^2 - 1)^2 + (u^m_{,x_2})^2] dx_1 dx_2 = 0 .$$

Hence the infimum of J subject to (21.6.5) is zero. It is not attained because any minimizer u would satisfy $u_{,x_2} \equiv 0$, which together with (21.6.5) implies that $u \equiv 0$ and hence that $J(u) = 1$, contradicting inf $J = 0$.

Let us now determine the Young measure $\nu_{(x_1,x_2)}$ corresponding to ∇u^m for any minimizing sequence $\{u^m\}_{m\in\mathbb{N}}$. We assume that a subsequence has already been extracted so that the Young measure is defined. In particular, $\nabla u^m \rightharpoonup \nabla u$ in $L^1(\Omega)$ for some u satisfying (21.6.5). Since obviously $\nabla u^m \to \mathcal{C} = \{(-1,0),(1,0)\}$ in measure, by Theorem 21.6.3 we have that supp $\nu_{(x_1,x_2)} \subset \mathcal{C}$ a.e., i.e.:

$$\nu_{(x_1,x_2)} = \lambda(x_\alpha)\delta_{(-1,0)} + (1 - \lambda(x_\alpha))\delta_{(1,0)} ,$$

where $0 \le \lambda(x_2) \le 1$ a.e. and $\delta_{(a,b)}$ stands for the Dirac delta concentrated at (a,b). Application of formula (21.6.2) yields

$$u_{,x_1}(x_\alpha) = \int_{\mathbb{R}^2} z_1[\lambda(x_\alpha)\delta_{(-1,0)} + (1 - \lambda(x_\alpha))\delta_{(1,0)}] = 1 - 2\lambda(x_\alpha) ,$$

$$u_{,x_2}(x_\alpha) = \int_{\mathbb{R}^2} z_2[\lambda(x_\alpha)\delta_{(-1,0)} + (1 - \lambda(x_\alpha))\delta_{(1,0)}] = 0 .$$

Hence, by (21.6.4) we conclude that $u = 0$ and consequently $\lambda(x_\alpha) = 1/2$ a.e. We have thus proved that

$$\nu_{(x_1,x_2)} = \frac{1}{2}\delta_{(-1,0)} + \frac{1}{2}\delta_{(1,0)} .$$

In particular, the Young measure is unique. $\qquad\square$

Bonnetier and Conca (1994) proved the converse part of the fundamental theorem for Young (parametrized) measures: given a parametrized measure $\{\nu_x\}_{x\in\Omega}$, find a sequence $\{u^j\}_{j\in\mathbb{N}}$ satisfying (26.1.3). Their construction can be sketched as follows. One considers a dense family of functions $\{\varphi_k\}_{k\in\mathbb{N}} \subset C(\mathcal{C})$, where $\mathcal{C}$ is a compact set in $\mathbb{R}^p$. For each l, one constructs a sequence of functions $u^{j,l} : \Omega \to \mathcal{C}$, that satisfy condition (21.6.3) for the first l φ_k. These functions are obtained using a result in measure theory, stated here as Theorem 21.6.6 (see below), which provides an approximation of the parametrized measure ν_x by convex sums of Dirac masses. A diagonal process, as l tends to infinity, yields the desired sequence $\{u^j\}$.

Let us pass to the formulation of the representation of the integral in (21.6.3). This result is referred to by Bonnetier and Conca (1994) as the "approximation theorem".

Let $\mathcal{C} = [0,1]^p$ and let $\{\varphi_k\}_{k\in\mathbb{N}}$ be a sequence of continuous functions defined on $\mathcal{C}$, that satisfy the following hypotheses:

(H_1) the functions $\{\varphi_k\}_{k\in\mathbb{N}}$ form a dense set of linearly independent functions in $C(\mathcal{C})$;

(H_2) the functions $\{\varphi_k\}_{k\in\mathbb{N}}$ are positive and bounded on $\mathcal{C}$, uniformly with respect to k, for example

$$\forall\, z \in \mathcal{C} , \qquad 0 \le \varphi_k(z) \le 1 ;$$

(H_3) there exists $N \geq 1$, such that for $j \leq N$, no point $(\varphi_1(z), \ldots, \varphi_j(z))$, with $z \in C$, can be written as a finite convex combination of points of the same form. That is, if

$$\begin{pmatrix} \varphi_1(z) \\ \vdots \\ \vdots \\ \varphi_j(z) \end{pmatrix} = \sum_{i=1}^{m} \theta_1 \begin{pmatrix} \varphi_1(z_i) \\ \vdots \\ \vdots \\ \varphi_j(z_i) \end{pmatrix}$$

with

$$z, z_i \in C , \qquad \theta_i \geq 0 , \qquad 1 \leq i \leq m , \qquad \sum_{i=1}^{m} \theta_i = 1 ,$$

then there exists i_0, $1 \leq i_0 \leq m$, such that

$$\begin{cases} \lambda_{i_0} = \lambda , \\ \theta_{i_0} = 1 . \end{cases}$$

Remark 21.6.5. Hypothesis (H_3) implies that the functions $\{\varphi_k\}_{k \in \mathbb{N}}$ separate the measures δ_z. When $p = 1$, and when the functions φ_k are the polynomials z^k, assumptions (H_1) — (H_3) are clearly satisfied.

Theorem 21.6.6. Let Ω be an open bounded set in $\mathbb{R}^n$, and $\{\mu_x\}_{x \in \Omega}$ be a family of positive Borel measures such that

$$\operatorname{supp} \mu_x \subset C , \qquad \langle \mu_x, 1 \rangle = 1 ,$$

for almost every $x \in \Omega$. Assume that $x \to \mu_x$ is measurable, i.e. that

$$\forall f \in C(C) , \qquad x \to \langle \mu_x, f \rangle$$

is Lebesgue measurable on Ω.

Then there exist $2(j + 1)$ measurable functions $\theta_d(x), a_d(x)$, defined on Ω,

$$\begin{cases} \theta_d(x) \in [0, 1] , \\ a_d(x) \in C , \\ \sum_{d=1}^{j+1} \theta_d(x) = 1 , \end{cases}$$

for a.e. $x \in \Omega$ such that

$$\forall 1 \leq l \leq j , \qquad \langle \mu_x, \varphi_k \rangle = \sum_{d=1}^{j+1} \theta_d(x) \varphi_l(a_d(x)) \quad \text{a.e. } x \in \Omega .$$

If, moreover, $p = 1$, and $\varphi_k = z^k$, then there exist $2j$ measurable functions $\theta_d(\boldsymbol{x})$, $a_d(\boldsymbol{x})$, defined on Ω, with values in $[0, 1]$, such that

$$\forall\, 1 \leq l \leq 2j - 1 , \qquad \langle \mu_{\boldsymbol{x}}, \lambda^l \rangle = \sum_{d=1}^{j} \theta_d(\boldsymbol{x}) a_d^l(\boldsymbol{x}) \ a.e. \ \boldsymbol{x} \in \Omega . \qquad \square$$

As an application of the last theorem, Bonnetier and Conca (1994) established the relative compactness for the weak-$*$ topology of some subsets of $L^\infty(\Omega)^r$.

Theorem 21.6.7. Let Ω be a bounded domain in $\mathbb{R}^n$, $\mathcal{C} = [0, 1]^p$, $p \geq 1$, and consider a family of r linearly independent functions $(\varphi_1, \ldots, \varphi_r) \in C(\mathcal{C})^r$, that satisfy (H_2) and (H_3).

Let $\mathcal{H}$ be the set of functions $\boldsymbol{h} = (h_1, \ldots h_r) \in L^\infty(\Omega)^r$, such that there exist

$$\begin{aligned} \theta_i &\in L^\infty(\Omega, [0, 1]) \\ a_i &\in L^\infty(\Omega, \mathcal{C})^p \end{aligned} \quad , \qquad 1 \leq i \leq r + 1 ,$$

with

$$\sum_{i=1}^{r+1} \theta_i(\boldsymbol{x}) \varphi_m(a_i(\boldsymbol{x})) = h_m(\boldsymbol{x}) , \quad 1 \leq m \leq r ,$$

$$\sum_{i=1}^{r+1} \theta_i = 1 .$$

Then $\mathcal{H}$ is compact for the weak-$*$ topology. $\qquad \square$

Corollary 21.6.8. The set

$$\mathrm{H}_3 = \{ (h, k, l) \in L^\infty(\Omega)^3 \mid h(\boldsymbol{x}) = \sum_{i=1}^{4} \theta_i(\boldsymbol{x}) h_i(\boldsymbol{x}) , \qquad k(\boldsymbol{x}) = \sum_{i=1}^{4} \theta_i(\boldsymbol{x}) h_i^3(\boldsymbol{x}) ,$$

$$l(\boldsymbol{x}) = \sum_{i=1}^{4} \theta_i(\boldsymbol{x}) h_i^{-3}(\boldsymbol{x}) \}$$

is compact for the weak-$*$ topology. $\qquad \square$

22.   Two-phase plate in bending. Hashin-Shtrikman bounds

Assume that a thin plate is made of a two-phase composite material. The basic cell Y will be viewed as a representative volume (area) element (RVE). Within the cell Y two isotropic materials are distributed and the area fractions m_1, m_2 are given; $m_1 + m_2 = 1$. The bending properties are characterized by isotropic bending stiffness tensors D_1, D_2. The perfect mixture of both materials leads to an isotropic effective plate characterized by two moduli: k_0 and μ_0. The problem of Hashin and Shtrikman is to find the lower and upper bounds for these moduli. The upper bounds will refer to the stiffest isotropic plate constructed from two given isotropic phases. The lower bounds refer to the softest two-phase composite plate. The bounds will be found by the translation method.

22.1.   Lower bound for the Kelvin modulus

Consider a balanced plate subject to bending loads. The plate is made of two materials distributed in a homogeneous manner in the transverse direction. Both materials are mixed at the local level. The distribution of the bending stiffness tensor D within the cell Y is given by

$$D(y) = \chi_1(y)D_1 + \chi_2(y)D_2 . \tag{22.1.1}$$

The cell Y is divided into two subdomains Y_α ($\alpha = 1, 2$) and χ_α is a characteristic function of the domain Y_α, i.e.

$$\chi_\alpha(y) = \begin{cases} 1, & \text{if } y \in Y_\alpha , \\ 0, & \text{if } y \notin Y_\alpha . \end{cases} \tag{22.1.2}$$

We have $\bar{Y}_1 \cup \bar{Y}_2 = \bar{Y}$, $Y_1 \cap Y_2 = \phi$. The domain Y_α is occupied by the α-th phase of the bending stiffness D_α. The isotropy assumption means that D_α is determined by two independent moduli k_α and μ_α according to Eq. (21.1.26):

$$D_\alpha = 2k_\alpha I_1 + 2\mu_\alpha I_2 . \tag{22.1.3}$$

If D_α are referred to the basis $\mathbf{a}_i \otimes \mathbf{a}_j$, then, cf. (21.1.22) and (21.1.33)

$$[D_\alpha^{ij}] = \operatorname{diag} [2k_\alpha, 2\mu_\alpha, 2\mu_\alpha] . \tag{22.1.4}$$

The area fraction of the α-th phase is defined by

$$m_\alpha = \langle \chi_\alpha(y) \rangle , \tag{22.1.5}$$

hence $m_1 + m_2 = 1$. Integration of $D(y)$ over Y gives

$$\langle D \rangle = m_1 D_1 + m_2 D_2 . \tag{22.1.6}$$

The bending compliance tensor d equals D^{-1}. We have

$$d(y) = \chi_1(y)d_1 + \chi_2(y)d_2 , \tag{22.1.7}$$

where $d_\alpha = D_\alpha^{-1}$. Moreover

$$\langle d \rangle = m_1 d_1 + m_2 d_2 . \tag{22.1.8}$$

The effective bending stiffness tensor is given by Eq. (3.4.12), i.e.:

$$\kappa_{\alpha\beta}^h D_0^{\alpha\beta\lambda\mu} \kappa_{\lambda\mu}^h = \min\{\langle \kappa_{\alpha\beta} D^{\alpha\beta\lambda\mu}(y)\kappa_{\lambda\mu}\rangle | \kappa \in \mathcal{K}_{\boldsymbol{\kappa}}^{per}(Y)\,, \langle \kappa \rangle = \kappa^h\} \qquad (22.1.9)$$

where $\mathcal{K}_{\boldsymbol{\kappa}}^{per}(Y)$ was given in Sec. 3.4.

We assume that the mixture is isotropic or

$$\boldsymbol{D}_0 = 2k_0\boldsymbol{I}_1 + 2\mu_0\boldsymbol{I}_2\,, \qquad (22.1.10)$$

where k_0 and μ_0 are the Kelvin and Kirchhoff moduli of the isotropic mixture. The aim of this section is to estimate k_0 from below. To this end we use the translation method.

To facilitate further computations we refer the entities of (22.1.9) to the vectorial basis $(\boldsymbol{a}_i)$:

$$\sum_{i,j=1}^{3} \kappa_h^i D_0^{ij} \kappa_h^j = J(\boldsymbol{\kappa}^h)\,, \qquad (22.1.11)$$

where

$$J(\boldsymbol{\kappa}^h) = \min\{\langle \sum_{i,j=1}^{3} \kappa^i D^{ij} \kappa^j\rangle | \kappa \in \mathcal{K}_{\boldsymbol{\kappa}}^{per}(Y), \langle \kappa \rangle = \kappa^h\} \qquad (22.1.12)$$

and rearrange $J(\boldsymbol{\kappa}^h)$ as follows

$$J(\boldsymbol{\kappa}^h) = \min\{\underline{\langle \boldsymbol{\kappa}^T(\boldsymbol{D}-\boldsymbol{T})\boldsymbol{\kappa}\rangle} + \langle \boldsymbol{\kappa}^T\boldsymbol{T}\boldsymbol{\kappa}\rangle | \kappa \in \mathcal{K}_{\boldsymbol{\kappa}}^{per}(Y), \langle \kappa \rangle = \kappa^h\}\,. \qquad (22.1.13)$$

Here $\boldsymbol{\kappa} = (\kappa^1, \kappa^2, \kappa^3)^T$, $\boldsymbol{T} = -t\,\overset{o}{\boldsymbol{T}}$, $t \in \mathbb{R}$ and $\overset{o}{\boldsymbol{T}}$ is defined by (21.1.19). Assume that t is chosen such that $\boldsymbol{D} - \boldsymbol{T} \geq 0$. The idea of the translation method lies in neglecting the differential constraints in estimating the underlined term in (22.1.13) and applying the equality $(21.4.7)_1$ to the second term. Thus the first term can be estimated according to (21.5.4). We find

$$J(\boldsymbol{\kappa}^h) \geq (\boldsymbol{\kappa}^h)^T[\langle(\boldsymbol{D}-\boldsymbol{T})^{-1}\rangle^{-1} + \boldsymbol{T}]\boldsymbol{\kappa}^h\,. \qquad (22.1.14)$$

To make the estimate possibly optimal we maximize its r.h.s.:

$$J(\boldsymbol{\kappa}^h) \geq \max_{\substack{\boldsymbol{T}=t\boldsymbol{T}^0, t\in\mathbb{R} \\ \boldsymbol{D}-\boldsymbol{T}\geq 0}} (\boldsymbol{\kappa}^h)^T[\langle(\boldsymbol{D}-\boldsymbol{T})^{-1}\rangle^{-1} + \boldsymbol{T}]\boldsymbol{\kappa}^h\,. \qquad (22.1.15)$$

Note that

$$\boldsymbol{D}_\alpha - \boldsymbol{T} = \text{diag}\,[2k_\alpha - t, 2\mu_\alpha + t, 2\mu_\alpha + t]\,; \quad \alpha = 1, 2\,. \qquad (22.1.16)$$

Thus the conditions $\boldsymbol{D}_\alpha - \boldsymbol{T} \geq 0$ imply

$$t \leq 2k_{min}\,, \qquad t \geq -2\mu_{min}\,. \qquad (22.1.17)$$

The formulae (22.1.11), (22.1.15) yield

$$\boldsymbol{D}_0 \geq \langle(\boldsymbol{D}-\boldsymbol{T})^{-1}\rangle^{-1} + \boldsymbol{T}\,, \qquad (22.1.18)$$

and by Lemma 21.2.3 this is equivalent to estimating

$$Y(\boldsymbol{D}_0) + \boldsymbol{T} \geq \boldsymbol{0} , \qquad (22.1.19)$$

with $Y(\cdot)$ given by (21.2.6). Since the matrices $\boldsymbol{D}_0$, $\boldsymbol{D}_\alpha$ and $\boldsymbol{T}$ are diagonal if referred to the basis $\boldsymbol{a}_i \otimes \boldsymbol{a}_j$, Remark 21.2.1 applies and the condition (22.1.19) is equivalent to the set of conditions (21.2.13). They now assume the form

$$y(2k_0, 2k_1, 2k_2, m_1, m_2) + t \geq 0 ,$$
$$y(2\mu_0, 2\mu_1, 2\mu_2, m_1, m_2) - t \geq 0 . \qquad (22.1.20)$$

Assume that

$$k_2 > k_1 , \qquad \mu_2 > \mu_1 , \qquad (22.1.21)$$

which is called the ordered case. Then $\mu_{min} = \mu_1$ and $k_{min} = k_1$. Hence $t \in [-2\mu_1, 2k_1]$. For the best estimate of k_0 one should take the greatest admissible $(-t) = 2\mu_1$. By $(22.1.20)_1$ we find

$$y(2k_0, 2k_1, 2k_2, m_1, m_2) \geq 2\mu_1 . \qquad (22.1.22)$$

By (21.2.18) this estimate is equivalent to

$$y(k_0, k_1, k_2, m_1, m_2) \geq \mu_1 . \qquad (22.1.23)$$

Hence $k_0 \geq \underline{k}_{HS}$, where $\underline{k}_{HS}$ is given by

$$y(\underline{k}_{HS}, k_1, k_2, m_1, m_2) = \mu_1 . \qquad (22.1.24)$$

By (21.2.22) the expression above is equivalent to the formula

$$\frac{m_2}{\underline{k}_{HS} - k_1} = \frac{1}{k_2 - k_1} + \frac{m_1}{k_1 + \mu_1} . \qquad (22.1.25)$$

The quantity $\underline{k}_{HS}$ is the lower Hashin-Shtrikman bound for k_0. The estimate $(22.1.20)_2$ does not lead to the lower estimate of μ_0 and that is why will not be analyzed.

The above derivation was based on the formula (22.1.9) concerning the case of a periodic composite. By the theorem of Kohn and Dal Maso, the periodicity assumption imposes no restriction on the final result Eq. (22.1.25), cf. Sec. 1.3.2 (Denseness of periodic composites) and Sec. 26.2.

22.2. Upper bound for the Kelvin modulus

Assume that the materials are ordered, cf. (22.1.21). The upper bound: $k_0 \leq \bar{k}_{HS}$ follows from (22.1.25) by replacing the material (1) with material (2). We find

$$\frac{m_1}{\bar{k}_{HS} - k_2} = \frac{1}{k_1 - k_2} + \frac{m_2}{k_2 + \mu_2} . \qquad (22.2.1)$$

The same result follows from estimating the effective complementary energy, cf. Sec. 3.6,

$$M_h^{\alpha\beta} d^0_{\alpha\beta\lambda\mu} M_h^{\lambda\mu} = \min\{\langle m^{\alpha\beta} d_{\alpha\beta\lambda\mu} m^{\lambda\mu}\rangle | m \in \mathcal{S}_2^{per}(Y), \langle m \rangle = M_h\}. \quad (22.2.2)$$

We decompose the averaged energy:

$$\langle m^T d\, m \rangle = \langle m^T(d - T)m \rangle + \langle m^T T m \rangle \quad (22.2.3)$$

and take $T = t\, \overset{o}{T}$, $t \geq 0$, $\overset{o}{T}$ given by (21.1.19). By using the estimate (21.5.4) to the first term and the estimate (21.4.15) to the second term one arrives at the inequality, equivalent to

$$Y(d_0) + T \geq 0.$$

Its analysis leads to (22.2.1).

22.3. *Lower bound for the Kirchhoff modulus*

To estimate μ_0 one should subject the cell Y to two simultaneous deformation states. The lower estimation of elastic energy stored in the cell leads to the optimal lower estimate of μ_0.

The functional to be estimated has the form

$$J(\kappa^h, \rho^h) = \min\{\langle \kappa^T D\kappa + \rho^T D\rho\rangle | \kappa, \rho \in \mathcal{K}_{\kappa}^{per}(Y),$$
$$\langle \kappa \rangle = \kappa^h, \langle \rho \rangle = \rho^h\}. \quad (22.3.1)$$

Here $\kappa = (\kappa^1, \kappa^2, \kappa^3)^T$, $\rho = (\rho^1, \rho^2, \rho^3)^T$, $\kappa^h = (\kappa_h^1, \kappa_h^2, \kappa_h^3)^T$, $\rho^h = (\rho_h^1, \rho_h^2, \rho_h^3)^T$. Let us introduce the notation

$$\mathcal{D} = \begin{bmatrix} D & 0 \\ 0 & D \end{bmatrix}, \quad (22.3.2)$$

$$\varepsilon = (\kappa, \rho)^T, \qquad \langle \varepsilon \rangle = (\langle \kappa \rangle, \langle \rho \rangle)^T, \qquad \varepsilon^h = (\kappa^h, \rho^h)^T.$$

Thus

$$J(\varepsilon^h) = \min\{\langle \varepsilon^T \mathcal{D}\varepsilon\rangle | \varepsilon \in \mathcal{K}_{\kappa}^{per}(Y) \times \mathcal{K}_{\kappa}^{per}(Y), \langle \varepsilon \rangle = \varepsilon^h\}. \quad (22.3.3)$$

Assume that $\varepsilon \in \mathcal{K}_{\kappa}^{per}(Y) \times \mathcal{K}_{\kappa}^{per}(Y)$ and $\langle \varepsilon \rangle = \varepsilon^h$. Then, by (21.4.12)

$$\langle \varepsilon^T \mathcal{D}\varepsilon\rangle = \langle \varepsilon^T(\mathcal{D} - \mathcal{T})\varepsilon\rangle + (\varepsilon^h)^T \mathcal{T}\varepsilon^h, \quad (22.3.4)$$

where $\mathcal{T}$ is defined by (21.4.10). Let us take $t_\alpha \in \mathbb{R}$ such that $\mathcal{D} - \mathcal{T} \geq 0$ and apply the inequality (21.5.4) for $A = \mathcal{D} - \mathcal{T}$, disregarding the differential constraints concealed in the definition of the set $\mathcal{K}_{\kappa}^{per}(Y)$. We find

$$J(\varepsilon^h) \geq (\varepsilon^h)^T[\langle(\mathcal{D} - \mathcal{T})^{-1}\rangle^{-1} + \mathcal{T}]\varepsilon^h. \quad (22.3.5)$$

Let us stipulate that the effective composite is isotropic or:

$$J(\varepsilon^h) = (\varepsilon^h)^T \mathcal{D}_0 \varepsilon^h \,, \tag{22.3.6}$$

with

$$[\mathcal{D}_0^{ij}] = \operatorname{diag}\left[2k_0, 2\mu_0, 2\mu_0, 2k_0, 2\mu_0, 2\mu_0\right] \,. \tag{22.3.7}$$

The condition

$$\mathcal{D}_0 \geq \langle(\mathcal{D} - \mathcal{T})^{-1}\rangle^{-1} + \mathcal{T} \tag{22.3.8}$$

can be written in the form

$$Y(\mathcal{D}_0) + \mathcal{T} \geq 0 \,. \tag{22.3.9}$$

The condition $\mathcal{D} - \mathcal{T} \geq 0$ means that the 6×6 matrix:

$$A = \begin{bmatrix} 2k_\alpha + t_1 & & & & & \\ & 2\mu_\alpha - t_1 & & & & -t_2 \\ & & 2\mu_\alpha - t_1 & & t_2 & \\ & & & 2k_\alpha + t_1 & & \\ & & t_2 & & 2\mu_\alpha - t_1 & \\ & -t_2 & & & & 2\mu_\alpha - t_1 \end{bmatrix} \tag{22.3.10}$$

is semipositive definite. The estimate (22.3.9) is equivalent to the condition of semipositive definiteness of the 6×6 matrix

$$B = \begin{bmatrix} a - t_1 & & & & & \\ & b + t_1 & & & t_2 & \\ & & b + t_1 & -t_2 & & \\ & & & a - t_1 & & \\ & & -t_2 & & b + t_1 & \\ & t_2 & & & & b + t_1 \end{bmatrix} \,, \tag{22.3.11}$$

where

$$a = y(2k_0, 2k_1, 2k_2, m_1, m_2) \,, \qquad b = y(2\mu_0, 2\mu_1, 2\mu_2, m_1, m_2) \,, \tag{22.3.12}$$

according to the Sylvester theorem the matrix A is semipositive definite, if the following determinants are non-negative

$$\det A = (2k_\alpha + t_1)^2 (2\mu_\alpha - t_1 - t_2)^2 (2\mu_\alpha - t_1 + t_2)^2 \,,$$

$$\det(A_{2 \leq i,j \leq 6}) = (2\mu_\alpha + t_2 - t_1)^2 (2\mu_\alpha - t_1 - t_2)^2 (2k_\alpha + t_1) \,, \tag{22.3.13}$$

$$\det(A_{3 \leq i,j \leq 6}) = (2\mu_\alpha - t_1)(2\mu_\alpha - t_1 + t_2)(2\mu_\alpha - t_1 - t_2)(2k_\alpha + t_1) \,,$$

and

$$2k_\alpha + t_1 \geq 0 \,, \qquad 2\mu_\alpha - t_1 \geq 0 \,. \tag{22.3.14}$$

Hence

$$(t_2)^2 \leq (2\mu_\alpha - t_1)^2 \,, \quad t_1 \geq -2k_\alpha \,. \tag{22.3.15}$$

If the ordering assumptions (22.1.21) hold, then

$$t_1 \geq -2k_1 \qquad t_2 \leq 2\mu_1 + 2k_1 \ . \tag{22.3.16}$$

Now let us consider the conditions of semipositive definiteness of the matrix $\boldsymbol{B}$. The following determinants should be non-negative

$$\det \boldsymbol{B} = (a - t_1)^2 (b + t_1 - t_2)^2 (b + t_1 + t_2)^2 \ ,$$

$$\det(\boldsymbol{B}_{2 \leq i,j \leq 6}) = (a - t_1)(b + t_1 - t_2)^2 (b + t_1 + t_2)^2 \ , \tag{22.3.17}$$

$$\det(\boldsymbol{B}_{3 \leq i,j \leq 6}) = (a - t_1)(b + t_1)(b + t_1 + t_2)(b + t_1 - t_2) \ ,$$

and

$$a - t_1 \geq 0 \ , \qquad b + t_1 \geq 0 \ . \tag{22.3.18}$$

Sylvester's conditions are satisfied provided that

$$(t_2)^2 \leq (b + t_1)^2 \ , \qquad t_1 \leq a \ . \tag{22.3.19}$$

Hence

$$b \geq t_2 - t_1 \ . \tag{22.3.20}$$

Of all admissible parameters t_1, t_2 one should choose those which maximize the expression $t_2 - t_1$. Thus we should take the largest t_2 and the smallest t_1 among their admissible values. Thus

$$\max(t_2 - t_1) = 2\mu_1 + 2k_1 - (-2k_1) = 4k_1 + 2\mu_1 \ .$$

The condition $b \geq 4k_1 + 2\mu_1$, equivalent to

$$y(2\mu_0, 2\mu_1, 2\mu_2, m_1, m_2) \geq 2(2k_1 + \mu_1) \ , \tag{22.3.21}$$

determines the estimate $\mu_0 \geq \underline{\mu}_{HS}$ with $\underline{\mu}_{HS}$ given by

$$y(2\underline{\mu}_{HS}, 2\mu_1, 2\mu_2, m_1, m_2) = 2(2k_1 + \mu_1) \ .$$

By (21.2.18) we find

$$y(\underline{\mu}_{HS}, \mu_1, \mu_2, m_1, m_2) = 2k_1 + \mu_1 \ , \tag{22.3.22}$$

and by (21.2.22) we have

$$\frac{m_2}{\underline{\mu}_{HS} - \mu_1} = \frac{1}{\mu_2 - \mu_1} + \frac{m_1}{2(k_1 + \mu_1)} \ . \tag{22.3.23}$$

The quantity $\underline{\mu}_{HS}$ represents the lower Hashin-Shtrikman estimate of μ_0.

22.4. Upper bound for the Kirchhoff modulus

By interchanging the role of both materials in the formula (22.3.23) one finds $\mu_0 \leq \bar{\mu}_{HS}$ where $\bar{\mu}_{HS}$ is determined by

$$\frac{m_1}{\underline{\mu}_2 - \bar{\mu}_{HS}} = \frac{1}{\mu_2 - \mu_1} - \frac{m_2}{2(k_2 + \mu_2)} \ . \tag{22.4.1}$$

The same expression can be found by estimating the sum of complementary energies. This derivation is left to the reader.

22.5. *Attainability of Hashin-Shtrikman bounds. The Francfort – Murat construction*

Consider a ribbed plate of third rank constructed by coating the second rank ribbed plate of stiffness tensor $\boldsymbol{D}_{hh}$ (Sec. 3.8, cf. Fig. 3.8.2) with layers of the stronger material (2). The layering direction is $\boldsymbol{p}$ and the area fractions of materials (hh) and (2) are β_1 and β_2 respectively, see Fig. 22.5.1.

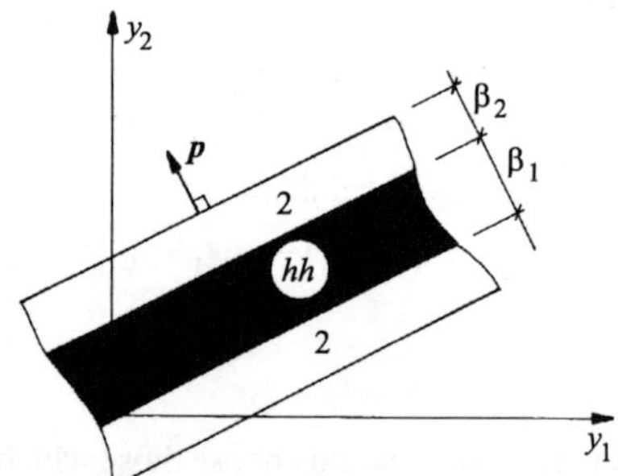

Fig. 22.5.1. A ribbed plate of third rank of effective bending stiffnesses D_{hhh}

Let us assume the ordered case (22.1.21). The effective stiffness tensor $\boldsymbol{D}_{hhh}$ of such a third-rank ribbed plate is implicitly given by formula (3.8.33), or

$$\beta_1(\boldsymbol{D}_2 - \boldsymbol{D}_{hhh})^{-1} = (\boldsymbol{D}_2 - \boldsymbol{D}_{hh})^{-1} - \frac{\beta_2}{s_2}\boldsymbol{\Gamma}_p \,, \tag{22.5.1}$$

where

$$\boldsymbol{\Gamma}_p = \boldsymbol{p} \otimes \boldsymbol{p} \otimes \boldsymbol{p} \otimes \boldsymbol{p} \,, \quad s_2 = k_2 + \mu_2 \,. \tag{22.5.2}$$

Multiplication of both sides by $\alpha_1\theta_1$ and taking into account (3.8.35) gives

$$\beta_1\theta_1\alpha_1(\boldsymbol{D}_2 - \boldsymbol{D}_{hhh})^{-1} = (\boldsymbol{D}_2 - \boldsymbol{D}_1)^{-1} - \frac{1}{s_2}\boldsymbol{\Gamma}_{n,m,p} \,, \tag{22.5.3}$$

with

$$\boldsymbol{\Gamma}_{n,m,p} = \theta_2\boldsymbol{\Gamma}_n + \theta_1\alpha_2\boldsymbol{\Gamma}_m + \theta_1\alpha_1\beta_2\boldsymbol{\Gamma}_p \,. \tag{22.5.4}$$

Note that

$$m_1 = \beta_1\theta_1\alpha_1 \,, \tag{22.5.5}$$

represents the area fraction of material (1). Let us compute the sum of the coefficients appearing in (22.5.4)

$$\theta_2 + \theta_1\alpha_2 + \theta_1\alpha_1\beta_2 = 1 - \theta_1 + \theta_1(1 - \alpha_1) + \theta_1\alpha_1(1 - \beta_1)$$
$$= 1 - \theta_1\alpha_1\beta_1 = 1 - m_1 = m_2 \,. \tag{22.5.6}$$

Thus the formula (22.5.3) can be put in the form

$$m_1(\boldsymbol{D}_2 - \boldsymbol{D}_{hhh})^{-1} = (\boldsymbol{D}_2 - \boldsymbol{D}_1)^{-1} - \frac{1}{s_2}(\rho_1 \boldsymbol{\Gamma}_n + \rho_2 \boldsymbol{\Gamma}_m + \rho_3 \boldsymbol{\Gamma}_p)\,, \qquad (22.5.7)$$

where

$$\sum_{i=1}^{3} \rho_i = m_2\,. \qquad (22.5.8)$$

The plate material (hhh) can further be coated by layers of material (2) and this process can be continued. The soft material (1) plays the role of a core and the strong material (2) is used for subsequent coatings. The subsequent layering directions are: $n_1 = n$, $n_2 = m$, $n_3 = p$, n_4, n_5, ..., n_N. The resulting stiffness tensor $\boldsymbol{D}_N$ is given by

$$m_1(\boldsymbol{D}_2 - \boldsymbol{D}_N)^{-1} = (\boldsymbol{D}_2 - \boldsymbol{D}_1)^{-1} - \frac{1}{s_2}\sum_{i=1}^{N}\rho_i \boldsymbol{\Gamma}_{n_i}\,, \qquad \sum_{i=1}^{N}\rho_i = m_2\,. \qquad (22.5.9)$$

Such laminates are sometimes called "stiff laminates". The "soft laminates" are constructed by taking material (2) as a core and material (1) as a coating.

We show below that one can choose the vectors n, m, p and the area fractions θ_1, α_1, β_1 such that the effective plate of stiffness $\boldsymbol{D}_{hhh}$ becomes isotropic.

Let us take the versors n, m, p such that their vertices form an equilateral triangle, see Fig. 22.5.2.

Let us take $n = e_1$ and

$$m = \left(-\frac{1}{2}, \frac{\sqrt{3}}{2}\right)\,, \qquad p = \left(-\frac{1}{2}, -\frac{\sqrt{3}}{2}\right)\,. \qquad (22.5.10)$$

For simplicity, the tensor $\boldsymbol{\Gamma}_{n,m,p}$ defined by (22.5.4) will be denoted by $\boldsymbol{\Gamma}$. Under the above choice of versors n, m and p the components $\Gamma_{\alpha\beta\lambda\mu}$ assume the form

$$\Gamma_{1111} = \theta_2 + \frac{\theta_1}{16}(\alpha_2 + \beta_2\alpha_1)\,, \qquad \Gamma_{2222} = \left(\frac{\sqrt{3}}{2}\right)^4 \theta_1(\alpha_2 + \beta_2\alpha_1)\,,$$

$$\Gamma_{1122} = \frac{1}{4}\left(\frac{\sqrt{3}}{2}\right)^4 \theta_1(\alpha_2 + \beta_2\alpha_1)\,, \qquad \Gamma_{1112} = \frac{\sqrt{3}}{16}\theta_1(\beta_2\alpha_1 - \alpha_2)\,, \qquad (22.5.11)$$

$$\Gamma_{2221} = \Gamma_{1112}\,, \qquad \Gamma_{1212} = \frac{3}{16}(\theta_1\alpha_2 + \theta_2\beta_2\alpha_1)\,.$$

Let us assume conditions of isotropy

$$\Gamma_{1222} = \Gamma_{1112} = 0\,, \qquad \Gamma_{1111} = \Gamma_{2222}\,, \qquad \Gamma_{1212} = \frac{1}{2}(\Gamma_{1111} - \Gamma_{1122})\,. \qquad (22.5.12)$$

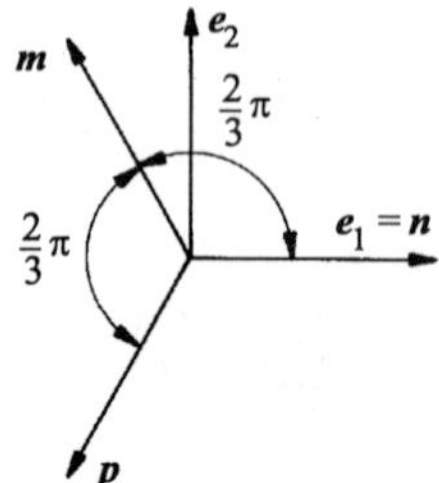

Fig. 22.5.2. Hexagonal choice of versors n, m, p

They are satisfied provided that

$$\beta_2 = \frac{\alpha_2}{\alpha_1}, \qquad \alpha_2 = \frac{\theta_2}{\theta_1}, \tag{22.5.13}$$

which makes the tensor Γ isotropic:

$$\Gamma = \frac{3}{4}\theta_2(2I_1 + I_2). \tag{22.5.14}$$

Let us assume that the tensor D_{hhh} is isotropic:

$$D_{hhh} = 2k_{hhh}I_1 + 2\mu_{hhh}I_2. \tag{22.5.15}$$

Then

$$m_1(D_2 - D_{hhh})^{-1} = \frac{m_1}{2(k_2 - k_{hhh})}I_1 + \frac{m_1}{2(\mu_2 - \mu_{hhh})}I_2. \tag{22.5.16}$$

On the other hand

$$(D_2 - D_1)^{-1} - \frac{1}{s_2}\Gamma = \left(\frac{1}{2\Delta k} - \frac{\frac{3}{2}\theta_2}{s_2}\right)I_1 + \left(\frac{1}{2\Delta\mu} - \frac{\frac{3}{4}\theta_2}{s_2}\right)I_2, \tag{22.5.17}$$

where $\Delta f = f_2 - f_1$, $f \in \{k, \mu\}$.

By equating (22.5.16) with (22.5.17) we arrive at two independent equations for k_{hhh} and μ_{hhh}

$$\frac{m_1}{k_2 - k_{hhh}} = \frac{1}{\Delta k} - \frac{3\theta_2}{s_2}, \qquad \frac{m_1}{\mu_2 - \mu_{hhh}} = \frac{1}{\Delta\mu} - \frac{3\theta_2}{2s_2}. \tag{22.5.18}$$

By (22.5.13) we find the relations

$$\theta_1 = 1 - \theta_2, \qquad \alpha_1 = \frac{1 - 2\theta_2}{1 - \theta_2}, \qquad \beta_1 = \frac{1 - 3\theta_2}{1 - 2\theta_2}. \tag{22.5.19}$$

By (22.5.5) $m_1 = 1 - 3\theta_2$, $m_2 = 1 - m_1$ and we can express θ_2, α_2 and β_2 in terms of m_2

$$\theta_2 = \frac{m_2}{3} , \qquad \alpha_2 = \frac{m_2}{3 - m_2} , \qquad \beta_2 = \frac{m_2}{3 - 2m_2} . \qquad (22.5.20)$$

Substituting $3\theta_2 = m_2$ into (22.5.18) gives

$$\frac{m_1}{k_2 - k_{hhh}} = \frac{1}{k_2 - k_1} - \frac{m_2}{k_2 + \mu_2} , \qquad \frac{m_1}{\mu_2 - \mu_{hhh}} = \frac{1}{\mu_2 - \mu_1} - \frac{m_2}{2(k_2 + \mu_2)} . \qquad (22.5.21)$$

According to (22.2.1) and (22.4.1) we recognize that $k_{hhh} = \bar{k}_{HS}$ and $\mu_{hhh} = \bar{\mu}_{HS}$.

The above derivation shows that:

(i) there exists an isotropic plate constructed by three subsequent layerings (by introducing ribs) with the stronger material as a coating,

(ii) this plate is the stiffest possible plate constructed from two isotropic materials taken in a given proportion,

(iii) the Hashin-Shtrikman upper bounds are attainable by the same composite plate. Thus the isotropic plate of moduli $\bar{k}_{HS}, \bar{\mu}_{HS}$ can effectively be constructed.

We observe that the function $\bar{k}_{HS}(m_2)$ increases from k_1 to k_2 monotonically if m_2 runs from 0 to 1. Behavior of the function $\bar{\mu}_{HS}(m_2)$ is similar, it increases from μ_1 to μ_2.

The construction of the softest isotropic plate from materials (1) and (2) with given area fractions m_1 and m_2 is performed similarly. One should take material (2) as a core and material (1) as a coating in the three subsequent layerings along versors $n = e_1$ and m, p given by (22.5.10). An isotropic plate, with moduli $\underline{k}_{HS}$ and $\underline{\mu}_{HS}$ given by Eqs. (22.1.25) and (22.3.23) is thus constructed. This procedure should be based on the formula (3.8.58) for compliances.

Remark 22.5.1. The analysis of Sec. 5.6 has shown that the formula (3.2.32) due to Duvaut applies only to plates made of periodicity cells of shapes of thin plates. Consequently the formula (3.8.33) applies to the same class of periodic plates. The construction of higher-rank ribbed plates (see Eq. (22.5.9)) is completely abstract, since the condition of thinness of the cell contradicts the condition of its small in-plane dimensions. This contradiction concerns the first-rank ribbed plates and is aggravated by further layering process. Thus one should abandon all hope of manufacturing the strongest plate by following the method of this section. This construction is nevertheless practical from another viewpoint. It proves that Hashin-Shtrikman bounds are optimal in the sense that they cannot be tightened.

23.   Two-phase plate. Hashin-Shtrikman bounds for the in-plane problem

The subject of our consideration will be in-plane (or membrane) deformation of a thin plate made of a two-phase composite material. The membrane properties are characterized by isotropic membrane stiffness tensors A_1 and A_2. The composite is viewed as isotropic on the macroscopic level. The effective membrane stiffness tensor A_0 is determined by two moduli, still denoted by k_0 and μ_0. The aim of this section is to find the lower and upper bounds for these moduli and to prove that these bounds cannot be tightened.

23.1.   *The upper and lower bounds for the Kelvin modulus*

The setting of the problem is similar to that of Sec. 22.1. Instead of the bending stiffness tensors D_0, D_α we deal here with the membrane stiffness tensors A_0 and A_α, $\alpha = 1, 2$. The compliances are denoted by $a_0 = A_0^{-1}$ and $a_\alpha = A_\alpha^{-1}$. Let us decompose

$$A_\alpha = A_\alpha^{ij} \mathbf{a}_i \otimes \mathbf{a}_j , \qquad a_\alpha = a_\alpha^{ij} \mathbf{a}_i \otimes \mathbf{a}_j . \tag{23.1.1}$$

Due to isotropy, the representations are diagonal:

$$[A_\alpha^{ij}] = \operatorname{diag}[2k_\alpha, 2\mu_\alpha, 2\mu_\alpha] , \qquad [a_\alpha^{ij}] = \operatorname{diag}[2\mathcal{K}_\alpha, 2\mathcal{L}_\alpha, 2\mathcal{L}_\alpha] , \tag{23.1.2}$$

where $\mathcal{K}_\alpha = k_\alpha^{-1}, \mathcal{L}_\alpha = \mu_\alpha^{-1}$. The moduli k_α, μ_α here have different meaning from those of Sec. 22, which should not lead to misunderstandings, because the membrane and bending problems are considered in the present chapter separately.

Within the cell Y the compliance $a(y)$ is distributed according to the rule

$$a(y) = \chi_1(y)a_1 + \chi_2(y)a_2 , \tag{23.1.3}$$

cf. Eq. (22.1.1). The effective compliances $a^0_{\alpha\beta\lambda\mu}$ are determined by the formula, cf. Sec. 3.6

$$N_h^{\alpha\beta} a^0_{\alpha\beta\lambda\mu} N_h^{\lambda\mu} = \min\{\langle n^{\alpha\beta}(y) a_{\alpha\beta\lambda\mu}(y) n^{\lambda\mu}(y)\rangle | n \in \mathcal{S}_1^{per}(Y) , \langle n \rangle = N_h\} \tag{23.1.4}$$

where $N_h \in \mathbb{E}_2^s$. We shall further refer to the basis $\mathbf{a}_i$ and define

$$J(N_h) = \sum_{i,j=1}^{3} N_h^i a_{ij}^0 N_h^j ,$$

$$J(N_h) = \min\{\langle n^i a_{ij} n^j \rangle | n \in \mathcal{S}_1^{per}(Y), \langle n \rangle = N_h\} . \tag{23.1.5}$$

We now proceed as in Sec. 22.1. We rearrange $J(N_h)$ as follows

$$J(N_h) = \min\{\langle n^T(a - T)n\rangle + \langle n^T T n\rangle | n \in \mathcal{S}_1^{per}(Y), \langle n \rangle = N_h\} , \tag{23.1.6}$$

where $T = t\,\overset{o}{T}, t \in \mathbb{R}$, $\overset{o}{T}$ being defined by (21.1.18). By (21.4.17) we have

$$\langle n^T T n\rangle = N_h^T T N_h \qquad \forall n \in \mathcal{S}_1^{per}(Y) , \qquad \langle n \rangle = N_h . \tag{23.1.7}$$

Assuming that $a - T \geq 0$, applying the estimate (21.5.4) to the first term in (23.1.6) and using (23.1.7) one arrives at

$$J(N_h) \geq \max_{\substack{t \in \mathbb{R} \\ a-T \geq 0}} \{N_h^T[\langle (a - T)^{-1}\rangle^{-1} + T]N_h\} . \tag{23.1.8}$$

Further steps are similar to those of Sec. 22.1, but not exactly the same. Note that now the ordering conditions: $k_2 > k_1$, $\mu_2 > \mu_1$ imply $\mathcal{K}_2 < \mathcal{K}_1$ and $\mathcal{L}_2 < \mathcal{L}_1$. The conditions

$$a - T \geq 0 , \qquad Y(a^0) + T \geq 0 \tag{23.1.9}$$

imply

$$2\mathcal{K}_\alpha - t \geq 0 , \qquad 2\mathcal{L}_\alpha + t \geq 0 , \qquad y(2\mathcal{K}_0, 2\mathcal{K}_1, 2\mathcal{K}_2, m_1, m_2) \geq -t . \tag{23.1.10}$$

Note that $\max(-t) = 2\mathcal{L}_2$. We find the bound

$$\mathcal{K}_0 \geq \underline{\mathcal{K}}_{HS} , \tag{23.1.11}$$

where $\underline{\mathcal{K}}_{HS}$ is determined by the equation

$$y(2\underline{\mathcal{K}}_{HS}, 2\mathcal{K}_1, 2\mathcal{K}_2, m_1, m_2) = 2\mathcal{L}_2 . \tag{23.1.12}$$

Now we use the properties (21.2.18), (21.2.17) and find

$$y(\bar{k}_{HS}, k_1, k_2, m_1, m_2) = \mu_2 , \tag{23.1.13}$$

where $\bar{k}_{HS} = (\underline{\mathcal{K}}_{HS})^{-1}$. After using (21.2.23) we have

$$\frac{m_1}{\bar{k}_{HS} - k_2} = \frac{1}{k_1 - k_2} + \frac{m_2}{k_2 + \mu_2} . \tag{23.1.14}$$

In this manner the upper Hashin-Shtrikman bound $k_0 \leq \bar{k}_{HS}$ has been determined. Note that the form of the formula (23.1.14) is the same as for the plate bending problem, cf. Eq. (22.2.1).

To find lower bound for k_0 it is sufficient to change the role of the phases. Thus we arrive at the formula for $\underline{k}_{HS}$ that coincides with Eq. (22.1.25). The formulae for both upper and lower bounds of the Kelvin modulus are the same in the bending and membrane problems.

23.2. *The upper and lower bounds for the Kirchhoff modulus*

We proceed similarly to Sec. 22.3. The derivation is, however, worth reporting.

To estimate $\mathcal{L}_0$ of an isotropic composite plate one should subject the cell Y to two simultaneous stress fields τ and n of class $\mathcal{S}_1^{per}(Y)$. Estimation of the complementary energy stored in Y, associated with both the stress fields leads to the optimal lower Hashin-Shtrikman bound for $\mathcal{L}_0$.

Let us introduce the functional

$$J(s^h) = \min\{\langle s^T \mathcal{A}s\rangle | \tau, n \in \mathcal{S}_1^{per}(Y), \langle s\rangle = s^h\} , \tag{23.2.1}$$

where

$$\boldsymbol{\mathcal{A}} = \operatorname{diag}[\boldsymbol{a}, \boldsymbol{a}] \,, \qquad \boldsymbol{s} = [\tau^1, \tau^2, \tau^3, n^1, n^2, n^3] \qquad (23.2.2)$$

and $\boldsymbol{s}^h$ is defined similarly; the components of $\boldsymbol{s}$ are referred to the basis $\boldsymbol{a}_i$. We shall use the identity (21.4.23) with the translation matrix $\boldsymbol{\mathcal{T}}$ given by (21.4.10). The translation bound assumes the form

$$J(\boldsymbol{s}^h) \geq (\boldsymbol{s}^h)^T [\langle (\boldsymbol{\mathcal{A}} - \boldsymbol{\mathcal{T}})^{-1} \rangle^{-1} + \boldsymbol{\mathcal{T}}] \boldsymbol{s}^h \,, \qquad (23.2.3)$$

where $\boldsymbol{\mathcal{A}} - \boldsymbol{\mathcal{T}} \geq 0$. Now we postulate that the mixture is isotropic, hence

$$J(\boldsymbol{s}^h) = (\boldsymbol{s}^h)^T \boldsymbol{\mathcal{A}}_0 \boldsymbol{s}^h \,, \qquad (23.2.4)$$

with

$$\boldsymbol{\mathcal{A}}_0 = \operatorname{diag}[2\mathcal{K}_0, 2\mathcal{L}_0, 2\mathcal{L}_0, 2\mathcal{K}_0, 2\mathcal{L}_0, 2\mathcal{L}_0] \qquad (23.2.5)$$

and (23.2.3) leads to

$$\boldsymbol{\mathcal{A}}_0 \geq \langle (\boldsymbol{\mathcal{A}} - \boldsymbol{\mathcal{T}})^{-1} \rangle^{-1} + \boldsymbol{\mathcal{T}} \,. \qquad (23.2.6)$$

The condition $\boldsymbol{\mathcal{A}} - \boldsymbol{\mathcal{T}} \geq 0$ is satisfied provided that

$$t_1 \geq -2\mathcal{K}_2 \,, \qquad t_2 \leq 2\mathcal{L}_2 + 2\mathcal{K}_2 \,. \qquad (23.2.7)$$

The inequality (23.2.6) is fulfilled if, in particular

$$y(2\mathcal{L}_0, 2\mathcal{L}_1, 2\mathcal{L}_2, m_1, m_2) \geq \max(t_2 - t_1) \,. \qquad (23.2.8)$$

By (23.2.7) we have: $\max(t_2 - t_1) = 2\mathcal{L}_2 + 4\mathcal{K}_2$, which gives $\mathcal{L}_0 \geq \underline{\mathcal{L}}_{HS}$, where $\underline{\mathcal{L}}_{HS}$ is determined by

$$y(2\underline{\mathcal{L}}_{HS}, 2\mathcal{L}_1, 2\mathcal{L}_2, m_1, m_2) = 2\mathcal{K}_2 + \mathcal{L}_2 \,. \qquad (23.2.9)$$

By denoting $\bar{\mu}_{HS} = (\underline{\mathcal{L}}_{HS})^{-1}$ and using (21.2.17) we find

$$y(\bar{\mu}_{HS}, \mu_1, \mu_2, m_1, m_2) = \frac{k_2 \mu_2}{2\mu_2 + k_2} \,. \qquad (23.2.10)$$

Applying (21.2.23) we obtain

$$\frac{m_1}{\underline{\mu}_2 - \bar{\mu}_{HS}} = \frac{1}{\mu_2 - \mu_1} - \frac{m_2(k_2 + 2\mu_2)}{2\mu_2(k_2 + \mu_2)} \,. \qquad (23.2.11)$$

The equation above determines the upper Hashin-Shtrikman bound for $\mu_0 : \mu_0 \leq \bar{\mu}_{HS}$. To find the lower bound: $\mu_0 \geq \underline{\mu}_{HS}$ we change the role of the stiff and soft phase. Thus we put $k_1, \mu_1, \mu_2, m_1, m_2$ in (23.2.10) instead of $k_2, \mu_2, \mu_1, m_2, m_1$, respectively. Thus we find the formula for $\underline{\mu}_{HS}$

$$\frac{m_2}{\underline{\mu}_1 - \underline{\mu}_{HS}} = \frac{1}{\mu_1 - \mu_2} - \frac{m_1(k_1 + 2\mu_1)}{2\mu_1(k_1 + \mu_1)} \,. \qquad (23.2.12)$$

23.3. *Attainability of Hashin-Shtrikman bounds*

This section is aimed at constructing a microstructure that realizes the bounds: $k_0 = \bar{k}_{HS}$, $\mu_0 = \bar{\mu}_{HS}$ simultaneously. The construction is similar to that presented in Sec. 22.5. We perform three layerings by mixing

(i) material (2) with (1) in proportions θ_2 and θ_1 respectively, to find a material of the effective stiffness tensor $\boldsymbol{A}_h$ determined by Eq. (3.9.15). The tensor $\boldsymbol{\Gamma}_A$ involved in (3.9.15) and given by (3.9.18) depends on the lamination direction determined by the versor $\boldsymbol{n}$. We shall emphasize this relation by writing $\boldsymbol{\Gamma}_A(\boldsymbol{n})$. By (3.9.15) and (3.9.18) we have

$$\theta_1(\boldsymbol{A}_2 - \boldsymbol{A}_h)^{-1} = (\boldsymbol{A}_2 - \boldsymbol{A}_1)^{-1} - \frac{\theta_2}{\mu_2}\boldsymbol{\Psi}^{(n)} , \tag{23.3.1}$$

where

$$\boldsymbol{\Psi}^{(n)} = \sum_{\alpha,\beta,\lambda,\mu} \Psi^{(n)}_{\alpha\beta\lambda\mu}\boldsymbol{e}_\alpha \otimes \boldsymbol{e}_\beta \otimes \boldsymbol{e}_\lambda \otimes \boldsymbol{e}_\mu , \tag{23.3.2}$$

$$\Psi^{(n)}_{\alpha\beta\lambda\mu} = \frac{1}{4}(\delta_{\beta\lambda}n_\alpha n_\mu + \delta_{\beta\mu}n_\alpha n_\lambda + \delta_{\alpha\lambda}n_\beta n_\mu + \delta_{\alpha\mu}n_\beta n_\lambda) - \psi n_\alpha n_\beta n_\lambda n_\mu ,$$

$$\psi = \frac{k_2}{k_2 + \mu_2} . \tag{23.3.3}$$

Note that $\boldsymbol{\Gamma}_{(A)}(\boldsymbol{n}) = \dfrac{1}{\mu_2}\boldsymbol{\Psi}^{(n)}$. The material thus constructed will shortly be called (h) material.

(ii) Material (2) with material (h) in proportions α_2 and α_1 respectively; the lamination direction is given by a versor $\boldsymbol{m}$. The material thus obtained will be called (hh) material and its stiffness tensor $\boldsymbol{A}_{hh}$ is determined by

$$\alpha_1(\boldsymbol{A}_2 - \boldsymbol{A}_{hh})^{-1} = (\boldsymbol{A}_2 - \boldsymbol{A}_h)^{-1} - \frac{\alpha_2}{\mu_2}\boldsymbol{\Psi}^{(m)} . \tag{23.3.4}$$

(iii) Material (2) with material (hh) in proportions β_2 and β_1 respectively; the lamination direction is given by a versor $\boldsymbol{p}$. The material thus obtained, called (hhh) material, is characterized by the stiffness tensor $\boldsymbol{A}_{hhh}$, determined by the implicit equation

$$\beta_1(\boldsymbol{A}_2 - \boldsymbol{A}_{hhh})^{-1} = (\boldsymbol{A}_2 - \boldsymbol{A}_{hh})^{-1} - \frac{\beta_2}{\mu_2}\boldsymbol{\Psi}^{(p)} . \tag{23.3.5}$$

In the sequence of laminations (i) – (iii) the soft material (1) plays the role of a core and the stiffer material (2) is used as a coating. Such laminates are called "stiff laminates".

By combining (23.3.1) – (23.3.5) one finds

$$m_1(\boldsymbol{A}_2 - \boldsymbol{A}_{hhh})^{-1} = (\boldsymbol{A}_2 - \boldsymbol{A}_1)^{-1} - \frac{1}{\mu_2}\boldsymbol{\Psi} , \tag{23.3.6}$$

with $m_1 = \theta_1\alpha_1\beta_1$ and

$$\boldsymbol{\Psi} = \theta_2\boldsymbol{\Psi}^{(n)} + \theta_1\alpha_2\boldsymbol{\Psi}^{(m)} + \beta_2\alpha_1\theta_1\boldsymbol{\Psi}^{(p)} . \tag{23.3.7}$$

Now we choose the vectors $\boldsymbol{n}, \boldsymbol{m}, \boldsymbol{p}$ as in Sec. 22.5 (see Eq. (22.5.10) and Fig. 22.5.2) and compute

$$\Psi^{(n)}_{1111} = 1 - \psi\,, \qquad \Psi^{(n)}_{1212} = \frac{1}{4}\,,$$
$$\Psi^{(n)}_{\alpha\alpha\beta\beta} = 0\,, \qquad \text{for } \alpha = 2\,, \ \beta \in \{1,2\}\,; \tag{23.3.8}$$

$$\Psi^{(m)}_{1111} = \frac{1}{4} - \frac{\psi}{16}\,, \qquad \Psi^{(m)}_{2222} = \frac{3}{4} - \frac{9}{16}\psi\,,$$
$$\Psi^{(m)}_{1122} = -\frac{3}{16}\psi\,, \qquad \Psi^{(m)}_{1112} = -\frac{\sqrt{3}}{8}\left(1 - \frac{1}{2}\psi\right)\,, \tag{23.3.9}$$
$$\Psi^{(m)}_{2221} = -\frac{\sqrt{3}}{8}\left(1 - \frac{3}{2}\psi\right)\,, \qquad \Psi^{(m)}_{1212} = \frac{1}{4} - \frac{3}{16}\psi\,;$$

$$\Psi^{(p)}_{\alpha\alpha\beta\beta} = \Psi^{(m)}_{\alpha\alpha\beta\beta}\,, \qquad \Psi^{(p)}_{1212} = -\Psi^{(m)}_{1212}\,,$$
$$\Psi^{(p)}_{\alpha\alpha\alpha\beta} = -\Psi^{(m)}_{\alpha\alpha\alpha\beta}\,, \qquad \alpha, \beta \in \{1,2\}\,. \tag{23.3.10}$$

Let us impose isotropy conditions on the tensor $\boldsymbol{\Psi}$. In particular the conditions: $\Psi_{1112} = 0$, $\Psi_{2221} = 0$ imply $\beta_2 = \alpha_2/\alpha_1$. The condition $\Psi_{1111} = \Psi_{2222}$ gives $\alpha_2 = \theta_2/\theta_1$. We finally find

$$\boldsymbol{\Psi} = \frac{3}{2}\theta_2\left[(1-\psi)\boldsymbol{I}_1 + \left(1 - \frac{1}{2}\psi\right)\boldsymbol{I}_2\right]\,, \tag{23.3.11}$$

the tensors $\boldsymbol{I}_\alpha$ being defined by Eqs. (3.8.29). By (23.3.6) the isotropy of $\boldsymbol{\Psi}$ implies the isotropy of $\boldsymbol{A}_{hhh}$. Thus there exist moduli k_{hhh}, μ_{hhh} such that

$$\boldsymbol{A}_{hhh} = 2k_{hhh}\boldsymbol{I}_1 + 2\mu_{hhh}\boldsymbol{I}_2\,. \tag{23.3.12}$$

Thus the left-hand side of (23.3.6) assumes the form

$$\frac{m_1}{2(k_2 - k_{hhh})}\boldsymbol{I}_1 + \frac{m_1}{2(\mu_2 - \mu_{hhh})}\boldsymbol{I}_2\,. \tag{23.3.13}$$

Noting that

$$(\boldsymbol{A}_2 - \boldsymbol{A}_1)^{-1} = \frac{1}{2\Delta k}\boldsymbol{I}_1 + \frac{1}{2\Delta\mu}\boldsymbol{I}_2 \tag{23.3.14}$$

and using (23.3.6), (23.3.11), (23.3.13) we obtain

$$\frac{m_1}{k_2 - k_{hhh}} = \frac{1}{\Delta k} - \frac{3\theta_2(1-\psi)}{\mu_2}\,, \qquad \frac{m_1}{\mu_2 - \mu_{hhh}} = \frac{1}{\Delta\mu} - \frac{3\theta_2\left(1 - \frac{1}{2}\psi\right)}{\mu_2}\,. \tag{23.3.15}$$

The relation $m_2 = 3\theta_2$ known from Sec. 22.5 also holds here. Thus (23.3.15) is equivalent to

$$\frac{m_1}{k_2 - k_{hhh}} = \frac{1}{k_2 - k_1} - \frac{m_2}{k_2 + \mu_2}\,, \qquad \frac{m_1}{\mu_2 - \mu_{hhh}} = \frac{1}{\mu_2 - \mu_1} - \frac{m_2(k_2 + 2\mu_2)}{2\mu_2(k_2 + \mu_2)}\,. \tag{23.3.16}$$

We compare these formulae with (23.1.14) and (23.2.10) to conclude that

$$k_{hhh} = \bar{k}_{HS} , \qquad \mu_{hhh} = \bar{\mu}_{HS} . \qquad (23.3.17)$$

This proves that the upper Hashin-Shtrikman bounds are not only attained, but they can be attained simultaneously.

Similarly one can construct a composite that realizes the lower Hashin-Shtrikman bounds $\underline{k}_{HS}$, $\underline{\mu}_{HS}$. To this end one should take material (2) as a core and use material (1) as a coating. The choice of vectors n, m, p should be the same as previously.

23.4. Summary of the main results

For the two-phase plate in bending, characterized by stiffnesses D_2 and D_1 such that $D_2 > D_1$, the Hashin-Shtrikman bounds assume the form

$$\underline{k}_{HS} \leq k_0 \leq \bar{k}_{HS} , \qquad \underline{\mu}_{HS} \leq \mu_0 \leq \bar{\mu}_{HS} , \qquad (23.4.1)$$

where the bounds are determined by

$$\begin{aligned}
y(\underline{k}_{HS}, k_1, k_2, m_1, m_2) &= \mu_1 , \\
y(\bar{k}_{HS}, k_2, k_1, m_2, m_1) &= \mu_2 , \\
y(\underline{\mu}_{HS}, \mu_1, \mu_2, m_1, m_2) &= 2k_1 + \mu_1 , \\
y(\bar{\mu}_{HS}, \mu_2, \mu_1, m_2, m_1) &= 2k_2 + \mu_2 .
\end{aligned} \qquad (23.4.2)$$

For the in-plane (membrane) problem of two-phase plate, characterized by stiffnesses A_2 and A_1 such that $A_2 > A_1$, the Hashin-Shtrikman bounds assume the form

$$\underline{\mathcal{K}}_{HS} \leq \mathcal{K}_0 \leq \bar{\mathcal{K}}_{HS} , \qquad \underline{\mathcal{L}}_{HS} \leq \mathcal{L}_0 \leq \bar{\mathcal{L}}_{HS} , \qquad (23.4.3)$$

where the bounds are determined by

$$\begin{aligned}
y(\underline{\mathcal{K}}_{HS}, \mathcal{K}_1, \mathcal{K}_2, m_1, m_2) &= \mathcal{L}_2 , \\
y(\bar{\mathcal{K}}_{HS}, \mathcal{K}_2, \mathcal{K}_1, m_2, m_1) &= \mathcal{L}_1 , \\
y(\underline{\mathcal{L}}_{HS}, \mathcal{L}_1, \mathcal{L}_2, m_1, m_2) &= 2\mathcal{K}_2 + \mathcal{L}_2 , \\
y(\bar{\mathcal{L}}_{HS}, \mathcal{L}_2, \mathcal{L}_1, m_2, m_1) &= 2\mathcal{K}_1 + \mathcal{L}_1 .
\end{aligned} \qquad (23.4.4)$$

Here $\mathcal{K}_\alpha = (k_\alpha)^{-1}$, $\mathcal{L}_\alpha = (\mu_\alpha)^{-1}$. If written as above, the formulae (23.4.2) and (23.4.4) reveal an analogy. After expressing the latter in terms of k and μ the analogy is lost. Note that the known analogies between $\mathcal{K}'_\epsilon(\Omega)$ and $\mathcal{S}_2(\Omega)$ and between $\mathcal{K}'_\kappa(\Omega)$ and $\mathcal{S}_1(\Omega)$ (see Secs. 3.4 and 3.6) are so suggestive that one could think that finding (23.4.4) having (23.4.2) is an easy task. However, this can be misleading.

Note first that the analogies mentioned do not correlate ϵ with M and κ with N, but $\epsilon_{\alpha\beta}$ with $\varepsilon_{\alpha\gamma}\,\varepsilon_{\beta\delta}\,M^{\gamma\delta}$ and $\kappa_{\alpha\beta}$ with $\varepsilon_{\alpha\gamma}\,\varepsilon_{\beta\delta}\,N^{\gamma\delta}$; ε is a permutation symbol. Second, the change: $(k, \mu) \rightsquigarrow (\mathcal{K}, \mathcal{L})$ transforms the inequalities: $k_2 > k_1 \rightsquigarrow \mathcal{K}_1 > \mathcal{K}_2$ and $\mu_2 > \mu_1 \rightsquigarrow \mathcal{L}_1 > \mathcal{L}_2$. Thus despite some analogies an independent derivation of all bounds (23.4.1) – (23.4.4) is necessary.

24. Explicit formulae for effective bending stiffnesses and compliances of ribbed plates

The general formulae for effective stiffnesses of ribbed plates of first, second and higher rank have been derived in Secs. 3.7 and 3.8. The aim of this section is to specify these formulae to the case of the first rank and orthogonal second rank ribbed plates constructed from isotropic phases. Such explicit formulae are helpful in proving the extremal properties of the "stiff" and "soft" ribbed plates, which will be the subject of Sec. 26.

24.1. First rank ribbed structure

Consider a transversely symmetric ribbed plate of first rank discussed in Sec. 3.8, see Fig. 3.8.1. We assume here that both materials are isotropic of stiffness tensors D_α ($\alpha = 1, 2$) given by (3.8.28). The ordering assumption: $D_2 > D_1$ holds. The materials (1) and (2) are taken in proportion θ_1 and θ_2 respectively; $\theta_1 + \theta_2 = 1$. Our aim is to find the explicit formulae for the components $\overset{o}{D}{}_h^{\alpha\beta\lambda\mu}$ of the tensor D_h given by Eq. (3.8.2). These components are referred to the natural basis: $\overset{o}{e}_1$, $\overset{o}{e}_2$ such that $\overset{o}{e}_1$ coincides with the lamination direction: $\overset{o}{e}_1 = n$, see Fig. 24.1.1.

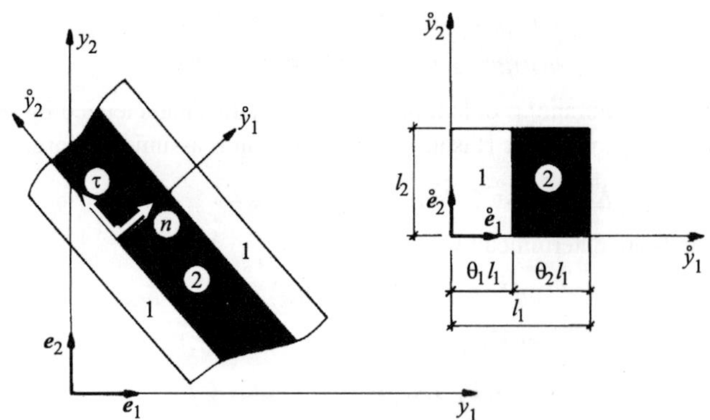

Fig. 24.1.1. First rank ribbed plate

Thus we decompose D_h as follows

$$D_h = \overset{o}{D}{}_h^{\alpha\beta\lambda\mu}\, \overset{o}{e}_\alpha \otimes \overset{o}{e}_\beta \otimes \overset{o}{e}_\lambda \otimes \overset{o}{e}_\mu ,\tag{24.1.1}$$

or

$$D_h = \overset{o}{D}{}_h^{ij}\, \overset{o}{a}_i \otimes \overset{o}{a}_j ,\tag{24.1.2}$$

where the tensors $\overset{o}{a}_i$ are defined by $\overset{o}{e}_\alpha$ according to (21.1.2). To find $\overset{o}{D}{}_h^{\alpha\beta\lambda\mu}$ one can use directly (3.7.7). It turns out, however, that simpler formulae can be found by using the formula (3.8.2) of Francfort and Murat.

Let us introduce the notation

$$s_\alpha = k_\alpha + \mu_\alpha , \qquad r_\alpha = k_\alpha - \mu_\alpha \qquad (24.1.3)$$

and operations:

$$\langle f \rangle_\theta = \theta_1 f_1 + \theta_2 f_2 , \quad [f]_\theta = \theta_1 f_2 + \theta_2 f_1 , \quad \{f\}_\theta = \frac{f_1 f_2}{[f]_\theta} , \quad \Delta f = f_2 - f_1 , \quad (24.1.4)$$

for $f \in \{k, \mu, s, r\}$.

The non-zero stiffnesses $\overset{o}{D}{}_h^{\alpha\beta\lambda\mu}$ are given by

$$\overset{o}{D}{}_h^{1111} = \langle s \rangle_\theta - t_\theta (\Delta s)^2 , \qquad \overset{o}{D}{}_h^{1122} = \langle r \rangle_\theta - t_\theta \Delta s \Delta r ,$$
$$\overset{o}{D}{}_h^{2222} = \langle s \rangle_\theta - t_\theta (\Delta r)^2 , \qquad \overset{o}{D}{}_h^{1212} = \langle \mu \rangle_\theta , \qquad (24.1.5)$$

and

$$\overset{o}{D}{}_h^{2211} = \overset{o}{D}{}_h^{1122} , \qquad \overset{o}{D}{}_h^{1221} = \overset{o}{D}{}_h^{2112} = \overset{o}{D}{}_h^{2121} = \overset{o}{D}{}_h^{1212} , \qquad (24.1.6)$$

$$t_\theta = \frac{\theta_1 \theta_2}{[s]_\theta} . \qquad (24.1.7)$$

The components $\overset{o}{D}{}_h^{ij}$ can now be found by using (21.1.24):

$$\frac{1}{2} \overset{o}{D}{}_h^{11} = \langle k \rangle_\theta - t_\theta (\Delta k)^2 , \qquad \frac{1}{2} \overset{o}{D}{}_h^{12} = -t_\theta \Delta k \Delta \mu ,$$
$$\frac{1}{2} \overset{o}{D}{}_h^{22} = \langle \mu \rangle - t_\theta (\Delta \mu)^2 , \qquad \frac{1}{2} \overset{o}{D}{}_h^{33} = \langle \mu \rangle_\theta , \qquad \overset{o}{D}{}_h^{21} = \overset{o}{D}{}_h^{12} , \qquad (24.1.8)$$

and other components are zero. The apparent simplicity of the formulae (24.1.8) is worth emphasizing.

Let us find now the components of the effective compliance tensor d_h, by using the formula (3.8.58). We decompose

$$d_h = \sum \overset{o}{d}{}_{ij}^{h} \, \overset{o}{\mathbf{a}}_i \otimes \overset{o}{\mathbf{a}}_j \qquad (24.1.9)$$

and introduce the following notation:

$$\mathcal{K}_\alpha = (k_\alpha)^{-1} , \quad \mathcal{L}_\alpha = (\mu_\alpha)^{-1} , \qquad \mathcal{S}_\alpha = \mathcal{K}_\alpha + \mathcal{L}_\alpha , \qquad T_\theta = \frac{\theta_1 \theta_2}{[\mathcal{S}]_\theta} . \qquad (24.1.10)$$

The non-zero components $\overset{o}{d}{}_{ij}^{h}$ are given by

$$2 \overset{o}{d}{}_{11}^{h} = \langle \mathcal{K} \rangle_\theta - T_\theta (\Delta \mathcal{K})^2 , \qquad 2 \overset{o}{d}{}_{12}^{h} = T_\theta \Delta \mathcal{K} \Delta \mathcal{L} ,$$
$$2 \overset{o}{d}{}_{22}^{h} = \langle \mathcal{L} \rangle_\theta - T_\theta (\Delta \mathcal{L})^2 , \qquad 2 \overset{o}{d}{}_{33}^{h} = \{ \mathcal{L} \}_\theta , \qquad \overset{o}{d}{}_{21}^{h} = \overset{o}{d}{}_{21}^{h} . \qquad (24.1.11)$$

Here $\Delta\mathcal{K} = \mathcal{K}_1 - \mathcal{K}_2 > 0$, $\Delta\mathcal{L} = \mathcal{L}_1 - \mathcal{L}_2 > 0$. Note three differences between (24.1.8) and (24.1.11):

i) the signs in the expressions for $\overset{o}{D}{}^{12}_{h}$ and $\overset{o}{d}{}^{h}_{12}$ are different,

ii) the arithmetic mean in the expression for $\overset{o}{D}{}^{33}_{h}$ transforms to the harmonic mean in the expressions for $\overset{o}{d}{}^{h}_{33}$,

iii) factor 2 transforms to $1/2$.

The composite of stiffnesses (24.1.11) will further be called material (h).

24.2. *Second rank ribbed structure with soft phase taken as a core*

Let us consider the ribbed plate of Fig. 3.8.2. Now we change notation: the area fractions of material (h) and material (2) are ω_1 and ω_2, respectively, see Fig. 24.2.1.

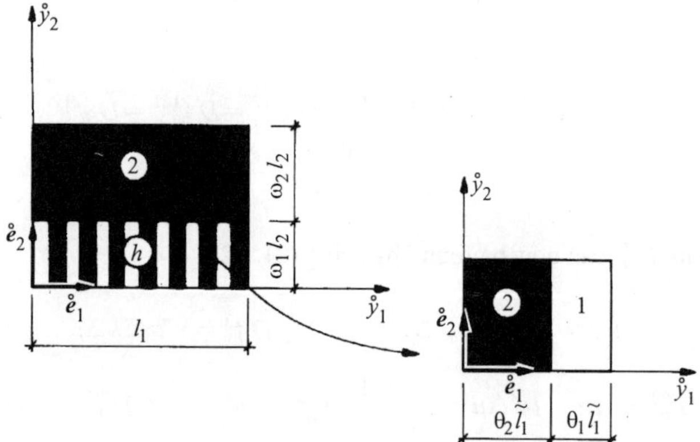

Fig. 24.2.1. Second rank ribbed plate. The stronger material (2) plays the role of a coating

The vector m is taken as $\overset{o}{e}_2$. The resulting homogenized material will be orthotropic, with orthotropy axes coinciding with the basis $(\overset{o}{e}_\alpha)$. The soft material (1) plays the role of a core, while material (2) is the coating. The resulting composite, called further (hh), of stiffness D_{hh}, belongs to the class of "stiff laminates". The resulting area fraction of material (1) equals $m_1 = \theta_1\omega_1$ and $m_2 = \omega_2 + \omega_1\theta_2$. Note that $m_1 + m_2 = 1$. The following formulae will be useful

$$\theta_2 - \theta_1\omega_2 = 1 + m_1 - 2\theta_1, \quad \theta_1\omega_2 = \theta_1 - m_1, \quad \theta_1\omega_2 + \theta_2 = m_2. \quad (24.2.1)$$

The condition $D_2 - D_1 > 0$ implies $D_2 > D_h$, D_h given by (24.1.8). To find D_{hh} we shall use Eq. (3.8.35) and refer all tensors to the basis $\overset{o}{a}_i \otimes \overset{o}{a}_j$. We decompose

$$D_{hh} = \overset{o}{D}{}^{ij}_{hh}\, \overset{o}{a}_i \otimes \overset{o}{a}_j. \quad (24.2.2)$$

The final formulae read as follows

$$
\begin{aligned}
\frac{1}{2}\overset{o}{D}{}^{11}_{hh} &= k_2 - \frac{m_1(k_2+[\mu]_m)\Delta k s_2}{f(\theta_1,m_1)}\,, \\[2mm]
\frac{1}{2}\overset{o}{D}{}^{12}_{hh} &= \frac{m_1(2\theta_1-1-m_1)\Delta k \Delta \mu s_2}{f(\theta_1,m_1)}\,, \\[2mm]
\frac{1}{2}\overset{o}{D}{}^{22}_{hh} &= \mu_2 - \frac{m_1(\mu_2+[k]_m)\Delta \mu s_2}{f(\theta_1,m_1)}\,, \\[2mm]
\frac{1}{2}\overset{o}{D}{}^{33}_{hh} &= \langle\mu\rangle_m\,, \qquad \overset{o}{D}{}^{12}_{hh} = \overset{o}{D}{}^{21}_{hh}\,,
\end{aligned}
\tag{24.2.3}
$$

where

$$
f(\theta_1,m_1) = 4(1-\theta_1)(\theta_1-m_1)\Delta\mu\Delta k + s_2[s]_m\,.
\tag{24.2.4}
$$

The notation (24.1.4) applies here. Consequently

$$
[g]_m = m_1 g_2 + m_2 g_1\,, \quad \langle g\rangle_m = m_1 g_1 + m_2 g_2\,, \quad \Delta g = g_2 - g_1\,,
\tag{24.2.5}
$$

for $g \in \{\mu, k, s\}$.

Let us sketch now the derivation of the formulae (24.2.3). According to (3.8.35) the tensor D_{hh} is given by

$$
m_1(D_2 - D_{hh})^{-1} = (D_2 - D_1)^{-1} - \frac{1}{s_2}(\theta_1\omega_2\Gamma_{e_2} + \theta_2\Gamma_{e_1})\,.
\tag{24.2.6}
$$

The tensors involved in (24.2.6) will be represented in the basis $\overset{o}{a}_i \otimes \overset{o}{a}_j$;

$$
[(D_2 - D_{hh})_{ij}] =
\begin{bmatrix}
2k_2 - \overset{o}{D}{}^{11}_{hh} & -\overset{o}{D}{}^{12}_{hh} & -\overset{o}{D}{}^{13}_{hh} \\[2mm]
-\overset{o}{D}{}^{12}_{hh} & 2\mu_2 - \overset{o}{D}{}^{22}_{hh} & -\overset{o}{D}{}^{23}_{hh} \\[2mm]
-\overset{o}{D}{}^{13}_{hh} & -\overset{o}{D}{}^{23}_{hh} & 2\mu_2 - \overset{o}{D}{}^{33}_{hh}
\end{bmatrix}\,,
$$

$$
[(D_2 - D_1)^{-1}_{ij}] = \operatorname{diag}\left[\frac{1}{2\Delta k}, \frac{1}{2\Delta\mu}, \frac{1}{2\Delta\mu}\right]\,,
$$

$$
[(\Gamma_{e_1})_{ij}] = \frac{1}{2}\begin{bmatrix} 1 & 1 & 0 \\ 1 & 1 & 0 \\ 0 & 0 & 0 \end{bmatrix}\,, \qquad
[(\Gamma_{e_2})_{ij}] = \frac{1}{2}\begin{bmatrix} 1 & -1 & 0 \\ -1 & 1 & 0 \\ 0 & 0 & 0 \end{bmatrix}\,.
$$

Equation (24.2.6) gives

$$
\left[\left[\frac{1}{m_1}(D_2 - D_{hh})\right]^{-1}_{ij}\right] = \frac{1}{2}[X_{ij}]\,,
$$

where

$$
[X_{ij}] =
\begin{bmatrix}
\dfrac{1}{\Delta k} - \dfrac{1}{s_2}(\theta_2 + \theta_1\omega_2) & -\dfrac{1}{s_2}(\theta_2 - \theta_1\omega_2) & 0 \\[2ex]
-\dfrac{1}{s_2}(\theta_2 - \theta_1\omega_2) & \dfrac{1}{\Delta\mu} - \dfrac{1}{s_2}(\theta_2 + \theta_1\omega_2) & 0 \\[2ex]
0 & 0 & \dfrac{1}{\Delta\mu}
\end{bmatrix} .
$$

The matrix inverse to X assumes the form

$$
[(X)_{ij}^{-1}] = \frac{s_2}{f(\theta_1, m_1)}
\begin{bmatrix}
(k_2 + [\mu]_m)\Delta k & (1 + m_1 - 2\theta_1)\Delta k\Delta\mu & 0 \\[2ex]
(1 + m_1 - 2\theta_1)\Delta k\Delta\mu & (\mu_2 + [k]_m)\Delta\mu & 0 \\[2ex]
0 & 0 & \dfrac{\Delta\mu s_2}{f(\theta_1, m_1)}
\end{bmatrix} .
$$

From the equation

$$
\frac{1}{2}D_{hh} = \frac{1}{2}D_2 - m_1 X^{-1}
$$

we arrive directly at (24.2.3).   □

Thus the formula (3.8.35) provides us with a convenient method to derive Eq. (24.2.3). Finding the same formulae by using two times (3.7.7) turns out to be surprisingly complex.

Substituting $\omega_2 = 0$ $(\theta_1 = m_1)$ into (24.2.23) leads to the previous formula (24.1.8) concerning the first rank ribbed plate, cf. Fig. 24.1.1.

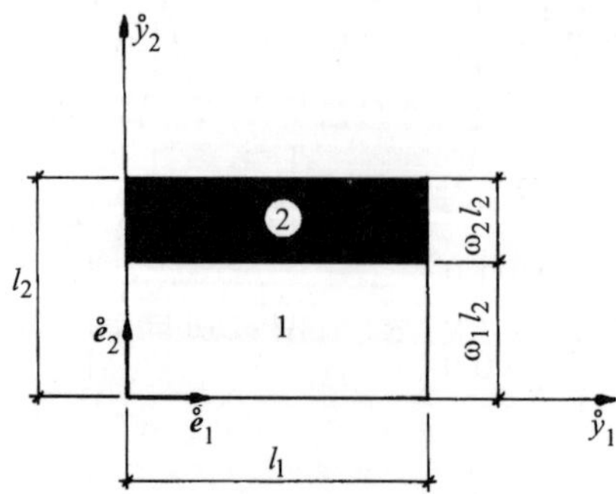

Fig. 24.2.2.

Substituting $\theta_1 = 1$ or $m_\alpha = \omega_\alpha$ yields

$$\frac{1}{2}\overset{o}{D}{}^{11}_{h} = \langle k\rangle_\omega - t_\omega(\Delta k)^2 , \qquad \frac{1}{2}\overset{o}{D}{}^{12}_{h} = t_\omega \Delta k \Delta \mu ,$$

$$\frac{1}{2}\overset{o}{D}{}^{22}_{h} = \langle \mu\rangle_\omega - t_\omega(\Delta \mu)^2 , \qquad \frac{1}{2}\overset{o}{D}{}^{33}_{h} = \langle \mu\rangle_\omega , \tag{24.2.7}$$

with $t_\omega = \omega_1\omega_2/[s]_\omega$. These formulae concern the first rank ribbed plate with the cell given in Fig. 24.2.2. Note the difference in sign in the formulae for $\overset{o}{D}{}^{12}_{h}$, cf. Eqs. $(24.1.8)_2$ and $(24.2.7)_2$. Thus the composites of Figs. 24.1.1 and 24.2.2 are characterized by the opposite values of this stiffness.

Let us find now the components of the effective compliance tensor $d_{hh} = (D_{hh})^{-1}$. Inversion of (24.2.3) gives

$$2\overset{o}{d}{}^{hh}_{11} = \mathcal{K}_2 + \frac{m_1(\mathcal{K}_2 + [\mathcal{L}]_m)\Delta \mathcal{K} S_2}{F(\theta_1, m_1)} ,$$

$$2\overset{o}{d}{}^{hh}_{12} = -\frac{m_1(2\theta_1 - 1 - m_1)\Delta \mathcal{K}\Delta \mathcal{L} S_2}{F(\theta_1, m_1)} ,$$

$$2\overset{o}{d}{}^{hh}_{22} = \mathcal{L}_2 + \frac{m_1(\mathcal{L}_2 + [\mathcal{K}]_m)\Delta \mathcal{L} S_2}{F(\theta_1, m_1)} , \tag{24.2.8}$$

$$2\overset{o}{d}{}^{hh}_{33} = \{\mathcal{L}\}_m ,$$

where

$$d_{hh} = \sum_{i,j} \overset{o}{d}{}^{hh}_{ij} \overset{o}{\mathbf{a}}_i \otimes \overset{o}{\mathbf{a}}_j ,$$

$$\mathcal{K}_\alpha = (k_\alpha)^{-1} , \quad \mathcal{L}_\alpha = (\mu_\alpha)^{-1} ,$$

$$\Delta \mathcal{K} = \mathcal{K}_1 - \mathcal{K}_2 , \quad \Delta \mathcal{L} = \mathcal{L}_1 - \mathcal{L}_2 , \quad S_2 = \mathcal{K}_2 + \mathcal{L}_2$$

and

$$F(\theta_1, m_1) = 4(1 - \theta_1)(\theta_1 - m_1)\Delta \mathcal{L}\Delta \mathcal{K} + S_2[S]_m . \tag{24.2.9}$$

To find the effective compliance of the first rank ribbed plate of Fig. 24.2.2 one should substitute $\theta_1 = 1$ or $m_\alpha = \omega_\alpha$ into (24.2.8). One finds

$$2\overset{o}{d}{}^{h}_{11} = \langle \mathcal{K}\rangle_\omega - T_\omega(\Delta \mathcal{K})^2 , \qquad 2\overset{o}{d}{}^{h}_{12} = -T_\omega \Delta \mathcal{K}\Delta \mathcal{L} ,$$

$$2\overset{o}{d}{}^{h}_{22} = \langle \mathcal{L}\rangle_\omega - T_\omega(\Delta \mathcal{L})^2 , \qquad 2\overset{o}{d}{}^{h}_{33} = \{\mathcal{L}\}_\omega , \tag{24.2.10}$$

with $T_\omega = \omega_1\omega_2([S]_\omega)^{-1}$,

Let us examine the second rank ribbed plate of compliances (24.2.8) once again. Assume that m_1 is fixed. Consider the quadratic form with respect to x and y:

$$H(\theta_1; x, y) = 2[\overset{o}{d}{}^{hh}_{11}(\theta_1)x^2 + 2\overset{o}{d}{}^{hh}_{12}(\theta_1)xy + \overset{o}{d}{}^{hh}_{22}(\theta_1)y^2] , \tag{24.2.11}$$

with $\overset{o}{d}{}^{hh}_{\alpha\beta}(\theta_1)$ given by (24.2.8). An extremal property of this form is expressed by

Lemma 24.1. Assume that $x \neq 0$, $y > 0$ and

$$|y/x| \leq \zeta_2 , \quad \zeta_2 = \frac{m_2 \Delta \mathcal{K}}{[\mathcal{K}]_m + \mathcal{L}_2} . \tag{24.2.12}$$

Then

$$\min_{0 \leq \theta_1 \leq 1} H(\theta_1; x, y) = H(\theta_1^*; x, y) , \tag{24.2.13}$$

where

$$\theta_1^* = \frac{1}{2} \left(1 + m_1 + m_2 \frac{y/x}{\zeta_2} \right) . \tag{24.2.14}$$

The function $H(\theta_1^*; x, y)$ has the form

$$H(\theta_1^*; x, y) = \check{K} x^2 + \mathcal{L}_2 y^2 , \tag{24.2.15}$$

where

$$\check{K} = \frac{\mathcal{K}_1 \mathcal{K}_2 + \mathcal{L}_2 \langle \mathcal{K} \rangle_m}{\mathcal{L}_2 + [\mathcal{K}]_m} . \tag{24.2.16}$$

Moreover

$$m_1 \leq \theta_1^* \leq 1 . \tag{24.2.17}$$

Proof. The necessary condition of minimum

$$\frac{\partial H}{\partial \theta_1}(\theta_1; x, y) = 0 \tag{24.2.18}$$

leads to an algebraic equation which can be factorized by the program MAPLE V, cf. Heal et al. (1996). The solutions of this equation have the form

$$\theta_1^{(1)} = \theta_1^* , \quad \theta_1^{(2)} = \frac{1}{2}(1 + m_1) + \frac{\mathcal{K}_2 + [\mathcal{L}]_m}{2 \Delta \mathcal{L}} \frac{x}{y} . \tag{24.2.19}$$

Using MAPLE V once again we check that

$$\text{sign} \left(\frac{\partial^2 H}{\partial (\theta_1)^2}(\theta_1^{(\alpha)}) \right) = (-1)^{\alpha+1} \text{sign} \, g \left(\frac{y}{x} \right) , \tag{24.2.20}$$

where

$$g \left(\frac{y}{x} \right) = \Delta \mathcal{K}([\mathcal{L}]_m + \mathcal{K}_2) - \left| \frac{y}{x} \right|^2 \Delta \mathcal{L}(\mathcal{L}_2 + [\mathcal{K}]_m) . \tag{24.2.21}$$

The condition $|y/x| \leq \zeta_2$ implies $g(y/x) > 0$. Thus the minimum is attained at $\theta_1 = \theta_1^{(1)} = \theta_1^*$. Substitution of $\theta_1 = \theta_1^*$ into (24.2.11) gives (24.2.15). To prove (24.2.17), assume first that $y/x > 0$. Then $0 < y/x < \zeta_2$. Hence $2\theta_1^* \leq 1 + m_1 + m_2 = 2$. On the other hand, $2\theta_1^* \geq 1 + m_1 \geq 2m_1$.

Assume now that $y/x < 0$. Then $2\theta_1^* = 1 + m_1 - m_2 \dfrac{|y/x|}{\zeta_2} \geq 1 + m_1 - m_2 = 2m_1$. On the other hand, we have: $2\theta_1^* \leq 1 + m_1 \leq 2$, which completes the proof. $\square$

Note that for $\theta_1 = \theta_1^*$ the orthotropic quadratic form (24.2.11) becomes isotropic, cf. (24.2.15). This fact will play a crucial role in proving the optimal properties of second rank ribbed plates (Sec. 26).

24.3.   *Second rank ribbed structure with the strong phase taken as a core*

Consider the second rank ribbed plate constructed as shown in Fig. 24.3.1. The strong material (2), taken as a core, is subsequently coated by the soft phase (1).

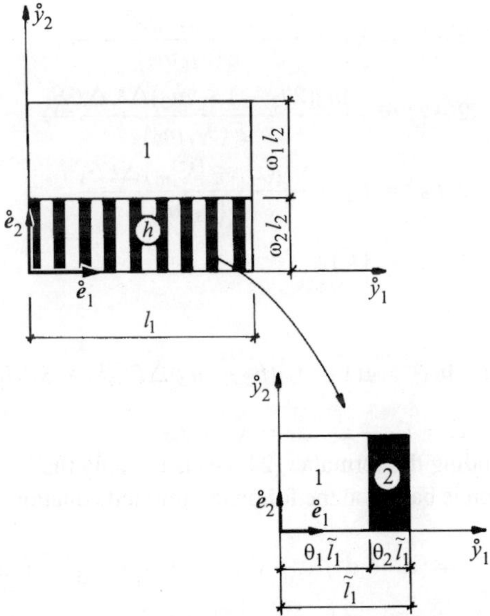

Fig. 24.3.1. Second rank ribbed plate. The weaker material (1) plays the role of a coating

Let us note that replacing

$$\theta_\alpha \rightsquigarrow \theta_\beta , \qquad \omega_\alpha \rightsquigarrow \omega_\beta , \qquad k_\alpha \rightsquigarrow k_\beta , \qquad \mu_\alpha \rightsquigarrow \mu_\beta , \qquad (24.3.1)$$

changes the structure of Fig. 24.3.1 into that of Fig. 24.2.1; here $\beta = 3 - \alpha$, $\alpha = 1, 2$. Performing these changes in (24.2.3) yields

$$\frac{1}{2} \overset{o}{D} \overset{11}{{}_{hh}} = k_1 + \frac{m_2(k_1 + [\mu]_m)\Delta k s_1}{f(\theta_2, m_2)} ,$$

$$\frac{1}{2} \overset{o}{D} \overset{12}{{}_{hh}} = \frac{m_2(2\theta_2 - 1 - m_2)\Delta k \Delta \mu s_1}{f(\theta_2, m_2)} ,$$

$$\frac{1}{2} \overset{o}{D} \overset{22}{{}_{hh}} = \mu_1 + \frac{m_2(\mu_1 + [k]_m)\Delta \mu s_1}{f(\theta_2, m_2)} , \qquad (24.3.2)$$

$$\frac{1}{2} \overset{o}{D} \overset{33}{{}_{hh}} = \langle \mu \rangle_m ,$$

where

$$f(\theta_2, m_2) = 4(1 - \theta_2)(\theta_2 - m_2)\Delta \mu \Delta k + s_1[s]_m . \qquad (24.3.3)$$

Inversion of the formulae (24.3.2) gives the effective compliances

$$2 \overset{o}{d} \overset{hh}{{}_{11}} = \mathcal{K}_1 - \frac{m_2(\mathcal{K}_1 + [\mathcal{L}]_m)\Delta \mathcal{K} S_1}{F(\theta_2, m_2)} ,$$

$$2 \overset{o}{d} \overset{hh}{{}_{12}} = -\frac{m_2(2\theta_2 - 1 - m_2)\Delta \mathcal{K} \Delta \mathcal{L} S_1}{F(\theta_2, m_2)} ,$$

$$2 \overset{o}{d} \overset{hh}{{}_{22}} = \mathcal{L}_1 - \frac{m_2(\mathcal{L}_1 + [\mathcal{K}]_m)\Delta \mathcal{L} S_1}{F(\theta_2, m_2)} , \qquad (24.3.4)$$

$$2 \overset{o}{d} \overset{hh}{{}_{33}} = \{\mathcal{L}\}_m ,$$

where

$$F(\theta_2, m_2) = 4(1 - \theta_2)(\theta_2 - m_2)\Delta \mathcal{L} \Delta \mathcal{K} + S_1[S]_m \qquad (24.3.5)$$

and $\Delta \mathcal{K} = \mathcal{K}_1 - \mathcal{K}_2$, $\Delta \mathcal{L} = \mathcal{L}_1 - \mathcal{L}_2$, $S_1 = \mathcal{K}_1 + \mathcal{L}_1$.

A natural way of finding the formulae (24.3.4) is to apply the Francfort-Murat equation (3.8.58). The derivation is based on the following implicit equation

$$m_2(d^1 - \overset{o}{d}{}^{hh})^{-1} = (d^1 - d^2)^{-1} + \Gamma , \qquad \Gamma = \theta_1 \Gamma_{\mathbf{d}}^{(1)} + \theta_2 \omega_1 \Gamma_{\mathbf{d}}^{(2)} , \qquad (24.3.6)$$

where the tensors $\Gamma_{\mathbf{d}}^{(\alpha)}$ are defined by (3.8.59); for $\alpha = 1$, $n = e_1$ and for $\alpha = 2$, $n = e_2$. The non-zero components of the tensors $\Gamma_{\mathbf{d}}^{(\alpha)}$ are

$$(\Gamma_{\mathbf{d}}^{(1)})_{2222} = -\frac{4}{S_1} , \qquad (\Gamma_{\mathbf{d}}^{(1)})_{1212} = -\frac{1}{\mathcal{L}_1} ,$$

$$(\Gamma_{\mathbf{d}}^{(2)})_{1111} = -\frac{4}{S_1} , \qquad (\Gamma_{\mathbf{d}}^{(2)})_{1212} = -\frac{1}{\mathcal{L}_1} . \qquad (24.3.7)$$

Thus we have

$$\Gamma_{1111} = -\frac{4\theta_2\omega_1}{S_1}\,, \quad \Gamma_{1122} = 0\,, \quad \Gamma_{2222} = -\frac{4\theta_1}{S_1}\,, \quad \Gamma_{1212} = -\frac{\theta_1 + \theta_2\omega_1}{\mathcal{L}_1}\,. \quad (24.3.8)$$

Now we represent Γ in the basis $\mathbf{a}_i \otimes \mathbf{a}_j$. The components Γ^{ij} are set up in the table:

$$[\Gamma^{ij}] = -2 \begin{bmatrix} \dfrac{\theta_2\omega_1 + \theta_1}{S_1} & \dfrac{\theta_2\omega_1 - \theta_1}{S_1} & 0 \\[2mm] \dfrac{\theta_2\omega_1 - \theta_1}{S_1} & \dfrac{\theta_2\omega_1 + \theta_1}{S_1} & 0 \\[2mm] 0 & 0 & \dfrac{\theta_2\omega_1 + \theta_1}{\mathcal{L}_1} \end{bmatrix}\,.$$

Further steps are similar to those performed in Sec. 24.2. They lead to Eq. (24.3.4).

25. Explicit formulae for effective membrane stiffnesses and compliances of ribbed plates

The relaxed formulation of the minimum compliance problem for a plate loaded in its plane involves orthogonal second rank ribbed composite plates. The aim of this section is to derive closed formulae for effective membrane stiffnesses of such structures. These formulae play a crucial role in the optimization procedure.

25.1. First rank ribbed plates

Consider the problem of finding explicit formulae for the effective stiffness and compliance tensors: A_h, a^h of the ribbed plate of Fig. 24.1.1. The easiest method is to use Eqs. (3.9.15) and (3.9.19). Here $n = e_1$, hence by Eq. (3.9.18)

$$(\mathbf{\Gamma}_A)_{1111} = \frac{1}{s_2}, \qquad (\mathbf{\Gamma}_A)_{1212} = \frac{1}{4\mu_2} \tag{25.1.1}$$

and the components with indices: 2211, 1122, 2222, 1222, 1112 vanish. By using (21.1.24) we can find the components of $\mathbf{\Gamma}_A$ referred to the basis $a_i \otimes a_j$. They form the matrix

$$[(\mathbf{\Gamma}_A)^{ij}] = \begin{bmatrix} \dfrac{1}{2s_2} & \dfrac{1}{2s_2} & 0 \\[2mm] \dfrac{1}{2s_2} & \dfrac{1}{2s_2} & 0 \\[2mm] 0 & 0 & \dfrac{1}{2\mu_2} \end{bmatrix}. \tag{25.1.2}$$

By (3.9.15) we have

$$\left[\left(\frac{A_2 - A_h}{\theta_1} \right)_{ij} \right]^{-1} = \frac{1}{2}X, \tag{25.1.3}$$

where

$$X = \begin{bmatrix} \dfrac{1}{\Delta k} - \dfrac{\theta_2}{s_2} & -\dfrac{\theta_2}{s_2} & 0 \\[2mm] -\dfrac{\theta_2}{s_2} & \dfrac{1}{\Delta \mu} - \dfrac{\theta_2}{s_2} & 0 \\[2mm] 0 & 0 & \dfrac{1}{\Delta \mu} - \dfrac{\theta_2}{\mu_2} \end{bmatrix}.$$

After inverting matrix X and using (25.1.3) one finds the stiffnesses

$$\frac{1}{2}\overset{o}{A}{}_{h}^{11} = \langle k \rangle_\theta - t_\theta(\Delta k)^2, \qquad \frac{1}{2}\overset{o}{A}{}_{h}^{12} = -t_\theta \Delta k \Delta \mu,$$

$$\frac{1}{2}\overset{o}{A}{}_{h}^{22} = \langle \mu \rangle_\theta - t_\theta(\Delta \mu)^2, \qquad \frac{1}{2}\overset{o}{A}{}_{h}^{33} = \{\mu\}_\theta. \tag{25.1.4}$$

The effective compliances are given by

$$2 \, \overset{o}{a} \, {}^{h}_{11} = \langle \mathcal{K} \rangle_\theta - T_\theta (\Delta \mathcal{K})^2 \, , \qquad 2 \, \overset{o}{a} \, {}^{h}_{12} = T_\theta \Delta \mathcal{K} \Delta \mathcal{L} \, ,$$

$$2 \, \overset{o}{a} \, {}^{h}_{22} = \langle \mathcal{L} \rangle_\theta - T_\theta (\Delta \mathcal{L})^2 \, , \qquad 2 \, \overset{o}{a} \, {}^{h}_{33} = \langle \mathcal{L} \rangle_\theta \, . \tag{25.1.5}$$

The notation of Sec. 24.1 also applies here.

25.2.　Second rank ribbed plates

Consider the ribbed second rank plate of Fig. 24.2.1. The membrane stiffnesses $\overset{o}{A} \, {}^{\alpha\beta}_{hh}$ and compliances $\overset{o}{a} \, {}^{hh}_{\alpha\beta}$ are expressed as (24.2.3) and (24.2.8) or as the same formulae as those for the bending stiffnesses $\overset{o}{D} \, {}^{\alpha\beta}_{hh}$ and bending compliances $\overset{o}{d} \, {}^{hh}_{\alpha\beta}; \alpha, \beta = 1, 2$. The difference concerns the shearing characteristics:

$$\frac{1}{2} \, \overset{o}{A} \, {}^{33}_{hh} = \{\mu\}_m \, , \qquad 2 \, \overset{o}{a} \, {}^{hh}_{33} = \langle \mathcal{L} \rangle_m \, . \tag{25.2.1}$$

For the ribbed plate of Fig. 24.3.1 the stiffnesses $\overset{o}{A} \, {}^{\alpha\beta}_{hh}$ and compliances $\overset{o}{a} \, {}^{hh}_{\alpha\beta}$ are given by (24.3.2) and (24.3.4) respectively. The shearing characteristics are given by Eq. (25.2.1).

26. Thin bending two-phase plates of minimum compliance

One of the most challenging optimization problem of plates in bending is formulated as
follows: given two materials of fixed total volume, find its optimal distribution within a
given domain such that the total compliance (or the work of the loading applied) assumes
a smallest possible value. This results in the stiffest plate design. Already in the late sixties
it was discovered that admitting ribs improves the design and that this process never stops,
cf. Kozłowski and Mróz (1969). This phenomenon has further been cleared up by Cheng
(1981) and Cheng and Olhoff (1981). If one uses the finite element method, the solutions
become mesh-dependent. Thus it turned out that one should abandon all hope that the
problem can be solved as originally posed.

Applications of the theory of G-convergence of operators and Γ-convergence of functionals have essentially helped in rearranging the optimization problem mentioned above
to its well-posed *relaxed form*. It turned out that optimizing the layout of two materials
should take into account its mixing at a microscale. Thus instead of filling up the given domain with both phases, one should admit a two-phase composite material made of the two
given materials. Moreover, it has been revealed that not all two-phase composites should be
taken into consideration. In the problem of minimizing the total compliance, it is sufficient
to admit so called orthogonal ribbed plates of the second rank, with stiffnesses depending
on three scalar unknowns.

The aim of the present section is to give a detailed justification of this relaxation formulation.

26.1. Ill-posedness of the initial formulation

The subject of our consideration is a thin transversely symmetric Kirchhoff plate subject
to a transverse loading $q = q(x)$, $x \in \Omega$, Ω being a mid-plane of the plate. The problem
reduces to the plate bending problem. The setting of the problem is fully two-dimensional:
deformation of the plate is determined by the deflection function $w = w(x)$, representing
transverse displacements of the plate mid-plane.

To fix the data, let us assume that the plate is clamped along $\Gamma_0 \subset \partial\Omega$:

$$w = 0 , \qquad \frac{\partial w}{\partial n} = 0 \qquad \text{on } \Gamma_0 , \tag{26.1.1}$$

and free along $\Gamma_1 = \partial\Omega\backslash\Gamma_0$. The bending stiffness tensor $D = D(x)$ is the only elastic
characteristic of the plate. Let us assume that distribution of the bending stiffness is piecewise constant: $D = D_1$ in Ω_1 and $D = D_2$ in Ω_2; the domains Ω_α satisfy the conditions:
$\overline{\Omega}_1 \cup \overline{\Omega}_2 = \overline{\Omega}$ and $\Omega_1 \cap \Omega_2 = \emptyset$. The division of Ω into Ω_α is arbitrary. The characteristic
function of the domain Ω_α will be denoted by $\chi_\alpha = \chi_\alpha(x)$. Thus one can write

$$D = \chi_1(x)D_1 + \chi_2(x)D_2 , \tag{26.1.2}$$

and $\chi_1(x) + \chi_2(x) = 1$, $x \in \Omega$. Assume, moreover, that both phases are isotropic, cf. Eq.
(21.1.26)

$$D_\alpha = 2k_\alpha I_1 + 2\mu_\alpha I_2 , \tag{26.1.3}$$

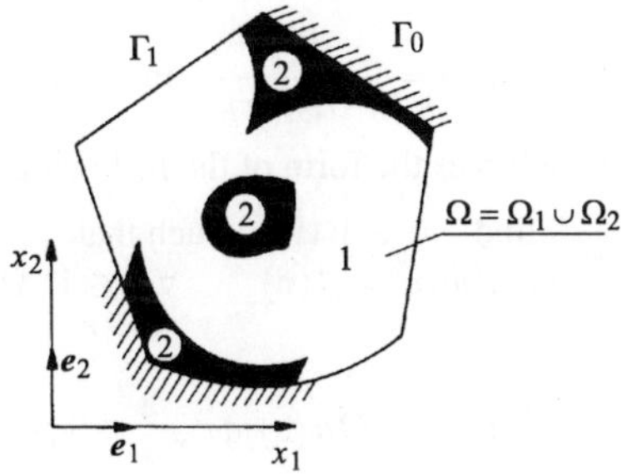

Fig. 26.1.1. Two-phase plate

$$d_\alpha = \frac{1}{2}\mathcal{K}_\alpha I_1 + \frac{1}{2}\mathcal{L}_\alpha I_2 \, , \tag{26.1.4}$$

where $\mathcal{K}_\alpha = (k_\alpha)^{-1}$, $\mathcal{L}_\alpha = (\mu_\alpha)^{-1}$ and the ordering assumption holds:

$$k_2 > k_1 \, , \qquad \mu_2 > \mu_1 \, , \qquad \mathcal{K}_1 > \mathcal{K}_2 \, , \qquad \mathcal{L}_1 > \mathcal{L}_2 \, . \tag{26.1.5}$$

The constitutive relations link the moment tensor M with the curvature tensor κ

$$M = D\kappa \, , \qquad \kappa = dM \, , \tag{26.1.6}$$

where $d = D^{-1}$. Referring these relations to the basis $\{e_\alpha\}$ one finds

$$M^{\alpha\beta} = D^{\alpha\beta\lambda\mu}\kappa_{\lambda\mu} \, , \qquad \kappa_{\lambda\mu} = d_{\lambda\mu\alpha\beta}M^{\alpha\beta} \, . \tag{26.1.7}$$

If we refer them to the basis a_i we have

$$M^i = \sum_{j=1}^{3} D^{ij}\kappa^j \, , \qquad \kappa^i = \sum_{j=1}^{3} d_{ij}M^j \, . \tag{26.1.8}$$

The components $D^{\alpha\beta\lambda\mu}$, $d_{\alpha\beta\lambda\mu}$, D^{ij} and d_{ij} are given by (21.1.30) – (21.1.32), where k, μ, $\mathcal{K}$ and $\mathcal{L}$ refer to the phase number (1) or (2). Note that the Greek indices at κ are at the lower level and the index i is put at the upper level. This mismatch is a consequence of not introducing the co-basis a^i, to simplify this formalism. Since the space is Euclidean, both notations are correct.

A moment field M is said to be statically admissible if $M \in \mathcal{S}_2(\Omega)$, with

$$\mathcal{S}_2(\Omega) := \{M = (M^{\alpha\beta}) \in L^2(\Omega, \mathbb{E}_2^s) | \int_\Omega M^{\alpha\beta}\kappa_{\alpha\beta}(v)dx = \int_\Omega qvdx \, \forall \, v \in V(\Omega)\} \, ,$$

where

$$V(\Omega) = \{v \in H^2(\Omega) | v = 0 \quad \text{and} \quad \frac{\partial v}{\partial n} = 0 \quad \text{on } \Gamma_0\} \, .$$

A strain field κ is said to be kinematically admissible if $\kappa \in \mathcal{K}_\kappa(\Omega)$, with $\mathcal{K}_\kappa(\Omega)$ defined in Sec. 3.4. Then there exists $w \in H^2(\Omega)$ such that

$$\kappa_{\alpha\beta}(w) = -\frac{\partial^2 w}{\partial x_\alpha \partial x_\beta} \quad \text{in} \quad \Omega \,. \tag{26.1.9}$$

The equilibrium problem assumes the form of the following problem:

$$(P) \quad \left|\begin{array}{l} \text{find} \quad w \in V(\Omega) \quad \text{such that} \\ a_{\boldsymbol{D}}(w,v) = f(v) \qquad \forall\, v \in V(\Omega) \,, \end{array}\right. \tag{26.1.10}$$

where

$$a_{\boldsymbol{D}}(w,v) = \int_\Omega \kappa(w) : (\boldsymbol{D}\kappa(v))dx \,, \qquad w,v \in H^2(\Omega) \tag{26.1.11}$$

and

$$f(v) = \int_\Omega qv dx \,. \tag{26.1.12}$$

The integrand of (26.1.11) is written in the invariant manner. It should be understood as follows:

$$\kappa(w) : (\boldsymbol{D}\kappa(v)) = \kappa_{\alpha\beta}(w) D^{\alpha\beta\lambda\mu}\kappa_{\lambda\mu}(v) = \sum_{i,j=1}^{3} \kappa^i(w) D^{ij}\kappa^i(v) \,. \tag{26.1.13}$$

The functional $J : L^\infty(\Omega) \to \mathbb{R}$

$$J(\chi_2) = f(w(\chi_2)) \,, \tag{26.1.14}$$

is called the plate compliance (or total compliance). Its argument is the function χ_2 which determines the layout of the materials (1) and (2) within Ω. Given χ_2, one can solve the problem (P) and, having $w(\chi_2)$, compute $J(\chi_2)$.

Let A be a positive constant. The minimum compliance problem is usually put in the form:

$$\inf\{J(\chi_2)|\, \chi_2 \in L^\infty(\Omega, \{0,1\}) \,, \int_\Omega \chi_2 dx = A \,,$$

$$\text{with } w \text{ being the solution to problem } (P)\} \,. \tag{26.1.15}$$

By Castigliano's theorem:

$$J(\chi_2) = \min\{\int_\Omega \boldsymbol{M} : (\boldsymbol{d}\boldsymbol{M})dx \mid \boldsymbol{M} \in \mathcal{S}_2(\Omega)\} \,. \tag{26.1.16}$$

Now we introduce the Lagrange multiplier λ for the isoperimetric constraint appearing in (26.1.15). Kohn and Strang (1986) proved that supremum over λ can be interchanged with infimum over χ_2. Thus for a fixed λ we put the problem (26.1.15) in the form:

$$\min_{\chi_2 \in L^\infty(\Omega,\{0,1\})} \min_{\boldsymbol{M} \in \mathcal{S}_2(\Omega)} \int_\Omega (\boldsymbol{M} : (\boldsymbol{d}\boldsymbol{M}) + \lambda\chi_2)dx \,. \tag{26.1.17}$$

The order of minima can be interchanged. Then we use Theorem 1.2.33 according to which the minimum over χ_2 can be put under the integral. We find

$$\min_{M \in \mathcal{S}_2(\Omega)} \left\{ \int_\Omega \min[M : (d_2 M) + \lambda, M : (d_1 M)] dx \right\}. \tag{26.1.18}$$

By (21.1.38) we have

$$M : (d_\sigma M) = \frac{1}{2} \mathcal{K}_\sigma I^2(M) + \frac{1}{2} \mathcal{L}_\sigma II^2(M), \tag{26.1.19}$$

where $I^2(M) = [I(M)]^2$ and $II^2(M) = [II(M)]^2$.

The integrand in (26.1.18) is a non-convex function of $I(M)$ and $II(M)$, see Fig. 26.1.2.

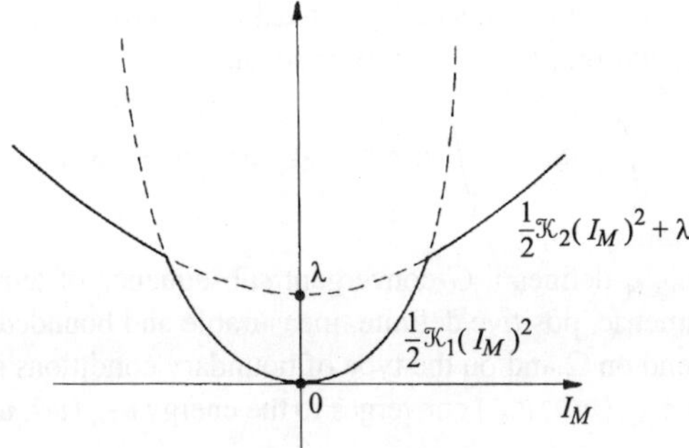

Fig. 26.1.2. The section: $II_M = II(M) = 0$; $I_M = I(M)$

We conclude that the problem (26.1.18) is ill-posed and requires relaxation. More precisely, the integral functional involved in this problem is not sequentially weakly lower semicontinuous.

26.2. Relaxation

The design variable in the problem (26.1.17) is the characteristic function $\chi = \chi_2$ which equals 1 in Ω_2 and 0 in Ω_1. Since this problem is ill-posed, one should extend the design space in such a manner that the weak-$*$ limit of a sequence of functions $\{\chi_n\}$ would belong to this space. Applying Corollary 1.1.4 we conclude that

$$\chi_n \rightharpoonup m_2 \quad \text{in} \quad L^\infty(\Omega, [0, 1]) \quad \text{weak-}* \quad \text{as} \quad n \to \infty. \tag{26.2.1}$$

Any function $m_2 \in L^\infty(\Omega, [0, 1])$ can be interpreted as a distribution of the area fraction of material (2).

Any sequence $\{\chi_n\}$ defines a sequence

$$D_n(x) = (1 - \chi_n(x)) D_1 + \chi_n(x) D_2 \tag{26.2.2}$$

of stiffness tensors and of compliance tensors $d_n(x) = (D_n(x))^{-1}$. They determine the sequence of the solutions $w_n \in V(\Omega)$ that satisfy

$$a_{\mathbf{D}_n}(w_n, v) = f(v) \qquad \forall\, v \in V(\Omega)\,. \tag{26.2.3}$$

The following theorem determines the properties of the limit function w^h to which the functions w_n converge.

Theorem 26.2.1. From the sequence $\{\chi_n\}_{n\in\mathbb{N}}$ such that the isoperimetric condition

$$\int_\Omega \chi_n dx = A\,, \qquad n \in \mathbb{N} \tag{26.2.4}$$

is fulfilled one can extract a subsequence $\{\chi_{n'}\}$ weakly-$*$ converging to $m_2 \in L^\infty(\Omega, [0, 1])$. Moreover, the area condition (26.2.4) is preserved, since

$$\int_\Omega \chi_{n'} dx \rightarrow \int_\Omega m_2 dx = A \quad \text{as} \quad n' \rightarrow \infty\,. \tag{26.2.5}$$

The sequence $\{\chi_{n'}\}_{n'\in\mathbb{N}}$ defines a G-convergent subsequence of tensors $D_{n'}(x)$. Its G-limit $D_h(\cdot)$ is a symmetric, positive definite, measurable and bounded tensor-valued function. It does not depend on Ω and on the type of boundary conditions imposed on $\partial\Omega$. The sequence of energies $a_{\mathbf{D}_{n'}}(w_{n'}, w_{n'})$ converges to the energy $a_{\mathbf{D}_h}(w^h, w^h)$, where w^h solves the problem

$$(P_h) \quad \left| \begin{array}{l} \text{find } w^h \in V(\Omega) \quad \text{such that} \\ a_{\mathbf{D}_h}(w^h, v) = f(v) \qquad \forall\, v \in V(\Omega)\,. \end{array} \right.$$

The sequence $w_{n'}$ converges to w^h weakly in $H^2(\Omega)$.

Proof. The convergence of $\chi_{n'}$ to m_2 and (26.2.5) result from (26.2.1). The sequence $D_{n'}(x)$ has the form (26.2.2). Thus the remaining part of the proof is a direct consequence of application of H-convergence results presented in Sec. 1.3.2. It is sufficient to take $\varepsilon = \dfrac{1}{n'}$. $\qquad\qquad\square$

Obviously, different $\{\chi_n\}$ satisfying (26.2.4) may tend to the same limit $m_2 \in L^\infty(\Omega, [0, 1])$ and produce different limiting (homogenized) tensors $D_h(x)$. Theorem 26.2.1 asserts that there exist tensor-valued functions $D_h(\cdot)$ associated with sequences $\{\chi_n\}_{n\in\mathbb{N}}$. According to Dal Maso and Kohn (see Sec. 1.3.2 and Allaire and Kohn (1993a)), these functions can be characterized pointwise for a.e. $x \in \Omega$ as follows. For each $\rho_2 \in [0, 1]$ there is a closed set of fourth-order tensors G_{ρ_2}, called the G-closure of D_2 and D_1 with area fractions ρ_2 and $\rho_1 = 1 - \rho_2$, with the following properties:
(i) if $m_2(x)$ and $D_h(x)$ are linked as in Th. 26.2.1, then $D_h(x) \in G_{m_2(x)}$ for a.e. $x \in \Omega$.

(ii) If $m_2(x)$ and $D_h(x)$ are measurable functions satisfying the condition $D_h(x) \in G_{m_2(x)}$ a.e. $x \in \Omega$, then there exists a sequence $\{\chi_n\}_{n\in\mathbb{N}}$ weakly-$*$ convergent in $L^\infty(\Omega, [0,1])$ to m_2 that determines the G-limit $D_h(x)$. $\qquad\square$

The following theorem is due to Dal Maso and Kohn (see Sec. 1.3.2 and Allaire and Kohn (1993a)) and reveals the role of periodic composites, cf. also Lipton (1994c).

Theorem 26.2.2. Let ρ_2 be a fixed number from $[0,1]$. Denote by $G^{per}_{\rho_2}$ the set of all effective stiffness tensors D_h of periodic plates, determined by, see (3.4.10),

$$\boldsymbol{\kappa} : (\boldsymbol{D}_h \boldsymbol{\kappa}) = \min\{\langle \boldsymbol{\kappa}^y : (\boldsymbol{D}\boldsymbol{\kappa}^y)\rangle \mid \boldsymbol{D} = \chi_1^Y(y)\boldsymbol{D}_1 + \chi_2^Y(y)\boldsymbol{D}_2 ,$$
$$\langle \chi_2^Y(y)\rangle = \rho_2 ,\ \boldsymbol{\kappa}^y \in \mathcal{K}^{per}_{\boldsymbol{\kappa}}(Y) ,\ \langle \boldsymbol{\kappa}^y\rangle = \boldsymbol{\kappa}\} . \qquad (26.2.6)$$

Here $\langle \cdot \rangle$ represents averaging over a certain rectangular cell Y. The periodicity cell Y is divided into Y_1 and Y_2; χ_α^Y represents the characteristic function of the domain Y_α.

Denote by $\overline{G^{per}_{\rho_2}}$ the closure of the set $G^{per}_{\rho_2}$. Then $G^{per}_{\rho_2} \subseteq G_{\rho_2}$ and $\overline{G^{per}_{\rho_2}} = G_{\rho_2}$. $\qquad\square$

Dual version of this theorem is formulated as follows.

Corollary 26.2.3. Let ρ_2 be a fixed number from $[0,1]$. The effective compliance tensors d_h of periodic plates are determined by, see Sec. 3.6

$$\boldsymbol{M} : (\boldsymbol{d}_h \boldsymbol{M}) = \min\{\langle \boldsymbol{m} : (\boldsymbol{d}\boldsymbol{m})\rangle \mid \boldsymbol{d} = \chi_1^Y(y)\boldsymbol{d}_1 + \chi_2^Y(y)\boldsymbol{d}_2 ,\ \langle \chi_2^Y\rangle = \rho_2 ,$$
$$\boldsymbol{m} = (m^{\alpha\beta}) \in \mathcal{S}^{per}_2(Y) ,\qquad \langle \boldsymbol{m}\rangle = \boldsymbol{M}\} . \qquad (26.2.7)$$

The set of all such tensors d_h is still $G^{per}_{\rho_2}$. As before $G^{per}_{\rho_2} \subseteq G_{\rho_2}$ and $\overline{G^{per}_{\rho_2}} = G_{\rho_2}$. $\qquad\square$

Let us pass to the relaxation of the problem (26.1.17). Consider the sequence $D_n(x)$ which defines the sequence of compliances $d_n(x) = (D_n(x))^{-1}$ and the sequence of moment fields M_n. Then

$$\int_\Omega M_n : (d_n M_n)dx \rightarrow \int_\Omega M_h : (d^h M_h)dx , \qquad (26.2.8)$$

where $d^h = (D_h)^{-1}$. The sequence M_n converges to M_h weakly in $L^2(\Omega, \mathbb{E}_2^s)$. Moreover, $M_h \in \mathcal{S}_2(\Omega)$.

To relax the problem (26.1.17) one replaces:

$$d \rightsquigarrow d_n ,\qquad M \rightsquigarrow M_n ,\qquad \chi_2 \rightsquigarrow \chi_n , \qquad (26.2.9)$$

and finds

$$\min_{\chi_n \in L^\infty(\Omega, \{0,1\})} \min_{M_n \in \mathcal{S}_2(\Omega)} \int_\Omega (M_n : (d_n M_n) + \lambda\chi_n)dx . \qquad (26.2.10)$$

Note that in the limit $(n \to \infty)$ the first minimum is replaced by the two minima: over $m_2(x) \in L^\infty(\Omega, [0, 1])$ and over $d^h(x) \in G_{m_2(x)}$. Thus we have

$$\min_{m_2 \in L^\infty(\Omega,[0,1])} \ \min_{d^h \in G_{m_2(x)}} \ \min_{M \in \mathcal{S}_2(\Omega)} \int_\Omega (M : (d^h(x)M) + \lambda m_2(x))dx \ . \qquad (26.2.11)$$

Let us compare the problem (26.1.17) with (26.2.11). We note that
(i) designs predicted within the initial formulation (26.1.17) are included in the class of generalized solutions to the relaxed formulation (26.2.11).
(ii) If the solution of the problem (26.2.11) lies within the class of the initial formulation (26.1.17), then the value of the compliance determined by the relaxed problem (26.2.11) coincides with the value of the compliance determined by the initial formulation.
(iii) A solution to the relaxed problem exists. The minimum is attained by the limits of conventional designs $\{\chi_n\}$.

The features (i) – (iii) mean that the formulation (26.2.11) is an appropriate relaxation of problem (26.1.17).

The features (i) and (ii) do not need proving. To prove (iii), let us assume that $\{\chi_n\}$ is a minimizing sequence of the problem (26.2.11). This sequence determines a sequence d_n of compliances. By Theorem (26.2.1) and Corollary (26.2.3) one can extract a subsequence, still denoted by $\{\chi_n\}$ such that $\chi_n(x) \rightharpoonup m_2(x)$ in $L^\infty(\Omega, [0, 1])$ weak-$*$ and $d_n(x)$ G-converges to $d^h(x)$. Then the functional in (26.1.17) tends to the value of the functional in (26.2.11) for those $m_2(x)$ and $d^h(x)$. On the other hand, the couple (m_2, d^h) is a minimizer of (26.2.11). This proves the property (iii). $\qquad\qquad\qquad\square$

The formulation (26.2.11) can be rearranged to the form

$$\min_{m_2 \in L^\infty(\Omega,[0,1])} \ \min_{M \in \mathcal{S}_2(\Omega)} \int_\Omega [2\mathcal{W}^*(M, m_2(x)) + \lambda m_2]dx \ , \qquad (26.2.12)$$

where

$$\mathcal{W}^*(M, m_2(x)) = \min_{d^h(x) \in G_{m_2(x)}} \frac{1}{2} M : (d^h(x)M) \ . \qquad (26.2.13)$$

By Corollary (26.2.3) one can write

$$2\mathcal{W}^*(M, m_2(x)) = \min_{d \in G^{per}_{m_2(x)}} (M : (dM)) \ , \qquad (26.2.14)$$

or

$$2\mathcal{W}^*(M, m_2(x)) = \inf_{d \in G^{per}_{m_2(x)}} M : (dM) \ . \qquad (26.2.15)$$

One can show that $2\mathcal{W}^*$ is only a function of M, m_2 and $d_\alpha(\alpha = 1, 2)$, while x is concealed in $m_2 = m_2(x)$. It turns out that this function can be explicitly calculated, which will be the subject of the next section.

26.3. Bounding the potential $\mathcal{W}^*$ by the translation method of Cherkaev-Gibianskii

By (26.2.7) and (26.2.15) we have

$$2\mathcal{W}^*(M, m_2) = \inf_{\chi_2^Y \in L^\infty(Y, \{0,1\})} \min\{\langle m : (dm)\rangle|$$

$$d = d_1 \chi_1^Y(y) + d_2 \chi_2^Y(y), d_\alpha \text{ of the form (26.1.4)},$$

$$\langle \chi_2^Y \rangle = m_2, m = (m^{\alpha\beta}) \in \mathcal{S}_2^{per}(Y), \langle m \rangle = M\} . \qquad (26.3.1)$$

Here m_2 represents a number from $[0, 1]$.

The expression above will be estimated by the translation method. To this end we refer the tensors d and m to the basis (a_i). Thus $m : (dm) = \widehat{m}^T \widehat{dm}$, where $\widehat{m} = (m^1, m^2, m^3)^T, \widehat{d} = [d^{ij}], d = d^{ij} a_i \otimes a_j$. In the sequel we shall neglect the sign "$\widehat{}$".

Let us take $\beta \geq 0$ such that

$$d - \frac{1}{2}\beta \overset{o}{T} \geq 0 , \qquad (26.3.2)$$

where $\overset{o}{T}$ is defined by (21.1.19), and rearrange (26.3.1) as follows

$$2\mathcal{W}^*(M, m_2) = \inf_{\substack{\chi_2^Y \\ \langle \chi_2^Y \rangle = m_2}} \min_{\substack{m \in \mathcal{S}_2^{per}(Y) \\ \langle m \rangle = M}} [\langle m^T(d - \frac{1}{2}\beta \overset{o}{T})m\rangle + \frac{\beta}{2}\langle m^T \overset{o}{T} m\rangle] . \qquad (26.3.3)$$

The first term will be estimated by disregarding the differential constraints involved in $\mathcal{S}_2^{per}(Y)$. The second term will be estimated by using the property of quasiconvexity of the function $F(m) = -\det m$, see (21.4.15) in Sec. 21.4.4. Thus we estimate the first term by making use of the estimate (21.5.4) for $A = d - \frac{1}{2}\beta \overset{o}{T}$, taking into account (26.3.2):

$$\langle m^T(d - \frac{1}{2}\beta \overset{o}{T})m\rangle \geq M^T \langle (d - \frac{1}{2}\beta \overset{o}{T})^{-1}\rangle^{-1}M . \qquad (26.3.4)$$

By using (21.4.15) we obtain

$$\langle m^T \overset{o}{T} m\rangle \geq M^T \overset{o}{T} M . \qquad (26.3.5)$$

Since the r.h.s. of (26.3.4) and (26.3.5) are independent of χ_2^Y, we arrive at the translation bound in the form

$$2\mathcal{W}^*(M, m_2) \geq M^T \mathbf{d}(\beta)M , \qquad (26.3.6)$$

with β such that (26.3.2) is satisfied and

$$\mathbf{d}(\beta) = \langle (d - \frac{1}{2}\beta \overset{o}{T})^{-1}\rangle^{-1} + \frac{1}{2}\beta \overset{o}{T} . \qquad (26.3.7)$$

To make the bound (26.3.6) possibly optimal we maximize the r.h.s. over feasible values of β

$$2\mathcal{W}^*(M, m_2) \geq \max_{\substack{\beta \geq 0 \\ d - \frac{1}{2}\beta \overset{o}{T} \geq 0}} M^T \mathbf{d}(\beta)M . \qquad (26.3.8)$$

According to (26.1.4) and (21.1.19) the tensors $d(y)$ and $\overset{o}{T}$ can be represented by

$$d(y) = \frac{1}{2}\mathcal{K}(y)\boldsymbol{I}_1 + \frac{1}{2}\mathcal{L}(y)\boldsymbol{I}_2 , \qquad \overset{o}{T} = -\boldsymbol{I}_1 + \boldsymbol{I}_2 , \tag{26.3.9}$$

where

$$\mathcal{K}(y) = \mathcal{K}_1\chi_1^Y(y) + \mathcal{K}_2\chi_2^Y(y) , \qquad \mathcal{L}(y) = \mathcal{L}_1\chi_1^Y(y) + \mathcal{L}_2\chi_2^Y(y) . \tag{26.3.10}$$

Now we can compute

$$\mathbf{d}(\beta) = \langle[\frac{1}{2}(\mathcal{K}(y) + \beta)\boldsymbol{I}_1 + \frac{1}{2}(\mathcal{L}(y) - \beta)\boldsymbol{I}_2]^{-1}\rangle^{-1} + \frac{1}{2}\beta(-\boldsymbol{I}_1 + \boldsymbol{I}_2) . \tag{26.3.11}$$

By using the rule

$$(a\boldsymbol{I}_1 + b\boldsymbol{I}_2)^{-1} = a^{-1}\boldsymbol{I}_1 + b^{-1}\boldsymbol{I}_2 , \tag{26.3.12}$$

valid for any $a, b \in \mathbb{R}\backslash\{0\}$, see Eq. (21.1.26), we find

$$\mathbf{d}(\beta) = \frac{1}{2}K(\beta)\boldsymbol{I}_1 + \frac{1}{2}L(\beta)\boldsymbol{I}_2 , \tag{26.3.13}$$

where

$$K(\beta) = \left(\frac{m_1}{\mathcal{K}_1 + \beta} + \frac{m_2}{\mathcal{K}_2 + \beta}\right)^{-1} - \beta , \qquad L(\beta) = \left(\frac{m_1}{\mathcal{L}_1 - \beta} + \frac{m_2}{\mathcal{L}_2 - \beta}\right)^{-1} + \beta . \tag{26.3.14}$$

The condition (26.3.2) implies

$$\mathcal{K}_1 + \beta \geq 0 , \qquad \mathcal{L}_1 - \beta \geq 0 ; \qquad \mathcal{K}_2 + \beta \geq 0 , \qquad \mathcal{L}_2 - \beta \geq 0 . \tag{26.3.15}$$

In view of the ordering assumption (26.1.5) and $\beta \geq 0$ we have

$$\beta \in [0, \mathcal{L}_2] . \tag{26.3.16}$$

According to the rule (21.1.38) we can write

$$\frac{1}{2}\boldsymbol{M}^T\mathbf{d}(\beta)\boldsymbol{M} = \frac{1}{4}K(\beta)I^2(\boldsymbol{M}) + \frac{1}{4}L(\beta)II^2(\boldsymbol{M}) . \tag{26.3.17}$$

Let us introduce a new invariant of $\boldsymbol{M}$:

$$\zeta_M = \frac{II(\boldsymbol{M})}{|I(\boldsymbol{M})|} , \qquad \zeta_M = \frac{[(M^2)^2 + (M^3)^2]^{1/2}}{|M^1|} , \tag{26.3.18}$$

or

$$\zeta_M = \frac{[(M^{11} - M^{22})^2 + (2M^{12})^2]^{1/2}}{|M^{11} + M^{22}|} , \tag{26.3.19}$$

provided that M is referred to the basis e_α. The potential (26.3.17) can be written in the form

$$M^T \mathbf{d}(\beta) M = \begin{cases} \dfrac{1}{2} I^2(M) U(\beta, \zeta_M) & \text{if } I(M) \neq 0 \\[2ex] \dfrac{1}{2} L(\beta) II^2(M) & \text{if } I(M) = 0 \, , \end{cases} \tag{26.3.20}$$

where

$$U(\beta, x) = K(\beta) + x^2 L(\beta) \, . \tag{26.3.21}$$

Let us maximize the expression (26.3.20) with respect to β satisfying (26.3.16). In the case of $I(M) \neq 0$ the problem reduces to finding β_0 such that

$$\mathcal{H}(\zeta_M) := U(\beta_0, \zeta_M) = \max_{\beta \in [0, \mathcal{L}_2]} U(\beta, \zeta_M) \, . \tag{26.3.22}$$

The extremality condition: $\partial U / \partial \beta = 0$ leads to the equation

$$|[\mathcal{L}]_m - \beta| \Delta \mathcal{K} = \Delta \mathcal{L} \zeta_M |[\mathcal{K}]_m + \beta| \, , \tag{26.3.23}$$

where $[f]_m = m_1 f_2 + m_2 f_1$, $\Delta f = f_1 - f_2$ for $f \in \{\mathcal{K}, \mathcal{L}\}$. The solution of Eq. (26.3.23) is denoted by β_2. The condition (26.3.16) implies

$$[\mathcal{K}]_m + \beta_2 \geq 0 \, , \qquad [\mathcal{L}]_m - \beta_2 \geq m_2(\mathcal{L}_1 - \mathcal{L}_2) > 0 \, . \tag{26.3.24}$$

Thus

$$\beta_2 = \frac{\Delta \mathcal{K} [\mathcal{L}]_m - \Delta \mathcal{L} [\mathcal{K}]_m \zeta_M}{\Delta \mathcal{K} + \zeta_M \Delta \mathcal{L}} \, . \tag{26.3.25}$$

The condition (26.3.16) leads to

$$\zeta_2 \leq \zeta_M \leq \zeta_1 \, , \tag{26.3.26}$$

where, cf. (24.2.12)$_2$

$$\zeta_1 = \frac{[\mathcal{L}]_m \Delta \mathcal{K}}{[\mathcal{K}]_m \Delta \mathcal{L}} \, , \qquad \zeta_2 = \frac{m_2 \Delta \mathcal{K}}{[\mathcal{K}]_m + \mathcal{L}_2} \, . \tag{26.3.27}$$

Substitution of (26.3.25) into (26.3.14) gives

$$K(\beta_2) = \frac{\mathcal{K}_1 \mathcal{K}_2 + [\mathcal{L}]_m \langle \mathcal{K} \rangle_m - m_1 m_2 \Delta \mathcal{K} \Delta \mathcal{L} \zeta_M}{[S]_m} \, ,$$

$$L(\beta_2) = \frac{\mathcal{L}_1 \mathcal{L}_2 + [\mathcal{K}]_m \langle \mathcal{L} \rangle_m - m_1 m_2 \Delta \mathcal{K} \Delta \mathcal{L} (\zeta_M)^{-1}}{[S]_m} \, , \tag{26.3.28}$$

where $[S]_m = [\mathcal{K} + \mathcal{L}]_m$ and $\langle f \rangle = m_1 f_1 + m_2 f_2$, $f \in \{\mathcal{K}, \mathcal{L}\}$. We conclude that $\beta_0 = \beta_2$ if $\zeta_M \in [\zeta_2, \zeta_1]$, see Fig. 26.3.1b.

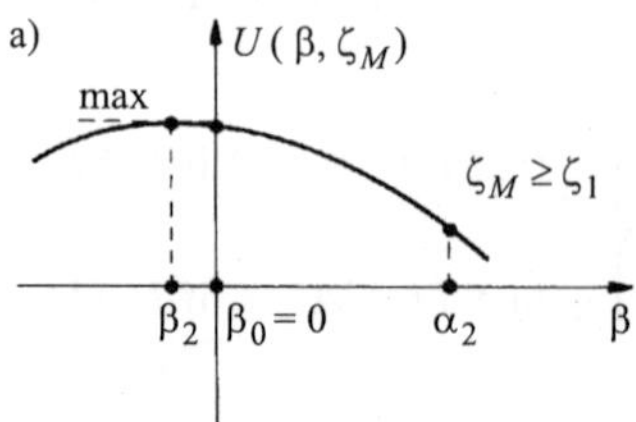

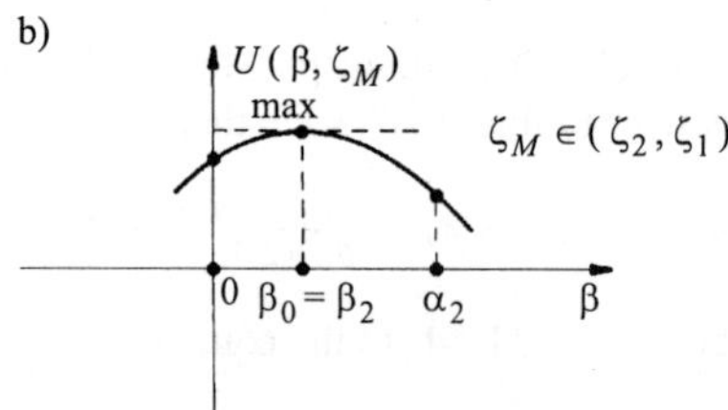

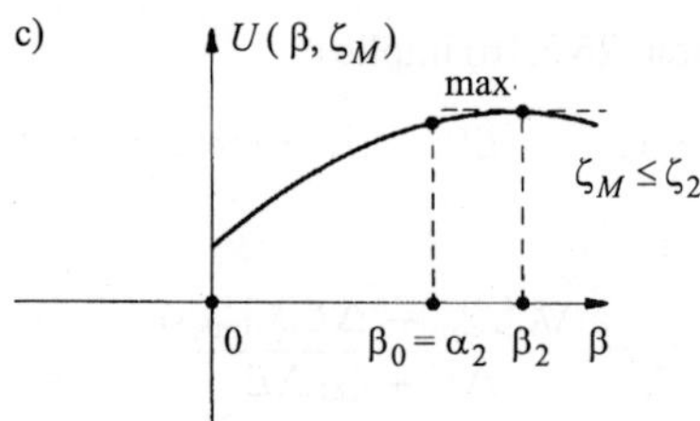

Fig. 26.3.1. Analysis of problem (26.3.22)

Assume that $\zeta_M > \zeta_1$. Then $\beta_2 < 0$ (cf. Fig. 26.3.1a) and $\beta_0 = 0$. We compute

$$K(0) = \{\mathcal{K}\}_m , \qquad L(0) = \{\mathcal{L}\}_m , \tag{26.3.29}$$

where $\{f\}_m = (m_1 f_1^{-1} + m_2 f_2^{-1})^{-1}$. Assume that $\zeta_M < \zeta_2$, see Fig. 26.3.1c. Then $\beta_2 > \mathcal{L}_2$ and $\beta_0 = \mathcal{L}_2$. We find

$$K(\mathcal{L}_2) = \check{K} , \qquad L(\mathcal{L}_2) = \mathcal{L}_2 , \tag{26.3.30}$$

where $\check{K}$ is determined by, cf. (24.2.16)

$$y(\check{K}, \mathcal{K}_1, \mathcal{K}_2, m_1, m_2) = \mathcal{L}_2 . \tag{26.3.31}$$

The above formula follows from $(26.3.14)_1$, (21.2.24) and (21.2.19). Comparing (26.3.31) with $(23.4.4)_1$ we conclude that

$$\check{K} = \underline{\mathcal{K}}_{HS} , \tag{26.3.32}$$

where $\underline{\mathcal{K}}_{HS}$ represents the lower Hashin-Shtrikman bound for the modulus $\mathcal{K}$.

The solution to the problem (26.3.22) can be summarized as follows

$$\beta_0 = \begin{cases} \mathcal{L}_2 & \text{if } \zeta_M \in [0, \zeta_2] \ \ (\text{regime 3}) , \\ \beta_2 & \text{if } \zeta_M \in [\zeta_2, \zeta_1] \ \ (\text{regime 2}) , \\ 0 & \text{if } \zeta_M \geq \zeta_1 \qquad (\text{regime 1}) , \end{cases} \qquad (26.3.33)$$

and

$$\mathcal{H}(\zeta) = \begin{cases} \mathcal{H}_L(\zeta) & \text{if } \zeta \in [0, \zeta_2] , \\ \mathcal{H}_i(\zeta) & \text{if } \zeta \in [\zeta_2, \zeta_1] , \\ \mathcal{H}_R(\zeta) & \text{if } \zeta \geq \zeta_1 , \end{cases} \qquad (26.3.34)$$

where

$$\mathcal{H}_L(\zeta) = a_L + c_L \zeta^2 , \qquad \mathcal{H}_R(\zeta) = a_R + c_R \zeta^2 ,$$
$$a_L = \check{K} , \quad c_L = \mathcal{L}_2 , \quad a_R = \{\mathcal{K}\}_m , \quad c_R = \{\mathcal{L}\}_m . \qquad (26.3.35)$$

Moreover, by using the identity:

$$\langle f \rangle_m [f]_m - f_1 f_2 = m_1 m_2 (\Delta f)^2 , \qquad (26.3.36)$$

one can express $\mathcal{H}_i(\zeta)$ as follows

$$\mathcal{H}_i(\zeta) = \langle \mathcal{K} \rangle_m - T_m (\Delta \mathcal{K})^2 - 2 T_m \Delta \mathcal{K} \Delta \mathcal{L} \zeta + [\langle \mathcal{L} \rangle_m - T_m (\Delta \mathcal{L})^2] \zeta^2 , \qquad (26.3.37)$$

with $T_m = m_1 m_2 ([S]_m)^{-1}$, cf. (24.1.10). One can prove that

$$\mathcal{H}_i(\zeta) = \mathcal{H}_L(\zeta) + A_L (\zeta - \zeta_2)^2 = \mathcal{H}_R(\zeta) + A_R (\zeta - \zeta_1)^2 , \qquad (26.3.38)$$

where

$$A_L = \frac{m_1 \Delta \mathcal{L} (\mathcal{L}_2 + [\mathcal{K}]_m)}{[S]_m} , \qquad A_R = \frac{m_1 m_2 (\Delta \mathcal{L})^2 [\mathcal{K}]_m}{[\mathcal{L}]_m [S]_m} . \qquad (26.3.39)$$

Thus we immediately see that

$$\mathcal{H}_i(\zeta) > \mathcal{H}_L(\zeta) \quad \text{and} \quad \mathcal{H}_i(\zeta) \geq \mathcal{H}_R(\zeta) , \qquad (26.3.40)$$

for all ζ. Moreover we have

$$\mathcal{H}_L(\zeta_2) = \mathcal{H}_i(\zeta_2) , \qquad \mathcal{H}_i(\zeta_1) = \mathcal{H}_R(\zeta_1) ,$$
$$\frac{d\mathcal{H}_L}{d\zeta}(\zeta_2) = \frac{d\mathcal{H}_i}{d\zeta}(\zeta_2) , \qquad \frac{d\mathcal{H}_i}{d\zeta}(\zeta_1) = \frac{d\mathcal{H}_R}{d\zeta}(\zeta_1) , \qquad (26.3.41)$$

which proves that the function $\mathcal{H}(\zeta)$ is smooth in the whole domain $\zeta \geq 0$, see Fig. 26.3.2. In the case of $I(\boldsymbol{M}) = 0$ one can show that

$$\max_{\beta \in [0, \mathcal{L}_2]} \frac{1}{2} L(\beta) II^2(\boldsymbol{M}) = \frac{1}{2} \{\mathcal{L}\}_m II^2(\boldsymbol{M}) . \qquad (26.3.42)$$

Finally, the r.h.s. of the estimate (26.3.8)

$$2\underline{\mathcal{W}}^*(\boldsymbol{M}, m_2) = \max_{\substack{\beta \geq 0 \\ \mathbf{d}(\beta) - \frac{1}{2}\beta\overset{o}{\mathbf{T}} \geq 0}} \boldsymbol{M}^T \mathbf{d}(\beta)\boldsymbol{M} \,, \tag{26.3.43}$$

is given by

$$2\underline{\mathcal{W}}^*(\boldsymbol{M}, m_2) = \begin{cases} \dfrac{1}{2}I^2(\boldsymbol{M})\mathcal{H}(\zeta_M) & \text{if } I(\boldsymbol{M}) \neq 0 \,, \\[2ex] \dfrac{1}{2}\{\mathcal{L}\}_m II^2(\boldsymbol{M}) & \text{if } I(\boldsymbol{M}) = 0 \,. \end{cases} \tag{26.3.44}$$

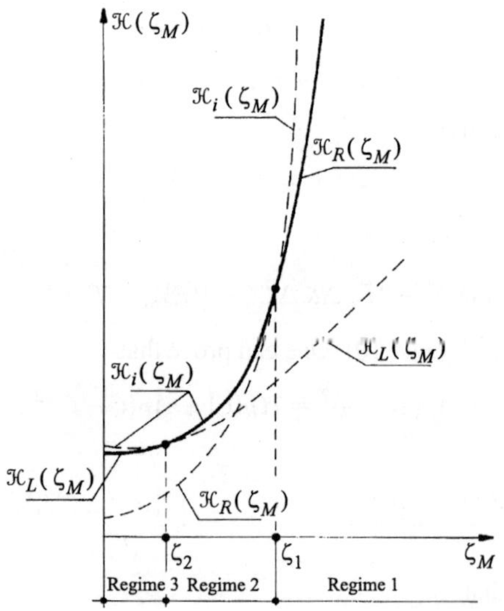

Fig. 26.3.2. Graph of the function $\mathcal{H}(\zeta_M)$

26.4. *Attainability of the translation bound*

We shall prove that there exists $\overline{\boldsymbol{d}} \in \overline{G_{m_2}^{per}}$ such that

$$\boldsymbol{M}^T \overline{\boldsymbol{d}} \boldsymbol{M} = 2\underline{\mathcal{W}}^*(\boldsymbol{M}, m_2) \,, \tag{26.4.1}$$

or $\mathcal{W}^*(\boldsymbol{M}, m_2) = \underline{\mathcal{W}}^*(\boldsymbol{M}, m_2)$, which means that the bound (26.3.8) is attained and the problem (26.2.12) can be explicitly formulated.

26.4.1. Regime (2): $\zeta_M \in [\zeta_2, \zeta_1]$

Let us compare the formula (24.2.10) for compliances of the first rank ribbed plate of Fig. 24.2.2 with the coefficients of the polynomial (26.3.37). We note that this polynomial can be written in the form:

$$\mathcal{H}_i(\zeta) = 2(\overset{o}{d}{}^h_{11} + 2\,\overset{o}{d}{}^h_{12}\zeta + \overset{o}{d}{}^h_{22}\zeta^2)\,, \tag{26.4.2}$$

where $\overset{o}{d}{}^h_{\alpha\beta}$ are given by (24.2.10) provided that m_α is substituted for ω_α. In the case of $I(M) \neq 0$, by (26.3.44) and (21.1.48) we have,

$$2\underline{W}^*(M, m_2) = \sum_{\alpha,\beta=1}^{2} d^M_{\alpha\beta} M^\alpha_M M^\beta_M\,, \tag{26.4.3}$$

where the moments $M^1_M = I(M)$, $M^2_M = II(M)$ and the compliances:

$$d^M_{\alpha\alpha} = \overset{o}{d}{}^h_{\alpha\alpha}\,, \qquad d^M_{12} = \operatorname{sgn}\left(I(M)\right) \overset{o}{d}{}^h_{12}\,, \tag{26.4.4}$$

are referred to the lines of principal moments.

In the case of $I(M) \geq 0$ the first rank ribbed plate of Fig. 24.2.2 realizes the bound $\underline{W}^*$. If $I(M) \leq 0$, the bound is realized by the first rank ribbed plate of Fig. 24.1.1, see Eq. (24.1.1). In the first case the ribs lie along the lines of the first principal moment M_I. In the second case the ribs follow the M_{II} direction. In the last case we have $M_{II} \leq M_I \leq 0$, hence in general the ribs follow the lines of the principal moment of greater absolute value. The case $I(M) = 0$ will be considered later on.

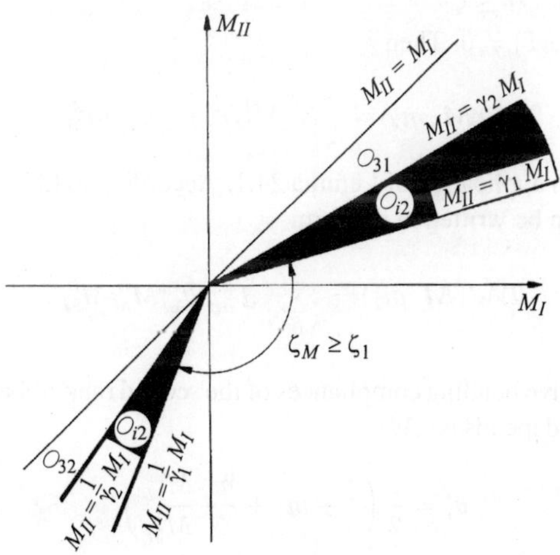

Fig. 26.4.1. The (M_I, M_{II}) plane. Regimes of the optimum solution

We conclude that the estimate (26.3.8) is attained in the class of the first rank ribbed plates whose ribs follow the directions of the greatest (in the sense of absolute value) principal moment.

Let us interpret the domain $\zeta_M \in [\zeta_2, \zeta_1]$ in the M_I, M_{II} plane. The domain of interest is defined by

$$O_i = \{(M_I, M_{II}) | M_I \geq M_{II}, \qquad \zeta_2 \leq \frac{M_I - M_{II}}{|M_I + M_{II}|} \leq \zeta_1 \},$$

since, by (26.3.18), we have

$$\zeta_M = \frac{M_I - M_{II}}{|M_I + M_{II}|}. \tag{26.4.5}$$

Let us define the parameters

$$\gamma_2 = \frac{1 - \zeta_2}{1 + \zeta_2}, \qquad \gamma_1 = \frac{1 - \zeta_1}{1 + \zeta_1}. \tag{26.4.6}$$

Note that $\gamma_1 - \gamma_2 < 0$, $\gamma_1 < 1$, $\gamma_2 \in (0, 1)$. One can show that $O_i = O_{i1} \cup O_{i2}$, where

$$O_{i1} = \{(M_I, M_{II}) | M_I \geq M_{II}, \qquad \gamma_1 M_I \leq M_{II} \leq \gamma_2 M_I \},$$
$$O_{i2} = \{(M_I, M_{II}) | M_I \geq M_{II}, \qquad \frac{1}{\gamma_1} M_I \leq M_{II} \leq \frac{1}{\gamma_2} M_I \}.$$

The domains $O_{i\alpha}$ ($\alpha = 1, 2$) form two cones joined at the origin, cf. Fig. 26.4.1. The case $\gamma_1 > 0$ or $\mathcal{K}_1 \mathcal{L}_2 < \mathcal{K}_2 \mathcal{L}_1$ is considered.

26.4.2.   Regime (3): $\zeta_M \leq \zeta_2$

Consider the case $I(M) \neq 0$. Then

$$2\mathcal{W}^*(M, m_2) = \frac{1}{2}\check{K}(M_M^1)^2 + \frac{1}{2}\mathcal{L}_2(M_M^2)^2. \tag{26.4.7}$$

Let us pass now to an application of Lemma 24.1. According to (24.2.13) and (24.2.11) the potential (26.4.7) can be written in the form

$$2\mathcal{W}^*(M, m_2) = \sum_{\alpha,\beta=1}^{2} \overset{o}{d}{}^{hh}_{\alpha\beta}(\theta_1^*) M_M^\alpha M_M^\beta, \tag{26.4.8}$$

where $\overset{o}{d}{}^{hh}_{\alpha\beta}$ are effective bending compliances of the second rank ribbed plate of Fig. 24.2.1. The area fraction θ_1^* depends on M:

$$\theta_1^* = \frac{1}{2}\left(1 + m_1 + \frac{m_2}{\zeta_2}\frac{M_M^2}{M_M^1}\right). \tag{26.4.9}$$

We observe that by (21.1.47) we have $M_M^2/M_M^1 = (\mathrm{sgn}\,\zeta_M)\zeta_M$. The expression (26.4.8) represents the complementary energy of the ribbed plate of Fig. 24.2.1 with the direction of ribs coinciding with the lines of principal moments ($M_M^3 = 0$). Note that there are two second rank ribbed microstructures realizing the potential (26.4.7): the vector $\overset{o}{e}_1$ of Fig. 24.2.1 can be tangent to the line of the first or the second principal moment. In the latter case the formula for θ_1^* is slightly different:

$$\theta_1^* = \frac{1}{2}\left(1 + m_1 - m_2 \mathrm{sgn}\,(M_M^1)\frac{\zeta_M}{\zeta_2}\right), \tag{26.4.10}$$

and the modulus $\overset{o}{d}{}_{12}^{hh}$ changes the sign, cf. the transformation formula (21.1.25) for $\alpha = \dfrac{\pi}{2}$.

Let us interpret the domain $\zeta_M \leq \zeta_2$ in the plane M_1, M_{II}. The domain

$$O_3 = \{(M_I, M_{II})|M_I \geq M_{II}\,,\quad \frac{M_I - M_{II}}{|M_I + M_{II}|} \leq \zeta_2\}$$

may be written as $O_3 = O_{31} \cup O_{32}$ with

$$O_{31} = \{(M_I, M_{II})|M_I + M_{II} > 0\,,\qquad M_I \geq M_{II}\,,\ M_{II} \geq \gamma_2 M_I\}\,,$$

$$O_{32} = \{(M_I, M_{II})|M_I + M_{II} < 0\,,\qquad M_I \geq M_{II}\,,\ M_{II} \geq \frac{1}{\gamma_2}M_I\}\,,$$

where γ_α are defined by (26.4.6), cf. Fig. 26.4.1.

Remark 26.4.1. The potential (26.4.7) can also be realized by the microstructure consisting of inclusions of special shape, discovered by Vigdergauz (1986, 1994). Thus the domain O_3 may be called the Vigdergauz domain. The inclusion of Vigdergauz have such shapes that the moments $M_{\tau\tau}$ around their boundaries are constant. This condition can be fulfilled if $\zeta_M \leq \zeta_2$. The first shapes of holes of equi-strength ($M_{\tau\tau} = $ const) boundaries have been found by Cherepanov (1974). Both Cherepanov's and Vigdergauz contours are given by elliptic functions.   □

26.4.3.   Regime (1): $\zeta_M \geq \zeta_1$

In the case where $\zeta_M \geq \zeta_1$ and $M_M^1 \neq 0$ the potential $\underline{\mathcal{W}}^*$ is given by the formula, cf. (26.3.44), (26.3.35)

$$\mathcal{W}^*(\boldsymbol{M}, m_2) = \frac{1}{4}\{\mathcal{K}\}_m(M_M^1)^2 + \frac{1}{4}\{\mathcal{L}\}_m(M_M^2)^2\,. \tag{26.4.11}$$

We shall prove below that the energy (26.4.11) can be stored in a first-rank ribbed plate with the ribs appropriately inclined, see Fig. 26.4.2.

Let φ_M and α represent the angles between the versors e_1^M and e_1 and between $\overset{o}{e}_1$ and e_1. The versor e_1^M determines the direction of the first principal moment M_I; the versor $\overset{o}{e}_1$ is the lamination direction cf. Fig. 26.4.2. Moreover, let $\phi = \varphi_M - \alpha$.

According to the transformation rules (21.1.40) applied to $\boldsymbol{M}$ we have, see Fig. 21.1.1

$$\overset{o}{M}{}^1 = M_M^1\,,\qquad \overset{o}{M}{}^2 = \cos 2\phi\,M_M^2\,,\qquad \overset{o}{M}{}^3 = \sin 2\phi\,M_M^2\,, \tag{26.4.12}$$

where $\boldsymbol{M} = \overset{o}{M}{}^i\,\overset{o}{\boldsymbol{a}}_i$; the basis $\overset{o}{\boldsymbol{a}}_i$ is defined by $\overset{o}{\boldsymbol{e}}_\alpha$. We claim that for each $\boldsymbol{M}$ such that $\zeta_M \geq \zeta_1$ one can find ϕ such that

$$\mathcal{W}^*(\boldsymbol{M}, m_2) = \frac{1}{2}[\overset{o}{d}{}_{\alpha\beta}^h\,\overset{o}{M}{}^\alpha\,\overset{o}{M}{}^\beta + \overset{o}{d}{}_{33}^h(\overset{o}{M}{}^3)^2]\,, \tag{26.4.13}$$

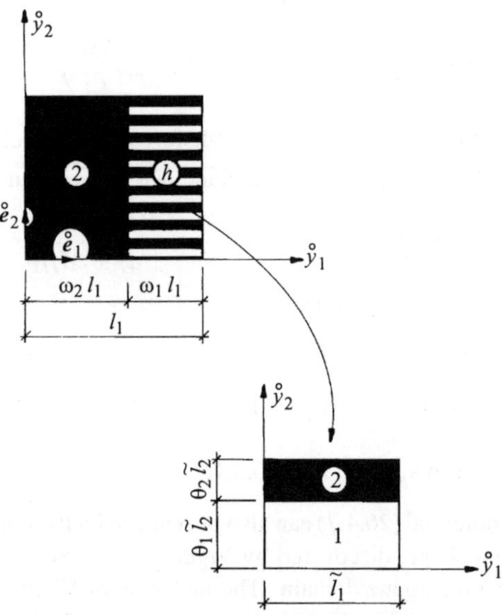

Fig. 26.4.2. Second rank ribbed plate

where $\overset{o}{d}{}^{\,h}_{\alpha\beta}$, $\overset{o}{d}{}^{\,h}_{33}$ represent the effective compliances of the first rank ribbed plate of Fig.
26.4.2, cf. Fig. 24.1.1 and Eq. (24.1.11); here $\theta_\alpha = m_\alpha$. Indeed, by substituting (26.4.12)
into (26.4.13) and equating with (26.4.11) one finds the equation:

$$[\langle \mathcal{K}\rangle_m - T_m(\Delta\mathcal{K})^2](M_M^1)^2 + 2T_m\Delta\mathcal{K}\Delta\mathcal{L}\cos 2\phi M_M^1 M_M^2$$
$$+[\langle \mathcal{L}\rangle_m - T_m(\Delta\mathcal{L})^2](\cos 2\phi M_M^2)^2 + \{\mathcal{L}\}_m(\sin 2\phi M_M^2)^2$$
$$= \{\mathcal{K}\}_m(M_M^1)^2 + \{\mathcal{L}\}_m(M_M^2)^2 \,, \tag{26.4.14}$$

where $T_M = m_1 m_2([\mathcal{K}+\mathcal{L}]_m)^{-1}$.

The condition $\zeta_M \geq \zeta_1$ implies $M_M^2 > 0$. Equation (26.4.14) is equivalent to

$$[\{\mathcal{K}\}_m - \langle \mathcal{K}\rangle_m](M_M^1)^2 + [\{\mathcal{L}\}_m - \langle \mathcal{L}\rangle_m](\cos 2\phi M_M^2)^2$$
$$+T_m[\Delta\mathcal{K}M_M^1 - \Delta\mathcal{L}\cos 2\phi M_M^2]^2 = 0 \,. \tag{26.4.15}$$

Applying the formula (26.3.36) for $f = \mathcal{K}$ and $f = \mathcal{L}$ we find

$$\frac{m_1 m_2}{[\mathcal{K}]_m}(\Delta\mathcal{K}M_M^1)^2 + \frac{m_1 m_2}{[\mathcal{L}]_m}(\cos 2\phi\Delta\mathcal{L}M_M^2)^2$$
$$= \frac{m_1 m_2}{[\mathcal{K}]_m + [\mathcal{L}]_m}(\Delta\mathcal{K}M_M^1 - \cos 2\phi\Delta\mathcal{L}M_M^2)^2 \,. \tag{26.4.16}$$

This equation is equivalent to

$$([\mathcal{L}]_m\Delta\mathcal{K}M_M^1 + [\mathcal{K}]_m\Delta\mathcal{L}\cos 2\phi M_M^2)^2 = 0 \,. \tag{26.4.17}$$

Hence we find

$$\cos 2\phi = -\frac{\zeta_1}{(M_M^2/M_M^1)} , \qquad (26.4.18)$$

or

$$\cos 2\phi = -\operatorname{sgn}(M_M^1)\frac{\zeta_1}{\zeta_M} , \qquad (26.4.19)$$

because $\zeta_M = M_M^2/|M_M^1|$ and $M_M^2 > 0$. Note that the condition: $\zeta_M \geq \zeta_1$ implies that the absolute value of the r.h.s. of (26.4.19) is smaller than 1, hence ϕ exists. The angle α assumes two values:

$$\alpha = \varphi_M \pm \frac{1}{2}\arccos\left(-\operatorname{sgn}(M_M^1)\frac{\zeta_1}{\zeta_M}\right) . \qquad (26.4.20)$$

Indeed, for both angles α the complementary energy density $\mathcal{W}^*$ given by (26.4.13) assumes the same value. Hence we conclude that there are two rank-one microstructures that realize the bound (26.4.11) for the regime $\zeta_M \geq \zeta_1$.

Consider now the case of $M_M^1 = M^1 = 0$. According to (26.3.44) we have

$$\mathcal{W}^*(\boldsymbol{M}, m_2) = \frac{1}{4}\{\mathcal{L}\}_m(M_M^2)^2 .$$

Then Eq. (26.4.14) is satisfied for $\phi = \pm\frac{\pi}{4}$. The half-line $M_{II} = -M_I$ lies within the region $\zeta_M \geq \zeta_1$. If referred to the (M_I, M_{II}) plane, this region forms a cone, see Fig. 26.4.1.

26.5.   Physical interpretation of the relaxed problem

For a given λ the problem (26.2.12) can be put in the form

$$\min_{m_2\in L^\infty(\Omega;[0,1])}[f(m_2) + \lambda\int_\Omega m_2(x)dx] , \qquad (26.5.1)$$

where

$$f(m_2) = 2\min_{\boldsymbol{M}\in\mathcal{S}_2(\Omega)}\int_\Omega \mathcal{W}^*(\boldsymbol{M}(x), m_2(x))dx , \qquad (26.5.2)$$

with $\mathcal{W}^*(\boldsymbol{M}, m_2(x))$ given by (26.3.44). The function $\mathcal{W}^*$ depends on the invariants $I(\boldsymbol{M})$ and $II(\boldsymbol{M})$ defined by (21.1.14) and (21.1.16), respectively.

The problem (26.5.2) can be viewed as an equilibrium problem for a hypothetic hyperelastic plate for which the inverted constitutive relationship assumes the form

$$\overline{\kappa}_{\alpha\beta} = \frac{\partial\mathcal{W}^*(\overline{\boldsymbol{M}}, m_2(x))}{\partial M^{\alpha\beta}} , \qquad (26.5.3)$$

where $\overline{\kappa}$ and $\overline{\boldsymbol{M}}$ are the solution of the problem (26.5.2). For simplicity, from now on the overbar will be omitted.

The equation (26.5.3) can be written explicitly:
(i) when $I(\boldsymbol{M}) = 0$

$$\kappa_{\alpha\beta} = \frac{1}{2}\{\mathcal{L}\}_m \cdot II(\boldsymbol{M})\frac{\partial II(\boldsymbol{M})}{\partial M^{\alpha\beta}} \ . \tag{26.5.4}$$

(ii) when $I(\boldsymbol{M}) \neq 0$

$$\kappa_{\alpha\beta} = \frac{\partial \mathcal{W}^*}{\partial I(\boldsymbol{M})}\frac{\partial I(\boldsymbol{M})}{\partial M^{\alpha\beta}} + \frac{\partial \mathcal{W}^*}{\partial \zeta_M}\frac{\partial \zeta_M}{\partial M^{\alpha\beta}} \ , \tag{26.5.5}$$

where

$$\frac{\partial \mathcal{W}^*}{\partial I(\boldsymbol{M})} = \frac{1}{2}I(\boldsymbol{M})\mathcal{H}(\zeta_M) \ , \qquad \frac{\partial \mathcal{W}^*}{\partial \zeta_M} = \frac{1}{4}I^2(\boldsymbol{M})\frac{d\mathcal{H}(\zeta)}{d\zeta}\bigg|_{\zeta=\zeta_M} \ . \tag{26.5.6}$$

The function $d\mathcal{H}/d\zeta$ is piecewise linear:

$$\frac{d\mathcal{H}}{d\zeta} = \begin{cases} 2c_L\zeta & \text{if } \zeta \in [0,\zeta_2] \ , \\ 2[(c_L + A_L)\zeta - A_L\zeta_2] = 2[(c_R + A_R)\zeta - A_R\zeta_1] & \text{if } \zeta \in [\zeta_2,\zeta_1] \ , \\ 2c_R\zeta & \text{if } \zeta \geq \zeta_1 \ . \end{cases} \tag{26.5.7}$$

For $\zeta \in [\zeta_2,\zeta_1]$ the line $(c_L + A_L)\zeta - A_L\zeta_2$ cuts the axis ζ at $\zeta_3 = \dfrac{A_L}{c_L + A_L}\zeta_2$. Note that $\zeta_3 \in (0,\zeta_2)$, see Fig. 26.5.1.

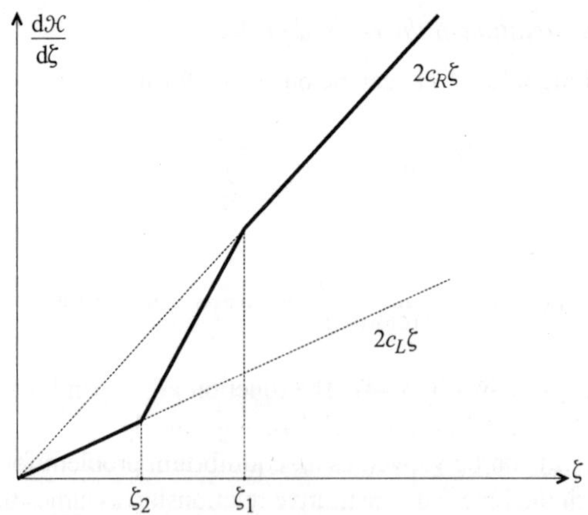

Fig. 26.5.1.

Thus the first term of (26.5.5) is smooth, while the second one is only continuous. Figure 26.5.1 suggests calling the sliding regime (ζ_2, ζ_1) – the regime of hardening.

We observe that problem (26.5.2) is solvable, see the next section.

26.6. *Primal formulation of the relaxed problem*

The compliance of the plate is equal to a minimum value of the complementary energy, cf. (26.1.16). This fact enables one to rearrange the minimum compliance problem to the minimization problem involving moments as dual variables, cf. (26.1.17) and (26.5.1). In this section we formulate the minimum compliance problem in terms of displacements.

Let us start with the problem (26.2.11). For fixed m_2 and $\boldsymbol{d}^h = \boldsymbol{d}$ the minimum over $\boldsymbol{M} \in \mathcal{S}_2(\Omega)$ is a standard minimization problem. Indeed, by applying Rockafellar's theory of duality with, see Sec. 1.2.5,

$$\Lambda \boldsymbol{M} = (\operatorname{div} \operatorname{div} \boldsymbol{M}, \ b_0(\boldsymbol{M}), \ b_1(\boldsymbol{M})) \,,$$

$$G(\Lambda \boldsymbol{M}) = I_{\mathcal{S}_2(\Omega)}(\Lambda \boldsymbol{M}) \,,$$

$$F(\boldsymbol{M}) = \int_{\Omega} \boldsymbol{M} : (\boldsymbol{d}(x)\boldsymbol{M})dx$$

we readily conclude

$$\min_{\boldsymbol{M} \in \mathcal{S}_2(\Omega)} \int_{\Omega} [\boldsymbol{M} : (\boldsymbol{d}(x)\boldsymbol{M}) + \lambda m_2(x)]dx$$

$$= \max_{v \in V(\Omega)} \int_{\Omega} [2qv - 2W(\boldsymbol{\kappa}(v)) + \lambda m_2(x)]dx \,, \tag{26.6.1}$$

where

$$W(\boldsymbol{\rho}) = \frac{1}{2}\boldsymbol{\rho} : (\boldsymbol{D}\boldsymbol{\rho}) \,, \qquad \rho \in \mathbb{E}_s^2 \,, \tag{26.6.2}$$

and $\boldsymbol{D} = \boldsymbol{d}^{-1}$. We recall that b_0 and b_1 are boundary operators.

Consequently we have

$$R_\lambda = \min_{m_2 \in L^\infty(\Omega, [0,1])} \ \min_{\boldsymbol{d} \in G_{m_2(x)}} \ \min_{\boldsymbol{M} \subset \mathcal{S}_2(\Omega)} \int_{\Omega} [\boldsymbol{M} : (\boldsymbol{d}(x)\boldsymbol{M}) + \lambda m_2(x)]dx$$

$$= \min_{m_2 \in L^\infty(\Omega, [0,1])} \ \min_{\boldsymbol{d} \in \overline{G_{m_2(x)}^{per}}} \ \min_{\boldsymbol{M} \in \mathcal{S}_2(\Omega)} \int_{\Omega} [\boldsymbol{M} : (\boldsymbol{d}(x)\boldsymbol{M}) + \lambda m_2(x)]dx$$

$$= \min_{m_2 \in L^\infty(\Omega, [0,1])} \ \min_{\boldsymbol{D} \in \overline{G_{m_2(x)}^{per}}} \ \max_{v \in V(\Omega)} \int_{\Omega} [2qv - \boldsymbol{\kappa}(v) : (\boldsymbol{D}(x)\boldsymbol{\kappa}(v)) + \lambda m_2(x)]dx \,. \tag{26.6.3}$$

Since $\overline{G_{m_2(x)}^{per}}$ is contained in $L^\infty(\Omega, \mathbb{E}_{2,4}^s)$, therefore, for a fixed $m_2 \in L^\infty(\Omega, [0,1])$ the minimax problem in (26.6.2) is not a standard one. By $\mathbb{E}_{2,4}^s$ we denote the space of matrices $(D^{\alpha\beta\lambda\mu})$ with usual symmetries: $D^{\alpha\beta\lambda\mu} = D^{\lambda\mu\alpha\beta} = D^{\beta\alpha\lambda\mu}$. To prove that the order of minimization over $\boldsymbol{D}$ and maximization over v may be interchanged one can use the Young measures. Such an approach has been followed by Lipton (1994b). However, the same aim can be achieved in a standard manner without recourse to the Young measures.

Indeed, by using Theorem 1.2.33 we write

$$
R_\lambda = \min_{m_2 \in L^\infty(\Omega,[0,1])} \ \min_{\mathbf{M} \in \mathcal{S}_2(\Omega)} \int_\Omega [\min_{\mathbf{d} \in \overline{G^{per}_{m_2(x)}}} \mathbf{M} : (\mathbf{d}(x)\mathbf{M}) + \lambda m_2(x)]dx
$$

$$
= \min_{m_2 \in L^\infty(\Omega,[0,1])} \ \min_{\mathbf{M} \in \mathcal{S}_2(\Omega)} \int_\Omega [2\mathcal{W}^*(\mathbf{M}, m_2(x)) + \lambda m_2(x)]dx \ . \tag{26.6.4}
$$

Consider now the following minimization problem

$$
(P_{m_2}) \qquad \min_{\mathbf{M} \in \mathcal{S}_2(\Omega)} \int_\Omega \mathcal{W}^*(\mathbf{M}, m_2(x))dx \ .
$$

To apply Rockafellar's theory of duality to the problem (P_{m_2}) we take Λ and $G(\Lambda \mathbf{M})$ as before; moreover we set

$$
F(\mathbf{M}) = \int_\Omega \mathcal{W}^*(\mathbf{M}, m_2(x))dx \ .
$$

For $v \in V(\Omega)$ we then have $\Lambda^*(v, -v, \frac{\partial v}{\partial \mathbf{n}}) = -\boldsymbol{\kappa}(v)$, where $\mathbf{n}$ stands for the outward unit normal vector to $\partial\Omega$.

The dual problem means evaluating:

$$
(P^*_{m_2}) \qquad \max\{-G^*(v) - F^*(\boldsymbol{\kappa}(v))|\ v \in V(\Omega)\} \ .
$$

Standard calculation yields:

$$
G^*(v) = -\int_\Omega qvdx \ ,
$$

$$
F^*(\boldsymbol{\kappa}(v)) = \sup_{\mathbf{M} \in \mathcal{S}_2(\Omega)} \{\int_\Omega \mathbf{M} : \boldsymbol{\kappa}(v)dx - \frac{1}{2}\int_\Omega \min_{\mathbf{d} \in \overline{G^{per}_{m_2(x)}}} \mathbf{M} : (\mathbf{d}(x)\mathbf{M})dx\}
$$

or

$$
F^*(\boldsymbol{\kappa}(v)) = \int_\Omega \mathcal{W}(\boldsymbol{\kappa}(v), m_2(x))dx \ ,
$$

with

$$
\mathcal{W}(\boldsymbol{\rho}, m_2(x)) = \frac{1}{2} \max_{\mathbf{D} \in \overline{G^{per}_{m_2(x)}}} \boldsymbol{\rho} : (\mathbf{D}\boldsymbol{\rho}) \ .
$$

According to Proposition 1.2.50 we have

$$
\min P_{m_2} = \max P^*_{m_2} \ . \tag{26.6.5}
$$

Substituting (26.6.5) into (26.6.4), one can write the relation (26.6.3) as follows

$$R_\lambda = \min_{m_2 \in L^\infty(\Omega,[0,1])} \ \min_{D \in \overline{G^{per}_{m_2(x)}}} \ \max_{v \in V(\Omega)} \int_\Omega [2qv - \kappa(v) : (D(x)\kappa(v)) + \lambda m_2(x)]dx$$

$$= \min_{m_2 \in L^\infty(\Omega,[0,1])} \ \min_{M \in \mathcal{S}_2(\Omega)} \int_\Omega [2\mathcal{W}^*(M, m_2(x)) + \lambda m_2(x)]dx$$

$$= \min_{m_2 \in L^\infty(\Omega,[0,1])} \ \max_{v \in V(\Omega)} \int_\Omega [2qv - 2\mathcal{W}(\kappa(v), m_2(x)) + \lambda m_2(x)]dx$$

$$= \min_{m_2 \in L^\infty(\Omega,[0,1])} \ \max_{v \in V(\Omega)} \int_\Omega [2qv - \max_{D \in G^{per}_{m_2(x)}} \kappa(v) : (D(x)\kappa(v)) + \lambda m_2(x)]dx$$

$$= \min_{m_2 \in L^\infty(\Omega,[0,1])} \ \max_{v \in V(\Omega)} \ \min_{D \in \overline{G^{per}_{m_2(x)}}} \int_\Omega [2qv - \kappa(v) : (D(x)\kappa(v)) + \lambda m_2(x)]dx.$$

$$(26.6.6)$$

The set $\overline{G^{per}_{m_2(x)}}$ is sequentially compact for the topology induced by G-convergence, cf. Sec. 1.3.2. The function $\mathcal{W}(\cdot, m_2(x))$ is convex and $\mathcal{W}(\rho, m_2(x))$ has quadratic growth in ρ independently of $m_2(x) \in [0,1]$. We conclude that R_λ is finite and the relaxed problem (26.6.6), or rather a chain of equivalent relaxed problems possesses a solution:

$$(\overline{v}, \overline{M}, \overline{m}_2, \overline{D}) \in V(\Omega) \times \mathcal{S}_2(\Omega) \times L^\infty(\Omega, [0,1]) \times \overline{G^{per}_{m_2(x)}}.$$

According to (26.4.1) we have $\overline{D} = (\overline{d})^{-1}$. The extremality relations (1.2.51), (1.2.52) yield:

$$\overline{M}(x) = \frac{\partial \mathcal{W}(\kappa(\overline{v}(x)), \overline{m}_2)}{\partial \kappa} \quad \text{or} \quad \kappa(\overline{v}) = \frac{\partial \mathcal{W}^*(\overline{M}(x), \overline{m}_2)(x)}{\partial M}. \qquad (26.6.7)$$

We recall that extremality relations hold also for nonconvex problems, provided that their solutions exist.

Note that by Rockafellar's Theorem 1.3.33 R_λ is also equal to:

$$R_\lambda = \min_{M \in \mathcal{S}_2(\Omega)} \int_\Omega \Gamma_\lambda(M)dx, \qquad (26.6.7b)$$

where

$$F_\lambda(M) = \min_{0 \le \rho_2 \le 1} \ \min_{d \in G_{\rho_2}} [M : (dM) + \lambda \rho_2].$$

In the sequel the bar over the quantities involved in (26.6.7) will be omitted. We observe that the principal directions of the tensors κ and M coincide. Indeed, let us refer the constitutive relationship (26.5.3) to the basis $\mathbf{a}_i^M$:

$$\kappa_M^i = \frac{\partial \mathcal{W}^*}{\partial M_M^i}. \qquad (26.6.8)$$

According to (26.3.44), the potential $\mathcal{W}^*$ depends on M_M^1 and M_M^2 but is independent of M_M^3. Hence $\kappa_M^3 = 0$. This yields

$$\boldsymbol{\kappa} = \kappa_M^1 \mathbf{a}_1 + \kappa_M^2 \mathbf{a}_2^M \ . \tag{26.6.9}$$

Thus $\Pi\boldsymbol{\kappa} = II(\boldsymbol{\kappa})\mathbf{a}_2^M$, where Π is the operator of projection onto the plane $\mathbf{a}_2^M$, $\mathbf{a}_3^M$. By definition $\Pi M = II(M)\mathbf{a}_2^M$, hence $\Pi\boldsymbol{\kappa}$ and ΠM are collinear. Consequently $\mathbf{a}_i^M = \mathbf{a}_i^\kappa$ and

$$\begin{aligned}
\kappa_M^1 &= \kappa_\kappa^1 \ , & \kappa_M^2 &= \kappa_\kappa^2 \ , & \kappa_M^3 &= \kappa_\kappa^3 = 0 \ , \\
M_M^1 &= M_\kappa^1 \ , & M_M^2 &= M_\kappa^2 \ , & M_M^3 &= M_\kappa^3 = 0 \ .
\end{aligned} \tag{26.6.10}$$

Let us proceed now to find the potential $\mathcal{W}$. The simplest way of finding the explicit form of $\mathcal{W}$ is inverting the relations (26.5.3) and then reconstructing $\mathcal{W}$ such that $(26.6.7)_1$ holds.

Regime (3): $\zeta_M \leq \zeta_2$

Substitution of (26.4.7) into (26.6.8) implies

$$\kappa_M^1 = \frac{1}{2}\check{K}M_M^1 \ , \qquad \kappa_M^2 = \frac{1}{2}\mathcal{L}_2 M_M^2 \ , \qquad \kappa_M^3 = 0 \ . \tag{26.6.11}$$

Taking into account (26.6.10) we have

$$M_\kappa^1 = 2\check{k}\kappa_\kappa^1 \ , \qquad M_\kappa^2 = 2\mu_2\kappa_\kappa^2 \ , \qquad M_\kappa^3 = 0 \ , \tag{26.6.12}$$

where $\mu_2 = (\mathcal{L}_2)^{-1}$, $\check{k} = (\check{K})^{-1}$ and

$$\check{k} = \frac{k_1 k_2 + \mu_2\langle k\rangle_m}{\mu_2 + [k]_m} \tag{26.6.13}$$

or $\check{k} = \overline{k}_{HS}$, where $\overline{k}_{HS}$ represents the upper Hashin-Shtrikman bound for k_0, see (22.2.1). The condition $\zeta_M \leq \zeta_2$ assumes the form

$$\zeta_M \leq \frac{m_2\mu_2\Delta k}{k_1 k_2 + \langle k\rangle_m} \ . \tag{26.6.14}$$

By (26.6.12) we have $\zeta_M = (\mu_2/\check{k})\zeta_\kappa$, with $\zeta_\kappa = \kappa_\kappa^2/|\kappa_\kappa^1|$. Thus the condition (26.6.14) yields

$$\zeta_\kappa \leq \check{\zeta}_2 \ , \qquad \check{\zeta}_2 = \frac{m_2\Delta k}{\mu_2 + [k]_m} \ , \tag{26.6.15}$$

where, as usual, $\Delta k = k_2 - k_1$. The elastic potential $\mathcal{W}$ assumes the form

$$\mathcal{W} = \check{k}(\kappa_\kappa^1)^2 + \mu_2(\kappa_\kappa^2)^2 \ . \tag{26.6.16}$$

Regime (1): $\zeta_M \geq \zeta_1$

Substitution of (26.4.11) into (26.6.8) yields

$$\kappa_M^1 = \frac{1}{2}\{\mathcal{K}\}_m M_M^1 , \qquad \kappa_M^2 = \frac{1}{2}\{\mathcal{L}\}_m M_M^2 , \qquad \kappa_M^3 = 0 . \qquad (26.6.17)$$

By inverting these relations and taking into account (26.6.10) we find

$$M_\kappa^1 = 2\langle k \rangle_m \kappa_\kappa^1 , \qquad M_\kappa^2 = 2\langle \mu \rangle_m \kappa_\kappa^2 , \qquad M_\kappa^3 = 0 . \qquad (26.6.18)$$

The inequality $\zeta_M \geq \zeta_1$ transforms into

$$\zeta_\kappa \geq \check{\zeta}_1 , \qquad \check{\zeta}_1 = \frac{\Delta k}{\Delta \mu} , \qquad (26.6.19)$$

where $\Delta \mu = \mu_2 - \mu_1$. Let us compute

$$\check{\zeta}_1 - \check{\zeta}_2 = \frac{\Delta k}{\Delta \mu} \frac{[k + \mu]_m}{\mu_2 + [k]_m} . \qquad (26.6.20)$$

Thus $\check{\zeta}_1 > \check{\zeta}_2$ and we conclude that the domains $(0, \check{\zeta}_2)$ and $(\check{\zeta}_1, \infty)$ are disjoint. The relations (26.6.18) determine the potential

$$W = \langle k \rangle_m (\kappa_\kappa^1)^2 + \langle \mu \rangle_m (\kappa_\kappa^2)^2 . \qquad (26.6.21)$$

Regime (2): $\zeta_M \in [\zeta_2, \zeta_1]$

According to (26.4.3) we find

$$\kappa_M^1 = \overset{o}{d}{}^{\,h}_{11} M_M^1 + \mathrm{sgn}\,(M_M^1)\, \overset{o}{d}{}^{\,h}_{12} M_M^2 ,$$
$$\kappa_M^2 = \mathrm{sgn}\,(M_M^1)\, \overset{o}{d}{}^{\,h}_{12} M_M^1 + \overset{o}{d}{}^{\,h}_{22} M_M^2 , \qquad \kappa_M^3 = 0 , \qquad (26.6.22)$$

where $\overset{o}{d}{}^{\,h}_{\alpha\beta}$ are given by (24.2.10) with $\omega_\alpha = m_\alpha$. By inverting the equations (26.6.22) we obtain

$$\frac{1}{2}M_M^1 = A\kappa_M^1 + \mathrm{sgn}\,(M_M^1)B\kappa_M^2 , \qquad \frac{1}{2}M_M^2 = \mathrm{sgn}\,(M_M^1)B\kappa_M^1 + C\kappa_M^2 , \qquad (26.6.23)$$

where

$$A = \langle k \rangle_m - t_m(\Delta k)^2 , \qquad B = t_m \Delta k \Delta \mu , \qquad C = \langle \mu \rangle_m - t_m(\Delta \mu)^2 . \qquad (26.6.24)$$

Here $t_m = m_1 m_2 ([k + \mu]_m)^{-1}$. Note that $A > 0, B > 0, C > 0$. Moreover, by algebraic manipulations one can prove the inequalities

$$A\zeta_\alpha - B > 0 , \qquad C - B\zeta_\alpha > 0 , \qquad \alpha = 1, 2 . \qquad (26.6.25)$$

(i) Case $M_M^1 > 0$

The condition $\zeta_M = M_M^2 / M_M^1 \leq \zeta_1$ implies

$$\kappa_M^1(A\zeta_1 - B) \geq (C - B\zeta_1)\kappa_M^2 . \qquad (26.6.26)$$

Since $\kappa_M^2 > 0$ by definition, we conclude that $\kappa_M^1 > 0$. The condition $\zeta_M \geq \zeta_2$ implies

$$\kappa_M^2(C - B\zeta_2) \geq (A\zeta_2 - B)\kappa_M^1 . \tag{26.6.27}$$

By (26.6.10) the conditions (26.6.26), (26.6.27) can be rearranged to the form

$$\check{\zeta}_2 \leq \zeta_\kappa \leq \check{\zeta}_1 , \tag{26.6.28}$$

where now $\zeta_\kappa = \kappa_\kappa^2/|\kappa_\kappa^1| = \kappa_\kappa^2/\kappa_\kappa^1$ and $\check{\zeta}_\alpha$ are defined by (26.6.15) and (26.6.19). The elastic potential assumes the form

$$\mathcal{W} = A(\kappa_\kappa^1)^2 + 2B\kappa_\kappa^1\kappa_\kappa^2 + C(\kappa_\kappa^2)^2 . \tag{26.6.29}$$

(ii) Case $M_M^1 < 0$

The condition: $\zeta_M = -M_M^2/M_M^1 \leq \zeta_1$ implies

$$(A\zeta_1 - B)(-\kappa_M^1) \geq (C - B\zeta_1)\kappa_M^2 . \tag{26.6.30}$$

Hence $-\kappa_M^1 > 0$. By (26.6.10) we have $\zeta_\kappa = -\kappa_\kappa^2/\kappa_\kappa^1$ and (26.2.30) implies

$$\zeta_\kappa \leq \frac{A\zeta_1 - B}{C - B\zeta_1} . \tag{26.6.31}$$

On the other hand, the condition $\zeta_M \geq \zeta_2$ gives

$$\zeta_\kappa \geq \frac{A\zeta_2 - B}{C - B\zeta_2} . \tag{26.6.32}$$

The inequalities (26.6.31) and (26.6.32) can be rearranged to the form (26.6.28). Thus (26.6.28) holds for arbitrary M_M^1. The potential assumes the form

$$\mathcal{W} = A(\kappa_\kappa^1)^2 - 2B\kappa_\kappa^1\kappa_\kappa^2 + C(\kappa_\kappa^2)^2 . \tag{26.6.33}$$

In general, we have

$$\mathcal{W} = A(\kappa_\kappa^1)^2 + 2B|\kappa_\kappa^1|\kappa_\kappa^2 + C(\kappa_\kappa^2)^2 . \tag{26.6.34}$$

Obviously the potential $\mathcal{W}$ is a non-differentiable function of κ_κ^1.

Final form of the potential $\mathcal{W}$:

$$\mathcal{W}(\boldsymbol{\kappa}, m_2) = \begin{cases} (I(\boldsymbol{\kappa}))^2 \mathcal{F}(\zeta_\kappa) & \text{if} \quad I(\boldsymbol{\kappa}) \neq 0 , \\ \langle\mu\rangle_m (II(\boldsymbol{\kappa}))^2 & \text{if} \quad I(\boldsymbol{\kappa}) = 0 . \end{cases} \tag{26.6.35}$$

The function $\mathcal{F}(\cdot)$ is defined as follows

$$\mathcal{F}(\zeta) = \begin{cases} \mathcal{F}_L(\zeta) & \text{if} \quad \zeta \in [0, \check{\zeta}_2] , \\ \mathcal{F}_i(\zeta) & \text{if} \quad \zeta \in [\check{\zeta}_2, \check{\zeta}_1] , \\ \mathcal{F}_R(\zeta) & \text{if} \quad \zeta \geq \check{\zeta}_1 , \end{cases} \tag{26.6.36}$$

with

$$\mathcal{F}_L(\zeta) = \breve{a}_L + \breve{c}_L \zeta^2 , \qquad \mathcal{F}_R(\zeta) = \breve{a}_R + \breve{c}_R \zeta^2 ,$$
$$\mathcal{F}_i(\zeta) = A + 2B\zeta + C\zeta^2 . \tag{26.6.37}$$

The coefficients involved in (26.6.37) are defined by

$$\breve{a}_L = \breve{k} , \qquad \breve{c}_L = \mu_2 , \qquad \breve{a}_R = \langle k \rangle_m , \qquad \breve{c}_R = \langle \mu \rangle_m , \tag{26.6.38}$$

and A, B, C are given by (26.6.24). One can prove that

$$\mathcal{F}_i(\zeta) = \mathcal{F}_L(\zeta) - \breve{A}_L(\zeta - \breve{\zeta}_2)^2 = \mathcal{F}_R(\zeta) - \breve{A}_R(\zeta - \breve{\zeta}_1)^2 , \tag{26.6.39}$$

where

$$\breve{A}_L = \frac{m_1 \Delta\mu(\mu_2 + [k]_m)}{[k + \mu]_m} , \qquad \breve{A}_R = \frac{m_1 m_2 (\Delta\mu)^2}{[k + \mu]_m} . \tag{26.6.40}$$

Consequently

$$\mathcal{F}_i(\zeta) \leq \mathcal{F}_L(\zeta) \quad \text{and} \quad \mathcal{F}_i(\zeta) \leq \mathcal{F}_R(\zeta) \tag{26.6.41}$$

for all $\zeta \geq 0$. The relations similar to (26.3.41) hold here, which proves that the function $\mathcal{F}(\zeta)$ is smooth in the whole domain see Fig. 26.6.1. Moreover, we observe that $\breve{a}_L < \breve{a}_R$ and $\breve{c}_L > \breve{c}_R$.

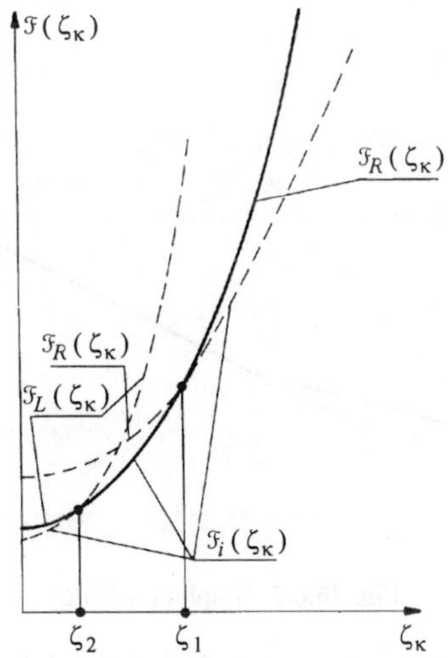

Fig. 26.6.1. Graph of $\mathcal{F}(\zeta_\kappa)$

Constitutive relationship

Substitution of (26.6.35) into (26.6.4) gives
i) case of $I(\kappa) = 0$

$$M^{\alpha\beta} = 2\langle\mu\rangle_m II(\kappa)\frac{\partial II(\kappa)}{\partial\kappa_{\alpha\beta}} \,, \tag{26.6.42}$$

ii) case of $I(\kappa) \neq 0$

$$M^{\alpha\beta} = \frac{\partial W}{\partial I(\kappa)}\frac{\partial I(\kappa)}{\partial\kappa_{\alpha\beta}} + \frac{\partial W}{\partial\zeta_\kappa}\frac{\partial\zeta_\kappa}{\partial\kappa_{\alpha\beta}} \,, \tag{26.6.43}$$

where

$$\frac{\partial W}{\partial I(\kappa)} = 2I(\kappa)\mathcal{F}(\zeta_\kappa) \,, \qquad \frac{\partial W}{\partial\zeta_\kappa} = (I(\kappa))^2\frac{d\mathcal{F}(\zeta)}{d\zeta}\bigg|_{\zeta=\zeta_\kappa} \,. \tag{26.6.44}$$

The function $d\mathcal{F}(\zeta)/d\zeta$ is piecewise linear

$$\frac{d\mathcal{F}}{d\zeta} = \begin{cases} 2\check{c}_L\zeta & \text{if } \zeta \in [0,\check{\zeta}_2] \,, \\[4pt] \begin{aligned} &2[(\check{c}_L - \breve{A}_L)\zeta + \breve{A}_L\check{\zeta}_2] \\ &= 2(\check{c}_R - \breve{A}_R)\zeta + \breve{A}_R\check{\zeta}_1] = 2[C\zeta + B] \end{aligned} & \text{if } \zeta \in [\check{\zeta}_2,\check{\zeta}_1] \,, \\[4pt] 2\check{c}_R\zeta & \text{if } \zeta \geq \check{\zeta}_1 \,. \end{cases} \tag{26.6.45}$$

For $\zeta \in [\check{\zeta}_2,\check{\zeta}_1]$, we have $d\mathcal{F}/d\zeta = 0$ at $\check{\zeta}_3 = -B/C < 0$, see Fig. 26.6.2

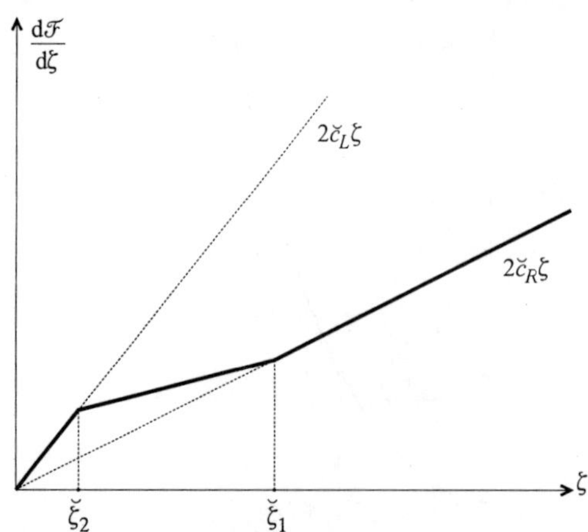

Fig. 26.6.2. Graph of $d\mathcal{F}/d\zeta$

In contrast to the dual formulation (see Fig. 26.5.1) the sliding regime can be called here the regime of softening.

26.7. On the shape design

The aim of this section is to formulate a shape design problem.

Assume that the plate occupying a domain Ω, clamped at Γ_0 is subject to bending moments M_n^0 and transverse forces Q^0 along the edge $\Gamma_1 \subset \partial\Omega$, or

$$f(v) = \int_{\Gamma_1} [M_n^0 \left(-\frac{\partial v}{\partial n}\right) + Q^0 v]ds \ . \tag{26.7.1}$$

The loading q is absent. Moreover, assume that one plate material of moduli (k_2, μ_2) is at our disposal. This material occupies the domain $\Omega_2 \subset \Omega$ such that $\Gamma_1 \subset \partial\Omega_2$. The boundary $\partial\Omega_1 = \partial\Omega\backslash\partial\Omega_2$ is free of loading. The minimum compliance problem has the form (26.1.15) with the difference that now the right-hand side of the equilibrium equation is given by (26.7.1) and the integral in (26.1.11) is taken over the domain Ω_2. Just the shape of Ω_2 is unknown.

This shape optimization problem needs a relaxation. In the context of the plane-stress problem, Allaire and Kohn (1993a) proved that the relaxed problem can be obtained by passing to zero: $k_1 \to 0$, $\mu_1 \to 0$ in the relaxed problem concerning the optimal distribution of two materials. Applying the same arguments one conclude that the same method holds for the Kirchhoff plate bending problem. Thus the relaxed form of the optimum shape design for thin plates can be found by passing to zero with k_1 and μ_1 in the formulation (26.5.1).

Let us substitute

$$k_1 = 0 \ , \qquad \mu_1 = 0 \tag{26.7.2}$$

into (26.3.44). We find that

$$W^*(\boldsymbol{M}, m_2) = \frac{\mathcal{L}_2}{4m_2}(II(\boldsymbol{M}))^2 \tag{26.7.3a}$$

if $I(\boldsymbol{M}) = 0$ and

$$W^*(\boldsymbol{M}, m_2) = \frac{1}{4}(I(\boldsymbol{M}))^2 \mathcal{H}_0(\zeta_M) \tag{26.7.3b}$$

for $I(\boldsymbol{M}) \neq 0$. Here

$$\mathcal{H}_0(\zeta) = \begin{cases} \dfrac{1}{m_2}(\mathcal{K}_2 + m_1\mathcal{L}_2) + \mathcal{L}_2\zeta^2 & \text{if} \quad \zeta \in [0,1] \ , \\[2mm] \dfrac{1}{m_2}(\mathcal{K}_2 + \mathcal{L}_2\zeta^2) & \text{if} \quad \zeta \geq 1 \ . \end{cases} \tag{26.7.4}$$

Since by (21.1.14) and (21.1.15)

$$\zeta_M = \left(1 - \frac{2\det \boldsymbol{M}}{(I(\boldsymbol{M}))^2}\right)^{1/2} , \tag{26.7.5}$$

the conditions $\zeta_M \leq 1$ and $\zeta_M \geq 1$ are satisfied provided that $\det \boldsymbol{M} \geq 0$ or $\det \boldsymbol{M} \leq 0$, respectively. The function $\mathcal{H}_0(\zeta_M)$ has a cusp at $\det \boldsymbol{M} = 0$ (or $\zeta_M = 1$), cf. Fig. 26.7.1.

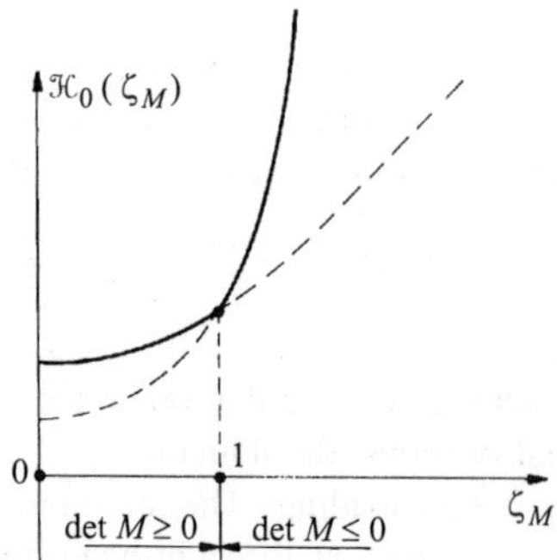

Fig. 26.7.1. Case of shape design. Graph of $\mathcal{H}_0(\zeta_M)$

The constitutive relationship (26.5.3) assumes the form

$$
\kappa_{\alpha\beta} = \begin{cases} 0 & \text{if } tr\,M = 0\,, \\ \dfrac{1}{2}I(M)\mathcal{H}_0(\zeta_M)\dfrac{\partial I}{\partial M^{\alpha\beta}} + \dfrac{1}{4}I^2(M)\dfrac{d\mathcal{H}_0(\zeta_M)}{d\zeta}\dfrac{\partial \zeta_M}{\partial M^{\alpha\beta}} & \text{if } tr\,M \neq 0\,, \end{cases} \qquad (26.7.6)
$$

where

$$
\frac{d\mathcal{H}_0}{d\zeta} = \begin{cases} \dfrac{2}{\mu_2}\zeta & \text{if } \zeta \leq 1\,, \\ \dfrac{2}{m_2\mu_2}\zeta & \text{if } \zeta \geq 1\,, \end{cases} \qquad (26.7.7)
$$

the last function having a jump at $\zeta = 1$, see Fig. 26.7.2.

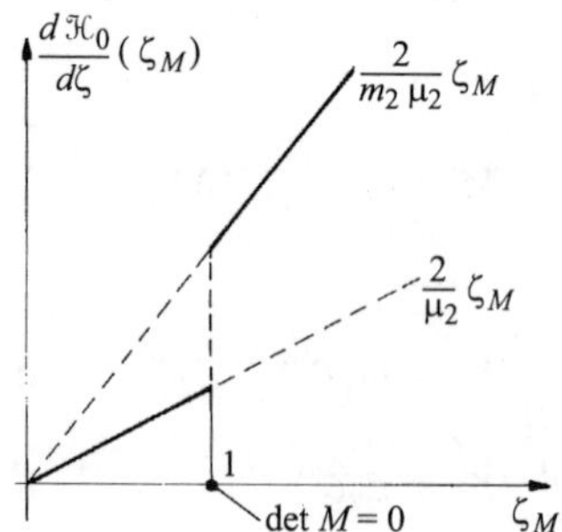

Fig. 26.7.2. Case of shape design. Graph of $d\mathcal{H}_0/d\zeta$

Thus in the case of (26.7.2) the sliding regime reduces to one point $\zeta_M = 1$, which corresponds to the domain in which $\det M = 0$. At this point one of the principal moments equals zero. In the final design the plate domain is divided into three subdomains in which $\det M$ is negative, zero, or positive.

Let us simplify notation as follows: $\mathcal{K}_2 = \mathcal{K}$, $\mathcal{L}_2 = \mathcal{L}$, $m_2 = \theta$. By (26.7.3a,b) the potential $\mathcal{W}^*$ assumes the form

$$\mathcal{W}^*(\boldsymbol{M},\theta) = \mathcal{W}_0^*(\boldsymbol{M}) + \frac{1-\theta}{4\theta}[\mathcal{K}(I(\boldsymbol{M}))^2 + \mathcal{L}u(\boldsymbol{M})]\,, \tag{26.7.8}$$

with

$$u(\boldsymbol{M}) = \begin{cases} (I(\boldsymbol{M}))^2 & \text{if} \quad \det \boldsymbol{M} \geq 0 \\ (II(\boldsymbol{M}))^2 & \text{if} \quad \det \boldsymbol{M} \leq 0\,, \end{cases} \tag{26.7.9}$$

and, see (21.1.38)

$$\mathcal{W}_0^*(\boldsymbol{M}) = \frac{1}{4}\mathcal{K}(I(\boldsymbol{M}))^2 + \frac{1}{4}\mathcal{L}(II(\boldsymbol{M}))^2 \tag{26.7.10}$$

represents the complementary energy of the plate material.

One can prove by inspection that the function $u(\boldsymbol{M})$ can be expressed by one formula:

$$u(\boldsymbol{M}) = \frac{1}{2}(|M_I| + |M_{II}|)^2\,. \tag{26.7.11}$$

Consequently the formula (26.7.8) can be rearranged as follows

$$\mathcal{W}^*(\boldsymbol{M},\theta) = \mathcal{W}_0^*(\boldsymbol{M}) + \frac{1-\theta}{2\theta}g(\boldsymbol{M})\,, \tag{26.7.12}$$

where

$$g(\boldsymbol{M}) = \frac{1}{4}\mathcal{K}(M_I + M_{II})^2 + \frac{1}{4}\mathcal{L}(|M_I| + |M_{II}|)^2\,. \tag{26.7.13}$$

The optimization problem (26.6.4) simplifies to the form (26.6.7b) with

$$F_\lambda(\boldsymbol{M}) = \min_{0 \leq \theta \leq 1}[2\mathcal{W}^*(\boldsymbol{M},\theta) + \lambda\theta]\,. \tag{26.7.14}$$

Minimization over θ can be easily done explicitly. We find

$$F_\lambda(\boldsymbol{M}) = 2\mathcal{W}_0^*(\boldsymbol{M}) + \begin{cases} 2\sqrt{\lambda g(\boldsymbol{M})} - g(\boldsymbol{M}) & \text{if} \quad g(\boldsymbol{M}) \leq \lambda \\ \lambda & \text{if} \quad g(\boldsymbol{M}) \geq \lambda\,. \end{cases} \tag{26.7.15}$$

If $\lambda \neq 0$ one can write

$$F_\lambda(\boldsymbol{M}) = 2\mathcal{W}_0^*(\boldsymbol{M}) + \lambda - \lambda\begin{cases} (1 - \sqrt{g(\boldsymbol{M})/\lambda})^2 & \text{if} \quad \sqrt{g(\boldsymbol{M})/\lambda} \leq 1\,, \\ 0 & \text{if} \quad \sqrt{g(\boldsymbol{M})/\lambda} \geq 1\,, \end{cases} \tag{26.7.16}$$

and this expression shows that the function $F_\lambda(\boldsymbol{M})$ is of C^1 class, the stitching along $g(\boldsymbol{M}) - \lambda = 0$ being smooth.

Theorem 26.7.1. The shape optimization problem given by (26.6.7b) with $F_\lambda(M)$ defined by (26.7.14) is well-posed. Its solution exists.

Proof. By using the definition of the norm of $M \in \mathbb{E}_2^s$

$$||M||^2 = (M_I)^2 + (M_{II})^2 \tag{26.7.17}$$

one can rearrange the potential (26.7.10) to the form

$$\mathcal{W}_0^*(M) = \frac{1}{8}(\mathcal{K} + \mathcal{L})||M||^2 + \frac{1}{4}(\mathcal{K} - \mathcal{L})\det M . \tag{26.7.18}$$

Let us define the increasing and convex function

$$f(t) = \begin{cases} 2t & \text{if } t \in [0, 1] \\ 1 + t^2 & \text{if } t \geq 1 . \end{cases} \tag{26.7.19}$$

and the function η and k

$$\eta(N) = f([|\,||N||^2 + 2\alpha \det N + 2\beta|\det N|]^{\frac{1}{2}}) - 2\beta|\det N| , \tag{26.7.20}$$
$$k(N) = \eta(N) - 2\alpha \det N$$

with arguments $(N) \in \mathbb{E}_2^s$ and

$$\alpha = \frac{\mathcal{K}}{\mathcal{K} + \mathcal{L}} , \qquad \beta = 1 - \alpha . \tag{26.7.21}$$

One can easily check that

$$F_\lambda(M) = \lambda G(M/\sqrt{\lambda}) , \tag{26.7.22}$$

where

$$G(N) = \frac{1}{2}(\mathcal{K} - \mathcal{L})\det N + k(\frac{1}{2}\sqrt{\mathcal{K} + \mathcal{L}}N) . \tag{26.7.23}$$

Note that $F_\lambda(M)$ has a quadratic growth. Thus to prove the assertion, it is sufficient to prove that $F_\lambda(M)$ is polyconvex, cf. Th. 2.3 in Dacorogna (1989) and Ciarlet (1988). Thus it is sufficient to prove polyconvexity of $\eta(N)$.

Let us define the functions

$$t_\sigma(N) = [||N||^2 + 2\alpha \det N + 2\sigma\beta(-\det N)]^{\frac{1}{2}} , \tag{26.7.24}$$
$$g_\sigma(N, z) = f(t_\sigma(N)) + 2\sigma\beta z ,$$

indexed by $\sigma \in \{-1, 1\}$; $z \in \mathbb{R}$. Note that $(t_\sigma(N))^2$ is a non-negative quadratic form of N. Indeed, for $\sigma = -1$ we have $(t_{-1}(N))^2 = (tr N)^2$ and for $\sigma = 1$

$$(t_1(N))^2 = (N_I)^2 + (N_{II})^2 + 2(\alpha - \beta)N_I N_{II} .$$

Since $\alpha - \beta < 1$ the form above is positive definite. Recall, moreover, that the Euclidean norm is convex. Thus the functions $t_\sigma(N)$ are convex. On the other hand, the function $f(t)$ is increasing and convex. Thus by Theorem 5.1 of Rockafellar (1970) the functions $g_\sigma(N, z)$ are convex with respect to N and z.

Moreover,

$$\eta(N) = \max_{\sigma \in \{-1,1\}} g_\sigma(N, \det\ N)\,. \tag{26.7.25}$$

This equality follows from the inequality

$$2\sqrt{r+s} - s \geq 2\sqrt{r-s} + s \tag{26.7.26}$$

which is valid for

$$r + s \leq 1\,, \qquad 0 < s < r\,, \qquad r, s \in \mathbb{R}\,. \tag{26.7.27}$$

Indeed, it is sufficient to choose

$$r = \|N\|^2 + 2\alpha\ \det\ N\,, \qquad s = 2\beta|\det\ N|$$

and then the conditions (26.7.27) hold irrespective of the sign of $\det\ N$. The maximum in (26.7.25) is attained for $\sigma = -\mathrm{sgn}(\det\ N)$. Thus (26.7.25) holds, which proves that $\eta(N)$ is polyconvex. $\qquad\qquad\qquad\qquad\qquad\qquad\qquad\qquad\qquad\qquad\qquad\qquad\square$

Interestingly enough, the assumption (26.7.2) does not lead to any degeneracy of the primal constitutive relationship (26.6.4). The potential (26.6.35) assumes the form

$$\mathcal{W}(\kappa, m_2) = \begin{cases} (I(\kappa))^2 \mathcal{F}_0(\zeta_\kappa) & \text{if}\quad I(\kappa) \neq 0\,, \\[2mm] m_2 \mu_2 II^2(\kappa) & \text{if}\quad I(\kappa) = 0\,, \end{cases} \tag{26.7.28}$$

where

$$\mathcal{F}_0(\zeta) = \begin{cases} \mu_2(\zeta_2^{(0)} + \zeta^2) & \text{if}\quad \zeta \in [0, \zeta_2^{(0)}]\,, \\[2mm] m_2 \dfrac{k_2 \mu_2}{k_2 + \mu_2}(1 + \zeta)^2 & \text{if}\quad \zeta \in [\zeta_2^{(0)}, \zeta_1^0]\,, \\[2mm] m_2 \mu_2(\zeta_1^{(0)} + \zeta^2) & \text{if}\quad \zeta \geq \zeta_1^{(0)}\,, \end{cases} \tag{26.7.29}$$

and

$$\zeta_1^{(0)} = \frac{k_2}{\mu_2}\,, \qquad \zeta_2^{(0)} = \frac{m_2 k_2}{\mu_2 + m_1 k_2}\,. \tag{26.7.30}$$

The constitutive relationship (26.6.4) assumes the form (26.6.42) – (26.6.45) with $\mathcal{F} = \mathcal{F}_0$. In particular, the function

$$\frac{d\mathcal{F}}{d\zeta} = \begin{cases} 2\mu_2\zeta & \text{if}\quad \zeta \in [0, \zeta_2^{(0)}]\,, \\[2mm] 2m_2 \dfrac{k_2 \mu_2}{k_2 + \mu_2}(1 + \zeta) & \text{if}\quad \zeta \in [\zeta_2^{(0)}, \zeta_1^0]\,, \\[2mm] 2m_2 \mu_2 \zeta & \text{if}\quad \zeta \geq \zeta_1^{(0)}\,, \end{cases}$$

forms a broken line, cf. Fig. 26.6.2 with $\check{\zeta}_3 = -1$. The sliding regime does not degenerate to a point. The length of the interval $(\zeta_2^{(0)}, \zeta_1^0)$ is positive. The function $\mathcal{W}(\cdot, m_2)$ is smooth and strictly convex.

Note that if $\zeta_M = 1$ the constitutive relationships (26.6.23) have the form

$$\begin{bmatrix} M_M^1 \\ M_M^2 \end{bmatrix} = m_2 \frac{k_2 \mu_2}{k_2 + \mu_2} \begin{bmatrix} 1 & \operatorname{sgn}\left(M_M^1\right) \\ \operatorname{sgn}\left(M_M^1\right) & 1 \end{bmatrix} \begin{bmatrix} \kappa_M^1 \\ \kappa_M^2 \end{bmatrix} . \tag{26.7.31}$$

Hence we conclude that

$$M_I = 2m_2 \frac{k_2 \mu_2}{k_2 + \mu_2} \kappa_I \,, \qquad M_{II} = 0 \,, \qquad \text{if} \quad M_I + M_{II} > 0 \,,$$

$$M_I = 0 \,, \qquad M_{II} = 2m_2 \frac{k_2 \mu_2}{k_2 + \mu_2} \kappa_{II} \,, \qquad \text{if} \quad M_I + M_{II} < 0 \,. \tag{26.7.32}$$

26.8. *Square clamped plates of minimum compliance*

Consider a square plate clamped at its edge, subject to a transverse loading $q = q(x)$, $x \in \Omega$. The plate is to be constructed of pieces of thicknesses h_1 and h_2. Thus the "plate materials" are characterized by the moduli

$$k_\alpha = \frac{E(h_\alpha)^3}{24(1-\nu)} \,, \qquad \mu_\alpha = \frac{E(h_\alpha)^3}{24(1+\nu)} \,, \tag{26.8.1}$$

where E and ν represent the Young modulus and the Poisson ratio, common for both phases. The assumption: $h_2 > h_1$ implies the ordering condition (26.1.5). The area occupied by the first phase is given by:

$$\int_\Omega m_1(x) dx = C \,. \tag{26.8.2}$$

Let us represent C as follows

$$C = (h_2 - h_1)^{-1}(-Vol + h_2 \cdot |\Omega|) \,, \tag{26.8.3}$$

where $|\Omega|$ stands for the area of the middle plane and Vol is a constant. Thus the condition (26.8.2) can be rearranged to the form

$$\int_\Omega (m_1(x)h_1 + m_2(x)h_2) dx = Vol \tag{26.8.4}$$

which means that the plate volume equals Vol and both conditions (26.8.2) and (26.8.4) are equivalent. The constant Vol should be chosen such that

$$h_1|\Omega| < Vol < h_2|\Omega| \,. \tag{26.8.5}$$

To find the plate of minimum compliance we assume that the plate is endowed with a second rank ribbed microstructure in which the stronger phase is chosen as a coating, cf. Fig. 24.2.1. Thus

$$m_1 = (1 - \theta_2)(1 - \omega_2) , \qquad m_2 = 1 - m_1 \qquad (26.8.6)$$

and the condition (26.8.4) is replaced with

$$\int_\Omega [h_1 + (\omega_2 + \theta_2 - \omega_2\theta_2)(h_2 - h_1)]dx = Vol , \qquad (26.8.7)$$

if θ_2 and ω_2 are taken as design variables. At each point $x \in \Omega$ the ribs can be differently inclined to the reference axis x_1, the inclination angle being denoted by $\alpha = \alpha(x)$. Thus the effective bending stiffnesses $D^{\alpha\beta\lambda\mu}_{\theta_2,\omega_2,\alpha}(x)$ depend on the values of the design variables: $\omega_2(x)$, $\theta_2(x)$, $\alpha(x)$ at this point. We do not report these formulae, since they are determined by (24.2.3), (21.1.25) and the relation inverse to (21.1.24). The design variables are subject to the constraints

$$-\theta_2 \leq 0 , \qquad \theta_2 - 1 \leq 0 ,$$
$$-\omega_2 \leq 0 , \qquad \omega_2 - 1 \leq 0 , \qquad (26.8.8)$$
$$-\pi \leq \alpha \leq \pi .$$

The deflection $w = w(x)$ of the plate satisfies the variational equilibrium equation

$$\int_\Omega qv dx - \int_\Omega \kappa_{\sigma\beta}(v) D^{\sigma\beta\lambda\mu}_{\theta_2,\omega_2,\alpha}\kappa_{\lambda\mu}(w)dx = 0 \qquad \forall v \in H^2_0(\Omega) . \qquad (26.8.9)$$

The minimum compliance problem amounts to finding:

$$(P_{\min}) \quad \left| \begin{array}{l} \displaystyle \min_{\theta_2,\omega_2,\alpha} J , \qquad J = \int_\Omega qw dx , \\[2ex] \text{where } w \text{ satisfies (26.8.9) and } \theta_2, \omega_2, \alpha \text{ are subject to (26.8.7) and (26.8.8).} \end{array} \right.$$

There are two ways of proceeding further. We can rearrange this problem to the form (26.5.1) or its dual counterpart (26.6.5). Then we are faced with the static problem of a hypothetic hyperelastic plate of nonlinear constitutive relations (26.6.7). The other way is to apply directly a standard optimality criteria method to the problem $(P_{\min})$. This method is described in Sec. 1.2 of the book of Bendsøe (1995). Just with the help of this algorithm the optimal layouts of m_2 for clamped plates subject to a uniform loading have been found in Lewiński and Othman (1997b) and four of these layouts are presented here in Figs. 26.8.1 – 26.8.4. They concern the same case of $h_2/h_1 = 2$ and different values of the volume assumed. The compliance of a homogeneous plate of a given volume is denoted by J_0. In all cases the ratios J/J_0 are smaller than 1. The profit of the optimization is greater for greater prescribed volumes.

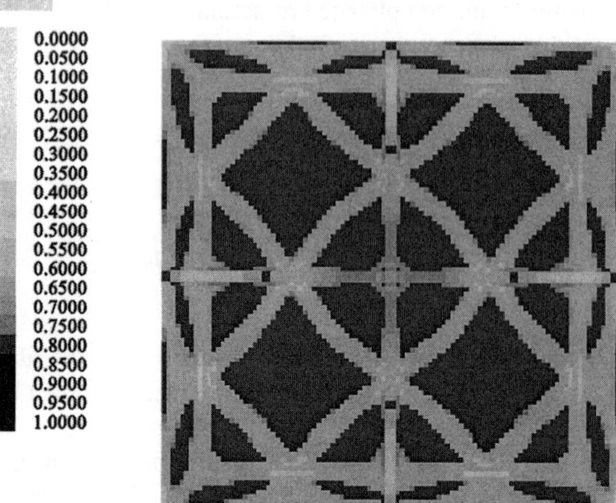

Fig. 26.8.1. Optimum layout of m_2. Case of $h_2/h_1 = 2$ and $Vol = 18000cm^3$. The relative compliance: $J/J_0 = 0.56$

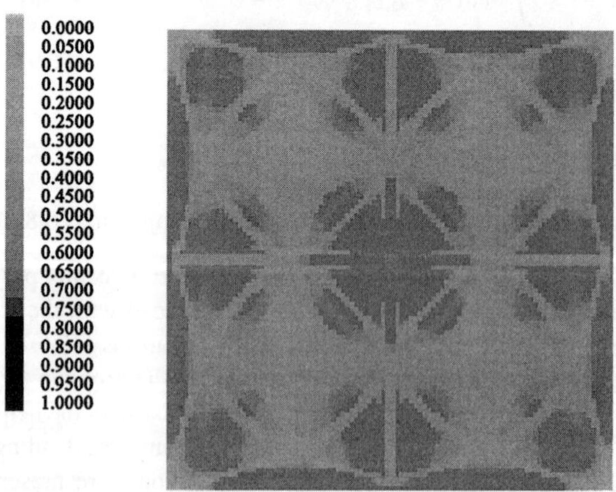

Fig. 26.8.2. Optimum layout of m_2. Case of $h_2/h_1 = 2$ and $Vol = 14000cm^3$. The relative compliance: $J/J_0 = 0.72$

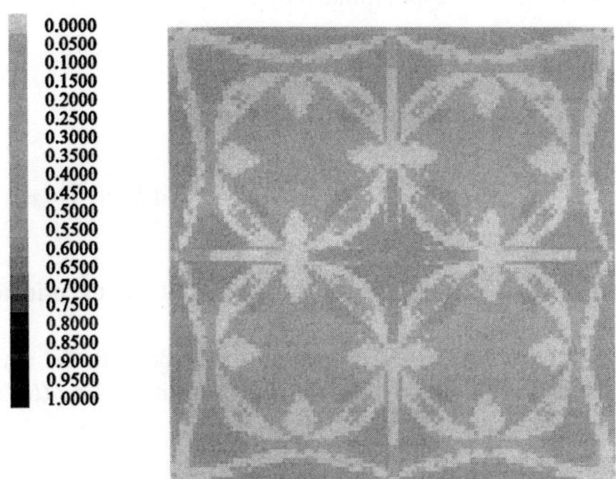

Fig. 26.8.3. Optimum layout of m_2. Case of $h_2/h_1 = 2$ and $Vol = 12000cm^3$. The relative compliance: $J/J_0 = 0.89$

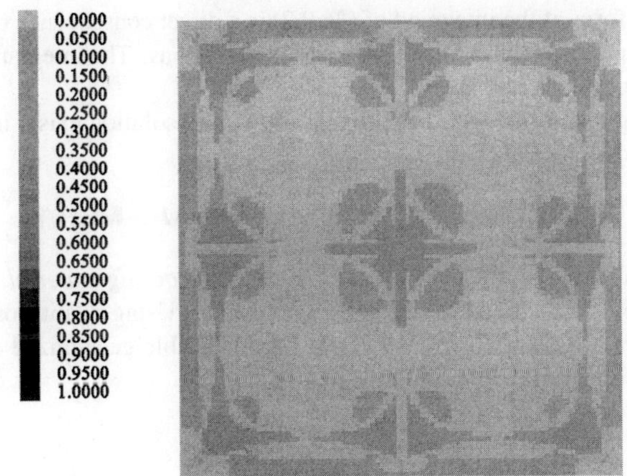

Fig. 26.8.4. Optimum layout of m_2. Case of $h_2/h_1 = 2$ and $Vol = 11000cm^3$. The relative compliance: $J/J_0 = 0.99$

26.9. *Optimal perforated plates of small volume*

Consider the shape optimization problem of Sec. 26.7 for the special case when the fixed volume condition imposes a very small value of the volume. Formally speaking the constant A in (26.1.15) is a very small number. Consequently the relevant Lagrangian multiplier λ must be very large. This implies that in (26.7.15) the condition $g(M) \leq \lambda$ prevails and the function $F_\lambda(\boldsymbol{M})$ can be approximated by

$$F_\lambda(\boldsymbol{M}) = \sqrt{\lambda}[\mathcal{K}(tr\ \boldsymbol{M})^2 + \mathcal{L}(|M_I| + |M_{II}|)^2]^{\frac{1}{2}} \ . \tag{26.9.1}$$

The problem (26.6.7b) assumes the degenerate form

$$\inf_{\boldsymbol{M}\in\mathcal{S}_2(\Omega)} \int_\Omega [\mathcal{K}(tr\ \boldsymbol{M})^2 + \mathcal{L}(|\lambda_1(\boldsymbol{M})| + |\lambda_2(\boldsymbol{M})|)^2]^{\frac{1}{2}} dx \ . \tag{26.9.2}$$

Here, the notation $\lambda_1(\boldsymbol{M}) \equiv M_I$, $\lambda_2(\boldsymbol{M}) \equiv M_{II}$ indicates that the principal moments are certain functions of $\boldsymbol{M} \in \mathbb{E}_2^s$.

Let us note that the problem (26.9.2) does not coincide with the analogous problem of optimal grillages, as formulated by Rozvany. According to Eq. (6.3) of the book by Rozvany (1976) the optimization problem of the form:

$$\min_{\boldsymbol{M}\in\mathcal{S}_2(\Omega)} \int_\Omega [|\lambda_1(\boldsymbol{M})| + |\lambda_2(\boldsymbol{M})|]dx \ , \tag{26.9.3}$$

determines optimal layout of perfectly plastic grillages. A mismatch between both the formulations follows from the known difference between the behavior of independent beams and a plate. The form of the integrand of (26.9.2) is a direct consequence of the Kirchhoff kinematic assumptions which link the slopes with defections. The integrand of (26.9.3) is not affected by such constraints.

Since the integrand in (26.9.2) has a linear growth its solution exists in a *large* space $\mathcal{Z}(\Omega, \mathbb{E}_2^s)$ defined by

$$\mathcal{Z}(\Omega, \mathbb{E}_2^s) = \{\boldsymbol{M} \in \mathbb{M}^1(\Omega, \mathbb{E}_2^s)|\ \text{div}\,\text{div}\boldsymbol{M} \in \mathbb{M}^1(\Omega)\} \ . \tag{26.9.4}$$

The plate of small volume is a plate made of a *perfectly locking material*. Consequently, in (26.9.1) – (26.9.3) $\boldsymbol{M}$ denotes the *rate of moments*. Using duality one can find the *locking condition*, which imposes restrictions on admissible generalized strains κ. The dual problem is solvable in $H^2(\Omega)$.

27. Minimum compliance problem for thin plates of varying thickness: application of Young measures

In this section we study the minimum compliance problem provided that h, the half-thickness of the plate, is a function of $x_1 \in (a, b)$. As in Example 1.3.5, (a, b) denotes the interval $\{x_1 | \ \exists x_2 \text{ with } (x_\alpha) \in \Omega\}$.

The given load q is always assumed in $L^2(\Omega)$, and the loading functional is given by (26.1.12). In contrast to Sec. 26, the only assumptions imposed on the bending stiffness tensor are those specified in Example 1.3.5.

We analyze two choices for the set of admissible half-thicknesses, $\mathcal{H}$,

$$\mathcal{H}_1 = \left\{ h \in L^\infty(a, b) \mid h(x_1) \in \{h_{min}, h_{max}\} \ , \ \int_\Omega h dx = C_0 \right\} , \qquad (27.1)$$

$$\mathcal{H}_2 = \left\{ h \in L^\infty(a, b) \mid h_{min} \le h(x_1) \le h_{max} \ , \ \int_\Omega h dx = C_0 \right\} , \qquad (27.2)$$

where h_{min}, h_{max}, C_0 are positive constants that satisfy

$$area(\Omega)h_{min} < C_0 < area(\Omega)h_{max} \ .$$

Concerning (27.1), the thickness function h is only allowed to take on the two values h_{min} and h_{max}. The minimum compliance problems may have no solutions in $\mathcal{H}_1$, $\mathcal{H}_2$. To overcome this difficulty it is necessary to introduce sets of generalized thicknesses $\widetilde{\mathcal{H}}$ and to perform some relaxation of f. More precisely, $\widetilde{\mathcal{H}}$ and an extension $\widetilde{f}$ of $f_{|_\mathcal{H}}$ to $\widetilde{\mathcal{H}}$ are to be such that:

(A_1) for each $\widetilde{h} \in \widetilde{\mathcal{H}}$, the generalized compliance $\widetilde{f}(\widetilde{h})$ is realizable by a limit of ordinary, admissible plates. In other words, there exists a sequence $\{h_m\}_{m \in M} \subset \mathcal{H}$ for which

$$\widetilde{f}(\widetilde{h}) = \lim_{m \to \infty} f(h_m) \ . \qquad (27.3)$$

(A_1) The functional $\widetilde{f}$ attains its minimum value on $\widetilde{\mathcal{H}}$.

Bonnetier and Vogelius (1987) considered the following candidates for the "generalized plate-thicknesses":

$$\widetilde{\mathcal{H}}_1 = \left\{ \theta \in L^\infty(a, b) \mid 0 \le \theta(x_1) \le 1, \ \int_\Omega [\theta h_{max} + (1 - \theta)h_{min}]dx = C_0 \right\} , \qquad (27.4)$$

$$\widetilde{\mathcal{H}}_2 = \Big\{ (h_s, \theta) \in L^\infty(a, b)^2 \mid h_{min} \le h_s(x_1) \le h_{max} \ , \ 0 \le \theta(x_1) \le 1 \ ,$$
$$\int_\Omega [\theta h_{max} + (1 - \theta)h_s]dx = C_0 \Big\} \ . \qquad (27.5)$$

Here θ plays the role of density of fine scale *stiffeners* parallel to the x_2-axis, and in the second case h_s represents the variable *minimum height of the stiffeners* (the maximum height is always h_{max}). The generalized compliance functional corresponding to $\widetilde{\mathcal{H}}_1$ is

$$\widetilde{f}_1(\theta) = \int_\Omega q\widetilde{w}^0 dx \,, \tag{27.6}$$

where $\widetilde{w}^0$ solves

$$\widetilde{w}^0 \in H_0^2(\Omega) \,, \qquad (\widetilde{D}_0^{\alpha\beta\lambda\mu}\widetilde{w}^0_{,\lambda\mu})_{,\alpha\beta} = q \quad \text{in } \Omega \,. \tag{27.7}$$

The nonzero elements of $\widetilde{D}_0^{\alpha\beta\lambda\mu}$ are

$$\widetilde{D}_0^{1111}(x_1) = \frac{2}{3}c(x_1)\frac{E}{1-\nu^2} \,,$$

$$\widetilde{D}_0^{2222}(x_1) = \frac{2}{3}m(x_1)E + \frac{2}{3}c(x_1)\frac{E\nu^2}{1-\nu^2} \,,$$

$$\widetilde{D}_0^{1122}(x_1) = \widetilde{D}_0^{2211}(x_1) = \frac{2}{3}c(x_1)\frac{E\nu}{1-\nu^2} \,, \tag{27.8}$$

$$\widetilde{D}_0^{1212}(x_1) = \widetilde{D}_0^{1221}(x_1) = \widetilde{D}_0^{2112}(x_1) = \widetilde{D}_0^{2121}(x_1) = \frac{1}{3}m(x_1)\frac{E}{1+\nu} \,,$$

with m denoting the average

$$m(x_1) = \theta(x_1)h_{max}^3 + (1 - \theta(x_1))h_{min}^3(x_1) \,, \tag{27.9}$$

and c the harmonic average

$$c^{-1}(x_1) = \theta(x_1)h_{max}^{-3} + (1 - \theta(x_1))h_{min}^{-3}(x_1) \,. \tag{27.10}$$

Bonnetier and Conca (1987) proved that $(\widetilde{f}_1, \widetilde{\mathcal{H}}_1)$ represents a *full relaxation* of $(f, \mathcal{H}_1)$. It means that both (A_1) and (A_2) are satisfied. Particularly, the functional $\widetilde{f}_1$ is sequentially weak-$*$ continuous.

Introduce now the generalized compliance functional corresponding to $\widetilde{\mathcal{H}}_2$:

$$\widetilde{f}_2(h_s, \theta) = \int_\Omega q\widetilde{w}dx \,, \tag{27.11}$$

where $\widetilde{w}$ solves problem (27.7) with $(\widetilde{D}_0^{\alpha\beta\lambda\mu})$ being replaced by $(\widetilde{D}^{\alpha\beta\lambda\mu})$. The last stiffness tensor has the form (27.8), where the average and harmonic average are given by

$$m(x_1) = \theta(x_1)h_{max}^3 + (1 - \theta(x_1))h_s^3(x_1) \,,$$
$$c^{-1}(x_1) = \theta(x_1)h_{max}^{-3} + (1 - \theta(x_1))h_s^{-3}(x_1) \,. \tag{27.12}$$

Bonnetier and Conca (1987) pointed out that $(\widetilde{f}_2, \widetilde{\mathcal{H}}_2)$ is only a *partial relaxation* of $(f, \mathcal{H}_2)$, i.e., in general only (A_1) is satisfied.

We observe that $\widetilde{D}_0$ is obtained from $\widetilde{D}$ by taking $h_s = h_{min}$. The proof of (27.8) in the general case, thus also for $h_s = h_{min}$, is based on the following result proved in Bonnetier and Conca (1987).

Lemma 27.1. Let $\theta \in L^\infty(a, b)$ with $0 \leq \theta \leq 1$ and let $g_\alpha \in L^\infty(a, b)$, $\alpha = 1, 2$. Define

$$g_\varepsilon(x_1) = \begin{cases} g_1(x_1) & \text{for } x_1/\varepsilon - [x_1/\varepsilon] \leq \theta(x_1) \,, \\ g_2(x_1) & \text{for } x_1/\varepsilon - [x_1/\varepsilon] > \theta(x_1) \,, \end{cases}$$

where $[\,]$ denotes the integer part. Under these assumptions

$$g_\varepsilon \rightharpoonup \theta g_1 + (1 - \theta)g_2 \qquad \text{in} \quad L^\infty(a, b) \text{ weak-} * \text{ as } \varepsilon \to 0 \,. \qquad \square$$

To apply this lemma we take $h_s \in L^\infty(a, b)$ and $\theta \in L^\infty(a, b)$ with $h_{min} \leq h_s \leq h_{max}$ and $0 \leq \theta \leq 1$. Next we define

$$h_\varepsilon(x_1) = \begin{cases} h_{max} & \text{for } x_1/\varepsilon - [x_1/\varepsilon] \leq \theta(x_1) \,, \\ h_s(x_1) & \text{for } x_1/\varepsilon - [x_1/\varepsilon] > \theta(x_1) \,. \end{cases}$$

Let $(D_\varepsilon^{\alpha\beta\lambda\mu})$ denote the bending stiffness tensor for an orthotropic plate with $h = h_\varepsilon$. Then

$$(D_\varepsilon^{1111})^{-1} = \left(\frac{2}{3}h_\varepsilon^3 \frac{E}{1 - \nu^2}\right)^{-1} \,, \qquad D_\varepsilon^{1122}(D_\varepsilon^{1111})^{-1} = \nu \,,$$
$$D_\varepsilon^{2222} - (D_\varepsilon^{1122})^2(D_\varepsilon^{1111})^{-1} = \frac{2}{3}h_\varepsilon^3 E \,, \qquad D_\varepsilon^{1212} = \frac{1}{3}h_\varepsilon^3 \frac{E}{1 + \nu} \,. \tag{27.13}$$

Applying Proposition 1.3.6 and Lemma 27.1 we get

$$(\widetilde{D}^{1111})^{-1} = \left(\frac{2}{3}c\frac{E}{1 - \nu^2}\right)^{-1} \,, \qquad \widetilde{D}^{1122}(\widetilde{D}^{1111})^{-1} = \nu \,,$$
$$\widetilde{D}^{2222} - (\widetilde{D}^{1122})^2(\widetilde{D}^{1111})^{-1} = \frac{2}{3}mE \,, \qquad \widetilde{D}^{1212} = \frac{1}{3}m\frac{E}{1 + \nu} \,, \tag{27.14}$$

with m and c given by (27.12). The last formula yields (27.8), provided that $h_s = h_{min}$. We observe that

$$h_\varepsilon^3 \rightharpoonup m = \theta h_{max}^3 + (1 - \theta)h_s^3 \,, \qquad h_\varepsilon^{-3} \rightharpoonup c^{-1} = \theta h_{max}^{-3} + (1 - \theta)h_s^3 \,, \tag{27.15}$$

in $L^\infty(a, b)$ weak-$*$ as $\varepsilon \to 0$. The first statement follows from Lemma 27.1 with $g_1 = h_{max}^3$ and $g_1 = h_s^3$; to obtain the second, the same lemma is used this time with $g_1 = h_{max}^3$ and $g_2 = h_s^{-3}$. Based on (27.13), (27.14) and (27.15) we get

$$(D_\varepsilon^{1111})^{-1} \rightharpoonup (D^{1111})^{-1} \,, \qquad D_\varepsilon^{1122}(D_\varepsilon^{1111})^{-1} \rightharpoonup \widetilde{D}^{1122}(\widetilde{D}^{1111})^{-1}$$
$$D_\varepsilon^{2222} - (D_\varepsilon^{1122})^2(D_\varepsilon^{1111})^{-1} \rightharpoonup \widetilde{D}^{2222} - (\widetilde{D}^{1122})^2(\widetilde{D}^{1111})^{-1} \,,$$
$$D_\varepsilon^{1212} \rightharpoonup \widetilde{D}^{1212} \,,$$

in $L^\infty(a, b)$ weak-$*$ as $\varepsilon \to 0$.

Bonnetier and Conca (1993, 1994) extended $(\widetilde{f}_2, \widetilde{\mathcal{H}}_2)$ to a full relaxation. Indeed, by Corollary 21.6.8 one considers

$$\widetilde{\widetilde{\mathcal{H}}}_2 = \{(\theta_1, \theta_2, \theta_3, \theta_4; h_1, h_2, h_3, h_4) \in L^\infty(a,b)^8 \mid \theta_M(x_1) \in [0,1],$$

$$h_{min} \le h_M(x_1) \le h_{max} \text{ for a.e. } x \in (a,b), 1 \le M \le 4, \sum_{M=1}^{4} \theta_M = 1,$$

$$\sum_M \int_\Omega \theta_M h_M dx = C_0 \} \; .$$

The generalized compliance functional $\widetilde{\widetilde{f}}$ is defined by

$$\widetilde{\widetilde{f}}(\theta_M, h_M) = \int_\Omega q\widetilde{\widetilde{w}}dx \; ,$$

where $\widetilde{\widetilde{w}}$ is a solution to

$$(\widetilde{\widetilde{D}}^{\alpha\beta\lambda\mu}\widetilde{\widetilde{w}}_{,\lambda\mu})_{,\alpha\beta} = q \text{ in } \Omega \; , \qquad \widetilde{\widetilde{w}} \in H_0^2(\Omega) \; . \tag{27.16}$$

The tensor $\widetilde{\widetilde{D}}$ is defined as in (27.14), but m and c are respectively replaced by

$$m(x_1) = \sum_{M=1}^{4} \theta_M(x_1)h_M^3(x_1) \; , \qquad c^{-1}(x_1) = \sum_{M=1}^{4} \theta_M(x_1)h_M^{-3}(x_1) \; . \tag{27.17}$$

It is now easy to prove that $(\widetilde{\widetilde{f}}, \widetilde{\widetilde{\mathcal{H}}}_2)$ represents a full relaxation of $(f, \mathcal{H}_2)$. Indeed, the property of partial relaxation (A_1) is obtained by using Lemma 27.1, considering elements of $\mathcal{H}_2$ of the form

$$h_\varepsilon(x_1) = \begin{cases} h_1(x_1) & \text{if } x_1/\varepsilon - [x_1/\varepsilon] \le \theta_1(x_1) \; , \\[2mm] h_2(x_1) & \text{if } \theta_1(x_1) < x_1/\varepsilon - [x_1/\varepsilon] \le \theta_1(x_1) + \theta_2(x_1) \; , \\[2mm] h_3(x_2) & \text{if } \theta_1(x_1) + \theta_2(x_1) < x_1/\varepsilon - [x_1/\varepsilon] \le \theta_1(x_1) + \theta_2(x_1) + \theta_3(x_1) \; , \\[2mm] h_4(x_1) & \text{if } \theta_1(x_1) + \theta_2(x_1) + \theta_3(x_1) < x_1/\varepsilon - [x_1/\varepsilon] \le 1 \; . \end{cases}$$

For (A_2), one shows that there exists an element of $\widetilde{\widetilde{\mathcal{H}}}_2$, at which $\widetilde{\widetilde{f}}$ attains the infimum. To this end, one takes a minimizing sequence $\{\theta_M^k, h_M^k\}_{k \in \mathbb{N}} \subset \widetilde{\widetilde{\mathcal{H}}}_2$ for $\widetilde{\widetilde{f}}$. Applying Proposition 1.3.6 and Lemma 27.1 we conclude that

$$\widetilde{\widetilde{w}}(\theta_M^{k'}, h_M^{k'}) \rightharpoonup \widetilde{\widetilde{w}}(\theta_M, h_M) \qquad \text{in } H_0^2(\Omega) \text{ weakly as } k' \to \infty$$

for a subsequence $\{k'\}$ of $\{k \in \mathbb{N}\}$, where $(\theta_M, h_M) \in \widetilde{\widetilde{\mathcal{H}}}_2$. Hence

$$\widetilde{\widetilde{f}}(\theta_M^{k'}, h_M^{k'}) \to \widetilde{\widetilde{f}}(\theta_M, h_M) = \inf_{\widetilde{\widetilde{\mathcal{H}}}_2} \widetilde{\widetilde{f}} \; .$$

An alternative point of view to the full relaxation of $(f, \mathcal{H}_2)$ was adopted by Muñoz and Pedregal (1998). These authors examine the problem of relaxation *in terms of parametrized measures* associated to sequences of half-thicknesses and define a new compliance functional $\overline{f}$ for such families of probability measures, which is an extension and reduces to f when the parametrized measure is trivial, i.e., a Dirac mass. In fact, the approach used by Muñoz and Pedregal (1998) develops the results due to Bonnetier and Conca (1993). In the last paper it is shown that it is sufficient to take $M = \alpha = 1, 2$. More precisely, $(^2f, \,^2\mathcal{H}_2)$ represents a full relaxation $(f, \mathcal{H}_2)$, where

$$^2\mathcal{H}_2 = \{ \sum_{\alpha=1}^{2} \theta_\alpha(x_1)(h_\alpha^3(x_1), h_\alpha^{-3}(x_1), h_\alpha(x_1)) | \ (\theta_\alpha(x_1), h_\alpha(x_1)) \in [0,1] \times [h_{min}, h_{max}],$$

$$\sum_{\alpha=1}^{2} \theta_\alpha(x_1) = 1 \ , \quad \int_\Omega (\sum_{\alpha=1}^{2} \theta_\alpha h_\alpha) dx = C_0 \} \ , \tag{27.18}$$

and

$$^2f(h_1, h_2; \theta) = \int_\Omega qw(h_1, h_2; \theta) dx \ . \tag{27.19}$$

Here $w(h_1, h_2; \theta)$ is the solution to (27.16) where m and c^{-1} are still given by (27.17); now, however, $M = \alpha = 1, 2$.

Let us pass to the full relaxation proposed by Muñoz and Pedregal (1998). Let $\overline{\mathcal{H}}_2$ be the set of parametrized measures associated to sequences $\{h_k\}_{k \in \mathbb{N}}$ of half-thicknesses:

$$\overline{\mathcal{H}}_2 = \{\mu = \{\mu_{x_1}\}_{x_1 \in (a,b)} | \ supp \ \mu_{x_1} \subset C \subset [h_{min}, h_{max}],$$

$$\text{a.e. } x_1 \in (a, b), \int_\Omega \int_C z d\mu_{x_1}(z) dx = C_0 \} \ .$$

According to Theorem 21.6.6 we can find a sequence $\{h_k\}_{k \in \mathbb{N}}$ taking values in C and whose associated parametrized measure is precisely μ. It might not be true, however, that

$$\int_\Omega h_k(x_1) dx = C_0 \qquad \text{for all } k \ .$$

We only know that

$$\int_\Omega h_k(x_1) dx \to C_0 \qquad \text{as } k \to \infty \ .$$

Further, let

$$m(x_1) = \int_C z^3 d\mu_{x_1}(z) \ , \qquad c^{-1}(x_1) = \int_C z^{-3} d\mu_{x_1}(z) \ . \tag{27.20}$$

The bending stiffness tensor $(\overline{D}^{\alpha\beta\lambda\mu})$ depends on μ through (27.20); i.e., in (27.14) m and c are given by (27.20).

The compliance functional $\overline{f}$ is defined by

$$\overline{f}(\mu) = \int_\Omega q\overline{w}dx \, , \qquad (27.21)$$

where $\overline{w}$ is the solution to (27.7) with the stiffness tensor $\overline{D}$.

Theorem 27.2. The pair $(\overline{f}, \overline{\mathcal{H}}_2)$ is a full relaxation for $(f, \mathcal{H}_2)$.

Proof. For $h \in \mathcal{H}$ we have

$$\mu = \delta_{h(x_1)} \in \overline{\mathcal{H}} \, ,$$

and, moreover, $f(h) = \overline{f}(\mu)$, so that

$$\inf_{\overline{\mathcal{H}}_2} \overline{f} \leq \inf_{\mathcal{H}_2} f \, . \qquad (27.22)$$

On the other hand, given any $\mu \in \overline{\mathcal{H}}_2$ we can find a sequence $\{h_k\}_{k \in \mathbb{N}} \subset \mathcal{H}_2$ whose parametrized measure is μ. Applying Proposition 1.3.6 we conclude that

$$\overline{f}(\mu) = \lim_{k \to \infty} f(h_k) \, .$$

The arbitrariness of μ yields the equality in (27.22).

To show existence of minimizers for $(\overline{f}, \overline{\mathcal{H}}_2)$, we take any minimizing sequence for f in $\mathcal{H}_2$. Then parametrized measure generated by such sequence μ belongs obviously to $\overline{\mathcal{H}}_2$ and by definition of $\overline{f}$ we have as before

$$\overline{f}(\mu) = \lim_{k \to \infty} f(h_k) \, .$$

Thus μ is a minimizer, and the proof is complete. $\square$

Now the following question arises: what is the form of parametrized measure rendering the functional $\overline{f}$ the minimal value?

By using rather simple convexity arguments it can be shown that $(\overline{f}, \overline{\mathcal{H}}_2)$ admits a minimizer of the form

$$\mu_{x_1} = \theta_1(x_1)\delta_{h_{min}} + \theta_2(x_1)\delta_{h_{max}} + \theta_3(x_1)\delta_{h(x_1)} \, , \qquad (27.23)$$

where $\theta_i(x_1) \in [0, 1]$, $\theta_1 + \theta_2 + \theta_3 = 1$ and $h \in \mathcal{H}_2$. Similarly, it can be shown that there are minimizers for $(\overline{f}, \overline{\mathcal{H}}_2)$ of the form

$$\mu_{x_1} = \theta_1(x_1)\delta_{h_1(x_1)} + (1 - \theta(x_1))\delta_{h_2(x_1)} \, , \qquad (27.24)$$

with $\theta_1(x_1) \in [0, 1]$ and $h_\alpha \in \mathcal{H}_2$.

Muñoz and Pedregal (1998) also argued that minimizers depending on two design variables: $\theta_1(x_1) \in [0,1]$ and $h_\alpha \in \mathcal{H}_2$ fail to exist in general. What we are missing is the volume constraint. To adjust this constraint, we introduce an additional parameter and include both h_{min} and h_{max} in the description of the minimizer. In this manner we find the *optimal full relaxation*. Indeed, the following theorem was established by Muñoz and Pedregal (1998).

Theorem 27.3. There exist $\theta_1(x_1) \in [0,1]$, $x_1 \in (a,b)$, $h \in \mathcal{H}_2$ and $z \in [a,b]$ such that the family of probability measures

$$\mu_{x_1} = \begin{cases} \theta(x_1)\delta_{h_{max}} + (1-\theta(x_1))\delta_{h(x_1)} & \text{for } x_1 \in (a,z) \\ \theta(x_1)\delta_{h_{min}} + (1-\theta(x_1))\delta_{h(x_1)} & \text{for } x_1 \in (z,b) , \end{cases}$$

is a minimizer for $(\overline{f}, \overline{\mathcal{H}}_2)$. $\square$

28.   Thin shells of minimum compliance

This section discusses the following layout problem: given a middle surface of a shell and an external loading and given two isotropic materials of a given volume, find their optimal distribution such that the shell made in this manner realizes the smallest compliance. Such layout problem requires relaxation, cf. Sec. 26. The optimal shell must be constructed from a composite *shell material*, its effective stiffnesses being determined by the homogenization formulae of Sec. 17. A natural relaxed formulation is expressed in terms of the dual variables. One of the aims of the present section is to put this formulation in its primal form, involving displacements and strain measures.

28.1.   *Setting of the problem*

Consider a shell with a middle surface S, an image of a plane reference domain Ω; $S = \Phi(\Omega)$, see Fig. 17.1.1. Assume that the plane subdomains Ω is a sum of two subdomains Ω_α, $\alpha = 1, 2$, such that $\Omega_1 \cap \Omega_2 = \emptyset$ and $\overline{\Omega}_1 \cup \overline{\Omega}_2 = \overline{\Omega}$. Two isotropic materials are at our disposal. Their moduli, reduced according to the plane-stress assumption, are denoted by $\widetilde{C}_\alpha$. Due to isotropy one can write

$$\widetilde{C}_\alpha = 2\widetilde{k}_\alpha \boldsymbol{I}_1 + 2\widetilde{\mu}_\alpha \boldsymbol{I}_2 \,, \tag{28.1.1a}$$

where

$$I_1^{\alpha\beta\lambda\mu} = \frac{1}{2}a^{\alpha\beta}a^{\lambda\mu} \,, \qquad I_2^{\alpha\beta\lambda\mu} = \frac{1}{2}(a^{\alpha\lambda}a^{\beta\mu} + a^{\alpha\mu}a^{\beta\lambda} - a^{\alpha\beta}a^{\lambda\mu}) \,, \tag{28.1.1b}$$

cf. Eqs. (3.8.29), (16.1.3), and $\widetilde{k}_\alpha, \widetilde{\mu}_\alpha$ represent the Kelvin and Kirchhoff moduli, respectively. They can be expressed in terms of the Young and Poisson moduli as follows

$$2\widetilde{k}_\alpha = \frac{E_\alpha}{1 - \nu_\alpha} \,, \qquad 2\widetilde{\mu}_\alpha = \frac{E_\alpha}{1 + \nu_\alpha} \,. \tag{28.1.2}$$

The material of moduli $(\widetilde{k}_\alpha, \widetilde{\mu}_\alpha)$ is placed transversely symmetrically within a shell of the mid-surface $\Phi(\Omega)$. The αth material forms the shell of thickness h_α. Thus the two-component shell has the piecewise constant thickness. The membrane stiffnesses of the αth part of the shell are

$$\boldsymbol{A}_\alpha = 2\widetilde{k}_\alpha h_\alpha \boldsymbol{I}_1 + 2\widetilde{\mu}_\alpha h_\alpha \boldsymbol{I}_2 \,, \tag{28.1.3}$$

while the bending stiffnesses have the form

$$\boldsymbol{D}_\alpha = 2\widetilde{k}_\alpha \frac{(h_\alpha)^3}{12}\boldsymbol{I}_1 + 2\widetilde{\mu}_\alpha \frac{(h_\alpha)^3}{12}\boldsymbol{I}_2 \,. \tag{28.1.4}$$

The piecewise distribution of the membrane and bending stiffnesses of the shell can be represented by the formulae

$$\boldsymbol{A}(\xi) = \chi_1(\xi)\boldsymbol{A}_1 + \chi_2(\xi)\boldsymbol{A}_2 \,, \qquad \boldsymbol{D}(\xi) = \chi_1(\xi)\boldsymbol{D}_1 + \chi_2(\xi)\boldsymbol{D}_2 \,, \tag{28.1.5}$$

with χ_α being a characteristic function of the domain Ω_α. The membrane and bending flexibilities are

$$\boldsymbol{A}^{-1}(\xi) = \chi_1(\xi)\boldsymbol{A}_1^{-1} + \chi_2(\xi)\boldsymbol{A}_2^{-2}\,, \quad \boldsymbol{D}^{-1}(\xi) = \chi_1(\xi)\boldsymbol{D}_1^{-1} + \chi_2(\xi)\boldsymbol{D}_2^{-1}\,, \qquad (28.1.6)$$

where

$$\boldsymbol{A}_\alpha^{-1} = \frac{1}{2\widetilde{k}_\alpha h}\boldsymbol{I}_1 + \frac{1}{2\widetilde{\mu}_\alpha h}\boldsymbol{I}_2\,, \quad \boldsymbol{D}_\alpha^{-1} = \frac{12}{2\widetilde{k}_\alpha (h_\alpha)^3}\boldsymbol{I}_1 + \frac{12}{2\widetilde{\mu}_\alpha (h_\alpha)^3}\boldsymbol{I}_2\,. \qquad (28.1.7)$$

Assume that the shell equilibrium is considered within the Koiter shell model, see Sec. 17.1. Thus the deformation of the shell is fully determined by the position of the shell middle surface, measured with respect to the local co-basis $\boldsymbol{a}^1$, $\boldsymbol{a}^2$, $\boldsymbol{a}^3$, see Sec. 16.1. The tangent and normal displacements of a point $\boldsymbol{\Phi}(\xi)$ of the middle surface are denoted by $u_\alpha(\xi)$ and $w(\xi)$, respectively. These displacements form a vector field $\boldsymbol{U}(\xi)$

$$\boldsymbol{U}(\xi) = [u_1(\xi), u_2(\xi), w(\xi)]^T\,. \qquad (28.1.8)$$

Assume that the shell is clamped along $\boldsymbol{\Phi}(\Gamma_0) \subset \partial S$

$$u_\alpha = 0, \qquad w = 0\,, \qquad \frac{\partial w}{\partial \boldsymbol{n}} = 0 \qquad \text{on } \boldsymbol{\Phi}(\Gamma_0)\,, \qquad (28.1.9)$$

where $\boldsymbol{n}$ is the outward unit vector normal to $\boldsymbol{\Phi}(\Gamma_0) \subset \partial S$. The space of admissible displacements has the form

$$V(\Omega) = \{u_\alpha \in H^1(\Omega), w \in H^2(\Omega) | \boldsymbol{u}, w \text{ satisfy } (28.1.9) \text{ for } \xi \in \Gamma_0\}\,.$$

The shell is subject to the surface loading q^α, q. The compliance (or rather total compliance) of the shell is defined as a functional of χ_2

$$I(\chi_2) = f(\boldsymbol{U})\,, \qquad (28.1.10)$$

with

$$f(\boldsymbol{U}) = \int_\Omega (q^\alpha u_\alpha + qw)\sqrt{a(\xi)}\,d\xi\,. \qquad (28.1.11)$$

The solution $\overline{U} = (\overline{\boldsymbol{u}}, \overline{w})$ to the static problem is determined by the layout of the second material. Hence I depends on χ_2 and $\chi_1 = 1 - \chi_2$.

The generalized strains associated with $\boldsymbol{U}$ are written in the form

$$\boldsymbol{\varepsilon}(\boldsymbol{U}) = [\epsilon(\boldsymbol{U}), \rho(\boldsymbol{U})]^T\,, \qquad (28.1.12)$$

where ϵ and ρ are defined by (16.2.7). The generalized stresses

$$\sigma = [\boldsymbol{N}, \boldsymbol{M}]^T\,, \qquad (28.1.13)$$

are interrelated with ε by the constitutive relations (16.2.15) which can be put in the form

$$\sigma = \mathcal{D}\varepsilon\,, \quad \text{or} \quad \varepsilon = \boldsymbol{d}\sigma \qquad (28.1.14)$$

where

$$\mathcal{D} = \operatorname{diag}(A, D), \quad d = \operatorname{diag}(A^{-1}, D^{-1}), \tag{28.1.15}$$

hence

$$\mathcal{D}(\xi) = \chi_1(\xi)\mathcal{D}_1 + \chi_2(\xi)\mathcal{D}_2, \quad d(\xi) = \chi_1(\xi)d_1 + \chi_2(\xi)d_2, \tag{28.1.16}$$

with

$$\mathcal{D}_\alpha = \operatorname{diag}(A_\alpha, D_\alpha), \quad d_\alpha = \operatorname{diag}(A_\alpha^{-1}, D_\alpha^{-1}). \tag{28.1.17}$$

The bilinear form (16.2.16) can be rewritten as follows

$$a_{\mathcal{D}}(U, Z) = \int_\Omega \varepsilon(U) : (\mathcal{D}\varepsilon(Z))\sqrt{a(\xi)}\, d\xi, \tag{28.1.18}$$

with $Z = (v, v)$.

The equilibrium problem reads:

$$(P) \quad \left| \begin{array}{l} \text{find } \overline{U} \in V(\Omega) \text{ such that} \\[2mm] a_{\mathcal{D}}(U, Z) = f(Z) \quad \forall Z \in V(\Omega). \end{array} \right. \tag{28.1.19}$$

Consider a family of shells of constant volume

$$\int_\Omega [\chi_1(\xi)h_1 + \chi_2(\xi)h_2]\sqrt{a(\xi)}\, d\xi = C. \tag{28.1.20}$$

Assume that $h_2 > h_1$. Then the quantity C must be chosen such that

$$h_1|\Omega| < C < h_2|\Omega|. \tag{28.1.21}$$

This condition determines a constant A such that

$$\int_\Omega \chi_2(\xi)\sqrt{a(\xi)}\, d\xi = A. \tag{28.1.22}$$

Then $0 < A < |\Omega|$.

The minimum compliance problem has the form:

$$\inf\{I(\chi_2)|\ \chi_2 \in L^\infty(\Omega, \{0, 1\}),\ \text{with } \chi_2 \text{ satisfying (28.1.22)}$$
$$\text{and } \overline{U} \text{ being the solution to problem } (P)\}. \tag{28.1.23}$$

Let us express this problem in terms of the generalized stresses σ. This quantity is said to be statically admissible if $\sigma \in S(\Omega)$, where, cf. also (17.2.4),

$$S(\Omega) = \{\sigma \in L^2(\Omega, \mathbb{E}_2^s) \times L^2(\Omega, \mathbb{E}_2^s)|\ N^{\alpha\beta}{}_{||\beta} \in L^2(S),\ M^{\alpha\beta}{}_{||\alpha\beta} \in L^2(S) \quad \text{and}$$
$$\int_\Omega \sigma : \varepsilon(Z)\sqrt{a(\xi)}\, d\xi = f(Z) \quad \forall Z \in V(\Omega)\}. \tag{28.1.24}$$

Here $\sigma : \varepsilon = N^{\alpha\beta}\epsilon_{\alpha\beta} + M^{\alpha\beta}\rho_{\alpha\beta}$.

Castigliano's theorem or the principle of the complementary energy for the Koiter shell yields

$$\inf_{\sigma \in S(\Omega)} \int_{\Omega} \sigma : (d\sigma)\sqrt{a}\, d\xi \; . \tag{28.1.25}$$

The theory of duality (see Sec. 1.2.5) implies that this minimal value of the functional appearing in (28.1.25) is equal to $f(\overline{U})$. Thus the layout problem (28.1.23) can be put in the form

$$\min_{\chi \in L^{\infty}(\Omega;\{0,1\})} \; \min_{\sigma \in S(\Omega)} \int_{\Omega} [\sigma : (d\sigma) + \lambda\chi]\sqrt{a}\, d\xi \; , \tag{28.1.26}$$

where $\chi = \chi_2$.

By repeating the arguments of Sec. 26.1 we conclude that the above problem is ill-posed and requires a relaxation.

28.2. Relaxation

The method of relaxation is similar to the case of plates in bending, cf. Sec. 26.2. Thus we extend the design space by admitting the weak-$*$ limits of sequences $\{\chi_n\}_{n\in\mathbb{N}}$, where $\chi_n = (\chi_2)_n$. These weak-$*$ limits are denoted by m_2, $m_2 \in L^{\infty}(\Omega, [0,1])$. Any sequence $\{\chi_n\}$ defines a sequence of the stiffness tensors $\{\mathcal{D}_n\}$ and the compliance tensors $\{d_n\}$. By the equilibrium problem (P) a sequence $U_n = ((u_\alpha)_n, w_n)$ is determined. Its limit $U^h = (u^h, w^h)$ satisfies the problem (P) with $\mathcal{D} = \mathcal{D}_h$ being a G-limit of the tensors $\{\mathcal{D}_n\}$.

Let us introduce the isoperimetric condition

$$\int_{\Omega} \chi_n \sqrt{a(\xi)}\, d\xi = A \; , \quad n \in \mathbb{N} \; , \tag{28.2.1}$$

where $\chi_n = (\chi_2)_n$. If $\chi_n \rightharpoonup m_2$ weak-$*$ in $L^{\infty}(\Omega, [0,1])$, then the condition (28.2.1) is preserved, or

$$\int_{\Omega} m_2(\xi) \sqrt{a(\xi)}\, d\xi = A \; . \tag{28.2.2}$$

Such sequences $\{\chi_n\}$ determine sequences of $\{\mathcal{D}_n\}$ and their G-limits. The set of all such limits will be denoted by $\mathcal{G}_{m_2}$. By the homogenization results of Sec. 17 we know that the elements $\mathcal{D}_h \in \mathcal{G}_{m_2}$ have the form

$$\mathcal{D}_h = \begin{bmatrix} A_h & E_h \\ F_h & D_h \end{bmatrix} \tag{28.2.3}$$

and $E_h \neq 0, F_h \neq 0$ in general. We conjecture here that Theorem 26.2.2 due to Dal Maso and Kohn still holds. Consequently the set $\mathcal{G}_{m_2}$ consists of all tensors $\mathcal{D}_h$ characterizing the effective stiffnesses of periodic shells. The relevant formulae are derived in Sec. 17. Obviously, the tensors $d^h = (\mathcal{D}_h)^{-1}$ constitute the same set $\mathcal{G}_{m_2}$.

The relaxation of the problem (28.1.26) leads to the following problem:

$$\min_{m_2 \in L^\infty(\Omega;[0,1])} \min_{d \in \mathcal{G}_{m_2}(\xi)} \min_{\sigma \in \mathcal{S}(\Omega)} \int_\Omega [\sigma : (d(\xi)\sigma) + \lambda m_2(\xi)]\sqrt{a}\, d\xi \ . \tag{28.2.4}$$

By interchanging the two last minima we find

$$\min_{m_2 \in L^\infty(\Omega;[0,1])} \min_{\sigma \in \mathcal{S}(\Omega)} \int_\Omega [2\mathcal{W}^*(\sigma, m_2(\xi)) + \lambda m_2(\xi)]\sqrt{a}\, d\xi \tag{28.2.5}$$

with

$$\mathcal{W}^*(\sigma, m_2) = \min\{\frac{1}{2}\sigma : (d\sigma)|d \in \mathcal{G}_{m_2}\} \ . \tag{28.2.6}$$

28.3.   Primal formulation of the relaxed problem

The procedure will be similar to that of Sec. 26.6. Note first that by the Castigliano theorem

$$\min_{\sigma \in \mathcal{S}(\Omega)} \int_\Omega \sigma : (d\sigma)\sqrt{a}\, d\xi = \max_{Z \in V(\Omega)} \int_\Omega [2(q^\alpha v_\alpha + qv) - 2W(\varepsilon(Z))]\sqrt{a}\, d\xi \tag{28.3.1}$$

where $Z = (v_1, v_2, v)$ and

$$W(\varepsilon) = \frac{1}{2}\varepsilon : (\mathcal{D}\varepsilon) \ , \quad \varepsilon \in \mathbb{E}_s^2 \times \mathbb{E}_s^2 \ . \tag{28.3.2}$$

Let us denote the value of the expression (28.2.4) by R_λ. By (28.3.1) we have

$$R_\lambda = \min_{m_2 \in L^\infty(\Omega;[0,1])} \min_{\mathcal{D} \in \mathcal{G}_{m_2}(\xi)} \max_{Z \in V(\Omega)} \int_\Omega [2(q^\alpha v_\alpha + qv)$$

$$- \varepsilon(Z) : (\mathcal{D}(\xi)\varepsilon(Z)) + \lambda m_2(\xi)]\sqrt{a}\, d\xi \ . \tag{28.3.3}$$

We shall prove below that the order of minimization over $\mathcal{D}$ and maximization over Z may be interchanged.

Let us start with (28.2.5) and consider the problem

$$(P_{m_2}) \qquad \min_{\sigma \in \mathcal{S}(\Omega)} \int_\Omega \mathcal{W}^*(\sigma, m_2(\xi))\sqrt{a}\, d\xi \ , \tag{28.3.4}$$

or

$$(P_{m_2}) \qquad \min_{\sigma \in \mathcal{S}(\Omega)} F(\sigma) \ , \tag{28.3.5}$$

with

$$F(\sigma) = \int_\Omega \mathcal{W}^*(\sigma, m_2(\xi))\sqrt{a(\xi)}\, d\xi \ . \tag{28.3.6}$$

To apply Rockafellar's theory of duality we take $\Lambda(Z) = \varepsilon(Z)$ and set

$$G(\Lambda\boldsymbol{\sigma}) = I_{S(\Omega)}(\Lambda\boldsymbol{\sigma}) \, .$$

The derivation of the explicit form of Λ^* is left to the reader as an exercise. cf. Sec. 17.2.
The dual problem means then evaluating

$$(P^*_{m_2}) \qquad \max\{-G^*(Z) - F^*(\varepsilon(Z))|Z \in V(\Omega)\} \, .$$

By standard arguments we find

$$G^*(Z) = -f(Z) \, , \tag{28.3.7}$$

$$F^*(\varepsilon(Z)) = \sup_{\boldsymbol{\sigma}\in\mathcal{S}(\Omega)} \{\int_\Omega \boldsymbol{\sigma} : \varepsilon(Z)\sqrt{a}\,d\xi - \frac{1}{2}\int_\Omega \min_{d\in\mathcal{G}_{m_2(\xi)}} \boldsymbol{\sigma} : (d(\xi)\boldsymbol{\sigma})\sqrt{a}\,d\xi\} \tag{28.3.8}$$

or

$$F^*(\varepsilon(Z)) = \int_\Omega \mathcal{W}(\varepsilon(Z), m_2(\xi))\sqrt{a}\,d\xi \, , \tag{28.3.9}$$

with

$$\mathcal{W}(\varepsilon, m_2(\xi)) = \frac{1}{2}\max_{\boldsymbol{D}\in\mathcal{G}_{m_2(\xi)}} \varepsilon : (\boldsymbol{D}\varepsilon) \, . \tag{28.3.10}$$

The equality (26.6.5) holds here. Thus

$$\min_{\boldsymbol{\sigma}\in\mathcal{S}(\Omega)} \int_\Omega \mathcal{W}^*(\boldsymbol{\sigma}, m_2(\xi))\sqrt{a}\,d\xi$$

$$= \max\{f(Z) - \int_\Omega \mathcal{W}(\varepsilon(Z), m_2(\xi))\sqrt{a}\,d\xi|Z \in V(\Omega)\} \, . \tag{28.3.11}$$

The l.h.s. of (28.2.5) has been denoted by R_λ. Taking into account (28.3.11) we have

$$R_\lambda = \min_{m_2\in L^\infty(\Omega;[0,1])} \max_{Z\in V(\Omega)} \{2f(Z)$$

$$- \int_\Omega [\max_{\boldsymbol{D}\in\mathcal{G}_{m_2(\xi)}} \varepsilon(Z) : (\boldsymbol{D}(\xi)\varepsilon(Z)) + \lambda m_2(\xi)]\sqrt{a}\,d\xi\} \, , \tag{28.3.12}$$

or

$$R_\lambda = \min_{m_2\in L^\infty(\Omega;[0,1])} \max_{Z\in V(\Omega)} \min_{\boldsymbol{D}\in\mathcal{G}_{m_2(\xi)}} \{\int_\Omega [2(q^\alpha v_\alpha + qv)$$

$$- \varepsilon(Z) : (\boldsymbol{D}(\xi)\varepsilon(Z)) + \lambda m_2(\xi)]\sqrt{a}\,d\xi\} \, , \tag{28.3.13}$$

which proves that the minimum over $\mathcal{D}$ and maximum over Z in (28.3.3) may be interchanged.

The primal formulation of the minimum compliance problem has the form (28.3.12). Note that the set $\mathcal{G}_{m_2(\xi)}$ is sequentially compact for the topology induced by G-convergence. By (28.3.10) the function $\mathcal{W}(\cdot, m_2(\xi))$ is convex and $\mathcal{W}(\varepsilon, m_2)$ has quadratic growth in ε, for all $m_2 \in [0,1]$. Consequently, the primal problem (28.3.12) is solvable, its solution being denoted by

$$(\overline{U}, \overline{\sigma}, \overline{m}_2, \overline{\mathcal{D}}) \in V(\Omega) \times \mathcal{S}(\Omega) \times L^\infty(\Omega; [0,1]) \times \mathcal{G}_{m_2(\xi)} .$$

Assume that the minimum in (28.2.6) is realized for $d = \overline{d}$ and maximum in (28.3.10) is attained for $\mathcal{D} = \overline{\mathcal{D}}$. Then $\overline{\mathcal{D}} = (\overline{d})^{-1}$. The extremality relation (1.2.55) give

$$\overline{\sigma}(\xi) = \frac{\partial \mathcal{W}(\varepsilon(\overline{U}), \overline{m}_2(\xi))}{\partial \varepsilon} , \tag{28.3.14}$$

or

$$\varepsilon(\overline{U}) = \frac{\partial \mathcal{W}^*(\overline{\sigma}(\xi), \overline{m}_2(\xi))}{\partial \sigma} . \tag{28.3.15}$$

To be more precise, these relations are written in terms of stress and couple resultants

$$\overline{N} = \frac{\partial \mathcal{W}(\epsilon(\overline{U}), \rho(\overline{U}), \overline{m}_2(\xi))}{\partial \epsilon} , \quad \overline{M} = \frac{\partial \mathcal{W}(\epsilon(\overline{U}), \rho(\overline{U}), \overline{m}_2(\xi))}{\partial \rho} . \tag{28.3.16}$$

The inverse relations are

$$\epsilon(\overline{U}) = \frac{\partial \mathcal{W}^*(\overline{N}(\xi), \overline{M}(\xi), \overline{m}_2(\xi))}{\partial N} , \quad \rho(\overline{U}) = \frac{\partial \mathcal{W}^*(\overline{N}(\xi), \overline{M}(\xi), \overline{m}_2(\xi))}{\partial M} . \tag{28.3.17}$$

The explicit form of the potentials $\mathcal{W}$ and $\mathcal{W}^*$ is still unknown. Since the curvature tensor $(b_\alpha{}^\beta.)$ is involved in the equilibrium equations of the shell, it is clear that just this tensor should effect the effective potentials $\mathcal{W}$ and $\mathcal{W}^*$. Moreover, it is not known whether the tensors $\overline{\mathcal{D}}$ and $\overline{d}$ can be realized by laminates of finite rank. It can only be taken for granted that orthogonal ribbed shells of second rank will not suffice.

28.4. On the in-plane minimum compliance problem of two-phase plates

The special case of the optimization problem (28.1.26) for plates in bending has been considered in Sec. 26. The analogous optimization problem for plates loaded in the plane does not require a detailed derivation, due to some similarities with the bending case. Thus the present section is aimed at giving a short outline of the in-plane problem.

In the in-plane (membrane) problem considered, we have $q = 0$, $w = 0$, $M = 0$. Thus the relaxed problem assumes a special form of (28.2.5)

$$\min_{m_2 \in L^\infty(\Omega;[0,1])} \min_{N \in \mathcal{S}_1(\Omega)} \int_\Omega [2\mathcal{W}^*(N, m_2(x)) + \lambda m_2(x)] dx . \tag{28.4.1}$$

The set $\mathbf{S}_1(\Omega)$ is defined by

$$\mathbf{S}_1(\Omega) = \{\mathbf{N} \in L^2(\Omega, \mathbb{E}_2^s) | \ N_{,\beta}^{\alpha\beta} \in L^2(\Omega) \ ,$$

$$\int_\Omega N^{\alpha\beta} e_{\alpha\beta}(\mathbf{v})dx = \int_\Omega q^\alpha v_\alpha dx \quad \forall \, \mathbf{v} \in V_1(\Omega)\} \ ,$$

where

$$V_1(\Omega) = \{\mathbf{v} \in H^1(\Omega)^2 | \ \mathbf{v} = \mathbf{0} \text{ on } \Gamma_0\}$$

and

$$W^*(\mathbf{N}, m_2(x)) = \min\{\frac{1}{2}\mathbf{N} : (\mathbf{A}^{-1}\mathbf{N}) | \mathbf{A} \in G_{m_2(x)}^{per}\} \ . \tag{28.4.2}$$

The above potential can be explicitly found by the translation method with the same translation matrix as used in Sec. 26.3. But here the differential conditions concealed in the set $\mathcal{S}_1(\Omega)$ are different, which makes the interval of admissible parameters β larger.

More specifically, the condition (26.3.2) with $\mathbf{d}$ replaced by $\mathbf{A}^{-1}$ is the only condition imposed on β. This point is crucial since instead of (26.3.16) we have now $\beta \in [-\mathcal{K}_2, \mathcal{L}_2]$, where $\mathcal{K}_\alpha$ and $\mathcal{L}_\alpha$ define the tensors $\mathbf{A}_\alpha$ by: $\mathbf{A}_\alpha^{-1} = \frac{1}{2}\mathcal{K}_\alpha \mathbf{I}_1 + \frac{1}{2}\mathcal{L}_\alpha \mathbf{I}_2$. Consequently the condition (26.3.26) still holds; here ζ_M is replaced with ζ_N. The definition of ζ_2 remains valid but the formula for ζ_1 is now different. It reads

$$\zeta_1 = \frac{\mathcal{K}_2 + [\mathcal{L}]_m}{m_2 \Delta \mathcal{L}} \ . \tag{28.4.3}$$

The potential $W^*(\mathbf{N}, m_2)$ is given by

$$W^*(\mathbf{N}, m_2) = \begin{cases} \dfrac{1}{2}I^2(\mathbf{N})\mathcal{H}(\zeta_N) & \text{if} \quad I(\mathbf{N}) \neq 0 \\[2ex] \dfrac{1}{2}\check{L}\, II^2(\mathbf{N}) & \text{if} \quad I(\mathbf{N}) = 0 \ , \end{cases} \tag{28.4.4}$$

where

$$\check{L} = \frac{\mathcal{L}_1\mathcal{L}_2 + \mathcal{K}_2\langle\mathcal{L}\rangle_m}{\mathcal{K}_2 + [\mathcal{L}]_m} \tag{28.4.5}$$

and the function $\mathcal{H}(\zeta)$ is given by (26.3.34) with $\mathcal{H}_L(\zeta)$ and $\mathcal{H}_i(\zeta)$ defined by (26.3.35) and (26.3.37), respectively. However, the function $\mathcal{H}_R(\zeta)$ is defined differently, as follows

$$\mathcal{H}_R(\zeta) = \mathcal{K}_2 + \check{L}\zeta^2 \ . \tag{28.4.6}$$

With the use of the formula in Sec. 25 one can show that the regimes $\zeta_N \leq \zeta_2$ and $\zeta_N \geq \zeta_1$ are realized by second rank ribbed plates and the sliding regime $\zeta_N \in [\zeta_2, \zeta_1]$ is realized by first rank ribbed plates. *In all regimes the directions of principal stress resultants, principal strains and directions of ribs coincide.* The characteristics of these microstructures can be identified by the methods of Sec. 26.4 with the use of the formulae in Sec. 25.

Remark 28.4.1. The results (26.3.44) and (28.4.4) differ in the regime $\zeta \geq \zeta_1$; recall that ζ_1 is defined differently for $\zeta = \zeta_M$ and $\zeta = \zeta_N$. In this regime the torsion (or shear) prevails. The origin of this discrepancy lies in the Kirchhoff assumptions which reduce the number of kinematic variables and, consequently, the number of equilibrium equations. The translation method of bounding the potential $\mathcal{W}^*$, as a method sensitive to the type of differential equations governing the problem, yields different results. Thus let it be emphasized here that the final layouts of the two-phase plates in bending are incomparable with the optimal layouts for the in-plane problems of two-phase plates, not because of differences in the loading, but because of essential differences in their optimal behavior. $\square$

Remark 28.4.2. The bending-membrane analogy (discussed in Sec. 23.4) interrelates the maximum compliance designing of bending plates with the minimum compliance design- ing of plates subject to in-plane loads, and vice versa. Thus the result (28.4.4) can be wrung from the upper bound of the potential of a two-phase plate in bending. Such an upper bound can be found by using the translation method for the plate potential expressed in terms of strains, see Gibianskii and Cherkaev (1984). On the other hand, Eq. (26.3.44) can be read off from the lower bound of the potential of a two-phase plate loaded in the plane, as found by Allaire and Kohn (1993b, 1993c, 1994). Let us acknowledge Kohn's private communication: Kohn (1993).

However, the bending-membrane analogy may be misleading (see comments in Sec. 23.4) and that is why the reader is encouraged to rederive all the energy bounds directly, by using the translation method. $\square$

Let us outline briefly the in-plane shape optimization problem. As in Sec. 26.7, one should confine to the case of a boundary loading. We have at our disposal one material of flexibilities $\mathcal{K}_2 = \mathcal{K}$ and $\mathcal{L}_2 = \mathcal{L}$ and of area fraction $m_2 = \theta$. The potential $\mathcal{W}^*(N, \theta)$ assumes the form

$$\mathcal{W}^*(N, \theta) = \mathcal{W}_0^*(N) + \frac{1 - \theta}{2\theta} g(N) , \tag{28.4.7}$$

with

$$g(N) = \frac{1}{4}(\mathcal{K} + \mathcal{L})(|\lambda_1(N)| + |\lambda_2(N)|)^2 , \tag{28.4.8}$$

where $\lambda_1(N) = N_I$, $\lambda_2(N) = N_{II}$. The optimization problem assumes the form (26.6.7b) with $F_\lambda(N)$ given by (26.7.15). Only the function $g(N)$ is defined differently here. Let us formulate now a counterpart of Theorem 26.7.1.

Theorem 28.4.1. (Allaire and Kohn (1993a)) The shape optimization problem formulated above admits at least one solution.

Proof. Polyconvexity of $F_\lambda(N)$ defined here follows from polyconvexity of the function $\eta(N)$ given by (26.7.20) in the case of $\alpha = 0$ and $\beta = 1$. $\square$

Let us consider now the shape design associated with very small values of the volume imposed. Then the relevant Lagrangian multiplier λ attains great values and the main term of F_λ assumes the form

$$F_\lambda(\boldsymbol{N}) = \sqrt{\lambda}\sqrt{\mathcal{K} + \mathcal{L}}(|\lambda_1(\boldsymbol{N})| + |\lambda_2(\boldsymbol{N})|) , \qquad (28.4.9)$$

known from the theory of Michell continua, see Sec. 29 and Strang and Kohn (1983). The optimization problem has the form (26.9.3) with $\mathcal{S}_2(\Omega)$ replaced by $\mathcal{S}_1(\Omega)$. Thus the Michell problem for the in-plane problem is similar to the bending problem of optimal grillages. The optimal Michell structures are composed of infinitely thin rods. That is why they can be alternatively derived from the discrete formulation or within the theory of rod structures. This approach is presented in Sec. 29.1, where the integrand (28.4.9) is obtained in a completely different manner.

29.   Truss-like Michell continua

Consider the following fundamental problems of engineering design:

(P_a): given two point supports A and B, one of which sliding, find the lightest frame structure capable of carrying a concentrated force P (perpendicular to the line AB) applied to the mid-point C, cf. Fig. 29.1.

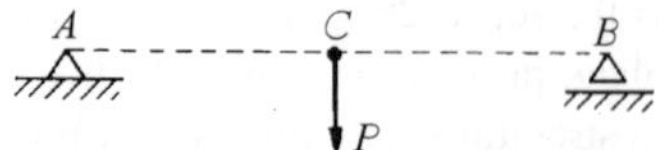

Fig. 29.1. How to transmit the force P to the point supports at A and B?

(P_b): solve problem (P_a) under the condition that the whole structure lies on the upper side of the line AB.

First, one can prove that the solutions do not take forms of frames (or trusses) of finite number of bars. If we add a bar and assume smaller cross sections, the weight can diminish. To figure out further consideration it is helpful to look at the exact solutions to the problem (P_a) and (P_b), see Fig. 29.2. They were found by Michell (1904).

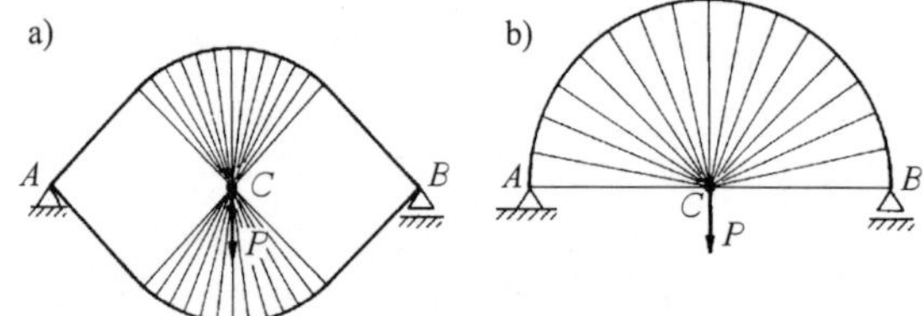

Fig. 29.2. Michell's solutions (1904). Feasible domains: a) Ω-the whole plane
b) Ω-the upper half plane.

Note that the optimum structures consist of bars of finite cross sections (thicker lines) and of fans of infinitely thin bars which resemble spokes of wheels of a bicycle.

Although 94 years elapsed since Michell discovered these solutions, the theory of this fundamental optimum design problem is far from being completely resolved.

29.1.   *Structures of minimum weight. Discrete versus continuum formulations*

The most effective bar structures are trusses since in each member, the cross-sections are uniformly loaded. In the uniformly stressed trusses we have

$$\sigma_p = \frac{|N_1|}{A_1} = \frac{|N_2|}{A_1} = \ldots \frac{|N_m|}{A_m} , \tag{29.1.1}$$

where N_j represents the axial force in the jth member; A_j stands for the area of the jth cross section. The critical stress σ_p is assumed to have the same absolute value for tension and compression. The axial forces in members equilibrate a given set of forces (P_I), $I =$

$1, \ldots, N$. The equilibrium equations have the form

$$\sum_{j=1}^{m} G_{jI} L_j N_j = P_I , \qquad (29.1.2)$$

where L_j represents the length of the jth member and (G_{jI}) are components of the geometric matrix of the truss.

The weight of the truss is given by

$$\Phi_T = \varrho \sum_{j=1}^{m} A_j L_j , \qquad (29.1.3)$$

where ϱ is the material density. Taking into account (29.1.1) one can express the weight in terms of (N_j)

$$\Phi_T = k \sum_{j=1}^{m} |N_j| L_j , \qquad (29.1.4)$$

where $k = \varrho / \sigma_p$. The problem of finding the lightest truss to be placed in a given domain Ω assumes the form

$$(P_D) \quad | \ \min\{\Phi_T| \ N_j \ \text{satisfy (29.1.2) and the joints lie within } \Omega\}$$

The constitutive relations are not postulated. Thus the problem (P_D) concerns the statically determined trusses for which the matrix (G_{jI}) is quadratic and invertible.

Numerical experiments show that increasing the number of bars usually makes the weight smaller. This suggests admitting from the very beginning that the bars can be infinitely thin and form a dense network. For simplicity, the two-dimensional or plane problems will be dealt with. Because of dense distribution of bars we treat them as embedded into a two-dimensional continuum that can be identified with a thin in-plane loaded plate of thickness $2c$ in which a plane-stress state prevails. The membrane stress resultants $N_h^{\alpha\beta}$ equal $2c\sigma_h^{\alpha\beta}$ and $\sigma_h^{\alpha\beta}$ are stresses in the plate. The subscript h indicates that these are homogenized stresses which should not be misinterpreted with stresses in bars, since the latter are constant and equal σ_p. In the continuum description the equilibrium equations (29.1.2) are replaced with

$$\sigma_h^{\alpha\beta}{}_{,\beta} = 0 \quad \text{in } \Omega , \qquad \sigma_h^{\alpha\beta} n_\beta = p^\alpha \quad \text{on } \Gamma = \partial\Omega . \qquad (29.1.5)$$

Here n is a unit vector normal to Γ. Instead of concentrated loads we deal with the densities p^α of a boundary loading.

The bars can only be stressed or compressed, hence they must inevitably lie along trajectories of principal stresses $\lambda_1(\sigma_h), \lambda_2(\sigma_h)$, see Fig. 29.1.1. The notation $\lambda_\alpha(\sigma_h)$ is here better than $(\sigma_h)_I, (\sigma_h)_{II}$, usually used. The trajectories of the principal stress form an orthogonal coordinate system $\xi^1 = \alpha, \xi^2 = \beta$ with the local basis a_1, a_2. The Lamé parameters of the system: $A(\alpha, \beta)$ and $B(\alpha, \beta)$ are given by

$$A(\alpha, \beta) = (a_{11})^{1/2} , \quad B(\alpha, \beta) = (a_{22})^{1/2} , \qquad (29.1.6)$$

where $a_{\alpha\beta} = a_\alpha \cdot a_\beta$. The elementary arc lengths are $dL_1 = A d\alpha$ and $dL_2 = B d\beta$.

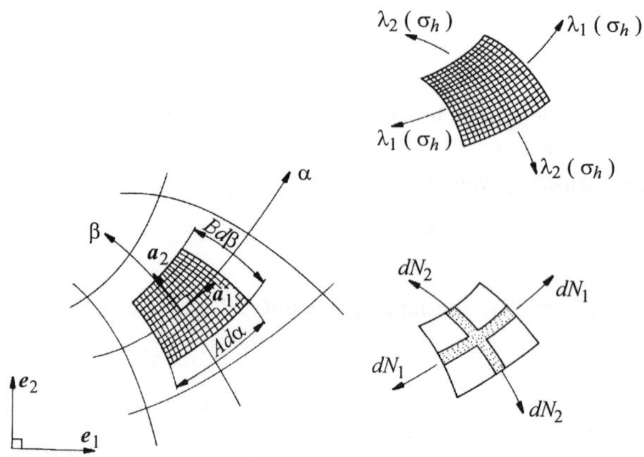

Fig. 29.1.1. From discrete to continuous description of a framework

The relations between the elementary forces dN_α and the averaged stresses $\lambda_\alpha(\sigma_h)$ assume the form

$$dN_1 = 2c\lambda_1(\sigma_h)Bd\beta , \qquad dN_2 = 2c\lambda_2(\sigma_h)Ad\alpha . \tag{29.1.7}$$

The weight of the truss equals to

$$\Phi = \int 2\varrho c(dA_1 dL_1 + dA_2 dL_2) , \tag{29.1.8}$$

where $2cdA_1$ and $2cdA_2$ are elementary cross sections of bars and the integral is taken over all such bars. From the conditions of uniform stress

$$|dN_1| = 2cdA_1\sigma_p , \quad |dN_2| = 2cdA_2\sigma_p \tag{29.1.9}$$

we express $2cdA_\alpha$ in terms of $|dN_\alpha|$ and rearrange Φ to the form

$$\Phi = k\int (|dN_1|dL_1 + |dN_2|dL_2) . \tag{29.1.10}$$

By (29.1.7) we express the weight by the integral over the domain Ω as follows

$$\Phi = 2ck\int (|\lambda_1(\sigma_h)| + |\lambda_2(\sigma_h)|)d\Omega , \tag{29.1.11}$$

where $d\Omega = ABd\alpha d\beta$. The function Φ thus derived represents the weight of a truss made from densely distributed bars formed along the trajectories of principal averaged stresses $\sigma_h^{\alpha\beta}$. The plane domain Ω will be called a feasible domain.

The counterpart of problem (P_D) reads

$$(P_C) \qquad \min\{\Phi|\ \sigma_h \text{ satisfies } (29.1.5)\} .$$

The passage from (P_D) (discrete formulation) to (P_C) (continuum formulation) may be viewed as convincing, but it is by no means rigorous. The future research should show that (P_C) is a correct relaxation of (P_D), the relaxation to be understood in an appropriate meaning, still weird.

Note that in the problem (P_b), the domain Ω is a half-plane, hence unbounded. The problem (P_a) exceeds the framework of the formulation (P_C), since there $\Omega = \mathbb{R}^2$ and the loading is concentrated inside Ω. In practical problems Ω is a given bounded domain, which renders the solution suboptimal.

Let us emphasize that the constitutive relations of the truss material have not been invoked.

29.2. Dual formulation

A formulation involving stresses is usually viewed as dual. In our case the situation is different: the dual formulation for (P_C) will involve Lagrange multipliers interpreted further as trial or virtual displacements. Most authors assume that $2c = 1$ and $k = 1$ and we shall follow this simplification. Moreover, the subscript "h" at σ_h will be omitted.

Let us rewrite the problem (P_C) in the form

$$(P_C) \qquad \min\{\Phi(\sigma)|\ \sigma \in \mathcal{S}_p(\Omega)\}\,,$$

where

$$\Phi(\sigma) = \int_\Omega (|\lambda_1(\sigma)| + |\lambda_2(\sigma)|)d\Omega\,, \qquad (29.2.1)$$

and

$$\mathcal{S}_p(\Omega) = \{\sigma|\ \mathrm{div}\,\sigma = 0\,,\ \sigma n = p \text{ on } \Gamma\}\,.$$

If $\sigma, \tau \in \mathcal{S}_p(\Omega)$, then $\widetilde{\sigma} = \theta\sigma + (1 - \theta)\tau$ satisfies $\mathrm{div}\widetilde{\sigma} = 0$ and $\widetilde{\sigma}n = p$ and hence $\widetilde{\sigma} \in \mathcal{S}_p(\Omega)$. Thus $\mathcal{S}_p(\Omega)$ is convex. Moreover, the integrand of Φ is convex, which renders the problem (P_C) convex.

Despite this nice property the problem (P_C) is highly nontrivial, since the integrand of Φ is: i) nondifferentiable, ii) has a linear growth, cf. Sec. 26.9. These two properties imply singularities in the optimal solutions: there can appear bars of finite cross-sections, embracing some regions of densely distributed infinitely thin bars. Along the bars of finite cross-sections, the principal stresses become concentrated to form axial forces. The discrete background (P_D) is thus partly revealed.

Prior to passing to the dual formulation let us prove the following lemma.

Lemma 29.2.1 Let $\tau \in \mathbb{E}_2^s$ and γ_1, γ_2 be arbitrary numbers. Let

$$I(\gamma_1, \gamma_2) := \min_{\tau \in \mathbb{E}_2^s} \{|\lambda_1(\tau)| + |\lambda_2(\tau)| - \gamma_1\tau^{11} - \gamma_2\tau^{22}\}\,, \qquad (29.2.2)$$

where $(\tau^{\alpha\beta})$ refer to the Cartesian basis $e_\alpha \otimes e_\beta$. Then

$$I(\gamma_1, \gamma_2) = \begin{cases} 0 & \text{if } |\gamma_1| \leq 1 \text{ and } |\gamma_2| \leq 1\,, \\ -\infty & \text{otherwise.} \end{cases} \qquad (29.2.3)$$

In case of $|\gamma_\alpha| \leq 1$ the minimum is attained at

$$\tau^{11} = \tau^{22} = 0 \qquad \text{if} \quad |\gamma_\alpha| < 1 \,,$$
$$\tau^{11} = K_1\gamma_1 \,, \qquad \tau^{22} = K_2\gamma_2 \,, \qquad \text{if} \quad |\gamma_\alpha| = 1 \,, \ \alpha = 1,2 \,,$$

where $K_\alpha \geq 0$.

Proof. Take $\boldsymbol{\tau} = K\gamma_1 \boldsymbol{e}_1 \otimes \boldsymbol{e}_1$, $K \geq 0$. Then $|\lambda_1(\boldsymbol{\tau})| = K|\gamma_1|$, $\gamma_1\tau^{11} = K|\gamma_1|^2$.
Thus

$$I(\gamma_1, \gamma_2) \leq \min_{K \geq 0} K|\gamma_1|(1 - |\gamma_1|) \,,$$

which implies

$$I(\gamma_1, \gamma_2) \leq \begin{cases} 0 & \text{if } |\gamma_1| \leq 1 \,, \\ -\infty & \text{otherwise.} \end{cases}$$

Similarly we prove that

$$I(\gamma_1, \gamma_2) \leq \begin{cases} 0 & \text{if } |\gamma_2| \leq 1 \,, \\ -\infty & \text{otherwise.} \end{cases}$$

On the other hand, the expression

$$|\lambda_1(\boldsymbol{\tau})| + |\lambda_2(\boldsymbol{\tau})| - \gamma_1\tau^{11} - \gamma_2\tau^{22}$$

is non-negative if $|\gamma_\alpha| \leq 1$, which completes the proof. $\qquad\square$

Let us pass now to the dual formulation of (P_C). We introduce a Lagrangian multiplier $\boldsymbol{u} = (u_1, u_2)$

$$\Phi = \min_{\boldsymbol{\sigma}\boldsymbol{n}=\boldsymbol{p}} \ \max_{\boldsymbol{u}} \int_\Omega [|\lambda_1(\boldsymbol{\sigma})| + |\lambda_2(\boldsymbol{\sigma})| + u_\alpha\sigma^{\alpha\beta}_{,\beta}]d\Omega \,.$$

Hence

$$\Phi = \min_{\boldsymbol{\sigma}} \ \max_{\boldsymbol{u}}[\int_\Omega [|\lambda_1(\boldsymbol{\sigma})| + |\lambda_2(\boldsymbol{\sigma})| - \sigma^{\alpha\beta}e_{\alpha\beta}(\boldsymbol{u})]d\Omega + \int_\Gamma p^\alpha u_\alpha d\Gamma] \,, \qquad (29.2.4)$$

where $e_{\alpha\beta}(\boldsymbol{u}) = \dfrac{1}{2}(u_{\alpha,\beta} + u_{\beta,\alpha})$ and $(\ )_{,\alpha} = \partial/\partial x_\alpha$. By convexity we can interchange min and max to find

$$\Phi = \max_{\boldsymbol{u}}[\int_\Omega g(e(\boldsymbol{u}))d\Omega + \int_\Gamma p^\alpha u_\alpha d\Gamma] \,, \qquad (29.2.5)$$

where

$$g(\boldsymbol{\epsilon}) = \min_{\boldsymbol{\sigma}\in\mathbf{E}^s_2}(|\lambda_1(\boldsymbol{\sigma})| + |\lambda_2(\boldsymbol{\sigma})| - \sigma^{\alpha\beta}\epsilon_{\alpha\beta}) \,. \qquad (29.2.6)$$

Let us now refer $\boldsymbol{\sigma}$ to the new orthonormal basis: $\boldsymbol{\sigma} = \sigma^{\alpha'\beta'}\boldsymbol{e}'_\alpha \otimes \boldsymbol{e}'_\beta$ such that $\boldsymbol{e}'_\alpha$ coincide with the principal directions of $\boldsymbol{\epsilon}$. Then

$$g(\epsilon) = \min_{\sigma \in \mathbf{E}_2^s}\{|\lambda_1(\boldsymbol{\sigma})| + |\lambda_2(\boldsymbol{\sigma})| - \lambda_1(\epsilon)\sigma^{1'1'} - \lambda_2(\epsilon)\sigma^{2'2'}\} \,. \tag{29.2.7}$$

Now we use Lemma 29.2.1 and find

$$g(\epsilon) = \begin{cases} 0 & \text{if } |\lambda_1(\epsilon)| \le 1 \,,\ |\lambda_2(\epsilon)| \le 1 \\ -\infty & \text{otherwise.} \end{cases} \tag{29.2.8}$$

In case of $|\lambda_1(\epsilon)| \le 1$ and $|\lambda_2(\epsilon)| \le 1$ the minimum is attained at

$$\begin{aligned} \sigma^{1'1'} = \sigma^{2'2'} = 0 \quad &\text{if } |\lambda_1(\epsilon)| < 1 \,,\ |\lambda_2(\epsilon)| < 1 \\ \sigma^{1'1'} = K_1\lambda_1(\epsilon) \,, \qquad \sigma^{2'2'} = K_2\lambda_2(\epsilon) \,, \quad &\text{if } |\lambda_\alpha(\epsilon)| = 1 \,, \end{aligned} \tag{29.2.9}$$

and $K_\alpha \ge 0$. Taking account of these results in (29.2.5) we arrive at the problem dual to (P_C)

$$(P_C^*) \quad \max\{\int_\Gamma p^\alpha u_\alpha d\Gamma |\ |\lambda_1(\epsilon(\boldsymbol{u}))| \le 1 \,,\ |\lambda_2(\epsilon(\boldsymbol{u}))| \le 1\} \,.$$

The reader is advised to derive this dual problem by using Rockafellar's theory of duality. The principal strains are roots of the equation

$$\lambda^2 - \lambda \operatorname{tr} e(\boldsymbol{u}) + \det e(\boldsymbol{u}) = 0 \,.$$

We know from algebra that the conditions $|\lambda_\alpha| \le 1$ are satisfied if and only if

$$\det e(\boldsymbol{u}) \le 1 \text{ and } |\operatorname{tr} e(\boldsymbol{u})| \le 1 + \det e(\boldsymbol{u}) \,, \tag{29.2.10}$$

which renders the problem (P_C^*) formulated in terms of $\operatorname{tr} e(\boldsymbol{u})$ and $\det e(\boldsymbol{u})$. When only some part of Γ is loaded and some of it serves as a support, then $\boldsymbol{u}$ should vanish somewhere on Γ. In the problem considered the part of Γ on which the structure is supported is a priori unknown.

According to (29.2.9) the bars lie along the lines $|\lambda_\alpha(e(\boldsymbol{u}))| = 1$, since there the stresses do not vanish. Thus there are two families of bars that form a curvilinear orthogonal network.

The structures related to the condition:

$$\lambda_1(e(\boldsymbol{u})) = \lambda_2(e(\boldsymbol{u})) = \pm 1$$

were first discussed by Maxwell, cf. Prager (1978). Such structures can be made from a finite number of bars.

Other family of structures related to the case:

$$\lambda_1(e(u)) = 1 , \quad \lambda_2(e(u)) = -1 \tag{29.2.11}$$

were first discovered by Michell (1904), two of them being given in Fig. 29.2, cf. Hemp (1973). These structures as two-dimensional discrete-continuous bodies go beyond the framework of both the truss theory and the plane-stress theory. The Michell structures are formed partly from infinitely thin bars and partly from bars of finite cross-sections. Note, however, that the formulations (P_C) and (P_C^*) exclude the latter case. Indeed, some particular Michell structures are known, but this knowledge did not help up till now to find a rigorous relaxation of the problem (P_D). The continuum formulations (P_C) and (P_C^*) are too narrow to encompass the case of stress concentrations along the lines (or the sets of dimension lower than the dimension of Ω).

Consider a special case of Ω being an exterior of a convex domain Ω_1. Let a force P be prescribed at a point $A \in \Omega$, cf. Fig. 29.2.1. Let us require that a structure is supported on a part of $\partial\Omega_1$ and subject to the force P at point A. Then one can show (cf. Hegemier and Prager, 1969) that at most one field u satisfying (29.2.11) exists. Thus, according to (P_C^*), the minimum weight of such an optimal cantilever is equal to $P \cdot u(A)$. The problem (P_C^*) is reduced to finding a kinematically admissible field u.

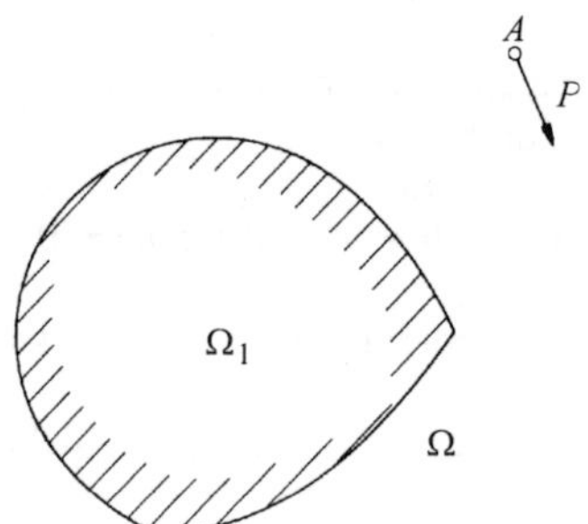

Fig. 29.2.1. How to transmit the force P to the boundary $\partial\Omega_1$?

29.3.   *A symmetric cantilever problem*

Assume that Ω is a strip of given width h. Our task is to find the lightest cantilevers capable of carrying the vertical force P taking a position along the symmetry line. The cantilever should be fully or partly supported along its left lateral edge, cf. Fig. 29.3.1a. Our aim is to report the shapes of the lightest cantilevers for various ratios of $\xi = x/h$, where now x represents the distance of the force to the support.

If $\xi \le 0.5$, the optimal cantilever consists of two bars, see Fig. 29.3.2b,c. If $0.5 < \xi < 1.82196$ three new regions occur, two of them being fans and the third being a Michell truss formed of two families of bars subject to compression and tension, see Fig. 29.3.2d,e.

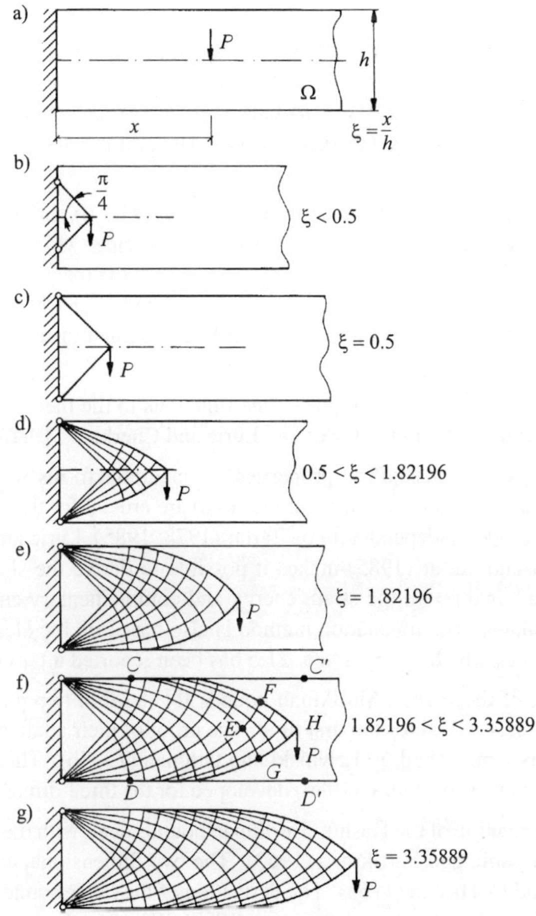

Fig. 29.3.1. Family of optimal cantilevers for various ratios of $\xi = x/h$.
After Lewiński et al. (1994a)

If $1.82196 < \xi \le 3.35889$, three new domains occur. In regions DEG, CEF, one family of bars is tangent to the lines DD' and CC' respectively. The solutions (d) and (e) were found by A.S.L. Chan (1960). The regions CEF and DEG were first analyzed by H.S.Y. Chan (1967). Extension to the $EFHG$ region has been developed in Lewiński et al. (1994a), where all details of the solutions $(b) - (g)$ can be found.

30. Comments and bibliographical notes

Many complicated computations involving the second rank tensors can be simplified by referring the tensors to the basis (21.1.2). The first applications of this basis can be found in the papers by Lurie et al. (1982), Rykhlevskii (1984) and Gibianskii and Cherkaev (1984, 1987), see also Lurie and Cherkaev (1986). The preliminary Secs. 21.1 – 21.5 have been inspired by these works.

The importance of the fraction-linear transformation (21.2.6), called Y-transformation, was stressed for the first time by Milton (1991). In this paper Milton also gives credit to an unpublished work of Gibianskii and Cherkaev in which the inequality (21.2.5) appeared for the first time. The proof of Lemma 21.2.2 is repeated after Cherkaev and Gibiansky (1993). Y-transformation is closely related to the Stieltjes function and Padé approximants, see Tokarzewski and Telega (1997).

Applications of quasiconvex and quasiaffine functions to the theory of optimum design can be found in Kohn and Strang (1986) and Lurie and Cherkaev (1986).

The quasiconvex and quasiaffine properties of quadratic forms involving stresses or strains (and similary, moments or curvature tensors) are crucial for the translation method. This method, developed independently by Tartar (1978, 1985), Lurie and Cherkaev (1984, 1986) and Murat and Tartar (1985) makes it possible to derive the sharp bounds (or the bounds that are attainable) on the strain energy and complementary energy of the elastic composites and plates. The translation method is also based on the elementary inequality (21.5.4), the proof of which given in Sec. 21.5 has been reported after Gibiansky (1993).

The derivation of the Hashin-Shtrikman bounds for the thin two-phase elastic plate is reported in Secs. 22.1 – 22.4 for the first time. The proof of their attainability given in Sec. 22.5 was previously published by Lewiński and Othman (1997b). This proof is based on the method of Francfort and Murat (1986) developed for the three-dimensional composites.

The original derivation of the Hashin-Shtrikman bounds dealt with the three-dimensional composites, see Hashin and Shtrikman (1963). The two-dimensional counterpart of these bounds were found by Hashin (1965). The derivation of the latter bounds presented in Sec. 23 is reported after Cherkaev and Gibiansky (1993). The proof of attainability given in Sec. 23.3 is inspired by the method of Francfort and Murat (1986) and was published in Lewiński and Othman (1997a).

Cherkaev and Gibiansky (1993) showed that by estimating the sums of strain and complementary energies, associated with independent strain and stress fields, one can find new coupled bounds for the bulk and shear moduli of the effective isotropic two-phase mixture. The explicit results concern the two-dimensional case. They turn out to be sharper than the coupled bounds found previously by Milton and Phan-Thien (1982) and by Berryman and Milton (1988).

A mixture of isotropic phases can have anisotropic properties. The bounds for these properties were reported by Milton and Kohn (1988) and Jikov et al. (1994, Chap. 13).

The counterparts of the Hashin-Shtrikman bounds for the thermal expansion moduli were found by Schapery (1968) and by Rosen and Hashin (1970). These bounds were recently substantially tightened by using the translation method by Gibiansky and Torquato (1997). These results were confirmed numerically by Sigmund and Torquato (1997).

The whole Sec. 26 concerns the minimum compliance problem for thin elastic two-phase plates in bending. By similar methods one can analyze the same optimization problem for the plates subject to an in-plane loading. Although both these problems have recently found their solutions some fundamental aspects remain unclear up till now. Before commenting these points let us recall much simpler second order problem which is now fully resolved:

$$(P_1) \qquad \max\{J(\chi_2) = \int_\Omega w(x)\,dx\}\,,$$
$$\chi_2 \in L^\infty[\Omega, \{0,1\}]$$

where w is a solution to the problem:

$$\left| \begin{aligned} &\text{find } w \in H_0^1(\Omega) \text{ such that} \\ &\int_\Omega D^{\alpha\beta}(x)w_{,\alpha}v_{,\beta}\,dx = \int_\Omega v\,dx \qquad \forall v \in H_0^1(\Omega)\,, \end{aligned} \right.$$

with $D^{\alpha\beta}(x) = D(x)\delta^{\alpha\beta}$, $D(x) = \chi_1(x)D_1 + \chi_2(x)D_2$, D_α are positive constants and

$$\int_\Omega \chi_2\,dx = C\,,$$

where C is prescribed. One of possible interpretation is that J represents a torsional stiffness of a two-phase bar of cross section Ω; the amount of both phases is fixed, cf. also Lavrov et al. (1980).

The problem (P_1) is ill-posed since the maximizing sequences $(\chi_2)_n$ tend to functions of the space $L^\infty(\Omega, [0,1])$. Consequently, the boundary between the domains in which $\chi_1 = 1$ and $\chi_2 = 0$ becomes a generalized curve of highly oscillating shape. To make the solution regular one can impose constraints on a Lipschitz constant of this curve, see Pironneau (1984, Chap. 3), or reformulate the problem to a relaxed form.

The relaxation can be done by two equivalent but methodologically different methods. In the first method the problem is directly convexified, see Kohn and Strang (1986) and Goodman et al. (1986). In the second method, developed e.g. by Lurie and Cherkaev (1986) the homogenization arguments are invoked. This last method consists in admitting all composite mixtures of the phases. This method is constructive since the G-closure of all effective tensors $D = (D^{\alpha\beta})$ characterizing such composites is known. This set was determined by Lurie and Cherkaev (1984), Tartar (1985) and Murat and Tartar (1985) and can be explicity expressed in terms of the two eigenvalues of D. Having the complete characterization of all effective tensors D one can put the relaxed formulation in the explicit form involving the microstructural characteristics as new design variables. In the problem considered it is sufficient to admit the laminates of first rank. If we perform maximization over the parameters characterizing such laminates we recover the formulation due to Goodman et al. (1986). The numerical procedures can start from this formulation or from the previous one, involving microstructural design variables. The second method has become popular later, when applied to elasticity problems, see the detailed historical comments in the monograph by Bendsøe (1995) and the review paper by Lewiński (1993).

The case $D_1 = 0$ refers to a shape optimization problem, The shape optimization problem can be in general put in the form

$$\min\{F(A)|\ A \in \mathcal{A}\}\,,$$

where $\mathcal{A}$ is a family of sets and F is a function defined on $\mathcal{A}$. To make such problems well-posed we either impose constrains on Lipschitz constants of ∂A (see Pironneau (1984, Chap. 3)) or enlarge the class $\mathcal{A}$. Assume that $\mathcal{A}$ is a family of open subsets A of a given bounded set $\Omega \subset \mathbb{R}^n$ such that $|A| = C$ and $F(A)$ depends on the solution of an elliptic second order partial differential equation on A. Buttazzo and Dal Maso (1993) proved that the problem is properly relaxed if the set $\mathcal{A}$ is augmented with the so-called quasi-open subsets A of Ω such that $|A| = C$. The generalized solution exists, if F is decreasing with respect to set inclusions and that F is lower semicontinuous with respect to γ-convergence, or convergence of resolvents of the Laplace operator with Dirichlet boundary conditions.

Recently, Bucur et al. (1998) extended the results due to Buttazzo and Dal Maso (1993) to the following optimization problem:

$$\min\{J(A_1,\ldots,A_k)|\ A_i \in \mathcal{A}(\Omega)\,,\qquad A_i \cap A_j = \varnothing\ \ \text{for}\ \ i \neq j\}\,,$$

where k is a fixed positive integer, Ω is a given bounded open set of $\mathbb{R}^n$ and $\mathcal{A}(\Omega)$ is the class of admissible domains. Obviously, $J\colon \mathcal{A}(\Omega)^k \to [0,\infty]$ is the cost functional. For $k = 1$ one recovers the general shape optimization problem investigated by Buttazzo and Dal Maso (1993).

In contrast to the second order problem (P_1) the fourth order plate bending problem (26.1.15) discussed in Sec. 26 has been less investigated since up till now the G-closure of all effective tensors $\boldsymbol{D} = (D^{\alpha\beta\lambda\mu})$ is unknown. There is only one known method of relaxation and it is based on the homogenization approach. Although the complete characterization of the G-closure is unknown, one can prove that the optimum is realized by orthogonal second rank laminates. This proof was first reported by Gibiansky and Cherkaev (1984); a similar proof is presented in Secs. 26.3, 26.4 with all necessary details. To prove attainability of the translation bound (26.3.43) we have used the closed formulae for effective stiffnesses of second rank ribbed plates, derived in Sec. 24 by using the two-dimensional counterpart of the formula of Francfort and Murat (1986), see Eq. (3.8.35). The translation method does not require the ordering assumption (26.1.5), which was admitted for the sake of simplicity.

The attainability of the energy bound (26.3.43) can be proved more systematically by using the Hashin-Shtrikman variational principle. This has been shown in Allaire and Kohn (1993b, 1993c, 1994) in the context of the two-dimensional elasticity problem.

Let us repeat after Remark 26.4.1 that the regime $\zeta_M \leq \zeta_2$ (see the domains O_{31} and O_{32} in Fig. 26.4.1) can be realized by both the second rank microstructure and the Vigdergauz inclusions, see Vigdergauz (1986, 1994) and Grabovsky and Kohn (1995). Whether other microstructures than second rank ribbed plates saturate the bound (26.4.1) for $\zeta_M > \zeta_2$ is an open problem. Up till now we know only that certain curved quadrilateral inclusions are optimal if their distribution is dilute, see the paper by Cherkaev et al. (1998), concerning the case of plane strain.

In its final form the relaxed formulation (26.5.1) is similar to an equilibrium problem of a physically nonlinear plate, expressed in terms of moments or dual variables. The passage to primal variables can be performed by applying Rockafellar's theory of duality, which is accomplished in Sec. 26.6. We prove there that the primal formulation can be found without using the theory of Young measures, which was suggested by Lipton (1994b). On the other hand, the inversion of the nonlinear constitutive relation (26.5.3) can be performed directly, by making use of the fact that the principal directions of the strain tensor and moment tensor coincide, see Sec. 26.6.

The optimal designs of two-phase plates formed on a square, reported in Sec. 26.8, were found by the numerical algorithm developed by Lewiński and Othman (1997 b). Further details and examples can be found in Othman (1997).

The classical, yet rigorous techniques of solving the shape optimization problems have been presented in the monograph by Sokolowski and Zolesio (1992).

It has turned out recently that the layout problem of mixing two materials can be rearranged to the shape optimization problem by passing to zero with the moduli of the weaker phase. The crucial problem is whether two operations: relaxation and degenerating the weaker phase commute. The affirmative answer was given by Allaire and Kohn (1993a). This justifies a formal derivation of Sec. 26.7, based on the assumption (26.7.1) which transforms the weaker phase into voids. The recent papers by Allaire et al. (1996, 1997) generalize the setting of the two-dimensional shape optimization problem to the three-dimensional shape problem for the elastic bodies.

The elastic structures and, in particular, the plates are usually subjected to various types of loading. A design optimal with respect to one loading is usually impractical for a different loading. The simplest method of taking into account that the plate should be capable of carrying various surface loadings $p_i(x)$, $i = 1, \ldots, N$, $x \in \Omega$ is to demand that the weighted sum of the compliances is minimal. Let ρ_i be weighted factors. Thus we have the multiple load minimum compliance problem (see Krog and Olhoff, 1997; Lipton, 1994b)

$$\min \sum_{i=1}^{N} \rho_i J_i(\chi_2) , \qquad J_i(\chi_2) = \int_\Omega p_i(x) w_i(x) \, dx ,$$

with $w_i \in V(\Omega)$ being a solution to the problem (see (26.1.10))

$$a_{\boldsymbol{D}}(w_i, v) = \int_\Omega p_i v \, dx , \qquad \forall \, v \in V(\Omega) ,$$

with

$$\int_\Omega \chi_2 \, dx = C \,.$$

The tensor D is given by (26.1.2).

Avellaneda and Milton (1989) proved that the relaxed formulation for the multiple load case should involve composite regions generated by the ribbed plates of third rank. The higher rank ribbed composites are redundant. This result was recently generalized to the three dimensional case by Francfort et al. (1995), where the proof is given that the ribbed (or laminated) plates of sixth rank should be introduced. Note that these numbers: 3 and 6 appear in the theorem by Francfort and Murat (1986): 3 (in the plane case) and 6 (in the spatial case) subsequent laminations suffice to construct isotropic composites attaining the Hashin-Shtrikman extreme values of the effective moduli. Thus the condition of the higher-order laminate being able to achieve isotropic properties is decisive here.

We have stressed above that relaxing the minimum compliance problem in the shape design means accepting infinitely small holes in the final design. If we prevent the final design from having such degeneracy, we confine ourselves with suboptimal solutions. Recently two independent methods of finding such solutions have been developed. The first one is based on putting a constraint on the perimeter of all holes, see Ambrosio and Buttazzo (1993), Haber et al. (1996) and Beckers (1997). The second one admits appearing of finite voids and is called a bubble method, see Eschenauer et al. (1994). It has turned out recently, that the characteristic function of the bubble method can be rigorously derived by using the concept of the topological derivative, introduced by Sokołowski and Żochowski (1997). Its first applications in the optimization of shells formed on a sphere can be found in Lewiński and Sokołowski (1998).

The minimum compliance problem for elastic plates can be based on other plate models than the thin plate model. One can use, for instance, the Reissner-Hencky plate theory described in Sec. 5.1. The effective stiffness of composite domains can be determined by the in-plane scaling based homogenization (Sec. 5.2). Such a relaxed formulation was used by Bendsøe and Diaz (1993), Diaz et al. (1995) and Krog and Olhoff (1997), cf. also Lipton (1994c) and Lipton and Diaz (1997). Some comments on using the formulae for effective stiffnesses derived in Sec. 2 for making the minimum compliance problem correctly posed can be found in Kohn and Vogelius (1986b). This problem as well as the problem of possible applications of the formulae of Secs. 5.3 and 5.4 for relaxing the optimization problems of plates are not fully resolved up till now.

The optimization problem of Sec. 26 can be viewed as designing a plate whose thickness assumes two possible values: h_1 and h_2. A natural question is how to regularize the problem in which the thickness is allowed to assume all the values from the interval $[h_1, h_2]$. This topic is discussed in Lurie and Cherkaev (1986). The one dimensional case when $h(x) = h(x_1)$ has been solved by Bonnetier and Conca (1993, 1994) and Muñoz and Pedregal (1998). These results are discussed in Sec. 27 and involve in an essential manner Young measures. The theory of Young measures is sketched in Sec. 21.6. For more details the reader is referred to the papers by Ball (1989), Ball and Knowles (1990), Roubiček (1997), Valadier (1994) and to the nice book by Pedregal (1997).

The problems of minimizing other functionals than the plate compliance are less investigated. One can minimize, for instance, the virtual work

$$J_1(w) = \int_\Omega pw\,dx\,,$$

where p is not a loading. One can also minimize

$$J_2(w) = \sup_{x \in \Omega} |w(x)|\,, \qquad J_3(w) = \int_\Omega w^2\,dx\,.$$

For such merit functions we conjecture that the relaxed formulation should involve more general composite plates: the orthogonal second rank ribbed plates will not suffice. At least such ribbed plates should be non-orthogonal, see Cherkaev (1994) and Lurie (1994).

On the other hand, instead of enlarging the design space one can shrink it to make the problem well posed. This concept was used by Myśliński and Sokołowski (1985) to minimize $J_2(w)$ among the plates of thickness $h(x)$ varying between h_1 and h_2.

The method of restricting the design space for the minimum weight design problem of plates of varying thickness h was recently proposed by Alvarez-Vázgez and Viaño (1997). There the following conditions on h are enforced:

$$\|\partial_\alpha h\|_{\infty,\Omega} \leq c_1\,, \qquad \|\partial_{\alpha\beta} h\|_{\infty,\Omega} \leq c_2$$

to make the design problem well-posed. For older references on this topic the reader is referred to Bendsøe (1995).

Theory of optimization of shells is less developed. The method based on the derivative of shape functionals is discussed in Khludnev and Sokołowski (1997) and Sokołowski (1996). The layout problem is considered in Tenek and Hagiwara (1994a, 1994b), where the formulae for effective stiffnesses are simplified for numerical aims. A general formulation of the minimum compliance problem for two-phase thin shells was presented for the first time by Lewiński and Telega (1997a) and is developed in Sec. 28. A mathematical justification of the dual homogenization formulae for thin shells is given in Telega and Lewiński (1998a, 1998b).

Consider the shape design of the plates loaded in-plane. If the volume diminishes, the designs tend to special forms called Michell trusses (or Michell continua). These optimal structures were discovered by Michell (1904). Their classical constructions are presented in Hemp (1973). In 1983, Strang and Kohn put forward a functional the minimization of which yields Michell's layouts. A mathematical justification of this functional has been recently presented by Allaire and Kohn (1993a) and Bendsøe and Haber (1993). This functional is derived in Sec. 29.1 by applying to the theory of gridworks. The dual formulation of Sec. 29.2 repeats the arguments of Strang and Kohn (1983). Since the classical framework does not allow for concentrated loads, Bouchitté et al. (1997a) proposed a general form of shape optimization problems, where for a load a measure is taken. The existence of an optimal measure for the total energy cost functional was shown.

The solution to the problem of transmitting a force to a straight rigid support (rigid wall) depends on the distance between the point at which the force is applied and the wall. For a short distance the optimal truss has only two bars. For longer distances we find the structures of Michell type, elaborated by A.S.L. Chan and H.S.Y. Chan, see Chan (1967). The longer cantilevers were found by Lewiński et al. (1994a, 1994b). Michell-type trusses of finite number of joints were found by Prager (1978), see also Sankaranarayanan et al. (1994) and Kołakowski and Holnicki-Szulc (1997), where the methods developed in the book by Holnicki-Szulc and Gierliński (1995) were used.

The Michell structures have their counterparts for the bending problem; then they are called Michell grillages. Their theory was developed by Rozvany and Gollub (1990).

The concept of shape design by admitting voids at a microscale was used and justified by Allaire and Kohn (1993a). This idea, along with the passage to the Michell continua is briefly outlined in Sec. 28.4. The counterpart of this problem concerning the bending case is considered in Sec. 26.7 with all necessary details because most of these results have not yet been published. The proof of Theorem 26.7.1 is constructed by the methods proposed by Kohn and Strang (1986, Part. I, Lemma 3.4), see also Allaire and Kohn (1993a, Remark 4.2) and Allaire et al. (1997, Eqs. (68), (69)).

In Sec. 26.9 a proof is given that admitting voids in a plate subjected to bending does not lead to the optimization problem of the least weight grillages, developed in the book by Rozvany (1976). The formulation (26.9.2) is new. One can solve the problem (26.9.2) by extending the results due to Demengel (1985). In the last paper the problem of existence of stresses in locking bodies was solved, see also Telega and Jemioło (1998). In our case, the problem dual to (26.9.2) will involve the locking condition imposed on $(\kappa_{\alpha\beta}(w))$, i.e., on the tensor of changes of curvature.

References

Aboudi J. (1987) Stiffness reduction of cracked solids, *Engrg. Fract. Mech.*, **26**, 637-650.

Acerbi E., Chiadò Piat V., Dal Maso G. and Percivale D. (1992) An extension theorem from connected sets and homogenization in general periodic domains, *Nonlinear Anal., Theory, Meth. Appl.*, **18**, 481-496.

Adams R.A. (1975) Sobolev Spaces, *Academic Press*, New York.

Aganović I., Marušić-Paloka E. and Tutek Z. (1996) Slightly wrinkled plate, *Asympt. Anal.*, **13**, 1-29.

Aganović I., Jurak M., Marušić-Paloka E. and Tutek Z. (1998) Moderately wrinkled plate, *Asympt. Anal.*, **16**, 273-297.

Alexandre R. (1997) Homogenization and θ-2 convergence, *Proc. R. Soc. Edinburgh*, **127A**, 441-455.

Alexiewicz A. (1969) Functional Analysis, *PWN*, Warszawa, in Polish.

Allaire G. (1992) Homogenization and two-scale convergence, *SIAM J. Math. Anal.*, **23**, 1482-1518.

Allaire G. (1997) Mathematical approaches and methods, Appendix A to: U. Hornung (ed.), *Homogenization and Porous Media, Springer*, New York, pp. 225-250.

Allaire G. and Kohn R.V. (1993a) Optimal design for minimum weight and compliance in plane stress using extremal microstructures, *Eur. J. Mech., A/Solids*, **12**, 839-878.

Allaire G. and Kohn R.V. (1993b) Optimal bounds on the effective behavior of a mixture of two well-ordered elastic materials, *Quart. Appl. Math.*, **51**, 643-674.

Allaire G. and Kohn R.V. (1993c) Explicit optimal bounds on the elastic energy of a two-phase composite in two space dimensions, *Quart. Appl. Math.*, **51**, 675-699.

Allaire G. and Kohn R.V. (1994) Optimal lower bounds on the elastic energy of a composite made from two non well-ordered isotropic materials, *Quart. Appl. Math.*, **52**, 311-333.

Allaire G., and Murat F. (1993) Homogenization of the Neumann problem with nonisolated holes, *Asympt. Anal.*, **7**, 81-95.

Allaire G., Belhachmi Z. and Jouve F. (1996) The homogenization method for topology and shape optimization. Single and multiple loads case, *Revue Europenne des Elements Finis.*, Vol. **5**, 649-672.

Allaire G., Bonnetier E., Francfort G. and Jouve F. (1997) Shape optimization by the homogenization method, *Numer. Math.*, **76**, 27-68.

Allen D. H. (1994) Damage evolution in laminates, in: *Damage Mechanics of Composite Materials*, ed. by R. Tarleja, *Elsevier*, Amsterdam, pp. 79-116.

Allen D., Harris C.E. and Groves S.E. (1987) A thermomechanical constitutive theory for elastic composites with distributed damage. Part I. Theoretical development. Part II. Application to matrix cracking in laminated composites, *Int. J. Solids Structures*, **23**, 1301-1318; 1319-1338.

Alvarez-Vazquez L.J. and Quintela-Estevez P. (1992) The effect of different scalings in the modelling of nonlinearly elastic plates with rapidly varying thickness, *Comp. Meth. Appl. Mech. Engrg.*, **96**, 1-24.

Alvarez-Vázquez L.J. and Viaño J.M. (1997) Modeling and optimization of non-symmetric plates, *Math. Modelling and Num. Anal.*, **31**, 733-763.

Amar M. (1998) Two-scale convergence and homogenization on $BV(\Omega)$, *Asympt. Anal.* **16**, 65-84.

Ambrosio L. and Buttazzo G. (1993) An optimal design problem with perimeter penalization, *Calcul. Variat. Partial. Diff. Eqs.*, **1**, 55-69.

Anzellotti G. (1986) On the minima of functionals with linear growth, *Rend. Sem. Mat. Univ. Padova*, **75**, 91-109.

Anzellotti G. and Giaquinta M. (1978) Funzioni BV e tracce, *Rend. Sem. Mat. Univ. Padova*, **60**, 1-21.

Anzellotti G., Baldo S. and Percivale D. (1994) Dimension reduction in variational problems, asymptotic development in Γ-convergence and thin structures in elasticity, *Asympt. Anal.*, **9**, 61-100.

Arnold D.N. and Falk R.S. (1987) Well-posedness of the fundamental boundary value problems for constrained anisotropic elastic materials, *Arch. Rat. Mech. Anal.*, **98**, 143-165.

Artola M. and Duvaut G. (1977) Homogénéisation d'une plaque renforcée, *C. R. Acad. Sci. Paris, Série A*, **284**, 707-710.

Artola M. and Duvaut G. (1978) Homogénéisation d'une plaque renforcée par un système périodique de barres curvilignes, *C. R. Acad. Sci. Paris, Série A*, **286**, 659-662.

Attouch H. (1984) Variational Convergence for Functions and Operators, *Pitman Adv. Publ. Program*, Boston-London-Melbourne.

Attouch H. and Murat F. (1985) Homogenization of fissured materials, *Publ. AVAMAC, Université de Perpignan*, No 85-03.

Aubin J.P. and Cellina A. (1984) Differential Inclusions, *Springer-Verlag*, Berlin.

Aubin J.P., Frankowska H. (1990) Set-Valued Analysis, *Birkhäuser*, Boston.

Azé D. (1984) Epi-convergence et dualité. Applications à la convergence des variables primales et duals pour der suites convexe, *Publications AVAMAC, Université de Perpignan*, No 84-12.

Azé D. (1986) Convergence des variables duals dans des problmes de transmission à travers des couches minces par des méthodes d'epi-convergence, *Ric. di Mat.*, **35**, 125-159.

Avellaneda M. and Milton G.W. (1989) Bounds on the effective elasticity tensor of composite based on two-point correlations, in: *Proc. ASME Energy Technology Conference and Exposition*, ed. by D. Hui, Houston, *ASME Press*, New York.

Babŭska I. and Li L. (1992) The problem of plate modelling: theoretical and computational results, *Comp. Meth. Appl. Mech. Eng.*, **100**, 249-273.

Baiocchi C. and Capelo A. (1984) Variational and Quasivariational Inequalities: Applications to Free Boundary Problems, *J. Wiley and Sons*, Chichester.

Bakhvalov N.S. and Panasenko G.P. (1984) Homogenization of Processes in Periodic Media (in Russian), *Nauka*, Moscow; English transl. by *Kluwer Publ.*, Dordrecht-Boston, London, 1989.

Ball J.M. (1989) A version of the fundamental theorem for Young measures, in: *Lecture Notes in Physics*, vol. 344, pp. 207-215, Berlin.

Ball J. M. and Knowles G. (1990) Young measures and minimization problems of mechanics, in: *Elasticity: Mathematical Methods and Applications*, ed. by G. Eason and R. W. Ogden, pp. 1-20, *Ellis Horwood*, Chichester.

Ball J.M. and Zhang K.-W. (1990) Lower semicontinuity of multiple integrals and the Biting Lemma, *Proc. R. Soc. Edinburgh*, **114A**, 367-379.

Ball J.M., Curie J.C. and Olver P.J. (1981) Null Lagrangians, weak continuity, and variational problems of arbitrary order, *J. Funct. Anal.*, **41**, 135-174.

Beckers M. (1977) Méthodes du perimetre et des filtres pour l'optimisation topologique en variables discrètes, *LTAS Rapport OF-45, Université de Liège, Struct. Aerospatiales*.

Benaouda M.K.E. and Telega J.J. (1997) On the existence of minimizers for Saint-Venant Kirchhoff bodies: placement boundary condition, *Bull. Polish Acad. Sci., Technical Sci.*, **45**, 211-223.

Bendsøe M.P. (1995) Optimization of Structural Topology, Shape and Material, *Springer* Berlin.

Bendsøe M.P. and Diaz A.R. (1993) Optimization of material properties for Mindlin plate design, *Struct. Optimiz.*, **6**, 268-270.

Bendsøe M.P. and Haber R.B. (1993) The Michell layout problem as a low volume fraction limit of the perforated plate topology optimization problem: an asymptotic study, *Struct. Optimiz.*, **6**, 263-267.

Bensoussan A., Lions J.-L. and Papanicolaou G. (1978) Asymptotic Analysis of Periodic Structures, *North-Holland*, Amsterdam.

Berezhnitskii L.T., Delyavskii M.V. and Panasyuk V.V. (1979) Bending of Thin Plates with Crack-Type Defects, *Naukova Dumka*, Kiev, in Russian.

Berger M.S. (1967) On Von Kármán's equations and the buckling of a thin elastic plate. I. The clamped plate, *Comm. Pure Appl. Math.*, **20**, 687-719.

Bernadou M. (1996) Finite Element Methods for Thin Shell Problems, *John Wiley & Sons*, Chichester; *Masson*, Paris.

Bernadou M. and Ciarlet P.G. (1976) Sur l'ellipticité du modèle linéaire de coques de W. T. Koiter, in: *Lecture Notes in Appl. Sci. and Eng.*, vol. 134, pp. 89-136, *Springer*, Berlin.

Bernadou M. and Oden J.T. (1988) An existence theorem for a class of nonlinear shallow shell problems, *J. Math. Pures Appl.*, **60**, 285-308.

Bernadou M., Ciarlet P.G. and Miara B. (1994) Existence theorems for two-dimensional linear shell theories, *J. Elasticity*, **34**, 111-138.

Bernadou M., Ciarlet P.G. and Viaño J.M. (eds., 1997) Shells: Mathematical Modelling and Scientitic Computing, *Universidade de Santiago de Compostela*, Cursos e Congresos No 105.

Berryman J.G. and Milton G.W. (1980) Microgeometry of random composites and porous media, *J. Phys., D. Appl. Phys.* **21**, 87-94.

Beyerlein I.J. and Phoenix S.L. (1997) Stress profiles and energy release rates around fiber breaks in a lamina with propagating zones of matrix yielding and debonding, *Compos. Sci. Technol.*, **57**, 869-885.

Beyerlein I.J., Phoenix S.L. and Sastry A.M. (1996) Comparison of shear-lag theory and continuum fracture mechanics for modelling fiber and matrix stresses in an elastic cracked composite lamina, *Int. J. Solids Structures*, **33**, 2543-2574.

Bielski W.R. and Telega J.J. (1988) On existence of solutions for geometrically nonlinear shells and plates, *ZAMM*, **68**, T155-T157.

Bielski W.R. and Telega J.J. (1996) A non-linear elastic plate model of moderate thickness: existence, uniqueness and duality, *J. Elasticity*, **42**, 243-273.

Bielski W. and Telega J.J. (1997) Non-linear moderately thick plates and homogenization, *Bull. Pol. Acad. Sci., Tech. Sci.*, **45**, 197-209.

Bielski W.R. and Telega J.J. (1998) Existence of solutions to obstacle problems for linear and nonlinear elastic plates, *Mathl. Comput. Modelling*, **28**, 55-66.

Bielski W.R., Telega J.J. (1999) Homogenization of a class of nonconvex functionals, *Bull. Pol. Acad. Sci.*, submitted.

Birman M.Sz. and Solomyak M.Z. (1967) Piecewise-polynomial approximation of functions of classes W_p^α, *Mat. Sbornik*, **73** (115), 331-355, in Russian.

Blanchard D. and Francfort G.A. (1987) Asymptotic thermoelastic behavior of flat plates, *Quart. Appl. Math.*, **44**, 645-667.

Blanchard D. and Xiang Y. (1992) Clamped plates as limits of genuinely clamped plates in linear and nonlinear elasticity, *Asympt. Anal.*, **5**, 495-516.

Bleich F. and Melan E. (1927) Die gewöhnlichen und partiellen Differenzengleichungen der Baustatik, Berlin.

Bojarski J. and Telega J.J. (1994) Existence of solutions for perfectly plastic shells, *C. R. Acad. Sci. Paris*, Série I, **319**, 1009-1014.

Bonnetier E. and Conca C. (1993) Relaxation totale d'un problème d'optimisation de plaques, *C. R. Acad. Sci. Paris*, Sér. I., **317**, 931-936.

Bonnetier E. and Conca C. (1994) Approximation of Young measures by functions and application to a problem of optimal design for plates with variable thickness, *Proc. R. Soc. Edinburgh.*, **124A**, 399-422.

Bonnetier E., Vogelius M. (1987) Relaxation of a compliance functional for a plate optimization problem, in: *Applications of Multiple Scaling in Mechanics*, ed. by P.G. Ciarlet and Sanchez-Palencia, pp. 31-53, *Masson*, Paris.

Bouchitté G. (1986-1987) Convergence et relaxation de fonctionnelles du calcul des variations à croissance linéaire, *Annales Fac. Sci. Toulouse*, **8**, 7-36.

Bouchitté G. (1987) Représentation intégrale de functionnelles convexes sur un espace de mesures. II. – Cas de l'epi-convergence, *Ann. Univ. Ferrara, Sez. VII, Sc. Mat.*, **33**, 113-156.

Bouchitté G. and Suquet P. (1991) Homogenization, plasticity and yield design, in: *Composite Media and Homogenization Theory*, ed. by G. Dal-Maso and G.F. Dell'Antonio, pp. 107-133, *Birkhäuser*, Boston.

Bouchitté G. and Valadier M. (1988) Integral representation of convex functionals on a space of measures, *J. Functional Anal.*, **80**, 398-420.

Bouchitté G. and Valadier M. (1989) Multifonctions s.c.i. et regularisée s.c.i. essentielle, in: *Analyse Non Linéaire*, ed. by H. Attouch, J.-P. Aubin, F. Clarke and I. Ekeland, pp. 123-149, *Gauthier-Villars*, Paris.

Bouchitté G., Buttazzo G. and Seppecher P. (1997a) Shape optimization via Monge-Kantorovich equation, *C. R. Acad. Sci. Paris*, Série I, **324**, 1185-1191.

Bouchitté G., Buttazzo G. and Seppecher P. (1997b) Energies with respect to a measure and applications to low dimensional structures, *Calc. Var.*, **5**, 37-54.

Bourgat J.F. (1978) Numerical experiments of the homogenization method for operators with periodic coefficients, *Rapport de Recherche*, 277, INRIA Rocquencourt.

Bourgat J.F. and Dervieux A. (1978) Méthode d'homogénéisation des opérateurs à coefficients périodiques: étude des correcteurs provenant du développement asymptotique, *Rapport de Recherche*, 278, INRIA Rocquencourt.

Bourgeat A. and Tapiéro R. (1983) Homogénéisation d'une plaque mince, thermoélastique, perforée transversalement, de structure non uniformement périodique, dans le modèle de la théorie naturelle, *C. R. Acad. Sci. Paris.*, Sér. I., **297**, 213-216.

Bourgeat A. and Tapiéro R. (1985) Homogenization of a thick plate model with inclusions or openings periodically distributed in the thickness, *Applicable Anal.*, **19**, 101-116.

Bourgeois S., Cartraud P. and Débordes O. (1997) Homogenization of periodic sandwiches. Numerical and analytical approaches. In: Proc. *Euromech 360. Mechanics of Sandwich Structures: Modelling, Numerical Simulation and Experimental Identification*, Sain-Etienne, 13-15 May 1997, France.

Bourquin F., Ciarlet P.G., Geymonat G. and Raoult A. (1992) Γ-convergence et analyse asymptotique des plaques minces, *C.R. Acad. Sci*, Série I, **315**, 1017-1024.

Braides A. (1983) Omogeneizzazione di integrali non coercivi, *Ric. di Matematica*, **32**, 347-368.

Braides A. and Chiadò Piat V. (1995) A derivation formula for convex integral functionals defined on $BV(\Omega)$, *J. Convex Anal.*, **2**, 69-85.

Brézis H. (1973) Operateurs Maximaux Monotones, *North-Holland*, Amsterdam.

Brézis H. (1983) Analyse Foncionnelle: Théorie et Applications, *Masson*, Paris.

Brezzi F. (1974) On the existence, uniqueness and approximation of saddle – point problems arising from Lagrangian multipliers, *RAIRO Anal. Numér.*, **8**, 129-151.

Bucur D., Buttazzo G. and Henrot A. (1998) Existence results for some optimal partition problems, Preprint.

Budiansky B. and O'Connell R.J. (1976) Elastic moduli of a cracked solid, *Int. J. Solids Structures*, **12**, 81-97.

Budiansky B. and Sanders J.L. (1963) On the "best" first-order linear shell theory, in: *Progress in Applied Mechanics. The Prager Anniversary Volume*, pp. 129-140, *The Macmillan Co.*, New York.

Busse S., Ciarlet P.G. and Miara B. (1997) Justification d'un modèle linéaire bi-dimensionnel de coques „faiblement courbées" en coordonnées curvilignes, *Math. Modelling Numer. Anal.*, **31**, 409-434.

Buttazzo G. (1989) Semicontinuity, Relaxation and Integral Representation in the Calculus of Variations, *Pitman Research Notes in Mathematics*, London, Harlow.

Buttazzo G. and Dal Maso G. (1980) Γ-limits of integral functionals, *J. Anal. Math.*, **37**, 145-185.

Buttazzo, G. and Dal Maso G. (1993) An existence result for a class of shape optimization problems, *Arch. Rat. Mech. Anal.*, **122**, 183-195.

Caillerie D. (1982) Plaques élastiques minces à structure périodique de période et d'épaisseur comparables, *C. R. Acad. Sci. Paris*, **294**, Série II, 159-162.

Caillerie D. (1984) Thin elastic and periodic plates, *Math. Meth. Appl. Sci.*, **6**, 159-191.

Caillerie D. (1987) Non homogeneous plate theory and conduction in composite, in: *Homogenization Techniques for Composite Media, Lecture Notes in Physics*, vol. 272, pp. 1-64, *Springer-Verlag*, Berlin.

Castaing C. and Valadier M. (1977) Convex Analysis and Measurable Multifunctions, *Lecture Notes in Mathematics*, vol. 580, *Springer-Verlag*, Berlin.

Cauchy A. (1828) Sur l'equilibre et le mouvement d'une plaque solide, *Exercises de Mathématique*, **3**, 381-411.

Cea J. (1971) Optimisation: Théorie et Algorithme, *Herrmann*, Paris.

Chacha D. and Sanchez-Palencia E. (1992) Overall behavior of elastic plates with periodically distributed fissures, *Asympt. Anal.*, **5**, 381-396.

Chan H.S.Y. (1967) Half-plane slip-line fields and Michell structures, *Quart. Appl. Math.*, **20**, 453-469.

Chavent G. and Kunisch K. (1997) Regularization of linear least squares problems by total bounded variation, *ESAIM: Control. Optim. Calculus Var.*, **2**, 359-376.

Cheng G. (1981) On non-smoothness in optimal design of solid elastic plates, *Int. J. Solids Structures*, **17**, 795-810.

Cheng G. and Olhoff N. (1981) An investigation concerning optimal design of solid elastic plates, *Int. J. Solids Structures*, **16**, 305-323.

Cherepanov G.P. (1974) Inverse problems of plane elasticity, *Prikl. Mat. Mekh.*, **38**, 963-979, in Russian.

Cherkaev A.V. (1994) Relaxation of problems of optimal structural design, *Int. J. Solids Structures*, **31**, 2251-2280.

Cherkaev A.V. and Gibianskii L.V. (1993) Coupled estimates for the bulk and shear moduli of a two-dimensional isotropic elastic composite, *J. Mech. Phys. Solids.*, **41**, 937-980.

Cherkaev A.V., Grabovsky Y., Movchan A.B., Serkov S.K. (1998) The cavity of the optimal shape under the shear stresses, *Int. J. Solids. Structures*, **35**, 4391-4410.

Chou T.W. (1992) Microstructural Design of Fiber Composites, *Cambridge University Press*, New York.

Christensen R.M. (1979) Mechanics of Composite Materials, *Wiley*, New York.

Christiansen E. (1986) On the collapse solution in limit analysis, *Arch. Rat. Mech. Anal.*, **91**, 119-135.

Ciarlet P.G. (1988) Mathematical Elasticity, vol I: Three-Dimensional Elasticity, *North-Holland*, Amsterdam.

Ciarlet P.G. (1990) Plates and Junctions in Elastic Multi-Structures, *Masson*, Paris; *Springer-Verlag*, Berlin.

Ciarlet P.G. (1997) Mathematical Elasticity, vol. II: Elastic Plates, *Elsevier*, Amsterdam.

Ciarlet P.G. and Destuynder P. (1979) A justification of the two-dimensional linear plate model, *J. Méc.*, **18**, 315-344.

Ciarlet P.G. and Lods V. (1996a) Asymptotic analysis of linearly elastic shells. I. Justification of membrane shell equations; II. Justification of flexural shell equations; III. Justification of Koiter's shell equations. *Arch. Rat. Mech. Anal.*, **136**, 119-161; 163-190; 191-200.

Ciarlet P.G. and Lods V. (1996b) On the ellipticity of linear membrane shell equations, *J. Math. Pures Appl.*, **75**, 107-124.

Ciarlet P.G. and Lods V. (1996c) Asymptotic analysis of linearly elastic shells: "generalized membrane" shells, *J. Elasticity.*, **43**, 147-188.

Ciarlet P.G. and Rabier P. (1980) Les Equations de von Kármán, *Springer-Verlag*, Berlin.

Ciarlet P.G. and Sanchez-Palencia E. (1996) An existence and uniqueness theorem for the two-dimensional linear membrane shell equations, *J. Math. Pures Appl.*, **75**, 51-67.

Cioranescu D. and Saint Jean Paulin J. (1979) Homogenization in open sets with holes, *J. Math. Anal. Appl.*, **71**, 590-607.

Cioranescu D. and Saint Jean Paulin J. (1986) Reinforced and honeycomb structures, *J. Math. Pure Appl.*, **65**, 403-422.

Cioranescu D. and Saint Jean Paulin J. (1988) Elastic behavior of very thin structures, in: *Material Instabilities in Continuum Mechanics*, ed. by J.M. Ball, pp. 64-75, *Clarendon Press*, Oxford.

Collard C. and Miara B. (1997) Analyse asymptotique formelle des coques non linéairement élastiques: calcul explicite des contraintes limites, *C. R. Acad. Sci. Paris*, Série I, **325**, 223-226.

Constanda C. (1995) On the bending of plates with transverse shear deformation and mixed periodic boundary conditions, *Math. Meth. Appl. Sci.*, **18**, 337-344.

Coutris N. and Monavon A. (1986) Application de la méthode des developpements asymptotiques raccordés à la théorie des plaques, 1-ère partie: Cas du bord encastré, 2-ème partie, *J. de Méc. théor. appl.*, **5**, 181-215; 853-896.

Dacorogna B. (1982) Weak Continuity and Weak Lower Semicontinuity of Non Linear Functionals, *Springer-Verlag*, Berlin (Russian edition: *Uspekhi Mat. Nauk.*, **44**, 35-98).

Dacorogna B. (1989) Direct Methods in the Calculus of Variations, *Springer*, Berlin.

Dal Maso G. (1993) An Introduction to Γ-Convergence, *Birkhäuser*, Boston.

Dal Maso G. and Kohn R.V. (1991) H-limits, *International Centre for Theoretical Physics*, Trieste, January.

Damlamian A. and Vogelius M. (1985) Homogenization limits of the equations of elasticity in thin domains, *IMA Preprint Series, No 170*, University of Minnesota.

Dauge M. and Gruais I. (1996) Asymptotics of arbitrary order in thin elastic plates and optimal estimates for the Kirchhoff-Love model, *Asympt. Anal.*, **13**, 167-197.

Dauge M. and Gruais I. (1998a) Edge layers in thin elastic plates, *Comp. Math. Appl. Mech. Eng.*, **157**, 335-347.

Dauge M. and Gruais I. (1998b) Asymptotics of arbitrary order for a thin elastic clamped plates; II. Analysis of the boundary layer terms, *Asympt. Anal.*, **16**, 99-124.

Dauge M., Djurdjevic I. and Rössle A. (1998) Higher order bending and membrane responses of thin linearly elastic plates, *C.R. Acad. Sci. Paris, Série I*, **326**, 519-524.

Dauge M., Gruais I. and Rössle A. (1997a) The influence of lateral boundary conditions on the asymptotics in thin elastic plates, I. *Institut de Recherche Mathématique de Rennes*, Prépublication, No 97-28.

Dauge M., Gruais I. and Rössle A. (1997b) The influence of lateral boundary conditions on the asymptotics in thin elastic plates, II. *Institut de Recherche Mathématique de Rennes*, Prépublication, No 97-29.

Dautray R. and Lions J.-L. (1990) Mathematical Analysis and Numerical Methods for Science and Technology, vol. 2, *Springer-Verlag*, Berlin.

Davet J.L. and Destuynder Ph. (1985) Singularité logarithmiques dans les effets de bord d'une plaque en matériaux composites, *J. Méc. Théorique et Appl.*, **4**, 357-373.

Davet J.L. and Destuynder Ph (1986) Free-edge stress concentration in composite laminate: a boundary layer approach, *Comp. Math. Appl. Mech. Eng.*, **59**, 129-140.

Davet J.L., Destuynder Ph. and Nevers Th. (1985) Some theoretical aspects in the modelling of delamination for multilayered plates, in: *Local Effects in the Analysis of Structures*, ed. by P. Ladevèze, pp. 181-197, *Elsevier Science Publisher*, Amsterdam.

Davis P.J. and Rabinowitz P. (1967) Numerical Integration, *Ginn (Blaisdell)*, Boston.

Davis P.J. and Rabinowitz P. (1975) Methods of Numerical Integration, *Academic Press*, New York.

Demengel F. (1982) Problemes variationnels en plasticité parfaite des plaques, *Thèse de 3-ème cycle, Université Paris-Sud*; abbreviated version: *Numer. Funct. Analysis Optimiz.*, **6**, 73-119 (1983).

Demengel F. (1984) Fonctions à hessien borné, *Annales Inst. Fourier Univ. Sci. Med. Grenoble*, **34**, 155-190.

Demengel F. (1985) Relaxation et existence pour le problème des matériaux à blocage, *Math. Modelling and Numer. Anal.*, **19**, 351-395.

Demengel F. (1989) Compactness theorems for spaces of functions with bounded derivatives and applications to limit analysis problems in plasticity, *Arch. Rat. Mech. Anal.*, **105**, 123-161.

Demengel F. and Tang Qi (1990) Convex function of a measure obtained by homogenization, *SIAM J. Math. Anal.*, **21**, 409-435.

Descloux J. (1990) Un problème d'approximation, Ecole Polytechnique Fédérale de Lausanne, unpublished manuscript.

Destuynder Ph. (1983) On existence theorem for a nonlinear shell model in large displacements analysis, *Math. Meth. in the Appl. Sci.*, **5**, 68-83.

Destuynder Ph. (1985) A classification of thin shell theories, *Acta Appl. Math.*, **4**, 15-63.

Destuynder P. (1986) Une Théorie Asymptotique des Plaques Minces en Elasticité Linéaire, *Masson*, Paris.

Destuynder Ph. and Gruais I. (1995) Error estimation for the linear three-dimensional elastic plate model, in: *Asymptotic Methods for Elastic Structures*, ed. by P.G. Ciarlet, L. Trabucho and J.M. Viaño, pp. 75-88, *Walter de Gruyter*, Berlin.

Destuynder Ph. and Theodory C. (1986) Homogénéisation de structures minces en béton armé, *Math. Model. Numer. Anal. (M²AN)*, **20**, 47-74.

Diaz A.R., Lipton R. and Soto C.A. (1995) A new formulation of the problem of optimum reinforcement of Reissner-Mindlin plates, *Comp. Meth. Appl. Mech. Engrg.*, **123**, 121-139.

DiPerna R.J. (1983) Convergence of approximate solutions to conservation laws, *Arch. Rat. Mech. Anal.*, **82**, 27-70.

Dumontet H. (1985a) Homogénéisation et effets de bord, in: *Actes du Troisième Colloque: Tendance Actuelles en Calcul de Structures*, t. 2., ed. by J. P. Grellier and G. M. Campel, pp. 1025-1042, *Pluralis*, Paris.

Dumontet H. (1985b) Boundary layers stresses in elastic composites, in: *Local Effects in the Analysis of Structures*, ed. by P. Ladevèze, pp. 215-232, *Elsevier Science Publisher*, Amsterdam.

Dumontet H. (1986) Study of a boundary layer problem in elastic composite material, *Math. Modelling Numer. Anal.*, **20**, 265-286.

Duvaut G. (1976) Analyse fonctionelle et mécanique des milieux continue. Application à l'étude des matériaux composites elastiques à structure périodique-homogénéisation, in: *Theoretical and Applied Mechanics*, ed. by W. T. Koiter, pp. 119-132., *North-Holland*, Amsterdam.

Duvaut G. (1977a) Homogénéisation des plaques à structure périodique en théorie non linéaire de von Kármán, in: *Journées d'Analyse Non Linéaire, Lecture Notes in Mathematics*, vol. 665, pp. 56-69, *Springer*, Berlin.

Duvaut G. (1977b) Comportement macroscopique d'une plaque perforée périodiquement, in: *Singular Perturbations and Boundary Layer Theory, Lecture Notes in Mathematics*, pp. 131-145, *Springer-Verlag*, Berlin.

Duvaut G. (1979) Cours sur les methodes variationnelles et la dualité, in: *Duality and Complementarity in Mechanics of Solids*, ed. by A. Borkowski, pp. 173-272, *Ossolineum*, Wrocław.

Duvaut G., Lions J.-L. (1974) Problèmes unilatéraux dans la théorie de la flexion forte des plaques, Part I. Le cas stationnaire, *J. Méc.*, **13**, 57-74.

Duvaut G. and Lions J.-L. (1976) Inequalities in Mechanics and Physics, *Springer-Verlag*, Berlin.

Duvaut G. and Metellus A.-M. (1976) Homogénéisation d'une plaque mince en flexion des structure périodique et symmétrique, *C. R. Acad. Sci.*, Paris, **A283**, 947-950.

Edwards R.E. (1965) Functional Analysis: Theory and Applications, *Holt, Rinehart and Winston*, New York.

Ekeland I. (1974) On the variational principle, *J. Math. Anal. Appl.*, **47**, 324-353.

Ekeland I. (1979) Nonconvex minimization problems, *Bull. Amer Math. Soc.*, **1**, 443-474.

Ekeland I. (1990) The variational principle revisited, in: *Methods of Nonconvex Analysis, Lecture Notes in Mathematics*, vol. 1446, pp. 1-15, *Springer-Verlag*, Berlin.

Ekeland I. and Temam R. (1976) Convex Analysis and Variational Problems, *North Holland*, Amsterdam.

El Otmani S. (1994) Étude de quelques problèmes d'homogénéisation en fonction des valeurs relatives de leurs differents paramètres, Thèse, *Université de Metz*.

El Otmani S., Sac-Épée and Saint Jean Paulin J. (1995) Study of a perforated thin plate according to the relative sizes of its different parameters, *Math. Meth. in the Appl. Sci.*, **18**, 571-589.

Eschenauer H.A., Kobelev V.V. and Schumacher A. (1994) Bubble method for topology and shape optimization of structures, *J. Struct. Optimiz.*, **8**, 42-51.

Folias E.S. and Reuter W.G. (1990) On the equilibrium of a linear elastic layer, *Comput. Mech.*, **5**, 459-468.

Francfort G.A. and Milton G.W. (1987) Optimal bounds for conduction in two-dimensional, multiphase, polycrystalline media, *J. Statistical Physics*, **46**, 161-177.

Francfort G.A. and Murat F. (1986) Homogenization and optimal bounds in linear elasticity, *Arch. Rat. Mech. Anal.*, **94**, 307-334.

Francfort G.A., Murat F. and Tartar L. (1995) Fourth-order moments of nonnegative measures on S^2 and applications, *Arch. Rat. Mech. Anal.*, **131**, 305-333.

Frąckiewicz H. (1970) Mechanics of Latticed Media, *PWN*, Warsaw, in Polish.

Friedrichs K.O. and Dressler R.F. (1961) A boundary layer theory for elastic plates, *Comm. Pure. Appl. Math.*, **14**, 1-33.

Fung Y.C. (1965) Foundations of Solid Mechanics, *Prentice-Hall*, Englewood Cliffs, New Jersey.

Gambin B. and Gałka A. (1995) Boundary layer problem in a piezoelectric composite, in: *IUTAM Symp. on Anisotropy, Inhomogeneity and Nonlinearity in Solid Mechanics*, ed. by D.F. Parker and A.H. England, pp. 327-332, *Kluwer Academic Publishers*.

Gamby D. and Rebiere J.L. (1993) A two-dimensional analysis of multiple matrix cracking in a laminated composite close to its characteristic damage state, *Compos. Structures*, **25**, 325-337.

Garrett K.W. and Bailey J.E. (1977) Multiple transverse fracture in 90° cross-ply laminates of a glass fibre-reinforced polyester, *J. Mater. Sci.*, **12**, 157-168.

Genevey K. (1996a) Sur quelques questions liées à la théorie mathématiques des coques minces, *Thèse de doctorat de l' Université Paris VI.*

Genevey K. (1996b) A regularity result for a linear membrane shell problem, *Math. Modelling Numer. Anal.*, **30**, 467-488.

Genevey K. (1997) Remarks on nonlinear membrane shell problems, *Math. Mech. Solids*, **2**, 215-237.

Genevey K. (1998) Asymptotic analysis of shells via Γ-convergence, *J. Comp. Math.*, in press.

Gibiansky L.V. (1993) Bounds on the effective moduli of composite materials, Lectures given during the School on Homogenization, *ICTP*, Trieste, Sept. 6-17, 1993, 134-164.

Gibianskii L.V. and Cherkaev A.V. (1984) Designing composite plates of extremal rigidity, *Fiz. Tekhn. Inst. im. A. F. Ioffe. AN SSSR*, Preprint No 914, Leningrad, in Russian; English translation in: Cherkaev, A. and Kohn, R. V. (eds.) *Topics in the Mathematical Modelling of Composite Materials, Birkhäuser*, Boston, 1997.

Gibianskii L.V. and Cherkaev A.V. (1987) Microstructures of elastic composites of extremal stiffness and the sharp bounds of the energy stored in them, *Fiz. Tekhn. Inst. im. A. F. Ioffe. AN SSSR*; Preprint No 1115, Leningrad, 1-52, in Russian.

Gibiansky L.V. and Torquato S. (1997) Thermal expansion of isotropic multiphase composites and polycrystals, *J. Mech. Phys. Solids*, **45**, 1223-1252.

Giusti E. (1984) Minimal Surfaces and Functions of Bounded Variations, *Birkhäuser*, Boston.

Glowinski R. and Le Tallec P. (1989) Augmented Lagrangian and Operator-Splitting Methods in Nonlinear Mechanics, *SIAM*, Philadelphia.

Goldenveizer A.L. (1962) Derivation of an approximate theory of bending of a plate by the method of asymptotic integration of the equations of the theory of elasticity, *Prikl. Mat. Mekh.*, **26**, 668-686, in Russian.

Goldenveizer A.L. and Kolos A.V. (1965) On a construction of two-dimensional equations of thin, elastic plates, *Prikl. Mat. Mekh.*, **29**, 141-155, in Russian.

Goodman J., Kohn R.V. and Reyna L. (1986) Numerical study of a relaxed variational problem from optimal design, *Comp. Meth. Appl. Mech. Engrg.*, **57**, 107-127.

Grabovsky Y. and Kohn R.V. (1995) Microstructures minimizing the energy of a two phase elastic composite in two space dimensions, I. The confocal ellipse construction, II. The Vigdergauz microstructure, *J. Mech. Phys. Solids.*, **43**, 933-947; 949-972.

Gregory R.D. and Wan F.Y.M. (1984) Decaying states of plane strain in a semi-infinite strip and boundary conditions for plate theory, *J. Elasticity*, **14**, 27-64.

Grigoliuk E.I. and Filshtinskii L.A. (1970) Perforated Plates and Shells, *Nauka*, Moscow, in Russian.

Groves S.E. (1986) Study of damage mechanics in continuous fibre composite laminates with matrix cracking and interply delamination, *Dissertation, Texas A & M University*.

Groves S.E., Harris C.E., Highsmith A.L., Allen D.H. and Norvell R.G. (1987) An experimental and analytical treatment of matrix cross-ply laminates, *Exper. Mech.*, **27**, 73-79.

Gudmundson P. and Östlund S. (1992a) First order analysis of stiffness reduction due to matrix cracking, *J. Comp. Mater.*, **26**, 1009-1030.

Gudmundson P. and Östlund S. (1992b) Numerical verification of a procedure for calculation of elastic constants in microcracking composite laminates, *J. Comp. Mater.*, **26**, 2480-2492.

Gudmundson P. and Zang W. (1993) An analytic model for thermoelastic properties of composite laminates containing transverse matrix cracks, *Int. J. Solids Structures*, **30**, 3211-3231.

Gutkowski W. (1973) Regular Skeletal Structures, *PWN*, Warsaw, in Polish.

Haber R.B., Jog Ch. and Bendsøe M.P. (1996) A new approach to variable-topology shape design using a constraint on perimeter, *TAM Report No 815, Univ. Illinois. UILU-ENG-96-6002*, 1-31.

Hadhri T. (1985a) Fonction convexe d'une mesure, *C. R. Acad. Sci. Paris*, Série I, **301**, 687-690.

Hadhri T. (1985b) Etude dans $HB \times BD$ d'un modèle de plaques élastoplastiques comportant une non-linéarité géometrique, *Math. Modelling Numer. Anal.*, **19**, 235-283.

Han Y.M. and Hahn H.T. (1989) Ply cracking and property degradations of symmetric balanced laminates under general in-plane loading, *Compos. Sci. Techn.*, **35**, 377-397.

Han Y.M., Hahn H.T. and Croman R.B. (1988) A simplified analysis of transverse ply cracking in cross-ply laminates, *Compos. Sci. Techn.*, **31**, 165-177.

Haryadi S.G., Kapania R.K. and Haftka R.T.(1998) Global/local analysis of composite plates with cracks, *Composites*, **29B**, 271-276.

Hashin Z. (1965) On elastic behaviour of fibre reinforced materials of arbitrary transverse phase geometry, *J. Mech. Phys. Solids.*, **13**, 119-134.

Hashin Z. (1985) Analysis of cracked laminates: a variational approach, *Mech. Mater.*, **4**, 121-136.

Hashin Z. (1987) Analysis of orthogonally cracked laminates under tension, *J. Appl. Mech. Trans. ASME* , **54**, 872-879.

Hashin Z. (1996) Finite thermoelastic fracture criterion with application to laminate cracking analysis, *J. Mech. Phys. Solids.*, **44**, 1129-1145.

Hashin Z. and Shtrikman, S. (1963) A variational approach to the theory of the elastic behaviour of multiphase materials, *J. Mech. Phys. Solids*, **11**, 127-140.

Heal K.M., Hansen M.L. and Rickard K.M. (1998) Maple V, Learning Guide, *Springer*, New York.

Hegemier G.A. and Prager W. (1969) On Michell trusses, *Int. J. Mech. Sci.*, **11**, 209-215.

Hemp W.S. (1973) Optimum Structures, *Clarendon Press*, Oxford.

Hencky H. (1947) Über die Berücksichtigung der Schubverzerrung in ebenen Platten, *Ing. Archiv.*, **16**, 72-76.

Highsmith A.L. and Reifsnider K.L. (1982) Stiffness reduction mechanisms in composite materials, in: *Damage in Composite Materials. ASTM STP*, ed. by K.L. Reifsnider, **775**, pp. 103-117.

Hill R. (1963) Elastic properties of reinforced solids: some theoretical principles, *J. Mech. Phys. Solids*, **11**, 357-372.

Hiriart-Urruty J.-B. and Lemaréchal C. (1996) Convex Analysis and Minimization Algorithms, vol. I. Fundamentals, vol. II. Advanced Theory and Bundle Methods, *Springer-Verlag*, Berlin.

Holnicki-Szulc J. and Gierliński J.T. (1995) Structural Analysis, Design and Control by the Virtual Distortion Method, *Wiley*, Chichester.

Hornung U. (ed., 1997) Homogenization and Porous Media, *Springer*, New York.

Hutchinson J.R. (1987) A comparison of Mindlin and Levinson plate theories, *Mech. Res. Comm.*, **14**, 165-170.

Ioffe A.D. and Tihomirov V.M. (1979) Theory of Extremal Problems, *North-Holland*, Amsterdam.

Ito K. and Kunisch K. (1990) An augmented Lagrangian technique for variational inequalities, *Appl. Math. Optim.*, **21**, 223-241.

Ito K. and Kunisch K. (1996) Augmented Lagrangian methods for nonsmooth convex optimization in Hilbert spaces, Preprint, Institut für Mathematik, *Technische Universität Graz*, Austria.

Iyengar K.T.S.R., Murthy M.V.V. and Rao M.N.B. (1988) Three-dimensional elastic analysis of cracked thick plates under bending fields, *Int. J. Solids Structures.*, **24**, 683-703.

Janas M., König J.A. and Sawczuk A. (1972) Plastic Analysis of Structures, *Ossolineum*, Wrocław, in Polish.

Jemielita G. (1991) On the winding paths of the theory of plates, Prace Naukowe, Budownictwo, Zeszyt **117**, *Wydaw. Politechniki Warszawskiej*, in Polish.

Jikov V.V., Kozlov S.M. and Oleinik O.A. (1994) Homogenization of Differential Operators and Integral Functionals, *Springer*, Berlin.

Joseph P.F. and Erdogan F. (1987) Surface crack problems in plates, *The National Aeronautics and Space Administration, Grant NAG-1-713. Dept. Mech. Eng. & Mech. Lehigh Univ. Bethlehem* USA, 1-46.

Kachanov L.M. (1971) Foundations of the Theory of Plasticity, *North-Holland*, Amsterdam.

Kalamkarov A.L. (1992) Composite and Reinforcement Elements of Construction, *Wiley*, New York.

Kalamkarov A.L. and Kolpakov A.G. (1996) On the analysis and design of fiber-reinforced composite shells, *J. Appl. Mech.*, **63**, 939-945.

Kalamkarov A.L. and Kolpakov A.G. (1997) Analysis, Design and Optimization of Composite Structures, *Wiley*, Chichester.

Kalamkarov A.L., Kudryavtsev B.A. and Parton Z. (1987) A problem of a curved composite layer with wavy faces of periodic geometry, *Prikl. Mat. Mekh.*, **51**, 68-75, in Russian.

Kączkowski Z. (1968) Plates. Statical Analysis, *Arkady*, Warsaw, in Polish.

Kesavan S. (1979) La méthode de Kikuchi appliquée aux equtions de von Kármán, *Numer. Math.*, **32**, 209-232.

Khludnev A.M. and Sokolowski J. (1997) Modelling and Control in Solid Mechanics, *Birkhäuser*, Basel.

Kikuchi F. (1976) An iterative finite element scheme for bifurcation analysis of semi-linear elliptic equations, Institute of Space and Aeronautical Science, *University of Tokyo*, Report No. 542, 203-231.

Kinderlehrer D. and Stampacchia G. (1980) An Introduction to Variational Inequalities and Their Applications, *Academic Press*, New York.

Kirchhoff G. (1850) Über das Gleichgewicht und die Bewegung einer elastischen Scheibe, *J. für Reine und Angew. Math.*, **40**, 55-88.

Kohn R.V. and Strang G. (1986) Optimal design and relaxation of variational problems *Comm. Pure Appl. Math.*, **39**, 113-137; 139-183; 353-379.

Kohn R.V. and Vogelius M. (1984) A new model for thin plates with rapidly varying thickness, *Int. J. Solids Structures*, **20**, 333-350.

Kohn R.V. and Vogelius M. (1985) A new model for thin plates with rapidly varying thickness. II: a convergence proof, *Quart. Appl. Math.*, **43**, 1-22.

Kohn R.V. and Vogelius M. (1986a) A new model for thin plates with rapidly varying thickness. III: comparison of different scalings, *Quart. Appl. Math.*, **44**, 35-48.

Kohn R.V. and Vogelius M. (1986b) Thin plates of rapidly varying thickness and their relation to structural optimization, in: *Homogenization and Effective Moduli of Materials and Media*, ed. by J. L. Ericksen, D. Kinderlehrer, R. Kohn, J.-L. Lions, pp. 126-149, *Springer*, Berlin.

Koiter W.T. (1960) A consistent first approximation in the general theory of thin elastic shells, in: *Proc. IUTAM Symp, Theory of Thin Elastic Shells*, ed. by W. T. Koiter, pp. 12-33, *North-Holland Publ. Co.*, Amsterdam.

Kolpakov A.G. (1991) Thin elastic plates with periodic structure and internal unilateral contact conditions, *Prikl. Mekh. Tekh. Fizika*, **5**, 136-142, in Russian.

Kolpakov A.G. (1992) On the thermoelasticity problem of non-uniform plates, *J. Appl. Math. Mech.*, **56**, 402-409; Supplement: ibid., **59**, 1995, 827-828.

Kolpakov A.G. (1997) Design of the reinforced plates using the homogenization method, in: *Proc. 2nd World Congress of Structural and Multidisciplinary Optimization*, ed. by W. Gutkowski and Z. Mróz, May 26-30, 1997, Zakopane, Poland, 761-766.

Kołakowski P. and Holnicki-Szulc J. (1997) Optimal remodelling of truss structures (simulation by virtual distortions), *Comp. Ass. Mech. Eng. Sci.*, **4**, 257-281.

Kozłowski W. and Mróz Z. (1969) Optimal design of solid plates, *Int. J. Solids Structures*, **5**, 781-794.

Krasnosel'ski M.A. (1964) Topological Methods in the Theory of Non Linear Integral Equations, *Pergamon Press*, New York.

Krog L.A. and Olhoff N. (1997) Topology and reinforcement layout optimization of disk, plate and shell structures, in: *Topology Optimization in Structural Mechanics, CISM Courses and Lectures*, No 374, ed. by G.I.N. Rozvany, pp. 237-322, *Springer*, Wien-New York.

Kubik J. (1993) Mechanics of Layered Structures, *Wydawnictwo TiT*, Opole, in Polish.

Kufner A., John O. and Fucik S. (1977) Function Spaces, *Noordhoff*, Leyden.

Landau L. and Lifchitz E. (1967) *Théorie de l' Élasticité*, Editions Mir, Moscow.

Laurent P.-J. (1972) Approximation et Optimisation, *Hermann*, Paris

Laws N. and Brockenbrough J.R. (1987) The effect of micro-crack systems on the loss of stiffness of brittle solids, *Int. J. Solids Structures*, **23**, 1247-1268.

Lavrov N.A., Lurie K.A. and Cherkaev A.V. (1980) Nonuniform rod of extremal torsional stiffness, *Mekh. Tverd. Tela*, **15**, 74-80, in Russian.

Le Dret H. and Raoult A. (1996) The membrane shell model in nonlinear elasticity: a variational asymptotic derivation, *J. Nonlinear Sci.*, **6**, 59-84.

Lee J.H. and Hong C.S. (1993) Refined two-dimensional analysis of cross-ply laminates with transverse cracks based on the assumed crack opening deformation, *Compos. Sci. Technol.*, **46**, 157-166.

Lee J.-W., Allen D.H. and Harris C.E. (1989) Internal state variable approach for predicting stiffness reductions in fibrous laminated composites with matrix cracks, *J. Compos. Mater.*, **23**, 1273-1291.

Lefik M. (1995) Finite element model for 3-D analysis of composite plates, *Eng. Trans.*, **43**, 225-244.

Leguillon D. and Sanchez-Palencia E. (1982) On the behavior of a cracked elastic body with (or without) friction, *J. Méc. Theor. Appl.*, **1**, 195-209.

Leissa A.W. (1969) Vibration of Plates, NASA SP-160, US Governement Printing Office, Washington DC, Reprinted by the Acoustical Society of America, 1993.

Leissa A.W. (1993) Vibration of Shells, Acoustical Society of America, American Institute of Physics.

Leissa A.W. (1998) The plate and shell vibration monographs, *Appl. Mech. Rev.*, **51**, R19-R28.

Lené F. (1978) Comportement macroscopique de matériaux élastiques comportant des inclusions rigides ou des trous répartis périodiquement, *C. R. Acad. Sci. Paris*, Série A, **286**, 75-78.

Lené F. (1984) Contribution à l'étude des matériaux composites et de leur endommagement, Thèse de Doctorat d'Etat es Sciences Mathématiques (Mécanique), *Université Paris VI*.

Levinson M. (1980) An accurate, simple theory of the statics and dynamics of elastic plates, *Mech. Res. Comm.*, **7**, 343-350.

Lewiński T. (1980) An analysis of various descriptions of state of strain in the linear Kirchhoff-Love shell theory, *Eng. Trans.*, **28**, 635-652.

Lewiński T. (1984a) Two versions of Woźniak's continuum model of hexagonal-type grid plates, *Mech. Teoret. Stos.*, **22**, 389-405.

Lewiński T. (1984b) Differential models of hexagonal-type grid plates, *Mech. Teoret. Stos.*, **22**, 407-421.

Lewiński T. (1986a) A note on averaging stiffnesses of thin elastic periodic plates, *Eng. Trans.*, **34**, 337-352.

Lewiński T. (1986b) A note on recent developments in the theory of elastic plates with moderate thickness, *Eng. Trans.*, **34**, 531-542.

Lewiński T. (1987) On refined plate models based on kinematical assumptions, *Ing. Archiv.*, **57**, 133-146.

Lewiński T. (1991a) Effective models of composite periodic plates. I- Asymptotic solution, *Int. J. Solids Structures*, **27**, 1155-1172.

Lewiński T. (1991b) Effective models of composite periodic plates. II- Simplifications due to symmetries, *Int. J. Solids Structures*, **27**, 1173-1184.

Lewiński T. (1991c) Effective models of composite periodic plates. III- Two-dimensional approaches, *Int. J. Solids Structures*, **27**, 1185-1203.

Lewiński T. (1991d) An algorithm for computing effective stiffnesses of plates with periodic structure, *ZAMM*, **71**, T330-T332.

Lewiński T. (1991e) On displacement-based theories of sandwich plates with soft core, *J. Eng. Math.*, **25**, 223-241.

Lewiński T. (1991f) Effective stiffnesses of cylindrical shells of periodic structure, *Mech. Res. Comm.*, **18**, 245-252.

Lewiński T. (1992) Homogenizing stiffnesses of plates with periodic structure, *Int. J. Solids Structures*, **29**, 309-326.

Lewiński T. (1993) On recent developments in the homogenization theory of elastic plates and their application to optimal design: Part I, *Struct. Optim.*, **6**, 59-64.

Lewiński T. (1995) Effective stiffnesses of transversely nonhomogeneous plates with unidirectional periodic structure, *Int. J. Solids Structures*, **32**, 3261-3287.

Lewiński T. (1997) Conformal scaling-based approach to evaluation of effective properties of thermoelastic non-homogeneous media, *Bull. Polon. Acad. Sci. Ser. Tech.*, **45**, 487-494.

Lewiński T. and Othman A.M. (1997a) On attainability of Hashin-Shtrikman bounds by iterative hexagonal layering. Plane elasticity problems, *Arch. Mech.*, **49**, 513-523.

Lewiński T. and Othman A.M. (1997b) Optimum design of two-phase thin plates, *3rd EUROMECH Solid Mechanics Conference* Stockholm, August, 18-22, Book of Abstracts, p. 233.

Lewiński T., Sokołowski J. (1998) Optimal shells formed on a sphere, The topological derivative method, INRIA Rapport de Recherche No 3495.

Lewiński T. and Telega J.J. (1985) On homogenization of fissured elastic plates, *Mech. Res. Comm.*, **12**, 271-281.

Lewiński T. and Telega J.J. (1988a) Asymptotic method of homogenization of two models of elastic shells, *Arch. Mech.*, **40**, 705-723.

Lewiński T. and Telega J.J. (1988b) Asymptotic method of homogenization of fissured elastic plates, *J. Elasticity*, **19**, 37-62.

Lewiński T. and Telega J.J. (1988c) Homogenization of fissured Reissner-like plates. Part I- Method of two-scale asymptotic expansions, *Arch. Mech.*, **40**, 97-117.

Lewiński T. and Telega J.J. (1988d) Homogenization of fissured Reissner-like plates. Part III- Some particular cases and an illustrative example, *Arch. Mech.*, **40**, 295-303.

Lewiński T. and Telega J.J. (1989) Overall properties of plates with partially penetrating fissures, *C. R. Acad. Sci. Paris*, **309**, Série II, 951-956.

Lewiński T. and Telega J.J. (1991a) Homogenization and effective properties of plates weakened by partially penetrating fissures: asymptotic analysis, *Int. J. Eng. Sci.*, **29**, 1129-1155.

Lewiński T. and Telega J.J. (1991b) Stiffness loss of cracked laminates, in: *Proc. Xth Polish Conference on Computer Methods in Mechanics*, ed. by E. Bielewicz, Świnoujście, Poland, 14-17 th May 1991, pp. 112-115, *Politechnika Szczecińska Press*.

Lewiński T. and Telega J.J. (1991c) Non-uniform homogenization and effective properties of a class of non-linear elastic shells. in: *Trends and Applications of Mathematics to Mechanics, Proc. 8th Symp. on Trends in Applications of Mathematics to Mechanics*, ed. by H. Troger, W. Schneider, F. Ziegler, pp. 248-253, *Longman Scientific and Technical*, London.

Lewiński T. and Telega J.J. (1992) Assessing stiffness loss of cracked laminates. The homogenization method, *ZAMM* , **72**, T161-T164.

Lewiński T. and Telega J.J. (1993) Effective properties of cracked cross-ply laminates, *Materials Science Forum*, vol. 123-125, pp. 515-524, *Proc. 7th Symp. on Continuum Models of Discrete Systems*, Paderborn, Germany, June 1992, ed. by K.H. Anthony and H.-J. Wagner, *Trans. Tech. Publ.*, Aedermannsdorf.

Lewiński T. and Telega J.J. (1996a) Stiffness loss in laminates with intralaminar cracks, Part I. Two-dimensional modelling, *Arch. Mech.*, **48**, 143-161.

Lewiński T. and Telega J.J. (1996b) Stiffness loss in laminates with intralaminar cracks, Part II. Periodic distribution of cracks and homogenization, *Arch. Mech.*, **48**, 163-190.

Lewiński T. and Telega J.J. (1996c) Stiffness loss of laminates with aligned intralaminar cracks, Part I. Macroscopic constitutive relations, *Arch. Mech.*, **48**, 245-264.

Lewiński T. and Telega J.J. (1996d) Stiffness loss of laminates with aligned intralaminar cracks, Part II. Comparisons, *Arch. Mech.*, **48**, 265-280.

Lewiński T. and Telega J.J. (1997a) Elastic plates and shells of minimal compliance, in: *Proc. 2nd World Congress of Structural and Multidisciplinary Optimization*, May 26-30, ed. by W. Gutkowski and Z. Mróz, vol. 2, pp. 841-846, Institute of Fundamental Technological Research, Polish Academy of Sciences, Warsaw.

Lewiński T. and Telega J.J. (1997b) Reduction of stiffness characteristics of balanced laminates with intralaminar cracks, in: *Proc. 1st Conference: Damage and Failure of Interfaces (DFI-1)*, ed by Rossmanith H., Wien, 22-24 September 1997, pp. 179-186, *Balkema Publ.*, Rotterdam.

Lewiński T. and Telega J.J. (1998) Stiffness reduction and stress analysis in cracked $[0^0_m/90^0_n]_s$ laminates, *Acta Mech.*, **131**, 177-201.

Lewiński T., Zhou M. and Rozvany G.I.N. (1994a) Extended exact solutions for least-weight truss layouts-Part I: Cantilever with a horizontal axis of symmetry, *Int. J. Mech. Sci.*, **36**, 375-398.

Lewiński T., Zhou M. and Rozvany G.I.N. (1994b) Extended exact solutions for least-weight truss layouts-Part II: Unsymmetric cantilevers, *Int. J. Mech. Sci.*, **36**, 399-419.

Li L., Babuška I. and Chen J. (1997) The boundary layer for p-model problems, Part I. Asymptotic analysis, Part II. Boundary layer behavior, *Acta Mech.*, **122**, 181-201; 203-216.

Li S., Reid S.R. and Soden P.D. (1994) A finite strip analysis of cracked laminates, *Mech. Mater.*, **18**, 289-311.

Li S., Reid S.R. and Soden P.D. (1997) Modelling transverse matrix cracking in laminated fibre-reinforced composite structures. in: *Damage and Failure of Interfaces (DFI-1)*, ed. by H.P. Rossmanith, pp. 131-138, *Balkema Publ.*, Rotterdam.

Lions J.-L. (1969) Quelques Méthodes de Résolution des Problèmes aux Limites Non Linéaires, *Dunod*, Paris.

Lions J.-L. and Magenes E. (1968) Problèmes aux Limites Non Homogènes et Applications, vol. I, Dunod, Paris.

Lipton R. (1994a) Optimal design and relaxation for reinforced plates subject to random transverse loads, *Probab. Eng. Mech.*, **9**, 167-177.

Lipton R. (1994b) A saddle-point theorem with application to structural optimization, *J. Optimiz. Theory Appl.*, **81**, 549-568.

Lipton R. (1994c) On optimal reinforcement of plates and choice of design parameters, *Control and Cybernetics*, **23**, 481-493.

Lipton R. and Diaz A. (1997) Reinforced Mindlin plates with extremal stiffness, *Int. J. Solids Structures*, **28**, 3691-3704.

Loboda V.V. (1981) Application of averaging method to calculation of plate reinforced with ribs, *Prikl. Mat. Mekhanika*, **45**, 867-875, in Russian.

Lods V. and Miara B. (1998) Nonlinearly elastic shell models: a formal asymptotic approach. II: The flexural model, Preprint, *Laboratoire d'Analyse Numérique, Université Pierre-et-Marie-Curie*, Paris.

Love A.E.H. (1889) The small vibrations and deformation of a thin elastic shell, *Phil. Trans. Roy. Soc. Lond.*, **179**, 491-546.

Lurie K.A. (1994) Direct relaxation of optimal layout problems for plates, *J. Optimiz. Theory Appl.*, **80**, 93-116.

Lurie K.A. and Cherkaev A.V. (1984) Exact estimates of conductivity of composites formed by two isotropically conducting media taken in prescribed proportion, *Proc. Roy. Soc. Edinburgh*, **A99**, 71-87.

Lurie K.A. and Cherkaev A.V. (1986) Effective characteristics of composite materials and optimum design of structural members, *Adv. Mech.*, **9**, 3-81, in Russian. English translation in: Cherkaev A. and Kohn R.V. (eds.) Topics in the Mathematical Modelling of Composite Materials, pp. 175-258, *Birkhäuser*, Boston 1997.

Lurie K.A., Cherkaev A.V. and Fedorov A.V. (1982) Regularization of optimal design problems for bars and plates, *J. Optimiz. Theory Appl.*, **37**, Part I. pp. 499-522, Part II. pp. 523-543.

Lurie K.A., Cherkaev A.V. and Fedorov A.V. (1984) On the existence of solutions to some problems of optimal design for bars and plates, *J. Optimiz. Theory. Appl.*, **42**, 247-281.

Lutoborski A. and Telega J.J. (1982) Effective moduli for an elastic arch, *Bull. Acad. Pol. Sci.*, Série Sci. Tech., **30**, 165-170.

Lutoborski A. and Telega J.J. (1984) Homogenization of a plane elastic arch., *J. Elasticity*, **14**, 65-77.

Matysiak S. and Nagórko W. (1989) Microlocal parameters in a modelling of microperiodic multilayered elastic plates, *Ing. Archiv.*, **59**, 434-444.

McCartney L.N. (1992) Theory of stress transfer in $0° - 90° - 0°$ cross-ply laminate containing a parallel array of transverse cracks, *J. Mech. Phys. Solids*, **40**, 27-68.

Messaoudi K. and Michaille G. (1994) Stochastic homogenization of nonconvex integral functionals, *Math. Modelling Numer. Anal.*, **28**, 329-356.

Miara B. (1994) Justification of the asymptotic analysis of elastic plates, I. The linear case, *Asympt. Anal.*, **8**, 259-276.

Miara B. (1998) Nonlinearly elastic shell models: a formal asymptotic approach, I. The membrane model, Preprint, Département de Sciences Mathématiques et Physiques, *Ecol. Supér. d'Ingén. en Electrotech. et Electron., Noisy-le-Grand*, France.

Miara B. and Sanchez-Palencia E. (1996) Asymptotic analysis of linearly elastic shells, *Asymptotic Anal.*, **12**, 41-54.

Michell A.G.M. (1904) The limits of economy of material in frame-structures, *Phil. Mag.*, **8**, No 47, 589-597.

Mignot F. and Puel J.-P. (1978) Homogénéisation d'un problème de bifurcation, in: *Proc of the Int. Meeting on Recent Methods in Non Linear Analysis*, ed. by E. de Giorgi, E. Magenes and U. Mosco, pp. 281-310, *Pitagora Editrice*, Bologna.

Mignot F., Puel J.-P. and Suquet P.M. (1980) Homogenization and bifurcation of perforated plates, *Int. J. Eng. Sci.*, **18**, 409-414.

Mignot F., Puel J.-P. and Suquet P.M. (1981) Flambage de plaques élastiques multiper-forées, *Annales Fac. Sci. Toulouse*, **3**, 1-57.

Milton G.W. (1991) The field equation recursion method, in: *Composite Media and Homogenization Theory*, ed by G. Dal Maso and G. Dal Antonio, *Birkhäuser*, Boston, 223-245.

Milton G.W. and Kohn R. V. (1988) Variational bounds on the effective moduli of anisotropic composites, *J. Mech. Phys. Solids*, **36**, 597-629.

Milton G.W. and Phan-Thien N. (1982) New bounds on the effective elastic moduli of two-component materials, *Proc. Roy. Soc. Lond.*, **A380**, 305-331.

Moreau J.J. (1974) On unilateral constraints, friction and plasticity, in: *New Variational Techniques in Mathematical Physics*, ed. by G. Capriz and G. Stampacchia, pp. 173-322, *Edizioni Cremonese*, Roma.

Moreau J.J. (1976) Champs et distributions de tenseurs deformation sur un ouvert de convexité quelconque, in: *Séminaire d'Analyse Convexe*, Montpellier, Exposé no 5.

Morel J.-M. and Solimini S. (1995) Variational Methods in Image Segmentation with Seven Image Processing Experiment, *Birkhäuser*, Boston.

Morrey C.B. (1966) Multiple Integrals in the Calculus of Variations, *Springer*, Berlin.

Movchan A.B. and Movchan N.V. (1995) Mathematical Modelling of Solids with Nonregular Boundaries, *CRC Press*, Boca Raton.

Muñoz J. and Pedregal P. (1998) On the relaxation of an optimal design problem for plates, *Asymptotic Anal.*, **16**, pp. 125-140.

Murat F. (1977/78) H-convergence, Seminaire d'Analyse Fonctionelle et Numérique, *Université d'Alger*, mimeographed lectures.

Murat F. (1987) A survey on compensated compactness, in: Contributions to Modern Calculus of Variations, ed. by L. Cesari, *Pitman Research Notes in Mathematics*, Series 148, pp. 145-183, Longman, Harlow.

Murat F. and Tartar L. (1985) Calcul des variations et homogénéisation, in: *Les Méthodes de l'Homogénéisation: Théorie et Applications en Physique, Coll. de la Dir. des Etudes et Recherches de Electr. de France*, pp. 319-370, *Eyrolles*, Paris.

Murat F. and Tartar L. (1997) H-convergence, in: *Topics in the Mathematical Modelling of Composite Materials*, ed. by A. Cherkaev and R. Kohn, pp. 21-43, *Birkhäuser*, Boston.

Müller S. (1987) Homogenization of non-convex integral functionals and cellular elastic materials, *Arch. Rat. Mech. Anal.*, **99**, 189-212.

Myśliński A. and Sokołowski J. (1985) Nondifferentiable optimization problems for elliptic systems, *SIAM J. Control. Optimiz.*, **23**, 632-648.

Naghdi P.M. (1963) Foundation of Elastic Shell Theory, in: *Progress in Solid Mechanics*, ed. by I. N. Sneddon and R. Hill, vol. 4, pp. 2-90, *North-Holland*, Amsterdam.

Nairn J.A. (1989) The strain energy release rate of composite microcracking: a variational approach, *J. Compos. Mater.*, **23**, 1106-1129.

Nairn J.A. and Hu S. (1994) Matrix microcracking, in: *Damage Mechanics of Composite Materials*, ed. by R. Talreja, pp. 187-243, *Elsevier Sc.B.V.*, Amsterdam.

Nazarov S.A., Plamenevskii B.A. (1991) Elliptic Problems in Domains with Piecewise Smooth Boundaries, *Nauka*, Moskva, in Russian.

Nečas J. (1967) Les Méthodes Directes en Théorie des Equations Elliptiques, *Masson*, Paris.

Nečas J. and Hlaváček I. (1981) Mathematical Theory of Elastic and Elasto-Plastic Bodies: An Introduction, *Elsevier*, Rotterdam.

Nemat-Nasser S. and Hori M. (1993) Micromechanics: Overall Properties of Heterogeneous Materials, *North-Holland*, Amsterdam.

Neuman E. and Schmidt J.W. (1983) On the convergence of quadratic spline interpolants, Institute of Computer Science, *Wrocław University*, Poland, Report No. N-125.

Nguetseng G. (1989) A general convergence result for a functional related to the theory of homogenization, *SIAM J. Math. Anal.*, **20**, 608-623.

Niordson F.I. (1985) Shell Theory, *North-Holland*, Amsterdam.

Noor A.K. (1988) Continuum modelling for repetitive lattice structures, *Appl. Mech. Rev.*, **41**, 285-296.

Noor A.K. and Burton W.S. (1989) Assessment of shear deformation theories for multilayered composite plates, *Appl. Mech. Rev.*, **42**, 1-13.

Norris A.N. (1989) The effective moduli of layered media- a new look at an old problem, in: *Micromechanics and Inhomogeneity – The Toshio Mura Ann.*, ed. by G.J. Weng and H. Abe, pp. 321-339, *Springer*, New York.

Novozhilov V.V. (1970) Thin Shell Theory, *Walters-Nordhoff*, Groningen.

Ogin S.L., Smith P.A. and Beaumont P.W.R. (1985) Matrix cracking reduction during the fatigue of a $(0^0/90^0)$, GFRP laminate, *Compos. Sci. Technol.*, **22**, 23-31.

Oleinik O.A. (1987) Homogenization problems in elasticity. Spectra of singularly perturbed operators, in: *Non-Classical Continuum Mechanics*, ed. by R.J. Knops and A.A. Lacey, pp. 53-95, *Cambridge University Press*, Cambridge.

Oleinik O.A., Shamaev S.A. and Yosifian G.A. (1992) Mathematical Problems in Elasticity and Homogenization, *North-Holland*, Amsterdam.

Olszak W., Perzyna P. and Sawczuk A. (1965) Theory of Plasticity, *Polish Scientific Publishers (PWN)*, Warszawa, in Polish.

Olszewski S. (1990) Effective stiffnesses of thin plates of periodic layout, *Warsaw University of Technology, Civil Eng. Faculty, Institute of Struct. Mech. Master's Thesis*, in Polish.

Orlov V.N. (1978) Piecewise-polynomial approximation in W_2^1 of functions from W_p^2, $1 \leq p \leq 2$, Zhurnal *Vychisl. Mat. Mat. Fiziki*, **18**, 935-942, in Russian.

Othman A.M. (1997) Thin plates of extremal compliance. Ph. D. thesis, *Warsaw University of Technology*, Warsaw.

Panagiotopoulos P.D. (1993) Hemivariational Inequalities: Aplications in Mechanics and Engineering, *Springer-Verlag*, Berlin.

Panasenko G.P. and Reztsov M.V. (1987) Averaging of three-dimensional elasticity problem in inhomogeneous media, *Doklady AN SSSR*, **294**, 1061-1065, in Russian.

Papadopoulos P. and Taylor R.L. (1990) Elasto-plastic analysis of Reissner-Mindlin plates, *Appl. Mech. Reviews*, **43**, Part 2, S40-S50.

Parvizi A., Garrett K.W. and Bailey J.E. (1978) Constrained cracking in glass fibre-reinforced epoxy cross-ply laminates, *J. Mater. Sci.*, **13**, 195-201.

Pécastaings F. (1985) Sur le principe de Saint-Venant pour les plaques, *Thèse de Doctorat d'Etat es Sciences Mathématiques, Univ. Pierre et Marie Curie, Laboratoire de Mécanique et Technologie*, Cachan (France).

Pedregal P. (1997) Parametrized Measures and Variational Principles, *Birkhäuser*, Basel.

Pedregal P. (1999) Optimization, relaxation and Young measures, *Bull. Amer. Math. Soc., New Series*, **36**, 27-58.

Percivale D. (1990) Perfectly plastic plates: a variational definition, *J. Reine Angew. Math.*, **411**, 39-58.

Persson L.-E., Persson L., Svanstedt N. and Wyller J. (1993) The Homogenization Method: An Introduction, Studentlitteratur, Lund.

Pietraszkiewicz W. (1979) Finite Rotations and Lagrangian Description in the Non-Linear Theory of Shells, *PWN*, Warszawa-Poznań.

Pironneau, O. (1984) Optimal Shape Design for Elliptic Systems, *Springer-Verlag*, New York.

Plantema F.J. (1966) Sandwich Construction. The Bending and Buckling of Sandwich Beams, Plates and Shells, *Wiley*, New York.

Poisson S. (1829) Mémoir sur l'équilibre et le mouvement des corps solides. §6 Equations de l'équilibre et le mouvement d'une plaque élastique, *Mém. de l'Ac. d. Sci. à Paris*, **8**, 523-545.

Ponte Castañeda P. and Suquet P. (1998) Nonlinear composites, in: *Advances in Applied Mechanics*, **34**, 171-302.

Powell M. J.D. and Sabin M.A. (1977) Piecewise quadratic approximations on triangles, *ACM Trans. Math. Software*, **3**, 316-325.

Prager W. (1978) Optimal layout of trusses with finite number of joints, *J. Mech. Phys. Solids*, **26**, 241-250.

Prat P.C. and Bazant Z.P. (1997) Tangential stiffness of elastic materials with systems of growing or closing cracks, *J. Mech. Phys. Solids*, **45**, 611-636.

Pruchnicki E. (1998) Overall properties of thin hyperelastic plate at finite strain with edge effects using asymptotic method, *Int. J. Eng. Sci.*, **36**, 973-1000.

Quintela-Estevez P. (1989) A new model for nonlinear elastic plates with rapidly varying thickness, *Appl. Anal.*, **32**, 107-127.

Raoult A. (1985) Construction d'un modèle d'évolution de plaques avec terme d'inertie de rotation, *Annal. Mat. Pura ed Applicata*, **139**, 361-400.

Reddy J. N. (1984) A refined nonlinear theory of plates with tranverse shear deformation, *Int. J. Solids. Structures*, **20**, 881-896.

Reissner E. (1944) On the theory of bending of elastic plates, *J. Math. Phys.*, **23**, 184-191.

Reissner E. (1945) The effect of transverse shear deformation on the bending of elastic plates, *J. Appl. Mech.*, **12**, A69-A77.

Reissner E. (1947) On bending of elastic plates, *Quart. Appl. Math.*, **5**, 55-68.

Reissner E. (1950) On a variational theorem in elasticity, *J. Math. Phys.*, **29**, 90-95.

Reissner E. (1985) Reflections on the theory of elastic plates, *Appl. Mech. Rev.*, **38**, 1453-1464.

Reztsov M.V. (1990) On effective moduli of composite plates, *Journal Vychisl. Mat. i Mat. Fiziki*, **30**, 1741-1743, in Russian.

Rockafellar R.T. (1970) Convex Analysis, *Princeton University Press*, Princeton.

Rockafellar R.T. (1976) Integral functionals, normal integrands and measurable selections, in: *Nonlinear Operators and Calculus of Variations, Lecture Notes in Mathematics*, vol. 543, pp. 157-207, *Springer-Verlag*, Berlin.

Rockafellar R.T. and Wets R.J.-B. (1998) Variational Analysis, *Springer*, Berlin.

Rogacheva N.N. (1994) The Theory of Piezoelectric Shells and Plates, *CRC Press*, 1994.

Rosen B. W. and Hashin Z. (1970) Effective thermal expansion and specific heat of composite materials, *Int. J. Eng. Sci.*, **8**, 157-173.

Roubiček T. (1997) Relaxation in Optimization Theory and Variational Calculus, *Walter de Gruyter*, Berlin.

Rozvany G.I.N. (1976) Optimal Design of Flexural Systems: Beams, Grillages, Slabs, Plates and Shells, *Pergamon Press*, Oxford.

Rozvany G.I.N. and Gollub W. (1990) Michell layouts for various combinations of line supports-I, *Int. J. Mech. Sci.*, **32**, 1021-1043.

Rychter Z. (1987) A sixth-order plate theory – derivation and error estimates, *J. Appl. Mech.*, **54**, 275-279.

Rykhlevskii Y. (Rychlewski J.) (1984) On Hooke's law, *Prikl. Mat. Mekh.*, **48**, 420-435, in Russian.

Sab K. (1994) Homogenization of non-linear random media by a duality method. Application to plasticity, *Asymptotic Anal.*, **9**, 311-336.

Sablonnière P. (1989) Espaces des fonctions splines polynomiales sur des domaines triangulaires du plan, Institut National des Sciences Appliquées de Rennes, Laboratoire Logiciels, Analyse Numérique et Statistiques, *Publications LANS No 16*.

Sac-Épée J.M. (1994) Etude de problémes d'homogénéisation définis sur des plaques minces perforées périodiquement, Thèse, *Université de Metz*.

Sanchez-Palencia E. (1980) Non-Homogeneous Media and Vibration Theory, *Springer*, Berlin.

Sanchez-Palencia E. (1987) Boundary layers and edge effects in composites, in: *Homogenization Techniques for Composite Media, Lecture Notes in Physics*, **272**, 121-192.

Sanchez-Hubert J. and Sanchez-Palencia E. (1992) Introduction aux Méthodes Asymptotiques et a' l'Homogénéisation, *Masson*, Paris.

Sanchez-Hubert J. and Sanchez-Palencia J. (1993) Exercices sur les Méthodes Asymptotiques et l'Homogénéisation, *Masson*, Paris.

Sanchez-Palencia E. and Zaoui A., (eds., 1987) Homogenization Techniques for Composite Media, *Springer-Verlag*, Berlin.

Sankaranarayanan R.T., Haftka R.T. and Kapania R.K. (1994) Truss topology optimization with simultaneous analysis and design, *AIAA Journal*, **32**, 420-424.

Santosa F. and Vogelius M. (1993) First-order corrections to the homogenized eigenvalues of a periodic composite structure, *SIAM J. Appl. Math.*, **53**, 1636-1668.

Save M.A. and Massonet C.E. (1972) Plastic Analysis and Design of Plastic Plates, Shells and Disks, *North-Holland*, Amsterdam.

Sawczuk A. and Duszek M. (1963) A note on the interaction of shear and bending in plastic plates, *Arch. Mech. Stos.*, **15**, 411-426.

Schapery R. A. (1968) Thermal expansion coefficients of composite materials based on energy principles, *J. Compos. Mater.*, **2**, 380-404.

Schwab C. and Wright S. (1995) Boundary layers of hierarchical beam and plate models, *J. Elasticity*, **38**, 1-40.

Schwab C. (1996) A-posteriori modelling error estimation for hierarchic plate models, *Numer. Math.*, **74**, 221-259.

Shi G. and Tong P. (1995) The derivation of equivalent constitutive equations of honeycomb structures by a two-scale method, *Comput. Mech.*, **15**, 395-407.

Sigmund O. and Torquato S. (1997) Design of materials with extreme thermal expansion using a three-phase topology optimization method, *J. Mech. Phys. Solids*, **45**, 1037-1067.

Sih G. C. (1971) A review of the three-dimensional stress problem for a cracked plate, *Int. J. Fracture Mech.*, **7**, 39-61.

Sławianowska A. and Telega J.J. (1999) Asymptotic analysis of anisotropic nonlinear elastic membranes, *Bull. Polish Acad. Sci., Tech. Sci.*, **47**, 115-126.

Smith P. (1985) Convexity Methods in Variational Calculus, Research Studies Press, *John Wiley and Sons*, New York.

Smith P.A. and Wood J.R. (1990) Poisson's ratio as a damage parameter in the static tensile loading of simple crossply laminates, *Compos. Sci. Techn.*, **38**, 85-93.

Sokołowski J. (1996) Displacement derivatives in shape optimization of thin shells, *INRIA, Rapport de Recherche*, No. 2995.

Sokołowski J. and Zolesio J. P. (1992) Introduction to Shape Optimization. Shape Sensitivity Analysis, *Springer*, New York.

Sokołowski J., Żochowski A. (1997) On topological derivative in shape optimization. IN-RIA, Rapport de Recherche No 3170, to appear in *SIAM J. Control. Cyber.*

Spagnolo S. (1968) Sulla convergenza delle soluzioni di equazioni paraboliche ed ellitiche, *Ann. Scuola Norm. Sup. Pisa. U. Sci.*, **3**, 571-597.

Stamm K. and Witte H. (1974) Sandwichkonstruktionen. Berechnung, Fertigung, Ausführung, *Springer*, Wien.

Stein E.M. (1970) Singular Integrals and Differentiability Properties of Functions, *Princeton University Press*, Princeton.

Strang G. and Kohn R. V. (1983) Hencky-Prandtl nets and constrained Michell trusses, *Comp. Meth. Appl. Mech. Enrg.*, **36**, 207-222.

Struwe M. (1990) Variational Methods, *Springer-Verlag*, Berlin.

Suquet P. (1981) Prévision par homogénéisation du flambement de plaques élastiques rein-forcées, *La Recherche Aérospatiale*, **2**, 99-107.

Suquet P.M. (1988) Discontinuities and plasticity, in: *Nonsmooth Mechanics and Applications*, ed. by J.J.Moreau and P.D. Panagiotopoulos, pp. 279-340, *Springer-Verlag*, Wien-New York.

Šverak V. (1992) Rank-one convexity does not imply quasiconvexity, *Proc. R. Soc. Edinburgh*, **120**, 185-189.

Synge J.L. (1957) The Hypercircle Method in Mathematical Physics, *Univ. Press*, New York.

Tadlaoui A. and Tapiero R. (1988) Calcul par homogénéisation des microcontraintes dans une plaque hétérogéne dans son epaisseur, *J. de Méc. Théor. Appl.*, **7**, 573-595.

Taghite M.B. and Lanchon-Ducauquis H. (1993) Determination of thermoelastic stresses in the plates which maintain the tube bundle of a heat exchanger, *Comp. Meth. Appl. Mech. Engrg.*, **104**, 261-290.

Taghite M.B., Rahmattulla A., Lanchon-Ducauquis H. and Taous K. (1997) Homogenization of a thermal problem in the plate of a heat exchanger, *Comp. Meth. Appl. Mech. Engrg.*, **145**, 381-402.

Talreja R. (1985) A continuum mechanics characterization of damage in composite materials, *Proc. Roy. Soc. London*, **A 399**, 195-216.

Talreja R. (1986) Stiffness properties of composite laminates with matrix cracking and interior delamination, *Engrg. Fract. Mech.*, **25**, 751-762.

Tang Qi (1990) Justification of elastoplastic plate models, *Asymptotic Anal.*, **3**, 57-76.

Tartar L. (1977) Cours Peccot, College de France.

Tartar L. (1978) Estimation de coefficients homogénéises, in: *Computing Methods in Applied Sciences and Engineering. Lecture Notes in Mathematics*, **704**, pp. 364-373, *Springer*, Berlin.

Tartar L. (1979) Compensated compactness and applications to partial differential equations, in: *Nonlinear Analysis and Mechanics, Heriot-Watt Symposium*, vol. IV, ed. by R. Knops, pp. 136-212, *Pitman Research Notes in Mathematics*, London.

Tartar L. (1985) Estimations fines de coefficients homogénéises, in: *Ennio De Giorgi Colloquium. Pitman Res. Notes in Math. V.*, ed. by P. Kree, **125**, pp. 168-187, Boston.

Telega J.J. (1985) Limit analysis theorems in the case of Signorini's conditions and friction, *Arch. Mech.*, **37**, 549-562.

Telega J.J. (1987) Variational methods in contact problems of the mechanics, *Advances in Mech.*, **10**, 3-95, in Russian.

Telega J.J. (1988) Topics on unilateral contact problems of elasticity and inelasticity, in: *Nonsmooth Mechanics and Applications*, ed. by J.J. Moreau and P.D. Panagiotopoulos, pp. 341-462, *Springer-Verlag*, Wien-New York.

Telega J.J. (1989) On the complementary energy principle in non-linear elasticity. Part I: Von Kármán plates and three-dimensional solids, *C. R. Acad. Sci. Paris*, Série II, **308**, 1193-1198; Part II: Linear elastic solid and non-convex boundary condition, ibid. 1313-1317.

Telega J.J. (1990a) Variational methods and convex analysis in contact problems and homogenization, IFTR Reports 39/90, in Polish.

Telega J.J. (1990b) Homogenization of fissured elastic solids in the presence of unilateral condtition and friction, *Comput. Mech.*, **6**, 109-127.

Telega J.J. (1990c) Some results of homogenization in plasticity: plates and fissured solids, in: *Inelastic Solids and Structures*, ed. by M. Kleiber and J.A. König, pp. 343-360. *Pineridge Press*, Swansea.

Telega J.J. (1992) Justification of a refined scaling of stiffnesses of Reissner plates with fine periodic structure, *Math. Models Meth. Appl. Sci.*, **2**, 375-406.

Telega J.J. (1991) Perfectly plastic plates loaded by boundary bending moments: relaxation and homogenization. *Arch. Mech.*, **43**, 715-741.

Telega J.J. (1993) Homogenization and effective properties of plates weakened by partially penetrating fissures: convergence and duality, *Math. Modell. Numer. Anal.*, **27**, 421-456.

Telega J.J. (1995) Epi-limit on HB and homogenization of heterogeneous plastic plates, *Nonlinear Analysis, Theory, Meth. Appl.*, **25**, 499-529.

Telega J.J. and Bielski W.R. (1987) A contribution to existence problems for a geometrically nonlinear shell model, in: *Selected Problems of Modern Continuum Theory*, ed. by W. Kosiński, T. Manacorda, A. Morro, T. Ruggeri, pp. 177-193, *Pitagora Editrice*, Bologna.

Telega J.J. and Gałka A. (1998) Augmented Lagrangian methods and applications to contact problems, in: Polish-Ukrainian Seminar "Theoretical Foundations of Civil Engineering", ed. by W. Szcześniak, pp. 335-348, *Of. Wyd. P.W.*, Warszawa.

Telega J.J. and Gałka A. (1999) Augmented Langrangian methods for contact problems, optimal control and image restoration, in: *From Convexity to Nonconvexity*, ed. by P.D. Panagiotopoulos, Kluwer, in press.

Telega J.J. and Gambin B. (1996) Effective properties of an elastic body damaged by random distribution of microcracks, in: *Continuum Models and Discrete Systems*, ed. by K.Z. Markov, pp. 300-307, *World Scientific*, Singapore.

Telega J.J. and Jemioło S. (1998) Macroscopic behavior of locking materials with microstructure, Part I. Primal and dual elastic-locking potential, relaxation, *Bull. Polish Acad. Sci., Tech. Sci.*, **46**, 265-276.

Telega J.J. and Lewiński T. (1988) Homogenization of fissured Reissner-like plates. Part II- Convergence, *Arch. Mech.*, **40**, 119-134.

Telega J.J. and Lewiński T. (1993) Stiffness loss of cross-ply laminates with intralaminar cracks, in: *MECAMAT 93, International Seminar on Micromechanics of Materials*, pp. 317-326, *Editions Eyrolles*, Paris.

Telega J.J. and Lewiński T. (1994) Mathematical aspects of modelling the microscopic behaviour of cross-ply laminates with intralaminar cracks, *Control and Cybernetics*, **23**, 773-792.

Telega J.J. and Lewiński T. (1998a) Homogenization of linear elastic shells: Γ-convergence and duality, Part I. Formulation of the problem, *Bull. Polon. Acad. Sci., Tech. Sci.*, **46**, 1-9.

Telega J.J. and Lewiński T. (1998b) Homogenization of linear elastic shells: Γ-convergence and duality, Part II. Dual homogenization, *Bull. Polon. Acad. Sci., Tech. Sci.*, **46**, 11-21.

Telega J.J. and Wojnar R. (1996) Main Polish historical and modern sources on applied mechanics, *Appl. Mech. Reviews*, **49**, 401-432.

Temam R. (1985) Mathematical Problems in Plasticity, *Gauthier-Villars*, Paris.

Tenek L. and Hagiwara I. (1994a) Optimal rectangular plate and shallow shell topologies using thickness distribution or homogenization, *Comp. Meth. Appl. Mech. Eng.*, **115**, 111-124.

Tenek L.H. and Hagiwara I. (1994b) Eigenfrequency maximization of plates by optimization of topology using homogenization and mathematical programming, *JSME Int. J.*, Series C, **37**, 667-677.

Tokarzewski S. and Telega J.J. (1997) Bounds on effective moduli by analytical continuation of the Stieltjes function expanded at zero and infinity, *Zeitschrift Ang. Math. Physik*, **48**, 1-20.

Tsai C.-L. and Daniel I. M. (1992) The behaviour of cracked cross-ply composite laminates under shear loading, *Int. J. Solids Structures*, **29**, 3251-3267.

Tsai C.-L., Daniel I.M. and Lee J.-W. (1990) Progressive matrix cracking of crossply composite laminates under biaxial loading, in: *Microcracking Induced Damage in Composites*, ed. by G.J.Dvorak and D.C. Lagoudas, *AMD* vol. 111, *MD* vol. 22, New York.

Urbański A.J. (1998) Finite element formulation of the homogenization problem for plates with periodic microstructure; in: *Shell Structures. Theory and Applications*, ed. by J. Chróścielewski and W. Pietraszkiewicz, pp. 269-270, *Technical University of Gdańsk*.

Valadier M. (1994) A course on Young measures, Rend. Istituto Mat. Università di Trieste, *Supplemento*, **26**, 349-394.

Vekua I.N. (1982) Some General Methods of Construction of Various Shell Theories, *Nauka*, Moscow, in Russian.

Vigdergauz S.B. (1986) Effective elastic parameters of a plate with a regular system of equi-strength openings, *Mekh. Tverd. Tela*, **2**, 162-166, in Russian.

Vigdergauz S.B. (1994) Two-dimensional grained composites of extreme rigidity, *J. Appl. Mech.*, **61**, 390-394.

Vlasov B.F. (1957) On equations of plates bending, *Izv. Akad. Nauk. SSSR O. M. N.*, **12**, 57-60, in Russian.

Vorovich I.I. (1989) Mathematical Problems of Nonlinear Theory of Shallow Shells, *Nauka*, Moskva, in Russian.

Woźniak C. (1970) Latticed Plates and Shells, *PWN*, Warsaw, in Polish.

Yang W. and Boehler J.P. (1992) Micromechanics modelling of anisotropic damage in cross-ply laminates, *Int. J. Solids. Structures*, **29**, 1303-1328.

Yosida K. (1987) Functional Analysis, *Springer-Verlag*, Berlin.

Young L.C. (1937) Generalized curves and the existence of an attained absolute minimum in the calculus of variations, *Comptes Rendus de la Societé des Sciences et des Lettres de Varsovie*, classe III, **30**, 212-234.

Young L.C.(1969) Lectures on the Calculus of Variations and Optimal Control Theory, *Saunders*, New York (reprinted by *Chelsea*, 1980).

Zocher M.A., Allen D.H. and Groves S.E. (1997) Stress analysis of a matrix-cracked viscoelastic laminate, *Int. J. Solids Structures*, **25**, 3235-3257.

Index

Series on Advances in Mathematics for Applied Sciences

Editorial Board

N. Bellomo
Editor-in-Charge
Department of Mathematics
Politecnico di Torino
Corso Duca degli Abruzzi 24
10129 Torino
Italy
E-mail: bellomo@polito.it

F. Brezzi
Editor-in-Charge
Istituto di Analisi Numerica del CNR
Via Abbiategrasso 209
I-27100 Pavia
Italy
E-mail: brezzi@dragon.ian.pv.cnr.it

M. A. J. Chaplain
Department of Mathematics
University of Dundee
Dundee DD1 4HN
Scotland

C. M. Dafermos
Lefschetz Center for Dynamical Systems
Brown University
Providence, RI 02912
USA

S. Kawashima
Department of Applied Sciences
Engineering Faculty
Kyushu University 36
Fukuoka 812
Japan

M. Lachowicz
Department of Mathematics
University of Warsaw
Ul. Banacha 2
PL-02097 Warsaw
Poland

S. Lenhart
Mathematics Department
University of Tennessee
Knoxville, TN 37996–1300
USA

P. L. Lions
University Paris XI-Dauphine
Place du Marechal de Lattre de Tassigny
Paris Cedex 16
France

B. Perthame
Laboratoire d'Analyse Numerique
University Paris VI
tour 55–65, 5ieme etage
4, place Jussieu
75252 Paris Cedex 5
France

K. R. Rajagopal
Department of Mechanical Engrg.
Texas A&M University
College Station, TX 77843-3123
USA

R. Russo
Dipartimento di Matematica
Università degli Studi Napoli II
81100 Caserta
Italy

V. A. Solonnikov
Institute of Academy of Sciences
St. Petersburg Branch of V. A. Steklov Mathematical
Fontanka 27
St. Petersburg
Russia

J. C. Willems
Mathematics & Physics Faculty
University of Groningen
P. O. Box 800
9700 Av. Groningen
The Netherlands

Series on Advances in Mathematics for Applied Sciences

Aims and Scope

This Series reports on new developments in mathematical research relating to methods, qualitative and numerical analysis, mathematical modeling in the applied and the technological sciences. Contributions related to constitutive theories, fluid dynamics, kinetic and transport theories, solid mechanics, system theory and mathematical methods for the applications are welcomed.

This Series includes books, lecture notes, proceedings, collections of research papers. Monograph collections on specialized topics of current interest are particularly encouraged. Both the proceedings and monograph collections will generally be edited by a Guest editor.

High quality, novelty of the content and potential for the applications to modern problems in applied science will be the guidelines for the selection of the content of this series.

Instructions for Authors

Submission of proposals should be addressed to the editors-in-charge or to any member of the editorial board. In the latter, the authors should also notify the proposal to one of the editors-in-charge. Acceptance of books and lecture notes will generally be based on the description of the general content and scope of the book or lecture notes as well as on sample of the parts judged to be more significantly by the authors.

Acceptance of proceedings will be based on relevance of the topics and of the lecturers contributing to the volume.

Acceptance of monograph collections will be based on relevance of the subject and of the authors contributing to the volume.

Authors are urged, in order to avoid re-typing, not to begin the final preparation of the text until they received the publisher's guidelines. They will receive from World Scientific the instructions for preparing camera-ready manuscript.

SERIES ON ADVANCES IN MATHEMATICS FOR APPLIED SCIENCES

SERIES ON ADVANCES IN MATHEMATICS FOR APPLIED SCIENCES